ANATOMY & PHYSIOLOGY

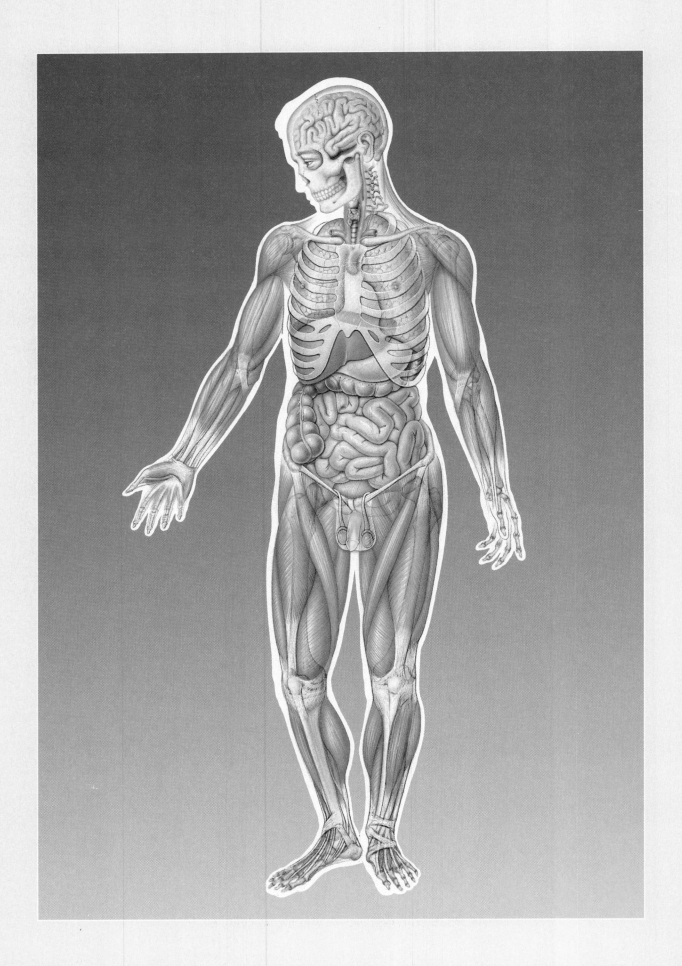

ANATOMY & PHYSIOLOGY

SECOND EDITION

Gary A. Thibodeau, Ph.D.

Chancellor and Professor of Biology,
University of Wisconsin—River Falls
River Falls, Wisconsin

Kevin T. Patton, Ph.D.

Professor, Department of Life Sciences
St. Charles County Community College
St. Peters, Missouri

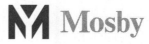 Mosby

St. Louis Baltimore Boston Chicago London Philadelphia Sydney Toronto

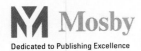

Dedicated to Publishing Excellence

Publisher: Alison Miller
Editor-in-Chief: James Smith
Editor: Sally Schrefer
Associate Developmental Editor: Laura J. Edwards
Project Manager: Carol Sullivan Wiseman
Senior Production Editor: Florence Achenbach
Senior Book Designer: Jeanne Wolfgeher
Part Opening Illustrations: Christy Krames
Cover Photos: © Chris Lane/Tony Stone Worldwide
            © Guy Motil/West Light

Credits for all illustrations and photos used by permission
appear before the index.

Second Edition

Printed in the United States of America

Mosby-Year Book, Inc.
11830 Westline Industrial Drive
St. Louis, MO. 63146

Library of Congress Cataloging in Publication Data

Thibodeau, Gary A.
    Anatomy & physiology / Gary A. Thibodeau, Kevin T. Patton. --2nd
ed.
        p.   cm.
    Includes bibliographical references and index.
    ISBN 0–8016–5005–4
    1. Human physiology.   2. Human anatomy.   I. Patton, Kevin T.
II. Title. III. Title: Anatomy and physiology
    [DNLM: 1. Anatomy.   2. Physiology.   QS 4 T427a]
QP34.5.T49   1992
612--dc20
DNLM/DLC
for Library of Congress   92-48787
                                                                        CIP

93 94 95 96 CL/CX/VH 9 8 7 6 5 4 3 2

About the Authors

Gary A. Thibodeau has been teaching anatomy and physiology for over 25 years. Since 1975, *Anatomy & Physiology* has been a logical extension of his interest and commitment to education. Gary's teaching style encourages active interaction with students, and he uses a wide variety of teaching methodologies—a style that has been incorporated into every aspect of this edition. While earning master's degrees in both zoology and pharmacology, as well as a Ph.D. in physiology, Gary says that he became "fascinated by the 'connectedness' of the life sciences." That fascination has led to this edition's unifying themes that focus on how each concept fits into the "big picture" of the human body.

Kevin T. Patton's decade of experience of teaching anatomy and physiology to community and junior college students from a variety of backgrounds has helped him produce a text that will be easier to understand for all students. "One thing I've learned," says Kevin, "is that most of us learn scientific concepts more easily when we can *see* what's going on." His talent for using imagery to teach is evident throughout this edition. Like Gary, Kevin found that the work that led him to a Ph.D. in vertebrate anatomy and physiology instilled in him an appreciation for the "big picture" of human structure and function.

Contributors

We gratefully acknowledge the following individuals for their contributions to this text:

Kara Lee Rusy, RN, MSN, CCRN
Critical Care Clinical Nurse Specialist
Assistant Professor
University of Utah Hospital
University of Utah College of Nursing
Salt Lake City, Utah

Sharon L. Sims, RN, PhD
Assistant Professor
Indiana School of Nursing
Indianapolis, Indiana

Robert R. Smith
Forest Park Community College
St. Louis, Missouri

Editorial Consultants

S. Paige Canfield
Timothy M. Griswold
Barbara Kelley
Linda Wendling

Preface to the Instructor

Anatomy & Physiology is both a teaching and a learning tool. It is intended to help teacher and student as they together face the difficult task of dealing effectively with a large and ever-growing body of scientific knowledge. This second edition of the text has been extensively revised. New pedagogical features, major full-color illustration revisions, and the addition of carefully selected new information in both anatomy and physiology provide an accurate and up-to-date presentation.

This new edition continues the long tradition of excellence in teaching about these complex subjects pioneered by Catherine Parker Anthony. Almost 50 years ago Professor Anthony created her classic *Textbook of Anatomy and Physiology*. It was a brilliant and innovative text designed to introduce students to the wonders of human form and function. We have retained the basic philosophy of personal and interactive teaching and learning championed by Professor Anthony in her text. In addition, essential, accurate, and current information continues to be presented without wordiness; emphasis is placed on concepts rather than descriptions, and the "connectedness" of human structure and function is repeatedly reinforced by unifying themes.

UNIFYING THEMES

Anatomy and physiology encompass a body of knowledge that is large and complex. Students are faced with the need to know and understand a multitude of individual structures and functions that constitute a bewildering array of seemingly disjointed information. Ultimately, the student of anatomy and physiology must be able to "pull together" this information to view the body as a whole. If a textbook is to be successful as a teaching tool in such a complex learning environment, it must help unify information, stimulate critical thinking, and motivate students to master a new vocabulary.

To accomplish this synthesis of information, unifying themes are required. In addition, a mechanism to position and implement these themes must be an integral part of each chapter. Unit One begins with "Seeing the Big Picture," an overview that encourages students to place individual structures or functions into an integrated and multifunctional framework. Then, throughout the book, the specific information presented is highlighted in a special "Big Picture" section so that it can be viewed as an integral component of a single multifaceted organism.

Anatomy & Physiology is dominated by two major unifying themes: the complementarity of normal structure and function and homeostasis. In every chapter of the book the student is shown how organized anatomical structures of a particular size, shape, form, or placement serve unique and specialized functions. The integrating principle of homeostasis is used to show how the "normal" interaction of structure and function is achieved and maintained by dynamic counterbalancing forces within the body. Repeated emphasis of these principles encourages students to integrate otherwise isolated factual information into a cohesive and understandable whole. The Big Picture summarizes the larger interaction between structures and functions of the different body systems. As a result, anatomy and physiology emerge as living and dynamic topics of personal interest and importance to the student.

AIMS OF THE REVISION

Our revision efforts focused on identifying the need for new or revised information and for visuals that might better serve to clarify important, yet sometimes difficult, content areas. For example, explanations of many complex physiological phenomena were expanded and illustrated, and the unifying themes needed to integrate form with function were reinforced.

During the revision effort, each change in the selection, sequencing, or method of presentation of material was evaluated by the authors and also by many other anatomy and physiology teachers working at every level of instruction—experienced and enthusiastic teachers currently assisting students to learn about human structure and function. The positive suggestions and feedback received from colleagues using or reviewing the first edition was of inestimable value in the revision effort. Their input was coupled with extensive information provided by Mosby-Year Book professionals and our own in-depth evaluation of the entire integrated teaching package. This package includes the text and an array of supplements, manuals, testbank materials, and visuals. As a result of this joint effort, we developed a shared vision for the new edition and were able to establish and meet specific goals that make this text and its accompanying supplements extremely effective teaching tools.

As teachers of anatomy and physiology we know that, to be effective, a text must be readable, and it must challenge and excite the student. This text remains one that students will read—one designed to help the teacher teach and the student learn. To accomplish this end requires improving

the comprehension of difficult material for students with thorough, consistent, and unintimidating explanations that are free of unnecessary terminology and extraneous information. We have also added many examples of highly effective illustrated tables and beautiful full-color artwork to provide sensory reinforcement of written material.

Illustrations and Design

A major strength of this text is the exceptional quality, accuracy, and beauty of the illustration program. The truest test of any illustration is how effectively it can complement and strengthen written information in the text and how successfully it can be used by the student as a learning tool. Each illustration is carefully referred to in the text and is designed to support the text discussion. Careful attention has been paid to placement and sizing of the illustrations to maximize usefulness and clarity. Each figure in the text was carefully reviewed for accuracy and all labels are relevant to and consistent with the text discussion. Each illustration has a boldface title and, where appropriate, small orientation diagrams or directional rosettes are included as part of the illustration to help students locate a structure with reference to the body as a whole or orient a small structure in a larger view.

The beauty of full-color artwork is aesthetically pleasing and functional. Color is used to highlight specific structures in drawings to help organize or highlight complex material in illustrated tables or conceptual flow charts. The text is also filled with dissection photographs, exceptional light micrographs, and scanning (SEM) and transmission (TEM) electron micrographs. In addition, examples of medical imagery, including CT scans, PET scans, MRIs, and x-rays, are also used to show structural detail, explain medical procedures, or enhance the understanding of differences that distinguish pathologic conditions from normal structure and function.

The most critically important illustrations in each chapter are identified with a key logo, or symbol. However, all illustrations used in the text are an integral part of the learning process and should be carefully studied by the student.

Organization and Content

The 33 chapters of *Anatomy & Physiology* are organized into six units. The sequencing of units and chapters in the book follows a course organization preferred by many instructors. However, each chapter is a self-contained learning module and may be assigned in any sequence. Thus instructors have the flexibility to alter the sequence of material to fit personal teaching preferences or the special content or time constraints of their particular courses or students.

Significant reorganization of material within or between chapters was implemented to provide students with better insight and understanding. To cite one example, the nervous system chapters (Chapters 11, 12, and 13) were completely revised and reorganized. The reflex arc was introduced earlier in this edition (Chapter 11) to provide a unifying conceptual framework to undergird material presented in the remaining chapters of the unit. Revision efforts within chapters also often focused on introduction of recent research results and theories and on improving the presentation and clarity of complex material. An example in Chapter 11 would be the thoroughly updated discussion of membrane potentials emphasizing the role of gated channels.

Unit One: The Body as a Whole

This four-chapter unit begins with an overview of the body as a unified and integrated organism. In Chapter 1 the concept of organizational levels is introduced and examples are provided. The unifying themes used throughout the book are fully explained in this chapter, and the terms used to describe anatomic structures and their relationships to each other and the body as a whole are defined and used in an appropriate context. In the case of homeostasis, the concept is first defined and, then, the control or feedback systems required to regulate the process are presented in a succinct and up-to-date treatment. Chapter 2 has been completely rewritten to present students with a nonthreatening and concise treatment of the principles required to understand physiological control mechanisms discussed in subsequent chapters of the text. The coverage of cells in Chapter 3 has been thoroughly revised to include the most current information on overall function. Discussion of cell transport mechanisms, DNA replication, and protein synthesis have been expanded. New micrographs, a carefully explained section on tissue repair, and a clarification of connective tissue classification are included in Chapter 4.

Unit Two: Support and Movement

The six chapters in Unit 2 represent a substantive change in both the content and organization of material from the first edition. Skin and Its Appendages (Chapter 5) now includes important and very current information on wound healing. Skeletal Tissues (Chapter 6) now includes a carefully explained and expanded discussion on both cell types and development of bone. The Skeletal System (Chapter 7) has been reorganized to include many new and less intimidating tables. To improve the clarity of some explanations, each table is now immediately juxtaposed to the relevant discussion. Articulations (Chapter 8) has been revised to accurately and understandably explain the most current classification of joints and joint movements.

New coverage of muscle anatomy and physiology has been divided into two chapters in this edition. As a result of the reorganization, thorough and up-to-date treatment of structure and function of muscle tissue is possible in chapters that are less lengthy and more "approachable" from a student perspective. Chapter 9, on muscle physiology, now includes a highly effective and thorough discussion of muscle contraction and new material on types of muscle fibers and effects of exercise on muscle function. Chapter 10, Anatomy of the Muscular System, now includes complete coverage of lever systems.

viii   Preface to the Instructor

Unit Three: Communication, Control, and Integration

The five chapters of Unit 3 provide exceptionally complete and understandable coverage of the nervous system (Chapters 11 through 13), the sense organs of the body (Chapter 14), and the endocrine system (Chapter 15). Significant content and organizational changes in coverage of the nervous system have occurred in this new edition. For example, the reflex arc is introduced as a unifying concept in Chapter 11. It "sets the stage" to help clarify and illuminate many of the complex and integrated control processes that are discussed in subsequent chapters. New and thorough treatment of membrane potentials emphasizing the role of gated channels is now an integral part of Chapter 11, as is discussion of neurotransmitters. In addition, convergence and divergence are explained in this chapter in the context of current research information. Chapter 12, Central Nervous System, has been updated with many new illustrations and reorganized for clarity. Treatment of the autonomic nervous system has been combined with coverage of somatic motor nerves and moved to Chapter 13, which also now includes discussion of the cranial and spinal nerves and their branches. Coverage of the sense organs in Chapter 14 has been refined and reworked to include expanded coverage of receptor physiology. In addition, new and updated treatment of the special senses, including innovative and full-color functional art, has been added to this chapter. The Endocrine System (Chapter 15) now includes the most up-do-date terminology of hormones and hormone producing glands and comprehensive coverage of steroid and nonsteroid hormone mechanisms of action.

Unit Four: Transportation and Defense

Coverage of subject matter related to transportation and defense is included in Chapters 16 through 21. The chapter on blood, Chapter 16, has been revised to include expanded coverage of RBC and WBC functions, as well as new material on RBC formation and destruction. Chapters dealing with the anatomy and physiology of the cardiovascular system (Chapters 17 and 18) have been reorganized to improve the sequencing of information. In addition, illustrations in both chapters were extensively revised, and numerous new full-color figures have been added to illustrate and improve the clarity of discussion related to physiological concepts. New material on the skeleton of the heart, expanded coverage of cardiac muscle physiology, and discussion of the anatomy of the blood vessel wall are part of the new revision. A number of new illustrations were also included as part of expanded coverage of the lymphatic system (Chapter 19). Revision of The Immune System (Chapter 20) posed significant challenges. This area is experiencing explosive change, and new theories are being advanced at a rapid rate. Therefore great selectivity was required in the addition of new material. Coverage of the basic concepts of immunity was updated, and a more detailed and current explanation of nonspecific vs. specific immunity was added.

Unit Five: Respiration, Nutrition, and Excretion

Information contained in the eight chapters of Unit 5 (Chapters 22 through 29) provides students with unique insights into the complexities of homeostasis. Each chapter in the unit has been updated as to content, and new illustrations have been added. Because of the dramatic impact on homeostasis, these chapters provide multiple opportunities to highlight for students the coupling of structural information with important functional concepts. Examples include an improved discussion of the structure of alveoli (Chapter 22) and a more thorough and carefully explained discussion of respiratory control mechanisms (Chapter 23). The digestive system (Chapter 24) has been reworked and revised to permit more visual reinforcement of the structural components. Coverage of the physiology of the digestive system (Chapter 25) begins with a new overview followed by carefully explained and accurate descriptions of the major digestive functions. Terms related to hormonal regulation of GI activity have been updated and expanded, and discussion of peristalsis and segmentation have been added. The chapter on nutrition and metabolism (Chapter 26) has been extensively revised from coverage in the first edition. Appropriate new information on nutrition has been added, including a thorough discussion of vitamins and minerals and their role in the body. The Urinary System (Chapter 27) has been reorganized so that the content material is more logically sequenced. In addition, up-to-date information has been added on hormonal regulation of urine formation and the role of urea in this complex process. Discussion of the basic regulatory mechanisms involved in both fluid and electrolyte balance (Chapter 28) and acid-base balance (Chapter 29) have been expanded, and new illustrations have been added to both chapters.

Unit Six: Reproduction and Development

Coverage of material in Unit Six (Chapters 30 through 33) has been significantly expanded from information included in the first edition. Chapters covering the anatomy and physiology of the reproductive system in both sexes (Chapters 30 and 31) are followed by two chapters new to this edition, Growth and Development (Chapter 32) and Genetics and Heredity (Chapter 33). Each chapter in this unit is richly illustrated with full-color artwork that provides accurate and innovative visual reinforcement of basic structure and physiological concepts. New micrographs and microdissection photographs are also included. Content and organizational changes to the male reproductive system text (Chapter 30) includes the addition of updated and more thorough coverage of capacitation and the supporting reproductive structures. In Chapter 31, Female Reproductive System, new easy-to-follow concept illustrations permit students to integrate the female regulatory cycles with appropriate anatomic structures, functional events, and time frames. Growth and Development (Chapter 32) covers the "cycle of life" from fertilization to death with illustrations and highly understandable explanations

found in no other text. The chapter on genetics and heredity (Chapter 33) is an important new addition to the text. It provides a highly accurate and concise treatment of genetics, including a discussion of the mechanisms of gene function and expression. The relationship between normal genetic expression and disease states provides students with insight into such areas of current interest as genetic counseling and gene therapy.

◢ LEARNING AIDS

Anatomy & Physiology is a student-oriented text. Written in a readable style, the text was designed with numerous pedagogical aids to maintain interest and motivation. The special features and learning aids listed below are intended to facilitate learning and retention of information in the most effective and efficient manner. No textbook can replace the direction and stimulation provided by an enthusiastic teacher to a curious and involved student. However, a full complement of innovative pedagogical aids that are carefully planned and implemented can contribute a great deal to the success of a text as a learning tool. An excellent textbook can and should be enjoyable to read and should be helpful to both student and teacher. We hope you agree that the learning aids in *Anatomy & Physiology* successfully meet the high expectations we have set.

Chapter Learning Aids
Chapter Outline

An overview outline introduces each chapter and enables the student to preview the content and direction of the chapter at the major concept level before beginning the detailed reading. Page references enable students to quickly locate topics in the chapter.

Chapter Objectives

Each chapter opening page lists measurable objectives for the student. These objectives clearly identify, before the student reads the chapter, what the key goals should be and what information should be mastered.

Key Terms

The most important key terms are presented and defined at the beginning of the chapter with a guide to their correct pronunciation. These key terms, with additional new terms, are also defined in the text body and are identified in **boldface** type to highlight their importance.

Quick Check Questions

Short objective-type questions are located immediately following major topic discussion areas throughout the body of the text. These questions cover important information presented in the preceding section. Students unable to answer the questions should reread the topic before proceeding.

Chapter Summary

Detailed end-of-chapter summaries provide excellent guides for students as they review the text materials before examinations. Many students also find the summaries to be useful as a chapter preview in conjunction with the chapter outline.

Review Questions

Review questions at the end of each chapter allow students to use a narrative format to discuss concepts and synthesize important chapter information for review by the instructor. The answers to these review questions are available in the Instructor's Resource Manual that accompanies the text.

Cycle of Life

In many body systems, changes in structure and function are frequently related to age or state of development. In appropriate chapters of the text these changes are highlighted in this special section.

The Big Picture

This special "whole body" section, located near the end of the chapter, helps students relate information about body structures or functions that are discussed in the chapter to the body as whole.

Mechanisms of Disease

Pathological examples are included at the end of many chapters of the book to stimulate student interest and to help students understand that the disease process is a disruption in homeostasis, a breakdown of the normal integration of form and function. The intent of the Mechanisms of Disease sections is to reinforce the normal mechanisms of the body while highlighting the general causes of disorders for that body system.

Case Studies

Case studies are found near the end of many chapters. The case study questions are critical thinking exercises that encourage students to apply information learned in the chapter, as well as seek information from outside sources, to answer questions in a clinical context. Answers to these questions are found in the Instructor's Resource Manual.

Boxed Information

Every effort has been made to update factual information and to incorporate the most current anatomy and physiology research findings in this new edition. However, although there has been an incredible explosion of knowledge in medicine and the life sciences, some of that information is appropriate for inclusion in a fundamental text and some is not. Therefore we have been highly selective in choosing new clinical, pathological, or special interest material for inclusion in this edition. This text remains focused on normal anatomy and physiology. Addition of new sports and fitness, special interest, or disease-related material is intended to stimulate student

interest and provide examples that reinforce the immediate personal relevance of anatomy and physiology as important disciplines for study.

Each chapter of the text contains information set off in color-highlighted boxes with an easily recognized symbol, or logo.

Health Matters

These boxes contain information related to health issues or clinical applications. In some instances, examples of structural anomaly or pathophysiology are presented. Information of this type is often useful in helping students understand the mechanisms involved in maintaining the "normal" interaction of structure and function.

Diagnostic Studies

These boxes deal with specific diagnostic tests used in clinical medicine or research. Specific blood chemistry analysis, medical imagery, and ECG interpretation are examples.

FYI

Topics of current interest such as new advances in anatomy and physiology research are covered in these boxes.

Sports and Fitness

Exercise physiology, sports injury, and physical education applications are highlighted in these special boxes.

Focus On. . . .

Many chapters contain longer essays, often clinical in nature, that relate to material covered in the chapter. Inflammation, diabetes mellitus, and skin cancer are examples of subjects covered in these boxes.

Unit Introductions

Each of the six major units of the text begins with a brief overview statement. The general content of the unit is discussed, and the chapters and their topics are listed. Before beginning the study of material in a new unit, students are encouraged to scan the introduction and each of the chapter outlines in the unit to understand the relationship and "connectedness" of the material to be studied.

Glossary

A comprehensive glossary of terms is located at the end of the text. Accurate, concise definitions and a phonetic pronunciation guide are provided for each entry.

Mini-Atlas of Human Anatomy

A full-color mini-atlas of human anatomy containing cadaver dissections, osteology, organ casts, and surface anatomy photographs is located at the end of the text.

SUPPLEMENTS

The supplements package was carefully planned and developed to assist instructors and to enhance their use of the text. Supplements were constructed and written to coordinate specifically with this second edition of *Anatomy & Physiology*. The authors and numerous other instructors carefully reviewed the supplements package for accuracy and relevancy.

Quick Reference To Anatomy and Physiology

This handy pocket reference summarizes key information using figures and tables. Included are directional terms and planes of the body, muscular and skeletal systems, organization of the nervous system, principal arteries and veins of the body, digestive functions, and much more!

The Human Body Videodisc

Comprehensive, versatile, and easy to use, this new resource provides unprecedented visual reinforcement for teaching anatomy and physiology. The outstanding features include:

- An overview of the structure and function of each system in the human body, including anatomical artwork, photographs of gross anatomy, micrographs of relevant tissues, and moving sequences to show the system in action or to explain how a process works.
- Animations prepared specifically for this videodisc to enable students to visualize such complex processes as movement across a cell membrane, muscle contraction, the mechanics of breathing, and the formation of urine.
- More than 1000 still images include cell structure and activity, epithelium and connective tissue, nervous system, sensory system, lymphoid system, integumentary system, respiratory system, digestive system, excretory system, male and female reproductive system, and development.
- More than 2 hours of video segments feature control, movement and circulation, defense, reproduction and development, and maintenance.
- A Print Directory lists all of the images available with their frame references and a copy of the narration for each of the moving sequences.
- A Bar Code Manual lists bar codes provided for each image to allow immediate access with the use of a bar code reader.

Instructor's Resource Manual

This invaluable resource includes a chapter synopsis and outline, ideas for presenting chapter content, teaching tips for difficult material, ideas for clinical applications, and

learning activities and assignments. The manual also includes answers to chapter review questions and case study questions, approximately 25 transparency masters of flow charts, and a directory of resources and suppliers for audiovisual materials.

Test Bank

This 3000-item test bank helps make test preparation easier and more efficient. Multiple-choice, true/false, matching, and short-answer questions are included. All questions are referenced to the text and rated for difficulty. Answers are included.

Computest Computerized Test Bank

This advanced-feature test generator allows you to add and edit questions; save and reload tests; create up to 99 different versions of each test; attach graphics to questions; import and export ASCII files; and select questions based on type, level of difficulty, or key word. This software can run on a network and enables you to save and reload tests at any time to create your own test library. Additional user-friendly features of this state-of-the-art software include:

♦ On-line testing, allowing students to sit at a computer terminal to take tests and quizzes
♦ On-screen control of page formatting for printing
♦ Six question types with answers—true/false, matching, multiple-choice, short-answer, fill-in, and essay
♦ Ability to scramble the order of questions for printing
♦ Multiple criteria for question selection—level of difficulty, type of question, author or user created as standard criteria, two optional author-defined criteria
♦ Full-function editing for adding new questions or modifying existing ones
♦ Two levels of password protection to protect the database from unauthorized users
♦ On-line, context-sensitive help accesses the relevant "help" screen from any point in the program
♦ Pull-down menu allows quick access to all program features
♦ On-screen status bars indicate number of questions selected and breakdowns by type, level of difficulty, and chapter range covered

Transparency Acetates

Featuring key illustrations from the text, 150 full-color transparencies, including paintings and line drawings, are an excellent supplement for lectures. Lettered callouts are consistently large and bold so they can be viewed easily, even from the back of a large lecture hall.

Human Body Systems Software

This interactive software helps beginning students explore applications of anatomy and physiology. Individual modules introduce the eleven body systems, and a **new** module on the cell features cell structure and transport across the membrane. Each module contains an introduction, a tutorial with practice review questions, practical applications, and a final quiz.

Human Cadaver Dissection Video

This 60-minute video takes the student through a dissection of the musculature of the human body, as well as the internal organs of the thorax and abdomen. The outstanding coverage includes a vivid, close-up detailed explanation of the dissection procedure with clear, precise, commentary.

Quiz Art

This unique study tool serves as a valuable, interactive exercise for learning key material from *Anatomy and Physiology*. Approximately 100 pieces of art from the text have been reproduced for students to label. Instructors may copy the artwork to use as handouts during lectures or as labeling quizzes.

Slides

A set of 40 slides carefully chosen from Erlandsen's *Color Atlas of Histology* will enhance your classroom and lab teaching.

Study and Review Guide

This valuable guide reinforces material in the text through Concept Reviews, organized by objectives and referenced to the text; Wordbits, a review of terminology through word roots; Clinical Challenges that apply the material to "real-life" situations; and Name-It figure labeling exercises. The preface offers study tips tailored specifically to an anatomy and physiology course.

Laboratory Manual

An invaluable resource for students, this manual contains 55 well-integrated exercises providing a hands-on learning experience to help students acquire a thorough understanding of the human body.

Extensively illustrated, the manual includes full-color reference sections on histology specimens and cat and fetal pig anatomy. Other features include boxed hints on handling specimens and managing lab activities, boxed safety tips, and coverage of cat, fetal pig, and laboratory rat dissection. Each exercise concludes with a lab report that may also serve as a self-test.

The accompanying *Instructor's Guide* offers detailed information to help the instructor prepare for the lab. Answers for all questions on the lab reports in the Student Laboratory Manual are also included.

◥ ACKNOWLEDGMENTS

Many people have contributed to the development and success of *Anatomy & Physiology*. We extend our thanks and deep appreciation to the various students and class-

room instructors who have provided us with helpful suggestions for this second edition of the text.

A specific "thank you" goes to the following instructors who critiqued in detail the first edition of this text or various drafts of the revision. Their invaluable comments were instrumental in the development of this new edition.

Charles T. Brown
Barton County Community College

Ed Calcaterra
DeSmet Jesuit High School

Laurence Campbell
Florida Southern College

Harry W. Colvin, Jr.
University of California—Davis

Dorwin Coy
University of North Florida

Douglas M. Dearden
General College of University of Minnesota

Gloria El Kammash
Wake Technical Community College

Norman Goldstein
California State University—Hayward

Charles J. Grossman
Xavier University

Rebecca Halyard
Clayton State College

Ann T. Harmer
Orange Coast College

Linden C. Haynes
Hinds Community College

Lee E. Henderson
Prairie View A&M University

Paula Holloway
Lewis and Clark Community College

Patricia Humphrey
Ohio University

Gayle Dranch Insler
Adelphi University

Carolyn Jaslow
Rhodes College

Murray L. Kaplan
Iowa State University

Clifton Lewis
Wayne County Community College

Donald Misumi
Los Angeles Trade-Technical Center

Robert Earl Olson
Briar Cliff College

Juanelle Pearson
Spalding University

Gerry Silverstein
University of Vermont—Burlington

Kathleen Tatum
Iowa State University

Judith B. Van Liew
State University of New York College at Buffalo

Clarence C. Wolfe
Northern Virginia Community College

We would also like to thank Stanley L. Erlandsen and Jean E. Magney for the use of a number of wonderful histology photos from their *Color Atlas of Histology* (Mosby-Year Book, Inc., 1992). In addition, many of the cadaver-dissection photos (including most of those found in the Mini-atlas) were borrowed from the *Color Atlas of Human Anatomy* by McMinn and Hutchings (Wolfe Medical Publications, 1990), for which we are very grateful.

At Mosby-Year Book, Inc., thanks are due all who have worked with us on this new edition. We are deeply indebted, in a very special way, to Virgil Mette, Executive Vice President, Editorial and Production, and Alison Miller, Senior Vice President, Nursing and College Editorial, for their faith in and support of this project and its authors. In addition, we wish to acknowledge the support and effort of Jim Smith, editor-in-chief, College; Sally Schrefer, editor; Laura J. Edwards, associate developmental editor; Carol Sullivan Wiseman, project manager; Florence Achenbach, production editor; and Jeanne Wolfgeher, our designer, all of whom were instrumental in bringing this edition to successful completion.

Gary A. Thibodeau
Kevin T. Patton

Preface to the Student

The purposes of a textbook are to facilitate the process of learning and to serve as a reference resource. We have worked hard to make this edition of *Anatomy & Physiology* the best learning and reference tool available. In addition to complete, up-to-date text that is written in an easy-to-read conversational style, we have included useful themes and features that make learning about human anatomy and physiology fun and easy.

Numerous features, from the short passages that introduce each unit to *The Big Picture* sections at the end of many chapters, help you to draw together isolated facts and concepts so that you can see the overall structure and function of the body. We have paid careful attention to the

illustration of key concepts so that you can gain a clearer understanding. Some of the best medical and scientific illustrators available have helped us in this effort. A variety of boxed asides, tables, charts, and special sections not only help you see the practical applications of the principles of anatomy and physiology; they also help you understand the "big picture" of the human body.

A few of the major features of this book are shown on the next few pages. We have also included a brief list of careers that challenge you to apply the principles of human anatomy and physiology in ways that help improve the quality of people's lives.

Objectives identify learning goals for each chapter.

Color-coded Tabs help you quickly find the information you need.

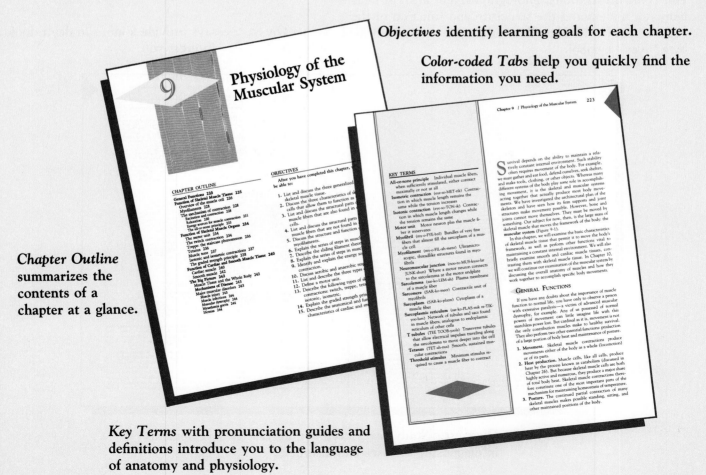

Chapter Outline summarizes the contents of a chapter at a glance.

Key Terms with pronunciation guides and definitions introduce you to the language of anatomy and physiology.

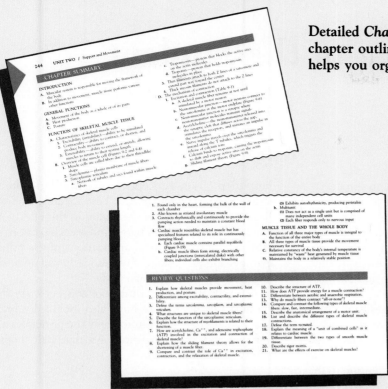

Detailed *Chapter Summaries* at the end of each chapter outline essential information in a way that helps you organize your study.

Review Questions help you determine whether you have mastered the important concepts of each chapter.

Full-color illustrations, photographs, and medical images help you understand the structure and function of the body more clearly. Key illustrations are easily identified by a "Key" symbol.

Focus On . . . essays provide a more in-depth look at subjects that may interest you.

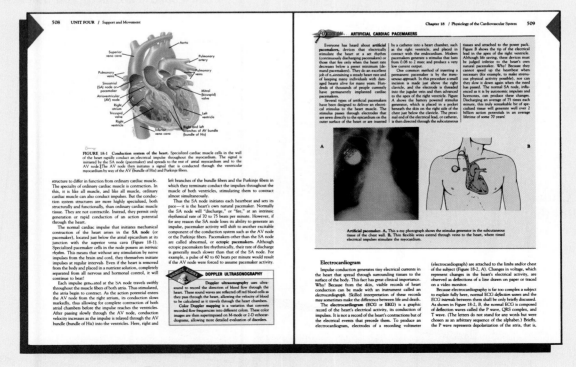

Diagnostic Study boxes keep you abreast of developments in diagnosing diseases and disorders.

Quick Check questions test your knowledge of material you just read.

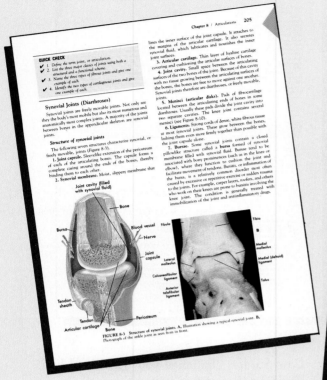

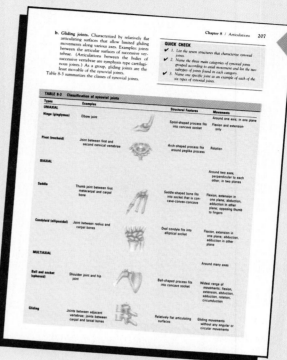

Unique illustrated tables combine text and graphics so you can more easily synthesize key information.

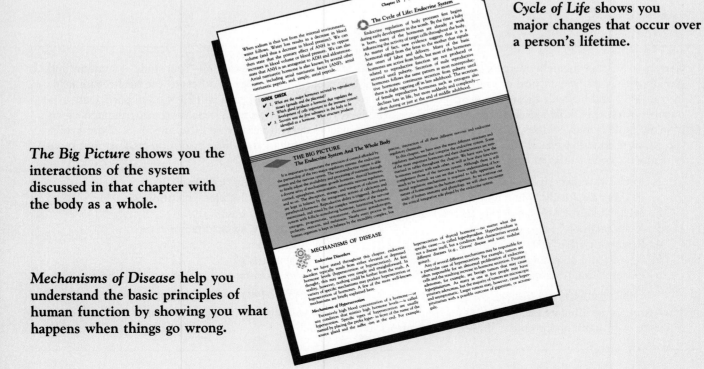

Cycle of Life shows you major changes that occur over a person's lifetime.

The Big Picture shows you the interactions of the system discussed in that chapter with the body as a whole.

Mechanisms of Disease help you understand the basic principles of human function by showing you what happens when things go wrong.

Health Matters boxes present current information on diseases, disorders, treatments, and other health issues related specifically to the normal structure and function discussed.

FYI boxes give you more in-depth information on interesting topics mentioned in the text.

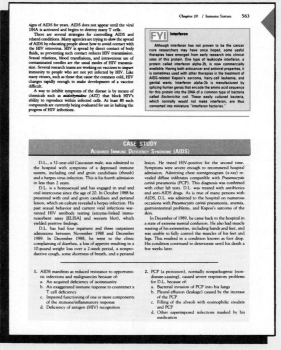

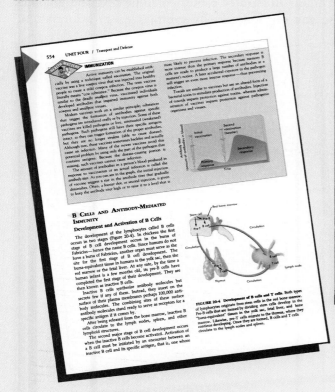

Case Studies challenge you with "real-life" clinical situations so you can creatively apply what you have learned.

Careers in health care are numerous, exciting, and personally and professionally rewarding. Your study of anatomy and physiology is a gateway to further study and eventual entry into a career of service to others in the health care field. In the 1990s and beyond, those who work in this complex and rapidly changing area will most often participate as a member of a highly skilled professional team. Successful delivery of health care services requires the coordinated efforts of many professionals. Nurses, physicians, dentists, and pharmacists depend heavily on other members of the health care team to function effectively and efficiently. A partial listing of some of the allied health professions that serve critically important roles in this team effort would include the following:

1. Medical, nursing, pharmacy, and dental assistants in every health discipline and specialty area
2. Medical and dental laboratory technologists
3. X-ray technologists
4. Respiratory therapists
5. Speech and communicative disorders professionals
6. Occupational therapists
7. Physical therapists
8. Medical transcriptionists
9. Holistic health care practitioners
10. Athletic trainers
11. Emergency medical technicians
12. Nutritionists
13. Optometrists
14. Psychologists and other mental health care professionals

This listing of health care occupations is far from complete, and new initiatives in the medical sciences will continue to create even more opportunities for rewarding careers in the future.

Contents in Brief

Detailed Contents

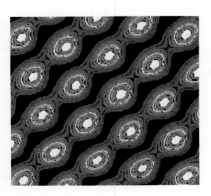

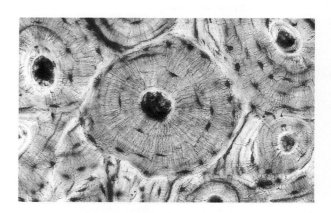

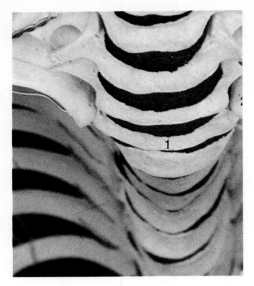

8 Articulations 201

9 Physiology of the Muscular System 223

10 Anatomy of the Muscular System 248

UNIT THREE
Communication, Control, and Integration

11 Nervous System Cells 280

UNIT FOUR
Transportation and Defense

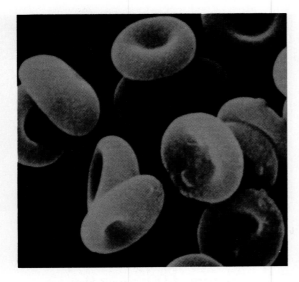

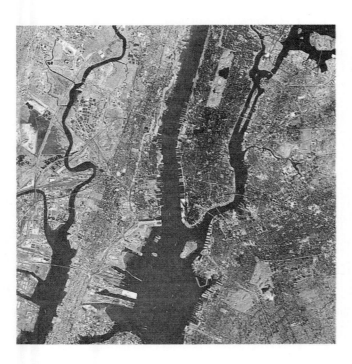

UNIT FIVE
Respiration, Nutrition, and Excretion

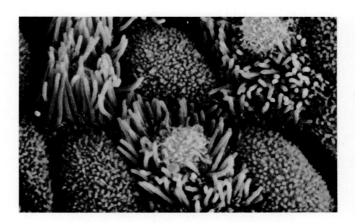

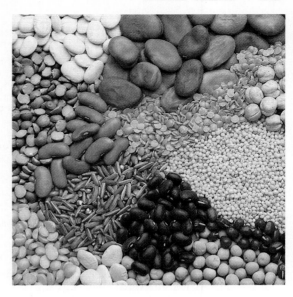

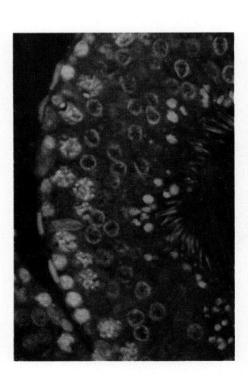

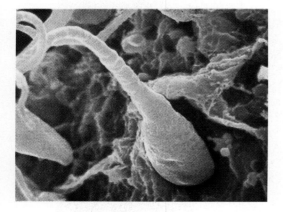

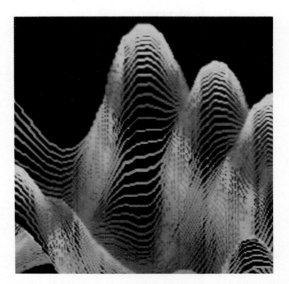

Boxes and Special Features

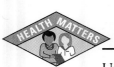

HEALTH MATTERS

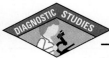

DIAGNOSTIC STUDIES

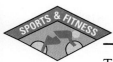

SPORTS AND FITNESS

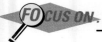

FOCUS ON . . .

FYI

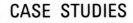

CASE STUDIES

CYCLE OF LIFE

THE BIG PICTURE

MECHANISMS OF DISEASE

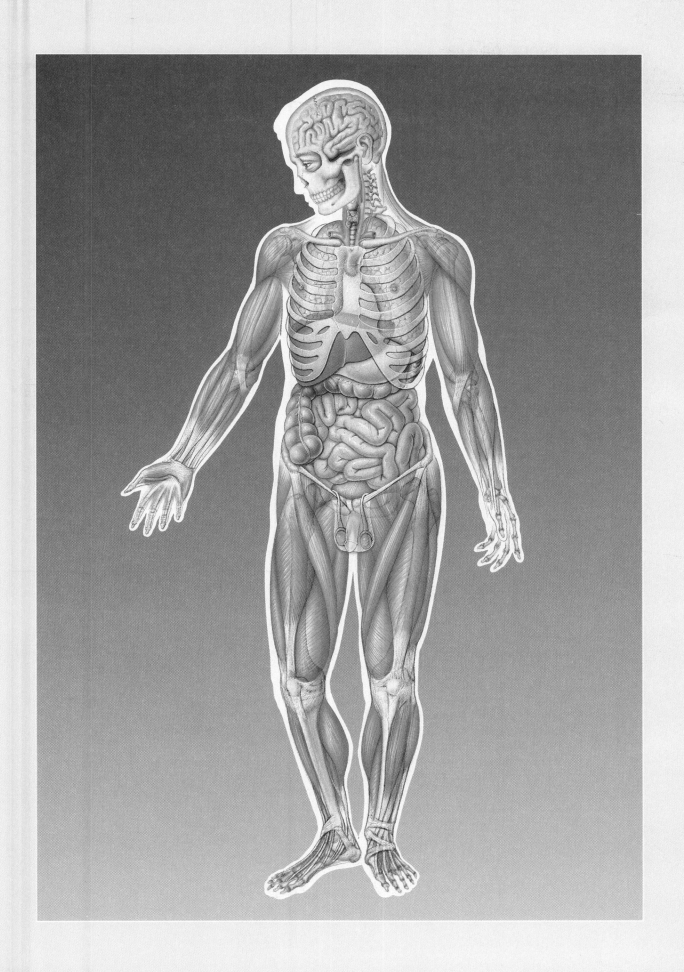

UNIT ONE

The Body as a Whole

The four chapters in Unit I "set the stage" for study of human anatomy and physiology. They provide the unifying information required to understand the "connectedness" of human structure and function and to understand how organized anatomical structures of a particular size, shape, form, or placement serve unique and specialized functions. Note that the illustration selected to introduce this unit shows the body not as an assemblage of isolated parts but as an integrated whole.

In Chapter 1 the concept of levels of organization in the body is presented, and the unifying theme of homeostasis is introduced to explain how interaction of structure and function at the chemical, organelle, cellular, tissue, organ, and system level is achieved and maintained by dynamic counterbalancing forces within the body. The material presented in Chapter 2—The Chemical Basis of Life—provides an understanding of the basic chemical interactions that influence the control, integration, and regulation of these counterbalancing forces in every organ system of the body.

Unit I concludes with information that builds on the organizational and biochemical information presented in the first two chapters. The structure and function of cells presented in Chapter 3 explains why physiologists often state that "all body functions are cellular functions." Grouping similar cells into functioning tissues is accomplished in Chapter 4. Subsequent chapters of the text will focus on the remaining organ systems of the body.

Seeing the Big Picture

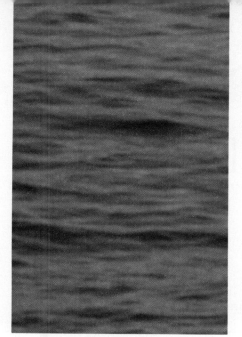

Water—essential for the whole body

Before beginning to read this introduction, you probably spent a few minutes flipping through this book. Naturally, you are curious about your course in human anatomy and physiology and you wanted to see what lay ahead. But it is more than that. You are intensely curious about the human body—about yourself, really. We all have that desire to learn more about how our bodies are put together and how all the parts work. Unlike many other people, though, you now have the opportunity to gain an understanding of the underlying scientific principles of human structure and function.

To truly understand the nature of the human body requires an ability to appreciate "the parts" and "the whole" at the same time. As you flipped through this book for the first time, you probably looked at many different body parts. Some were microscopic—like muscle cells—and some were very large—like arms and legs. In looking at these parts, however, you gained very little insight as to how they worked together to allow you to sit here, alive and breathing, and read and comprehend these words.

Think about it for a moment. What does it take to be able to read these words and understand them? You might begin by thinking about the eye. How do all its many intricate parts work together to form an image? But the eye is not the only organ you are using right now. What about the bones, joints, and muscles you are using to hold the book, turn the pages, and to move your eyes as they scan this paragraph? Let's not forget the nervous system. The brain, spinal cord, and nerves are receiving information from the eyes, evaluating it, and using it to coordinate the muscle movements. The squiggles we call letters are being interpreted near the top of the brain to form complex ideas. In short, you are *thinking* about what you are reading.

But we still have not covered everything. How are you getting the energy to operate your eyes, muscles, brain, and nerves? Energetic chemical reactions inside each cell of these organs require oxygen and nutrients captured by the lungs and digestive tract and delivered by the heart and blood vessels. These chemical reactions produce wastes that are handled by the liver, kidneys, and other organs. All of these functions must be coordinated, a feat accomplished by regulation of body organs by hormones, nerves, and other mechanisms.

Learning to name the various body parts, to describe their detailed structure, and to explain the mechanisms underlying their functions are all essential steps that lead to the goal of understanding of the human body. To actually reach that goal, however, you must be able to draw together isolated facts and concepts. In other words, understanding the nature of individual body parts is meaningless if you do not understand how those parts work together in a living, *whole* person.

Many textbooks are written like reference books—like encyclopedias, for example. They provide detailed descriptions of the structure and function of individual body parts, often in logical groupings, while rarely stopping to step back and look at the whole person. In this book, however, we have incorporated the "whole body" aspect into our discussion of every major topic. In chapter and unit introductions, in appropriate paragraphs within each section, and in specific sections near the end of most chapters, we have stepped back from the topic at hand and refocused attention to the broader view.

We are confident that our "whole body" approach will help you put each new fact or concept you learn into its proper place within a larger framework of understanding. You may also better appreciate why it is important to learn some detailed facts that may at first seem to have no practical value to you. When you have finished learning the many details covered in this course, however, you will have also gained a more complete understanding of the essential nature of the human body.

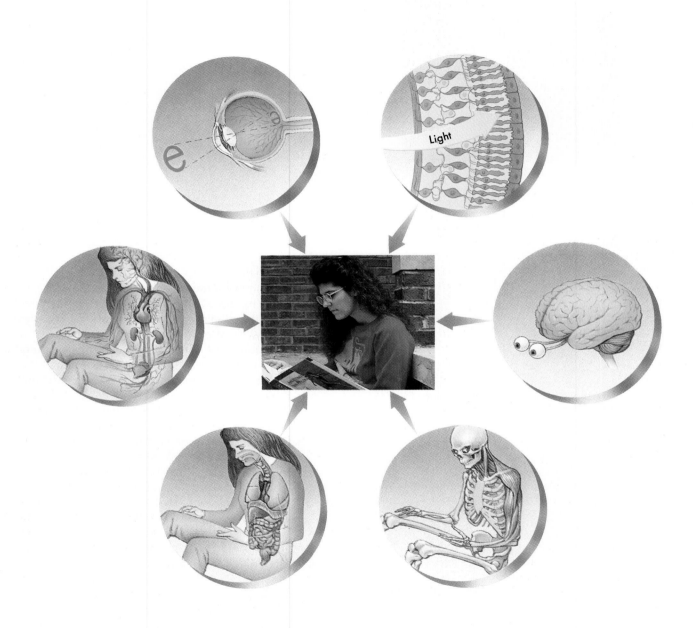

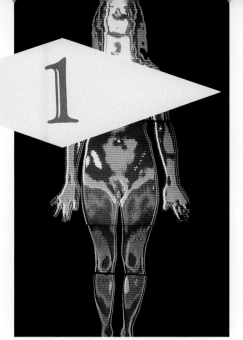

1

Organization of the Body

Thermograph of a woman

OBJECTIVES

After you have completed this chapter, you should be able to:

1. Define the terms *anatomy* and *physiology*.
2. Identify the classic "characteristics of life."
3. List and discuss in order of increasing complexity the levels of organization of the body.
4. List and briefly discuss the major organ systems of the body and tell the functions of each.
5. Explain the interaction between structure and function.
6. Discuss the concept of body type (somatotype).
7. Define homeostasis.
8. Explain the importance of homeostatic control mechanisms and and operation of negative and positive feedback loops.
9. Describe the anatomical position.
10. Discuss and contrast the axial and appendicular subdivisions of the body by identifying the specific anatomical regions in each area.
11. List the nine abdominal regions and the four abdominal quadrants.
12. List and define the principal directional terms and body planes employed in describing the body and the relationship of its parts.
13. Name the cavities of the body and identify the major organs found in each.

You are about to begin the study of one of nature's most wondrous structures—the human body. Anatomy (ah-NAT-o-me) and physiology (fiz-ee-OL-o-je) are branches of biology concerned with the form and functions of the body. Anatomy is the study of body structure, whereas physiology deals with body function—how the body parts work to support life. As you learn about the complex interdependence of structure and function in the human body, you will become in a very real sense the subject of your own study.

Regardless of your field of study or future career goals, acquiring and using information about your body structure and functions will enable you to live a more knowledgeable, involved, and healthy life in this science-conscious age. Your study of anatomy and physiology will provide a unique and fascinating understanding of self and allow for more active and informed participation in your own personal health care decisions.

If you are pursuing a health-related career your study of anatomy and physiology takes on added significance. It will provide the necessary foundation for an understanding of the clinical sciences. Knowledge of normal structure and function is required to understand both the nature of disease and how best to treat the sick and injured. Individuals who have a solid understanding of body structure and function are able to better understand and interact with health care providers. Results of medical diagnostic tests will have more meaning and convey greater understanding. You will be better able to judge the risks and the benefits of differing medical treatments or prescribed drugs. In addition, such preventive health care measures as breast self-examination, stress reduction, and adjustments in the nature of diet and exercise will be more meaningful and more easily incorporated into your personal life-style.

It is important that your study of human anatomy and physiology be based on knowledge gained from actual observation and experimentation. Information in each chapter of the text is the result of such work by scientists in each discipline. This was not always the case. For over 1500 years, writings of the Greek physician Galen (c. 130-200) served as the classic source of knowledge about the human body. Unfortunately, Galen lived in an age when actual dissection of human cadavers was prohibited. His knowledge about human anatomy and physiology was based on studies of monkeys and pigs. Only when direct and systematic study of the *human* body was undertaken by visionary scientists such as Andreas Vesalius (1514-1564) did accurate and more useful information become available.

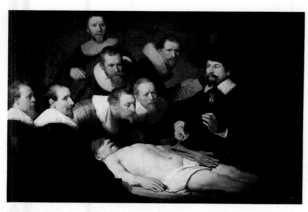

FIGURE 1-1 The anatomy lesson of Dr. Tulp. This famous painting by Rembrandt shows the importance of human dissection in teaching anatomy to 17th century medical students.

ANATOMY AND PHYSIOLOGY

Anatomy

Anatomy is often defined as the study of the structure of an organism and the relationships of its parts. The word anatomy is derived from two Greek words (*ana*, "up," and *temos* or *tomos*, "cutting"). Students of anatomy still learn about the structure of the human body by literally cutting it apart. This process, called dissection, remains a principal technique used to isolate and study the structural components or parts of the human body.

Biology is defined as the study of life. Both anatomy and physiology are subdivisions of that very broad area of inquiry. Just as biology can be subdivided into specific areas for study, so can anatomy and physiology. For example, the term *gross anatomy* is used to describe the study of those body parts visible to the naked eye. Before the discovery of the microscope, anatomists had to study human structure using only the eye during dissection. These early anatomists could make only a *gross*, or whole, examination (Figure 1-1). With the use of modern microscopes, many anatomists now specialize in *microscopic anatomy*, including the study of cells, called *cytology* (sye-TOL-o-jee), and tissues, called *histology* (his-TOL-o-jee). Other branches of anatomy include the study of human growth and development (*developmental anatomy*) or the study of diseased body structures (*pathological anatomy*). In the chapters that follow you will study the body by systems—a process called *systemic anatomy*. Systems are groups of organs that have a common function, such as the bones in the skeletal system and the muscles in the muscular system.

Physiology

Physiology is the science that treats the functions of the living organism and its parts. The term is a combination of two Greek words (*physis*, "nature," and *logos*, "science or study"). Simply stated, it is the study of physiology that helps us understand how the body works. Physiologists attempt to discover and understand, through active experimentation, the intricate control systems and regulatory mechanisms that permit the body to operate and survive, in an often hostile environment. As a scientific discipline, physiology can be subdivided according to (1) the type of organism involved, such as human physiology or plant physiology; (2) the organizational level studied, such as molecular or cellular physiology; or (3) a specific or *systemic* function being studied, such as neurophysiology, respiratory physiology, or cardiovascular physiology.

In the chapters that follow, we will study both anatomy and physiology by specific organ systems. This unit begins with an overview of the body as a whole. In subsequent chapters, the body will be dissected and studied, both structurally (anatomy) and functionally (physiology), into "levels of organization" so that its component parts can be more easily understood and then "fit together" into a living and integrated whole. It is knowledge of anatomy and physiology that allows us to understand how nerve impulses travel from one part of the body to another, how muscles contract, how light energy can be transformed into visual images, how we breathe, digest food, reproduce, excrete wastes, sense changes in our environment, and even how we think and reason.

> **QUICK CHECK**
> ✔ 1. Define anatomy and physiology.
> ✔ 2. List the three ways in which physiology can be subdivided as a scientific discipline.
> ✔ 3. What name is used to describe the study of the body that considers groups of organs that have a common function?

CHARACTERISTICS OF LIFE

Anatomy and physiology are important disciplines in biology—the study of life. But what is life? What is the quality that distinguishes a vital and functional being from a dead body? We know that a living organism is endowed with certain characteristics not associated with inorganic matter. However, there is no short and very specific definition of life because no single criterion adequately describes life. Instead of a single "difference" that separates living and nonliving things, scientists most frequently define life by a listing of attributes that, taken together, are often called *characteristics of life*.

A listing of characteristics of life cited by physiologists may differ, depending on the type of organism being studied and the way in which life functions are grouped and defined. Attributes that characterize life in bacteria, plants, or animals may vary. Those that are considered most important in humans are described as follows.

Responsiveness. Responsiveness or irritability is that characteristic of life that permits an organism to sense, monitor, and respond to changes in its external environment. Withdrawing from a painful stimulus, such as a pin prick, is an example of responsiveness.

Conductivity. Conductivity refers to the capacity of living cells and tissues to selectively transmit or propagate a wave of excitation from one point to another within the

body. Responsiveness and conductivity are highly developed in both nerve and muscle cells in living organisms.

Growth. Growth occurs as a result of a normal increase in size or number of cells. In most instances, it produces an increase in size of the individual, or of a particular organ or part, but little change in the shape of the organism as a whole or of the part affected.

Respiration. Respiration involves those processes that result in the absorption, transport, utilization, or exchange of respiratory gases (oxygen and carbon dioxide) between an organism and its environment. The exchange of gases may occur between the blood and individual body cells (internal respiration) as the cells utilize nutrients to produce energy or between the blood and air in the lungs (external respiration).

Digestion. Digestion is the process by which complex food products are broken down into simpler substances that can be absorbed and used by individual body cells.

Absorption. Absorption refers to the movement of digested nutrients through the wall of the digestive tube and into the body fluids for transport to cells for use.

Secretion. Secretion is the production and delivery of specialized substances such as digestive juices and hormones for diverse body functions.

Excretion. Excretion refers to removal of waste products produced during many body functions, including the breakdown and use of nutrients in the cell. Carbon dioxide is a gaseous waste that is excreted during respiration.

Circulation. Circulation refers to the movement of body fluids and many other substances such as nutrients, hormones, and waste products from one body area to another.

Reproduction. Reproduction involves the formation of a new individual and also the formation of new cells (via cell division) in the body to permit growth, wound repair, and replacement of dead or aging cells on a regular basis.

Each "characteristic of life" is related to the sum total of all the physical and chemical reactions occurring in the body. The term *metabolism* is used to describe these various processes. They include the steps involved in the breakdown of nutrient materials to produce energy and the transformation of one material into another. For example, if we eat and absorb more sugar than needed for the body's immediate energy requirements, it is converted into an alternate form, such as fat, that can be stored in the body. Metabolic reactions are also required for making complex compounds out of simpler ones, as in tissue growth, wound repair, or manufacture of body secretions.

Each characteristic of life—its functional manifestation in the body, its integration with other body functions and structures, and its mechanism of control—will be the subject of study in subsequent chapters of the text.

QUICK CHECK

✔ 1. *List the characteristics of life in humans.*
✔ 2. *Define the term "metabolism" as it applies to the characteristics of life.*

LEVELS OF ORGANIZATION

Before you begin the study of the structure and function of the human body and its many parts, it is important to think about how the parts are organized and how they might logically fit together and function effectively. Figure 1-2 illustrates the differing levels of organization that influence body structure and function.

Chemical Level—Basis for Life

Note in Figure 1-2 that organization of the body begins at the chemical level. There are over 100 different chemical building blocks of nature, called atoms—tiny spheres of matter so small that they are invisible. Every material thing in our universe, including the human body, is composed of atoms. Combinations of atoms form larger chemical aggregates, called molecules. Molecules, in turn, often combine with other atoms and molecules to form larger and more complex chemicals, called macromolecules.

The unique and complex relationships that exist between atoms, molecules, and macromolecules in living material form a semifluid matrix-type material called *cytoplasm* (SYE-toe-plazm)—the essential material of human life. Unless proper relationships between chemical elements are maintained, death results. Maintaining the type of chemical organization in cytoplasm required for life requires the expenditure of energy. In Chapter 2, important information related to the chemistry of life will be discussed in more detail.

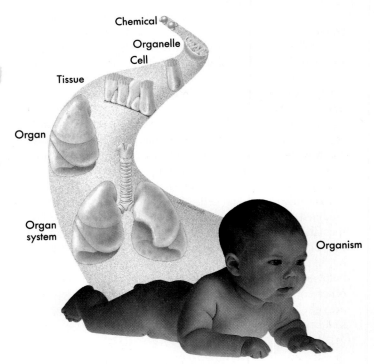

FIGURE 1-2 Levels of organization. Structural levels of organization in the body.

Organelle Level

Chemical constituents may be organized within larger units called cells to form a variety of structures called organelles (or-gan-ELZ), the next level of organization (see Figure 1-2). Organelles may be defined as collections of molecules organized in such a way that they can perform an individual function. It is the sum property of these structures that allows each cell to live. Organelles cannot survive outside the cell, but without organelles a cell could not survive either.

More than two dozen organelles have been identified. A few examples include: mitochondria (my-toe-KON-dree-ah), the "power house" of cells, which provide the energy needed by the cell to carry on day-to-day functioning, growth, and repair (see Figure 1-2); Golgi (GOL-jee) apparatus, which provides a "packaging" service to the cell by storing material for future internal use or for export from the cell; and endoplasmic reticulum, the transport channels within the cell, which act as "highways" for the movement of chemicals. Chapter 3 contains a complete discussion of organelles and their functions.

Cellular Level

The characteristics of life ultimately result from a hierarchy of structure and function that begins with the organization of atoms, molecules, and macromolecules. Continued organization resulting in organelles is the next organizational step. However, when viewed from the perspective of the anatomist, the most important function of the chemical and organelle levels of organization is to furnish the basic building blocks and specialized structures required for the next higher level of body structure—the **cellular level.** Cells are the smallest and most numerous structural units that possess and exhibit the basic characteristics of living matter. How many cells are there in the body? One estimate places the number of cells in a 150-pound adult human body at 100,000,000,000,000. In case you cannot translate this number—1 with 14 zeroes after it—it is 100 trillions! or 100,000 billions! or 100 million millions!

Each cell is surrounded by a membrane and is characterized by a single nucleus surrounded by cytoplasm that contains the numerous organelles required for specialized activity. Although all cells have certain features in common, they specialize or *differentiate* in order to perform unique functions. Fat cells, for example, are structurally modified to permit the storage of lipid material. The specialized cells shown in Figure 1-2 line the tubes of the respiratory tract. They secrete mucus and are covered by hairlike projections called cilia (SIL-ee-ah). These cells help protect the respiratory tract from inhaled dust and other contaminants. Muscle, bone, nerve, and blood cells are other examples of structurally and functionally specialized cells.

Tissue Level

As you can see in Figure 1-2, the next higher level of organization beyond the cell is the **tissue level.** Tissues represent another step in the progressive or hierarchical organization of living matter. By definition, a *tissue* is an organization of a great many similar cells that are specialized to perform a certain function. Tissue cells are surrounded by varying amounts and kinds of nonliving, intercellular substances, or the matrix.

There are four major or principal tissue types: *epithelial, connective, muscle,* and *nervous.* Considering the complex nature of the human body, this is a surprisingly short list of major tissues. Each of the four major tissues, however, can be subdivided into a number of specialized subtypes. Together, the body tissues are able to meet all of the structural and functional needs of the body.

The tissue used as an example in Figure 1-2 is a specialized type of epithelium that lines the tubes of the respiratory tract. There are several types of epithelial tissues. Some are specialized to form sheets that cover the body surface, whereas other types line body cavities or protect the passageways of the respiratory or digestive tracts. The details of tissue structure and function will be covered in Chapter 4.

Organ Level

Organs are more complex units than tissues. An organ is an organization of several different kinds of tissues so arranged that together they can perform a special function. For example, each lung is an example of organization at the **organ level:** muscle and specialized connective tissues form the many tubes that convey air, epithelial tissues line the microscopic air sacs, and nervous tissues permit control of air flow and muscular contraction.

Tissues seldom exist in isolation. Instead, joined together, they form organs that represent discrete but functionally complex operational units. Each organ has a unique shape, size, appearance, and placement in the body, and each can be identified by the pattern of tissues that forms it. The lungs, heart, brain, kidneys, liver, and spleen are all examples of organs.

System Level

Systems are the most complex of the component organizational units of the body. The **system level** of organization involves varying numbers and kinds of organs so arranged that together they can perform complex functions for the body—functions designed to meet specialized needs. Eleven major systems compose the human body: integumentary, skeletal, muscular, nervous, endocrine, circulatory, lymphatic/immune, respiratory, digestive, urinary, and reproductive. Systems that work together to accomplish the general needs of the body will be described next.

Outer protection

The skin, or **integumentary system,** is crucial to survival. Its primary function is protection. It protects underlying tissue against invasion by harmful microorganisms, bars entry of most chemicals, and minimizes mechanical injury of underlying structures. In addition, the skin

serves to regulate body temperature, synthesizes important chemicals and hormones, and functions as a sophisticated sense organ. The integumentary system includes the skin and its appendages: hair, nails, and specialized glands (Figure 1-3). Study of the integument and its diseases is called *dermatology*.

Support and movement

The skeletal and muscular systems work together to support and move the body (Figure 1-4).

The **skeletal system** consists of bones and related tissues such as cartilage and ligaments that together provide the body with a rigid framework for support and protection. In addition, the skeletal system, through joints, or *articulations*, makes movement possible. Bones also serve as reservoirs for mineral storage and function in *hematopoiesis*, or blood cell formation.

Individual muscles are the organs of the **muscular system.** In addition to *voluntary*, or *skeletal*, muscles, which have the ability to contract when stimulated and are under conscious control, the muscular system also contains *smooth*, or *involuntary*, muscles and the *cardiac* muscle of the heart. Muscles not only produce movement (or maintain body posture) but are also responsible for generating heat required for maintenance of a constant core temperature.

Communication, control, and integration

For the body to function as a whole, its various structures must be coordinated and regulated. The nervous and endocrine systems accomplish this vital task (Figure 1-5).

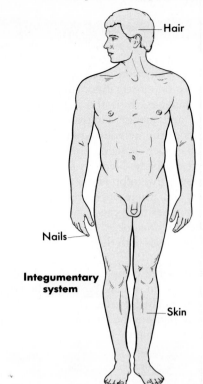

FIGURE 1-3 System that covers and protects the body.

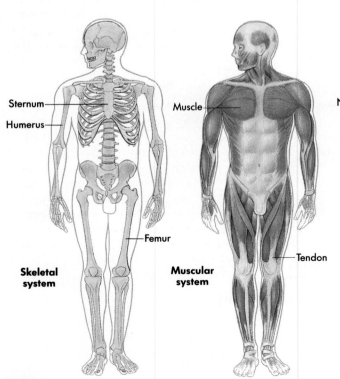

FIGURE 1-4 Systems that provide support and movement.

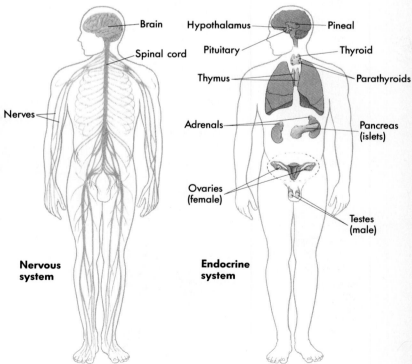

FIGURE 1-5 Systems that provide communication, control, and integration.

The brain, spinal cord, and nerves are the organs of the **nervous system.** The primary functions of this complex system include communication, integration, and control of body functions. These functions are accomplished by the generation, transmission, integration, and recognition of specialized nerve impulses. It is the nerve impulse that permits the rapid and precise control of diverse body functions. Elements of the nervous system serve to recognize certain stimuli—such as light, pressure, or temperature—which affect the body. Nervous impulses may then be generated to convey this information to the brain where it can be analyzed and where appropriate action can be initiated. Nerve impulses also cause muscles to contract and glands to secrete. *Neurology* or *neurobiology* is the branch of science that deals with the nervous system and its disorders.

The **endocrine system** is composed of specialized glands that secrete chemicals known as *hormones* directly into the blood. Sometimes called *ductless glands*, the organs of the endocrine system perform the same general functions as the nervous system—namely, communication, integration, and control. The nervous system provides rapid, brief control by fast-traveling nerve impulses. The endocrine system provides slower but longer-lasting control by secretion of hormones. The organs that are acted on and respond in some way to a particular hormone are referred to descriptively as *target organs*.

Hormones are the main regulators of metabolism, growth and development, reproduction, and many other body activities. They play roles of the utmost importance in such areas as fluid and electrolyte balance, acid-base balance, and energy metabolism.

The pituitary gland, pineal gland, hypothalamus, thyroid, parathyroids, adrenals, pancreas, ovaries, testes, thymus, and placenta all function as endocrine glands. The study of endocrine glands and their hormones is called *endocrinology.*

Transportation and defense

For all the body's cells to derive the benefits of specialized organ functions—intake of food and output of wastes, for example—a network must exist to permit distribution of substances back and forth between the organ, or system, level and the cellular level. Thus hormones produced by the endocrine system, for example, could reach the cells of the skeletal system as needed. The body's distribution network is the cardiovascular system.

The immune response, the body's defense mechanism, uses the cardiovascular and lymphatic systems of the body (Figure 1-6).

The **cardiovascular system** consists of the heart and a closed system of vessels called arteries, veins, and capillaries. As the name implies, blood contained in the circulatory system is pumped by the heart around a closed circle or circuit of vessels as it passes through the body.

The primary function of the cardiovascular system is transportation. The need for an efficient transportation system in the body is obvious. Critical transportation needs include movement of oxygen and carbon dioxide, nutrients, hormones, waste products, and other important substances on a continuing basis.

The **lymphatic system** is composed of lymph, lymphatic vessels, lymph nodes, and specialized lymphatic organs such as the thymus and spleen.

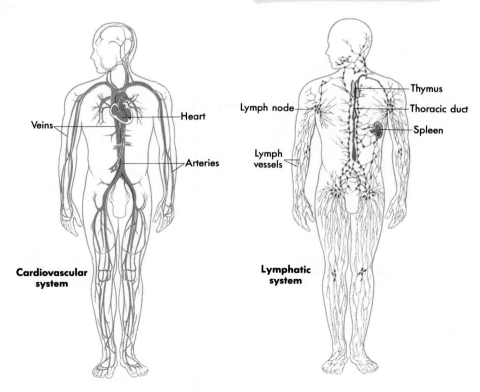

FIGURE 1-6 Systems that provide transportation and defense.

The functions of the lymphatic system include movement of fluids and certain large molecules, such as proteins, from the tissue spaces surrounding the cells to the blood and movement of fat-related nutrients from the digestive tract where they are absorbed back to the general circulation.

The **immune response** is intimately involved with lymphatic tissue. However, instead of a grouping of specialized organs the immune system is said to be a *functional system*. It is composed of specialized cells and molecules that confer protection and resistance to disease. Our ability to resist infections, to fight cancer and many other types of disease, is based on the proper functioning of our lymphatic system and immunity. The study of the immune response functions is called *immunology*.

Processing, regulation, and maintenance

The respiratory, digestive, and urinary systems all contribute to the maintenance of a stable environment for the body's cells. Foodstuffs are broken down into nutrients. Nutrients and oxygen are delivered to cells in exchange for waste products, which are eliminated. These processes are regulated by mechanisms that constantly monitor and respond to the body's internal conditions (Figure 1-7).

The organs of the **respiratory system** include the nose, pharynx, larynx, trachea, bronchi, and lungs. Together these organs permit the movement of air into the tiny, thin-walled sacs of the lungs, called *alveoli*. It is in the alveoli that oxygen from the air is exchanged for the waste product carbon dioxide, which is carried to the lungs by the blood so it can be eliminated from the body.

The main organs of the **digestive system** include the mouth, pharynx (throat), esophagus, stomach, small intestine, large intestine, rectum, and anal canal. These organs form a tube, open at both ends, called the *gastrointestinal*, or *GI, tract*. Accessory organs of digestion contribute to the proper functioning of the system but are not a part of the GI tract itself. The accessory digestive organs include teeth, tongue, salivary glands, liver, gallbladder, and pancreas. Food that enters the tract is digested, nutrients are absorbed, and the undigested residue is eliminated from the body as waste material, called *feces*. The scientific study of the GI tract and its diseases is called *gastroenterology*.

The organs of the **urinary system** include the two kidneys, the ureters, and the bladder and urethra. The kidneys function to "clear" or clean the blood of the many waste products that are continually produced by metabolism of foodstuffs in the body cells. The kidneys also play an important role in maintaining the electrolyte, water, and acid-base balance in the body.

The waste product produced by the kidneys is called *urine*. Once produced it flows out of the kidneys through the two ureters into the urinary bladder where it is stored. Urine passes from the bladder to the outside of the body through the urethra.

Reproduction and development

The **reproductive systems** of males and females combine to ensure the conception and development of offspring (Figure 1-8). The importance of normal reproductive system function is notably different from the end result of "normal function" as measured in any other organ system of

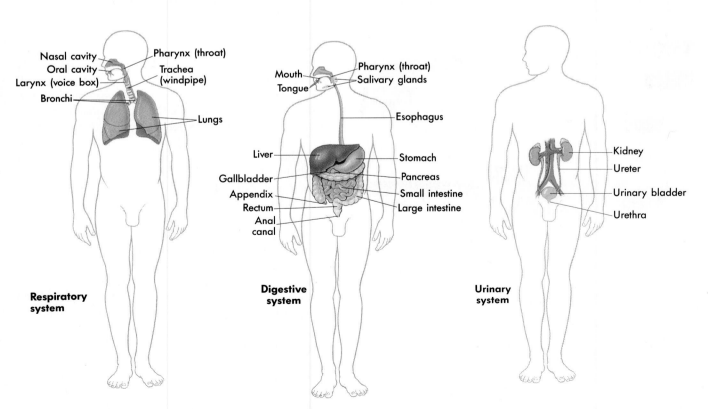

FIGURE 1-7 Systems involved with processing, regulation, and maintenance.

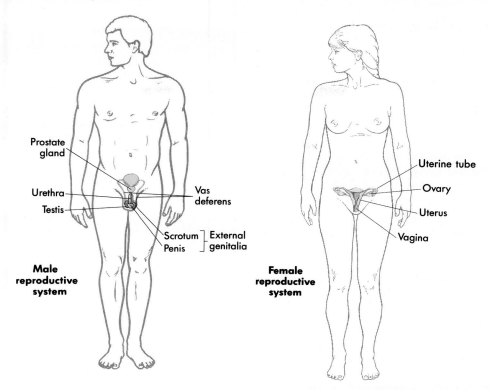

FIGURE 1-8 **Systems involved with reproduction and development.**

the body. The proper functioning of the reproductive system ensures survival, not of the individual but of the genetic code. In addition, production of hormones that permits development of the sexual characteristics occurs as a result of normal reproductive system activity.

In the male, the reproductive system organs include the *gonads* (testes), which produce sex cells, or sperm; a number of genital ducts, including the vas deferens and urethra; *accessory glands,* which contribute secretions important to reproductive function, such as the prostate gland; and the supporting structures, including the penis and scrotum (*genitalia*). Functioning together, these structures produce, transfer, and ultimately introduce sperm into the female reproductive tract, where fertilization can occur.

The female gonads are the ovaries. The accessory organs include the uterus, uterine (Fallopian) tubes, and vagina. The breasts, or mammary glands, are also classified as external accessory sex organs. The reproductive organs in the female are intended to produce sex cells or ova, receive the male sex cells, permit fertilization and transfer of the sex cells to the uterus, and allow for development, birth, and nourishment of the offspring. Study of the female reproductive system is called *gynecology*.

Organism Level

The living human organism is certainly more than the sum of its parts. It is a marvelously integrated assemblage of interactive structures that is able to survive and flourish in an often hostile environment. The human body can not only reproduce itself and effect ongoing repair and replacement of many damaged or ageing parts, it can maintain in a constant and predictable way an incredible number of variables required for us to lead healthy, productive lives.

We are able to maintain a "normal" body temperature and fluid balance under widely varying environmental extremes; we maintain constant blood levels of many important chemicals and nutrients; we experience effective protection against disease, elimination of waste products, and coordinated movement; and we correctly and quickly interpret sound, visual images, and other external stimuli with great regularity. These are but a few examples of how the different levels of organization in the human organism permit the expression of the characteristics associated with life.

As you study the structure and function of the human body it is too easy to think of each part or function in isolation from the body as a whole. Always remember that you are ultimately dealing with information related to the entire human organism—not information limited to an understanding of the structure and function of a single organelle, cell, tissue, organ, or organ system. Do not limit your learning to the memorization of facts. Instead, integrate and conceptualize factual information so that your understanding of human structure and function is related not to a part of the body but to the body as a whole.

QUICK CHECK

✔ 1. *List the seven levels of organization.*
✔ 2. *Identify three organelles.*
✔ 3. *List the four major tissue types.*
✔ 4. *List the 11 major organ systems.*

INTERACTION OF STRUCTURE AND FUNCTION

One of the most unifying and important concepts of the study of anatomy and physiology is the principle of *complementarity of structure and function*. In the chapters that follow, you will note again and again that anatomical structures seem "designed" to perform specific functions. Each structure has a particular size, shape, form, or placement in the body that makes it especially efficient at performing a unique and specialized activity.

The relationships between the levels of structural organization shown in Figure 1-2 will take on added meaning as you study the respiratory system in Chapter 22. For example, you will learn about a highly specialized chemical substance secreted by cells in the lungs that helps to keep tiny air sacs in these organs from collapsing during respiration. Hereditary material called DNA (a macromolecule) "directs" the differentiation of specialized cells in the lungs during development so that they can effectively contribute to respiratory function. As a result of DNA activity, special chemicals are produced, cells are modified, and tissues appear that are uniquely suited to this organ system. The cilia (organelles), which cover the exposed surface of cells that form the tissues that line the respiratory passageways, help trap and eliminate inhaled contaminants such as dust. The structures of the respiratory tubes and of the lungs assist in the efficient and rapid movement of air and also make possible the exchange of critical respiratory gases such as oxygen and carbon dioxide between the air in the lungs and the blood. Working together as the respiratory system, specialized chemicals, organelles, cells, tissues, and organs supply every cell of our body with necessary oxygen and constantly remove carbon dioxide.

Structure not only determines function, but function itself influences the actual anatomy of an organism over time. Understanding this fact helps students better understand the mechanisms of disease and the structural abnormalities often associated with pathology. Current research in the study of human biology is now focused in large part on integration, interaction, development, modification, and control of functioning body structures.

By applying the principle of complementarity of structure and function as you study the structural and functional levels of the body's organization in each organ system, you will be able to integrate otherwise isolated factual information into a cohesive and understandable whole. A memorized set of individual and isolated facts is soon forgotten—the parts of an anatomical structure that can be related to its function are not.

QUICK CHECK

✔ 1. Define "complementarity of structure and function."
✔ 2. Give an example of how the chemical macromolecule, DNA, can have an influence on body structure.
✔ 3. Discuss how structure relates to function at the tissue level of organization in the respiratory system.

BODY TYPE

The term **somatotype** is used to describe a particular category of body build, or *physique*. Although the human body comes in many sizes and shapes, every individual can be classified as belonging to one of three basic body types, or somatotypes. The names used to describe these body types are *endomorph, mesomorph,* and *ectomorph.* In Figure 1-9, extreme examples of the three somatotypes are shown. By carefully studying the body build of numerous individuals, scientists have found that the basic components that determine the different categories of physique occur in varying degrees in every person—both men and women. Only in very rare instances does an individual show almost total dominance by a single somatotype component.

Endomorphy

Endomorphic persons tend to be fat. Typically, they have a heavy torso with a protruding abdomen and slightly smaller chest. Smoothness of contours caused by accumulation of fat under the skin all but eliminates muscular definition. In the extreme endomorph the neck is short, and the head is almost always large and spherical. The limbs are relatively short in most endomorphs, with tapering and rounding of the thighs and upper arms.

Mesomorphy

Mesomorphs are heavily muscled and have large, prominent bones. In most individuals with this body type the distal segments of the arms and legs are prominent and massive. The shoulders of a mesomorphic person are usually well defined and tend to project outward from the torso. In addition, the chest segment of the trunk predominates over the abdominal segment, and the waist is low.

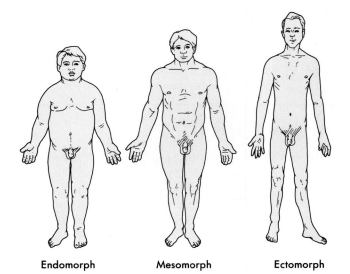

Endomorph   Mesomorph   Ectomorph

FIGURE 1-9 Types of body build, or physique. Examples of extreme somatotypes. Note the smooth, soft contours of the endomorph; the defined muscularity of the mesomorph; and the dominance of height over fat or muscle in the ectomorph.

FOCUS ON... **BODY TYPE AND DISEASE**

Until recently the concept of somatotype was considered largely "historical" and of relatively little practical importance. However, new research findings have rekindled interest in this area. We now know, for example, that knowledge of physique can provide health care professionals and educators with vital information useful in such areas as disease screening procedures, programs designed to identify individuals who may be "at risk" for developing certain diseases, and for predicting performance capability in selected physical education programs.

Researchers have discovered that individuals (especially endomorphs) who have large waistlines and are "apple-shaped," or fattest in the abdomen, have a greater risk for heart disease, stroke, high blood pressure, and diabetes than individuals with a lower "pear-shaped" distribution pattern of fat in the hips, thighs, and buttocks. Breast cancer in postmenopausal women is also associated with the storage of fat in the abdomen and upper body area (apple shape). En-domorphic individuals of the same height and weight but with a lower, or pear-shaped, body fat distribution pattern develop these diseases in larger numbers than mesomorphs and ectomorphs but less frequently than endomorphs with an apple-shaped, or high body fat, distribution pattern. This information explains why a growing number of clinical screening procedures include both somatotype identification procedures and analysis of body fat distribution patterns. Individuals "at high risk" can be advised to watch closely for signs of diseases now known to be associated with body shape.

Gender, exercise, diet, and heredity all play a role in determining body shape and fat content. Although body shape is not gender specific, men are more likely than women to be apple-shaped and are therefore more at risk for diseases now associated with this higher body fat distribution pattern. This suggests that the male sex hormone, testosterone, influences the fat deposition pattern. In apple-shaped individuals, abdominal fat is stored deep in abdominal tissues, whereas in the pear-shaped physique, most excess fat is located just below the skin near the body surface.

The *waist-to-hip ratio* is now used with assessment of somatotype to evaluate the risk of individuals developing diseases we know are related to body fat distribution patterns. Endomorphs with an "apple-shape" are at highest risk. To determine the waist-to-hip ratio, the hips are measured at their widest point around the buttocks, and the waist is measured at the level of the umbilicus. To obtain the ratio simply divide the waist measurement by the hip size. A ratio greater than 1.0 for men and 0.8 for women places the person "at risk" for diseases associated with a high body, or apple-shaped, fat distribution pattern. In men the risk factor appears if the waist size exceeds the hip measurement, and in women, risk increases significantly if the waist size exceeds 80% of the hip measurement.

Ectomorphy

In ectomorphs there is a relative predominance of height over fat or muscle. These persons tend to be tall and thin. Typically, they have a relatively short trunk, long limbs, and poorly developed musculature. A shoulder drop is common in ectomorphs. The distal segments of the arms and legs tend to be relatively long and thin. In addition, the neck tends to be long and slender, and the face is small, with sharp, fragile features.

QUICK CHECK

✔ 1. *Name the three basic types of body build or physique.*
✔ 2. *List five characteristics typical of each basic somatotype.*
✔ 3. *What somatotype would best characterize most Olympic gymnasts, swimmers, and wrestlers?*

HOMEOSTASIS

More than a century ago a great French physiologist, Claude Bernard (1813–1878), made a remarkable observation. He noted that body cells survived in a healthy condition only when the temperature, pressure, and chemical composition of their fluid environment remained relatively constant. He called the environment of cells the internal environment, or *milieu interieur*. Bernard realized that although many elements of the external environment in which we live are in a constant state of change, important elements of the internal environment, such as body temperature, remain remarkably stable. For example, the traveler who flies from Alaska to Florida in January will be exposed to dramatic changes in air temperature. Fortunately, in a healthy individual, body temperature will remain at or very near the normal of 37° C (98.6° F) regardless of temperature changes that may occur in the external environment. Just as the external environment surrounding the body as a whole is subject to change, so too is the fluid environment surrounding each body cell. The specialized fluid that bathes each cell contains literally dozens of different substances. Good health, indeed life itself, is dependent on the correct and constant amount of each substance in the blood and other body fluids. The precise and constant chemical composition of the internal environment must be maintained within very narrow limits ("normal ranges") or sickness and death will result.

In 1932 a famous American physiologist, Walter B. Cannon, suggested the name **homeostasis** (ho-me-o-STA-sis) for the relatively constant states maintained by the body. Homeostasis is a key word in modern physiology. It comes from two Greek words (*homoios*, "the same", and *stasis*, "standing"). "Standing or staying the same" then is the literal meaning of homeostasis. In his classic publication entitled *The Wisdom of the Body*, Cannon advanced one of the most unifying and important themes of physiology. He suggested that every regulatory mechanism of the body existed to maintain homeostasis or constancy of the body's internal environment. However, as Cannon emphasized, homeostasis does not mean something set and immobile that stays exactly the same all the time. In his words, homeostasis "means a condition that may vary, but which is relatively constant." It is the maintenance of

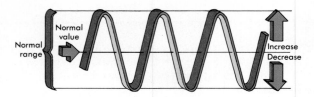

FIGURE 1-10 Homeostasis of blood glucose. Range over which a given value, such as blood glucose concentration, is maintained through homeostasis. Note that the concentration of glucose fluctuates above and below a normal value (90 mg/ml) within a normal range (80 to 100 mg/ml).

relatively constant internal conditions despite changes in the external environment that characterizes homeostasis. For example, even if external temperatures vary, homeostasis of body temperature means that it remains relatively constant at about 37° C, although it may vary slightly above or below that point and still be "normal." The fasting concentration of blood glucose, an important nutrient, can also vary somewhat and still remain within normal limits (Figure 1-10). A value of between 80 to 100 mg per 100 ml of blood, depending on dietary intake and timing of meals, is typical. Although levels of the important gases, oxygen and carbon dioxide, also vary with respiratory rate, these substances, like body temperature and blood glucose levels, must be maintained within very narrow limits.

Specific regulatory mechanisms are responsible for adjusting body systems to maintain homeostasis. This ability of the body to "self-regulate," or "return to normal" to maintain homeostasis, is a critically important concept in modern physiology and also serves as a basis for understanding mechanisms of disease. Each cell of the body, each tissue, and each organ system plays an important role in homeostasis. Each of the diverse regulatory systems described in subsequent chapters of the text will be explained as a function of homeostasis. You will learn how specific regulatory activities such as temperature control or carbon dioxide elimination are accomplished. In addition, an understanding of the relationship of homeostasis to healthy survival will explain why such mechanisms are necessary.

◤ HOMEOSTATIC CONTROL MECHANISMS

Maintaining homeostasis means that the cells of the body are in an environment that meets their needs and permits them to function normally under changing external conditions. Devices for maintaining or restoring homeostasis are known as **homeostatic control mechanisms.** They involve virtually all of the body's organs and systems. If circumstances occur that require changes or more active regulation in some aspect of the internal environment, the body must have appropriate control mechanisms available that will respond to these changing needs and then restore and maintain a healthy internal environment. For example, exercise will increase the need for oxygen and result in accumulation of the waste product carbon dioxide. By increasing our breathing rate above its average of about 17

breaths per minute we can maintain an adequate blood oxygen level and also increase the elimination of carbon dioxide. When exercise stops, the need for an increased respiratory rate no longer exists, and the frequency of breathing will return to normal.

To accomplish this self-regulation a highly complex and integrated communication control system or network is required. This type of network is called a **feedback control loop.** Different networks in the body control such diverse functions as blood carbon dioxide levels, temperature, heart rate, sleep cycles, and thirst. Information may be transmitted in these control loops by nervous impulses or by specific chemical messengers called hormones, which are secreted into the blood. Regardless of the body function being regulated or the mechanism of information transfer (nerve impulse or hormone secretion), these feedback control loops have the same basic components and work in the same way.

Basic Components of Control Mechanisms

There are a minimum of three basic components in every feedback control loop. They include:

1. Sensor mechanism
2. Integrating, or control, center
3. Effector mechanism

The process of regulation and the concept of "return to normal" first requires that the body be able to "sense" or identify the variable being controlled. Specialized nerve cells or hormone producing (endocrine) glands frequently act as homeostatic sensors. To function in this way a specialized sensor must be able to identify the element being controlled. It must also be able to respond to any changes that may occur from the normal set point range. If deviations from the normal set point range occur the sensor generates a signal (nerve impulse or hormone) to transmit that information to the second component of the feedback loop—the integration or control center.

When the integration or control center of the feedback loop (often a discrete area of the brain) receives input from a homeostatic sensor, that information is analyzed and integrated with input from other sensors, and then some specific action is initiated, if necessary, to maintain homeostasis. First, the level or magnitude of the variable being measured by the sensor is compared to the normal "set point" level that must be maintained for homeostasis. If significant deviation from that predetermined level exists the integration/control center will send its own specialized signal to the third component of the control loop—the effector mechanism.

Effectors are organs such as muscles or glands that directly influence controlled physiological variables. For example, it is effector action that increases or decreases variables such as body temperature, heart rate, blood pressure, or blood sugar concentration, to keep them within their normal range. The activity of effectors is ultimately regulated by feedback of information regarding their own effects on a controlled variable.

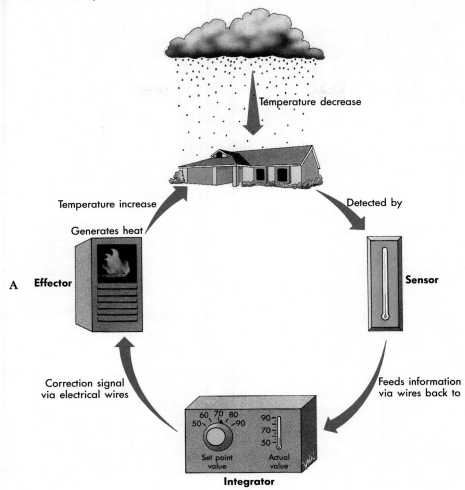

Temperature decrease

Temperature increase

Detected by

Generates heat

A   **Effector**

Sensor

Correction signal
via electrical wires

Feeds information
via wires back to

60 70 80
50   90

90
70
50

Set point
value

Actual
value

Integrator

FIGURE 1-11   Basic components of homeostatic control mechanisms. A, Heat regulation by a furnace controlled by a thermostat.

Many instructors use the example of a furnace controlled by a thermostat to explain how feedback control systems work. This analogy is a good one because it parallels the homeostatic mechanism used to control body temperature. Note that in Figure 1-11, A, changes in room temperature (the controlled variable) are detected by a thermometer (sensor) attached to the thermostat (integrator). The thermostat contains a switch that controls the furnace (effector). When cold weather causes a decrease in room temperature, the change is detected by the thermometer and relayed to the thermostat. The thermostat compares the actual room temperature to the set point temperature. After the integrator determines that the actual temperature is too low, it sends a "correction" signal by switching on the furnace. The furnace produces heat and thus increases room temperature back toward normal. As the room temperature increases above normal, feedback information from the thermometer causes the thermostat to switch off the furnace. Thus by intermittently switching the furnace off and on, a relatively constant room temperature can be maintained.

Figure 1-11, B, shows how body temperature can be regulated in much the same way as room temperature is regulated by the furnace system just described. Here, sensory receptors in the skin and blood vessels act as sensors by monitoring body temperature. When cold weather causes the body temperature to decrease, feedback information is relayed via nerves to the "thermostat" in a part of the brain called the **hypothalamus** (hi-po-THAL-ah-mus). The hypothalamic integrator compares the actual body temperature to the "built-in" set point body temperature and subsequently sends a nerve signal to effectors. In this example, the skeletal muscles act as effectors by shivering and thus producing heat. Shivering increases body temperature back to normal, when it stops as a result of feedback information that causes the hypothalamus to shut off its stimulation of the skeletal muscles. More specifics of body temperature control will be discussed in Chapter 5.

The impact of effector activity on sensors may be positive or negative in nature. Therefore homeostatic control mechanisms are categorized as either negative or positive feedback systems. By far the most important and numerous of the homeostatic control mechanisms are negative feedback systems.

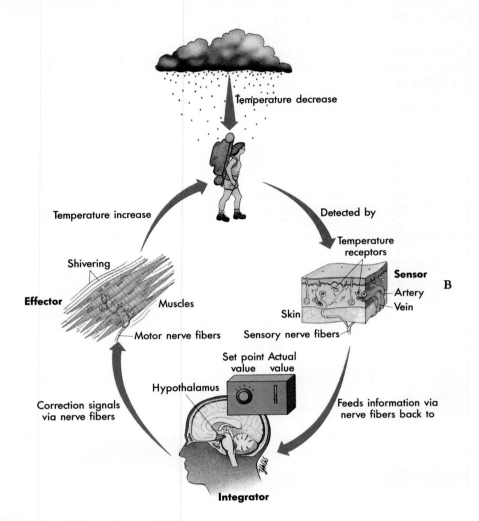

FIGURE 1-11 **Basic components of homeostatic control mechanisms. cont'd, B,** Homeostasis of body temperature. Note that in both examples a stimulus (drop in temperature) activates a sensor mechanism (thermostat or body temperature receptor) that sends input to an integrating or control center (on-off switch or hypothalamus), which then sends input to an effector mechanism (furnace or contracting muscle). The resulting heat that is produced maintains the temperature in a "normal range." Feedback of effector activity to the sensor mechanism completes the control loop.

Negative Feedback Control Systems

The example of temperature regulation by action of a thermostatically regulated furnace is a classic example of *negative feedback.* **Negative feedback control systems are inhibitory.** They oppose a change (such as a drop in temperature) by creating a response (production of heat) that is opposite in direction to the initial disturbance (fall in temperature below a normal set point). All negative feedback mechanisms in the body respond in this way regardless of the variable being controlled. They produce an action that is opposite to the change that activated the system. It is important to emphasize that negative feedback control systems *stabilize* physiological variables. They keep variables from straying too far outside of their normal ranges. Negative feedback systems are responsible for maintaining a constant internal environment.

Positive Feedback Control Systems

Positive feedback is also possible in control systems. However, because positive feedback does not operate to help the body maintain a stable or homeostatic condition it is often harmful, even disastrous, to survival. **Positive feedback control systems are stimulatory.** Instead of opposing a change in the internal environment and causing a "return to normal," positive feedback tends to amplify or reinforce the change that is occurring. In the example of the furnace controlled by a thermostat, a positive feedback loop would continue to increase the temperature. It would do so by stimulating the furnace to produce more and more heat. Each increase in heat production would be followed by a positive stimulation to increase the temperature even more. Typically, such responses result in instability and disrupt homeostasis. Why? Because the variable in question

Positive Feedback During Childbirth

One of the mechanisms that operates during delivery of a newborn illustrates the concept of positive feedback. As delivery begins, the baby is pushed from the *womb*, or *uterus*, into the birth canal, or *vagina*. Stretch receptors in the wall of the reproductive tract detect the increased stretch caused by the movement of the baby. Information regarding increased stretch is fed back to the brain, which triggers the pituitary gland to secrete a hormone called *oxytocin*. Oxytocin travels through the bloodstream to the uterus, where it stimulates stronger contractions. Stronger contractions push the baby farther along, increasing stretch, and thus stimulating release of more oxytocin. Uterine contractions quickly get stronger and stronger until finally the baby is pushed out of the body and the positive feedback loop is broken.

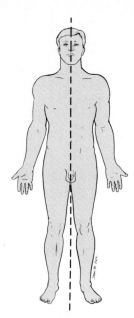

FIGURE 1-12   Anatomical position and bilateral symmetry. In the anatomical position, the body is in an erect, or standing, posture with the arms at the sides and palms forward. The head and feet are also pointing forward. The dotted line shows the body's bilateral symmetry. As a result of this organizational feature, the right and left sides of the body are mirror images of each other.

would continue to deviate further and further away from its normal range.

Only a few examples of positive feedback operate in the body under normal conditions. In each case, positive feedback accelerates the process in question. The feedback causes an ever-increasing rate of events to occur until something stops the process. Events that lead to a simple "sneeze," the birth of a baby, or the formation of a blood clot are examples of positive feedback.

In summary, most homeostatic mechanisms operate on the negative feedback principle. They are activated, or turned on, by changes in the environment that surrounds every body cell. Negative feedback systems are inhibitory. They reverse the change that initially activated the homeostatic mechanism. By reversing the initial change, a homeostatic mechanism tends to maintain or restore internal constancy. Occasionally, a positive (stimulatory) feedback mechanism helps to promote survival. Such positive or stimulatory feedback systems may be required to bring specific body functions to swift completion.

QUICK CHECK

✔ 1. *Define the term* homeostasis.
✔ 2. *List the three basic components of every feedback control system.*
✔ 3. *Explain the mechanism of action of negative and positive feedback control systems.*

ANATOMICAL POSITION

Discussions about the body, how it moves, its posture, or the relationship of one area to another, assume that the body as a whole is in a specific position called the **anatomical position.** In this reference position (Figure 1-12) the body is in an erect, or standing, posture with the arms at the sides and palms turned forward. The head and feet are also pointing forward. The anatomical position is a reference position that gives meaning to the directional

terms used to describe the body parts and regions.

Bilateral symmetry is one of the most obvious of the external organizational features in humans. The figure shown in Figure 1-12 is divided by a line into bilaterally symmetrical sides. To say that humans are bilaterally symmetrical simply means that the right and left sides of the body are mirror images of each other, and only one plane can divide the body into left and right sides. One of the most important features of bilateral symmetry is balanced proportions. There is a remarkable correspondence in size and shape when comparing similar anatomical parts or external areas on opposite sides of the body.

The terms *ipsilateral* and *contralateral* are often used to identify placement of one body part with respect to another on the same or opposite side of the body. These terms are used most frequently in describing injury to an extremity. Ipsilateral simply means "on the same side," and contralateral means "on the opposite side." Injuries to an arm or leg require careful comparison of the injured with the noninjured side. Minimal swelling or deformity on one side of the body is often apparent only to the trained observer who compares a suspected area of injury with its corresponding part on the opposite side of the body. If the right knee were injured, for example, the left knee would be designated as the contralateral knee.

BODY CAVITIES

The body, contrary to its external appearance, is not a solid structure. It contains two major cavities that are, in

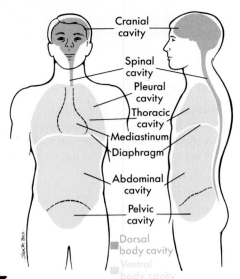

FIGURE 1-13   Body cavities. Location and subdivisions of the major body cavities.

TABLE 1-1	Organs in ventral body cavities
Area	**Organs**
THORACIC CAVITY	
Right pleural cavity	Right lung (in pleural cavity)
Mediastinum	Heart (in pericardial cavity) Trachea Right and left bronchi Esophagus Thymus gland Aortic arch and thoracic aorta Venae cavae Various lymph nodes and nerves Thoracic duct
Left pleural cavity	Left lung (in pleural cavity)
ABDOMINOPELVIC CAVITY	
Abdominal cavity	Liver Gallbladder Stomach Pancreas Intestines Spleen Kidneys Ureters

Pelvic cavity	Urinary bladder
	Female reproductive organs
	  Uterus
	  Uterine tubes
	  Ovaries
	Male reproductive organs
	  Prostate gland
	  Seminal vesicles
	  Parts of vas deferens
	Part of large intestine, namely,
	  sigmoid colon and rectum

turn, subdivided and contain compact, well-ordered arrangements of internal organs. The two major cavities are called the *ventral* and *dorsal* body cavities. The location and outlines of the body cavities are illustrated in Figure 1-13.

The **ventral cavity** consists of the *thoracic* or *chest cavity* and the *abdominopelvic cavity*. The thoracic cavity consists of a right and a left *pleural cavity* and a midportion called the *mediastinum*. Fibrous tissue forms a wall around the mediastinum, completely separating it from the right pleural cavity, in which the right lung lies, and from the left pleural cavity, in which the left lung lies. Thus the only organs in the thoracic cavity that are not located in the mediastinum are the lungs. Organs that are located in the mediastinum are the following: the heart (enclosed in its pericardial cavity), the trachea and right and left bronchi, the esophagus, the thymus, various blood vessels (e.g., thoracic aorta, superior vena cava), the thoracic duct and other lymphatic vessels, various lymph nodes, and nerves (such as the phrenic and vagus nerves). The **abdominopelvic cavity** consists of an upper portion, the *abdominal cavity*, and a lower portion, the *pelvic cavity*. The abdominal cavity contains the liver, gallbladder, stomach, pancreas, intestines, spleen, kidneys, and ureters. The bladder, certain reproductive organs (uterus, uterine tubes, and ovaries in the female; prostate gland, seminal vesicles, and part of the vas deferens in the male), and part of the large intestine (namely, the sigmoid colon and rectum) lie in the pelvic cavity (Table 1-1).

The **dorsal cavity** consists of the *cranial* and *spinal* cavities. The canial cavity lies in the skull and houses the brain. The spinal cavity lies in the spinal column and houses the spinal cord (Figure 1-13).

The thin filmy membranes that line body cavities or cover the surface of organs within body cavities also have special names. The term **parietal** refers to the actual wall of a body cavity or the lining membrane that covers its surface. **Visceral** refers not to the wall or lining of a body cavity but to the thin membrane that covers the organs, or *viscera*, within a cavity.

The membrane lining the inside of the abdominal cavity, for example, is called the *parietal peritoneum*. The membrane that covers the organs within the abdominal cavity is called the *visceral peritoneum*. You can see in Figure 1-18 that there is a space or opening between the two membranes in the abdomen. This is called the *peritoneal cavity*. Body membranes will be discussed in greater detail in Chapter 4.

BODY REGIONS

Identification of an object begins with overall or generalized recognition of its structure and form. Initially, it is in this way that the human form can be distinguished from other creatures or objects. Recognition occurs as soon as you can identify overall shape and basic outline. For more specific identification to occur, details of size, shape, and appearance of individual body areas must be described. Individuals differ in overall appearance because specific body areas such as the face or torso have unique identifying characteristics. Detailed descriptions of the human

form require that specific regions be identified and appropriate terms be employed to describe them (Figure 1-14 and Table 1-2).

The body as a whole can be subdivided into two major portions or components: **axial** and **appendicular.** The axial portion of the body consists of the head, neck, and torso, or trunk; the appendicular portion consists of the upper and lower extremities and their connections to the axial portion. Each major area is subdivided as shown in Figure 1-14. Note, for example, that the torso is composed of thoracic, abdominal, and pelvic areas and the upper extremity is divided into arm, forearm, wrist, and hand components. Although most terms used to describe gross

body regions are familiar, misuse is common. The term *leg* is a good example. To an anatomist, "leg" refers to the area of the lower extremity between the knee and ankle and *not* to the entire lower extremity.

Abdominal Regions

For convenience in locating abdominal organs, anatomists divide the abdomen into nine imaginary regions. The following is a list of the nine regions (Figure 1-15) identified from right to left and from top to bottom:

1. Right hypochondriac region
2. Epigastric region
3. Left hypochondriac region

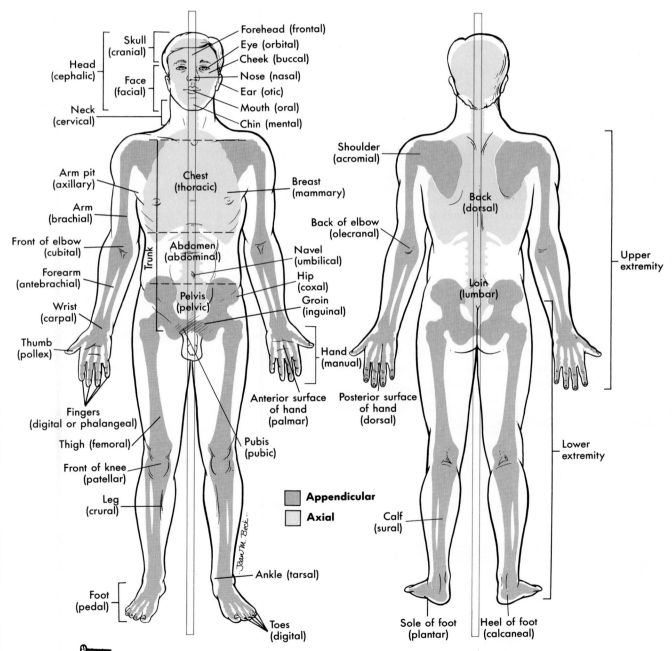

FIGURE 1-14   Specific body regions. Note that the body as a whole can be subdivided into two major portions or components: axial and appendicular.

4. Right lumbar region
5. Umbilical region
6. Left lumbar region
7. Right iliac (inguinal) region
8. Hypogastric region
9. Left iliac (inguinal) region

The most superficial organs located in each of the nine abdominal regions are shown in Figure 1-15. In the right hypochondriac region the right lobe of the liver and the gallbladder are visible. In the epigastric area, parts of the right and left lobes of the liver and a large portion of the stomach can be seen. Viewed superficially, only a small portion of the stomach and large intestine is visible in the left hypochondriac area. The right lumbar region in Figure 1-15 shows part of the large and small intestine. The superficial organs seen in the umbilical region include a portion of the transverse colon and loops of the small intestine. Addi-

tional loops of the small intestine and a part of the colon can be seen in the left lumbar region. The right iliac region contains the cecum and parts of the small intestine. Only loops of the small intestine, the urinary bladder, and the appendix are seen in the hypogastric region. The left iliac region in Figure 1-15 shows portions of the colon and the small intestine.

Abdominopelvic Quadrants

Physicians and other health professionals frequently divide the abdomen into four quadrants (Figure 1-16) to describe the site of abdominopelvic pain or locate some type of internal pathology such as a tumor or abscess. As you can see in Figure 1-16, a horizontal and vertical line passing through the umbilicus (navel) divides the abdomen into *right* and *left upper quadrants* and *right* and *left lower quadrants*.

TABLE 1-2   Descriptive terms for body regions

Body Region	Area or Example	Body Region	Area or Example
Abdominal (ab-DOM-in-al)	Anterior torso below diaphragm	**Gluteal** (GLOO-tee-al)	Buttock
		Inguinal (ING-gwi-nal)	Groin
Antebrachial (an-tee-BRAY-kee-al)	Forearm	**Lumbar** (LUM-bar)	Lower back between ribs and pelvis
Antecubital (an-tee-KYOO-bi-tal)	Depressed area just in front of elbow	**Mammary** (MAM-er-ee)	Breast
		Manual (MAN-yoo-al)	Hand
Axillary (AK-si-lair-ee)	Armpit	**Navel** (NAY-val)	Area around navel, or umbilicus
Brachial (BRAY-kee-al)	Upper arm		
Calcaneal (cal-CANE-ee-al)	Heel of foot	**Occipital** (ok-SIP-i-tal)	Back of lower skull
		Olecranal (o-LECK-ra-nal)	Back of elbow
Carpal (KAR-pal)	Wrist	**Palmar** (PAHL-mar)	Palm of hand
Cephalic (se-FAL-ik)	Head	**Patellar** (pa-TELL-er)	Front of knee
Cervical (SER-vi-kal)	Neck	**Pedal** (PED-al)	Foot
Coxal (COX-al)	Hip	**Pelvic** (PEL-vik)	Lower portion of torso
Cranial (KRAY-nee-al)	Skull	**Perineal** (pair-i-NEE-al)	Area (perineum) between anus and genitals
Crural (KROOR-al)	Leg		
Cubital (KYOO-bi-tal)	Front of the elbow		
Cutaneous (kyoo-TANE-ee-us)	Skin (or body surface)	**Plantar** (PLAN-tar)	Sole of foot
		Pollex (POL-ex)	Thumb
Digital (DIJ-i-tal)	Fingers or toes	**Popliteal** (pop-li-TEE-al)	Area behind knee
Dorsal (DOR-sal)	Back or top	**Pubic** (PYOO-bik)	Pubis
Facial (FAY-shal)	Face	**Supraclavicular** (soo-pra-cla-VIK-yoo-lar)	Area above clavicle
Buccal (BUK-al)	Cheek (inside)		
Frontal (FRON-tal)	Forehead	**Sural** (SUR-al)	Calf
Nasal (NAY-zal)	Nose	**Tarsal** (TAR-sal)	Ankle
Oral (OR-al)	Mouth	**Temporal** (TEM-por-al)	Side of skull
Orbital or **ophthalmic** (OR-bi-tal or op-THAL-mik)	Eyes	**Thoracic** (tho-RAS-ik)	Chest
		Zygomatic (zye-go-MAT-ik)	Cheek
Otic (O-tick)	Ear		
Femoral (FEM-or-al)	Thigh		

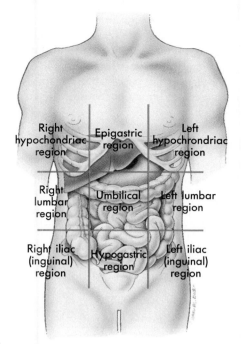

FIGURE 1-15   Nine regions of abdominopelvic cavity. The nine regions of the abdominopelvic cavity showing the most superficial organs.

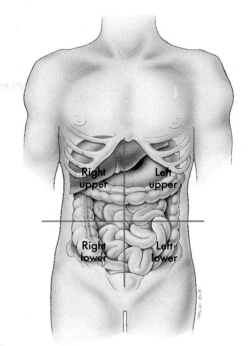

FIGURE 1-16   Division of the abdomen into four quadrants. Diagram shows relationship of internal organs to the four abdominopelvic quadrants: *1*, right upper quadrant (RUQ); *2*, left upper quadrant (LUQ); *3*, right lower quadrant (RLQ); *4*, left lower quadrant (LLQ).

HEALTH MATTERS

UMBILICUS

The umbilicus, or navel, is perhaps the "landmark" most frequently used when describing the surface anatomy of the abdomen. In most individuals the umbilicus is located at the level of the fourth lumbar vertebra. Because it is such a prominent and universal landmark, location of the umbilicus, used either alone or in combination with other body structures, is important in many clinical and diagnostic procedures. For example, a point on the left side of the abdominal wall, called *Munro's point,* is located midway between the umbilicus and a bony prominence that can be felt through the skin over a portion of the hip or pelvic bone. This is a common point for a surgeon to perform an abdominal puncture for insertion of instruments used to view the internal abdominal organs. On the right side of the body a similar line can be drawn between the umbilicus and the bony pelvic prominence on that side. If tenderness exists at a point about two thirds of the distance from the umbilicus to this pelvic prominence (*McBurney's point),* appendicitis is suspected. The umbilicus also serves as an external point of reference for placement of many types of surgical incisions to expose internal abdominal or pelvic organs.

QUICK CHECK

✔ 1. Define the term anatomical position.
✔ 2. Name the two major subdivisions of the body as a whole.
✔ 3. Identify the two major body cavities and the subdivisions of each.

TERMS USED IN DESCRIBING BODY STRUCTURE

Directional Terms

To minimize confusion when discussing the relationship between body areas or the location of a particular anatomical structure, specific terms must be used. When the body is in the anatomical position, the following directional terms can be used to describe the location of one body part with respect to another (Figure 1-17).

Superior and **inferior. Superior** means "toward the head," and **inferior** means "toward the feet." Superior also means "upper" or "above," and inferior means "lower" or "below." For example, the lungs are located superior to the diaphragm, whereas the stomach is located inferior to it.

Anterior and **posterior. Anterior** means "front" or "in front of"; **posterior** means "back" or "in back of." In humans who walk in an upright position, **ventral** (toward the belly) can be used in place of anterior, and **dorsal** (toward the back) can be used for posterior. For example, the nose is on the anterior surface of the body, and the shoulder blades are on its posterior surface.

Medial and **lateral. Medial** means "toward the midline of the body"; **lateral** means "toward the side of the body, or away from its midline." For example, the great toe is at the medial side of the foot, and the little toe is at its lateral side. The heart lies medial to the lungs, and the lungs lie lateral to the heart.

Proximal and **distal. Proximal** means "toward or nearest

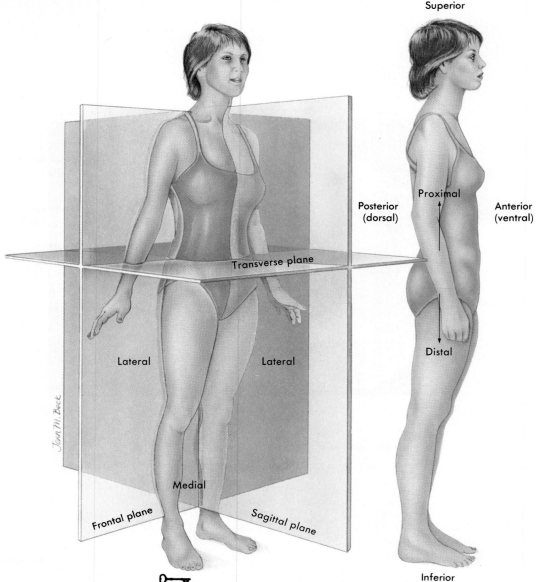

FIGURE 1-17   Directions and planes of the body.

the trunk of the body, or nearest the point of origin of one of its parts"; **distal** means "away from or farthest from the trunk or the point of origin of a body part." For example, the elbow lies at the proximal end of the lower arm, whereas the hand lies at its distal end.

Superficial and **deep.** **Superficial** means "nearer the surface"; **deep** means "farther away from the body surface." For example, the skin of the arm is superficial to the muscles below it, and the bone of the upper arm is deep to the muscles that surround and cover it.

▶ BODY PLANES AND SECTIONS

The transparent glass-like plates in Figure 1-17 dividing the body into parts represent cuts or *sections* that can be made along a particular axis or line of orientation called a

plane. There are three major planes that lie at right angles to each other. They are called the *sagittal, coronal (kuh-RO-nul),* and *transverse* (or *horizontal*) *planes.* Literally hundreds of sections can be made in each plane, and each section made is named after the particular plane along which it occurs. For example, the transverse plane in Figure 1-17 is shown dividing the individual into upper and lower parts at about the level of the umbilicus. Many other transverse sections are possible in parallel transverse planes. A transverse section through the knee would amputate the lower extremity at that joint, and a transverse section through the neck would result in decapitation.

Read the following definitions and identify each term in Figure 1-17:

Sagittal. A lengthwise plane running from front to back; divides the body or any of its parts into right and left sides.

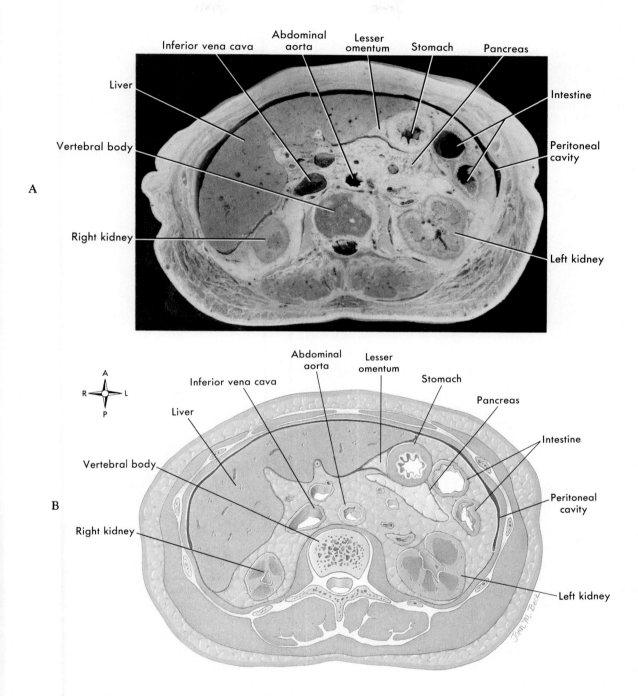

FIGURE 1-18   Transverse section of abdomen. A, Transverse, or horizontal, plane through the abdomen shows the position of various organs within the cavity. **B,** Drawing of the photograph helps to clarify the photo.

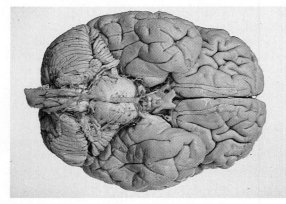

A

FIGURE 1-19   **Three views of the brain. A,** Inferior aspect of brain. **B,** Lateral view of right half of brain. **C,** Medial view of left half of brain (visible only after midsagittal section). Direction "rosettes" indicate anterior, right, superior, and inferior, as relates to anatomical position. Can you see why, if the section in **A** were viewed from above, the "L" and "R" would be reversed?

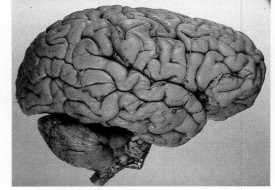

B

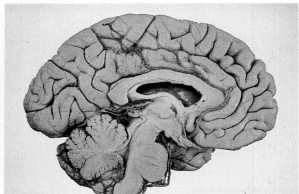

C

If a sagittal section is made in the exact midline, resulting in equal and symmetrical right and left halves, the plane is called a *midsagittal* or *median sagittal* plane.

Coronal. A lengthwise plane running from side to side; divides the body or any of its parts into anterior and posterior portions; also called a *frontal* plane.

Transverse. A crosswise plane; divides the body or any of its parts into upper and lower parts; also called a horizontal plane.

Figure 1-18 shows the organs of the abdominal cavity as they would appear in the transverse, or horizontal, plane or "cut" through the abdomen represented in Figure 1-17. In addition to the actual photograph a simplified line diagram helps in identifying the primary organs. Note that organs near the bottom of the photo or line drawing are in a posterior position. The cut vertebra of the spine, for example, can be identified in its position behind or *posterior* to the stomach. The kidneys are located on either side of the vertebra—they are *lateral* and the vertebra is *medial*. To make the reading of anatomical figures a little easier, an *anatomical compass* is used throughout this book. On many figures, you will notice a small compass rosette similar to those on geographical maps. Rather than being labeled N, S, E, and W, the anatomical rosette is labeled with abbreviated anatomical directions:

A, Anterior
I, Inferior
L (opposite R), Left

L (opposite M), Lateral
M, Medial
P, Posterior
R, Right
S, Superior

The brain is an internal organ that shows bilateral symmetry. If cut into halves using a midsagittal section, the right and left sides will be almost identical (mirror images) in their appearance. In Figure 1-19, a series of photographs begins with a view of the brain as seen from below (Figure 1-19, A). You are looking up at the bottom of the most *inferior* surface of the organ after its removal from the skull.

Figure 1-19, B, shows the right half of the brain as viewed from the side, or *laterally*. The *lateral surface* is the external surface. In Figure 1-19, C, the same right half is viewed showing its *medial surface*. The medial surface, of course, can only be seen if the organ is cut in half. As you can see, an understanding of directional terms and body planes or sections is required to fully appreciate and interpret the anatomical illustrations.

QUICK CHECK

✔ 1. *List and define the three major planes that are used to divide the body into parts.*
✔ 2. *List the nine abdominal regions and four abdominopelvic quadrants.*

MEDICAL IMAGING OF THE BODY

Cadavers (preserved human bodies used for scientific study) can be cut into sagittal, frontal, or transverse sections for easy viewing of internal structures, but living bodies cannot. This fact has been troublesome for medical professionals who must determine whether internal organs are injured or diseased. In some cases, the only sure way to detect a *lesion* or variation from normal is extensive *exploratory surgery*. Fortunately, recent advances in medical imaging allow physicians to visualize internal structures of the body without risking the trauma or other complications associated with extensive surgery. Some of the more widely used techniques are briefly described here.

Radiography

Radiography, or x-ray photography, is the oldest and still most widely used method of noninvasive imaging of internal body structures. In this method, energy in the x band of the radiation spectrum is beamed through the body to photographic film (Figure A). The x-ray photograph shows the outlines of bones and other dense structures that partially absorb the x rays. In *fluoroscopy*, a phosphorescent screen sensitive to x rays is used instead of photographic film. A visible image is formed on the screen as x rays passing through the subject cause the screen to glow. Fluoroscopy allows a medical professional to view the internal structures of the subject's body as it moves. One way to make soft, hollow structures such as blood vessels or digestive organs more visible is to use *radiopaque* contrast media. Substances such as barium sulfate, which absorb x rays, are injected or swallowed to fill the hollow organ of interest. As the screen in Figure A shows, the hollow organ shows up as distinctly as a dense bone.

Computed Tomography

A newer variation of traditional x-ray photography is called **computed tomography (CT)** or computed axial tomography (CAT) scanning. In this method, a device with an x-ray source on one side of the body and an x-ray detector on the other side is rotated around the central axis of the subject's body (Figure B). Information from the x-ray detectors is interpreted by a computer, which generates a video image of the body as if it were cut into anatomical sections. The term *computed tomography* literally means "picturing a cut using a computer." Because CT scanning and other recent advances in diagnostic imaging produce images of the body as if it were actually cut into sections, it has become especially important for students of the health sciences to be familiar with *sectional anatomy*. Sectional anatomy is the study of the structural relationships visible in anatomical sections.

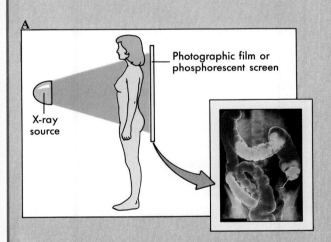

A

Photographic film or phosphorescent screen

X-ray source

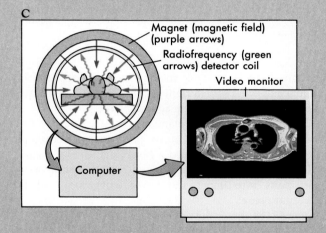

C

Magnet (magnetic field) (purple arrows)

Radiofrequency (green arrows) detector coil

Video monitor

Computer

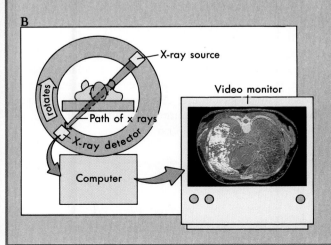

B

X-ray source

rotates

Path of x rays

X-ray detector

Computer

Video monitor

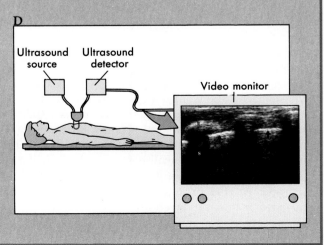

D

Ultrasound source

Ultrasound detector

Video monitor

Magnetic Resonance Imaging

Magnetic resonance imaging (MRI) is a type of scanning that uses a magnetic field to induce tissues to emit radio frequency (RF) waves. An RF detector coil senses the waves and sends the information to a computer that constructs sectional images similar to those produced in CT scanning (Figure C). Different tissues can be distinguished because each emits different radio signals. MRI, also called *nuclear magnetic resonance (NMR)* imaging, avoids the use of potentially harmful x-radiation and often produces sharper images of soft tissues than other imaging methods.

Ultrasonography

In **ultrasonography,** high-frequency (ultrasonic) waves are reflected off internal tissues to produce an image called a *sonogram.* Because it does not involve x-radiation, and because it is relatively inexpensive and easy to use, ultrasonography has been used extensively—especially in studying maternal or fetal structures in pregnant women. However, the image produced is not nearly as clear or sharp as in MRI, CT scanning, or traditional radiography (Figure D).

Variations of these and other technological advances that have improved the ability to study the structure and functions of the human body will be discussed in later chapters.

 The Cycle of Life: *Life span considerations*

An important generalization about body structure is that every organ, regardless of location or function, undergoes change over the years. In general, the body performs its functions least well at both ends of life—in infancy and in old age. Organs develop and grow during the years before maturity, and body functions gradually become more and more efficient and effective. In the healthy young adult all body systems are mature and fully operational. Homeostatic mechanisms tend to function most effectively during this period of life to maintain the constancy of one's internal environment. After maturity, effective repair and replacement of the body's structural components often decrease. The term *atrophy* is used to describe the wasting effects of advancing age. In addition to structural atrophy, the functioning of many physiological control mechanisms also decreases and become less precise with advancing age. The changes in functions that occur during the early years are called developmental processes. Those that occur during the late years are called aging processes. The study of aging processes and other changes that occur in our lives as we get older is called *gerontology.* Many specific age changes will be noted in the chapters that follow.

 MECHANISMS OF DISEASE

Some General Considerations

Clearer understanding of the normal function of the body often comes from our study of disease. **Pathophysiology** is the organized study of the underlying physiological processes associated with disease. Pathophysiologists attempt to understand the mechanisms of a disease and its course of development, or *pathogenesis*. Near the end of each chapter of this book we will briefly describe some important disease mechanisms that illustrate the normal functions described earlier in that chapter.

Many diseases are best understood as disturbances to homeostasis, the relative constancy of the body's internal environment. If homeostasis is disturbed, various negative-feedback mechanisms usually return the body to normal. When a disturbance goes beyond the normal fluctuation of everyday life, we can say that a disease condition exists. In acute conditions, the body recovers its homeostatic balance quickly. In chronic diseases, a normal state of balance may never be restored. If the disturbance keeps the body's internal environment too far from normal for too long, death may result (Figure 1-20).

Basic Mechanisms of Disease

Disturbances to homeostasis and the body's responses are the basic mechanisms of disease. Because of their variety, disease mechanisms can be categorized for easier study:

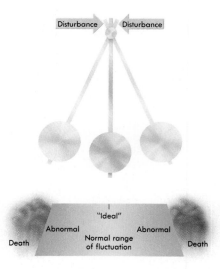

FIGURE 1-20 Model of homeostatic balance. Movement of the pendulum represents the fluctuation of a physiological variable, such as body temperature or blood pressure, around an "ideal" value. Disturbance in either direction could move the variable into the "abnormal" range, the disease state. If the disturbance is so extreme that it goes "off the scale" or outside the tolerable range, death results.

1. Genetic mechanisms. Altered, or *mutated*, genes can cause abnormal proteins to be made. These abnormal proteins often do not perform their intended function—resulting in the absence of an essential function. On the other hand, such proteins may actually perform an abnormal, disruptive function. Either case poses a potential threat to the constancy of the body's internal environment. The action of genes is first discussed in Chapter 3, and the mechanisms by which genes are inherited are discussed in Chapter 33.

2. Pathogenic organisms. Many important disorders are caused by *pathogenic* (disease-causing) organisms that

damage the body in some way (Figure 1-21). Any organism that lives in or on another organism to obtain its nutrients is called a *parasite*. The presence of microscopic or larger parasites may interfere with normal body functions of the *host*, causing disease. Besides parasites, there are organisms that poison or otherwise damage the human body to cause disease. Some of the major pathogenic organisms include:

♦ **Viruses** are intracellular parasites that consist of a DNA or RNA core surrounded by a protein coat

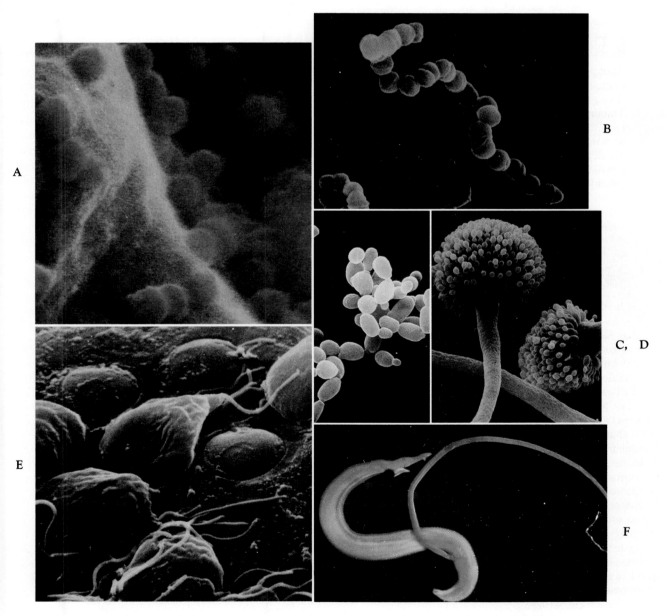

FIGURE 1-21   Examples of pathogenic organisms. **A,** Viruses (the human immunodeficiency virus, or HIV, that causes AIDS). **B,** Bacteria (streptococcus bacteria that cause "strep throat" and other infections). **C,** Fungi (yeast cells that commonly infect the urinary and reproductive tracts). **D,** Fungi (the mold that causes aspergillosis). **E,** Protozoans (the flagellated cells that cause "traveler's diarrhea." **F,** Pathogenic animals (the parasitic worms that cause "snail fever").

and, sometimes, a lipoprotein envelope. They invade human cells and cause them to produce viral components.

♦ **Bacteria** are tiny, primitive cells that lack nuclei. They cause infection by parasitizing tissues or otherwise disrupting normal function.

♦ **Fungi** are simple organisms similar to plants but lack the chlorophyll pigments that allow plants to make their own food. Since they cannot make their own food, fungi must parasitize other tissues, including those of the human body.

♦ **Protozoa** are *protists,* one-celled organisms larger than bacteria and whose DNA is organized into a nucleus. Many types of protozoa parasitize human tissues.

♦ **Pathogenic animals** are large, multicellular organisms such as insects and worms. Such animals can parasitize human tissues, bite or sting, or otherwise disrupt normal body function.

Examples of infections or other conditions caused by pathogenic organisms are given in many chapters throughout this book.

3. Tumors and cancer. Abnormal tissue growths, or *neoplasms,* can cause various physiological disturbances, as described later in Chapter 4.

4. Physical and chemical agents. Agents such as toxic or destructive chemicals, extreme heat or cold, mechanical injury, and radiation can each affect the normal homeostasis of the body. Examples of healing of tissues damaged by physical agents is discussed in Chapters 4, 5, 6, and other chapters.

5. Malnutrition. Insufficient or imbalanced intake of nutrients causes various diseases; these are outlined in Chapter 26.

6. Autoimmunity. Some diseases result from the immune system attacking one's own body (*autoimmunity*) or from other mistakes or overreactions of the immune response. Autoimmunity, literally "self-immunity," is discussed in Chapter 20 with other disturbances of the immune system.

7. Inflammation. The body often responds to disturbances with the *inflammatory response.* The inflammatory response, which is described in Chapter 4, is a normal mechanism that usually speeds recovery from an infection or injury. However, when the inflammatory response occurs at inappropriate times or is abnormally prolonged or severe, normal tissues may become damaged. Thus some disease symptoms are *caused by* the inflammatory response.

8. Degeneration. By means of many still unknown processes, tissues sometimes break apart or *degenerate.* Although a normal consequence of aging, degeneration of one or more tissues resulting from disease can occur at any time. The degeneration of tissues associated with aging is discussed in nearly every chapter of this book.

HEALTH MATTERS — DISEASE TERMINOLOGY

Everyone is interested in *pathology*—the study of disease. Researchers want to know the scientific basis of abnormal conditions. Health practitioners want to know how to prevent and treat various diseases. Every one of us, when we suffer from the inevitable head cold or something more serious, want to know what is going on and how best to deal with it. Pathology has its own terminology, as in any specialized field. Most of these terms are derived from Latin and Greek word parts. For example, *patho-* comes from the Greek word for disease (*pathos*) and is used in many terms, including "pathology" itself.

Disease conditions are usually *diagnosed* or identified by signs and symptoms. **Signs** are objective abnormalities that can be seen or measured by someone other than the patient, whereas **symptoms** are the subjective abnormalities that are felt only by the patient. Although *sign* and *symptom* are distinct terms, we often use them interchangeably. A **syndrome** is a collection of different signs and symptoms that occur together. When signs and symptoms appear suddenly, persist for a short time, then disappear, we say that the disease is **acute.** On the other hand, diseases that develop slowly and last for a long time (perhaps for life) are labeled **chronic** diseases. The term *subacute* refers to diseases with characteristics somewhere between acute and chronic.

The study of all the factors involved in causing a disease is its **etiology.** The etiology of a skin infection often involves a cut or abrasion and subsequent invasion and growth of a bacterial colony. Diseases with undetermined causes are said to be **idiopathic. Communicable** diseases are those that can be transmitted from one person to another.

The term *etiology* refers to the theory of a disease's cause, but the actual pattern of a disease's development is called its **pathogenesis.** The common cold, for example, begins with a *latent* or "hidden" stage during which the cold virus establishes itself in the patient. No signs of the cold are yet evident. In infectious diseases, the latent stage is also called **incubation.** The cold may then manifest itself as a mild nasal drip, triggering a few sneezes. It then progresses to its full fury and continues for a few days. After the cold infection has run its course, a period of *convalescence,* or recovery, occurs. During this stage, body functions return to normal. Some chronic diseases, such as cancer, exhibit a temporary reversal that seems to be a recovery. Such reversal of a chronic disease is called a **remission.** If a remission is permanent, we say that the person is "cured."

Epidemiology is the study of the occurrence, distribution, and transmission of diseases in human populations. A disease that is native to a local region is called an **endemic** disease. If the disease spreads to many individuals at the same time, the situation is called an **epidemic. Pandemics** are epidemics that affect large geographic regions, perhaps spreading worldwide. Because of the speed and availability of modern air travel, pandemics are more common then they once were. Almost every flu season, we see a new strain of influenza virus quickly spreading from continent to continent.

Risk Factors

Other than direct causes or disease mechanisms, certain *predisposing conditions* may exist that make the development of a disease more likely to occur. Usually called **risk factors,** they often do not actually cause a disease but just put one "at risk" for developing it. Some of the major types of risk factors are listed below:

1. Genetic factors. There are several types of genetic risk factors. Sometimes, an inherited trait puts one at a greater-than-normal risk for developing a specific disease. For example, light-skinned people are more at risk for developing certain forms of skin cancer than dark-skinned people. This occurs because light-skinned people have less pigment in their skin to protect them from cancer-causing ultraviolet radiation (see Chapter 5). Membership in a certain ethnic group, or *gene pool,* involves the "risk" of inheriting a disease-causing gene that is common in that gene pool. For example, certain Africans and their descendants are at a greater-than-average risk of inheriting *sickle-cell anemia*—a deadly blood disorder.
2. Age. Biological and behavioral variations during different phases of the human life cycle put us at greater risk for developing certain diseases at certain times in our life. For example, middle ear infections are more common in infants than in adults because of the difference in ear structure at different ages.
3. Life-style. The way we live and work can put us at risk for some diseases. People whose work or personal activity puts them in direct sunlight for long periods have a greater chance of developing skin cancer, because this puts them in more frequent contact with ultraviolet radiation from the sun. Some researchers believe that the high-fat, low-fiber diet common among people in the "developed" nations increases the risk of developing cancer.
4. Stress. Physical, psychological, or emotional stress can put one at risk of developing problems such as chronic high blood pressure (hypertension), peptic ulcers, and headaches. Conditions caused by psychological factors are sometimes called *psychogenic* (mind-caused) disorders. Chapter 21 discusses the concept of stress and its impact on health.
5. Environmental factors. Although environmental factors such as climate and pollution can actually cause injury or disease, some environmental situations simply put us at greater risk for getting certain diseases. For example, because some parasites survive only in tropical environments, we are not at risk if we live in a temperate climate.
6. Preexisting conditions. A preexisting disease, such as an infection, can adversely affect our capacity to defend ourselves against further attack. Thus the *primary* (preexisting) condition can put a person at risk of developing a *secondary* condition. For example, blisters from a preexisting burn may break open and thus increase the risk of a bacterial infection of the skin.

Risk factors can combine, increasing a person's chances of developing a specific disease even more. For example, a light-skinned person can add to the genetic risk of developing skin cancer by spending a large amount of time in the sun without skin protection—a life-style risk. Many of these categories of risk factors overlap. For example, stress can be a component of life-style—or it could be considered a preexisting condition. Sometimes a high-risk group is identified by epidemiologists but the exact risk mechanism is uncertain. For example, high incidence of heart disease in a small ethnic group may point to a genetic risk factor but could also result from some aspect of a shared life-style.

CHAPTER SUMMARY

ANATOMY AND PHYSIOLOGY

A. Anatomy and physiology are branches of biology concerned with the form and functions of the body
B. Anatomy—science of the structure of an organism and the relationship of its parts
  1. Gross anatomy—study of the body and its parts using only the naked eye
  2. Microscopic anatomy—study of body parts using a microscope
    a. Cytology—study of cells
    b. Histology—study of tissues
  3. Developmental anatomy—study of human growth and development
  4. Pathological anatomy—study of diseased body structures
  5. Systemic anatomy—study of body by systems
C. Physiology—science of the functions of organisms; subdivisions named according to:
  1. Organism involved—human or plant physiology
  2. Organizational level—molecular or cellular physiology
  3. Systemic function—respiratory, neuro-, or cardiovascular physiology

CHARACTERISTICS OF LIFE

A. No single criterion adequately describes life
B. Characteristics of life considered most important in humans:
  1. Responsiveness
  2. Conductivity
  3. Growth
  4. Respiration
  5. Digestion
  6. Absorption
  7. Secretion
  8. Excretion
  9. Circulation
  10. Reproduction
C. Metabolism—sum total of all physical and chemical reactions occurring in the living body

LEVELS OF ORGANIZATION (Figure 1-2)

A. Chemical level—basis for life
  1. Organization of chemical constituents separates living from nonliving material

2. Organization of atoms, molecules, and macromolecules results in living matter—a semifluid matrix-type material called cytoplasm
B. Organelle level
   1. Chemical constituents organized to form organelles that perform individual functions
   2. It is the sum property of these organelles that allows the cell to live
   3. More than two dozen organelles have been identified, including:
      a. Mitochondria
      b. Golgi apparatus
      c. Endoplasmic reticulum
C. Cellular level
   1. Cells—smallest and most numerous units that possess and exhibit characteristics of life
   2. Cell—nucleus surrounded by cytoplasm within a limiting membrane
   3. Cells differentiate to perform unique functions
D. Tissue level
   1. Tissue—an organization of similar cells specialized to perform a certain function
   2. Tissue cells surrounded by nonliving matrix
   3. Four major tissue types
      a. Epithelial tissue
      b. Connective tissue
      c. Muscle tissue
      d. Nervous tissue
E. Organ level
   1. Organ—organization of several different kinds of tissues to perform a special function
   2. Organs represent discrete and functionally complex operational units
   3. Each organ has a unique size, shape, appearance, and placement in the body
F. System level
   1. Systems—most complex organizational units of the body
   2. System level involves varying numbers and kinds of organs arranged to perform complex functions:
      a. Outer protection
      b. Support and movement
      c. Communication, control, and integration
      d. Transportation and defense
      e. Processing, regulation, and maintenance
      f. Reproduction and development
   3. Outer protection (Figure 1-3)
      a. Skin—includes layers of skin proper and appendages:
         (1) Hair
         (2) Nails
         (3) Specialized glands
      b. Crucial to survival; primary function is protection
      c. Also serves to regulate body temperature, synthesize important chemicals and hormones, and function as a sense organ
   4. Support and movement (Figure 1-4)
      a. Skeletal system
         (1) Consists of bones and related tissues such as cartilage and ligaments that provide rigid framework for support and protection
         (2) Articulations, or joints, make movement possible
         (3) Bones serve as reservoirs for mineral storage and function in blood cell formation, or hemopoiesis
      b. Muscular system
         (1) Individual muscles of the body are the organs of the system

(2) Muscle types
   (a) skeletal (voluntary muscle)
   (b) smooth (involuntary muscle)
   (c) cardiac (heart muscle)
(3) Function is to produce movement, maintain posture, and generate heat
5. Communication, control, and integration (Figure 1-5)
   a. Nervous system
      (1) Composed of brain, spinal cord, and nerves
      (2) Primary functions
         (a) communication
         (b) control
         (c) integration
      (3) Functions occur rapidly and are generally of short duration
      (4) Functions are accomplished using the nerve impulse
      (5) System generates and interprets nerve impulses
   b. Endocrine system
      (1) Composed of ductless glands that secrete chemical substances, called hormones, directly into the blood
      (2) Primary functions
         (a) communication
         (b) control
         (c) integration
      (3) Functions occur slowly and are generally of long duration
      (4) Hormones act on specific, or target, organs
      (5) Hormones regulate:
         (a) metabolism
         (b) growth
         (c) reproduction (and many other body functions)
      (6) Examples of endocrine glands include pituitary, hypothalamus, thyroid, pancreas, thymus, and gonads
6. Transportation and defense (Figure 1-6)
   a. Circulatory system
      (1) Consists of the heart and a closed system of vessels called arteries, veins, and capillaries
      (2) Primary function is transportation
   b. Lymphatic system and immunity
      (1) Composed of lymph, lymphatic vessels, lymph nodes, thymus, and spleen
      (2) Functions:
         (a) movement of fat-related nutrients, fluid, and large molecules from tissue spaces back to blood
         (b) immune response confers protection and resistance to disease via specialized cells and molecules
7. Processing, regulation, and maintenance (Figure 1-7)
   a. Respiratory system
      (1) Respiratory organs include nose, pharynx, larynx, trachea, bronchi, and lungs
      (2) Organs permit movement of air into and out of lungs and the exchange of respiratory gases between air in lungs and the blood
   b. Digestive system
      (1) Composed of mouth, pharynx, esophagus, stomach, small and large intestine, rectum, and anal canal
      (2) Accessory organs include teeth, tongue, salivary glands, liver, gallbladder, and pancreas
      (3) Ingested food in digested, nutrients are absorbed, and undigested residue is eliminated as feces

c. Urinary system
   (1) Urinary organs include:
       (a) Kidneys
       (b) Ureters
       (c) Bladder
       (d) Urethra
   (2) Kidneys "clean" or clear blood of wastes and form urine
   (3) Ureters transfer urine to urinary bladder for storage
   (4) Urine passes to exterior of body through urethra
8. Reproduction and development (Figure 1-8)
   a. Normal function ensures survival of species and the development of secondary sexual characteristics
   b. Male components
      (1) Gonads, or testes
      (2) Genital ducts
          (a) Vas deferens
          (b) Urethra
      (3) Accessory glands—contribute secretions important to reproductive function
      (4) Supporting structures
          (a) genitalia (penis and scrotum)
   c. Functions—male
      (1) Production of sperm
      (2) Maturation of sperm and transfer to exterior
   d. Female components
      (1) Gonads—ovaries
      (2) Accessory organs
          (a) Uterus
          (b) Uterine (Fallopian) tubes, or oviducts
          (c) Vagina
      (3) Supporting structures
          (a) Breasts
   e. Functions—female
      (1) Production of ova
      (2) Reception of male sex cells and promotion of fertilization
      (3) Transfer of fertilized ovum to uterus
      (4) Development of fertilized ovum
      (5) Allow birth of young
      (6) Provide for nourishment of young after birth
G. Organism level
   1. The living human organism is greater than the sum of its parts
   2. All the components interact to allow the human to survive and flourish

INTERACTION OF STRUCTURE AND FUNCTION

A. Complementarity of structure and function is an important and unifying concept in the study of anatomy and physiology
B. Anatomical structures often seem "designed" to perform specific functions because of their unique size, shape, form, or body location
C. Understanding the interaction of structure and function assists in integration of otherwise isolated factual information

BODY TYPE (FIGURE 1-9)

A. Endomorphy
   1. Persons are fat with smooth contours
   2. Heavy torso with protruding abdomen and smaller chest
   3. Neck short and head large and spherical
   4. Limbs relatively short with tapering and rounding of thighs and upper arms
B. Mesomorphy
   1. Persons are heavily muscled with large prominent bones
   2. Distal segments of arms and legs are prominent and massive
   3. Well-defined shoulders
   4. Chest predominates over abdominal segment
   5. Waist is low
C. Ectomorphy
   1. Persons have predominance of height over fat or muscle
   2. Tendency to be tall and thin
   3. Generally short trunk, long limbs, and poorly developed musculature
   4. Distal segments of arms and legs relatively long and thin
   5. Neck long and slender; face small
   6. Fragile facial features and shoulder droop are common

HOMEOSTASIS (FIGURE 1-10)

A. Term homeostasis coined by American physiologist, Walter B. Cannon
B. Homeostasis is the term used to describe the relatively constant states maintained by the body—internal environment around body cells remains constant
C. Body adjusts important variables from a normal "set point" in an acceptable or normal range
D. Examples of homeostasis
   1. Temperature regulation
   2. Regulation of blood carbon dioxide level
   3. Regulation of blood glucose level

HOMEOSTATIC CONTROL MECHANISMS
(FIGURE 1-11)

A. Devices for maintaining or restoring homeostasis by self-regulation via feedback control loops
B. Basic components of control mechanisms
   1. Sensor mechanism—specific sensors detect and react to any changes from normal
   2. Integrating or control center—information is analyzed, integrated, and then, if needed, a specific action is initiated
   3. Effector mechanism—effectors directly influence controlled physiological variables
C. Negative feedback control systems
   1. Are inhibitory
   2. Stabilize physiological variables
   3. Produce an action that is opposite to the change that activated the system
   4. Are responsible for maintaining homeostasis
   5. Much more common than positive feedback control systems
D. Positive feedback control systems
   1. Are stimulatory

2. Amplify or reinforce the change that is occurring
3. Tend to produce destabilizing effects and disrupt homeostasis
4. Bring specific body functions to swift completion

ANATOMICAL POSITION (FIGURE 1-12)

A. Reference position
B. Body erect with arms at sides and palms forward
C. Head and feet pointing forward
D. Bilateral symmetry is term meaning that right and left sides of body are mirror images
   1. Bilateral symmetry confers balanced proportions
   2. Remarkable correspondence of size and shape between body parts on opposite sides of the body

BODY CAVITIES (FIGURE 1-13; TABLE 1-1)

A. Ventral body cavity
   1. Thoracic cavity
      a. Right and left pleural cavities
      b. Mediastinum
   2. Abdominopelvic cavity
      a. Abdominal cavity
      b. Pelvic cavity
B. Dorsal body cavity
   1. Cranial cavity
   2. Spinal cavity

BODY REGIONS (FIGURE 1-14; TABLE 1-2)

A. Axial subdivision
   1. Head and shoulders
   2. Neck
   3. Torso, or trunk, and its subdivisions
B. Appendicular subdivision
   1. Upper extremity and subdivisions
   2. Lower extremity and subdivisions
C. Abdominal Regions (Figure 1-15)
   1. Right hypochondriac region
   2. Epigastric region
   3. Left hypochondriac region
   4. Right lumbar region
   5. Umbilical region
   6. Left lumbar region
   7. Right iliac (inguinal) region
   8. Hypogastric region
   9. Left iliac (inguinal) region
D. Abdominopelvic Quadrants (Figure 1-16)
   1. Right upper quadrant
   2. Left upper quadrant
   3. Right lower quadrant
   4. Left lower quadrant

TERMS USED IN DESCRIBING BODY STRUCTURE

A. Directional terms (Figure 1-17)
   1. Superior
   2. Inferior
   3. Anterior (ventral)
   4. Posterior (dorsal)
   5. Medial
   6. Lateral
   7. Proximal
   8. Distal
   9. Superficial
   10. Deep

BODY PLANES AND SECTIONS (FIGURE 1-17)

A. Planes are lines of orientation along which cuts or sections can be made to divide the body, or a body part, into smaller pieces
B. There are three major planes, which lie at right angles to each other:
   1. Sagittal—plane runs front to back so that sections through this plane divide body (or body part) into right and left sides
      a. If section divides body (or part) into symmetrical right and left *halves*, the plane is called midsagittal or median sagittal
   2. Frontal (coronal)—plane runs lengthwise (side to side) and divides body (or part) into anterior and posterior portions
   3. Transverse (horizontal)—plane is a "crosswise" plane—it divides body (or part) into upper and lower parts

CYCLE OF LIFE: LIFE SPAN CONSIDERATIONS

A. Structure and function of body undergo changes over the early years (developmental processes) and late years (aging processes)
B. Infancy and old age are periods of time when the body functions least well
C. Young adulthood is period of greatest homeostatic efficiency
D. Atrophy—term to describe the wasting effects of advancing age

MECHANISMS OF DISEASE: SOME GENERAL CONSIDERATIONS

A. Pathophysiology is the organized study of physiological processes associated with disease
B. Categories of basic mechanisms of disease:
   1. Genetic mechanisms
   2. Pathogenic organisms, including:
      a. Viruses
      b. Bacteria
      c. Fungi
      d. Protozoa
      e. Pathogenic animals
   3. Tumors and cancer
   4. Physical and chemical agents
   5. Malnutrition
   6. Autoimmunity
   7. Inflammation
   8. Degeneration
C. Risk factors—predisposing factors that make the development of disease more likely to occur
   1. Genetic factors
   2. Age
   3. Life-style
   4. Stress
   5. Environmental factors
   6. Preexisting conditions

REVIEW QUESTIONS

1. Define the terms *anatomy* and *physiology*.
2. Identify the "characteristics of life."
3. Discuss the concept of organization in living things.
4. List and briefly describe the levels of organization that relate the structure of an organism to its function. Give examples characteristic of each level.
5. Define *exercise physiology*.
6. Give examples of each system level of organization in the body and briefly discuss the function of each.
7. What does the term *somatotype* mean?
8. List the three major somatotype categories and briefly describe the generalized characteristics of each.
9. Do characteristics of body build, or physique, predispose an individual to injury or disease? If so, give an example.
10. What is meant by the term *anatomical position*? How do the specific anatomical terms of position or direction relate this body orientation?
11. What is bilateral symmetry? What terms are used to identify placement of one body part with respect to another on the same or opposite sides of the body?
12. Identify the two major subdivisions of the body as a whole and list the primary anatomical areas or components of each.
13. List by name the nine abdominal regions, the four abdominal quadrants, and identify the major organs located in each.
14. Define briefly each of the following terms: *anterior, distal, sagittal plane, medial, dorsal, coronal plane, organ, parietal peritoneum, superior, tissue*.
15. Name the two major body cavities and the subdivisions of each.
16. Locate the mediastinum.
17. What does the term *homeostasis* mean? Illustrate some generalizations about body function using homeostatic mechanisms as examples.
18. Define homeostatic control mechanisms and feedback control loops.
19. Identify the three basic components of a control loop.
20. Distinguish between positive and negative feedback control loops.
21. Discuss in general terms the principle of complementarity of structure and function.

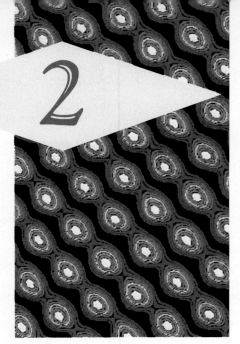

Individual silicon atoms

2 The Chemical Basis of Life

CHAPTER OUTLINE

OBJECTIVES

After you have completed this chapter, you should be able to:

1. Explain why an understanding of basic chemistry is important in the study of life processes.
2. Explain the relationship between elements, compounds, atoms, and molecules.
3. List the major elements and major mineral elements found in cytoplasm.
4. Discuss atomic structure and explain how an atom's electron shells influence its ability to enter into chemical reactions.
5. Compare and contrast the three major types of chemical bonds.
6. List and describe the three basic types of chemical reactions that occur in living material.
7. Discuss the properties that make water such an important inorganic molecule in living organisms.
8. Discuss the concept of pH and its relationship to acids, bases, and salts in the body.
9. List the four major groups of organic substances in the body and give examples and functions of specific types in each group.
10. Define the term *bioenergy* and identify the most important of the bioenergy molecules.
11. Define or explain the following terms or phrases: *atomic number, octet rule, isotope, polymer, electrolyte, polarity, nucleotide, base pair, high energy bond.*

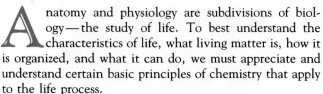

Anatomy and physiology are subdivisions of biology—the study of life. To best understand the characteristics of life, what living matter is, how it is organized, and what it can do, we must appreciate and understand certain basic principles of chemistry that apply to the life process.

Life itself depends on proper levels and proportions of chemical substances in the cytoplasm of cells. The various structural levels of organization described in Chapter 1 are based, ultimately, on the existence and interrelationships of atoms and molecules. Chemistry, like biology, is a very broad scientific discipline. It deals with structure, arrangement and composition of substances, and the reactions they undergo. Just as biology may be subdivided into many subdisciplines or branches, like anatomy and physiology, chemistry may also be divided into specialized areas. **Biochemistry** is that specialized area of chemistry that deals with living organisms and life processes. It deals directly with the chemical composition of living matter and the processes that underlie such life activities as growth, muscle contraction, and transmission of nervous impulses.

Modern biochemistry is in reality many disciplines. It is closely related to the other life sciences and to modern medicine. Biochemists use many different chemical, physical, biological, nutritional, and immunological techniques to probe life processes at every level of organization. An understanding of homeostatic processes and control mechanisms is in many cases dependent on a knowledge of basic chemistry and on selected facts and concepts in the specialized area of biochemistry.

BASIC CHEMISTRY

Elements and Compounds

Chemists use the term *matter* to describe in a general sense all of the materials or substances around us. Anything that has mass and occupies space is matter.

Substances are either **elements** or **compounds.** An element is said to be "pure." That is, it cannot be broken down or decomposed into two or more different substances. Pure oxygen is a good example of an element. In most living material, elements do not exist alone in their pure state. Instead, two or more elements are joined to form chemical combinations called compounds. Compounds can be broken down or decomposed into the elements that are contained within them. Water is a compound (H_2O). It can be broken down into atoms of hydrogen and atoms of oxygen in a 2:1 ratio.

Other examples of elements include phosphorus, cop-

per, and nitrogen. For convenience in writing chemical formulas and in other types of notation, chemists assign to each element a symbol, usually the first letter or two of the English or Latin name of the element: P, phosphorus; Cu, copper (Latin *cuprum*); N, nitrogen. Note in Table 2-1 that 26 elements are listed as being present in the human body. Although all are important, 11 are called *major elements*. Four of these major elements—carbon, oxygen, hydrogen, and nitrogen—make up about 96% of the material in the human body. The 15 remaining elements are present in amounts less than 0.1% of body weight and are called *trace elements*. It is important to note, however, that the unique "aliveness" of a living organism does not depend on a single element or mixture of elements but on the complexity, organization, and interrelationships of all elements required for life.

QUICK CHECK

✔ 1. *What is biochemistry?*
✔ 2. *What is the difference between an element and a compound?*
✔ 3. *What elements make up 96% of the material in the human body?*

TABLE 2-1   Elements in the human body

Element	Symbol	Human Body Weight (%)	Importance or Function
MAJOR ELEMENTS			
Oxygen	O	65.0	Necessary for cellular respiration; component of water
Carbon	C	18.5	Backbone of organic molecules
Hydrogen	H	9.5	Component of water and most organic molecules; necessary for energy transfer and respiration
Nitrogen	N	3.3	Component of all proteins and nucleic acids
Calcium	Ca	1.5	Component of bones and teeth; triggers muscle contraction
Phosphorus	P	1.0	Principal component in backbone of nucleic acids; important in energy transfer
Potassium	K	0.4	Principal positive ion within cells; important in nerve function
Sulfur	S	0.3	Component of most proteins
Sodium	Na	0.2	Important positive ion surrounding cells; important in nerve function
Chlorine	Cl	0.2	Important negative ion surrounding cells
Magnesium	Mg	0.1	Component of many energy-transferring enzymes
TRACE ELEMENTS			
Silicon	Si	<0.1	—
Aluminum	Al	<0.1	—
Iron	Fe	<0.1	Critical component of hemoglobin in the blood
Manganese	Mn	<0.1	—
Fluorine	F	<0.1	—
Vanadium	V	<0.1	—
Chromium	Cr	<0.1	—
Copper	Cu	<0.1	Key component of many enzymes
Boron	B	<0.1	—
Cobalt	Co	<0.1	—
Zinc	Zn	<0.1	Key component of some enzymes
Selenium	Se	<0.1	—
Molybdenum	Mo	<0.1	Key component of many enzymes
Tin	Sn	<0.1	—
Iodine	I	<0.1	Component of thyroid hormone

Atoms

The most important of all chemical theories was advanced in 1805 by the English chemist John Dalton. He proposed the concept that matter is composed of **atoms** (from the Greek *atomos,* "indivisible"). His idea was revolutionary and yet simple—that all matter, regardless of the form it may assume (liquid, gas, or solid), is composed of units he called atoms.

Dalton conceived of atoms as solid, indivisible particles, and for about 100 years this was believed to be true. We now know that atoms are divisible into even smaller or *subatomic* particles, some of which exist in a "cloud" surrounding a dense central core called a nucleus. Over 100 million atoms of even very dense and heavy substances if lined up would measure barely an inch and would consist mostly of empty space! Our knowledge about the number and nature of the subatomic particles and the central nucleus around which they move continues to grow as a result of ongoing research.

Atomic structure

Atoms contain several different kinds of smaller or subatomic particles that are found in either a central *nucleus* or its surrounding *electron cloud* or field. Figure 2-1, A shows an atomic model of carbon illustrating the most important types of subatomic particles:

♦ Protons (p$^+$)
♦ Neutrons (n^o)
♦ Electrons (e$^-$)

Note that the carbon atom in Figure 2-1 has a central corelike *nucleus.* It is located deep inside the atom and is made up of six positively charged *protons* (+ or *p)* and six uncharged *neutrons* (n). Note also that the nucleus is surrounded by a "cloud" or "field" of six negatively charged electrons (− or *e*). Because protons are positively charged and neutrons are neutral, the nucleus of an atom bears a positive electrical charge equal to the number of protons that are present in it. Electrons move around the atom's nucleus in what can be represented as an electron "cloud" or field (Figure 2-1, *B*). The number of negatively charged electrons moving around an atom's nucleus equals the number of positively charged protons in the nucleus. The opposite charges therefore cancel or neutralize each other, and atoms are electrically neutral particles.

Atomic number and atomic weight

Elements differ in their chemical and physical properties because of differences in the number of protons in their atomic nuclei. The number of protons in an atom's nucleus, called its *atomic number,* is therefore critically important—it identifies the kind of element it is. Look again at the elements important in living organisms listed in Table 2-1. Each element is identified by its symbol and atomic number. Hydrogen, for example, has an atomic number of 1; this means that all hydrogen atoms—and only hydrogen atoms—have one proton in their nucleus. All carbon atoms, and only carbon atoms, contain six protons and have an atomic number of 6. All oxygen atoms, and only oxygen atoms, have eight protons and an atomic number of 8. In short, each element is identified by its own unique number of protons, that is, by its own unique atomic number. If two atoms contain a different number of protons, they necessarily have different atomic numbers and are different elements.

There are 92 elements that occur naturally on earth. Since each element is characterized by the number of protons in its atoms (atomic number), there are atoms that contain from 1 to 92 protons.

The term *atomic weight* refers to the mass of a single atom. It equals the number of protons plus the number of neutrons in the atom's nucleus. The weight of electrons is for practical purposes negligible. Since protons and neutrons weigh almost exactly the same amount, the equation for determining atomic weight is as follows:

$$\text{Atomic Weight} = (p + n)$$

The largest naturally occurring atom is uranium. It has an atomic weight of 238, with a nucleus containing 92 protons and 146 neutrons. In contrast, hydrogen, which has only one proton and no neutrons in its nucleus, has an atomic weight of 1.

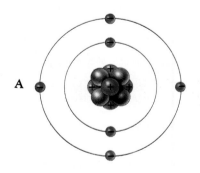

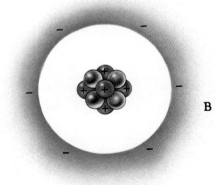

FIGURE 2-1 Models of the atom. The nucleus—protons (+) and neutrons—is at the core. Electrons inhabit outer regions called electron shells **(A)** or "clouds" **(B).** This is a carbon atom, a fact that is determined by the number of its protons. All carbon atoms (and only carbon atoms) have six protons. (Not all of the protons in the nucleus are visible in this illustration.)

Electron shells

The total number of electrons in an atom equals the number of protons in its nucleus (Figure 2-1). These electrons are known to exist in a cloudlike envelope surrounding the atom's nucleus.

The "cloud" suggests that any one electron cannot be exactly located at a specific point at any particular time. Earlier models suggested that electrons moved in regular patterns around the nucleus much like the planets in our solar system move around the sun. The so-called "Bohr model" (Figure 2-1, A) is perhaps most useful in visualizing the structure of atoms as they enter into chemical reactions. It is named after a Danish physicist, Niels Bohr, who contributed much to our understanding of atomic structure.

In this type of model the electrons are shown in shells or concentric circles, showing relative distances of the electrons from the nucleus. The electrons surrounding the atom's nucleus are seen in the Bohr model as existing in simple two-dimensional concentric rings. Each ring or shell represents a specific **energy level,** and each shell can hold only a certain maximum number of electrons. The number and arrangement of electrons orbiting in an atom's energy shells have great significance; they determine whether or not the atom is chemically active.

In chemical reactions between atoms it is the electrons in the outermost shell that participate in the formation of chemical bonds. In each shell, electrons tend to group in pairs. As a rule of thumb, an atom can be listed as chemically inert and unable to react with another atom if its outermost shell has four pairs, or eight, electrons. Such an atom is said to have a stable electron configuration. Pairing of electrons is important. If the outer shell contains single, unpaired electrons, the atom will be chemically active. Atoms with fewer or more than eight electrons in the outer shell will attempt to lose, gain, or share electrons with other atoms to achieve stability. This tendency is called the **octet rule.** This rule holds true except for atoms that are limited to a single shell that is filled by a maximum of two electrons. For example, hydrogen has but one electron in its single shell. It therefore has an incomplete shell with an unpaired electron. The result is a highly reactive tendency of hydrogen to enter into many chemical reactions. Helium, however, has two electrons in its single shell. Since this is the maximum number for this shell, no chemical activity is possible, and no naturally occurring compound containing helium exists.

The atoms shown in Figure 2-2 illustrate several of the most important facts related to electron shells. Note that even in the hydrogen atom with its very basic structure, positive and negative charges balance. However, its single energy shell contains only one electron because the hydrogen nucleus contains only one proton. As a result of the unpaired electron, hydrogen is chemically active. In contrast, the helium atom has a full outer shell and is therefore inactive, or *inert,* as is neon. With only four electrons and six electrons in the outer shells of carbon and oxygen, respectively, these elements will react chemically, since they do not satisfy the octet rule.

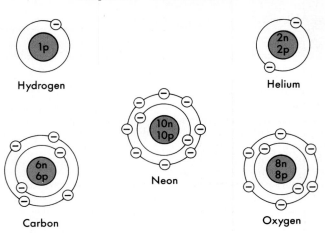

FIGURE 2-2 Electron shells of five common elements. All atoms are balanced with respect to positive and negative charges. In atoms with a single shell, two electrons are required for stability. Hydrogen with its single electron is reactive, whereas helium with its full shell is not. In atoms with more than one shell, eight electrons in the outermost shell are required for stability. Neon is stable because its outer shell has eight electrons. Oxygen and carbon with six and four electrons, respectively, in their outer shells are chemically active.

QUICK CHECK

✔ 1. List and define the three most important types of subatomic particles.
✔ 2. How are the atomic number and atomic weight of an atom defined?
✔ 3. What is an electron shell?
✔ 4. Explain what is meant by the "octet rule."

Isotopes

All atoms of the same element contain the same number of protons but do not necessarily contain the same number of neutrons. **Isotopes** of an element contain the same number of protons, but different numbers of neutrons. Isotopes have the same basic chemical properties as any other atom of the same element, and they also have the same atomic number, but because they have a different number of neutrons they differ in atomic weight. Usually a hydrogen atom has only one proton and no neutrons (atomic number, 1; atomic weight, 1). Figure 2-3 illustrates this most common type of hydrogen and two of its isotopes. Note that the isotope of hydrogen, called *deuterium,* has one proton and one neutron (atomic weight, 2). *Tritium* is the isotope of hydrogen that has one proton and two neutrons (atomic weight, 3).

The atomic nuclei of over 99% of all carbon atoms in nature have six protons and six neutrons (atomic number, 6; atomic weight, 12). An important isotope of carbon has eight neutrons instead of six. It is called *carbon 14.* Carbon 14 is an example of a special type of isotope that is unstable and undergoes nuclear breakdown—it is designated as a **radioactive isotope, or radioisotope.** (Tritium, incidentally, is also a radioisotope.) During breakdown, radioactive isotopes emit nuclear particles and radiation—a process called "decay."

Radioactivity is the emission of radiations from an atom's nucleus. Alpha particles, beta particles, and gamma rays are the three kinds of radiations. **Alpha particles** are relatively heavy particles consisting of two protons plus two neutrons. They shoot out of a radioactive atom's nucleus at a reported speed of 18,000 miles per second. **Beta particles** are electrons formed in a radioactive atom's nucleus by one of its neutrons breaking down into a proton and an electron. The proton remains behind in the nucleus, and the electron is ejected from it as a beta particle. Beta particles, since they are electrons, are much smaller than alpha particles, which consist of two protons and two neutrons. Also, beta particles travel at a much greater speed than alpha particles. **Gamma rays** are electromagnetic radiations, a form of light energy.

How does radioactivity change an atom? To find out how the emission of an alpha particle changes the nucleus of a radium atom, examine Figure A. Note that after the ejection of an alpha particle, the atom's nucleus contains fewer protons and neutrons. Basic principles about atoms, you will recall, are these: All atoms of the same element contain the same number of protons; atoms that contain different numbers of protons are therefore different elements. After an alpha particle is ejected from the nucleus of a radium atom, the atom contains 86 instead of 88 protons and has been changed into an atom of radon. Figure B shows how the emission of a beta particle changes an atom's nucleus: iodine loses a neutron and gains a proton to become xenon. Radioactivity changes the chemical identity of an atom. It transforms an atom of one element into an atom of a different element by changing the number of protons in the atom's nucleus.

Our bodies are continually exposed to low levels of radiation present in the environment. When alpha or beta particles or gamma rays score direct hits on atoms present in living cells, they ionize the atoms by knocking electrons out of their outer shells. The effect of ionization may injure, kill, or change cells. Knowledge of this fact underlies the use of radiation therapy to kill cancer cells. But radiation can also have an opposite effect. It can lead to the development of cancer cells. Abundant evidence indicates that leukemia and other types of cancer may result from exposure to very high levels of radiation or to lower levels of radiation for prolonged periods of time. For example, many more survivors of the atomic bombings of Hiroshima and Nagasaki developed leukemia than did individuals not exposed to radiation caused by the bombs.

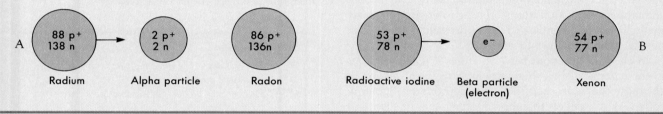

A   Radium (88 p+ 138 n) → Alpha particle (2 p+ 2 n)   Radon (86 p+ 136n)   Radioactive iodine (53 p+ 78 n) → Beta particle (electron) (e−)   Xenon (54 p+ 77 n)   B

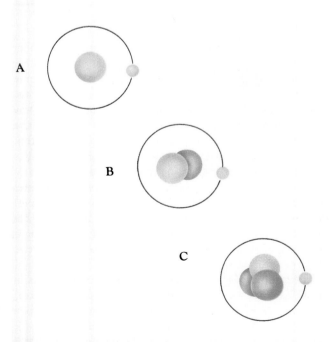

A

B

C

FIGURE 2-3   Structure of hydrogen and two of its isotopes. A, The most common form of hydrogen. **B,** An isotope of hydrogen called *deuterium.* **C,** The hydrogen isotope *tritium.* Note that isotopes of an element differ only in the number of its neutrons.

Interactions Between Atoms —Chemical Bonds

Interactions between two or more atoms occur largely as a result of activity between electrons in their outermost shell. The result, called a **chemical reaction,** most often involves unpaired electrons.

Ultimately, in atoms with fewer or more than eight electrons in the outer shell, reactions will occur that result in the loss, gain, or sharing of one atom's unpaired electrons with those of another atom to satisfy the octet rule for both atoms. The result of such reactions between atoms is the formation of a **molecule.** For example, two atoms of oxygen can combine with one carbon atom to form molecular carbon dioxide, or CO_2. If atoms of more than one element combine, the result, as defined earlier, is a compound. In other words, oxygen exists as a molecule (O_2) and is an element. Water exists as a molecule (H_2O) and is a compound. Reactions that hold atoms together do so by the formation of **chemical bonds.** There are two types of chemical bonds that unite atoms into molecules: ionic, or electrovalent, bonds and covalent bonds.

Ionic, or electrovalent, bonds

A chemical bond formed by the transfer of electrons from one atom to another is called an **ionic,** or **electrova-**

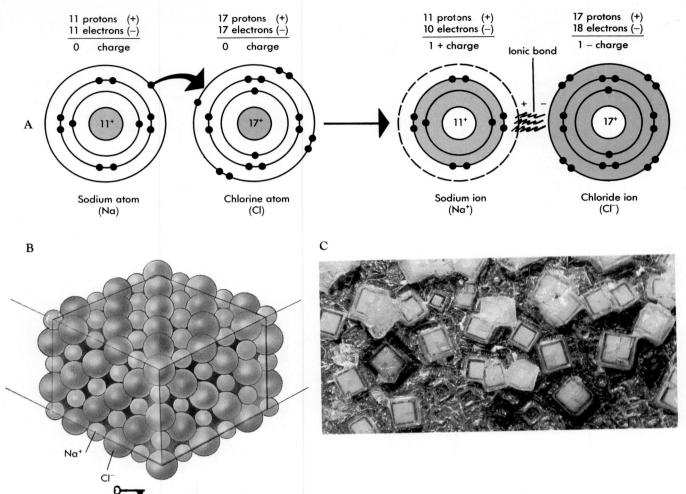

11 protons (+)
11 electrons (−)
0   charge

17 protons (+)
17 electrons (−)
0   charge

11 protons (+)
10 electrons (−)
1 + charge

Ionic bond

17 protons (+)
18 electrons (−)
1 − charge

A

Sodium atom
(Na)

Chlorine atom
(Cl)

Sodium ion
(Na⁺)

Chloride ion
(Cl⁻)

B

C

Na⁺

Cl⁻

FIGURE 2-4   Example of an ionic bond. A, Steps involved in forming an ionic bond between atoms of sodium and chlorine (see text for explanation). **B,** Molecules of sodium chloride (table salt) in typical cube shape formation. **C,** Photomicrograph showing crystals of sodium chloride.

lent, bond. Such a bond occurs as a result of the attraction between atoms that have become electrically charged by the loss or gain of electrons. Such atoms (Figure 2-4) are called **ions.** It is important to remember that ions can be positively or negatively charged and that ions with opposite charges are attracted to each other.

Note in Figure 2-4 A, that in the outer shell of the sodium atom there is a single unpaired electron. If this electron were "lost" the outer ring would be stable because it would have a full outer octet (four pairs of electrons). The loss of the electron would result in the formation of a sodium ion (Na^+) with a positive charge. This is because there is now one more proton (+) than electron (−). The chlorine atom, in contrast, has one unpaired electron plus three paired electrons, or a total of seven electrons, in its outer shell. By the addition of another electron, chlorine would satisfy the octet rule—its outer energy shell would have a full complement of four paired electrons. Addition of another electron would result in the formation of a negatively charged chloride ion (Cl^-). A chemical reaction is set to occur. Sodium transfers or donates its one unpaired electron to chlorine and becomes a positively

charged sodium ion (Na^+). Chlorine accepts the electron from sodium and pairs it with its one unpaired electron, thereby filling its outer shell with the maximum of four electron pairs and becomes a negatively charged chlorine (Cl^-) ion. The positively charged sodium ion (Na^+) is attracted to the negatively charged chloride ion (Cl^-), and the formation of NaCl, ordinary table salt, results. This chemical reaction illustrates ionic or electrovalent bonding. The electron transfer changed the two atoms of the elements sodium and chlorine into ions. The ionic bond is simply the strong electrostatic force that binds the positively and negatively charged ions together.

Covalent bonds

Just as atoms can be held together by ionic bonds formed when atoms gain or lose electrons, atoms can also be bonded together by *sharing* electrons. A chemical bond formed by the sharing of one or more pairs of electrons between the outer shells of two atoms is called a **covalent bond.** This type of chemical bonding is of great significance in physiology. The major elements of the body (carbon, oxygen, hydrogen, and nitrogen) almost always share

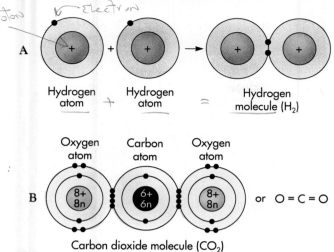

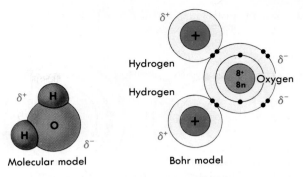

FIGURE 2-6   Water is a polar molecule. Diagram shows the polar nature of water. The two hydrogen atoms are nearer one end of the molecule, giving that end a partial positive charge. The opposite end of the molecule has a partial negative charge.

FIGURE 2-5   Types of covalent bonds. A, A single covalent bond forms by the sharing of one electron pair between two atoms of hydrogen, resulting in a molecule of hydrogen gas. **B,** A double covalent bond (double bond) forms by the sharing of *two* pairs of electrons between two atoms. In this case, two double bonds form—one between carbon and *each* of the two oxygen atoms.

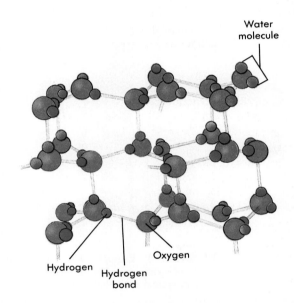

FIGURE 2-7   Hydrogen bonds between water molecules. Hydrogen bonds serve to weakly attach the negative (*oxygen*) side of one water molecule with the positive (*hydrogen*) side of an adjacent water molecule.

electrons to form covalent bonds. For example, if two atoms of hydrogen are bound together by the sharing of one electron pair, a *single* covalent bond is said to exist, and a molecule of hydrogen gas results (Figure 2-5, A). Covalent bonds that bind atoms together by sharing *two* pairs of electrons are called *double* covalent bonds, or simply, **double bonds** (Figure 2-5, B). The example shown illustrates two atoms of oxygen, each sharing two electrons with a carbon atom to acquire a complete outer shell of eight electrons and thus satisfy the octet rule. The result is a molecule of carbon dioxide.

Hydrogen bonds

In addition to ionic and covalent bonds, which actually form molecules, a third type of chemical bond, called a **hydrogen bond,** can exist within or between biologically important molecules. Hydrogen bonds are much weaker than ionic or covalent bonds because they require less energy to break. Instead of forming as a result of transfer or sharing of electrons, hydrogen bonds result from unequal charge distribution on a molecule. Such molecules, like water, are said to be **polar.** Note in Figure 2-6 that although an atom of water is electrically neutral (the number of negative charges equals the number of positive charges), it has a partial positive charge (the hydrogen side) and a partial negative charge (the oxygen side). Hydrogen bonds serve to weakly attach the negative (oxygen) side of one water molecule with the positive (hydrogen) side of an adjacent water molecule. Figure 2-7 illustrates hydrogen bonding between water molecules. This ability of water molecules to form hydrogen bonds with each other accounts for many of the unique properties of water, which make it an ideal medium for the chemistry of life. Hydrogen

bonds are also important in maintaining the three-dimensional structure of proteins and nucleic acids, also described later in the chapter.

Chemical Reactions

Chemical reactions involve interactions between atoms and molecules that, in turn, involve the formation or breaking of chemical bonds. Three basic types of chemical reactions that you will learn to recognize as you study physiology are:

1. Synthesis reactions
2. Decomposition reactions
3. Exchange reactions

To the chemist, reactions can be symbolized by variations on a simple formula. In **synthesis** (from the Greek *syn,* "together," and *thesis,* "putting") **reactions,** two or more substances called *reactants* combine to form a different

more complex substance called a *product.* The process can be summarized by the formula:

$$A + B \xrightarrow{\text{Energy}} AB$$
$$\text{(Reactants)} \qquad\qquad \text{(Product)}$$

Synthesis reactions result in the formation of new bonds, and energy is required for the reaction to occur and the new product to form. Many such reactions occur in the body. Every cell, for example, combines amino acid molecules as reactants to form complex protein compounds as products. The ability of the body to synthesize new tissue in wound repair is a good example of this type of reaction.

Decomposition reactions result in breakdown of a complex substance into two or more simpler substances. In this type of reaction, chemical bonds are broken and energy is released. Energy can be released in the form of heat, or it can be captured for storage and future use. Decomposition reactions can be summarized by the formula:

$$AB \rightarrow A + B + \text{Energy}$$

Decomposition reactions occur when a complex nutrient is broken down in a cell to release energy for other cellular functions. The products of such a reaction are ultimately waste products. Decomposition and synthesis are opposites. Synthesis builds up; decomposition breaks down. Synthesis forms chemical bonds; decomposition breaks chemical bonds. Decomposition and synthesis reactions are often coupled to one another in such a way that the energy released by a decomposition reaction can be used to drive a synthesis reaction.

The nature of **exchange reactions** permits two different reactants to exchange components and, as a result, form two new products. An exchange reaction is often symbolized by the following:

$$AB + CD \rightarrow AD + CB$$

Exchange reactions break down or decompose two compounds and, in exchange, synthesize two new compounds. Certain exchange reactions take place in the blood. One example is the reaction between lactic acid and sodium bicarbonate. The decomposition of both substances is exchanged for the synthesis of sodium lactate and carbonic acid. These changes can be seen more easily in an equation.

$$\text{H} \cdot \text{Lactate} + \text{NaHCO}_3 \rightarrow \text{Na} \cdot \text{Lactate} + \text{H} \cdot \text{HCO}_3$$

The formula "H · lactate" represents lactic acid; "NaHCO$_3$" is the formula for sodium bicarbonate; "Na · lactate" represents sodium lactate; and "H · HCO$_3$" represents carbonic acid.

Reversible reactions, as the name suggests, proceed in both directions. A great many synthesis, decomposition, or exchange reactions are reversible, and a number of them are cited in later chapters of the book. An arrow pointing in both directions is used to denote a reversible reaction:

$$A + B \leftrightarrow AB$$

QUICK CHECK

✔ 1. *List the three types of chemical bonds and explain how they are formed.*

✔ 2. *Diagram the three basic types of chemical reactions.*

INORGANIC MOLECULES

In living organisms, there are two kinds of compounds: *organic* and *inorganic.* Organic compounds are generally defined as compounds composed of molecules that contain carbon–carbon (C–C) covalent bonds or carbon–hydrogen (C–H) covalent bonds—or both kinds of bonds. Few inorganic compounds have carbon atoms in them, and none have C–C or C–H bonds. Organic molecules are generally larger and more complex than inorganic molecules. The human body has both kinds of compounds because both are equally important to the chemistry of life. We will discuss the chemistry of inorganic compounds first, then move on to some of the more important types of organic compounds.

Water

Water has been called the "cradle of life" because all living organisms require water to survive. Each body cell is bathed in fluid, and it is only in this precisely regulated and homeostatically controlled environment that cells can function. In addition to being surrounded by water, the basic substance of each cell, *cytoplasm,* is itself largely water. Water is certainly the body's most abundant and important compound. It makes up almost 70% of body weight and serves a host of vital functions. It is because of its pervasive importance in all living organisms that an understanding of the basics of water chemistry is so important. In a very real sense, "water chemistry" forms the basis for the chemistry of life.

Properties of water

The chemist views water as a simple compound. It has an atomic structure that results from the combination of two covalent bonds between a single oxygen atom and two hydrogen atoms.

Recall that water molecules are polar (see Figure 2-6); they have a positively charged end and a negatively charged end. This simple chemical property, called *polarity,* allows water to act as a very effective *solvent.* The proper functioning of a cell requires the presence of many chemical substances. Many of these compounds are quite large and must be broken into smaller and more reactive particles (ions) for reactions to occur. Because of its polar nature, water has a tendency to ionize substances in solution (Figure 2-8). The fact that so many substances dissolve in water is of the utmost importance in the life process.

The critical role that water plays as a solvent permits the *transportation* of many essential materials within the body. By dissolving oxygen and food substances in the blood, for instance, water enables these materials to enter and leave the blood capillaries in the lungs and digestive organs and

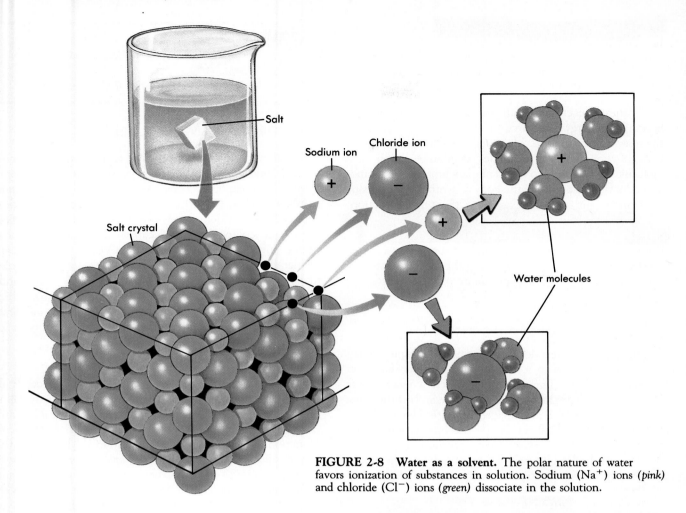

FIGURE 2-8   Water as a solvent. The polar nature of water favors ionization of substances in solution. Sodium (Na⁺) ions (*pink*) and chloride (Cl⁻) ions (*green*) dissociate in the solution.

eventually enter cells in every area of the body. In turn, waste products are transported from where they are produced to excretory organs for elimination from the body.

Another important function of water stems from the fact that water both absorbs and gives up heat slowly. These properties of water enable it to maintain a relatively constant temperature. This allows the body, which has a large water content, to resist sudden changes in temperature. Chemists describe this property by saying that water has a high *specific heat,* that is, water can lose and gain large amounts of heat with little change in temperature. As a result, excess body heat produced by contraction of muscles during exercise, for example, can be transported by the blood to the body surface and dissipated into the environment with little actual change in core temperature.

Both chemists and biologists recognize water's high *heat of vaporization* as another important physical quality. This characteristic requires absorption of significant amounts of heat to change water from a liquid to a gas. The energy is required to break the many hydrogen bonds that hold adjacent water molecules together in the liquid state. Thus, the body can dissipate excess heat and maintain a normal temperature by evaporation of water (sweat) from the skin surface whenever excess heat is being produced.

An understanding and appreciation of the importance of water in the life process is critical. Water does more than act as a solvent, produce ionization, and facilitate chemical reactions. It has essential chemical roles of its own in addition to the many important physical qualities it brings to body function (see Table 2-2). It plays a key role in such processes as cell permeability, active transport of materials, secretion, and membrane potential, to name a few. The requirement of the body to maintain homeostasis will be stressed in each chapter of the text. As you progress from system to system in the chapters that follow, the importance of water in almost every regulatory and control mechanism studied will be a constant that should not go unnoticed.

Oxygen and Carbon Dioxide

Oxygen (O_2) and carbon dioxide (CO_2) are important inorganic substances that are closely related to cellular respiration. Molecular oxygen in the body is present as two oxygen atoms joined by a double covalent bond. Oxygen is required to complete the decomposition reactions required for the release of energy from nutrients burned by the cell. Carbon dioxide is considered one of a group of very simple carbon-containing inorganic compounds. It is an important exception to the "rule of thumb" that inorganic substances

TABLE 2-2   Properties of water

Property	Description	Example of Benefit to the Body
Strong Polarity	Polar water molecules attract ions and other polar compounds, causing them to dissociate	Many kinds of molecules can dissolve in cells, permitting a variety of chemical reactions and allowing many substances to be transported
High Specific Heat	Hydrogen bonds absorb heat when they break and release heat when they form, minimizing temperature changes	Body temperature stays relatively constant
High Heat of Vaporization	Many hydrogen bonds must be broken for water to evaporate	Evaporation of water in perspiration cools the body
Cohesion	Hydrogen bonds hold molecules of water together	Water works as lubricant or cushion to protect against damage from friction or trauma

do not contain carbon. Like oxygen, CO_2 is involved in cellular respiration. It is produced as a waste product during the breakdown of complex nutrients and also serves an important role in maintaining the appropriate acid-base balance in the body.

Electrolytes

Other inorganic substances include acids, bases, and salts. These substances belong to a large group of compounds called **electrolytes** (e-LEK-tro-lites). Electrolytes are substances that break up, or *dissociate*, in solution to form charged particles, or ions. Ions with a positive charge are called *cations*, and those with a negative charge are called *anions*. Figure 2-8 shows the way in which water molecules work to dissociate a typical electrolyte, sodium chloride (NaCl), into Na^+ cations and Cl^- anions.

Acids and bases

Acids and bases are common and very important chemical substances in the body. Early chemists categorized acids and bases using such characteristics as taste or the ability to change the color of certain dyes. Acids, for example, taste sour and bases taste bitter. The dye *litmus* will turn blue in the presence of bases and red when exposed to an acid. These and other observations illustrate a fundamental point, namely, that acids and bases are chemical opposites. Although acids and bases both dissociate in solution they release different types of ions. The unique chemical properties of acids and bases when they are in solution is perhaps the best way to differentiate them.

Acids. By definition, an **acid** is any substance that will release a hydrogen ion (H^+) when in solution. A hydrogen ion is simply a bare proton—the nucleus of a hydrogen atom. Therefore acids are frequently called "proton donors." It is the concentration of hydrogen ions that accounts for the chemical properties of acids. The level of "acidity" of a solution depends on the number of hydrogen ions a particular acid will release.

One point should be understood about water. Water molecules dissociate continually in a reversible reaction to form hydrogen ions (H^+) and hydroxide ions (OH^-):

$$H_2O \rightleftharpoons H^+ + OH^-$$

Recall from our discussion of ionic bonds (pp. 40 to 41) that having a single unpaired electron in the outer shell makes an atom unstable and that losing that electron results in a more stable structure. This is precisely the reason why dissociation of water occurs. In pure water, the balance between these two ions is equal. However, when an acid such as hydrochloric acid (HCl) dissociates into H^+ and Cl^-, it shifts the H^+/OH^- balance in favor of excess H^+ ions, thus increasing the level of acidity.

A *strong acid* is an acid that completely, or almost completely, dissociates to form H^+ ions. A *weak acid*, on the other hand, dissociates very little and therefore produces few excess H^+ ions in solution. There are many important acids in the body, and they perform many functions. Hydrochloric acid, for example, is the acid produced in the stomach to aid the digestive process.

Bases. Bases, or **alkaline** compounds, are electrolytes that, when dissociated in solution, shift the H^+/OH^- balance in favor of OH^-. This can be accomplished by increasing the number of hydroxide (OH^-) ions in solution or decreasing the number of H^+ ions present. The fact that bases will combine with or accept hydrogen ions (protons) is the reason the term *proton acceptor* is used to describe these substances. The dissociation of a common base, sodium hydroxide, yields the cation Na^+ and the OH^- anion.

Like acids, bases are classified as strong or weak, depending on how readily and completely they dissociate into ions. Important bases in the body such as the bicarbonate ion (HCO_3^-) play critical roles in the transportation of respiratory gases and in the elimination of waste products from the body.

The pH scale: measuring acidity and alkalinity. The term *pH* is a symbol used to mean the hydrogen ion (H^+) concentration of a solution (Figure 2-9). Actually, pH stands for the negative logarithm of the hydrogen ion concentration. pH indicates the degree of **acidity** or **alkalinity** of a solution. As the concentration of hydrogen ions increases, the pH goes down and the solution becomes more acidic; a decrease in hydrogen ion concentration makes the solution more alkaline and the pH goes up. A pH of 7 indicates neutrality (equal amounts of H^+ and OH^-), a pH of less than 7 indicates acidity (more H^+ than

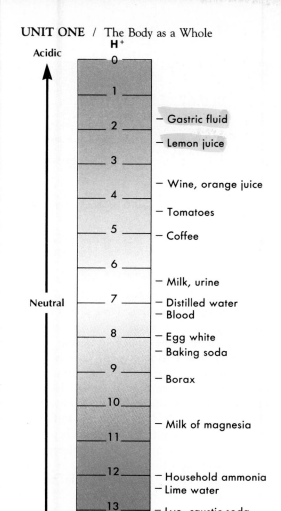

FIGURE 2-9   The pH range. Note that as concentration of H$^+$ increases, the solution becomes increasingly acidic and the pH value decreases. As OH$^-$ concentration increases, the pH value also increases, and the solution becomes more and more basic, or alkaline.

OH$^-$), and a pH greater than 7 indicates alkalinity (more OH$^-$ than H$^+$). The overall pH range is often expressed numerically on a logarithmic scale of 1 to 14. Keep in mind that a change of 1 pH unit on this type of scale represents a tenfold difference in actual concentration of hydrogen ions.

Buffers

The normal pH range of blood and other body fluids is extremely narrow. For example, venous blood (pH 7.36) is only slightly more acidic than arterial blood (pH 7.41). The difference results primarily from carbon dioxide entering venous blood as a waste product of cellular metabolism. Carbon dioxide is carried as carbonic acid (H$_2$CO$_3$) and therefore lowers the pH of venous blood. Over 30 *liters* of carbonic acid are transported in the venous blood each day and eliminated as CO$_2$ by the lungs, and yet

1 liter of venous blood contains only about 1/100,000,000 g more hydrogen ions than does 1 liter of arterial blood! The incredible constancy of the pH homeostatic mechanism is due to the presence of substances, called **buffers,** that minimize changes in the concentrations of H$^+$ and OH$^-$ ions in our body fluids. Buffers are said to act as a "reservoir" for hydrogen ions. They donate or remove hydrogen ions to a solution if that becomes necessary to maintain a constant pH. The specifics of buffer action will be discussed in Chapter 29.

Salts

A salt is any compound that results from the chemical interaction of an acid and a base. Salts, like acids and bases, are electrolyte compounds and dissociate in solution to form positively and negatively charged ions. When mixed and allowed to react, the positive ion (cation) of a base and the negative ion (anion) of an acid will join to form a salt and water in the manner of a typical exchange reaction. The reaction between an acid and base to form a salt and water is called a **neutralization reaction:**

(A B	+	C D	$\longrightarrow$	C B	+	A D)
HCl	+	NaOH	$\longrightarrow$	NaCl	+	H$_2$O
Acid		Base		Salt		Water
(hydrochloric acid)		(sodium hydroxide)		(sodium chloride)		

Note that the sodium and the chloride join to form the salt, whereas the hydroxyl ion "accepts" or combines with a hydrogen ion to form water.

The source of many of the major and trace mineral elements listed in Table 2-1 are inorganic salts, which are common in many body fluids and specialized tissues such as bone. These elements often exert their full physiological effects only when present as charged atoms or ions in solution.

The proper amount and concentration of such mineral salt electrolytes as potassium (K$^+$), calcium (Ca^{++}), and sodium (Na$^+$) are required for proper functioning of nerves and for contraction of muscle tissue. See Chapter 28 for specific homeostatic control mechanisms that regulate electrolyte balance in blood and other body fluids. Table 2-3 lists several inorganic salts, which on dissociation in body fluids contribute important electrolytes required for numerous body functions.

TABLE 2-3   Inorganic salts important in body functions

Inorganic Salt	Chemical Formula	Electrolytes
Sodium chloride	NaCl	Na^+ + Cl^-
Calcium chloride	$CaCl_2$	Ca^{++} + $2Cl^-$
Magnesium chloride	$MgCl_2$	Mg^{++} + $2Cl^-$
Sodium bicarbonate	$NaHCO_3$	Na^+ + HCO_3^-
Potassium chloride	KCl	K^+ + Cl^-
Sodium sulfate	Na_2SO_4	$2 Na^+$ + $SO_4^=$
Calcium carbonate	$CaCO_3$	Ca^{++} + $CO_3^=$
Calcium phosphate	$Ca_3(PO_4)_2$	$3 Ca^{++}$ + $2PO_4^{\equiv}$

TABLE 2-3   Inorganic salts important in body functions

Inorganic Salt	Chemical Formula	Electrolytes
Sodium chloride	NaCl	Na^+ + Cl^-
Calcium chloride	$CaCl_2$	Ca^{++} + $2Cl^-$
Magnesium chloride	$MgCl_2$	Mg^{++} + $2Cl^-$
Sodium bicarbonate	$NaHCO_3$	Na^+ + HCO_3^-
Potassium chloride	KCl	K^+ + Cl^-
Sodium sulfate	Na_2SO_4	$2 Na^+$ + $SO_4^=$
Calcium carbonate	$CaCO_3$	Ca^{++} + $CO_3^=$
Calcium phosphate	$Ca_3(PO_4)_2$	$3 Ca^{++}$ + $2PO_4^{\equiv}$

QUICK CHECK

✔ 1. *Discuss the properties of water that make it so important in living organisms.*
✔ 2. *What is an electrolyte?*
✔ 3. *How do acids and bases react with each other when in solution?*
✔ 4. *What is pH?*

ORGANIC MOLECULES

The term *organic* is used to describe the enormous number of compounds that contain carbon—specifically C–C or C–H bonds. Recall that carbon atoms have only four electrons in their outer shell (see Figure 2-1, A); it requires four electrons to satisfy the octet rule. As a result, each carbon atom can join with four other atoms to form literally thousands of molecules of varying sizes and shapes. In the human body, four major groups of organic substances are most important:

1. Carbohydrates
2. Proteins
3. Lipids
4. Nucleic acids

Figure 2-10 shows examples of the four major organic substances represented by three-dimensional models.

Carbohydrates

All carbohydrate compounds contain the elements carbon, hydrogen, and oxygen with the carbon atoms linked to one another to form chains of varying lengths. Carbohydrates include the substances commonly called sugars and starches and represent the primary source of

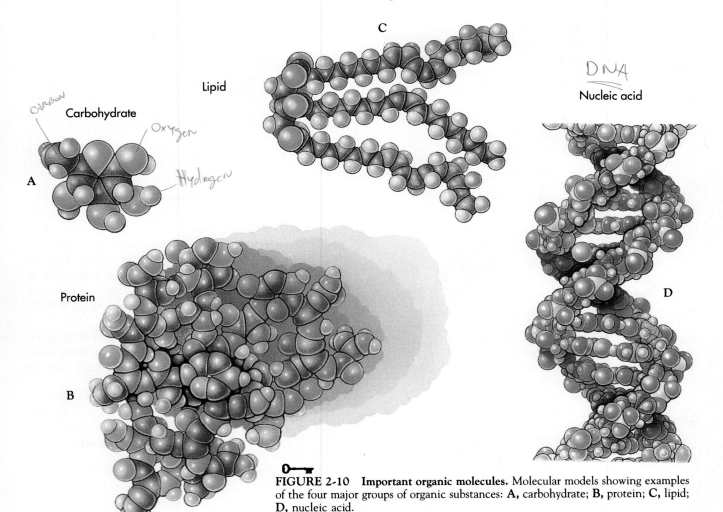

FIGURE 2-10   Important organic molecules. Molecular models showing examples of the four major groups of organic substances: **A,** carbohydrate; **B,** protein; **C,** lipid; **D,** nucleic acid.

chemical energy needed by every body cell. In addition, carbohydrates serve a structural role as components of such critically important molecules as RNA and DNA, which are involved in cell reproduction and protein synthesis.

As a group, carbohydrates are divided into three types or classes, which are characterized by the length of their carbon chains. The three types are called:

1. Monosaccharides (simple sugars)
2. Disaccharides (double sugars)
3. Polysaccharides (complex sugars)

Monosaccharides

Monosaccharides, or simple sugars, have short carbon chains. The most important simple sugar is *glucose*. It is a six-carbon sugar with the formula $C_6H_{12}O_6$. The chemical formula indicates that each molecule of glucose contains six atoms of carbon, twelve atoms of hydrogen, and six atoms of oxygen. Because it has six carbon atoms it is called a **hexose** (*hexa,* "six"). Glucose is present in the dry state as a "straight chain" but forms a "cyclic" compound when dissolved in water. In Figure 2-11 the straight chain and cyclic arrangements are shown with a three-dimensional model of the molecule. It is important to remember that all forms of glucose, however represented in models or illustrations, are the same molecule.

In addition to glucose, other important hexoses, or six-carbon simple sugars, include fructose and galactose. Not all monosaccharides, however, are hexoses. Some are **pentoses** (*penta,* "five"), so-named because they contain five carbon atoms. *Ribose* and *deoxyribose* are pentose

monosaccharides of great importance in the body—they will be covered further when we discuss nucleic acids. Like all monosaccharides, ribose and deoxyribose are sugars—but strange sugars in that they are not sweet.

Disaccharides and polysaccharides

Substances classified as disaccharides (double sugars) or polysaccharides (complex sugars) are carbohydrates composed of two or more simple sugars that are bonded together through a synthesis reaction that involves the removal of water. Sucrose (table sugar), maltose, and lactose are all disaccharides. Each consists of two monosaccharides linked together. Figure 2-12 shows the formation of sucrose from glucose and fructose. Note that a hydrogen atom from the glucose molecule combines with a hydroxyl group (OH) from the fructose molecule to form water, leaving an oxygen atom to bind the two subunits together.

Polysaccharides consist of many monosaccharides chemically joined to form straight or branched chains. Once again, water is removed as the many monosaccharide subunits are joined. Any large molecule made up of many identical small molecules is called a *polymer.* Polysaccharides are polymers of monosaccharides. *Glycogen* is sometimes referred to as animal starch. It is the main polysaccharide in the body and has an estimated molecular weight of several million—truly a macromolecule.

QUICK CHECK

✔ 1. List the four major groups of organic substances.
✔ 2. Identify the most important monosaccharide, or simple sugar.
✔ 3. Identify a carbohydrate polymer and explain how it is formed.

Proteins

All **proteins** are characterized by the presence of four elements: carbon, oxygen, hydrogen, and nitrogen. A number of more specialized proteins also contain sulfur, iron, and phosphorus. *Proteins* (from the Greek *proteios,* "of the first rank") are the most abundant of the carbon-containing, or organic, compounds in the body, and as their name implies, their functions are of first-rank importance (Table 2-4). Protein molecules are giant sized; that is, they are macromolecules.

Compared to water with a molecular weight of 18, giant protein molecules may weigh in at several million! However, all protein molecules, regardless of size, have a similar

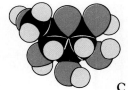

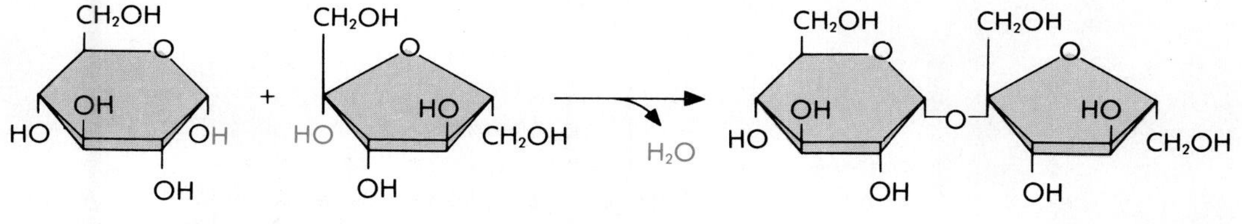

FIGURE 2-11 Structure of glucose. A, Straight chain, or linear model, of glucose. **B,** Ring model representing glucose in solution. **C,** Three-dimensional, or space filling, model of glucose.

FIGURE 2-12 Formation of sucrose. Glucose and fructose are joined in a synthesis reaction that involves the removal of water.

TABLE 2-4   Major functions of human protein compounds

Functions	Examples
Provide Structure	Structural proteins include keratin of skin, hair, and nails; parts of cell membranes; tendons
Catalyze Chemical Reactions	Lactase (enzyme in intestinal digestive juice) catalyzes chemical reaction that changes lactose to glucose and galactose
Transport Substances in Blood	Proteins classified as albumins combine with fatty acids to transport them in form of lipoproteins
Communicate Information to Cells	Insulin, a protein hormone, serves as chemical message from islet cells of pancreas to cells all over the body
Act as Receptors	Binding sites of certain proteins on surfaces of cell membranes serve as receptors for insulin and various other hormones
Defend Body Against any Harmful Agents	Proteins called antibodies or immunoglobulins combine with various harmful agents to render them harmless
Provide Energy	Proteins can be metabolized for energy

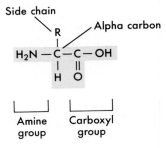

FIGURE 2-13 Basic structural formula for an amino acid. Note relationship of the side chain (R), amine group, and carboxyl group to the alpha carbon.

Individual amino acids are often compared to the letters of the alphabet. Just as combinations of individual letters form word combinations, different amino acids form protein chains. Think of amino acids as the alphabet of proteins.

The ability of amino acids to "link up" in all possible combinations allows the body to build or synthesize an almost infinite variety of different protein "words" or chains that may contain a dozen, several hundreds, or even thousands of amino acids.

Amino acids frequently become joined by peptide bonds. A **peptide bond** is one that binds the carboxyl group of one amino acid to the amino group of another amino acid. OH from the carboxyl group of one amino acid and H from the amino group of another amino acid split off to form water plus a new compound called a peptide. A peptide made up of only two amino acids linked by a peptide bond is a *dipeptide*. A *tripeptide* consists of three amino acids linked by two bonds. A long sequence or chain of amino acids—usually 100 or more—linked by peptide bonds constitutes a *polypeptide*. When the length of the polymer chain exceeds about 100 amino acids, the molecule is called a protein rather than a polypeptide.

Do you see a similarity between the formation of a disaccharide, such as sucrose, from simple sugar "building blocks" and the formation of a dipeptide from amino acid "building blocks"? In both processes, two subunits are joined together, resulting in the loss of a water molecule. Such processes are called **dehydration synthesis reactions** and are very common in living organisms. Repetitive dehydration synthesis reactions involving the addition of simple sugars will result in the formation of a polysaccharide polymer. If amino acids are linked together in this fashion by formation of peptide bonds, the resulting polymer is a polypeptide.

Levels of protein structure

Protein molecules are highly organized and show a very definite relationship between their structural appearance and their function. For example, the strong, inelastic **structural proteins** found in tendons and ligaments are linear or threadlike, insoluble, and very stable molecules. In contrast, **functional proteins** such as antibody molecules are globular, soluble, and chemically active molecules.

basic structure. They are chainlike polymers composed of multiple subunits or building blocks linked end to end. The building blocks of all proteins are called *amino acids*.

Amino acids

The elements that make up a protein molecule are bonded together to form chemical units called **amino acids.** Proteins are composed of 20 commonly occurring amino acids, and nearly all of the 20 amino acids are usually present in every protein. Of these 20, eight are known as **essential amino acids.** These cannot be produced by the body and must be included in the diet. The 12 remaining **nonessential amino acids** can be produced from other amino acids or from simple organic molecules readily available to the body cells. The basic structural formula for an amino acid is shown in Figure 2-13. As you can see, it consists of a *carbon atom* (called the alpha carbon) to which are bonded an *amino group* (NH_2), a *carboxyl group* (COOH), a *hydrogen atom*, and a *side chain*, or group of elements designated by the letter R. It is this side chain that constitutes the unique, identifying part of an amino acid. Several representative amino acids are shown in Figure 2-14.

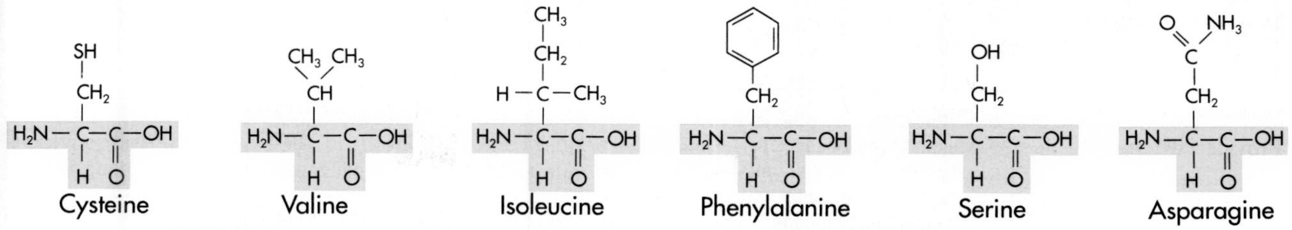

FIGURE 2-14 Representative amino acids. These structural formulas show that each amino acid has the same chemical backbone but differs from the others in the side, or R, group that it possesses.

Biochemists often describe four levels of protein organization:

1. Primary
2. Secondary
3. Tertiary
4. Quaternary

The levels of protein structure are illustrated in Figure 2-15.

The *primary structure* of a protein refers simply to the number, kind, and sequence of amino acids that make up the polypeptide chain. The hormone of the human parathyroid gland, parathyroid hormone (PTH), is a primary structure protein—it consists of only one polypeptide chain of 84 amino acids.

Most polypeptides do not exist as a straight chain. Instead, they show a *secondary structure* in which the chains are coiled or bent into pleated sheets. The most common type of coil takes a clockwise direction and is called an "alpha helix." In this type of secondary structure, the coils of the protein chain resemble a spiral staircase with the coils stabilized by hydrogen bonds between successive turns of the spiral. This stabilizing function of hydrogen bonding in protein structure is critical.

Just as a primary structure polypeptide chain can pleat or bend into a helical secondary structure, so too can a secondary structure protein chain undergo other contortions and be further twisted, resulting in a globular shaped *tertiary structure* protein. In this structure, the polypeptide chain is so twisted that its coils touch one another in many places, and "spot welds," or interlocking connections, occur. These linkages result from strong covalent bonds between amino acid units that exist in the same chain. In addition, hydrogen bonds also help stabilize the twisted and convoluted loops of the structure. The specialized muscle protein, myoglobin, which will be discussed in Chapter 9, is an example of a tertiary structure protein.

A *quaternary structure* protein is one that contains clusters of more than one polypeptide chain. Antibody molecules that protect us from disease (Chapter 20) and hemoglobin molecules in red blood cells (Chapter 16) are examples.

QUICK CHECK

✔ 1. *What element is present in all proteins but not in carbohydrates?*
✔ 2. *Identify the "building blocks" of proteins and explain what common chemical features they all share.*
✔ 3. *Explain the four levels of protein structure.*

 Disulfide Linkages

Hair contains a fibrous threadlike protein called *keratin* that is rich in the sulfur-containing amino acid, cysteine. The protein chains in keratin are linked in numerous places by sulfur to sulfur bonds, which form between sulfur-containing amino acids, such as cysteine, in adjacent hair fibrils in each hair shaft. The number and placement of these bonds, called **disulfide linkages** are determined by heredity and partly explains the "natural" straightness or curl of hair seen in every individual.

The object of a "permanent wave" (at least to a chemist!) is to change the arrangement of these bonds by breaking the naturally occurring disulfide linkages and then causing them to re-form in another pattern. In a permanent wave, strong chemicals are applied to the hair that break the existing or "natural" disulfide linkages. The hair is then curled on some type of roller or straightened, and another chemical is applied that causes the disulfide bridges to become reestablished in the new or reoriented configuration. Such treatment must be repeated as new hair growth occurs.

Lipids

Lipids, according to one definition, are water-insoluble organic biomolecules. Although insoluble in water, most of these compounds, many with an oil-like consistency and greasy feel, dissolve readily in such nonpolar organic solvents as ether, alcohol, or benzene. Like the carbohydrates, lipids are composed largely of carbon, hydrogen, and oxygen. However, the proportion of oxygen in lipids is much lower than that in carbohydrates. Many lipids also contain other elements such as nitrogen and phosphorus. As a group, lipids include a large assortment of compounds that have been classified in several ways. Classification of lipids includes triglycerides or fats, phospholipids, steroids, and prostaglandins.

Lipids are critically important biological compounds and play a number of major roles in the body (Table 2-5). Many are used for energy purposes, whereas others serve a structural role and function as integral parts of cell membranes. Other important lipid compounds serve as

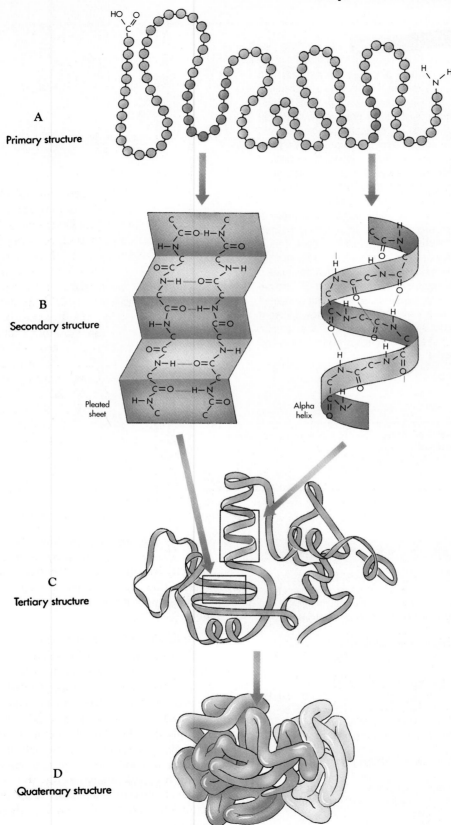

FIGURE 2-15   **Structural levels of protein. A,** *Primary structure:* determined by number, kind, and sequence of amino acids in the chain; **B,** *Secondary structure:* hydrogen bonds stabilize folds or helical spirals; **C,** *Tertiary structure:* globular shape maintained by strong (covalent) intramolecular bonding and by stabilizing hydrogen bonds; **D,** *Quaternary structure:* results from bonding between more than one polypeptide unit.

How Enzymes Function

Enzymes, the largest group of proteins in the body, are **chemical catalysts.** This means that they help a chemical reaction occur but are not reactants or products themselves. They participate in chemical reactions but are not changed by the reactions. Enzymes act to speed up the rate at which metabolic reactions occur. Nearly 2,000 enzymes are known and each is responsible for speeding up the rate of a very particular and unique chemical reaction— sometimes by a factor of 10 and often by a million or more. No reaction in the body occurs fast enough unless the specific enzymes needed for that reaction are present. These important proteins are able to accomplish their function even when present in very small quantities. This is possible because they are not "used up" in a reaction but remain unchanged to be used again and again as needed. Note how shape is important to the function of enzyme molecules. Each enzyme, by means of its uniquely shaped binding sites, binds to a very specific substance, called a **substrate,** that it works on—much as a key fits a specific lock. This explanation of enzyme action is sometimes called the **lock-and-key model.**

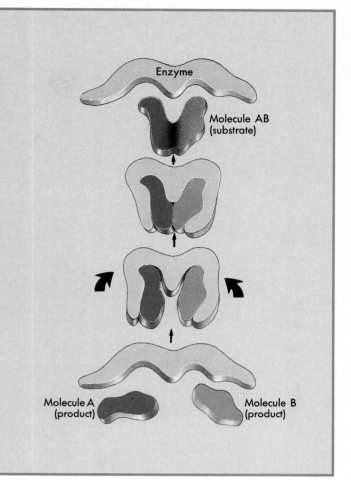

Enzyme

Molecule AB
(substrate)

Molecule A
(product)

Molecule B
(product)

Enzymes are functional proteins whose molecular shape allows them to catalyze chemical reactions. Subtrates (molecules *A* and *B*) are brought together by the enzyme to form a larger molecule (*AB*).

TABLE 2-5	Major functions of human lipid compounds
Function	**Example**
Energy	Lipids can be stored and broken down later for energy; they yield more energy per unit of weight than carbohydrates or proteins
Structure	Phospholipids and cholesterol are required components of cell membranes
Vitamins	Fat-soluble vitamins: vitamin A forms retinol (necessary for night vision); vitamin D increases calcium uptake; vitamin E promotes wound healing; and vitamin K is required for the synthesis of blood clotting proteins
Protection	Fat surrounds and protects organs
Insulation	Fat under the skin minimizes heat loss; fatty tissue (myelin) covers nerve cells and electrically insulates them
Regulation	Steroid hormones regulate many physiological processes. Examples: estrogen and testosterone are responsible for many of the differences between females and males; prostaglandins help regulate inflammation and tissue repair

vitamins or protect vital organs by serving as "fat pads," or shock-absorbers, in certain body areas. A specialized lipid material actually serves as an "insulator material" around nerves, thus serving to prevent "short circuits" and speed nervous impulse transmissions.

Triglycerides or fats

Triglycerides (triacylglycerols) or fats are the most abundant lipids, and they function as the body's most concentrated source of energy. Two types of building blocks are needed to synthesize or build a fat molecule: *glycerol* and *fatty acids.* Each glycerol unit is joined to three fatty acids. The glycerol building block is the same in each fat molecule. Therefore it is the specific type of fatty acid molecule or component that identifies and determines the chemical nature of any fat.

Types of fatty acids. Fatty acids vary in the length of their carbon chains (number of carbon atoms) and in the number of hydrogen atoms that are attached to, or "saturate," the available bonds around each carbon in the chain. Figure 2-16 shows a structural formula and three-dimensional model for a saturated (palmitic) and unsaturated (linolenic) fatty acid.

By definition a *saturated fatty acid* is one in which all available bonds of its hydrocarbon chain are filled, that is,

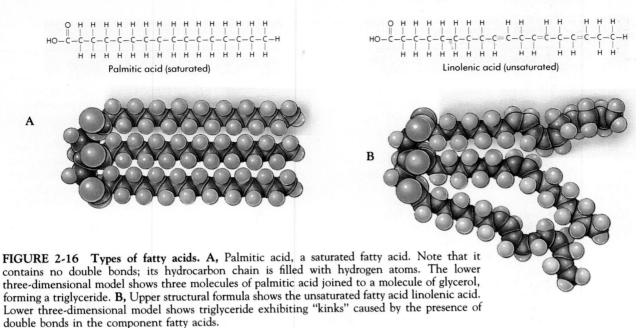

Palmitic acid (saturated)

Linolenic acid (unsaturated)

A

B

FIGURE 2-16 Types of fatty acids. A, Palmitic acid, a saturated fatty acid. Note that it contains no double bonds; its hydrocarbon chain is filled with hydrogen atoms. The lower three-dimensional model shows three molecules of palmitic acid joined to a molecule of glycerol, forming a triglyceride. **B,** Upper structural formula shows the unsaturated fatty acid linolenic acid. Lower three-dimensional model shows triglyceride exhibiting "kinks" caused by the presence of double bonds in the component fatty acids.

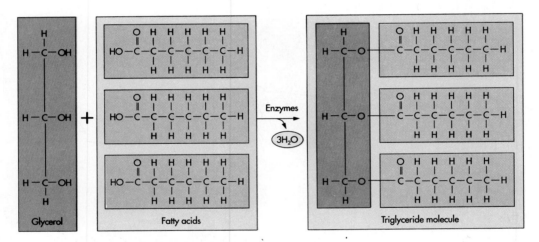

FIGURE 2-17   Formation of triglyceride. Glycerol tristerate is a composite molecule, made up of three molecules of stearic acid (a fatty acid) coupled in a dehydration synthesis reaction to a single glycerol backbone. In addition to the triglyceride, this process results in the formation of three molecules of water.

saturated, with hydrogen atoms. The chain contains no double bonds (Figure 2-16, A). In contrast, an *unsaturated fatty acid* has one or more double bonds in its hydrocarbon chain because not all of the chain carbon atoms are saturated with hydrogen atoms (Figure 2-16, B). The degree of saturation is the most important factor in determining the physical, as well as the chemical, properties of fatty acids. For example, animal fats such as tallow and lard are solids at room temperature, whereas vegetable oils are liquids. The difference lies in the extent of unsaturation—animal fats are saturated, whereas vegetable oils are not. Note in Figure 2-16, B, that the presence of double bonds in a fatty acid molecule will cause the chain to "kink" or bend.

Fats become more oily and liquid as the number of unsaturated double bonds increases. The "kinks" and

"bends" in the unsaturated molecules keep them from fitting closely together. In contrast, the lack of kinks in saturated fatty acids allows the molecules to fit tightly together to form a solid mass at higher temperatures.

Formation of triglycerides. Figure 2-17 shows the formation of a triglyceride. Its name, *glycerol tristearate,* suggests that it contains three molecules of stearic acid attached to a glycerol molecule. Note that the three stearic acid "building blocks" attach by their carboxyl groups (COOH) to the three hydroxyl groups (OH) of the glycerol molecule, forming the triglyceride and three molecules of water. The process is one you are now familiar with—it is a dehydration synthesis reaction. Keep in mind that although some fats, like glycerol tristearate, contain three molecules of the same fatty acid, others may have two or three different fatty acids attached to glycerol.

BLOOD LIPOPROTEINS

A lipid such as cholesterol can travel in the blood only after it has attached to a protein molecule—forming a lipoprotein. Some of these molecules are called high-density lipoproteins (HDLs) because they have a high density of protein (more protein than lipid). Another type of molecule contains less protein (and more lipid), so it is called low-density lipoprotein (LDL).

The cholesterol in LDLs is often called "bad" cholesterol because high blood levels of LDL are associated with atherosclerosis, a life-threatening blockage of arteries. LDLs carry cholesterol to cells, including the cells that line blood vessels. HDLs, on the other hand, carry so-called "good" cholesterol away from cells and toward the liver for elimination from the body. A high proportion of HDL in the blood is associated with a low risk of developing atherosclerosis. Factors such as cigarette smoking decrease HDL levels and thus contribute to risk of atherosclerosis. Factors such as exercise increase HDL levels and thus decrease the risk of atherosclerosis.

The 1985 Nobel Prize in Physiology or Medicine was awarded to Drs. Michael Brown and Joseph Goldstein for their research on specialized receptor sites on LDL molecules, which are elevated in the blood of individuals with certain types of heart disease. Lipid metabolism will be discussed in detail in Chapter 26.

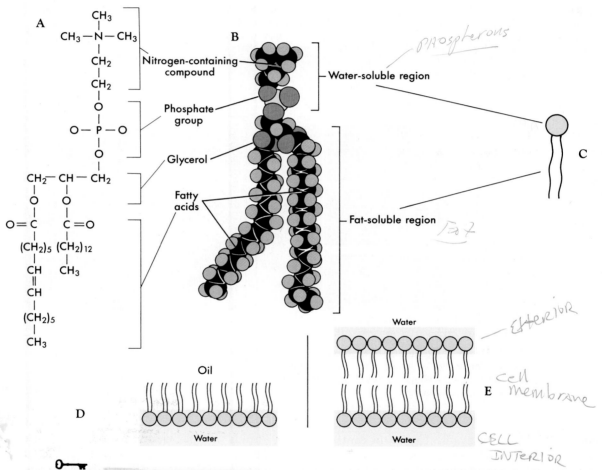

FIGURE 2-18 Phospholipid molecule. A, Chemical formula of a phospholipid molecule. **B,** Molecular model showing water- and fat-soluble regions. **C,** The way phospholipids are often depicted. **D,** Orientation of phospholid molecules in an oil-water interface. **E,** Orientation of phospholid molecules when surrounded by water.

Phospholipids

Phospholipids are fat compounds similar to triglycerides. They are modified, however, in that one of the three fatty acids attached to glycerol in a triglyceride is replaced in a phospholipid by another type of chemical structure containing phosphorus and nitrogen. The structural formula of a phospholipid is shown in Figure 2-18. Observe that the phospholipid molecule contains glycerol. Joined to the glycerol at one end of the molecule are two fatty acids. Attached to glycerol but extending in the opposite direction is the phosphate group, which is attached to a nitrogen-containing compound.

The end of the molecule containing the phosphorus group in a phospholipid is water soluble, and the end

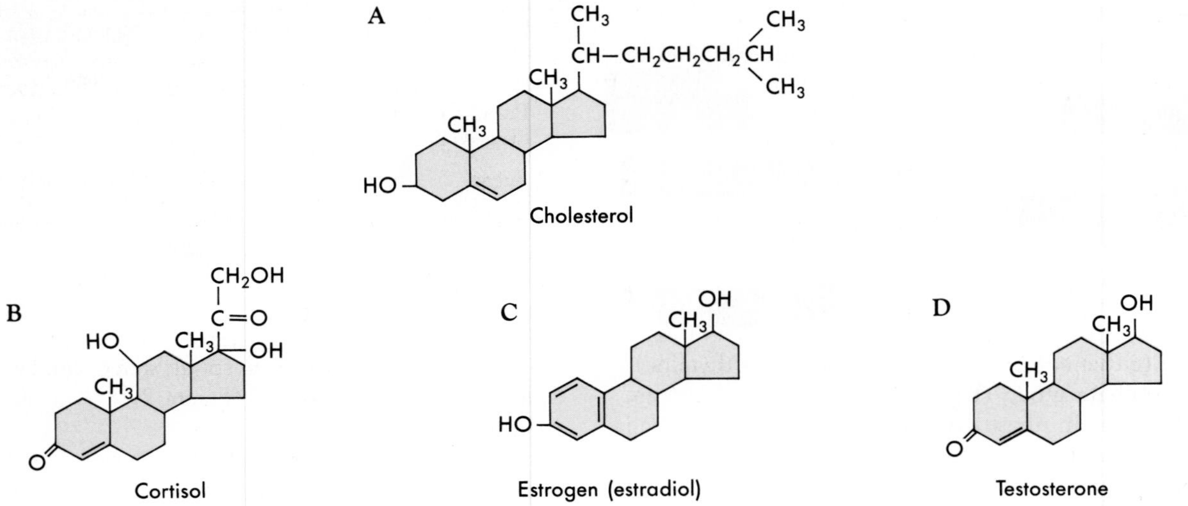

FIGURE 2-19   The steroid nucleus. The steroid nucleus—highlighted in yellow—found in cholesterol **(A)** forms the basis for many other important compounds such as cortisol **(B)**, estradiol (an estrogen) **(C)**, and testosterone **(D)**.

formed by the two fatty acids is fat soluble. This unique property means that phospholipid molecules can bridge or join two different chemical environments—a water environment on one side and a lipid environment on the other. For this reason, phospholipids are a primary component of cell membranes and will be discussed further in Chapter 3.

Steroids

Steroids are an important and large group of compounds whose molecules have as their principal component the *steroid nucleus* (Figure 2-19). Steroids are widely distributed in the body and are involved in many important structural and functional roles. Cholesterol is a steroid found in the plasma membrane surrounding every body cell (see Chapter 3). Its presence helps stabilize this important cellular structure and is required for many reactions that cells must perform to survive. Cholesterol and other steroid compounds are also required for the synthesis of such hormones as estrogens, testosterone, and cortisol.

Prostaglandins

Prostaglandins, often called "tissue hormones," are lipids composed of a 20-carbon unsaturated fatty acid that contains a 5-carbon ring. Many different kinds of prostaglandins exist in the body. We now classify 16 prostaglandin types (PGs) into nine broad categories, called PGA to PGI. Each major grouping of prostaglandins can be further subdivided based on chemical structure and function.

Prostaglandins were first associated with prostate tissue and were named accordingly. Subsequent discoveries, however, have shown that these biologically powerful chemical substances are produced by cell membranes located in almost every body tissue. They are formed and then released from cell membranes in response to a particular stimulus. Once released, they have a very local effect and are then inactivated.

FYI   **Aspirin and Prostaglandins**

In the presence of an appropriate stimulus such as irritation or injury, fatty acids required for prostaglandin synthesis are released by cell membranes. If a specific enzyme, *cyclooxygenase,* is present to interact with these fatty acids, prostaglandins will be synthesized and released from the cell membrane into the surrounding tissue fluid.

Prostaglandins serve as inflammatory agents. They cause local dilation of blood vessels with resulting heat (fever), swelling, redness, and pain. Aspirin works to relieve these symptoms by blocking the activity of the enzyme cyclooxygenase. If this enzyme cannot function properly, prostaglandin synthesis will be inhibited, and symptoms will be relieved.

The effects of prostaglandins in the body are many and varied. They play a crucial role in regulating the effects of several hormones, influence blood pressure and secretion of digestive juices, enhance the body immune system and inflammatory response, and play an important role in blood clotting and respiration, to name a few. The use of prostaglandins and prostaglandin inhibitors as drugs is an exciting and rapidly growing area in clinical medicine. Treatment of specific disease states, symptoms, or medical conditions using prostaglandins or drugs that inhibit prostaglandin action ranges from their use to relieve menstrual cramps to treatment of asthma, high blood pressure, and ulcers.

QUICK CHECK

✔ 1. *What are the building blocks of a triglyceride or fat?*
✔ 2. *Give an example of a dehydration synthesis reaction.*
✔ 3. *What is a phospholipid, and why is it an important molecule?*
✔ 4. *Identify an important steroid.*

Nucleic Acids

Survival of humans as a species—and survival of every other species—depends largely on two kinds of **nucleic acid compounds.** Their abbreviated names, DNA and RNA, almost everyone has heard or seen, but their full names are much less familiar. They are deoxyribonucleic and ribonucleic acids. Nucleic acid molecules are polymers of thousands and thousands of smaller molecules called nucleotides—deoxyribonucleotides in DNA molecules and ribonucleotides in RNA molecules. A *deoxyribonucleotide* consists of the pentose sugar named deoxyribose, a nitrogenous base (either adenine, cytosine, guanine, or thymine), and a phosphate group (Figure 2-20). *Ribonucleotides* are similar but contain ribose instead of deoxyribose and uracil instead of thymine (Table 2-6.) Two of the bases in a deoxyribonucleotide, specifically adenine and guanine, are called **purine bases** because they derive from *purine.* Purines have a double ring structure. Cytosine and thymine derive from *pyrimidine,* so they are known as **pyrimidine bases.** Pyrimidines have a single ring structure. The pyramidine base, uracil, replaces thymine in RNA. Differences between DNA and RNA will be discussed in Chapter 3.

DNA molecules, the largest molecules in the body, are very large polymers composed of many nucleotides. Two long polynucleotide chains compose a single DNA molecule. The chains coil around each other to form a

double helix. A helix is a spiral shape similar to the shape of a wire in a spring. Figure 2-20 is a diagram of the double helix DNA.

Each helical chain in a DNA molecule has its phosphate-sugar backbone toward the outside and its bases pointing inward toward the bases of the other chain. More than that, each base in one chain is joined to a base in the other chain by means of hydrogen bonds to form what is known as a **base pair.** The two polynucleotide chains of a DNA molecule are thus held together by hydrogen bonds between the two members of each base pair (Figure 2-20). One important principle to remember is that only two kinds of base pairs are present in DNA. What are they? Symbols used to represent them are A-T and G-C. Although a DNA molecule contains only these two kinds

TABLE 2-6	Comparison of DNA and RNA structure	
	DNA	**RNA**
Polynucleotide Strands	2	1
Sugar	Deoxyribose	Ribose
Base Pairs	Adenine-thymine Guanine-cytosine	Adenine-uracil Guanine-cytosine

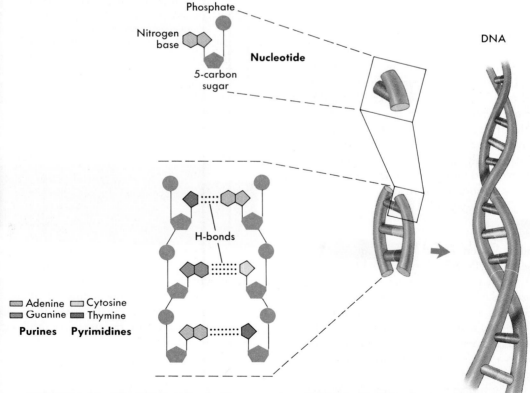

FIGURE 2-20   The DNA molecule. Representation of DNA double helix showing the general structure of a nucleotide and the two kinds of "base pairs": adenine (A) (blue) with thymine (T) (purple), and guanine (G) (green) with cytosine (C) (yellow). Note that the G-C base pair has three hydrogen bonds and an A-T base pair, two. Hydrogen bonds are extremely important in maintaining the structure of this molecule.

FIGURE 2-21 Metabolic reactions. Hydrolysis is a catabolic reaction that adds water to break down large molecules into smaller molecules, or subunits. Dehydration synthesis is an anabolic reaction that operates in the reverse fashion: small molecules are assembled into large molecules by removing water. Note that specific examples of dehydration synthesis were shown in Figures 2-12 and 2-17.

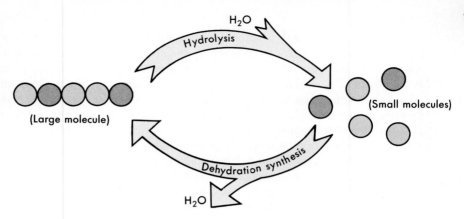

of base pairs, it contains millions of them—over 100 million pairs estimated in one human DNA molecule! Two other impressive facts are these: the millions of base pairs occur in the same sequence in all the millions of DNA molecules in one individual's body but in a different sequence in the DNA of all other individuals. In short, the base pair sequence in DNA is unique to each individual. This fact has momentous significance, but what it is we shall save to reveal in Chapter 3. For now, we shall merely state that DNA functions as the heredity molecule. It carries out a weighty responsibility, that of passing the traits of one generation on to the next. How it accomplishes this will be told in other portions of this text.

Metabolism

The term **metabolism** is used to describe all of the chemical reactions that occur in body cells. The important topics of nutrition and metabolism will be discussed fully in Chapter 26. Nutrition and metabolism are described together because the total of all of the chemical reactions or *metabolic activity* occurring in cells is associated with the use the body makes of foods after they have been digested,

absorbed, and circulated to cells. The terms *catabolism* and *anabolism* are used to describe the two major types of metabolic activity. **Catabolism** describes those chemical reactions—usually hydrolysis reactions— that break down larger food molecules into smaller chemical units and in so doing *release energy*. **Anabolism** involves the many chemical reactions—usually dehydration synthesis reactions— that build larger and more complex chemical molecules from smaller subunits (Figure 2-21). Anabolic chemical reactions *require energy*—energy in the form of **adenosine triphosphate, ATP.**

ATP is composed of three building blocks: the first is the pentose sugar *ribose*, which serves as the point of attachment for the nitrogen-containing molecule, *adenine*, and a unique grouping of three phosphate subunits (Figure 2-22). The "squiggle" lines indicate covalent bonds between the phosphate groups. These bonds are called **high-energy bonds,** because when they are broken during catabolism-type chemical reactions, they release large amounts of energy. The energy stored in ATP is used to do the body's work—the work of muscle contraction and movement, of active transport, and of biosynthesis.

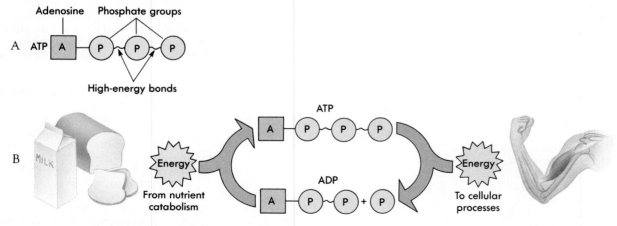

FIGURE 2-22 Adenosine triphosphate (ATP). A, Structure of ATP. A single adenosine group *(A)* has three attached phosphate groups *(P)*. High-energy bonds between the phosphate groups can release chemical energy to do cellular work. **B,** General scheme of ATP energy cycle. ATP stores energy in its last high-energy phosphate bond. When that bond is later broken, energy is released to do cellular work. The ADP and phosphate groups that result can be resynthesized into ATP, capturing additional energy from nutrient catabolism.

Catabolism

Catabolism consists of those chemical reactions that not only break down relatively complex compounds into simpler ones but also release energy from them. This breakdown process represents a type of chemical reaction called *hydrolysis* (see Figure 2-21). As a result of hydrolysis occurring during catabolism, a water molecule is added to break a larger compound into smaller subunits. For example, hydrolysis of a fat molecule would break it down into its subunits—glycerol and fatty acid molecules; a disaccharide such as sucrose would be broken down into its monosaccharide subunits—glucose and fructose; the subunits of protein hydrolysis would be amino acids. Ultimately, catabolism reactions will further degrade these building blocks of food compounds—glycerol, fatty acids, monosaccharides, and amino acids—into the end products carbon dioxide, water, and other waste products. During this process, energy is released. Some of the energy released by catabolism is heat energy, the heat that keeps our bodies warm. However, more than half of the released energy is immediately recaptured and put back into storage in ATP.

Anabolism

Anabolism is the term used to describe chemical reactions that join simple molecules together to form more complex biomolecules, notably, carbohydrates, lipids, proteins, and nucleic acids. Literally thousands of anabolic reactions take place continually in the body. The type of chemical reaction responsible for this joining together of smaller units to form larger molecules is called *dehydration synthesis* (see Figure 2-21). It is a key reaction during anabolism. As a result of dehydration synthesis, water is removed as smaller subunits are fused together. The process requires energy, which is supplied by the breakdown of ATP. It is anabolism-type reactions that join monosaccharide units to form larger carbohydrates, fuse amino acids into peptide chains, and form fat molecules from glycerol and fatty acid subunits.

QUICK CHECK

✔ 1. *Name two important nucleic acids.*
✔ 2. *What is a nucleotide?*
✔ 3. *What is meant by the term "base pair"?*
✔ 4. *How is energy stored in the ATP molecule?*

CHAPTER SUMMARY

BASIC CHEMISTRY

A. Elements and compounds
   1. Matter—anything that has mass and occupies space
   2. Element—simple form of matter, a substance that cannot be broken down into two or more different substances
      a. There are 26 elements in the human body
      b. There are 11 "major elements," four of which (carbon, oxygen, hydrogen, and nitrogen) make up 96% of the human body
      c. There are 15 "trace elements" that make up less than 2% of body weight
   3. Compound—atoms of two or more elements joined to form chemical combinations
B. Atoms (Figure 2-1)
   1. The concept of an atom was proposed by the English chemist John Dalton
   2. Atomic structure—atoms contain several different kinds of subatomic particles; the most important are:
      a. Protons (+ or p)—positively charged subatomic particles found in the nucleus
      b. Neutrons (n)—neutral subatomic particles found in the nucleus
      c. Electrons (− or e)—negatively charged subatomic particles found in the electron cloud
   3. Atomic number and atomic weight
      a. Atomic number
         (1) The number of protons in an atom's nucleus
         (2) The atomic number is critically important; it identifies the kind of element
      b. Atomic weight
         (1) The mass of a single atom
         (2) It is equal to the number of protons plus the number of neutrons in the nucleus (p + n)
   4. Electron shells (Figure 2-2)
      a. The total number of electrons in an atom equals the number of protons in the nucleus

   b. The electrons form a "cloud" around the nucleus
   c. "Bohr model"—two-dimensional model useful in visualizing the structure of atoms
      (1) Exhibits electrons in concentric circles showing relative distances of the electrons from the nucleus
      (2) Each ring, or shell, represents a specific energy level and can only hold a certain number of electrons
      (3) The number and arrangement of electrons determine if an atom is chemically stable
      (4) An atom with eight, or four pairs of, electrons in the outermost shell is chemically inert
      (5) An atom without a full outermost shell is chemically active
   d. Octet rule—atoms with fewer or more than eight electrons in the outer shell will attempt to lose, gain, or share electrons with other atoms to achieve stability
   5. Isotopes (Figure 2-3)
      a. Isotopes of an element contain the same number of protons but contain different numbers of neutrons
      b. Isotopes have the same atomic number, and therefore the same basic chemical properties, as any other atom of the same element but have a different atomic weight
      c. Radioactive isotope—an unstable isotope that undergoes nuclear breakdown and emits nuclear particles and radiation
C. Interactions between atoms—chemical bonds
   1. Chemical reaction—interaction between two or more atoms that occurs as a result of activity between electrons in their outermost shells
   2. Molecule—two or more atoms joined together
   3. Compound—consists of molecules formed by atoms of two or more elements

4. Chemical bonds—two types unite atoms into molecules
   a. Ionic, or electrovalent, bond (Figure 2-4)—formed by transfer of electrons; strong electrostatic force that binds positively and negatively charged ions together
   b. Covalent bond (Figure 2-5)—formed by sharing of electron pairs between atoms; the more significant bond in physiology
3. Hydrogen bond (Figure 2-7)
   a. Much weaker than ionic or covalent bonds
   b. Results from unequal charge distribution on molecules
D. Chemical reactions
   1. Involve the formation or breaking of chemical bonds
   2. There are three basic types of chemical reactions involved in physiology:
      a. Synthesis reaction—combining of two or more substances to form a more complex substance; formation of new chemical bonds; A + B → AB
      b. Decomposition reaction—breaking down of a substance into two or more simpler substances; breaking of chemical bonds; AB → A + B
      c. Exchange reaction—decomposition of two substances and, in exchange, synthesis of two new compounds from them; AB + CD → AD + CB
      d. Reversible reactions—occur in both directions

INORGANIC MOLECULES

A. Inorganic compounds—few have carbon atoms and none have C—C or C—H bonds
B. Water
   1. The body's most abundant and important compound
   2. Properties of water (Table 2-2)
      a. Polarity—allows water to act as an effective solvent; ionizes substances in solution
      b. The solvent allows transportation of essential materials throughout the body
      c. High specific heat—water can lose and gain large amounts of heat with little change in its own temperature; enables the body to maintain a relatively constant temperature
      d. High heat of vaporization—water requires absorption of significant amounts of heat to change water from a liquid to a gas, allowing the body to dissipate excess heat
C. Oxygen and carbon dioxide—closely related to cellular respiration
   1. Oxygen—required to complete decomposition reactions necessary for the release of energy in the body
   2. Carbon dioxide—produced as a waste product and also helps maintain the appropriate acid-base balance in the body
D. Electrolytes
   1. Large group of inorganic compounds, which includes acids, bases, and salts
   2. Substances that dissociate in solution to form ions
   3. Positively charged ions are cations; negatively charged ions are anions
   4. Acids and bases—common and important chemical substances that are chemical opposites
      a. Acids
         (1) Any substance that releases a hydrogen ion (H^+) when in solution; "proton donor"
         (2) Level of "acidity" depends on the number of hydrogen ions a particular acid will release
      b. Bases
         (1) Electrolytes that dissociate to yield hydroxyl ions (OH^-) or other electrolytes that combine with hydrogen ions (H^+)
         (2) Described as "proton acceptors"

E. pH scale—measuring acidity and alkalinity (Figure 2-9)
   1. pH indicates the degree of acidity or alkalinity of a solution
   2. pH of 7 indicates neutrality (equal amounts of H^+ and OH^-); a pH of less than 7 indicates acidity; a pH of more than 7 indicates alkalinity
F. Buffers
   1. Maintain the constancy of the pH
   2. Minimize changes in the concentrations of H^+ and OH^- ions
   3. Act as a "reservoir" for hydrogen ions
G. Salts (Table 2-3)
   1. Compound that results from chemical interaction of an acid and a base
   2. Reaction between an acid and a base to form a salt and water is called a neutralization reaction

ORGANIC MOLECULES (FIGURE 2-10)

A. "Organic" describes compounds that contain carbon-carbon or carbon-hydrogen bonds
B. Carbohydrates—organic compounds containing carbon, hydrogen, and oxygen; commonly called sugars and starches
   1. Monosaccharides—simple sugars with short carbon chains; those with six carbons are hexoses (e.g., glucose), whereas those with five are pentoses (e.g., ribose, deoxyribose)
   2. Disaccharides and polysaccharides—two (di-) or more (poly-) simple sugars that are bonded together through a synthesis reaction
C. Proteins
   1. Most abundant organic compounds
   2. Chainlike polymers
   3. Amino acids—building blocks of proteins
      a. Essential amino acids—eight amino acids that cannot be produced by the human body
      b. Nonessential amino acids—12 amino acids can be produced from molecules available in the human body
      c. Amino acids consist of a carbon atom, an amino group, a carboxyl group, a hydrogen atom, and a side chain
   4. Levels of protein structure (Figure 2-15)
      a. Protein molecules are highly organized and show a definite relationship between structure and function
      b. There are four levels of protein organization
         (1) Primary structure—refers to the number, kind, and sequence of amino acids that make up the polypeptide chain
         (2) Secondary structure—polypeptide is coiled or bent into pleated sheets stabilized by hydrogen bonds
         (3) Tertiary structure—a secondary structure can be further twisted, resulting in a globular shape; the coils touch in many places and are "welded" by covalent and hydrogen bonds
         (4) Quaternary structure—highest level of organization occurring when protein contains more than one polypeptide chain
D. Lipids
   1. Water-insoluble organic molecules that are critically important biological compounds
   2. Major roles:
      a. Energy source
      b. Structural role
      c. Integral parts of cell membranes
   3. Triglycerides, or fats (Figures 2-16 and 2-17)
      a. Most abundant lipids and most concentrated source of energy

b. The building blocks of triglycerides are glycerol (the same for each fat molecule) and fatty acids (different for each fat and determines the chemical nature)
    (1) Types of fatty acids—saturated fatty acid (all available bonds are filled) and unsaturated fatty acid (has one or more double bonds)
    (2) Triglycerides are formed by a dehydration synthesis

4. Phospholipids (Figure 2-18)
    a. Fat compounds similar to triglyceride
    b. One end of the phospholipid is water soluble; the other end is fat soluble
    c. Phospholipids can join two different chemical environments

5. Steroids (Figure 2-19)
    a. Main component is steroid nucleus
    b. Involved in many structural and functional roles

6. Prostaglandins
    a. Commonly called "tissue hormones"; produced by cell membranes throughout the body
    b. Effects are many and varied; however, they are released in response to a specific stimulus and are then inactivated

E. Nucleic acids
1. DNA (deoxyribonucleic acid)
    a. Composed of deoxyribonucleotides; that is, structural units composed of the pentose sugar (deoxyribose), phosphate group, and nitrogenous base (cytosine, thymine, guanine, or adenine)
    b. DNA molecule consists of two long chains of deoxyribonucleotides coiled into double-helix shape (Figure 2-20)
    c. Alternating deoxyribose and phosphate units form backbone of the chains
    d. Base pairs hold the two chains of DNA molecule together
    e. Specific sequence of over 100 million base pairs constitute one human DNA molecule; all DNA molecules in one individual are identical and different from all other individuals
    f. DNA functions as hereditary molecule

2. RNA (ribonucleic acid)
    a. Composed of the pentose sugar (ribose), phosphate group, and a nitrogenous base
    b. Nitrogenous bases for RNA are adenine, uracil, guanine, or cytosine (uracil replaces thymine)

F. Metabolism—all the chemical reactions that occur in body cells (Figure 2-21)
1. Catabolism
    a. Chemical reactions that break down complex compounds into simpler ones and release energy; hydrolysis is a common catabolic reaction
    b. Ultimately, the end products of catabolism are carbon dioxide, water, and other waste products
    c. More than half the energy released is put back into storage as ATP (adenosine triphosphate), which is then used to do cellular work (Figure 2-22)

2. Anabolism
    a. Chemical reactions that join simple molecules together to form more complex molecules
    b. Chemical reaction responsible for anabolism is dehydration synthesis

REVIEW QUESTIONS

1. Differentiate between the terms *chemistry* and *biochemistry*.
2. Define the following terms: *element, compound, atom, molecule.*
3. Compare early nineteenth century and present-day concepts of atomic structure.
4. Name and define three kinds of subatomic particles.
5. Are atoms electrically charged particles? Give reason for answer.
6. What four elements make up approximately 96% of the body's weight?
7. Define and contrast meanings of the terms *atomic number* and *atomic weight.*
8. Explain the general rule where an atom can be listed as chemically inert and unable to react with another atom.
9. Explain and give an example of the octet rule.
10. Define and give example of an isotope.
11. Explain what the term *radioactivity* means.
12. How does radioactivity differ from chemical activity?
13. Define the terms: *alpha particles, beta particles,* and *gamma rays.*
14. Explain how radioactive atoms become transformed into atoms of a different element.
15. Explain what the term *chemical reaction* means.
16. Differentiate between ionic, covalent, and hydrogen bonds. Give another name for ionic bonds.
17. Identify and differentiate between the three basic types of chemical reactions.
18. Define the term *inorganic.*
19. What percent of body weight is due to water?
20. Explain why water is said to be polar and list at least four functions of water that are crucial to survival.
21. What are electrolytes and how are they formed?
22. What is a cation? an anion? an ion? Give an example of each.
23. Define the terms *acid, base, salt,* and *buffer.*
24. Explain how pH indicates the degree of acidity or alkalinity of a solution.
25. What are the structural units, or building blocks, of proteins? of carbohydrates? of triglycerides? of DNA?
26. Explain what a protein molecule's binding site is. What function does it serve in enzymes?
27. What is the "lock and key model"?
28. Describe some of the functions proteins perform.
29. Proteins, carbohydrates, lipids—which of these are insoluble in water? contain nitrogen? include prostaglandins? include phosphoglycerides?
30. Differentiate between saturated and unsaturated fatty acids.
31. What groups compose a nucleotide?
32. What pentose sugar is present in a deoxyribonucleotide?
33. Describe the size, shape, and chemical structure of the DNA molecule.
34. What base is thymine always paired with in the DNA molecule? What other two bases are always paired?
35. What is the function of DNA?
36. What is catabolism. What function does it serve?
37. Compare catabolism, anabolism, and metabolism.
38. ATP is the abbreviation for what important biomolecule? Why is it important?

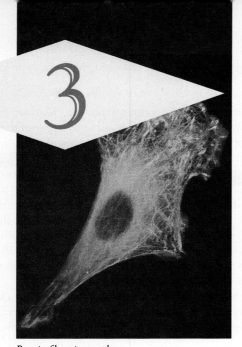

3 ▷ Cells

Protein fibers in cytoplasm

OBJECTIVES

After you have completed this chapter, you should be able to:

1. Discuss the structure of a "typical" cell.
2. Describe the molecular structure and function of cell membranes.
3. Identify by name the membranous and nonmembranous organelles of the cell.
4. Discuss the structure and function of the following cell structures: endoplasmic reticulum, ribosomes, Golgi apparatus, mitochondria, lysosomes, peroxisomes, cytoskeleton, cell fibers, centrosome, centrioles, cell extensions, nucleus, and cell connections.
5. Discuss the organization and generalized function of chromatin material in the nucleus.
6. Compare the processes of diffusion, dialysis, facilitated diffusion, osmosis, and filtration.
7. Discuss and compare the factors that determine potential osmotic pressure of electrolyte and nonelectrolyte solutions.
8. Discuss the "active" cell transport mechanisms responsible for movement of some materials through cell membranes.
9. Describe the molecular structure of DNA.
10. Discuss how genes control protein synthesis and determine hereditary characteristics.
11. Discuss the four distinct phases of mitosis.
12. Compare and contrast mitosis and meiosis.
13. Discuss and give examples of how cells adapt to changing conditions and what kinds of changes may be harmful to the body.

In the 1830s, two German scientists, Mathias Schleiden and Theodor Schwann, advanced one of the most important and unifying concepts in biology—the **cell theory.** It states simply that the cell is the fundamental organizational unit of life. Although earlier scientists had seen cells, Schleiden and Schwann were the first to suggest that all living things are composed of cells. Some 100 trillion of them make up the human body. Actually, the study of cells has captivated the interest of countless scientists for over 300 years. But as yet these small structures have not yielded all their secrets, not even to the probing tools of present-day researchers.

To introduce you to the world of cells, this chapter begins by describing common cell structures and then discusses some important and representative cell functions.

CELL STRUCTURE AND FUNCTION

The principle of complementarity of structure and function was introduced in Chapter 1 and is evident in the relationships that exist between cell size, shape, and function. Almost all human cells are microscopic in size. Their diameters range from 7.5 micrometers (μm) (red blood cells) to 300 μm (female sex cell or ovum). The period at the end of this sentence measures about 100 μm—roughly 13 times as large as our smallest cells and one third the size of the human ovum. Like other anatomic structures, cells exhibit a particular size or form because they are intended to perform a specialized activity. An individual nerve cell, for example, may have threadlike extensions over a meter in length! Such a cell is ideally suited to transmit nervous impulses from one area of the body to another. Muscle cells are specialized to contract or shorten, whereas other types of cells may serve protective or secretory functions (Table 3–1).

The Typical Cell

Regardless of their distinctive anatomic characteristics and specialized functions, cells have many similarities. There is no cell that truly represents or contains all of the specialized components found in the many types of body cells. As a result, students are often introduced to cell structure and function by studying a so-called *typical* or **composite cell**—one that exhibits the most important characteristics of many distinctive cell types. Such a generalized cell is illustrated in Figure 3–1. Keep in mind that no such "typical" cell actually exists in the body; it is a composite structure created for study purposes. Refer to Figure 3–1 and Table 3–2 often as you learn about the principal cell structures described in the paragraphs that follow.

TABLE 3–1 Examples of cell types

Type	Example	Structural Features	Functions
Nerve Cell		Surface that is sensitive to stimuli Long extensions	Detects changes in internal or external environment Transmits nerve impulses from one part of the body to another
Muscle Cell		Elongated, threadlike Contains tiny fibers that slide together forcefully	Contracts (shortens), allowing movement of body parts
Red Blood Cells		Contains hemoglobin, a red pigment that attracts, then releases, oxygen	Transports oxygen in the bloodstream (from the lungs to other parts of the body)
Gland Cells		Contains sacs that release a secretion to the outside of the cell	Releases substances such as hormones, enzymes, mucus, and sweat
Immune Cells		Some have outer membranes able to engulf other cells Some have systems that manufacture antibodies Some are able to destroy other cells	Recognize and destroy "nonself" cells such as cancer cells and invading bacteria

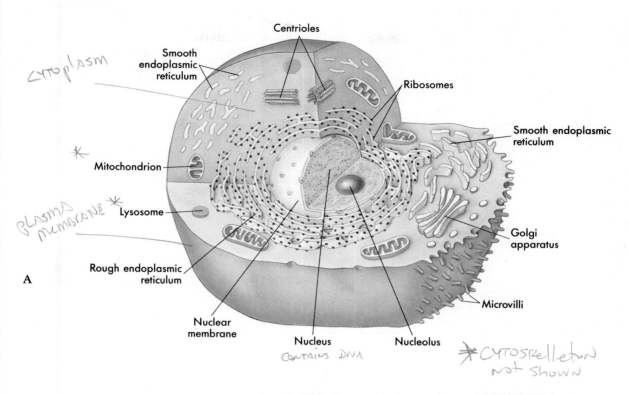

ORGANELLES

cytoplasm

PLASMA MEMBRANE

CONTAINS DNA

CYTOSKELLETON not shown

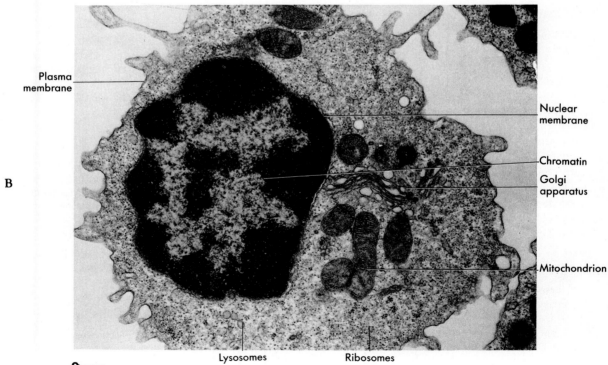

FIGURE 3-1 Typical or composite cell. A, Artist's interpretation of cell structure. **B,** Color-enhanced electron micrograph of a cell. Both show the many mitochondria, popularly known as the "power plants of the cell." Note, too, the innumerable dots bordering the endoplasmic reticulum. These are ribosomes, the cell's "protein factories."

TABLE 3–2   Some major cell structures and their functions

Cell Structure	Functions
MEMBRANOUS	
Plasma membrane	Serves as the boundary of the cell, maintaining its integrity; protein molecules on outer surface of plasma membrane perform various functions; for example, they serve as markers that identify cells of each individual, as receptor molecules for certain hormones and other molecules, and as transport mechanisms
Endoplasmic reticulum (ER)	Ribosomes attached to rough ER synthesize proteins that leave cells via the Golgi complex; smooth ER synthesizes lipids incorporated in cell membranes, steroid hormones, and certain carbohydrates used to form glycoproteins
Golgi apparatus	Synthesizes carbohydrate, combines it with protein, and packages the product as globules of glycoprotein
Lysosomes	A cell's "digestive system"
Peroxisomes	Contain enzymes that detoxify harmful substances
Mitochondria	Catabolism; ATP synthesis; a cell's "power plants"
Nucleus	Dictates protein synthesis, thereby playing essential role in other cell activities, namely, cell transport, metabolism, and growth
NONMEMBRANOUS	
Ribosomes	Synthesize proteins; a cell's "protein factories"
Cytoskeleton	Acts as a framework to support the cell and its organelles; functions in cell movement; forms cell extensions (microvilli, cilia, flagella)
Cilia and Flagella	Hairlike cell extensions that serve to move substances over the cell surface (cilia) or propel sperm cells (flagella)
Nucleolus	Plays an essential role in the formation of ribosomes

Cell Structures

Ideas about cell structure have changed considerably over the years. Early biologists saw cells as simple, fluid-filled bubbles. Today's biologists know that cells are far more complex than this. Each cell is surrounded by a plasma membrane that separates the cell from its surrounding environment. The inside of the cell is composed largely of a thick fluid called cytoplasm (literally, "cell substance"). Suspended in the cytoplasm are various organelles, including a central nucleus. Each different organelle is structurally suited to perform a specific function within the cell—much as each of your organs is suited to a specific function within your body. In short, the main cell structures are (1) the plasma membrane, (2) cytoplasm, and (3) the organelles (Figure 3-1).

> **QUICK CHECK**
> ✔ 1. What important concept in biology was proposed by Schleiden and Schwann?
> ✔ 2. Give an example of how cell structure relates to its function.
> ✔ 3. List the three main structural components of a typical cell.

◣ CELL MEMBRANES

Figure 3-1 shows that a typical cell contains a variety of membranes. The outer boundary of the cell or **plasma** membrane is just one of these membranes. Each cell also has various *membranous organelles*. Membranous organelles are sacs and canals made of the same membrane material as the plasma membrane. This membrane material is a very thin sheet—only about 75 Å or .0000003 inch thick—made of lipid, protein, and other molecules.

Figure 3-2 shows the current model of how cell membranes are constructed. This concept of cell membranes is called the **fluid mosaic model.** Like the tiles in an art mosaic, the molecules that comprise a cell membrane are arranged in a sheet. Unlike art mosaics, however, this mosaic of molecules is *fluid.* That is, the molecules are able to slowly float around the membrane like icebergs. The fluid mosaic model shows us that the molecules of a cell membrane are bound tightly enough to form a continuous sheet but loosely enough that the molecules can slip past one another.

What are the forces that hold together a cell membrane? The short answer to that question is: chemical attractions. The primary structure of a cell membrane is a double layer of phospholipid molecules. Recall from Chapter 2 that phospholipid molecules have "heads" that are water soluble and double "tails" that are lipid soluble (see Figure 2-18 on p. 54). Because their heads are **hydrophilic** (water-loving) and their tails are **hydrophobic** (water-fearing), phospholipid molecules naturally arrange themselves into double layers or *bilayers* in water. This allows all the hydrophilic heads to face water and all the hydrophobic tails to face away from the water. Because the internal environment of

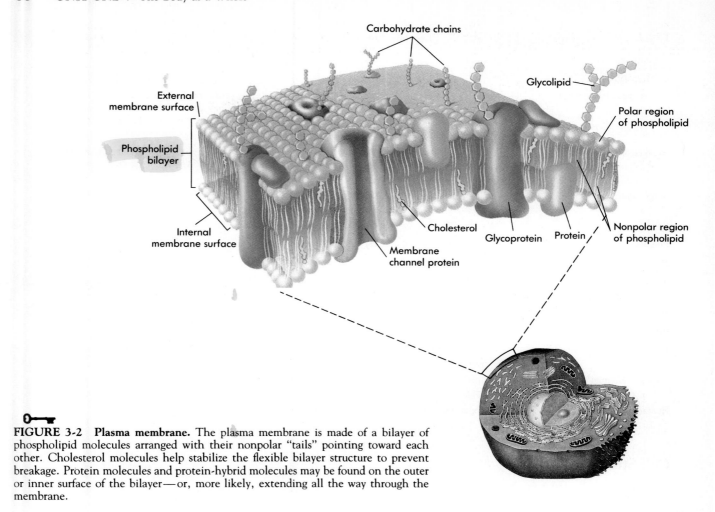

FIGURE 3-2   Plasma membrane. The plasma membrane is made of a bilayer of phospholipid molecules arranged with their nonpolar "tails" pointing toward each other. Cholesterol molecules help stabilize the flexible bilayer structure to prevent breakage. Protein molecules and protein-hybrid molecules may be found on the outer or inner surface of the bilayer—or, more likely, extending all the way through the membrane.

the body is simply a water-based solution, phospholipid bilayers appear wherever phospholipid molecules are scattered among the water molecules. Cholesterol is a steroid lipid that mixes with phospholipid molecules to form a blend of lipids that stays just fluid enough to function properly at body temperature. Without cholesterol, cell membranes would break far too easily.

Each human cell manufactures phospholipid and cholesterol molecules, which then arrange themselves into a bilayer to form a natural "fencing" material that can be used throughout the cell. This "fence" allows lipid-soluble molecules to pass through easily—just like a picket fence allows air and water to pass through easily. However, because most of the phospholipid bilayer is hydrophobic, cell membranes do not allow water or water-soluble molecules to pass through easily. This characteristic of cell membranes is ideal, because most of the substances in the internal environment are water soluble. What good is a membrane if it allows just about everything to pass through it?

A cell can control what moves through any section of membrane by means of proteins imbedded in the phospholipid bilayer (see Figure 3-2). Many of these *membrane proteins* have openings that, like gates in a fence, allow water-soluble molecules to pass through the membrane.

Specific kinds of transport proteins allow only certain kinds of molecules to pass through—and the cell can determine whether these "gates" are open or closed at any particular time. We will consider this function of membrane proteins again when we discuss transport mechanisms in the cell.

Membrane proteins also serve a variety of other functions (Table 3-3). Some membrane proteins have carbohydrates attached to their outer surface—forming *glycoprotein* molecules—that act as identification markers. Such identification markers, which are recognized by other molecules, act as signs on a fence that identify the enclosed area. Cells and molecules of the immune system can thus distinguish between "self" cells and "nonself" cells. This mechanism not only allows us to attack cancer or bacterial cells, it also prevents us from receiving blood donations from people without similar cell markers. Other membrane proteins are *receptors* that can react to the presence of hormones or other regulatory chemicals and thereby trigger metabolic changes in the cell. Some membrane proteins are enzymes that catalyze cellular reactions. Some membrane proteins bind to other membrane proteins to form connections between cells or bind to support filaments within the cell to anchor them.

Table 3-3 summarizes some of the essential functions of cell membranes.

TABLE 3–3   Essential functions of cell membranes

Function		Structure
Maintain wholeness (integrity) of a cell or membranous organelle		Sheet (bilayer) of phospholipids stabilized by cholesterol
Controlled transport of water-soluble molecules from one compartment to another		Membrane proteins that act as channels or carriers of molecules
Sensitivity to hormones and other regulatory chemicals		Receptor molecules that trigger metabolic changes in membrane (or on other side of membrane)
Regulation of metabolic reactions		Enzyme molecules that catalyze specific chemical reactions
Form connections between cells		Membrane proteins that bind to one another
Support and maintain the shape of a cell or membranous organelle		Membrane proteins that bind to support filaments within the cytoplasm
Identification of cells or organelles		Glycoproteins or proteins in the membrane that act as markers

CYTOPLASM AND ORGANELLES

Cytoplasm is the thick internal fluid of cells in which are suspended many tiny structures. Early cell scientists believed cytoplasm to be a rather uniform fluid with only one object suspended within it: the nucleus. We now know that the cytoplasm of each cell contains hundreds or even thousands of "little organs" or **organelles**—each with specific functions (see Table 3-2). Until relatively recently, when newer microscope technology became available, many of these organelles simply could not be seen. If they were able to be seen, they were simply called *inclusions* because their roles as integral parts of the cell were not yet recognized. Undoubtedly, some of the cellular structures we now call inclusions will eventually be recognized as organelles and given specific names.

There are many types of organelles that have been identified in various cells of the body. To make it easier to study them, they have been classified into two major groups: the *membranous organelles* and the *nonmembranous organelles.* The membranous organelles are the organelles that are specialized sacs or canals made of cell membrane. The nonmembranous organelles are not made of membrane but are made of microscopic filaments or other nonmembranous materials. As you read through the following sections, be sure to note whether the organelle is membranous or not (see Table 3-2).

Endoplasmic Reticulum

Endoplasm means the cytoplasm located toward the center of a cell. Reticulum means network. Therefore the name **endoplasmic reticulum (ER)** means literally a network located deep inside the cytoplasm. And when first seen, it appeared to be just that. Later on, however, more highly magnified views under the electron microscope showed the endoplasmic reticulum distributed throughout the cytoplasm. It consists of membranous-walled canals and flat, curving sacs arranged in parallel rows.

There are two types of endoplasmic reticulum: *rough* and *smooth.* Innumerable small granules—ribosomes by name—dot the outer surfaces of the membranous walls of the rough type and give it its "rough" appearance. Ribosomes are themselves organelles. The canals of the endoplasmic reticulum wind tortuously through the cytoplasm, extending all the way from the plasma membrane to the nucleus. The membrane forming the walls of the endoplasmic reticulum has essentially the same molecular structure as the plasma membrane.

The structural fact that the endoplasmic reticulum is an interconnected system of canals suggests that it might function as a miniature circulatory system for the cell. And in fact, proteins do move through the canals. The ribosomes attached to the rough endoplasmic reticulum synthesize proteins. These proteins enter the canals, move through them toward the Golgi apparatus, and eventually leave the cell. Thus the rough endoplasmic reticulum functions in both protein synthesis and intracellular transportation.

No ribosomes border the membranous wall of the smooth endoplasmic reticulum—hence its smooth appearance and its name. Its functions are less well established and probably more varied than those of the rough type. The smooth endoplasmic reticulum is now believed to synthesize certain lipids and carbohydrates. Included among these are the steroid hormones and some of the carbohydrates used to form glycoproteins. Lipids that form cell membranes are also synthesized here. As these membrane lipids are made, they simply become part of the smooth ER's wall. Bits of the ER break off from time to time and travel to other membranous organelles—even the plasma membrane—and thus add to their membrane. The smooth ER, then, is the organelle that makes membrane for use throughout the cell.

Ribosomes

Every cell contains thousands of **ribosomes.** Many of them are attached to the rough endoplasmic reticulum and many of them lie free, scattered through the cytoplasm. Find them in both locations in Figure 3-1. Because ribosomes are too small to be seen with a light microscope, no one knew they existed until the electron microscope revealed them. Research has now yielded information about their molecular structure. Each ribosome is a nonmembranous structure made of two pieces, a large subunit and a small subunit (Figure 3-3). Each subunit is composed of ribonucleic acid (RNA) bonded to protein.

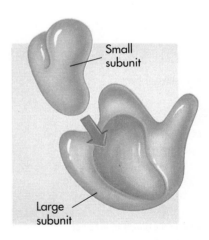

Small subunit

Large subunit

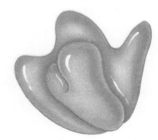

Ribosome

FIGURE 3-3 Ribosome. A ribosome is composed of a small subunit and a large subunit. The subunits fit together to form a complete ribosome.

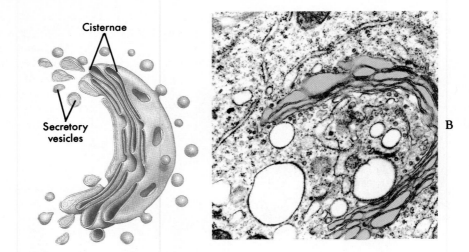

FIGURE 3-4   Golgi apparatus. A, Artist's interpretation of the structure of the Golgi apparatus, showing a stack of flattened sacs, or *cisternae,* and numerous small membranous bubbles, or *secretory vesicles.* **B,** Transmission electron micrograph showing the Golgi apparatus highlighted with color.

Ribosomal RNA is often abbreviated as rRNA. Other types of ribonucleic acid in the cell are messenger RNA (mRNA) and transfer RNA (tRNA) (see pp. 86 to 87).

The function of ribosomes is protein synthesis. Ribosomes are the molecular machines that make proteins. Or to use a popular term, they are the cell's "protein factories." The ribosomes attached to the endoplasmic reticulum synthesize proteins mainly for "export." The ribosomes that are free in the cytoplasm make proteins for the cell's own domestic use. They make both its structural and its functional proteins (enzymes). Working ribosomes, those that are actually making proteins, appear to function in groups called *polyribosomes.* Under the electron microscope, polyribosomes look like short strings of beads.

Golgi Apparatus

The **Golgi apparatus** is a membranous organelle consisting of tiny sacs or *cisternae* stacked on one another and located near the nucleus (Figure 3-4; see also Figure 3-1). Like the endoplasmic reticulum, the Golgi apparatus processes molecules within its membranes. As a matter of fact, the Golgi apparatus seems to be part of the same mechanism that prepares protein molecules for export from the cell.

The role of the Golgi apparatus in processing and packaging protein molecules for export from the cell is summarized in Figure 3-5. First, proteins synthesized by ribosomes and transported to the end of an ER canal are packaged into tiny membranous bubbles, or *vesicles,* that break away from the ER. The vesicles then move to the Golgi apparatus and fuse with the first cisterna. Protein molecules thus released into the cisterna are then chemically altered by enzymes present there. For example, the enzymes may attach carbohydrate molecules synthesized in the Golgi apparatus to form glycoproteins. The molecules are then "pinched off" in another vesicle, which moves to the next cisterna for further processing. The proteins and

glycoproteins eventually end up in the last cisterna, from which vesicles pinch off and move to the plasma membrane. There, they release their contents outside the cell in a process called *secretion.* Some protein and glycoprotein molecules may instead be incorporated into the membrane of a Golgi vesicle. This means that these molecules eventually become part of the plasma membrane, as seen in Figure 3-2.

Lysosomes

Like the endoplasmic reticulum and the Golgi apparatus, **lysosomes** have membranous walls—indeed, they are vesicles that have pinched off from the Golgi apparatus. The size and shape of lysosomes change with the stage of their activity. In their earliest, inactive stage, they look like mere granules. Later, as they become active, they take on the appearance of small vesicles or sacs (see Figure 3-1) and often contain tiny particles such as fragments of membranes or pigment granules. The interior of the lysosome contains various kinds of enzymes capable of breaking down all the main components of cells. These enzymes can, and under some circumstances actually do, destroy cells by digesting them. It is little wonder that these powerful and dangerous substances are usually kept sealed up in lysosomes. However, lysosomal enzymes more often protect than destroy cells. Large molecules and large particles (for example, bacteria) that find their way into cells enter lysosomes, and their enzymes dispose of them by digesting them. Therefore "digestive bags" and even "cellular garbage disposals" are nicknames for lysosomes. White blood cells serve as scavenger cells for the body, engulfing bacteria and destroying them in their lysosomes.

Peroxisomes

In addition to lysosomes, another type of smaller membranous sac containing enzymes is also present in the cytoplasm of some cells. These organelles, called **peroxi-**

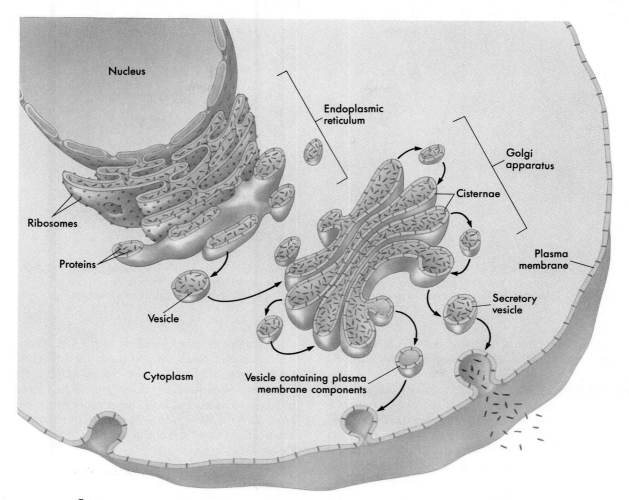

FIGURE 3-5 **Function of the Golgi apparatus.** The Golgi apparatus processes and packages protein molecules delivered from the ER by small vesicles. After entering the first cisterna of the Golgi apparatus, a protein molecule undergoes a series of chemical modifications, is sent (by means of a vesicle) to the next cisterna for further modification, and so on, until it is ready to exit the last cisterna. When it is ready to exit, a molecule is packaged in a membranous secretory vesicle that migrates to the surface of the cell and "pops open" to release its contents into the space outside the cell. Some vesicles remain inside the cell for some time, serving as storage vessels for the substance to be secreted.

somes, serve to detoxify harmful substances which may enter cells. They are often seen in kidney and liver cells, which serve important detoxification functions in the body. Peroxisomes contain the enzymes *peroxidase* and *catalase,* which are important in metabolic reactions involving hydrogen peroxide—a chemical toxic to cells.

Mitochondria

Note the **mitochondria** shown in Figures 3-1, 3-6, and 3-7. Magnified thousands of times, as they are there, they look like small, partitioned sausages—if you can imagine sausages only 1.5 µm long and one half as wide. (In case you can visualize inches better than micrometers, 1.5 µm equals about 3/50,000 of an inch.) Yet, like all organelles, and tiny as they are, mitochondria have a highly organized molecular structure. Their membranous walls consist of not one but two delicate membranes. They form a sac within a

sac. The inner membrane is thrown into folds called **cristae.** Imbedded in the inner membrane are enzymes that are essential for the making of one of the most important chemicals in the world. Without this compound, life cannot exist. Its long name, *adenosine triphosphate,* and its abbreviation, *ATP,* were mentioned in Chapter 2. Chapter 26 presents more detailed information about this vital substance.

Both inner and outer membranes of the mitochondrion have essentially the same molecular structure as the cell's plasma membrane. All evidence so far indicates that the proteins in the membranes of the cristae are arranged precisely in the order of their functioning. This is another example, but surely an impressive one, of the principle stressed in Chapter 1 that organization is a foundation stone and a vital characteristic of life.

Enzymes in the mitochondrial inner membrane catalyze

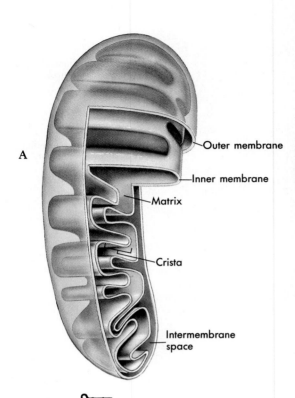

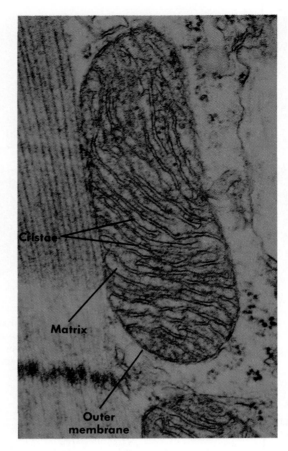

FIGURE 3-6   Mitochondrion. A, Cutaway sketch showing outer and inner membranes. Note the many folds (cristae) of the inner membrane. **B,** Transmission electron micrograph of a mitochondrion. Although some mitochondria have the capsule shape shown here, many are round or oval.

oxidation reactions. These are the chemical reactions that provide cells with most of the energy that does all of the many kinds of work that keeps them and the body alive. Thus do mitochondria earn their now familiar title, the "power plants" of cells.

The fact that mitochondria generate most of the power for cellular work suggests that the number of mitochondria in a cell might be directly related to its amount of activity. This principle does seem to hold true. In general, the more work a cell does, the more mitochondria its cytoplasm contains. Liver cells, for example, do more work and have more mitochondria than sperm cells. A single liver cell contains 1,000 or more mitochondria, whereas only about 25 mitochondria are present in a single sperm cell.

CYTOSKELETON

As its name implies, the **cytoskeleton** is the cell's internal supporting framework. Like the bony skeleton of the body, the cytoskeleton is made up of rather rigid, rodlike pieces that not only provide support but also allow movement. In this section, we will discuss the basic characteristics of the cell's skeleton, as well as several organelles associated with it.

Cell Fibers

No one knew much about **cell fibers** until the development of two new research methods; one uses fluorescent molecules and the other uses stereomicroscopy, that is, three-dimensional pictures of whole, unsliced cells made with high-voltage electron microscopes. Using these techniques, investigators discovered intricate arrangements of fibers of varying widths. The smallest fibers seen have a width of about 3 to 6 nm — less than one millionth of an inch! Particularly striking about these fine fibers is their arrangement. They form a three-dimensional, irregularly shaped lattice, a kind of scaffolding in the cell. These fibers appear to support parts of the cell formerly thought to float free in the cytoplasm — the endoplasmic reticulum, mitochondria, and "free" ribosomes (Figure 3-7).

The smallest cell fibers are called *microfilaments.* Microfilaments often serve as "cellular muscles." Microfilaments are made of thin, twisted strands of protein molecules (Figure 3-8, A) and usually form bundles that lie parallel to the long axis of a cell. The different proteins in the microfilaments can slide past one another, causing shortening of the cell. The most obvious example of this occurs in muscle cells, where many bundles of microfilaments work together to shorten the cells with great force.

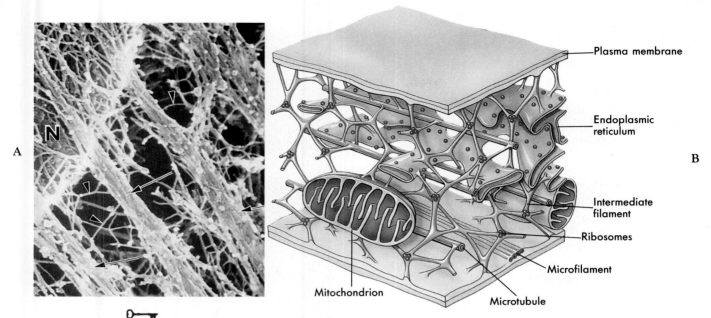

FIGURE 3-7 **The cytoskeleton. A,** Color-enhanced electron micrograph of a portion of the cell's internal framework. The letter *N* marks the nucleus, the arrowheads mark the intermediate filaments, and the complete arrows mark the microtubules. **B,** Artist's interpretation of the cell's internal framework. Notice that the "free" ribosomes and other organelles are not really free at all.

Cell fibers called *intermediate filaments* are twisted protein strands that are slightly thicker than microfilaments (Figure 3-8, *B*). Intermediate filaments are thought to form much of the supporting framework in many types of cells. For example, the protective cells in the outer layer of skin are filled with a dense arrangement of tough intermediate filaments.

The thickest of the cell fibers are tiny, hollow tubes called *microtubules*. As Figure 3-8, *C*, shows, microtubules are made of protein subunits arranged in a spiral fashion. Microtubules are sometimes called the "engines" of the cell because they often move things around in the cell—or even cause movement of the entire cell. For example, the movement of vesicles within the cell and the movement of chromosomes during cell division are both thought to be accomplished by microtubules.

Centrosome

The **centrosome** is an area of the cytoplasm near the nucleus that coordinates the building and breaking of microtubules in the cell. For this reason, this nonmembranous structure has sometimes been called the *microtubule-organizing center*. The centrosome plays an important role during cell division, when a special "spindle" of microtubules is constructed for the purpose of moving chromosomes around the cell.

The boundaries of the centrosome are rather indistinct because it lacks a membranous wall. However, the general location of the centrosome is easy to find because of a pair of cylindrical structures called *centrioles*.

Under the light microscope, **centrioles** appear as two

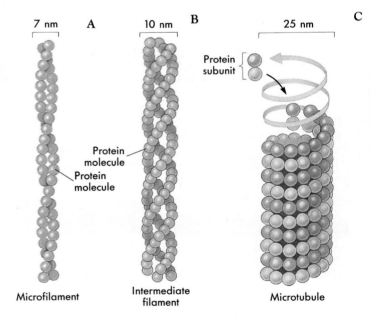

FIGURE 3-8 **Cell fibers. A,** *Microfilaments* are thin, twisted strands of protein molecules. **B,** *Intermediate filaments* are thicker, twisted protein strands. **C,** *Microtubules* are hollow fibers made of a spiral arrangement of protein subunits.

dots located near the nucleus. The electron microscope, however, reveals them not as mere dots but as tiny cylinders (see Figure 3-1). The walls of the cylinders consist of nine bundles of microtubules, with three tubules in each bundle. A curious fact about these two tubular-walled

cylinders is their position at right angles to each other. This special arrangement occurs when the centrioles separate in preparation for cell division (see p. 88 in this chapter). Before separating, a daughter centriole is formed perpendicularly to each of the original pair, so that a complete pair may be distributed to each new cell. In addition to their involvement in forming the spindle that appears during cell division (see Table 3-5), centrioles are essential for the formation of microtubular cell extensions, discussed in the next section.

Cell Extensions

In some cells, the cytoskeleton forms projections that extend the plasma membrane outward to form tiny, fingerlike processes. These processes, *microvilli, cilia,* and *flagella,* are present only in certain types of cells—depending, of course, on a cell's specialized functions. **Microvilli,** for example, are found in epithelial cells that line the intestines and other areas where absorption is important (Figure 3-9, A). Like tiny fingers crowded against each other, microvilli cover part of the surface of a cell (see Figure 3-1). A single microvillus measures about 0.5 μm long and only 0.1 μm or less across. Since one cell has hundreds of these projections, the surface area of the cell is increased manyfold—a structural feature that enables the cell to perform its function of absorption at a faster rate.

Cilia and **flagella** are cell processes that both have cylinders made of microtubules at their core. This cylinder is composed of nine double microtubules arranged around two single microtubules in the center—a slightly different arrangement than in centrioles. This particular arrangement is suited to movement. Additional short microtubules at the base of the cylinder react with the longer microtubules, causing them to slip back and forth to produce a "wiggling" movement.

What distinguishes cilia from flagella are their size and number. Cilia are shorter and more numerous than flagella (Figure 3-9, B). Under low magnification, cilia look like tiny hairs. In the lining of the respiratory tract, movement of cilia keep contaminated mucus moving toward the throat where it can be swallowed. In the lining of the female reproductive tract, cilia keep the ovum moving toward the uterus. Flagella are single, long structures in the only type of human cell that has this feature: the human sperm cell (Figure 3-9, C). A sperm cell's flagellum moves like the tail of an eel, allowing the cell to "swim" toward the female sex cell (ovum).

FIGURE 3-9 Cell processes. A, *Microvilli* are numerous, fingerlike projections that increase the surface area of absorptive cells. This electron micrograph shows a cross section of microvilli from a cell lining the small intestine. Note the bundles of microfilaments that support the microvilli. **B,** *Cilia* are numerous, fine processes that transport fluid across the surface of the cell. This electron micrograph shows cilia (long projections) and microvilli (small bumps) on cells lining the lung airways. **C,** In humans, *flagella* are single, elongated processes on sperm cells that enable these cells to "swim."

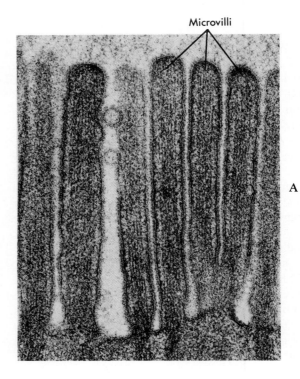

Microvilli

A

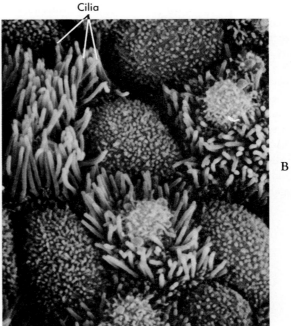

Cilia

B

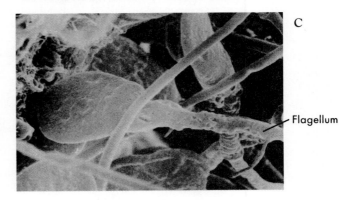

C

Flagellum

NUCLEUS

The **nucleus,** one of the largest cell structures (see Figure 3-1), occupies the central portion of the cell. Both the shape of the nucleus and the number of nuclei present in a cell vary. One spherical nucleus per cell, however, is common. Electron micrographs show that two membranes perforated by openings or pores enclose the *nucleoplasm* (nuclear fluid) (Figure 3-10).

These nuclear membranes, together called the *nuclear envelope,* have essentially the same structure as other cell membranes and appear to be extensions of the membranous walls of the endoplasmic reticulum. Probably the most important fact to remember about the nucleus is this: it contains DNA molecules, the famous heredity molecules. In nondividing cells they appear as granules or threads, which are named *chromatin* (from the Greek *chroma,* "color") because they readily take the color of dyes. When the process of cell division begins, DNA molecules become tightly coiled. They then look like short, rodlike structures and are called **chromosomes.** All normal human cells (except mature sex cells) contain 46 chromosomes, and each chromosome consists of one DNA molecule plus some protein molecules.

The functions of the nucleus are actually functions of DNA molecules. In general, DNA molecules dictate both the structure and the function of cells, and they determine heredity. The most prominent structure visible in the nucleus is a small nonmembrane body that stains densely and is called the **nucleolus** (Figure 3-10, A). Like chromosomes, it consists chiefly of a nucleic acid, but the nucleic acid is *not* DNA. It is RNA, ribonucleic acid (see p. 86).

The nucleolus functions to synthesize ribosomal RNA (rRNA) and combine it with protein to form ribosomes, the protein synthesizers of cells. You might guess, therefore, and correctly so, that the more protein a cell makes, the larger its nucleolus will appear. Cells of the pancreas, to cite just one example, make large amounts of protein and have large nucleoli.

CELL CONNECTIONS

The tissues and organs of the body must be held together, so they must of course be connected in some way. Some cells are held together by fibrous nets that surround groups of cells. Certain muscle cells are held together this way. More often, though, the cells form direct connections with each other. Such connections not only hold the cells together, they sometimes also allow direct communication between the cells. The major types of direct cell connections are summarized in Figure 3-11.

Desmosomes are like small "spot welds" that hold adjacent cells together. Adjacent skin cells are held together this way. Notice in Figure 3-11 that fibers on the outer surface of each desmosome interlock with each other. This arrangement resembles Velcro, a product that holds things together tightly when tiny plastic hooks become interlocked with fabric loops. Notice also that the desmosomes are anchored internally by intermediate filaments of the cytoskeleton.

Gap junctions are formed when membrane channels of adjacent plasma membranes adhere to each other. As Figure 3-11 shows, such junctions have two effects: (1) they form gaps or "tunnels" that join the cytoplasm of two cells

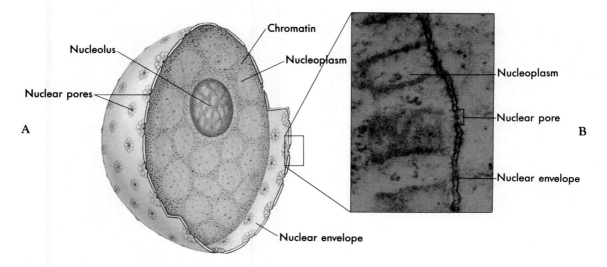

FIGURE 3-10   Nucleus. A, Artist's rendering and **B,** electron micrograph show that the nuclear envelope is composed of two separate membranes and is perforated by large openings, or nuclear pores.

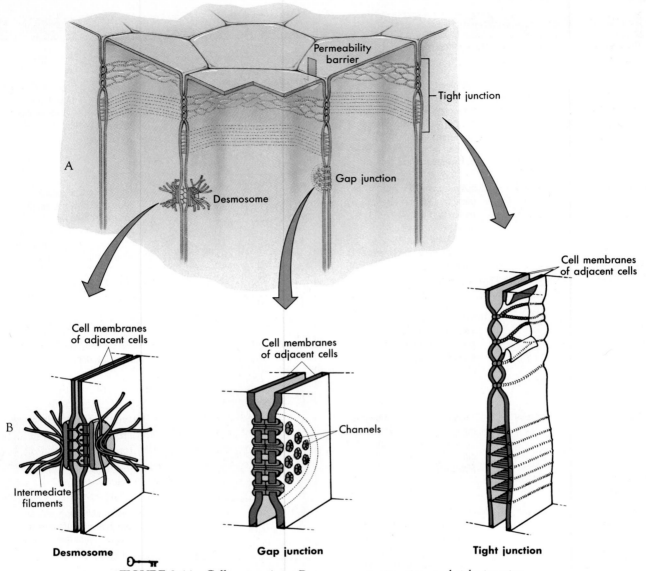

FIGURE 3-11   Cell connections. Desmosome, gap junction, and tight junction.

and (2) they fuse the two plasma membranes into a single structure. One advantage of this arrangement is that certain molecules can pass directly from one cell to another. Another advantage is that impulses traveling along a membrane can travel over many cell membranes in a row without stopping because they have "run out of membrane." Heart muscle cells are joined by gap junctions so that a single impulse can travel to, and thus stimulate, many cells at the same time.

Tight junctions occur in cells that are joined by "collars" of tightly fused membrane. As you can see in Figure 3-11, rows of membrane proteins that extend all the way around a cell fuse with similar membrane proteins in neighboring cells. An entire sheet of cells can be bound together the way soft drink cans are held in a six-pack by plastic collars—only more tightly. When tight junctions hold a sheet of cells together, molecules cannot *permeate,* or spread through, the cracks between the cells. Tight junctions occur in the lining of the intestines and other

parts of the body, where it is important to control what gets past a sheet of cells. The only way for a molecule to get past the intestinal lining is through controlled channel or carrier molecules in the plasma membranes of the cells.

MOVEMENT OF SUBSTANCES THROUGH CELL MEMBRANES

If a cell is to survive, it must be able to move substances to places where they are needed. We already know one way that cells move organelles within the cytoplasm: pushing or pulling by the cytoskeleton. A cell must also be able to move various molecules in and out through the plasma membrane, as well as from compartment to compartment within the cell. In this part of the chapter, we will briefly describe the basic mechanisms a cell uses to move substances across its membranes. As you read through this section refer to Table 3-4, which summarizes the various transport mechanisms mentioned in the text.

TABLE 3-4   Some important transport processes

Process	Type	Description		Examples
Simple Diffusion	Passive	Movement of particles through the phospholipid bilayer or through channels from an area of high concentration to an area of low concentration—that is, down the concentration gradient		Movement of carbon dioxide out of all cells; movement of sodium ions into nerve cells as they conduct an impulse
Dialysis	Passive	Diffusion of small solute particles, but not larger solute particles, through a selectively permeable membrane; results in separation of large and small solutes		During procedure called *peritoneal dialysis,* small solutes diffuse from blood vessels but blood proteins do not (thus removing only small solutes from the blood)
Osmosis	Passive	Diffusion of water through a selectively permeable membrane in the presence of at least one impermeant solute		Diffusion of water molecules into and out of cells to correct imbalances in water concentration
Facilitated Diffusion	Passive	Diffusion of particles through a membrane by means of carrier molecules; also called *carrier-mediated passive transport*		Movement of glucose molecules into most cells
Active Transport	Active	Movement of solute particles from an area of low concentration to an area of high concentration (up the concentration gradient) by means of a carrier molecule		In muscle cells, pumping of nearly all calcium ions to special compartments—or out of the cell
Phagocytosis	Active	Movement of cells or other large particles into a cell by trapping it in a section of plasma membrane that pinches off to form an intracellular vesicle; type of *endocytosis*		Trapping of bacterial cells by phagocytic white blood cells
Pinocytosis	Active	Movement of fluid and dissolved molecules into a cell by trapping them in a section of plasma membrane that pinches off to form an intracellular vesicle; type of *endocytosis*		Trapping of large protein molecules by some body cells
Exocytosis	Active	Movement of proteins or other cell products out of the cell by fusing a secretory vesicle with the plasma membrane		Secretion of the hormone, prolactin, by pituitary cells

Before beginning a discussion of individual processes, we must point out that membrane transport processes can be labeled as *passive* or *active*. Passive processes do not require any energy expenditure or "activity" of the cell membrane—the particles move by using energy that they already have. Active processes, on the other hand, do require the expenditure of metabolic energy by the cell. In active processes, the transported particles are actively "pulled" across the membrane. Keep this distinction in mind as we explore the basic mechanisms of cell membrane transport.

Passive Transport Processes

Diffusion

Often molecules simply spread or *diffuse* through the membranes. The term *diffusion* refers to a natural phenomenon caused by the tendency of small particles to spread out evenly within any given space. All molecules in a solution bounce around in short, chaotic paths. As they collide with one another, they tend to spread out or diffuse. Think of the example of a lump of sugar dissolving in water (Figure 3-12). After the lump is placed in the water, the sugar molecules are very close to one another—there is a very high sugar concentration in the lump. As the sugar molecules dissolve, they begin colliding with one another and thus push each other away. Given enough time, the sugar molecules will eventually diffuse evenly through the water.

Notice that during diffusion, molecules move from an area of high concentration to an area of low concentration. It is not surprising, then, that molecules tend to move from the side of the membrane with high concentration to the side of the membrane with a lower concentration of that molecule. Another way of stating this principle is to say that diffusion occurs down a *concentration gradient*. A concentration gradient is simply a measurable difference in concentration from one area to another. Since molecules spread from the area of *high* concentration to the area of

low concentration, they spread *down* the concentration gradient.

Perhaps the best way to learn the principle of diffusion across a membrane is to look at the example illustrated in Figure 3-13. Suppose a 10% glucose solution is separated from a 20% glucose solution by a membrane. Suppose further that the membrane has pores in it that allow glucose molecules to pass through, as well as pores that allow water molecules to pass. Glucose molecules and water molecules darting about the solution collide with each other and with the membrane. Some inevitably hit the membrane pores from the 20% glucose side, and some hit the membranes pores from the 10% side. Just as inevitably, some pass through the pores in both directions. For a while, more glucose molecules enter the pores from the 20% side simply because they are more numerous there than on the 10% side. More of these particles, therefore, move through the membrane from the 20% glucose solution into the 10% glucose solution than diffuse through it in the opposite

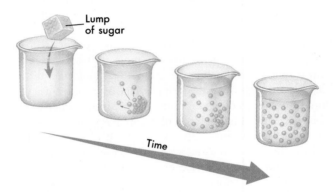

FIGURE 3-12   Diffusion. The molecules of a lump of sugar are very densely packed when they enter the water. As sugar molecules collide frequently in the area of high concentration, they eventually spread away from each other—toward the area of lower concentration. Eventually, the sugar molecules are evenly distributed.

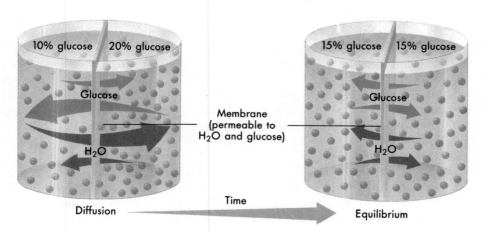

FIGURE 3-13   Diffusion through a membrane. Note that the membrane allows glucose and water to pass and that it separates a 10% glucose solution from a 20% glucose solution. The container on the left shows the two solutions separated by the membrane at the start of diffusion. The container on the right shows the result of diffusion after time.

direction. In other words, the overall direction of diffusion is from the side where concentration is higher (20%) to the side where concentration is lower (10%).

During the time that diffusion of glucose molecules is taking place, diffusion of water molecules is also going on. Remember, the direction of diffusion of any substance is always down that substance's concentration gradient. Water molecules are more concentrated on the 10% glucose solution side because the solution is more dilute—or watery—on that side. Thus water molecules move from the 10% glucose solution side to the 20% glucose solution side. As Figure 3-13 shows, diffusion of both kinds of molecules eventually produces an equilibrium in which both solutions have equal concentrations. We say that *equilibration* has occurred.

Equilibrium is not a static state with no movement of molecules across the membrane. Instead, it is a balanced state in which the number of molecules of a substance bouncing to one side of the membrane exactly equals the number of molecules of that substance that are bouncing to the other side. Once equilibration has occurred, overall diffusion may have stopped, but balanced diffusion of small numbers of molecules may continue.

Now that we know that concentration gradients drive diffusion, we can explore how the molecules actually find a way through a cell membrane. Sometimes molecules diffuse directly through the bilayer of phospholipid molecules that forms most of a cell membrane. As discussed earlier in this chapter, lipid-soluble molecules can pass through easily. Such molecules simply dissolve in the phospholipid fluid, diffuse through this fluid, and then move into the water solution on the other side of the membrane. For a long time, biologists thought that this was the only way that molecules could diffuse through a cell membrane. They found that water-soluble molecules such as sodium ions (Na^+) and glucose molecules could not pass through an artificial phospholipid bilayer easily. Yet they also found that such small, water-soluble molecules could pass through living cell membranes quickly. Even water molecules, which are small enough to pass through the thin phospholipid film with only moderate ease, diffuse rapidly through most cell membranes. It was not until the presence of various transport proteins, such as membrane channels, was discovered that we understood how these molecules diffuse across cell membranes.

As you already know, cell membranes possess protein "tunnels" better known as *membrane channels* (see Table 3-3 and Figure 3-14). Membrane channels are pores through which specific ions or other small, water-soluble molecules can pass. For example, sodium ions (Na^+) pass only through *sodium channels* and chloride ions (Cl^-) pass only through *chloride channels*. When molecules are allowed to cross a membrane, they are said to *permeate* the membrane. Thus a particular membrane is *permeable* to sodium only if sodium channels are present in that membrane. The membrane is said to be *impermeable* to sodium if there are no channels or other transport proteins that allow sodium ions to pass. Thus living membranes can

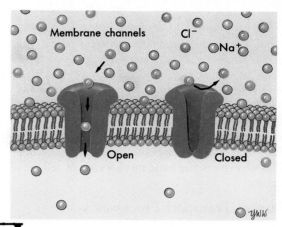

FIGURE 3-14 Membrane channels. Gated channel proteins form tunnels through which specific molecules may pass—as long as the "gates" are open. Notice that the transported molecules move from an area of high concentration to an area of low concentration.

be permeable to some molecules but not to others, depending on the type of channels present. As Figure 3-14 shows, the permeability of a membrane can also be affected by the opening and closing of specific channels. Channels that operate this way are called *gated channels*. Because a living cell membrane can limit the diffusion of some molecules this way, we say the membrane is *selectively permeable*.

Diffusion is a passive process. In other words, the energy for transport through a membrane does not come from the membrane but from the energy of collision already possessed by the moving molecule. The only requirement of the cell is that it be permeable to the type of molecule in question.

Dialysis

Under certain circumstances, a type of diffusion called **dialysis** may occur. Dialysis is a form of diffusion in which the selectively permeable nature of a membrane causes the separation of smaller solute particles from larger solute particles. *Solutes* are the particles dissolved in a *solvent* such as water. Together, the solutes and solvents form a mixture called a *solution*. Figure 3-15 illustrates the principle of dialysis. A bag made of dialysis membrane—material with microscopic pores—is filled with a solution containing glucose, water, and albumin (protein) molecules and immersed in a container of pure water. Both water and glucose molecules are small enough to pass through the pores in the dialysis membrane. Albumin molecules, like all protein molecules, are very large and do not pass through the membrane's pores. Because of differences in concentration, glucose molecules diffuse out of the bag as small water molecules diffuse into the bag. Despite a concentration gradient, the albumin molecules do not diffuse out of the bag. Why not? Because they simply will not fit through the tiny pores in the membrane. After some time has passed, the large solutes are still trapped within the bag, but most of the smaller solutes are outside it.

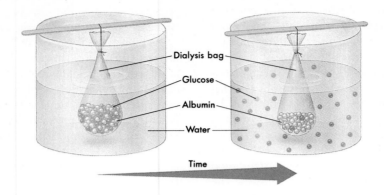

FIGURE 3-15 Dialysis. A dialysis bag containing glucose, water, and albumin (protein) molecules is suspended in pure water. Over time, the smaller solute molecules (glucose) diffuse out of the bag. The larger solute molecules (albumin) remain trapped in the bag because the bag is impermeable to them. Thus dialysis is diffusion that results in separation of small and large solute particles.

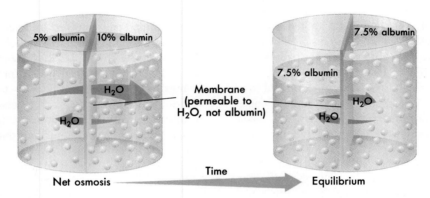

FIGURE 3-16 Osmosis. Osmosis is the diffusion of water through a selectively permeable membrane. The membrane shown is this diagram is permeable to water but not to albumin. Because there are relatively more water molecules in 5% albumin than 10% albumin, more water molecules osmose from the more dilute into the more concentrated solution (as indicated by the larger arrow in the left-hand diagram) than osmose in the opposite direction. The overall direction of osmosis, in other words, is toward the more concentrated solution. Net osmosis produces the following changes in these solutions: their concentrations equilibrate, both the volume and the pressure of the originally more concentrated solution increase, and the volume and the pressure of the other solution decrease proportionately.

Osmosis

Another special case of diffusion is called **osmosis.** Osmosis is the diffusion of water through a selectively permeable membrane. Often, water is able to diffuse across a living membrane that does not allow diffusion of one or more other substances. Thus water can equilibrate its concentration on both sides of the membrane but the *impermeant* solutes cannot. This can have very important consequences in cells, as we shall see.

First, let us look at an example of osmosis. Imagine that you have a 10% albumin solution separated by a membrane from a 5% albumin solution (Figure 3-16). Assume that the membrane is freely permeable to water but impermeable to albumin. The water molecules diffuse or *osmose* through the membrane from the area of high water concentration to the area of low water concentration. That is, the water moves from the more dilute 5% albumin solution to the less dilute 10% albumin solution. Although equilibrium is eventually

reached, the albumin does not diffuse across the membrane. Only the water diffuses. Because of this osmosis, one solution loses volume, and the other solution gains volume (see Figure 3-16).

Unlike the open container pictured in Figure 3-16, cells are closed containers. They are enclosed by their plasma membranes. Actually, most of the body is composed of closed compartments such as cells, blood vessels, tubes, and bladders. In closed compartments, such as a toy water-balloon, changes in volume also mean changes in pressure. Adding volume to a cell by osmosis increases its pressure, just as adding volume to a water-balloon increases its pressure. Water pressure that develops in a solution as a result of osmosis into that solution is called **osmotic pressure.** Taking this principle a step further, we can state that osmotic pressure develops in the solution that originally has the higher concentration of impermeant solute.

Potential osmotic pressure is the maximum osmotic

pressure that *could develop* in a solution when it is separated from pure water by a selectively permeable membrane. Actual osmotic pressure, on the other hand, is pressure that *already has developed* in a solution by means of osmosis. Actual osmotic pressure is easy to measure because it is already there. Since potential osmotic pressure is a *prediction* of what the actual osmotic would be, it cannot be measured directly. What determines a solution's potential osmotic pressure? The answer, simply put, is the concentration of particles of impermeant solutes dissolved in the solution. Thus one can predict the direction of osmosis and the amount of pressure it will produce by knowing the concentrations of impermeant solutes in two solutions. A method for determining the potential osmotic pressure of a solution is outlined in Appendix B.

The concept of osmosis and osmotic pressure has very important practical consequences in human physiology and medicine. Homeostasis of volume and pressure is necessary to maintain the healthy functioning of human cells. The volume and pressure of body cells tend to remain fairly constant because intracellular fluid (fluid inside the cell) is maintained at about the same potential osmotic pressure as the extracellular fluid (fluid outside the cell). Two fluids that have the same potential osmotic pressure are said to be **isotonic** to each other (Figure 3-17, *B*). Isotonic comes from the word parts *iso-*, meaning "same," and *-tonic*, meaning "pressure." Isotonic solutions have the same

potential osmotic pressure because they have the same concentration of impermeant solutes.

A human cell placed in a concentrated solution of impermeant solutes will shrivel up. Look at the example of a red blood cell in Figure 3-17, *C*. The pictured cell is placed in a solution with a higher concentration of impermeant solutes than the cell and, therefore, a higher potential osmotic pressure. The extracellular solution is said to be **hypertonic** (higher pressure) than the intracellular solution. Cells placed in solutions that are hypertonic to intracellular fluid always shrivel. If cells shrivel too much they may become permanently damaged—or even die. Obviously, this fact is of great medical importance. Large amounts of solutes cannot be introduced into the body without considering the effect they will have on the concentration of impermeant solutes in the extracellular fluid. If a treatment or procedure causes extracellular fluid to become hypertonic to the cells of the body, serious damage may occur.

If a human cell is placed in a very dilute solution, such as pure water, the cell may swell. If it swells enough, the cell may burst, or *lyse*. Look at the example of a red blood cell in Figure 3-17, *A*. This cell is placed in a solution that is **hypotonic** (lower pressure) to the intracellular fluid. Hypotonic solutions tend to lose pressure because they have a lower concentration of impermeant solutes, and thus a higher water concentration, than the opposite solution. In

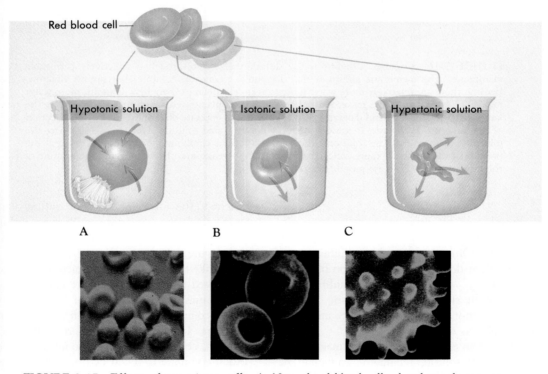

FIGURE 3-17 Effects of osmosis on cells. A, Normal red blood cells placed in a hypotonic solution may swell (as the scanning electron micrograph shows) or even burst (as the drawing shows). This change results from the inward diffusion of water (osmosis). **B,** Cells placed in an isotonic solution maintain a constant volume and pressure because the potential osmotic pressure of the intracellular fluid matches that of the extracellular fluid. **C,** Cells placed in a solution that is hypertonic to the intracellular fluid lose volume and pressure as water osmoses out of the cell and into the hypertonic solution. The "spikes" seen in the scanning electron micrograph are rigid microtubles of the cytoskeleton. These supports become visible as the cell "deflates."

this example, the extracellular fluid is *hypotonic* to intracellular fluid, which in turn is *hypertonic* to the extracellular fluid. Water always osmoses from the hypotonic solution to the hypertonic solution.

In summary, we can make several generalizations about osmosis. First, osmosis is the diffusion of water across a membrane that limits the diffusion of at least some of the solute molecules. That is, at least one impermeant solute must be present. Second, osmosis results in the gain of volume (and thus, pressure) on one side of the membrane and loss of volume (and pressure) on the other side of the membrane. Third, the direction of osmosis and the resulting changes in pressure can be predicted if you know the potential osmotic pressures or *tonicity* of each solution separated by the membrane.

Facilitated diffusion

Another special kind of diffusion is called **facilitated diffusion.** This term is used when movement of molecules is facilitated or made more efficient by the action of carrier mechanisms in a cell membrane. Figure 3-18 shows that a carrier molecule attracts a solute to a binding site, changes shape, then releases the solute to the other side of the membrane. This mechanism differs from channel-mediated transport, which does not involve binding the solute molecule and changing shape to release the bound solute. Carrier-mediated diffusion is also faster than other types of diffusion. However, facilitated diffusion is similar to simple diffusion in that it also transports substances down a concentration gradient (that is, from high to low concentration). The energy required for either simple or facilitated diffusion comes from the collision energy of the solute—not from the cell. Thus facilitated diffusion is simply carrier-mediated passive transport.

Filtration

Another important passive transport process is **filtration.** This form of transport involves the passing of water and permeable solutes through a membrane by the force of hydrostatic pressure. *Hydrostatic pressure* is the force, or weight, of a fluid pushing against a surface. Filtration is movement of molecules through a membrane from an area of high hydrostatic pressure to an area of low hydrostatic pressure—that is, down a hydrostatic pressure gradient. Actually, filtration most often transports substances through a sheet of cells. The force of pressure pushes the molecules through or between the cells that form the sheet. Because the filtration membrane does not allow larger particles through, filtration results in the separation of large and small particles. This is similar to dialysis, except that dialysis is driven by a *concentration gradient* whereas filtration is driven by a *hydrostatic pressure gradient.*

A simple model of filtration is found in many drip-type coffee makers. A mixture of water and ground coffee beans is placed in a bowl made of filter paper. Gravity pulls downward on the mixture, generating hydrostatic pressure against the bottom of the filter. The pores in the paper filter are large enough to let water molecules and other small particles pass through to a coffee pot below the filter. Most of the coffee grounds are too large to pass through the filter and thus cannot be filtered. The coffee in the pot below is called the *filtrate.*

How and where does filtration occur in the body? Most often, in tiny blood vessels called *capillaries,* which are found throughout the body. Hydrostatic pressure of the blood (blood pressure) generated by heart contractions, gravity, and other forces pushes water and small solutes out of the capillaries and into the interstitial spaces of a tissue.

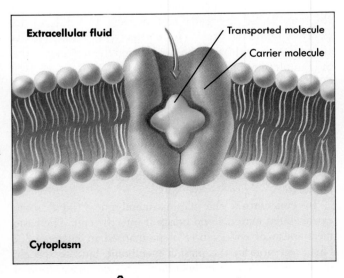

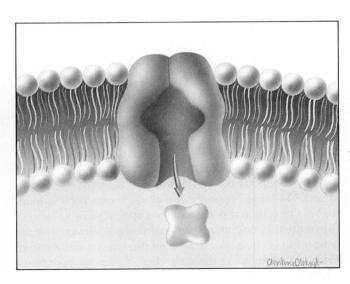

A

B

FIGURE 3-18 Carrier-mediated transport. In carrier-mediated transport, a membrane-bound carrier protein attracts a solute molecule to a binding site **(A)** and changes shape in a manner that allows the solute to move to the other side of the membrane **(B).** This type of transport can be passive (driven by a concentration gradient) or active (driven by cellular energy needed to overcome a concentration gradient).

Blood cells and large blood proteins are too large to fit through pores in the capillary wall; therefore they cannot be filtered out of the blood. Capillary filtration allows the blood vessels to supply tissues with water and other essential substances quickly and easily without losing its cells and blood proteins. Capillary filtration is also the first step used by the kidney to form urine.

Active Transport Processes

All the transport processes that we have discussed so far are passive processes. The force of movement comes from the concentration gradient or the hydrostatic pressure gradient—that is, from the physical forces of nature. The driving force for active transport processes, on the other hand, comes from the cell itself. Energy of metabolism must be used by cells to force particles across a membrane that otherwise would not move across.

Active transport

The term **active transport** refers to a carrier-mediated process in which cellular energy is used to move molecules "uphill" through a cell membrane. By "uphill," we mean that the substance moves from an area of low concentration to an area of higher concentration. An actively transported substance moves *against* its concentration gradient. This is exactly the opposite of diffusion, in which a substance is transported *down* its concentration gradient—or "down-hill." It is important to remember that molecules will not travel "uphill" on their own, anymore than a ball will roll uphill by itself. Molecules will travel uphill only when they are forced by carrier mechanisms powered by cellular energy.

Active transport is an extremely important process. It allows cells to move certain ions or other water-soluble particles to specific areas. For example, active calcium carriers or *calcium pumps* in the membranes of muscle cells allows the cell to force nearly all of the intracellular calcium ions (Ca^{++}) into special compartments—or out of the cell entirely. This is important, because a muscle cell cannot operate properly unless the intracellular Ca^{++} concentration is kept low during rest. Other cells use active transport carriers, or *pumps,* for similar purposes.

One type of active transport pump, the **sodium-potassium pump,** operates in the plasma membrane of all human cells (Figure 3-19). It is essential for healthy cell survival. As its name suggests, the sodium-potassium pump actively transports both sodium ions (Na^+) and potassium ions (K^+)—but in opposite directions. It transports sodium ions *out* of cells and potassium ions *into* cells. By so doing, the sodium-potassium pump maintains a lower sodium concentration in intracellular fluid than in the surrounding extracellular fluid. At the same time, this pump maintains a higher potassium concentration in the intracellular fluid than in the surrounding extracellular fluid. Both ions bind to the same carrier, a molecule known as *sodium-potassium ATPase.* Figure 3-19 shows that three Na^+ ions bind to sodium-binding sites on the carrier's inner face. At the same time, an energy-containing ATP molecule produced

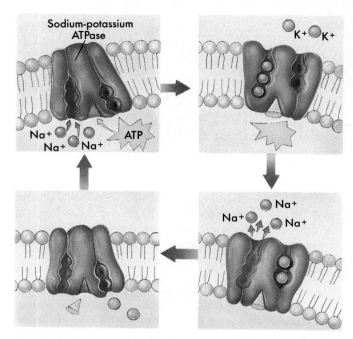

FIGURE 3-19   Sodium-potassium pump. Three Na^+ ions bind to sodium-binding sites on the carrier's inner face. At the same time, an energy-containing ATP molecule produced by the cell's mitochondria binds to the carrier. The ATP breaks apart, transferring its stored energy to the carrier. The carrier then changes shape, releases the three Na^+ ions to the outside of the cell, and attracts two K^+ ions to its potassium-binding sites. The carrier then returns to its original shape, releasing the two K^+ ions and the remnant of the ATP molecule to the inside of the cell. The carrier is now ready for another pumping cycle.

by the cell's mitochondria binds to the carrier. The ATP breaks apart, transferring its stored energy to the carrier. The carrier then changes shape, releases the three Na^+ to the outside of the cell, and attracts two K^+ to its potassium-binding sites. The carrier then returns to its original shape, releasing the two K^+ and the remnant of the ATP molecule to the inside of the cell.

Endocytosis and exocytosis

Like active transport pumps, *endocytosis* and *exocytosis* are mechanisms that require expenditure of metabolic energy by the cell. They differ from pump mechanisms in that they allow substances to enter or leave the interior of a cell without actually moving through its plasma membrane.

In **endocytosis** the plasma membrane "traps" some extracellular material and brings it into the cell. The basic mechanism of endocytosis is summarized in Figure 3-20. First, receptors in the plasma membrane bind to specific molecules in the extracellular fluid. This causes a portion of the plasma membrane to be pulled inward by the cytoskeleton, forming a small pocket around the material to be moved into the cell. The edges of the membranous pocket eventually fuse, forming a vesicle. The vesicle is then pulled inward—away from the plasma membrane—by the

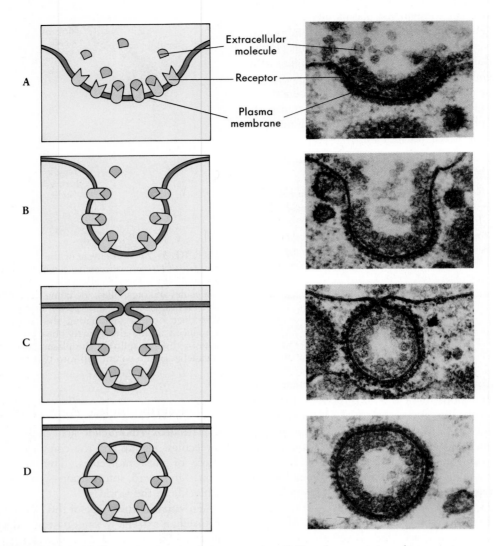

FIGURE 3-20   Endocytosis. An artist's interpretation *(left)* and transmission electron micrographs *(right)* show the basic steps of endocytosis: **A,** Membrane receptors bind to specific molecules in the extracellular fluid. **B,** A portion of the plasma membrane is pulled inward by the cytoskeleton, forming a small pocket around the material to be moved into the cell. **C,** The edges of the pocket eventually fuse, forming a vesicle. **D,** The vesicle is then pulled inward—away from the plasma membrane—by the cytoskeleton. In this example, only the receptor-bound molecules enter the cell. In some cases, some free molecules or even entire cells may also be trapped within the vesicle and transported inward.

cytoskeleton. Sometimes endocytosis picks up a variety of molecules and other particles along with receptor-bound molecules. On the other hand, sometimes only the molecules that bind to membrane receptors are brought into the cell.

There are two basic forms of endocytosis: *phagocytosis* and *pinocytosis.* In phagocytosis, microorganisms or other large particles are engulfed by the plasma membrane and enter the cell in vesicles that have pinched off from the membrane. Once inside, they fuse with the membranous walls of the lysosomes. Enzymes from the lysosome then digest the particles into their component molecules. The products of digestion may then diffuse through the mem-

branous wall of the vesicle into the cytoplasm. The term *phagocytosis* means "condition of the cell eating" (from the word parts *phago-,* meaning "eat," *-cyto-,* meaning "cell," and *-osis,* meaning "condition"). *Pinocytosis,* or "condition of the cell drinking," is a similar process in which fluid and the substances dissolved in it enter a cell. How does a cell know what to "eat" or "drink"? A current hypothesis suggests that cells usually consume material that contains molecules that bind with membrane receptors. In a way, this is similar to how we choose what to eat or drink—we recognize appropriate substances when they react with taste receptors in the taste buds of the tongue.

Exocytosis is the process by which large molecules,

notably proteins, can leave the cell even though they are too large to move out through the plasma membrane (see Figure 3-5). They are first enclosed in membranous vesicles. The vesicles are then pulled out to the plasma membrane by the cytoskeleton. The vesicles then fuse with the plasma membrane and release their contents outside the cell. Some gland cells secrete their products by exocytosis. Besides providing a mechanism of transport, exocytosis also provides a way for new membrane material to be added to the plasma membrane.

QUICK CHECK

✔ 1. *Name three different passive processes that transport substances across a cell membrane. How are they alike? How are they different?*
✔ 2. *What causes osmotic pressure to develop in a cell?*
✔ 3. *Describe three different active processes that transport substances across a cell membrane. What distinguishes them from passive processes?*

▶ GROWTH AND REPRODUCTION OF CELLS

Cell growth and reproduction are the most fundamental of all living functions. These two processes together constitute the **cell life cycle** (Figure 3-21). On these processes depend the continued survival of all organisms already living and the creation of all new organisms. Cell growth depends on using genetic information in DNA to make the structural and functional proteins needed for cell survival. Cell reproduction ensures that the genetic information is passed from one generation of cells to the next and from one generation of organisms to the next. Mistakes in these processes can cause lethal genetic disorders, cancer, and other conditions. Recent advances in our ability to manipulate genetic information, and thus cell growth and reproduction, now present us with ethical implications more far-reaching than those accompanying the birth of the atomic age. All this makes the cell life cycle a worthy and fascinating topic of study.

Four decades ago, an American, James Watson, and three British scientists, Francis Crick, Maurice Wilkins, and Rosalind Franklin, won the race to solve the puzzle of DNA's molecular structure (Figure 3-22). Watson and Crick announced their discovery with an understated, one-page report in the journal, *Nature*, in 1953. Nine years later, Watson, Crick, and Wilkins received the Nobel Prize for their brilliant and significant work—hailed as the greatest biological discovery of our time. (Franklin had passed away before the Nobel prize was awarded.) Watson tells his insider's view of how the discovery was made in *The Double Helix* (1969), a book that is now a classic for both its science and its intrigue. We begin our outline of the cell life cycle with a discussion of DNA because it is, as Watson called it in his book, "the most golden of all molecules."

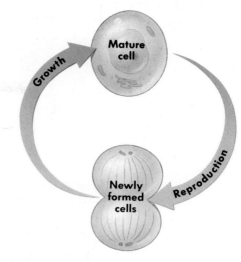

FIGURE 3–21 Life cycle of the cell. The processes of growth and reproduction of successive generations of cells exhibit a cyclic pattern. Newly formed cells grow to maturity by synthesizing new molecules and organelles, including the replication of an extra set of DNA molecules in anticipation of reproduction. Mature cells reproduce by first distributing the two identical sets of DNA (produced during the growth phase) in the orderly process of *mitosis*, then by splitting the plasma membrane, cytoplasm, and organelles of the parent cell into two distinct daughter cells.

Deoxyribonucleic Acid (DNA)

The **deoxyribonucleic acid** molecule is a giant among molecules. Its size and the complexity of its shape exceed those of most molecules. The importance of its function—in a word, information—surpasses that of any other molecule in the world.

To visualize the shape of the DNA molecule, picture to yourself an extremely long, narrow ladder made of a pliable material (see Figure 3-22). Now see it twisting round and round on its axis and taking on the shape of a steep spiral staircase millions of turns long. This is the shape of the DNA molecule—a double spiral or *double helix*.

The DNA molecule is a *polymer*. This means that it is a large molecule made up of many smaller molecules joined together in sequence. DNA is a polymer of millions of pairs of nucleotides. A nucleotide is a compound formed by combining phosphoric acid with a sugar and a nitrogenous base. In the DNA molecule there are four different kinds of nucleotides. Each nucleotide consists of a phosphate group that attaches to the sugar deoxyribose that attaches to one of four bases. Nucleotides differ, therefore, in their nitrogenous base component—containing either adenine or guanine (purine bases) or cytosine or thymine (pyrimidine bases). (Deoxyribose is a sugar that is not sweet and one whose molecules contain only five carbon atoms.) Notice what the name *deoxyribonucleic acid* tells you—that this compound contains deoxyribose, that it occurs in nuclei, and that it is an acid.

Figure 3-22 reveals additional and highly significant facts about DNA's molecular structure. First, observe which compounds form the sides of the DNA spiral

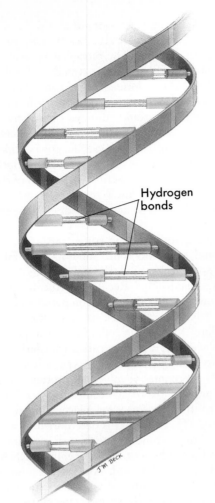

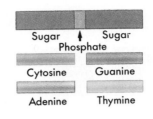

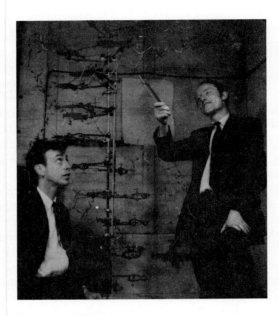

FIGURE 3–22 Watson-Crick model of the DNA molecule. The DNA structure illustrated here is based on that published by James Watson (*photograph, left*) and Francis Crick (*photograph, right*) in 1953. Note that each side of the DNA molecule consists of alternating sugar and phosphate groups. Each sugar group is united to the sugar group opposite it by a pair of nitrogenous bases (adenine-thymine or cytosine-guanine). The sequence of these pairs constitutes a genetic code that determines the structure and function of a cell.

staircase—a long line of phosphate and deoxyribose units joined alternately one after the other. Look next at the stair steps. Notice two facts about them: that two bases join (loosely bound by hydrogen bonds) to form each step, and that only two combinations of bases occur. The same two bases invariably pair off with each other in a DNA molecule. Adenine always goes with thymine (or vice versa, thymine with adenine), and guanine always goes with cytosine (or vice versa). This fact about DNA's molecular structure is called **obligatory base-pairing.** Pay particular attention to it, for it is the key to understanding how a DNA molecule is able to duplicate itself. DNA duplication, or replication as it is usually called, is one of the most important of all biological phenomena because it is an essential and crucial part of the mechanism of genetics.

Another fact about DNA's molecular structure that has great functional importance is the sequence of its base pairs. Although the base pairs in all DNA molecules are the same, the sequence of these base pairs is not the same in all DNA molecules. For instance, the sequence of the base pairs composing the seventh, eighth, and ninth steps of one DNA molecule might be cytosine-guanine, adenine-thymine, and thymine-adenine. Such a sequence of three bases forms a code word or "triplet" called a **codon.** In another DNA molecule the coding sequence of the base pairs making up these same steps might be entirely different, perhaps thymine-adenine, guanine-cytosine, and cytosine-guanine. Perhaps these seem to be minor details, but nothing could be further from the truth, since it is the sequence of the base pairs in the nucleotides composing DNA molecules that identifies each gene. Hence it is the sequence of base pairs that determines all hereditary traits.

A human **gene** is a segment of a DNA molecule. One gene consists of a chain of about 1,000 pairs of nucleotides joined one after the other in a precise sequence. One gene controls the production within the cell of one polypeptide chain. One or more polypeptides make up each of a cell's

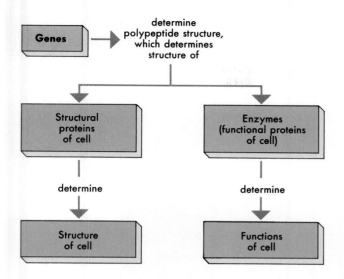

FIGURE 3-23 How genes determine a cell's structural and functional characteristics.

structural proteins and enzymes. Each enzyme catalyzes one chemical reaction. Therefore, as Figure 3-23 indicates, the millions of genes constituting a cell's DNA determine the cell's structure and its functions.

Cell Growth

Recall that one of the two major phases of the cell life cycle is the growth phase (see Figure 3-21). It is during this phase that a newly formed cell produces new molecules, from which it constructs the additional cell membrane, cell fibers, and other structures necessary for growth. As we have just stated, all of the structural proteins, plus the enzymes needed to make lipids, carbohydrates, and other substances, are made by the cell using information contained in the genes of DNA molecules. In this brief section on cell growth, we will first present a simplified account of how these proteins are made. Later, we will discuss how the cell replicates its DNA molecules in anticipation of reproduction.

Protein synthesis

Protein synthesis is an anabolic process, meaning small molecules are joined together to form large molecules. Amino acids are strung together in a specific sequence to form polypeptide chains. Two or more polypeptide chains are then linked to form even longer chains of amino acids—whole protein molecules. In the following few paragraphs, we will briefly describe how the cell "copies" information from genes and interprets it to form specific polypeptides. As you read through the steps of protein synthesis, refer often to Figure 3-24.

First, a strand of RNA (ribonucleic acid) forms along a segment of one strand of a DNA molecule. RNA differs from DNA in certain respects. Its molecules are smaller than those of DNA, and RNA contains ribose instead of deoxyribose. Also, one of the four bases in RNA is uracil instead of thymine. As a strand of RNA is forming along a strand of DNA, uracil attaches to adenine, and guanine attaches to cytosine. The process is known as **complementary pairing.** Thus a single-strand molecule of *messenger RNA (mRNA)* is formed. The name "messenger RNA" describes its function. As soon as it is formed, it separates from the DNA strand, diffuses out of the nucleus, and carries a "message" to a ribosome in the cell's cytoplasm, directing its synthesis of a specific polypeptide. Synthesis of a mRNA molecule is often called **transcription** because it actually copies or "transcribes" a portion of the DNA code. Cell biologists now know that after a mRNA molecule is formed, its message may be "edited" before it reaches a ribosome.

In the cytoplasm, the edited mRNA molecule attracts large and small ribosome subunits (see Figure 3-3) that come together around the mRNA molecule to form a "mRNA sandwich." Recall that the two subunits of the now-complete ribosome are composed largely of ribosomal RNA (rRNA). The cell is now ready to interpret or "translate" the genetic code, forming a specific sequence of amino acids, in a process called **translation.**

In translation, yet another type of RNA—*transfer RNA (tRNA)*—becomes involved in protein synthesis. As the name implies, tRNA molecules carry or "transfer" amino acids to the ribosome for placement in the prescribed sequence. This function is determined by a unique molecular structure: a binding site for a specific amino acid at one end and a binding site for a specific mRNA codon (base triplet) at the other end. After picking up its amino acid from a pool of 20 different types of amino acids floating free in the cytoplasm, a tRNA molecule moves to the ribosome. There, it attaches to a complementary mRNA codon—the codon that signifies the specific amino acid carried by that tRNA molecule. After the next tRNA brings its amino acid into place, the two amino acids form a peptide bond. The ribosome then moves down the mRNA strand, making the next mRNA codon available for a complementary tRNA molecule bearing the appropriate amino acid. More tRNA molecules, one after the other in rapid sequence, bring more amino acids to the ribosome and fit them into their proper positions in the growing chain of amino acids. As Figure 3-24 shows, the ribosome is all the while moving down the mRNA strand—until it reaches the end, where the ribosome subunits fall away. Each of the tools needed for translation—mRNA, tRNA, and ribosome subunits—can be reused again and again to form copies of the same polypeptide.

Once the proper polypeptides are formed, enzymes in the ER, Golgi apparatus, or cytoplasm link them together to form whole protein molecules. Protein anabolism is now complete. It is one of the major kinds of cellular work. One human cell is estimated to synthesize perhaps 2,000 different enzymes! In addition to this staggering workload, it produces many different protein compounds that help form its own structures, and many cells also synthesize special proteins. Liver cells are an example; they synthesize prothrombin, fibrinogen, albumins, and globulins.

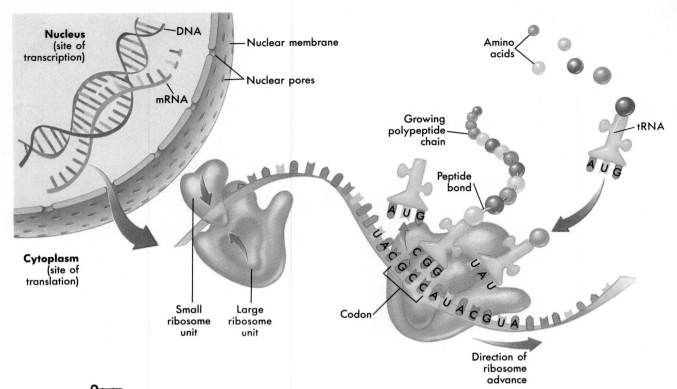

FIGURE 3-24   Protein synthesis. Protein synthesis begins with *transcription*, a process in which a messenger RNA (mRNA) molecule forms along one gene sequence of a DNA molecule within the cell's nucleus. As it is formed, the mRNA molecule separates from the DNA molecule and leaves the nucleus through the large nuclear pores. Outside the nucleus, ribosome subunits attach to the beginning of the mRNA molecule and begin the process of *translation.* In translation, transfer RNA (tRNA) molecules bring specific amino acids—encoded by each mRNA codon—into place at the ribosome site. As the amino acids are brought into the proper sequence, they are joined together by peptide bonds to form long strands called polypeptides. Several polypeptide chains may be needed to make a complete protein molecule.

DNA *replication*

As you may have noticed by now, nucleic acids such as RNA are synthesized directly on the DNA molecule. Earlier in the chapter, we discussed the fact that rRNA is made and formed into ribosome subunits in the nucleolus. In the previous section, we described how mRNA molecules form during transcription by the complementary pairing of RNA bases with DNA bases. Transfer RNA is formed on DNA molecules in a similar fashion, then remain in the cytoplasm to be used over and over. As a cell becomes larger, mechanisms that are just now beginning to be understood trigger the synthesis of a complete copy of the nucleus's entire set of DNA molecules. Replication of the entire set of DNA molecules, or **genome,** prepares the cell for reproduction, when one set will go to one daughter cell and the other set to the other daughter cell. The mechanics of DNA replication resemble those of RNA synthesis—as we shall see.

In the first step of DNA replication, the tightly coiled DNA molecules uncoil except for small segments. (Since these remaining tight little coils are denser than the thin elongated sections, they absorb more stain and appear as chromatin granules under the microscope. The thin uncoiled sections, in contrast, are invisible because they absorb so little stain.) As the DNA molecule uncoils, its two strands come apart. Then, along each of the two separated strands of nucleotides, a complementary strand forms. Intracellular fluid contains many DNA nucleotides. By the mechanism of obligatory base-pairing and with the work of specific enzymes, nucleotides become attached at their correct places along each DNA strand (Figure 3-25). Interpreted, this means that new thymine, that is, from the intracellular fluid, attaches to the "old" adenine in the original DNA strand. Conversely, new adenine attaches to old thymine. Also, new guanine joins old cytosine and, vice versa, new cytosine joins old guanine. By the end of the growth phase, each of the two DNA strands of the original DNA molecule has a complete new complementary strand attached to it. Each half of the DNA molecule or strand, in other words, has duplicated itself to create a whole new DNA molecule. Thus two new chromosomes now replace each original chromosome. However, at this stage (before cell reproduction has actually begun), they are called *chromatids* instead of chromosomes. The two chro-

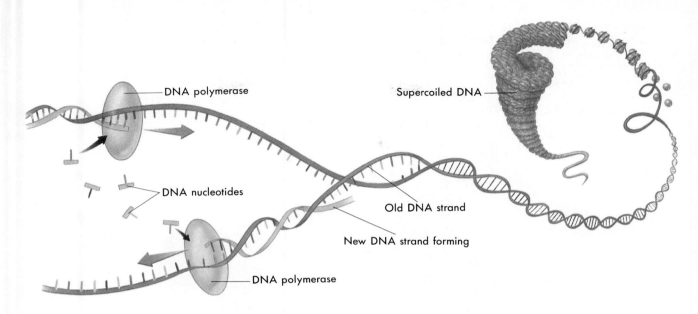

FIGURE 3-25   DNA replication. When a DNA molecule makes a copy of itself, it "unzips" to expose its nucleotide bases. Through the mechanism of obligatory base-pairing, coordinated by the enzyme *DNA polymerase*, new nucleotides bind to the exposed bases. This forms a new "other half" to each half of the original molecule. After all the bases have new nucleotides bound to them, two identical DNA molecules will be ready for distribution to the two daughter cells.

matids formed from each original chromosome contain duplicate copies of DNA and, therefore, the same genes as the chromosome from which they were formed. Chromatids are present as attached pairs. *Centromere* is the name of their point of attachment.

Cell Reproduction

Now that we have briefly looked at the growth phase, it is time to turn our attention to the other major phase of the cell life cycle: cell reproduction. Simply put, cells reproduce by splitting themselves into two separate cells. One *parent cell* thus becomes two smaller *daughter cells.* Splitting the plasma membrane and cytoplasm into two is called **cytokinesis** (meaning "cell movement"). This is an apt name because the cell's cytoskeleton moves the plasma membrane and internal structures in a way that pinches it in half, forming two equivalent daughter cells. Of course, each daughter cell must possess all the resources necessary for survival if the cycle of life is to remain unbroken. That means that beside sufficient cytoplasm, mitochondria, and other organelles, each cell must also have a complete set of genetic information (DNA) needed to run a cell properly. This aspect of cell reproduction or *cell division* is accomplished by a process called **mitosis.** During mitosis, the cell organizes replicated DNA into two identical sets and then distributes one complete set to each daughter cell.

Mitosis

Mitosis, the process of organizing and distributing nuclear DNA during cell division, is a continuous process (Table 3-5) consisting of four distinct phases:

1. *Prophase*   3. *Anaphase*
2. *Metaphase*   4. *Telophase*

When the cell is not experiencing mitosis—during the growth phase between cell divisions—it is said to be in **interphase** (meaning "between-phase"). A cell is not actively reproducing during interphase, but it is actively *preparing for* reproduction. Not only are the DNA molecules being replicated, but the centrioles (the only easily visible part of the centrosome) are also being replicated. Additional cytoplasm, membrane, and other cell structures are also being constructed in anticipation of cell division.

The cell enters **prophase** when it begins to divide, usually before cytokinesis, or "pinching in half," becomes apparent. During prophase, which literally means "before-phase," the nuclear envelope falls apart as the paired chromatids coil up to form dense, compact **chromosomes** (Figure 3-26). By the end of prophase, each chromosome consists of a pair of short, thick bodies joined together at a centromere. At the same time that chromosomes are forming, the centriole pairs move toward opposite ends, or "poles," of the parent cell. As they move apart, a parallel arrangement of microtubules, or *spindle fibers,* is constructed between them.

The term **metaphase** literally means "position-changing-phase." This name is appropriate because during this phase the chromosomes, no longer trapped within a nucleus, are moved by the cytoskeleton into an orderly pattern. The chromosomes are aligned along a plane at the "equator" of the cell about midway between the centriole pairs at opposite poles of the cell (the *equatorial plate*). One

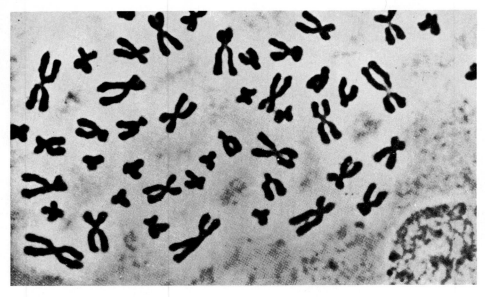

FIGURE 3-26   Chromosomes of a normal human cell prior to cell division.

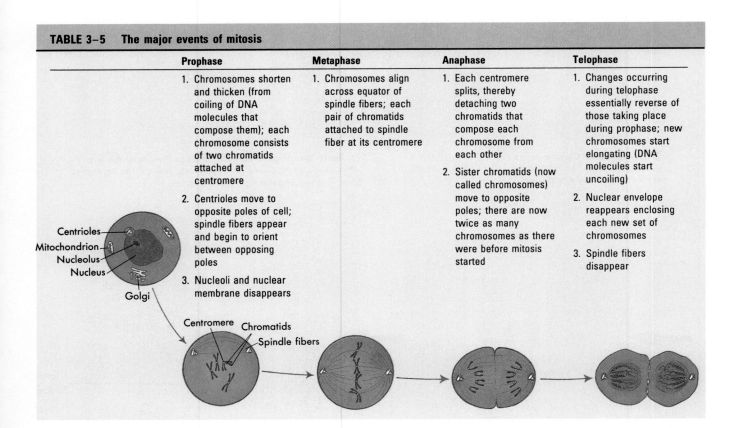

TABLE 3–5   The major events of mitosis

	Prophase	Metaphase	Anaphase	Telophase
	1. Chromosomes shorten and thicken (from coiling of DNA molecules that compose them); each chromosome consists of two chromatids attached at centromere	1. Chromosomes align across equator of spindle fibers; each pair of chromatids attached to spindle fiber at its centromere	1. Each centromere splits, thereby detaching two chromatids that compose each chromosome from each other	1. Changes occurring during telophase essentially reverse of those taking place during prophase; new chromosomes start elongating (DNA molecules start uncoiling)
	2. Centrioles move to opposite poles of cell; spindle fibers appear and begin to orient between opposing poles		2. Sister chromatids (now called chromosomes) move to opposite poles; there are now twice as many chromosomes as there were before mitosis started	2. Nuclear envelope reappears enclosing each new set of chromosomes
	3. Nucleoli and nuclear membrane disappears			3. Spindle fibers disappear

Centrioles
Mitochondrion
Nucleolus
Nucleus
Golgi

Centromere   Chromatids
Spindle fibers

chromatid of each chromosome faces one pole of the cell, and its identical sister chromatid faces the opposite pole. Each chromatid then attaches to a spindle fiber.

Anaphase, or the "apart-phase," begins as soon as all the chromosomes have aligned along the cell's equator. During this phase of mitosis, the centromere of each chromosome splits to form two chromosomes, each consisting of a single DNA molecule. Each chromosome is pulled toward the nearest pole (that is, toward the nearest centrosome) by a spindle fiber. The effect of this movement is that one set of DNA molecules reaches one end of the cell, and another set reaches the other end of the cell—forming two separate, but identical, pools of genetic information. By the time a dividing cell has reached anaphase, cytokinesis has usually become apparent. The cell has not completely split yet, but cleavage along the equator may be visible.

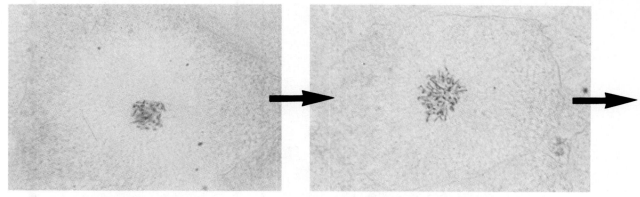

A, Early prophase. Chromosomes are beginning to form from the nuclear chromatin material. Chromosomes have already been duplicated.

B, Later prophase. Chromosomes are now fully formed and easily visible in the cell nucleus.

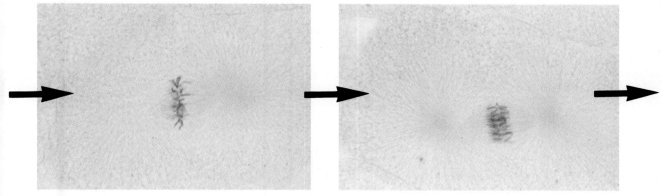

C, Metaphase. Chromosomes are lined up at the equatorial plate. Spindle fibers (microtubules) of the mitotic apparatus are distinct.

D, Early anaphase. Spindle fibers pull each of the two chromatids (now called chromosomes) toward the centrioles. have divided, separating the chromatids.

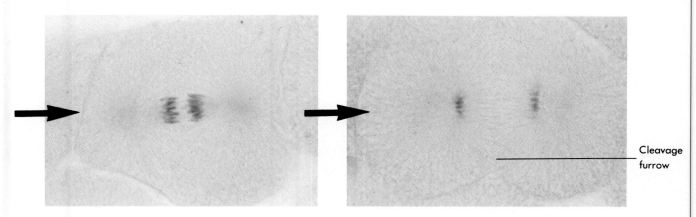

Cleavage furrow

E, Later anaphase. The chromatids are moving toward the two poles of the mitotic apparatus.

F, Telophase. Chromosomes have completed movement to opposite poles. The cleavage furrow (of cytokinesis) that will eventually constrict the cell into two daughter cells is evident.

FIGURE 3-27 Mitotic cell division. Series of photomicrographs showing animal cells undergoing cell division from **(A)** prophase to **(F)** telophase.

FOCUS ON... CHANGES IN CELL GROWTH AND REPRODUCTION

Cells have the ability to adapt to changing conditions. Cells may alter their size, reproductive rate, or other characteristics to adapt to changes in the internal environment. Such adaptations usually allow cells to work more efficiently. However, sometimes cells alter their characteristics abnormally—decreasing their efficiency and threatening the health of the body. Common types of changes in cell growth and reproduction are summarized below.

Cells may respond to changes in function, hormone signals, or availability of nutrients by increasing or decreasing in size. The term *hypertrophy* (hye-PER-tro-fee) refers to an increase in cell size, and the term *atrophy* (AT-ro-fee) refers to a decrease in cell size. Either type of adaptive change can occur easily in muscle tissue. When a person continually uses muscle cells to pull against heavy resistance, as in weight training, the cells respond by increasing in size. Body build-

ers thus increase the size of their muscles by hypertrophy—increasing the size of muscle cells. Atrophy often occurs in underused muscle cells. For example, when a broken arm is immobilized in a cast for a long period, muscles that move the arm often atrophy. Because the muscles are temporarily out of use, muscle cells decrease in size. Atrophy may also occur in tissues whose nutrient or oxygen supply is diminished.

Sometimes cells respond to changes in the internal environment by increasing their rate of reproduction—a process called *hyperplasia* (hye-per-PLAY-zha). The word *-plasia* comes from a Greek word that means "formation"—referring to formation of new cells. Because *hyper-* means "excessive," *hyperplasia* means excessive cell reproduction. Like hypertrophy, hyperplasia causes an increase in the size of a tissue or organ. However, hyperplasia is an increase in the *number of cells* rather than an increase in the size of each

cell. A common example of hyperplasia occurs in the milk-producing glands of the female breast during pregnancy. In response to hormone signals, the glandular cells reproduce rapidly, preparing the breast for milk production and nursing.

If the body loses its ability to control mitosis, abnormal hyperplasia may occur. The new mass of cells thus formed is a tumor or *neoplasm* (NEE-o-plazm). Many neoplasms also exhibit a characteristic called *anaplasia* (an-ah-PLAY-zha). Anaplasia is a condition in which cells change in orientation to each other and fail to mature normally; that is, they fail to *differentiate* into a specialized cell type. Neoplasms may be relatively harmless growths called *benign* (be-NINE) tumors. If tumor cells can break away and travel through the blood or lymphatic vessels to other parts of the body, the neoplasm is a *malignant tumor,* or cancer.

Telophase is the "end-phase," or "completion-phase," of mitosis. It is during this phase that the DNA is returned to its original form and location within the cell. The cell rebuilds the nuclear envelope, trapping the DNA molecules within a membranous basket once again. At the same time, the chromosomes elongate back into the chromatin form. Recall that DNA cannot participate in protein synthesis unless at least some of its length is uncoiled. Spindle fibers, which are no longer needed, disappear during telophase. Cytokinesis is usually completed during, or just after, the telophase portion of mitosis. Each genetically identical daughter cell thus formed is now in interphase. It will grow and develop into a mature cell, perhaps becoming a parent itself.

A series of photographs summarizing the events of mitotic cell division (mitosis and cytokinesis) is shown in Figure 3-27.

Meiosis

Meiosis is the type of cell division that occurs only in primitive sex cells during the process of their becoming mature sex cells. As a result of meiosis the primitive sex cells (spermatogonia in the male and oogonia in the female) become mature sex cells called **gametes.** Male gametes are named spermatozoa but are usually called sperm. Female gametes are named ova (singular, ovum). In humans, all somatic cells contain 46 chromosomes. This total of 46 chromosomes per cell is known as the **diploid** number of chromosomes. Diploid comes from the Greek *diploos,* meaning "two" or "pair." In somatic cells the 46 chromosomes are present in 22 homologous pairs—the remaining two being the sex chromosomes XY (male) or

XX (female). During meiosis, or **reduction division,** the diploid chromosome number (46) of the primitive spermatogonium or oogonium is reduced to the **haploid** number of 23 found in the mature sex cells or gametes. Figure 3-28 shows that meiotic division occurs in two steps: *meiosis I,*

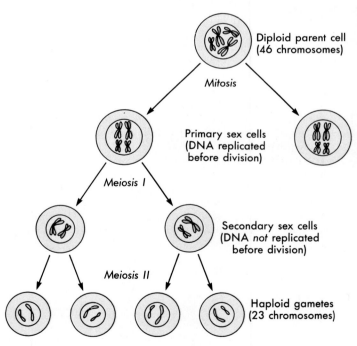

FIGURE 3-28 **Meiosis.** Meiotic cell division takes place in two steps: *meiosis I* and *meiosis II.* Meiosis is called *reduction division* because the number of chromosomes is reduced by half (from the diploid number to the haploid number).

during which the number of chromosomes is halved but the chromatid pairs remain together, and *meiosis II*, during which the chromatids finally split apart. The end result of fertilization is the fusion of two gametes, each containing the haploid number (23) of chromosomes. Fertilization results in formation of a zygote that is a diploid cell having 46 chromosomes, 23 chromosomes being contributed by each parent. The zygote is the first cell of the human offspring. It will then undergo mitotic division to form two cells, then four, and so on, until 100 trillion cells are eventually formed. During development, the new daughter cells will specialize, or *differentiate*, to become specific cell types, which in turn will form specific organs. More information about the formation of sex cells, fertilization, and development is given in Chapters 30 through 32.

QUICK CHECK

✔ 1. *What are the two major phases of the cell life cycle? During which of these phases does mitosis occur?*

✔ 2. *How does DNA act as the "master molecule" of a cell?*

✔ 3. *What is the difference between mitotic cell division and meiotic cell division?*

Cycle of Life: *Cells*

Different types of cells have highly variable life cycles. The active life span of a single cell may vary from a few minutes to years, depending on its function and level of activity. Some cells may remain dormant or inactive for years. Then, when they are "activated" by some biological need and become functional, their life span may be shortened dramatically.

One example involves cells in the immune system that are programmed to produce antibodies against a specific disease. Other examples include the female sex cells, or ova, which are present from birth. They mature throughout the reproductive life span of the individual so that each month at least one cell will become fully developed and provide an opportunity for fertilization to occur.

Structure follows function at every level of organization in the body. Frequently, function decreases with advancing age, and resulting changes occur in cell numbers and in their ability to function effectively. As a result, we lose functional capacity in every body organ system. Our muscles may atrophy, the skin will lose its elasticity, and our respiratory, cardiovascular, and skeletal systems will become affected because of cellular changes that accompany aging.

THE BIG PICTURE
Cells and the Whole Body

When exploring the microscopic world of cells, it is easy to get caught up in the intricate mechanisms that operate in each specific organelle. Once you feel comfortable with these mechanisms, try to put them together into a bigger picture of cell function. For example, most of the processes that we discussed in this chapter are going on at about the same time within each and every cell of your body. Each cell is transcribing genes and synthesizing polypeptides, which are then dumped into the ER and transported to a Golgi apparatus for processing and packaging before being sent off to become a lysosome or being secreted by exocytosis. At the same time, energy for this and other cell work is being transferred from food molecules to ATP molecules—which act as energy-storage batteries for the cell. The cytoskeleton and the cell membrane are transporting materials into, out of, and around the cytoplasm. Studying cell structure and function is like looking at the score of a symphony for the first time. All the

individual parts look unrelated and somewhat confusing, but with a little effort you can combine them to form a coherent whole. The "symphony" of normal cell function results from a coordinated combination of many processes dictated by the cell's "musical score"—the genetic code.

After considering the cell as a whole living unit, take another step back from your mental image of a cell. Picture a huge "society" of trillions of cells—the human body. Each cell contributes to the survival of its society (the body)—and itself—by specializing in functions that help maintain the relative constancy of the internal environment. When we think of that relative constancy, homeostasis, we should appreciate that it is all accomplished by the action of many individual cells. How are individual cells grouped together? How do they function as groups to promote homeostasis? These questions will be answered, at least in part, in Chapter 4.

MECHANISMS OF DISEASE

Cellular Disease

The more that we learn about the mechanisms of disease, the more apparent it becomes that most diseases known to medicine involve abnormalities of cells. Even a few abnormal cells can so disrupt the internal environment of the body that a person's health can be in immediate danger. The following paragraphs list several important categories of disease that are caused by cell abnormalities. Specific diseases belonging to these categories are discussed further in later chapters, but this sampler will give you a basic understanding of cellular mechanisms of disease.

Disorders Involving Cell Transport

Several very severe diseases result from damage to cell transport mechanisms. *Cystic fibrosis (CF)*, for example, is an inherited condition in which chloride ion (Cl^-) pumps in the plasma membrane are missing. Because Cl^- transport is altered, secretions such as sweat, mucus, and pancreatic juice are very salty—and often very thick. Abnormally thick mucus in the lungs impairs normal breathing and often leads to recurring lung infections. Thick pancreatic secretions can plug ducts that carry important enzymes to the digestive tract. Figure 3-29 shows a child with CF next to a normal child of the same age. Because of the breathing, digestive, and other problems caused by the disease, the affected child has not developed normally. Recent evidence suggests that *Duchenne muscular dystrophy (DMD)*, another severe inherited condition, results from "leaky" membranes in muscle cells. Calcium (Ca^{++}) enters affected muscle cells through leaky membranes and triggers chemical reactions that destroy the muscle, causing life-threatening paralysis.

Disorders Involving Cell Membrane Receptors

Disorders involving cell membrane receptors are just now being investigated by biologists. One particularly common disorder that involves cell membrane receptors is an adult-onset form of *diabetes mellitus,* called *type II diabetes* or *non-insulin-dependent diabetes mellitus (NIDDM)*. This condition is produced by a cellular response to obesity, which triggers a reduction in the number of membrane receptors for the hormone *insulin.* Cells throughout the body thus become less sensitive to insulin, the hormone that allows glucose molecules to enter the plasma membrane. Without sufficient stimulation by insulin, the cells literally starve for energy-rich glucose, even though it is available outside the cell.

Disorders Involving Cell Reproduction

As mentioned earlier in this chapter (see essay, p. 91), abnormalities in mitotic division can cause tumors to arise. *Cancers* are tumors that tend to spread, often disrupting vital functions and eventually killing the victim. Even noncancerous tumors can cause significant health impairment—or death—depending on their size and location.

Disorders Involving DNA and Protein Synthesis

Genetic disorders are pathological conditions caused by mistakes, or *mutations*, in a cell's genetic code. Abnormal genes cause the production of abnormal enzymes or other proteins. Abnormal proteins, in turn, cause abnormalities in cellular function—producing a specific disease. Many diseases are known to be caused by this mechanism, including some of the diseases in the other categories. Another example is *sickle-cell anemia*, a blood disease caused by the production of abnormal hemoglobin (the protein in red blood cells that carries oxygen). Mistakes in enzyme production can cause a whole group of metabolic disorders called *inborn errors of metabolism.*

Infections

Bacteria and *viruses* can infect cells and thus damage them in ways that produce disease. Bacteria, tiny one-celled organisms, may parasitize cells directly and thus destroy them, or the bacteria may produce toxins that damage cells or elicit violent reactions of the immune system. Viruses are microscopic particles that contain DNA or RNA. Viruses cause disease by taking over the genetic apparatus of a cell and thus using the cell's resources to produce viral products—or even new viruses. If enough cells are damaged during a bacterial or viral infection to disrupt vital functions, death may result.

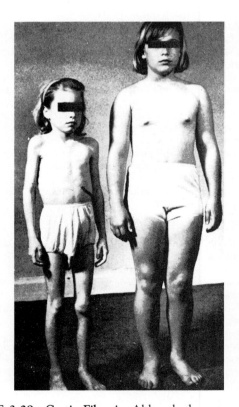

FIGURE 3-29 Cystic Fibrosis. Although the same age, the child with cystic fibrosis (*left*) is smaller and thinner than the unaffected child (*right*). In cystic fibrosis, the absence of chloride ion pumps causes thickening of some glandular secretions. Because thickened secretions block airways and digestive ducts, children born with this disease become weakened—often dying before adulthood.

CHAPTER SUMMARY

CELL STRUCTURE AND FUNCTION

A. The typical cell (Figure 3-1)
B. Cell structures
  1. Plasma membrane—separates the cell from its surrounding environment
  2. Cytoplasm—thick fluid filling the inside of the cell
  3. Organelles—a variety of structures (including the nucleus) suspended in the cytoplasm, with each structure suited to perform a specific function

CELL MEMBRANES

A. Each cell contains a variety of membranes
  1. Plasma membrane (Figure 3-2)
  2. Membranous organelles—sacs and canals made of the same material as the plasma membrane
B. Fluid mosaic model—theory explaining how cell membranes are constructed
  1. Molecules of the cell membrane are arranged in a sheet
  2. The mosaic of molecules is fluid, that is, the molecules are able to float around slowly
  3. This model illustrates that the molecules of the cell membrane form a continuous sheet
C. Chemical attractions are the forces that hold membranes together
D. Primary structure of a cell membrane is a double layer of phospholipid molecules
  1. Heads are hydrophilic (water-loving)
  2. Tails are hydrophobic (water-fearing)
  3. Arrange themselves in bilayers in water
  4. Cholesterol molecules are scattered among the phospholipids to allow the membrane to function properly at body temperature
  5. Most of the bilayer is hydrophobic, therefore, water or water-soluble molecules do not pass through easily
E. Membrane proteins (Table 3-1)
  1. A cell controls what moves through the membrane by means of membrane proteins imbedded in the phospholipid bilayer
  2. Some membrane proteins have carbohydrates attached to them, forming glycoproteins that act as identification markers
  3. Some membrane proteins are receptors that react to specific chemicals

CYTOPLASM AND ORGANELLES

A. Cytoplasm—thick internal fluid of cells that contains many organelles
B. Two major groups of organelles (Table 3-2)
  1. Membranous organelles are specialized sacs or canals made of cell membranes
  2. Nonmembranous organelles are made of microscopic filaments or other nonmembranous materials
C. Endoplasmic reticulum
  1. Made of membranous-walled canals and flat, curving sacs arranged in parallel rows throughout the cytoplasm; extend from the plasma membrane to the nucleus
  2. Proteins move through the canals
  3. Two types of endoplasmic reticulum
    a. Rough endoplasmic reticulum
      (1) Ribosomes dot the outer surface of the membranous walls
      (2) Ribosomes synthesize proteins, which move toward the Golgi apparatus and then eventually leave the cell
      (3) Function in both protein synthesis and intracellular transportation
    b. Smooth endoplasmic reticulum
      (1) No ribosomes border the membranous wall
      (2) Functions are less well established and probably more varied than for the rough endoplasmic reticulum
      (3) Synthesizes certain lipids and carbohydrates and creates membranes for use throughout the cell
D. Ribosomes (Figure 3-3)
  1. Many are attached to the rough endoplasmic reticulum and many lie free, scattered through the cytoplasm
  2. Each ribosome is a nonmembranous structure made of two pieces, a large subunit and a small subunit; each subunit is composed of rRNA
  3. Ribosomes in the endoplasmic reticulum make proteins for "export"; free ribosomes make proteins for the cells domestic use
E. Golgi apparatus
  1. Membranous organelle consisting of cisternae stacked on one another and located near the nucleus (Figure 3-4)
  2. Processes protein molecules from the endoplasmic reticulum (Figure 3-5)
  3. Processed proteins leave the final cisterna in a vesicle; contents may then be secreted to outside the cell
F. Lysosomes
  1. Made of microscopic membranous sacs that have "pinched off" from Golgi apparatus
  2. The cell's own digestive system; enzymes in lysosomes digest particles or large molecules that enter them; under some conditions, digest and thereby destroy cells
G. Peroxisomes
  1. Small membranous sacs containing enzymes that detoxify harmful substances that enter the cells
  2. Often seen in kidney and liver cells
H. Mitochondria (Figure 3-6)
  1. Made up of microscopic sacs; wall composed of inner and outer membranes separated by fluid; thousands of particles make up enzyme molecules attached to both membranes
  2. The "power plants" of cells; mitochondrial enzymes catalyze series of oxidation reactions that provide about 95% of cell's energy supply

CYTOSKELETON

A. The cell's internal supporting framework made up of rigid, rodlike pieces that provide support and allow movement
B. Cell fibers
  1. Intricately arranged fibers of varying lengths that form a three-dimensional, irregularly shaped lattice
  2. Fibers appear to support the endoplasmic reticulum, mitochondria, and "free" ribosomes (Figure 3-7)
  3. Smallest cell fibers are microfilaments
    a. "Cellular muscles"
    b. Made of thin, twisted strands of protein molecules that lie parallel to the long axis of the cell
    c. Microfilaments can slide past each other, causing shortening of the cell
  4. Intermediate filaments are twisted protein strands slightly thicker than microfilaments; form much of the supporting framework in many types of cells
  5. Microtubules are tiny, hollow tubes that are the thickest of the cell fibers; they are made of protein subunits arranged in a spiral fashion; their function is to move things around in the cell

C. Centrosome
1. An area of the cytoplasm near the nucleus that coordinates the building and breaking of microtubules in the cell
2. Nonmembranous structure has sometimes been called the microtubule-organizing center
3. Plays an important role during cell division
4. The general location of the centrosome is identified by the centrioles
D. Cell extensions
1. Cytoskeleton forms projections that extend the plasma membrane outward to form tiny, fingerlike processes
2. There are three types of these processes, which have specific functions (Figure 3-9)
   a. Microvilli—found in epithelial cells that line the intestines and other areas where absorption is important; they help to increase the surface area manyfold
   b. Cilia and flagella—cell processes that have cylinders made of microtubules at their core; cilia are shorter and more numerous than flagella; flagella are only found on human sperm cells

NUCLEUS

A. Definition—spherical body in center of cell; enclosed by pore-filled membrane
B. Structure
1. Consists of nuclear envelope (composed of two membranes each with essentially the same molecular structure as plasma membrane) surrounding nucleoplasm; nuclear envelope contains pores (Figure 3-10)
2. Contains DNA (heredity molecules), which appear as:
   a. Chromatin threads or granules in nondividing cells
   b. Chromosomes in early stages of cell division
C. Functions of nucleus are functions of DNA molecules; DNA determines both structure and function of cells and heredity

CELL CONNECTIONS

A. Cells are held together by fibrous nets that surround groups of cells (e.g., muscle cells) or, more frequently, cells have direct connections
B. There are three types of direct cell connections (Figure 3-11)
1. Desmosomes—"spot welds" that hold adjacent cells together; fibers on the outer surface of each desmosome interlock with each other; anchored internally by intermediate filaments of the cytoskeleton
2. Gap junctions—membrane channels of adjacent plasma membranes adhere to each other; have two effects:
   a. Form gaps or "tunnels" that join the cytoplasm of two cells
   b. Fuse two plasma membranes into a single structure
3. Tight junctions—occur in cells that are joined by "collars" of tightly fused material; molecules cannot permeate through the cracks of tight junctions; occur in the lining of the intestines and other parts of the body, where it is important to control what gets through a sheet of cells

MOVEMENT OF SUBSTANCES THROUGH CELL MEMBRANES

A. Passive transport processes—do not require any energy expenditure of the cell membrane
1. Diffusion—a passive process (Figures 3-12 and 3-13)
   a. Molecules spread through the membranes

b. Molecules move from an area of high concentration to an area of low concentration, down a concentration gradient
c. As molecules diffuse, a state of equilibrium will occur
d. Membrane channels are pores in cell membranes through which specific ions or other small, water-soluble molecules can pass
2. Dialysis—a form of diffusion in which the selectively permeable nature of a membrane causes the separation of smaller solute particles from larger solute particles
3. Osmosis (Figure 3-16)
   a. Diffusion of water through a selectively permeable membrane; limits the diffusion of at least some of the solute particles
   b. Water pressure that develops as a result of osmosis is called osmotic pressure
   c. Potential osmotic pressure is the maximum pressure that could develop in a solution when it is separated from pure water by a selectively permeable membrane; knowledge of potential osmotic pressure allows the prediction of the direction of osmosis and the resulting change of pressure
   d. Isotonic—when two fluids have the same potential osmotic pressure
   e. Hypertonic—"higher pressure"; cells placed in solutions that are hypertonic to intracellular fluid always shrivel as water flows out of cell; this has great medical importance: if medical treatment causes the extracellular fluid to become hypertonic to the cells of the body, serious damage may occur
   f. Hypotonic—"lower pressure"; cells placed in a hypotonic solution may swell as water flows into cell; water always osmoses from the hypotonic solution to the hypertonic solution
   g. Osmosis results in the gain of volume on one side of the membrane and loss of volume on the other side of the membrane
4. Facilitated diffusion
   a. A special kind of diffusion when movement of molecules is made more efficient by the action of carrier mechanisms in a cell membrane
   b. Transports substances down a concentration gradient
   c. Energy required comes from the collision energy of the solute, therefore it is simply carrier-mediated passive transport
5. Filtration
   a. Passage of water and permeable solutes through a membrane by the force of hydrostatic pressure
   b. Small molecules travel down a hydrostatic pressure gradient and through a sheet of cells
   c. Filtration results in the separation of large and small particles
   d. Filtration occurs most often in the capillaries
B. Active transport processes—require the expenditure of metabolic energy by the cell
1. Active transport
   a. Carrier-mediated process that moves substances against its concentration gradient (Figure 3-18)
   b. Opposite of diffusion
   c. Substances are moved by "pumps" (e.g., calcium pumps and sodium-potassium pumps)
2. Endocytosis and exocytosis—allow substances to enter or leave the interior of a cell without actually moving through its plasma membrane
   a. Endocytosis—the plasma membrane "traps" some

extracellular material and brings it into the cell in a vesicle; there are two basic types of endocytosis (Figure 3-20)

(1) Phagocytosis—"condition of cell eating"; large particles are engulfed by the plasma membrane and enter the cell in vesicles; vesicles fuse with lysosomes where the particles are digested

(2) Pinocytosis—"cell drinking"; fluid and the substances dissolved in it enter the cell

b. Exocytosis

(1) Process by which large molecules, notably proteins, can leave the cell even though they are too large to move out through the plasma membrane

(2) Large molecules are enclosed in membranous vesicles and then pulled to the plasma membrane by the cytoskeleton, where the contents are released

(3) Exocytosis also provides a way for new material to be added to the plasma membrane

GROWTH AND REPRODUCTION OF CELLS

A. Cell growth and reproduction of cells are the most fundamental of all living functions and together constitute the cell life cycle (Figure 3-21)

B. Cell growth—depends on using genetic information in DNA to make the structural and functional proteins needed for cell survival

C. Cell reproduction—ensures that genetic information is passed from one generation to the next

D. Deoxyribonucleic acid (DNA) (Figure 3-22)

1. A double helix polymer (composed of nucleotides) that has a most important function; the transfer of information; information, coded in genes, directs synthesis of all cell's proteins

2. Gene—a segment of a DNA molecule that consists of approximately 1,000 pairs of nucleotides

E. Cell growth—a newly formed cell produces a variety of molecules and other structures necessary for growth using the information contained in the genes of DNA molecules; this stage is known as interphase

1. Protein synthesis

a. An anabolic process—small molecules join to form large molecules

b. Steps of protein synthesis (Figure 3-24)

(1) mRNA forms along a segment of one strand of DNA; process is called transcription

(2) mRNA carries "message" to a ribosome, where the ribosome will "translate" the genetic code to form a specific sequence of amino acids; process is called translation

(3) tRNA becomes involved in protein synthesis during translation; tRNA carries amino acids to the ribosome for placement in the prescribed sequence

(4) Once the proper polypeptides are formed, enzymes in the ER, Golgi apparatus, or cytoplasm link them together to form whole protein molecules

2. DNA replication

a. Replication of the genome prepares the cell for reproduction; mechanics are similar to RNA synthesis

b. DNA replication (Figure 3-25)

(1) DNA strand uncoils and strands come apart

(2) Along each separate strand, a complementary strand forms

(3) The two new strands are called chromatids, instead of chromosomes

(4) Chromatids are attached pairs, and the centromere is the name of their point of attachment

F. Cell reproduction—cells reproduce by splitting themselves into two smaller daughter cells (Table 3-5)

1. Mitotic cell division—the process of organizing and distributing nuclear DNA during cell division has four distinct phases

a. Prophase—"before-phase"

(1) After the cell has prepared for reproduction during interphase, the nuclear envelope falls apart as the chromatids coil up to form chromosomes, which are joined at the centromere

(2) As chromosomes are forming, the centriole pairs move toward the poles of the parent cell and spindle fibers are constructed between them

b. Metaphase—"position-changing phase"

(1) Chromosomes move so that one chromatid of each chromosome faces its respective pole

(2) Each chromatid attaches to a spindle fiber

c. Anaphase—"apart phase"

(1) Centromere of each chromosome has split to form two chromosomes, each consisting of a single DNA molecule

(2) Each chromosome is pulled toward the nearest pole, forming two separate but identical pools of genetic information

d. Telophase—"end phase"

(1) DNA returns to its original form and location within the cell

(2) After completion of telophase each daughter cell begins interphase to develop into a mature cell

2. Meiosis (Figure 3-28)

CYCLE OF LIFE: CELLS

A. Diferent types of cells have different life cycles

B. Advancing age creates changes in cell numbers and in their ability to function effectively

1. Examples of decreased functional ability include muscle atrophy, loss of elasticity of the skin, and changes in the cardiovascular, respiratory, and skeletal systems

THE BIG PICTURE: CELLS AND THE WHOLE BODY

A. Most cell processes are occurring at the same time in all of the cells throughout the body

B. The processes of normal cell function result from the coordination dictated by the genetic code

MECHANISMS OF DISEASE: CELLULAR DISEASE

A. Most diseases involve abnormalities of cells

B. Disorders involving cell transport

1. Damage to cell transport mechanisms can result in diseases such as cystic fibrosis and Duchenne muscular dystrophy (Figure 3-29)

C. Disorders involving cell membrane receptors

D. Disorders involving cell reproduction

1. Tumors arise from abnormalities in mitotic division

E. Disorders involving DNA and protein synthesis

1. Genetic disorders are caused by mutations in a cell's genetic code

2. Abnormal genes cause the production of abnormal enzymes, which causes abnormalities in cellular function, resulting in a specific disease such as sickle-cell anemia

F. Infections—bacteria and viruses damage cells

REVIEW QUESTIONS

1. What is the range in human cell diameters?
2. List the three main cell structures.
3. Describe the location, molecular structure, and width of the plasma membrane.
4. Explain the communication function of the plasma membrane, its transportation function, and its identification function.
5. Briefly describe the structure and function of these cellular structures/organelles: endoplasmic reticulum, ribosomes, Golgi apparatus, mitochondria, lysosomes, peroxisomes, cytoskeleton, cell fibers, centrosome, centrioles, and cell extensions.
6. Describe the three types of intercellular junctions.
7. Describe briefly the functions of the nucleus and the nucleoli.
8. Contrast passive and active transport mechanisms.
9. Define the terms *diffusion, dialysis, facilitated diffusion, osmosis,* and *filtration.*
10. Explain how a concentration gradient relates to the process of diffusion.
11. Describe and give an example of a membrane channel.
12. Differentiate between actual osmotic pressure and potential osmotic pressure.
13. What factor directly determines the potential osmotic pressure of a solution?
14. Explain the terms *isotonic, hypotonic,* and *hypertonic.*
15. State the principle about the solution that develops an osmotic pressure, given appropriate conditions.
16. State the principle about the direction of active transport.
17. Name and describe the active transport pump that operates in the plasma membrane of all human cells.
18. Explain the processes of endocytosis and exocytosis.
19. Differentiate between the processes of phagocytosis and pinocytosis.
20. Describe the size and shape of a DNA molecule.
21. Where is DNA located and what is its function?
22. Explain the steps of DNA replication.
23. What are the steps involved in DNA replication and when does it occur?
24. Watson called DNA "the most golden of all molecules," primarily because of its function. What is it?
25. Define mitosis.
26. Briefly describe the four distinct phases of mitosis.
27. Define meiosis.
28. When does the reduction of chromosomes from the diploid to haploid number take place?
29. List several ways cells can adapt to changing conditions.
30. Differentiate between hypertrophy and atrophy.
31. Give examples of normal and abnormal hyperplasia.

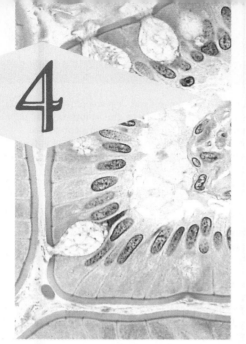

4 Tissues

Simple columnar epithelium

OBJECTIVES

After you have completed this chapter, you should be able to:

1. Define the term *tissue*.
2. List the four major categories of tissues and discuss the basic function of each type.
3. List and discuss some important structural and functional generalizations that apply to epithelium as a principal tissue type.
4. Classify membranous epithelium using cell shape and cell layers as criteria; discuss each type in terms of its structure, function, and location in the body.
5. Discuss glandular epithelium and compare endocrine and exocrine glands in terms of generalized function.
6. Discuss the structural classification of exocrine glands.
7. Explain how apocrine, holocrine, and merocrine glands differ in their method of secretion.
8. List the major types of connective tissues and contrast their important structural and functional differences.
9. Discuss the major types of connective tissue fibers, cells, and matrix in terms of structure and function.
10. Compare bone and cartilage in terms of generalized function, cell types, organizational structure, and blood supply.
11. Discuss blood as a tissue.
12. Compare characteristics of neurons and neuroglia in terms of nervous system function.
13. Explain the process of regeneration as it relates to tissue repair.
14. Discuss and give examples of the two major categories or types of body membranes.

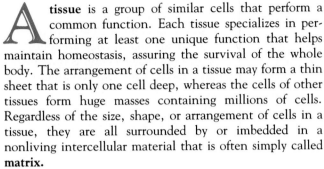

KEY TERMS

Connective tissue Most abundant tissue in the body; it connects, supports, transports, and defends

Epithelial tissue (ep-i-THEE-lee-al TISH-yoo) Tissue that covers or lines the body and some of its parts (membranous) or forms secretory units (glandular)

Gland Epithelium specialized for secretory activity

Histogenesis (his-toe-JEN-e-sis) Process of the primary germ layers differentiating into different kinds of tissues

Inflammation Reaction of cells and tissues to injury

Matrix (MAY-triks) Nonliving intercellular material surrounding the cells of a tissue

Mucous membrane (MYOO-kus MEM-brane) Epithelial membrane that lines body surfaces opening directly to the exterior

Nervous tissue Consists of two basic kinds of cells, neurons and neuroglia; it processes and transmits information

Regeneration (ree-jen-er-AY-shun) Growth of new tissue

Serous membrane (SE-rus MEM-brane) Epithelial membrane that lines body cavities and covers the surfaces of organs

Tissue Group of similar cells that performs a common function

A tissue is a group of similar cells that perform a common function. Each tissue specializes in performing at least one unique function that helps maintain homeostasis, assuring the survival of the whole body. The arrangement of cells in a tissue may form a thin sheet that is only one cell deep, whereas the cells of other tissues form huge masses containing millions of cells. Regardless of the size, shape, or arrangement of cells in a tissue, they are all surrounded by or imbedded in a nonliving intercellular material that is often simply called **matrix.**

Tissues differ as to the amount and kind of intercellular, or "between-the-cells," matrix. It is the unique and specialized nature of the matrix in bone and cartilage, for example, that contributes to the strength and resiliency of the body. Some tissues contain almost no intercellular matrix. Other tissues are almost entirely matrix—with only a few cells present. Some types of tissue matrix contain fibers that make them flexible or elastic, some contain mineral crystals that make them rigid, and others are very fluid.

It is primarily the intercellular junctions, such as the desmosomes and tight junctions described in the previous chapter, that hold groups of cells together to form tissues found in sheets or other continuous masses of cells. The tissue that forms the outer layer of skin is held together this way. In other tissues, it is the matrix that holds the cells together—if they are held together at all. For example, the fibers and crystals of bone matrix hold bone tissue together, whereas the fluid nature of blood matrix (plasma) does not hold blood tissue in a solid mass at all.

The four major types of human tissue that were introduced in Chapter 1 are described in more detail in this chapter. An understanding of the major tissue types will help you understand the next higher levels of organization in the body—organs and organ systems. Eventually, your knowledge of **histology** (the biology of tissues) will give you a better appreciation for the nature of the whole body.

◢ PRINCIPAL TYPES OF TISSUE

Although a number of subtypes are present in the body, all tissues can be classified by their structure and function into four principal types:

1. **Epithelial tissue** covers and protects the body surface, lines body cavities, specializes in moving substances into and out of the blood (secretion, excretion, and absorption), and forms many glands.

2. **Connective tissue** is specialized to support the body and its parts, to connect and hold them together, to transport substances through the body, and to protect it from foreign invaders. The cells in connective tissue are often relatively far apart and separated by large quantities of nonliving matrix.

3. **Muscle tissue** produces movement; it moves the body and its parts. Muscle cells are specialized for contractility and produce movement by the shortening of contractile units found in the cytoplasm.

4. **Nervous tissue** is the most complex tissue in the body. It specializes in communication between the various parts of the body and in integration of their activities. This tissue's major function is the generation of complex messages for the coordination of body functions.

EMBRYONIC DEVELOPMENT OF TISSUES

The four major tissues of the body appear early in the *embryonic* period of development (first 2 months after conception). After fertilization has occurred, repeated cell divisions soon convert the single-celled zygote into a hollow ball of cells called a **blastocyst.** The blastocyst implants in the uterus, and within 2 weeks the cells move and regroup in an orderly way into three **primary germ layers** called **endoderm, mesoderm,** and **ectoderm** (Figure 4-1). The process by which blastocyst cells move and then differentiate into the three primary germ layers is called **gastrulation.** During this process the cells in each germ layer become increasingly differentiated to form specific tissues and eventually give rise to the structures listed in

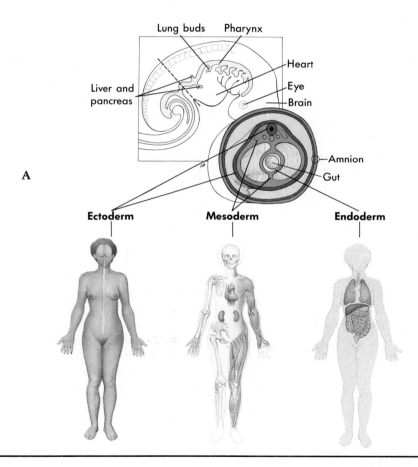

ECTODERM

Epithelium (epidermis) of skin
Lining of mouth, anus, nostrils
Sweat and sebaceous glands
Epidermal derivatives (hair, enamel of teeth)
Nervous system (brain and spinal cord)
Epithelial (sensory) parts of eyes, nose, ear

MESODERM

Muscles
Skeleton (bones and cartilage)
Blood
Epithelial lining of blood vessels
Dermis of skin and dentin of teeth
Organs (except lining) of excretory and
  reproductive systems
Connective tissue

ENDODERM

Epithelium (lining) of digestive and respiratory
  systems
Secretory parts of liver and pancreas
Urinary bladder
Epithelial lining of urethra
Thyroid, parathyroid, thymus

FIGURE 4-1   Primary germ layers. A, Illustration shows the primary germ layers and the body systems into which they develop. **B,** Structures derived from primary germ layers.

Figure 4-1, *B*. In summary, some epithelial tissues develop from each of the primary germ layers, whereas connective and muscle tissues arise from mesoderm, and nerve tissue develops from ectoderm. The process of the primary germ layers differentiating into the different kinds of tissues is called **histogenesis.** Chapter 32 provides additional details of human development, including a discussion of differentiation of organs and body systems.

QUICK CHECK

✔ 1. *Name the four basic tissue types and give the major function of each.*

✔ 2. *What is a primary germ layer?*

EPITHELIAL TISSUE

Types and Locations

Epithelial tissue, or **epithelium,** is often subdivided into two types: (1) **membranous** (covering or lining) epithelium and (2) **glandular** epithelium. Membranous epithelium covers the body and some of its parts and lines the serous cavities (pleural, pericardial, and peritoneal), the blood and lymphatic vessels, and the respiratory, digestive, and genitourinary tracts. Glandular epithelium is grouped in solid cords or specialized follicles that form the secretory units of endocrine and exocrine glands.

Functions

Epithelial tissues have a widespread distribution throughout the body and serve several important functions:

♦ **Protection.** Generalized protection is the most important function of membranous epithelium. It is the relatively tough and impermeable epithelial covering of the skin that protects the body from mechanical and chemical injury and also from invading bacteria and other disease-causing microorganisms.

♦ **Sensory functions.** Epithelial structures specialized for sensory functions are found in the skin, nose, eye, and ear.

♦ **Secretion.** Glandular epithelium is specialized for secretory activity. Secretory products include hormones, mucus, digestive juices, and sweat.

♦ **Absorption.** The lining epithelium of the gut and respiratory tracts allows for the absorption of nutrients from the gut and the exchange of respiratory gases between air in the lungs and the blood.

♦ **Excretion.** It is the specialized epithelial lining of kidney tubules that makes the excretion and concentration of excretory products in the urine possible.

Generalizations About Epithelial Tissue

♦ Most epithelial tissues are characterized by extremely limited amounts of intercellular, or matrix, material. This explains their characteristic appearance, when viewed under a light microscope, of a continuous sheet of cells packed tightly together. With the electron microscope, however, narrow spaces—about one millionth of an inch (20 nm) wide—can

be seen around the cells. These spaces, like other intercellular spaces, contain interstitial fluid.

♦ Sheets of epithelial cells compose the surface layer of skin and of mucous and serous membranes. The epithelial tissue attaches to an underlying layer of connective tissue by means of a thin noncellular layer of adhesive, permeable material called the **basement membrane** (see Figure 4-2). Both epithelial and connective tissue cells synthesize the basement membrane, which is made up of glycoprotein material secreted by the epithelial components and a fine mesh of fibers produced by the connective tissue cells. Histologists refer to the glycoprotein material secreted by the epithelial cells as the **basal lamina** and to the connective tissue fibers as the **reticular lamina.** The union of basal and reticular lamina forms the basement membrane.

♦ Epithelial tissues contain no blood vessels. As a result, epithelium is said to be **avascular** (*a,* "without"; *vascular,* "vessels"). Hence oxygen and nutrients must diffuse from capillaries in the underlying connective tissue through the permeable basement membrane to reach epithelial cells.

♦ At intervals between adjacent epithelial cells, their plasma membranes are modified so as to hold the cells together. These specialized intercellular structures, such as **desmosomes** and **tight junctions,** are described in Chapter 3.

♦ Epithelial cells can reproduce themselves. They frequently go through the process of cell division. Since epithelial cells in many locations meet considerable wear and tear, this fact has practical importance. It means, for example, that new cells can replace old or destroyed epithelial cells in the skin or in the lining of the gut or respiratory tract.

CLASSIFICATION OF EPITHELIAL TISSUE

Membranous (Covering or Lining) Epithelium
Classification based on cell shape

The shape of membranous epithelial cells may be used for classification purposes. Four cell shapes, called **squamous, cuboidal, columnar,** and **pseudostratified columnar,** are used in this classification scheme (Figure 4–2). Squamous (Latin, "scaly") cells are flat and platelike. Cuboidal cells, as the name implies, are cube-shaped and have more cytoplasm than the scalelike squamous cells. Columnar epithelial cells are higher than they are wide and appear narrow and cylindrical. Pseudostratified columnar epithelium has only one layer of oddly shaped columnar cells. Although each cell touches the basement membrane, the tops of some pseudostratified cells do not fully extend to the surface of the membrane. Also, some nuclei are near the "top" of the cell and some near the "bottom" of the cell—rather than all nuclei being near the bottom. The result is a false (*pseudo*) appearance of layering, or stratification, when only a single layer of cells is present.

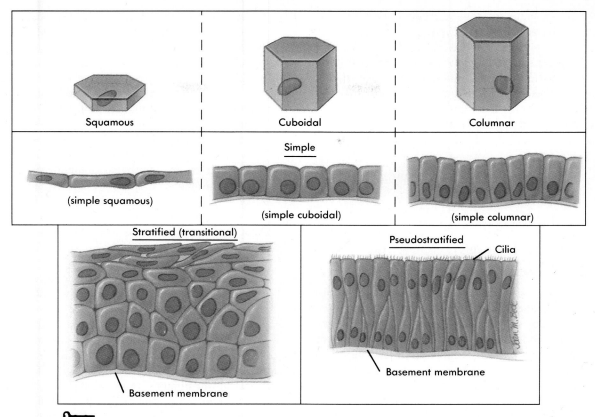

FIGURE 4-2 **Classification of epithelial tissues.** The tissues are classified according to the shape and arrangement of cells.

Classification based on layers of cells

In most cases the location and function of membranous epithelium determine whether or not its cells will be stacked and layered or arranged in a sheet one-cell-layer thick. Arrangement of epithelial cells in a single layer is called **simple epithelium.** If epithelial cells are layered one on another, the tissue is called **stratified epithelium.** **Transitional epithelium** (described later) is a unique arrangement of differing cell shapes in a stratified, or layered, epithelial sheet.

If membranous or covering epithelium is classified both by the shape and layering of its cells, the specific types listed in Table 4–1 are possible. Notice that stratified tissue types are named for the shape of cells in their top layer only. Each type is described in the paragraphs that follow, and selected examples are illustrated in Figures 4–3 to 4–9.

Simple epithelium.

Simple squamous epithelium. Simple squamous epithelium consists of only one layer of flat, scale-like cells (Figure 4-3). Consequently, substances can readily diffuse or filter through this type of tissue. The microscopic air sacs (alveoli) of the lungs, for example, are composed of this kind of tissue, as are the linings of blood and lymphatic vessels and the surfaces of the pleura, pericardium, and peritoneum. (Blood and lymphatic vessel linings are called

TABLE 4-1	Classification scheme for membranous epithelial tissues	
	Shape of Cells*	**Tissue Type**
One layer	Squamous	Simple squamous
	Cuboidal	Simple cuboidal
	Columnar	Simple columnar
	Pseudostratified columnar	Pseudostratified columnar
Several layers	Squamous	Stratified squamous
	Cuboidal	Stratified cuboidal
	Columnar	Stratified columnar
	(Varies)	Transitional

*In the top layer (if more than one layer in the tissue).

endothelium, and the surfaces of the pleura, pericardium, and peritoneum are called **mesothelium.** Some histologists classify these as connective tissue.)

Simple cuboidal epithelium. Simple cuboidal epithelium is composed of one layer of cuboidal-shaped cells resting on a basement membrane (Figure 4-4). This type of epithelium is seen in many types of glands and their ducts. It is also found in the ducts and tubules of other organs, such as the kidney.

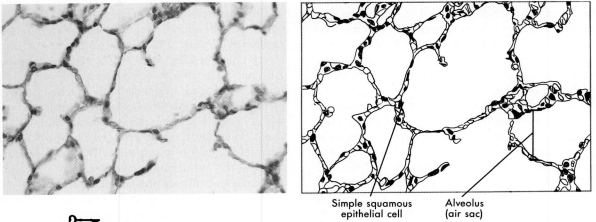

FIGURE 4-3 **Simple squamous epithelium. A,** Photomicrograph of lung tissue shows thin simple squamous epithelium lining the alveolar air sacs. **B,** Sketch of micrograph.

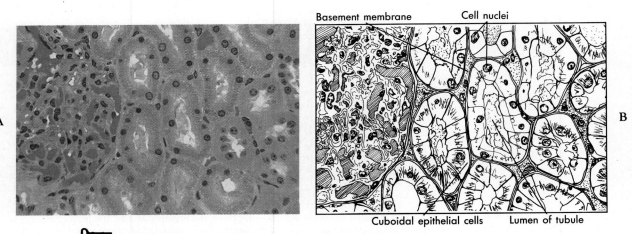

FIGURE 4-4 **Simple cuboidal epithelium. A,** Photomicrograph of kidney tubules shows the single layer of cuboidal cells touching a basement membrane. **B,** Sketch of the micrograph. Note the cuboidal cells that enclose the tubule (lumen) opening.

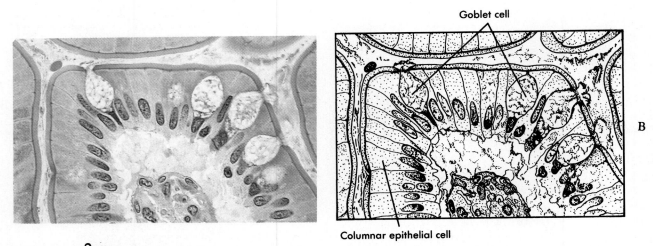

FIGURE 4-5 **Simple columnar epithelium. A,** Photomicrograph. **B,** Sketch of the photomicrograph. Note the goblet, or mucus-producing, cells present.

Simple columnar epithelium. Simple columnar epithelium composes the surface of the mucous membrane that lines the stomach, intestine, uterus, uterine tubes, and parts of the respiratory tract (Figure 4-5). It consists of a single layer of cells, many of which have a modified structure. Three common modifications are goblet cells, cilia, and microvilli. Goblet cells have large, secretory vesicles, which give them the appearance of a goblet. The vesicles contain mucus, which goblet cells produce in great quantity and secrete onto the surface of the epithelial membrane. Mucus is a solution of water, electrolytes, and glycoproteins. In the intestine the plasma membranes of many columnar cells extend out in hundreds and hundreds of microscopic fingerlike projections called **microvilli.** By greatly increasing the surface area of the intestinal mucosa, microvilli make it especially well suited for absorbing nutrients and fluids from the intestine.

Pseudostratified columnar epithelium. Pseudostratified columnar epithelium is found lining the air passages of the respiratory system and certain segments of the male reproductive system such as the urethra (Figure 4-6). Although appearing to be stratified, only a single layer of irregularly shaped columnar cells touches the basement membrane. The cells are of differing heights, and many are not tall enough to reach the upper surface of the epithelial sheet. This fact, coupled with placement of cell nuclei at odd and irregular levels in the cells, gives a false (pseudo) impression of stratification. Mucus-secreting goblet cells are numerous and cilia are present. In the respiratory system air passages, uniform motion of the cilia causes a thin layer of tacky mucus to move in one direction over the free surface of the epithelium. As a result, dust particles in the air are trapped and moved toward the mouth and away from the delicate lung tissues.

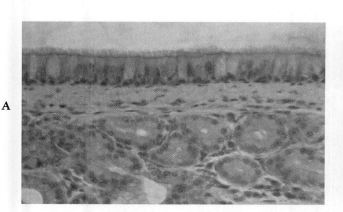
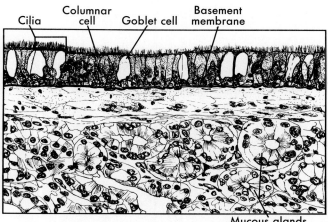

FIGURE 4-6   Pseudostratified ciliated epithelium. A, This photomicrograph of the trachea shows that each irregularly shaped columnar cell touches the underlying basement membrane. **B,** Sketch of the photomicrograph. Placement of cell nuclei at irregular levels in the cells gives a false (pseudo) impression of stratification.

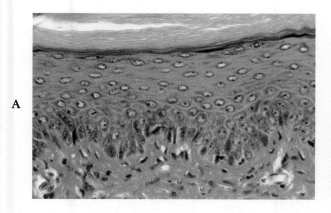
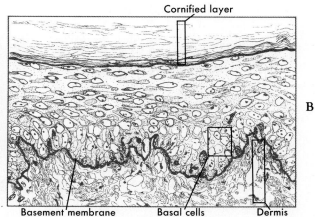

FIGURE 4-7   Stratified squamous (keratinized) epithelium. A, Photomicrograph of the skin shows cells becoming progressively flattened and scale-like as they approach the surface and are lost. **B,** Sketch of the photomicrograph. The outer surface of this epithelial sheet contains many flattened cells, which have lost their nuclei.

Stratified epithelium.

Stratified squamous (keratinized) epithelium. Stratified squamous epithelium is characterized by multiple layers of cells with typical flattened squamous cells at the free or outer surface of the sheet (Figure 4-7). The presence of keratin in these cells contributes to the protective qualities of skin covering the body surface. Details of the histology of this type of epithelium are presented in Chapter 5.

Stratified squamous (nonkeratinized) epithelium. Nonkeratinized stratified squamous epithelium is found lining the vagina, mouth, and esophagus (Figure 4-8). Its free surface is moist, and the outer epithelial cells, unlike those found in the skin, do not contain keratin. This type of epithelium serves a protective function.

Stratified cuboidal epithelium. The cuboidal variety of stratified epithelium also serves a protective function. Typically, two or more rows of low cuboidal-shaped cells

are arranged randomly over a basement membrane. Stratified cuboidal epithelium can be located in the sweat gland ducts, in the pharynx, and over parts of the epiglottis.

Stratified columnar epithelium. Although this protective epithelium has multiple layers of columnar cells, only the most superficial cells are truly columnar in appearance. Epithelium of this type is rare. It is located in segments of the male urethra and in the mucous layer near the anus.

Stratified transitional epithelium. Transitional epithelium is a stratified tissue that is typically found in body areas, such as the wall of the urinary bladder, that are subjected to stress and tension changes (Figure 4-9). In many instances 10 or more layers of cuboidal-like cells of varying shape are present in the absence of stretching or tension. As tension increases, the epithelial sheet is expanded, the number of observable cell layers will decrease, and cell shape will change from cuboidal to

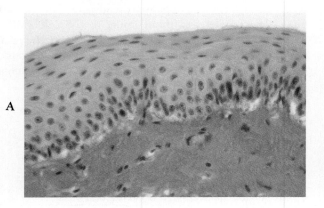

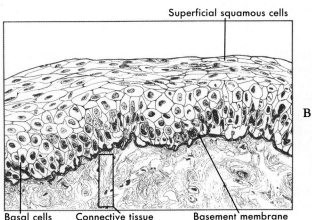

FIGURE 4-8   Stratified squamous (nonkeratinized) epithelium. A, Photomicrograph of vaginal tissue. Each cell in the layer is flattened near the surface and attached to the sheet. No flaking of dead cells from the surface occurs. **B,** Sketch of the photomicrograph. All cells have nuclei. Compare to Figure 4-7.

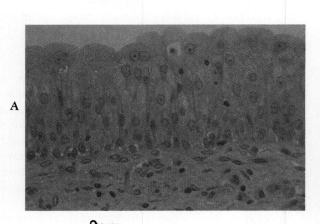

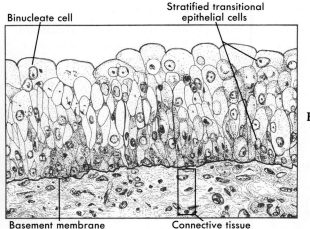

FIGURE 4-9   Transitional epithelium. A, Photomicrograph of the urinary bladder shows that cell shape is variable from cuboidal to squamous. Several layers of cells are present. Intermediate and surface cells do not touch basement membrane. **B,** Sketch of photomicrograph.

squamous in appearance. This ability of transitional epithelium to stretch protects the bladder wall and other distensible structures that it lines from tearing when stretched with great force.

Glandular Epithelium

Epithelium of the glandular type is specialized for secretory activity. Regardless of the secretory product produced, glandular activity is dependent on complex and highly regulated cellular activities requiring the expenditure of stored energy.

Unlike the single or layered cells of membranous epithelium that are typically found in protective coverings or linings, glandular epithelial cells may function singly as **unicellular glands** or in clusters, solid cords, or specialized follicles as **multicellular glands.** Glandular secretions may be discharged into ducts, directly into the blood, into the lumen of hollow visceral structures, or onto the body surface.

All glands in the body can be classified as either exocrine or endocrine glands. **Exocrine glands,** by definition, discharge their secretion products into ducts. The salivary glands are typical exocrine glands. The secretion product (saliva) is produced in the gland and then discharged into a duct that transports it to the mouth. **Endocrine glands** are often called **ductless glands** because they discharge their secretion products (hormones) directly into the blood or interstitial fluid. The pituitary, thyroid, and adrenal glands are typical endocrine glands.

Structural classification of exocrine glands

Multicellular exocrine glands are most often classified by structure, using the shape of their ducts and the complexity (branching) of their duct systems as distinguishing characteristics. Shapes include **tubular** and **alveolar** (saclike). **Simple** exocrine glands have only one duct leading to the surface, and **compound** exocrine glands have two or more ducts. Table 4-2 describes some of the major structural types of exocrine glands.

Functional classification of exocrine glands

In addition to structural differences, exocrine glands also differ in the method by which they discharge their secretion products from the cell. Using these functional criteria, three types of exocrine glands can be identified (Figure 4-10):

1. Apocrine
2. Holocrine
3. Merocrine

Apocrine glands collect their secretory products near the apex, or tip, of the cell and then release it into a duct by pinching off the distended end. This process results in some loss of cytoplasm and damage to the cell. Recovery and repair of cells are rapid, however, and continued secretion occurs. The milk-producing mammary glands are examples of apocrine-type glands.

Holocrine glands—such as the sebaceous glands that produce oil to lubricate the skin—collect their secretory

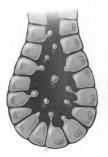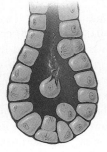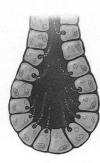

Apocrine gland   Holocrine gland   Merocrine gland

FIGURE 4-10 Exocrine glands. Here exocrine glands are classified by method of secretion.

product inside the cell and then rupture completely to release it. These cells literally self-destruct to complete their function.

Merocrine glands discharge their secretion product directly through the cell or plasma membrane. This discharge process is completed without injury to the cell wall and without loss of cytoplasm. Only the secretion product passes from the glandular cell into the duct. Most secretory cells are of this type. The salivary glands are good examples of merocrine-type exocrine glands.

QUICK CHECK

✔ 1. List at least three functions of epithelial tissue.
✔ 2. What are the three basic shapes of epithelial cells?
✔ 3. Distinguish between a simple epithelial tissue and a stratified epithelial tissue.
✔ 4. How do exocrine glands secrete their products?

CONNECTIVE TISSUE

Connective tissue is one of the most widespread tissues in the body, found in or around nearly every organ of the body. It exists in more varied forms than the other three basic tissues: delicate tissue-paper webs, tough resilient cords, rigid bones, and a fluid, namely, blood, are all forms of connective tissue.

Functions, Characteristics, and Types

Connective tissue connects, supports, transports, and defends. It connects tissues to each other, for example. It also connects muscles to muscles, muscles to bones, and bones to bones. It forms a supporting framework for the body as a whole and for its organs individually. One kind of connective tissue—blood—transports a large array of substances between parts of the body. And finally, several kinds of connective tissue cells defend us against microbes and other invaders.

Connective tissue consists predominantly of intracellular material called **matrix.** Imbedded in the matrix are relatively few cells, varying numbers and kinds of fibers, fluid, and perhaps other material called **ground substance.** The qualities of the matrix and fibers largely determine the

TABLE 4-2   Structural Classification of Multicellular Exocrine Glands

Shape*	Complexity †		Type	Example
Tubular (single, straight)	Simple		Simple tubular	Intestinal glands
Tubular (coiled)	Simple		Simple coiled tubular	Sweat glands
Tubular (multiple)	Simple		Simple branched tubular	Gastric (stomach) glands
Alveolar (single)	Simple		Simple alveolar	Sebaceous (skin oil) glands
Alveolar (multiple)	Simple		Simple branched alveolar	Sebaceous glands
Tubular (multiple)	Compound		Compound tubular	Mammary glands
Alveolar (multiple)	Compound		Compound alveolar	Mammary glands
Some tubular; some alveolar	Compound		Compound tubuloalveolar	Salivary glands

*Shape of the distal, secreting units of the gland.
†Number of ducts *reaching the surface*.

structural characteristics of each type of connective tissue. The matrix of blood, for example, is a fluid (plasma). It contains numerous blood cells but no fibers, except when it coagulates. Some connective tissues have the consistency of a soft gel, some are firm but flexible, some hard and rigid, some tough, others delicate—and in each case it is their matrix and extracellular fibers that make them so.

A connective tissue's matrix contains one or more of the following kinds of fibers: collagenous (or white), reticular, or elastic. Fibroblasts and some other cells produce these protein fibers. Collagenous fibers are tough and strong, reticular fibers are delicate, and elastic fibers are extensible and elastic. Collagenous or white fibers are made of *collagen* and often occur in bundles—an arrangement that provides great tensile strength. Reticular fibers, in contrast, occur in networks and, although delicate, support small structures such as capillaries and nerve fibers. Reticular fibers are made of a specialized type of collagen called *reticulin.* Collagen in its hydrated form you know as gelatin. Of all the hundreds of different protein compounds in the body, collagen is the most abundant. Biologists now estimate that it constitutes somewhat over one fourth of all the protein in the body. And interestingly, one of the most basic factors in the aging process, according to some researchers, is the change in the molecular structure of collagen that occurs gradually with the passage of the years.

Elastic fibers are made of a protein called *elastin,* which returns to its original length after being stretched. Elastic fibers are found in "stretchy" tissues, such as the cartilage of the external ear.

Connective tissues have been classified by histologists in several different ways. Usually, they are placed into different categories or types according to the structural characteristics of the intercellular material. The classification scheme we have adopted here is widely used and includes most of the major types:

1. Fibrous
   a. Loose (areolar)
   b. Adipose
   c. Reticular
   d. Dense
2. Bone
3. Cartilage
   a. Hyaline
   b. Fibrocartilage
   c. Elastic
4. Blood

Fibrous tissues (areolar, adipose, reticular, and dense) have extracellular fibers as their predominant feature. The type and arrangement of the fibers is what distinguishes members of the group from each other. Bone is considered a separate category because it has both fibers and a hard mineral ground substance. Cartilage is yet another category because besides fibers, it has a specialized ground substance that traps water to form a firm gel. Blood, the last category listed, is characterized by the lack of fibers in its matrix. These major types of connective tissue are described further in the following pages and in Table 4-3.

TABLE 4–3 Tissues		
Tissue	**Location**	**Function**
EPITHELIAL		
Simple squamous	Alveoli of lungs	Absorption by diffusion of respiratory gases between alveolar air and blood
	Lining of blood and lymphatic vessels (called *endothelium;* classified as connective tissue by some histologists)	Absorption by diffusion, filtration, osmosis
	Surface layer of pleura, pericardium, peritoneum (called *mesothelium;* classified as connective tissue by some histologists)	Absorption by diffusion and osmosis; also, secretion
Stratified squamous	Surface of mucous membrane lining mouth, esophagus, and vagina	Protection
	Surface of skin (epidermis)	Protection
Transitional	Surface of mucous membrane lining urinary bladder and ureters	Permits stretching
Simple columnar	Surface layer of mucous lining of stomach, intestines, and part of respiratory tract	Protection; secretion; absorption; moving of mucus (by ciliated columnar epithelium)

TABLE 4-3   Tissues—cont'd

Tissue	Location	Function
Pseudostratified	Surface of mucous membrane lining trachea, large bronchi, nasal mucosa, and parts of male reproductive tract (epididymis and vas deferens); lines large ducts of some glands (e.g., parotid)	Protection
Glandular	Glands	Secretion
CONNECTIVE		
Fibrous		
Loose, ordinary (areolar)	Between other tissues and organs	Connection
	Superficial fascia	Connection
Adipose (fat)	Under skin	Protection
	Padding at various points	Insulation
		Support
		Reserve food
Reticular	Inner framework of spleen, lymph nodes, bone marrow	Support
		Filtration
Dense fibrous		
Regular	Tendons	Flexible but strong connection
	Ligaments	
	Aponeuroses	
Irregular	Deep fascia	Connection
	Dermis	Support
	Scars	
	Capsule of kidney, etc.	
Bone	Skeleton	Support
		Protection
		Calcium reservoir
Cartilage		
Hyaline	Part of nasal septum	Firm but flexible support
	Covering articular surfaces of bones	
	Larynx	
	Rings in trachea and bronchi	
Fibrocartilage	Disks between vertebrae	
	Symphysis pubis	
Elastic	External ear	
	Eustachian tube	
Blood	In blood vessels	Transportation
		Protection
MUSCLE		
Skeletal (striated voluntary)	Muscles that attach to bones	Movement of bones
	Extrinsic eyeball muscles	Eye movements
	Upper third of esophagus	First part of swallowing
Smooth (nonstriated, involuntary, or visceral)	In walls of tubular viscera of digestive, respiratory, and genitourinary tracts	Movement of substances along respective tracts
	In walls of blood vessels and large lymphatic vessels	Change diameter of blood vessels, thereby aiding in regulation of blood pressure
	In ducts of glands	Movement of substances along ducts
	Intrinsic eye muscles (iris and ciliary body)	Change diameter of pupils and shape of lens
	Arrector muscles of hairs	Erection of hairs (gooseflesh)
Cardiac (striated involuntary)	Wall of heart	Contraction of heart
NERVOUS	Brain	Excitability
	Spinal cord	Conduction
	Nerves	

Fibrous Connective Tissues
Loose connective tissue (areolar)

Loose connective tissue, shown in Figure 4-11, is often called *loose, ordinary connective tissue,* or *areolar tissue.* It is loose because it is stretchable and ordinary because it is one of the most widely distributed of all tissues. It is common and ordinary, not special like some kinds of connective tissue (e.g., bone and cartilage) that help form comparatively few structures. "Areolar" was the early name for the loose, ordinary connective tissue that connects many adjacent structures of the body. It acts like a glue spread between them—but an elastic glue that permits movement. The word **areolar** means "like a small space" and refers to the bubbles that appear as areolar tissue is pulled apart during dissection.

The matrix of areolar tissue is a soft, viscous gel mainly because it contains hyaluronic acid. An enzyme, hyaluronidase, can change the matrix from its viscous gel state to a watery consistency. Physicians have made use of this knowledge for many years. They frequently inject a commercial preparation of hyaluronidase with drugs or fluids. By de-creasing the viscosity of intercellular material, the enzyme not only hastens diffusion and absorption of the injected material, but it also lessens tissue tension and pain. Some bacteria, notably pneumococci and streptococci, spread through connective tissues by secreting hyaluronidase.

The matrix of areolar tissue contains numerous fibers and cells. Typically, there are many interwoven collagenous and elastic fibers and a half dozen or so kinds of cells. **Fibroblasts** are usually present in the greatest numbers in areolar tissue, and **macrophages** are second. Fibroblasts synthesize both the gel-like ground substance and the fibers present in it. Macrophages (also known by several other names, for example, "histiocytes" and "resting wandering cells") carry on phagocytosis and so are classified as phagocytes. Phagocytosis is part of the body's vital complex of defense mechanisms. Other kinds of cells found in loose, ordinary connective tissue are mast cells, some wandering white blood cells (leukocytes), an occasional fat cell, and some plasma cells. Both macrophages and mast cells are derived from white blood cells.

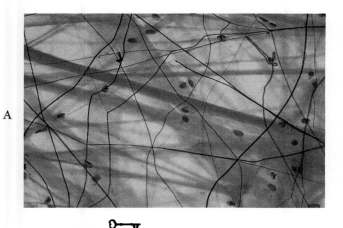

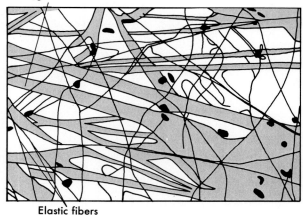

Bundle of collagen fibers

Elastic fibers

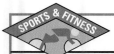

FIGURE 4-11 Loose, ordinary (areolar) connective tissue. **A,** Photomicrograph. **B,** Sketch of photomicrograph. Note the loose arrangement of fibers compared to fibers in Figures 4-15 and 4-17.

SPORTS & FITNESS

Tissues and Fitness

Achieving and maintaining an ideal body weight is a health-conscious goal. However, a better indicator of health and fitness is **body composition.** Exercise physiologists assess body composition to identify the percentage of the body made of lean tissue and the percentage made of fat. Body-fat percentage is often determined by using calipers to measure the thickness of skin folds at certain locations on the body. The thickness measurements, which reflect the volume of adipose tissues under the skin, are then used to estimate the percentage of fat in the entire body. A much more accurate method is to weigh a subject totally immersed in a tank of water. Fat has a very low density and thus increases the buoyancy of the body. Thus the lower a person's weight while immersed, the higher the body-fat percentage.

A person with low body weight may still have a high ratio of fat to muscle, an unhealthy condition. In this case the individual is "underweight" but "overfat." In other words, fitness depends more on the percentage and ratio of specific tissue types than the overall amount of tissue present. Therefore one goal of a good fitness program is a desirable body-fat percentage. For men, the ideal is 15% to 18%, and for women, the ideal is 20% to 22%.

Because fat contains stored energy (measured in calories), a low fat percentage means a low energy reserve. High body-fat percentages are associated with several life-threatening conditions, including cardiovascular disease. A balanced diet and an exercise program ensure that the ratio of fat to muscle tissue stays at a level appropriate for maintaining homeostasis.

Adipose tissue

Adipose tissue differs from loose, ordinary connective tissue mainly in that it contains predominantly fat cells and many fewer fibroblasts, macrophages, and mast cells (Figure 4-12). Adipose tissue forms supporting, protective pads around the kidneys and various other structures. It also serves two other functions—it constitutes a storage depot for excess food, and it acts as an insulating material to conserve body heat. See Figure 4-13 for the location of the main fat storage areas.

Reticular tissue

A three-dimensional web, that is, a reticular network, identifies reticular type tissue (Figure 4-14). Slender, branching reticular fibers with reticular cells overlying them compose the reticular meshwork. Branches of the cytoplasm of reticular cells follow the branching reticular fibers.

Reticular tissue forms the framework of the spleen, lymph nodes, and bone marrow. It functions as part of the body's complex mechanism for defending itself against microbes and injurious substances. The reticular meshwork filters injurious substances out of the blood and lymph, and the reticular cells phagocytose (engulf and destroy) them. Another function of reticular cells is to make reticular fibers.

Dense fibrous tissue

Dense fibrous tissue consists mainly of fibers packed densely in the matrix. It contains relatively few fibroblast cells. Some dense fibrous tissues are designated as *regular* and others as *irregular*, depending on the arrangement of fibers. In dense fibrous (regular) tissues, the bundles of fibers are arranged in regular, parallel rows (Figure 4-15). A structure composed of dense fibrous (regular) tissue is predominately bundles of collagenous fibers and is flexible but possesses great tensile strength. These characteristics are desirable in structures that anchor muscle to bone, such

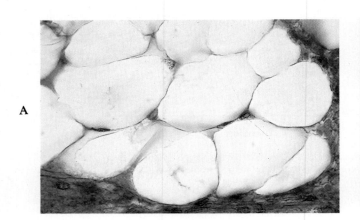

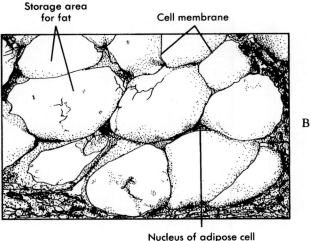

Storage area for fat

Cell membrane

Nucleus of adipose cell

A

B

FIGURE 4-12 Adipose tissue. A, Photomicrograph. **B,** Sketch of photomicrograph. Note the large storage spaces for fat inside the adipose tissue cells.

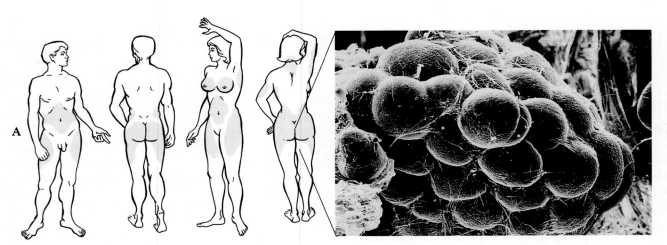

A

B

FIGURE 4-13 Fat storage areas. A, The different distribution of fat in male and female bodies. **B,** Electron micrograph of a cluster of adipose cells held together by a network of fine reticular fibers (×150.).

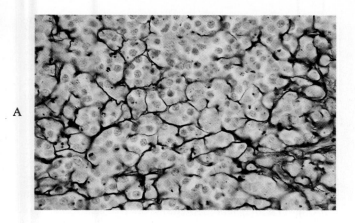

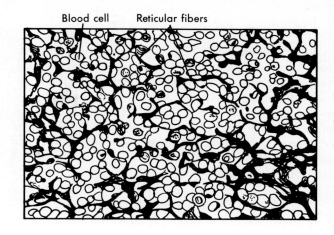

Blood cell   Reticular fibers

A

B

FIGURE 4-14 **Reticular connective tissue. A,** The supporting framework of reticular fibers are stained black in this section of spleen tissue. **B,** Sketch of photomicrograph.

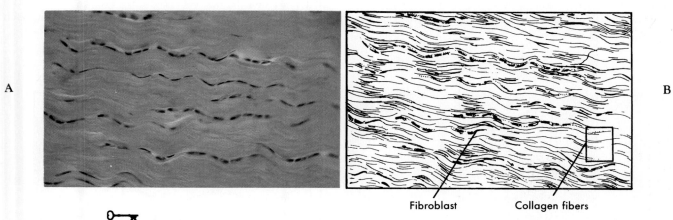

A

B

Fibroblast   Collagen fibers

FIGURE 4-15 **Dense fibrous (regular) connective tissue. A,** Photomicrograph of tissue in a tendon. **B,** Sketch of the photomicrograph. Note the multiple (regular) bundles of collagenous fibers arranged in parallel rows.

as tendons (Figure 4-16). Ligaments (which connect bone to bone) instead have a predominance of elastic fibers. Hence ligaments exhibit some degree of elasticity. In dense fibrous (irregular) tissues, the bundles of fibers are not arranged in parallel rows (Figure 4-17). Instead, the fibers intertwine with one another to form a thick mat of strong connective tissue that can withstand stresses applied from any direction. Dense fibrous (irregular) tissue forms the strong inner layer of the skin called the dermis. It also forms the outer capsule of such organs as the kidney and the spleen.

Bone tissue

Bone, or *osseus tissue,* is one of the most highly specialized forms of connective tissue. The mature cells of bone, **osteocytes,** are imbedded in a unique matrix material containing both organic collagen material and mineral salts. The inorganic (bone salt) portion makes up about 65% of the total matrix material and is responsible for the hardness of bone.

Bones are the organs of the skeletal system. They provide support and protection for the body and serve as

points of attachment for muscles. In addition, the calcified matrix of bones serves as a mineral reservoir for the body.

The basic organizational or structural unit of bone is the microscopic **haversian system** (Figure 4-18)). Osteocytes, or bone cells, are located in small spaces, or **lacunae,** which are arranged in concentric layers of bone matrix called **lamellae.** Small canals called **canaliculi** connect each lacuna and osteocyte with nutrient blood vessels found in the central haversian canal.

In addition to mature osteocytes, there are two additional types of bone cells: **osteoblasts,** or bone-forming cells, and **osteoclasts,** or bone-destroying cells. Mature bone can grow and be reshaped by the simultaneous activity of osteoblasts laying down new bone as the osteoclasts break down and remove existing bone tissue.

Certain bones called **membrane bones** (e.g., flat bones of the skull) are formed within membranous tissue, whereas others (e.g., long bones such as the humerus) are formed indirectly through replacement of cartilage in a process called **endochondral ossification.** The details of bone formation will be presented in Chapter 6.

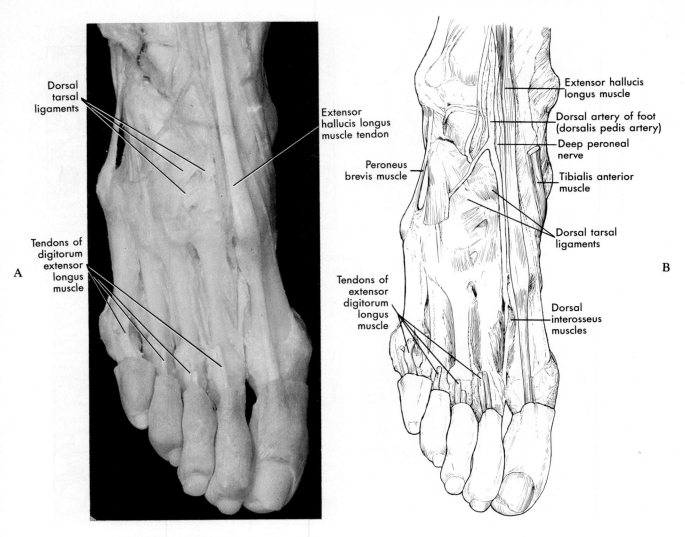

FIGURE 4-16   Tendons and ligaments. A, Tendons and ligaments of the foot are examples of dense fibrous connective tissue. **B,** Sketch of cadaver dissection. Note the cordlike tendons passing to the toes and the flat, shiny tarsal ligaments on the dorsum (*top*) of the foot.

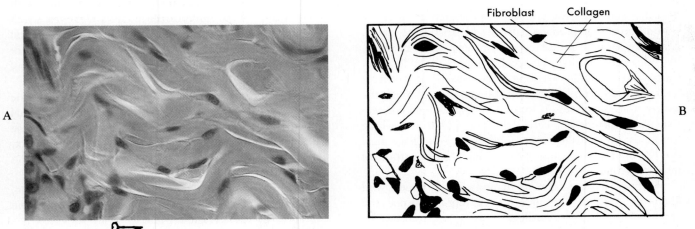

FIGURE 4-17   Dense fibrous (irregular) connective tissue. A, Section of skin (dermis) showing arrangements of collagenous fibers (pink) and purple-staining fibroblast cell nuclei. **B,** Sketch of photomicrograph.

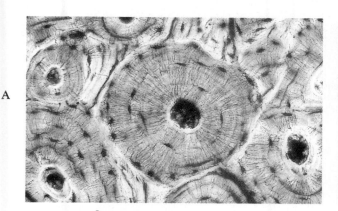

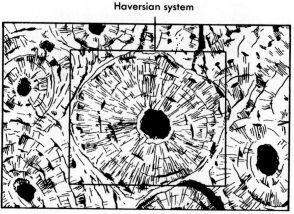

Haversian system

FIGURE 4-18   Bone tissue. A, Photomicrograph of dried, ground bone. **B,** Sketch of photomicrograph. Many wheel-like structural units of bone, known as *haversian systems,* are apparent in this section.

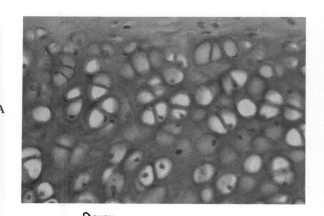

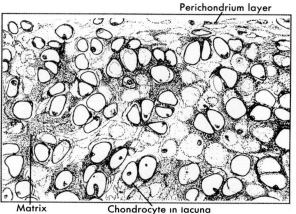

Perichondrium layer

Matrix   Chondrocyte in lacuna

FIGURE 4-19   Hyaline cartilage. A, Photomicrograph of trachea. Note the many spaces, or lacunae, in the gel-like matrix. **B,** Sketch of the photomicrograph.

Cartilage

Cartilage differs from other connective tissues in that only one cell type, the **chondrocyte,** is present. It is the chondrocytes that produce both the fibers and the tough, gristlelike ground substance of cartilage. Chondrocytes, like bone cells, are found in small openings called lacunae. Cartilage is avascular (lacking blood vessels) so nutrients must reach the cells by diffusion. Movement is through the matrix from blood vessels located in a specialized connective tissue membrane, called the **perichondrium,** which surrounds the cartilage mass. Injuries to cartilage heal slowly, if at all, because of this inefficient method of nutrient delivery.

Hyaline cartilage takes its name from the Greek word *hyalos* or "glass." The name is appropriate, since the appearance of hyaline cartilage is shiny and translucent. This is the most prevalent type of cartilage and is found in the support rings of the respiratory tubes and covering the ends of bones that articulate at joints (Figure 4-19).

Fibrocartilage is the strongest and most durable type of cartilage (Figure 4-20). The matrix is rigid and is filled with strong white fibers. Fibrocartilage disks serve as shock absorbers between adjacent vertebrae (intervertebral disks) and in the knee joint. Damage to the fibrocartilage pads or joint menisci in the knee occurs frequently as a result of sport-related injuries.

Elastic cartilage contains large numbers of very fine elastic fibers that give the matrix material additional strength and flexibility (Figure 4-21). This type of cartilage is found in the external ear and in the voice box, or larynx.

Blood

Blood is perhaps the most unusual connective tissue, since it exists in a liquid state and contains neither ground substance nor fibers (Figure 4-22).

Whole blood is often divided into a **liquid fraction** called **plasma** and **formed elements** or cells that can be divided into three classes: red cells, or **erythrocytes;** white

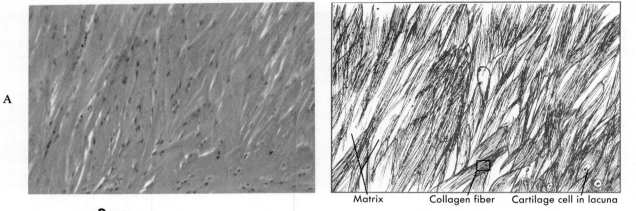

FIGURE 4-20   Fibrocartilage. A, Photomicrograph of pubic symphysis joint. The strong dense fibers that fill the matrix convey shock-absorbing qualities. **B,** Sketch of the photomicrograph.

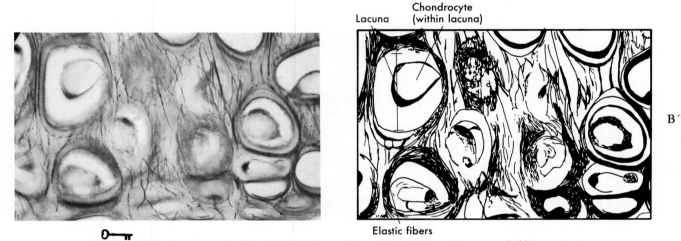

FIGURE 4-21   Elastic cartilage. A, Note the cartilage cells in the lacunae surrounded by matrix and dark staining elastic fibers. **B,** Sketch of photomicrograph.

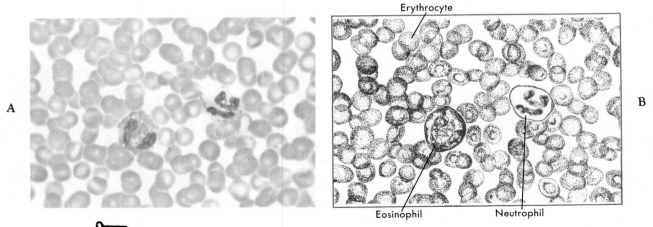

FIGURE 4-22   Blood. A, Photomicrograph of human blood smear shows two white blood cells, or leukocytes, surrounded by numerous smaller red blood cells. **B,** Sketch of the photomicrograph.

FOCUS ON... INFLAMMATION

The terms **inflammation** or **inflammatory response** are used to describe the complex way in which cells and tissues react to injury. Many of the events, which are now identified as steps in the inflammatory response, are so dramatic that for centuries they were often thought to be a primary disease. It was in the first century AD that the Roman physician Celsus first established inflammation as an entity by describing its four cardinal signs: **rubor** (redness), **calor** (heat), **tumor** (swelling), and **dolor** (pain). His accurate and detailed description of the visible signs that signal the body's response to injury is considered a classic in the annals of medicine.

The inflammatory response can best be described as a series of sequenced events that occur as a result of an inflammatory stimulus or "insult." Heat, physical pressure, caustic chemicals, toxins released by harmful bacteria, or any other type of noxious stimulus will initiate an inflammatory response. Early studies with rabbits, using a device known as a **transparent ear chamber,** permitted scientists to view changes in living tissue after an injury for prolonged periods. Skin over an animal's external ear was subjected to very slight injury and then viewed through the chamber under a microscope.

Immediately after an injury occurs, there is a very brief constriction of surrounding blood vessels that lasts but a moment. Then, almost immediately, blood vessels dilate, or open, and blood flow increases.

Injured tissues release a number of chemicals that affect blood vessels. These chemicals include **histamine, serotonin,** and a group of chemically related compounds called **kinins.** All of these substances result in both vasodilation and an increase in the permeability of blood vessels so that components that would normally be retained in the blood are permitted to leak out into the tissue spaces.

In the absence of injury, blood flow through a small vessel is such that the cells tend to pass in large measure within the central two thirds of the lumen with a thin layer of plasma flowing closest to the outer walls. This is called **axial flow.** After an injury, blood cells no longer pass in a central stream. Microscopic examination of vessels near an injured site shows that white cells begin to accumulate in the vessel near the point of injury and then stick, or **marginate,** to the wall. This **margination of leukocytes** continues until the endothelial surface of the vessel is covered with adherent white cells. Within minutes these cells begin to pass through the endothelial lining and out of the vessels into the interstitial spaces near the injury. One of the important functions of many white blood cells is **phagocytosis**— the process of engulfing and destroying bacteria.

Movement of white cells into the area of injury or infection is called **diapedesis.** The term **chemotaxis** describes the attraction of leukocytes, especially neutrophils, into the interstitial spaces. The attractive force is produced by the release of kinins and other chemicals by injured tissue. **Leukocytosis** means an increase in the number of leukocytes in the blood. A substance called **leukocytosis-promoting (LP) factor** is also released by injured tissue. It stimulates the release of white cells from storage areas and increases the number of circulating white blood cells.

The accumulation of dead leukocytes and tissue debris may lead to the formation of **pus** at the focal point of infection.

Should this occur, an **abscess,** or cavity, formed by the disintegration of tissues, may fill with pus and require surgical drainage.

Increased permeability of blood vessels, increased blood flow, and the migration and accumulation of white blood cells all contribute to the formation of **inflammatory exudate,** which accumulates in the interstitial spaces in the area of injury. The result is often swelling, or **edema,** and pain. In addition to white blood cells and tissue debris, inflammatory exudate contains the "leaked" substances normally retained in the blood but allowed to escape into the interstitial spaces because of increased capillary permeability. One such substance is a soluble protein that is soon converted into **fibrin** in the interstitial spaces. Fibrin formation results in development of a clot, which helps to seal off the infected area and decrease the spread of bacteria or other infectious material.

The cardinal signs of inflammation "make sense" when examined in the light of our understanding of the process.

- The redness (rubor) is also caused by increased blood flow and pooling of blood following injury.
- The heat (calor) is largely the result of increased blood flow to the area of injury.
- Swelling (tumor) results because of edema and accumulation of inflammatory exudate and clot formation in the affected tissue spaces.
- Pain (dolor) is caused by chemicals such as the kinins (especially **bradykinin**) and other chemical mediators that are released following tissue injury and cellular death.

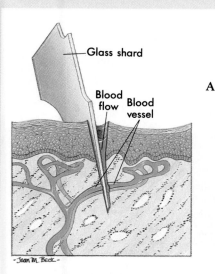

A

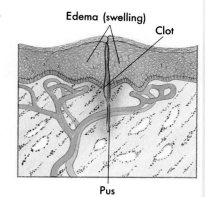

B

Inflammation in a section of traumatized skin. A, Injury or insult to tissue resulting in dilation of blood vessels and increase in blood flow. **B,** Inflammatory response— the body attempts to seal off an area of insult to limit bacterial invasion.

cells, or **leukocytes;** and **thrombocytes,** or platelets. The liquid fraction makes up about 55% of whole blood, and the formed elements compose about 45%.

Blood performs many body transport functions, including movement of respiratory gases (oxygen and carbon dioxide), nutrients, and waste products. In addition, blood plays a critical role in maintaining a constant body temperature and in regulating the pH of body fluids. The white blood cells function in destroying harmful microorganisms.

Circulating blood tissue is formed in the *red marrow* of bones and in other tissues by a process of differentiation called *hematopoiesis.* This blood-forming tissue is sometimes given the status of a separate connective tissue type: **hematopoietic tissue.**

Blood and its formation are described in detail in Chapter 16.

MUSCLE TISSUE

Three types of muscle tissue are present in the body—skeletal muscle, visceral muscle, and cardiac muscle. Their names suggest their locations. **Skeletal muscle tissue** (Figure 4-23) composes muscles attached to bones; these are the organs that we think of as our muscles. **Smooth muscle tissue,** also sometimes called *visceral muscle tissue* (Figure 4-24), is found in the walls of the viscera (hollow

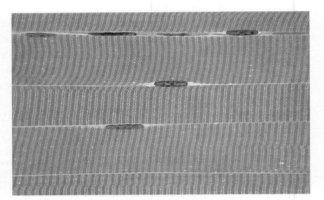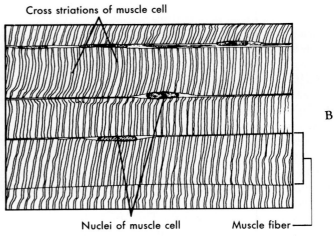

Cross striations of muscle cell

Nuclei of muscle cell   Muscle fiber

FIGURE 4-23   Skeletal muscle. A, Photomicrograph. **B,** Sketch of the photomicrograph. Note the striations of the muscle cell fibers in longitudinal section.

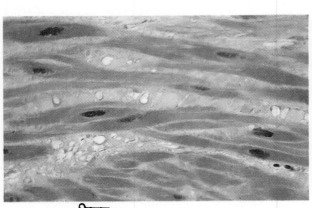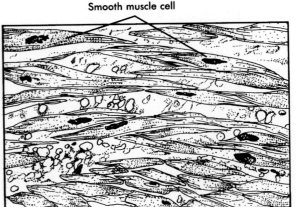

Smooth muscle cell

FIGURE 4-24   Smooth muscle. A, Photomicrograph, longitudinal section. **B,** Sketch of the photomicrograph. Note the central placement of nuclei in the spindle-shaped smooth muscle fibers.

internal organs—e.g., the stomach, intestines, and blood vessels). **Cardiac muscle tissue** makes up the wall of the heart (Figure 4-25). Another name for skeletal muscle is *striated voluntary* muscle. The term *striated* refers to cross striations (stripes) visible in microscopic slides of the tissue. The term *voluntary* indicates that voluntary or willed control of skeletal muscle contractions is possible. Another name for smooth muscle is *nonstriated involuntary*. It has no cross striations and cannot ordinarily be controlled by the will. Another name for cardiac muscle is *striated involuntary* muscle. Like skeletal muscle, cardiac muscle has cross striations, and, like smooth muscle, its contractions cannot ordinarily be controlled by the will.

Look now at Figure 4-23 and observe the following structural characteristics of skeletal muscle cells: many cross striations, many nuclei per cell, and long, narrow, threadlike shape of the cells. Skeletal muscle cells may have a length of more than 3.75 cm, but they have diameters of only 10 to 100 μm. Because this gives them a threadlike appearance, muscle cells are often called muscle fibers. Chapter 9 gives more detailed information about the structure of skeletal muscle tissue.

Smooth muscle cells are also long, narrow fibers but not nearly as long as striated fibers. One can see the full length of a smooth muscle fiber in a microscopic field but only a small part of a striated fiber. According to one estimate, the longest smooth muscle fibers measure about 500 μm and the longest striated fibers about 40,000 μm. As Figure 4-24 shows, smooth muscle fibers have only one nucleus per fiber and are nonstriated or smooth in appearance.

Under the light microscope, cardiac muscle fibers (Figure 4-25) have cross striations and unique dark bands (intercalated disks). They also seem to be incomplete cells that branch into each other to form a big continuous mass of cytoplasm (a syncytium). The electron microscope, however, has revealed that the intercalated discs are actually places where the plasma membranes of two cardiac fibers abut. Cardiac fibers do branch in and out, but a complete plasma membrane encloses each cardiac fiber—around its end (at intercalated discs), as well as its sides.

Muscle cells are the movement specialists of the body. Because their cytoskeletons include bundles of microfilaments specialized for movement, they have a higher degree of contractility (ability to shorten or contract) than cells of any other tissue.

NERVOUS TISSUE

The basic function of the nervous system is to rapidly regulate, and thereby integrate, the activities of the different parts of the body. Functionally, rapid communication is possible because nervous tissue has much more developed excitability and conductivity characteristics than any other type of tissue.

The organs of the nervous system are the brain, the spinal cord, and the nerves. Actual nerve tissue is ectodermal in origin and consists of two basic kinds of cells: nerve cells, or **neurons,** which are the conducting units of the system, and special connecting and supporting cells called **neuroglia** (Figure 4-26).

All neurons are characterized by a cell body called the **soma** and, generally, at least two processes: one **axon,** which transmits nerve impulses away from the cell body, and one or more **dendrites,** which carry impulses toward the cell body. Most neurons are located within the organs of the central nervous system.

The anatomy and physiology of the nervous system are presented in Chapters 11 through 13.

QUICK CHECK

✔ *1. Name the two types of involuntary muscle. Where is each found in the body?*

✔ *2. What are the two principal types of cell in nervous tissue? What is the function of each?*

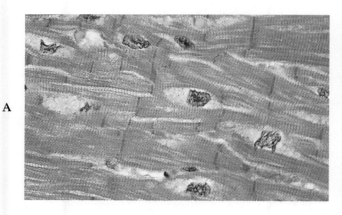

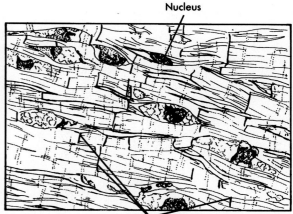

Nucleus

Intercalated disk

FIGURE 4-25 Cardiac muscle. A, Photomicrograph. **B,** Sketch of the photomicrograph. The dark bands, called *intercalated discs,* which are characteristic of cardiac muscle, are easily identified in this tissue section.

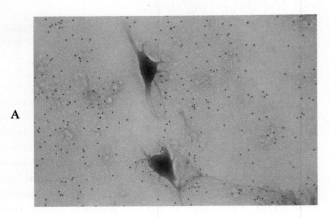
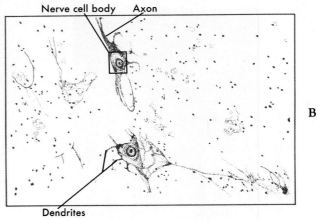

FIGURE 4-26 **Nervous tissue. A,** Photomicrograph. Multipolar neurons in smear of spinal cord. Both neurons in this slide show characteristic soma or cell bodies and multiple cell processes. **B,** Sketch of photomicrograph.

Tissue Repair

When damaged by mechanical or other injuries, tissues have varying capacity to repair themselves. Damaged tissue will regenerate or be replaced by tissue we know as scars. Tissues usually repair themselves by allowing the phagocytic cells to remove dead or injured cells, then filling in the gaps that are left. This growth of functional new tissue is called **regeneration.**

Epithelial and connective tissues have the greatest capacity to regenerate (see Figure 4-27). When a break in an epithelial membrane occurs, as in a cut, cells quickly divide to form daughter cells that fill the wound. In connective tissues, cells that form collagen fibers become active after an injury and fill in a gap with an unusually dense mass of fibrous connective tissue. If this dense mass of fibrous tissue is small, it may be replaced by normal tissue later. If the mass is deep or large, or if cell damage was extensive, it may remain as a dense fibrous mass, called a **scar.** An unusually thick scar that develops in the lower layer of the skin, such as that shown in Figure 4-28, is called a **keloid.**

Muscle tissue, on the other hand, has a very limited capacity to regenerate and thus heal itself. Damaged muscle is often replaced with fibrous connective tissue instead of muscle tissue. When this happens, the organ involved loses some or all of its ability to function.

Like muscle tissue, nerve tissue also has a limited capacity to regenerate. Neurons outside the brain and spinal cord can sometimes regenerate, but very slowly, and only if certain neuroglia are present to "pave the way." In the normal adult brain and spinal cord, neurons do not

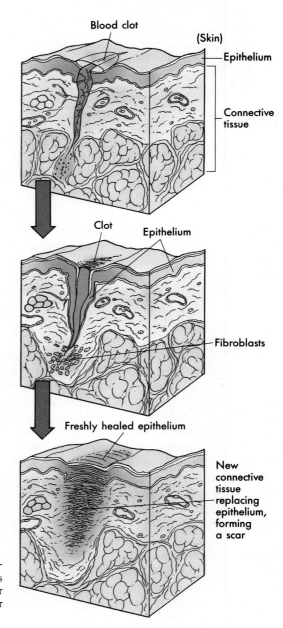

FIGURE 4-27 **Healing of a minor wound.** When a minor injury damages a layer of epithelium and the underlying connective tissue (as in a minor skin cut), both the epithelial tissue and the connective tissue can repair itself.

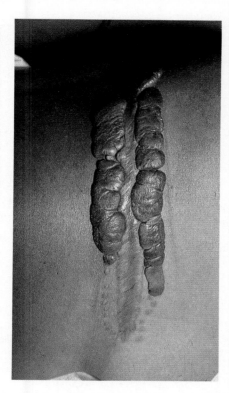

FIGURE 4-28 Keloid. Keloids are thick scars that form in the lower layer of the skin in predisposed individuals. This photograph shows keloids that formed at suture marks after surgery.

grow back when injured. Thus brain and spinal cord injuries nearly always result in permanent damage. Fortunately, the discovery of *nerve growth factors* produced by neuroglia offers the promise of treating brain damage by stimulating release of these factors.

BODY MEMBRANES

The term **membrane** refers to a thin, sheetlike structure that may have many important functions in the body. Membranes cover and protect the body surface, line body cavities, and cover the inner surfaces of the hollow organs such as the digestive, reproductive, and respiratory passageways. Some membranes anchor organs to each other or to bones, and others cover the internal organs. In certain areas of the body, membranes secrete lubricating fluids that reduce friction during organ movements such as the beating of the heart or lung expansion and contraction. Membrane lubricants also decrease friction between bones and joints. There are two major categories, or types, of body membranes:

1. Epithelial membranes, composed of epithelial tissue and an underlying layer of specialized connective tissue

2. Connective tissue membranes composed exclusively of various types of connective tissue; no epithelial cells are present in this type of membrane

Epithelial Membranes

There are three types of epithelial tissue membranes in the body: (1) cutaneous membrane, (2) serous membranes, and (3) mucous membranes.

Cutaneous membranes

The **cutaneous membrane, or skin,** is the primary organ of the integumentary system. It is one of the most important and certainly one of the largest and most visible organs. In most individuals the skin composes some 16% of the body weight. It fulfills the requirements necessary for an epithelial tissue membrane in that it has a superficial layer of epithelial cells and an underlying layer of supportive connective tissue. Its structure is uniquely suited to its many functions. The skin will be discussed in depth in Chapter 5.

Serous membranes

Like all epithelial membranes, a **serous** (SE-rus) **membrane** is composed of two distinct layers of tissue. The epithelial sheet is a thin layer of simple squamous epithelium. The connective tissue layer forms a very thin sheet that holds and supports the epithelial cells.

The serous membrane that lines body cavities and covers the surfaces of organs in those cavities is in reality a single, continuous sheet covering two different surfaces. The *parietal membrane* is the portion that lines the wall of the cavity like wallpaper; the *visceral membrane* covers the surface of the viscera (organs within the cavity). Two important serous membranes are identified in Figure 4-29: the *pleura*, which surrounds a lung and lines the thoracic cavity, and the *peritoneum*, which covers the abdominal viscera and lines the abdominal cavity. Another example is the *pericardium*, which surrounds the heart.

Serous membranes secrete a thin, watery fluid that lubricates organs as they rub against one another and against the walls of cavities.

Mucous membranes

Mucous (MYOO-kus) **membranes** are epithelial membranes that line body surfaces opening directly to the exterior. Examples of mucous membranes include those lining the respiratory, digestive, urinary, and reproductive

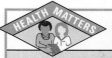

INFLAMMATION OF SEROUS MEMBRANES

Pleurisy (PLOOR-i-see) (also called pleuritis) is a very painful pathological condition characterized by inflammation of the serous membranes (pleurae) that line the chest cavity and cover the lungs. Pain is caused by irritation and friction as the lungs rub against the walls of the chest cavity. In severe cases the inflamed surfaces of the pleura fuse, and permanent damage may develop. The term **peritonitis** (pair-i-toe-NYE-tis) is used to describe inflammation of the serous membranes in the abdominal cavity. Peritonitis is sometimes a serious complication of an infected appendix.

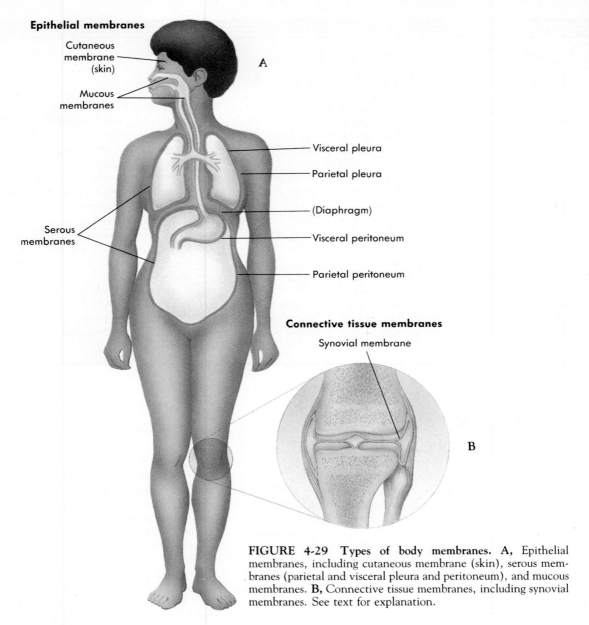

Epithelial membranes

Cutaneous membrane (skin)

Mucous membranes

Serous membranes

A

Visceral pleura

Parietal pleura

(Diaphragm)

Visceral peritoneum

Parietal peritoneum

Connective tissue membranes

Synovial membrane

B

FIGURE 4-29 Types of body membranes. A, Epithelial membranes, including cutaneous membrane (skin), serous membranes (parietal and visceral pleura and peritoneum), and mucous membranes. **B,** Connective tissue membranes, including synovial membranes. See text for explanation.

tracts. The epithelial component of the mucous membrane varies, depending on its location and function. In the esophagus, for example, a tough, abrasion-resistant stratified squamous epithelium is found. A thin layer of simple columnar epithelium covers the walls of the lower segments of the digestive tract.

Mucous membranes get their name from the fact that they produce a film of mucus that coats and protects the underlying cells. Besides protection, mucus also serves other purposes. For example, in the digestive tract, mucus acts as a lubricant for food as it moves along the digestive tract. In the respiratory tract, it serves as a sticky trap for contaminants.

Connective Tissue Membranes

Unlike cutaneous, serous, and mucous membranes, connective tissue membranes do not contain epithelial components. The **synovial membranes** lining the spaces between bones and joints that move are classified as connective tissue membranes. These membranes are smooth and slick and secrete a thick and colorless lubricating fluid called **synovial fluid.** The membrane itself, with its specialized fluid, helps reduce friction between the opposing surfaces of bones in movable joints. Synovial membranes also line the small, cushionlike sacs called **bursae** that are found between some moving body parts.

QUICK CHECK

✔ 1. Which two of the four major tissue types have the greatest capacity to regenerate after an injury?

✔ 2. Name the four principal types of body membranes. Which are epithelial membranes?

THE BIG PICTURE

Tissues, Membranes, and the Whole Body

Tissues and body membranes are sometimes called "the fabric of the body." Like the pieces of fabric in a garment, tissues and body membranes are portions of a larger integrated structure. Just as each type of fabric in a complex garment has a different functional role determined by its structural characteristics, so does each type of tissue within the body. One of the ultimate functional goals of most tissues and membranes is maintenance of relative constancy in the body: homeostasis.

How do the major tissue types help maintain homeostasis? Epithelial tissues promote constancy of the body's internal environment in several ways. They form membranes that contain and protect the internal fluid environment, they absorb nutrients and other substances needed to maintain an optimum concentration in the body, and they secrete various products that regulate body functions involved in homeostasis. Connective tissues hold organs and systems together to form a whole, connected body. They also form structures that support the body and permit

movement, such as the components of the skeleton. Some connective tissues, such as blood, transport nutrients, wastes, and other substances within the internal environment. Some blood cells help protect the internal environment by participating in the body's immune system. Muscle tissues work in conjunction with connective tissues (bones, tendons, etc.) to permit movement (a function needed to avoid injury); communicate; and to find food, shelter, and other requirements. Nervous tissue works along with glandular epithelial tissue to regulate various body functions in a way that maintains homeostatic balance.

Now that you have a basic knowledge of the various types of body "fabric"—the tissues and body membranes—you are ready to study the structure and function of specific organs and systems. As you take this next step in your studies, pay close attention to the tissue types that comprise each organ. If you do, you will find it easier to understand the characteristics of a particular organ, and you will also improve your understanding of the integrated nature of the whole body.

MECHANISMS OF DISEASE

Tumors and Cancer

Neoplasms

The term **neoplasm** literally means "new matter" and refers to any abnormal growth of cells. Also called **tumors,** neoplasms can be distinct lumps of abnormal cells or, in blood tissue, can be diffuse. Neoplasms are often classified as **benign** or **malignant.** Benign tumors are called that because they do not spread to other tissues, and they usually grow very slowly. Their cells are often well differentiated, unlike the undifferentiated cells typical of malignant tumors. Cells in a benign tumor tend to stay together, and they are often surrounded by a capsule of dense tissue. Benign tumors are usually not life-threatening but can be if they disrupt the normal function of a vital organ. Malignant tumors or **cancers,** on the other hand, are not encapsulated and tend to spread to other regions of the body. For example, cells from malignant breast tumors usually form new (secondary) tumors in bone, brain, and lung tissues. The cells migrate by way of lymphatic or blood vessels. This manner of spreading is called **metastasis.** Cells that do not metastasize can spread another way: they grow rapidly and extend the tumor into nearby tissues (Figure 4-30). Malignant tumors may replace part of a vital organ with abnormal, undifferentiated tissue—a life-threatening situation.

Neoplasms are further classified into subgroups depending on the tissue in which they originate. Both benign and malignant tumors can be divided into three types: epithelial tumors, connective tissue tumors, and miscellaneous tumors. Benign tumor types that arise from epithelial tissues include: *papilloma* (a fingerlike projection), *adenoma* (glandular tumor), and *nevus* (small, pigmented skin tumors). Benign tumor types that arise from connective tissues include: *lipoma* (adipose

tumor), *osteoma* (bone tumor), and *chondroma* (cartilage tumor). Malignant tumors that arise from epithelial tissues are generally called **carcinomas.** Examples include *melanoma* (cancer of skin-pigment cells), and *adenocarcinoma* (glandular cancer). Malignant tumors that arise from connective tissues are generally called **sarcomas.** Examples include: *lymphoma* (lymphatic cancer), *osteosarcoma* (bone cancer), and *fibrosarcoma* (cancer of fibrous connective tissue). Miscellaneous tumors are those that do not fit either of the previous categories. For example, an *adenofibroma* is a benign neoplasm formed by both epithelial and connective tissues. Another example is *neuroblastoma,* a malignant tumor that arises from nerve tissue.

Causes of Cancer

The etiologies (origins) of various forms of cancer puzzle medical science no less today than a hundred years ago. We do know that it involves uncontrolled cell division: **hyperplasia** (too many cells) and/or **anaplasia** (abnormal, undifferentiated cells). Thus the mechanism of all cancers is a mistake or problem in cell division. What we are uncertain of is the cause of the abnormal cell division. Currently, several factors are known to play a role:

◆ **Genetic factors.** Over a dozen forms of cancer are known to be directly inherited, perhaps involving abnormal "cancer genes" called **oncogenes.** Another type of gene, called a **tumor suppressor gene,** may fail to operate—allowing cancer to develop. Exactly how these genes work is still being investigated. It is thought

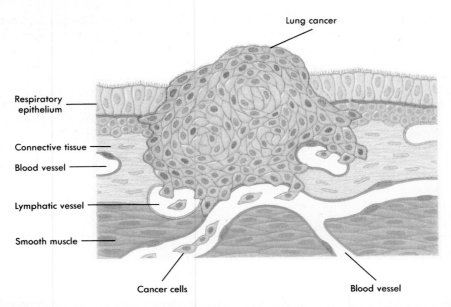

Lung cancer

Respiratory epithelium

Connective tissue

Blood vessel

Lymphatic vessel

Smooth muscle

Cancer cells

Blood vessel

FIGURE 4-30   Cancer. This abnormal mass of proliferating cells in the lining of lung airways is a malignant tumor—lung cancer. Notice how some cancer cells are leaving the tumor and entering the blood and lymph vessels.

that many cancers involve a genetic predisposition (risk factor) coupled with other cancer-causing mechanisms. Cancers with known genetic risk factors include basal cell carcinoma (a type of skin cancer), breast cancer, and neuroblastoma (a cancer of nerve tissue).

♦ **Carcinogens.** Carcinogens (cancer makers) are agents that affect genetic activity in some way, causing abnormal cell reproduction. Some carcinogens are **mutagens** (mutation makers) that cause changes or mutations in a cell's DNA structure. Although many industrial products are known to be carcinogens, a wide variety of natural mineral, vegetable, and animal materials are also carcinogenic. Exposure to damaging types of radiation or other physical injuries can be carcinogenic. For example, sunburns or chronic exposure to sunlight can cause skin cancer. Even viruses have been known to cause cancer, perhaps by altering the genetic code of cells during an infection or by damaging the body's ability to suppress cancer. Papilloma (wart) viruses have been blamed for some cases of cervical cancer in women.

♦ **Age.** Certain cancers are found primarily in young people (e.g., leukemia) and others primarily in older adults (e.g., colon cancer). The age factor may result from changes in the genetic activity of cells over time or from accumulated effects of cell damage.

Detection and Treatment of Cancer

Signs of cancer are those one would expect of a malignant neoplasm: the appearance of abnormal, rapidly growing tissue. Cancer specialists, or *oncologists,* have stressed that early detection of cancer is important because it is in the early stages of development of primary tumors, before metastasis and the development of secondary tumors have begun, that cancer is most treatable. Some methods currently used to detect the presence of cancer include:

♦ **Self-examination** for the early signs of cancer previously described is one method for detection of cancer. For example, women are encouraged to perform a monthly breast self-examination. Likewise, males are encouraged to perform a monthly testicular self-examination. If an abnormality is found, it can be further investigated with one of the following described methods.

♦ **Medical imaging** techniques that visualize deep tissues for medical study are often used to detect cancers (see pages 26 to 27). Radiography (x-ray photography) is often used to detect the presence of tumors. *Mammography,* x-ray photography of a breast, is considered an important detection tool for this type of cancer. *Computed tomography (CT)* (x-ray scanning), *magnetic resonance imaging (MRI)* (electromagnetic scanning), and *ultrasonography* (ultrasound scanning) produce cross-sectional images of body regions suspected of having tumors.

♦ **Blood tests** to determine the concentration of ions, enzymes, or other blood components are useful in detecting cancer when the results show abnormalities associated with particular forms of cancer. Cancer cells may also produce or trigger production of substances often referred to as *tumor markers.* For example, tests for a prostate cancer marker are now being used in conjunction with other diagnostic tests.

♦ **Biopsy** is the removal and examination of living tissue. Microscopic examination of tumor tissue removed surgically or through a needle sometimes reveals whether it is malignant or benign.

The information gained from these techniques can be used to *stage* and *grade* malignant tumors. Staging involves classifying a tumor based on its size and the extent of its spread. Grading is an assessment of what the tumor is likely to do based on the degree of abnormality of the cells—a useful basis for making a

Continued

prognosis (statement of probable outcome).

Without treatment, cancer usually results in death. The progress of a particular type of cancer depends on the type of cancer and its location. Many cancer patients suffer from *cachexia,* a syndrome involving loss of appetite, weight loss, and general weakness. Various anatomical or functional abnormalities may arise as a result of damage to particular organs. The ultimate causes of death in cancer patients include secondary infection by pathogenic microbes, organ failure, hemorrhage (blood loss), and, in some cases, undetermined factors.

Of course, once cancer has been identified, every effort is made to treat it and thus prevent or delay its development. Surgical removal of cancerous tumors is sometimes done, but the probability that malignant cells have been left behind must be addressed. **Chemotherapy,** or "chemical therapy," using *cytotoxic* (cell-killing) compounds, or *antineoplastic* drugs, can be used after surgery to destroy remaining malignant cells. **Radiation therapy,** also called *radiotherapy,* using destructive x-ray or gamma radiation may be used alone or with chemotherapy to destroy remaining cancer cells. **Laser therapy,** in which an intense beam of light destroys a tumor, is also sometimes coupled with chemotherapy or radiation therapy. **Immunotherapy,** a newer type of cancer treatment, bolsters the body's own defenses against cancer cells. Since viruses cause some types of cancer, oncologists hope that vaccines against certain forms of cancer will be developed. Although new and different approaches to cancer treatment are being investigated, many researchers are concentrating on improving existing methods.

CHAPTER SUMMARY

INTRODUCTION

A. Tissue—group of similar cells that perform a common function
B. Matrix—nonliving intercellular material

PRINCIPAL TYPES OF TISSUE

A. Epithelial tissue
B. Connective tissue
C. Muscle tissue
D. Nervous tissue

EMBRYONIC DEVELOPMENT OF TISSUES

A. Primary germ layers (Figure 4-1)
  1. Endoderm
  2. Mesoderm
  3. Ectoderm
B. Gastrulation—process of cell movement and differentiation, which results in development of primary germ layers
C. Histogenesis—the process of the primary germ layers differentiating into different kinds of tissue

EPITHELIAL TISSUE

A. Types and locations
  1. Epithelium is divided into two types:
    a. Membranous (covering or lining) epithelium
    b. Glandular epithelium
  2. Locations
    a. Membranous epithelium—covers the body and some of its parts; lines the serous cavities, blood and lymphatic vessels, and respiratory, digestive, and genitourinary tracts
    b. Glandular epithelium—secretory units of endocrine and exocrine glands
B. Functions
  1. Protection
  2. Sensory functions
  3. Secretion
  4. Absorption
  5. Excretion
C. Generalizations about epithelial tissue
  1. Limited amount of matrix material
  2. Membranous type attached to a basement membrane
  3. Avascular

4. Cells are in close proximity, with many desmosomes and tight junctions
5. Capable of reproduction

CLASSIFICATION OF EPITHELIAL TISSUE

A. Membranous (covering or lining) epithelium
  1. Classification based on cell shape
    a. Squamous
    b. Cuboidal
    c. Columnar
    d. Pseudostratified columnar
  2. Classifications based on layers of cells (Table 4-1)
    a. Simple epithelium
      (1) Simple squamous epithelium (Figure 4-3)
        (a) One-cell layer of flat cells
        (b) Permeable to many substances
        (c) Examples: endothelium—lines blood vessels; mesothelium—pleura
      (2) Simple cuboidal epithelium (Figure 4-4)
        (a) One-cell layer of cuboidal-shaped cells
        (b) Found in many glands and ducts
      (3) Simple columnar epithelium (Figure 4-5)
        (a) Single layer of tall, column-shaped cells
        (b) Cells often modified for specialized functions such as goblet cells (secretion), cilia (movement), microvilli (absorption)
        (c) Often lines hollow visceral structures
      (4) Pseudostratified columnar epithelium (Figure 4-6)
        (a) Columnar cells of differing heights
        (b) All cells rest on basement membrane but may not reach the free surface above
        (c) Cell nuclei at odd and irregular levels
        (d) Found lining air passages and segments of male reproductive system
        (e) Motile cilia and mucus are important modifications
    b. Stratified epithelium
      (1) Stratified squamous (keratinized) epithelium
        (a) Multiple layers of flat, squamous cells (Figure 4-7)
        (b) Cells filled with keratin
        (c) Covering outer skin on body surface
      (2) Stratified squamous (nonkeratinized) epithelium (Figure 4-8)

           (a) Lining vagina, mouth, and esophagus
           (b) Free surface is moist
           (c) Primary function is protection
         (3) Stratified cuboidal epithelium
           (a) Two or more rows of cells is typical
           (b) Basement membrane is indistinct
           (c) Located in sweat gland ducts and pharynx
         (4) Stratified columnar epithelium
           (a) Multiple layers of columnar cells
           (b) Only most superficial cells are typical in
               shape
           (c) Rare
           (d) Located in segments of male urethra and
               near anus
         (5) Stratified transitional epithelium (Figure 4-9)
           (a) Located in lining of hollow viscera subjected
               to stress (e.g., urinary bladder)
           (b) Often 10 or more layers thick
           (c) Protects organ walls from tearing
  B. Glandular epithelium
     1. Specialized for secretory activity
     2. Exocrine glands—discharge secretions into ducts
     3. Endocrine glands—"ductless" glands; discharge
         secretions directly into the blood or interstitial fluid
     4. Structural classification of exocrine glands
       a. Multicellular exocrine glands are classified by the
           shape of their ducts and the complexity of their duct
           system
       b. Shapes include: tubular and alveolar
       c. Simple exocrine glands—only one duct leads to the
           surface
       d. Compound exocrine glands—have two or more
           ducts
     5. Functional classification of exocrine glands (Figure 4-10)
       a. Apocrine glands
         (1) Secretory products collect near apex of cell and
             are secreted by pinching off the distended end
         (2) Secretion process results in some damage to cell
             wall and some loss of cytoplasm
         (3) Mammary glands are good examples of this
             secretory type
       b. Holocrine glands
         (1) Secretion products, when released, cause rupture
             and death of the cell
         (2) Sebaceous glands are holocrine
       c. Merocrine glands
         (1) Secrete directly through cell membrane
         (2) Secretion proceeds with no damage to cell wall
             and no loss of cytoplasm
         (3) Most numerous gland type

CONNECTIVE TISSUE

A. Functions, characteristics, and types
   1. General function—connects, supports, transports, and
       protects
   2. General characteristics—matrix predominates in most
       connective tissues and determines its physical
       characteristics; consists of fluid, gel, or solid matrix,
       with or without extracellular fibers (collagenous,
       reticular, and elastic)
   3. Four main types (Table 4-3)
     a. Fibrous
       (1) Loose (areolar)
       (2) Adipose
       (3) Reticular
       (4) Dense
     b. Bone
     c. Cartilage
       (1) Hyaline

         (2) Fibrocartilage
         (3) Elastic
       d. Blood
  B. Fibrous connective tissue
     1. Loose (areolar) connective tissue (Figure 4-11)
       a. One of the most widely distributed of all tissues
       b. Intercellular substance is prominent and consists of
           collagenous and elastic fibers loosely interwoven and
           embedded in soft viscous ground substance
       c. Several kinds of cells present, notably, fibroblasts
           and macrophages, also mast cells, plasma cells, fat
           cells, and some white blood cells
       d. Function—connection
     2. Adipose tissue (Figure 4-12)
       a. Similar to loose connective tissue but contains
           mainly fat cells
       b. Functions—protection, insulation, support, and food
           reserve
     3. Reticular tissue (Figure 4-14)
       a. Forms framework of spleen, lymph nodes, and bone
           marrow
       b. Consists of network of branching reticular fibers with
           reticular cells overlying them
       c. Functions—defense against microbes and other
           injurious substances; reticular meshwork filters out
           injurious particles and reticular cells phagocytose
           them
     4. Dense fibrous tissue (Figures 4-15 and 4-17)
       a. Matrix consists mainly of fibers packed densely and
           relatively few fibroblast cells
         (1) Regular—bundles of fibers are arranged in regular
             parallel rows
         (2) Irregular—fibers intertwine to form a thick mat
       b. Locations—composes structures that need great
           tensile strength, such as tendons and ligaments; also
           dermis and the outer capsule of kidney and spleen
       c. Function—furnishes flexible but strong connection
  C. Bone tissue
     1. Highly specialized connective tissue type (Figure 4-18)
       a. Cells—osteocytes—imbedded in a calcified matrix
       b. Inorganic component of matrix accounts for 65% of
           total bone tissue
     2. Functions
       a. Support
       b. Protection
       c. Point of attachment for muscles
       d. Reservoir for minerals
     3. Haversian system
       a. Structural unity of bone
       b. Spaces for osteocytes called lacunae
       c. Matrix present in concentric rings called lamellae
       d. Canaliculi are canals that join lacunae with the
           central haversian canal
     4. Cell types
       a. Osteocyte—mature bone cell
       b. Osteoblast—bone-forming cell
       c. Osteoclast—bone-destroying cell
     5. Formation (ossification)
       a. In membranes—e.g., flat bones of skull
       b. From cartilage (endochondral)—e.g., long bones,
           such as the humerus
  D. Cartilage
     1. Chondrocyte is only cell type present
     2. Lacunae house cells as in bone
     3. Avascular—therefore nutrition of cells depends on
         diffusion of nutrients through matrix
     4. Heals slowly after injury because of slow nutrient
         transfer to the cells
     5. Perichondrium is membrane that surrounds cartilage

6. Types
   a. Hyaline (Figure 4-19)
      (1) Appearance is shiny and translucent
      (2) Most prevalent type of cartilage
      (3) Located on the ends of articulating bones
   b. Fibrocartilage (Figure 4-20)
      (1) Strongest and most durable type of cartilage
      (2) Matrix is semirigid and filled with strong white fibers
      (3) Found in intervertebral disks and pubic symphysis
      4. Serves as shock-absorbing material between bones at the knee (menisci)
   c. Elastic (Figure 4-21)
      (1) Contains many fine elastic fibers
      (2) Provides strength and flexibility
      (3) Located in external ear and larynx
E. Blood
   1. A liquid tissue (Figure 4-22)
   2. Contains neither ground substance nor fibers
   3. Composition of whole blood
      a. Liquid fraction, or plasma, is 55% of total blood
      b. Formed elements contribute 45% of total blood
         (1) Red blood cells, erythrocytes
         (2) White blood cells, leukocytes
         (3) Platelets, thrombocytes
   4. Functions
      a. Transportation
      b. Regulation of body temperature
      c. Regulation of body pH
      d. White blood cells destroy bacteria
   5. Circulating blood tissue is formed in the red bone marrow by a process called hematopoiesis; the blood-forming tissue is sometimes called hematopoietic tissue

MUSCLE TISSUE

A. Types
   1. Skeletal, or striated voluntary (Figure 4-23)
   2. Smooth, or nonstriated involuntary, or visceral (Figure 4-24)
   3. Cardiac, or striated involuntary (Figure 4-25)
B. Microscopic characteristics
   1. Skeletal muscle—threadlike cells with many cross striations and many nuclei per cell
   2. Smooth muscle—elongated narrow cells, no cross striations, one nucleus per cell
   3. Cardiac muscle—branching cells with intercalated disks (formed by abutment of plasma membranes of two cells)

NERVOUS TISSUE

A. Functions—rapid regulation and integration of body activities
B. Specialized characteristics
   1. Excitability
   2. Conductivity
C. Organs
   1. Brain
   2. Spinal cord
   3. Nerves
D. Cell types
   1. Neuron—conducting unit of system (Figure 4-26)
      a. Cell body, or soma
      b. Processes
         (1) Axon (single process)—transmits nerve impulse away from the cell body
         (2) Dendrites (one or more)—transmits nerve impulse toward the cell body
   2. Neuroglia—special connecting and supporting cells

TISSUE REPAIR

A. Tissues have a varying capacity to repair themselves; damaged tissue will regenerate or be replaced by scar tissue
B. Regeneration—growth of new tissue (Figure 4-27)
C. Scar—dense fibrous mass; unusually thick scar is a keloid (Figure 4-28)
D. Epithelial and connective tissues have the greatest ability to regenerate
E. Muscle and nervous tissues have a limited capacity to regenerate

BODY MEMBRANES

A. Thin tissue layers that cover surfaces, line cavities, and divide spaces or organs (Figure 4-29)
B. Epithelial membranes are most common type
   1. Cutaneous membrane (skin)
      a. Primary organ of integumentary system
      b. One of the most important organs
      c. Comprises approximately 16% of body weight
   2. Serous membrane
      a. Parietal membranes—line closed body cavities
      b. Visceral membranes—cover visceral organs
      c. Pleura—surrounds a lung and lines the thoracic cavity
      d. Peritoneum—covers the abdominal viscera and lines the abdominal cavity
   3. Mucous membrane
      a. Lines and protects organs that open to the exterior of the body
      b. Found lining ducts and passageways of respiratory and digestive tracts
C. Connective tissue membranes
   1. Do not contain epithelial components
   2. Synovial membranes—line the spaces between bone in joints
   3. Have smooth and slick membranes that secrete synovial fluid
   4. Help reduce friction between opposing surfaces in a moveable joint
   5. Synovial membranes also line bursae

THE BIG PICTURE: TISSUES, MEMBRANES, AND THE WHOLE BODY

A. Tissues and membranes maintain homeostasis
   1. Epithelial tissues
      a. Form membranes that contain and protect the internal fluid environment
      b. Absorb nutrients
      c. Secrete products that regulate functions involved in homeostasis
   2. Connective tissues
      a. Hold organs and systems together
      b. Form structures which support the body and permit movement
   3. Muscle tissues
      a. Work with connective tissues to permit movement
   4. Nervous tissues
      a. Work with glandular epithelial tissue to regulate body functions

MECHANISMS OF DISEASE: TUMORS AND CANCER

A. Tumors and cancer
   1. Neoplasm—any abnormal growth of cells
      a. Benign tumor
      b. Malignant tumor (cancer) (Figure 4-30)
         (1) Metastasis—lymphatic or blood vessel cell migration

  2. Neoplasm subgroups
    a. Epithelial tumors
    b. Connective tissue tumors
    c. Miscellaneous tumors
B. Causes of cancer
  1. Hyperplasia—too many cells
  2. Anaplasia—abnormal, undifferentiated cells
  3. Factors in abnormal cell division
    a. Genetic factors
      (1) Oncogenes
      (2) Tumor suppressor gene
    b. Carcinogens

    c. Age
C. Detection and treatment of cancer
  1. Methods
    a. Self-examination
    b. Medical imaging
    c. Blood tests
    d. Biopsy
  2. Staging—classification of tumor by size and extent of spread
  3. Grading—assessment of tumor from degree of cell abnormality

REVIEW QUESTIONS

1. Define the term *tissue* and identify the four principal tissue types.

2. List at least three structures derived from each of the primary germ layers.

3. What are the five most important functions of epithelial tissue?

4. Discuss the structure of merocrine, epocrine, and holocrine glands and discuss the method of secretion.

5. Which of the following best describes the number of blood vessels in epithelial tissue: none, very few, very numerous?

6. Explain how the shape of epithelial cells is used for classification purposes. Identify the four types of epithelium described in this classification process.

7. Classify epithelium according to the layers of cells present.

8. List the types of simple and stratified epithelium and give examples of each.

9. What is glandular epithelium? Give examples.

10. Discuss the structural classification of exocrine glands. Give examples of each type.

11. Compare merocrine, apocrine, and holocrine glands by identifying the method by which they discharge their secretion products from the cells. Give an example of each type.

12. List the types of connective tissue and briefly discuss the function and location of each variety.

13. Describe loose connective tissue.

14. How is adipose tissue different from most other types of connective tissue? Is fat distributed similarly or differently in male and female bodies?

15. Differentiate between regular and iregular dense fibrous tissues.

16. Discuss and compare the microscopic anatomy of bone and cartilage tissue.

17. Compare the structure of the three major types of cartilage tissue. Locate and give an example of each type.

18. List the components of whole blood and discuss the basic function of each fraction or cell type.

19. List the three major types of muscle tissue.

20. Identify the two basic types of cells in nervous tissue.

21. What are the four cardinal signs of inflammation? Discuss the process by which tissues respond to injury.

22. Differentiate between a scar and a keloid.

23. Describe the regenerative capacity of both muscle and nerve tissues.

24. List and give examples of the three types of epithelial tissue membranes.

25. Compare and contrast epithelial and connective tissue membranes.

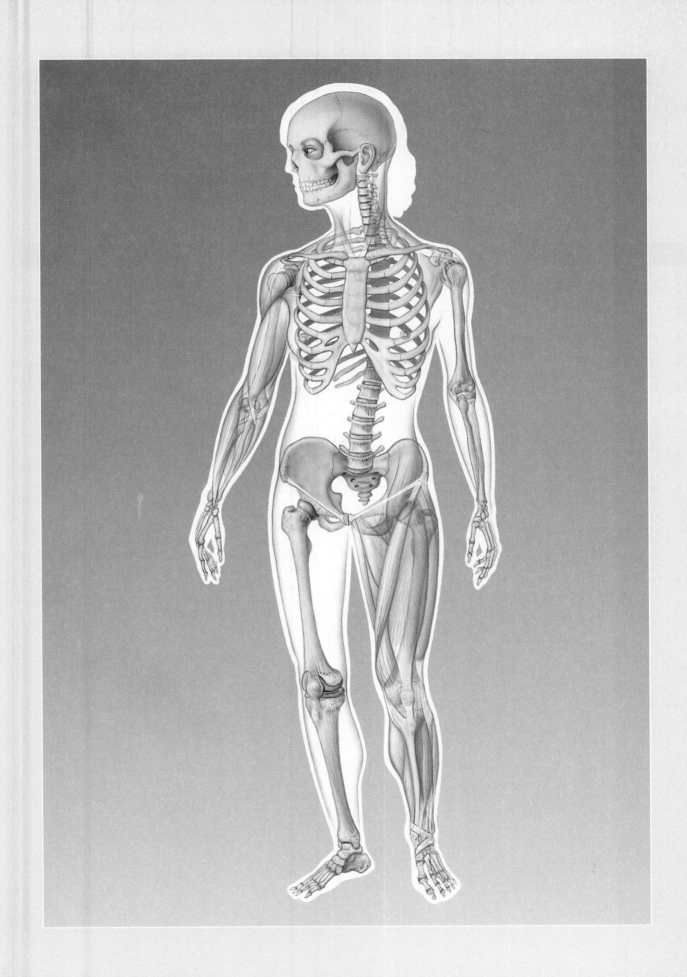

UNIT TWO

Support and Movement

The six chapters in Unit Two describe the outer covering of the body, as well as the bones, muscles, and articulations, or joints, of the body. The skin (and its appendages) is selected in Chapter 5 as the first organ system to be studied. Chapter 6—Skeletal Tissues—provides information on the types of skeletal tissues, how they are formed, grow, repair if injured, and how they function. Skeletal tissues serve not only to protect and support body structures, but also as storage sites for important mineral elements vital to many body functions. Blood cell formation also occurs within the marrow of bones.

The organs of the skeletal system, bones, are identified by size, shape, and specialized markings and then are organized or grouped into major subdivisions in Chapter 7. Movement between bones occurs at joints, or articulations, which are classified in Chapter 8 according to both structure and potential for movement.

Muscle physiology and anatomy of the major muscle groups are discussed in Chapters 9 and 10. The microscopic and molecular structure of muscle cells and tissues is related to function, as is the gross structure of individual muscles and muscle groups. Movement and heat production constitute the two most important functions of muscle. Discussion of muscle groups in Chapter 10 focuses on how muscles function and on how they attach to bones, how they are named, and how they are integrated functionally with other body organ systems.

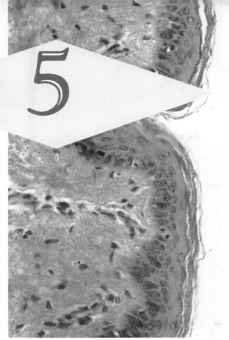

5 Skin and its Appendages

Cross-section of thin skin
Erlandsen/Magney: Color Atlas of Histology

OBJECTIVES

After you have completed this chapter, you should be able to:

1. Define the terms *integument* and *integumentary system.*
2. Discuss the generalized functions of the skin as an organ system.
3. Describe the cell types and cell layers of the epidermis in thick skin and give the function of each.
4. Discuss epidermal growth and repair.
5. Describe the layers, structural components, and functions of the dermis.
6. Discuss factors that influence skin color.
7. Describe the formation, structure, and growth of hair and nails.
8. Discuss and compare the structure and function of sweat (sudoriferous), sebaceous, and ceruminous glands.
9. Discuss the composition and function of skin surface film.
10. Explain how the skin functions in homeostasis of body temperature.
11. Explain the classification of burns into first, second, and third degree.

nails, and skin glands—as an organ system. Ideally, by studying the skin and its appendages before you proceed to the more complex organ systems in the chapters that follow, you will improve your understanding of how structure is related to function. **Integument** is another name for the skin. **Integumentary system** is a term used to denote the skin and its appendages.

SKIN FUNCTIONS

Skin functions are crucial to maintenance of homeostasis and thus to survival itself. They are also diverse. They include such different processes as protection, temperature regulation, synthesis of important chemicals and hormones (such as vitamin D), and excretion of water and salts. Also, certain substances, it is now known, can be absorbed through the skin. These include the fat-soluble vitamins (A, D, E, and K), estrogens and other sex hormones, corticoid hormones, and certain drugs such as nicotine, nitroglycerine, and dimethyl sulfoxide (DMSO). In addition, sensory receptors in the skin enable it to function as a sophisticated sense organ. They serve as antennas that detect stimuli leading to sensations of heat, cold, pressure, touch, and pain.

The skin also produces melanin—the pigment that serves as an extremely effective screen to potentially harmful ultraviolet light—and keratin—one of nature's most flexible yet enduring protective proteins. The keratinized stratified squamous epithelial tissue that composes the epidermis makes it a formidable barrier. It protects underlying tissues against invasion by unconquerable hordes of microorganisms, bars entry of most harmful chemicals, and minimizes mechanical injury of underlying structures.

The skin plays a very important role in maintaining homeostasis of body temperature. (We shall discuss this in some detail later in the chapter.) Briefly, blood vessels in the dermis dilate and sweat secretion increases if body temperature rises above normal. More heat is therefore lost by radiation from the larger volume of blood in the skin and by evaporation of sweat on the skin's surface. Together these changes tend to decrease blood temperature back to its normal level. If blood temperature decreases below normal, skin blood vessels constrict and sweat secretion decreases.

STRUCTURE OF THE SKIN

The skin is a thin, relatively flat organ classified as a membrane—the **cutaneous membrane.** Two main layers compose it: an outer, thinner layer called the **epidermis** and an inner, thicker layer named the **dermis** (Figure 5-1). The cellular epidermis is an epithelial layer derived from the ectodermal germ layer of the embryo. By the seventeenth week of gestation, the epidermis of the developing baby has all the essential characteristics of the adult. The deeper dermis is a relatively dense and vascular connective tissue layer that may average just over 4 mm in thickness in some body areas. The specialized area where the cells of the

---

Vital, diverse, complex, extensive—these adjectives describe the body's largest, thinnest, and one of its most important organs, the skin. It forms a self-repairing and protective boundary between the internal environment of the body and an often hostile external world. The skin surface is as large as the body itself, an area in average-sized adults of roughly 1.6 to 1.9 m² (17 to 20 square feet). Its thickness varies from slightly less than 0.05 cm (1/50 inch) to slightly more than 0.3 cm (1/8 inch).

As you know, the body is characterized by a "nested" or hierarchical type of organization. Complexity progresses from cells to tissues and then to organs and organ systems. This chapter discusses the skin and its appendages—hair,

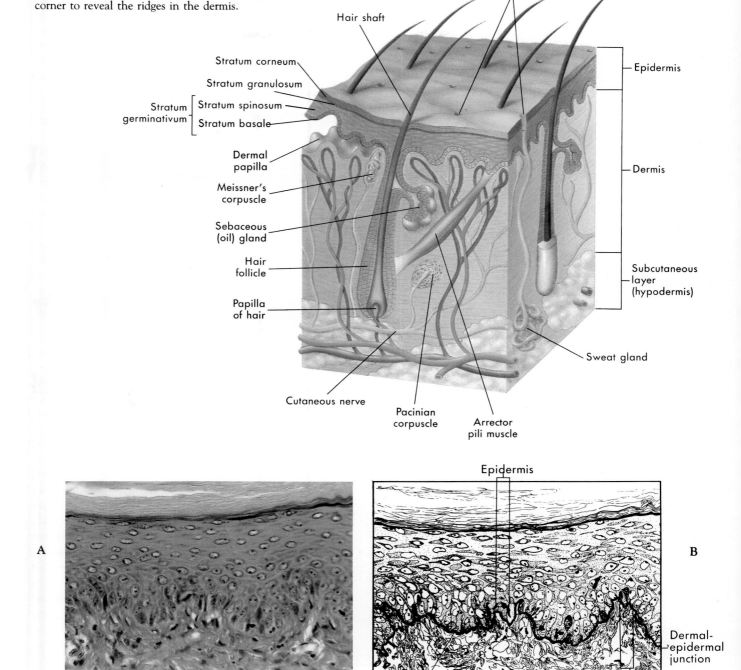

Openings of sweat ducts

Hair shaft

Epidermis

Stratum corneum

Stratum granulosum

Stratum germinativum — Stratum spinosum

Stratum basale

Dermal papilla

Dermis

Meissner's corpuscle

Sebaceous (oil) gland

Hair follicle

Papilla of hair

Subcutaneous layer (hypodermis)

Sweat gland

Cutaneous nerve

Pacinian corpuscle

Arrector pili muscle

A

B

Epidermis

Dermal-epidermal junction

Basement membrane

Dermis

FIGURE 5-2 Photomicrograph of skin. A, Note the many-layered epidermis separated from the dermis below by a distinct basement membrane. **B,** Sketch of photomicrograph.

epidermis meet the connective tissue cells of the dermis is called the **dermal-epidermal junction** (Figure 5-2). Beneath the dermis lies a loose **subcutaneous layer** rich in fat and areolar tissue. It is sometimes called the **hypodermis,** or **superficial fascia.** The fat content of the hypodermis varies with the state of nutrition and in obese individuals may

exceed 10 cm in thickness in certain areas. The density and arrangement of fat cells and collagen fibers in this area determine the relative mobility of the skin. When skin is removed from an animal by blunt dissection, separation occurs in the "cleavage plane" that exists between the superficial fascia and the underlying tissues.

SUBCUTANEOUS INJECTION

Although the subcutaneous layer is not part of the skin itself, it carries the major blood vessels and nerves to the skin above. The rich blood supply and loose spongy texture of this area make it an ideal site for the rapid and relatively pain-free absorption of injected material. Liquid medicines, such as insulin, and pelleted implant materials are often administered by *subcutaneous injection* into this spongy and porous layer beneath the skin.

Epidermis

The epidermis of the skin is composed of stratified squamous epithelium. In "thin skin," which covers most of the body surface and has a total depth of 1 to 3 mm, the outer epidermis is much thinner than most of us would probably guess—less than 0.17 mm thick (1/200 inch) in most areas. Exceptions are body surfaces chronically exposed to pressure or friction, such as the soles of the feet and the palms of the hands. Here, in the "thick skin," which has a total thickness of 4 to 5 mm, the epidermis is appreciably thicker (1 to 1.3 mm) than in the thin skin that covers most of the body.

Cell types

The epidermis is composed of several types of epithelial cells. **Keratinocytes** become filled with a tough, fibrous protein called **keratin.** These cells, arranged in distinct *strata*, or layers, are by far the most important cells in the epidermis. They comprise over 90% of the epidermal cells and form the principal structural element of the outer skin. **Melanocytes** contribute color to the skin and serve to filter ultraviolet light. Although they may comprise over 5% of the epidermal cells, melanocytes may be completely absent from the skin in certain nonlethal conditions. Another cell type, *Langerhans' cells,* are thought to play a limited role in immunological reactions that affect the skin and may serve as a defense mechanism for the body.

Cell layers

The cells of the epidermis are found in up to five distinct layers, or strata. Each stratum (meaning "layer") is named for its structural or functional characteristics.

1. **Stratum corneum** (horny layer). The stratum corneum is the most superficial layer of the epidermis. It is composed of very thin squamous (flat) cells that at the skin surface are dead and are continually being shed and replaced. The cytoplasm in these cells has been replaced by a water-repellent protein called *keratin.* In addition, the cell membranes become thick and chemically resistant. Specialized junctions (desmosomes) that hold adjacent keratinocytes together strengthen this layer even more and permit it to withstand considerable wear and tear. The process by which cells in this layer are formed from cells in deeper layers of the epidermis and are then filled with keratin and moved to the surface is called **keratinization.**

The stratum corneum is sometimes called the **barrier area** of the skin because it functions as a barrier to water loss and to many environmental threats ranging from microorganisms and harmful chemicals to physical trauma. Once this barrier layer is damaged, the effectiveness of the skin as a protective covering is greatly reduced, and most contaminants can easily pass through the lower layers of the cellular epidermis. Certain diseases of the skin cause the stratum corneum layer of the epidermis to thicken far beyond normal limits—a condition called **hyperkeratosis.** The result is a thick, dry, scaly skin that is inelastic and subject to painful fissures.

2. **Stratum lucidum** (clear layer). The keratinocytes in the stratum lucidum are closely packed and clear. Typically, the nuclei are absent, and the cell outlines are indistinct. These cells are filled with a soft gel-like substance called **eleiden,** which will eventually be transformed into keratin. Eleidin is rich in protein-bound lipids and serves to block water penetration or loss. This layer is absent in thin skin but is quite apparent in sections of thick skin from the soles of the feet or palms of the hands.

3. **Stratum granulosum** (granular layer). The process of keratinization begins in the stratum granulosum of the epidermis. Cells are arranged in a sheet two to four layers deep and are filled with intensely staining granules called **keratohyalin,** which is required for keratin formation. Cells in the stratum granulosum have started to degenerate. As a result, high levels of lysosomal enzymes are present in the cytoplasm, and the nuclei are missing or degenerate. Like the stratum lucidum, this layer of the epidermis may also be missing in some regions of thin skin.

4. **Stratum spinosum** (spiny layer). The stratum spinosum layer of the epidermis is formed from eight to ten layers of irregularly shaped cells with very prominent intercellular bridges or desmosomes. When viewed under a microscope, the desmosomes joining adjacent cells give the layer a spiny or prickly appearance (Latin *spinosus,* "spinelike"). Cells in this epidermal layer are rich in ribonucleic acid (RNA) and are therefore well equipped to initiate protein synthesis required for production of keratin.

5. **Stratum basale** (base layer). The stratum basale is a single layer of columnar cells. Only the cells in this deepest stratum of the epithelium undergo mitosis. As a result of this regenerative activity, cells transfer or migrate from the basal layer through the other layers until they are shed from the skin surface.

The term **stratum germinativum** (growth layer) is sometimes used to describe the stratum spinosum and the stratum basale together.

QUICK CHECK

✔ *1. Name three functions of the skin.*
✔ *2. What are the two principal layers of the skin? What tissue type dominates each layer?*
✔ *3. What is the function of keratin?*

Epidermal Growth and Repair

The most important function of the integument—protection—largely depends on the special structural features of the epidermis and its ability to create and repair itself following injury or disease. *Turnover* and *regeneration time* are terms used to describe the time period required for a population of cells to mature and reproduce. Obviously, as the surface cells of the stratum corneum are lost, replacement of keratinocytes by mitotic activity must occur. New cells must be formed at the same rate that old keratinized cells flake off from the stratum corneum to maintain a constant thickness of the epidermis. Cells push upward from the stratum basale into each successive layer, die, become keratinized, and eventually desquamate (fall away), as did their predecessors. This fact nicely illustrates a physiological principle: while life continues, the body's work is never done. Even at rest it is producing millions upon millions of new cells to replace old ones.

HEALTH MATTERS — BLISTERS

Blisters (see figure) may result from injury to cells in the epidermis or from separation of the dermal-epidermal junction. Regardless of cause, they represent a basic reaction of skin to injury. Any irritant that damages the physical or chemical bonds that hold adjacent skin cells or layers together will initiate blister formation. The specialized junctions (desmosomes) that serve to hold adjacent cells in the epidermis together are essential for integrity of the skin. If these intercellular bridges, sometimes described as "spot-welds" between adjacent cells, are weakened or destroyed, the skin literally falls apart and away from the body. Damage to the dermal-epidermal junction will produce similar results. Blister formation follows burns, friction injuries, exposure to primary irritants, or accumulation of toxic breakdown products following cell injury or death in the layers of the skin. Typically, chemical agents that break disulfide linkages or hydrogen bonds cause blisters. Since both types of these chemical bonds are the functional connecting links in intercellular bridges (or desmosomes), their involvement in blister formation serves as a good example of the relationship between structure and function at the chemical level of organization.

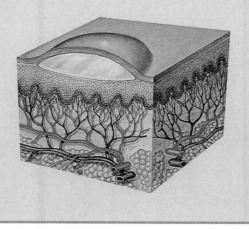

Current research suggests that the regeneration time required for completion of mitosis, differentiation, and movement of new keratinocytes from the stratum basale to the surface of the epidermis is about 35 days. The process can be accelerated by abrasion of the skin surface, which tends to peel off a few of the cell layers of the stratum corneum. The result is an intense stimulation of mitotic activity in the stratum basale and a shortened turnover period. If abrasion continues over a prolonged period of time, the increase in mitotic activity and shortened turnover time will result in an abnormally thick stratum corneum and the development of **calluses** at the point of friction or irritation. Although callus formation is a normal and protective response of the skin to friction, several skin diseases are also characterized by an abnormally high mitotic activity in the epidermis. In such conditions the thickness of the corneum is dramatically increased. As a result, scales accumulate, and development of skin lesions often occurs.

Normally about 10% to 12% of all cells in the stratum basale enter mitosis each day. Cells migrating to the surface proceed upward in vertical columns from discrete groups of eight to ten of these basal cells that are undergoing mitosis. Each group of active basal cells, together with its vertical columns of migrating keratinocytes, is called an *epidermal proliferating unit* or *EPU*. Keratinization proceeds as the cells migrate toward the stratum corneum. As mitosis continues and new basal cells enter the column and migrate upward, fully cornified "dead" cells are sloughed off at the skin surface.

Dermal-epidermal Junction

Electron microscopy and histochemical studies have demonstrated the existence of a definite basement membrane, specialized fibrous elements, and a unique polysaccharide gel that together serve to cement the superficial epidermis to the dermis below. The area of contact between the two skin layers forms a specialized junction. It functions to "glue" the two layers together and to provide a mechanical support for the epidermis, which is attached to its upper surface. In addition, it serves as a partial barrier to the passage of some cells and large molecules. Certain dyes, for example, if injected into the dermis cannot passively diffuse upward into the epidermis unless the junctional barrier is damaged by heat, enzymes, or other chemicals that change its permeability characteristics. Although the junction is remarkably effective in preventing separation of the two skin layers even when they are subjected to relatively high shear forces, this barrier is thought to play only a limited role in preventing passage of harmful chemicals or disease-causing organisms through the skin from the external environment.

Dermis

The dermis, or *corium*, is sometimes called the "true skin." It is composed of a thin **papillary** and a thicker **reticular** layer. The dermis is much thicker than the epidermis and may exceed 4 mm on the soles and palms. It

is thinnest on the eyelids and penis, where it seldom exceeds 0.5 mm. As a rule of thumb, the dermis on the ventral surface of the body and over the appendages is generally thinner than on the dorsal surface. The mechanical strength of the skin is in the dermis. In addition to serving a protective function against mechanical injury and compression, this layer of the skin provides a reservoir storage area for water and important electrolytes. A specialized network of nerves and nerve endings in the dermis also serves to process sensory information such as pain, pressure, touch, and temperature. At various levels of the dermis, there are muscle fibers, hair follicles, sweat and sebaceous glands, and many blood vessels. It is the rich vascular supply of the dermis that plays a critical role in regulation of body temperature—a function to be described later in the chapter.

Papillary layer

Note in Figure 5-1 that the thin superficial layer of the dermis forms bumps, called **dermal papillae,** which project into the epidermis. The papillary layer takes its name from these papillae arranged in rows on its surface. Between the sculptured surface of the papillary layer and the stratum basale lies the important dermal-epidermal junction. The papillary layer and its papillae are composed essentially of loose connective tissue elements and a fine network of thin collagenous and elastic fibers. The thin epidermal layer of the skin conforms tightly to the ridges of dermal papillae. As a result, the epidermis also has characteristic ridges on its surface. Epidermal ridges are especially well defined on the tips of the fingers and toes. In each of us they form a unique pattern—an anatomical fact made famous by the art of fingerprinting. These ridges perform a function very important to human survival: they allow us to grip surfaces well enough to walk upright on slippery surfaces and to grasp and use tools.

Reticular layer

The thick reticular layer of the dermis consists of a much more dense *reticulum,* or network of fibers, than is seen in the papillary layer above it. It is this dense layer of tough and interlacing white collagenous fibers that, when commercially processed from animal skin, results in leather. Although most of the fibers in this layer are of the collagenous type, which give toughness to the skin, elastic fibers are also present. These make the skin stretchable and elastic (able to rebound).

The dermis contains skeletal (voluntary) and smooth (involuntary) muscle fibers. A number of skeletal muscles are located in the skin of the face and scalp. These muscles permit a wide variety of facial expressions and are also responsible for voluntary movement of the scalp. The distribution of smooth muscle fibers in the dermis is much more extensive than the skeletal variety. Each hair follicle has a small bundle of involuntary muscles attached to it. These are the **arrector pili muscles.** Contraction of these muscles makes the hair "stand on end"—as in extreme fright, for example, or from cold. As the hair is pulled into

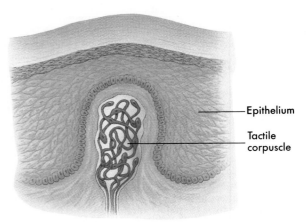

FIGURE 5-3 Skin receptors. Receptors are specialized nerve endings that make it possible for the skin to act as a sense organ. This *Meissner's corpuscle* is capable of detecting light touch (slight pressure). (See also Figure 5-1.)

an upright position, it raises the skin around it into what we commonly call "goose pimples." In the dermis of the skin of the scrotum and in the pigmented skin called the areolae surrounding the nipples, smooth muscle cells form a loose network. Contraction of these smooth muscle cells will wrinkle the skin and cause elevation of the testes or erection of the nipples.

Millions of specialized nerve endings called **receptors** are located in the dermis of all skin areas (Figure 5-3 and see Figure 5-1). They permit the skin to serve as a sense organ transmitting sensations of pain, pressure, touch, and temperature to the brain. The sensory receptors of the skin are discussed in Chapter 14.

Dermal Growth and Repair

Unlike the epidermis, the dermis does not continually shed and regenerate itself. It does maintain itself, but rapid regeneration of connective tissue in the dermis occurs only during unusual circumstances, as in the healing of wounds (see Figure 4-27, p. 119). In the healing of a wound such as a surgical incision, fibroblasts in the dermis quickly reproduce and begin forming an unusually dense mass of new connective tissue fibers. If this dense mass is not replaced by normal tissue, it remains as a *scar.* The dense bundles of white collagenous fibers that characterize the reticular layer of the dermis tend to orient themselves in patterns that differ in appearance from one body area to another. The result is formation of patterns called **Langer's lines,** or **cleavage lines** (Figure 5-4). If surgical incisions are made parallel to the cleavage, or Langer's lines, the resulting wound will have less tendency to gape open and will tend to heal with a thin and less noticeable scar.

If the elastic fibers in the dermis are stretched too much—for example, by a rapid increase in the size of the abdomen during pregnancy or as a result of great obesity—these fibers will weaken and tear. The initial result is formation of pinkish or slightly bluish depressed furrows with jagged edges. These tiny linear markings (*stretch*

FOCUS ON . . . SKIN CANCER

Skin cancers are the most common neoplasms, or abnormal growths, seen in humans. They account for almost one fourth of all cancer in men and nearly 14% of reported cancers in women. Of the many types of skin cancer, the most common forms are (1) **squamous cell carcinoma,** (2) **basal cell carcinoma,** and (3) **malignant melanoma.** All three types are epithelial cancers and apparently result from cell changes in the epidermis.

1. Squamous cell carcinoma. This slow-growing malignant tumor of the epidermis is the most common type of skin cancer. Lesions typical of this form of skin cancer are hard, raised nodules that are usually painless *(A)*. If not treated, squamous cell carcinoma will metastasize, invading other organs.

2. Basal cell carcinoma. Usually occurring on the upper face, this type of skin cancer is much less likely to metastasize than other types. This malignancy begins in cells at the base of the epidermis (stratum basale). Basal cell carcinoma lesions typically begin as *papules* (firm, raised lesions) that erode in the center to form a bleeding, crusted crater *(B)*.

3. Melanoma. Malignant melanoma, the fastest increasing cancer in the United States today, is the most serious form of skin cancer; it causes death in about one in every four cases. This type of cancer sometimes develops from a pigmented *nevus* (mole) to become a dark, spreading lesion *(C)*. Benign moles should be checked regularly for warning signs of melanoma because early detection and removal is essential in treating this rapidly spreading cancer. The "ABCD" rule of self-examination of moles is summarized here:

Asymmetry
Benign moles are *symmetrical;* their halves are mirror images of each other. Melanoma lesions are asymmetrical, or lopsided.
Border
Benign moles are outlined by a distinct border, but malignant melanoma lesions are often irregular or indistinct.
Color
Benign moles may be any shade of brown but are relatively evenly colored. Melanoma lesions tend to be unevenly colored, exhibiting a mixture of shades or colors.
Diameter
By the time a melanoma lesion exhibits characteristics A, B, and C, it is also probably larger than 6 mm (¼ inch).

Although genetic predisposition also plays a role, many pathophysiologists believe that exposure to the sun's ultraviolet (UV) radiation is the most important factor in causing the common skin cancers. UV radiation damages the DNA in skin cells, causing the mistakes in mitosis that produce cancer. Skin cells have a natural ability to repair UV damage to the DNA, but in some people, this mechanism may not be able to deal with a massive amount of damage. People with the rare, inherited condition *xeroderma* (zee-roh-DERM-ah) *pigmentosa* cannot repair UV damage at all and almost always develop skin cancer.

One of the rarer skin cancers, **Kaposi's** (ka-PO-sees) **sarcoma,** has increased recently in some parts of the world. Once associated mainly with certain ethnic groups, a form of this cancer now appears in many cases of AIDS and other immune deficiencies. Kaposi's sarcoma, first appearing as purple papules *(D)*, quickly spreads to the lymph nodes and internal organs. Some researchers believe that a virus or other agent, perhaps transmitted along with the HIV, is a possible cause of this cancer.

A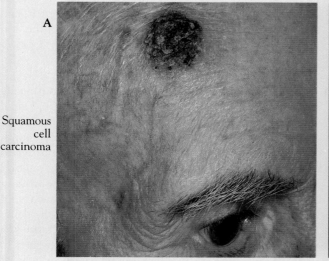

Squamous cell carcinoma

B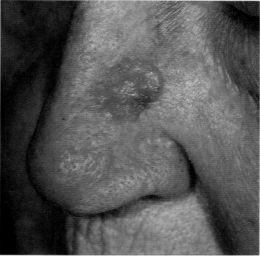

Basal cell carcinoma

C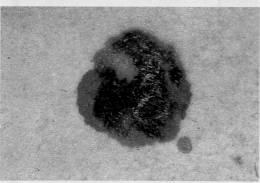

Malignant melanoma

D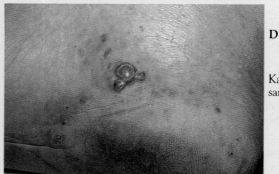

Kaposi's sarcoma

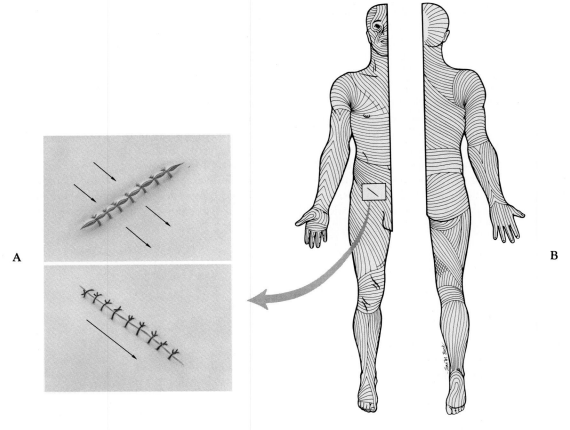

FIGURE 5-4   Langer's cleavage lines. A, If an incision "cuts across" cleavage lines, stress tends to pull the cut edges apart and may retard healing. **B,** Surgical incisions that are parallel to cleavage lines are subjected to less stress and tend to heal more rapidly.

marks) are really tiny tears. When they heal and lose their color, the *striae* (Latin, "furrows") that remain appear as glistening silver-white scar lines.

QUICK CHECK

✔ 1. *What is the name of the gluelike layer separating the dermis from the epidermis?*

✔ 2. *Which layer of the dermis forms the bumps that produce ridges on the palms and soles?*

✔ 3. *Which layer is vascular: the epidermis or dermis?*

◢ Skin Color

Human skin, as everyone knows, comes in a wide assortment of colors. The basic determinant of skin color is the quantity of **melanin** deposited in the cells of the epidermis. The number of pigment-producing melanocytes that are scattered throughout the stratum basale of the epidermis in most body areas is about the same in all races. It is the amount of melanin pigment actually produced by these cells that accounts for a majority of skin color variations. Melanocytes are highly specialized and unique.

HEALTH MATTERS   **VITILIGO**

An acquired condition called **vitiligo** results in loss of pigment in certain areas of the skin. The patches of depigmented white skin that characterize this condition contain melanocytes, but for unknown reasons they no longer produce pigment.

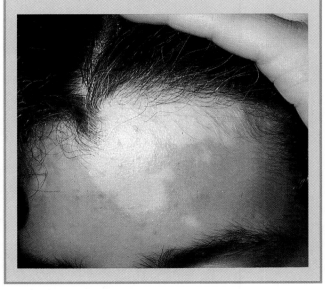

Of all body cells, only melanocytes have the ability to routinely convert the amino acid *tyrosine* into the dark brown melanin pigment. The pigment granules are then transferred to the other epidermal cells. The pigment-producing process is regulated by the enzyme *tyrosinase*. But this conversion process depends on several factors (Figure 5-5). Heredity is first of all. Geneticists tell us that four to six pairs of genes exert the primary control over the amount of melanin formed by melanocytes. If the enzyme tyrosinase is absent from birth because of a congenital defect, the melanocytes cannot form melanin, and a condition called **albinism** results. Albino individuals have a characteristic absence of pigment in their hair, skin, and eyes. Thus heredity determines how dark or light one's skin color will

be (see Chapter 33). Other factors can modify the genetic effect. Sunlight is an obvious example. Prolonged exposure to sunlight causes melanocytes to increase melanin production and darken skin color. So, too, does excess secretion of both adrenocorticotropic hormone (ACTH) or melanocyte-stimulating hormone (MSH) by the anterior pituitary gland. Increasing age may also influence melanocyte activity. In many individuals, decreasing tyrosinase activity is evident in graying of the hair. In addition to melanin, other pigments, such as the yellow pigment *carotene*, also contribute to skin color.

An individual's basic skin color changes, as we have just observed, whenever its melanin content changes appreciably. But skin color can also change without any change in

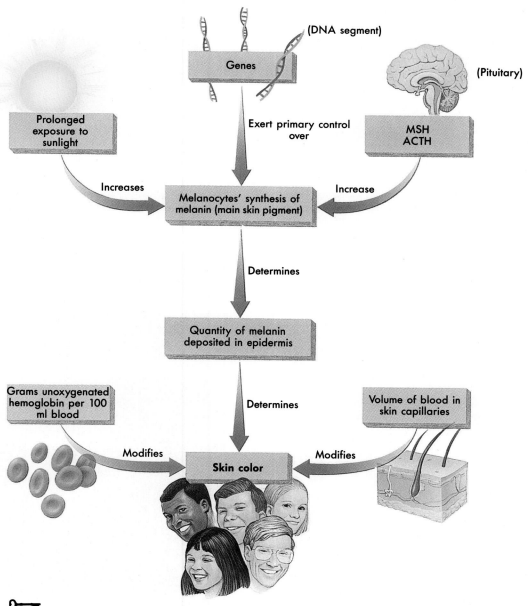

FIGURE 5-5 **How genes affect skin color.** Genes determine an individual's basic skin color by controlling the amount of melanin synthesized and deposited in the epidermis. But, as the diagram shows, other factors may modify the basic skin color.

melanin. In this case the change is usually temporary and most often stems from a change in the volume of blood flowing through skin capillaries. If skin blood vessels constrict, for example, skin blood volume decreases, and the skin may turn pale. Or if skin blood vessels dilate, as they do in blushing, the skin appears to be pinker.

In general, the sparser the pigments in the epidermis, the more transparent the skin is and therefore the more vivid the change in skin color will be with a change in skin blood volume. Conversely, the richer the pigmentation, the more opaque the skin, resulting in less skin color changes with a change in skin blood volume.

In some abnormal conditions, skin color changes due to an excess amount of unoxygenated hemoglobin in the skin capillary blood. If skin contains relatively little melanin, it will appear bluish, that is, *cyanotic,* when its blood has a high proportion of unoxygenated hemoglobin. In general, the darker the skin pigmentation, the greater the amount of unoxygenated hemoglobin that must be present before **cyanosis** ("condition of blueness") becomes visible.

◣ APPENDAGES OF THE SKIN

Appendages of the skin consist of the hair, nails, and skin glands.

Hair

Only a few areas of the skin are hairless—notably the palms of the hands and the soles of the feet. Hair is also absent from lips, nipples, and some areas of the genitalia.

Many months before birth, hair follicles begin to develop in most parts of the skin. By about the sixth month of pregnancy the developing fetus is all but covered by an extremely fine and soft hair coat, called **lanugo.** Most of the lanugo hair is lost before birth. Soon after birth any lanugo hair that remains is lost and then replaced by new hair that is more pigmented and strong. Replacement hair growth appears first on the scalp, eyelids, and eyebrows. The coarse pubic and axillary hair that develops at puberty is called *terminal hair.*

Hair growth begins when cells of the epidermis spread down into the dermis, forming a small tube, the **follicle.** The stratum germinativum develops into the follicle's innermost layer and forms at the bottom of the follicle a cap-shaped cluster of cells known as the **germinal matrix.** Protruding into the germinal matrix is a small mound of the dermis, called the **hair papilla**—a highly important structure, since it contains the blood capillaries that nourish the germinal matrix (Figure 5-6). Cells of the germinal matrix are responsible for forming hairs. They undergo repeated mitosis, push upward in the follicle, and become keratinized to form a hair. As long as cells of the germinal matrix remain alive, hair will regenerate even though it is cut or plucked or otherwise removed.

Part of the hair, namely, the root, lies hidden in the follicle. The visible part of a hair is called the **shaft.** The inner core of a hair is known as the **medulla** and the outer portion around it is called the **cortex.** Layers of keratinized cells make up the cortex. Deposited in these cells are varying amounts of melanin, the pigment responsible for brown or black hair. Besides adding color, melanin also imparts strength to the hair. Whether hair is straight or wavy depends mainly on the shape of the shaft. Straight hair has a round, cylindrical shaft. Wavy hair, in contrast, has a flat shaft that is not as strong. As a result it is more

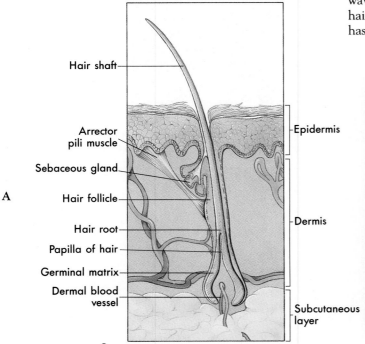

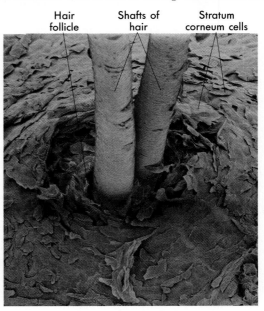

A
Hair shaft
Arrector pili muscle
Sebaceous gland
Hair follicle
Hair root
Papilla of hair
Germinal matrix
Dermal blood vessel
Epidermis
Dermis
Subcutaneous layer

B
Hair follicle
Shafts of hair
Stratum corneum cells

FIGURE 5-6 Hair follicle. A, Relationship of a hair follicle and related structures to the epidermal and dermal layers of the skin. **B,** This scanning electron micrograph shows shafts of hair extending from their follicles.

easily broken or damaged than straight hair. Two or more small *sebaceous glands* secrete **sebum,** an oily substance, into each hair follicle. The sebaceous gland secretions lubricate hair and keep it from becoming dry, brittle, and easily damaged.

Hair alternates between periods of growth and rest. On the average, hair on the head grows a little less than 12 mm (½ inch) per month, or about 5 inches a year. Body hair grows more slowly. Head hairs reportedly live between 2 and 6 years, then die and are shed. Normally, however, new hairs replace those lost. But baldness can develop—a fact we all know. The common type of baldness occurs only when two requirements are met: genes for baldness must be inherited and the male sex hormone, testosterone, must be

present. When the right combination of causative factors exist, common baldness or **male-pattern baldness** (Figure 5-7) inevitably results. The only known treatment that actually reverses male-pattern baldness, rather than simply covering it up or transplanting follicles, uses the drug Rogaine *(minoxidil).* Unfortunately, this treatment is extremely expensive, must be continued for life if new hair growth is to be retained, and does not always produce as dramatic an effect as hoped.

Contrary to what many people believe, hair growth is not stimulated by frequent cutting or shaving. In addition, stories about hair or beard growth continuing after death are also false. What may appear to be continuing beard growth after death is really caused by dehydration and shrinkage of the skin over the face. As a result, a "five o'clock shadow" may appear, or the beard may become more noticeable 2 or 3 days after death even if the face was carefully shaved at the time the body was initially prepared for viewing. In these cases what appears to be continued hair growth is in reality only a more visible beard in a skin surface dehydrated by environmental conditions or the embalming process.

Nails

Heavily keratinized epidermal cells compose fingernails and toenails. The visible part of each nail is called the **nail body.** The rest of the nail, namely, the **root,** lies in a groove hidden by a fold of skin called the **cuticle.** The nail body nearest the root has a crescent-shaped white area known as the **lunula,** or "little moon." Under the nail lies a layer of epithelium called the **nail bed** (Figure 5-8). Because it contains abundant blood vessels, it appears pink in color through the translucent nail bodies. Nails grow by mitosis of cells in the stratum germinativum beneath the lunula. On the average, nails grow about 0.5 mm a week. Fingernails, however, as you may have noticed, grow faster than toenails, and both grow faster in the summer than in the winter.

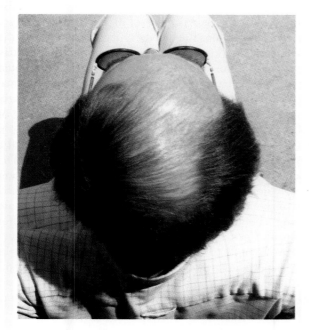

FIGURE 5-7   Male pattern baldness.

QUICK CHECK

✔ *1. List three factors that affect the color of the skin.*
✔ *2. What substance comprises most of the hair and nails?*

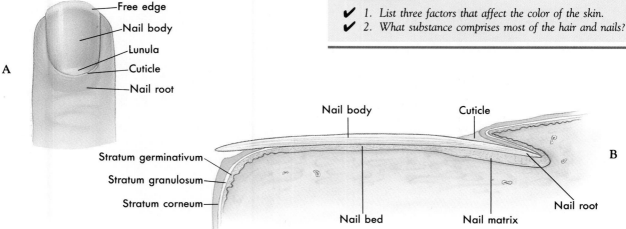

FIGURE 5-8   **Structure of nails. A,** Fingernail viewed from above. **B,** Sagittal section of fingernail and associated structures.

Skin Glands

The skin glands include three kinds of microscopic glands: sweat, sebaceous, and ceruminous (Figure 5-9).

Sweat glands

Sweat glands are the most numerous of the skin glands. They can be classified into two groups—*eccrine* and *apocrine*—based on type of secretion, location, and nervous system connections.

Eccrine sweat glands are by far the most numerous, important, and widespread sweat glands in the body. They are quite small, with a secretory portion less than 0.4 mm in diameter, and are distributed over the total body surface with the exception of the lips, ear canal, glans penis, and nail beds. Eccrine sweat glands are a simple, coiled, tubular type of gland. They function throughout life to produce a transparent watery liquid (*perspiration,* or *sweat*) rich in salts, ammonia, uric acid, urea, and other wastes. In addition to elimination of waste, sweat plays a critical role in helping the body maintain a constant core temperature. Histologists estimate that a single square inch of skin on the palms of the hands contains about 3,000 sweat glands. Eccrine sweat glands are also very numerous on the soles of the feet, forehead, and upper torso. With a good magnifying glass you can locate the openings of these sweat gland ducts on the skin ridges of the palms and on the skin of the palmar surfaces of the fingers.

Apocrine sweat glands are located deep in the subcutaneous layer of the skin in the armpit (axilla), the areola of the breast, and the pigmented skin areas around the anus. They are much larger than eccrine glands and often have secretory units that reach 5 mm or more in diameter. They are connected with hair follicles and are classified as simple, branched tubular glands. Apocrine glands enlarge and begin to function at puberty, producing a more viscous and colored secretion than eccrine glands. In the female, apocrine gland secretions show cyclic changes that are linked to the menstrual cycle. Odor often associated with apocrine gland secretion is not caused by the secretion itself. Instead, it is caused by contamination and decomposition of the secretion by skin bacteria.

Sebaceous glands

Sebaceous glands secrete oil for the hair and skin. Wherever hairs grow from the skin, there are sebaceous glands, at least two for each hair. The oil, or **sebum,** keeps the hair supple and the skin soft and pliant. It serves as nature's own protective skin cream by preventing excessive water loss from the epidermis. Because sebum is rich in chemicals such as triglycerides, waxes, fatty acids, and cholesterol that have an antifungal effect, it also contributes to reducing fungal activity on the skin surface. This property of sebum helps protect the skin from numerous types of fungal infections. Sebaceous glands are simple branched glands of varying size that are found in the dermis, except in the skin of the palms and soles. Although almost always associated with hair follicles, some specialized sebaceous glands do open directly on the skin surface in such areas as the glans penis, lips, and eyelids. Sebum secretion increases during adolescence, stimulated by increased blood levels of the sex hormones. Frequently sebum accumulates in and enlarges some of the ducts of the sebaceous glands, forming white pimples. With oxidation this accumulated sebum darkens, forming a **blackhead.**

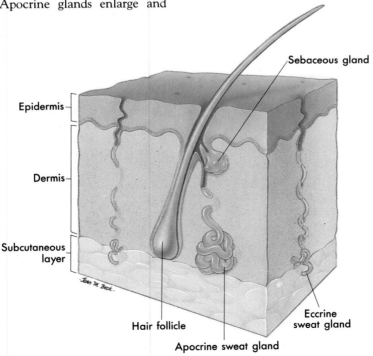

FIGURE 5-9   Skin glands.

ACNE

Common *acne*, or *acne vulgaris*, occurs most frequently in the adolescent years as a result of overactive secretion by the sebaceous glands, with blockage and inflammation of their ducts. There is an increase of more than fivefold in the rate of sebum secretion between 10 and 19 years of age. As a result sebaceous gland ducts may become plugged with sloughed skin cells and sebum contaminated with bacteria. The inflamed plug is called a *comedo* and is the most characteristic sign of acne. Pus-filled *pimples* or *pustules* result from secondary infections within or beneath the epidermis, often in a hair follicle or sweat pore.

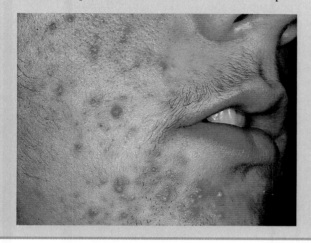

Ceruminous glands

Ceruminous glands are a special variety or modification of apocrine sweat glands. Histologically, they appear as simple coiled tubular glands with excretory ducts that open onto the free surface of the skin in the external ear canal or with sebaceous glands into the necks of hair follicles in this area. The mixed secretions of sebaceous and ceruminous glands form a brown waxy substance called **cerumen.** Although it serves a useful purpose in protecting the skin of the ear canal from dehydration, excess cerumen can harden and cause blockage in the ear, resulting in loss of hearing.

SURFACE FILM

The ability of the skin to act as a protective barrier against a wide array of potentially damaging assaults from the environment begins with the proper functioning of a thin film of emulsified material spread over its surface. The **surface film** is produced by the mixing of residue and secretions from sweat and sebaceous glands with epithelial cells constantly being cast off from the epidermis. The shedding of epithelial elements from the skin surface is called **desquamation.** Functions of surface film include:

- ◆ Antibacterial and antifungal activity
- ◆ Lubrication
- ◆ Hydration of the skin surface
- ◆ Buffering of caustic irritants
- ◆ Blockade of many toxic agents

Chemical composition of the surface film includes (1) amino acids, sterols, and complex phospholipids from the breakdown of sloughed epithelial cells; (2) fatty acids, triglycerides, and waxes from sebum; and (3) water, ammonia, lactic acid, urea, and uric acid from sweat. The specific chemical composition of surface film is quite variable, and samples taken from skin covering one body area will often have a different "mix" of chemical components than film covering skin in another area. This difference helps explain the unique and localized distribution patterns of certain skin diseases and why the skin covering one area of the body is sometimes more susceptible to attack by certain bacteria or fungi.

QUICK CHECK

✔ 1. *What are the two types of sweat glands? How do they differ?*
✔ 2. *List two functions of sebum.*
✔ 3. *What substances make up the skin's surface film?*

HOMEOSTASIS OF BODY TEMPERATURE

Warm-blooded animals such as humans maintain a remarkably constant temperature despite sizable variations in environmental temperatures.

Normally, in most people, body temperature moves up and down very little in the course of a day. It hovers close to a set point of about 37° C, increasing perhaps to 37.6° C by late afternoon and decreasing to around 36.2° C by early morning. This homeostasis of body temperature is of the utmost importance. Why? Because healthy survival depends on biochemical reactions taking place at certain rates. And these rates in turn depend on normal enzyme functioning, which depends on body temperature staying within the narrow range of normal.

To maintain an even temperature, the body must, of course, balance the amount of heat it produces with the amount it loses. This means that if extra heat is produced in the body, this same amount of heat must then be lost from it. Obviously if this does not occur, if increased heat loss does not closely follow increased heat production, body temperature will climb steadily upward. If body temperature increases above normal for any reason, the skin plays a critical role in heat loss by the physical phenomena of evaporation, radiation, conduction, and convection.

Heat Production

Heat is produced by one means—metabolism of foods. Because the muscles and glands (especially the liver) are the most active tissues, they carry on more metabolism and therefore produce more heat than any of the other tissues. So the chief determinant of how much heat the body produces is the amount of muscular work it does. During exercise and shivering, for example, metabolism and heat

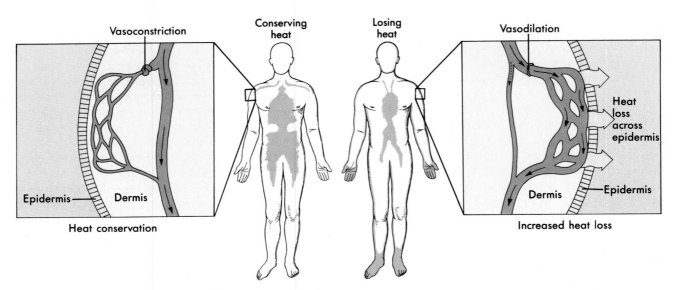

FIGURE 5-10   The skin as a thermoregulatory organ. When homeostasis requires that the body conserve heat, blood flow in the warm organs of the body's core increases *(left)*. When heat must be lost to maintain the stability of the internal environment, flow of warm blood to the skin increases *(right)*. Heat can be lost from the blood and skin by means of radiation, conduction, convection, and evaporation.

production increase greatly. But during sleep, when very little muscular work is being done, metabolism and heat production decrease.

Heat Loss

As already stated, one mechanism the body uses to maintain relative constancy of internal temperature is to regulate the amount of heat loss. Some 80% or more of this transfer of heat occurs through the skin; the remainder takes place in mucous membranes. As Figure 5-10 shows, heat loss can be regulated by altering the flow of blood in the skin. If heat must be conserved to maintain a constant body temperature, then dermal blood vessels constrict (vasoconstriction), keeping most of the warm blood circulating deeper in the body. If heat loss must be increased to maintain a constant temperature, then dermal blood vessels widen (vasodilation), increasing the skin's supply of warm blood from deeper tissues. Heat transferred from the warm blood to the epidermis can then be lost to the external environment through the physical processes of evaporation, radiation, conduction, and convection.

Evaporation

Heat energy must be expended to evaporate any fluid. Evaporation of water constitutes one method by which heat is lost from the body, especially from the skin. Evaporation is especially important at high environmental temperatures when it is the only method by which heat can be lost from the skin. A humid atmosphere necessarily retards evaporation and therefore lessens the cooling effect derived from it—the explanation for the fact that the same degree of temperature seems hotter in humid climates than in dry ones. At moderate temperatures, evaporation accounts for about half as much heat loss as does radiation.

SPORTS & FITNESS   ### EXERCISE AND THE SKIN

Excess heat produced by the skeletal muscles during exercise increases the core body temperature far beyond the normal range. Because blood in vessels near the skin's surface dissipates heat well, the body's control centers adjust blood flow so that more warm blood from the body's core is sent to the skin for cooling. During exercise, blood flow in the skin can be so high that the skin takes on a redder coloration.

To help dissipate even more heat, sweat production increases to as high as 3 L per hour during exercise. Although each sweat gland produces very little of this total, over 3 million individual sweat glands are found throughout the skin. Sweat evaporation is essential to keeping body temperature in balance, but excessive sweating can lead to a dangerous loss of fluid. Because normal drinking may not replace the water lost through sweating, it is important to increase fluid consumption during and after any type of exercise to avoid *dehydration.*

Radiation

Radiation is the transfer of heat from the surface of one object to that of another without actual contact between the two. Heat radiates from the body surface to nearby objects that are cooler than the skin and radiates to the skin from those that are warmer than the skin. This is, of course, the principle of heating and cooling systems. In cool environmental temperatures, radiation accounts for a greater percentage of heat loss from the skin than both conduction and evaporation combined. However, in hot environments, no heat is lost by radiation but instead may be gained by radiation from warmer surfaces to the skin.

Conduction

Conduction means the transfer of heat to any substance actually in contact with the body—to clothing or jewelry, for example, or even to cold foods or liquids ingested. This process accounts for a relatively small amount of heat loss.

Convection

Convection is the transfer of heat away from a surface by movement of heated air or fluid particles. Usually convection causes very little heat loss from the body's surface. But, under certain conditions, it can account for considerable heat loss—as you know if you have ever stepped from your bath into even slightly moving air from an open window.

Homeostatic Regulation of Heat Loss

The operation of the skin's blood vessels and sweat glands must be coordinated carefully and must take into account moment-by-moment fluctuations in body temperature. Like most homeostatic mechanisms, heat loss by the skin is controlled by a negative-feedback loop (Figure 5-11). Temperature receptors in a part of the brain, called the *hypothalamus*, detect changes in the body's internal temperature. If some disturbance, such as exercise, increases body temperature above the set point value of 37° C, then the hypothalamus acts as an integrator and sends a nervous signal to the sweat glands and blood vessels of the skin. The sweat glands and blood vessels (the effectors) respond to the signal by acting in ways that promote heat loss. That is, sweat glands increase their output of sweat (increasing heat loss by evaporation), and blood vessels increase their diameter (increasing heat loss by radiation and other means). This continues until the set point is reached.

FIGURE 5-11 Role of skin in homeostasis of body temperature. Body temperature is continually monitored by nerve receptors in the skin and other parts of the body. These receptors feed information back to the hypothalamus of the brain, which compares the actual temperature to the set point temperature and then sends out an appropriate correction signal to effectors. If actual body temperature is above the set point temperature, sweat glands in the skin are signaled to increase their secretion and thus promote evaporation and cooling. At the same time, blood vessels in the dermis are signaled to dilate and thus promote radiation of heat away from the skin's surface.

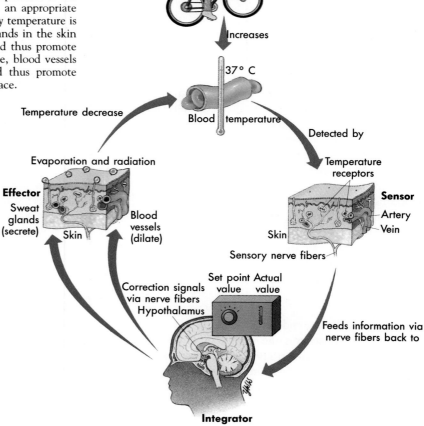

BURNS

Typically we think of a burn as a thermal injury or lesion caused by contact of the skin with some hot object or fire. In addition, overexposure to ultraviolet light (sunburn) or contact with an electric current or corrosive chemicals will also cause injury or death to skin cells. The injuries that result can all be classified as **burns.**

Estimating Body Surface Area

When burns involve large areas of the skin, treatment and prognosis for recovery depend in large part on total area involved and severity of the burn. The severity of a burn is determined by the depth of the lesion, as well as the extent (percent of body surface area burned). There are several ways to estimate the extent of body surface area burned. One method is called the **"rule of palms"** and is based on the assumption that the palm size of the burn victim is about 1% of the total body surface area. Therefore estimating the number of "palms" that are burned will approximate the percentage of body surface area involved.

The **"rule of nines"** (Figure 5-12) is another and more accurate method of determining the extent of a burn injury. Using this technique the body is divided into 11 areas of 9%, with the area around the genitals, called the *perineum,* representing the additional 1% of body surface area. As Figure 5-12 shows, 9% of the skin covers the head and each upper extremity, including front and back surfaces. Twice as much, or 18%, of the total skin area covers the front and back of the trunk and each lower extremity, including front and back surfaces. The rule of nines works well with adults but does not reflect the differences in body surface area seen in small children. Special tables called *Lund-Browder Charts,* which take the large surface area of certain body areas (such as the head) in the growing child into account, are used by physicians to estimate burn percentages in children.

The depth of a burn injury depends on the tissue layers of the skin that are involved. A **first-degree burn** (typical sunburn) will cause minor discomfort and some reddening of the skin. Although the surface layers of the burned area may peel in 1 or 2 days, no blistering occurs and actual tissue destruction is minimal. First- and second-degree burns are called **partial-thickness burns.**

Second-degree burns involve the deep epidermal layers and always cause injury to the upper layers of the dermis. In deep second-degree burns, damage to sweat glands, hair follicles, and sebaceous glands may occur but tissue death is not complete. Blisters, severe pain, generalized swelling, and edema characterize this type of burn. Scarring is common.

Third-degree, or **full-thickness, burns** are characterized by destruction of both the epidermis and dermis. Tissue death extends below the hair follicles and sweat glands. Burning may involve underlying muscles, fasciae, or even bone. A distinction between second- and third-degree burning is the fact that the third-degree lesion is insensitive to pain immediately after injury because of the destruction of nerve endings. Scarring is a serious problem.

QUICK CHECK

✔ 1. *What is the one means of heat production in the body? In what type of organs does most heat production occur?*

✔ 2. *Name three of the four physical processes in which heat is lost from the body.*

✔ 3. *Distinguish between the three classes of burns.*

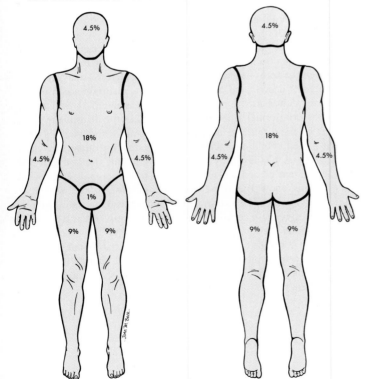

FIGURE 5-12 "Rule of nines." "Rule of nines" is one method used to estimate amount of skin surface burned in an adult.

HEALTH MATTERS

SUNBURN AND SKIN CANCER

Burns caused by exposure to harmful UV radiation in sunlight are commonly called *sunburns.* As with any burn, serious sunburns can cause tissue damage and lead to secondary infections and fluid loss. Cancer researchers have recently theorized that blistering (second-degree) sunburns during childhood may trigger the development of malignant melanoma later in life. Some epidemiological studies show that adults who had more than two blistering sunburns before the age of 20 have a greater risk of developing melanoma than someone who experienced no such burns. If this theory is true, it could explain the dramatic increase in skin cancer rates in the United States in recent years. Those who grew up as sunbathing and "suntans" became popular in the 1950s and 1960s are now, as adults, exhibiting melanoma at a much higher rate than in previous generations.

Cycle of Life: *Skin*

Everyone is aware of dramatic changes in skin that each person experiences from birth and through the mature years. Infants and young children have relatively smooth and unwrinkled skin characterized by the elasticity and flexibility associated with extreme youth. Because the skin tissues are in an active phase of new growth, healing of skin injuries is often rapid and efficient. Young children have fewer sweat glands than adults, so their bodies rely more on increased blood flow to maintain a normal body temperature. This explains why preschoolers often become "red-faced" while playing outdoors on a warm day.

As adulthood begins, at puberty, hormones stimulate the development and activation of sebaceous glands and sweat glands. After the sebaceous glands become active, especially during the initial years, they often overproduce sebum and thus give the skin an unusually oily appearance. Sebaceous ducts may become clogged or infected and form acne pimples or other blemishes on the skin. Activation of apocrine sweat glands during puberty causes increased sweat production—an ability needed to maintain an adult body

properly—and also the possibility of increased "body odor." Body odor is caused by wastes produced by bacteria that feed on the organic compounds found in apocrine sweat and on the surface of the skin.

As one continues past early adulthood, the sebaceous and sweat glands become less active. Although this can provide a welcome relief to those who suffer from acne or other problems associated with overactivity of these glands, it can affect normal function of the body. For example, the reduction of sebum production can cause the skin and hair to become less resilient and therefore more likely to wrinkle or crack. Wrinkling can also be caused by an overall degeneration in the skin's ability to maintain itself as efficiently as it did during the early years of development. Loss of function in sweat glands as adulthood advances adversely affects the body's ability to cool itself during exercise or when the external temperature is high. Thus elderly individuals are more likely to suffer severe problems during hot weather than young adults.

THE BIG PICTURE
The Skin and the Whole Body

The skin is one of the major components of the body's structural framework. It is continuous with the connective tissues that hold the body together, including those of fascia, bones, tendons, and ligaments. The integumentary, skeletal, and muscular systems work together to protect and support the whole body.

As stated several times in this chapter, the skin is a barrier that separates the internal environment from the external environment. Put another way, the skin *defines* the internal environment of the body. The barrier formed by the skin is a formidable one indeed, possessing innumerable mechanisms for protecting internal structures from the sometimes harsh external environment. Without these protective mechanisms, the internal environment could not maintain a relative constancy that is independent of the external environment.

First, the dermis and epidermis work together to form a tough, waterproof envelope that protects us from drying out and from the dangers of chemical or microbial contamination. The sebaceous secretions of the skin, along with other components of the skin's surface film, enhance the skin's ability to protect the internal environment. Protection against mechanical injuries is provided by hair, calluses, and the layers of the skin itself. Pigmentation in

the skin, and our ability to regulate its concentration, protect us from the harmful effects of solar radiation.

Although its primary functions are support and protection, the skin has other important roles in maintaining homeostasis. For example, the skin is also an important agent in the regulation of body temperature—serving as a sort of "radiator" that can be activated or deactivated as needed. It helps maintain a constant level of calcium in the body by producing vitamin D, which is necessary for normal absorption of calcium in the digestive tract. Ridges on the palms and fingers allow us to make and use tools for getting food, building shelters, and conducting other survival tasks. The skin's flexibility and elasticity permit the free movement required to perform such tasks. Sensory nerve receptors in the dermis allow the skin to be a "window on the world." Information about the external environment is relayed from skin receptors to nervous control (integration) centers, where it is used to coordinate the function of other organs.

As you continue your study of the various organ systems of the body, keep in mind that none of them could operate properly without the structural and functional assistance of the integumentary system.

MECHANISMS OF DISEASE

Skin Disorders

Any disorder of the skin can be called a **dermatosis,** which simply means "skin condition." Many dermatoses involve inflammation of the skin, or **dermatitis.** Various disorders involving the skin have already been discussed in this chapter. A few more representative disorders are described here (see essay on skin cancer, p. 136).

Skin Infections

The skin is the first line of defense against microbes that might invade the body's internal environment. It is no wonder that the skin is a common site of infection. In adults, the antimicrobial characteristics of sebum in the skin's surface film often inhibit skin infections. In children, the lack of sebum in the surface film makes the skin less resistant to infection.

Many different viruses, bacteria, and fungi cause skin conditions. Here are a few examples of skin infections caused by different types of pathogenic (disease-causing) organisms:

1. Impetigo. This highly contagious bacterial condition results from *Staphylococcus* or *Streptococcus* infection and occurs most often in young children. Impetigo starts as a reddish discoloration, or *erythema,* but soon develops into vesicles (blisters) and yellowish crusts. Occasionally, the infection becomes systemic (body-wide) and thus becomes life-threatening.

2. Tinea. Tinea is the general name for many different *mycoses* (fungal infections) of the skin. Ringworm, jock itch, and athlete's foot are all classified as tinea. Signs of tinea include erythema, scaling, and crusting. Occasionally, fissures, or cracks, in the epidermis develop at creases in the epidermis. Figure 5-13 shows a case of ringworm, a tinea infection that typically forms a round rash that heals in the center to form a ring. Antifungal agents usually stop the acute infection but are unable to completely destroy the fungus. Recurrence of tinea can be avoided by keeping the skin dry, since fungi require a moist environment to grow.

3. Warts. Caused by papillomaviruses, warts are nipple-like neoplasms of the skin. Although they are usually benign, some warts transform to become malignant. Transmission of warts generally occurs through direct contact with warts on the skin of an infected person. Warts can be removed by freezing, drying, laser therapy, or application of chemicals.

4. Boils. Also called *furuncles,* boils are local *Staphylococcus* infections of hair follicles characterized by large, inflamed pus-filled lesions. A group of untreated boils may fuse into even larger lesions called *carbuncles.*

Vascular and Inflammatory Skin Disorders

Everyone who might ever be called on to provide care to a bedridden or otherwise immobilized individual should be aware of the causes and nature of pressure sores, or **decubitus ulcers** (Figure 5-14). Decubitus means "lying down," a name that hints at a common cause of pressure sores: lying in one position for long periods. Also called *bedsores,* these lesions appear after blood flow to a local area of skin slows because of pressure on skin covering bony prominences such as the ankles. Ulcers form and infections develop as lack of blood flow causes tissue damage. Frequent changes in body position and soft support cushions help prevent decubitus ulcers.

A common type of skin disorder that involves blood vessels is **urticaria,** or *hives.* This condition is characterized by raised red lesions, called *wheals,* caused by leakage of fluid from the

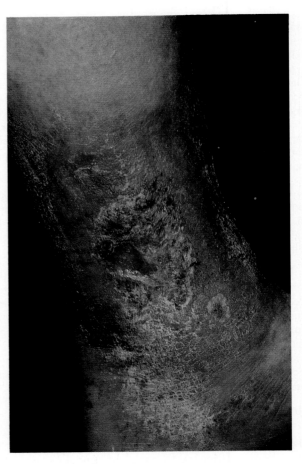

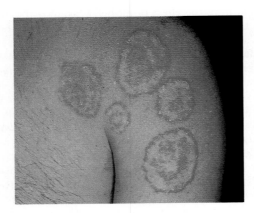

FIGURE 5-13 Tinea.

FIGURE 5-14 Decubitus ulcer.

Continued

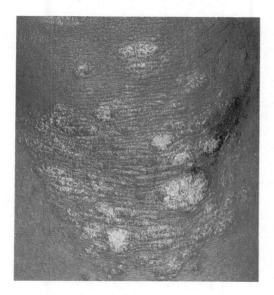

FIGURE 5-15   Psoriasis.

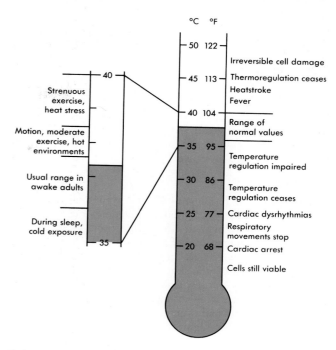

FIGURE 5-16   Body temperature. Diagram, modeled after a thermometer, shows some physiological consequences of an abnormal body temperature. The normal range of body temperature under various conditions is shown in the inset.

skin's blood vessels. Urticaria is often associated with severe itching. Hypersensitivity or allergic reactions, physical irritants, and systemic diseases are common causes.

Scleroderma is an autoimmune disease that affects the blood vessels and connective tissues of the skin. The name scleroderma comes from the word parts *sclera-*, which means "hard," and *derma*, which means "skin." Hard skin is a good description of the lesions characteristic of scleroderma. Scleroderma begins as a mild inflammation that later develops into a patch of yellowish, hardened skin. Scleroderma most commonly remains a mild, localized condition. Very rarely, localized scleroderma progresses to a systemic form, affecting large areas of the skin and other organs. Persons with advanced systemic scleroderma seem to be wearing a mask because skin hardening prevents them from moving their mouths freely. Both forms of scleroderma occur more commonly in women than in men.

Psoriasis is a chronic inflammatory disorder of the skin thought to have a genetic basis. This common skin problem is characterized by cutaneous inflammation accompanied by scaly lesions that develop from an excessive rate of epithelial cell growth (Figure 5-15).

Eczema is the most common inflammatory disorder of the skin. This condition is characterized by inflammation often accompanied by papules (bumps), vesicles (blisters), and crusts. Eczema is not a distinct disease but rather a sign or symptom of an underlying condition. For example, an allergic reaction called *contact dermatitis* can progress to become eczematous. Poison ivy is a form of contact dermatitis—occurring on *contact* with chemicals produced by the poison ivy plant.

Abnormal Body Temperature

Maintenance of a body temperature within a narrow range is necessary for normal functioning of the body. As the figure shows, straying too far out of the normal range of body temperatures can have very serious physiological consequences (Figure 5-16). A few important conditions related to body temperature follow:

1. Fever. A fever or *febrile* state is an unusually high body temperature associated with a systemic inflammation response. In the case of infections, chemicals called *pyrogens* (literally "fire-makers") cause the thermostatic control centers of the hypothalamus to produce a fever. Because the body's "thermostat" is reset to a higher setting, a person feels a need to warm up to this new temperature and often experiences "chills" as the febrile state begins. The high body temperature associated with infectious fever is thought to enhance the body's immune responses, eliminating the pathogen. Strategies aimed at reducing the temperature of a febrile person are normally counteracted by the body's heat-generating mechanisms and have the effect of further weakening the infected person. Under ordinary circumstances, it is best to let the fever "break" on its own after the pathogen is destroyed.

2. Malignant hyperthermia (MH). An inherited condition characterized by abnormally increased body temperature (hyperthermia) and muscle rigidity when exposed to certain anesthetics or muscle relaxants (succinylcholine, for example). The drug *dantrolene (Dantrium)*, which inhibits heat-producing muscle contractions, has been used to prevent or relieve effects of this condition.

3. Heat exhaustion. Occurs when the body loses a large amount of fluid resulting from heat-loss mechanisms. This usually happens when environmental temperatures are high. Although a normal body temperature is maintained, the loss of water and electrolytes can cause weakness, vertigo (dizziness), nausea, and possibly loss of consciousness. Heat exhaustion

may also be accompanied by skeletal muscle cramps that are often called *heat cramps*. Heat exhaustion is treated with rest (in a cool environment) accompanied by fluid replacement.

4. Heat stroke, or *sunstroke*. A severe, sometimes fatal, condition resulting from the inability of the body to maintain a normal temperature in an extremely warm environment. Such thermoregulatory failure may result from factors such as old age, disease, drugs that impair thermoregulation, or simply overwhelming elevated environmental temperatures. Heat stroke is characterized by body temperatures of 41° C (105° F) or higher, tachycardia (rapid heart rate), headache, and hot, dry skin. Confusion, convulsions, or loss of consciousness may occur. Unless the body is cooled and body fluids replaced immediately, death may result.

5. Hypothermia. The inability to maintain a normal body temperature in extremely cold environments. Hypothermia is characterized by body temperatures lower than 35° C (95° F), shallow and slow respirations, and a faint, slow pulse. Hypothermia is usually treated by slowly warming the affected person's body.

6. Frostbite. Local damage to tissues caused by extremely low temperatures. Damage to tissues results from formation of ice crystals accompanied by a reduction in local blood flow. *Necrosis* (tissue death) and even *gangrene* (decay of dead tissue) can result from frostbite.

CASE STUDY
MALIGNANT MELANOMA

Jane M. is a 45-year-old woman who has come to see her nurse practitioner because she is conerned about a mole (nevus) on her arm. Jane has been healthy all her life and has been very active in outdoor activities. She has been a swimmer, and when she was in her teens, she worked many summers as a lifeguard at the local swimming pool. Jane never protected her skin from the sun, as it was considered attractive to tan as deeply as possible. She does remember having blistering sunburns at least twice.

The mole on her arm has been present since her late teens, but lately it has looked different to Jane.

Originally it was flat with a definite border, dark brown in color, and about one half centimeter in diameter. Now it seems to have taken on more than one color—some areas appear to be reddish, and Jane has noted a grayish spot as well. The mole is not painful, but it does itch.

The nurse practitioner refers Jane to a surgeon who excises the mole and several surrounding lymph nodes. The surgical specimens are sent to the laboratory for pathological analysis. The pathology report confirms the diagnosis of malignant melanoma without metastasis to the surrounding lymph nodes.

---

1. Which of the following is thought to be the strongest causative factor in the development of melanoma?
   A. Heredity
   B. Solar radiation
   C. Presence of moles
   D. Location of moles

2. Which of the following characteristics of changes in appearance in a mole are warning signs of possible melanoma?
   A. Color change
   B. Irregular border
   C. Increased size
   D. All of the above

3. Which of the following best explains the pathological mechanism in malignant melanoma?

   A. DNA damage leads to mistakes in mitosis that produce cancer.
   B. Loss of melanin from the cells makes the skin more susceptible to cancer.
   C. Hormonal changes of puberty trigger malignant activity in the cells.
   D. The skin is unable to repair damage, and this predisposes it to cancer.

4. Which of the following locations of lesions has the best prognosis for survival with malignant melanoma?
   A. Head
   B. Neck
   C. Estremities
   D. Trunk

CHAPTER SUMMARY

INTRODUCTION

A. Skin (integument) is body's largest organ
B. Approximately 1.6 to 1.9 m² in average-sized adult
C. Integumentary system describes the skin and its appendages—the hair, nails, and skin glands

SKIN FUNCTIONS

A. Crucial to maintenance of homeostasis
B. Protection (physical barrier to microorganisms)
C. Important role in maintaining body temperature
D. Synthesis of important chemicals (e.g., vitamin D) and hormones
E. Excretion of water, wastes, and salts
F. Absorbs fat-soluble vitamins (A, D, E, K), estrogens, and certain chemicals
G. Receptors allow skin to function as a sense organ (heat, cold, pressure, touch, and pain)
H. Produces melanin (screens out ultraviolet light)
I. Produces keratin for protection (water-repellent)

STRUCTURE OF THE SKIN

A. Skin classified as a cutaneous membrane
B. Two primary layers—epidermis and dermis; joined by dermal-epidermal junction (Figure 5-2)
C. Subcutaneous layer (hypodermis, or superficial fascia) lies beneath dermis
D. Epidermis
  1. Thin outer layer of skin
    a. "Thin skin"—covers most of body surface (1 to 3 mm thick)
    b. "Thick skin"—soles and palms (4 to 5 mm thick)
  2. Cell types
    a. Keratinocytes—constitute over 90% of cells present; principal structural element of the outer skin
    b. Melanocytes—pigment-producing cells (5% of the total); contribute to skin color; filter ultraviolet light
    c. Langerhans' cells—play a role in immune response
  3. Cell layers
    a. Stratum corneum (horny layer)—most superficial layer; dead cells filled with keratin (barrier area)
    b. Stratum lucidum (clear layer)—cells filled with keratin precursor called eleiden; absent in thin skin
    c. Stratum granulosum (granular layer)—cells arranged in two to four layers and filled with keratohyalin granules; contain high levels of lysosomal enzymes
    d. Stratum spinosum (spiny layer)—cells arranged in eight to ten layers with prominent desmosomes; cells rich in RNA
    e. Stratum basale (base layer)—single layer of columnar cells; only these cells undergo mitosis and then migrate through the other layers until they are shed
    f. Stratum germinativum (growth layer)—describes the stratum spinosum and stratum basale together
E. Epidermal growth and repair
  1. Turnover or regeneration time refers to time required for epidermal cells to form in the stratum basale and migrate to the skin surface—about 35 days
  2. Shortened turnover time will increase the thickness of the stratum corneum and result in callus formation
  3. Normally 10% to 12% of all cells in stratum basale enter mitosis daily
  4. Each group of eight to ten basal cells in mitosis with their vertical columns of migrating keratinocytes is called an epidermal proliferating unit, or EPU
F. Dermal-epidermal junction
  1. A definite basement membrane, specialized fibrous

elements, and a polysaccharide gel serve to "glue" the epidermis to the dermis below
  2. The junction serves as a partial barrier to the passage of some cells and large molecules
G. Dermis
  1. Called "true skin"—much thicker than the epidermis and lies beneath it
  2. Gives strength to the skin
  3. Serves as a reservoir storage area for water and electrolytes
  4. Contains specialized sensory receptors (Figure 5-3), muscle fibers, hair follicles, sweat and sebaceous glands, and many blood vessels
  5. Rich vascular supply plays a critical role in temperature regulation
  6. Layers of dermis
    a. Papillary layer—composed of dermal papillae that project into the epidermis; contains fine collagenous and elastic fibers; contains the dermal-epidermal junction; forms a unique pattern that gives individual fingerprints
    b. Reticular layer—contains dense, interlacing white collagenous fibers and elastic fibers to make the skin tough yet stretchable; when processed from animal skin produces leather
H. Dermal growth and repair
  1. The dermis does not continually shed and regenerate itself as does the epidermis
  2. During wound healing, the fibroblasts begin forming an unusually dense mass of new connective fibers; if not replaced by normal tissue, this mass remains a scar
  3. Langer lines (cleavage lines) (Figure 5-4)—patterns formed by the collagenous fibers of the reticular layer of the dermis

SKIN COLOR

A. Basic determinant is quantity of melanin
B. Melanin formed in melanocytes from tyrosine
C. Albinism—congenital absence of melanin
D. Process regulated by tyrosinase, exposure to sunlight, and hormones MSH and ACTH
E. Carotene (yellowish color) can also contribute to skin color
F. Color changes also occur as a result of changes in blood flow to skin and circulating levels of unoxygenated hemoglobin

APPENDAGES OF THE SKIN

A. Hair (Figure 5-6)
  1. Distribution—over entire body except palms of hands and soles of feet and a few other small areas
  2. Fine and soft hair coat existing before birth called lanugo
  3. Coarse pubic and axillary hair that develops at puberty called terminal hair
  4. Hair follicles and hair develop from epidermis; stratum germinativum forms innermost layer of follicle and germinal matrix; mitosis of cells of germinal matrix forms hairs
  5. Papilla—cluster of capillaries under germinal matrix
  6. Root—part of hair embedded in follicle in dermis
  7. Shaft—visible part of hair
  8. Medulla—inner core of hair; cortex—outer portion
  9. Color—result of different amounts of melanin in cortex of hair
  10. Growth—hair growth and rest periods alternate; hair on head averages 5 inches of growth per year

11. Sebaceous glands—attach to and secrete sebum into follicle
12. Male pattern baldness results from combination of genetic tendency and male sex hormones (Figure 5-7)

B. Nails (Figure 5-8)
   1. Consist of epidermal cells converted to hard keratin
   2. Nail body—visible part of each nail
   3. Root—part of nail in groove hidden by fold of skin, the cuticle
   4. Lunula—moon-shaped white area nearest root
   5. Nail bed—layer of epithelium under nail body; contains abundant blood vessels
   6. Growth—nails grow by mitosis of cells in stratum germinativum beneath the lunula; average growth about 0.5 mm per week, or slightly over 1 inch per year

C. Skin glands (Figure 5-9)
   1. There are two types of sweat glands
      a. Eccrine glands
         (1) Most numerous sweat glands; quite small
         (2) Distributed over total body surface with exception of a few small areas
         (3) Simple, coiled, tubular glands
         (4) Function throughout life
         (5) Secrete perspiration or sweat; eliminate wastes and help maintain a constant core temperature
      b. Apocrine glands
         (1) Located deep in subcutaneous layer
         (2) Limited distribution—axilla, areola of breast, and around anus
         (3) Large in size (often over 5 mm in diameter)
         (4) Simple, branched, tubular glands
         (5) Begin to function at puberty
         (6) Secretion shows cyclic changes in female with menstrual cycle
   2. Sebaceous glands
      a. Secrete sebum—oily substance that keeps hair and skin soft and pliant; prevents excessive water loss from the skin
      b. Lipid components have antifungal activity
      c. Simple, branched glands
      d. Found in dermis except in palms and soles
      e. Secretion increases in adolescence; may lead to formation of pimples and blackheads
   3. Ceruminous glands
      a. Modified apocrine sweat glands
      b. Simple, coiled, tubular glands
      c. Empty contents into external ear canal either alone or with sebaceous glands
      d. Mixed secretions of sebaceous and ceruminous glands called cerumen (wax)
      e. Function of cerumen to protect area from dehydration; excess secretion can cause blockage of ear canal and loss of hearing

SURFACE FILM

A. Emulsified protective barrier formed by mixing of residue and secretions of sweat and sebaceous glands with sloughed epithelial cells from skin surface; shedding of epithelial elements is called desquamation
B. Functions
   1. Antibacterial, antifungal activity
   2. Lubrication
   3. Hydration of skin surface
   4. Buffer of caustic irritants
   5. Blockade of toxic agents
C. Chemical composition
   1. From epithelial elements—amino acids, sterols, and complex phospholipids
   2. From sebum—fatty acids, triglycerides, and waxes
   3. From sweat—water and ammonia, urea, and lactic acid and uric acid

HOMEOSTASIS OF BODY TEMPERATURE

A. To maintain homeostasis of body temperature, heat production must equal heat loss; skin plays a critical role in this process
B. Heat production
   1. By metabolism of foods in skeletal muscles and liver
   2. Chief determinant of heat production is the amount of muscular work being performed
C. Heat loss—approximately 80% of heat loss occurs through the skin; remaining 20% occurs through the mucosa of the respiratory, digestive, and urinary tracts (Figure 5-10)
   1. Evaporation—to evaporate any fluid, heat energy must be expended; this method of heat loss is especially important at high environmental temperatures when it is the only method by which heat can be lost from the skin
   2. Radiation—transfer of heat from one object to another without actual contact; important method of heat loss in cool environmental temperatures
   3. Conduction—transfer of heat to any substance actually in contact with the body; accounts for relatively small amounts of heat loss
   4. Convection—transfer of heat away from a surface by movement of air; usually accounts for a small amount of heat loss
D. Homeostatic regulation of heat loss (Figure 5-11)
   1. Heat loss by the skin is controlled by a negative-feedback loop
   2. Receptors in the hypothalamus monitor the body's internal temperature
   3. If the body temperature is increased, the hypothalamus sends a nervous signal to the sweat glands and blood vessels of the skin
   4. The hypothalamus continues to act until the body's temperature returns to normal

BURNS

A. Defined as injury or death to skin cells caused by heat, ultraviolet light, electric current, or corrosive chemicals
B. Severity of burn injury is determined by depth of lesion and percent of body surface burned
C. Estimating body surface area
   1. "Rule of palms"—based on assumption that palm size of burn victim is about 1% of body surface, therefore estimating the number of "palms" burned will approximate the percentage of body surface involved
   2. "Rule of nines"—9% of total skin area covers head and each upper extremity, including front and back surfaces, 18% of total skin area covers each of the following: front of trunk, back of trunk, and each lower extremity, including front and back surfaces (Figure 5-12)
   3. Lund-Browder charts—make allowances for the large percent of surface area in certain body regions in children (such as the head); permit more accurate estimates of burned surface area in children
D. First-degree burn
   1. Minor pain—no real tissue destruction
   2. Some reddening of the skin
   3. No blistering, but some peeling of surface occurs
   4. No scarring
E. Second-degree burn
   1. Severe pain
   2. Damage or destruction of epidermis and upper dermal layers
   3. Blisters form with swelling and edema
   4. Dermal tissue death not complete but scarring common

F. Third-degree burn
  1. Total destruction of both epidermis and upper dermal layers
  2. Tissue death extends below level of hair follicles and sweat glands
  3. No immediate pain; nerve endings are destroyed
  4. Burning may involve deep tissues, including muscle and bone
  5. Scarring is a serious problem
G. Classification of burns by depth of injury
  1. Partial-thickness burn—includes both first- and second-degree burns
  2. Full-thickness burn—refers to third-degree burns only

CYCLE OF LIFE: SKIN

A. Children
  1. Skin is smooth, unwrinkled, and characterized by elasticity and flexibility
  2. Few sweat glands
  3. Rapid healing
B. Adults
  1. Development and activation of sebaceous and sweat glands
  2. Increased sweat production
    a. Body odor
  3. Increased sebum production
    a. Acne
C. Old age
  1. Decreased sebaceous and sweat gland activity
    a. Wrinkling
    b. Decrease of body's ability to cool itself

THE BIG PICTURE: SKIN AND THE WHOLE BODY

A. Skin is a major component of the body's structural framework
B. Skin defines the internal environment of the body
C. Primary functions are support and protection

MECHANISMS OF DISEASE

A. Skin disorders (dermatosis)
  1. Dermatitis—inflammation of the skin
  2. Skin infections
    a. Adults—antimicrobial characteristics of sebum inhibit infection
    b. Children—lack of sebum makes skin less resistant to infection
    c. Examples of skin infections
      (1) Impetigo
      (2) Tinea—fungal infections (mycoses) (Figure 5-13)
      (3) Warts
      (4) Boils (furuncles)
  3. Vascular and inflammatory skin disorders
    a. Decubitus ulcers (bedsores) (Figure 5-14)
    b. Urticaria, or hives, characterized by wheals
    c. Scleroderma
    d. Psoriasis (Figure 5-15)
    e. Eczema—most common inflammatory skin disorder
B. Abnormal body temperature
  1. Maintenance of temperature within a small range is necessary for normal function (Figure 5-16)
    a. Fever—usually high body temperature associated with a systemic inflammation response
    b. Malignant hypothermia—inherited condition characterized by hypothermia and muscle rigidity
    c. Heat exhaustion—loss of fluid caused by high environmental temperatures
    d. Heat stroke (sunstroke)—severe condition resulting from the body's inability to maintain normal temperature in extremely warm environment
    e. Hypothermia—inability to maintain body temperature in extremely cold environments
    f. Frostbite—damage to tissue caused by extremely cold temperatures

REVIEW QUESTIONS

1. How do the terms *integument* and *integumentary system* differ in meaning?
2. List and briefly discuss several of the different functions of the skin.
3. Identify the two main layers of the skin from superficial to deep. How are these layers related to the dermal-epidermal junction and the subcutaneous layer?
4. List three cell types found in the epidermis.
5. List and describe the cell layers of the epidermis from superficial to deep.
6. What layer of the epidermis is sometimes called the barrier area?
7. Discuss the process of epidermal growth and repair.
8. What is keratin? Where is it found and how is it formed?
9. What part of the skin contains blood vessels?
10. What is the dermal-epidermal junction and how does it function?
11. Why is the process of blister formation a good example of the relationship between the skin's structure and function?
12. List the two layers of the dermis. Which layer helps make the skin stretchable and able to rebound?
13. What are arrector pili muscles?
14. What are Langer's cleavage lines? Why are these lines important to surgeons?
15. What is melanin? Where is it found in the skin?
16. List the appendages of the skin.
17. Identify each of the following: hair papilla, germinal matrix, hair root, hair shaft, follicle.
18. List the three primary types of skin glands.
19. What is the difference between eccrine and apocrine sweat glands?
20. Discuss the importance of the surface film of the skin.
21. What is the chemical composition of skin surface film? How is the chemical composition related to its protective function?
22. How is heat lost from the body? Discuss the homeostatic regulation of heat loss.
23. How is the "rule of nines" used in determining the extent of a burn injury?
24. What are the differences between first-, second-, and third-degree burns?

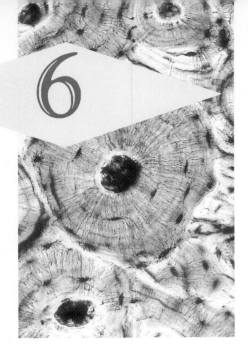

Dried, ground bone

6 Skeletal Tissues

OBJECTIVES

After you have completed this chapter, you should be able to:

1. List the four types of bones and give examples of each.
2. Identify the six major structures of a typical long bone.
3. Identify each of the major constituents of bone as a tissue and discuss how structural organization contributes to function.
4. Identify by name and discuss each of the major components of a haversian system.
5. List and describe the function of the three major types of cells found in bones.
6. List and discuss the five homeostatic functions of bones.
7. Compare and contrast the development of intramembranous and endochondral bone.
8. Describe steps involved in bone fracture repair.
9. Compare the basic structural units of bone and cartilage.
10. Identify the three specialized types of cartilage, give examples of each, and summarize the structural and functional differences between them.
11. Compare the mechanism of growth in bone and cartilage.

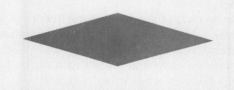

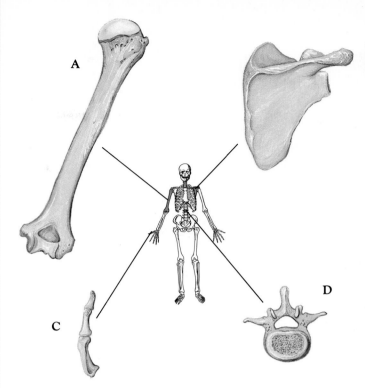

FIGURE 6-1 **Types of bones.** Examples of bone types include **A,** long bones (humerus); **B,** flat bones (scapula); **C,** short bones (phalanx); and **D,** irregular bones (vertebra).

This chapter focuses on two highly specialized types of connective tissues—**bone** and **cartilage.** In addition, the functional characteristics of cartilage and a comparison of cartilage and bone will be discussed later in the chapter.

Other types of tissue in the skeletal system include fibrous and loose connective tissue, blood, nervous tissue, epithelium, lymphatic tissue, myeloid tissue (bone marrow), and adipose, or fat, tissue.

Individual bones, which are considered separate, discrete organs will be discussed in Chapter 7. *Articulations,* or joints, are points of contact between bones that make movement possible and will be considered in Chapter 8.

TYPES OF BONES

Structurally, we can name four types of bones. Their names suggest their shapes: long bones, short bones, flat bones, and irregular bones. Figure 6-1 gives an illustration of each type. Bones serve differing needs and their size, shape, and appearance will vary to meet those needs. For example, some bones must bear great weight, whereas others serve a protective function or serve as delicate support structures for the fingers and toes. Bones not only differ in size and shape but also in the amount and proportion of two different types of bone tissue that comprise them. **Compact bone** is dense and "solid" in appearance, whereas **cancellous,** or **spongy, bone** is characterized by open space partially filled by an assemblage of needlelike structures. Both types will be discussed when the microscopic structure of bone is described later in the chapter.

QUICK CHECK

✔ *1. Name the two major types of connective tissue found in the skeletal system.*
✔ *2. Name the two different types of bone tissue.*

Long Bones

A long bone consists of the following structures visible to the naked eye: diaphysis, epiphyses, articular cartilage, periosteum, medullary (marrow) cavity, and endosteum. Identify each part in Figure 6-2 as you read about it.

1. Diaphysis. Main shaftlike portion. Its hollow, cylindrical shape and the thick compact bone that composes it adapt the diaphysis well to its function of providing strong support without cumbersome weight.

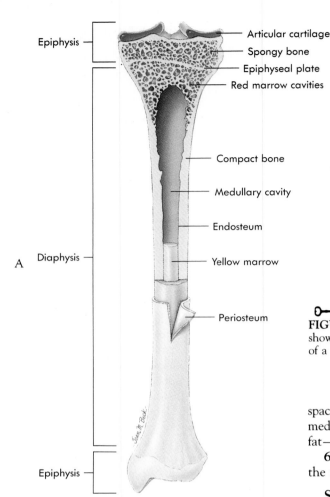

Epiphysis

Articular cartilage
Spongy bone
Epiphyseal plate
Red marrow cavities

Compact bone

Medullary cavity

Endosteum

A   Diaphysis

Yellow marrow

Periosteum

Epiphysis

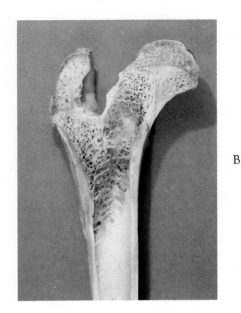

B

FIGURE 6-2   Long bone. A, Longitudinal section of long bone showing both cancellous and compact bone. **B,** Cutaway section of a long bone.

space in the diaphysis of a long bone. In the adult, the medullary cavity is filled with connective tissue rich in fat—a substance called *yellow marrow.*

6. Endosteum. A thin epithelial membrane that lines the medullary cavity of long bones.

Short, Flat, and Irregular Bones

Short bones, flat bones, and irregular bones all have an inner portion of cancellous bone covered over on the outside with compact bone. Red marrow fills the spaces in the cancellous bone inside a few irregular and flat bones—for example, in the vertebrae and sternum. To help in the diagnosis of leukemia and certain other diseases, a physician may decide to perform a needle puncture of one of these bones. In this type of diagnostic procedure, a needle is inserted through the skin and compact bone into the red marrow, and a small amount of the marrow is then aspirated and examined under the microscope for normal or abnormal blood cells.

◢ Bone Tissue

Bone (osseous) tissue is perhaps the most distinctive form of connective tissue in the body. It is typical of other connective tissues in that it consists of cells, fibers, and extracellular material, or *matrix.* However, its extracellular components are hard and *calcified.* In bone the extracellular material, or matrix, predominates. It is much more abundant than the bone cells, and it contains many fibers of collagen (the body's most abundant protein). The rigidity of bone enables it to serve supportive and protective functions.

As a tissue, bone is ideally suited to its functions, and

2. Epiphyses. Both ends of a long bone. Epiphyses have a bulbous shape that provides generous space near joints for muscle attachments and also gives stability to joints. Look at Figure 6-2 to note the innumerable small spaces in the bone of the epiphysis. They make this kind of bone look a little like a sponge—hence its name, spongy, or cancellous, bone. A specialized type of soft connective tissue, called **red marrow,** fills the spaces within this spongy bone. Early in development, epiphyses are separated from the diaphysis by a layer of cartilage, the *epiphyseal plate.*

3. Articular cartilage. Thin layer of hyaline cartilage that covers articular or joint surfaces of epiphyses. Resiliency of this material cushions jars and blows.

4. Periosteum. Dense, white fibrous membrane that covers bone except at joint surfaces, where articular cartilage forms the covering. Many of the periosteum's fibers penetrate the underlying bone, welding these two structures to each other.

Muscle tendon fibers interlace with periosteal fibers, thereby anchoring muscles firmly to bone. The periosteum contains many small blood vessels that send branches into the bone. This important membrane is essential for bone cell survival and for bone formation, a process that continues throughout life.

5. Medullary (or marrow) cavity. A tubelike hollow

Bone-seeking Isotopes

The disastrous nuclear accident that occurred in Chernobyl in the former Soviet Union early in 1986 released large quantities of what are described as **bone-seeking isotopes** into the environment. Nuclear reactors generate numerous radioactive elements in the fission of uranium or plutonium. Radioactive strontium is one of the most hazardous bone-seeking isotopes produced in nuclear fission reactors. Once ingested, it will quickly substitute for calcium in the apatite crystals of bone and damage the red marrow and other body tissues by radioactive emissions.

the concept that structure and function are interrelated is apparent in this highly specialized tissue. It has a tensile strength nearly equal to cast iron but at less than one third the weight. Bone is organized so that its great strength and minimal weight result from the interrelationships of its structural components. The relationship of structure to function is apparent in its chemical, cellular, tissue, and organ levels of organization.

Composition of Bone Matrix

The extracellular **bone matrix** can be subdivided into two principal chemical components: *inorganic salts* and *organic matrix*.

Inorganic salts

The calcified nature and thus the hardness of bone result from the deposition of highly specialized chemical crystals of calcium and phosphate, called *apatite*. The needlelike apatite crystals are about 300 Å in length by 20 Å in thickness. They are oriented in the organic components of the bone so that they can most effectively resist stress and mechanical deformation. In addition to calcium and phosphate, other mineral constituents such as magnesium and sodium are also found in bone.

Organic matrix

The organic matrix of bone is a composite of collagenous fibers and an amorphous mixture of protein and polysaccharides called *ground substance*. Connective tissue cells secrete the gel-like and homogeneous ground substance that surrounds the fibers found in bone matrix.

The organic components not only add to overall strength but also give bone some degree of plastic-like resilience so that applied stress—within reasonable limits—does not result in frequent crush or fracture injuries.

QUICK CHECK

✔ 1. *List the six structural components of a typical long bone that are visible to the naked eye.*

✔ 2. *Identify the two principal chemical components of bone matrix.*

MICROSCOPIC STRUCTURE OF BONE

The basic structural components and cell types of bone were described briefly in Chapter 4. In the paragraphs that follow, additional information about bone structure and cell types will serve as a basis for learning the functional characteristics of this important tissue. Understanding how a bone forms and grows, how it repairs itself following injury, and how it interacts with other tissues and organs in maintaining various important homeostatic mechanisms is based on a knowledge of its basic structure—a structure as unique as its chemical composition.

Compact Bone

Compact bone contains many cylinder-shaped structural units called **osteons,** or **haversian systems** (in honor of Clopton Havers, a seventeenth century English anatomist who first described them). Note in Figure 6-3 that each osteon surrounds a canal that runs lengthwise through the bone. Living bone cells in these units are literally cemented together to constitute the structural framework of compact bone. The unique structure of the osteon permits delivery of nutrients and removal of waste products from metabolically active but imprisoned bone cells.

Four types of structures make up each osteon, or haversian system: lamellae, lacunae, canaliculi, and a haversian canal. As you read the following definitions, identify each structure in Figure 6-3.

Lamellae. Concentric, cylinder-shaped layers of calcified matrix.

Lacunae (Latin for "little lakes"). Small spaces containing tissue fluid in which bone cells lie imprisoned between the hard layers of the lamellae.

Canaliculi. Ultrasmall canals radiating in all directions from the lacunae and connecting them to each other and into a larger canal, the haversian canal.

Haversian canal. Extends lengthwise through the center of each haversian system; contains blood vessels, lymphatic vessels, and nerves from the haversian canal; nutrients and oxygen move through canaliculi to the lacunae and their bone cells—a short distance of about 0.1 mm or less.

Lengthwise running haversian canals are connected to each other by transverse **Volkmann's canals.** These communicating canals contain nerves and vessels that carry blood and lymph from the exterior surface of the bone to the osteons.

Cancellous Bone

Cancellous, or spongy, bone differs in microscopic structure from compact bone. As you recall, the structural unit of compact bone is the highly organized osteon, or haversian system. There are no osteons in cancellous bone. Instead, it consists of needlelike bony spicules called **trabeculae.** Bone cells are found within the trabeculae. Nutrients are delivered to the cells and waste products are removed by diffusion through tiny canaliculi that extend to the surface of the very thin spicules. The scanning electron micrograph of cancellous bone shown in Figure 6-4

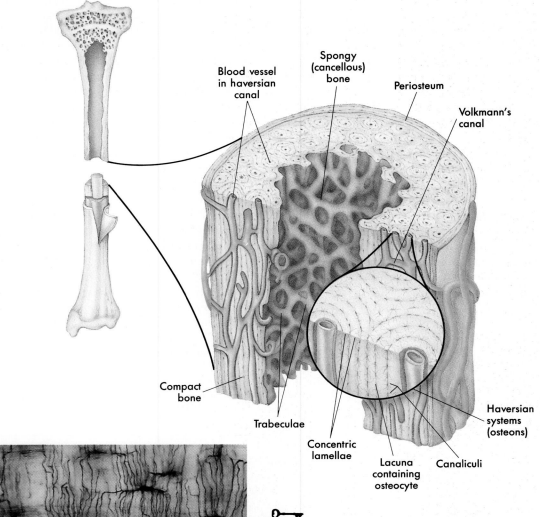

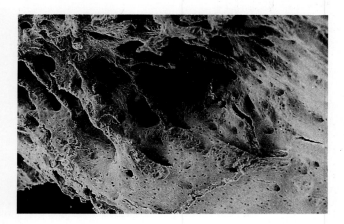

FIGURE 6-3 Microscopic structure of bone. Longitudinal section of a long bone cut away to show microscopic structure of compact bone. Note that the hard shell of the bone is constructed of cylindrical units called *haversian systems*. Spongy bone is constructed of bony projections called *trabeculae*. Inset shows a highly magnified haversian system where *lacunae* joined by numerous *canaliculi* can be easily identified. (*Inset, Courtesy Erlandsen/Magney: Color Atlas of Histology.*)

FIGURE 6-4 Trabeculae. Scanning electron micrograph (SEM) of trabeculae in cancellous bone shows the openings of many canaliculi on the surface of the trabeculae. (Courtesy Erlandsen/Magney: Color Atlas of Histology.)

illustrates the many openings of canaliculi on the surface of trabeculae.

The placement of trabeculae in the spongy bone shown in Figure 6-4 is not as random and unorganized as it might first appear. The bony spicules are actually arranged along lines of stress, and their orientation will therefore differ between individual bones according to the nature and magnitude of the applied load. This feature greatly enhances a bone's strength and is yet another example of the relationship between structure and function.

Locked within a seemingly lifeless calcified matrix, bone cells are active metabolically. They must, like all living cells, continually receive food and oxygen and excrete their wastes. So blood supply to bone is both important and abundant. One or more arteries supply the bone marrow in the internal medullary cavity and provide nutrients to areas of cancellous bone. In addition, blood vessels from the periosteum penetrate bone and then, by way of Volkmann's canals, connect with vessels in the haversian canals, ultimately serving the needs of cells in compact bone.

QUICK CHECK

✔ 1. Identify the four structures that form the osteon, or haversian system.

✔ 2. Name the transverse canals that connect blood vessels between adjacent haversian systems.

✔ 3. Name the needlelike spicules present in cancellous bone.

Types of Bone Cells

Three major types of cells are found in bone: *osteoblasts* (bone-forming cells), *osteoclasts* (bone-reabsorbing cells), and *osteocytes* (mature bone cells). All bone surfaces are covered with a continuous layer of cells that is critical to the survival of bone. This layer is composed of relatively large numbers of osteoblasts interspersed with a much smaller population of osteoclasts. **Osteoblasts** are small

cells that synthesize and secrete a specialized organic matrix, called *osteoid*, that is an important part of the ground substance of bone. Collagen fibrils line up in regular arrays in the osteoid and serve as a framework for the deposition of calcium and phosphate. The process ultimately results in accumulation of mineralized bone.

Osteoclasts are giant multinucleate cells (Figure 6-5) that are responsible for the active erosion of bone minerals. They are formed by fusion of several precursor cells and contain large numbers of mitochondria and lysosomes. Each 24-hour day sees an alternation of primarily osteoblast, then osteoclast, activity. For this reason, bone is a highly active, dynamic tissue that undergoes continuous change and remodeling.

Osteocytes are mature, nondividing osteoblasts that have become surrounded by matrix and now lie within lacunae. Figure 6-6 is a scanning electron micrograph showing a mature osteocyte within a lacuna. Note that a cytoplasmic process from the cell is extending into a canaliculus below. Numerous collagen fibers are seen in the ground substance and mineralized bone surrounding the osteocyte.

The way in which these cell types work together to produce bone will be described in detail when the development of bone is discussed later in this chapter.

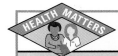

BONE MARROW TRANSPLANTS

Red marrow can be removed for use in **marrow transplants.** In certain types of cancer treatments or following accidental radiation poisoning, the red marrow is destroyed and must be replaced for survival. Highly specialized techniques are required to remove large quantities of living marrow from a donor and then to successfully reinject this delicate tissue into a recipient without destroying the tissue's ability to function.

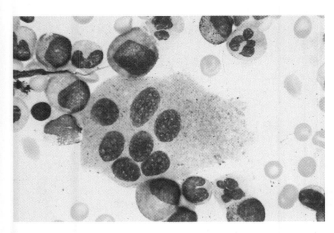

FIGURE 6-5   Osteoclast in bone marrow. Osteoclasts are giant multinucleate cells responsible for erosion of bone. (Courtesy Erlandsen/Magney: Color Atlas of Histology.)

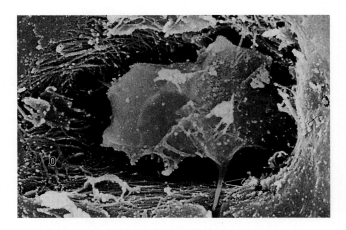

FIGURE 6-6   Osteocyte. Scanning electron micrograph (SEM) showing an osteocyte within a lacuna. Note the cytoplasmic process (*arrowhead*) extending into canaliculi below. The cell is surrounded by collagen fibers and mineralized bone. (Courtesy Erlandsen/Magney: Color Atlas of Histology.)

BONE MARROW

Bone marrow is a specialized type of soft, diffuse connective tissue called *myeloid tissue*. It serves as the site for production of blood cells and is found in the medullary cavities of long bones and in the spaces of spongy bone.

During the lifetime of an individual, two types of marrow occur. In an infant's or child's body, virtually all of the bones contain *red marrow*. It is named for its function in the production of red blood cells. As an individual ages, the red marrow is gradually replaced by *yellow marrow*. In yellow marrow, the marrow cells have become saturated with fat and, as a result, are inactive in blood cell production.

The main bones in an adult that still contain red marrow include the ribs, bodies of the vertebrae and the ends of the humerus in the upper arm, and the femur, or thigh, bone. During times of a decreased blood supply, yellow marrow in an adult can alter to become red marrow. Such a transition may occur during periods of prolonged anemia caused by chronic blood loss, exposure to radiation or toxic chemicals, and certain diseases.

FUNCTIONS OF BONE

Bones perform five functions for the body. Each one is important for maintaining homeostasis and for optimal body function.

1. Support. Bones serve as the supporting framework of the body, much as steel girders are the supporting framework of our modern buildings. They contribute to the shape, alignment, and positioning of the body parts.

2. Protection. Hard, bony "boxes" serve to protect the delicate structures they enclose. For example, the skull protects the brain, and the rib cage protects the lungs and the heart.

3. Movement. Bones with their joints constitute levers. Muscles are anchored firmly to bones. As muscles contract and shorten, they pull on bones, thereby producing movement at a joint. This process will be discussed further in Chapter 8.

4. Mineral storage. Bones serve as the major reservoir for calcium, phosphorus, and certain other minerals. Homeostasis of blood calcium concentration—essential for healthy survival—depends largely on changes in the rate of calcium movement between the blood and bones. If, for example, blood calcium concentration increases above normal, calcium moves more rapidly out of the blood into bones and more slowly in the opposite direction. Result? Blood calcium concentration decreases—usually to its homeostatic level.

5. Hematopoiesis. *Hematopoiesis*, or blood cell formation, is a vital process carried on by red bone marrow, or *myeloid tissue*. Myeloid tissue, in the adult, is located primarily in the ends or epiphyses of certain long bones, in the flat bones of the skull, in the pelvis, and in the sternum and ribs.

DEVELOPMENT OF BONE

When the skeleton begins to form in a baby before its birth, it consists not of bones but of cartilage and fibrous structures shaped like bones. Gradually these cartilage "models" become transformed into real bones when the cartilage is replaced with calcified bone matrix. This process of constantly "remodeling" a growing bone as it changes from a small cartilage model to the characteristic shape and proportion of the adult bone requires continuous activity by the bone-forming osteoblasts and bone-resorbing osteoclasts. The laying down of calcium salts in the gel-like matrix of the forming bones is an ongoing process. This calcification process is what makes bones as "hard as bone." The combined action of the osteoblasts and osteoclasts sculpts bones into their adult shapes. The term **osteogenesis** is used to describe this process.

"Sculpting" by the bone-forming and bone-resorbing cells allows bones to respond to stress or injury by changing size, shape, and density. The stresses placed on certain bones during exercise increase the rate of bone deposition. For this reason, athletes or dancers may have denser, stronger bones than less active people.

Most bones of the body are formed from cartilage models in a process called **endochondral ossification**, meaning "formed in cartilage." A few flat bones are formed within fibrous membrane, rather than cartilage, in a process known as **intramembranous ossification.**

Intramembranous Ossification

Intramembranous ossification takes place, as its name implies, within a connective tissue membrane. The flat bones of the skull, for example, begin to take shape when groups of cells within the membrane differentiate into **osteoblasts.** These clusters of osteoblasts are called *centers of ossification.* They secrete matrix material and collagenous fibrils. The Golgi apparatus in an osteoblast specializes in synthesizing and secreting carbohydrate compounds of the type called *mucopolysaccharides,* and its endoplasmic reticulum makes and secretes collagen, a protein. In time, relatively large amounts of the mucopolysaccharide substance, or *ground substance,* accumulate around each individual osteoblast. Numerous bundles of collagenous fibers then become imbedded in the ground substance. Together, the ground substance and collagenous fibers constitute the organic bone matrix. Calcification of the organic bone matrix occurs when complex calcium salts are deposited in it.

As calcification of bone matrix continues, the **trabeculae** appear and join in a network to form **spongy bone.** In time the core layer of spongy bone will be covered on each side by plates of compact, or dense, bone. Once formed, a flat bone grows in size by the addition of osseous tissue to its outer surface. The process is called *appositional growth.* Flat bones cannot grow by interior expansion as is the case with endochondral bone growth described in the following section.

Endochondral Ossification

Most bones of the body are formed from cartilage models, with bone formation spreading essentially from the center to the ends. The steps of endochondral ossification are illustrated in Figure 6-7. The cartilage model of a typical long bone, such as the tibia, can be identified early in embryonic life (Figure 6-7, A). The cartilage model then develops a periosteum (Figure 6-7, B) that soon enlarges and produces a ring, or collar, of bone. Bone is deposited by osteoblasts, which differentiate from cells on the inner surface of the covering periosteum. Soon after the appearance of the ring of bone, the cartilage begins to calcify (Figure 6-7, C) and a **primary ossification center** forms when a blood vessel enters the rapidly changing cartilage model at the midpoint of the diaphysis (Figure 6-7, D). Endochondral ossification progresses from the diaphysis toward each epiphysis (Figure 6-7, E) and the bone grows in length. Eventually, **secondary ossification centers** appear in the epiphyses (Figure 6-7, F), and bone growth proceeds toward the diaphysis from each end (Figure 6-7, G).

Until bone growth in length is complete a layer of the cartilage, known as the **epiphyseal plate,** remains between each epiphysis and the diaphysis. During periods of growth, proliferation of epiphyseal cartilage cells brings about a thickening of this layer. Ossification of the additional cartilage nearest the diaphysis then follows—that is, osteoblasts synthesize organic bone matrix, and the matrix undergoes calcification. As a result, the bone becomes longer. It is the epiphyseal plate that allows the diaphysis of a long bone to increase in length.

The epiphyseal plate shown in Figure 6-8 is composed of four layers of cells. The top layer of cells closest to the epiphysis is composed of "resting" cartilage cells. These cells are not proliferating or undergoing change. This layer serves as a point of attachment firmly joining the outer end, or epiphysis, of a bone to the shaft.

The second layer of cells shown in figure 6-8 is called the *zone of proliferation.* It is composed of cartilage cells, which are undergoing active mitosis. As a result of mitotic division and increased cellular activity the layer thickens and the plate as a whole increases in length.

The third layer of cells, called the *zone of hypertrophy,* is composed of older, enlarged cells, which are undergoing degenerative changes associated with calcium deposition.

The fourth layer, closest to the diaphysis, is called the *zone of calcification.* It is a thin layer composed of dead or dying cartilage cells undergoing rapid calcification. As the process of calcification progresses, this layer becomes fragile and disintegrates. The resulting space is soon filled with new bone tissue, and the bone as a whole grows in length.

When epiphyseal cartilage cells stop multiplying and the cartilage has become completely ossified, bone growth has ended. Radiographs can reveal any epiphyseal cartilage still present. When bones have grown their full length, the epiphyseal cartilage disappears. Bone has replaced it and is

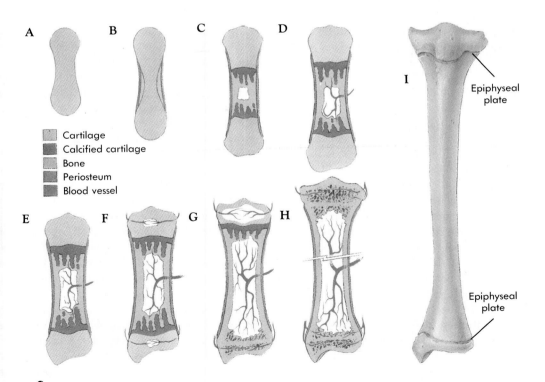

□ Cartilage
■ Calcified cartilage
■ Bone
■ Periosteum
■ Blood vessel

FIGURE 6-7 Endochondral bone formation. A, Cartilage model. **B,** Subperiosteal bone collar formation. **C,** Development of primary ossification center. **D,** Entrance of blood vessel. **E,** Prominent marrow cavity, with thickening and lengthening of collar. **F,** Development of secondary ossification centers in epiphyseal cartilage. **G,** With cessation of bone growth, lower, then upper, epiphyseal plates disappear. **H,** Appearance of mature bone showing continuous marrow cavity and residual epiphyseal line. **I,** External view of epiphyseal plates on a juvenile tibia bone.

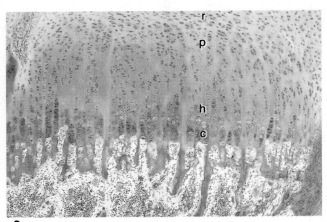

FIGURE 6-8 **Epiphyseal plate.** Note the zone of resting cartilage (r) seen at top and the marrow cavity seen at bottom of the micrograph. Calcified cartilage matrix with thin depositions of newly formed bone on their surfaces projects into the marrow cavity (zones of, p, proliferation; h, hypertrophy; c, calcification). (Courtesy Erlandsen/Magney: Color Atlas of Histology.)

then continuous between epiphysis and diaphysis. The point of articulation between the epiphysis and diaphysis of a growing long bone, however, is susceptible to injury if overstressed—especially in a young child or preadolescent athlete. In these individuals, the epiphyseal plate can be separated from the diaphysis or epiphysis, causing an **epiphyseal fracture** (Figure 6-9).

QUICK CHECK

✔ 1. *Name the three major types of bone cells.*
✔ 2. *Name the two types of bone marrow.*
✔ 3. *What are the five functions of bone?*
✔ 4. *Identify the two types of bone formation.*

BONE GROWTH AND RESORPTION

Bones grow in diameter by the combined action of two of the three bone cell types: osteoblasts and osteoclasts. Osteoclasts enlarge the diameter of the medullary cavity by eating away the bone of its walls. At the same time, osteoblasts from the periosteum build new bone around the outside of the bone. By this dual process a bone with a larger diameter and larger medullary cavity has been produced from a smaller bone with a smaller medullary cavity.

The formation of bone tissue continues long after bones have stopped growing in size. Throughout life, bone formation (ossification) and bone destruction (resorption) go on concurrently. These opposing processes balance each other during adulthood's early to middle years. The rate of bone formation equals the rate of bone destruction. Bones, therefore, neither grow nor shrink. They stay constant in size. Not so, in the earlier years. During childhood and adolescence, ossification occurs at a faster rate than bone

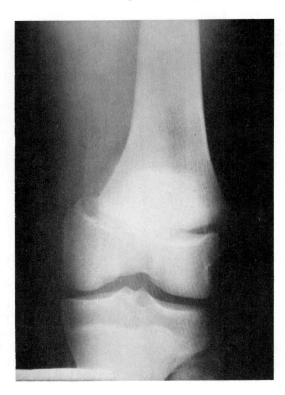

FIGURE 6-9 **Epiphyseal fracture.** X-ray shows epiphyseal fracture of the distal femur in a young athlete. Note the separation of the diaphysis and epiphysis at the level of the growth plate.

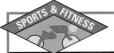

SPORTS & FITNESS

EXERCISE AND BONE DENSITY

Walking, jogging, and other forms of exercise subject bones to stress. They respond by laying down more collagen fibers and mineral salts in the bone matrix. This, in turn, makes bones stronger. But inactivity and lack of exercise tend to weaken bones because of decreased collagen formation and excessive calcium withdrawal. To prevent these changes, as well as many others, astronauts perform special exercises regularly in space.

---

resorption. Bone gain outstrips bone loss, and bones grow larger. But between the ages of 35 and 40 years, the process reverses, and from that time on bone loss exceeds bone gain. Bone gain occurs slowly at the outer, or periosteal, surfaces of bones. Bone loss, on the other hand, occurs at the inner, or endosteal, surfaces and occurs at a somewhat faster pace. More bone is lost on the inside than gained on the outside, and inevitably bones become remodeled as the years go by.

REPAIR OF BONE FRACTURES

The term *fracture* is defined as a break in the continuity of a bone. Types of bone fractures are discussed in Chapter 7, p. 196. *Fracture healing* is considered the prototype of bone repair. The complex bone tissue repair process that follows a fracture is apparently initiated by bone death or by damage to periosteal and haversian system blood vessels.

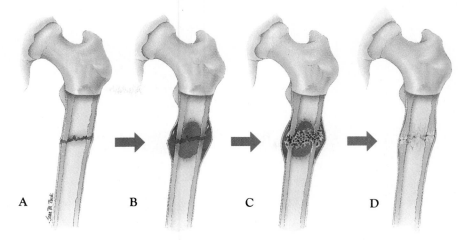

FIGURE 6-10 **Bone fracture and repair. A,** Fracture of femur. **B,** Formation of fracture hematoma. **C,** Formation of internal and external callus. **D,** Bone remodeling complete.

A bone fracture invariably tears and destroys blood vessels that carry nutrients to osteocytes. It is this vascular damage that initiates the repair sequence. Eventually, dead bone is either removed by osteoclastic resorption or serves as a scaffolding or framework for the deposition of a specialized repair tissue called *callus.*

The process of fracture healing is shown in Figure 6-10, A to D. Vascular damage occurring immediately after a fracture results in hemorrhage and the pooling of blood at the point of injury. The resulting blood clot is called a *fracture hematoma* (Figure 6-10, B). As the hematoma is resorbed, the formation of specialized callus tissue occurs. It serves to bind the broken ends of the fracture on both the outside surface and along the marrow cavity internally. The rapidly growing callus tissue effectively "collars" the broken ends and stabilizes the fracture so that healing can proceed (Figure 6-10, C). If the fracture is properly aligned and immobilized and if complications do not develop, callus tissue will be actively "modeled" and eventually replaced with normal bone as the injury heals completely (Figure 6-10, D).

CARTILAGE

Types of Cartilage

Cartilage is classified as connective tissue and consists of three specialized types called **hyaline, elastic,** and **fibrocartilage.** As a tissue, cartilage both resembles and differs from bone. Innumerable collagenous fibers reinforce the matrix of both tissues, and, like bone, cartilage consists more of extracellular substance than of cells. However, in cartilage the fibers are embedded in a firm gel instead of in a calcified cement substance. Hence cartilage has the flexibility of a firm plastic material, whereas bone has the rigidity of cast iron. Another difference is that no canal system and no blood vessels penetrate the cartilage matrix. Cartilage is avascular and bone is abundantly vascular. Nevertheless, cartilage cells, like bone cells, lie in lacunae.

However, because no canals and blood vessels interlace cartilage matrix, nutrients and oxygen can reach the scattered, isolated **chondrocytes** (cartilage cells) only by diffusion. They diffuse through the matrix gel from capillaries in the fibrous covering of the cartilage—the **perichondrium**—or from synovial fluid in the case of articular cartilage.

The three cartilage types differ from one another largely by the amount of matrix material that is present and also by the relative amounts of elastic and collagenous fibers that are embedded in them. *Hyaline* is the most abundant type, and both elastic and fibrocartilage varieties are considered as modifications of the hyaline type. Collagenous fibers are present in all three types but are most numerous in *fibrocartilage.* Hence it has the greatest tensile strength. *Elastic* cartilage matrix contains elastic fibers, as well as collagenous fibers, and so has elasticity, as well as firmness.

Cartilage is an excellent skeletal support tissue in the developing embryo. It forms rapidly and yet retains a significant degree of rigidity, or stiffness. A majority of the bones that eventually form both the axial and the appendicular skeleton described in Chapter 7 first appear as cartilage models. Skeletal maturation involves replacement of the cartilage models with bone.

After birth there is a decrease in the total amount of cartilage tissue present in the body. However, it continues to play an important role in the growth of long bones until skeletal maturity and is found throughout life as the material that covers the articular surfaces of bones in joints. The three types of cartilage also serve numerous specialized functions throughout the body.

Hyaline cartilage

Hyaline, in addition to being the most common type of cartilage, serves numerous specialized functions. It resembles milk glass in appearance (Figure 6-11, A). In fact, its name is derived from the Greek word meaning "glassy." It is semitransparent and has a bluish, opalescent cast.

A

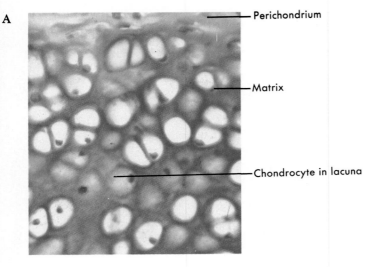

Perichondrium

Matrix

Chondrocyte in lacuna

B

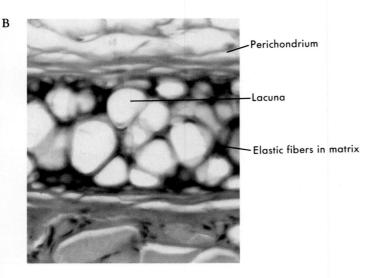

Perichondrium

Lacuna

Elastic fibers in matrix

C

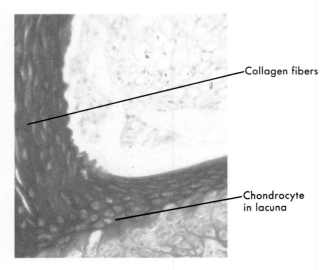

Collagen fibers

Chondrocyte in lacuna

FIGURE 6-11 Types of cartilage. A, Hyaline cartilage of trachea. **B,** Elastic cartilage of epiglottis. Note black elastic fibers in cartilage matrix and perichondrium layers on both surfaces. **C,** Fibrocartilage of intervertebral disk.

In the embryo, hyaline cartilage forms from differentiation of specialized mesenchymal cells that become crowded together in so-called **centers of chondrification.** As the cells enlarge, they secrete matrix material that surrounds the delicate fibrils. Eventually, the continued production of matrix separates and isolates the cells, or **chondrocytes,** into compartments, which, like in bone, are called *lacunae.* The two principal chemical components of matrix material (ground substance) are a mucoprotein called **chondroitin-sulfate** and a gel-like **polysaccharide.** Both substances are secreted from chondrocytes in much the same way as protein and carbohydrates are secreted from glandular cells.

In addition to covering the articular surfaces of bones, hyaline cartilage forms the costal cartilages that connect the anterior ends of the ribs with the sternum, or breastbone. It also forms the cartilage rings in the trachea, bronchi of the lungs, and the tip of the nose.

Elastic cartilage

Elastic cartilage gives form to the external ear, the epiglottis that covers the opening of the respiratory tract when swallowing, and the eustachian, or auditory, tubes that connect the middle ear and nasal cavity. The collagenous fibers of hyaline cartilage are also present—but in fewer numbers—in elastic cartilage. Large numbers of easily stained elastic fibers confer the elasticity and resiliency typical of this form of cartilage. In most stained sections, elastic cartilage has a yellowish color and has a greater opacity than the hyaline variety (Figure 6-11, *B*).

Fibrocartilage

Fibrocartilage is characterized by small quantities of matrix and abundant fibrous elements (Figure 6-11, C). It is strong, rigid, and most often associated with regions of dense connective tissue in the body. It occurs in the symphysis pubis, intervertebral disks, and near the points of attachment of some large tendons to bones.

Histophysiology of Cartilage

The gristle-like nature of cartilage permits it to sustain great weight when covering the articulating surfaces of bones or when serving as a shock-absorbing pad between articulating bones in the spine. In other areas, such as the external ear, nose, or respiratory passages, cartilage provides a strong yet pliable support structure that resists deformation or collapse of tubular passageways. It is cartilage that permits growth in length of long bones and that is largely responsible for their adult shape and size.

Growth of Cartilage

The growth of cartilage occurs in two ways:

1. **Interstitial** growth
2. **Appositional** growth

During interstitial growth, cartilage cells within the substance of the tissue mass divide and begin to secrete additional matrix. Internal division of chondrocytes is possible because of the soft, pliable nature of cartilage

tissue. This form of growth is most often seen during childhood and early adolescence, when a majority of cartilage is still soft and capable of expansion from within.

Appositional growth occurs when chondrocytes in the deep layer of the perichondrium begin to divide and secrete additional matrix. The new matrix is then deposited on the surface of the cartilage, causing it to increase in size. Appositional growth is unusual in early childhood but, once initiated, continues beyond adolescence and throughout life.

QUICK CHECK

✔ 1. *Name three specialized types of cartilage.*
✔ 2. *Identify the primary type of cartilage cell.*
✔ 3. *List the two mechanisms of cartilage growth.*

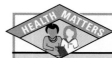

CARTILAGE AND NUTRITIONAL DEFICIENCIES

Certain nutritional deficiencies and other metabolic disturbances have an immediate and very visible impact on cartilage. It is for this reason that changes in cartilage often serve as indicators of inadequate vitamin, mineral, or protein intake. Vitamin A and protein deficiency, for example, will decrease the thickness of epiphyseal plates in the growing long bones of young children—an effect immediately apparent on x-ray examination. The opposite effect occurs in vitamin D deficiencies. As the epiphyseal cartilages increase in thickness but fail to calcify, the growing bones become deformed and bend under weight bearing. The bent long bones are a sign of **rickets.**

 ## Cycle of Life: *Skeletal tissues*

This chapter focused on the changes that occur in bone and cartilage tissue from the time before birth to advanced old age. For instance, we have outlined in some detail the process by which the soft cartilage and membranous skeleton becomes ossified over a period of years. By the time a person is a young adult in the midtwenties, the skeleton has become fully ossified. A few areas of soft tissue—the cartilaginous areas of the nose and ears, for example—may continue to grow and ossify very slowly throughout adulthood, so that by advanced old age, some structural changes are apparent.

Changes in skeletal tissue that occur during adulthood usually result from specific conditions. For example, the mechanical stress of weight-bearing exercise can trigger dramatic increases in the density and strength of bone tissue. Pregnancy, nutritional deficiencies, and illness can all cause a loss of bone density accompanied by a loss of structural strength.

In advanced adulthood, degeneration of bone and cartilage tissue become apparent. Replacement of hard bone matrix by softer connective tissue result in a loss of strength that increases susceptibility to injury. This is especially true in older women who suffer from osteoporosis. Fortunately, recent experiments show that even very light exercise by elderly individuals can counteract some of the skeletal tissue degeneration associated with old age.

 ## MECHANISMS OF DISEASE

Diseases of Skeletal Tissues

Neoplasms

There are many different types of neoplasms affecting skeletal tissues. They can be primary bone tumors or metastatic tumors from primary cancers elsewhere in the body (e.g., lung, prostate, breast). Slow-growing benign tumors are the most common type of skeletal neoplasm.

Osteochondroma is the most common type of skeletal tissue tumor. This condition develops early in bone growth, causing projections (spurs) at the ends of affected long bones. These tumors tend to originate within the periosteum near the epiphyseal plate and are asymptomatic unless irritating to surrounding tissue. Osteochondromas have the potential to become metastatic. However, the prognosis is good if complete surgical removal is accomplished.

Osteosarcoma (osteogenic sarcoma) is the most common primary malignant tumor of skeletal tissue and is often the most fatal. It appears more frequently in men, with its peak age of incidence between 10 and 25 years. Common sites of involvement are the tibia, femur, and humerus. Roughly 10% of patients experience metastases to the lungs, and if left untreated, the course can involve widespread metastases and death within 1 year. Therapy commonly involves surgery followed with chemotherapy.

Chondrosarcoma is a malignant tumor of hyaline cartilage arising from chondroblasts. It is a large, bulky, slow-growing tumor occurring most frequently in middle-aged persons. Common sites of involvement include the femur, spine, pelvis, ribs, or scapula. Large excisions or amputation of the affected extremity can improve survival rates. Chemotherapy has not been proven to be effective.

Metabolic Bone Diseases

Metabolic bone diseases are disorders of bone remodeling. *Osteoporosis* is one of the most common and serious of all bone diseases. It is characterized by excessive loss of calcified matrix, bone mineral, and collagenous fibers, causing a reduction in total bone mass.

Since both estrogen and testosterone serve important roles in stimulating osteoblast activity after puberty, decreasing levels of these hormones in the elderly reduces new bone growth and maintenance of existing bone mass. In women, decreasing estrogen levels associated with menopause cause accelerated bone resorption. Inadequate intake of calcium or vitamin D, necessary for normal bone mineralization, over a period of years can also result in decreased bone mass and the development of osteoporosis.

In osteoporosis, bones become porous, brittle, and fragile, fracturing easily under stress. As a result, they are often characterized by pathological changes in the mass or chemical composition of skeletal tissue. The result is a dangerous pathological condition resulting in increased susceptibility to "spontaneous fractures" and pathological curvature of the spine. Osteoporosis occurs most frequently in elderly white women.

Osteomalacia is a metabolic bone disease characterized by inadequate mineralization of bone. A large amount of osteoid (organic bone matrix) does not calcify in patients with this disease. Risk factors for development of osteomalacia include malabsorption problems, vitamin D and calcium deficiencies, chronic renal failure, and inadequate exposure to sunlight. Symptoms, though subtle, may include muscle weakness, fractures, generalized bone pain and tenderness in the extremities, and lower back pain. Treatment includes dietary supplements of vitamin D and calcium. In addition, exposure to sunlight may be instituted to promote vitamin D synthesis in the body.

Paget's disease, also known as *osteitis deformans,* is a disorder affecting older adults. It is characterized by proliferation of osteoclasts and compensatory increased osteoblastic activity. The result is rapid and disorganized bone remodeling. The bones formed are poorly constructed and weakened. It commonly affects the skull, femur, vertebra, and pelvic bones. Clinical manifestations may include bone pain, tenderness, and fractures. However, the majority of patients experience minimal changes and never know they have the disease. No treatment is recommended in the asymptomatic patient.

Osteomyelitis is a bacterial infection of the bone and marrow tissue. Infections of the bone are often more difficult to treat than soft-tissue infections because of decreased blood supply and density of the bone. Bacteria, viruses, fungi, and other pathogens may cause osteomyelitis. *Staphylococcus* bacteria are the most common pathogens. Osteomyelitis is associated with extension of another infection (e.g., bacteremia, urinary tract infection, vascular ulcer) or direct bone contamination (e.g., gunshot wound, open fracture). Patients who are elderly, poorly nourished, or diabetic are also at risk. Thrombosis of blood vessels in osteomyelitis often results in ischemia and bone necrosis. As a result, infection can extend under the periosteum and spread to adjacent soft tissues and joints. Signs and symptoms may include an area swollen, warm, tender to touch, and painful. Early recognition of infection and antimicrobial management are required. Sometimes patients may require 6 weeks of antibiotics.

CHAPTER SUMMARY

TYPES OF BONES

A. Structurally, there are four types of bones (Figure 6-1)
  1. Long bones
  2. Short bones
  3. Flat bones
  4. Irregular bones
B. Bones serve a variety of needs, and their size, shape, and appearance will vary to meet those needs
C. Bones vary in proportion of compact and cancellous (spongy) bone; compact bone is dense and solid in appearance, whereas cancellous bone is characterized by open space partially filled with needle-like structures
D. Long bones (Figure 6-2)
  1. Diaphysis
    a. Main shaft of long bone
    b. Hollow, cylindrical shape and thick compact bone
    c. Function is to provide strong support without cumbersome weight
  2. Epiphyses
    a. Both ends of a long bone, made of cancellous bone filled with marrow
    b. Bulbous shape
    c. Function is to provide attachments for muscles and give stability to joints
  3. Articular cartilage
    a. Layer of hyaline cartilage that covers the articular surface of epiphyses
    b. Function is to cushion jars and blows
  4. Periosteum
    a. Dense, white fibrous membrane that covers bone
    b. Attaches tendons firmly to bones
    c. Contains blood vessels that send branches into bone
    d. Essential for bone cell survival and bone formation
  5. Medullary (or marrow) cavity
    a. Tubelike, hollow space in diaphysis
    b. Filled with yellow marrow
  6. Endosteum—thin epithelial membrane that lines medullary cavity
E. Short, flat, and irregular bones
  1. Inner portion is cancellous bone covered on the outside with compact bone
  2. Spaces inside cancellous bone of a few irregular and flat bones are filled with red marrow

BONE TISSUE

A. Most distinctive form of connective tissue
B. Extracellular components are hard and calcified
    a. Apatite—highly specialized chemical crystals of calcium and phosphate contribute to bone hardness
    b. Slender needlelike crystals are oriented to most effectively resist stress and mechanical deformation
    c. Magnesium and sodium are also found in bone
  2. Organic matrix
    a. Composite of collagenous fibers and an amorphous mixture of protein and polysaccharides called ground substance
    b. Ground substance is secreted by connective tissue cells
    c. Adds to overall strength of bone and gives some degree of resilience to the bone

MICROSCOPIC STRUCTURE OF BONE (Figure 6-3)

A. Compact bone
1. Contains many cylinder-shaped structural units called osteons, or haversian systems
2. Osteons surround canals that run lengthwise through bone and are connected by transverse Volkmann's canals
3. Living bone cells are located in these units that constitute the structural framework of compact bone
4. Osteons permit delivery of nutrients and removal of waste products
5. Four types of structures make up each osteon
   a. Lamella—concentric, cylinder-shaped layers of calcified matrix
   b. Lacunae—Small spaces containing tissue fluid in which bone cells are located between hard layers of the lamella
   c. Canaliculi—ultrasmall canals radiating in all directions from the lacunae and connecting them to each other and to the haversian canal
   d. Haversian canal—extends lengthwise through the center of each osteon and contains blood vessels and lymphatic vessels
B. Cancellous bones
1. No osteons in cancellous bone, instead it has trabeculae
2. Nutrients are delivered and waste products removed by diffusion through tiny canaliculi
3. Bony spicules are arranged along lines of stress, enhancing the bone's strength
C. Blood supply
1. Bone cells are metabolically active and need a blood supply, which comes from the bone marrow in the internal medullary cavity of cancellous bone
2. Compact bone, in addition to bone marrow and blood vessels from the periosteum penetrate bone and then, by way of Volkmann's canals, connect with vessels in the haversian canals
D. Types of bone cells
1. Osteoblasts
   a. Bone-forming cells found in all bone surfaces
   b. Small cells synthesize and secrete osteoid, an important part of the ground substance
   c. Collagen fibrils line up in osteoid and serve as a framework for the deposition of calcium and phosphate
2. Osteoclasts (Figure 6-5)
   a. Giant multinucleate cells
   b. Responsible for the active erosion of bone minerals
   c. Contain large numbers of mitochondria and lysosomes
3. Osteocytes—mature, nondividing osteoblast surrounded by matrix, lying within lacunae (Figure 6-6)

BONE MARROW

A. Specialized type of soft, diffuse connective tissue; called myeloid tissue
B. Site for the production of blood cells
C. Found in medullary cavities of long bones and in the spaces of spongy bone
D. Two types of marrow occur during a person's lifetime
1. Red marrow
   a. Found in virtually all bones in an infant's or child's body
   b. Functions to produce red blood cells
2. Yellow marrow
   a. As an individual ages, red marrow is replaced by yellow marrow
   b. Marrow cells become saturated with fat and are no longer active in blood cell production
E. The main bones in an adult that still contain red marrow include the ribs, bodies of the vertebrae, the humerus, and the femur
F. Yellow marrow can alter to red marrow during times of decreased blood supply, such as anemia, exposure to radiation, and certain diseases

FUNCTIONS OF BONE

A. Support—bones form the framework of the body and contribute to the shape, alignment, and positioning of the body parts
B. Protection—bony "boxes" protect the delicate structures they enclose
C. Movement—bones with their joints constitute levers that move as muscles contract
D. Mineral storage—bones are the major reservoir for calcium, phosphorus, and other minerals
E. Hematopoiesis—blood cell formation is carried out by myeloid tissue

DEVELOPMENT OF BONE

A. Osteogenesis—development of bone from small cartilage model to an adult bone
B. Intramembranous ossification
1. Occurs within a connective tissue membrane
2. Flat bones begin when groups of cells differentiate into osteoblasts
3. Osteoblasts are clustered together in centers of ossification
4. Osteoblasts secrete matrix material and collagenous fibrils
5. Large amounts of ground substance accumulate around each osteoblast
6. Collagenous fibers become embedded in the ground substance and constitute the bone matrix
7. Bone matrix calcifies when calcium salts are deposited
8. Trabeculae appear and join in a network to form spongy bone
9. Apposition growth occurs by adding osseous tissue
C. Endochondral ossification (Figure 6-7)
1. Most bones begin as a cartilage model with bone formation spreading essentially from the center to the ends
2. Periosteum develops and enlarges, producing a collar of bone
3. Primary ossification center forms
4. Blood vessel enters the cartilage model at the midpoint of the diaphysis
5. Bone grows in length as endochondral ossification progresses from the diaphysis towards each epiphysis
6. Secondary ossification centers appear in the epiphysis, and bone growth proceeds toward the diaphysis
7. Epiphyseal plate remains between diaphysis and each epiphysis until bone growth in length is complete
8. Epiphyseal plate is composed of four layers (Figure 6-8)
   a. "Resting" cartilage cells—point of attachment joining the epiphysis to the shaft
   b. Zone of proliferation—cartilage cells undergoing active mitosis, causing the layer to thicken and the plate to increase in length
   c. Zone of hypertrophy—older, enlarged cells undergoing degenerative changes associated with calcium deposition
   d. Zone of calcification—dead or dying cartilage cells undergoing rapid calcification

BONE GROWTH AND RESORPTION

A. Bones grow in diameter by the combined action of osteoclasts and osteoblasts
B. Osteoclasts enlarge the diameter of the medullary cavity

C. Osteoblasts from the periosteum build new bone around the outside of the bone

REPAIR OF BONE FRACTURES

A. Fracture—break in the continuity of a bone
B. Fracture healing (Figure 6-10)
C. Fracture healing
   1. Fracture tears and destroys blood vessels that carry nutrients to osteocytes
   2. Vascular damage initiates repair sequence
   3. Callus—specialized repair tissue that binds the broken ends of the fracture together
   4. Fracture hematoma—blood clot occurring immediately after the fracture, is then resorbed and replaced by callus

CARTILAGE

A. Characteristics
   1. Avascular connective tissue
   2. Fibers of cartilage are embedded in a firm gel
   3. Has the flexibility of firm plastic
   4. No canal system or blood vessels
   5. Chondrocytes receive oxygen and nutrients by diffusion
   6. Perichondrium—fibrous covering of the cartilage
   7. Cartilage types differ due to the amount of matrix present and the amounts of elastic and collagenous fibers
B. Types of cartilage (Figure 6-11)
   1. Hyaline cartilage
      a. Most common type
      b. Covers the articular surfaces of bones
      c. Forms the costal cartilages, cartilage rings in the trachea, bronchi of the lungs, and the tip of the nose
      d. Forms from specialized cells in centers of chondrification, which secrete matrix material
      e. Chondrocytes are isolated into lacunae
   2. Elastic cartilage
      a. Forms external ear, epiglottis, and eustachian tubes
      b. Large number of elastic fibers confer elasticity and resiliency
   3. Fibrocartilage
      a. Occurs in symphysis pubis and intervertebral disks
      b. Small quantities of matrix and abundant fibrous elements
      c. Strong and rigid
C. Histophysiology of cartilage
   1. Gristlelike nature permits cartilage to sustain great weight or serve as a shock absorber

2. Strong yet pliable support structure
   3. Permits growth in length of long bones
D. Growth of cartilage
   1. Interstitial or endogenous growth
      a. Cartilage cells divide and secrete additional matrix
      b. Seen during childhood and early adolescence while cartilage is still soft and capable of expansion from within
   2. Appositional or exogenous growth
      a. Chondrocytes in the deep layer of the perichondrium divide and secrete matrix
      b. New matrix is deposited on the surface, increasing its size
      c. Unusual in early childhood, but, once initiated, continues throughout life

CYCLE OF LIFE: SKELETAL TISSUES

A. Skeleton fully ossified by midtwenties
   1. Soft tissue may continue to grow—ossifies more slowly
B. Adults—changes occur from specific conditions
   1. Increased density and strength from exercise
   2. Decreased density and strength from pregnancy, nutritional deficiencies, and illness
C. Advanced adulthood—apparent degeneration
   1. Hard bone matrix replaced by softer connective tissue
   2. Exercise can counteract degeneration

MECHANISMS OF DISEASE: DISEASES OF SKELETAL TISSUES

A. Neoplasms
   1. Osteochondroma—most common skeletal tissue tumor; causes spurs at the ends of long bones
   2. Osteosarcoma—most common malignant tumor
   3. Chondrosarcoma—malignant tumor of hyaline cartilage from chondroblasts
B. Metabolic bone disease
   1. Osteoporosis—most common bone disease
      a. Reduces total bone mass from loss of calcified matrix, bone mineral, and collagenous fibers
      b. Porous, brittle, and fragile bones fracture easily under stress
   2. Osteomalacia—inadequate mineralization of bone
   3. Paget's disease (osteitis deformans)—proliferations of osteoclasts and increased osteoblastic activity resulting in rapid and disorganized bone remodeling
   4. Osteomyelitis—bacterial infection of bone and marrow tissue

REVIEW QUESTIONS

1. Describe the microscopic structure of bone and cartilage.
2. Describe the structure of a long bone.
3. Explain the functions of the periosteum.
4. Describe the two principal chemical components of extracellular bone.
5. List and discuss each of the major anatomical components that together constitute an osteon.
6. Compare and contrast the three major types of cells found in bone.
7. What are the homeostatic functions that bone performs for the body?

8. Compare and contrast bone formation in intramembranous and endochondral ossification.
9. Discuss the sequence of steps characteristic of fracture healing.
10. Compare and contrast the basic structural elements of bone and cartilage.
11. Compare the structure and function of the three types of cartilage.
12. How does the mechanism of cartilage growth differ from bone growth?

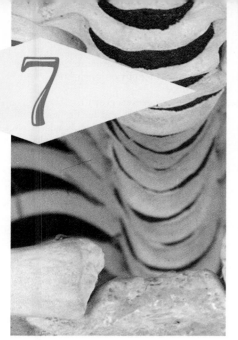

The Skeletal System

Thoracic inlet, viewed from above

OBJECTIVES

After you have completed this chapter, you should be able to:

1. Identify the two main subdivisions of the skeleton.
2. List the primary subdivisions of the axial skeleton.
3. Distinguish between the bones of the skull and those of the face.
4. List the sutures and fontanels of the skull.
5. Discuss the clinical significance of the cribriform plate of the ethmoid bone.
6. Name the regions of the vertebral column and give the number of vertebrae in each segment.
7. Discuss the bony components of the rib cage, or chest.
8. List the primary subdivisions of the appendicular skeleton.
9. List the bony components of the shoulder and pelvic girdles.
10. Compare the structure and function of the wrist and hand with the ankle and foot.
11. Discuss the structural components and functional significance of the arches of the foot.
12. List the skeletal differences between men and women.

Appendicular skeleton Bones (126) that form the appendages to the axial skeleton; the bones of the upper and lower extremities of the body

Axial skeleton Bones (80) that make up the head, neck, and torso

Cranium (KRAY-nee-um) Bones (8) of the skull that make up the brain case

Fontanel (FON-tah-nel) Unossified area in an infant's skull; "soft spots"

Pelvic girdle Group of bones that forms a stable base for the trunk and connect the legs to the trunk

Shoulder girdle Portion of skeleton where the upper extremity joins the trunk

Sinus (SYE-nus) Mucosa-lined, air-filled cavity found inside some cranial bones

Suture (SOO-chur) An immovable joint

Thorax (THOR-aks) Chest

Vertebra (VER-te-bra) One of the bones that make up the spinal column

Just as skeletal tissues are organized to form bones, the bones in turn are organized or grouped to form the major subdivisions of the skeletal system described below. The rigid bones lie buried within the muscles and other soft tissues, thus providing support and shape to the body as a whole. An understanding of the relationship of bones to each other and to other body structures provides a basis for understanding the function of many other organ systems. Coordinated movement, for example, is only possible because of the way bones are joined to one another in joints and the way muscles are attached to those bones. In addition, knowledge of the placement of bones within the soft tissues assists in locating and identifying other body structures.

The adult skeleton is composed of 206 separate bones. Variations in the total number of bones in the body may occur as a result of certain anomalies such as extra ribs or from failure of certain small bones to fuse in the course of development.

In Chapter 6 the basic types of skeletal tissue, including bone and cartilage, were discussed. Comparisons between the structural and functional characteristics of dense (compact) and cancellous (spongy) bone set the stage for study in this chapter of individual bones and their interrelationships in the skeleton. Chapter 8 takes your studies one step farther, by considering articulations—that is, how the bones form joints.

◢ DIVISIONS OF SKELETON

The human skeleton consists of two main divisions—the axial skeleton and the appendicular skeleton (Figure 7-1). Eighty bones make up the **axial skeleton.** This includes 74 bones that form the upright axis of the body and 6 tiny middle ear bones. The **appendicular skeleton** consists of 126 bones—more than half again as many as in the axial skeleton. Bones of the appendicular skeleton form the appendages to the axial skeleton, that is, the shoulder girdles, arms, wrists, and hands and the hip girdles, legs, ankles, and feet.

One of the first things you should do in studying the skeleton is to familiarize yourself with the names of individual bones listed in Table 7-1. Next, look at Table 7-2, which lists some terms that are often used to name or describe *bone markings*—specific features on an individual bone. After thus preparing yourself, begin a step-by-step exploration of the skeletal system by studying the illustrations, text, and tables that constitute the rest of this

TABLE 7-1 Bones of skeleton (206 total)*

AXIAL SKELETON (80 bones total)		AXIAL SKELETON—cont'd	
Part of Body	Name of Bone	Part of Body	Name of Bone
Skull (28 bones total)		**Sternum and ribs** (25 bones total)	Sternum (1)
Cranium (8 bones)	Frontal (1)		True ribs (14)
	Parietal (2)		False ribs (10)
	Temporal (2)	**APPENDICULAR SKELETON (126 bones total)**	
	Occipital (1)	Part of Body	Name of Bone
	Sphenoid (1)	**Upper extremities**	
	Ethmoid (1)	(including shoulder girdle) (64 bones total)	Clavicle (2)
Face (14 bones)	Nasal (2)		Scapula (2)
	Maxillary (2)		Humerus (2)
	Zygomatic (malar)((2)		Radius (2)
	Mandible (1)		Ulna (2)
	Lacrimal (2)		Carpals (16)
	Palatine (2)		Metacarpals (10)
	Inferior conchae (turbinates) (2)		Phalanges (28)
	Vomer (1)	**Lower extremities**	
Ear bones (6 bones)	Malleus (hammer) (2)	(including hip girdle) (62 bones total)	Coxal bones (2)
	Incus (anvil) (2)		Femur (2)
	Stapes (stirrup) (2)		Patella (2)
Hyoid bone (1)			Tibia (2)
Spinal column (26 bones total)	Cervical vertebrae (7)		Fibula (2)
	Thoracic vertebrae (12)		Tarsals (14)
	Lumbar vertebrae (5)		Metatarsals (10)
	Sacrum (1)		Phalanges (28)
	Coccyx (1)		

*An inconstant number of small, flat, round bones known as **sesamoid bones** (because of their resemblance to sesame seeds) are found in various tendons in which considerable pressure develops. Because the number of these bones varies greatly between individuals, only two of them, the patellae, have been counted among the 206 bones of the body. Generally, two of them can be found in each thumb (in flexor tendon near metacarpophalangeal and interphalangeal joints) and great toe plus several others in the upper and lower extremities. **Wormian bones,** the small islets of bone frequently found in some of the cranial sutures, have not been counted in this list of 206 bones because of their variable occurrence.

TABLE 7-2 Terms used to describe bone markings

Term	Meaning	Term	Meaning
Angle	A corner	Margin	Edge of a flat bone or flat portion or edge of a flat area
Body	The main portion of a bone	Meatus	Tubelike opening or channel (pl. *meati*)
Condyle	Rounded bump; usually fits into a fossa on another bone, forming a joint	Neck	A narrowed portion, usually at the base of a head
Crest	Moderately raised ridge; generally a site for muscle attachment	Notch	A V-like depression in the margin or edge of a flat area
Epicondyle	Bump near a condyle; often gives the appearance of a "bump on a bump"; for muscle attachment	Process	A raised area or projection
		Ramus	Curved portion of a bone, like a ram's horn (pl. *rami*)
Facet	Flat surface that forms a joint with another facet or flat bone	Sinus	Cavity within a bone
Fissure	Long, cracklike hole for blood vessels and nerves	Spine	Similar to a crest but raised more; a sharp, pointed process; for muscle attachment
Foramen	Round hole for vessels and nerves (pl. *foramina*)	Sulcus	Groove or elongated depression (pl. *sulci*)
Fossa	Depression; often receives an articulating bone (pl. *fossae*)	Trochanter	Large bump for muscle attachment (larger than tubercle or tuberosity)
Head	Distinct epiphysis on a long bone, separated from the shaft by a narrowed portion (or neck)	Tuberosity	Oblong, raised bump, usually for muscle attachment; small tuberosity is called a *tubercle*
Line	Similar to a crest but not raised as much (is often rather faint)		

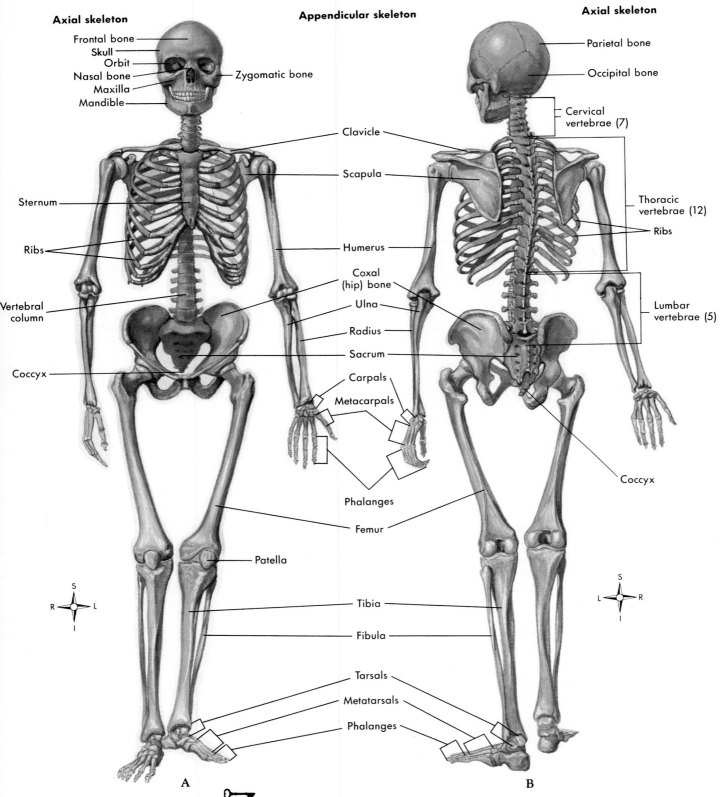

Axial skeleton

Frontal bone
Skull
Orbit
Nasal bone
Maxilla
Mandible
Zygomatic bone

Appendicular skeleton

Clavicle
Scapula
Humerus
Coxal (hip) bone
Ulna
Radius
Carpals
Metacarpals
Phalanges
Femur
Patella

Sternum
Ribs
Vertebral column
Coccyx

Tibia
Fibula

Tarsals
Metatarsals
Phalanges

Axial skeleton

Parietal bone
Occipital bone
Cervical vertebrae (7)
Thoracic vertebrae (12)
Ribs
Lumbar vertebrae (5)
Sacrum
Coccyx

A
B

FIGURE 7-1   Skeleton. A, Anterior view. **B,** posterior view.

chapter. One picture is said to be worth a thousand words. If this is true and if you refer often to the illustrations, you should find it easy—and perhaps even fun—to learn the names of the bones and to identify their markings.

AXIAL SKELETON
Skull

Twenty-eight irregularly shaped bones form the skull (Figures 7-2 to 7-8). It consists of two major divisions: the **cranium,** or brain case, and the **face.** The cranium is formed by eight bones, namely, frontal, two parietal, two temporal, occipital, sphenoid, and ethmoid (Table 7-3). The 14 bones that form the face are: two maxilla, two zygomatic (malar), two nasal, mandible, two lacrimal, two palatine, two inferior nasal conchae (turbinates), and vomer (Table 7-4). Note that all the face bones are paired except for the mandible and vomer. All cranial bones, on the other hand, are single (unpaired) except for the parietal and temporal bones, which are paired. The frontal and ethmoid bones of the skull help shape the face but are not numbered among the facial bones.

Cranial bones

The **frontal bone** forms the forehead and the anterior part of the top of the cranium. It contains mucosa-lined air-filled spaces, or **sinuses**—the *frontal sinuses.* The frontal sinuses, along with similar sinuses in the sphenoid, ethmoid, and maxillae, are often called *paranasal sinuses* because they have narrow channels that open into the nasal cavity. Paranasal sinuses are also discussed in Chapter 22,

page 583. A portion of the frontal bone forms the upper part of the orbits. It unites with the two parietal bones posteriorly in an immovable joint, or **suture,** the *coronal suture.* Several of the more prominent frontal bone markings are described in Table 7-3.

The two **parietal bones** give shape to the bulging topside of the cranium. They form immovable joints with several bones: the *lambdoidal suture* with the occipital bone, the *squamous suture* with the temporal bone and part of the sphenoid, and the *coronal suture* with the frontal bone.

The lower sides of the cranium and part of its floor are fashioned from two **temporal bones.** They house the middle and inner ear structures and contain the *mastoid sinuses,* notable because of the occurrence of *mastoiditis,* an inflammation of the mucous lining of these spaces. For a description of several other temporal bone markings, see Table 7-3.

MASTOIDITIS

Mastoiditis (mas-toy-DYE-tis), or inflammation of the air spaces within the mastoid portion of the temporal bone, can produce very serious medical problems unless treated promptly. Infectious material frequently finds its way into the mastoid air cells from middle ear infections. The mastoid air cells do not drain into the nose, as do the paranasal sinuses. As a result, infectious material that accumulates may erode the thin bony partition that separates the air cells from the cranial cavity. Should this occur, the inflammation may spread to the brain or its covering membranes.

TABLE 7-3   Cranial bones and their markings

Bones and Markings	Description	Bones and Markings	Description
FRONTAL	Forehead bone; also forms most of roof of orbits (eye sockets) and anterior part of cranial floor	**SPHENOID**	Keystone of cranial floor; forms its midportion; resembles bat with wings outstretched and legs extended downward posteriorly; lies behind and slightly above nose and throat; forms part of floor and sidewalls of orbit
Supraorbital margin	Arched ridge just below eyebrow, forms upper edge of orbit		
Frontal sinuses	Cavities inside bone just above supraorbital margin; lined with mucosa; contain air	**Body**	Hollow, cubelike central portion
Frontal tuberosities	Bulge above each orbit; most prominent part of forehead	**Greater wings**	Lateral projections from body, form part of outer wall of orbit
Superciliary ridges	Ridges caused by projection of frontal sinuses; eyebrows lie superficial to these ridges	**Lesser wings**	Thin, triangular projections from upper part of sphenoid body; form posterior part of roof of orbit
Supraorbital foramen (sometimes notch)	Foramen or notch in supraorbital margin slightly medial to its midpoint; transmits supraorbital nerve and blood vessels	**Sella turcica (or Turk's saddle)**	Saddle-shaped depression on upper surface of sphenoid body; contains pituitary gland
		Sphenoid sinuses	Irregular air-filled mucosa-lined spaces within central part of sphenoid
Glabella	Smooth area between superciliary ridges and above nose	**Pterygoid processes**	Downward projections on either side where body and greater wing unite; comparable to extended legs of bat if entire bone is likened to this animal; form part of lateral nasal wall
PARIETAL	Prominent, bulging bones behind frontal bone; forms top sides of cranial cavity		

TABLE 7-3   Cranial bones and their markings—cont'd

Bones and Markings	Description	Bones and Markings	Description
Optic foramen	Opening into orbit at root of lesser wing; transmits optic nerve	Stylomastoid foramen	Opening between styloid and mastoid processes where facial nerve emerges from cranial cavity
Superior orbital fissure	Slitlike opening into orbit; lateral to optic foramen; transmits third, fourth, and part of fifth cranial nerves	Jugular fossa	Depression on undersurface of petrous portion; dilated beginning of internal jugular vein lodged here
Foramen rotundum	Opening in greater wing that transmits maxillary division of fifth cranial nerve	Jugular foramen	Opening in suture between petrous portion and occipital bone; transmits lateral sinus and ninth, tenth, and eleventh cranial nerves
Foramen ovale	Opening in greater wing that transmits mandibular division of fifth cranial nerve		
Foramen lacerum	Opening at the junction of the sphenoid, temporal, and occipital bones; transmits branch of the ascending pharyngeal artery	Carotid canal (or foramen)	Channel in petrous portion; best seen from undersurface of skull; transmits internal carotid artery
Foramen spinosum	Opening in greater wing that transmits the middle meningeal artery to supply meninges	**OCCIPITAL**	Forms posterior part of cranial floor and walls
		Foramen magnum	Hole through which spinal cord enters cranial cavity
TEMPORAL	Form lower sides of cranium and part of cranial floor; contain middle and inner ear structures	Condyles	Convex, oval processes on either side of foramen magnum; articulate with depressions on first cervical vertebra
Squamous portion	Thin, flaring upper part of bone	External occipital protuberance	Prominent projection on posterior surface in midline short distance above foramen magnum; can be felt as definite bump
Mastoid portion	Rough-surfaced lower part of bone posterior to external auditory meatus	Superior nuchal line	Curved ridge extending laterally from external occipital protuberance
Petrous portion	Wedge-shaped process that forms part of center section of cranial floor between sphenoid and occipital bones; name derived from Greek word for stone because of extreme hardness of this process; houses middle and inner ear structures	Inferior nuchal line	Less well-defined ridge paralleling superior nuchal line a short distance below it
		Internal occipital protuberance	Projection in midline on inner surface of bone; grooves for lateral sinuses extend laterally from this process and one for sagittal sinus extends upward from it
Mastoid process	Protuberance just behind ear		
Mastoid air cells	Air-filled mucosa-lined spaces within mastoid process	**ETHMOID**	Complicated irregular bone that helps make up anterior portion of cranial floor, medial wall of orbits, upper parts of nasal septum, and sidewalls and part of nasal roof; lies anterior to sphenoid and posterior to nasal bones
External auditory meatus (or canal)	Tube extending into temporal bone from external ear opening to tympanic membrane		
Zygomatic process	Projection that articulates with malar (or zygomatic) bone	Horizontal (cribriform) plate	Olfactory nerves pass through numerous holes in this plate
Internal auditory meatus	Fairly large opening on posterior surface of petrous portion of bone; transmits eighth cranial nerve to inner ear and seventh cranial nerve on its way to facial structures	Crista galli	Meninges (membranes around the brain) attach to this process
		Perpendicular plate	Forms upper part of nasal septum
Mandibular fossa	Oval-shaped depression anterior to external auditory meatus; forms socket for condyle of mandible	Ethmoid sinuses	Honeycombed, mucosa-lined air spaces within lateral masses of bone
		Superior and middle conchae (turbinates)	Help to form lateral walls of nose
Styloid process	Slender spike of bone extending downward and forward from undersurface of bone anterior to mastoid process; often broken off in dry skull; several neck muscles and ligaments attach to styloid process	Lateral masses	Compose sides of bone; contain many air spaces (ethmoid cells or sinuses); inner surface forms superior and middle conchae

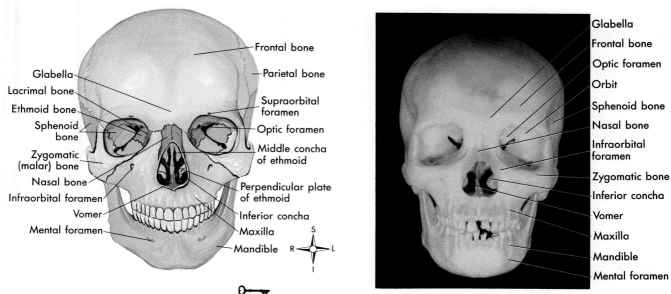

FIGURE 7-2   Anterior view of the skull.

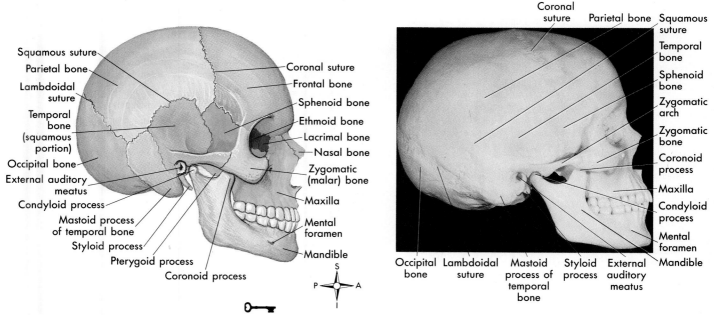

FIGURE 7-3   Skull viewed from the right side.

The **occipital bone** creates the framework of the lower, posterior part of the skull. It forms immovable joints with three other cranial bones—the parietal, temporal, and sphenoid—and a movable joint with the first cervical vertebra. Table 7-3 lists a description of some of its markings.

The shape of the **sphenoid bone** resembles a bat with its wings outstretched and legs extended down and back. Note in Figure 7-4 the location of the sphenoid bone in the central portion of the cranial floor. Here it serves as the keystone in the architecture of the cranium, anchoring the frontal, parietal, occipital, and ethmoid bones. The sphenoid bone also forms part of the lateral wall of the cranium and part of the floor of each orbit (Figures 7-2 and 7-3). The sphenoid bone contains fairly large mucosa-lined

air-filled spaces, the *sphenoid sinuses* (see Figure 7-6). Several prominent sphenoid markings are described in Table 7-3 (Figures 7-4 to 7-6).

The **ethmoid,** a complicated, irregular bone, lies anterior to the sphenoid but posterior to the nasal bones. It helps fashion the anterior part of the cranial floor (Figure 7-4), the medial walls of the orbits (Figures 7-2 and 7-7), the upper parts of the nasal septum (Figure 7-2) and of the sidewalls of the nasal cavity (Figure 7-8), and the part of the nasal roof (the cribriform plate) perforated by small foramina through which olfactory nerve branches reach the brain. The lateral masses of the ethmoid bone are honeycombed with sinus spaces (Figure 7-8). For more ethmoid bone markings, see Table 7-3.

TABLE 7-4   Facial bones and their markings

Bones and Markings	Description	Bones and Markings	Description
PALATINE	Form posterior part of hard palate, floor, and part of sidewalls of nasal cavity and floor of orbit	**MAXILLA**	Upper jaw bones; form part of floor of orbit, anterior part of roof of mouth, and floor of nose and part of sidewalls of nose
Horizontal plate	Joined to palatine processes of maxillae to complete part of hard palate	**Alveolar process**	Arch containing teeth
		Maxillary sinus (antrum of Highmore)	Large air-filled mucosa-lined cavity within body of each maxilla; largest of sinuses
MANDIBLE	Lower jawbone; largest, strongest bone of face	**Palatine process**	Horizontal inward projection from alveolar process; forms anterior and larger part of hard palate
Body	Main part of bone; forms chin	**Infraorbital foramen**	Hole on external surface just below orbit; transmits vessels and nerves
Ramus	Process, one on either side, that projects upward from posterior part of body	**Lacrimal groove**	Groove on inner surface; joined by similar groove on lacrimal bone to form canal housing nasolacrimal duct
Condyle (or head)	Part of each ramus that articulates with mandibular fossa of temporal bone	**NASAL**	Small bones forming upper part of bridge of nose
Neck	Constricted part just below condyles		
Alveolar process	Teeth set into this arch	**ZYGOMATIC**	Cheekbones; form part of floor and sidwall of orbit
Mandibular foramen	Opening on inner surface of ramus; transmits nerves and vessels to lower teeth	**LACRIMAL**	Thin bones about size and shape of fingernail; posterior and lateral to nasal bones in medial wall of orbit; help form sidewall of nasal cavity, often missing in dry skull
Mental foramen	Opening on outer surface below space between two bicuspids; transmits terminal branches of nerves and vessels that enter bone through mandibular foramen; dentists inject anesthetics through these foramina	**INFERIOR NASAL CONCHAE (turbinates)**	Thin scroll of bone forming kind of shelf along inner surface of sidewall of nasal cavity; lies above roof of mouth
Coronoid process	Projection upward from anterior part of each ramus; temporal muscle inserts here	**VOMER**	Forms lower and posterior part of nasal septum; shaped like ploughshare
Angle	Juncture of posterior and inferior margins of ramus		

THE CRIBRIFORM PLATE

Separation of the nasal and cranial cavities by the **cribriform plate** of the ethmoid bone has great clinical significance. The cribriform plate is perforated by many small openings, which permit branches of the olfactory nerve responsible for the special sense of smell to enter the cranial cavity and reach the brain. Separation of these two cavities by a thin, perforated plate of bone presents real hazards. If the cribriform plate is damaged as a result of trauma to the nose, it is possible for potentially infectious material to pass directly from the nasal cavity into the cranial fossa. If fragments of a fractured nasal bone are pushed through the cribriform plate they may tear the coverings of the brain or enter the substance of the brain itself.

Facial bones

The two **maxillae** together serve as the keystone in the architecture of the face just as the sphenoid bone acts as the keystone of the cranium. Each maxilla articulates not only with the other maxilla but also with a nasal, a zygomatic, an inferior concha, and a palatine bone. Of all the facial bones, only the mandible does not articulate with the maxillae. The maxillae form part of the floor of the orbits, part of the roof of the mouth, and part of the floor and sidewalls of the nose. Each maxilla contains a mucosa-lined space, the maxillary sinus (Figure 7-8). This sinus is the largest of the paranasal sinuses, that is, sinuses connected by channels to the nasal cavity (see Figure 22-5, p. 583). For other markings of the maxillae, see Table 7-4.

Unlike the upper jaw, which is formed by the articula-

tion of the two maxillae, the lower jaw, because of fusion of its halves during infancy, consists of a single bone, the **mandible** (see Figure 7-6). It is the largest, strongest bone of the face. It articulates with the temporal bone in the only movable joint of the skull. Its major markings are identified in Table 7-4.

The cheek is shaped by the underlying **zygomatic,** or malar, bone. This bone also forms the outer margin of the orbit and, with the zygomatic process of the temporal bone, makes the zygomatic arch. It articulates with four other facial bones: the maxillary, temporal, frontal, and sphenoid bones.

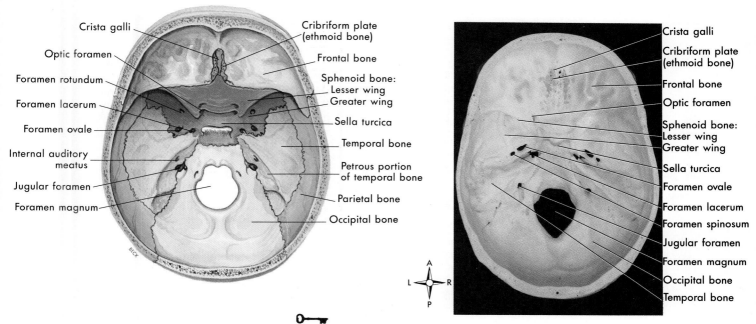

FIGURE 7-4   Floor of the cranial cavity.

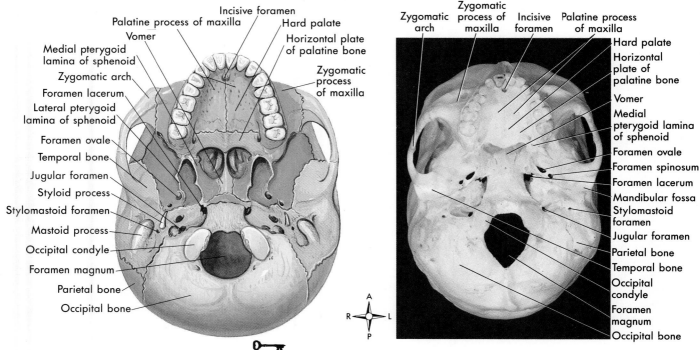

FIGURE 7-5   Skull viewed from below.

Shape is given to the nose by the two **nasal bones,** which form the upper part of the bridge of the nose, and by *cartilage,* which forms the lower part. Although small in size, the nasal bones enter into several articulations: with the perpendicular plate of the ethmoid bone, the cartilaginous part of the nasal septum, the frontal bone, the maxillae, and each other.

An almost paper-thin bone, shaped and sized about like a fingernail, lies just posterior and lateral to each nasal bone. It helps form the sidewall of the nasal cavity and the medial wall of the orbit. Because it contains a groove for the nasolacrimal (tear) duct, this bone is called the **lacrimal bone** (see Figures 7-3 and 7-7). It joins the maxilla, frontal bone, and ethmoid bone.

The two **palatine bones** join to each other in the midline like two Ls facing each other. Their united horizontal portions form the posterior part of the hard palate (see Figure 7-5). The vertical portion of each palatine bone forms the lateral wall of the posterior part of each nasal cavity. The palatine bones articulate with the maxillae and the sphenoid bone.

There are two **inferior nasal conchae** (turbinates). Each

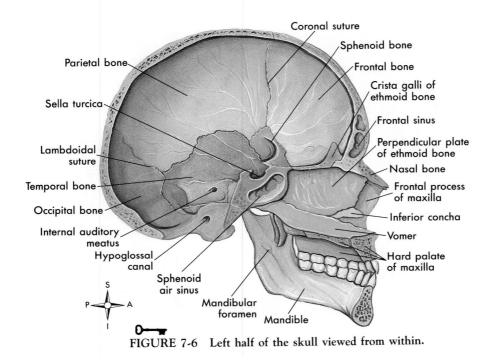

FIGURE 7-6 Left half of the skull viewed from within.

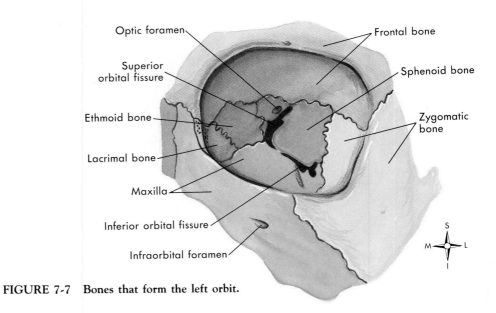

FIGURE 7-7 Bones that form the left orbit.

one is scroll-like in shape and forms a kind of ledge projecting into the nasal cavity from its lateral wall. In each nasal cavity there are three such ledges. The superior and middle conchae (which are projections of the ethmoid bone) form the upper and middle ledges. The inferior concha (which is a separate bone) forms the lower ledge. They are mucosa covered and divide each nasal cavity into three narrow, irregular channels, the *nasal meati*. The inferior nasal conchae form immovable joints with the ethmoid, lacrimal, maxilla, and palatine bones.

Two structures that enter into the formation of the nasal septum have already been mentioned—the perpendicular plate of the ethmoid bone and the septal cartilage. One other structure, the **vomer bone,** completes the septum posteriorly (see Figure 7-6 and Table 7-5). It is usually described as being shaped like a ploughshare. It forms immovable joints with four bones: the sphenoid, ethmoid, palatine, and maxillae.

Special features of the skull include sutures, *fontanels* (Figure 7-9), sinuses, orbits, nasal septum, wormian bones, and the *auditory ossicles;* all of which are described in Table 7-5.

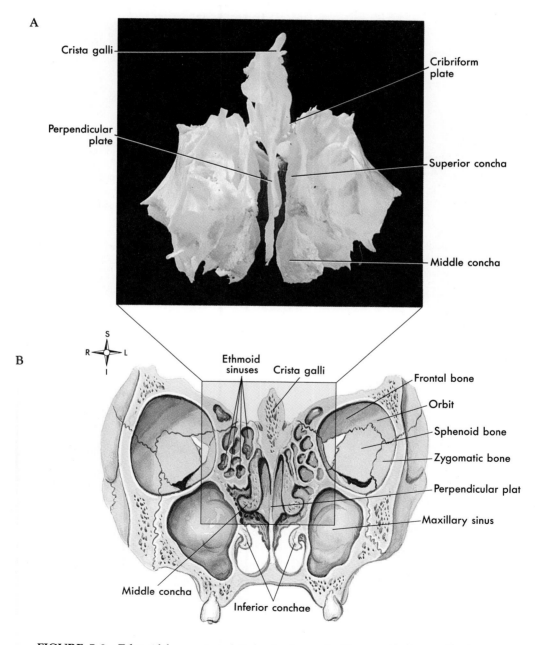

FIGURE 7-8 Ethmoid bone, viewed from the front. A, Photograph of an isolated ethmoid bone. **B,** The ethmoid as seen in a frontal section of the skull.

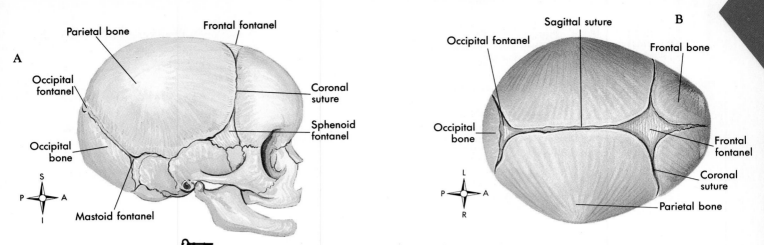

FIGURE 7-9 Skull at birth. A, Viewed from the side (lateral). **B,** Viewed from above (superior).

TABLE 7-5   Special features of skull

Feature	Description	Feature	Description
SUTURES	Immovable joints between skull bones	**AIR SINUSES**	Spaces or cavities within bones; those that communicate with nose called *paranasal sinuses* (frontal, sphenoidal, ethmoidal, and maxillary); mastoid cells communicate with middle ear rather than nose, therefore not included among paranasal sinuses
Squamous	Line of articulation along top curved edge of temporal bone		
Coronal	Joint between parietal bones and frontal bone		
Lambdoidal	Joint between parietal bones and occipital bone		
Sagittal	Joint betwen right and left parietal bones	**ORBITS FORMED BY**	
		Frontal	Roof of orbit
FONTANELS	"Soft spots" where ossification is incomplete at birth; allow some compression of skull during birth; also important in determining position of head before delivery; six such areas located at angles of parietal bones	Ethmoid	Medial wall
		Lacrimal	Medial wall
		Sphenoid	Lateral wall
		Zygomatic	Lateral wall
		Maxillary	Floor
		Palatine	Floor
Frontal (or anterior)	At intersection of sagittal and coronal sutures (juncture of parietal bones and frontal bone); diamond shaped; largest of fontanels; usually closed by 1½ years of age	**NASAL SEPTUM FORMED BY**	Partition in midline of nasal cavity; separates cavity into right and left halves
		Perpendicular plate of ethmoid bone	Forms upper part of septum
Occipital (or posterior)	At intersection of sagittal and lambdoidal sutures (juncture of parietal bones and occipital bone); triangular in shape; usually closed by second month	Vomer bone	Forms lower, posterior part
		Cartilage	Forms anterior part
		WORMIAN BONES	Small islets of bones within suture
Sphenoid (or anterolateral)	At juncture of frontal, parietal, temporal, and sphenoid bones	**MALLEUS, INCUS, STAPES**	Tiny bones, referred to as auditory ossicles, in middle ear cavity in temporal bones; resemble, respectively, miniature hammer, anvil, and stirrup
Mastoid (or posterolateral)	At juncture of parietal, occipital, and temporal bones; usually closed by second year		

TABLE 7-6 Hyoid, vertebrae, and thoracic bones and their markings

Bones and markings	Description	Bones and markings	Description
HYOID	U-shaped bone in neck between mandible and upper part of larynx; distinctive as only bone in body not forming a joint with any other bone; suspended by ligaments from styloid processes of temporal bones	**VERTEBRAL COLUMN—cont'd**	
		Superior articulating processes	Project upward from laminae
		Inferior articulating processes	Project downward from laminae; articulate with superior articulating processes of vertebrae below
VERTEBRAL COLUMN	Not actually a column but a flexible, segmented curved rod; forms axis of body; head balanced above, ribs and viscera suspended in front, and lower extremities attached below; encloses spinal cord	**Spinal foramen**	Hole in center of vertebra formed by union of body, pedicles, and laminae; spinal foramina, when vertebrae, superimposed one on other, form spinal cavity that houses spinal cord
General features	Anterior part of each vertebra (except first two cervical) consists of body; posterior part of vertebrae consists of neural arch, which, in turn, consists of two pedicles, two laminae, and seven processes projecting from laminae	**Intervertebral foramina**	Opening between vertebrae through which spinal nerves emerge
Body	Main part; flat, round mass located anteriorly; supporting or weight-bearing part of vertebra	**CERVICAL VERTEBRAE**	First or upper seven vertebrae; foramen in each transverse process for transmission of vertebral artery, vein, and plexus of nerves; short bifurcated spinous processes except on seventh vertebrae, where it is extra long and may be felt as protrusion when head bent forward; bodies of these vertebrae small, whereas spinal foramina large and triangular
Pedicles	Short projections extending posteriorly from body		
Lamina	Posterior part of vertebra to which pedicles join and from which processes project	**Atlas**	First cervical vertebra; lacks body and spinous process; superior articulating processes concave ovals that act as rockerlike cradles for condyles of occipital bone; named *atlas* because it supports the head as Atlas supports the world in Greek mythology
Neural arch	Formed by pedicles and laminae; protects spinal cord posteriorly; congenital absence of one or more neural arches is known as *spina bifida* (cord may protrude right through skin)		
		Axis (epistropheus)	Second cervical vertebra, so named because atlas rotates about this bone in rotating movements of head; *dens,* or odontoid process, peglike projection upward from body of axis, forming pivot for rotation of atlas
Spinous process	Sharp process projecting inferiorly from laminae in midline		
Transverse processes	Right and left lateral projections from laminae		

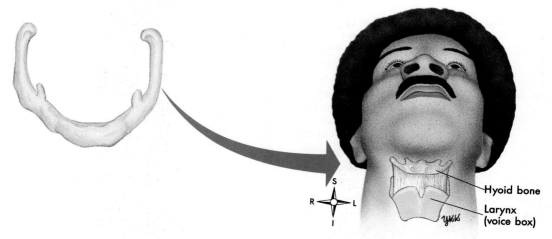

FIGURE 7-10 Hyoid bone.

TABLE 7-6   Hyoid, vertebrae, and thoracic bones and their markings—cont'd

Bones and markings	Description	Bones and markings	Description
THORACIC VERTEBRAE	Next 12 vertebrae; 12 pairs of ribs attached to these; stronger, with more massive bodies than cervical vertebrae; no transverse foramina; two sets of facets for articulations with corresponding rib: one on body, second on transverse process; upper thoracic vertebrae with elongated spinous process.	**CURVES—cont'd**	
		Secondary	Concavities in *cervical* and *lumbar* regions; cervical concavity results from infant's attempts to hold head erect (2 to 4 months); lumbar concavity, from balancing efforts in learning to walk (10 to 18 months)
LUMBAR VERTEBRAE	Next 5 vertebrae; strong, massive; superior articulating processes directed medially instead of upward; inferior articulating processes, laterally instead of downward; short, blunt spinous process	**STERNUM**	Breastbone; flat dagger-shaped bone; sternum, ribs, and thoracic vertebrae together form bony cage known as *thorax*
		Body	Main central part of bone
		Manubrium	Flaring, upper part
SACRUM	Five separate vertebrae until about 25 years of age; then fused to form one wedge-shaped bone	Xiphoid process	Projection of cartilage at lower border of bone
Sacral promontory	Protuberance from anterior, upper border of sacrum into pelvis; of obstetrical importance because its size limits anteroposterior diameter of pelvic inlet	**RIBS**	
		True ribs	Upper seven pairs; fasten to sternum by costal cartilages
COCCYX	Four or five separate vertebrae in child but fused into one in adult	*False ribs*	False ribs do not attach to sternum directly; upper three pairs of false ribs attach by means of costal cartilage of seventh ribs; last two pairs do not attach to sternum at all, therefore called "floating" ribs
CURVES	Curves have great structural importance because they increase carrying strength of vertebral column, make balance possible in upright position (if column were straight, weight of viscera would pull body forward), absorb jars from walking (straight column would transmit jars straight to head), and protect column from fracture	Head	Projection at posterior end of rib; articulates with corresponding thoracic vertebra and one above, except last three pairs, which join corresponding vertebrae only
		Neck	Constricted portion just below head
		Tubercle	Small knob just below neck; articulates with transverse process of corresponding thoracic vertebra; missing in lowest 3 ribs
Primary	Column curves at birth from head to sacrum with convexity posteriorly; after child stands, convexity persists only in *thoracic* and *sacral* regions, which therefore are called *primary curves*	Body or shaft	Main part of rib
		Costal cartilage	Cartilage at sternal end of true ribs; attaches ribs (except floating ribs) to sternum

Hyoid Bone

The hyoid bone is a single bone in the neck—a part of the axial skeleton (Table 7-6). Its U shape may be felt just above the larynx (voice box) and below the mandible where it is suspended from the styloid processes of the temporal bones (Figure 7-10). Several muscles attach to the hyoid bone. Among them are an extrinsic tongue muscle and certain muscles of the floor of the mouth. The hyoid claims the distinction of being the only bone in the body that articulates with no other bones.

QUICK CHECK

✔ 1. Name the eight bones of the cranium and describe how they fit together.
✔ 2. Name the 14 bones of the face and describe how they fit together.
✔ 3. Which bone is the only bone that normally does not form a joint with any other bone of the skeleton?

Vertebral Column

The vertebral, or spinal, column forms the longitudinal axis of the skeleton. It is a flexible rather than a rigid column because it is segmented. As Figure 7-11 shows, the vertebral column consists of 24 vertebrae plus the sacrum and coccyx. Joints between the vertebrae permit forward, backward, and sideway movement of the column. Consider too these further facts about the vertebral column. The head is balanced on top, the ribs and viscera are suspended in front, the lower extremities are attached below, and the spinal cord is enclosed within. It is indeed the "backbone" of the body.

The seven **cervical vertebrae** constitute the skeletal framework of the neck (Figure 7-11). The next 12 vertebrae are called **thoracic vertebrae** because of their location in the posterior part of the chest or thoracic region. The next five, the **lumbar vertebrae,** support the small of the back. Below the lumbar vertebrae lie the *sacrum* and *coccyx.* In the adult the sacrum is a single bone that has resulted from the fusion of five separate vertebrae, and the coccyx is a single bone that has resulted from the fusion of four or five vertebrae.

All the vertebrae resemble each other in certain features and differ in others. For example, all except the first cervical vertebra have a flat, rounded body placed anteriorly and centrally, plus a sharp or blunt *spinous process* projecting inferiorly in the posterior midline and two transverse processes projecting laterally (Figures 7-11 to 7-13). All but the sacrum and coccyx have a central opening, the *vertebral foramen.* An upward projection (the *dens*) from the body of the second cervical vertebra furnishes an axis for rotating the head. A long, blunt spinous process that can be felt at the back of the base of the neck characterizes the seventh cervical vertebra. Each thoracic vertebra has articular facets for the ribs. More detailed descriptions of separate vertebrae are given in Table 7-6. The vertebral column, as a whole, articulates with the head, ribs, and iliac bones. Individual vertebrae articulate with each other in joints between their bodies and between their articular processes.

To increase the carrying strength of the vertebral column and to make balance possible in the upright position, the vertebral column is curved. At birth there is a continuous posterior convexity from head to coccyx. Later, as the child learns to sit and stand, secondary posterior concavities necessary for balance develop in the cervical and lumbar regions (see Figures 32-18 and 32-19).

Sternum

The medial part of the anterior chest wall is supported by the **sternum,** a somewhat dagger-shaped bone consisting of three parts: the upper handle part, the *manubrium;* the middle blade part, the *body;* and a blunt cartilaginous lower tip, the *xiphoid process.* The last ossifies during adult life.

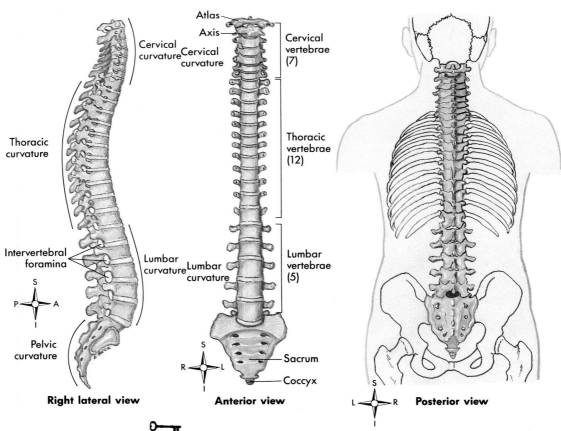

Atlas
Axis
Cervical curvature
Cervical curvature
Cervical vertebrae (7)
Thoracic curvature
Thoracic vertebrae (12)
Intervertebral foramina
Lumbar curvature
Lumbar curvature
Lumbar vertebrae (5)
Pelvic curvature
Sacrum
Coccyx

Right lateral view   **Anterior view**   **Posterior view**

FIGURE 7-11 The vertebral column (three views.)

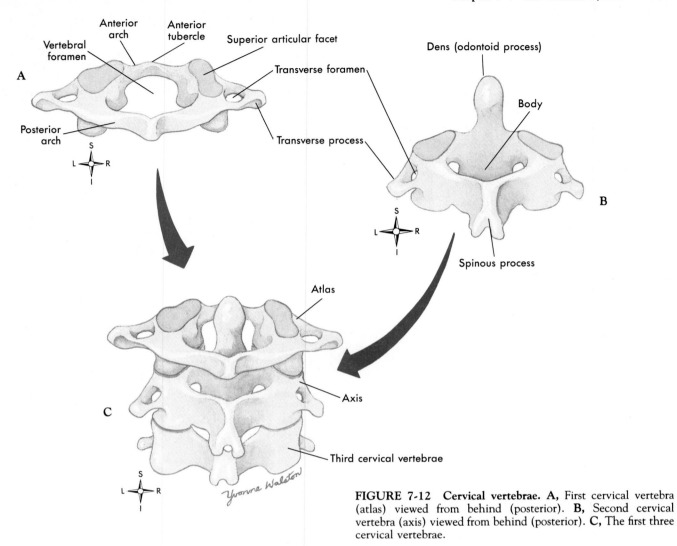

FIGURE 7-12 Cervical vertebrae. A, First cervical vertebra (atlas) viewed from behind (posterior). **B,** Second cervical vertebra (axis) viewed from behind (posterior). **C,** The first three cervical vertebrae.

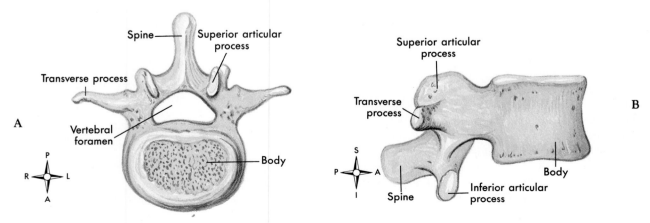

FIGURE 7-13 Lumbar vertebra. A, Third lumbar vertebra viewed from above (superior). **B,** Third lumbar vertebra viewed from the side (lateral).

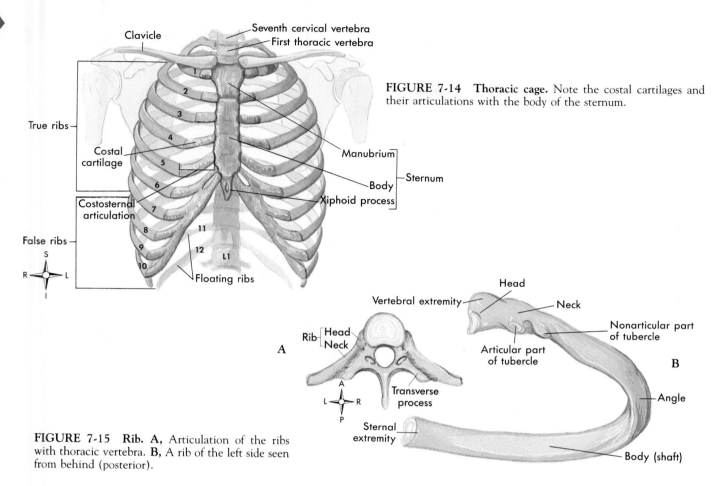

FIGURE 7-14 Thoracic cage. Note the costal cartilages and their articulations with the body of the sternum.

FIGURE 7-15 Rib. A, Articulation of the ribs with thoracic vertebra. **B,** A rib of the left side seen from behind (posterior).

The manubrium articulates with the clavicle and first rib, whereas the next nine ribs join the body of the sternum, either directly or indirectly, by means of the *costal cartilages* (Figure 7-14).

Ribs

Twelve pairs of ribs, together with the vertebral column and sternum, form the bony cage known as the *thoracic cage,* or simply, the **thorax.** Each rib articulates with both the body and the transverse process of its corresponding thoracic vertebra. The head of each rib articulates with the body of the corresponding thoracic vertebra, and the tubercle of each rib articulates with the vertebra's transverse process (Figure 7-15). In addition, the second through the ninth ribs articulate with the body of the vertebra above. From its vertebral attachment, each rib curves outward, then forward and downward (Figures 7-1 and 7-15), a mechanical fact important for breathing. Anteriorly, each rib of the first seven pairs joins a costal cartilage that attaches to the sternum. For this reason, these ribs are often called the *true ribs.* Ribs of the remaining five pairs, the *false ribs,* do not attach directly to the sternum. Instead, each costal cartilage of pairs eight, nine, and ten attaches to the costal cartilage of the rib above it—indirectly attaching it to the sternum. Ribs of the last two pairs of false ribs are designated as *floating ribs* because they do not attach even indirectly to the sternum (Figure 7-14).

APPENDICULAR SKELETON
Upper Extremity

The upper extremity consists of the bones of the shoulder girdle, upper arm, lower arm, wrist, and hand. Two bones, the **clavicle** and **scapula,** compose the **shoulder girdle.** Contrary to appearances, this girdle forms only one bony joint with the trunk: the sternoclavicular joint between the sternum and clavicle. At its outer end, the clavicle articulates with the scapula, which attaches to the ribs by muscles and tendons, not by a joint. All shoulder movements therefore involve the sternoclavicular joint. Various markings of the scapula are described in Table 7-7 (see also Figure 7-16).

QUICK CHECK
1. Name the three types of vertebrae and how many of each type are found in the vertebral column. Besides the vertebrae, what other bones are in the vertebral column?
2. What bones make up the bony cage known as the thorax? How do these bones fit together to form this structure?
3. What is a floating rib?

TABLE 7-7   Upper extremity bones and their markings

Bones and markings	Description	Bones and markings	Description
CLAVICLE	Collar bones; shoulder girdle joined to axial skeleton by articulation of clavicles with sternum (scapula does not form joint with axial skeleton)	**HUMERUS—cont'd**	
		Trochlea	Projection with deep depression through center similar to shape of pulley; articulates with ulna
SCAPULA	Shoulder blades; scapulae and clavicles together comprise shoulder girdle	Olecranon fossa	Depression on posterior surface just above trochlea; receives olecranon process of ulna when lower arm extends
Borders			
Superior	Upper margin	Coronoid fossa	Depression on anterior surface above trochlea; receives coronoid process of ulna in flexion of lower arm
Vertebral	Margin toward vertebral column		
Axillary	Lateral margin	**RADIUS**	Bone of thumb side of forearm
Spine	Sharp ridge running diagonally across posterior surface of shoulder blade	Head	Disk-shaped process forming proximal end of radius; articulates with capitulum of humerus and with radial notch of ulna
Acromion process	Slightly flaring projection at lateral end of scapular spine; may be felt as tip of shoulder; articulates with clavicle		
		Radial tuberosity	Roughened projection on ulnar side, short distance below head; biceps muscle inserts here
Coracoid process	Projection on anterior surface from upper border of bone; may be felt in groove between deltoid and pectoralis major muscles, about 1 inch below clavicle	Styloid process	Protuberance at distal end on lateral surface (with forearm in anatomical position)
Glenoid cavity	Arm socket	**ULNA**	Bone of little finger side of forearm; longer than radius
HUMERUS	Long bone of upper arm	Olecranon process	Elbow
Head	Smooth, hemispherical enlargement at proximal end of humerus	Coronoid process	Projection on anterior surface of proximal end of ulna; trochlea of humerus fits snugly between olecranon and coronoid processes
Anatomical neck	Oblique groove just below head		
Greater tubercle	Rounded projection lateral to head on anterior surface	Semilunar notch	Curved notch between olecranon and coronoid process into which trochlea fits
Lesser tubercle	Prominent projection on anterior surface just below anatomical neck	Radial notch	Curved notch lateral and inferior to semilunar notch; head of radius fits into this concavity
Intertubercular groove	Deep groove between greater and lesser tubercles; long tendon of biceps muscle lodges here	Head	Rounded process at distal end; does not articulate with wrist bones but with fibrocartilaginous disk
Surgical neck	Region just below tubercles; so named because of its liability to fracture	Styloid process	Sharp protuberance at distal end; can be seen from outside on posterior surface
Deltoid tuberosity	V-shaped, rough area about midway down shaft where deltoid muscle inserts	**CARPALS**	Wrist bones; arranged in two rows at proximal end of hand: proximal row (from little finger toward thumb)—*pisiform, triquetrum, lunate,* and *scaphoid*; distal row—*hamate, capitate, trapezoid,* and *trapezium*
Radial groove	Groove running obliquely downward from deltoid tuberosity; lodges radial nerve		
Epicondyles (medial and lateral)	Rough projections at both sides of distal end	**METACARPALS**	Long bones forming framework of palm of hand; numbered I through V
Capitulum	Rounded knob below lateral epicondyle; articulates with radius; sometimes called *radial head* of humerus	**PHALANGES**	Miniature long bones of fingers, three (proximal, middle, distal) in each finger, two (proximal, distal) in each thumb

The **humerus** or upper arm bone, like other long bones, consists of a shaft, or diaphysis, and two ends, or epiphyses (Figures 7-17 and 7-18). The upper epiphysis bears several identifying structures: the head, anatomical neck, greater and lesser tubercles, intertubercular groove, and surgical neck. On the diaphysis are found the deltoid tuberosity and the radial groove. The distal epiphysis has four projections—the medial and lateral epicondyles, the capitulum, and the trochlea—and two depressions—the olecranon and coronoid fossae. For descriptions of all of these markings, see Table 7-7. The humerus articulates proximally with the scapula and distally with both the radius and the ulna.

Two bones form the framework for the forearm: the **radius** on the thumb side and the **ulna** on the little finger side. At the proximal end of the ulna the olecranon process projects posteriorly and the coronoid process projects anteriorly. There are also two depressions: the semilunar notch on the anterior surface and the radial notch on the lateral surface. The distal end has two projections: a rounded head and a sharper styloid process. For more detailed identification of these markings, see Table 7-7. The ulna articulates proximally with the humerus and radius and distally with a fibrocartilaginous disk, but not with any of the carpal bones.

The radius has three projections: two at its proximal end, the head and radial tuberosity, and one at its distal end, the styloid process (see Figures 7-17 and 7-18). There are two proximal articulations: one with the capitulum of the humerus and the other with the radial notch of the ulna. The three distal articulations are with the scaphoid and lunate carpal bones and with the head of the ulna.

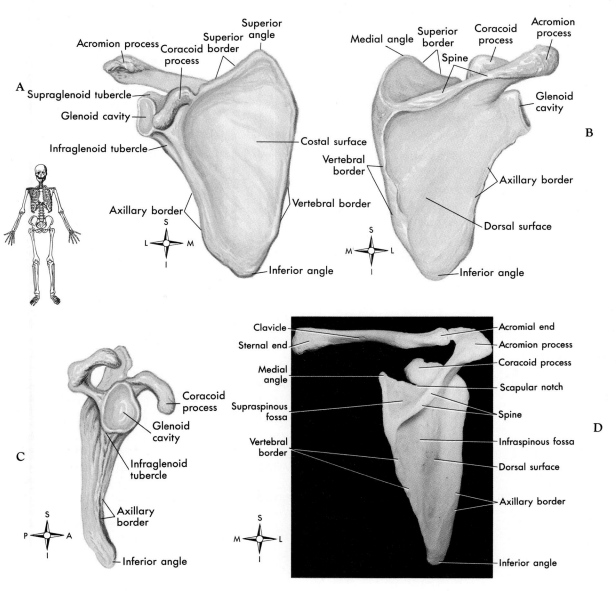

FIGURE 7-16 Right scapula. A, Anterior view. **B,** Posterior view. **C,** Lateral view. **D,** Posterior view showing articulation with clavicle.

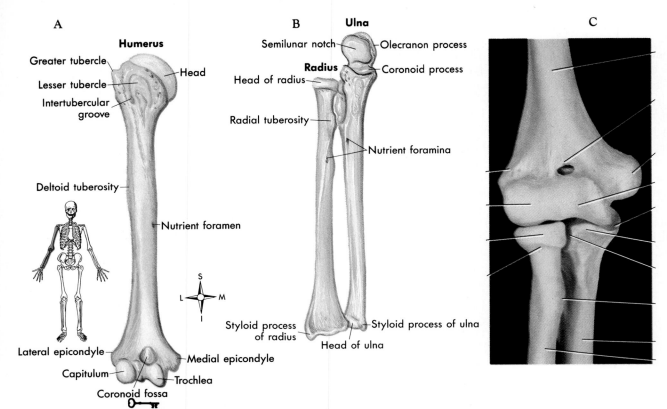

A

Humerus

Greater tubercle

Lesser tubercle

Intertubercular groove

Head

Deltoid tuberosity

Nutrient foramen

Lateral epicondyle

Capitulum

Coronoid fossa

Medial epicondyle

Trochlea

B

Ulna

Semilunar notch

Radius

Head of radius

Radial tuberosity

Olecranon process

Coronoid process

Nutrient foramina

Styloid process of radius

Styloid process of ulna

Head of ulna

C

FIGURE 7-17   **Bones of the arm** (right arm, anterior view). **A,** Humerus (upper arm). **B,** Radius and ulna (forearm). **C,** Elbow joint, showing how the distal end of the humerus joins the proximal ends of the radius and ulna. (The inset shows the relative position of the right arm bones within the entire skeleton.)

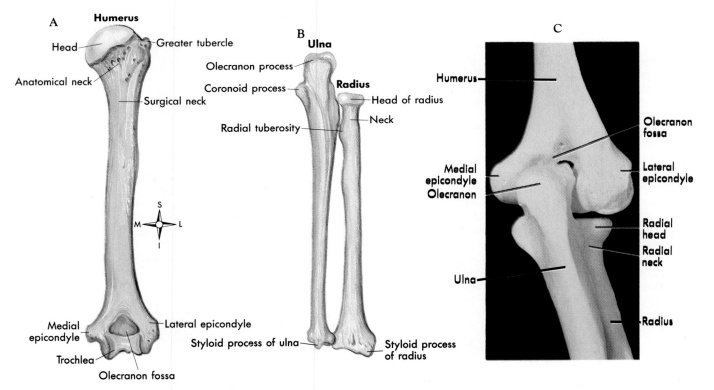

A

Humerus

Head

Anatomical neck

Greater tubercle

Surgical neck

Medial epicondyle

Trochlea

Olecranon fossa

Lateral epicondyle

B

Ulna

Olecranon process

Coronoid process

Radius

Radial tuberosity

Head of radius

Neck

Styloid process of ulna

Styloid process of radius

C

Humerus

Medial epicondyle

Olecranon

Ulna

Olecranon fossa

Lateral epicondyle

Radial head

Radial neck

Radius

FIGURE 7-18   **Bones of the arm** (right arm, posterior view). **A,** Humerus (upper arm). **B,** Radius and ulna (forearm). **C,** Elbow joint, showing how the distal end of the humerus joins the proximal ends of the radius and ulna.

The eight **carpal bones** (Figure 7-19) form what most people think of as the upper part of the hand but what, anatomically speaking, is the wrist. Only one of these bones is evident from the outside, the *pisiform bone*, which projects posteriorly on the little finger side as a small rounded elevation. Ligaments bind the carpals closely and firmly together in two rows of four each: proximal row (from little finger toward thumb)—pisiform, triquetrum, lunate, and scaphoid bones; distal row—hamate, capitate, trapezoid, and trapezium bones. The joints between the carpals and radius permit wrist and hand movements.

Of the five **metacarpal bones** that form the framework of the hand, the thumb metacarpal forms the most freely movable joint with the carpals. This fact has great significance. Because of the wide range of movement possible between the thumb metacarpal and the trapezium,

particularly the ability to oppose the thumb to the fingers, the human hand has much greater dexterity than the forepaw of any animal and has enabled humans to manipulate their environment effectively. The heads of the metacarpals, prominent as the proximal knuckles of the hand, articulate with the phalanges.

QUICK CHECK

✔ 1. *What bones make up the shoulder girdle? Where does the shoulder girdle form a joint with the axial skeleton?*

✔ 2. *What are the two bones of the forearm? In the anatomical position, which one is lateral?*

✔ 3. *Name the bones of the hand and wrist.*

 Palpable Bony Landmarks

Health professionals often identify externally palpable bony landmarks when dealing with the sick and injured. **Palpable** bony landmarks are bones that can be touched and identified through the skin. They serve as reference points in identifying other body structures.

There are externally palpable bony landmarks throughout the body. Many skull bones, such as the zygomatic bone, can be palpated. The medial and lateral epicondyles of the humerus, the olecranon process of the ulna, and the styloid process of the ulna and the radius at the wrist can be palpated on the upper extremity. The highest corner of the shoulder is the acromion process of the scapula.

When you put your hands on your hips, you can feel the superior edge of the ilium, called the *iliac crest*. The anterior end of the crest, called the *anterior superior iliac spine*, is a prominent landmark used often as a clinical reference. The medial malleolus of the tibia and the lateral malleolus of the fibula are prominent at the ankle. The calcaneus or heel bone is easily palpated on the posterior aspect of the foot. On the anterior aspect of the lower extremity, examples of palpable bony landmarks include the patella or kneecap; the anterior border of the tibia, or shin bone; and the metatarsals and phalanges of the toes. Try to identify as many of the externally palpable bones of the skeleton as possible on your own body. Using these as points of reference will make it easier for you to visualize the placement of other bones that cannot be touched or palpated through the skin.

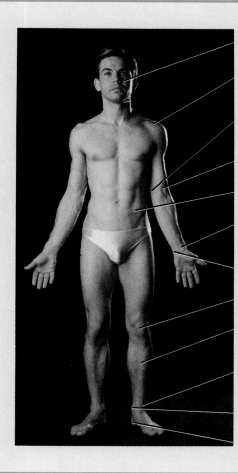

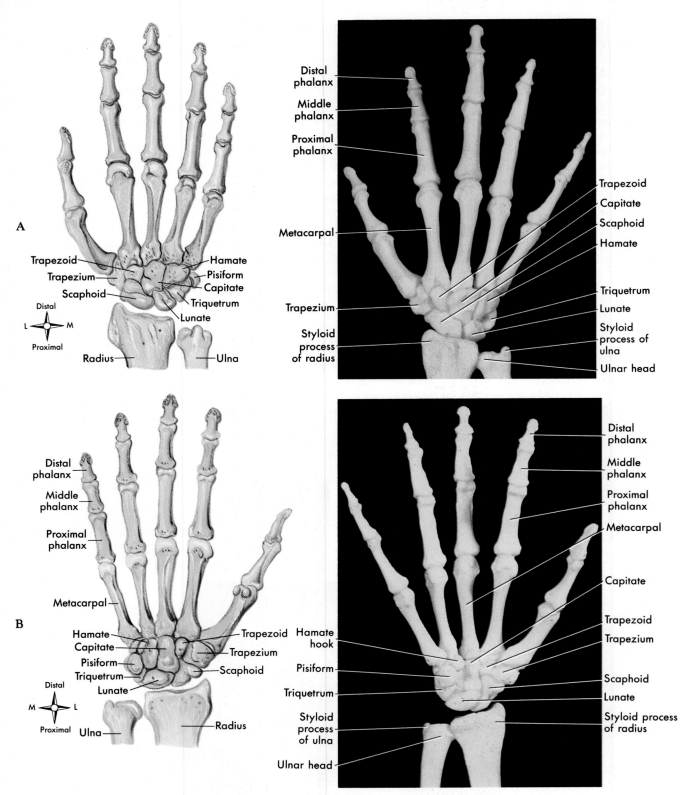

FIGURE 7-19 Bones of the hand and wrist. A, Dorsal view of the right hand and wrist. **B,** Palmar view of the right hand and wrist.

Lower Extremity

Bones of the hip, thigh, lower leg, ankle, and foot constitute the lower extremity (Table 7-8). Strong ligaments bind each **coxal bone** (*os coxae,* or *innominate bone*) to the sacrum posteriorly and to each other anteriorly to form the **pelvic girdle** (Figures 7-20 and 7-25), a stable, circular base that supports the trunk and attaches the lower extremities to it. In early life, each coxal bone is made up of three separate bones. Later on, they fuse into a single, massive irregular bone that is broader than any other bone in the body. The largest and uppermost of the three bones is the **ilium;** the strongest and lowermost, the **ischium;** and the anteriormost, the **pubis.** Numerous markings are present on the three bones (see Table 7-8 and Figures 7-20 and 7-21).

The pelvis can be divided into two parts by an imaginary plane, called the *pelvic inlet.* The edge of this plane, outlined in Figure 7-20, is called the *pelvic brim,* or *brim of the true pelvis.* The structure above the pelvic inlet, termed the **false pelvis,** is bordered by muscle in the front and bone along the sides and back. The structure below the pelvic inlet, the so-called true pelvis, creates the boundary of another imaginary plane, called the pelvic outlet. It is through the pelvic outlet that the digestive tract empties. The female reproductive tract also passes through the pelvic outlet; this is a fact of great importance in childbirth. The pelvic outlet is just large enough for the passage of a baby during delivery; however, careful positioning of the baby's head is required. Measurements such as those shown in Figure 7-20 are routinely made by obstetricians to ensure successful delivery. Despite its apparent rigidity, the joint between the pubic portions of each coxal bone, the *symphysis pubis,* softens prior to delivery. This allows the pelvic outlet to expand to accommodate the newborn's head as it passes out of the birth canal. The tiny coccyx bone, which protrudes into the pelvic outlet, sometimes breaks when the force of labor contractions pushes the newborn's head against it.

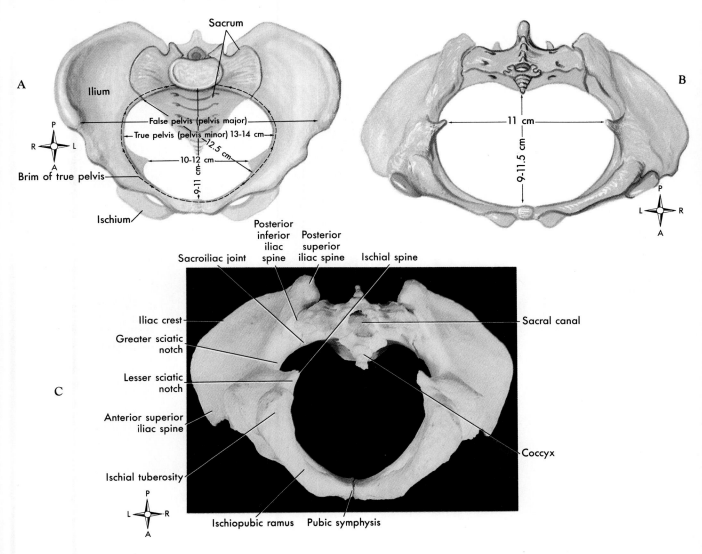

FIGURE 7-20 **The female pelvis. A,** Pelvis viewed from above. Note that the brim of the true pelvis (*dotted line*) marks the boundary between the superior false pelvis (*pelvis major*) and the inferior true pelvis (*pelvis minor*). **B** and **C,** Pelvis viewed from below.

FIGURE 7-21 Right coxal bone. The right coxal bone is disarticulated from the skeleton and viewed from the side with the bone turned so as to look directly into the acetabulum.

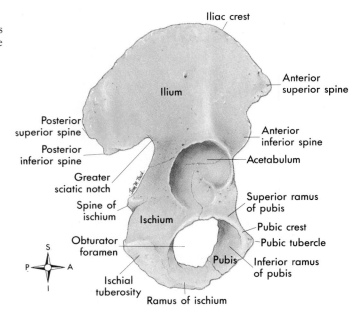

Bones and markings	Description	Bones and markings	Description
COXAL	Large hip bone; with sacrum and coccyx, forms basinlike pelvic cavity; lower extremities attached to axial skeleton by coxal bones	**COXAL—cont'd**	
		Superior ramus of pubis	Part of pubis lying between symphysis and acetabulum; forms upper part of obturator foramen
Ilium	Upper, flaring portion	**Inferior ramus**	Part extending down from symphysis; unites with ischium
Ischium	Lower, posterior portion		
Pubic bone (pubis)	Medial, anterior section	**Pubic arch**	Angle formed by two inferior rami
Acetabulum	Hip socket; formed by union of ilium, ischium, and pubis	**Pubic crest**	Upper margin of superior ramus
		Pubic tubercle	Rounded process at end of crest
Iliac crests	Upper, curving boundary of ilium	**Obturator foramen**	Large hole in anterior surface of os coxa; formed by pubis and ischium; largest foramen in body
Iliac spines			
Anterior superior	Prominent projection at anterior end of iliac crest; can be felt externally as "point" of hip	**Pelvic brim (or inlet)**	Boundary of aperture leading into true pelvis; formed by pubic crests, iliopectineal lines, and sacral promontory; size and shape of this inlet have obstetrical importance, since if any of its diameters too small, infant skull cannot enter true pelvis for natural birth
Anterior inferior	Less prominent projection short distance below anterior superior spine		
Posterior superior	At posterior end of iliac crest	**True pelvis (or pelvis minor)**	Space below pelvic brim; true "basin" with bone and muscle walls and muscle floor; pelvic organs located in this space
Posterior inferior	Just below posterior superior spine		
Greater sciatic notch	Large notch on posterior surface of ilium just below posterior inferior spine	**False pelvis (or pelvis major)**	Broad, shallow space above pelvic brim, or pelvic inlet; name "false pelvis" is misleading, since this space is actually part of abdominmal cavity, not pelvic cavity
Ischial tuberosity	Large, rough, quadrilateral process forming inferior part of ischium; in erect sitting position body rests on these tuberosities		
Ischial spine	Pointed projection just above tuberosity	**Pelvic outlet**	Irregular circumference marking lower limits of true pelvis; bounded by tip of coccyx and two ischial tuberosities *Continued.*
Symphysis pubis	Cartilaginous, amphiarthrotic joint between pubic bones		

TABLE 7-8   Lower extremity bones and their markings

TABLE 7-8 Lower extremity bones and their markings—cont'd

Bones and markings	Description	Bones and markings	Description
COXAL—cont'd		**TIBIA**	Shin bone
Pelvic girdle (or bony pelvis)	Complete bony ring; composed of two hip bones (ossa coxae), sacrum, and coccyx; forms firm base by which trunk rests on thighs and for attachment of lower extremities to axial skeleton	**Condyles**	Bulging prominences at proximal end of tibia; upper surfaces concave for articulation with femur
		Intercondylar eminence	Upward projection on articular surface between condyles
		Crest	Sharp ridge on anterior surface
FEMUR	Thigh bone; largest, strongest bone of body	**Tibial tuberosity**	Projection in midline on anterior surface
Head	Rounded, upper end of bone; fits into acetabulum	**Medial malleolus**	Rounded downward projection at distal end of tibia; forms prominence on medial surface of ankle
Neck	Constricted portion just below head		
Greater trochanter	Protuberance located inferiorly and laterally to head	**FIBULA**	Long, slender bone of lateral side of lower leg
Lesser trochanter	Small protuberance located inferiorly and medially to greater trochanter	**Lateral malleolus**	Rounded prominence at distal end of fibula; forms prominence on lateral surface of ankle
Intertrochanteric line	Line extending between greater and lesser trochanter	**TARSALS**	Bones that form heel and proximal or posterior half of foot
Linea aspera	Prominent ridge extending lengthwise along concave posterior surface	**Calcaneus**	Heel bone
Supracondylar ridges	Two ridges formed by division of linea aspera at its lower end; medial supracondylar ridge extends inward to inner condyle, lateral ridge to outer condyle	**Talus**	Uppermost of tarsals; articulates with tibia and fibula; boxed in by medial and lateral malleoli
		Longitudinal arches	Tarsals and metatarsals so arranged as to form arch from front to back of foot
Condyles	Large, rounded bulges at distal end of femur; one medial and one lateral	**Medial**	Formed by calcaneus, talus, navicular, cuneiforms, and three medial metatarsals
Epicondyles	Blunt projections from the sides of the condyles; one on the medial aspect and one on the lateral aspect	**Lateral**	Formed by calcaneus, cuboid, and two lateral metatarsals
Adductor tubercle	Small projection just above medial condyle; marks termination of medial supracondylar ridge	**Transverse (or metatarsal) arch**	Metatarsals and distal row of tarsals (cuneiforms and cuboid) so articulated as to form arch across foot; bones kept in two arched positions by means of powerful ligaments in sole of foot and by muscles and tendons
Trochlea	Smooth depression between condyles on anterior surface; articulates with patella		
Intercondyloid fossa (notch)	Deep depression between condyles on posterior surface; cruciate ligaments that help bind femur to tibia lodge in this notch	**METATARSALS**	Long bones of feet
		PHALANGES	Miniature long bones of toes; two in each great toe; three in other toes
PATELLA	Kneecap; largest sesamoid bone of body; embedded in tendon of quadriceps femoris muscle		

The two thigh bones, or **femurs,** have the distinction of being the longest and heaviest bones in the body. Several prominent markings characterize them. For example, three projections are conspicuous at each epiphysis: the head and greater and lesser trochanters proximally and the medial and lateral condyles and adductor tubercle distally (Figure 7-22). Both condyles and the greater trochanter may be felt externally. For a description of the various femur markings, see Table 7-8.

The largest sesamoid bone in the body, and the one that is almost universally present, is the **patella,** or kneecap, located in the tendon of the quadriceps femoris muscle as a projection to the underlying knee joint. Although some individuals have sesamoid bones in tendons of other muscles, lists of bone names usually do not include them because they are not always present, are not found in any particular tendons, and are seemingly of no importance. (See the footnote in Table 7-1.) When the knee joint is extended, the patellar outline may be distinguished through the skin, but as the knee flexes, it sinks into the

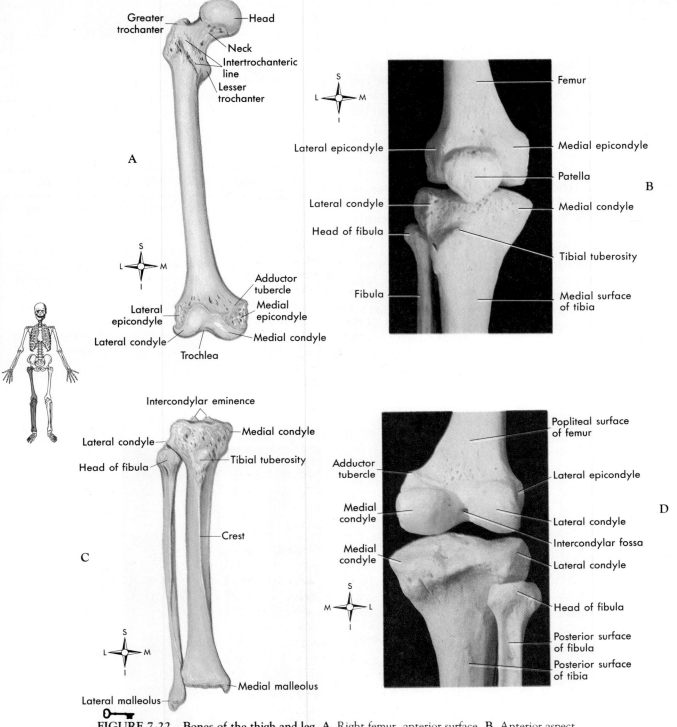

FIGURE 7-22   Bones of the thigh and leg. A, Right femur, anterior surface. **B,** Anterior aspect of the right knee skeleton. **C,** Right tibia and fibula, anterior surface. **D,** Posterior aspect.

intercondylar notch of the femur and can no longer be easily distinguished.

The **tibia** is the larger and stronger and more medially and superficially located of the two leg bones. The **fibula** is smaller and more laterally and deeply placed. At its proximal end it articulates with the lateral condyle of the tibia. The proximal end of the tibia, in turn, articulates with the femur to form the knee joint, the largest and one of the most stable joints of the body. Distally the tibia articulates with the fibula and also with the talus. The latter fits into a boxlike socket (ankle joint) formed by the medial and lateral malleoli, projections of the tibia and fibula, respectively. For other tibial markings, see Table 7-8 and Figure 7-22.

Structure of the **foot** is similar to that of the hand, with certain differences that adapt it for supporting weight (Figure 7-23). One example of this is the much greater solidity and the more limited mobility of the great toe compared to the thumb. Then, too, the foot bones are held together in such a way as to form springy lengthwise and crosswise arches (Figure 7-24). This is architecturally sound, since arches furnish more supporting strength per given amount of structural material than any other type of construction. Hence the two-way arch construction makes a highly stable base. The **longitudinal arch** has an inner, or medial, portion and an outer, or lateral, portion. Both are formed by the placement of tarsals and metatarsals. Specifically, some of the tarsals (calcaneus, talus, navicular, and cuneiforms) and the first three metatarsals (starting with the great toe) form the **medial longitudinal arch.** The calcaneus and cuboid tarsals plus the fourth and fifth metatarsals shape the **lateral longitudinal arch.** The transverse arch results from the relative placement of the distal row of tarsals and the five metatarsals. (See Table 7-8 for specific bones of different arches.) Strong ligaments and leg muscle tendons normally hold the foot bones firmly in their arched positions. Not infrequently, however, these weaken, causing the arches to flatten—a condition aptly called fallen arches, or flatfeet (Figure 7-24, *B*). Look at

Figure 7-24, *D* to see what high heels do to the position of the foot. They give a forward thrust to the body, which forces an undue amount of weight on the heads of the metatarsals. Normally the tarsals and metatarsals play the major role in the functioning of the foot as a supporting structure, with the phalanges relatively unimportant. The reverse is true for the hand. Here, manipulation is the main function rather than support. Consequently, the phalanges of the fingers are all important, and the carpals and metacarpals are subsidiary.

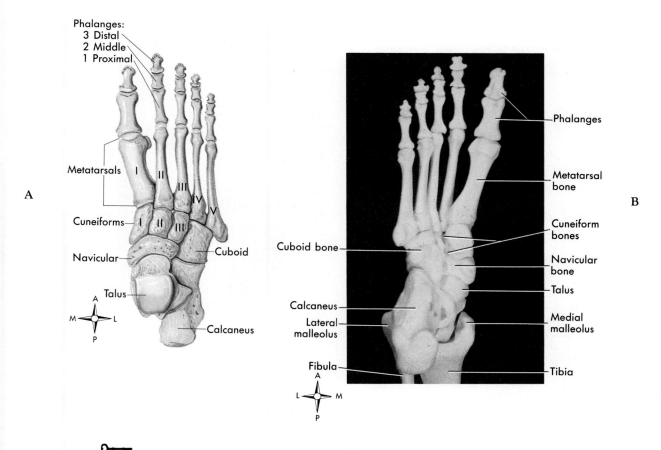

FIGURE 7-23 The foot. A, Bones of right foot viewed from above. Tarsal bones consist of cuneiforms, navicular, talus, cuboid, and calcaneus. **B,** Posterior aspect of the right ankle skeleton and inferior aspect of the right foot skeleton.

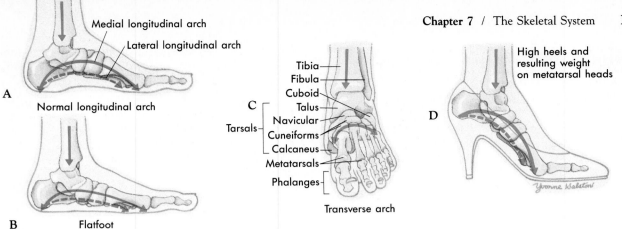

FIGURE 7-24 Arches of the foot. A, Longitudinal arch. Medial portion formed by calcaneus, talus, navicular, cuneiforms, and three metatarsals; lateral portion formed by calcaneus, cuboid, and two lateral metatarsals. **B,** "Flatfoot" results when there is a weakening of tendons and ligaments attached to the tarsal bones. Downward pressure by the weight of the body gradually flattens out the normal arch of the bones. **C,** Transverse arch in the metatarsal region of the right foot. **D,** High heels throw the weight forward, causing the heads of the metatarsals to bear most of the body's weight. (*Arrows* show direction of force.)

SKELETAL DIFFERENCES IN MEN AND WOMEN

Both general and specific differences exist between male and female skeletons. The general difference is one of size and weight, the male skeleton being larger and heavier. The specific differences concern the shape of the pelvic bones and cavity. Whereas the male pelvis is deep and funnel shaped with a narrow pubic arch (usually less than 90 degrees), the female pelvis, as Figure 7-25 shows, is shallow, broad, and flaring, with a wider pubic arch (usually greater than 90 degrees). The childbearing function obviously explains the necessity for these and certain other modifications of the female pelvis. These and other differences between the male and female skeleton are summarized in Table 7-9, p. 196.

> **QUICK CHECK**
> ✔ 1. Which three bones fuse together during skeletal development to form the coxal (hip) bone?
> ✔ 2. List the bones of the lower extremity, indicating their positions in the skeleton.
> ✔ 3. What is the functional advantage of foot arches?
> ✔ 4. Name two differences between typical male and female skeletons.

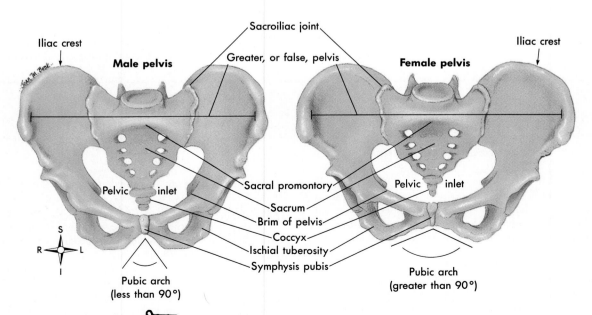

FIGURE 7-25 Comparison of male and female bony pelvis.

TABLE 7-9   Comparison of male and female skeletons

Portion of Skeleton	Male	Female
GENERAL FORM	Bones heavier and thicker Muscle attachment sites more massive Joint surfaces relatively large	Bones lighter and thinner Muscle attachment sites less distinct Joint surfaces relatively small
SKULL	Forehead shorter vertically Mandible and maxillae relatively larger Facial area more pronounced Processes more prominent	Forehead more elongated vertically Mandible and maxillae relatively smaller Facial area rounder, with less pronounced features Processes less pronounced
PELVIS		
Pelvic cavity	Narrower in all dimensions Deeper Pelvic outlet relatively small	Wider in all dimensions Shorter and Roomier Pelvic outlet relatively large
Sacrum	Long, narrow, with smooth concavity (sacral curvature); sacral promontory more pronounced	Short, wide, flat concavity more pronounced in a posterior direction; sacral promontory less pronounced
Coccyx	Less movable	More movable and follows posterior direction of sacral curvature
Pubic arch	Less than a 90° angle	Greater than a 90° angle
Symphysis pubis	Relatively deep	Relatively shallow
Ischial spine, ischial tuberosity, and anterior superior iliac spine	Turned more inward	Turned more outward and further apart
Greater sciatic notch	Narrow	Wide

Cycle of Life: *Skeletal system*

The changes that occur in the body's skeletal framework over the course of a life result primarily from structural changes in bone, cartilage, and muscle tissues. For example, the resilience of incompletely ossified bone in young children allows their bones to withstand the mechanical stresses of childbirth and learning to walk with relatively little risk of fracturing. The density of bone and cartilage in the young to middle-aged adult permit the carrying of great loads. Loss of bone density in later adulthood can make a person so prone to fractures that simply walking or lifting with moderate force can cause bones to crack or break. The loss of skeletal tissue density may result in a compression of weight-bearing bones that causes a loss of height and perhaps an inability to maintain a standard posture. Degeneration of skeletal muscle tissue in late adulthood may also contribute to postural changes and loss of height.

MECHANISMS OF DISEASE

Fractures and Abnormal Spinal Curvatures

Bone Fractures

A *bone fracture* is defined as a partial or complete break in the continuity of the bone that occurs under mechanical stress. The most common cause of a fracture is traumatic injury. Bone cancer or metabolic bone disorders can also cause fractures by weakening a bone to the point that it fractures under very little stress. An *open fracture*, also known as a *compound fracture*, is one in which broken bone projects through surrounding tissue and skin, inviting possibility of infection (Figure 7-26, A). A *closed fracture*, also known as a *simple fracture*, does not produce a break in the skin and therefore does not pose an immediate danger of bone infection (Figure 7-26, B). As Figure 7-26, C shows, fractures are also classified as "complete" or "incomplete." A *complete fracture* involves a break across the entire section of bone, whereas an *incomplete fracture* involves only a partial break where bone fragments are still partially joined.

Fractures are also described anatomically according to the bone involved (e.g., femur) and the region of bone in which the fracture occurs (e.g., distal). There are many different specific types of fractures. For example, a *greenstick fracture* is one in which one side of the bone is bent and the other side

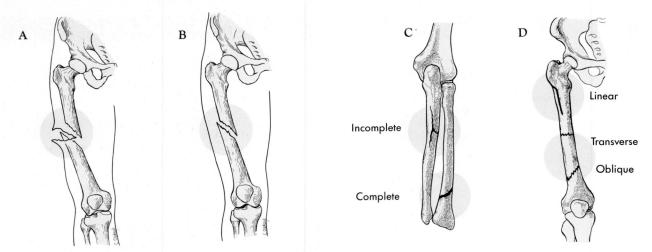

FIGURE 7-26 Bone Fractures. A, Open. **B,** Closed. **C,** Incomplete and complete. **D,** Linear, transverse, and oblique.

is broken. This type of fracture commonly occurs in children because their growing bones are less brittle than in the adult. A "dentate" fracture results in fragmented ends of the bone being jagged and opposing each other, fitting together like teeth on a gear. A "hangman's" fracture is a fracture of the posterior elements of the second vertebral bone of the spine.

Sometimes the angle of the fracture line or crack is used in labeling fracture types (Figure 7-26, *D*). A "linear" fracture involves a fracture line parallel to the bone's long axis. A fracture line at a right angle to bone's long axis is labeled a "transverse" fracture. "Oblique" fractures occur at slanted, or diagonal, angles to longitudinal axis of bone.

Clinical signs and symptoms of a fracture are loss of function or false motion, pain, soft tissue edema, and deformity. These vary with the type and location of the fracture. Treatment usually involves "reduction" or realignment of the bone, immobilization, and restoring function through rehabilitation. Repair of bone tissue is discussed in Skeletal Tissues (Chapter 6).

Abnormal Spinal Curvatures

The normal curvature of the spine is convex through the thoracic region and concave through the cervical and lumbar regions (see Figure 7-11). This gives the spine strength to support the weight of the rest of the body and balance necessary to stand and walk. A curved structure has more strength than a straight one of the same size and material. Poor posture or disease may cause the lumbar curve to be abnormally accentuated—a condition known as "sway back," or *lordosis* (Figure 7-27, *A*). This condition is frequently seen during pregnancy as the woman adjusts to changes in her center of gravity. It may also be idiopathic, secondary to traumatic injury, or a degenerative process of the vertebral bodies. *Kyphosis,* or "hunchback," is an abnormally increased roundness in the thoracic curvature (Figure 7-27, *B*). It is frequently seen in elderly people with osteoporosis or chronic arthritis, those with neuromuscular diseases, or compression fractures of the thoracic vertebrae. In a condition called Scheuermann's disease, kyphosis can develop in children at puberty. Abnormal side-to-side curvature is called *scoliosis* (Figure 7-27, *C*). This too may be idiopathic or a result of

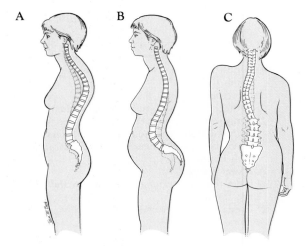

FIGURE 7-27 Abnormal spinal curvatures. A, Kyphosis. **B,** Lordosis. **C,** Scoliosis.

damage to the supporting muscles along the spine. It is a relatively common condition that appears before adolescence.

All three abnormal curvatures can interfere with normal breathing, posture, and other vital functions. The degree of curvature and resulting deformity of the vertebral column will determine the various treatments instituted. The traditional treatment for scoliosis is the use of a supportive brace, called the Milwaukee brace, that is worn on the upper body 23 hours per day for up to several years. A newer approach to straighten abnormal curvature is transcutaneous stimulation. In this method, muscles on one side of the vertebral column are electrically stimulated to contract and pull the vertebrae into a more normal position. If these methods fail, surgical intervention is used in which pieces of bone from elsewhere in the skeleton, or metal rods, are grafted to the deformed vertebrae to hold them in proper alignment. If treated early enough, kyphosis due to poor posture can be corrected with special exercises and instructions for appropriate posture. Kyphosis due to pathological causes may also require special braces or surgical intervention.

CHAPTER SUMMARY

INTRODUCTION

A. Skeletal tissues form bones—the organs of the skeletal system

B. The relationship of bones to each other and to other body structures provides a basis for understanding the function of other organ systems

C. The adult skeleton is composed of 206 separate bones

DIVISIONS OF SKELETON (FIGURE 7-1)

A. Axial skeleton—the bones of the head, neck, and torso; composed of 74 bones that form the upright axis of the body and six tiny middle ear bones

B. Appendicular skeleton—the 126 bones that form the appendages to the axial skeleton; the upper and lower extremities

AXIAL SKELETON

A. Skull—made up of 28 bones in two major divisions, cranial bones and facial bones (Figures 7-2 to 7-7)
  1. Cranial bones
    a. Frontal bone
      (1) Forms the forehead and anterior part of the top of the cranium
      (2) Contains the frontal sinuses
      (3) Forms the upper portion of the orbits
      (4) Forms the coronal suture with the two parietal bones
    b. Parietal bones
      (1) Form the bulging topside of the cranium
      (2) Form several sutures: lambdoidal suture with occipital bone; squamous suture with temporal bone and part of sphenoid; and coronal suture with frontal bone
    c. Temporal bones
      (1) Form the lower sides of the cranium and part of the cranial floor
      (2) Contain the inner and middle ears
    d. Occipital bone
      (1) Forms the lower, posterior part of the skull
      (2) Forms immovable joints with three other cranial bones and a movable joint with the first cervical vertebra
    e. Sphenoid bone
      (1) A bat-shaped bone located in the central portion of the cranial floor
      (2) Anchors the frontal, parietal, occipital, and ethmoid bones and forms part of the lateral wall of the cranium and part of the floor of each orbit
      (3) Contains the sphenoid sinuses
    f. Ethmoid bone (Figure 7-8)
      (1) A complicated, irregular bone that lies anterior to the sphenoid and posterior to the nasal bones
      (2) Forms the anterior cranial floor, medial orbit walls, upper parts of the nasal septum, and sidewalls of the nasal cavity
      (3) The cribriform plate is located in the ethmoid
  2. Facial bones
    a. Maxilla (upper jaw)
      (1) Two maxillae form the keystone of the face
      (2) Maxillae articulate with each other and with nasal, zygomatic, inferior concha, and palatine bones
      (3) Forms parts of the orbital floors, roof of the mouth, and floor and sidewalls of the nose
      (4) Contains maxillary sinuses
    b. Mandible (lower jaw)
      (1) Largest, strongest bone of the face
      (2) Forms the only movable joint of the skull with the temporal bone
    c. Zygomatic bone
      (1) Shapes the cheek and forms the outer margin of the orbit
      (2) Forms the zygomatic arch with the zygomatic process of the temporal bones
    d. Nasal bone
      (1) Both nasal bones form the upper part of the bridge of the nose, whereas cartilage forms the lower part
      (2) Articulates with the ethmoid, nasal septum, frontal, maxillae, and the other nasal bone
    e. Lacrimal bone
      (1) Paper-thin bone that lies just posterior and lateral to each nasal bone
      (2) Forms the nasal cavity and medial wall of the orbit
      (3) Contains groove for the nasolacrimal (tear) duct
      (4) Articulates with the maxilla, frontal, and ethmoid bones
    f. Palatine bone
      (1) Two bones form the posterior part of the hard palate
      (2) Vertical portion forms the lateral wall of the posterior part of each nasal cavity
      (3) Articulates with the maxillae and the sphenoid bone
    g. Inferior nasal conchae (turbinates)
      (1) Form lower edge projecting into the nasal cavity and form the nasal meati
      (2) Articulates with ethmoid, lacrimal, maxillary, and palatine bones
    h. Vomer bone
      (1) Forms posterior portion of the nasal septum
      (2) Articulates with the sphenoid, ethmoid, palatine, and maxillae

B. Hyoid bone (Figure 7-10)
  1. U-shaped bone located just above the larynx and below the mandible
  2. Suspended from the styloid processes of the temporal bone
  3. Only bone in the body that articulates with no other bones

C. Vertebral column (Figure 7-11)
  1. Forms the flexible longitudinal axis of the skeleton
  2. Consists of 24 vertebrae plus the sacrum and coccyx
  3. Segments of the vertebral column:
    a. Cervical vertebrae, 7
    b. Thoracic vertebrae, 12
    c. Lumbar vertebrae, 5
    d. Sacrum—in the adult, results from the fusion of five separate vertebrae
    e. Coccyx—in the adult, results from the fusion of four or five separate vertebrae
  4. Characteristics of the vertebrae (Figures 7-12 and 7-13)
    a. All vertebrae, except the first, have a flat, rounded body anteriorly and centrally, a spinous process posteriorly, and two transverse processes laterally
    b. All but the sacrum and coccyx have a vertebral foramen
    c. Second cervical vertebrae has an upward projection, the dens, to allow rotation of the head
    d. Seventh cervical vertebra has a long, blunt spinous process

e. Each thoracic vertebra has articular facets for the ribs

5. Vertebral column as a whole articulated with the head, ribs, and iliac bones

6. Individual vertebrae articulate with each other in joints between their bodies and between their articular processes

D. Sternum (Figure 7-14)
1. Dagger-shaped bone in the middle of the anterior chest wall made up of three parts:
   a. Manubrium—the upper handle part
   b. Body—middle blade part
   c. Xiphoid process—blunt cartilaginous lower tip, which ossifies during adult life
2. Manubrium articulates with the clavicle and first rib
3. Next nine ribs join the body of the sternum, either directly or indirectly, by means of the costal cartilages

E. Ribs (Figures 7-14 and 7-15)
1. Twelve pairs of ribs, with the vertebral column and sternum, form the thorax
2. Each rib articulates with both the body and transverse process of its corresponding thoracic vertebra
3. Ribs 2 through 9 articulate with the body of the vertebra above
4. From its vertebral attachment, each rib curves outward, then forward and downward
5. Rib attachment to the sternum:
   a. Ribs 1 through 7 join a costal cartilage that attaches it to the sternum
   b. Costal cartilage of ribs 8 through 10 joins the cartilage of the rib above to be indirectly attached to the sternum
   c. Ribs 11 and 12 are floating ribs, since they do not attach even indirectly to the sternum

APPENDICULAR SKELETON

A. Upper extremity
1. Consists of the bones of the shoulder girdle, upper arm, lower arm, wrist, and hand
2. Shoulder girdle (Figure 7-16)
   a. Made up of the scapula and clavicle
   b. Clavicle forms the only bony joint with the trunk, the sternoclavicular joint
   c. At its distal end, the clavicle articulates with the acromion process of the scapula
3. Humerus (Figures 7-17 and 7-18)
   a. The long bone of the upper arm
   b. Articulates proximally with the glenoid fossa of the scapula and distally with the radius and ulna
4. Ulna
   a. The long bone found on the little finger side of the forearm
   b. Articulates proximally with the humerus and radius and distally with a fibrocartilaginous disk
5. Radius
   a. The long bone found on the thumb side of the forearm
   b. Articulates proximally with the capitulum of the humerus and the radial notch of the ulna; articulates distally with the scaphoid and lunate carpals and with the head of the ulna
6. Carpal bones (Figure 7-19)
   a. Eight small bones that form the wrist
   b. Carpals are bound closely and firmly by ligaments and form two rows of four carpals each
      (1) Proximal row is made up of the pisiform, triquetrum, lunate, and scaphoid
      (2) Distal row is made up of the hamate, capitate, trapezoid, and trapezium
   c. The joints between the radius and carpals allow wrist and hand movements

7. Metacarpal bones
   a. Form the framework of the hand
   b. The thumb metacarpal forms the most freely movable joint with the carpals
   c. Heads of the metacarpals (the knuckles) articulate with the phalanges

B. Lower extremity
1. Consists of the bones of the hip, thigh, lower leg, ankle, and foot
2. Pelvic girdle is made up of the sacrum and os coxae bound tightly by strong ligaments (Figure 7-20)
   a. A stable circular base that supports the trunk and attaches the lower extremities to it
   b. Each coxal bone is made up of three bones that fuse together:
      1. Ilium—largest and uppermost
      2. Ischium—strongest and lowermost
      3. Pubis—anteriormost
3. Femur—longest and heaviest bone in the body (Figure 7-22)
4. Patella—largest sesamoid bone in the body
5. Tibia
   a. The larger, stronger, and more medially and superficially located of the two leg bones
   b. Articulates proximally with the femur to form the knee joint
   c. Articulates distally with the fibula and talus
6. Fibula
   a. The smaller, more laterally and deeply placed of the two leg bones
   b. Articulates with the tibia
7. Foot (Figures 7-23 and 7-24)
   a. Structure is similar to that of the hand with adaptations for supporting weight
   b. Foot bones are held together to form spring arches
      (1) Medial longitudinal arch is made up of the calcaneus, talus, navicular, cuneiforms, and the medial three metatarsals
      (2) Lateral longitudinal arch is made up of the calcaneus, cuboid, and fourth and fifth metatarsals

SKELETAL DIFFERENCES IN MEN AND WOMEN

A. Male skeleton is larger and heavier than female skeleton
B. Pelvic differences (Figure 7-25)
1. Male pelvis—deep and funnel shaped with a narrow pubic arch
2. Female pelvis—shallow, broad, and flaring with a wider pubic arch

CYCLE OF LIFE: SKELETAL SYSTEM

A. Changes in the skeletal framework result from changes in bone, cartilage, and muscle tissue
B. Older Adults
1. Loss of bone density
   a. Prone to fractures
2. Loss of skeletal tissue density
   a. Compression of weight-bearing bones
      (1) Loss of height
      (2) Postural changes
3. Degeneration of skeletal muscle tissue
   a. Loss of height
   b. Postural changes

MECHANISMS OF DISEASE: FRACTURES AND ABNORMAL SPINAL CURVATURES

A. Bone fracture—partial or complete break in the bone occurring under stress (Figure 7-26)
1. Open fracture (compound)—bone projects through skin, potential for infection
2. Closed fracture (simple)—bone does not break skin

3. Complete fracture—break across the entire bone
4. Incomplete fracture—bone fragments remain partially joined
5. Fractures also described by bone and region of break
6. Symptoms
   a. Loss of function or false motion
   b. Pain
   c. Soft tissue edema
   d. Deformity
7. Treatment
   a. Realignment of bone
   b. Immobilization
   c. Rehabilitation

B. Abnormal Spinal Curvatures (Figure 7-27)
1. Normal spine—convex through thoracic region and concave through cervical and lumbar regions
2. Lordosis (swayback)—abnormal lumbar curve
3. Kyphosis (hunchback)—abnormal roundness in thoracic curvature
4. Scoliosis—abnormal side-to-side curvature
5. Abnormal curvatures interfere with breathing, posture, and other functions
6. Treatment
   a. Braces
   b. Transcutaneous stimulation
   c. Surgical intervention
   d. Exercise

REVIEW QUESTIONS

1. Describe the skeleton as a whole and identify its two major subdivisions.
2. Identify the bones in the cranium and face.
3. Name and locate the fontanels and sutures of the skull.
4. Name the five pairs of bony sinuses in the skull.
5. Discuss the clinical (medical) importance of the cribriform plate of the ethmoid bone and the mastoid air cells in the temporal bone.
6. Identify and discuss the normal primary and secondary curves of the spine.
7. Describe the different kinds of bone fractures and discuss symptoms and treatments for broken bones.

8. Identify the bony components of the thorax.
9. Identify the bones of the shoulder and pelvic girdles.
10. Identify and compare the bones of the arm, forearm, wrist, and hand with those of the thigh, lower leg, ankle, and foot.
11. Discuss the arches of the foot.
12. Describe the skeletal variations between men and women.
13. Describe some common age changes in the skeleton.
14. List and explain the three most important pathological curvatures of the spine.

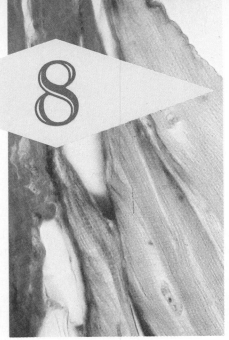

Junction between muscle and tendon
Erlandsen/Magney: Color Atlas of Histology

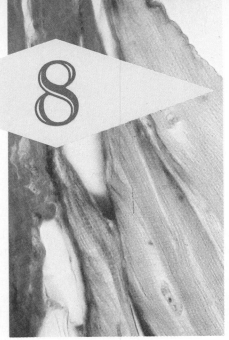
In the diamond at the top: **8** **Articulations**

CHAPTER OUTLINE

OBJECTIVES

After you have completed this chapter, you should be able to:

1. Define the term *articulation*.
2. Compare the classification of joints according to both structure and function.
3. List the types of fibrous joints and give an example of each.
4. Discuss the seven structures that characterize diarthrotic joints.
5. Explain the functional significance of bursae.
6. Discuss the structural characteristics of uniaxial, biaxial, and multiaxial diarthroses and give an example of each.
7. Identify the types of movement at synovial joints and give examples of specific joints where each occurs.
8. Discuss the knee joint as a typical synovial joint.
9. Explain why knee injuries are among the most common type of athletic injury.

Amphiarthrosis (am-fee-ar-THRO-sis) Slightly movable joint

Articulation (ar-tik-yoo-LAY-shun) Joint

Bursa (BER-sah) Cushionlike sac found between some moving body parts

Diarthrosis (dye-ar-THRO-sis) Freely movable joint

Menisci (me-NI-sky) Articular disks located within the knee joint

Suture (SOO-chur) Immovable joint found between flat bones of the skull

Symphysis (SIM-fe-sis) Joint formed by fibrocartilage

Synarthrosis (sin-ar-THRO-sis) Fibrous joint allowing little or no movement

An **articulation,** or joint, is a point of contact between bones. Although most joints in the body allow considerable movement, others are completely immovable or permit only limited motion or motion in only one plane or direction. In the case of immovable joints, such as the sutures of the skull, adjacent bones are bound together into a strong and rigid protective plate. In some joints, movement is possible but highly restricted. For example, joints between the bodies of the spinal vertebrae perform two seemingly contradictory functions. They help to firmly bind the components of the spine to each other and yet permit normal but restricted movement to occur. Most joints in the body allow considerable movement to occur as a result of skeletal muscle contractions. It is the existence of such joints that permits us to execute complex, highly coordinated, and purposeful movements. Functional articulations between bones in the extremities such as the shoulder, elbow, hip, and knee contribute to controlled and graceful movement and provide a large measure of our enjoyment of life. This chapter begins by classifying joints and describing their identifying features. Coverage of joint classification and structure is followed by a discussion of body movements and a description of selected major joints. The chapter concludes by describing life cycle changes and some common joint diseases.

CLASSIFICATION OF JOINTS

Joints may be classified into three major categories using either a structural or a functional scheme. If a *structural classification* is employed, joints are named according to the type of connective tissue that joins the bones together (fibrous or cartilaginous joints) or by the presence of a fluid-filled joint capsule (synovial joints). If a *functional classification* scheme is used, joints are divided into three classes according to the degree of movement they permit: synarthroses (immovable); amphiarthroses (slightly movable); and diarthroses (freely movable). Table 8-1 classifies joints according to structure, function, and range of movement. Refer often to this table, and to the illustrations that follow, as you read about each of the major joint types.

Fibrous Joints (Synarthroses)

The articulating surfaces of bones that form fibrous joints fit closely together. The different types and amounts of connective tissue joining bones in this group may permit very limited movement in some fibrous joints, but most are

fixed. There are three subtypes of fibrous joints, called syndesmoses, sutures, and gomphoses.

Syndesmoses

Syndesmoses (SIN-dez-MO-ses) are joints in which fibrous bands (ligaments) connect two bones. The joint between the distal ends of the radius and ulna is joined by the *radioulnar interosseous ligament* (Figure 8-1). Although this joint is classified as a fibrous joint, some movement is possible because of ligament flexibility.

Sutures

Sutures are found only in the skull. In most sutures, teethlike projections jut out from adjacent bones and interlock with each other with only a thin layer of fibrous tissue between them. Sutures become ossified in older adults, forming extremely strong lines of fusion between opposing skull bones (Figure 8-1).

Gomphoses

Gomphoses (gom-FO-ses) are unique joints that occur between the root of a tooth and the alveolar process of the mandible or maxilla (Figure 8-1). The fibrous tissue between the tooth's root and the alveolar process is called the *periodontal membrane*.

Cartilaginous Joints (Amphiarthroses)

The bones that articulate to form cartilaginous joints are joined together by either hyaline cartilage or fibrocartilage. Joints characterized by the presence of hyaline cartilage between articulating bones are called *synchondroses*, and those joined by fibrocartilage are called *symphyses*. Cartilaginous joints permit only very limited movement between articulating bones in certain circumstances. During childbirth, for example, slight movement at the symphysis pubis facilitates the baby's passage through the pelvis.

Synchondroses

Note that joints identified as **synchondroses** (SIN-kon-DRO-ses) have hyaline cartilage between articulating bones. Examples include the articulation between the first rib and sternum (Figure 8-2) and the joint present during the growth years between the epiphyses of a long bone and its diaphysis.

Symphyses

A **symphysis** (SIM-fa-sis) is a joint in which a pad or disk of fibrocartilage connects two bones. The tough fibrocartilage disks in these joints may permit slight movement when pressure is applied between the bones. Most symphyses are located in the midline of the body.

TABLE 8-1	Primary joint classifications		
Functional Name	**Structural Name**	**Degree of Movement Permitted**	**Example**
SYNARTHROSES	Fibrous	Immovable	Sutures of skull
AMPHIARTHROSES	Cartilaginous	Slightly movable	Symphysis pubis
DIARTHROSES	Synovial	Freely movable	Shoulder joint

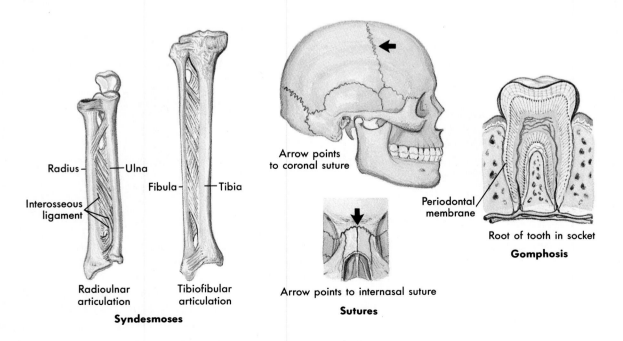

FIGURE 8-1 Fibrous joints. Examples of the types of fibrous joints.

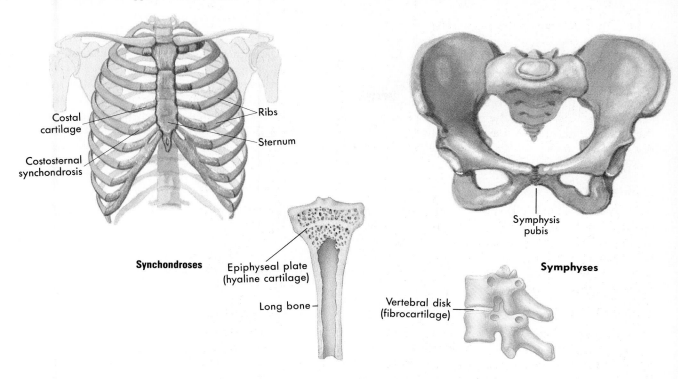

FIGURE 8-2 **Cartilaginous joints.** Examples of the types of cartilaginous joints.

TABLE 8-2	Classification of fibrous and cartilaginous joints		
Types	**Examples**	**Structural Features**	**Movements**
FIBROUS JOINTS			
Syndesmoses	Joints between distal ends of radius and ulna	Fibrous bands (ligaments) connect articulating bones	Slight
Sutures	Joints between skull bones	Teethlike projections of articulating bones interlock with thin layer of fibrous tissue connecting them	None
Gomphoses	Joints between roots of teeth and jaw bones	Fibrous tissue connects roots of teeth to alveolar processes	None
CARTILAGINOUS JOINTS			
Synchondroses	Costal cartilage attachments of first rib to sternum; epiphyseal plate between diaphysis and epiphysis of growing long bone	Hyaline cartilage connects articulating bones	Slight
Symphyses	Symphysis pubis; joints between *bodies* of vertebrae	Fibrocartilage between articulating bones	Slight

Examples of symphyses include the symphysis pubis (Figure 8-2) (see Figure 7-12 and the articulation between the *bodies* of adjacent vertebrae). The intervertebral disk in these joints is composed of tough and resilient fibrocartilage, which serves to absorb shock and permit limited movement. The bones of the vertebral column have numerous points of articulation between them. Collectively, these joints permit limited motion of the spine in a very restricted range. It is the articulation between the *bodies* of adjacent vertebrae that is classified as a cartilaginous joint. The points of contact between the *articular facets* of adjacent vertebrae are considered synovial joints and will be described later in the chapter.

Table 8-2 summarizes the different kinds of fibrous and cartilaginous joints.

Synovial Joints (Diarthroses)

Synovial joints are freely movable joints. Not only are they the body's most mobile but also its most numerous and anatomically most complex joints. A majority of the joints between bones in the appendicular skeleton are synovial joints.

Structure of synovial joints

The following seven structures characterize synovial, or freely movable, joints (Figure 8-3).

1. Joint capsule. Sleevelike extension of the periosteum of each of the articulating bones. The capsule forms a complete casing around the ends of the bones, thereby binding them to each other.

2. Synovial membrane. Moist, slippery membrane that lines the inner surface of the joint capsule. It attaches to the margins of the articular cartilage. It also secretes synovial fluid, which lubricates and nourishes the inner joint surfaces.

3. Articular cartilage. Thin layer of hyaline cartilage covering and cushioning the articular surfaces of bones.

4. Joint cavity. Small space between the articulating surfaces of the two bones of the joint. Because of this cavity with no tissue growing between the articulating surfaces of the bones, the bones are free to move against one another. Synovial joints therefore are diarthroses, or freely movable, joints.

5. Menisci (articular disks). Pads of fibrocartilage located between the articulating ends of bones in some diarthroses. Usually these pads divide the joint cavity into two separate cavities. The knee joint contains two menisci (see Figure 8-10).

6. Ligaments. Strong cords of dense, white fibrous tissue at most synovial joints. These grow between the bones, lashing them even more firmly together than possible with the joint capsule alone.

7. Bursae. Some synovial joints contain a closed pillowlike structure called a **bursa** formed of synovial membrane filled with synovial fluid. Bursae tend to be associated with bony prominences (such as in the knee or elbow), where they function to cushion the joint and facilitate movement of tendons. *Bursitis,* or inflammation of the bursa, is a relatively common disorder most often caused by excessive or repetitive exercise or sudden trauma to the joints. For example, carpet layers, roofers, and others who work on their knees are prone to bursitis involving the knee joint. The condition is generally treated with immobilization of the joint and antiinflammatory drugs.

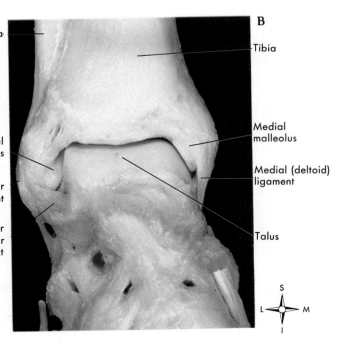

FIGURE 8-3   Structure of synovial joints. A, Illustration showing a typical synovial joint. **B,** Photograph of the ankle joint as seen from in front.

Types of Synovial Joints

Synovial joints are divided into three main groups—uniaxial, biaxial, and multiaxial. Each of these is subdivided into two subtypes as follows:

1. **Uniaxial joints.** Synovial joints that permit movement around only one axis and in only one plane. Hinge and pivot joints are types of uniaxial joints (Figure 8-4, A and B).

   a. **Hinge joints.** Those in which the articulating ends of the bones form a hinge-shaped unit. Like a common door hinge, hinge joints permit only back and forth movements, namely, flexion and extension. If you have access to an articulated skeleton, examine the articulating end of the humerus (the trochlea) and of the ulna (the semilunar notch). Observe their interaction as you flex and extend the forearm. Do you see why you can flex and extend your forearm but cannot move it in any other way at this joint? The shapes of the trochlea and of the semilunar notch (see Figures 7-17 and 7-18, p. 187) permit only the uniaxial, horizontal plane movements of flexion and extension at the elbow. Other hinge joints include the knee and interphalangeal joints.

   b. **Pivot joints.** Those in which a projection of one bone articulates with a ring or notch of another bone. Examples: a projection (dens) of the second cervical vertebra articulates with a ring-shaped portion of the first cervical vertebra; the head of the radius articulates with radial notch of the ulna.

2. **Biaxial joints.** Diarthroses that permit movements around two perpendicular axes in two perpendicular planes. Saddle and condyloid joints are types of biaxial joints (Figure 8-4, C and D).

   a. **Saddle joints.** Those in which the articulating ends of the bones resemble reciprocally shaped miniature saddles. Only two saddle joints—one in each thumb—are present in the body. The thumb's metacarpal bone articulates in the wrist with a carpal bone (trapezium). The saddle-shaped articulating surfaces of these bones make it possible for the thumb to move over to touch the tips of the fingers—that is, to oppose the fingers. How important is this? To answer this for yourself, consider the following. Opposing the thumb to the fingers enables us to grip small objects. Were it not for this movement, we would have no manual dexterity. A surgeon could not grasp a scalpel or suture needle. None of us could hold a pen or pencil for writing.

   b. **Condyloid (ellipsoidal) joints.** Those in which a condyle fits into an elliptical socket. Examples: condyles of the occipital bone fit into elliptical depressions of the atlas; the distal end of the radius fits into depressions of the carpal bones (scaphoid, lunate, and triquetrum).

3. **Multiaxial joints.** Joints that permit movements around three or more axes and in three or more planes (Figure 8-4, E and F).

   a. **Ball and socket joints** (spheroid joints). Our most movable joints. A ball-shaped head of one bone fits into a concave depression on another, thereby allowing the first bone to move in many directions. Examples: shoulder and hip joints.

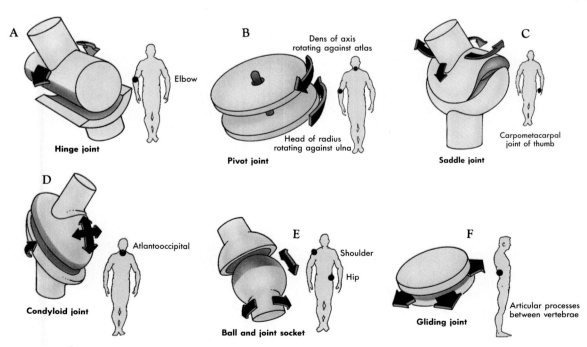

FIGURE 8-4 **Synovial joint types.** Uniaxial: **A,** hinge and **B,** pivot; Biaxial: **C,** saddle and **D,** condyloid; Multiaxial: **E,** ball and socket and **F,** gliding.

b. Gliding joints. Characterized by relatively flat articulating surfaces that allow limited gliding movements along various axes. Examples: joints between the articular surfaces of successive vertebrae. (Articulations between the *bodies* of successive vertebrae are symphysis type cartilaginous joints.) As a group, gliding joints are the least movable of the synovial joints.

Table 8-3 summarizes the classes of synovial joints.

QUICK CHECK

✔ 1. List the seven structures that characterize synovial joints.
✔ 2. Name the three main categories of synovial joints grouped according to axial movement and list the two subtypes of joints found in each category.
✔ 3. Name one specific joint as an example of each of the six types of synovial joints.

TABLE 8-3   Classification of synovial joints

Types	Examples	Structural Features	Movements
UNIAXIAL			Around one axis; in one plane
Hinge	Elbow joint	Spool-shaped process fits into concave socket	Flexion and extension only
Pivot	Joint between first and second cervical vertebrae	Arch-shaped process fits around peglike process	Rotation
BIAXIAL			Around two axes, perpendicular to each other; in two planes
Saddle	Thumb joint between first metacarpal and carpal bone	Saddle-shaped bone fits into socket that is concave-convex-concave	Flexion, extension in one plane; abduction, adduction in other plane; opposing thumb to fingers
Condyloid (ellipsoidal)	Joint between radius and carpal bones	Oval condyle fits into elliptical socket	Flexion, extension in one plane; abduction, adduction in other plane
MULTIAXIAL			Around many axes
Ball and socket	Shoulder joint and hip joint	Ball-shaped process fits into concave socket	Widest range of movements; flexion, extension, abduction, adduction, rotation, circumduction
Gliding	Joints between adjacent vertebrae; joints between carpal and tarsal bones	Relatively flat articulating surfaces	Gliding movements without any angular or circular movements

JOINT MICE

The thin layer of cartilage that covers the articulating surfaces of bones in synovial joints may become cracked as a result of traumatic injury—especially to the knee. Such injuries are called *chondral fractures*. They result in fissuring of the articular cartilage, which can be likened to the cracking of an egg shell. If such an injury occurs, small fragments may become detached from the damaged layer of cartilage and enter the joint cavity. These fragments are called "loose bodies" or **joint mice** because they often cause clicking sounds or catching and locking sensations during joint movement. Symptoms often vary over time as cartilage fragments move about the joint cavity in the synovial fluid. Pain and loss of motion will occur if they become lodged between opposing bone surfaces.

TYPES OF MOVEMENT AT SYNOVIAL JOINTS

The types of movement possible at synovial joints depend on the shapes of the articulating surfaces of the bones and on the positions of the joints' ligaments and nearby muscles and tendons. All synovial joints, however, permit one or more of the following types of movements: angular, circular, gliding, and special movements. Illustrations in Figures 8-5 through 8-7 provide examples of each type of movement at selected synovial joints. Refer to these illustrations frequently as you read about the various types of movement possible at synovial joints.

You'll notice that the movements can often be classified as opposites: flexion is the opposite of extension, protraction is the opposite of retraction, and so on. It will be good

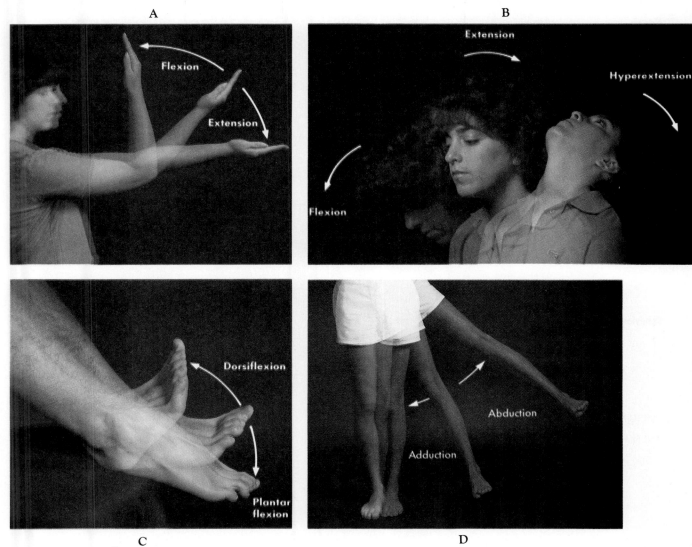

A
B
C
D

FIGURE 8-5   Angular Movements. A, Flexion and extension at the elbow. **B,** Flexion, extension, and hyperextension of the neck. **C,** Dorsiflexion and plantar flexion. **D,** Adduction and abduction of the thigh.

to keep this concept in mind when the mechanism of antagonistic muscle groups is discussed in Chapter 10.

Angular Movements

Angular movements change the size of the angle between articulating bones. Flexion, extension, abduction, and adduction are some of the different types of angular movements.

Flexion

Flexion decreases the angle between bones. It bends or folds one part on another. For example, if you bend your head forward on your chest, you are flexing it. If you bend your arm at the elbow, you are flexing your lower arm (Figure 8-5, A and B). Flexion, in short, is bending, folding, or withdrawing a part.

Extension and hyperextension

Extension increases the angle between bones. It returns a part from its flexed position to its anatomical position (Figure 8-5, A). Extensions are straightening or stretching movements. Stretching an extended part beyond its anatomical position is called **hyperextension.** Hyperextension of the neck is illustrated in Figure 8-5, B.

Plantar flexion and dorsiflexion

Plantar flexion occurs when the foot is stretched down and back. This movement increases the angle between the top of the foot and the front of the leg. **Dorsiflexion** occurs when the foot is tilted upward, thus decreasing the angle between the top of the foot and front of the leg (Figure 8-5, C).

Abduction and adduction

Abduction moves a part away from the median plane of the body, as in moving the leg straight out to the side or the fingers away from the midline of the hand. **Adduction** moves a part toward the median plane. Examples: bringing the leg back to the side; moving the fingers toward the midline of the hand. These opposing movements can be seen in Figure 8-5, D.

Circular Movements

Circular movements result in arclike rotation of a structure around an axis. The primary circular movements are rotation, circumduction, supination, and pronation.

Rotation and circumduction

Rotation consists of pivoting a bone on its own axis. For example, moving the head from side to side as in indicating "no" (Figure 8-6, A). **Circumduction** moves a part so its distal end moves in a circle (Figure 8-6, B). When a pitcher winds up to throw a ball, he circumducts his arm.

Supination and pronation

Supination and **pronation** of the hand are shown in Figure 8-6, C. Pronation turns the palm of the hand down, whereas supination turns the hand palm side up.

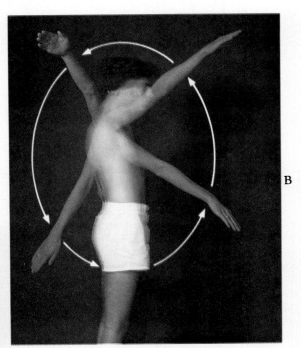

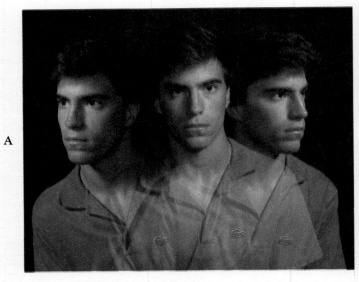

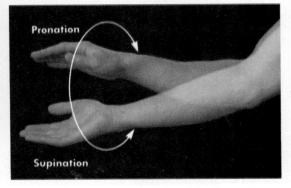

FIGURE 8-6 Circular Movements. A, Rotation of the head. **B,** Circumduction of the arm. **C,** Pronation and supination of the hand.

JOINT REPLACEMENT

Total or partial replacement of many different types of diseased joints in the body is now a common surgical procedure. Total hip replacement (THR) is the most common orthopedic operation performed on older persons, done over 200,000 times in the United States in 1992. In addition, large numbers of knee, elbow, shoulder, and finger joint replacement procedures are also performed annually. Osteoarthritis and other types of degenerative bone disease frequently destroy or severely damage joints and cause unremitting pain. Total joint replacement is often the only effective treatment option available.

The THR procedure involves replacement of both the femoral head by a metallic alloy prosthesis and the acetabular socket with a high-density polyethylene cup. In the past, both sides were generally cemented into place with methyl methacrylate adhesive. Many of the newer prostheses, however, are "porous-coated" to permit natural ingrowth of bone for stability and retention. The advantage of bone ingrowth is prevention of loosening that occurs if the cement bond weakens. With the newer prostheses and advanced surgical techniques, long-term THR success rates have risen to approximately 85%. The x-ray illustration shows placement of a noncemented, porous-coated femoral prosthesis made of titanium alloy and a cemented acetabular socket made of high-density plastic.

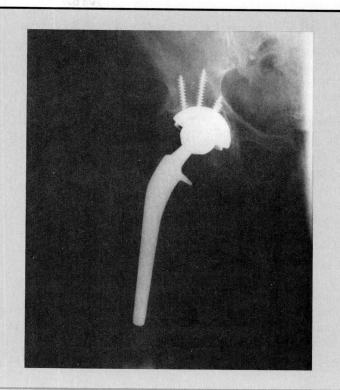

MEASURING RANGE OF MOTION

Measuring range of motion (ROM) is often one of the first assessment techniques employed by a health care provider to determine the degree of damage in an injured joint. In the absence of disease or injury, major synovial joints should function within a normal ROM. Joint ROM can be measured actively or passively. In active movements the individual moves the joint or body part through its ROM, and in passive movements the physician or other health care provider moves the part with the muscles relaxed. Normally, both active and passive ROM should be about equal. If a joint has an obvious increase or limitation in its range of motion, a specialized instrument, called a **goniometer,** is used to measure the angle. A goniometer consists of two rigid shafts that intersect at a hinge joint. A protractor is fixed to one shaft so that motion can be read directly from the scale in degrees. The starting position is defined as the point at which the movable segment is at zero degrees (usually anatomical position). Measuring joint ROM provides a physician, nurse, athletic trainer, or therapist with information required to assess normal joint function, accurately measure dysfunction, or gauge treatment and rehabilitative progress following injury or disease.

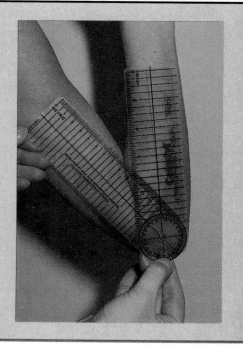

Gliding Movements

Gliding movements are the simplest of all movements. The articular surface of one bone moves over the articular surface of another without any angular or circular movement. Gliding movements occur between the carpal and tarsal bones and between the articular facets of adjoining spinal vertebrae.

Special Movements

Special movements are often unique or unusual type movements that occur only in a very limited number of joints. These specialized movements do not fit well into other movement categories and are generally described separately. Special movements include inversion, eversion, protraction, retraction, elevation, and depression.

Inversion and eversion

Inversion turns the sole of the foot inward, whereas **eversion** turns it outward (Figure 8-7, A).

Protraction and retraction

Protraction moves a part forward, whereas **retraction** moves it back. For instance, if you stick out your jaw, you protract it, and if you pull it back, you retract it (Figure 8-7, B).

Elevation and depression

Elevation moves a part up, as in closing the mouth (Figure 8-7, C). **Depression** lowers a part, moving it in the opposite direction from elevation.

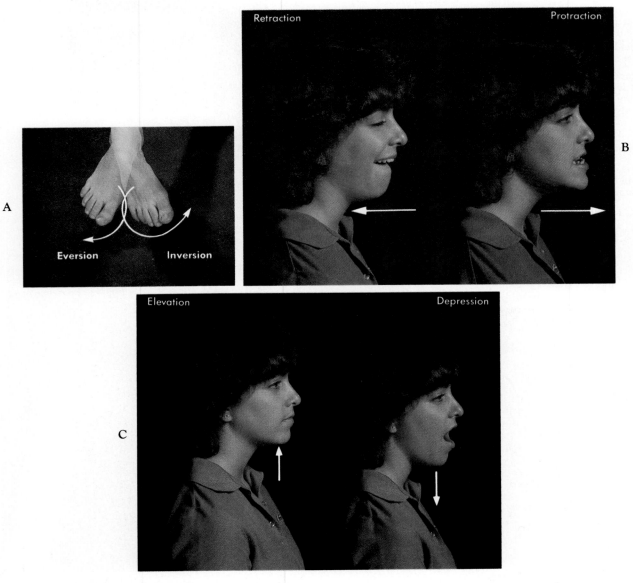

FIGURE 8-7   Special Movements. A, Eversion and inversion of the foot. **B,** Protraction and retraction of the jaw. **C,** Elevation and depression of the jaw.

REPRESENTATIVE SYNOVIAL JOINTS

Humeroscapular Joint

The joint between the head of the humerus and the glenoid cavity of the scapula is the one we usually refer to as the shoulder joint (Figure 8-8). It is our most mobile joint. One anatomical fact, the shallowness of the glenoid cavity, largely accounts for this. The shallowness offers little interference to movements of the head of the humerus. Were it not for the *glenoidal labrum* (Figure 8-8, C), a narrow rim of fibrocartilage around the glenoid cavity, it would have scarcely any depth at all. Some structures strengthen the shoulder joint and give it a degree of stability, notably several ligaments, muscles, tendons, and bursae. Note in Figure 8-8, A, for example, the superior, medial, and inferior glenohumeral ligaments. Each of these is a thickened portion of the joint capsule. Also in this figure, identify the subscapularis muscle, the tendon of the long head of the biceps brachii muscle, and a bursa. Shoulder muscles and tendons form a cufflike arrangement around the joint. It is called the *rotator cuff*. Baseball pitchers frequently injure the rotator cuff in the shoulder of their pitching arms. The main bursa of the shoulder joint is the *subdeltoid bursa*. It lies wedged between the inferior surface of the deltoid muscle and the superior surface of the joint capsule. Other bursae of the shoulder joint are the subscapular, subacromial, and subcoracoid bursae. All in all, the shoulder joint is more mobile than stable. Dislocations of the head of the humerus from the glenoid cavity occur frequently.

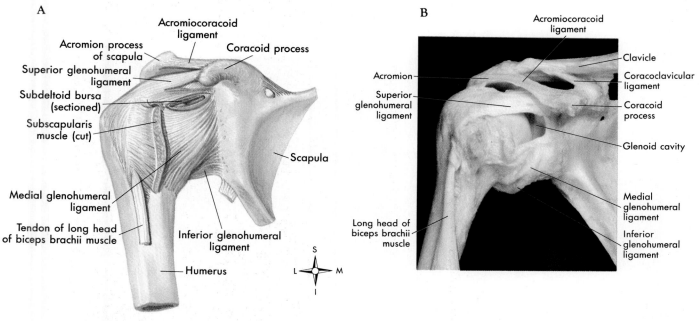

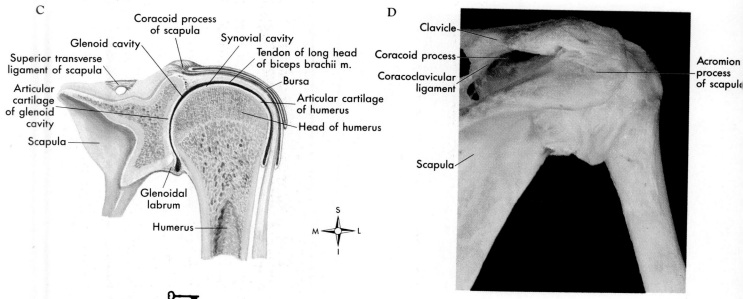

FIGURE 8-8   The shoulder joint. A and **B,** Anterior view. **C** and **D,** Viewed from behind through shoulder joint.

Hip Joint

The first characteristic to remember about the hip joint is stability; the second is mobility (Figure 8-9). The stability of the hip joint derives largely from the shapes of the head of the femur and of the acetabulum, the socket of the hip bone into which the femur head fits. Turn to Figure 7-21 and note the deep, cuplike shape of the acetabulum, and then observe the ball-like head of the femur in Figure 7-22, A. Compare these with the shallow, almost saucer-shaped glenoid cavity (Figure 7-16, C) and the head of the humerus (Figure 7-17). From these observations, do you see why the hip joint necessarily has a somewhat more limited range of movement than the shoulder joint? Both joints, however, allow multiaxial movements. Both permit flexion, extension, abduction, adduction, rotation, and circumduction.

A joint capsule and several ligaments hold the femur and hip bones together and contribute to the hip joint's stability. The iliofemoral ligament connects the ilium with the femur, and the ischiofemoral and pubofemoral ligaments join the ischium and pubic bone to the femur. The iliofemoral ligament is one of the strongest ligaments in the body.

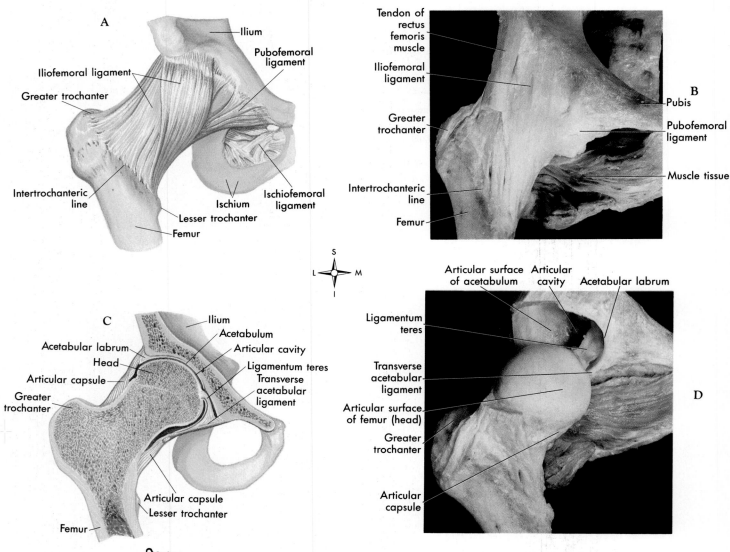

FIGURE 8-9   The hip joint. A and **B,** Anterior view. **C** and **D,** Frontal section through the hip joint.

Knee Joint

The knee, or tibiofemoral, joint is the largest and one of the most complex and most frequently injured joints in the body (Figures 8-10 to 8-12). The condyles of the femur articulate with the flat upper surface of the tibia. Although this is a precariously unstable arrangement, there are counteracting forces supplied by a joint capsule, cartilages, and numerous ligaments and muscle tendons. Note, for example, in Figure 8-10, the shape of the two cartilages labeled *medial meniscus* and *lateral meniscus.* They attach to the flat top of the tibia and, because of their concavity, form a kind of shallow socket for the condyles of the femur. Of the many ligaments that hold the femur bound to the tibia, five can be seen in Figure 8-10. The anterior cruciate

ligament attaches to the anterior part of the tibia between its condyles, then crosses over and backward and attaches to the posterior part of the lateral condyle of the femur. The posterior cruciate ligament attaches posteriorly to the tibia and lateral meniscus, then crosses over and attaches to the front part of the femur's medial condyle. The ligament of Wrisberg attaches posteriorly to the lateral meniscus and extends up and over to attach to the medial condyle behind the attachment of the posterior cruciate ligament (see Figure 8-10, C). The transverse ligament connects the anterior margins of the two menisci. Strong ligaments, the fibular and tibial collateral ligaments, located at the sides of the knee joint can be seen in Figure 8-10, C.

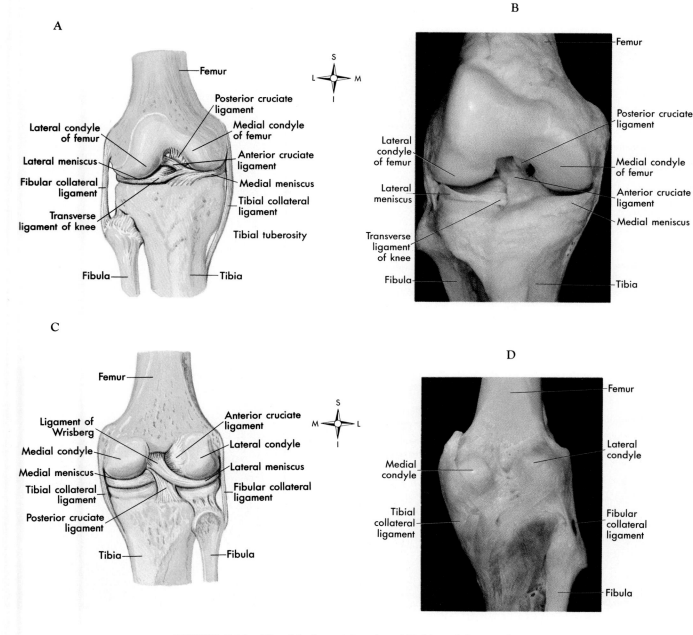

FIGURE 8-10 The right knee joint. A and **B,** Viewed from in front. **C** and **D,** Viewed from behind.

A baker's dozen of bursae serve as pads around the knee joint: four in front, four located laterally, and five medially. Of these, the largest is the prepatellar bursa (Figure 8-11) inserted in front of the patellar ligament, between it and the skin. The painful ailment called "housemaid's knee" is prepatellar bursitis.

Compared to the hip joint, the knee joint is relatively unprotected by surrounding muscles. Consequently the knee, more often than the hip, is injured by blows or sudden stops and turns. Athletes, for example, frequently

tear a knee cartilage, that is, one of the menisci (Figure 8-12).

The structure of the knee joint permits the hingelike movements of flexion and extension. Also, with the knee flexed, some internal and external rotation can occur. In most of our day-to-day activities—even such ordinary ones as walking, going up and down stairs, and getting into and out of chairs—our knees bear the brunt of the load; they are the main weight bearers. Knee injury or disease, therefore, can be badly crippling.

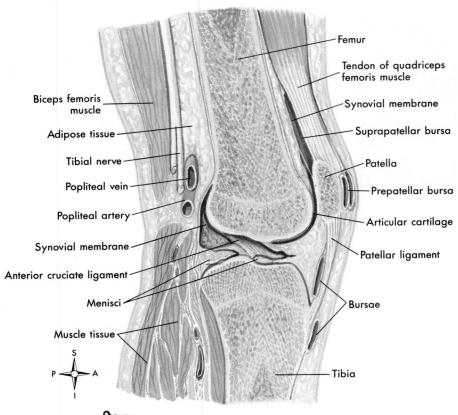

FIGURE 8-11   Sagittal section through knee joint.

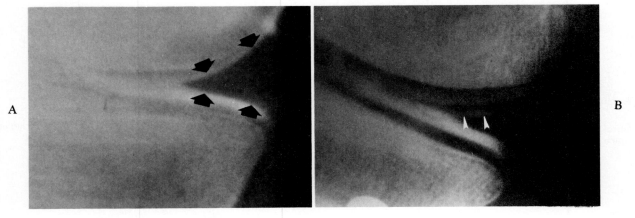

FIGURE 8-12   Knee arthrogram. A, Normal medial meniscus. Spot film shows normal triangular shape of meniscus (*arrows*). **B,** Linear tear of medial meniscus (*arrowheads*).

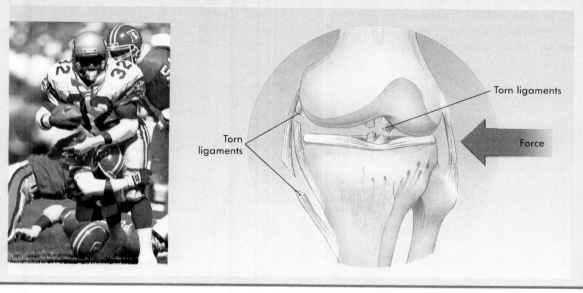

THE KNEE JOINT

The knee is the largest and most vulnerable joint. Because the knee is often subjected to sudden, strong forces during athletic activity, knee injuries are among the most common type of athletic injury. Sometimes, the articular cartilages on the tibia become torn when the knee twists while bearing weight. The ligaments holding the tibia and femur together can also be injured in this way. Knee injuries may also occur when a weight-bearing knee is hit by another person.

Vertebral Joints

One vertebra connects to another by several joints—between their bodies; laminae; and articular, transverse, and spinous processes. These joints hold the vertebrae firmly together so they are not easily dislocated, but also these joints form a flexible column. Consider how many ways you can move the trunk of your body. You can flex it forward or laterally, you can extend it, and you can circumduct or rotate it. The bodies of adjacent vertebrae are connected by intervertebral disks and strong ligaments. Fibrous tissue and fibrocartilage form a disk's outer rim (called the *annulus fibrosus*). Its central core (the *nucleus*

pulposus), in contrast, consists of a pulpy, elastic substance (Figure 8-13, *A*). With age the nucleus loses some of its resiliency. It may then be suddenly compressed by exertion or trauma and pushed through the annulus, with fragments protruding into the spinal canal and pressing on spinal nerves or the spinal cord itself. Severe pain results. In medical terminology, this is called a *herniated disk* or *herniated nucleus pulposus* (HNP); in popular language, it is a "slipped disk" (Figure 8-13, *B*).

In Figure 8-14 the following ligaments that bind the vertebrae together can be identified: The *anterior longitudinal ligament,* a strong band of fibrous tissue, connects the

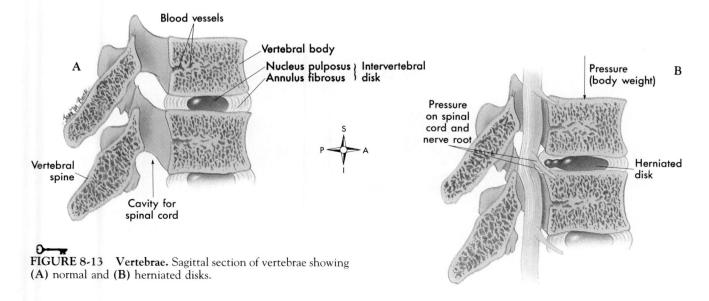

FIGURE 8-13   Vertebrae. Sagittal section of vertebrae showing **(A)** normal and **(B)** herniated disks.

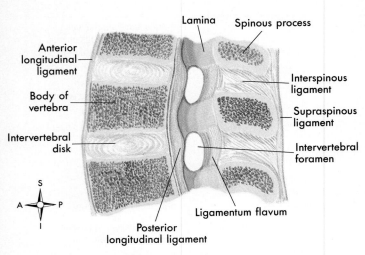

Anterior longitudinal ligament

Body of vertebra

Intervertebral disk

Lamina Spinous process

Interspinous ligament

Supraspinous ligament

Intervertebral foramen

Ligamentum flavum

Posterior longitudinal ligament

S
A—P
I

FIGURE 8-14 Vertebrae and their ligaments. Sagittal section of two lumbar vertebrae and their ligaments.

anterior surfaces of the vertebral bodies from the atlas down to the sacrum. Connecting the posterior surfaces of the bodies is the *posterior longitudinal ligament*. The *ligamenta flava* bind the laminae of adjacent vertebrae firmly together. Spinous processes are connected by *interspinous ligaments*. In addition, the tips of the spinous processes of the cervical vertebrae are connected by the *ligamentum nuchae*; its extension, the *supraspinous ligament*, connects the tips of the rest of the vertebrae down to the sacrum. And finally, *intertransverse ligaments* connect the transverse processes of adjacent vertebrae.

Table 8-4 summarizes the whole chapter with descriptions of most of the individual joints of the body.

QUICK CHECK

✔ 1. *Name one example of each of the following types of movement at a synovial joint: angular, circular, gliding, and special.*
✔ 2. *Which joint is the largest, most complex, and most frequently injured in the body?*
✔ 3. *List the two anatomic components of an intervertebral disk.*

TABLE 8-4 Description of individual joints

Name	Articulating Bones	Type	Movements
Atlantoepistropheal	Anterior arch of atlas rotates about dens of axis (epistropheus)	Synovial (pivot)	Pivoting or partial rotation of head
Vertebral	Between bodies of vertebrae	Cartilaginous (symphyses)	Slight movement between any two vertebrae but considerable motility for column as whole
	Between articular processes	Synovial (gliding)	Gliding
Sternoclavicular	Medial end of clavicle with manubrium of sternum	Synovial (gliding)	Gliding
Acromioclavicular	Distal end of clavicle with acromion of scapula	Synovial (gliding)	Gliding; elevation, depression, protraction, and retraction
Thoracic	Heads of ribs with bodies of vertebrae	Synovial (gliding)	Gliding
	Tubercles of ribs with transverse processes of vertebrae	Synovial (gliding)	Gliding
Shoulder	Head of humerus in glenoid cavity of scapula	Synovial (ball and socket)	Flexion, extension, abduction, adduction, rotation, and circumduction of upper arm
Elbow	Trochlea of humerus with semilunar notch of ulna; head of radius with capitulum of humerus	Synovial (hinge)	Flexion and extension
	Head of radius in radial notch of ulna	Synovial (pivot)	Supination and pronation of lower arm and hand; rotation of lower arm on upper
Wrist	Scaphoid, lunate, and triquetral bones articulate with radius and articular disk	Synovial (condyloid)	Flexion, extension, abduction, and adduction of hand
Carpal	Between various carpals	Synovial (gliding)	Gliding

Continued

TABLE 8-4   **Description of individual joints—cont'd**

Name	Articulating Bones	Type	Movements
Hand	Proximal end of first metacarpal with trapezium	Synovial (saddle)	Flexion, extension, abduction, adduction, and circumduction of thumb and opposition to fingers
	Distal end of metacarpals with proximal end of phalanges	Synovial (hinge)	Flexion, extension, limited abduction, and adduction of fingers
	Between phalanges	Synovial (hinge)	Flexion and extension of finger sections
Sacroiliac	Between sacrum and two ilia	Synovial (gliding)	None or slight
Symphysis Pubis	Between two pubic bones	Cartilaginous (symphysis)	Slight, particularly during pregnancy and delivery
Hip	Head of femur in acetabulum of os coxae	Synovial (ball and socket)	Flexion, extension, abduction, adduction, rotation, and circumduction
Knee	Between distal end of femur and proximal end of tibia	Synovial (hinge)	Flexion and extension; slight rotation of tibia
Tibiofibular (Proximal)	Head of fibula with lateral condyle of tibia	Synovial (gliding)	Gliding
Ankle	Distal ends of tibia and fibula with talus	Synovial (hinge)	Flexion (dorsiflexion) and extension (plantar flexion)
Foot	Between tarsals	Synovial (gliding)	Gliding; inversion and eversion
	Between metatarsals and phalanges	Synovial (hinge)	Flexion, extension, slight abduction, and adduction
	Between phalanges	Synovial (hinge)	Flexion and extension

 Cycle of Life: *Articulations*

This chapter discusses the types of articulations that occur between bones and presents information on the common types of diseases that affect joints during the life cycle.

Changes in the way bones develop and in the sequence of ossification that occurs between birth and skeletal maturity also affects our joints. Fontanels, which exist between bones of the cranium, disappear with increasing age, and the epiphyseal plates, which serve as points of articulation between the epiphyses and diaphyses of growing long bones, ossify at skeletal maturity.

Range of motion at synovial joints is typically greater early in life. With advancing age, joints are often described as being "stiff," range of motion decreases, and changes in gait are common. Difficulty with locomotion that affects many elderly individuals may result from disease conditions that influence the functional unit comprised of muscles, bones, and joints.

Some skeletal diseases have profound effects that are manifested as joint problems. Abnormal bone growth, or "lipping," which results in bone spurs, or sharp projections, on the articular surfaces of bones, dramatically influence joint function. Many of these disease conditions are associated with specific developmental periods, including early childhood, adolescence, and adulthood.

MECHANISMS OF DISEASE

Joint Disorders

Joint disorders may be classified as either **noninflammatory joint disease** or **inflammatory joint disease.** Serving as fulcrums, the joints permit smooth and precise movement to occur when a muscle contracts and pulls on bones. Because joints, bones, and muscles act together as a functional unit, joint disorders, regardless of type or classification, have profound effects on body mobility.

Noninflammatory Joint Disease

Osteoarthritis also known more accurately as *degenerative joint disease* (DJD), is the most common noninflammatory disorder of movable joints. It is characterized by deterioration and atrophy of articular cartilage and formation of new bone at the joint surfaces. As the joint undergoes repeated mechanical stress, the cartilage that covers surfaces of articulating bones thins and the joint space narrows. The articular plate no longer acts as a "shock absorber." Bony spurs form at the edges of the joint, the synovial membranes thicken, and the ligaments calcify. Osteoarthritis occurs most often in weight-bearing joints, such as the hips, lumbar spine, and knees. Symptoms include stiffness, pain on movement, and limited joint motion. The cause of osteoarthritis is unknown and, unfortunately, there is no treatment available to stop the degenerative joint disease process. In advanced cases, partial or total joint replacements may be necessary.

Traumatic injuries are a frequent cause of joint injury. *Dislocation* (subluxation) of a joint due to trauma can be an emergency because of associated damage to important blood vessels and nerves. In a dislocation the articular surface of bones forming the joint are no longer in proper contact. This displacement of bone alters the normal contour of the joint and can tear surrounding ligaments. Pain and soft tissue swelling are associated with dislocation injuries. Effective treatment involves early reduction or re-alignment.

One of the most common traumatic joint injuries associated with athletic activity involves damage to the cartilaginous menisci of the knee. Because the menisci act as shock absorbers and stabilize the knee, severe tears may produce edema, pain, instability, and limited motion. Arthroscopic surgical procedures are frequently used in both diagnosis and treatment of this type of knee injury (see box).

A *sprain* is an acute musculoskeletal injury to the ligamentous structures surrounding a joint, disrupting continuity of the synovial membrane. A common cause is a twisting or wrenching movement often associated with "whiplash" type injuries. Blood vessels may be ruptured and bruising and swelling may occur. Limitation of joint motion is common in all sprain injuries.

Inflammatory Joint Disease

Arthritis is a general name for many different inflammatory joint diseases. One type, *rheumatoid arthritis,* is a chronic and

---

HEALTH MATTERS   **ARTHROSCOPY**

Arthroscopy is an imaging technique that allows a physician to examine the internal structure of a joint without the use of extensive surgery. **A,** a narrow tube with lenses and a fiberoptic light source is inserted into the joint space through a small puncture. Isotonic saline (salt) solution is injected through a needle to expand the volume of the synovial space. This spreads the joint structures and makes viewing easier **(B).**

Although arthroscopy is often used as a diagnostic procedure, it can be used to perform joint surgery. While the surgeon views the internal structure of the joint through the arthroscope or on an attached video monitor, instruments can be inserted through puncture holes and used to repair or remove damaged tissue. Arthroscopic surgery is much less traumatic than previous methods in which the joint cavity was completely opened.

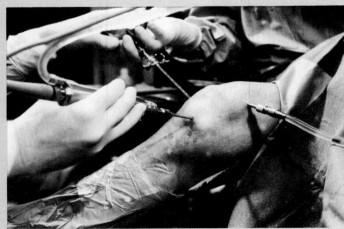

A

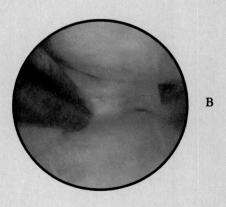

B

systemic inflammatory connective tissue disease of unknown cause. It is characterized by destruction of the synovial membrane, resulting in joint deformity. Immunological mechanisms appear to play an important role. Swelling of the synovial membrane leads to destruction of cartilage and erosion of bone. Loss of articular surface and joint motion is common, and progressive crippling and deformity may occur. Swelling, joint pain, tenderness, and stiffness or limited motion are common signs and symptoms. Small joints of the hand, wrist, and feet are often involved first, but the disease often progresses to larger joints. Because it is a systemic disease, fever, weight loss, fatigue, and possible pericarditis (inflammation of the membranous sac around the heart) may develop.

Juvenile rheumatoid arthritis is more severe than the adult form but exhibits similar deterioration and deformity of the joints. Onset occurs during childhood and is more common in girls. The joint inflammatory process often destroys growth of cartilage, and growth of long bones is arrested.

Gouty arthritis is another type of inflammatory arthritis. It is a metabolic disorder in which excess blood levels of uric acid, a nitrogenous waste, deposit as sodium urate crystals within the synovial fluid of joints and in other tissues. These crystals trigger the chronic inflammation and tissue damage seen with the disease. Swelling, tenderness, or pain typically appears in the fingers, wrists, elbows, ankles, and knees. Allopurinol (Zyloprim) is a drug used to treat the disease. It acts to inhibit the synthesis of uric acid.

CASE STUDY
TRAUMATIC INJURY TO THE KNEE

Joe is a 15-year-old boy who is a running back on his high school football team. Today during practice, he was running down the field when he was tackled from the side at knee level. Joe felt his knee "pop" and was unable to walk off the field unassisted. Ice is applied to his injured knee, and he is taken to the school health room. The school athletic trainer elevates Joe's knee and performs an initial evaluation of the injury. She notes that the affected knee is not swollen, but it has joint line tenderness on the medial side of the knee. Joe has limited flexion in the affected knee and is unable to fully extend his leg. His knee gives way when he tries to bear weight. He is referred to the local orthopedist for treatment.

The orthopedist examines the knee, takes a series of x-rays and an MRI (magnetic resonance imaging). He then schedules Joe for arthroscopic surgery. The arthroscopy shows a tear in the medial meniscus (cartilage). The surgeon is able to repair the tear with minimal removal of miniscal tissue. Joe is placed in a hinged splint and begins a program of rehabilitation with the sports medicine clinic.

1. Which of the following represent normal movements of the knee joint?
   A. Flexion
   B. Extension
   C. Supination
   D. Hyperextension

2. Which of the following factors makes the knee joint more vulnerable to injuries such as Joe's?
   A. It is relatively unprotected by muscle.
   B. The ligaments are very fragile.
   C. It is a weight-bearing joint.
   D. It cannot be pronated.

3. Why is Joe unable to fully extend his leg on the affected side?
   A. He may have a mechanical obstruction from the displaced meniscus.
   B. Muscle spasm in the thigh may prevent full extension.
   C. Joint effusion may prevent full extension.
   D. All of the above.

4. Which types of movement during the actual blow to the knee cause the meniscus to tear?
   A. Valgus movement
   B. Flexion
   C. Hyperextension
   D. All of the above

CHAPTER SUMMARY

INTRODUCTION

A. Articulation—point of contact between bones
B. Joints are mostly very movable, but others are immovable or allow only limited motion
C. Movable joints allow complex, highly coordinated and purposeful movements to be executed

CLASSIFICATION OF JOINTS

A. Joints may be classified using a structural or functional scheme (Table 8-1)
  1. Structural classification—joints are named according to:
    a. Type of connective tissue that joins bones together (fibrous or cartilaginous joints)
    b. Presence of a fluid-filled joint capsule (synovial joint)
  2. Functional classification—joints are named according to degree of movement allowed
    a. Synarthroses—immovable joint
    b. Amphiarthroses—slightly movable
    c. Diarthroses—freely movable
B. Fibrous joints (synarthroses)—bones of joints fit together closely, allowing little or no movement (Figure 8-1)
  1. Syndesmoses—joints in which ligaments connect two bones
  2. Sutures—found only in skull; teeth-like projections from adjacent bones interlock with each other
  3. Gomphoses—between root of a tooth and the alveolar process of the mandible or maxilla
C. Cartilaginous joints (amphiarthroses)—bones of joints are joined together by either hyaline cartilage or fibrocartilage; allow very little motion (Figure 8-2)
  1. Synchondroses—hyaline cartilage present between articulating bones
  2. Symphyses—joints in which a pad or disk of fibrocartilage connects two bones
D. Synovial joints (diarthroses)—freely movable joints (Figure 8-3)
  1. Structures of synovial joints
    a. Joint capsule—sleevelike casing around the ends of the bones, binding them together
    b. Synovial membrane—membrane that lines the joint capsule and also secretes synovial fluid
    c. Articular cartilage—hyaline cartilage covering articular surfaces of bones
    d. Joint cavity—small space between the articulating surfaces of the two bones of the joint
    e. Menisci (articular disks)—pads of fibrocartilage located between articulating bones
    f. Ligaments—strong cords of dense white fibrous tissue that hold the bones of synovial joint more firmly together
    g. Bursae—synovial membranes filled with synovial fluid; cushion joints and facilitates the movement of tendons
  2. Types of synovial joints (Figure 8-4)
    a. Uniaxial joints—synovial joints that permit movement around only one axis and in only one plane
      1. Hinge joints—articulating ends of bones form a hinge-shaped unity that allows only flexion and extension
      2. Pivot joints—a projection of one bone articulates with a ring or notch of another bone
    b. Biaxial joints—synovial joints that permit movements around two perpendicular axes in two perpendicular planes

      1. Saddle joints—synovial joints in which the articulating ends of the bones resemble reciprocally shaped miniature saddles; only example in body is in the thumbs
      2. Condyloid (ellipsoidal) joints—synovial joints in which a condyle fits into an elliptical socket
    c. Multiaxial joints—synovial joints that permit movements around three or more axes in three or more planes
      1. Ball and socket (spheroid) joints—most movable joints; ball-shaped head of one bone fits into a concave depression
      2. Gliding joints—relatively flat articulating surfaces that allow limited gliding movements along various axes

TYPES OF MOVEMENT AT SYNOVIAL JOINTS

A. Angular movements—change the size of the angle between articulating bones (Figure 8-5)
  1. Flexion—decreases the angle between bones; bends or folds one part on another
  2. Extension and hyperextension
    a. Extension—increases the angle between bones, returns a part from its flexed position to its anatomical position
    b. Hyperextension—stretching or extending part beyond its anatomical position
  3. Plantar flexion and dorsiflexion
    a. Plantar flexion—increases the angle between the top of the foot and the front of the leg
    b. Dorsiflexion—decreases the angle between the top of the foot and the front of the leg
  4. Abduction and adduction
    a. Abduction—moves a part away from the median plane of the body
    b. Adduction—moves a part toward the median plane of the body
B. Circular movements (Figure 8-6)
  1. Rotation and circumduction
    a. Rotation—pivoting a bone on its own axis
    b. Circumduction—moves a part so its distal end moves in a circle
  2. Supination and pronation
    a. Supination—turns the hand palm side up
    b. Pronation—turns the hand palm side down
C. Gliding movements—simplest of all movements; articular surface of one bone moves over the articular surface of another without any angular or circular movement
D. Special movements (Figure 8-7)
  1. Inversion and eversion
    a. Inversion—turning the sole of the foot inward
    b. Eversion—turning the sole of the foot outward
  2. Protraction and retraction
    a. Protraction—moves a part forward
    b. Retraction—moves a part backward
  3. Elevation and depression
    a. Elevation—moves a part up
    b. Depression—lowers a part

REPRESENTATIVE SYNOVIAL JOINTS

A. Humeroscapular joint (Figure 8-8)
  1. Shoulder joint
  2. Most mobile joint due to the shallowness of the glenoid cavity
  3. Glenoid labrum—narrow rim of fibrocartilage around the glenoid cavity that lends depth to the glenoid cavity

4. Structures that strengthen the shoulder joint are ligaments, muscles, tendons, and bursae
B. Hip joint (Figure 8-9)
   1. Stable joint due to the shape of the head of the femur and of the acetabulum
   2. A joint capsule and ligaments contribute to the joint's stability
C. Knee joint (Figures 8-10 and 8-11)
   1. Largest and one of the most complex and most frequently injured joints
   2. Tibiofemoral joint is supported by a joint capsule, cartilage, and numerous ligaments and muscle tendons
   3. Permits flexion and extension and, with the knee flexed, some internal and external rotation
D. Vertebral joints (Figures 8-13 and 8-14)
   1. Vertebrae are connected to one another by several joints to form a strong flexible column
   2. Bodies of adjacent vertebrae are connected by intervertebral disks and ligaments
   3. Intervertebral disks are made up of two parts
      a. Annulus fibrosus—disk's outer rim, made of fibrous tissue and fibrocartilage
      b. Nucleus pulposus—disk's central core, made of a pulpy, elastic substance

THE CYCLE OF LIFE: ARTICULATIONS

A. Bone development and the sequence of ossification between birth and skeletal maturity affects joints
   1. Fontanels between cranial bones disappear
   2. Epiphysial plates ossify at maturity
B. Older Adults
   1. Range of motion decreases
   2. Changes in gait
C. Skeletal diseases manifest as joint problems
   1. Abnormal bone growth (lipping)—influences joint motion
   2. Disease conditions can be associated with specific developmental periods

MECHANISMS OF DISEASE: JOINT DISORDERS

A. Joint disorders affect mobility
B. Noninflammatory joint disease
   1. Osteoarthritis—degenerative joint disease
      a. Deterioration and atrophy of articular cartilage and formation of new bone at joint surfaces
      b. Occurs in weight-bearing joints
      c. Cause unknown
      d. Treatment will not stop process
         (1) Joint replacement
   2. Traumatic injuries
      a. Dislocation (subluxation)
      b. Sprain—musculoskeletal injury to ligamentous structures surrounding joint
C. Inflammatory joint disease
   1. Arthritis—general name
      a. Rheumatoid arthritis—chronic and systemic inflammatory joint disease
         (1) Unknown cause
         (2) Destroys synovial membrane, resulting in joint deformation
      b. Juvenile rheumatoid arthritis
         (1) Occurs in childhood—more common in girls
         (2) More severe than in adults
         (3) Destroys growth of cartilage and arrests growth of long bones
      c. Gouty arthritis—metabolic disorder
         (1) Excess uric acid deposits sodium urate crystals in synovial fluid in joints and tissue
         (2) Zyloprim—drug used to inhibit synthesis of uric acid

REVIEW QUESTIONS

1. Classify joints according to function and according to structure.
2. Define the terms *fibrous joints*, *cartilaginous joints*, and *synovial joints*.
3. Define the terms *synarthroses*, *amphiarthroses*, and *diarthroses*.
4. Name and define three types of fibrous joints. Give an example of each.
5. Name and define two types of cartilaginous joints. Give an example of each.
6. Describe the characteristic structural features of synovial joints.
7. List and describe the different types of synovial joints. Give an example of each.
8. Name and define four kinds of angular movements permitted by some synovial joints.
9. Define and give an example of the following: *rotation, circumduction, pronation, supination.*
10. Describe the function of a goniometer.
11. What joint makes possible much of the dexterity of the human hand? Describe it and the movements it permits.
12. Describe vertebral joints.
13. Why does the range of movement vary at different joints? What factors determine this?
14. Describe and differentiate between the following joints: humeroscapular, hip, knee.
15. How do loose bodies, or cartilaginous "joint mice," differ from menisci?
16. Discuss the surgical procedure of total hip replacement (THR).
17. Why are knee injuries a common type of athletic injury?

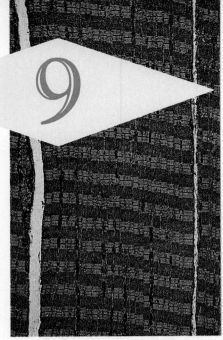

TEM of skeletal muscle fiber

9 ▶ Physiology of the Muscular System

OBJECTIVES

After you have completed this chapter, you should be able to:

1. List and discuss the three generalized functions of skeletal muscle tissue.
2. Discuss the three characteristics of skeletal muscle cells that allow them to function as they do.
3. List and discuss the structural parts of skeletal muscle fibers that are also found in other types of cells.
4. List and discuss the structural parts of skeletal muscle fibers that are not found in other cells.
5. Discuss the structure and function of myofilaments.
6. Explain the series of steps in muscle contraction.
7. Describe the sliding filament theory.
8. Explain the series of steps in muscle relaxation.
9. Identify and explain the energy sources for muscle contraction.
10. Discuss aerobic and anaerobic respiration.
11. Define a motor unit.
12. Describe the following types of skeletal muscle contractions: twitch, treppe, tetanus, tonic, isotonic, isometric.
13. Explain the graded strength principle.
14. Describe the anatomical and functional characteristics of cardiac and smooth muscle.

KEY TERMS

All-or-none principle   Individual muscle fibers, when sufficiently stimulated, either contract maximally or not at all

Isometric contraction (eye-so-MET-rik)   Contraction in which muscle length remains the same while the tension increases

Isotonic contraction (eye-so-TON-ik)   Contraction in which muscle length changes while the tension remains the same

Motor unit   Motor neuron plus the muscle fiber it innervates

Myofibril (my-o-FYE-bril)   Bundles of very fine fibers that almost fill the sarcoplasm of a muscle cell

Myofilament (my-o-FIL-ah-ment)   Ultramicroscopic, threadlike structures found in myofibrils

Neuromuscular junction (noo-ro-MUS-kyoo-lar JUNK-shun)   Where a motor neuron connects to the sarcolemma at the motor endplate

Sarcolemma (sar-ko-LEM-ah)   Plasma membrane of a muscle fiber

Sarcomere (SAR-ko-meer)   Contractile unit of myofibrils

Sarcoplasm (SAR-ko-plazm)   Cytoplasm of a muscle fiber

Sarcoplasmic reticulum (sar-ko-PLAS-mik re-TIK-yoo-lum)   Network of tubules and sacs found in muscle fibers; analogous to endoplasmic reticulum of other cells

T tubules   Transverse tubules that allow electrical impulses traveling along the sarcolemma to move deeper into the cell

Tetanus (TET-uh-nus)   Smooth, sustained muscular contractions

Threshold stimulus   Minimum stimulus required to cause a muscle fiber to contract

Survival depends on the ability to maintain a relatively constant internal environment. Such stability often requires movement of the body. For example, we must gather and eat food, defend ourselves, seek shelter, and make tools, clothing, or other objects. Whereas many different systems of the body play *some* role in accomplishing movement, it is the skeletal and muscular systems acting together that actually produce most body movements. We have investigated the architectural plan of the skeleton and have seen how its firm supports and joint structures make movement possible. However, bone and joints cannot move themselves. They must be moved by something. Our subject for now, then, is the large mass of skeletal muscle that moves the framework of the body: the **muscular system** (Figure 9-1).

In this chapter, we will examine the basic characteristics of skeletal muscle tissue that permit it to move the body's framework, as well as perform other functions vital to maintaining a constant internal environment. We will also briefly examine smooth and cardiac muscle tissues, contrasting them with skeletal muscle tissue. In Chapter 10, we will continue our examination of the muscular system by discussing the overall anatomy of muscles and how they work together to accomplish specific body movements.

◆ GENERAL FUNCTIONS

If you have any doubts about the importance of muscle function to normal life, you have only to observe a person with extensive paralysis—a victim of advanced muscular dystrophy, for example. Any of us possessed of normal powers of movement can little imagine life with this matchless power lost. But cardinal as it is, movement is not the only contribution muscles make to healthy survival. They also perform two other essential functions: production of a large portion of body heat and maintenance of posture.

1. **Movement.** Skeletal muscle contractions produce movements either of the body as a whole (locomotion) or of its parts.
2. **Heat production.** Muscle cells, like all cells, produce heat by the process known as catabolism (discussed in Chapter 26). But because skeletal muscle cells are both highly active and numerous, they produce a major share of total body heat. Skeletal muscle contractions therefore constitute one of the most important parts of the mechanism for maintaining homeostasis of temperature.
3. **Posture.** The continued partial contraction of many skeletal muscles makes possible standing, sitting, and other maintained positions of the body.

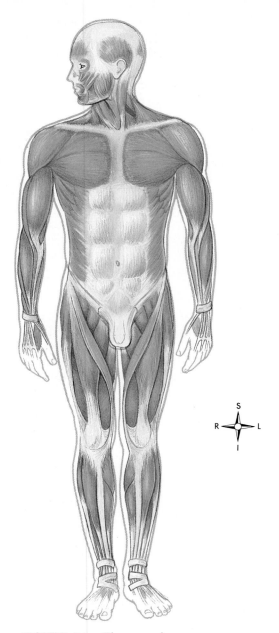

FIGURE 9-1   The muscular system.

FUNCTION OF SKELETAL MUSCLE TISSUE

Skeletal muscle cells have several characteristics that permit them to function as they do. One such characteristic is the ability to be stimulated, often called **excitability** or *irritability*. Because skeletal muscle cells are excitable, they can respond to regulatory mechanisms such as nerve signals. **Contractility** of muscle cells, the ability to contract or shorten, allows muscle tissue to pull on bones and thus produce body movement. **Extensibility,** the ability to extend or stretch, allows muscles to return to their resting length after having contracted. These characteristics are related to the microscopic structure of skeletal muscle cells. In the following passages, we will first discuss the basic structure of a muscle cell. We will then explain how a muscle cell's structural components allow it to perform its specialized functions.

Overview of the Muscle Cell

Look at Figure 9-2. As you can see there, a skeletal muscle is composed of bundles of skeletal muscle fibers that generally extend the entire length of the muscle. They are called *fibers,* instead of *cells,* because of their threadlike shape (1 to 40 mm long but with a diameter of only 10 to 100 μm). Skeletal muscle fibers have many of the same structural parts as other cells. Several of them, however, bear different names in muscle fibers. For example, **sarcolemma** is the plasma membrane of a muscle fiber. **Sarcoplasm** is its cytoplasm. Muscle cells contain a network of tubules and sacs known as the **sarcoplasmic reticulum (SR)**—a structure analogous, but not identical, to the endoplasmic reticulum of other cells. Muscle fibers contain many mitochondria, and, unlike most other cells, they have several nuclei.

Certain structures not found in other cells are present in skeletal muscle fibers. For instance, bundles of very fine fibers—**myofibrils**—extend lengthwise along skeletal muscle fibers and almost fill their sarcoplasm. Myofibrils, in turn, are made up of still finer fibers, called *thick* and *thin myofilaments* (Figure 9-2, *D*). Find the label *sarcomere* in this drawing. Note that a **sarcomere** is a segment between two successive Z lines. Each myofibril consists of a lineup of many sarcomeres, each of which functions as a contractile unit. The A bands of the sarcomeres appear as relatively wide, dark stripes (cross striae) under the microscope, and they alternate with narrower, lighter colored stripes formed by the I bands (Figure 9-2, *D*). Because of its cross striae, skeletal muscle is also called *striated muscle.* Electron microscopy of skeletal muscle (Figure 9-3) has revolutionized our concept of both its structure and function.

FIGURE 9-2 Structure of skeletal muscle. A, Skeletal muscle organ, composed of bundles of contractile muscle fibers. **B,** Greater magnification of single fiber showing smaller fibers—myofibrils—in the sarcoplasm. Note sarcoplasmic reticulum and T tubules forming triad. **C,** Myofibril magnified further to show sarcomere between successive Z lines. Cross striae are visible. **D,** Molecular structure of myofibril showing thick myofilaments and thin myofilaments.

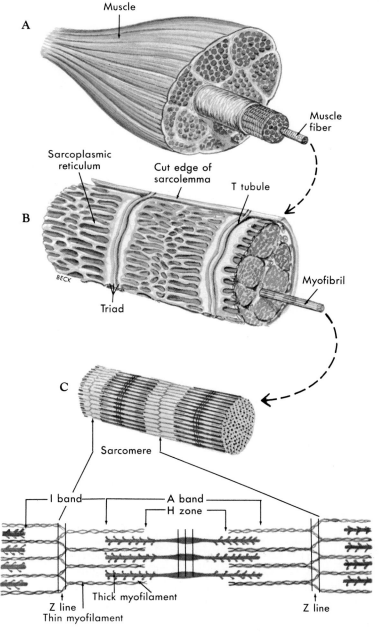

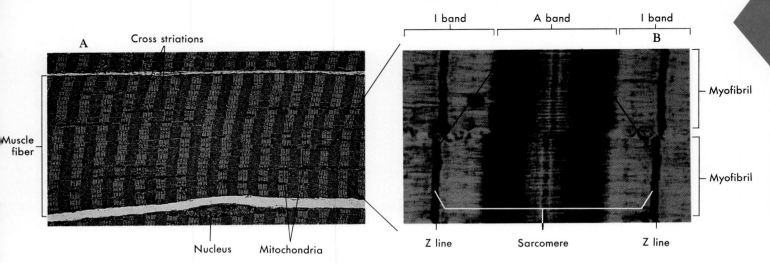

FIGURE 9-3 **Electron micrographs of striated muscle.** B shows detail of A at greater magnification. Note that the myofilaments of each myofibril form a pattern that, when viewed together, produce the striated pattern typical of skeletal muscle.

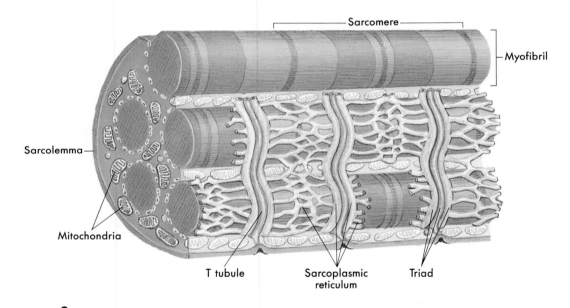

FIGURE 9-4 **Unique features of the skeletal muscle cell.** Notice especially the *T tubules,* which are extensions of the plasma membrane, or *sarcolemma,* and the *sarcoplasmic reticulum (SR),* which forms networks of tubular canals and sacs. A *triad* is a triplet of adjacent tubules: a sac of the SR, a T tubule, and another sac of the SR.

Another structure unique to muscle cells is a system of transverse tubules, or **T tubules.** This name derives from the fact that these tubules extend transversely across the sarcoplasm, at a right angle to the long axis of the cell. As Figures 9-2, *B,* and 9-4 show, the T tubules are formed by inward extensions of the sarcolemma. The chief function of T tubules is to allow electrical signals, or *impulses,* traveling along the sarcolemma to move deeper into the cell.

The sarcoplasmic reticulum is also a system of membranous tubules in a muscle fiber. It is separate from the T tubule system, forming extensive networks of connected canals and sacs. The membrane of the sarcoplasmic reticulum continually pumps calcium ions (Ca^{++}) from the sarcoplasm and stores them within its sacs. Notice in Figures 9-2, *B,* and 9-4 that a tubular sac of the sarcoplasmic reticulum butts up against each side of every T tubule in a muscle fiber. This triplet of tubules (a T tubule sandwiched between sacs of the sarcoplasmic reticulum) is called a **triad.** The triad is an important feature of the muscle cell because it allows an electrical impulse traveling along a T tubule to stimulate the membranes of adjacent sacs of the sarcoplasmic reticulum.

Myofilaments

Each muscle fiber contains a thousand or more parallel subunits, called myofibrils, that are only about 1 μm thick. Lying side by side in each myofibril are thousands of **thick** and **thin myofilaments.** Over the years, a clear picture of the molecular structure of myofilaments has emerged. This picture has revealed the mechanism of how muscle fibers contract and do so powerfully. It is wise, therefore, to take a moment to study the molecular structure of myofilaments before discussing the detailed mechanism of muscle contraction.

First of all, there are four different kinds of protein molecules that comprise myofilaments: *myosin, actin, tropomyosin,* and *troponin.* The thin filaments are made of a combination of three different proteins. Figure 9-5, *A,* shows that globular actin molecules are strung together like beads to form two fibrous strands that twist around each other to form the bulk of each thin filament. Actin and myosin molecules have a chemical attraction for one another, but, at rest, the active sites on the actin molecules are covered up by long tropomyosin molecules. The tropomyosin molecules seem to be held in this blocking position by troponin molecules spaced at intervals along the thin filament.

As Figure 9-5, *B,* shows, the thick filaments are made almost entirely of myosin molecules. Notice that the myosin molecules are shaped like golf clubs, with their long

shafts bundled together to form a thick filament and their "heads" sticking out from the bundle. The myosin heads are chemically attracted to the actin molecules of the nearby thin filaments, so they angle toward the thin filaments. Because they attempt to bridge the gap between adjacent myofilaments, the myosin heads are usually called **cross bridges.**

Within a myofibril the thick and thin filaments alternate, as shown in Figure 9-2, *D.* This arrangement is crucial for contraction. Another fact important for contraction is that the thin filaments attach to both Z lines of a sarcomere and that they extend in from the Z lines part way toward the center of the sarcomere. When the muscle fiber is relaxed, the thin filaments terminate at the outer edges of the H zones. In contrast, the thick myosin filaments do not attach to the Z lines, and they extend only the length of the A bands of the sarcomeres.

The Mechanism of Contraction

To accomplish the powerful shortening, or *contraction,* of a muscle fiber, several processes must be coordinated in a stepwise fashion. These steps are summarized in the following and in Table 9-1.

Excitation and contraction

Under normal circumstances, a skeletal muscle fiber remains "at rest" until it is stimulated by a signal from a special type of nerve cell called a **motor neuron.** As Figure 9-6 shows, motor neurons connect to the sarcolemma of a muscle fiber at a folded **motor endplate** to form a junction called a **neuromuscular junction.** A neuromuscular junction is a type of connection called a *synapse,* characterized by a narrow gap, or synaptic cleft, across which *neurotransmitter* molecules transmit signals. When nerve impulses reach the end of a motor neuron fiber, small vesicles release a neurotransmitter—**acetylcholine**—into the synaptic cleft. Diffusing swiftly across this microscopic gap, acetylcholine molecules contact the sarcolemma of the adjacent muscle fiber. There, they stimulate acetylcholine receptors and thus initiate an impulse in the sarcolemma.

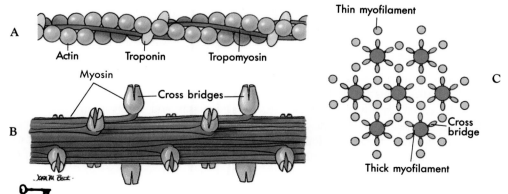

FIGURE 9-5 Structure of myofilaments. A, Thin myofilament. **B,** Thick myofilament. **C,** Cross section of several thick and thin myofilaments, showing the arrangements of myofilaments and cross bridges.

TABLE 9-1	Major events of muscle contraction and relaxation

EXCITATION AND CONTRACTION

1 A nerve impulse reaches the end of a motor neuron, triggering the release of the neurotransmitter *acetylcholine.*

2 Acetylcholine diffuses rapidly across the gap of the neuromuscular junction and binds to acetylcholine receptors on the motor endplate of the muscle fiber.

3 Stimulation of acetylcholine receptors initiates an impulse that travels along the sarcolemma, through the T tubules, to sacs of the sarcoplasmic reticulum (SR).

4 Ca^{++} is released from the SR into the sarcoplasm, where it binds to troponin molecules in the thin myofilaments.

5 Tropomyosin molecules in the thin myofilaments shift, exposing actin's active sites.

6 Energized myosin cross bridges of the thick myofilaments bind to actin and use their energy to pull the thin myofilaments toward the center of each sarcomere. This cycle repeats itself many times per second, as long as ATP is available.

7 As the thin filaments slide past the thick myofilaments, the entire muscle fiber shortens.

RELAXATION

1 After the impulse is over, the SR begins actively pumping Ca^{++} back into its sacs.

2 As Ca^{++} is stripped from troponin molecules in the thin myofilaments, tropomyosin returns to its position, blocking actin's active sites.

3 Myosin cross bridges are prevented from binding to actin and thus can no longer sustain the contraction.

4 Since the thick and thin myofilaments are no longer connected, the muscle fiber may return to its longer, resting length.

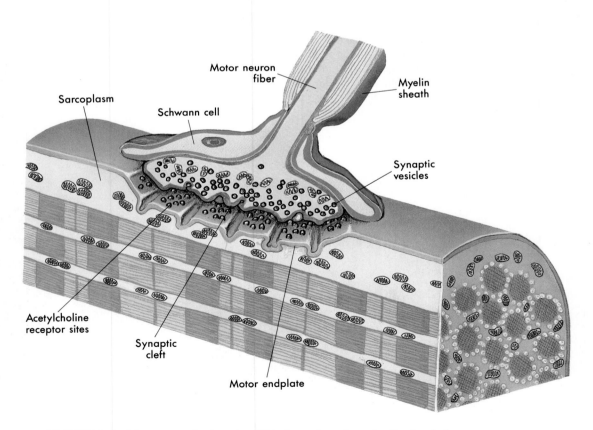

FIGURE 9-6   Neuromuscular junction. This figure shows how the distal end of a motor neuron fiber forms a synapse, or "chemical junction," with an adjacent muscle fiber. Neurotransmitters (specifically, acetylcholine) are released from the neuron's synaptic vesicles and diffuse across the synaptic cleft. There, they stimulate receptors in the motor endplate region of the sarcolemma.

The impulse, a temporary electrical imbalance, is conducted over the muscle fiber's sarcolemma and inward along the T tubules (Figure 9-7). The impulse in the T tubules triggers the release of a flood of calcium ions from the adjacent sacs of the SR. In the sarcoplasm, the calcium ions combine with troponin molecules in the thin filaments of the myofibrils. Recall that troponin normally holds tropomyosin strands in a position that blocks the chemically active sites of actin. When calcium binds to troponin, however, the tropomyosin shifts to expose active sites on the actin molecules (Figure 9-8). Once the active sites are exposed, energized myosin cross bridges of the thick filaments bind to actin molecules in the nearby thin filaments. The myosin heads bend with great force, literally pulling the thin filaments past them. Each head then releases itself, binds to the next active site, and pulls again. Figure 9-9 shows how the sliding of the thin filaments

toward the center of each sarcomere quickly shortens the entire myofibril—and thus the entire muscle fiber. This model of muscle contraction has been called the *sliding filament theory.*

Relaxation

Almost immediately after the sarcoplasmic reticulum releases its flood of calcium ions into the sarcoplasm, it begins actively pumping them back into its sacs once again. Within a few milliseconds, much of the calcium is recovered. Because the active transport carriers of the SR have a greater affinity to calcium than troponin molecules do, the calcium ions are stripped off the troponin molecules and returned to the sacs of the SR. As you might suspect, this shuts down the whole process of contraction. Troponin without its bound calcium allows the tropomyosin to once again block actin's active sites. Myosin cross bridges reaching for the next active site on actin are blocked, and thus the thin filaments are no longer being held—or pulled—by the thick filaments. The muscle fiber may remain at its contracted length, but forces outside the muscle fiber are likely to pull it back to its longer resting length. In short, the contraction process in a skeletal muscle fiber automatically shuts itself off within a small fraction of a second after the initial stimulation.

FYI **Rigor Mortis**

The term **rigor mortis** is a Latin phrase that means "stiffness of death." In a medical context, the term *rigor mortis* refers to the stiffness of skeletal muscles sometimes observed shortly after death. What causes rigor mortis? At the time of death, stimulation of muscle cells ceases. However, some muscle fibers may have been in midcontraction at the time of death—when the myosin-actin cross bridges are still intact. ATP is required to release the cross bridges and "energize" them for their next attachment. Because the last of a cell's ATP supply is used up at the time it dies, many cross bridges may be left "stuck" in the contracted position. Thus muscles in a dead body may be stiff because individual muscle fibers ran out of the ATP required to "turn off" a muscle contraction.

QUICK CHECK

✔ 1. *Describe the structure of thin myofilaments and thick myofilaments, naming the kinds of proteins that comprise them.*

✔ 2. *What is a neuromuscular junction? How does it work?*

✔ 3. *What is the role of calcium ions (Ca^{++}) in muscle contraction?*

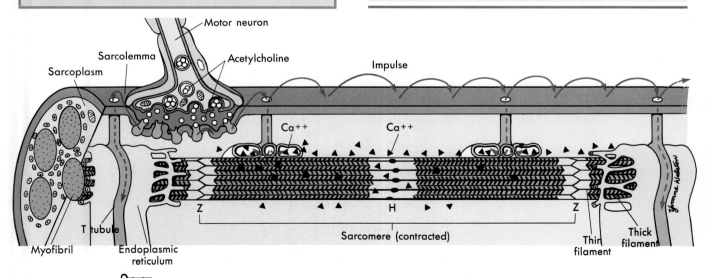

FIGURE 9-7 **Effects of excitation on a muscle fiber.** Excitation of the sarcolemma by a nerve impulse initiates an impulse in the sarcolemma. The impulse travels across the sarcolemma and through the T tubules, where it triggers adjacent sacs of the sarcoplasmic reticulum to release a flood of calcium ions (Ca^{++}) into the sarcoplasm. The Ca^{++} are then free to bind to troponin molecules in the thin filaments. This binding, in turn, initiates the chemical reactions that produce a contraction.

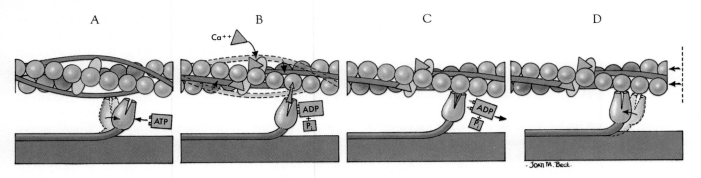

FIGURE 9-8   The molecular basis of muscle contraction. A, Each myosin cross bridge in the thick filament moves into a resting position after an ATP binds and transfers its energy. **B,** Calcium ions released from the sarcoplasmic reticulum bind to troponin in the thin filament, allowing tropomyosin to shift from its position blocking the active sites of actin molecules. **C,** Each myosin cross bridge then binds to an active site on a thin filament, displacing the remnants of ATP hydrolysis—adenosine diphosphate (ADP) and inorganic phosphate (P_i). **D,** The release of stored energy from step **A** provides the force needed for each cross bridge to move back to its original position, pulling actin along with it. Each cross bridge will remain bound to actin until another ATP binds to it and pulls it back into its resting position **(A).**

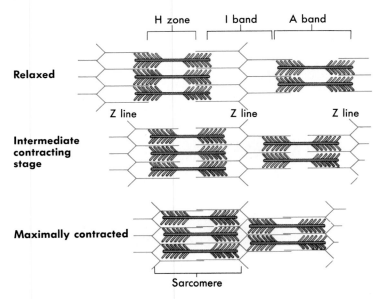

FIGURE 9-9   Sliding filament theory. During contraction, myosin cross bridges pull the thin filaments toward the center of each sarcomere, thus shortening the myofibril and the entire muscle fiber.

Energy sources for muscle contraction

The energy required for muscular contraction is obtained by hydrolysis of a nucleotide called **adenosine triphosphate,** or ATP. Recall from Chapter 2 (Figure 2-22, p. 57) that this molecule has an adenine and ribose group (together called adenosine) attached to three phosphate groups. Two of the three phosphate groups in ATP are attached to the molecule by *high-energy bonds.* It is the breaking of these high-energy bonds that provides the energy necessary to pull the thin myofilaments during muscle contraction. As Figure 9-8, A, shows, before contraction occurs, each myosin cross bridge moves into a resting position when an ATP molecule binds to it. The ATP molecule breaks its outermost high-energy bond,

releasing the inorganic phosphate (P_i) and transferring the energy to the myosin cross bridge. In a way, this is like pulling back the elastic band of a sling-shot—the apparatus is "at rest" but ready to spring. When myosin binds to actin, the stored energy is released, and the cross bridge does indeed spring back to its original position. Thus, the energy transferred from ATP is used to do the work of pulling the thin filaments during contraction. Another ATP molecule then binds to the myosin cross bridge, which then releases actin and moves into its resting position again—all set for the next "pull." This cycle repeats itself over and over again, as long as ATP is available and actin's active sites are unblocked.

Muscle fibers must continually resynthesize ATP be-

cause they can store only small amounts of it. Immediately after ATP breaks down, energy for its resynthesis can be supplied by the breakdown of another high-energy compound, creatine phosphate (CP), which is also present in small amounts in muscle fibers (Figure 9-10). Ultimately, energy for both ATP and CP synthesis comes from the catabolism of foods.

Note that continued, efficient nutrient catabolism by muscle fibers requires two essential ingredients: glucose and oxygen. Glucose is a nutrient molecule that contains many chemical bonds. The potential energy stored in these chemical bonds is released during catabolic reactions in the sarcoplasm and mitochondria and transferred to ATP or CP molecules. Some muscle fibers ensure an uninterrupted supply of glucose by storing it in the form of *glycogen.* Oxygen, which is needed for a catabolic process known as *aerobic respiration,* can also be stored by cells. During rest, excess oxygen molecules in the sarcoplasm are bound to a large protein molecule called **myoglobin.** Myoglobin is a reddish pigment similar to the pigment *hemoglobin* that gives blood its red color. Like hemoglobin, myoglobin contains iron (Fe) groups that attract oxygen molecules and hold them temporarily. When the oxygen concentration inside a muscle fiber decreases rapidly—as it does during exercise—it can be quickly resupplied from the myoglobin.

Muscle fibers that contain large amounts of myoglobin take on a deep red appearance and are often called *red fibers.* Muscles fibers with little myoglobin in them appear to be light pink and are often called *white fibers.* Most muscle tissues contain a mixture of red and white fibers.

Aerobic (oxygen-requiring) respiration is a catabolic process that produces the maximum amount of energy available from each glucose molecule. When oxygen concentration is low, however, muscle fibers can shift toward an increased use of another catabolic process: *anaerobic respiration.* As its name implies, anaerobic respiration does not require the immediate use of oxygen. Besides its ability to produce ATP without oxygen, anaerobic respiration has the added advantage of being very rapid. Muscle fibers having difficulty getting oxygen—or fibers that generate a great deal of force very quickly—may rely on anaerobic respiration to resynthesize their ATP molecules.

Anaerobic respiration may allow the body to avoid the use of oxygen in the short term, but not in the long term. Anaerobic respiration results in the formation of an incompletely catabolized molecule called **lactic acid.** Lactic acid may accumulate in muscle tissue during exercise and cause a burning sensation. Some of the lactic acid eventually diffuses into the blood and is delivered to the

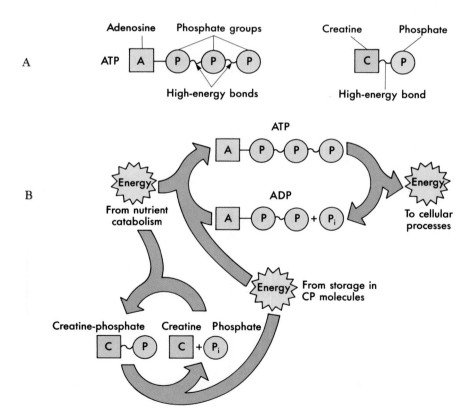

FIGURE 9-10 Energy sources for muscle contraction. A, The basic structure of two high-energy molecules in the sarcoplasm: adenosine triphosphate (ATP) and creatine phosphate (CP). **B,** This diagram shows how energy released during the catabolism of nutrients can be transferred to the high-energy bonds of ATP directly, or instead stored temporarily in the high-energy bond of the CP. During contraction, ATP is hydrolyzed and the energy of the broken bond transferred to a myosin cross bridge.

liver, where an oxygen-consuming process later converts it back into glucose. This is one of the reasons that after heavy exercise, when the lack of oxygen in some tissues caused the production of lactic acid, a person may still continue to breathe heavily. The body is repaying the so-called *oxygen debt* by using the extra oxygen gained by heavy breathing to process the lactic acid produced during exercise.

Because the catabolic processes of cells are never 100% efficient, some of the energy released is lost as heat. Because skeletal muscle tissues produce such a massive amount of heat—even when they are hardly doing any work—they have a great effect on body temperature. Recall from Chapter 5 that various heat-loss mechanisms of the skin can be employed to cool the body when it becomes overheated (see Figure 5-10, p. 143). Skeletal muscle tissues can likewise be employed when the body's temperature falls below the set point value determined by the "thermostat" in the hypothalamus of the brain. As Figure 9-11 shows, a low external temperature can reduce body

temperature below the set point. Temperature sensors in the skin and other parts of the body feed this information back to the hypothalamus, which compares the actual value to the set point value (usually about 37° C). The hypothalamus responds to a decrease in body temperature by signalling skeletal muscles to contract. The shivering contractions that result produce enough waste heat to warm the body back to the set point temperature—and homeostatic balance is maintained.

We will discuss the subject of energy metabolism more thoroughly in Chapter 26.

The all-or-none principle

Individual muscle fibers operate according to the **all-or-none principle.** This means that when they are sufficiently stimulated, they either contract with all the force possible under existing conditions, or they do not contract at all. The minimal level of stimulation required to cause a fiber to contract is called the **threshold stimulus.** However, if conditions at the time of stimulation change, the force of

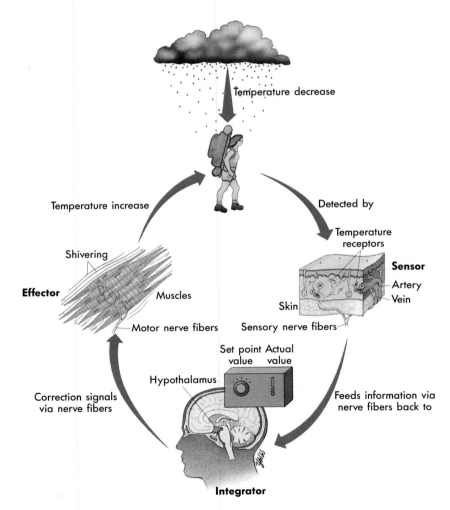

FIGURE 9-11   The role of skeletal muscle tissues in maintaining a constant body temperature. This diagram shows that a drop in body temperature caused by cold weather can be corrected by a negative feedback mechanism that triggers shivering (muscle contraction), which in turn produces enough heat to warm the body.

FOCUS ON... TYPES OF MUSCLE FIBERS

Skeletal muscle fibers can be classified into three types according to their structural and functional characteristics: (1) *slow* (red) *fibers,* (2) *fast* (white) *fibers,* and (3) *intermediate fibers.* Each type is best suited to a particular type or style of muscular contraction. Although each muscle organ contains a mix of all three fiber types, different organs have these fibers in different proportions, depending on the types of contraction that they most often perform.

Slow fibers are also called red fibers because they contain a high concentration of myoglobin, the reddish pigment used by muscle cells to store oxygen. They are called slow fibers because their thick myofilaments are made of a type of myosin that reacts at a slow rate. Because they contract so slowly, slow fibers are usually able to produce ATP quickly enough to keep pace with the energy needs of the myosin and thus avoid fatigue. This effect is enhanced by a larger number of mitochondria than other fiber types and the rich oxygen store provided by the myoglobin. The slow, nonfatiguing characteristics of slow fibers make them especially well suited to the sustained contractions exhibited by postural muscles. Postural muscles containing a high proportion of slow fibers can hold the skeleton upright for long periods without fatigue.

Fast fibers are also called white fibers because they contain very little myoglobin. Fast fibers can contract much more rapidly than slow fibers because they have a faster type of myosin and because their system of T tubules and sarcoplasmic reticulum is more efficient at quickly delivering Ca^{++} to the sarcoplasm. The price of a rapid contraction mechanism is the rapid depletion of ATP. Despite the fact that fast fibers typically contain a high concentration of glycogen, they have few mitochondria and so must rely primarily on anaerobic respiration to regenerate ATP. Because anaerobic respiration produces relatively small amounts of ATP, fast fibers cannot produce enough ATP to sustain a contraction for very long. Because they can generate great force very quickly but not for a long duration, fast fibers are best suited for muscles that move the fingers and eyes in darting motions.

Intermediate fibers have characteristics somewhere in between the two extremes of fast and slow fibers. They are more fatigue resistant than fast fibers and can generate more force more quickly than slow fibers. This type of muscle fiber predominates in muscles that both provide postural support and are occasionally required to generate rapid, powerful contractions. One example is the *soleus,* or calf muscle, that helps to support the leg but is also used in walking, running, and jumping.

---

the cell's contraction changes. For example, if a muscle fiber is slightly warmer than usual, calcium ions will diffuse through the sarcoplasm more efficiently and more actin-myosin reactions will occur during a single contraction. Thus, up to a point, a warm fiber contracts more strongly than a cool fiber.

QUICK CHECK

✔ 1. *Where does the energy stored in ATP come from?*
✔ 2. *Contrast aerobic respiration and anaerobic respiration in muscle fibers.*
✔ 3. *What is the role of myoglobin in muscle fibers?*
✔ 4. *What is meant by the statement that "muscle fibers contract all-or-none"?*

FUNCTION OF SKELETAL MUSCLE ORGANS

Although each skeletal muscle fiber is distinct from all other fibers, it operates as part of the large group of fibers that form a skeletal muscle organ. Skeletal muscle organs, often simply called *muscles,* are composed of bundle upon bundle of muscle fibers that are held together by fibrous connective tissues (see Figure 9-2, A). The details of muscle organ anatomy are discussed in Chapter 10. For now, we will turn our attention to the matter of how skeletal muscle organs function as a single unit.

The Motor Unit

Recall that each muscle fiber receives its stimulus from a motor neuron. This motor neuron, often called a *somatic motor neuron,* plus the muscle fibers to which it attaches, constitutes a functional unit called a **motor unit** (Figure 9-12). The single fiber of a somatic motor neuron divides into a variable number of branches on entering the skeletal muscle. The neuron branches of some motor units terminate in only a few muscle fibers, whereas others terminate in numerous fibers. Consequently, impulse conduction by one motor unit may stimulate only a half dozen or so muscle fibers to contract at one time, whereas conduction by another motor unit may activate a hundred or more fibers simultaneously. This fact bears a relationship to the function of the muscle as a whole. As a rule, the fewer the number of fibers supplied by a skeletal muscle's individual motor units, the more precise the movements that muscle can produce. For example, in certain small muscles of the hand, each motor unit includes only a few muscle fibers, and these muscles produce precise finger movements. In contrast, motor units in large abdominal muscles that do not produce precise movements are reported to include more than a hundred muscle fibers each.

The Twitch Contraction

Many experimental methods have been used to study the contractions of skeletal muscle organs. They vary from relatively simple procedures, such as observing or palpating muscles in action, to the more complicated method of *electromyography* (recording electrical impulses from muscles as they contract). One method of studying muscle contraction that is particularly useful for the purposes of our

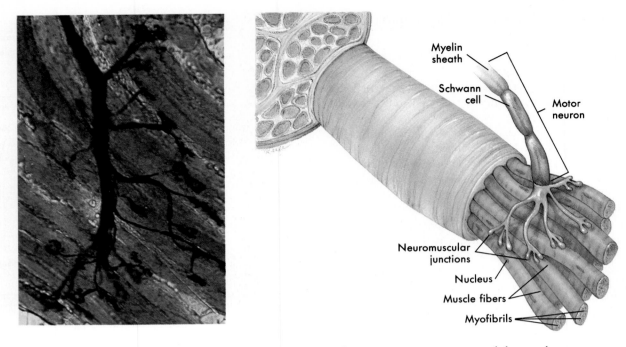

FIGURE 9-12   **Motor unit.** A motor unit consists of one somatic motor neuron and the muscle fibers supplied by its branches.

discussion is called, simply, **myography.** Myography, a term that means "muscle graphing," is a procedure in which the contraction of an isolated muscle is recorded as a line that rises and falls as the muscle contracts and relaxes. To get the muscle to contract, an electrical stimulus of sufficient intensity (the threshold stimulus) is applied to the muscle. A single, brief threshold stimulus produces a quick jerk of the muscle, called a **twitch contraction.**

The myogram of a twitch contraction shown in Figure 9-13 shows that the muscle does not begin to contract at the instant of stimulation but rather a fraction of a second later. The muscle then increases its tension (or shortens) until a peak is reached, after which it gradually returns to its resting state. These three phases of the twitch contraction are called, respectively, the *latent period,* the *contraction phase,* and the *relaxation phase.* The entire twitch usually lasts less than one tenth of a second.

Because the entire muscle is stimulated at once, the phases of the twitch contraction represent the events occurring in each fiber of each motor unit. During the latent period, the impulse initiated by the stimulation travels through the sarcolemma and T tubules to the SR, where it triggers the release of calcium ions into the sarcoplasm. It is not until the calcium binds to troponin and the sliding of the myofilaments begins that contraction is observed. After a few milliseconds, the forceful sliding of the myofilaments ceases and relaxation begins. By the end of the relaxation phase, all of the myosin-actin reactions in all the fibers have ceased.

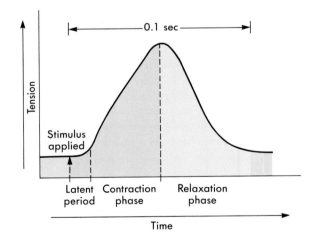

FIGURE 9-13   **The twitch contraction.** Three distinct phases are apparent: (1) the latent period, (2) the contraction phase, and (3) the relaxation phase.

Twitch contractions rarely happen in the body. Motor units are controlled by separate somatic motor neurons that normally do not all "fire" at the same time. Only when an electrical stimulus is applied, or when overactivity of the nervous system stimulates all the motor neurons in a muscle, do such contractions occur. However, knowledge of the twitch contraction gives us important insights about the mechanisms of more typical types of muscle contractions.

Treppe: The Staircase Phenomenon

One interesting effect that can be seen in myographic studies of the twitch contraction is called **treppe,** or the *staircase phenomenon.* Treppe is a gradual, steplike increase in the strength of contraction that can be observed in a series of twitch contractions that occur about 1 second apart (Figure 9-14, B).

In other words, a muscle contracts more forcefully after it has contracted a few times than when it first contracts—a principle used by athletes when they warm up, but one not yet satisfactorily explained. Presumably, it relates partly to the rise in temperature of active muscles and partly to their accumulation of metabolic products. After the first few stimuli, muscle responds to a considerable number of successive stimuli with maximal contractions. Eventually, it will respond with less and less strong contractions. The relaxation phase becomes shorter and finally disappears entirely. In other words, the muscle stays partially contracted—an abnormal state of prolonged contraction called *contracture.*

Repeated stimulation of muscle in time lessens its excitability and contractility and may result in **muscle fatigue,** a condition in which the muscle does not respond to the strongest stimuli. Complete muscle fatigue can be readily induced in an isolated muscle but very seldom occurs in the body.

Tetanus

The concept of the simple twitch can help us understand the smooth, sustained types of contraction that are commonly observed in the body. Such smooth, sustained contractions are called *tetanic contractions,* or, simply, **tetanus.** Figure 9-14, C, shows that if a series of stimuli come in a rapid enough succession, the muscle does not have time to relax completely before the next contraction phase begins. Muscle physiologists describe this effect as *multiple wave summation*—so named because it seems as if multiple twitch waves have been added together to sustain

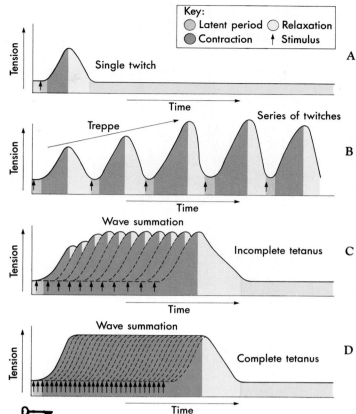

FIGURE 9-14 Myograms of various types of muscle contractions. A, A single twitch contraction. **B,** The treppe phenomenon, or "staircase effect," is a steplike increase in the force of contraction over the first few in a series of twitches. **C,** Incomplete tetanus occurs when a rapid succession of stimuli produces "twitches" that seem to add together (wave summation) to produce a rather sustained contraction. **D,** Complete tetanus is a smoother sustained contraction, produced by the summation of "twitches" that occur so close together that the muscle cannot relax at all.

muscle tension for a longer time. The type of tetanus produced when very short periods of relaxation occur between peaks of tension is called *incomplete tetanus.* It is "incomplete" because the tension is not sustained at a completely constant level. Figure 9-14, D, shows that when the frequency of stimuli increases, the distance between peaks of tension decrease to a point at which they seem to fuse into a single, sustained peak. This produces a very smooth type of tetanic contraction, called *complete tetanus.* Tetanus is the kind of contraction exhibited by normal skeletal muscles most of the time.

SPORTS & FITNESS

MUSCLE FATIGUE

Broadly defined, **muscle fatigue** is simply a state of exhaustion (a loss of strength or endurance) produced by strenuous muscular activity. *Physiological muscle fatigue* is caused by a relative lack of ATP, rendering the myosin cross bridges incapable of producing the force required for further muscle contractions. The low levels of ATP that produce fatigue may result from a depletion of oxygen or glucose in muscle fibers or from the inability to regenerate ATP quickly enough. High levels of lactic acid or other metabolic waste products also contribute to physiological fatigue. Under ordinary circumstances, however, physiological fatigue seldom occurs. It is usually *psychological fatigue* that produces the exhausted feeling that stops us from continuing a muscular activity. Thus in physiological muscle fatigue, we *cannot* contract our muscles, but in psychological muscle fatigue, we simply *will not* contract our muscles because we feel tired.

QUICK CHECK

✔ 1. *What are the three phases of a twitch contraction? What molecular events occur during each of these phases?*

✔ 2. *What is the difference between a twitch contraction and a tetanic contraction?*

✔ 3. *How does the treppe effect relate to the warm-up exercises of athletes?*

Muscle Tone

A **tonic contraction** (*tonus,* "tone") is a continual, partial contraction. At any one moment a small number of the total fibers in a muscle contract, producing a tautness of the muscle rather than a recognizable contraction and movement. Different groups of fibers scattered throughout the muscle contract in relays. Tonic contraction, or **muscle tone,** is characteristic of the muscles of normal individuals when they are awake. It is particularly important for maintaining posture. A striking illustration of this fact is the following: when a person loses consciousness, muscles lose their tone, and the person collapses in a heap, unable to maintain a sitting or standing posture. Posture is discussed in more detail in Chapter 10. Muscles with less than normal tone are described as *flaccid muscles,* and those with more than normal tone are called *spastic.*

Muscle tone is maintained by negative feedback mechanisms centered in the nervous system, specifically in the spinal cord. Stretch sensors in the muscles and tendons detect the degree of stretch in a muscle organ and feed this information back to an integrator mechanism in the spinal cord. When the actual stretch (detected by the stretch receptors) deviates from the set point stretch, signals sent via the somatic motor neurons adjust the strength of tonic contraction. This type of subconscious mechanism is often called a spinal reflex (discussed further in Chapters 11 to 13).

Isotonic and Isometric Contractions

An **isotonic contraction** is a contraction in which the tone or tension within a muscle remains the same, but the length of the muscle changes (Figure 9-15, A). The term *isotonic* literally means "same tension." Since there is little or no resistance (load) placed on the muscle in an isotonic contraction, all of the energy of contraction is used to pull on the thin myofilaments and thus decrease the length of a fiber's sarcomeres. Put another way, in isotonic contractions the myosin cross bridges "win" the tug-of-war against a light load and are thus able to pull the thin myofilaments along.

An **isometric contraction,** on the other hand, is a contraction in which muscle length remains the same but in which muscle tension increases (Figure 9-15, B). The term *isometric* means "same length." You can observe isometric contraction by pushing your arms against a wall and feeling the tension increase in your arm muscles. Isometric contractions can do work by "tightening" to resist a force, but they do not produce movements. In isometric contractions, the tension produced by the "power stroke" of the myosin cross bridges cannot overcome the load placed on the muscle. Using the tug-of-war analogy, we can say that in isometric contractions the myosin cross bridges reach a "draw"—they hold their own against the load placed on the muscle but do not make any progress in sliding the thin myofilaments.

Although most muscles can produce either isotonic or isometric contractions, most body movements are a mixture of the two. Can you explain how walking and running are examples of actions produced by the combination of isotonic and isometric contractions?

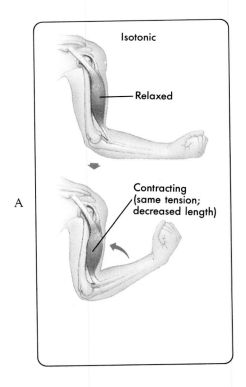
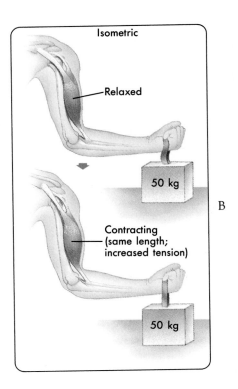

FIGURE 9-15 Isotonic and isometric contraction. A, In isotonic contraction the muscle shortens, producing movement. **B,** In isometric contraction the muscle pulls forcefully against a load but does not shorten.

FO(CUS ON)... EFFECTS OF EXERCISE ON SKELETAL MUSCLES

Most of us believe that exercise is good for us, even if we have no idea what or how many specific benefits can come from it. Some of the good consequences of regular, properly practiced exercise are greatly improved muscle tone, better posture, more efficient heart and lung function, less fatigue, and looking and feeling better.

Skeletal muscles undergo changes that correspond to the amount of work that they normally do. During prolonged inactivity, muscles usually shrink in mass, a condition called **disuse atrophy.** Exercise, on the other hand, may cause an increase in muscle size called **hypertrophy.**

Muscle hypertrophy can be enhanced by **strength training,** which involves contracting muscles against heavy resistance. Isometric exercises and weight lifting are common strength-training activities. This type of training results in increased numbers of myofilaments in each muscle fiber. Although the number of muscle fibers stays the same, the increased number of myofilaments greatly increases the mass of the muscle.

Endurance training, often called **aerobic training,** does not usually result in muscle hypertrophy. Instead, this type of exercise program increases a muscle's ability to sustain moderate exercise over a long period. Aerobic activities such as running, bicycling, or other primarily isotonic movements increase the number of blood vessels in a muscle without significantly increasing its size. The increased blood flow allows a more efficient delivery of oxygen and glucose to muscle fibers during exercise. Aerobic training also causes an increase in the number of mitochondria in muscle fibers. This allows production of more ATP as a rapid energy source.

The Graded Strength Principle

Skeletal muscle organs do not contract according to the all-or-none principle, as do the individual muscle cells composing them. In other words, skeletal muscles contract with varying degrees of strength at different times—a fact called the **graded strength principle.** Because muscle organs can generate different grades of strength, we can match the force of a movement to the demands of a specific task.

Various factors contribute to the phenomenon of graded strength. We have already discussed some of these factors. For example, we have stated that the metabolic condition of individual fibers influences their capacity to generate force. Thus if many fibers of a muscle organ are unable to maintain a high level of ATP and become fatigued, the entire muscle organ will suffer some loss in its ability to generate maximum force of contraction. On the other hand, the improved metabolic conditions that produce the treppe effect allow a muscle organ to increase its contraction strength.

Another factor that influences the grade of strength exhibited by a muscle organ is the number of fibers contracting simultaneously. Obviously, the more muscle fibers contracting at the same time, the stronger the contraction of the whole muscle organ. How large this number is depends on how many motor units are activated or **recruited.** Recruitment of motor units, in turn, depends on the intensity and frequency of stimulation. In general, the more intense and the more frequent a stimulus, the more motor units are recruited and the stronger the contraction. Figure 9-16 shows that increasing the strength of the stimulus beyond the threshold level of the most sensitive motor units causes an increase in strength of contraction. As the threshold level of each additional motor unit is crossed, the strength of contraction increases. This continues as the strength of stimulation increases until the maximal level of contraction is reached. At this point, the limits of the muscle organ to recruit new motor units have been reached. Even if stimulation increases above the maximal level, the muscle cannot contract any more

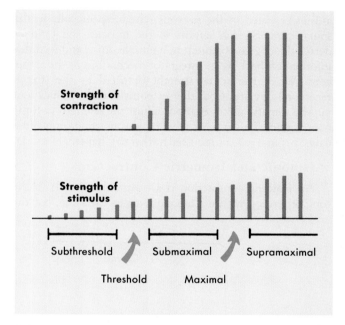

FIGURE 9-16   The strength of muscle contraction compared to the strength of the stimulus. After the threshold stimulus is reached, a continued increase in stimulus strength produces a proportional increase in muscle strength until the maximal level of contraction strength is reached.

strongly. As long as the supply of ATP holds out, the muscle organ can sustain a tetanic contraction at the maximal level when motor units contract and relax in overlapping "relays" (see Figure 9-14, *D*).

The maximal strength that a muscle can develop is directly related to the initial length of its fibers—this is the *length-tension relationship.* A muscle that begins a contraction from a short initial length cannot develop much tension because its sarcomeres are already compressed. Conversely, a muscle that begins a contraction from an overstretched initial length cannot develop much tension because the thick myofilaments are too far away from the thin myofilaments to effectively pull them and thus compress the sarcomeres. The strongest maximal contraction is possible only when the muscle organ has been

stretched to an optimal initial length. To illustrate this point, extend your elbow fully and try to contract the *biceps brachii* muscle on the ventral side of your upper arm. Now flex the elbow just a little and contract the biceps again. Try it a third time with the elbow completely flexed. The greatest tension—seen as the largest "bulge" of the biceps—occurs when the elbow is partly flexed and the biceps only moderately stretched.

Another factor that influences the strength of a skeletal muscle contraction is the amount of load imposed on the muscle. Within certain limits, the heavier the load, the stronger the contraction. Lift your hand with palm up in front of you and then put this book in your palm. You can feel your arm muscles contract more strongly as the book is placed in your hand. This occurs because of a *stretch reflex,*

a response in which the body tries to maintain a constancy of muscle length (Figure 9-17). An increased load threatens to stretch the muscle beyond the set point length that you are trying to maintain. Your body exhibits a negative feedback response when it detects the increased stretch caused by an increased load, feeds the information back to an integrator in the nervous system, and increases its stimulation of the muscle to counteract the stretch. This reflex maintains a relatively constant muscle length as load is increased up to a maximum sustainable level. When the load becomes too heavy, and thus threatens to cause injury to the muscle or skeleton, the body abandons this reflex and forces you to relax and drop the load.

The major factors involved in the graded strength principle are summarized in Figure 9-18.

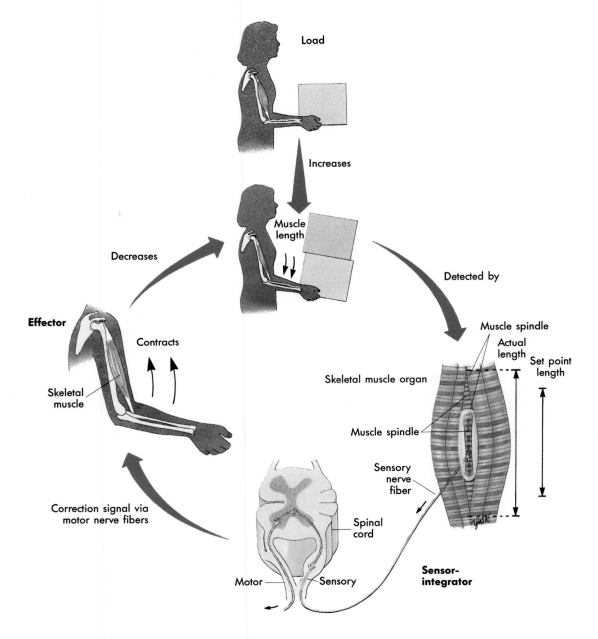

FIGURE 9-17   The stretch reflex. The strength of a muscle organ can be matched to the load imposed on it by a negative feedback response centered in the spinal cord.

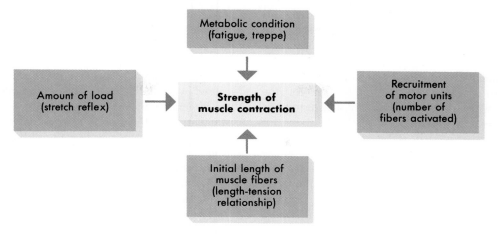

FIGURE 9-18 Factors that influence the strength of muscle contraction.

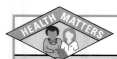

ABNORMAL MUSCLE CONTRACTIONS

Cramps are painful muscle spasms (involuntary twitches). Cramps often occur when a muscle organ is mildly inflamed, but they can be a symptom of any irritation or ion and water imbalance.

Convulsions are abnormal, uncoordinated tetanic contractions of varying groups of muscles. Convulsions may result from a disturbance in the brain or *seizure* in which the output along motor nerves increases and becomes disorganized.

Fibrillation is an abnormal type of contraction in which individual fibers contract asynchronously rather than at the same time. This produces a flutter of the muscle but no effective movement. Fibrillation can also occur in cardiac muscle, where it reduces the heart's ability to pump blood.

QUICK CHECK

✔ 1. *What is meant by the term muscle tone?*
✔ 2. *What is the difference between isotonic and isometric contractions?*
✔ 3. *Name four factors that influence the strength of a skeletal muscle contraction.*
✔ 4. *What is meant by the phrase "recruitment of motor units"?*

▶ FUNCTION OF CARDIAC AND SMOOTH MUSCLE TISSUE

Cardiac and smooth muscle tissues operate by mechanisms similar to those in skeletal muscle tissues. The detailed study of cardiac and smooth muscle function will be set aside until we discuss specific smooth and cardiac muscle organs in later chapters. However, it may be helpful to preview some of the basic principles of cardiac and smooth muscle physiology so that we can compare them with those that operate in skeletal muscle tissue. Table 9-2

summarizes the characteristics of the three major types of muscle.

Cardiac Muscle

Cardiac muscle, also known as *striated involuntary* muscle, is found in only one organ of the body: the heart. Forming the bulk of the wall of each heart chamber, cardiac muscle contracts rhythmically and continuously to provide the pumping action necessary to maintain a relative constancy of blood flow through the internal environment. As you shall see, its physiological mechanisms are well adapted to this function.

The functional anatomy of cardiac muscle tissue resembles that of skeletal muscle to a degree but exhibits specialized features related to its role in continuously pumping blood. As Figure 9-19 shows, each cardiac muscle fiber contains parallel myofibrils that are each comprised of sarcomeres that give the whole fiber a striated appearance. However, the cardiac muscle fiber does not taper like a skeletal muscle fiber but, instead, forms strong, electrically coupled junctions **(intercalated disks)** with other fibers. This feature, along with the branching exhibited by individual cells, allows cardiac fibers to form a continuous, electrically coupled mass, called a *syncytium* (meaning "unit of combined cells"). Cardiac muscles thus form a continuous, contractile band around the heart chambers that conducts a single impulse across a virtually continuous sarcolemma—features necessary for an efficient, coordinated pumping action.

Although the cardiac muscle fiber has T tubules and sarcoplasmic reticulum, they are arranged a little differently than in skeletal muscle fibers. The T tubules are larger, and they form *diads* (double structures) rather than triads (triple structures), with a rather sparse sarcoplasmic reticulum. Much of the calcium (Ca^{++}) that enters the sarcoplasm during contraction enters from the outside of the cells through the T tubules, rather than from storage in the SR. Because the sarcolemma of the cardiac muscle sustains each impulse longer than in skeletal muscle, Ca^{++} remains in the sarcolemma longer. This means that even though many

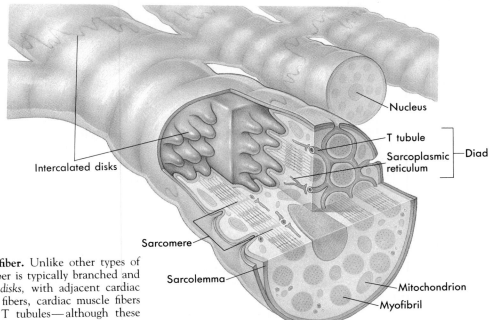

FIGURE 9-19   Cardiac muscle fiber. Unlike other types of muscle fibers, the cardiac muscle fiber is typically branched and forms junctions, called *intercalated disks*, with adjacent cardiac muscle fibers. Like skeletal muscle fibers, cardiac muscle fibers contain sarcoplasmic reticula and T tubules—although these structures are not as highly organized as in skeletal muscle fibers.

adjacent cardiac muscle cells contract simultaneously, they exhibit a prolonged contraction rather than a rapid twitch. It also means that impulses cannot come rapidly enough to produce tetanus. Because it cannot sustain long tetanic contractions, cardiac muscle does not normally run low on ATP and thus does not experience fatigue. Obviously, this characteristic of cardiac muscle is vital to keeping the heart continuously pumping.

Unlike skeletal muscle, in which a nervous impulse excites the sarcolemma to produce its own impulse, cardiac muscle is self-exciting. Cardiac muscle cells thus exhibit a continuing rhythm of excitation and contraction on their own, although the rate of self-induced impulses can be altered by nervous or hormonal input.

The structure and function of the heart are discussed further in Chapters 17 and 18.

TABLE 9-2 Characteristics of muscle tissues			
	Skeletal	**Cardiac**	**Smooth**
PRINCIPAL LOCATION	Skeletal muscle organs	Wall of heart	Walls of many hollow organs
PRINCIPAL FUNCTIONS	Movement of bones, heat production, posture	Pumping of blood	Movement in walls of hollow organs (peristalsis, mixing)
TYPE OF CONTROL	Voluntary	Involuntary	Involuntary
STRUCTURAL FEATURES			
Striations	Present	Present	Absent
Nucleus	Many, near sarcolemma	Single	Single, near center of cell
T tubules	Narrow, form triads with SR	Large diameter, form diads with SR, regulate Ca^{++} entry into sarcoplasm	Absent
Sarcoplasmic reticulum	Extensive, stores and releases Ca^{++}	Less extensive than in skeletal muscle	Very poorly developed
Cell junctions	No gap junctions	Intercalated disks	Visceral: many gap junctions Multiunit: few gap junctions
CONTRACTION STYLE	Rapid twitch contractions of motor units usually summate to produce sustained tetanic contractions; must be stimulated by a neuron	Syncytium of fibers compress heart chambers in slow, separate contractions (does not exhibit tetanus or fatigue); exhibits autorhythmicity	Visceral: electrically coupled sheets of fibers contract autorhythmically, producing peristalsis or mixing movements Multiunit: individual fibers contract when stimulated by neuron

Smooth Muscle

As we mentioned in Chapter 4, smooth muscle is comprised of small, tapered cells with single nuclei. Smooth muscle cells do not have T tubules and have only loosely organized sarcoplasmic reticula. The calcium required for contraction comes from outside the cell and binds to a protein called *calmodulin,* rather than to troponin, to trigger a contraction event.

The lack of striations in smooth muscle fibers can be accounted for by the fact that the thick and thin myofilaments are arranged quite differently than in skeletal or cardiac muscle fibers. As Figure 9-20 shows, thin arrangements of myofilaments crisscross the cell and attach at their ends to the cell's plasma membrane. When cross bridges pull the thin filaments together, the muscle "balls up" and thus contracts the cell. Because the myofilaments are not organized into sarcomeres, they have more freedom of movement and thus can contract a smooth muscle fiber to shorter lengths than in skeletal and cardiac muscle.

There are two types of smooth muscle tissue: *visceral* and *multiunit.* In **visceral,** or *single-unit,* muscle, gap junctions join individual smooth muscle fibers into large, continuous sheets—much like the syncytium of fibers observed in cardiac muscle. This is the most common type of smooth muscle, forming a muscular layer in the walls of many hollow structures such as the digestive, urinary, and reproductive tracts. Like cardiac muscle, this type of smooth muscle commonly exhibits a rhythmic self-excitation or *autorhythmicity* (meaning "self-rhythm") that spreads across the whole tissue. When these rhythmic, spreading waves of contraction become strong enough, they can push the contents of a hollow organ progressively along its lumen. This phenomenon, called **peristalsis,** moves food along the digestive tract, assists the flow of urine to the bladder, and pushes a baby out of the womb during labor. Such contractions can also be coordinated to produce mixing movements in the stomach and other organs.

Multiunit smooth muscle tissue does not act as a single unit (as in visceral muscle) but instead is comprised of many independent single-cell units. Each independent fiber does not usually generate its own impulse but rather responds only to nervous input. Although this type of smooth muscle can form thin sheets, as in the walls of large blood vessels, it is more often found in bundles (for example, the *arrector pili* muscles of the skin) or as single fibers (such as those surrounding small blood vessels).

The structure and function of smooth muscle organs are discussed in later chapters.

QUICK CHECK
1. *How do slow, separate, autorhythmic contractions of cardiac muscle make it well suited to its role in pumping blood?*
2. *What produces the striations in cardiac muscle?*
3. *How are myofilaments arranged in a smooth muscle fiber?*

A
Relaxed

B
Fully contracted

Plasma membrane

Plasma membrane

Thick myofilament

Thin myofiflament

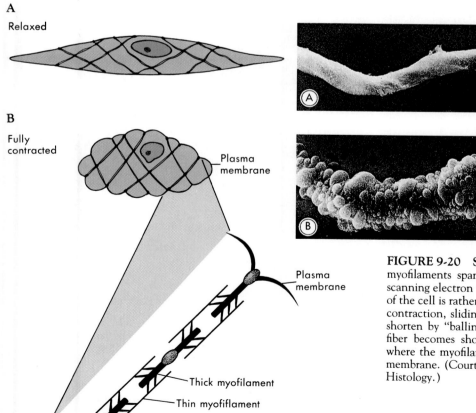

FIGURE 9-20 Smooth muscle fiber. A, Thin bundles of myofilaments span the diameter of a relaxed fiber. The scanning electron micrograph *(right)* shows that the surface of the cell is rather flat when the fiber is relaxed. **B,** During contraction, sliding of the myofilaments causes the fiber to shorten by "balling up." The micrograph shows that the fiber becomes shorter and thicker, exhibiting "dimples" where the myofilament bundles are pulling on the plasma membrane. (Courtesy of Erlandsen/Magney: Color Atlas of Histology.)

THE BIG PICTURE
Muscle Tissue and the Whole Body

The function of all three major types of muscle (skeletal, smooth, and cardiac) is integral to the function of the entire body. What does the function of muscle tissue contribute to the homeostasis of the whole body? First, all three types of muscle tissue provide the movement necessary for survival. Skeletal muscle moves the skeleton so that we can seek shelter, gather food, and defend ourselves. All three muscle types produce movements that power vital homeostatic mechanisms such as breathing, blood flow, digestion, and urine flow.

The relative constancy of the body's internal temperature could not be maintained in a cool external environment if not for the "waste" heat generated by muscle tissue—especially the large mass of skeletal muscle found throughout the body. Maintenance of a relatively stable body position—posture—is also a primary function of the skeletal muscular system. Posture, specific body movements, and other contributions of the skeletal muscular system to the homeostasis of the whole body is discussed further in Chapter 10. The homeostatic roles of smooth muscle organs and the cardiac muscle organ (the heart) are examined in later chapters.

Like all tissues of the body, muscle tissue both gives and takes. A number of systems support the function of muscle tissues. Without these systems, muscle would cease to operate. For example, the nervous system directly controls the contraction of skeletal muscle and multiunit smooth muscle. It also influences the rate of rhythmic contractions in cardiac muscle and visceral smooth muscle. The endocrine system produces hormones that assist the nervous system in regulation of muscle contraction throughout the body. The blood delivers nutrients and carries away waste products. Nutrients for the muscle are ultimately procured by the respiratory system (oxygen) and digestive system (glucose and other foods). The respiratory system also helps get rid of the waste of muscle metabolism, as does the urinary system. The liver processes lactic acid produced by muscles and converts it back to glucose. The immune system helps defend muscle tissue against infection and cancer—as it does for all body tissues. The fibers that comprise muscle tissues, then, are truly members of the large, interactive "society of cells" that forms the human body.

MECHANISMS OF DISEASE

Major Muscular Disorders

As you might expect, muscle disorders, or **myopathies** generally disrupt the normal movement of the body. In mild cases, these disorders vary from inconvenient to slightly troublesome. Severe muscle disorders, however, can impair the muscles used in breathing—a life-threatening situation.

Muscle Injury

Injuries to skeletal muscles resulting from overexertion or trauma usually result in a muscle **strain** (Figure 9-21). Muscle strains are characterized by muscle pain, or **myalgia** (my-AL-jee-ah), and involve overstretching or tearing of muscle fibers. If an injury occurs in the area of a joint and a ligament is damaged, the injury may be called a **sprain.** Any muscle inflammation, including that caused by a muscle strain, is termed **myositis** (my-o-SYE-tis). If tendon inflammation occurs with myositis, as in a *charley horse,* the condition is termed **fibromyositis** (fi-bro-my-o-SYE-tis). Although inflammation may subside in a few hours or days, it usually takes several weeks for damaged muscle fibers to repair themselves. Some damaged muscle cells may be replaced by fibrous tissue, forming scars. Occasionally, hard calcium is deposited in the scar tissue.

Cramps are painful muscle spasms (involuntary twitches). Cramps often result from mild myositis or fibromyositis, but they can be a symptom of any irritation or of an ion and water imbalance.

Minor trauma to the body, especially a limb, may cause a muscle *bruise,* or **contusion** (kon-TOO-zhun). Muscle contusions involve local internal bleeding and inflammation. Severe trauma to a skeletal muscle may cause a *crush injury.* Crush injuries not only greatly damage the affected muscle tissue, but

the release of muscle fiber contents into the bloodstream can be life-threatening. For example, the reddish muscle pigment *myoglobin* can accumulate in the blood and cause kidney failure.

Stress-induced muscle tension can result in myalgia and stiffness in the neck and back and is thought to be one cause of "stress headaches." Headache and back-pain clinics use

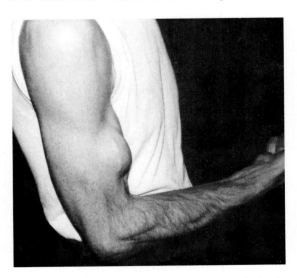

FIGURE 9-21 Muscle Strain. Severe strain of the biceps brachii muscle. In a severe muscle strain, a muscle may break in two pieces, causing a visible gap in muscle tissue under the skin. Notice how the broken ends of the muscle reflexively contract (spasm) to form a knot of tissue.

various strategies to treat stress-induced muscle tension. These treatments include massage, biofeedback, and relaxation training.

Muscle Infections

Several bacteria, viruses, and parasites are known to infect muscle tissue—often producing local or widespread myositis. For example, in trichinosis, widespread myositis is common. The muscle pain and stiffness that sometimes accompany influenza is another example.

Once a tragically common disease, **poliomyelitis** (po-lee-o-my-el-EYE-tis) is a viral infection of the nerves that control skeletal muscle movement. Although the disease can be asymptomatic, it often causes paralysis that may progress to death. Virtually eliminated in the United States as a result of a comprehensive vaccination program, it still affects millions in other parts of the world.

Muscular Dystrophy

Muscular dystrophy (DIS-tro-fee) is not a single disorder but a group of genetic diseases characterized by atrophy (wasting) of skeletal muscle tissues. Some, but not all, forms of muscular dystrophy can be fatal.

The common form of muscular dystrophy is **Duchenne's** (doo-SHENZ) **muscular dystrophy (DMD).** This form of the disease is also called *pseudohypertrophy* (meaning "false muscle growth") because the atrophy of muscle is masked by excessive replacement of muscle by fat and fibrous tissue. DMD is characterized by mild leg muscle weakness that progresses rapidly to include the shoulder muscles. The first signs of DMD are apparent at about 3 years of age, and the stricken child is usually severely affected within 5 to 10 years. Death from respiratory or cardiac muscle weakness often occurs by the time the victim is 21 years old.

Many pathophysiologists believe that DMD is caused by a missing fragment in an X chromosome, although other factors may be involved. DMD occurs primarily in boys. Because girls have two X chromosomes and boys only one, genetic diseases involving X chromosome abnormalities are more likely to occur in boys than in girls. This is true because girls with one damaged X chromosome may not exhibit an "X-linked" disease if their other X chromosome is normal (see Chapter 33).

Myasthenia Gravis

Myasthenia gravis (my-es-THEE-nee-ah GRA-vis) is a chronic disease characterized by muscle weakness, especially in the face and throat. Most forms of this disease begin with mild weakness and chronic muscle fatigue in the face, then progress to wider muscle involvement. When severe muscle weakness causes immobility in all four limbs, a *myasthenic crisis* is said to have occurred. A person in myasthenic crisis is in danger of dying from respiratory failure because of weakness in the respiratory muscles.

Myasthenia gravis is an autoimmune disease in which the immune system attacks muscle cells at the neuromuscular junction. Nerve impulses from motor neurons are then unable to fully stimulate the affected muscle.

Hernias

Weakness of abdominal muscles can lead to a *hernia*, or protrusion, of an abdominal organ (commonly the small intestine) through an opening in the abdominal wall. There are several types of hernias. The most common one, *inguinal hernia* (Figure 9-22), occurs when the hernia extends down the inguinal canal, often into the scrotum or labia. Males experience this most often, and it can occur at any age. Women may experience a *femoral hernia* below the groin because of changes during pregnancy.

Hernia is referred to as "reducible" when the protruding organ is manipulated back into the abdominal cavity, either naturally by laying down or manual reduction through a surgical opening in the abdomen. A "strangulated" hernia occurs when the mass is not reducible and blood flow to the affected organ (i.e., intestine) is stopped. Obstruction and gangrene can occur. Pain and vomiting are usually experienced and emergency surgical intervention is required.

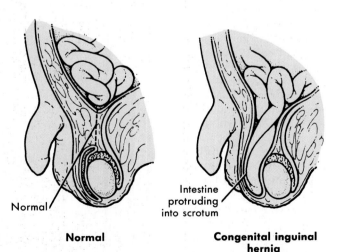

Normal

Normal

Congenital inguinal hernia

Intestine protruding into scrotum

FIGURE 9-22   Inguinal hernia.

CHAPTER SUMMARY

INTRODUCTION

A. Muscular system is responsible for moving the framework of the body

B. In addition to movement, muscle tissue performs various other functions

GENERAL FUNCTIONS

A. Movement of the body as a whole or of its parts

B. Heat production

C. Posture

FUNCTION OF SKELETAL MUSCLE TISSUE

A. Characteristics of skeletal muscle cells

   1. Excitability (irritability)—ability to be stimulated

   2. Contractility—ability to contract, or shorten, and produce body movement

   3. Extensibility—ability to extend, or stretch, allowing muscles to return to their resting length

B. Overview of the muscle cell (Figures 9-2 and 9-4)

   1. Muscle cells are called fibers due to their threadlike shape

   2. Sarcolemma—plasma membrane of muscle fibers

   3. Sarcoplasmic reticulum

      a. Network of tubules and sacs found within muscle fibers

      b. Membrane of the sarcoplasmic reticulum continually pumps calcium ions from the sarcoplasm and stores the ions within its sacs

   4. Muscle fibers contain many mitochondria and several nuclei

   5. Myofibrils—numerous fine fibers packed close together in sarcoplasm

   6. Sarcomere

      a. Segment of myofibril between two successive Z lines

      b. Each myofibril consists of many sarcomeres

      c. Contractile unit of muscle fibers

   7. Striated muscle

      a. Dark stripes called A bands; light H zone runs across midsection of each dark A band

      b. Light stripes called I bands; dark Z line extends across center of each light I band

   8. T tubules

      a. Transverse tubules extend across the sarcoplasm at right angles to the long axis of the muscle fiber

      b. Formed by inward extensions of the sarcolemma

      c. Allow electrical impulses traveling along the sarcolemma to move deeper into the cell

   9. Triad

      a. Triplet of tubules; a T tubule sandwiched between two sacs of sarcoplasmic reticulum

      b. Allows an electrical impulse traveling along a T tubule to stimulate the membranes of adjacent sacs of the sarcoplasmic reticulum

C. Myofilaments (Figure 9-5)

   1. Each myofibril contains thousands of thick and thin myofilaments

   2. Four different kinds of protein molecules comprise myofilaments

      a. Myosin

         (1) Makes up almost all of the thick filament

         (2) Myosin "heads" are chemically attracted to actin molecules

         (3) Myosin "heads" are known as cross bridges

      b. Actin—globular protein that forms two fibrous strands twisted around each other to form the bulk of the thin filament

      c. Tropomyosin—protein that blocks the active sites on the actin molecules

      d. Troponin—protein that holds tropomyosin molecules in place

   3. Thin filaments attach to both Z lines of a sarcomere and extend part way toward the center

   4. Thick myosin filaments do not attach to the Z lines

D. The mechanism of contraction

   1. Excitation and contraction (Table 9-1)

      a. A skeletal muscle fiber remains at rest until stimulated by a motor neuron

      b. Neuromuscular junction—motor neurons connect to the sarcolemma at the motor endplate (Figure 9-6)

      c. Neuromuscular junction is a synapse where neurotransmitter molecules transmit signals

      d. Acetylcholine—the neurotransmitter released into the synaptic cleft that diffuses across the gap, stimulates the receptors, and initiates an impulse in the sarcolemma

      e. Nerve impulse travels over the sarcolemma and inward along the T tubules, which triggers the release of calcium ions

      f. Calcium binds to troponin, causing the tropomyosin to shift and expose active sites on the actin

      g. Sliding filament theory (Figure 9-8)

         (1) When active sites on the actin are exposed, myosin cross bridges will bind

         (2) Myosin heads bend, pulling the thin filaments past them

         (3) Each head releases, binds to the next active site, and pulls again

         (4) The entire myofibril shortens

   2. Relaxation

      a. Immediately after the Ca^{++} are released, the sarcoplasmic reticulum begins actively pumping them back into the sacs

      b. Ca^{++} are removed from the troponin molecules, shutting down the contraction

   3. Energy sources for muscle contraction (Figure 9-10)

      a. Hydrolysis of ATP yields the energy required for muscular contraction

      b. An ATP binds to the myosin cross bridge and then transfers its energy to the cross bridge to perform the work of pulling the thin filament during contraction

      c. Muscle fibers continually resynthesize ATP from the breakdown of creatine phosphate

      d. Catabolism by muscle fibers requires glucose and oxygen

      e. At rest, excess O_2 in the sarcoplasm is bound to myoglobin

         (1) Red fibers—muscle fibers with high levels of myoglobin

         (2) White fibers—muscle fibers with little myoglobin

      f. Aerobic respiration occurs when adequate O_2 is available

      g. Anaerobic respiration occurs when low levels of O_2 are available and results in the formation of lactic acid

   4. The all-or-none principle

      a. Muscle fibers, when stimulated, either contract with all possible force or not at all

      b. Threshold stimulus—minimum level of stimulation needed to cause a fiber to contract

FUNCTION OF SKELETAL MUSCLE ORGANS

A. Muscles are composed of bundles of muscle fibers that are held together by fibrous connective tissue
B. Motor unit (Figure 9-12)
   1. Motor unit—motor neuron plus the muscle fibers to which it attaches
   2. Some motor units consist of only a few muscle fibers, whereas others consist of numerous fibers
   3. Generally, the smaller the number of fibers in a motor unit, the more precise the movements available; the larger the number of fibers in a motor unit, the more powerful a contraction is available
C. Twitch contraction
   1. A quick jerk of a muscle that is produced as a result of a single, brief threshold stimulus (generally occurs only in experimental situations)
   2. The twitch contraction has three phases (Figure 9-13)
      a. Latent phase—nerve impulse travels to the sarcoplasmic reticulum to trigger release of Ca^{++}
      b. Contraction phase—Ca^{++} bind to troponin and sliding of filaments occurs
      c. Relaxation phase—sliding of filaments ceases
D. Treppe: the staircase phenomenon (Figure 9-14, B)
   1. Gradual, steplike increase in the strength of contraction that is seen in a series of twitch contractions that occur 1 second apart
   2. Eventually, the muscle will respond with less forceful contractions and relaxation phase will become shorter
   3. If relaxation phase disappears completely, a contracture occurs
E. Tetanus—smooth, sustained contractions
   1. Multiple wave summation—multiple twitch waves are added together to sustain muscle tension for a longer time
   2. Incomplete tetanus—very short periods of relaxation occur between peaks of tension (Figure 9-14, C)
   3. Complete tetanus—the stimulation is such that twitch waves fuse into a single, sustained peak (Figure 9-14, D)
F. Muscle tone
   1. Tonic contraction—continual, partial contraction of a muscle
   2. At any one time, a small number of muscle fibers within a muscle contract, producing a tightness or muscle tone
   3. Muscles with less tone than normal are flaccid
   4. Muscles with more tone than normal are spastic
   5. Muscle tone is maintained by negative feedback mechanisms
G. Isotonic and isometric contractions (Figure 9-15)
   1. Isotonic contraction
      a. Contraction in which the tone or tension within a muscle remains the same but the length of the muscle shortens
      b. Isotonic—literally means "same tension"
      c. All of the energy of contraction is used to pull on thin myofilaments and thereby decrease the length of a fiber's sarcomeres
   2. Isometric contraction
      a. Contraction in which muscle length remains the same while the muscle tension increases
      b. Isometric—literally means "same length"
   3. Most body movements occur as a result of both types of contractions
H. Graded strength principle
   1. Muscle cells contract according to the all-or-none principle, but entire skeletal muscles do not
   2. Graded strength principle—skeletal muscles contract with varying degrees of strength at different times

3. Factors that contribute to the phenomenon of graded strength
   a. Metabolic condition of individual fibers
   b. Number of muscle fibers contracting simultaneously; the greater the number of fibers contracting, the stronger the contraction
   c. Number of motor units recruited
   d. Intensity and frequency of stimulation
4. Length-tension relationship
   a. Maximal strength that a muscle can develop bears a direct relationship to the initial length of its fibers
   b. A shortened muscle's sarcomeres are compressed, therefore the muscle cannot develop much tension
   c. An overstretched muscle cannot develop much tension because the thick myofilaments are too far from the thin myofilaments
   d. Strongest maximal contraction is possible only when the skeletal muscle has been stretched to its optimal length
5. Stretch reflex (Figure 9-17)
   a. The load imposed on a muscle influences the strength of a skeletal contraction
   b. Stretch reflex—the body tries to maintain a constancy of muscle length in response to increased load
   c. Maintains a relatively constant length as load is increased up to a maximum sustainable level

FUNCTION OF CARDIAC AND SMOOTH MUSCLE TISSUE

A. Cardiac muscle
   1. Found only in the heart, forming the bulk of the wall of each chamber
   2. Also known as striated involuntary muscle
   3. Contracts rhythmically and continuously to provide the pumping action needed to maintain a constant blood flow
   4. Cardiac muscle resembles skeletal muscle but has specialized features related to its role in continuously pumping blood
      a. Each cardiac muscle contains parallel myofibrils (Figure 9-19)
      b. Cardiac muscle fibers form strong, electrically coupled junctions (intercalated disks) with other fibers; individual cells also exhibit branching
      c. Syncytium—continuous, electrically coupled mass
      d. Cardiac muscle fibers form a continuous, contractile band around the heart chambers that conducts a single impulse across a virtually continuous sarcolemma
      e. T tubules are larger and form diads with a rather sparse sarcoplasmic reticulum
      f. Cardiac muscle sustains each impulse longer than in skeletal muscle, therefore impulses cannot come rapidly enough to produce tetanus
      g. Cardiac muscle does not run low on ATP and does not experience fatigue
      h. Cardiac muscle is self-stimulating
B. Smooth muscle
   1. Smooth muscle is comprised of small, tapered cells with single nuclei (Figure 9-20)
   2. No T tubules are present, and only a loosely organized sarcoplasmic reticulum is present
   3. Ca^{++} comes from outside the cell and binds to calmodulin instead of troponin to trigger a contraction
   4. No striations, since thick and thin myofilaments are arranged quite differently than in skeletal or cardiac

muscle fibers; myofilaments are not organized into sarcomeres

5. Two types of smooth muscle tissue
   a. Visceral muscle (single-unit)
      (1) Gap junctions join smooth muscle fibers into large, continuous sheets
      (2) Most common type, forms a muscular layer in the walls of hollow structures such as the digestive, urinary, and reproductive tracts
      (3) Exhibits autorhythmicity, producing peristalsis
   b. Multiunit
      (1) Does not act as a single unit but is comprised of many independent cell units
      (2) Each fiber responds only to nervous input

THE BIG PICTURE: MUSCLE TISSUE AND THE WHOLE BODY

A. Function of all three major types of muscle is integral to the function of the entire body
B. All three types of muscle tissue provide the movement necessary for survival
C. Relative constancy of the body's internal temperature is maintained by "waste" heat generated by muscle tissue
D. Maintains the body in a relatively stable position

MECHANISMS OF DISEASE: MAJOR MUSCULAR DISORDERS

A. Major Muscular Disorders
   1. Myopathies—muscle disorders; can range from mild to life threatening
   2. Muscle injury
      a. Strain—injury from overexertion or trauma; involves stretching or tearing of muscle fibers
         (1) Often accompanied by myalgia (muscle pain)
         (2) May result in inflammation of muscle (myositis) or of muscle and tendon (fibromyositis)
         (3) If injury is near a joint and involves ligament damage, it may be called a *sprain*

b. Cramps are painful muscle spasms (involuntary twitches)
c. Crush injuries result from severe muscle trauma and may release cell contents that ultimately cause kidney failure
d. Stress-induced muscle tension can cause headaches and back pain
   3. Infections
      a. Several bacteria, viruses, and parasites can infect muscles
      b. Poliomyelitis is a viral infection of motor nerves that ranges from mild to life threatening
   4. Muscular dystrophy
      a. A group of genetic disorders characterized by muscle atrophy
      b. Duchenne's (pseudohypertrophic) muscular dystrophy is the most common type
         (1) Characterized by rapid progression of weakness and atrophy, often resulting in death by age 21
         (2) X-linked inherited disease, affecting mostly boys
   5. Myasthenia gravis—autoimmune muscle disease characterized by weakness and chronic fatigue
B. Hernias—protrusion of an abdominal organ through an opening in the abdominal wall (Figure 9-22)
   1. Inguinal hernia—when organ extends through inguinal canal into scrotum or labia (most common); occurs more often in males
   2. Reducible hernia—protruding organ can be manipulated back into the abdominal cavity
   3. Strangulated hernia—organ is not reducible, and blood flow to the organ may be stopped; obstruction and gangrene are complications

REVIEW QUESTIONS

1. Explain how skeletal muscles provide movement, heat production, and posture.
2. Differentiate among excitability, contractility, and extensibility.
3. Define the terms *sarcolemma, sarcoplasm,* and *sarcoplasmic reticulum.*
4. What structures are unique to skeletal muscle fibers?
5. Describe the function of the sarcoplasmic reticulum.
6. Explain how the structure of myofilaments is related to their function.
7. How are acetylcholine, Ca^{++}, and adenosine triphosphate (ATP) involved in the excitation and contraction of skeletal muscle?
8. Explain how the sliding filament theory allows for the shortening of a muscle fiber.
9. Compare and contrast the role of Ca^{++} in excitation, contraction, and relaxation of skeletal muscle.
10. Describe the structure of ATP.
11. How does ATP provide energy for a muscle contraction?
12. Differentiate between aerobic and anaerobic respiration.
13. Why do muscle fibers contract "all-or-none"?
14. Compare and contrast the following types of skeletal muscle fibers: slow, fast, intermediate.
15. Describe the anatomical arrangement of a motor unit.
16. List and describe the different types of skeletal muscle contractions.
17. Define the term *recruited.*
18. Explain the meaning of a "unit of combined cells" as it relates to cardiac muscle.
19. Differentiate between the two types of smooth muscle tissue.
20. Describe rigor mortis.
21. What are the effects of exercise on skeletal muscles?

Detail of muscle pattern in neck

10 ▶ Anatomy of the Muscular System

CHAPTER OUTLINE

OBJECTIVES

After you have completed this chapter, you should be able to:

1. List the major connective tissue elements related to skeletal muscle.
2. Discuss the attachment of muscles.
3. Explain the functional classification of muscles based on movement pattern.
4. Identify six features that may be used to name a muscle.
5. Identify major muscles, their points of attachment, and their function in the following areas:
   a. Muscles of facial expression
   b. Muscles of mastication
   c. Muscles that move the head
   d. Muscles that move the abdominal and chest wall
   e. Muscles of the pelvic floor
   f. Muscles acting on the shoulder girdle
   g. Muscles that move the upper and lower arm
   h. Muscles that move the wrist, hand, and fingers
   i. Muscles that move the thigh and lower leg
   j. Muscles that move the ankle and foot
6. Define posture and discuss its importance to the body as a whole.

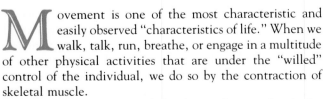

M ovement is one of the most characteristic and easily observed "characteristics of life." When we walk, talk, run, breathe, or engage in a multitude of other physical activities that are under the "willed" control of the individual, we do so by the contraction of skeletal muscle.

There are over 600 skeletal muscles in the body. Collectively, they constitute 40% to 50% of our body weight. And, together with the scaffolding provided by the skeleton, muscles also determine the form and contours of our body.

Contraction of individual muscle cells is ultimately responsible for purposeful movement. In Chapter 9, the physiology of muscular contraction was discussed. In this chapter, we will learn how contractile units are grouped into unique functioning organs—or muscles. The manner in which muscles are grouped, the relationship of muscles to joints, and how muscles attach to the skeleton determine purposeful body movement. A discussion of muscle shape and how muscles attach to and move bones will be followed by information on specific muscles and muscle groups. The chapter will end with a review of the concept of posture.

SKELETAL MUSCLE STRUCTURE

Connective Tissue Components

The highly specialized skeletal muscle fibers described in Chapter 9 are covered by a delicate connective tissue membrane called the **endomysium.** Groups of skeletal muscle fibers, called *fascicles,* are then bound together by a tougher connective tissue envelope called the **perimysium.** The muscle as a whole is covered by a coarse sheath called the **epimysium.** Because all three of these structures are continuous with the fibrous structures that attach muscles to bones or other structures, muscles are firmly harnessed to the structures they pull on during contraction. The epimysium, perimysium, and endomysium of a muscle, for example, may be continuous with fibrous tissue that extends from the muscle as a **tendon,** a strong tough cord continuous at its other end with the fibrous periosteum covering a bone. Or the fibrous wrapping of a muscle may extend as a broad, flat sheet of connective tissue called an **aponeurosis,** which usually merges with the fibrous wrappings of another muscle. So tough and strong are tendons and aponeuroses that they are not often torn, even by injuries forceful enough to break bones or tear muscles. They are, however, occasionally pulled away from bones.

Tube-shaped structures of fibrous connective tissue called **tendon sheaths** enclose certain tendons, notably those of the wrist and ankle. Like the bursae, tendon sheaths have a lining of synovial membrane. Its moist, smooth surface enables the tendon to move easily, almost frictionlessly, in the tendon sheath.

Size, Shape, and Fiber Arrangement

The structures called skeletal muscles are organs. They consist mainly of skeletal muscle tissue plus important connective and nervous tissue components. Skeletal muscles vary considerably in size, shape, and arrangement of fibers. They range from extremely small strands, such as the stapedius muscle of the middle ear, to large masses, such as the muscles of the thigh. Some skeletal muscles are broad in shape and some are narrow. Some are long and tapering and some are short and blunt. Some are triangular, some quadrilateral, and some irregular. Some form flat sheets and others form bulky masses.

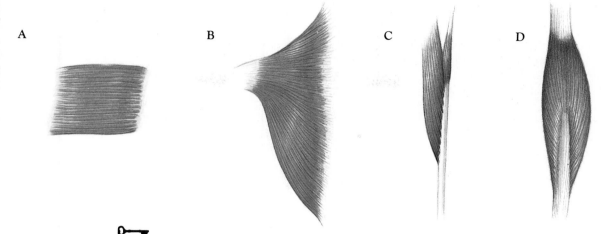

FIGURE 10-1 **Muscle shape and fiber arrangement. A,** Parallel. **B,** Convergent. **C,** Pennate. **D,** Bipennate.

Arrangement of fibers varies in different muscles. In some muscles the fibers are parallel to the long axis of the muscle (Figure 10-1, *A*). In some they converge to a narrow attachment (Figure 10-1, *B*), and in some they are oblique and either pennate (Figure 10-1, *C*) like the feathers in an old-fashioned plume pen or bipennate (double-feathered) (Figure 10-1, *D*)). Fibers may even be curved, as in the sphincters of the face, for example. The direction of the fibers composing a muscle is significant because of its relationship to function. For instance, a muscle with the bipennate fiber arrangement can produce a stronger contraction than a muscle having a parallel fiber arrangement.

QUICK CHECK

✔ 1. *Identify the connective tissue membrane that: (1) covers individual muscle fibers, (2) surrounds groups of skeletal muscle fibers (fascicles), and (3) covers the muscle as a whole.*
✔ 2. *Name the tough connective tissue cord that serves to attach a muscle to a bone.*
✔ 3. *Name three types of fiber arrangement seen in skeletal muscle.*

Attachment of Muscles

Most of our muscles span at least one joint and attach to both articulating bones. When contraction occurs, one bone usually remains fixed and the other moves. The points of attachment are called the origin and insertion. The **origin** is that point of attachment that does *not* move when the muscle contracts. Therefore the origin bone is the more stationary of the two bones at a joint when contraction occurs. The **insertion** is the point of attachment that moves when the muscle contracts (Figure 10-2). The insertion bone therefore moves toward the origin bone when the muscle shortens. In case you are wondering why both bones do not move, since both are pulled on by the contracting muscle, the reason is that one of them is normally stabilized by isometric contractions of other

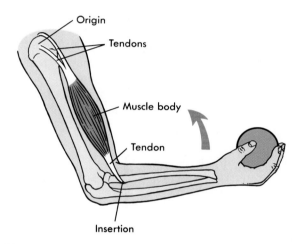

FIGURE 10-2 **Attachments of a skeletal muscle.** A muscle originates at a relatively stable part of the skeleton (origin) and inserts at the skeletal part that is moved when the muscle contracts (insertion).

muscles or by certain features of its own that make it less mobile.

The terms *origin* and *insertion* provide us with useful points of reference. Many muscles have multiple points of origin or insertion. Understanding the functional relationship of these attachment points during muscle contraction helps in deducing muscle actions. Attachment points of the biceps brachii shown in Figure 10-2 help provide functional information. Distal insertion on the radius of the lower arm causes flexion to occur at the elbow when contraction occurs. It should be realized, however, that origin and insertion are points that may change under certain circumstances. For example, not only can you grasp an object above your head and pull it down, you can also pull yourself up to the object. Although origin and insertion are convenient terms, they do not always provide the necessary information to understand the full functional potential of muscle action.

Muscle Actions

Skeletal muscles almost always act in groups rather than singly. As a result, most movements are produced by the coordinated action of several muscles. Some of the muscles in the group contract while others relax. The result is a movement pattern that allows for the functional classification of muscles or muscle groups. Several terms are used to describe muscle action during any specialized movement pattern. The terms *prime mover, antagonist, synergist,* and *fixator* are especially important and are discussed in the following paragraphs. Each term suggests an important concept that is essential to an understanding of such functional muscle patterns as flexion, extension, abduction, adduction, and other movements discussed in Chapter 8.

The term **prime mover** is used to describe a muscle or group of muscles that directly performs a specific movement. The movement produced by a muscle acting as a prime mover is described as the "action" or "function" of that muscle. For example, the biceps brachii shown in Figure 10-2 is acting as a prime mover during flexion of the forearm.

Antagonists are muscles that, when contracting, directly oppose prime movers. They are relaxed while the prime mover is contracting to produce movement. Simultaneous contraction of both a prime mover and its antagonist muscle results in rigidity and lack of motion. The term *antagonist* is perhaps unfortunate, since muscles cooperate, rather than oppose, in normal movement patterns. Antagonists are important in providing precision and control during contraction of prime movers.

Synergists are muscles that contract at the same time as the prime mover. They facilitate or complement prime mover actions so that the prime mover produces a more effective movement.

Fixator muscles generally function as joint stabilizers. They frequently serve to maintain posture or balance during contraction of prime movers acting on joints in the arms and legs.

Movement patterns are complex, and most muscles function not only as prime movers but also as antagonists, synergists, or fixators. A prime mover in a particular movement pattern, such as flexion, may be an antagonist during extension or a synergist or fixator in other types of movement.

QUICK CHECK

✔ 1. Identify the point of attachment of a muscle to a bone that: (1) does not move when the muscle contracts; (2) moves when the muscle contracts.

✔ 2. What name is used to describe a muscle that directly performs a specific movement?

✔ 3. What type of muscles help maintain posture or balance during contraction of muscles acting on joints in the arms and legs?

✔ 4. Name the type of muscles that generally function as joint stabilizers.

Lever Systems

When a muscle shortens, the central body portion, called the **belly,** contracts. The type and extent of movement is determined by the weight or resistance that is moved, the attachment of the tendinous extremities of the muscle to bone (origin and insertion), and by the particular type of joint involved. In almost every instance, muscles that move a part do not lie over that part. Instead, the muscle belly lies proximal to the part moved. Thus muscles that move the lower arm lie proximal to it, that is, in the upper arm.

Knowledge of **lever systems** is important in understanding muscle action. By definition, a *lever* is any rigid bar free to turn about a fixed point called its *fulcrum.* Bones serve as levers, and joints serve as fulcrums of these levers. A contracting muscle applies a pulling force on a bone lever at the point of the muscle's attachment to the bone. This causes the insertion bone to move about its joint fulcrum.

A lever system is a simple mechanical device composed of four component parts: (1) a rigid rod or bar (bone), (2) a fixed pivot, or fulcrum (F), around which the rod moves (joint), (3) a weight (W) or resistance that is moved, and

---

SPORTS & FITNESS   ASSESSING MUSCLE STRENGTH

Certified athletic trainers and other health care providers are often required to assess muscle strength in the evaluation of athletic injuries. A basic principle of muscle action in a lever system is called the **optimum angle of pull.** An understanding of this principle is required for correct assessment of muscle strength.

Generally, the optimum angle of pull for any muscle is a right angle to the long axis of the bone to which it is attached. When the angle of pull departs from a right angle and becomes more parallel to the long axis, the strength of contraction decreases dramatically. Contraction of the brachialis muscle demonstrates this principle very well. The brachialis crosses the elbow from humerus to ulna. In the anatomic position, the elbow is extended and the angle of pull of the brachialis is parallel to the long axis of the ulna (Figure 10-15, *D*). Contraction of the brachialis at this angle is very inefficient. As the elbow is flexed and the angle of pull approaches a right angle, the contraction strength of the muscle is greatly increased. Therefore to test brachialis muscle strength correctly, the forearm should be flexed at the elbow. Understanding the optimum angle of pull for any given muscle makes a rational approach to correct assessment of functional strength in that muscle possible.

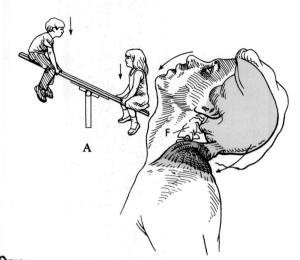

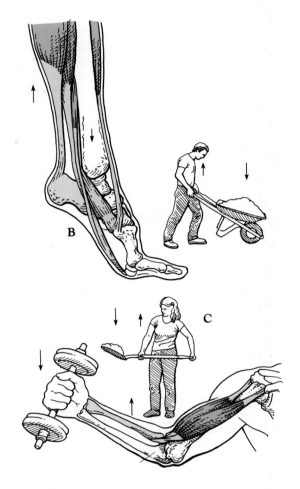

O═┳

FIGURE 10-3 **Lever classes. A,** Class I: fulcrum (F) between the weight (W) and force or pull (P); **B,** Class II: weight (W) between the fulcrum (F) and force or pull (P); **C,** Class III: force or pull (P) between the fulcrum (F) and the weight (W). The lever rod itself is colored yellow in each.

(4) a force, or pull (P), that produces movement (muscle contraction). Figure 10-3 shows the three different types of lever arrangements. All three types are found in the human body.

First-class levers

As you can see in Figure 10-3, the placement of the fulcrum in a first-class lever lies between the effort or pull (P) and the resistance or weight (W), as in a set of scales, a pair of scissors, or a child's seesaw. In the body, the head being raised or tipped backward on the atlas would be an example of a first-class lever in action. The facial portion of the skull is the weight, the joint between the skull and atlas is the fulcrum, and the muscles of the back produce the pull. In the human body, first-class levers are not abundant. They generally serve as levers of stability.

Second-class levers

In second-class levers the weight lies between the fulcrum and the joint at which the pull is exerted. The wheelbarrow is often used as an example. The presence of second-class levers in the human body is a controversial issue. Some authorities interpret the raising of the body on the toes as an example of this type of lever (Figure 10-3, B). In this example the point of contact between the toes and the ground would be the fulcrum, the weight would be located at the ankle, and pull would be exerted by the gastrocnemius muscle through the Achilles tendon. Opening the mouth against resistance (depression of the mandible) is also considered to be an example of a second-class lever.

Third-class levers

In a third-class lever the pull is exerted between the fulcrum and resistance or weight to be moved. Flexing of the forearm at the elbow joint is a frequently used example of this type of lever (Figure 10-3, C). Third-class levers permit rapid and extensive movement and are the most common type found in the body. They allow insertion of a muscle very close to the joint that it moves.

◢ HOW MUSCLES ARE NAMED

Muscle names seem more logical and therefore easier to learn when one understands the reasons for the names. The superficial muscles of the body are shown in Figures 10-4 and 10-5. Many are named using one or more of the following features.

- **Location.** Many muscles are named as a result of location. The *brachialis* (arm) muscle and *gluteus* (buttock) muscles are examples. Table 10-1 is a listing of muscles grouped by location.
- **Function.** The function of a muscle is frequently a part of its name. The *adductor* muscles of the thigh adduct, or move the leg toward the midline of the body. Table 10-2 lists selected muscles grouped according to function.

TABLE 10-1   Selected muscles grouped according to location

Location	Muscles	Location	Muscles
Neck	Sternocleidomastoid	**Thigh**	
Back	Trapezius	**Anterior surface**	Quadriceps femoris group
	Latissimus dorsi		Rectus femoris
Chest	Pectoralis major		Vastus lateralis
	Serratus anterior		Vastus medialis
			Vastus intermedius
Abdominal wall	External oblique	**Medial surface**	Gracilis
Shoulder	Deltoid		Adductor group (brevis, longus, magnus)
Upper arm	Biceps brachii	**Posterior surface**	Hamstring group
	Triceps brachii		Biceps femoris
	Brachialis		Semitendinosus
			Semimembranosus
Forearm	Brachioradialis	**Leg**	
	Pronator teres	**Anterior surface**	Tibialis anterior
Buttocks	Gluteus maximus	**Posterior surface**	Gastrocnemius
	Gluteus minimus		Soleus
	Gluteus medius	**Pelvic floor**	Levator ani
	Tensor fascia latae		Levator coccygeus
			Rectococcygeus

TABLE 10-2   Selected muscles grouped according to function

Part Moved	Example of Flexor	Example of Extensor	Example of Abductor	Example of Adductor
Head	Sternocleidomastoid	Semispinalis capitis		
Upper arm	Pectoralis major	Trapezius	Deltoid	Pectoralis major with latissimus dorsi
		Latissimus dorsi		
Forearm	With forearm supinated: biceps brachii	Triceps brachii		
	With forearm pronated: brachialis			
	With semisupination or semipronation: brachioradialis			
Hand	Flexor carpi radialis and ulnaris	Extensor carpi radialis, longus, and brevis	Flexor carpi radialis	Flexor carpi ulnaris
	Palmaris longus	Extensor carpi ulnaris		
Thigh	Iliopsoas	Gluteus maximus	Gluteus medius and gluteus minimus	Adductor group
	Rectus femoris (of quadriceps femoris group)			
Leg	Hamstrings	Quadriceps femoris group		
Foot	Tibialis anterior	Gastrocnemius	Evertors	Invertor
		Soleus	Peroneus longus	Tibialis anterior
			Peroneus brevis	
Trunk	Iliopsoas	Sacrospinalis		
	Rectus abdominis			

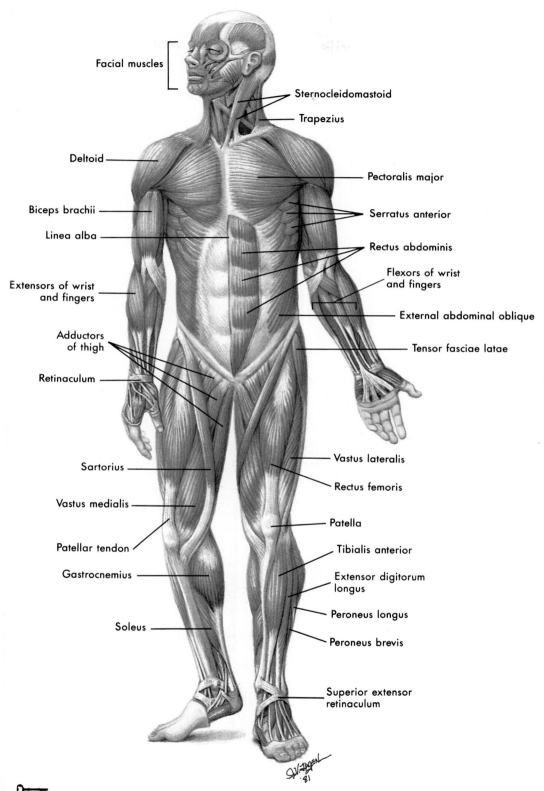

Facial muscles

Sternocleidomastoid

Trapezius

Deltoid

Pectoralis major

Biceps brachii

Serratus anterior

Linea alba

Rectus abdominis

Flexors of wrist
and fingers

Extensors of wrist
and fingers

External abdominal oblique

Adductors
of thigh

Tensor fasciae latae

Retinaculum

Vastus lateralis

Sartorius

Rectus femoris

Vastus medialis

Patella

Patellar tendon

Tibialis anterior

Gastrocnemius

Extensor digitorum
longus

Peroneus longus

Soleus

Peroneus brevis

Superior extensor
retinaculum

FIGURE 10-4 **General overview of the body musculature.** Anterior view.

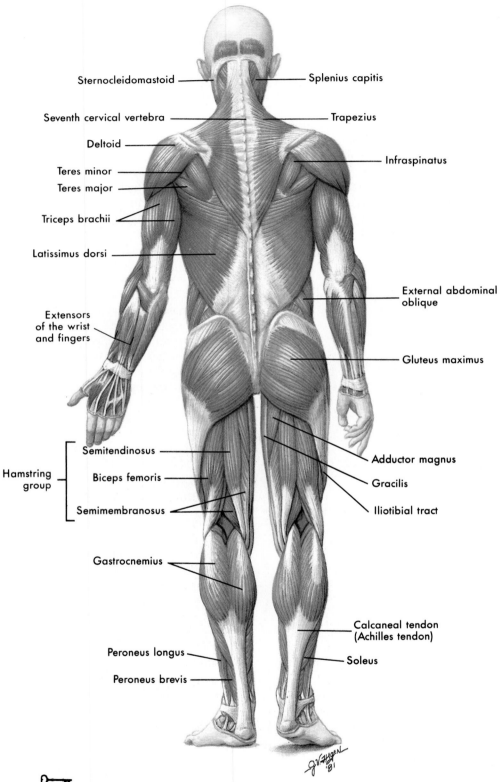

Sternocleidomastoid

Splenius capitis

Seventh cervical vertebra

Trapezius

Deltoid

Infraspinatus

Teres minor

Teres major

Triceps brachii

Latissimus dorsi

External abdominal oblique

Extensors of the wrist and fingers

Gluteus maximus

Hamstring group

Semitendinosus

Biceps femoris

Semimembranosus

Adductor magnus

Gracilis

Iliotibial tract

Gastrocnemius

Calcaneal tendon (Achilles tendon)

Peroneus longus

Soleus

Peroneus brevis

FIGURE 10-5 **General overview of the body musculature.** Posterior view.

♦ **Shape.** Shape is a descriptive feature used for naming many muscles. The *deltoid* (triangular) muscle covering the shoulder is delta, or triangular, in shape.

♦ **Direction of fibers.** Muscles may be named according to the orientation of their fibers. The term *rectus* means straight. The fibers of the *rectus abdominis* muscle run straight up and down and are parallel to each other.

♦ **Number of heads or divisions.** The number of divisions or heads (points of origin) may be used to name a muscle. The *biceps* (two); *triceps* (three); and *quadriceps* (four) refer to multiple points of origin. The *biceps brachii* is a muscle having two heads located in the arm.

♦ **Points of attachment.** Origin and insertion points may be used to name a muscle. For example, the *sternocleidomastoid* has its origin on the sternum and clavicle and inserts on the mastoid process of the temporal bone.

QUICK CHECK

✔ 1. *Name the four major components of any lever system.*

✔ 2. *Identify the three types of lever systems found in the human body and give one example of each.*

✔ 3. *What type of lever system permits rapid and extensive movement and is the most common type found in the body?*

✔ 4. *List five criteria that may determine a muscle's name and give an example of a specific muscle named using each criterion.*

Hints on How to Deduce Muscle Actions

To understand muscle actions, you need first to know certain anatomical facts such as which bones muscles attach to and which joints they pull across. Then if you relate these structural facts to functional principles, you may find your study of muscles more interesting and less difficult than you anticipate. Some specific suggestions for deducing muscle actions follow.

1. Start by making yourself familiar with the names, shapes, and general locations of the larger muscles, using Table 10-1 as a guide.

2. Try to deduce which bones the two ends of a muscle attach to from your knowledge of the shape and general location of the muscle. For example, look carefully at the deltoid muscle as illustrated in Figures 10-4 and 10-13. To what bones does it seem to attach? Check your answer with Table 10-9, p. 264.

3. Next, make a guess which bone moves when the muscle shortens. (The bone moved by a muscle's contraction is its insertion bone; the bone that remains relatively stationary is its origin bone.) In

many cases, you can tell by trying to move one bone and then another which is the insertion bone. In some cases, either bone may function as the insertion. Although not all muscle attachments can be deduced as readily as those of the deltoid, they can all be learned more easily by using this deduction method than by relying on rote memory alone.

4. Deduce a muscle's actions by applying the principle that its insertion moves toward its origin. Check your conclusions with the text. Here, as in steps 2 and 3, the method of deduction is intended merely as a guide and is not adequate by itself for determining muscle actions.

5. To deduce which muscle produces a given action (instead of which action a given muscle produces, as in step 4), start by inferring the insertion bone (bone that moves during the action). The body and origin of the muscle will lie on one or more of the bones toward which the insertion moves—often a bone, or bones, proximal to the insertion bone. Couple these conclusions about origin and insertion with your knowledge of muscle names and locations to deduce the muscle that produces the action.

For example, if you wish to determine the prime mover for the action of raising the upper arms straight out to the sides, you infer that the muscle inserts on the humerus, since this is the bone that moves. It moves toward the shoulder, that is, the clavicle and scapula, so that probably the muscle has its origin on these bones. Because you know that the deltoid muscle fulfills these conditions, you conclude, and rightly so, that it is the muscle that raises the upper arms sideways.

IMPORTANT SKELETAL MUSCLES

The major skeletal muscles of the body are listed, grouped, and illustrated in the tables and figures that follow. Begin your study with an overview of important superficial muscles, shown in Figures 10-4 and 10-5. The remaining figures in this chapter illustrate individual muscles or important muscle groups.

Basic information about many muscles is given in Tables 10-3 to 10-14. Each table has a description of a group of muscles that move one part of the body. The actions listed for each muscle are those for which it is a prime mover. Remember, however, that a single muscle acting alone rarely accomplishes a given action. Instead, muscles act in groups as prime movers, synergists, antagonists, and fixators to bring about movements.

Muscles of Facial Expression

The muscles of facial expression (Table 10-3) are unique in that at least one of their points of attachment is to the deep layers of the skin over the face or neck. Contraction of these muscles (Figure 10-6) produces a variety of facial expressions.

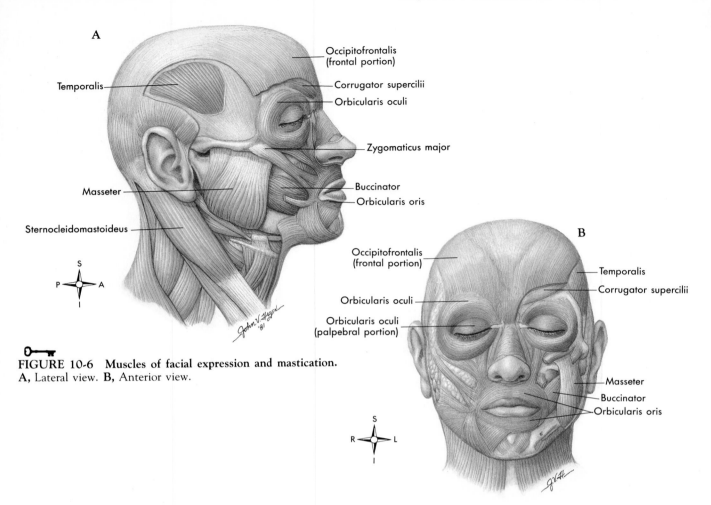

FIGURE 10-6 Muscles of facial expression and mastication.
A, Lateral view. **B,** Anterior view.

TABLE 10-3	**Muscles of facial expression and of mastication**			
Muscle	**Origin**	**Insertion**	**Function**	**Nerve Supply**
MUSCLES OF FACIAL EXPRESSION				
Epicranius (occipitofrontalis)	Occipital bone	Tissues of eyebrows	Raises eyebrows, wrinkles forehead horizontally	Cranial nerve VII
Corrugator supercilii	Frontal bone (superciliary ridge)	Skin of eyebrow	Wrinkles forehead vertically	Cranial nerve VII
Orbicularis oculi	Encircles eyelid		Closes eye	Cranial nerve VII
Zygomaticus major	Zygomatic bone	Angle of mouth	Laughing (elevates angle of mouth)	Cranial nerve VII
Orbicularis oris	Encircles mouth		Draws lips together	Cranial nerve VII
Buccinator	Maxillae	Skin of sides of mouth	Permits smiling	Cranial nerve VII
			Blowing, as in playing a trumpet	
MUSCLES OF MASTICATION				
Masseter	Zygomatic arch	Mandible (external surface)	Closes jaw	Cranial nerve V
Temporalis	Temporal bone	Mandible	Closes jaw	Cranial nerve V
Pterygoids (internal and external)	Undersurface of skull	Mandible (medial surface)	Grates teeth	Cranial nerve V

The **epicranius** (ep-i-KRA-ne-us), or *occipitofrontalis,* is in reality two muscles. One portion lies over the forehead, and the other covers the occipital bone in back of the head. The two muscular parts, or bellies, are joined by a connective tissue apponeurosis that covers the top of the skull. The frontal portion of the epicranius raises the eyebrows (surprise) and wrinkles the skin of the forehead horizontally. The **corrugator supercilii** (COR-u-GA-tor su-per-SIL-i) draws the eyebrows together, producing vertical wrinkles above the nose (frowning). The **orbicularis oculi** (or-BIC-u-LAR-us OK-u-li) encircles and closes the eye (blinking), whereas the **orbicularis oris** (OR-us) and **buccinator** (BUK-si-NA-tor) pucker the mouth (kissing) and press the lips and cheeks against the teeth. The **zygomaticus** (ZI-go-MAT-i-kus) **major** draws the corner of the mouth upward (laughing).

Muscles of Mastication

The muscles of **mastication** (mass-ti-KA-shun) shown in Figure 10-6 are responsible for chewing movements. These powerful muscles (Table 10-3) either elevate and retract the mandible (**masseter** [mas-SE-ter] and **temporalis** [tempo-RAL-is]) or open and protrude it while causing sideways movement (**pterygoids** [TER-i-goids]). The pull of gravity helps open the mandible during mastication, and the buccinator muscles play an important function by holding food between the teeth as the mandible moves up and down and from side to side.

Muscles That Move the Head

Paired muscles on either side of the neck are responsible for head movements (Figure 10-7). Note the points of attachment and functions of important muscles in this group listed in Table 10-4. When both **sternocleidomastoid** (STER-no-KLI-do-MAS-toyd) muscles (see Figure 10-4) contract at the same time, the head is flexed on the

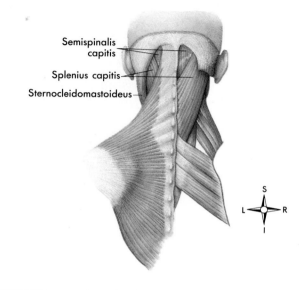

FIGURE 10-7 **Muscles that move the head.** Posterior view of muscles of the neck and the back.

TABLE 10-4	Muscles that move the head			
Muscle	**Origin**	**Insertion**	**Function**	**Nerve Supply**
Sternocleidomastoid	Sternum Clavicle	Temporal bone (mastoid process)	Flexes head (prayer muscle) One muscle alone, rotates head toward opposite side; spasm of this muscle alone or associated with trapezius called *torticollis* or *wryneck*	Accessory nerve
Semispinalis capitis	Vertebrae (transverse processes of upper six thoracic, articular processes of lower four cervical)	Occipital bone (between superior and inferior nuchal lines)	Extends head; bends it laterally	First five cervical nerves
Splenius capitis	Ligamentum nuchae Vertebrae (spinous processes of upper three or four thoracic)	Temporal bone (mastoid process) Occipital bone	Extends head Bends and rotates head toward same side as contracting muscle	Second, third, and fourth cervical nerves
Longissimus capitis	Vertebrae (transverse processes of upper six thoracic, articular processes of lower four cervical)	Temporal bone (mastoid process)	Extends head Bends and rotates head toward contracting side	Multiple innervation

thorax—hence the name "prayer muscle." If only one muscle contracts, the head and face are turned to the opposite side. The broad **semispinalis capitis** (sem-e-spi-NAL-is KAP-i-tis) is an extensor of the head and helps to bend it laterally. Acting together, the **splenius capitis** (SPLE-ne-us KAP-i-tis) muscles serve as strong extensors that return the head to the upright position following flexion. When either muscle acts alone, contraction results in rotation and tilting toward that side. The bandlike **longissimus capitis** (lon-JIS-i-mus KAP-i-tis) muscles are covered and not visible in Figure 10-7. They run from the neck vertebrae to the mastoid process of the temporal bone on either side and cause extension of the head when acting together. One contracting muscle will bend and rotate the head toward the contracting side.

◢ TRUNK MUSCLES
Muscles of the Thorax

The muscles of the thorax are of critical importance in respiration and will be discussed in detail in Chapter 23. Note in Figure 10-8 and Table 10-5 that the **internal** and **external intercostal** (IN-ter-KOS-tal) **muscles** attach to the ribs at different places and their fibers are oriented in different directions. As a result, the contraction of the external intercostals elevates and the internal intercostals depress the ribs—an important part of the breathing process. During inspiration the dome-shaped **diaphragm** (DI-a-fram) flattens out, thus increasing the size and volume of the thoracic cavity. As a result, air enters the lungs.

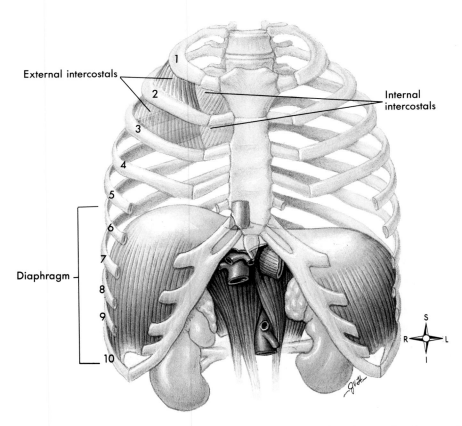

FIGURE 10-8   Muscles of the thorax. Anterior view. Note relationship of internal and external intercostal muscles and placement of diaphragm.

TABLE 10-5	Muscles of the thorax			
Muscle	**Origin**	**Insertion**	**Function**	**Nerve Supply**
External intercostals	Rib (lower border; forward fibers)	Rib (upper border of rib below origin)	Elevate ribs	Intercostal nerves
Internal intercostals	Rib (inner surface, lower border; backward fibers)	Rib (upper border of rib below origin)	Depress ribs	Intercostal nerves
Diaphragm	Lower circumference of thorax (of rib cage)	Central tendon of diaphragm	Enlarges thorax, causing inspiration	Phrenic nerves

Muscles of the Abdominal Wall

The muscles of the anterior and lateral abdominal wall (Figure 10-9 and Table 10-6) are arranged in three layers, with the fibers in each layer running in different directions much like the layers of wood in a sheet of plywood. The result is a very strong "girdle" of muscle that covers and supports the abdominal cavity and its internal organs.

The three layers of muscle in the anterolateral (side) abdominal walls are arranged as follows: the outermost layer, or **external oblique;** a middle layer, or **internal oblique;** and the innermost layer, or **transversus abdominis.** In addition to these sheetlike muscles, the band-shaped (or strap-shaped) **rectus abdominis** muscle runs down the midline of the abdomen from the thorax to the pubis. In addition to protecting the abdominal viscera, the rectus abdominis flexes the spinal column.

Muscles of the Pelvic Floor

Structures in the pelvic cavity are supported by a reinforced muscular floor that guards the outlet below. The muscular pelvic floor filling the diamond-shaped outlet is called the **perineum** (per-i-NE-um). Passing through the floor is the anal canal and urethra in both sexes and the vagina in the female.

The two **levator ani** and **coccygeus** muscles form most of the pelvic floor. They stretch across the pelvic cavity like a hammock. This diamond-shaped outlet can be divided into two triangles by a line drawn from side to side between the ischial tuberosities. The **urogenital triangle** is anterior (above) this line, extending to the symphysis pubis, and the **anal triangle** is posterior (behind it), ending at the coccyx. Note in Figure 10-10 that structures in the urogenital triangle include the **ischiocavernosus** and **bulbospongiosus** muscles associated with the penis in the male or vagina in the female. Constriction of muscles called **sphincter urethrae,** which encircle the urethra in both sexes, helps control urine flow. The anal triangle allows passage of the anal canal. The terminal portion of the canal is surrounded by the **external anal sphincter,** which regulates defecation. The origin, insertion, function, and innervation of important muscles of the pelvic floor are listed in Table 10-7. The coccygeus muscles lie behind the levator ani and are not visible in Figure 10-10.

TABLE 10-6	Muscles of the abdominal wall			
Muscle	**Origin**	**Insertion**	**Function**	**Nerve Supply**
External oblique	Ribs (lower eight)	Ossa coxae (iliac crest and pubis by way of inguinal ligament) Linea alba by way of an aponeurosis	Compresses abdomen Important postural function of all abdominal muscles is to pull front of pelvis upward, thereby flattening lumbar curve of spine; when these muscles lose their tone, common figure faults of protruding abdomen and lordosis develop	Lower seven intercostal nerves and iliohypogastric nerves
Internal oblique	Ossa coxae (iliac crest and inguinal ligament) Lumbodorsal fascia	Ribs (lower three) Pubic bone Linea alba	Same as external oblique	Last three intercostal nerves; iliohypogastric and ilioinguinal nerves
Transversus abdominis	Ribs (lower six) Ossa coxae (iliac crest, inguinal ligament) Lumbodorsal fascia	Pubic bone Linea alba	Same as external oblique	Last five intercostal nerves; iliohypogastric and ilioinguinal nerves
Rectus abdominis	Ossa coxae (pubic bone and symphysis pubis)	Ribs (costal cartilage of fifth, sixth, and seventh ribs) Sternum (xiphoid process)	Same as external oblique; because abdominal muscles compress abdominal cavity, they aid in straining, defecation, forced expiration, childbirth, etc.; abdominal muscles are antagonists of diaphragm, relaxing as it contracts and vice versa Flexes trunk	Last six intercostal nerves

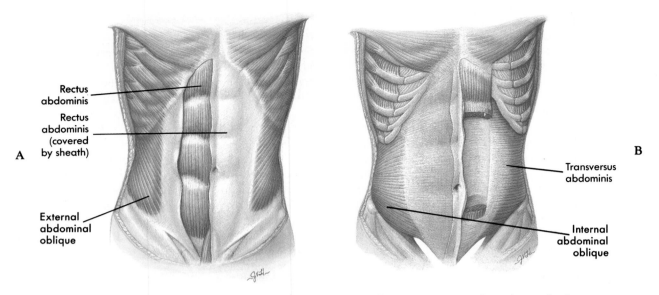

FIGURE 10-9 **Muscles of the trunk and abdominal wall. A,** Anterior view showing superficial muscles. **B,** Anterior view showing deeper muscles.

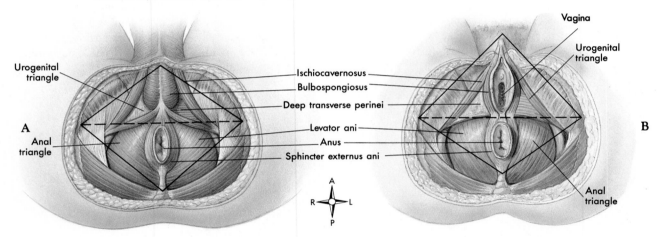

FIGURE 10-10 **Muscles of the pelvic floor. A,** Male, inferior view. **B,** Female, inferior view.

TABLE 10-7   Muscles of the pelvic floor

Muscle	Origin	Insertion	Function	Nerve Supply
Levator ani	Pubis and spine of ischium	Coccyx	Together with coccygeus muscles form floor of pelvic cavity and support pelvic organs	Pudendal nerve
Ischiocavernosus	Ischium	Penis or clitoris	Compress base of penis or clitoris	Perineal nerve
Bulbospongiosus				
Male	Bulb of penis	Perineum and bulb of penis	Constricts urethra and erects penis	Pudendal nerve
Female	Perineum	Base of clitoris	Erects clitoris	Pudendal nerve
Deep transverse perinei	Ischium	Central tendon (median raphe)	Support pelvic floor	Pudendal nerve
Sphincter urethrae	Pubic ramus	Central tendon (median raphe)	Constrict urethra	Pudendal nerve
Sphincter externus anii	Coccyx	Central tendon (median raphe)	Close anal canal	Pudendal and S4

UPPER LIMB MUSCLES

The muscles of the upper limb include those acting on the shoulder or pectoral girdle, as well as muscles located in the arm, forearm, and hand.

Muscles Acting on the Shoulder Girdle

Attachment of the upper extremity to the torso is by muscles that have either an anterior location (chest) or posterior placement (back and neck). Six muscles (Table 10-8 and Figure 10-11) that pass from the axial skeleton to the shoulder or pectoral girdle (scapula and clavicle) serve not only to "attach" the upper extremity to the body but do so in such a way that extensive movement is possible. The clavicle can be elevated and depressed, as well as moved forward and back. The scapula is capable of even a greater variety of movements.

The **pectoralis** (pek-to-RAL-is) **minor** lies under the larger pectoralis major muscle on the anterior chest wall. It helps "fix" the scapula against the thorax and also raises the ribs during forced inspiration. Another anterior chest wall muscle—the **serratus** (ser-RAY-tus) **anterior**—helps hold the scapula against the thorax to prevent "winging" and is a strong abductor that is useful in pushing or punching movements.

The posterior muscles acting on the shoulder girdle include the **levator scapulae** (le-VAY-tor SCAP-yoo-le), which elevates the scapula; the **trapezius** (trah-PEE-zee-us), which is used to "shrug" the shoulders; and the **rhomboideus** (rom-BOID-ee-us) **major** and **minor** muscles, which serve to adduct and elevate the scapula.

Muscles That Move the Upper Arm

The shoulder is a synovial joint of the ball-and-socket type. As a result, extensive movement is possible in every plane of motion. Muscles that move the upper arm can be grouped according to function as flexors, extensors, abductors, adductors, and medial and lateral rotators (Table 10-9; see Figure 10-13). The actions listed in Table 10-9 include both primary actions and important secondary functions.

The **deltoid** (DEL-toid) is a good example of a multi-function muscle. It has three groups of fibers and may act as three separate muscles. Contraction of anterior fibers will flex the arm while lateral fibers abduct and posterior fibers serve as extensors. Four other muscles serve as both a structural and functional cap or cuff around the shoulder joint and are referred to as the **rotator cuff muscles** (Figure 10-12). They include the **infraspinatus, supraspinatus, subscapularis,** and **teres minor.**

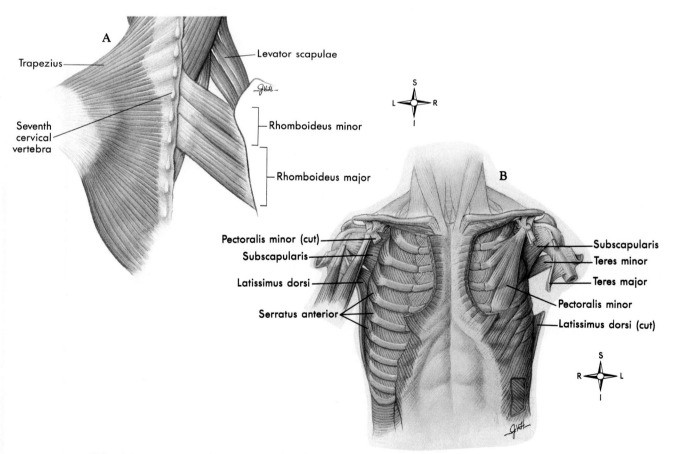

FIGURE 10-11   Muscles acting on the shoulder girdle. A, Posterior view. Trapezius has been removed on the right to reveal the deeper muscles. **B,** Anterior view. Pectoralis major has been removed on both sides. The pectoralis minor also has been removed on the right side.

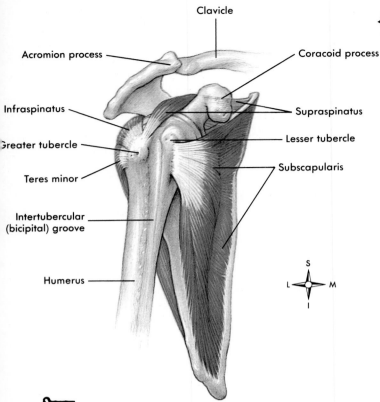

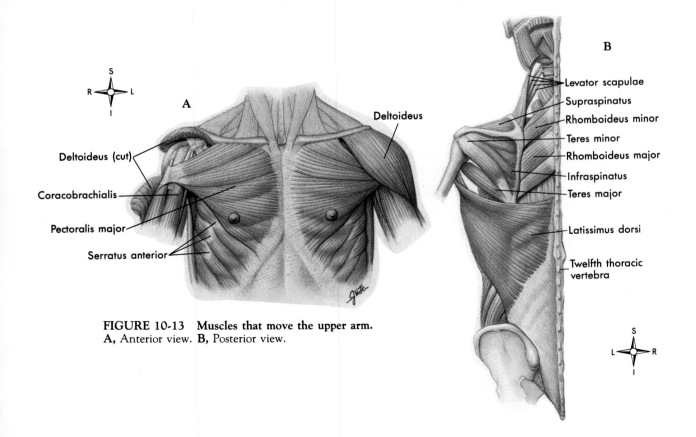

SHOULDER JOINT STABILITY

The disparity in size between the large and nearly hemispheric head of the humerus and the much smaller and shallow glenoid cavity of the scapula is of great clinical significance. Because the head of the humerus is over two times larger than the shallow glenoid concavity that receives it, only about one quarter of the articular surface of the humeral head is in contact with the fossa in any given position of the joint. This anatomical fact helps explain the inherent instability of the shoulder—our most mobile joint. The soft tissues surrounding the shoulder, such as the joint capsule, ligaments, and adjacent muscles, provide the primary restraint against excessive motion and potential dislocation.

Unfortunately, only a thin articular capsule surrounds the shoulder joint. It is extremely loose and does not function to keep the articulating bones of the joint in contact. This fact is obviously correlated with both the great range of motion (ROM) possible at this articulation and its tendency to dislocate as a result of athletic injury or other trauma. The tendons of the supraspinatus, infraspinatus, teres minor, and subscapularis muscles (called the **SITS** muscles) all blend with and strengthen the articular capsule. The musculotendinous cuff resulting from this fusion is called the **rotator cuff** (see Figure 10-12). The rotator cuff provides the necessary strength to help prevent anterior, superior, and posterior displacement of the humeral head during most types of activity.

FIGURE 10-12   Rotator cuff muscles. Note the tendons of the teres minor, infraspinatus, supraspinatus, and subscapularis muscles surrounding the head of the humerus.

FIGURE 10-13   Muscles that move the upper arm. A, Anterior view. **B,** Posterior view.

TABLE 10-8   Muscles acting on the shoulder girdle

Muscle	Origin	Insertion	Function	Nerve Supply
Trapezius	Occipital bone (protuberance)	Clavicle	Raises or lowers shoulders and shrugs them	Spinal accessory; second, third, and fourth cervical nerves
	Vertebrae (cervical and thoracic)	Scapula (spine and acromion)	Extends head when occiput acts as insertion	
Pectoralis minor	Ribs (second to fifth)	Scapula (coracoid)	Pulls shoulder down and forward	Medial and lateral anterior thoracic nerves
Serratus anterior	Ribs (upper eight or nine)	Scapula (anterior surface, vertebral border)	Pulls shoulder forward; abducts and rotates it upward	Long thoracic nerve
Levator scapulae	C1-C4 (transverse processes)	Scapula (superior angle)	Elevates and retracts scapula and abducts neck	Dorsal scapular nerve
Rhomboideus				
Major	T1-T4	Scapula (medial border)	Retracts, rotates, fixes scapula	Dorsal scapular nerve
Minor	C6-C7	Scapula (medial border)	Retracts, rotates, elevates, and fixes scapula	Dorsal scapular nerve

TABLE 10-9   Muscles that move the upper arm

Muscle	Origin	Insertion	Function	Nerve Supply
Pectoralis major	Clavicle (medial half)	Humerus (greater tubercle)	Flexes upper arm	Medial and lateral anterior thoracic nerves
	Sternum		Adducts upper arm anteriorly; draws it across chest	
	Costal cartilages of true ribs			
Latissimus dorsi	Vertebrae (spines of lower thoracic, lumbar, and sacral)	Humerus (intertubercular groove)	Extends upper arm	Thoracodorsal nerve
	Ilium (crest)		Adducts upper arm posteriorly	
	Lumbodorsal fascia			
Deltoid	Clavicle	Humerus (lateral side about halfway down—deltoid tubercle)	Abducts upper arm	Axillary nerve
	Scapula (spine and acromion)		Assists in flexion and extension of upper arm	
Coracobrachialis	Scapula (coracoid process)	Humerus (middle third, medial surface)	Adduction; assists in flexion and medial rotation of arm	Musculocutaneous nerve
Supraspinatus	Scapula (supraspinous fossa)	Humerus (greater tubercle)	Assists in abducting arm	Suprascapular nerve
Teres minor	Scapula (axillary border)	Humerus (greater tubercle)	Rotates arm outward	Axillary nerve
Teres major	Scapula (lower part, axillary border)	Humerus (upper part, anterior surface)	Assists in extension, adduction, and medial rotation of arm	Lower subscapular nerve
Infraspinatus	Scapula (infraspinatus border)	Humerus (greater tubercle)	Rotates arm outward	Suprascapular nerve
Subscapularis	Scapula (subscapular fossa)	Humerus (lesser tubercle)	Medial rotation	Subscapular nerve

Muscles That Move the Forearm

Selected superficial and deep muscles of the upper extremity are shown in Figure 10-14. Recall that most muscles acting on a joint lie proximal to that joint. Muscles acting directly on the forearm, therefore, are found proximal to the elbow and attach the bones of the forearm (ulna and radius) to the humerus or scapula above. Table 10-10 lists the muscles acting on the lower arm giving origin, insertion, function, and innervation of each. Figure 10-15 shows the detail of attachment of several important muscles in this group.

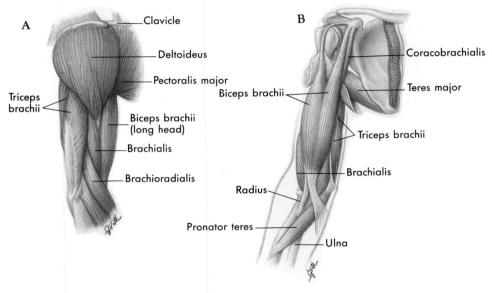

FIGURE 10-14 Muscles acting on the forearm. A, Lateral view of the right shoulder and arm. **B,** Anterior view of the right shoulder and arm (deep). Deltoid and pectoralis major muscles have been removed to reveal deeper structures.

TABLE 10-10   Muscles that move the forearm

Muscle	Origin	Insertion	Function	Nerve Supply
Biceps brachii	Scapula (supraglenoid tuberosity)	Radius (tubercle at proximal end)	Flexes supinated forearm	Musculocutaneous nerve
	Scapula (coracoid)		Supinates forearm and hand	
Brachialis	Humerus (distal half, anterior surface)	Ulna (front of coronoid process)	Flexes pronated forearm	Musculocutaneous nerve
Brachioradialis	Humerus (above lateral epicondyle)	Radius (styloid process)	Flexes semipronated or semisupinated forearm; supinates forearm and hand	Radial nerve
Triceps brachii	Scapula (infraglenoid tuberosity)	Ulna (olecranon process)	Extends lower arm	Radial nerve
	Humerus (posterior surface—lateral head above radial groove; medial head, below)			
Pronator teres	Humerus (medial epicondyle)	Radius (middle third of lateral surface)	Pronates and flexes forearm	Median nerve
	Ulna (coronoid process)			
Pronator quadratus	Ulna (distal fourth, anterior surface)	Radius (distal fourth, anterior surface)	Pronates forearm	Median nerve
Supinator	Humerus (lateral epicondyle)	Radius (proximal third)	Supinates forearm	Radial nerve
	Ulna (proximal fifth)			

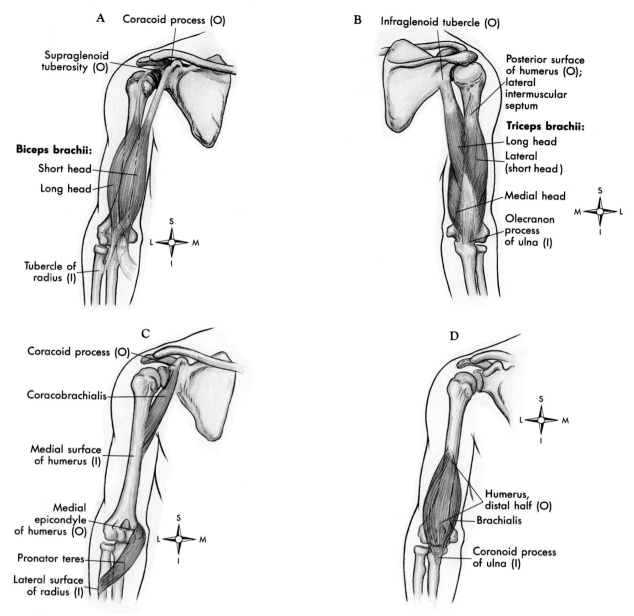

A
Coracoid process (O)
Supraglenoid tuberosity (O)
Biceps brachii:
Short head
Long head
Tubercle of radius (I)

B
Infraglenoid tubercle (O)
Posterior surface of humerus (O); lateral intermuscular septum
Triceps brachii:
Long head
Lateral (short head)
Medial head
Olecranon process of ulna (I)

C
Coracoid process (O)
Coracobrachialis
Medial surface of humerus (I)
Medial epicondyle of humerus (O)
Pronator teres
Lateral surface of radius (I)

D
Humerus, distal half (O)
Brachialis
Coronoid process of ulna (I)

FIGURE 10-15 **Muscles acting on the forearm. A,** Biceps brachii. **B,** Triceps brachii. **C,** Coracobrachialis and pronator teres. **D,** Brachialis. O, Origin. I, Insertion.

Muscles That Move the Wrist, Hand, and Fingers

Muscles acting on the wrist, hand, and fingers are located on either the anterior or the posterior surfaces of the forearm (Figure 10-16). In most instances, those muscles located on the anterior surface of the forearm are flexors, and those on the posterior surface are extensors of the wrist, hand, and fingers (Table 10-11).

A number of **intrinsic muscles** are responsible for precise movements of the hand and fingers. Examples include the **lumbrical** (LUM-bri-kal) and **interosseous** (in-ter-OS-ee-us) muscles, which originate from and fill the spaces between the metacarpal bones and then insert on the phalanges of the fingers. As a group the intrinsic muscles abduct and adduct the fingers and aid in flexing them. Eight additional muscles serve the thumb, enabling it to be placed in opposition to the fingers in tasks requiring grasping and manipulation. The **opponens pollicis** (o-PO-nenz POL-i-cis) is a particularly important thumb muscle. It allows the thumb to be drawn across the palm to touch the tip of any finger—a critical movement for many manipulative-type activities. Figure 10-17 shows the placement and points of attachment for various individual muscles acting on the wrist, hand, and fingers.

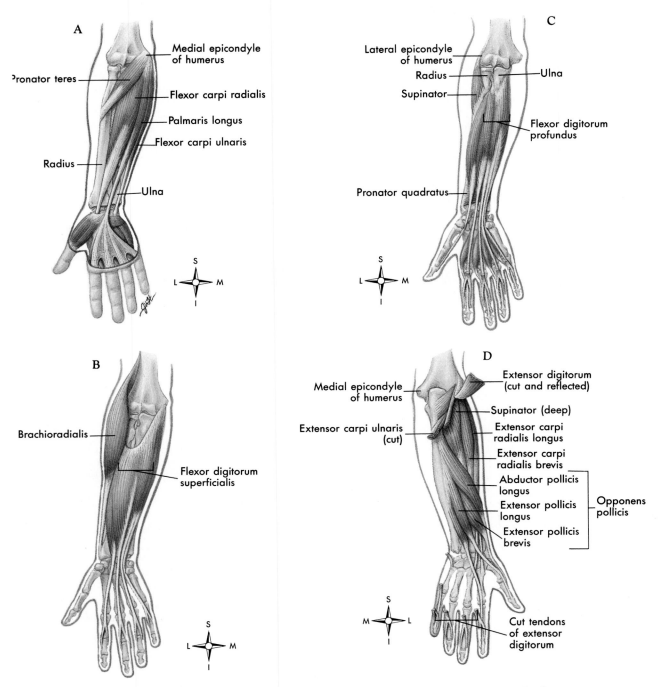

FIGURE 10-16   Muscles of the forearm. A, Anterior view shows right forearm (superficial). Brachioradialis muscle has been removed. **B,** Anterior view shows right forearm (deeper than **A**). Pronator teres, flexor carpi radialis and ulnaris, and palmaris longus muscles have been removed. **C,** Anterior view shows right forearm (deeper than **A** or **B**). Brachioradialis, pronator teres, flexor carpi radialis and ulnaris, palmaris longus, and flexor digitorum superficialis muscles have been removed. **D,** Posterior view shows deep muscles of the right forearm. Extensor digitorum, extensor digiti minimi, and extensor carpi ulnaris muscles have been cut to reveal deeper muscles.

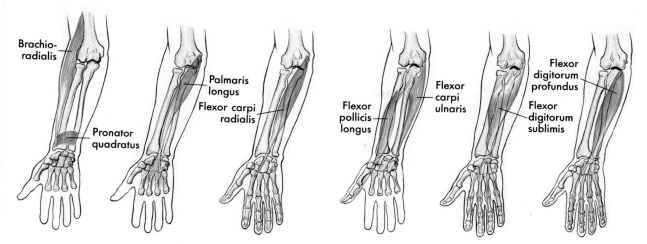

FIGURE 10-17   Some muscles of the anterior aspect of the right forearm.

TABLE 10-11   Muscles that move the wrist, hand, and fingers

Muscle	Origin	Insertion	Function	Nerve Supply
Flexor carpi radialis	Humerus (medial epicondyle)	Second metacarpal (base of)	Flexes hand Flexes forearm	Median nerve
Palmaris longus	Humerus (medial epicondyle)	Fascia of palm	Flexes hand	Median nerve
Flexor carpi ulnaris	Humerus (medial epicondyle) Ulna (proximal two thirds)	Pisiform bone Third, fourth, and fifth metacarpals	Flexes hand Adducts hand	Ulnar nerve
Extensor carpi radialis longus	Humerus (ridge above lateral epicondyle)	Second metacarpal (base of)	Extends hand Abducts hand (moves toward thumb side when hand supinated)	Radial nerve
Extensor carpi radialis brevis	Humerus (lateral epicondyle)	Second, third metacarpals (bases of)	Extends hand	Radial nerve
Extensor carpi ulnaris	Humerus (lateral epicondyle) Ulna (proximal three fourths)	Fifth metacarpal (base of)	Extends hand Adducts hand (moves toward little finger side when hand supinated)	Radial nerve
Flexor digitorum profundus	Ulna (anterior surface)	Distal phalanges (fingers 2 to 5)	Flexes distal interphalangeal joints	Median and ulnar nerves
Flexor digitorum superficialis	Humerus (medial epicondyle) Radius Ulna (coronoid process)	Tendons of fingers	Flexes fingers	Median nerve
Extensor digitorum	Humerus (lateral epicondyle)	Phalanges (fingers 2 to 5)	Extends fingers	Radial nerve
Opponens pollicis	Greater multangular	Thumb metacarpal	Opposes thumb to fingers	Median nerve

CARPAL TUNNEL SYNDROME

Some epidemiologists specialize in the field of occupational health, the study of health matters related to work or the workplace. Many problems seen by occupational health experts are caused by repetitive motions of the wrists or other joints. Word processors (typists) and meat cutters, for example, are at risk of developing conditions caused by repetitive motion injuries.

One common problem often caused by such repetitive motion is **tenosynovitis** (ten-o-sin-o-VYE-tis)—inflammation of a tendon sheath. Tenosynovitis can be painful, and the swelling characteristic of this condition can limit movement in affected parts of the body. For example, swelling of the tendon sheath around tendons in an area of the wrist known as the *carpal tunnel* can limit movement of the wrist, hand, and fingers. The figure shows the relative positions of the tendon sheath and median nerve within the carpal tunnel. If this swelling, or any other lesion in the carpal tunnel, presses on the *median nerve*, a condition called **carpal tunnel syndrome** may result. Because the median nerve connects to the palm and radial side (thumb side) of the hand, carpal tunnel syndrome is characterized by weakness, pain, and tingling in this part of the hand. The pain and tingling may also radiate to the forearm and shoulder. Prolonged or severe cases of carpal tunnel syndrome may be relieved by injection of antiinflammatory agents. A permanent cure is sometimes accomplished by surgical cutting or removal of the swollen tissue pressing on the median nerve.

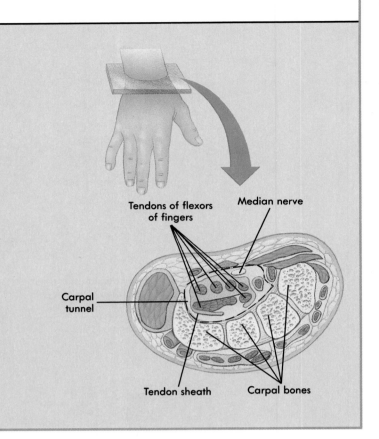

LOWER LIMB MUSCLES

The musculature, bony structure, and joints of the pelvic girdle and lower extremity function in locomotion and maintenance of stability. Powerful muscles at the back of the hip, at the front of the thigh, and at the back of the leg also serve to raise the full body weight from a sitting to a standing position. The muscles of the lower limb include those acting on the hip or pelvic girdle, as well as muscles located in the thigh, leg, and foot. Unlike the highly mobile shoulder girdle, the pelvic girdle is essentially fixed. Therefore our study of muscles in the lower extremity begins with those arising from the pelvic girdle and pass to the femur, producing their effects at the hip joint by moving the thigh.

Muscles That Move the Thigh and Lower Leg

Table 10-12 identifies muscles that move the thigh and lists the origin, insertion, function, and nerve supply of each (see Figure 10-19). Refer to Figures 10-4 and 10-5, as well as Figures 10-18 through 10-21, which show individual muscles, as you study the information provided in the table. Muscles acting on the thigh can be divided into three groups: (1) muscles crossing the front of the hip, (2) the three gluteal (GLOO-tee-al) muscles and the tensor fasciae latae (TEN-sor FASH-ee LAT-tee), and (3) the thigh adductors.

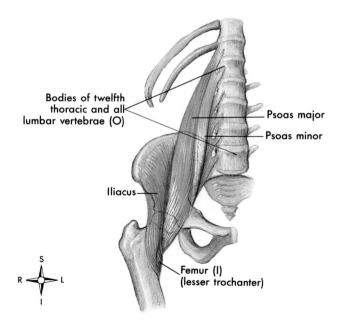

FIGURE 10-18 Iliopsoas muscle (iliacus, psoas major, and psoas minor muscles). O, Origin. I, Insertion.

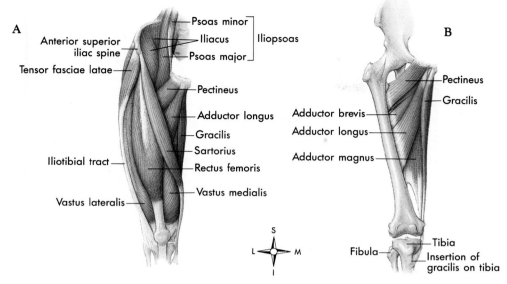

FIGURE 10-19 **Muscles of the anterior thigh. A,** Anterior view of the right thigh. **B,** Adductor region of the right thigh. Tensor fasciae latae, sartorius, and quadriceps muscles have been removed.

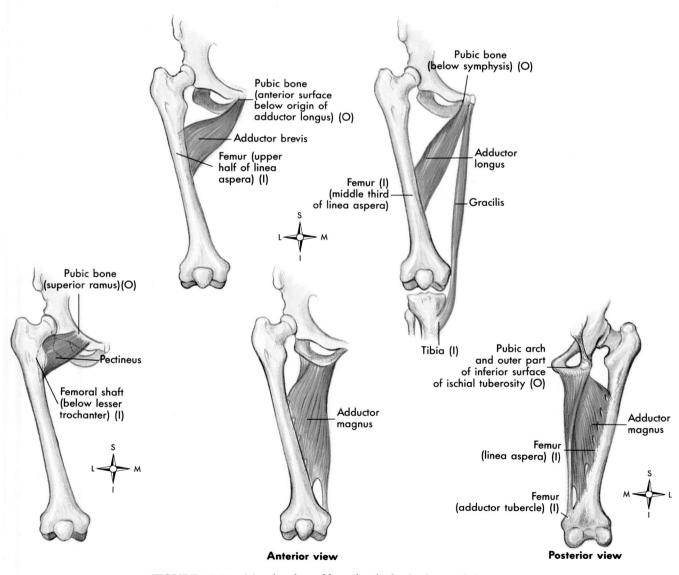

Anterior view

Posterior view

FIGURE 10-20 **Muscles that adduct the thigh.** O, Origin. I, Insertion.

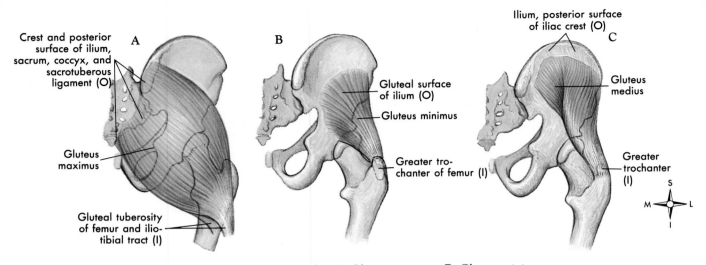

Crest and posterior surface of ilium, sacrum, coccyx, and sacrotuberous ligament (O)

Gluteus maximus

Gluteal tuberosity of femur and ilio-tibial tract (I)

A

B

Gluteal surface of ilium (O)

Gluteus minimus

Greater tro-chanter of femur (I)

Ilium, posterior surface of iliac crest (O)   C

Gluteus medius

Greater trochanter (I)

FIGURE 10-21   Gluteal muscles. A, Gluteus maximus. **B,** Gluteus minimus. **C,** Gluteus medius.

TABLE 10-12	Muscles that move the thigh			
Muscle	**Origin**	**Insertion**	**Function**	**Nerve Supply**
Iliopsoas (iliacus, psoas major, and psoas minor)	Ilium (iliac fossa)	Femur (lesser trochanter)	Flexes thigh	Femoral and second to fourth lumbar nerves
	Vertebrae (bodies of twelfth thoracic to fifth lumbar)		Flexes trunk (when femur acts as origin)	
Rectus femoris	Ilium (anterior, inferior spine)	Tibia (by way of patellar tendon)	Flexes thigh	Femoral nerve
			Extends lower leg	
Gluteal group				
Maximus	Ilium (crest and posterior surface)	Femur (gluteal tuberosity)	Extends thigh—rotates outward	Inferior gluteal nerve
	Sacrum and coccyx (posterior surface)	Iliotibial tract		
	Sacrotuberous ligament			
Medius	Ilium (lateral surface)	Femur (greater trochanter)	Abducts thigh—rotates outward; stabilizes pelvis on femur	Superior gluteal nerve
Minimus	Ilium (lateral surface)	Femur (greater trochanter)	Abducts thigh; stabilizes pelvis on femur	Superior gluteal nerve
			Rotates thigh medially	
Tensor fasciae latae	Ilium (anterior part of crest)	Tibia (by way of iliotibial tract)	Abducts thigh	Superior gluteal nerve
			Tightens iliotibial tract	
Adductor group				
Brevis	Pubic bone	Femur (linea aspera)	Adducts thigh	Obturator nerve
Longus	Pubic bone	Femur (linea aspera)	Adducts thigh	Obturator nerve
Magnus	Pubic bone	Femur (linea aspera)	Adducts thigh	Obturator nerve
Gracilis	Pubic bone (just below symphysis)	Tibia (medial surface behind sartorius)	Adducts thigh and flexes and adducts leg	Obturator nerve

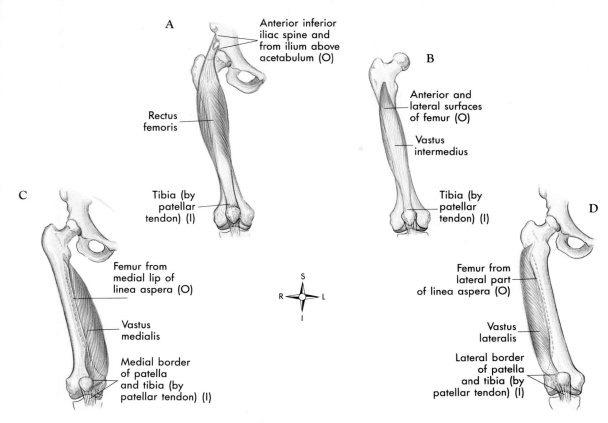

FIGURE 10-22 **Quadriceps femoris group of thigh muscles. A,** Rectus femoris. **B,** Vastus intermedius. **C,** Vastus medialis. **D,** Vastus lateralis. O, Origin. *I,* Insertion.

TABLE 10-13	Muscles that move the lower leg			
Muscle	**Origin**	**Insertion**	**Function**	**Nerve Supply**
Quadriceps femoris group				
Rectus femoris	Ilium (anterior inferior spine)	Tibia (by way of patellar tendon)	Flexes thigh Extends leg	Femoral nerve
Vastus lateralis	Femur (linea aspera)	Tibia (by way of patellar tendon)	Extends leg	Femoral nerve
Vastus medialis	Femur	Tibia (by way of patellar tendon)	Extends leg	Femoral nerve
Vastus intermedius	Femur (anterior surface)	Tibia (by way of patellar tendon)	Extends leg	Femoral nerve
Sartorius	Coxal (anterior, superior iliac spines)	Tibia (medial surface of upper end of shaft)	Adducts and flexes leg Permits crossing of legs tailor fashion	Femoral nerve
Hamstring group				
Biceps femoris	Ischium (tuberosity)	Fibula (head of)	Flexes leg	Hamstring nerve (branch of sciatic nerve)
	Femur (linea aspera)	Tibia (lateral condyle)	Extends thigh	Hamstring nerve
Semitendinosus	Ischium (tuberosity)	Tibia (proximal end, medial surface)	Extends thigh	Hamstring nerve
Semimembranosus	Ischium (tuberosity)	Tibia (medial condyle)	Extends thigh	Hamstring nerve

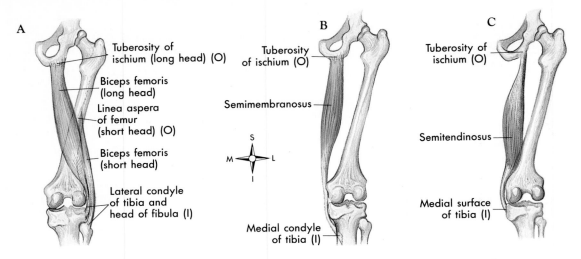

FIGURE 10-23   **Hamstring group of thigh muscles. A,** Biceps femoris. **B,** Semimembranosus. **C,** Semitendinosus. *O,* Origin. *I,* Insertion.

Table 10-13 identifies muscles that move the lower leg. Again, see Figures 10-4 and 10-5 and refer to Figures 10-22 and 10-23 as you study the table.

Muscles That Move the Ankle and Foot

Muscles listed in Table 10-14 and shown in Figure 10-24 are responsible for movements of the ankle and foot. These muscles, called **extrinsic foot muscles,** are located in the leg but exert their actions by pulling on tendons that insert on bones in the ankle and foot. Extrinsic foot muscles are responsible for such movements as dorsiflexion, plantar flexion, inversion, and eversion of the foot. Muscles located within the foot itself are called **intrinsic foot muscles.** They are responsible for flexion, extension, abduction, and adduction of the toes.

The leg muscles listed in Table 10-14 may be divided into four functional groups: (1) dorsal flexors, (2) plantar flexors, (3) invertors, and (4) evertors of the foot.

The superficial muscles located on the posterior surface of the leg form the bulging "calf." The common tendon of the **gastrocnemius** (gas-trok-NEE-me-us) and **soleus** is called the **calcaneal,** or *Achilles tendon.* It inserts into the calcaneus, or heel bone. By acting together, these muscles serve as powerful flexors (plantar flexion) of the foot.

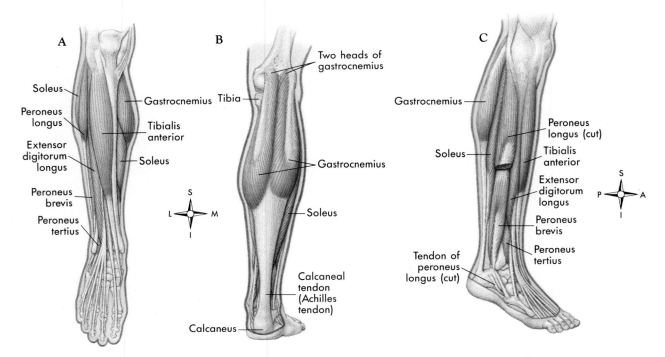

FIGURE 10-24   **Superficial muscles of the leg. A,** Anterior view. **B,** Posterior view. **C,** Lateral view.

TABLE 10-14   Muscles that move the foot

Muscle	Origin	Insertion	Function	Nerve Supply
Tibialis anterior	Tibia (lateral condyle of upper body)	Tarsal (first cuneiform)	Flexes foot	Common and deep peroneal nerves
		Metatarsal (base of first)	Inverts foot	
Gastrocnemius	Femur (condyles)	Tarsal (calcaneus by way of Achilles tendon)	Extends foot	Tibial nerve (branch of sciatic nerve)
			Flexes lower leg	
Soleus	Tibia (underneath gastrocnemius)	Tarsal (calcaneus by way of Achilles tendon)	Extends foot (plantar flexion)	Tibial nerve
	Fibula			
Peroneus longus	Tibia (lateral condyle)	First cuneiform	Extends foot (plantar flexion)	Common peroneal nerve
	Fibula (head and shaft)	Base of first metatarsal		
			Everts foot	
Peroneus brevis	Fibula (lower two thirds of lateral surface of shaft)	Fifth metatarsal (tubercle, dorsal surface)	Everts foot	Superficial peroneal nerve
			Flexes foot	
Peroneus tertius	Fibula (distal third)	Fourth and fifth metatarsals (bases of)	Flexes foot	Deep peroneal nerve
			Everts foot	
Extensor digitorum longus	Tibia (lateral condyle)	Second and third phalanges (four lateral toes)	Dorsiflexion of foot; extension of toes	Deep peroneal nerve
	Fibula (anterior surface)			

INTRAMUSCULAR INJECTIONS

Many drugs are administered by intramuscular injection. If the amount to be injected is 2 ml or less, the deltoid muscle is often selected as the site of injection. Note in Figure A that the needle is inserted into the muscle about two-fingers' breadth below the acromion process of the scapula and lateral to the tip of the acromion. If the amount of medication to be injected is 2 to 3 ml, the gluteal area shown in Figure B is often used. Injections are made into the gluteus medius muscle near the center of the upper outer quadrant, as shown in the illustration. Another technique of locating the proper injection site is to draw an imaginary diagonal line from a point of reference on the back of the bony pelvis (posterior superior iliac spine) to the greater trochanter of the femur. The injection is given about three-fingers' breadth above and one third of the way down the line. It is important to avoid the sciatic nerve and the superior gluteal blood vessels during the injection. Proper technique requires a knowledge of the underlying anatomy.

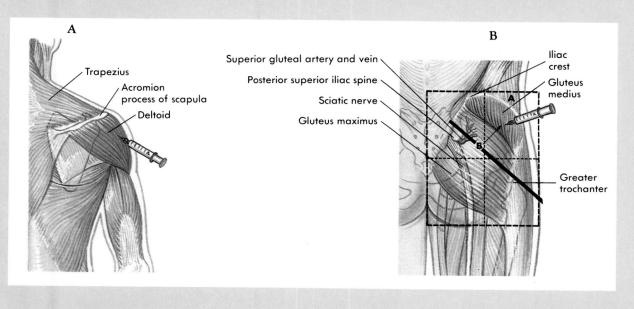

Dorsal flexors of the foot, located on the anterior surface of the leg, include the **tibialis** (tib-ee-AL-is) **anterior, peroneus tertius** (per-o-NEE-us TER-shus), and **extensor digitorum longus.** In addition to functioning as a dorsiflexor of the foot the extensor digitorum longus also everts the foot and extends the toes.

POSTURE

We have already discussed the major role muscles play in movement and heat production. We shall now turn our attention to a third way in which muscles serve the body as a whole—that of maintaining the posture of the body. Let us consider a few aspects of this important function.

The term *posture* means simply position or alignment of body parts. "Good posture" means many things. It means body alignment that most favors function; it means position that requires the least muscular work to maintain, which puts the least strain on muscles, ligaments, and bones; it means keeping the body's center of gravity over its base. Good posture in the standing position, for example, means head and chest held high, chin, abdomen, and buttocks pulled in, knees bent slightly, and feet placed firmly on the ground about 15 cm (6 inches) apart.

How Posture is Maintained

Since gravity pulls on the various parts of the body at all times, and since bones are too irregularly shaped to balance themselves on each other, the only way the body can be held upright is for muscles to exert a continual pull on bones in the opposite direction from gravity. Gravity tends to pull the head and trunk forward and downward; muscles (head and trunk extensors) must therefore pull backward and upward on them. For instance, gravity pulls the lower jaw downward; muscles must pull upward on it. Muscles exert this pull against gravity by virtue of their property of tonicity. Because tonicity is absent during sleep, muscle pull does not then counteract the pull of gravity. Hence, we cannot sleep standing up.

Many structures other than muscles and bones play a part in the maintenance of posture. The nervous system is responsible for the existence of muscle tone and also regulates and coordinates the amount of pull exerted by the individual muscles. The respiratory, digestive, circulatory, excretory, and endocrine systems all contribute something toward the ability of muscles to maintain posture. This is one of many examples of the important principle that all body functions are interdependent.

The importance of posture can perhaps be best evaluated by considering some of the effects of poor posture. Poor posture throws more work on muscles to counteract the pull of gravity and therefore leads to fatigue more quickly than good posture. Poor posture puts more strain on ligaments. It puts abnormal strains on bones and may eventually produce deformities. It also interferes with various functions such as respiration, heart action, and digestion.

 Cycle of Life: *Muscular system*

Acting together the muscular, skeletal, and nervous systems permit us to move in a coordinated and controlled way. However, it is the contraction, or shortening, of muscles that ultimately provides the actual movement necessary for physical activity. Dramatic changes occur in the muscular system throughout the cycle of life. Muscle cells may increase or decrease in number, size, and ability to shorten most effectively at different periods in life. In addition to age-related changes many pathologic conditions occurring at different ages may also affect the muscular system.

Because of the functional interdependence of the musculoskeletal and nervous systems, life cycle changes affecting the muscles are often manifested in other components of the functional unit. During infancy and childhood the ability to coordinate and control the strength of muscle contraction permits a sequential series of development steps to occur. A developing youngster learns to hold the head up, roll over, sit up, stand alone, and then walk and run as developmental changes permit better control and coordination of muscular contraction.

Degenerative changes associated with advancing age often result in replacement of muscle cells with non-functional connective tissue. As a result, the strength of muscular contraction diminishes. Pathological conditions associated with specific age ranges can also affect one or more components of the functional unit that permits us to move smoothly and effortlessly.

CHAPTER SUMMARY

INTRODUCTION

A. There are over 600 skeletal muscles in the body
B. Forty to fifty percent of our body weight is skeletal muscle
C. Muscles, along with the skeleton, determine the form and contour of our body

SKELETAL MUSCLE STRUCTURE

A. Connective tissue components
  1. Endomysium—delicate connective tissue membrane that covers specialized skeletal muscle fibers
  2. Perimysium—tough connective tissue binding together fascicles
  3. Epimysium—coarse sheath covering the muscle as a whole
  4. These three fibrous components may become a tendon, or an aponeurosis

B. Size, shape, and fiber arrangement (Figure 10-1)
  1. Skeletal muscles vary considerably in size, shape, and fiber arrangement
  2. Size—range from extremely small to large masses
  3. Shape—variety of shapes, such as broad, narrow, long, tapering, short, blunt, triangular, quadrilateral, irregular, flat sheets, or bulky masses
  4. Arrangement—variety of arrangements, such as parallel to long axis, converge to a narrow attachment, oblique, pennate, bipennate, or curved; the direction of fibers is significant due to its relationship to function

C. Attachment of muscles (Figure 10-2)
  1. Origin—point of attachment that does *not* move when the muscle contracts
  2. Insertion—point of attachment that moves when the muscle contracts

D. Muscle actions
  1. Most movements are produced by the coordinated action of several muscles; some muscles in the group contract while others relax
    a. Prime mover—a muscle or group of muscles that directly performs a specific movement
    b. Antagonist—muscles that, when contracting, directly oppose prime movers; relax while prime mover is contracting to produce movement; provide precision and control during contraction of prime movers
    c. Synergists—muscles that contract at the same time as the prime movers; they facilitate prime mover actions to produce a more efficient movement
    d. Fixator muscles—joint stabilizers

E. Lever systems
  1. In the human body, bones serve as levers and joints serve as fulcrums; contracting muscle applies a pulling force on a bone lever at the point of the muscle's attachment to the bone, causing the insertion bone to move about its joint fulcrum
  2. Lever system—composed of four component parts (Figure 10-3)
    a. Rigid bar (bone)
    b. Fulcrum (F) around which the rod moves (joint)
    c. Weight (W) that is moved
    d. Pull (P) that produces movement (muscle contraction)
  3. First-class levers
    a. Fulcrum lies between the effort and the resistance
    b. Not abundant in the human body; serve as levers of stability

  4. Second-class levers
    a. Weight lies between the fulcrum and the joint at which the pull is exerted
    b. Presence of these levers in the human body is a controversial issue
  5. Third-class levers
    a. Pull is exerted between the fulcrum and resistance
    b. Permit rapid and extensive movement
    c. Most common type of lever found in the body

HOW MUSCLES ARE NAMED

A. Muscles are named using one or more of the following features:
  1. Location
  2. Function
  3. Shape
  4. Direction of fibers—named according to fiber orientation
  5. Number of heads or divisions
  6. Points of attachment—origin and insertion points
B. Hints on how to deduce muscle actions

IMPORTANT SKELETAL MUSCLES

A. Muscles of facial expression—unique in that at least one point of attachment is to the deep layers of the skin over the face or neck (Table 10-3 and Figure 10-6)
B. Muscles of mastication—responsible for chewing movements (Table 10-3 and Figure 10-6)
C. Muscles that move the head—paired muscles on either side of the neck are responsible for head movements (Figure 10-7)

TRUNK MUSCLES

A. Muscles of the thorax—critical importance in respiration (Figure 10-8 and Table 10-5)
B. Muscles of the abdominal wall—arranged in three layers, with fibers in each layer running in different directions to increase strength (Figure 10-9 and Table 10-6)
C. Muscles of the pelvic floor—support the structures in the pelvic cavity (Figure 10-10 and Table 10-7)

UPPER LIMB MUSCLES

A. Muscles acting on the shoulder girdle—muscles that attach the upper extremity to the torso are located anteriorly (chest) or posteriorly (back and neck); these muscles also allow extensive movement (Figure 10-11 and Table 10-8)
B. Muscles that move the upper arm—the shoulder is a synovial joint allowing extensive movement in every plane of motion (Figure 10-12 and Table 10-9)
C. Muscles that move the forearm—found proximally to the elbow and attach to the ulna and radius (Figures 10-14 and 10-15 and Table 10-10)
D. Muscles that move the wrist, hand, and fingers—these muscles are located on the anterior or posterior surfaces of the forearm (Figure 10-16 and Table 10-11)

LOWER LIMB MUSCLES

A. The pelvic girdle and lower extremity function in locomotion and maintenance of stability
B. Muscles that move the thigh and lower leg (Figures 10-4, 10-5, 10-18 through 10-21, and Table 10-12)
C. Muscles that move the ankle and foot (Figure 10-24 and Table 10-14)

1. Extrinsic foot muscles are located in the leg and exert their actions by pulling on tendons that insert on bones in the ankle and foot; responsible for dorsiflexion, plantar flexion, inversion, and eversion
2. Intrinsic foot muscles are located within the foot; responsible for flexion, extension, abduction, and adduction of the toes

POSTURE

A. Maintaining the posture of the body is one of the major roles muscles play
B. "Good posture"—body alignment that most favors function, requires the least muscular work to maintain, keeping the body's center of gravity over its base
C. How posture is maintained
   1. Muscles exert a continual pull on bones in the opposite direction from gravity
   2. Structures other than muscle and bones play a role in maintaining posture
      a. Nervous system—responsible for the existence of muscle tone and also regulation and coordination of the amount of pull exerted by individual muscles
      b. Respiratory, digestive, excretory, and endocrine systems all contribute to maintain posture

CYCLE OF LIFE: MUSCULAR SYSTEM

A. Muscle cells—increase or decrease in number, size, and ability to shorten at different periods
B. Pathological conditions at different periods may affect the muscular system
C. Life cycle changes—manifested in other components of functional unit:
   1. Infancy and childhood— coordination and controlling of muscle contraction permits sequential development steps
D. Degenerative changes of advancing age result in replacement of muscle cells with nonfunctional connective tissue
   1. Diminished strength

REVIEW QUESTIONS

1. Define the terms *endomysium, perimysium,* and *epimysium.*
2. Differentiate between a tendon and an aponeurosis.
3. How do the origin and insertion of a muscle relate to each other in regard to actual movement?
4. Describe muscles that are classified as antagonists and synergists, including their relationship to the prime mover.
5. Identify and describe the most common type of lever system found in the body.
6. Give an example of a muscle named by: location, function, shape, fiber direction, number of heads, points of attachment.
7. Name the main muscles of the back, chest, abdomen, neck, shoulder, upper arm, lower arm, thigh, buttocks, leg, and pelvic floor.
8. Name the main muscles that flex, extend, abduct, and adduct the upper arm; that raise and lower the shoulder; that flex and extend the lower arm; that flex, extend, abduct, and adduct the thigh; that flex and extend the lower leg and thigh; that flex and extend the foot; that flex, extend, abduct, and adduct the head; that move the abdominal wall; that move the chest wall.
9. How is posture maintained?
10. Discuss the clinical significance regarding the difference in size between the large head of the humerus and the small and shallow glenoid cavity of the scapula.
11. Define carpal tunnel syndrome. What causes this condition?

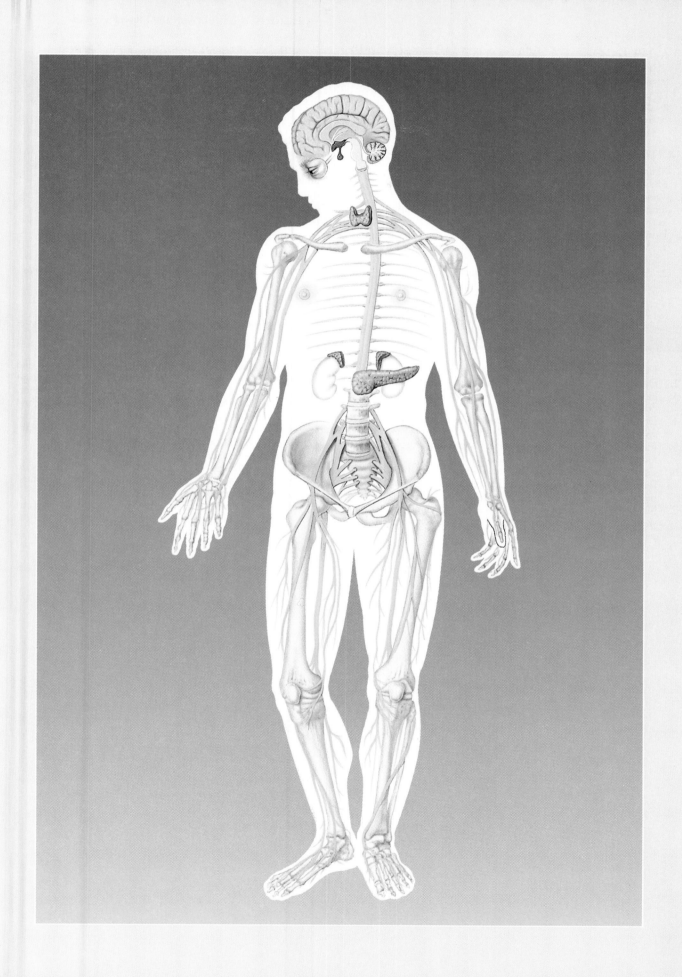

UNIT THREE

Communication, Control, and Integration

The anatomic structures and functional mechanisms that permit communication, control, and integration of bodily functions are discussed in the chapters of Unit 3. To maintain homeostasis, the body must have the ability to monitor and then respond appropriately to changes that may occur in either the internal or external environment. The nervous and endocrine systems provide this capability. Information originating in sensory nerve endings found in complex special sense organs such as the eye and in simple receptors located in skin or other body tissues provides the body with the necessary input. Nervous impulses traveling rapidly from the brain and spinal cord over nerves to muscles and glands initiate immediate coordinating and regulating responses. Slower acting chemical messengers, hormones produced by endocrine glands, serve to effect more long-term changes in physiological activities to maintain homeostasis.

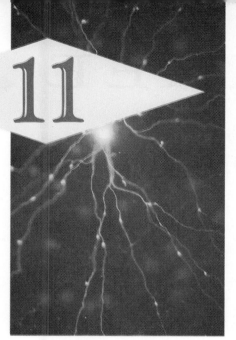

Neuron of the retina

11 Nervous System Cells

OBJECTIVES

**After you have completed this chapter, you should
be able to:**

1. Describe the generalized functions of the nervous
system.
2. Identify and describe the various subdivisions of
the nervous system.
3. Identify and describe the general structural and
functional characteristics of the two main types of
cells that compose the nervous system.
4. List and describe the structure and function of the
five types of neuroglia.
5. Identify the type of neuroglia cells found only in
the peripheral nervous system.
6. Classify neurons according to structural and
functional characteristics.
7. Discuss the structural and functional components
of a three-neuron ipsilateral reflex arc.
8. Differentiate between white and gray matter.
9. Describe the stages of the healing process after
injury to a peripheral motor neuron.
10. Identify the characteristics of resting membrane
potentials and local potentials.
11. Describe the sequence of events in an action
potential.
12. Compare and contrast continuous propagation of
an action potential with saltatory conduction.
13. List and describe the structural components of a
synapse.
14. Explain the mechanism of conduction of an
action potential across a synapse.
15. Compare spatial and temporal summation.
16. Identify and describe examples of a
neurotransmitter from each of the four main
chemical classes.

KEY TERMS

Action potential Membrane potential of an active neuron; nerve impulse

Axon (AK-son) Nerve cell process that transmits impulses away from the cell body

Central nervous system (CNS) Brain and spinal cord

Dendrite (DEN-drite) Nerve cell process that transmits impulses toward the cell body

Membrane potential The difference in electrical charge between the inside and outside of a plasma membrane

Myelin (MY-e-lin) White, fatty substance found in the myelin sheath around some nerve fibers

Nerve Bundle of peripheral nerve fibers

Neuroglia (noo-ROG-lee-ah) Supporting cells of nervous tissue

Neuron (NOO-ron) Nerve cell, including its axons and dendrites

Peripheral nervous system (PNS) Nerves connecting the brain and spinal cord to other parts of the body

Reflex arc Nerve pathway to and from the CNS that forms the basis for a reflex

Synapse (SIN-aps) Junction between adjacent neurons where nerve impulses are transmitted

The nervous system and the endocrine system together perform a vital function for the body—communication. Homeostasis and therefore survival depend on this function. Why? Because communication provides the means for controlling and integrating the many different functions performed by organs, tissues, and cells. Integrating means unifying. Unifying bodily functions means controlling them in ways that make them work together like parts of one machine to accomplish homeo-

stasis and thus survival. **Communication** makes possible control; control makes possible integration; integration makes possible homeostasis; homeostasis makes possible survival.

The **nervous system**—comprised of the *brain, spinal cord,* and *nerves*—is probably the most complex and least understood body system. Facts, theories, and questions about this system are as fascinating as they are abundant. We shall approach this somewhat daunting mass of information by considering in this chapter the cells of the nervous system and how they work together to accomplish their function. Then in Chapter 12 we shall discuss the brain and spinal cord. Chapter 13 presents the nerves of the body, and Chapter 14 continues the discussion by describing the structure and function of the sense organs.

◂ OVERVIEW OF THE NERVOUS SYSTEM

The nervous system is organized to detect changes in the internal and external environment, evaluate that information, and possibly respond by initiating changes in muscles or glands. To make this complex network of information lines and processing circuits easier to understand, biologists have subdivided the nervous system into the smaller "systems" listed here and illustrated in Figure 11-1.

I. Central nervous system (CNS)
II. Peripheral nervous system (PNS)
   A. Afferent (sensory) nervous system
   B. Efferent (motor) nervous system
      1. Somatic (motor) nervous system
      2. Autonomic nervous system (ANS)
         a. Sympathetic division
         b. Parasympathetic division

The **central nervous system (CNS)** is, as its name implies, the structural and functional center of the entire nervous system. Consisting of the brain and spinal cord, the CNS integrates incoming pieces of sensory information, evaluates the information, and initiates an outgoing response.

The **peripheral nervous system (PNS)** consists of the nerves that lie in the periphery, or "outer regions," of the nervous system. Nerves that originate from the brain are called *cranial nerves,* and nerves that originate from the spinal cord are called *spinal nerves.* The cranial and spinal nerves and all their branches consist of fibers that form incoming information pathways and outgoing pathways. For this reason, it is often convenient to say that the PNS is comprised of two main subdivisions. The **afferent nervous system** consists of all of the incoming *sensory* or *afferent* pathways. The **efferent nervous system** consists of all the outgoing *motor* or *efferent* pathways. The literal meanings of the terms *afferent* (carry toward) and *efferent* (carry away) may help you distinguish between these two divisions of the nervous system more easily.

The pathways of the efferent (motor) nervous system can be further subdivided according to the types of organs to which they travel. Pathways of the **somatic nervous**

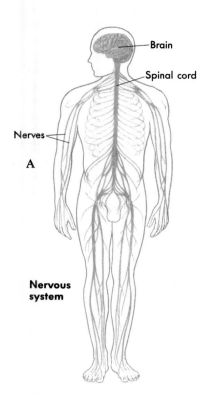

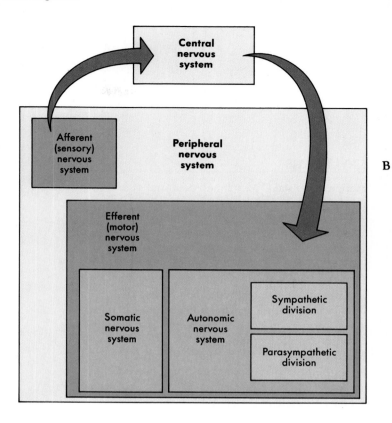

FIGURE 11-1 **The nervous system. A,** Major anatomical features of the human nervous system **B,** Diagram summarizes the organizational plan used by most biologists in studying the nervous system. The central nervous system processes information from the sensory portion of the peripheral nervous system and outputs information to the various motor divisions of the peripheral nervous system.

MULTIPLE SCLEROSIS (MS)

A number of diseases are associated with disorders of the oligodendrocytes. Because these neuroglial cells are involved in myelin formation, the diseases are called **myelin disorders.** The most common primary disease of the CNS is a myelin disorder called **multiple sclerosis,** or **MS.** It is characterized by myelin loss and destruction accompanied by varying degrees of oligodendrocyte injury and death. The result is demyelination throughout the white matter of the CNS. Hard plaquelike lesions replace the destroyed myelin, and affected areas are invaded by inflammatory cells. As the myelin surrounding nerve fibers is lost, nerve conduction is impaired and weakness, incoordination, visual impairment, and speech disturbances occur. Although the disease occurs in both sexes and all age groups, it is most common in women between 20 and 40 years of age.

The cause of multiple sclerosis is thought to be related to autoimmunity and to viral infections in some individuals. MS is characteristically relapsing and chronic in nature, but some cases of acute and unremitting disease have been reported. In most instances the disease is prolonged, with remissions and relapses occurring over a period of many years. There is no known cure.

system carry information to the *somatic effectors,* which are the skeletal muscles. Pathways of the **autonomic nervous system (ANS)** carry information to the *autonomic,* or *visceral, effectors,* which are the smooth muscles, cardiac muscle, and glands. As its name implies, the autonomic nervous system is autonomous of voluntary control—it seems to govern itself without our conscious knowledge. The pathways of the ANS can be divided into the **sympathetic division** and the **parasympathetic division.** The sympathetic division, comprised of pathways that exit the middle portions of the spinal cord, is involved in preparing the body to deal with immediate threats to the internal environment. It produces the "fight-or-flight" response. The parasympathetic pathways exit at the brain or lower portions of the spinal cord and coordinate the body's normal resting activities. The parasympathetic division is thus sometimes called the "rest-and-repair" division.

QUICK CHECK

1. *List the subdivisions of the human nervous system.*
2. *What two organs comprise the CNS?*
3. *Contrast the somatic nervous system with the autonomic nervous system.*

CELLS OF THE NERVOUS SYSTEM

Two main types of cells compose the nervous system, namely, *neurons* and *neuroglia*. **Neurons** are excitable cells that conduct the impulses that make possible all nervous system functions. In other words, they form the "wiring" of the nervous system's information circuits. **Neuroglia** or *neuroglial cells*, on the other hand, do not conduct information themselves but support the function of neurons in a variety of ways. Types of neuroglia and neurons are described in the following sections.

Neuroglia

The number of neuroglia in the human nervous system is beyond imagination. One estimate places the figure at a staggering 900 billion, or nine times the estimated number of stars in our galaxy! Unlike neurons, neuroglial cells retain their capacity for cell division throughout adulthood. Although this characteristic gives them the ability to replace themselves, it also makes them susceptible to abnormalities of cell division—such as cancer. Most benign and malignant tumors found in the nervous system originate in neuroglial cells.

As stated earlier, neuroglia serve various roles in supporting the function of neurons. To get a sense of this variety of function, we shall briefly examine five of the major types of neuroglia (Figure 11-2):

1. Astrocytes
2. Microglia
3. Ependymal cells
4. Oligodendrocytes
5. Schwann cells

The star-shaped neuroglia, **astrocytes** (Figure 11-2, A), derive their name from the Greek *astron*, "star." They are the largest and most numerous type of neuroglia. Webs of astrocytes form tight sheaths around the brain's blood capillaries. These sheaths and the tight junctions between the endothelial cells that form brain capillary walls constitute the so-called *blood-brain barrier (BBB)*. Small molecules (e.g., oxygen, carbon dioxide, water, alcohol) diffuse rapidly through the barrier to reach brain neurons. Larger molecules penetrate it slowly or not at all.

Microglia (Figure 11-2, B) are small, usually stationary cells. In inflamed or degenerating brain tissue, however, microglia enlarge, move about, and carry on phagocytosis. In other words, they engulf and destroy microbes and cellular debris.

Ependymal cells (Figure 11-2, C) are neuroglia that resemble epithelial cells, forming thin sheets that line fluid-filled cavities in the brain and spinal cord. Some ependymal cells take part in producing the fluid that fills these spaces. Other ependymal cells have cilia that help keep the fluid circulating within the cavities.

Oligodendrocytes (Figure 11-2, D) are smaller than astrocytes and have fewer processes. The name, oligodendrocytes, literally means "cell with few branches" (*oligo-*, few; *-dendro-*, branch; *-cyte*, cell). Some oligodendrocytes lie clustered around nerve cell bodies; some are arranged in rows between nerve fibers in the brain and cord. They help hold nerve fibers together and also serve another and probably more important function—they produce the fatty *myelin sheath* around nerve fibers in the CNS. Notice in Figure 11-2, D, how their processes wrap around surrounding nerve fibers to form this sheath.

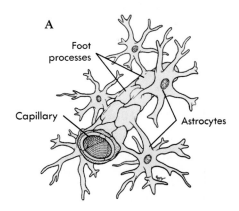

A

Foot processes

Capillary

Astrocytes

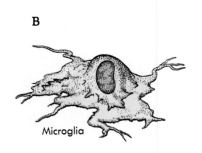

B

Microglia

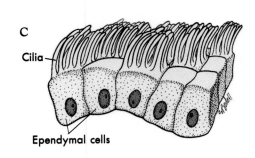

C

Cilia

Ependymal cells

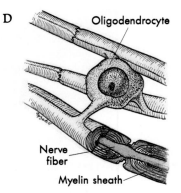

D

Oligodendrocyte

Nerve fiber

Myelin sheath

FIGURE 11-2   Types of neuroglia. A, Astrocytes attached to the outside of a capillary blood vessel in the brain. **B,** A phagocytic microglial cell. **C,** Ciliated ependymal cells forming a sheet that usually lines fluid cavities in the brain. **D,** An oligodendrocyte with processes that wrap around nerve fibers in the CNS to form myelin sheaths.

Continued

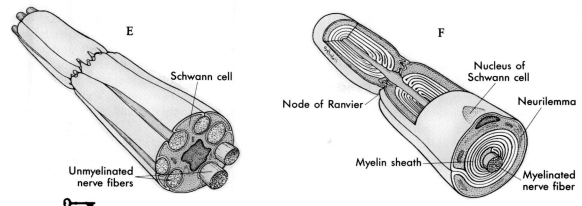

FIGURE 11-2, cont'd   Types of neuroglia. E, A Schwann cell supporting a bundle of nerve fibers in the PNS. **F,** Another type of Schwann cell wrapping around a peripheral nerve fiber to form a thick myelin sheath.

Unlike the other types of neuroglia, **Schwann cells** (Figure 11-2, *E* and *F*) are found only in the peripheral nervous system. Here they serve as the functional equivalent of the oligodendrocytes, supporting nerve fibers and forming a myelin sheath around them. As Figure 11-2, *F*, shows, many Schwann cells can wrap themselves around a single nerve fiber. The myelin sheath is formed by layers of Schwann cell membrane containing the white, fatty substance, **myelin.** Microscopic gaps in the sheath, between adjacent Schwann cells, are called **nodes of Ranvier.** The myelin sheath and its tiny gaps are important in the proper conduction of impulses along nerve fibers in the PNS. As each Schwann cell wraps around the nerve fiber, its nucleus and cytoplasm are squeezed to the perimeter to form the **neurilemma,** or sheath of Schwann. The neurilemma is essential to the regeneration of injured nerve fibers. Figure 11-2, *E,* shows that some Schwann cells do not wrap around nerve fibers to form a thick myelin sheath but simply hold fibers together in a bundle. Nerve fibers with many Schwann cells forming a thick myelin sheath are called *myelinated fibers,* or *white fibers.* When several nerve fibers are held by a single Schwann cell that does not wrap around them to form a thick myelin sheath, the fibers are called *unmyelinated fibers,* or *gray fibers.*

QUICK CHECK

✔ 1. *What are the five main types of neuroglia?*
✔ 2. *Describe the myelin sheath found on some nerve fibers.*
✔ 3. *What is a neurilemma?*

HEALTH MATTERS

THE BLOOD-BRAIN BARRIER

The blood-brain barrier (BBB) helps maintain the very stable environment required for normal functioning of the brain. The tight junctions between epithelial cells in the capillary wall, along with the covering formed by footlike extensions of the astrocytes, form a barrier that regulates the passage of most ions between the blood and the brain tissue. If they crossed to and from the brain freely, ions such as sodium (Na^+) and potassium (K^+) could disrupt the transmission of nerve impulses. Water, oxygen, carbon dioxide, and glucose can cross the barrier easily. Small, lipid-soluble molecules such as alcohol can also diffuse easily across the barrier.

The blood-brain barrier must be taken into consideration by researchers trying to develop new drug treatments for brain disorders. Many drugs and other chemicals simply will not pass through the barrier, although they might have therapeutic effects if they could get to the cells of the brain. For example, the abnormal control of muscle movements characteristic of *Parkinson's disease* can be alleviated by the substance dopamine, which is deficient in the brains of Parkinson victims. Since dopamine cannot cross the blood-brain barrier, dopamine injections or tablets are ineffective. Researchers found that the chemical used by brain cells to make dopamine, *levodopa (L-dopa),* can cross the barrier. Levodopa administered to patients with Parkinson's disease crosses the barrier and converts to dopamine, and the effects of the condition are reduced.

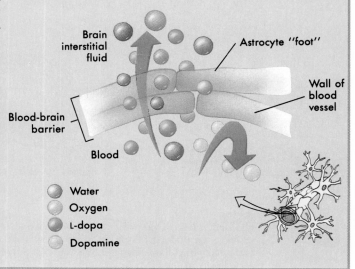

Neurons

The human brain is estimated to contain about 100 billion neurons, or about 10% of the total number of nervous system cells in the brain. All neurons consist of a cell body (also called the *soma* or *perikaryon*) and at least two processes: one *axon* and one or more *dendrites* (Figure 11-3). Because dendrites and axons are threadlike extensions from a neuron's cell body, they are often called *nerve fibers*.

In many respects the cell body, the largest part of a nerve cell, resembles other cells. It contains a nucleus, cytoplasm, and various organelles found in other cells, for example, mitochondria and a Golgi apparatus. A neuron's cytoplasm extends through its cell body and its processes. A plasma membrane encloses the entire neuron. Extending through the cytoplasm of each neuron are fine strands called **neurofibrils** (see Figure 11-4). Neurofibrils are bundles of thin microtubules and microfilaments formed by the neuron's cytoskeleton. Neurofibrils form a sort of "railway" for the rapid transport of molecules to and from the far ends of a neuron. Neurofibrils also separate the rough endoplasmic reticulum of the cell body into darkly staining structures called **Nissl bodies.** Nissl bodies provide protein molecules needed for transmission of nerve impulses from one neuron to another. They also provide proteins that are useful in maintaining and regenerating nerve fibers.

Dendrites branch extensively from the cell body—like tiny trees. In fact, their name derives from the Greek word for tree. The distal ends of dendrites of sensory neurons are called *receptors* because they receive the stimuli that initiate nerve impulses. Dendrites conduct impulses to the cell body of the neuron.

The **axon** of a neuron is a single process that extends from a tapered portion of the cell body called the *axon hillock.* Axons conduct impulses away from the cell body. Although a neuron has only one axon, that axon often has one or more side branches, called *axon collaterals.* Moreover, the distal tips of axons form branches called *telodendria* that each terminate in a *synaptic knob* (Figure 11-3). Each synaptic knob contains mitochondria and numerous vesicles.

Axons vary in both length and diameter. Some are a meter long. Some measure only a few millimeters. Axon diameters also vary considerably, from about 20 μm down to about 1 μm—a point of interest because axon diameter relates to velocity of impulse conduction. In general, the larger the diameter, the more rapid the conduction. A neuron's axon conducts impulses away from its cell body.

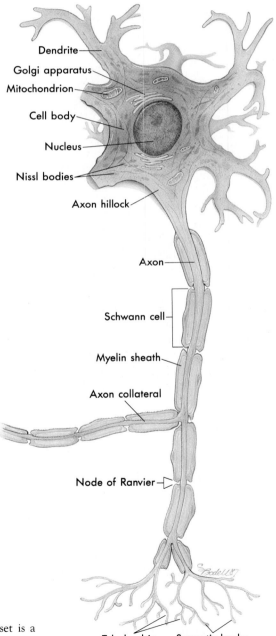

FIGURE 11-3 Structure of a typical neuron. The inset is a scanning electron micrograph of a neuron.

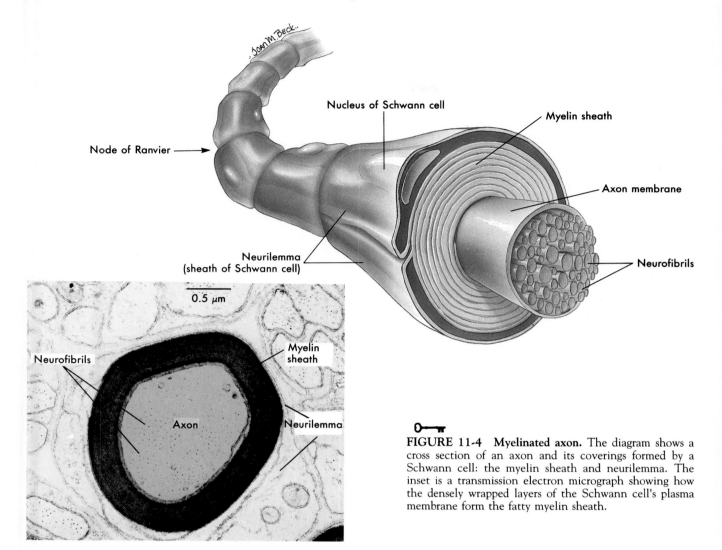

FIGURE 11-4 **Myelinated axon.** The diagram shows a cross section of an axon and its coverings formed by a Schwann cell: the myelin sheath and neurilemma. The inset is a transmission electron micrograph showing how the densely wrapped layers of the Schwann cell's plasma membrane form the fatty myelin sheath.

Whether an axon is myelinated or not also affects the speed of impulse conduction. Figure 11-4 shows a cross section of a typical myelinated axon. Notice how a series of Schwann cells have grown over the axon in a spiral fashion to form the myelin sheath and neurilemma. The role of the myelin sheath and nodes of Ranvier in impulse conduction will be discussed later.

Classification of Neurons
Structural classification

Classified according to the number of their processes, there are three types of neurons (Figure 11-5):

1. Multipolar
2. Bipolar
3. Unipolar

Multipolar neurons have only one axon but several dendrites. Most of the neurons in the brain and spinal cord are multipolar. *Bipolar neurons* have only one axon and also only one dendrite and are the least numerous kind of neuron. They are found in the retina of the eye, in the inner ear, and in the olfactory pathway. *Unipolar neurons* originate in the embryo as bipolar neurons, but in the course of development their two processes become fused into one for a short distance beyond the cell body. Then they separate into clearly distinguishable axon and dendrite. Sensory neurons are usually unipolar.

Functional classification

Classified according to the direction in which they conduct impulses, there are also three types of neurons:

1. Afferent neurons
2. Efferent neurons
3. Interneurons

Figure 11-6 shows one neuron of each of these types. *Afferent (sensory) neurons* transmit nerve impulses to the spinal cord or brain. *Efferent (motor) neurons* transmit nerve impulses away from the brain or spinal cord to or toward muscles or glands. *Interneurons* conduct impulses from afferent neurons toward or to motor neurons. Interneurons lie entirely within the central nervous system (brain and spinal cord).

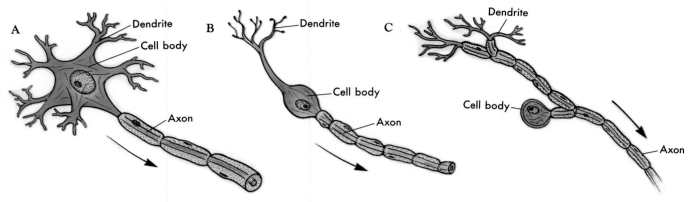

FIGURE 11-5 Structural classification of neurons. A, Multipolar neuron. **B,** Bipolar neuron. **C,** Unipolar neuron.

Reflex Arc

Notice in Figure 11-6 that neurons are often arranged in a semicircular pattern often called a *reflex arc*. Basically, a reflex arc is an impulse conduction route to and from the central nervous system (the brain and spinal cord). The most common form of reflex arc is the three-neuron arc (Figure 11-7, A). It consists of an afferent neuron, an interneuron, and an efferent neuron. Afferent or sensory neurons conduct impulses to the central nervous system from *sensory receptors* in the peripheral nervous system. Efferent neurons, or *motor neurons,* conduct impulses from the central nervous system to effectors. An effector is

muscle tissue or glandular tissue. Interneurons conduct impulses from afferent neurons toward or to motor neurons. In its simplest form, a reflex arc consists of an afferent neuron and an efferent neuron; this is called a *two-neuron arc.* In essence, a reflex arc is an impulse conduction route from receptors to the central nervous system and out to effectors. By now you should recognize that the reflex arc is simply an example of the information pathway described in Chapter 1 as a regulatory *feedback loop.* To confirm this point, compare the reflex arc (Figure 11-6) to the feedback loop illustrated in Figure 1-11, B, p. 17.

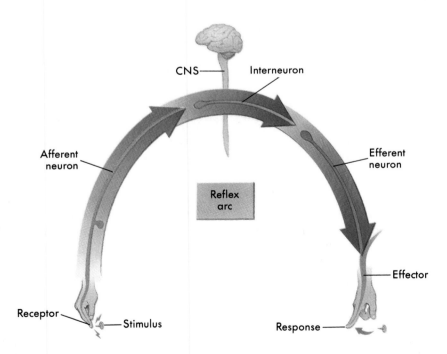

FIGURE 11-6 Functional classification of neurons. Neurons can be classified according to the direction in which they conduct impulses. Notice that the typical route of impulse conduction follows a pattern called the reflex arc.

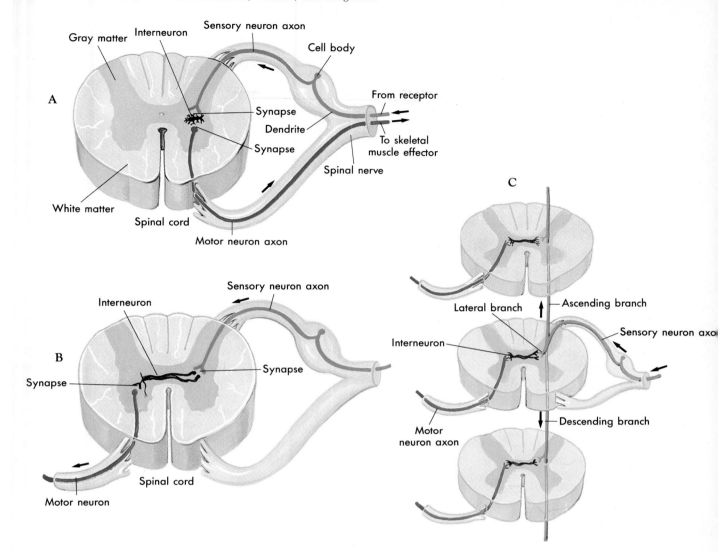

FIGURE 11-7 Examples of reflex arcs. A, Three-neuron ipsilateral reflex arc. Sensory information enters on the same side of the CNS as the motor information leaves the CNS. **B,** Three-neuron contralateral reflex arc. Sensory information enters on the opposite side of the CNS from the side that motor information exits the CNS. **C,** Intersegmental contralateral reflex arc. Divergent branches of a sensory neuron bring information to several segments of the CNS at the same time. Motor information leaves each segment on the opposite side of the CNS.

Now look again at Figure 11-7, A. Note the two labels for synapse. A **synapse** is the place where nerve impulses are transmitted from one neuron to another. Synapses are located between the synaptic knobs on one neuron and the dendrites or cell body of another neuron. For example, in Figure 11-7, A, the first synapse lies between the sensory neuron's synaptic knobs and the interneuron's dendrites. The second synapse lies between the interneuron's synaptic knobs and the motor neuron's dendrites. This reflex arc is called an *ipsilateral reflex arc* because the receptors and effectors are located on the same side of the body. Figure 11-7, B, shows a *contralateral reflex arc,* one whose receptors and effectors are located on opposite sides of the body.

Besides simple two-neuron and three-neuron arcs, intersegmental arcs (Figure 11-7, C)—even more complex multineuron, multisynaptic arcs—also exist. An important

principle is this: all impulses that start in receptors do not invariably travel over a complete reflex arc and terminate in effectors. Many impulses fail to be conducted across synapses. Moreover, all impulses that terminate in effectors do not invariably start in receptors. Many of them, for example, are thought to originate in the brain.

QUICK CHECK

✔ 1. *What is the difference between an axon and a dendrite?*
✔ 2. *What are the three structural categories of neurons?*
✔ 3. *What are the three main functional categories of neurons?*
✔ 4. *Describe the components of a reflex arc.*

NERVES AND TRACTS

Nerves are bundles of peripheral nerve fibers held together by several layers of connective tissues (Figure 11-8). Surrounding each nerve fiber is a delicate layer of fibrous connective tissue called the **endoneurium.** Bundles of fibers (each with its own endoneurium), called *fascicles,* are held together by a connective tissue layer called the **perineurium.** Numerous fascicles, along with the blood vessels that supply them, are held together to form a complete nerve by a fibrous coat called the **epineurium.** Within the central nervous system, bundles of nerve fibers are called *tracts* rather than nerves.

The creamy white color of myelin distinguishes bundles of myelinated fibers from surrounding unmyelinated tissues, which appear darker in comparison. Bundles of myelinated fibers make up the so-called *white matter* of the nervous system. In the PNS, white matter consists of myelinated nerves; in the CNS, white matter consists of myelinated tracts. Cell bodies and unmyelinated fibers make up the darker *gray matter* of the nervous system. Distinct regions of gray matter within the CNS are usually called *nuclei.* In peripheral nerves, similar regions of gray matter are more often called *ganglia.*

Most nerves in the human nervous system are *mixed nerves.* That is, they contain both sensory and motor neurons. Nerves that contain predominantly sensory neurons are often called *sensory nerves.* Likewise, nerves that contain mostly motor neurons are called *motor nerves.*

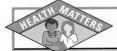

REDUCING DAMAGE TO NERVE FIBERS

Crushing and bruising cause most injuries to the spinal cord—often damaging nerve fibers irreparably. This usually results in *paralysis* or loss of function in the muscles normally supplied by the damaged fibers. Unfortunately, the inflammation of the injury site usually damages even more fibers and thus increases the extent of the paralysis. However, early treatment of the injury with the antiinflammatory drug *methylprednisolone* can reduce the inflammatory response in the damaged tissue and thus limit the severity of a spinal cord injury. Although early studies failed to confirm the effectiveness of standard doses of this steroid drug, later studies showed that very large doses administered within 8 hours of the injury reduced the extent of nerve cell damage dramatically. Since about 95% of the 10,000 Americans suffering spinal cord injuries each year are admitted for treatment well before the 8-hour limit, this drug may prove to be the first effective therapy for spinal cord injuries.

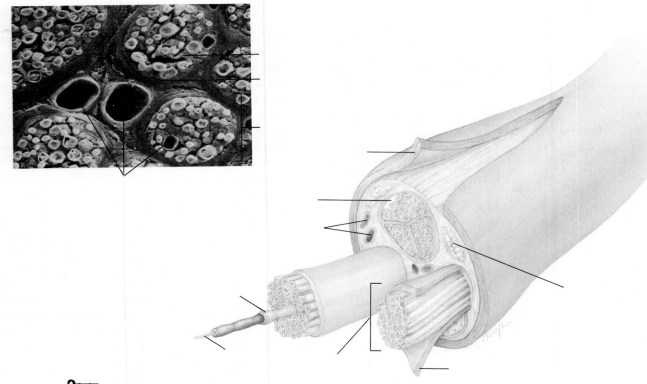

FIGURE 11-8 The nerve. Each nerve contains axons bundled into fascicles. A connective-tissue epineurium wraps the entire nerve. Perineurium surrounds each fascicle. Inset shows a scanning electron micrograph of a cross section of a nerve.

REPAIR OF NERVE FIBERS

Because mature neurons are incapable of cell division, damage to nervous tissue can be permanent. Since damaged neurons cannot be replaced, the only option for healing injured or diseased nervous tissue is by repairing the neurons that are already present. Unfortunately, neurons have a very limited capacity to repair themselves. Only if the damage is not extensive, when the cell body and neurilemma remain intact, and scarring has not occurred can nerve fibers be repaired. Figure 11-9 shows the stages of the healing process in the axon of a peripheral motor neuron. Immediately after the injury occurs, the distal portion of the axon degenerates as does its myelin sheath. Macrophages then move into the area and remove the debris. The remaining neurilemma and endoneurium forms a pathway or tunnel from the point of injury to the effector. New Schwann cells grow within this tunnel, maintaining a path for regrowth of the axon. Meanwhile, the cell body of the damaged neuron has reorganized its Nissl bodies to provide the proteins necessary to extend the remaining healthy portion of the axon. Several growing axon "sprouts" appear. When one of these growing fibers reaches the tunnel, it increases its growth rate—growing as much as 3 to 5 mm per day. If all goes well, the neuron's connection with the effector is quickly reestablished. Notice in Figure 11-9 that the skeletal muscle cell supplied by the damaged neuron atrophies during the absence of nervous input. Only after the nervous connection is reestablished and stimulation resumed does the muscle cell grow back to its original size.

In the central nervous system, similar repair of damaged nerve fibers is very unlikely. First of all, neurons in the CNS lack the neurilemma needed to form the guiding tunnel from the point of injury to the distal connection. Second, astrocytes quickly fill damaged areas and thus block regrowth of the axon with scar tissue. Despite great strides recently made by researchers looking for ways to stimulate the repair of neurons in the CNS, most injuries to the brain and spinal cord cause permanent damage.

QUICK CHECK

✔ 1. *What are the three layers of connective tissues that hold the fibers of a nerve together?*
✔ 2. *What is the difference between a nerve and a tract?*
✔ 3. *How does white matter differ from gray matter?*
✔ 4. *Under what circumstances can a nerve fiber be repaired?*

NERVE IMPULSES

Neurons are remarkable among cells because they initiate and conduct signals called "nerve impulses." Expressed differently, neurons exhibit both *excitability* and *conductivity*. What exactly *is* a nerve impulse? How is a neuron able to conduct this signal along its entire length—sometimes a full meter? These questions, and more, are answered in the paragraphs that follow.

Membrane Potentials

One way to describe a nerve impulse is as a wave of electrical fluctuation that travels along the plasma membrane. To understand this phenomenon more fully, however, requires some familiarity with the electrical nature of the plasma membrane. All living cells, including neurons, maintain a difference in the concentration of ions across their membranes. There is a slight excess of positive ions on the outside of the membrane and a slight excess of negative ions on the inside of the membrane. This, of course, results in a difference in electrical charge across their plasma membranes called the **membrane potential.** This difference in electrical charge is called a "potential" because it is a type of stored energy called potential energy. Whenever opposite electrical charges (in this case, opposite ions) are thus separated by a membrane, they have the potential to move toward one another if they are allowed to cross the membrane. When a membrane potential is maintained by a cell, opposite ions are held on opposite sides of the membrane like water behind a dam—ready to rush through with force when the proper membrane channels open.

A membrane that exhibits a membrane potential is said to be *polarized.* That is, its membrane has a negative pole (the side on which there is an excess of negative ions) and a positive pole (the side on which there is an excess of positive ions). The magnitude of potential difference between the two sides of a polarized membrane is measured in volts (V) or millivolts (mV). The sign of a membrane's voltage indicates the charge on the *inside* surface of a polarized membrane. For example, the value -70 mV indicates that the potential difference has a magnitude of 70 mV and that the inside of the membrane is negative

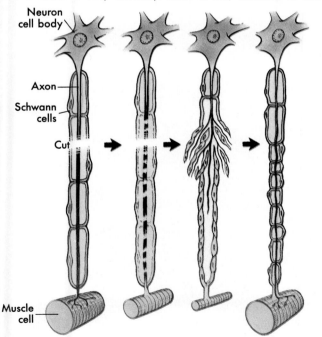

Neuron cell body

Axon

Schwann cells

Cut

Muscle cell

FIGURE 11-9 Repair of a peripheral nerve fiber. When cut, a damaged motor axon can regrow to its distal connection only if the neurilemma remains intact (to form a guiding tunnel) and if scar tissue does not block its way.

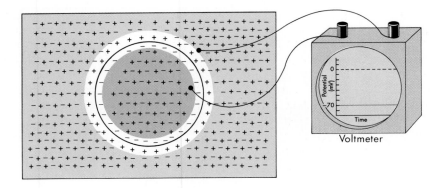

FIGURE 11-10 Membrane potential. The diagram on the left represents a cell maintaining a very slight difference in the concentration of oppositely charged ions across its plasma membrane. The voltmeter records the magnitude of electrical difference over time, which, in this case, does not fluctuate from −70 mV.

with respect to the outside surface (Figure 11-10). A value of +30 mV indicates a potential difference of 30 mV and that the inside of the membrane is positive (and thus the outside of the membrane is negative).

Resting Membrane Potentials

When a neuron is not conducting impulses, it is said to be "resting." At rest, a neuron's membrane potential is typically maintained at about −70 mV (Figure 11-10). The membrane potential maintained by a nonconducting neuron's plasma membrane is called the **resting membrane potential (RMP).**

The mechanisms that produce and maintain the RMP do so by promoting a slight ionic imbalance across the neuron's plasma membrane. Specifically, these mechanisms produce a slight excess of positive ions on its outer surface. This imbalance of ion concentrations is produced primarily by ion transport mechanisms in the neuron's plasma membrane.

The sodium-potassium pump is an active transport mechanism in the plasma membrane that transports sodium ions (Na^+) and potassium ions (K^+) in opposite directions and at different rates (Figure 11-11). It moves three sodium ions out of a neuron for every two potassium ions it moves into it. If, for instance, the pump transports 100 potassium ions into a neuron from the extracellular fluid, it concurrently transports 150 sodium ions out of the cell. The sodium-potassium pump thus creates an imbalance in the distribution of positive ions, resulting in a difference in electrical charge across the membrane. As this pump operates, the inside surface of the membrane becomes slightly less positive—that is, slightly negative—with respect to its outer surface.

There is more, however. Recall from Chapter 3 that the permeability characteristics of each cell's plasma membrane are determined in part by the presence of specific membrane transport channels. Many of these channels are *gated channels,* allowing specific molecules to diffuse across the membrane only when the "gate" of each channel is open (see Figure 3-14 on p. 78). In the neuron's plasma membrane, channels for the transport of the major anions

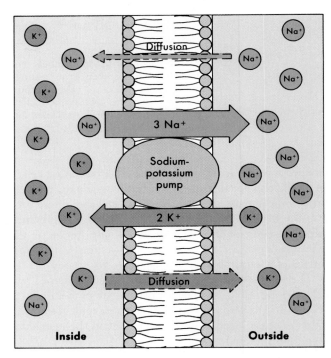

FIGURE 11-11 Sodium-potassium pump. This mechanism in the plasma membrane actively pumps sodium ions (Na^+) out of the neuron and potassium ions (K^+) into the neuron—at an unequal (3:2) rate. Because very little sodium reenters the cell via diffusion, this maintains an imbalance in the distribution of ions and thus maintains the resting potential.

(negative particles) are either nonexistent or closed. For example, there are no channels to allow the exit of the large anionic protein molecules that dominate the intracellular fluid. Chloride ions (Cl^-), the dominant extracellular anions, are likewise "trapped" on one side of the membrane because chloride channels are usually closed. This means that the only ions that can move efficiently across a neuron's membrane are the positive ions sodium and potassium. In a resting neuron, many of the potassium channels are open but most of the sodium channels are closed (Figure 11-12). This means that potassium ions

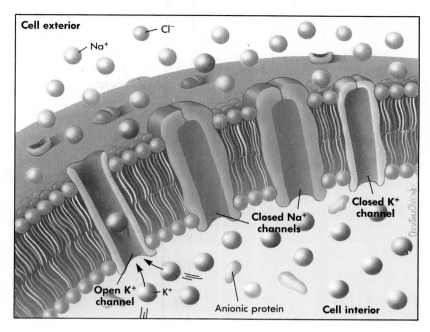

FIGURE 11-12   Role of ion channels in maintaining the resting membrane potential (RMP). Some K$^+$ channels are open in a "resting" membrane, allowing K$^+$ to diffuse down its concentration gradient (out of the cell) and thus add to the excess of positive ions on the outer surface of the plasma membrane. Diffusion of Na$^+$ in the opposite direction would counteract this effect but is prevented from doing so by closed Na$^+$ channels. Compare this figure to Figure 11-11.

pumped into the neuron can diffuse back out of the cell in an attempt to equalize its concentration gradient, but very little of the sodium pumped out of the cell can diffuse back into the neuron (see Figure 11-11). Thus the membrane's selective permeability characteristics also promote the maintenance of a slight excess of positive ions on the outer surface of the membrane.

The resting membrane potential can be maintained by a cell as long as its sodium-potassium pumps continue to operate and its permeability characteristics remain stable. If either of these mechanisms is altered, then the membrane potential is altered as well.

Local Potentials

In neurons, membrane potentials can fluctuate above or below the resting membrane potential in response to certain stimuli (Figure 11-13). A slight shift away from the RMP in a specific region of the plasma membrane is often called a **local potential.**

Excitation of a neuron occurs when a stimulus triggers the opening of additional Na$^+$ channels, which permits more Na$^+$ to enter the cell. As the excess of positive ions outside the plasma membrane decreases, the magnitude of the membrane potential is reduced. Such movement of the membrane potential toward zero is called **depolarization.** In *inhibition*, a stimulus triggers the opening of additional K$^+$ channels. As more K$^+$ diffuses out of the cell, the excess of positive ions outside the plasma membranes increases—increasing the magnitude of the membrane potential. Movement of the membrane potential away from zero (thus below the usual RMP) is called **hyperpolarization.**

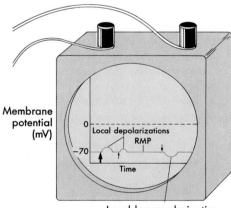

FIGURE 11-13   Local potentials. Excitatory stimuli (↑) cause depolarizations (movements toward 0 mV) in proportion to the strength of the stimuli. Inhibitory stimuli (↓) cause hyperpolarizations, local deviations away from 0 mV that cause the membrane potential to dip below the level of the RMP (resting membrane potential).

Local potentials are called *graded potentials* because the magnitude of deviation from the RMP is proportional to the magnitude of the stimulus. In short, local potentials can be large or small—they are not all-or-none events. Local potentials are called "local" because they are more or less isolated to a particular region of the plasma membrane. That is, local potentials do not spread all the way to the end of a neuron's axon.

QUICK CHECK

✔ 1. *What mechanisms are involved in producing the resting membrane potential (RMP)?*

✔ 2. *In a resting neuron, what positive ion is most abundant outside the plasma membrane? What positive ion is most abundant inside the plasma membrane?*

✔ 3. *How is depolarization of a membrane different from hyperpolarization?*

ACTION POTENTIAL

An **action potential** is, as the term itself suggests, the membrane potential of an active neuron, that is, one that is conducting an impulse. A synonym commonly used for action potential is *nerve impulse.* As mentioned earlier, the action potential, or nerve impulse, is an electrical fluctuation that travels along the surface of a neuron's plasma membrane. A step-by-step description of the mechanism that produces the action potential is given in the following paragraphs and in Table 11-1. Refer to Figures 11-14 and 11-15 as you read each step.

1. When an adequate stimulus is applied to a neuron, some of the Na^+ channels at the point of stimulation open. Na^+ diffuses rapidly into the cell because of the concentration gradient and electrical gradient, producing a local depolarization (Figure 11-14).

2. If the magnitude of the local depolarization surpasses a limit called the **threshold potential** (typically −59 mV), even more Na^+ channels are stimulated to open. As more Na^+ rushes into the cell, the membrane moves rapidly toward 0 mV, then continues in a positive direction to a peak of +30 mV (Figure 11-15). The positive value at the peak of the action potential indicates that there is an excess of positive ions *inside* the membrane. If the local depolarization fails to cross the threshold of −59 mV, the additional voltage-sensitive Na^+ channels do not open, and the membrane simply recovers back to the resting potential of −70 mV without producing an action potential.

3. Voltage-sensitive Na^+ channels stay open for only about 1 millisecond (ms) before they automatically close. This means that once they are stimulated, the Na^+ channels always allow sodium to rush in for the same amount of time, which in turn produces the same magnitude of action potential. In other words, the action potential is an *all-or-none* response. If the threshold potential is surpassed, the full peak of the action potential is always reached; if the threshold potential is not surpassed, no action potential will occur at all.

4. Once the peak of the action potential is reached, the membrane potential begins to move back toward the resting potential (−70 mV) in a process called **repolarization.** Surpassing the threshold potential not only triggers the opening of voltage-sensitive Na^+ channels but also voltage-sensitive K^+ channels. The voltage-sensitive K^+ channels are slow to respond, however, and thus do not begin opening until the inward diffusion of Na^+ ions has caused the membrane potential to reach + 30 mV. Once the K^+ channels open, K^+ rapidly diffuses out of the cell because of the concentration gradient and because it is repulsed by the now-positive interior of the cell. The outward rush of K^+ restores the original excess of positive ions on the outside surface of the membrane—thus repolarizing the membrane (Figure 11-14).

5. Because the K^+ channels often remain open as the membrane reaches its resting potential, too many K^+ may rush out of the cell. This causes a brief period of hyperpolarization before the resting potential is restored by the action of the sodium-potassium pump and the return of ion channels to their resting state.

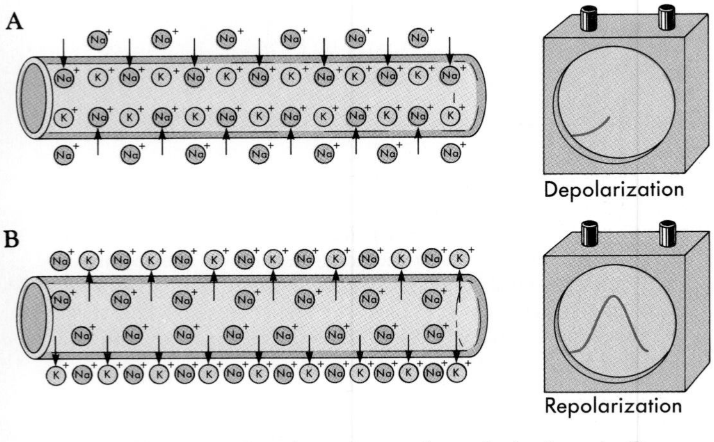

FIGURE 11-14 Depolarization and repolarization. A, Depolarization of a membrane occurs when Na^+ channels open, allowing Na^+ to move to an area of lower concentration (and more negative charge) *inside* the cell—reversing the polarity to an inside-positive state. **B,** Repolarization of a membrane occurs when K^+ channels then open, allowing K^+ to move to an area of lower concentration (and more negative charge) *outside* the cell—reversing the polarity back to an inside-negative state.

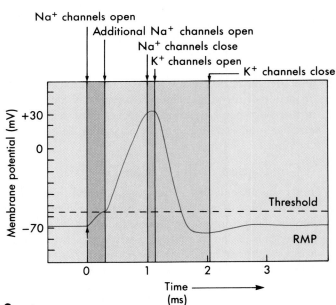

FIGURE 11-15 The action potential. Changes in membrane potential in a local area of a neuron's membrane result from changes in membrane permeability.

TABLE 11-1	Steps of the mechanism that produces an action potential
Step	**Description**
1	A stimulus triggers Na$^+$ channels to open and allow inward Na$^+$ diffusion. This causes the membrane to depolarize.
2	As the threshold potential is reached, additional Na$^+$ channels open and even more Na$^+$ enters the cell—causing the membrane to depolarize further.
3	The magnitude of the action potential peaks (at +30 mV) when Na$^+$ channels close.
4	Repolarization begins when K$^+$ channels open, allowing outward diffusion of K$^+$.
5	After a brief period of hyperpolarization, the resting potential is restored by the sodium-potassium pump and the return of ion channels to their resting state.

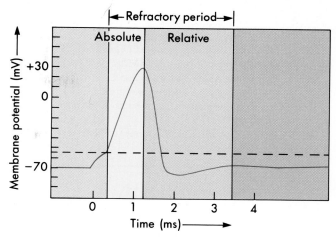

FIGURE 11-16 **Refractory period.** During the absolute refractory period, the membrane will not respond to any stimulus. During the relative refractory period, however, a very strong stimulus may elicit a response in the membrane.

Refractory Period

The refractory period is a brief period during which a local area of a neuron's membrane resists restimulation (Figure 11-16). For about half a millisecond after the membrane surpasses the threshold potential, it will not respond to any stimulus, no matter how strong. This is called the **absolute refractory period.** The **relative refractory period** is the few milliseconds after the absolute refractory period—the time during which the membrane is repolarizing and restoring the resting membrane potential. During the relative refractory period the membrane will respond only to very strong stimuli.

Because only very strong stimuli can produce an action potential during the relative refractory period, a series of closely spaced action potentials can occur only when the magnitude of the stimulus is great. The greater the magnitude of the stimulus, the earlier a new action potential can be produced, and thus the greater the frequency of action potentials. This means that although the magnitude of the stimulus does not affect the magnitude of the action potential, which is an all-or-none response, it does cause a proportional increase in the frequency of impulses. Thus the nervous system uses the frequency of nerve impulses to code for the strength of a stimulus—not changes in the magnitude of the action potential.

Conduction of the Action Potential

At the peak of the action potential, the inside of the neuron's plasma membrane is positive relative to the outside. That is, its polarity is now the *reverse* of that of the resting membrane potential. Such reversal in polarity causes electrical current to flow between the site of the action potential and the adjacent regions of membrane. This local current flow triggers voltage-sensitive Na$^+$ channels in the next segment of membrane to open. As Na$^+$ rushes inward, this next segment exhibits an action potential. The action potential thus has moved from one

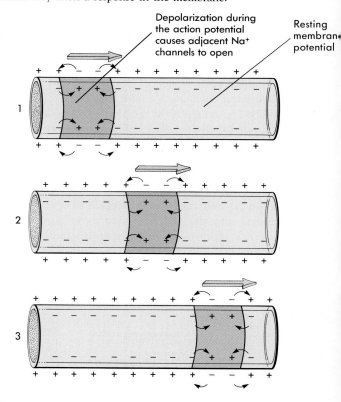

FIGURE 11-17 **Conduction of the action potential.** The reverse polarity characteristic of the peak of the action potential causes local current flow to adjacent regions of the membrane (*small arrows*). This stimulates voltage-sensitive Na$^+$ channels to open and thus create a new action potential. This cycle continues, producing wavelike conduction of the action potential from point to point along a nerve fiber.

point to the next along the neuron's membrane (Figure 11-17). This cycle goes on repeating itself because each action potential always causes enough local current flow to surpass the threshold potential for the next region of membrane. And because each action potential is an

all-or-none phenomenon, the fluctuation in membrane potential moves along the membrane without any decrement, or decrease, in magnitude.

The action potential never moves backward, restimulating the region from which it just came. It is prevented from doing so because the previous segment of membrane remains in a refractory period too long to allow such restimulation. This is the mechanism responsible for the one-way movement of action potentials along axons and dendrites.

In myelinated fibers, the insulating properties of the thick myelin sheath resists ion movement and the resulting local flow of current. Electrical changes in the membrane can only occur at gaps in the myelin sheath, that is, at the nodes of Ranvier. Figure 11-18 shows that when an action potential occurs at one node, current flows *across* the insulating myelin sheath to the next node. This stimulates an action potential at that node, which in turn stimulates the next node. Thus the action potential seems to "leap" from node to node along the myelinated fiber. This type of impulse conduction is called *saltatory conduction* (from the Latin verb *saltare*, "to leap").

How fast does a nerve fiber conduct impulses? It depends on its diameter and on the presence or absence of a myelin sheath. The speed of conduction of a nerve fiber is proportional to its diameter: the larger the diameter, the faster it conducts impulses. Myelinated fibers conduct impulses more rapidly than unmyelinated fibers. This is because saltatory conduction is more rapid than point-to-point conduction. The fastest fibers, such as those that innervate the skeletal muscles, can conduct impulses up to about 130 meters per second (close to 300 miles per hour). The slowest fibers, such as those from sensory receptors in the skin, may conduct impulses at only about 0.5 meter per second (little more than 1 mile per hour).

ANESTHETICS

HEALTH MATTERS

Anesthetics are substances that are administered to reduce or eliminate the sensation of pain—thus producing a state called *anesthesia*. Many anesthetics produce their effects by inhibiting the opening of sodium channels and thus blocking the initiation and conduction of nerve impulses. Anesthetics such as bupivacaine (Marcaine) are often used in dentistry to minimize pain involved in tooth extractions and other dental procedures. Procaine has likewise been used to block the transmission of impulses in sensory pathways of the spinal cord. Benzocaine and phenol, local anesthetics found in several over-the-counter products that relieve pain associated with teething in infants, sore throat pain, and other ailments, also produce their effect by blocking initiation and conduction of nerve impulses.

QUICK CHECK

✔ 1. List the events that lead to the initiation of an action potential.
✔ 2. What is meant by the term threshold potential?
✔ 3. How does impulse conduction in an unmyelinated fiber differ from impulse conduction in a myelinated fiber?

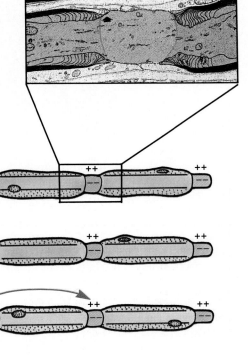

FIGURE 11-18 Saltatory conduction. This series of diagrams show that the insulating nature of the myelin sheath prevents ion movement everywhere but at the nodes of Ranvier. The action potential at one node triggers current flow (*arrows*) *across* the myelin sheath to the next node—producing an action potential there. The action potential thus seems to "leap" rapidly from node to node. The inset is a transmission electron micrograph showing a node of Ranvier in a myelinated fiber. (*Inset, Courtesy Erlandsen/Magney: Color Atlas of Histology.*)

SYNAPTIC TRANSMISSION

Structure of the Synapse

A **synapse** is the place where impulses are transmitted from one neuron, called the *presynaptic neuron,* to another neuron, called the *postsynaptic neuron.* Three structures make up a synapse:

1. A synaptic knob
2. A synaptic cleft
3. The plasma membrane of a postsynaptic neuron

A *synaptic knob* is a tiny bulge at the end of a terminal branch of a presynaptic neuron's axon (Figure 11-19). Each synaptic knob contains numerous small sacs or vesicles. Each vesicle contains about 10,000 molecules of a chemical compound called a **neurotransmitter.** A *synaptic cleft* is the space between a synaptic knob and the plasma membrane of a postsynaptic neuron. It is an incredibly narrow space—only 20 to 30 nanometers, or about one millionth of an inch in width. Identify the synaptic cleft in Figure 11-19. The plasma membrane of a *postsynaptic neuron* has protein molecules embedded in it opposite each synaptic knob. These serve as receptors to which neurotransmitter molecules bind.

Mechanism of Synaptic Transmission

An action potential that has traveled the length of a neuron stops at its axon terminals. Action potentials cannot cross synaptic clefts, miniscule barriers though they are. Instead, neurotransmitters are released from the synaptic knob, cross the synaptic cleft, and bring about a response by the postsynaptic neuron. *Excitatory* neurotransmitters cause depolarization of the postsynaptic membrane, whereas *inhibitory* neurotransmitters cause hyperpolarization of the postsynaptic membrane (see Figure 11-13).

The mechanism of synaptic transmission, summarized in Figure 11-20, consists of the following sequence of events:

1. When an action potential reaches a synaptic knob, voltage-sensitive calcium channels in its membrane open and allow calcium ions (Ca^{++}) to diffuse into the knob rapidly.

2. The increase in intracellular Ca^{++} concentration triggers the movement of neurotransmitter vesicles to the plasma membrane of the synaptic knob. Once there, they

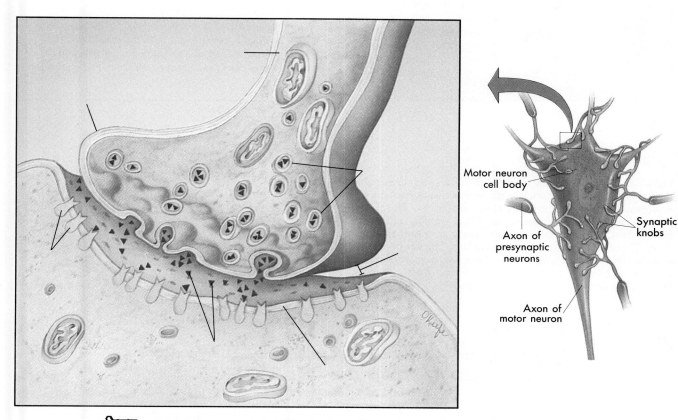

FIGURE 11-19 Structure of a synapse. Diagram shows synaptic knob, or axon terminal, of presynaptic neuron, the plasma membrane of a postsynaptic neuron, and a synaptic cleft. On the arrival of an action potential at a synaptic knob, neurotransmitter molecules are released from vesicles in the knob into the synaptic cleft. The combining of neurotransmitter and receptor molecules in the plasma membrane of the postsynaptic neuron initiates impulse conduction in the postsynaptic neuron.

Motor neuron cell body

Axon of presynaptic neurons

Axon of motor neuron

Synaptic knobs

fuse with the membrane and release their neurotransmitter via exocytosis. Thousands of neurotransmitter molecules spurt out of the open vesicles into the synaptic cleft.

3. The released neurotransmitter molecules almost instantaneously diffuse across the narrow synaptic cleft and contact the postsynaptic neuron's plasma membrane. Here, neurotransmitters bind to receptor molecules that are also gated channels or that are coupled to gated channels. Binding of neurotransmitters triggers the channels to open.

4. The opening of ion channels in the postsynaptic membrane may produce a local potential called a **postsynaptic potential.** Excitatory neurotransmitters cause both Na$^+$ channels and K$^+$ channels to open. Because Na$^+$ rushes inward faster than K$^+$ rushes outward, there is a temporary depolarization called an **excitatory postsynaptic potential (EPSP).** If the EPSP reaches the threshold potential, an action potential in the postsynaptic mem-

brane begins. Inhibitory neurotransmitters cause K$^+$ channels and/or Cl$^-$ channels to open. If K$^+$ channels open, K$^+$ rushes outward; if Cl$^-$ channels open, Cl$^-$ rushes inward. Either event makes the inside of the membrane even more negative than at the resting potential. This temporary hyperpolarization is called an **inhibitory postsynaptic potential (IPSP).**

5. Once a neurotransmitter binds to its postsynaptic receptors, its action is quickly terminated. Either one or both of two mechanisms bring this about. Some neurotransmitter molecules are transported back into the synaptic knobs, where they can be repacked into vesicles and used again. Other neurotransmitter molecules are metabolized into inactive compounds by synaptic enzymes.

The mechanism of synaptic transmission is summarized in Figure 11-20.

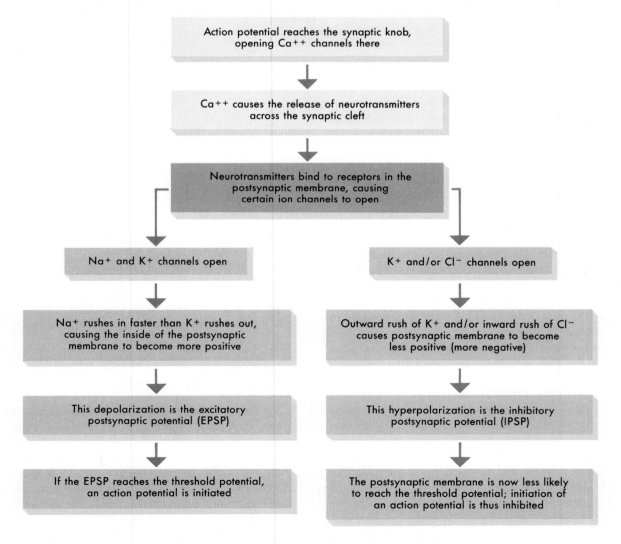

FIGURE 11-20 Summary of synaptic transmission.

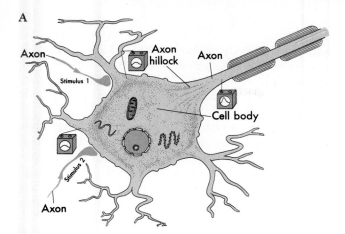

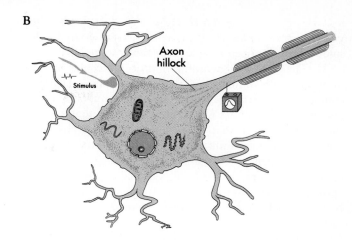

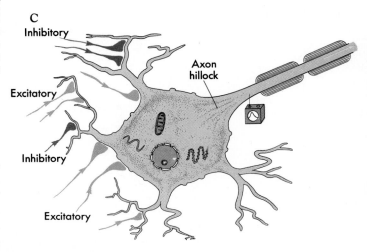

FIGURE 11-21 Summation. A, Spatial summation is the effect produced by simultaneous stimulation by a number of synaptic knobs on the same postsynaptic neuron. **B,** Temporal summation is the effect produced by a rapid succession of stimuli on a single postsynaptic neuron. **C,** Generation of an action potential at the axon hillock depends on the summation of effects produced by all the inhibitory and excitatory stimuli input to the postsynaptic membrane.

Summation

At least several, usually thousands, and in some cases more than 100,000, knobs synapse with a single postsynaptic neuron. The amount of excitatory neurotransmitter released by one knob is not enough to trigger an action potential. It may, however, *facilitate* initiation of an action potential by producing a local depolarization of the synaptic membrane—an excitatory postsynaptic potential (EPSP). When a number of knobs are activated simultaneously, neurotransmitters stimulate different locations on the postsynaptic membrane. The effects of the stimuli may add together, or *summate,* to produce an action potential (Figure 11-21, A). This phenomenon is called **spatial summation.** Likewise, when synaptic knobs stimulate a postsynaptic neuron in rapid succession, their effects can add up over a brief period of time to produce an action potential (Figure 11-21, B). This type of summation is called **temporal summation.**

Usually both excitatory and inhibitory transmitters are released at the same synapse. The excitatory neurotransmitters produce local EPSPs and the inhibitory neurotransmitters produce local IPSPs. Summation of these opposing local potentials occurs at the axon hillock, where many voltage-sensitive ion channels are located. If EPSPs predominate over the IPSPs enough to depolarize the membrane to the threshold potential, the voltage-sensitive channels will respond and produce an action potential (Figure 11-21, C). The action potential is then conducted without decrement along the axon's membrane. On the other hand, if IPSPs predominate over EPSPs, the membrane will not reach the threshold potential. The voltage-sensitive channels at the axon hillock will not respond, and no action potential will be conducted along the axon.

QUICK CHECK

✔ 1. *What are the three structural components of a synapse?*
✔ 2. *List the steps of synaptic transmission.*
✔ 3. *What is an EPSP? What is an IPSP?*
✔ 4. *How does temporal summation differ from spatial summation?*

◣ NEUROTRANSMITTERS

Neurotransmitters are the means by which neurons talk to one another. At billions, or more likely, trillions, of synapses throughout the body, presynaptic neurons release neurotransmitters that act to facilitate, stimulate, or inhibit postsynaptic neurons and effector cells. More than thirty compounds are known to be neurotransmitters. Dozens of other compounds are suspected of being neurotransmitters. For the most part, they are not distributed diffusely or at random throughout the nervous system. Instead, specific neurotransmitters are localized in discrete groups of neurons and thus released in specific nerve pathways.

Nervous pathways conduct information along complex series of neurons joined together by synapses. One thing that makes these pathways complex is that they often *converge* and *diverge*. **Convergence** occurs when more than one presynaptic axon synapses with a single postsynaptic neuron (A). Convergence allows information from several different pathways to be funneled into a single pathway. For example, the pathways that innervate the skeletal muscles may originate in several different areas of the CNS. Because pathways from each of these motor control areas converge on a single motor neuron, each area has an opportunity to control the same muscle. **Divergence** occurs when a single presynaptic axon synapses with many different postsynaptic neurons (B). Divergence allows information from one pathway to be "split" or "copied" and sent to different destinations in the nervous system. For example, a single bit of visual information may be sent to many different areas of the brain for processing.

Classification of Neurotransmitters

Neurotransmitters are commonly classified by their function or by their chemical structure, depending on the context in which they are discussed. You are already familiar with two major functional classifications: *excitatory neurotransmitters* and *inhibitory neurotransmitters*. Some neurotransmitters can have inhibitory effects at some synapses and excitatory effects at other synapses. For example, the neurotransmitter *acetylcholine*, discussed in Chapter 9, excites skeletal muscle cells but inhibits cardiac muscle cells. This illustrates an important point about the action of neurotransmitters: their function is determined by the postsynaptic receptors, not by the neurotransmitters themselves.

Another way to classify neurotransmitters by function is to identify the mechanism by which they cause a change in the postsynaptic neuron or effector cell. Some neurotransmitters trigger the opening or closing of ion channels directly, whereas others produce their effects by activating chemical messengers within the postsynaptic cell. An example of this "second messenger" mechanism occurs when the neurotransmitter *norepinephrine* binds to its receptors and activates the membrane-bound enzyme adenyl cyclase. The adenyl cyclase removes phosphate groups from ATP (adenosine triphosphate) to form cyclic AMP (adenosine monophosphate). Cyclic AMP is the "second messenger," triggering a series of chemical reactions that eventually cause Na^+ channels in the postsynaptic membrane to open (Figure 11-22, p. 300).

Because the functions of specific neurotransmitters vary by location it is often more useful to classify them according to their chemical structure. Neurotransmitters can thus be grouped into four main chemical classes:

1. Acetylcholine
2. Amines
3. Amino acids
4. Neuropeptides

Examples of neurotransmitters in each of these classes are given in the following discussion and in Table 11-2.

Acetylcholine

The neurotransmitter **acetylcholine (ACh)** is in a class of its own because it has a chemical structure unique among neurotransmitters. It is synthesized in neurons by combining an acetate (acetyl-coenzyme-A) with choline (one of the B complex vitamins). Postsynaptic membranes contain the enzyme *acetylcholinesterase*, which rapidly inactivates the acetylcholine bound to postsynaptic receptors. Choline molecules released by this reaction are transported back into the presynaptic neuron where they are combined with acetate to form more acetylcholine.

As Table 11-2 shows, acetylcholine is found in various locations in the nervous system. In many of these locations, it has an excitatory effect (for example, at the neuromuscular junctions of skeletal muscles). In others, it has an inhibitory effect (for example, at the neuromuscular junctions of cardiac muscle tissue).

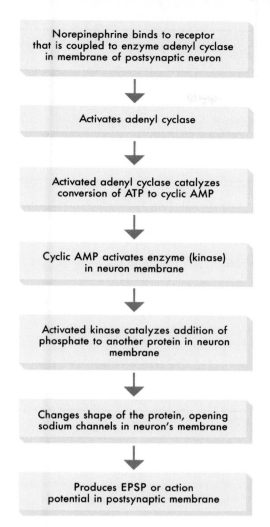

FIGURE 11-22 Second-messenger mechanism. Norepinephrine and many other neurotransmitters initiate nerve impulse conduction by using the chemical messenger within the postsynaptic cell (compare with Figure 11-20).

Amines

Amine neurotransmitters are synthesized from amino acid molecules such as tyrosine, tryptophan, or histidine. Amines include the neurotransmitters *serotonin* and *histamine.* Also included are neurotransmitters of the **catecholamine** subclass: *dopamine, epinephrine,* and *norepinephrine.*

The amine neurotransmitters are found in various regions of the brain, where they affect learning, emotions, motor control, and other activities. Dopamine, for example, is known to have an inhibitory effect on certain somatic motor pathways. When dopamine is deficient, the tremors and general overstimulation of muscles characteristic of *parkinsonism* occur. Epinephrine and norepinephrine are also involved in motor control, specifically in the sympathetic pathways of the autonomic nervous system. Some autonomic neurons in the adrenal gland do not terminate at a postsynaptic effector cell but instead release

their neurotransmitters directly into the bloodstream. When this occurs, epinephrine and norepinephrine are called *hormones* instead of neurotransmitters.

Amino Acids

Many biologists now believe that amino acids are among the most common neurotransmitters in the CNS. For example, it is thought that the amino acid *glutamate (glutamic acid)* is responsible for up to 75% of the excitatory signals in the brain. *Gamma-aminobutyric acid* (GABA), which is derived from glutamate, is the most common inhibitory neurotransmitter in the brain. In the spinal cord, the most widely distributed inhibitory neurotransmitter is the simple amino acid *glycine.*

Amino acids are found in all cells of the body, where they are used to synthesize various structural and functional proteins. In the nervous system, however, they are also stored in synaptic vesicles and used as neurotransmitters. Specialized receptors in the postsynaptic membrane are sensitive to high quantities of certain amino acids and thus trigger specific responses in the postsynaptic cell. It is believed that an imbalance of certain amino acids in the body could produce similar effects, and thus alter the function of the nervous system.

Neuropeptides

The **neuropeptide** neurotransmitters are short strands of amino acids called *polypeptides.* Polypeptides were first discovered to have regulatory effects in the digestive tract, where they act as hormones and regulate digestive function. In the 1970s some of these "gut" polypeptides such as *vasoactive intestinal peptide* (VIP), *cholecystokinin* (CCK), and *substance P* were also found to be acting as neurotransmitters in the brain. Researchers also found that receptors of many of the gut-brain polypeptides also bind morphine and other opium derivatives. For example, two subclasses of neuropeptides—**enkephalins** and **endorphins**—that bind to the opiate receptors serve as the body's own supply of opiates. Enkephalins and endorphins have important pain-relieving effects in the body.

Although neuropeptides may be secreted by a synaptic knob by themselves, some may be secreted along with a second, or even third, neurotransmitter. In such cases, the neuropeptide is thought to act as a **neuromodulator.** A neuromodulator is a "cotransmitter" that regulates or modulates the effects of the neurotransmitter(s) released along with it.

QUICK CHECK

✔ *1. How do excitatory neurotransmitters differ from inhibitory neurotransmitters?*

✔ *2. What are the four chemical classes of neurotransmitters?*

✔ *3. What are neuromodulators?*

TABLE 11-2   Examples of neurotransmitters

Neurotransmitter	Location*	Function*
ACETYLCHOLINE	Junctions with motor effectors (muscles, glands); many parts of brain	Excitatory or inhibitory; involved in memory
AMINES		
Serotonin	Several regions of the CNS	Mostly inhibitory; involved in moods and emotions, sleep
Histamine	Brain	Mostly excitatory; involved in emotions and regulation of body temperature and water balance
Dopamine	Brain; autonomic system	Mostly inhibitory; involved in emotions/moods and in regulating motor control
Epinephrine	Several areas of the CNS and in the sympathetic division of ANS	Excitatory or inhibitory; acts as a hormone when secreted by sympathetic neurons of the adrenal gland
Norepinephrine	Several areas of the CNS and in the sympathetic division of ANS	Excitatory or inhibitory; regulates sympathetic effectors; in brain, involved in emotional responses
AMINO ACIDS		
Glutamate (glutamic acid)	CNS	Excitatory; most common excitatory neurotransmitter in CNS
Gamma-aminobutyric acid (GABA)	Brain	Inhibitory; most common inhibitory neurotransmitter in brain
Glycine	Spinal cord	Inhibitory; most common inhibitory neurotransmitter in spinal cord
NEUROPEPTIDES		
Vasoactive intestinal peptide (VIP)	Brain; some ANS and sensory fibers; retina; gastrointestinal tract	Function in nervous system uncertain
Cholecystokinin (CCK)	Brain; retina	Function in nervous system uncertain
Substance P	Brain, spinal cord, sensory pain pathways; gastrointestinal tract	Mostly excitatory; transmits pain information
Enkephalins	Several regions of CNS; retina; intestinal tract	Mostly inhibitory; act like opiates to block pain
Endorphins	Several regions of CNS; retina; intestinal tract	Mostly inhibitory; act like opiates to block pain

*These are examples only; most of these neurotransmitters are also found in other locations, and many have additional functions.

HEALTH MATTERS   ANTIDEPRESSANTS

Severe psychic depression occurs when a deficit of norepinephrine, dopamine, serotonin, and other amines exists in certain brain synapses. This fact led to the development of antidepressant drugs. Certain of these antidepressants inhibit *catechol-O-methyl transferase (COMT)*, the enzyme that inactivates norepinephrine. When COMT is inhibited by an antidepressant drug, the amount of active norepinephrine in brain synapses increases—relieving the symptoms of depression. Antidepressants such as phenelzine (Nardil) block the action of *monoamine oxidase (MAO)*, the enzyme that inactivatesdopamine and serotonin. Widely used antidepressants such as imipramine (Tofranil) and amitriptyline (Elavil) increase amine levels at brain synapses by blocking their uptake into the axon terminals. Cocaine, which is often used in medical practice as a local anesthetic, produces a feeling of well-being in cocaine abusers by similarly blocking the uptake of dopamine. Unfortunately, cocaine and similar drugs can also adversely affect blood flow and heart function when taken in large amounts—leading to death in some individuals.

The Cycle of Life: *Nervous system cells*

The development of nerve tissue begins from the *ectoderm* during the first weeks after conception and exhibits many complicated stages before becoming mature by early adulthood. The most rapid and obvious development of nervous tissue occurs in the womb and for several years shortly after birth.

One of the most remarkable aspects of neural development involves the way in which nervous cells become organized to form a coordinated network spread throughout the body. Although we know very little about these processes, we do know that neurons require the coordinated actions of several agents to promote proper "wiring" of the nervous system. For example, we know that *nerve growth factors* released by neuroglia stimulate the growth of neuron processes and help direct them to the proper destination. Factors released by effector cells may also attract neuron processes, thus ensuring that they will be connected to the developing neural network.

During the first years of neural development, electrical synapses are made, broken, and reformed until the basic organization of the nervous system is intact. As rather permanent connections are made, electrical synapses are replaced by chemical synapses so that by the time a person reaches early adulthood the chemical type of synapse predominates. Neurobiologists believe that the sensory stimulation that serves as the essence of early learning in infants and children plays a critical role in directing the formation of synapses in the nervous system. The formation of new synapses is thought to be the primary physiological mechanism of learning and memory.

In advanced old age, degeneration of neurons, neuroglia, and the blood vessels that supply them may destroy certain portions of nervous tissue. This, coupled with age-related syndromes such as *Alzheimer's disease,* may produce a loss of memory, coordination, and other neural functions sometimes referred to as *senility.*

THE BIG PICTURE
Nervous System Cells and the Whole Body

Neurons, the conducting cells of the nervous system, act as the "wiring" that connects the structures needed to maintain the internal constancy that is so vital to our survival. They also form processing "circuits" that make decisions regarding appropriate responses to stimuli that threaten our internal constancy. Sensory neurons act as sensors or receptors that detect changes in our external and internal environment that may be potentially threatening. Sensory neurons then relay this information to integrator mechanisms in the central nervous system. There, the information is processed—often by one or more interneurons—and an outgoing response signal is relayed to effectors by way of motor neurons. At the effector, a chemical messenger or

neurotransmitter triggers a response that tends to restore homeostatic balance. Neurotransmitters released into the bloodstream, where they are called hormones, can enhance and prolong such homeostatic responses.

Of course, neurons do much more than simply respond to stimuli in a preprogrammed manner. Circuits of interneurons are capable of remembering, learning new responses, generating rational and creative thought, and other complex processes. The exact mechanisms of many of these complex integrative functions are yet to be discovered, but you will learn some of what we already know in Chapter 12, in our discussion of the central nervous system.

MECHANISMS OF DISEASE

Disorders of Nervous System Cells

Most disorders of nervous system cells involve neuroglia rather than neurons. *Multiple sclerosis (MS)*, one of the **myelin disorders** discussed on p. 282, is a good example of this principle. A few other important disorders involving neuroglia are described in the following paragraphs.

The general name for tumors arising in nervous system structures is **neuroma** (noo-RO-mah). Tumors do not usually develop directly from neurons but from neuroglia, membrane tissues, and blood vessels. A common type of brain tumor— **glioma** (glee-O-mah)—occurs in neuroglia. Gliomas are usually benign but may still be life-threatening. Because they often develop in deep areas of the brain, they are difficult to treat. Untreated gliomas may grow to a size that disrupts normal brain function—perhaps leading to death. Most malignant tumors of neuroglia and other tissues in the nervous system do not arise there but are secondary tumors resulting from metastasis of cancer cells from the breast, lung, or other organs.

Tumors in the central nervous system

Astrocytoma is a type of glioma that originates from astrocytes. It is a slow-growing, infiltrating tumor of the brain that usually appears during the fourth decade of life. Seizures, headaches, or neurological deficits indicative of the area of the brain involved are usually the presenting symptoms. **Glioblastoma multiforme,** a highly malignant form of astrocytic tumor, spreads throughout the white matter of the brain. Because of its invasive nature, surgical removal is difficult and the average survival is less than 1 year. **Ependymoma** is a neuroglial tumor arising from ependymal cells, which line the fluid-filled cavities (*ventricles*) of the brain and spinal cord. This is the most common glioma in children but can occur in adults. Because of its location, fluid pathways are obstructed, causing increased pressure in the brain, which in turn causes neurological damage. Surgical correction is possible and the average postoperative survival is roughly 5 years. Glioma of oligodendrocytes is called **oligodendroglioma.** This tumor commonly occurs in the anterior portion of the brain and has a peak

incidence at 40 years of age. The prognosis is better, with an average survival of 10 years after the onset of symptoms.

Tumors in the peripheral nervous system

Glial tumors can also develop in or on the cranial nerves. **Acoustic neuroma** is a lesion of the sheath of Schwann cells surrounding the eighth cranial nerve, responsible for hearing and balance. This tumor may be the size of a pea or walnut, but the person typically experiences difficulty deciphering speech through the affected ear, dizziness, tinnitus (ringing in the ear), and a slow, progressive hearing loss. With the use of microsurgical techniques, the tumor can be removed, but some nerve damage caused by the surgical procedure is common.

Glial tumors can also appear in other regions of the peripheral nervous system. **Multiple neurofibromatosis** (noo-ro-fye-bro-mah-TOE-sis) is an inherited disease characterized by numerous fibrous neuromas throughout the body (Figure 11-23). The tumors are benign, appearing first as small nodules in the Schwann cells of nerve fibers in the skin. In some cases, involvement spreads as large, disfiguring fibrous tumors in many areas of the body including muscles, bones, and internal organs. The disfigurement can be severe, as in the famous case of the "Elephant Man," who suffered crippling deformities in the skull, spinal column, and many other parts of the body.

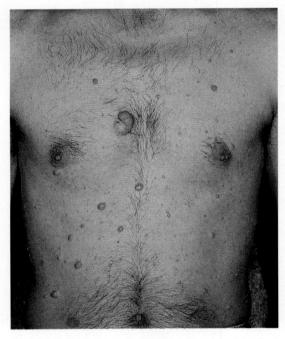

FIGURE 11-23 Multiple neurofibromatosis. Multiple tumors of Schwann cells in nerves of the skin that are characteristic of this inherited condition.

CHAPTER SUMMARY

INTRODUCTION

A. The function of the nervous system, along with the endocrine system, is to communicate

B. The nervous system is comprised of the brain, spinal cord, and nerves

OVERVIEW OF THE NERVOUS SYSTEM

A. Organized to detect changes in internal and external environments, evaluate the information, and initiate an appropriate response

B. Subdivided into smaller "systems" (Figure 11-1)
  1. Central nervous system (CNS)
     a. Structural and functional center of the entire nervous system
     b. Consists of the brain and spinal cord
     c. Integrates sensory information, evaluates it, and initiates an outgoing response
  2. Peripheral nervous system (PNS)
     a. Nerves that lie in the "outer regions" of the nervous system
     b. Cranial nerves—originate from the brain
     c. Spinal nerves—originate from the spinal cord
     d. Two major subdivisions
        (1) Afferent nervous system—consists of all incoming sensory pathways
        (2) Efferent nervous system—consists of all outgoing motor pathways
     e. Efferent nervous system—further subdivided according to the types of organs they innervate
        (1) Somatic nervous system—carries information to the somatic effectors (skeletal muscles)
        (2) Autonomic nervous system—carries information to the autonomic or visceral effectors (smooth and cardiac muscles and glands)

  (a) Sympathetic division—prepares the body to deal with immediate threats to the internal environment; produces "fight-or-flight" response
  (b) Parasympathetic division—coordinates the body's normal resting activities; sometimes called the "rest-and-repair" division

CELLS OF THE NERVOUS SYSTEM

A. Neuroglia
  1. Neuroglial cells support the neurons
  2. Five major types of neuroglia (Figure 11-2)
     a. Astrocytes
        (1) Star-shaped, largest, and most numerous type of neuroglia
        (2) Form tight sheaths around brain capillaries, which with tight junctions between capillary endothelial cells constitute the blood-brain barrier
     b. Microglia
        (1) Small, usually stationary cells
        (2) In inflamed brain tissue, they enlarge, move about, and carry on phagocytosis
     c. Ependymal cells
        (1) Resemble epithelial cells and form thin sheets that line fluid-filled cavities in the CNS
        (2) Some produce fluid; others aid in circulation of fluid
     d. Oligodendrocytes
        (1) Smaller than astrocytes with fewer processes
        (2) Hold nerve fibers together and produce the myelin sheath
     e. Schwann cells
        (1) Found only in peripheral nervous system

(2) Support nerve fibers and form myelin sheaths

(3) Gaps in the myelin sheath are called nodes of Ranvier

B. Neurons

1. Excitable cells that initiate and conduct impulses that make possible all nervous system functions

2. Components of neurons (Figure 11-3)

   a. Neurofibrils

   (1) "Bundles" of thin microtubules and microfilaments formed by the neuron's cytoskeleton

   (2) Allow the rapid transport of molecules to and from the far ends of a neuron

   b. Nissl bodies

   (1) Provide protein molecules needed for transmission of nerve impulses from one neuron to another

   (2) Provide proteins for maintaining and regenerating nerve fibers

   c. Dendrites

   (1) Each neuron has one or more dendrites, which branch from the cell body

   (2) Conduct nerve impulses to the cell body of the neuron

   (3) Distal ends of dendrites of sensory neurons are receptors

   d. Axon

   (1) A single process extending from the axon hillock, sometimes covered by fatty layer called myelin sheath (Figure 11-4)

   (2) Conduct nerve impulses away from the cell body of the neuron

   (3) Distal tips of axons are telodendria, which terminate in a synaptic knob

C. Classification of neurons

1. Structural classification—classified according to number of processes (Figure 11-5)

   a. Multipolar—one axon and several dendrites

   b. Bipolar—only one axon and one dendrite; least numerous kind of neuron

   c. Unipolar—one process comes off neuron cell body but divides almost immediately into one axon and one dendrite

2. Functional classification

   a. Afferent (sensory) neurons—conduct impulses to spinal cord or brain

   b. Efferent (motor) neurons—conduct impulses away from spinal cord or brain toward muscles or glandular tissue

D. Reflex arc

1. An impulse conduction route to and from the central nervous system with the impulse beginning in receptors and ending in effectors

2. Three-neuron arc—most common; consists of afferent neurons, interneurons, and efferent neurons (Figure 11-6)

   a. Afferent neurons—conduct impulses to the CNS from the receptors

   b. Efferent neurons—conduct impulses from the CNS to effectors (muscle or glandular tissue)

3. Two-neuron arc—simplest form; consists of afferent and efferent neurons

4. Synapse

   a. Where nerve impulses are transmitted from one neuron to another

   b. Located between the synaptic knob of one neuron and the dendrites or cell body of another neuron

NERVES AND TRACTS

A. Nerves—bundles of peripheral nerve fibers held together by several layers of connective tissue (Figure 11-8)

1. Endoneurium—delicate layer of fibrous connective tissue surrounding each nerve fiber

2. Perineurium—connective tissue holding together fascicles (bundles of fibers)

3. Epineurium—fibrous coat surrounding numerous fascicles and blood vessels to form a complete nerve

B. Within the CNS, bundles of nerve fibers are called tracts rather than nerves

C. White matter

1. Peripheral nervous system—myelinated nerves

2. Central nervous system—myelinated tracts

D. Gray matter

1. Made up of cell bodies and unmyelinated fibers

2. CNS—referred to as nuclei

3. PNS—referred to as ganglia

E. Mixed nerves

1. Contain both sensory and motor neurons

2. Sensory nerves—nerves with predominantly sensory nerves

3. Motor nerves—nerves with predominantly motor nerves

REPAIR OF NERVE FIBERS

A. Mature neurons are incapable of cell division, therefore damage to nervous tissue can be permanent

B. Neurons have limited capacity to repair themselves

C. If the damage is not extensive, the cell body and neurilemma are intact, and scarring has not occurred, nerve fibers can be repaired

D. Stages of repair of an axon in a peripheral motor neuron

1. Following injury, distal portion of axon and myelin sheath degenerates

2. Macrophages remove the debris

3. Remaining neurilemma and endoneurium form a tunnel from the point of injury to the effector

4. New Schwann cells grow in the tunnel to maintain a path for regrowth of the axon

5. Cell body reorganizes its Nissl bodies to provide the needed proteins to extend the remaining healthy portion of the axon

6. Axon "sprouts" appear

7. When "sprout" reaches tunnel, its growth rate increases

8. The skeletal muscle cell atrophies until the nervous connection is reestablished

E. In CNS, similar repair of damaged nerve fibers is unlikely

NERVE IMPULSES

A. Membrane potentials

1. All living cells maintain a difference in the concentration of ions across their membranes

2. Membrane potential—slight excess of positively charged ions on the outside of the membrane and slight excess of negatively charged ions on the inside of the membrane (Figure 11-10)

3. Difference in electrical charge is called potential because it is a type of stored energy

4. Polarized membrane—a membrane that exhibits a membrane potential

5. Magnitude of potential difference between the two sides of a polarized membrane is measured in volts (V) or millivolts (mV); the sign of a membrane's voltage indicates the charge on the *inside* surface of a polarized membrane

B. Resting membrane potential (RMP)

1. Membrane potential maintained by a nonconducting neuron's plasma membrane; typically −70 mV

2. The slight excess of positive ions on a membrane's outer surface is produced primarily by ion transport mechanisms

3. Sodium-potassium pump (Figure 11-11)
   a. Active transport mechanism in plasma membrane that transports Na$^+$ and K$^+$ in opposite directions and different rates
   b. Creates an imbalance in the distribution of positive ions, resulting in the inside surface becoming slightly negative with respect to its outer surface

4. The membrane's selective permeability characteristics also help maintain a slight excess of positive ions on the outer surface of the membrane (Figure 11-12)

C. Local potentials
   1. Local potentials—slight shift away from the resting membrane in a specific region of the plasma membrane (Figure 11-13)
   2. Excitation—when a stimulus triggers the opening of additional Na$^+$ channels, allowing the membrane potential to move toward zero (depolarization)
   3. Inhibition—when a stimulus triggers the opening of additional K$^+$ channels, increasing the membrane potential (hyperpolarization)
   4. Local potentials are called graded potentials because the magnitude of deviation from the resting membrane potential is proportional to the magnitude of the stimulus

ACTION POTENTIAL

A. Action potential—the membrane potential of a neuron that is conducting an impulse; also known as a nerve impulse

B. Mechanism that produces the action potential (Figure 11-15)
   1. When an adequate stimulus triggers Na$^+$ channels to open, allowing Na$^+$ to diffuse rapidly into the cell, producing a local depolarization
   2. As threshold potential is reached, additional Na$^+$ channels open and more Na$^+$ enters the cell, causing further depolarization
   3. The action potential is an all-or-none response
   4. Voltage-sensitive Na$^+$ channels stay open for only about 1 ms before they automatically close
   5. After action potential peaks, membrane begins to move back toward the resting membrane potential when K$^+$ channels open, allowing outward diffusion of K$^+$; process is known as repolarization
   6. Brief period of hyperpolarization occurs and then the resting membrane potential is restored by the sodium-potassium pumps

C. Refractory period (Figure 11-16)
   1. Absolute refractory period—brief period (lasting approximately half a millisecond) during which a local area of a neuron's membrane resists restimulation, will not respond to a stimulus, no matter how strong
   2. Relative refractory period—time during which the membrane is repolarized and restoring the resting membrane potential; the few milliseconds after the absolute refractory period; will respond only to a very strong stimuli

D. Conduction of the action potential
   1. At the peak of the action potential, the plasma membrane's polarity is now the reverse of the resting membrane potential
   2. The reversal in polarity causes electrical current to flow between the site of the action potential and the adjacent regions of membrane and triggers Na$^+$ channels in the next segment to open; this next segment exhibits an action potential (Figure 11-17)

3. This cycle continues to repeat
4. The action potential never moves backward due to the refractory period
5. In myelinated fibers, electrical charges in the membrane only occur at the nodes of Ranvier; this type of impulse conduction is called saltatory conduction (Figure 11-18)
6. Speed of nerve conduction depends on diameter and on the presence or absence of a myelin sheath

SYNAPTIC TRANSMISSION

A. Structure of the synapse (Figure 11-19)
   1. Synaptic knob—tiny bulge at the end of a terminal branch of a presynaptic neuron's axon that contains vesicles housing neurotransmitters
   2. Synaptic cleft—space between a synaptic knob and the plasma membrane of a postsynaptic neuron
   3. Plasma membrane of a postsynaptic neuron—has protein molecules that serve as receptors for the neurotransmitters

B. Mechanism of synaptic transmission—sequence of events are:
   1. Action potential reaches a synaptic knob, causing calcium ions to diffuse into the knob rapidly
   2. Increased calcium concentration triggers the release of neurotransmitter via exocytosis
   3. Neurotransmitter molecules diffuse across the synaptic cleft and bind to receptor molecules, causing ion channels to open
   4. Opening of ion channels produces a postsynaptic potential, either an excitatory postsynaptic potential (EPSP) or an inhibitory postsynaptic potential (IPSP)
   5. The neurotransmitter's action is quickly terminated by either neurotransmitter molecules being transported back into the synaptic knob and/or metabolized into inactive compounds

C. Summation (Figure 11-21)
   1. Spatial summation—adding together the effects of a number of knobs being activated simultaneously and stimulating different locations on the postsynaptic membrane, producing an action potential
   2. Temporal summation—when synaptic knobs stimulate a postsynaptic neuron in rapid succession, their effects can summate over a brief period of time to produce an action potential

NEUROTRANSMITTERS

A. Neurotransmitters—means by which neurons communicate with one another; there are more than 30 compounds known to be neurotransmitters, and dozens of others are suspected

B. Classification of neurotransmitters—commonly classified by:
   1. Function—the function of a neurotransmitter is determined by the postsynaptic receptor; two major functional classifications are excitatory neurotransmitters and inhibitory neurotransmitters
   2. Chemical structure—the mechanism by which neurotransmitters cause a change; there are four main classes; since the functions of specific neurotransmitters vary by location, they are usually classified according to chemical structure

C. Acetylcholine
   1. Unique chemical structure; acetate (acetyl coenzyme A) with choline
   2. Acetylcholine is deactivated by acetylcholinesterase, with the choline molecules being released and transported back to presynaptic neuron to combine with acetate
   3. Present at various locations, sometimes in an excitatory role, other times, inhibitory

D. Amines
   1. Synthesized from amino acid molecules
   2. Found in various regions of the brain, affecting learning, emotions, motor control, etc.
E. Amino acids
   1. Believed to be among the most common neurotransmitters of the CNS
   2. In the PNS, amino acids are stored in synaptic vesicles and used as neurotransmitters
F. Neuropeptides
   1. Made up of polypeptides
   2. May be secreted by themselves or in conjunction with a second or third neurotransmitter; in this case, neuropeptides act as a neuromodulator, a "cotransmitter" that regulates the effects of the neurotransmitter released along with it

CYCLE OF LIFE: NERVOUS SYSTEM CELLS

A. Nerve tissue development
   1. Begins in ectoderm
   2. Occurs most rapidly in womb and in first 2 years
B. Nervous cells organize into body network
C. Electrical synapses
   1. Form and reform until nervous system is intact
   2. Are replaced by chemical synapses as permanent connections are made
   3. Formation of new synapses stimulates learning and memory
D. Aging causes degeneration of the nervous system, which may lead to senility

THE BIG PICTURE: NERVOUS SYSTEM CELLS AND THE WHOLE BODY

A. Neurons act as the "wiring" that connects structures needed to maintain homeostasis
B. Sensory neurons—act as receptors to detect changes in the internal and external environment; relay information to integrator mechanisms in the CNS
C. Information is processed and a response is relayed to the appropriate effectors through the motor neurons
D. At the effector, neurotransmitter triggers a response to restore homeostasis
E. When neurotransmitters are released into the bloodstream, they are hormones
F. Neurons are responsible for more than just responding to stimuli; circuits are capable of remembering, learning new responses, generation of thought, etc.

MECHANISMS OF DISEASE: DISORDERS OF NERVOUS SYSTEM CELLS

A. Most disorders of the nervous system involve neuroglia rather than neurons
B. Neuroma—tumors arising in nervous system structures
   1. They develop from neuroglia, membrane tissues, and blood vessels
   2. Glioma—common brain tumor—usually benign; difficult to treat because they develop in deep brain areas
C. Most malignant tumors of the nervous system are secondary tumors resulting from metastasis of cancer cells from the breast, lung, or other organs
D. Tumors of the central nervous system
   1. Astrocytoma—glioma originating from astrocytes; a slow-growing infiltrating brain tumor usually appearing in the fourth decade of life
      a. Symptoms—seizures, headaches, or neurological deficits indicative of the area involved
   2. Glioblastoma multiforme—malignant form of astrocytic tumor that spreads throughout the white matter of the brain; surgical removal is difficult and average survival is less than 1 year
   3. Ependymoma—neurological tumor arising from ependymal cells lining the ventricles of the brain and spinal cord; most common glioma in children. Obstructs fluid pathways, causing increased pressure, which leads to neurological damage; surgical correction leads to an average postoperative survival of approximately 5 years
   4. Oligodendroglioma—glioma of oligodendrocytes, occurring in the anterior portion of the brain; peak incidence at 40 years of age; average survival, 10 years
E. Tumors in the peripheral nervous system
   1. Acoustic neuroma—lesion of the Schwann cells around the eighth cranial nerve; small tumor
      a. Symptoms—difficulty deciphering speech through the affected ear, dizziness, tinnitus, and slow, progressive hearing loss
      b. Tumor may be removed by microsurgery
F. Glial tumors can develop in or on the cranial nerves and in other regions of the peripheral nervous system
G. Multiple neurofibromatosis—inherited disease characterized by numerous fibrous neuromas throughout the body; tumors are usually benign but can cause severe disfigurement

REVIEW QUESTIONS

1. Briefly explain the general function the nervous system performs for the body.
2. Identify the other body system that performs the same general function.
3. Compare and contrast the general characteristics of the central nervous system and the peripheral nervous system.
4. Compare neurons and neuroglia as to numbers, types, structure, and function.
5. Differentiate between multipolar, bipolar, and unipolar neurons. Do the same with afferent neurons, efferent neurons, and interneurons.
6. How do white and gray matter differ from each other?
7. Explain why damage to the nervous system may be permanent.
8. How does the repair of nerve fibers differ in the peripheral nervous system and the central nervous system?
9. Define the term *potential* and differentiate from resting membrane.
10. Describe the function of sodium and potassium in the generation of an action potential.
11. Why is an action potential an all-or-none response?
12. How does the myelin sheath affect the speed of an action potential? What about the diameter of the nerve fiber?
13. Describe the structure of a synapse.
14. Describe the series of events that mediate conduction across synapses.
15. What is the difference between spatial and temporal summation?
16. List and describe the four main chemical classes of neurotransmitters.
17. Describe the actions of acetylcholine.
18. Why is the blood-brain barrier an important consideration in the treatment of brain disorders?
19. Describe one of the ways an anesthetic may relieve the sensation of pain.
20. Define the terms *convergence* and *divergence*. How do these events affect the functioning of the nervous system?

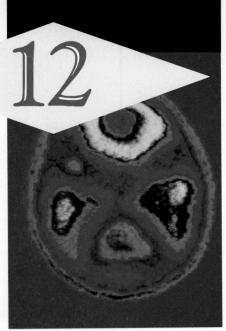

12

The Central Nervous System

PET scan of the brain

OBJECTIVES

**After you have completed this chapter, you should
be able to:**

1. List the major divisions of the nervous system.
2. Identify and locate the layers of the meninges.
3. Discuss the formation, circulation, and function of cerebrospinal fluid.
4. Discuss the location and generalized structure of the spinal cord.
5. List and give one primary function of the major ascending and descending tracts of the spinal cord.
6. List the three major divisions of the brain.
7. Identify the major divisions of the brain stem and describe the functions of each.
8. Describe the structure of the cerebellum. Explain how its function is related to the control of skeletal muscles.
9. Identify and discuss the primary functions of the two major components of the diencephalon.
10. Describe the structure of the cerebrum.
11. Discuss the sensory, motor, and integrative functions of the cortex.
12. Define electroencephalogram (EEG) and explain how it is produced.
13. Compare and contrast somatic sensory and somatic motor pathways.

Recall from Chapter 11 that the wonderfully complex nervous system is said to be comprised of two major divisions: the *central nervous system (CNS)* and the *peripheral nervous system (PNS)*. The reason for designating two distinct divisions is to make the study of the nervous system easier. In this chapter, we will discuss the part of the nervous system that lies at the center of the regulatory process: the central nervous system. Comprised of both the brain and the spinal cord, the central nervous system is the principal integrator of sensory input and motor output. Thus the central nervous system is capable of evaluating incoming information and formulating responses to changes that threaten our homeostatic balance.

This chapter begins with a description of the protective coverings of the brain and spinal cord. After that, we will briefly discuss the watery *cerebrospinal fluid (CSF)* and the spaces in which it is found. We will then outline the overall structure and function of the major organs of the central nervous system, beginning at the bottom with the spinal cord. Then our focus will move upward to the brain, beginning first with the narrow brain stem (Figure 12-1) and the roughly spherical cerebellum attached to its dorsal surface. Again shifting our attention upward, we will describe the structure and function of the diencephalon and then move on to a discussion of the cerebrum. As we move up the central nervous system you will notice that the complexity of function increases. The spinal cord mediates simple reflexes, whereas the brain stem and diencephalon are involved in the regulation of the more complex maintenance functions such as regulation of heart rate and breathing. The cerebral hemispheres, which together form the largest part of the brain, perform complex integrative functions such as conscious thought, learning, memory, language, and problem-solving. We will end the chapter with a discussion of the somatic sensory pathways and the somatic motor pathways. This will prepare us for Chapter 13, which covers the peripheral nervous system, and Chapter 14, which covers the sense organs.

◢ COVERINGS OF THE BRAIN AND SPINAL CORD

Because the brain and spinal cord are both delicate and vital, nature has provided them with two protective coverings. The outer covering consists of bone: cranial bones encase the brain; vertebrae encase the spinal cord. The inner covering consists of membranes known as **meninges.** Three distinct layers compose the meninges:

1. Dura mater
2. Arachnoid membrane
3. Pia mater

Observe their respective locations in Figures 12-2 and 12-3. The dura mater, made of strong white fibrous tissue, serves as the outer layer of the meninges and also as the inner periosteum of the cranial bones. The arachnoid membrane, a delicate, cobwebby layer, lies between the dura mater and the pia mater or innermost layer of the meninges. The transparent pia mater adheres to the outer surface of the brain and spinal cord and contains blood vessels.

The dura mater has three important inward extensions:
1. Falx cerebri
2. Falx cerebelli
3. Tentorium cerebelli

The *falx cerebri* projects downward into the longitudinal fissure to form a kind of partition between the two cerebral hemispheres. The *falx cerebelli* separates the two halves, or hemispheres, of the cerebellum. The *tentorium cerebelli* separates the cerebellum from the cerebrum. It is called a tentorium (meaning "tent") because it forms a tentlike covering over the cerebellum.

Figure 12-2 shows a large space within the dura, where the falx cerebri begins to descend between the left and right cerebral hemisphere. This space, called the *superior sagittal sinus,* is one of several dural sinuses. Dural sinuses function as veins, collecting blood from brain tissues for the return trip to the heart.

There are several spaces between and around the meninges (Figure 12-2). Three of these spaces are:

1. **Epidural space.** The epidural ("on the dura") space is immediately outside the dura mater but inside the bony coverings of the brain and spinal cord. It contains a supporting cushion of fat and other connective tissues.

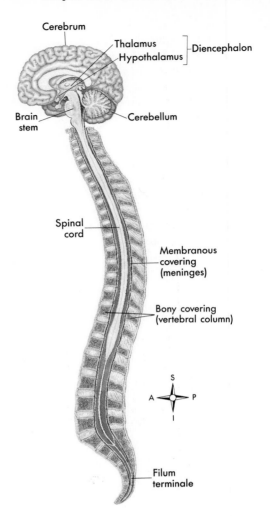

FIGURE 12-1   The central nervous system.

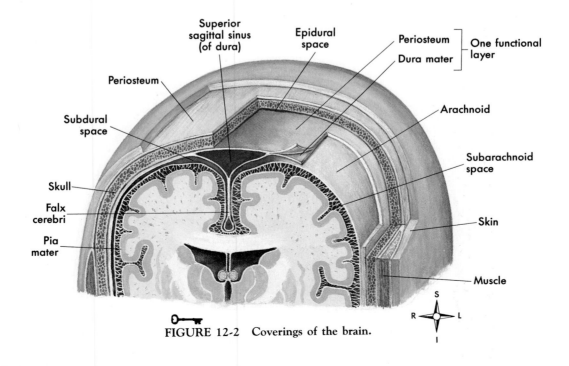

FIGURE 12-2   Coverings of the brain.

2. **Subdural space.** The subdural ("under the dura") space is between the dura mater and arachnoid membrane. The subdural space contains a small amount of lubricating serous fluid.

3. **Subarachnoid space.** As its name suggests, the subarachnoid space is under the arachnoid and outside the pia mater. This space contains a significant amount of cerebrospinal fluid.

The meninges of the cord (Figure 12-3) continue on down inside the spinal cavity for some distance below the end of the spinal cord. The pia mater forms a slender filament known as the *filum terminale* (see Figure 12-1). At the level of the third segment of the sacrum, the filum terminale blends with the dura mater to form a fibrous cord that disappears in the periosteum of the coccyx.

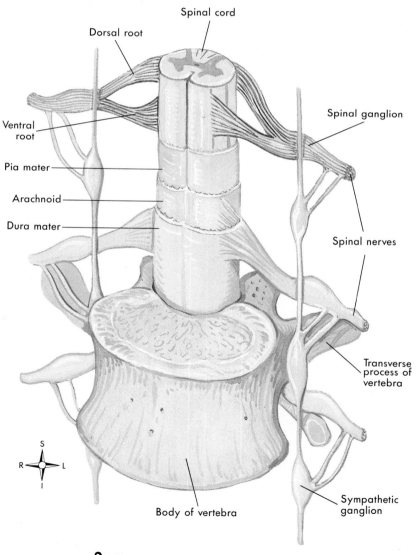

FIGURE 12-3 Coverings of the spinal cord.

CEREBROSPINAL FLUID

In addition to its bony and membranous coverings, nature has further protected the brain and spinal cord against injury by providing a cushion of fluid both around the organs and within them. This fluid is the **cerebrospinal fluid (CSF).** The cerebrospinal fluid does more than simply provide a supportive, protective cushion, however. It is also a reservoir of circulating fluid that, along with blood, the brain monitors for changes in the internal environment. For example, changes in the carbon dioxide (CO_2) content of CSF trigger homeostatic responses in the respiratory control centers of the brain stem that help regulate the overall CO_2 content and pH of the body.

Fluid Spaces

Cerebrospinal fluid is found in the subarachnoid space around the brain and spinal cord and within the cavities and canals of the brain and spinal cord.

The large, fluid-filled spaces within the brain are called **ventricles.** There are four of them. Two of them, the lateral (or first and second) ventricles, are located one in each hemisphere of the cerebrum. As you can see in Figure 12-4, the third ventricle is little more than a thin, vertical pocket of fluid below and medial to the lateral ventricles. The fourth ventricle is a tiny, diamond-shaped space where the cerebellum attaches to the back of the brain stem. Actually, the fourth ventricle is simply a slight expansion of the central canal extending up from the spinal cord.

LUMBAR PUNCTURE

The extension of the meninges beyond the cord is convenient for performing lumbar punctures without danger of injuring the spinal cord. A **lumbar puncture** is a withdrawal of some of the cerebrospinal fluid from the subarachnoid space in the lumbar region of the spinal cord. The physician inserts a needle just above or below the fourth lumbar vertebra, knowing that the spinal cord ends an inch or more above that level. The fourth lumbar vertebra can be easily located because it lies on a line with the iliac crest. Placing a patient on the side and arching the back by drawing the knees and chest together separates the vertebrae sufficiently to introduce the needle.

Cerebrospinal fluid removed through a lumbar puncture can be tested for the presence of blood cells, bacteria, or other abnormal characteristics that may indicate an injury or infection. A sensor called a manometer is sometimes attached to the needle to determine the pressure of the CSF within the subarachnoid space. The lumbar puncture can also be used to introduce diagnostic agents such as radiopaque dyes for x-ray photography into the subarachnoid space.

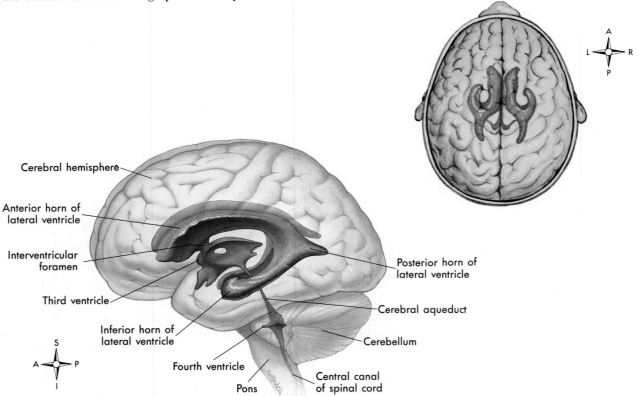

FIGURE 12-4 **Fluid spaces of the brain.** The large figure shows the ventricles highlighted within the brain in a left lateral view. The small figure shows the ventricles from above.

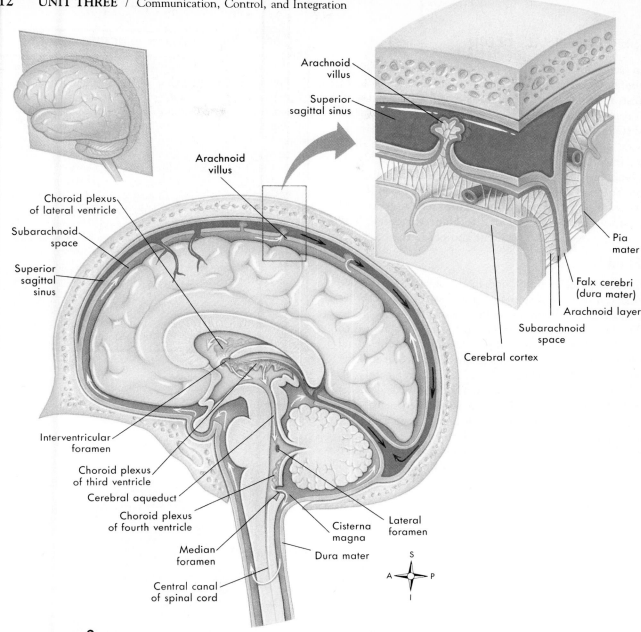

FIGURE 12-5 **Flow of cerebrospinal fluid.** The fluid produced by filtration of blood by the choroid plexus of each ventricle flows inferiorly through the lateral ventricles, interventricular foramen, third ventricle, cerebral aqueduct, fourth ventricle, and subarachnoid space and to the blood.

Formation and Circulation of Cerebrospinal Fluid

Formation of cerebrospinal fluid occurs mainly by separation of fluid from blood in the **choroid plexuses.** Choroid plexuses are networks of capillaries that project from the pia mater into the lateral ventricles and into the roofs of the third and fourth ventricles. From each lateral ventricle the fluid seeps through an opening, the interventricular foramen (of Monro), into the third ventricle, then through a narrow channel, the cerebral aqueduct (or

aqueduct of Sylvius), into the fourth ventricle (Figure 12-5). Some of the fluid moves from the fourth ventricle directly into the central canal of the cord. Some of it moves out of the fourth ventricle through openings in its roof, two lateral foramina (foramina of Luschka) and one median foramen (foramen of Magendie). These openings allow cerebrospinal fluid to move into the *cisterna magna,* a space behind the medulla that is continuous with the subarachnoid space around the brain and cord. The fluid circulates in the subarachnoid space, then is absorbed into venous

HYDROCEPHALUS

Occasionally, some condition interferes with circulation of cerebrospinal fluid. For example, a brain tumor may press against the cerebral aqueduct, shutting off the flow of fluid from the third to the fourth ventricle. In such an event the fluid accumulates within the lateral and third ventricles because it continues to form even though its drainage is blocked. This condition is known as **internal hydrocephalus.** If the fluid accumulates in the subarachnoid space around the brain, **external hydrocephalus** results. Subarachnoid hemorrhage, for example, may lead to formation of blood clots that block drainage of the cerebrospinal fluid from the subarachnoid space. With decreased drainage an increased amount of fluid remains in the space.

When internal hydrocephalus occurs in an infant whose skull has not completely ossified, the increasing fluid pressure in the ventricles causes the cranium to swell (A). This condition can be treated by surgical placement of a shunt, or tube, to drain the excess CSF (B). When this condition occurs in an older child or adult, the skull will not yield to the increasing fluid pressure. The pressure instead compresses the soft nervous tissue of the brain, potentially leading to coma or even death.

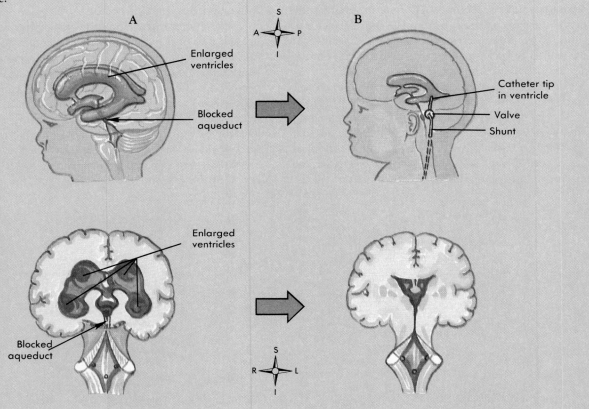

blood through the arachnoid villi (fingerlike projections of the arachnoid membrane into the brain's venous sinuses). Briefly, here is the circulation route of cerebrospinal fluid: it is formed by separation of fluid from blood in the choroid plexuses into the ventricles of the brain, circulates through the ventricles and into the central canal and subarachnoid spaces, and is absorbed back into blood.

The amount of cerebrospinal fluid in the average adult is about 140 ml (about 23 ml in the ventricles and 117 ml in the subarachnoid space of brain and cord).

QUICK CHECK
✔ 1. Name the three membranous coverings of the central nervous system in order, beginning with the outermost layer.
✔ 2. Trace the path of cerebrospinal fluid from its formation by a choroid plexus to its reabsorption into the blood.

SPINAL CORD

Structure of the Spinal Cord

The spinal cord lies within the spinal cavity, extending from the foramen magnum to the lower border of the first lumbar vertebra (Figure 12-6), a distance of about 45 cm (18 inches) in the average body. The spinal cord does not completely fill the spinal cavity—it also contains the meninges, cerebrospinal fluid, a cushion of adipose tissue, and blood vessels.

The spinal cord is an oval-shaped cylinder that tapers slightly from above downward and has two bulges, one in the cervical region and the other in the lumbar region (Figure 12-6). Two deep grooves, the *anterior median fissure* and the *posterior median sulcus,* just miss dividing the cord into separate symmetrical halves. The anterior fissure is the deeper and the wider of the two grooves—a useful factor to remember when you examine spinal cord diagrams. It enables you to tell at a glance which part of the cord is anterior and which is posterior.

Two bundles of nerve fibers called nerve roots project from each side of the spinal cord (see Figure 12-6). Fibers comprising the **dorsal nerve root** carry sensory information into the spinal cord. Cell bodies of these unipolar, sensory neurons comprise a small region of gray matter in the dorsal nerve root called the *dorsal root ganglion.* Fibers of the **ventral nerve root** carry motor information out of the

spinal cord. Cell bodies of these multipolar, motor neurons are in the gray matter that composes the inner core of the spinal cord. Numerous interneurons are also located in the spinal cord's gray matter core. On each side of the spinal cord, the dorsal and ventral nerve roots join together to form a single mixed nerve called, simply, a **spinal nerve.** Spinal nerves, components of the peripheral nervous system, will be considered in more detail in the next chapter.

Although the spinal cord's core of gray matter looks like a flat letter **H** in transverse sections of the cord, it actually has three dimensions, since the gray matter extends the length of the cord. The limbs of the **H** are called anterior, posterior, and lateral horns of gray matter, or gray columns (Figure 12-6). They consist predominantly of cell bodies of interneurons and motor neurons.

White matter surrounding the gray matter is subdivided in each half of the cord into three columns (or funiculi): the anterior, posterior, and lateral white columns. Each white column, or funiculus, consists of a large bundle of nerve fibers (axons) divided into smaller bundles called tracts, shown in Figure 12-7. The names of most spinal cord tracts indicate the white column in which the tract is located, the structure in which the axons that make up the tract originate, and the structure in which they terminate. Examples: the lateral corticospinal tract is located in the lateral white column of the cord, and the axons that

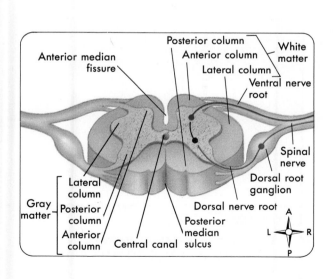

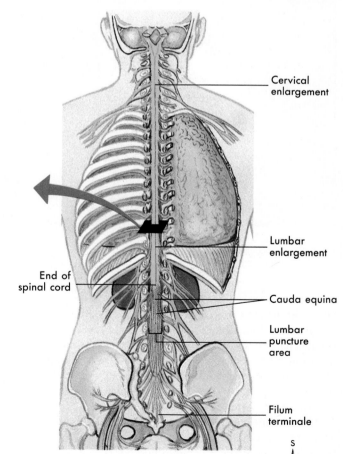

FIGURE 12-6 **Spinal cord.** The inset shows a transverse section of the spinal cord shown in the broader view.

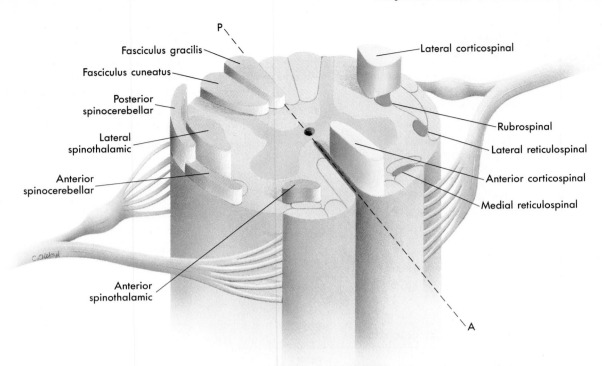

FIGURE 12-7 Major tracts of the spinal cord. The major ascending (sensory) tracts are highlighted in blue. The major descending (motor) tracts are highlighted in red. The broken line indicates the anterior/posterior orientation angle.

compose it originate from neuron cell bodies located in the spinal cortex (of the cerebrum) and terminate in the spinal cord. The anterior spinothalamic tract lies in the anterior white column, and the axons that compose it originate from neuron cell bodies located in the spinal cord and terminate in a portion of the brain called the thalamus.

Functions of the Spinal Cord

The spinal cord performs two general functions. Briefly, it provides the two-way conduction routes to and from the brain and serves as the integrator, or reflex center, for all spinal reflexes.

Spinal cord tracts provide two-way conduction paths to and from the brain. **Ascending tracts** conduct impulses up the cord to the brain. **Descending tracts** conduct impulses down the cord from the brain. Bundles of axons compose all tracts. Tracts are both structural and functional organizations of these nerve fibers. They are structural organizations in that all of the axons of any one tract originate from neuron cell bodies located in the same structure and all of the axons terminate in the same structure. For example, all the fibers of the spinothalamic tract are axons originating from neuron cell bodies located in the spinal cord and terminating in the thalamus. Tracts are functional organizations in that all the axons that compose one tract serve one general function. For instance, fibers of the spinothalamic tracts serve a sensory function. They transmit impulses that produce the sensations of crude touch, pain, and temperature.

Because so many different tracts make up the white columns of the cord, we shall mention only a few of the more important ones. Locate each tract in Figure 12-7.

Consult Tables 12-1 and 12-2 for a brief summary of information about these tracts.

Four important ascending or sensory tracts and their functions, stated very briefly, are as follows:

1. Lateral spinothalamic tracts. Crude touch, pain, and temperature
2. Anterior spinothalamic tracts. Crude touch and pressure
3. Fasciculi gracilis and cuneatus. Discriminating touch and conscious sensation of position and movement of body parts (kinesthesia)
4. Spinocerebellar tracts. Subconscious kinesthesia

Further discussion of the sensory neural pathways may be found on pp. 331-332.

Five important descending or motor tracts and their functions in brief are as follows:

1. Lateral corticospinal tracts. Voluntary movement; contraction of individual or small groups of muscles, particularly those moving hands, fingers, feet, and toes on opposite side of body
2. Anterior corticospinal tracts. Same as preceding except mainly muscles of same side of body
3. Lateral reticulospinal tracts. Transmits facilitatory impulses to anterior horn motor neurons to skeletal muscles
4. Medial reticulospinal tracts. Mainly inhibitory impulses to anterior horn motor neurons to skeletal muscles
5. Rubrospinal tracts. Transmit impulses that coordinate body movements and maintenance of posture

TABLE 12-1   Major ascending tracts of spinal cord

Name	Function	Location	Origin*	Termination†
Lateral spinothalamic	Pain, temperature, and crude touch opposite side	Lateral white columns	Posterior gray column opposite side	Thalamus
Anterior spinothalamic	Crude touch and pressure	Anterior white columns	Posterior gray column opposite side	Thalamus
Fasciculi gracilis and cuneatus	Discriminating touch and pressure sensations, including vibration, stereognosis, and two-point discrimination; also conscious kinesthesia	Posterior white columns	Spinal ganglia same side	Medulla
Anterior and posterior spinocerebellar	Unconscious kinesthesia	Lateral white columns	Anterior or posterior gray column	Cerebellum

*Location of cell bodies of neurons from which axons of tract arise.
†Structure in which axons of tract terminate.

TABLE 12-2   Major descending tracts of spinal cord

Name	Function	Location	Origin*	Termination†
Lateral corticospinal (or crossed pyramidal)	Voluntary movement, contraction of individual or small groups of muscles, particularly those moving hands, fingers, feet, and toes of opposite side	Lateral white columns	Motor areas of cerebral cortex opposite side from tract location in cord	Lateral or anterior gray columns
Anterior corticospinal (direct pyramidal)	Same as lateral corticospinal except mainly muscles of same side	Anterior white columns	Motor cortex but on same side as tract location in cord	Lateral or anterior gray columns
Lateral reticulospinal	Mainly facilitatory influence on motor neurons to skeletal muscles	Lateral white columns	Reticular formation, midbrain, pons, and medulla	Lateral or anterior gray columns
Medial reticulospinal	Mainly inhibitory influence on motor neurons to skeletal muscles	Anterior white columns	Reticular formation, medulla mainly	Lateral or anterior gray columns
Rubrospinal	Coordination of body movement and posture	Lateral white columns	Red nucleus (of midbrain)	Lateral or anterior gray columns

*Location of cell bodies of neurons from which axons of tract arise.
†Structure in which axons of tract terminate.

Further discussion of motor neural pathways may be found on pp. 332-334.

The spinal cord also serves as the reflex center for all spinal reflexes. The term *reflex center* means the center of a reflex arc or the place in the arc where incoming sensory impulses become outgoing motor impulses. They are structures that switch impulses from afferent to efferent neurons. In two-neuron arcs, reflex centers are merely synapses between neurons. In all other arcs, reflex centers consist of interneurons interposed between afferent and efferent neurons. Spinal reflex centers are located in the gray matter of the cord.

QUICK CHECK

✔ 1. What are spinal nerve roots? How does the dorsal root differ from the ventral root?

✔ 2. Name the regions of the white and gray matter seen in a horizontal section of the spinal cord.

✔ 3. Contrast ascending tracts and descending tracts of the spinal cord. Can you give an example of each?

The term *pain control area* means a place in the pain conduction pathway where impulses from pain receptors can be inhibited. The first pain control area suggested was a segment of the posterior gray horns of the spinal cord. *Substantia gelatinosa* is the name of this segment. Here, axon terminals of neurons that conduct from pain receptors to the spinal cord synapse with neurons that conduct pain impulses up the spinal cord to the brain. Several decades ago, researchers made a surprising discovery about these synapses in the substantia gelatinosa. They found they could inhibit pain conduction across them by stimulating skin touch receptors in a painful area. From this knowledge, a new theory about pain developed—the aptly named *gate-control theory of pain*. According to this theory, the substantia gelatinosa functions as a gate that can close and thus bar the entry of pain impulses into ascending paths to the brain. One way to close the gate is to stimulate skin touch receptors in a painful area. Today this is usually done by a device called the transcutaneous electrical nerve stimulation (TENS) unit. A patient uses the TENS unit to apply a low level of stimulation for a long period of time. This results in closure of the spinal cord pain gate and relief of pain.

In recent years, pain control areas have been identified in the brain, notably in the gray matter around the cerebral aqueduct and around the third ventricle. Neurons in these areas send their axons down the spinal cord to terminate in the substantia gelatinosa, where they release enkephalins. These chemicals act to prevent pain impulse conduction across synapses in the substantia gelatinosa. In short, enkephalins tend to close the spinal cord pain gate. Brief, intense transcutaneous stimulation at trigger or acupuncture points has been found to relieve pain in distant sites. The intense stimulation is postulated to activate brain pain control areas. They then send impulses down the cord to the substantia gelatinosa, closing the spinal cord pain gate.

THE BRAIN

The brain is one of the largest organs in adults. It consists, in round numbers, of 100 billion neurons and 900 billion neuroglia. In most adults, it weighs about 1.4 kg (3 pounds). Neurons of the brain undergo mitotic cell division only during the prenatal period and the first few months of postnatal life. Although they grow in size after that, they do not increase in number. Malnutrition during the crucial prenatal months of neuron multiplication is reported to hinder the process and result in fewer brain cells. The brain attains full size by about the eighteenth year but grows rapidly only during the first 9 years or so.

Six major divisions of the brain, named from below, upward, are as follows: *medulla oblongata, pons, midbrain, cerebellum, diencephalon,* and *cerebrum*. Very often the medulla oblongata, pons, and midbrain are referred to collectively as the brain stem. Look at these three structures in Figure 12-8. Do you agree that they seem to form a stem for the rest of the brain?

Structure of the Brain Stem

Three divisions of the brain make up the brain stem. The **medulla oblongata** forms the lowest part of the brain stem, the **midbrain** forms the uppermost part, and the **pons** lies between them, that is, above the medulla and below the midbrain.

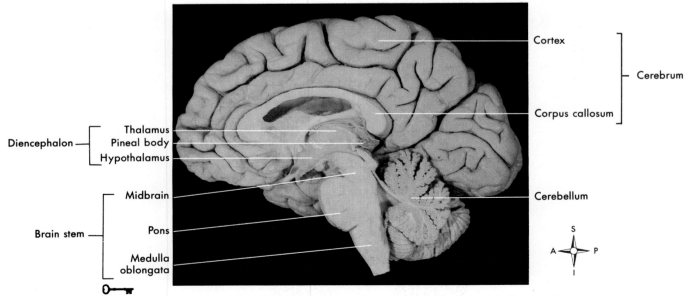

FIGURE 12-8   Divisions of the brain. A midsagittal section of the brain reveals features of its major divisions.

Medulla oblongata

The medulla oblongata is the part of the brain that attaches to the spinal cord. It is, in fact, an enlarged extension of the spinal cord located just above the foramen magnum. It measures only a few centimeters (about an inch) in length and is separated from the pons above by a horizontal groove. It is composed of white matter (projection tracts) and a network of gray and white matter called the *reticular formation* (see Figure 12-18).

The *pyramids* (Figure 12-9) are two bulges of white matter located on the ventral surface of the medulla. Fibers of the so-called pyramidal tracts form the pyramids.

The *olive* (Figure 12-9) is an oval projection appearing one on each side of the ventral surface of the medulla, lateral to the pyramids.

Located in the medulla's reticular formation are various *nuclei,* or clusters of neuron cell bodies. Some nuclei are called control centers—for example, the cardiac, respiratory, and vasomotor control centers.

Pons

Just above the medulla lies the pons, composed, like the medulla, of white matter and reticular formation. Fibers that run transversely across the pons and through the middle cerebellar peduncles into the cerebellum make up the external white matter of the pons and give it its arching, bridgelike appearance.

Midbrain

The midbrain (mesencephalon) is appropriately named. It forms the midsection of the brain, since it lies above the pons and below the cerebrum. Both white matter (tracts) and reticular formation compose the midbrain. Extending divergently through it are two ropelike masses of white matter named *cerebral peduncles* (Figure 12-9). Tracts in the peduncles conduct impulses between the midbrain and cerebrum. In addition to the cerebral peduncles, another landmark of the midbrain is the *corpora quadrigemina* (literally, "body of four twins"). The corpora quadrigemina are two *inferior colliculi* and two *superior colliculi*. Notice the location of the two sets of twin colliculi or, in other words, the corpora quadrigemina in Figure 12-9, *B.* They form the posterior, upper part of the midbrain, the part that lies just above the cerebellum. Certain auditory centers are located in the inferior colliculus. The superior colliculus contains visual centers. Two other midbrain structures are the *red nucleus* and the *substantia nigra*. Each of these consists of clusters of cell bodies of neurons involved in muscular control. The substantia nigra (literally, "black matter") gets its name from the dark pigment in some of its cells.

Functions of the Brain Stem

The brain stem, like the spinal cord, performs sensory, motor, and reflex functions. The spinothalamic tracts are important sensory tracts that pass through the brain stem on their way to the thalamus in the diencephalon. The fasciculi cuneatus and gracilis and the spinoreticular tracts are sensory tracts whose axons terminate in the gray matter of the brain stem. Corticospinal and reticulospinal tracts are two of the major tracts present in the white matter of the brain stem.

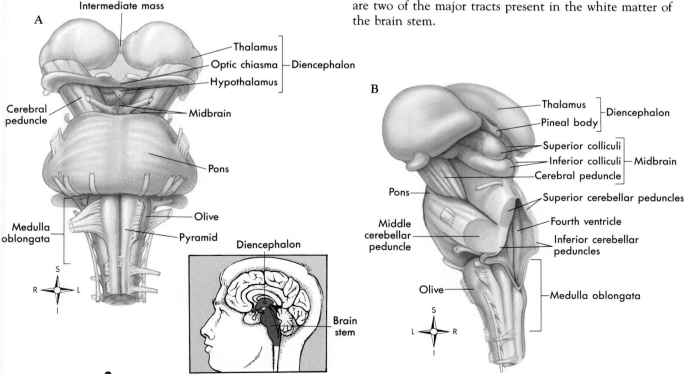

FIGURE 12-9 The brain stem and diencephalon. A, Anterior aspect. **B,** Posterior aspect (shifted slightly to lateral).

Nuclei in the medulla contain a number of reflex centers. Of first importance are the cardiac, vasomotor, and respiratory centers. Other centers present in the medulla are for various nonvital reflexes such as vomiting, coughing, sneezing, hiccupping, and swallowing.

The pons contains centers for reflexes mediated by the fifth, sixth, seventh, and eighth cranial nerves. The locations and functions of these peripheral nerves are discussed in Chapter 13. In addition, the pons contains the pneumotaxic centers that help regulate respiration.

The midbrain, like the pons, contains reflex centers for certain cranial nerve reflexes, for example, pupillary reflexes and eye movements, mediated by the third and fourth cranial nerves, respectively.

Structure of the Cerebellum

The cerebellum, the second largest part of the brain, is located just below the posterior portion of the cerebrum and is partially covered by it (Figure 12-10). A transverse fissure separates the cerebellum from the cerebrum. These two parts of the brain have several characteristics in common. For instance, gray matter makes up the outer portion, or *cortex*, of each. White matter predominates in

the interior of each. Observe the *arbor vitae*, that is, the internal white matter of the cerebellum in Figure 12-10. Note its distinctive pattern, similar to the veins of a leaf. Note, too, that the surfaces of both the cerebellum and the cerebrum have numerous grooves (*sulci*) and raised areas (*gyri*). The gyri of the cerebellum, however, are much more slender and less prominent than those of the cerebrum. Like the cerebrum, the cerebellum consists of two large lateral masses, the cerebellar hemispheres, and a central section called the *vermis*.

The internal white matter of the cerebellum is composed of some short and some long tracts. The shorter tracts conduct impulses from neuron cell bodies located in the cerebellar cortex to neurons whose dendrites and cell bodies compose nuclei located in the interior of the cerebellum. The longer tracts conduct impulses to and from the cerebellum. Fibers of the longer tracts enter or leave the cerebellum by way of its three pairs of peduncles (see Figure 12-9), as follows:

1. Inferior cerebellar peduncles. Composed chiefly of tracts into the cerebellum from the medulla and cord (notably, spinocerebellar, vestibulocerebellar, and reticulocerebellar tracts)
2. Middle cerebellar peduncles. Composed almost entirely of tracts into the cerebellum from the pons, that is, pontocerebellar tracts
3. Superior cerebellar peduncles. Composed principally of tracts from dentate nuclei in the cerebellum through the red nucleus of the midbrain to the thalamus

An important pair of cerebellar nuclei are the *dentate nuclei*, one of which lies in each hemisphere. Tracts connect these nuclei with the thalamus and with motor areas of the cerebral cortex. By means of these tracts, cerebellar impulses influence the motor cortex. Impulses also travel the reverse direction and enable the motor cortex to influence the cerebellum.

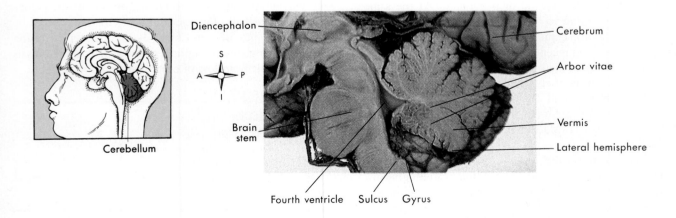

FIGURE 12-10 The cerebellum. This midsagittal section shows features of the cerebrum and surrounding structures of the brain.

Functions of the Cerebellum

The cerebellum performs three general functions, all of which have to do with the control of skeletal muscles:

1. It acts with the cerebral cortex to produce skilled movements by coordinating the activities of groups of muscles.
2. It helps control posture. It functions below the level of consciousness to make movements smooth instead of jerky, steady instead of trembling, and efficient and coordinated instead of ineffective, awkward, and uncoordinated.
3. It controls skeletal muscles to maintain equilibrium.

There have been many theories about cerebellar functions. One theory, based on comparative anatomy studies and substantiated by experimental methods, regards the cerebellum as three organs, each with a somewhat different function: coordinated control of muscle action, excitation and inhibition of postural reflexes, and maintenance of equilibrium.

Coordinated control of muscle action, which is ascribed to the neocerebellum (superior vermis and hemispheres), is closely associated with cerebral motor activity. Normal muscle action involves groups of muscles, the various members of which function together as a unit. In any given action, for example, the prime mover contracts and the antagonist relaxes but then contracts weakly at the proper moment to act as a brake, checking the action of the prime mover. Also, the synergists contract to assist the prime mover, and the fixation muscles of the neighboring joint contract. Through such harmonious, coordinated group action, normal movements are smooth, steady, and precise as to force, rate, and extent.

Achievement of such movements results from cerebellar activity added to cerebral activity. Impulses from the cerebrum may start the action, but those from the cerebellum synergize or coordinate the contractions and relaxations of the various muscles once they have begun. Figure 12-11 shows how the cerebrum and cerebellum work together. Impulses from the motor control areas of the cerebrum travel down the corticospinal tract and, through peripheral nerves, to skeletal muscle tissue. At the same time, the impulses go to the cerebellum. The cerebellum compares the motor commands of the cerebrum to information coming from receptors in the muscle. In effect, the cerebellum compares the intended movement to the actual movement. Impulses then travel from the cerebellum to both the cerebrum and the muscle tissue to adjust or coordinate the movements to produce the intended action. Most physiologists consider this the main function of the cerebellum.

One part of the cerebellum is also thought to be concerned with both exciting and inhibiting postural reflexes.

In addition, part of the cerebellum presumably discharges impulses important to the maintenance of equilibrium. Sensory impulses from equilibrium receptors in the ear reach the cerebellum. Here, connections are made with the proper motor fibers for contraction of the necessary muscles for stabilizing the body.

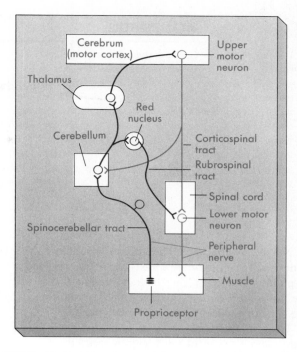

FIGURE 12-11 Coordinating function of the cerebellum. Impulses from the motor control areas of the cerebrum travel down to skeletal muscle tissue and to the cerebellum at the same time. The cerebellum, which also receives sensory information from the muscle tissue, compares the intended movement to the actual movement. It then sends impulses to both the cerebrum and the muscle tissue, thus coordinating and "smoothing" muscle activity.

Cerebellar disease (abscess, hemorrhage, tumors, trauma, etc.) produces certain characteristic symptoms. Predominant among them are ataxia (muscle incoordination), hypotonia, tremors, and disturbances of gait and equilibrium. One example of ataxia is overshooting a mark or stopping before reaching it when trying to touch a given point on the body (finger-to-nose test). Drawling, scanning, and singsong speech are also examples of ataxia. Tremors are particularly pronounced toward the end of the movements and with the exertion of effort. Disturbances of gait and equilibrium vary, depending on the muscle groups involved. The walk, for instance, is often characterized by staggering or lurching and by a clumsy manner of raising the foot too high and bringing it down with a clap. Paralysis does not result from loss of cerebellar function.

QUICK CHECK

✔ 1. Name the three major divisions of the brain stem and briefly describe the function of each.
✔ 2. What are gyri? What are sulci?
✔ 3. Describe how the cerebellum works with the cerebrum to coordinate muscle activity.

The Diencephalon

The diencephalon (literally, "between-brain") is the part of the brain located between the cerebrum and the

midbrain (mesencephalon). Although the diencephalon consists of several structures located around the third ventricle, the main ones are the **thalamus** and **hypothalamus.** The diencephalon also includes the **optic chiasma,** the **pineal body,** and several other small but important structures.

The thalamus

The thalamus is a dumbbell-shaped mass of gray matter made up of many nuclei. As Figures 12-9 and 12-12 show, each *lateral mass* of the thalamus forms one lateral wall of the third ventricle. Extending through the third ventricle, and thus joining the two lateral masses of the thalamus, is the *intermediate mass.* Two important groups of nuclei comprising the thalamus are the *geniculate bodies,* located in the posterior region of each lateral mass. The geniculate bodies play a role in processing auditory and visual input.

Large numbers of axons conduct impulses into the thalamus from the spinal cord, brain stem, cerebellum, basal ganglia, and various parts of the cerebrum. These axons terminate in thalamic nuclei, where they synapse with neurons whose axons conduct impulses out of the thalamus to virtually all areas of the cerebral cortex. Thus the thalamus serves as the major relay station for sensory impulses on their way to the cerebral cortex.

The thalamus performs the following primary functions:

1. Plays two parts in the mechanism responsible for sensations
   a. Impulses from appropriate receptors, on reaching the thalamus, produce conscious recognition of the crude, less critical sensations of pain, temperature, and touch
   b. Neurons whose dendrites and cell bodies lie in certain nuclei of the thalamus relay all kinds of sensory impulses, except possibly olfactory, to the cerebrum

2. Plays a part in the mechanism responsible for emotions by associating sensory impulses with feelings of pleasantness and unpleasantness
3. Plays a part in the arousal or alerting mechanism
4. Plays a part in mechanisms that produce complex reflex movements

The hypothalamus

The **hypothalamus** consists of several structures that lie beneath the thalamus and form the floor of the third ventricle and the lower part of its lateral walls. Prominent among the structures composing the hypothalamus are the supraoptic nuclei, the paraventricular nuclei, and the mamillary bodies. The supraoptic nuclei consist of gray matter located just above and on either side of the optic chiasma. The paraventricular nuclei of the hypothalamus are so named because of their location close to the wall of the third ventricle. The midportion of the hypothalamus gives rise to the **infundibulum,** the stalk leading to the posterior lobe of the *pituitary gland (neurohypophysis).* The posterior part of the hypothalamus consists mainly of the mamillary bodies (see Figure 12-19), which are involved with the olfactory sense (smell).

The hypothalamus is a small but functionally important area of the brain. It weighs little more than 7 g (¼ oz), yet it performs many functions of the greatest importance both for survival and for the enjoyment of life. For instance, it functions as a link between the psyche (mind) and the soma (body). It also links the nervous system to the endocrine system. Certain areas of the hypothalamus function as pleasure centers or reward centers for the primary drives such as eating, drinking, and sex. The following paragraphs give a brief summary of hypothalamic functions.

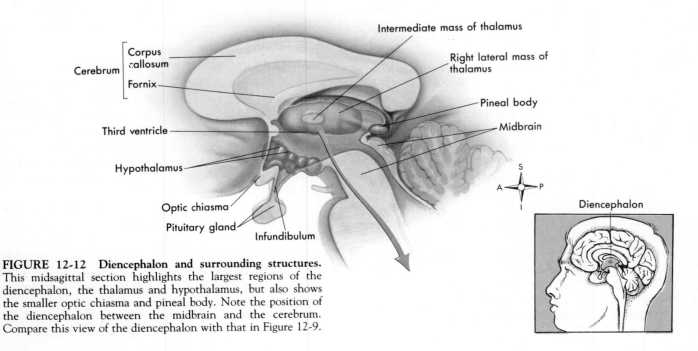

FIGURE 12-12 Diencephalon and surrounding structures. This midsagittal section highlights the largest regions of the diencephalon, the thalamus and hypothalamus, but also shows the smaller optic chiasma and pineal body. Note the position of the diencephalon between the midbrain and the cerebrum. Compare this view of the diencephalon with that in Figure 12-9.

1. The hypothalamus functions as a higher autonomic center or, rather, as several higher autonomic centers. By this we mean that axons of neurons whose dendrites and cell bodies lie in nuclei of the hypothalamus extend in tracts from the hypothalamus to both parasympathetic and sympathetic centers in the brain stem and cord. Thus impulses from the hypothalamus can simultaneously or successively stimulate or inhibit few or many lower autonomic centers. In other words the hypothalamus serves as a regulator and coordinator of autonomic activities. It helps control and integrate the responses made by autonomic (visceral) effectors all over the body.

2. The hypothalamus functions as the major relay station between the cerebral cortex and lower autonomic centers. Tracts conduct impulses from various centers in the cortex to the hypothalamus. Then, by way of numerous synapses in the hypothalamus, these impulses are relayed to other tracts that conduct them on down to autonomic centers in the brain stem and cord and also to spinal cord somatic centers (anterior horn motor neurons). Thus the hypothalamus functions as the link between the cerebral cortex and lower centers—hence between the psyche and the soma. It provides a crucial part of the route by which emotions can express themselves in changed bodily functions. It is the all-important relay station in the neural pathways that makes possible the mind's influence over the body—sometimes, unfortunately, even to the profound degree of producing "psychosomatic disease."

3. Neurons in the supraoptic and paraventricular nuclei of the hypothalamus synthesize the hormones released by the posterior pituitary gland (neurohypophysis). Because one of these hormones affects the volume of urine excreted, the hypothalamus plays an indirect but essential role in maintaining water balance (see Chapters 27 and 28).

4. Some neurons in the hypothalamus function as endocrine glands. Their axons secrete chemicals, *releasing hormones*, into blood, which circulate to the anterior pituitary gland. Releasing hormones control the release of certain anterior pituitary hormones—specifically growth hormone and hormones that control hormone secretion by sex glands, thyroid gland, and the adrenal cortex (discussed in Chapter 15). Thus indirectly the hypothalamus helps control the functioning of every cell in the body.

5. The hypothalamus plays an essential role in maintaining the waking state. Presumably it functions as part of an arousal or alerting mechanism. Clinical evidence of this is that somnolence (sleepiness) characterizes some hypothalamic disorders.

6. The hypothalamus functions as a crucial part of the mechanism for regulating appetite and therefore the amount of food intake. Experimental and clinical findings indicate the presence of an "appetite center" in the lateral part of the hypothalamus and a "satiety center" located medially. For example, an animal with an experimental lesion in the ventromedial nucleus of the hypothalamus will consume tremendous amounts of food. Similarly, a human being with a tumor in this region of the hypothalamus may eat insatiably and gain an enormous amount of weight.

7. The hypothalamus functions as a crucial part of the mechanism for maintaining normal body temperature. Hypothalamic neurons whose fibers connect with autonomic centers for vasoconstriction, dilation, and sweating and with somatic centers for shivering constitute heat-regulating centers. Marked elevation of body temperature frequently characterizes injuries or other abnormalities of the hypothalamus.

Pineal body

Although the thalamus and hypothalamus account for most of the tissue that makes up the diencephalon, there are several smaller structures of importance. For example, the *optic chiasma* is a region where the right and left *optic nerves* enter the brain and cross each other—exchanging fibers as they do so. The resulting bundles of fibers are called the *optic tracts*. Various small nuclei just outside the thalamus and hypothalamus, collectively referred to as the *epithalamus*, are also included among the structures of the diencephalon. One of the most intriguing of the epithalamic structures is the *pineal body*, or *epiphysis*.

As Figures 12-9 and 12-12 show, the pineal body is located just above the corpora quadrigemina of the midbrain. Its name comes from the fact that it resembles a tiny pine cone. The functions of the pineal body are still not well understood, but it seems to be involved in regulating the body's biological clock. It does produce some hormones, most notably *melatonin*, so it is also called the *pineal gland*. Melatonin is thought to synchronize various body functions with each other and with external stimuli. More discussion regarding the function of the pineal body appears in Chapter 15.

QUICK CHECK

✔ 1. *What are the two main components of the diencephalon? Where are they located?*
✔ 2. *Name three general functions of the thalamus.*
✔ 3. *Name three general functions of the hypothalamus.*

Structure of the Cerebrum
Cerebral cortex

The cerebrum, the largest and uppermost division of the brain, consists of two halves, the right and left **cerebral hemispheres.** The surface of the cerebrum—called the *cerebral cortex*—is made up of gray matter only 2 to 4 mm (roughly 1/12 to 1/6 inch) thick. But despite its thinness, the cortex has six layers, each composed of millions of axon terminals synapsing with millions of dendrites and cell bodies of other neurons.

If one uses a little imagination, the surface of the cerebral cortex looks like a group of small sausages. Each "sausage" represents a *convolution*, or gyrus. Names of some of these are the *precentral gyrus, postcentral gyrus, cingulate gyrus,* and *hippocampal gyrus.*

Between adjacent gyri lie either shallow grooves called sulci or deeper grooves called *fissures.* Fissures divide each cerebral hemisphere into five *lobes.* Four of them are named for the bones that lie over them: **frontal lobe, parietal lobe,**

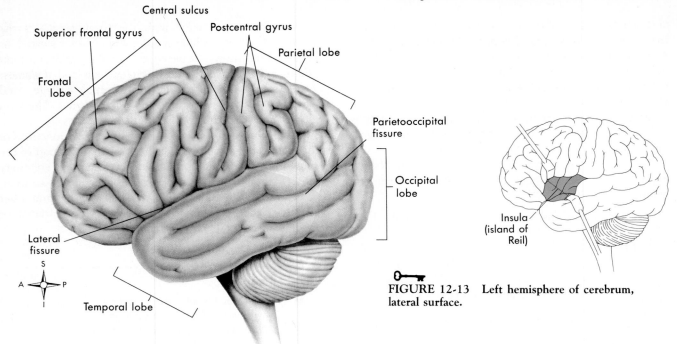

FIGURE 12-13   Left hemisphere of cerebrum, lateral surface.

temporal lobe, and **occipital lobe** (Figure 12-13). A fifth lobe, the **insula** (*island of Reil*), lies hidden from view in the lateral fissure. Names and locations of prominent cerebral fissures are the following (see Figure 12-13):

1. **Longitudinal fissure.** The deepest groove in the cerebrum; divides the cerebrum into two hemispheres
2. **Central sulcus (fissure of Rolando).** Groove between the frontal and parietal lobes
3. **Lateral fissure (fissure of Sylvius).** A deep groove between the temporal lobe below and the frontal and parietal lobes above; island of Reil lies deep in the lateral fissure
4. **Parietooccipital fissure.** Groove that separates the occipital lobe from the two parietal lobes

Cerebral tracts and basal ganglia

Beneath the cerebral cortex lies the large interior of the cerebrum. It consists mostly of white matter made up of numerous tracts. A few islands of gray matter, however, lie deep inside the white matter of each hemisphere. Collectively these are called **basal ganglia.** Since small regions of gray matter within the CNS are normally called *nuclei* instead of *ganglia*, this group of nuclei should be called the *cerebral nuclei.* As accurate as it might be, this term is still not used as commonly as the historical name, *basal ganglia.*

Tracts that make up the cerebrum's internal white matter are of three types: projection tracts, association tracts, and commissural tracts (Figure 12-14). *Projection*

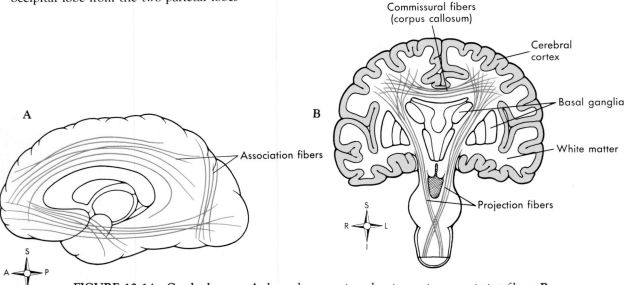

FIGURE 12-14   Cerebral tracts. A, Lateral perspective, showing various association fibers. **B,** Frontal (coronal) perspective, showing commissural fibers that make up the corpus callosum and the projection fibers that communicate with lower regions of the nervous system.

tracts are extensions of the ascending, or sensory, spinothalamic tracts and descending, or motor, corticospinal tracts. *Association tracts* are the most numerous of cerebral tracts; they extend from one convolution to another in the same hemisphere. *Commissural tracts*, in contrast, extend from one convolution to a corresponding convolution in the other hemisphere. Commissural tracts compose the **corpus callosum** (prominent white curved structure seen in Figure 12-12) and the anterior and posterior commissures.

Basal ganglia, seen in Figure 12-15, include the following masses of gray matter in the interior of each cerebral hemisphere:

1. **Caudate nucleus.** Observe the curving "tail" shape of this basal ganglion
2. **Lentiform nucleus.** So named because of its lenslike shape; note in Figure 12-15 that the lentiform nucleus consists of two structures, the *putamen* and *globus pallidus;* the putamen lies lateral to the globus pallidus (also called the pallidum)
3. **Amygdaloid nucleus.** Observe the location of this almond-shaped structure at the tip of the caudate nucleus

A structure associated with the basal ganglia is the *internal capsule.* It is a large mass of white matter located, as Figure 12-15 shows, between the caudate and lentiform nuclei and between the lentiform nucleus and thalamus. Together the caudate nucleus, internal capsule, and lentiform nucleus constitute the *corpus striatum.* The term means "striped body."

Researchers are still investigating the exact functions of the basal ganglia, but we already know that this part of the cerebrum plays an important role in regulating voluntary motor functions. For example, most of the muscle contractions involved in maintaining posture, walking, and other gross or repetitive movements seem to be initiated or modulated in the basal ganglia.

QUICK CHECK

✔ 1. *Name the five lobes that comprise each cerebral hemisphere. Where is each located?*

✔ 2. *Name the basal ganglia and describe where they are located within the cerebrum.*

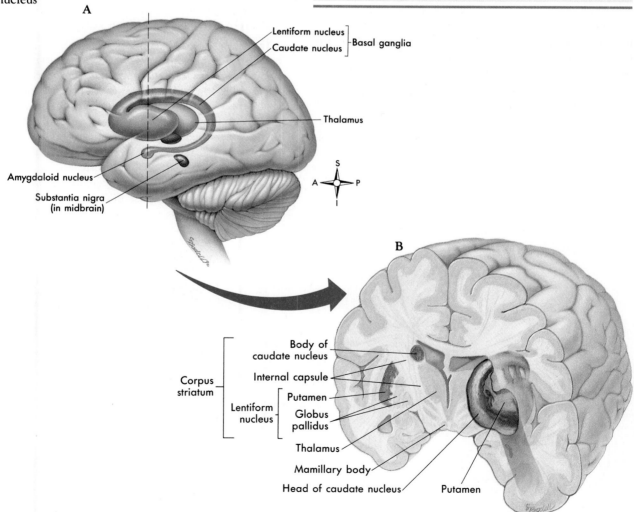

FIGURE 12-15 **Basal ganglia. A,** The basal ganglia seen through the cortex of the left cerebral hemisphere. **B,** The basal ganglia seen in a frontal (coronal) section of the brain.

PARKINSON'S DISEASE

The importance of the basal ganglia in regulating voluntary motor functions is made clear in cases of Parkinson's disease. Normally, neurons that lead from the substantia nigra to the basal ganglia secrete dopamine. Dopamine inhibits the excitatory effects of acetylcholine produced by other neurons in the basal ganglia. Such inhibition produces a balanced, restrained output of muscle-regulating signals from the basal ganglia. In Parkinson's disease, however, neurons leading from the substantia nigra degenerate and thus do not release normal amounts of dopamine. Without dopamine, the excitatory effects of acetylcholine are not restrained, and the basal ganglia produce an excess of signals that affect voluntary muscles in several areas of the body. Overstimulation of postural muscles in the neck, trunk, and upper limbs produces the syndrome of effects that typify this disease: rigidity and tremors of the head and limbs; an abnormal, shuffling gait; absence of relaxed arm-swinging while walking; and a forward tilting of the trunk.

Functions of the Cerebral Cortex
Functional areas of the cortex

During the past decade or so, research scientists in various fields—neurophysiology, neurosurgery, neuropsychiatry, and others—have added mountains of information to our knowledge about the brain. But questions come faster than answers, and clear, complete understanding of the brain's mechanisms still eludes us. Perhaps it forever will. Perhaps the capacity of the human brain falls short of the ability to fully understand its own complexity.

We do know that certain areas of the cortex in each hemisphere of the cerebrum engage predominantly in one particular function.

The function of each region of the cerebral cortex depends on the structures with which it communicates. For example, the postcentral gyrus (Figures 12-16 and 12-17) functions mainly as a general somatic sensory area. It receives impulses from receptors activated by heat, cold, and touch stimuli. The precentral gyrus, on the other hand, functions chiefly as the somatic motor area (Figures 12-16 and 12-17). Impulses from neurons in this area descend over motor tracts and eventually stimulate somatic effectors, the skeletal muscles. The transverse gyrus of the temporal lobe serves as the primary auditory area. The primary visual areas are in the occipital lobe. It is important to remember that no part of the brain functions alone. Many structures of the central nervous system must function together in order for any one part of the brain to function normally.

Sensory functions of the cortex

Various areas of the cerebral cortex are essential for normal functioning of the somatic, or "general," senses, as

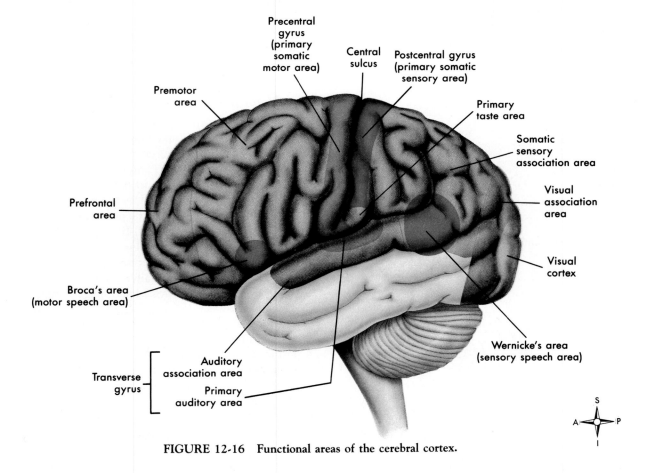

FIGURE 12-16 Functional areas of the cerebral cortex.

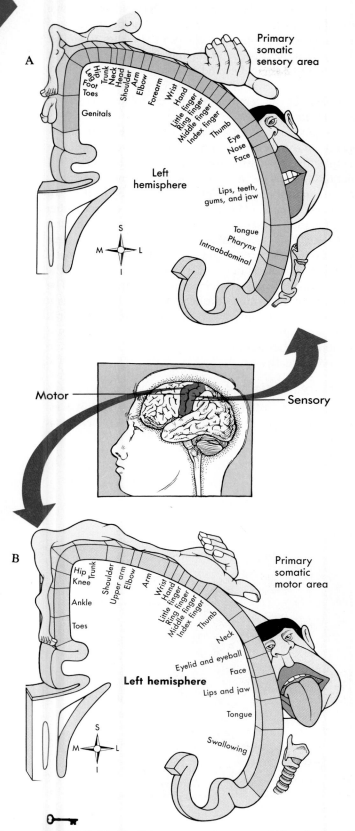

FIGURE 12-17 **Primary somatic sensory (A) and motor (B) areas of the cortex.**

well as the so-called "special senses." The somatic senses include sensations of touch, pressure, temperature, body position (proprioception), and similar perceptions that do not require complex sensory organs. The special senses include vision, hearing, and other types of perception that require complex sensory organs, for example, the eye and the ear.

As stated earlier, the postcentral gyrus serves as a primary area for the general somatic senses. As Figure 12-17, A, shows, sensory fibers carrying information from receptors in specific parts of the body terminate in specific regions of the somatic sensory area. In other words, the cortex contains a sort of "somatic sensory map" of the body. Areas such as the face and hand have a proportionally larger number of sensory receptors, so their part of the somatic sensory map is larger. Likewise, information regarding vision is mapped in the visual cortex, and auditory information is mapped in the primary auditory area (see Figure 12-16).

The cortex does more than just register separate and simple sensations, however. Information sent to the primary sensory areas is in turn relayed to the various sensory association areas, as well as to other parts of the brain. There, the sensory information is compared and evaluated. Eventually, the cortex integrates separate bits of information into whole perceptions.

Suppose, for example, that someone put an ice cube in your hand. You would, of course, see it and sense something cold touching your hand. But also, you would probably know that it was an ice cube because you would perceive a total impression compounded of many sensations such as temperature, shape, size, color, weight, texture, and movement and position of your hand and arm.

Discussion of somatic sensory pathways begins on p. 331. The special senses are discussed in Chapter 14.

Motor functions of the cortex

Mechanisms that control voluntary movements are extremely complex and imperfectly understood. It is known, however, that for normal movements to take place, many parts of the nervous system—including certain areas of the cerebral cortex—must function.

The precentral gyrus, that is, the most posterior gyrus of the frontal lobe constitutes the primary somatic motor area (see Figures 12-16 and 12-17, B). A secondary motor area lies in the gyrus immediately anterior to the precentral gyrus. Neurons in the precentral gyrus are said to control individual muscles, especially those that produce movements of distal joints (wrist, hand, finger, ankle, foot, and toe movements). Notice in Figure 12-17 that the primary somatic motor area is mapped according to the specific areas of the body it controls. Neurons in the premotor area just anterior to the precentral gyrus are thought to activate groups of muscles simultaneously.

Motor pathways descending from the cerebrum through the brain stem and spinal cord are discussed on pp. 332-334. Autonomic motor pathways are discussed in Chapter 13.

Integrative functions of the cortex

"Integrative functions" is a nebulous phrase. Even more obscure, however, are the neural processes it designates. They consist of all events that take place in the cerebrum between its reception of sensory impulses and its sending out of motor impulses. Integrative functions of the cerebrum include consciousness and mental activities of all kinds. Consciousness, use of language, emotions, and memory are the integrative cerebral functions that we shall discuss briefly.

Consciousness. Consciousness may be defined as a state of awareness of one's self, one's environment, and other beings. Very little is known about the neural mechanisms that produce consciousness. We do know, however, that consciousness depends on excitation of cortical neurons by impulses conducted to them by a network of neurons known as the reticular activating system. The **reticular activating system** consists of centers in the brain stem's *reticular formation* that receive impulses from the spinal cord and relay them to the thalamus and from the thalamus to all parts of the cerebral cortex (Figure 12-18). Both direct spinal reticular tracts and collateral fibers from the specialized sensory tracts (spinothalamic, lemniscal, auditory, and visual) relay impulses over the reticular activating system to the cortex. Without continual excitation of cortical neurons by reticular activating impulses, an individual is unconscious and cannot be aroused. Here, then, are two current concepts about the reticular activating system: (1) it functions as the arousal or alerting system for the cerebral

cortex, and (2) its functioning is crucial for maintaining consciousness. Drugs known to depress the reticular activating system decrease alertness and induce sleep. Barbiturates, for example, act this way. Amphetamine, on the other hand, a drug known to stimulate the cerebrum and to enhance alertness and produce wakefulness, probably acts by stimulating the reticular activating system.

Certain variations in the levels or state of consciousness are normal. All of us, for example, experience different levels of wakefulness. At times, we are highly alert and attentive. At other times, we are relaxed and nonattentive. All of us also experience different levels of sleep. Two of the best known stages are those called slow-wave sleep (SWS) and rapid eye movement (REM) sleep. Slow-wave sleep takes its name from the slow frequency, high-voltage brain waves that identify it. It is almost entirely a dreamless sleep. REM sleep, on the other hand, is associated with dreaming.

In addition to the various normal states of consciousness, altered states of consciousness also occur under certain conditions. Anesthetic drugs produce an altered state of consciousness, namely *anesthesia*. Disease or injury of the brain may produce an altered state called *coma*.

Peoples of various cultures have long been familiar with an altered state called *meditation*. Meditation is a waking state but differs markedly in certain respects from the usual waking state. According to some, meditation is a "higher" or "expanded" level of consciousness. This higher consciousness is accompanied, almost paradoxically, by a high degree of both relaxation and alertness. With training in meditation techniques and practice, an individual can enter the meditative state at will and remain in it for an extended period of time.

Language. Language functions consist of the ability to speak and write words and the ability to understand spoken and written words. Certain areas in the frontal, parietal, and temporal lobes serve as speech centers—as crucial areas, that is, for language functions. The left cerebral hemisphere contains these areas in about 90% of the population; in the remaining 10%, either the right hemisphere or both hemispheres contain them. Lesions in speech centers give rise to language defects called *aphasias*. For example, with damage to an area in the inferior gyrus of the frontal lobe (Broca's area, see Figure 12-16), a person becomes unable to articulate words but can still make vocal sounds and understand words heard and read.

Emotions. Emotions—both the subjective experiencing and objective expression of them—involve functioning of the cerebrum's **limbic system.** The name *limbic* (Latin for "border or fringe") suggests the shape of the cortical structures that make up the system. They form a curving border around the corpus callosum, the structure that connects the two cerebral hemispheres. Look now at Figure 12-19. Here on the medial surface of the cerebrum lie most of the structures of the limbic system. They are the cingulate gyrus and the hippocampus (the extension of the hippocampal gyrus that protrudes into the floor of the

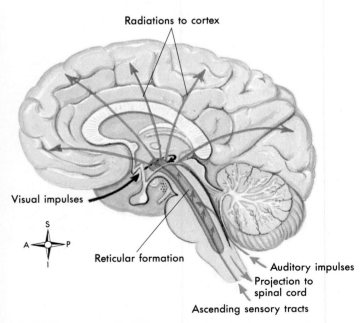

Radiations to cortex

Visual impulses

S
A — P
I

Reticular formation

Auditory impulses
Projection to spinal cord
Ascending sensory tracts

FIGURE 12-18 Reticular activating system. Consists of centers in the brain stem reticular formation plus fibers that conduct to the centers from below and fibers that conduct from the centers to widespread areas of the cerebral cortex. Functioning of the reticular activating system is essential for consciousness.

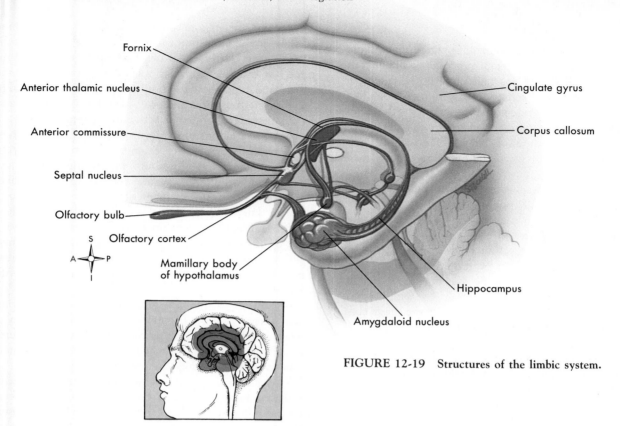

FIGURE 12-19   Structures of the limbic system.

inferior horn of the lateral ventricle). These limbic system structures have primary connections with various other parts of the brain, notably the thalamus, fornix, septal nucleus, amygdaloid nucleus (the tip of the caudate nucleus, one of the basal ganglia), and the hypothalamus. Some physiologists, therefore, include these connected structures as parts of the limbic system.

The limbic system (or to use its more descriptive name, the *emotional brain*) functions in some way to make us experience many kinds of emotions—anger, fear, sexual feelings, pleasure, and sorrow, for example. To bring about the normal expression of emotions, parts of the cerebral cortex other than the limbic system must also function. Considerable evidence exists that limbic activity without the modulating influence of the other cortical areas may bring on the attacks of abnormal, uncontrollable rage suffered periodically by some unfortunate individuals.

Memory. Memory is one of our major mental activities. The cortex is capable of storing and retrieving both *short-term memory* and *long-term memory*. Short-term memory involves the storage of information over a few seconds or minutes. Short-term memories can be somehow consolidated by the brain and stored as long-term memories that can be retrieved days—or even years—later.

It is an established fact that both short-term and long-term memory are functions of many parts of the cerebral cortex, especially of the temporal, parietal, and occipital lobes. Findings made by Dr. Wilder Penfield, a noted Canadian neurosurgeon, first gave evidence of this in the 1920s. He electrically stimulated the temporal lobes of epileptic patients undergoing brain surgery. They re-

sponded, much to his surprise, by recalling in the most minute detail songs and events from their past. Such long-term memories are believed to consist of some kind of structural traces—called *engrams*—in the cerebral cortex. Widely accepted today is the theory that an engram consists of some kind of permanent change in the synapses in a specific circuit of neurons. Repeated impulse conduction over a given neuronal circuit produces the synaptic change. What the change is is still a matter of speculation. Two suggestions are that it represents an increase in the number of presynaptic axon terminals or an increase in the number of receptor proteins in the postsynaptic neuron's membrane. Whatever the change is, it is such that it facilitates impulse transmission at the synapses.

A number of research findings indicate that the cerebrum's limbic system—the "emotional brain"—plays a key role in memory. To mention one role, when the hippocampus (part of the limbic system) is removed, the patient loses the ability to recall new information. Personal experience substantiates a relationship between emotion and memory.

Specialization of cerebral hemispheres

The right and left hemispheres of the cerebrum specialize in different functions. For example, as already noted, the left hemisphere specializes in language functions—it does the talking, so to speak. The left hemisphere also appears to dominate the control of certain kinds of hand movements, notably skilled and gesturing movements. Most people use their right hands for performing skilled movements, and the left side of the cerebrum controls the

muscles on the right side that execute these movements. The next time you are with a group of people who are talking, observe their gestures. The chances are about 9:1 that they will gesture mostly with their right hands—indicative of left cerebral control.

Evidence that the right hemisphere of the cerebrum specializes in certain functions has also been reported. It seems that one of the right hemisphere's specialties is the perception of certain kinds of auditory stimuli. For instance, some studies show that the right hemisphere perceives nonspeech sounds such as melodies, coughing, crying, and laughing better than does the left hemisphere. The right hemisphere may also function better at tactual perception and for perceiving and visualizing spatial relationships.

Despite the specializations of each cerebral hemisphere, both sides of a normal person's brain communicate with each other via the corpus callosum to accomplish the many complex functions of the brain.

The electroencephalogram (EEG)

Cerebral activity goes on as long as life itself. Only when life ceases (or moments before) does the cerebrum cease its functioning. Only then do all of its neurons stop conducting impulses. Proof of this has come from records of brain electrical potentials known as **electroencephalograms,** or **EEGs.** These records are usually made from a number of electrodes placed on different regions of the scalp, and they consist of waves—*brain waves*, as they are called (Figure 12-20).

Four types of brain waves are recognized based on frequency and amplitude of the waves. Frequency, or the number of wave cycles per second, is usually referred to as Hertz (Hz, from Hertz, a German physicist). Amplitude means voltage. Listed in order of frequency from fastest to slowest, brain wave names are *beta, alpha, theta,* and *delta. Beta waves* have a frequency of over 13 Hz and a relatively low voltage. *Alpha waves* have a frequency of 8 to 13 Hz and a relatively high voltage. *Theta waves* have both a relatively low frequency—4 to 7 Hz—and a low voltage. *Delta waves* have the slowest frequency—less than 4 Hz—but a high voltage. Brain waves vary in different regions of the brain, in different states of awareness, and in abnormal conditions of the cerebrum.

Fast, low-voltage beta waves characterize EEGs recorded from the frontal and central regions of the cerebrum when an individual is awake, alert, and attentive, with eyes open. Beta waves predominate when the cerebrum is busiest, that is, when it is engaged with sensory stimulation or mental activities. In short, beta waves are "busy waves." Alpha waves, in contrast, are "relaxed waves." They are moderately fast, relatively high-voltage waves that dominate EEGs recorded from the parietal lobe, occipital lobe, and posterior parts of the temporal lobes when the cerebrum is idling, so to speak. The individual is awake but has eyes closed and is in a relaxed, nonattentive state. This state is sometimes called the "alpha state." When drowsiness descends, moderately slow, low-voltage theta waves appear. Theta waves are "drowsy waves." "Deep sleep

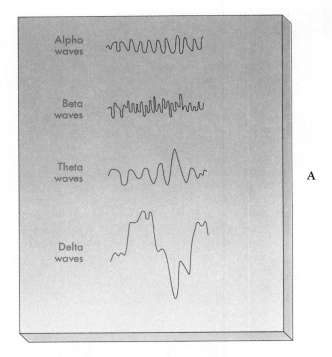

A

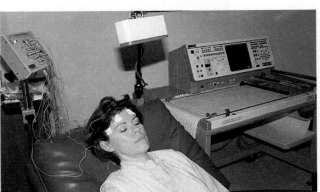

B

FIGURE 12-20 **The electroencephalogram (EEG). A,** Examples of alpha, beta, theta, and delta waves seen on an EEG. **B,** Photograph showing a person undergoing an EEG test. Notice the scalp electrodes that detect voltage fluctuations within the cranium.

waves," on the other hand, are delta waves. These slowest brain waves characterize the deep sleep from which one is not easily aroused. For this reason, deep sleep is referred to as slow-wave sleep (SWS).

QUICK CHECK

✔ 1. *Where is the primary somatic motor area of the cerebral cortex? Where is the primary somatic sensory area?*

✔ 2. *What does the reticular activating system have to do with alertness?*

✔ 3. *What is the function of the limbic system?*

✔ 4. *What kind of information can be gained from an EEG?*

BRAIN STUDIES

Neurobiologists have adapted many methods of medical imaging to the diagnosis of brain disorders without the trauma of extensive exploratory surgery. A few of the many approaches to studying the brain are listed here:

♦ X-ray photography. Traditional radiography (x-ray photography) of the head sometimes reveals tumors or injuries but does not show the detail of soft tissue necessary to diagnose many brain problems.

♦ Computed tomography (CT). This radiographic imaging technique involves scanning a person's head with a revolving x-ray generator. X-rays that pass through tissue hit x-ray sensors, which send the information to a computer that constructs an image that appears as a "slice of brain" on a video screen. CT scans can be "stacked" by the computer to give a three-dimensional view not possible with traditional radiography. Besides revealing the normal structure of the brain, CT scanning can often detect hemorrhages, tumors, and other abnormalities.

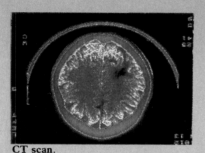

CT scan.

♦ Positron-emission tomography (PET). PET scanning is a variation of CT scanning in which a radioactive substance is introduced into the blood supply of the brain. The radioactive material shows up as a bright spot on the image. Different substances are taken up by the brain in different amounts, depending on the type of tissue and the level of activity—enabling radiologists to determine the *functional* characteristics of specific parts of the brain.

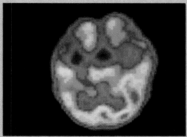

PET scan.

♦ Single-photon emission computed tomography (SPECT). SPECT is a method of scanning similar to PET, but using more stable substances and different detectors. SPECT is used to visualize blood flow patterns in the brain—making it useful in diagnosing cerebrovascular accidents (CVAs or strokes) and brain tumors.

♦ Ultrasonography. In this method, high-frequency sound (ultrasound) waves are reflected off anatomical structures to form images. This is similar to the way radar works. Because it does not use harmful radiation, ultrasonography is often used in diagnosing hydrocephalus or brain tumors in infants. When used on the brain, ultrasonography is often called *echoencephalography*.

♦ Magnetic resonance imaging (MRI). Also called *nuclear magnetic resonance (NMR)* imaging, this scanning method also has the advantage of avoiding the use of harmful radiation. In MRI, a magnetic field surrounding the head induces brain tissues to emit radio waves that can be used by a computer to construct a sectional image. MRI has the added advantage of producing sharper images than CT scanning and ultrasound. This makes it very useful in detecting small brain abnormalities.

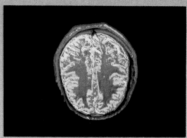

MRI scan.

♦ Evoked potential (EP). As discussed in this chapter, electroencephalography is the measurement of the electrical activity of the brain. The EP test is similar to the electroencephalogram (EEG), but the brain waves observed are caused (evoked) by specific stimuli such as a flash of light or a sudden sound. Recently, this information has been analyzed by a computer that then produces a color-coded graphic image of the brain generated on a video screen—a *brain electrical activity map (BEAM)*. Changes in color on the BEAM represent changes in brain activity evoked by each stimulus that is given. This technique is useful in diagnosing abnormalities of the visual or auditory systems because it reveals whether a sensory impulse is reaching the appropriate part of the brain.

♦ Magnetoencephalography (MEG). This new method of measuring brain activity uses a sensitive machine called a *biomagnetometer*, which detects the very small magnetic fields generated by neural activity. It can accurately pinpoint the part of the brain involved in a CVA (stroke), seizure, or other disorder or injury. Based on technology first developed by the military to locate submarines, MEG promises to improve our ability to diagnose epilepsy, Alzheimer's disease, and perhaps even certain types of addiction.

SOMATIC SENSORY PATHWAYS IN THE CENTRAL NERVOUS SYSTEM

For the cerebral cortex to perform its *sensory* functions, impulses must first be conducted to its sensory areas by way of relays of neurons referred to as sensory pathways. Most impulses that reach the sensory areas of the cerebral cortex have traveled over at least three pools of sensory neurons. We shall designate these as sensory neurons I, II, and III (Figure 12-21).

Sensory neurons I of the relay conduct from the periphery to the central nervous system. Sensory neurons II conduct from the cord or brain stem up to the thalamus. Their dendrites and cell bodies are located in spinal cord or brain stem gray matter. Their axons ascend in ascending tracts up the cord, through the brain stem, and terminate in the thalamus. Here they synapse with sensory neuron III dendrites or cell bodies (Figure 12-21). Sensory neurons III conduct from the thalamus to the postcentral gyrus of the parietal lobe, the somaticosensory area. Bundles of axons of

sensory neurons III form thalamocortical tracts. They extend through the portion of cerebral white matter known as the *internal capsule* to the cerebral cortex (see Figure 12-21).

For the most part, sensory pathways to the cerebral cortex are crossed pathways. This means that each side of the brain registers sensations from the opposite side of the body. Look again at Figure 12-21. The axons that *decussate* (cross from one side to the other) in these sensory pathways are which sensory neurons, I, II, or III? Usually it is the axon of sensory neuron II that decussates at some level in its ascent to the thalamus. General sensations of the right side of the body are predominantly experienced by the left somatic sensory area. General sensations of the left side of the body are predominantly experienced by the right somatic sensory area.

Two sensory pathways conduct impulses that produce sensations of touch and pressure, namely, the *medial lemniscal system* and the *spinothalamic pathway* (see Figure 12-21). The **medial lemniscal system** consists of the tracts that make up the posterior white columns of the cord (the

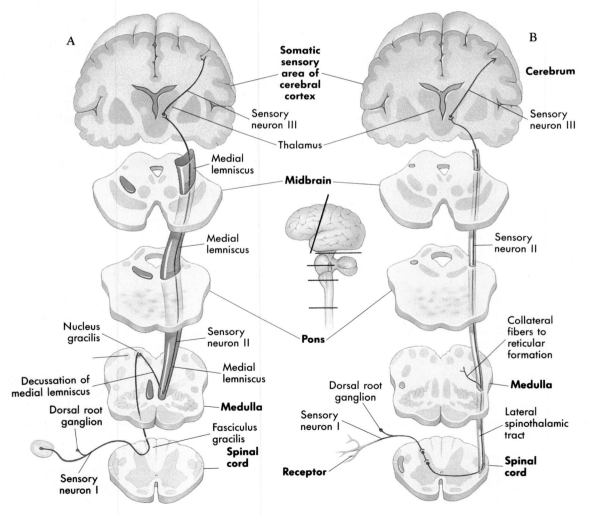

FIGURE 12-21   Examples of somatic sensory pathways. A, A pathway of the medial lemniscal system that conducts information about discriminating touch and kinesthesia. **B,** A spinothalamic pathway that conducts information about pain and temperature.

fasciculi cuneatus and gracilis) plus the *medial lemniscus*, a flat band of white fibers extending through the medulla, pons, and midbrain. (Derivation of the name *lemniscus* may interest you. It comes from the Greek word *lemniskos* meaning "woolen band." Apparently, to some early anatomist, the medial lemniscus looked like a band of woolen material running through the brain stem.)

The fibers of the medial lemniscus, like those of the spinothalamic tracts, are axons of sensory neurons II. They originate from cell bodies in the medulla, decussate, and then extend upward to terminate in the thalamus on the opposite side. The function of the medial lemniscal system is to transmit impulses that produce our more discriminating touch and pressure sensations, including *stereognosis* (awareness of an object's size, shape, and texture), precise localization, two-point discrimination, weight discrimination, and sense of vibrations. The sensory pathway for *kinesthesia* (sense of movement and position of body parts) is also part of the medial lemniscal system.

Crude touch and pressure sensations are functions of the **spinothalamic pathway.** Knowing that something touches the skin is a crude touch sensation, whereas knowing its precise location, size, shape, or texture involves the discriminating touch sensations of the medial lemniscal system.

QUICK CHECK

✔ 1. *Over how many afferent neurons does somatic sensory information usually pass?*
✔ 2. *Explain why stimuli on the left side of the body are perceived by the right side of the cerebral cortex.*

SOMATIC MOTOR PATHWAYS IN THE CENTRAL NERVOUS SYSTEM

For the cerebral cortex to perform its *motor* functions, impulses must be conducted from its motor areas to skeletal muscles by relays of neurons referred to as somatic motor pathways. Somatic motor pathways consist of motor neurons that conduct impulses from the central nervous system to somatic effectors, that is, skeletal muscles. Some motor pathways are extremely complex and not at all clearly defined. Others, notably spinal cord reflex arcs, are simple and well established. You read about these in Chapter 11. Look back now at Figure 12-11. From this diagram you can derive a cardinal principle about somatic motor pathways—the *principle of the final common path.* It is this: only one final common path, namely, the motor neuron from the anterior gray horn of the spinal cord, conducts *impulses* to skeletal muscles. Axons from the anterior gray horn are the only ones that terminate in skeletal muscle cells. This principle of the final common path to skeletal muscles has important practical implications. For example, it means that any condition that makes anterior horn motor neurons unable to conduct impulses also makes skeletal muscle cells supplied by these neurons

unable to contract. They cannot be willed to contract nor can they contract reflexively. They are, in short, paralyzed. Most famous of the diseases that produce paralysis by destroying anterior horn motor neurons is *poliomyelitis.* Numerous somatic motor paths conduct impulses from motor areas of the cerebrum down to anterior horn motor neurons at all levels of the cord.

Two methods are used to classify somatic motor pathways—one based on the location of their fibers in the medulla and the other on their influence on the lower motor neurons. The first method divides them into pyramidal and extrapyramidal tracts. The second classifies them as facilitatory and inhibitory tracts.

Pyramidal tracts are those whose fibers come together in the medulla to form the *pyramids,* hence their name. Because axons composing the pyramidal tracts originate from neuron cell bodies located in the cerebral cortex, they also bear another name—*corticospinal tracts* (Figure 12-22). About three fourths of their fibers decussate (cross over from one side to the other) in the medulla. After decussating, they extend down the spinal cord in the crossed corticospinal tract located on the opposite side of the spinal cord in the lateral white column. About one fourth of the corticospinal fibers do not decussate. Instead, they extend down the same side of the spinal cord as the cerebral area from which they came. One pair of uncrossed tracts lies in the anterior white columns of the cord, namely, the anterior corticospinal tracts. The other uncrossed corticospinal tracts form part of the lateral corticospinal tracts. About 60% of corticospinal fibers are axons that arise from neuron cell bodies in the precentral (frontal lobe) region of the cortex. About 40% of corticospinal fibers originate from neuron cell bodies located in postcentral areas of the cortex, areas classified as sensory; now, more accurately, they are often called sensorimotor areas.

Relatively few corticospinal tract fibers synapse directly with anterior horn motor neurons. Most of them synapse with interneurons, which in turn synapse with anterior horn motor neurons. All corticospinal fibers conduct impulses that facilitate resting negativity of anterior horn motor neurons. The effects of facilitatory impulses acting rapidly on any one neuron add up or summate. Each impulse, in other words, decreases the magnitude of the neuron's resting potential a little bit more. If sufficient numbers of impulses impinge rapidly enough on a neuron, its negativity decreases to threshold level. At that moment the neuron starts conducting impulses—it is stimulated. Stimulation of anterior horn motor neurons by corticospinal tract impulses results in stimulation of individual muscle groups (mainly of the hands and feet). Precise control of their contractions is, in short, the function of the corticospinal tracts. Without stimulation of anterior horn motor neurons by impulses over corticospinal fibers, willed movements cannot occur. This means that paralysis results whenever pyramidal corticospinal tract conduction is interrupted. For instance, the paralysis that so often follows cerebral vascular accidents (CVAs, or strokes) comes from pyramidal neuron injury—sometimes of their cell bodies in

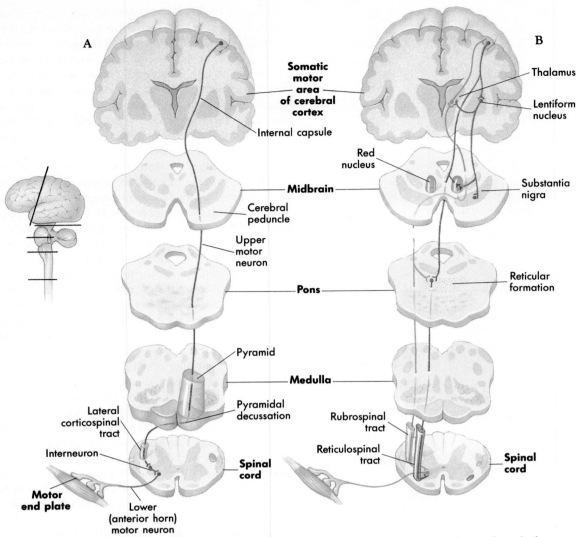

FIGURE 12-22   Examples of somatic motor pathways. A, A pyramidal pathway, through the lateral corticospinal tract. **B,** Extrapyramidal pathways through the rubrospinal and reticulospinal tracts.

the motor areas, sometimes of their axons in the internal capsule (see Figure 12-22).

Extrapyramidal tracts are much more complex than pyramidal tracts. They consist of all motor tracts from the brain to the spinal cord anterior horn motor neurons except the corticospinal (pyramidal) tracts. Within the brain, extrapyramidal tracts consist of numerous relays of motor neurons between motor areas of the cortex, basal ganglia, thalamus, cerebellum, and brain stem. In the cord, some of the most important extrapyramidal tracts are the reticulospinal tracts.

Fibers of the *reticulospinal tracts* originate from cell bodies in the reticular formation of the brain stem and terminate in gray matter of the spinal cord, where they synapse with interneurons that synapse with lower (anterior horn) motor neurons. Some reticulospinal tracts function as facilitatory tracts, others as inhibitory tracts. Summation of these opposing influences determines the lower motor neuron's response. It initiates impulse conduction only when facilitatory impulses exceed inhibitory impulses sufficiently to

decrease the lower motor neuron's negativity to its threshold level.

Conduction by extrapyramidal tracts plays a crucial part in producing our larger, more automatic movements because extrapyramidal impulses bring about contractions of groups of muscles in sequence or simultaneously. Such muscle action occurs, for example, in swimming and walking and, in fact, in all normal voluntary movements.

Conduction by extrapyramidal tracts plays an important part in our emotional expressions. For instance, most of us smile automatically at things that amuse us and frown at things that irritate us. It is extrapyramidal, not pyramidal, impulses that produce the smiles or frowns.

Axons of many different neurons converge on, that is, synapse, with each anterior horn motor neuron (see Figure 12-22). Hence many impulses from diverse sources—some facilitatory and some inhibitory—continually bombard this final common path to skeletal muscles. Together the added, or summated, effect of these opposing influences determines lower motor neuron functioning. Facilitatory

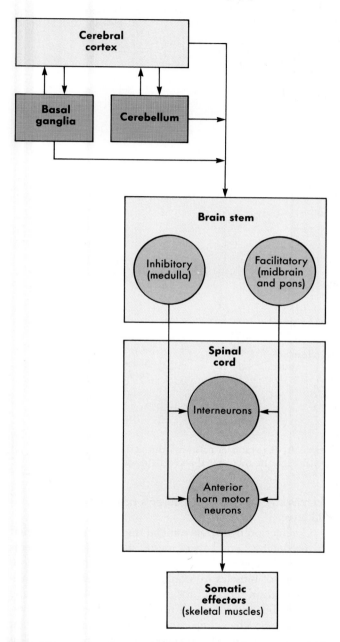

FIGURE 12-23 Concept of extrapyramidal motor control. Programmed movements result from a set of impulses called a *motor program*. The motor program organized in the basal ganglia and cerebellum in response to a command from the cortex is sent back to the primary motor control area of the cortex. From there, it is sent to the brain stem, then on through the spinal cord to the skeletal muscles. All along the motor pathway, the motor program can be refined by various components of this complex pathway.

impulses reach these cells via sensory neurons (whose axons, you will recall, lie in the posterior roots of spinal nerves), pyramidal (corticospinal) tracts, and extrapyramidal facilitatory reticulospinal tracts. Impulses over facilitatory reticulospinal fibers facilitate the lower motor neurons that supply extensor muscles. At the same time, they reciprocally inhibit the lower motor neurons that supply flexor muscles. Hence facilitatory reticulospinal impulses tend to increase the tone of extensor muscles and decrease the tone of flexor muscles.

Inhibitory impulses reach lower motor neurons mainly via inhibitory reticulospinal fibers that originate from cell bodies located in the *bulbar inhibitory area* in the medulla. They inhibit the lower motor neurons to extensor muscles (and reciprocally stimulate those to flexor muscles). Hence inhibitory reticulospinal impulses tend to decrease extensor muscle tone and increase flexor muscle tone—opposite effects from facilitatory reticulospinal impulses.

The set of coordinated commands that control the programmed muscle activity mediated by extrapyramidal pathways is often called a **motor program.** Traditionally, the primary somatic motor areas of the cerebral cortex was thought to be the principal organizer of motor programs sent along the extrapyramidal pathway. That view has been replaced by the concept illustrated in Figure 12-23. This newer concept holds that motor programs result from the interaction of several different centers in the brain. Apparently, many voluntary motor programs are organized in the basal ganglia and cerebellum—perhaps in response to a willed command by the cerebral cortex. Impulses that constitute the motor program are then channeled through the thalamus and back to the cortex, specifically to the primary motor area. From there, the motor program is sent down to the inhibitory and facilitatory regions of the brain stem. Signals from the brain stem then continue on down one or more spinal tracts and out to the muscles by way of the lower (anterior horn) motor neurons. All along the way, neural connections among the various motor control centers allow refinement and adjustment of the motor program. If all this sounds complicated and confusing, imagine what it must be like to be a neurobiologist trying to figure out how all this works! In fact, scientists working in this field admit that they have not worked out all the details of the extrapyramidal circuits—or exactly how they control muscle activity. However, the model shown in Figure 12-23 summarizes the current notion that it is a complex, interactive process.

QUICK CHECK

✔ 1. *What is the "principle of final common path" as it pertains to somatic motor pathways?*

✔ 2. *Distinguish between pyramidal and extrapyramidal pathways.*

SIGNS OF MOTOR PATHWAY INJURY

Injury of *upper motor neurons* (those whose axons lie in either pyramidal or extrapyramidal tracts) produces symptoms frequently referred to as "pyramidal signs," notably a spastic type of paralysis, exaggerated deep reflexes, and a positive Babinski reflex (p. 359). Actually, pyramidal signs result from interruption of both pyramidal and extrapyramidal pathways. The paralysis stems from interruption of pyramidal tracts, whereas the spasticity (rigidity) and exaggerated reflexes come from interruption of inhibitory extrapyramidal pathways.

Injury to *lower motor neurons* produces symptoms different from those of *upper motor neuron* injury. Anterior horn cells or lower motor neurons, you will recall, constitute the final common path by which impulses reach skeletal muscles. This means that if they are injured, impulses can no longer reach the skeletal muscles they supply. This in turn results in the absence of all reflex and willed movements produced by contraction of the muscles involved. Unused, the muscles soon lose their normal tone and become soft and flabby (flaccid). In short, absence of reflexes and flaccid paralysis are the chief "lower motor neuron signs."

Cycle of Life: *Central nervous system*

If the most obvious *structural* change over the life span is the overall growth, then degeneration, of the skeleton and other body parts, then the most obvious *functional* change is the development, then degeneration, of the complex integrative capacity of the central nervous system.

Although the development of the brain and spinal cord begins in the womb, further development is required by the time a baby is born. The lack of development of the central nervous system in a newborn is evidenced by lack of the more complex integrative functions such as language, complex memory, comprehension of spatial relationships, and complex motor skills such as walking. As childhood proceeds, one can easily see evidence of the increasing capacity of the central nervous system for complex function. A child learns to use language, to remember both concrete and abstract ideas, to walk, and even to behave in ways that conform to the norms of society. By the time a person reaches adulthood, most, if not all, of these complex functions have become fully developed. We use them throughout adult life to help us maintain internal stability in an unstable external world.

As we enter very late adulthood, the tissues of the brain and spinal cord degenerate to a degree that varies from one individual to the next. In some cases, the degeneration is profound—or it occurs in a critical part of the brain—and an elderly person becomes unable to communicate, unable to walk, or unable to perform some other complex function. In many cases, however, the degeneration produces milder effects such as temporary lapses in memory of fumbling with certain very complex motor tasks.

THE BIG PICTURE
The Central Nervous System and the Whole Body

The central nervous system is the ultimate regulator of the entire body. It serves as the anatomical and functional center of the countless feedback loops that maintain the relative constancy of the internal environment. The CNS directly or indirectly regulates, or at least influences, nearly every organ in the body.

The intriguing thing about the way in which the CNS regulates the whole body is that it is able to integrate, or bring together, literally millions of bits of information from all over the body and make sense of it all. Not only does the CNS make sense of all this information, it compares it to previously stored memories and makes decisions based on its own conclusions about the data. The complex integrative functions of human language, consciousness, learning, and memory enable us to adapt to situations that less complex organisms could not. Thus, our wonderfully complex central nervous system is essential to our survival.

MECHANISMS OF DISEASE

Disorders of the Central Nervous System

Destruction of brain tissue

Injury or disease can destroy neurons. A common example is the destruction of neurons of the motor area of the cerebrum that results from a **cerebrovascular accident (CVA)**. A CVA, or *stroke*, is a hemorrhage from or cessation of blood flow through cerebral blood vessels. When this happens, the oxygen supply to portions of the brain is disrupted and neurons cease functioning. If the lack of oxygen is prolonged, the neurons die. If the damage occurs in a motor control area of the brain (see Figures 12-16 and 12-17), a person can no longer voluntarily move the parts of the body controlled by the affected area(s). Because motor neurons cross over from side to side in the brainstem (see Figure 12-22), paralysis appears on the side of the body opposite the side of the brain on which the CVA occurred. The term **hemiplegia** (hem-i-PLEE-ja) refers to paralysis (loss of voluntary muscle control) of one whole side of the body.

One of the most common crippling diseases that appears during childhood, **cerebral palsy,** also results from damage to brain tissue. Cerebral palsy involves permanent, nonprogressive damage to motor control areas of the brain. Such damage is present at birth or occurs shortly after birth and remains throughout life. Possible causes of brain damage include prenatal infections or diseases of the mother; mechanical trauma to the head before, during, or after birth; nerve-damaging poisons; reduced oxygen supply to the brain; and other factors. The resulting impairment to voluntary muscle control can manifest itself in various ways. Many people with cerebral palsy exhibit **spastic paralysis,** a type of paralysis characterized by involuntary contractions of affected muscles. In cerebral palsy, spastic paralysis often affects one entire side of the body **(hemiplegia),** or both legs **(paraplegia),** both legs and one arm **(triplegia),** or all four extremities **(quadriplegia).**

Dementia

Various degenerative diseases can result in destruction of neurons in the brain. This degeneration can progress to adversely affect memory, attention span, intellectual capacity, personality, and motor control. The general term for this syndrome is **dementia** (de-MEN-sha).

Alzheimer's (ALZ-hye-merz) **disease** is characterized by dementia. Its characteristic lesions develop in the cortex during the middle to late adult years. Exactly what causes dementia-producing lesions to develop in the brains of individuals with Alzheimer's disease is not known. There is some evidence that this disease has a genetic basis—at least in some families. Other evidence indicates that environmental factors may have a role. Because the exact cause of Alzheimer's disease is still not known, development of an effective treatment has proven difficult. Currently, people diagnosed with this disease are treated by helping them maintain their remaining mental abilities and looking after their hygiene, nutrition, and other aspects of personal health management.

Huntington's disease (HD) is an inherited disease characterized by *chorea* (involuntary, purposeless movements) that progresses to severe dementia and death. The initial symptoms of this disease first appear between ages 30 and 40, with death generally occurring by age 55. Now that the gene responsible for Huntington's disease has been located, researchers are hopeful that an effective treatment will be found.

Acquired immune deficiency syndrome (AIDS), caused by the HIV virus, can also cause dementia. The immune deficiency characteristic of AIDS results from HIV infection of white blood cells that are critical to the proper function of the immune system (see Chapter 20). However, the HIV virus also infects neurons and can cause progressive degeneration of the brain—resulting in dementia.

Seizure disorders

Some of the most common nervous system abnormalities belong to the group of conditions called *seizure disorders*. These disorders are characterized by **seizures**—sudden bursts of abnormal neuron activity that result in temporary changes in brain function. Seizures may be very mild, causing subtle changes in level of consciousness, motor control, or sensory perception. On the other hand, seizures may be quite severe—resulting in jerky, involuntary muscle contractions called *convulsions* or even unconsciousness.

Recurring or chronic seizure episodes constitute a condition called **epilepsy.** Although some cases of epilepsy can be traced to specific causes such as tumors or chemical imbalances, most epilepsy is idiopathic (of unknown cause). Epileptics are often treated with anticonvulsive drugs such as *phenobarbital, phenytoin,* or *valproic acid* that block neurotransmitters in affected areas of the brain. By thus blocking synaptic transmission, such drugs inhibit the explosive bursts of neuron activity associated with seizures. With proper medication, many epileptics lead normal lives without the fear of experiencing uncontrollable seizures.

Diagnosis and evaluation of epilepsy or any seizure disorder often rely on *electroencephalography* (see Figure 12-20). As Figure 12-24 illustrates, a normal EEG shows the moderate rise and fall of voltage in various parts of the brain, but a seizure manifests itself as an explosive increase in the size and frequency of voltage fluctuations. Different classifications of epilepsy are based on the location(s) and the duration of these changes in brain activity.

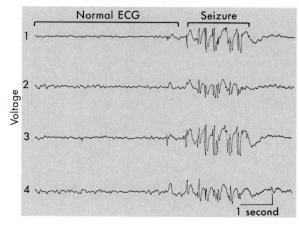

FIGURE 12-24 EEG of a seizure. An EEG tracing showing activity in four different places in the brain (obtained from four sets of electrodes). Compare the moderate, chaotic activity identified as "normal" with the explosive activity that occurs during a seizure.

CASE STUDY
SPINAL CORD INJURY

A.D. is a 19-year-old college student who was rock climbing and fell 30 feet to the ground. A.D. is 6 feet 7 inches tall and very active on his college basketball team. His medical history is relatively negative except for the usual childhood illnesses and minor accidents. On his basketball physical examination before school started, his vital signs were as follows: blood pressure 110/82; heart rate 88; respirations 18.

A.D. was picked up at the scene of the accident by paramedics who found him lying on his back, unable to move any extremities, and complaining of some neck discomfort. He appeared awake, alert, and oriented to his current location, the date and day of the week, and details of the fall. His responses to verbal questioning were appropriate. He complained that he could not feel or move his arms and legs. His pupils were equal and reactive to light. He showed no signs of other injury except for several scrapes on his arms. His vital signs revealed blood pressure of 110/72; heart rate of 86;

respirations of 18, unlabored and regular. The paramedics applied a cervical collar, placed him on a back board, immobilized his head, and transported him to the medical center by helicopter.

A.D. was found to exhibit no deep tendon reflexes of the extremities in the emergency room. His perception of sensory stimulus ended just above the nipple line of the chest. He had some sensory perception of the arms but was not able to demonstrate any consistent pattern of perception with repeated examinations. He had some ability to tighten the biceps but could not overcome gravity to raise his arms. He was unable to expand his chest wall. The remainder of his physical examination revealed blood pressure at 100/60; heart rate 68; respirations 24, somewhat shallow; oral temperature 99.8° F. His color was dusky. His skin was warm and dry to touch. Neurosurgeons were called in to evaluate A.D. They found on roentgenogram (x-ray photographs) that he had a cervical dislocated fracture of C5 and C6.

1. Which of the following best explains the changes in A.D.'s vital signs from those indicated in his past history?
   a. Lying on his back
   b. Spinal shock
   c. Anxiety
   d. Lack of mobility

2. A.D. asks why he can feel and move his shoulders and some parts of his arms. Your response would be based on the understanding that:
   a. The concussion is causing these transient signs
   b. His injury affected the autonomic nervous center in the cerebellum
   c. These findings are consistent with the C5 and C6 injury level
   d. The brain pathways are too complex for one to know the exact etiology

3. In considering A.D.'s respiratory status, you would be concerned if:
   a. He does not continue taking deep breaths
   b. His anxiety causes him to have episodes of shallow breathing
   c. He is breathing only with his diaphragm
   d. His breath sounds continue to be bronchovesicular

4. Monitoring A.D.'s gastrointestinal system is important because:
   a. He may develop a stress ulcer quickly
   b. His constipation will resolve easily with the use of medication
   c. A bowel program must begin as soon as possible
   d. Enemas will be needed every day to clear the bowel because it will not function on its own

CHAPTER SUMMARY

COVERINGS OF THE BRAIN AND SPINAL CORD
A. Two protective coverings (Figure 12-2)
   1. Outer covering is bone; cranial bones encase the brain, and vertebrae encase the spinal cord
   2. Inner covering is the meninges; the meninges of the cord continue inside the spinal cavity beyond the end of the spinal cord
B. Meninges—three membranous layers (Figure 12-3)
   1. Dura mater—strong, white fibrous tissue; outer layer of meninges and inner periosteum of the cranial bones; has three important extensions

   a. Falx cerebri
      (1) Projects downward into the longitudinal fissure between the two cerebral hemispheres
      (2) Dural sinuses—function as veins, collecting blood from brain tissues for return to the heart
      (3) Superior sagittal sinus—one of several dural sinuses
   b. Falx cerebelli—separates the two hemispheres of the cerebellum
   c. Tentorium cerebelli—separates the cerebellum from the cerebrum

2. Arachnoid membrane—delicate, cobwebby layer between the dura mater and pia mater
3. Pia mater—innermost, transparent layer; adheres to the outer surface of the brain and spinal cord; contains blood vessels; beyond the spinal cord, forms a slender filament called filum terminale, at level of sacrum, blends with dura mater to form a fibrous cord that disappears into the periosteum of the coccyx
4. Several spaces exist between and around the meninges
   a. Epidural space—located between the dura mater and inside the bony covering of the brain and spinal cord; contains a supporting cushion of fat and other connective tissues
   b. Subdural space—located between the dura mater and arachnoid membrane; contains lubricating serous fluid
   c. Subarachnoid space— located between the arachnoid and pia mater; contains a significant amount of cerebrospinal fluid

CEREBROSPINAL FLUID

A. Functions
   1. Provide a supportive, protective cushion
   2. Reservoir of circulating fluid, which is monitored by the brain to detect changes in the internal environment
B. Fluid spaces
   1. Cerebrospinal fluid—found within the subarachnoid space around the brain and spinal cord and within the cavities and canals of the brain and spinal cord
   2. Ventricles—fluid-filled spaces within the brain; four ventricles within the brain (Figure 12-4)
      a. First and second ventricles (lateral)—one located in each hemisphere of the cerebrum
      b. Third ventricle—thin, vertical pocket of fluid below and medial to the lateral ventricles
      c. Fourth ventricle—tiny, diamond-shaped space where the cerebellum attaches to the back of the brain stem
C. Formation and circulation of cerebrospinal fluid (Figure 12-5)
   1. Occurs by separation of fluid from blood in the choroid plexuses
      a. Fluid from the lateral ventricles seeps through the interventricular foramen (of Monro) into the third ventricle
      b. From the third ventricle goes through the cerebral aqueduct into the fourth ventricle
      c. From the fourth ventricle fluid goes to two different areas
         (1) Some fluid flows directly into the central canal of the spinal cord
         (2) Some fluid leaves the fourth ventricle through openings in its roof into the cisterna magna, a space that is continuous with the subarachnoid space
      d. Fluid circulates in the subarachnoid space and then is absorbed into venous blood through the arachnoid villi

SPINAL CORD

A. Structure of the spinal cord (Figure 12-6)
   1. Lies within the spinal cavity and extends from the foramen magnum to the lower border of the first lumbar vertebra
   2. Oval-shaped cylinder that tapers slightly from above downward
   3. Two bulges, one in the cervical region and one in the lumbar region
   4. Anterior median fissure and posterior median sulcus are two deep grooves; anterior fissure is deeper and wider
   5. Nerve roots
      a. Fibers of dorsal nerve root
         (1) Carry sensory information into the spinal canal
         (2) Dorsal root ganglion—cell bodies of unipolar, sensory neurons make up a small region of gray matter in the dorsal nerve root
      b. Fibers of ventral nerve root
         (1) Carry motor information out of the spinal cord
         (2) Cell bodies of multipolar, motor neurons are in the gray matter of the spinal cord
   6. Interneurons are located in the spinal cord's gray matter core
   7. Spinal nerve—a single mixed nerve on each side of the spinal cord where the dorsal and ventral nerve roots join together
   8. Gray matter
      a. Extends the length of the cord
      b. Consists predominantly of cell bodies of interneurons and motor neurons
      c. In transverse section, looks like an H with the limbs being called the anterior, posterior, and lateral horns of gray matter
   9. White matter
      a. Surrounds the gray matter and is subdivided in each half on the cord into three funiculi: anterior, posterior, and lateral white columns
      b. Each funiculus consists of a large bundle of axons divided into tracts
      c. Names of spinal tracts indicate the location of the tract, the structure in which the axons originate, and the structure in which they terminate
B. Functions of the spinal cord
   1. Provides two-way conduction routes to and from the brain
      a. Ascending tracts—conduct impulses up the cord to the brain
      b. Descending tracts—conduct impulses down the cord from the brain
      c. Bundles of axons compose all tracts
      d. Tracts are both structural and functional organizations of nerve fibers
         (1) Structural—all axons of any one tract originate in the same structure and terminate in the same structure
         (2) Functional—all axons that compose one tract serve one general function
      e. Important ascending (sensory) tracts (Figure 12-7)
         (1) Lateral spinothalamic tracts—crude touch, pain, and temperature
         (2) Anterior spinothalamic tracts—crude touch, pressure
         (3) Fasciculi gracilis and cuneatus—discriminating touch and conscious kinesthesia
         (4) Spinocerebellar tracts—subconscious kinesthesia
      f. Important descending (motor) tracts (Figure 12-7)
         (1) Lateral corticospinal tracts—voluntary movements on opposite side of the body
         (2) Anterior corticospinal tracts—voluntary movements on same side of body
         (3) Lateral reticulospinal tracts—transmits facilitatory impulses to anterior horn motor neurons to skeletal muscles
         (4) Medial reticulospinal tracts—inhibitory impulses to anterior horn motor neurons to skeletal muscles
         (5) Rubrospinal tracts—transmits impulses that coordinate body movements and maintenance of posture

g. Spinal cord—reflex center for all spinal reflexes; spinal reflex centers are located in the gray matter of the cord

THE BRAIN

A. Structures of the brain stem (Figure 12-9)
1. Medulla oblongata
   a. Lowest part of the brain stem
   b. Part of the brain that attaches to spinal cord, located just above the foramen magnum
   c. A few centimeters in length and separated from the pons above by a horizontal groove
   d. Composed of white matter and a network of gray and white matter called the reticular formation network
   e. Pyramids—two bulges of white matter located on the ventral side of the medulla; formed by fibers of the pyramidal tracts
   f. Olive—oval projection located lateral to the pyramids
   g. Nuclei—clusters of neuron cell bodies located in the reticular formation
2. Pons
   a. Located above the medulla and below the midbrain
   b. Composed of white matter and reticular formation
3. Midbrain
   a. Located above the pons and below the cerebrum; forms the midsection of the brain
   b. Composed of white tracts and reticular formation
   c. Extending divergently through the midbrain are cerebral peduncles; conduct impulses between the midbrain and cerebrum
   d. Corpora quadrigemina—landmark in midbrain
      (1) Made up of two inferior colliculi and two superior colliculi
      (2) Forms the posterior, upper part of the midbrain that lies just above the cerebellum
      (3) Inferior colliculus—contains auditory centers
      (4) Superior colliculus—contains visual centers
   e. Red nucleus and substantia nigra—clusters of cell bodies of neurons involved in muscular control
B. Functions of the brain stem
1. Performs sensory, motor, and reflex functions
2. Spinothalamic tracts—important sensory tracts that pass through the brain stem
3. Fasciculi cuneatus and gracilis and spinoreticular tracts—sensory tracts whose axons terminate in the gray matter of the brain stem
4. Corticospinal and reticulospinal tracts—two of the major tracts present in the white matter of the brain stem
5. Nuclei in medulla—contain reflex centers
   a. Of primary importance—cardiac, vasomotor, and respiratory centers
   b. Nonvital reflexes—vomiting, coughing, sneezing, etc.
6. Pons—contains reflexes mediated by fifth, sixth, seventh, and eighth cranial nerves and pneumotaxic centers that help regulate respiration
7. Midbrain—contains centers for certain cranial nerve reflexes
C. Structure of the cerebellum (Figure 12-10)
1. Second largest part of the brain
2. Located just below the posterior portion of the cerebrum; transverse fissure separates the two parts of the brain
3. Gray matter makes up the cortex, and white matter predominates in the interior
4. Arbor vitae—internal white matter of the cerebellum; distinctive pattern similar to the veins of a leaf

5. Cerebrum has numerous sulci and gyri
6. Consists of the cerebellar hemispheres and the vermis
7. Internal white matter—composed of short and long tracts
   a. Shorter tracts—conduct impulses from neuron cell bodies located in the cerebellar cortex to neurons whose dendrites and cell bodies compose nuclei located in the interior of the cerebellum
   b. Longer tracts—conduct impulses to and from the cerebellum; fibers enter or leave by way of three pairs of peduncles
      (1) Inferior cerebellar peduncles—composed chiefly of tracts into the cerebellum from the medulla and cord
      (2) Middle cerebellar peduncles—composed almost entirely of tracts into the cerebellum from the pons
      (3) Superior cerebellar peduncles—composed principally of tracts from dentate nuclei in the cerebellum through the red nucleus of the midbrain to the thalamus
8. Dentate nuclei
   a. Important pair of cerebellar nuclei, one of which is located in each hemisphere
   b. Nuclei connected with thalamus and with motor areas of the cerebral cortex by tracts
   c. By means of the tracts, cerebellar impulses influence the motor cortex, and the motor cortex influences the cerebellum
D. Functions of the cerebellum
1. Three general functions, all of which have to do with the control of skeletal muscles
   a. Acts with cerebral cortex to produce skilled movements by coordinating the activities of groups of muscles
   b. Controls skeletal muscles to maintain equilibrium
   c. Controls posture; operates at subconscious level to smooth movements and make movements efficient and coordinated
2. Cerebellum compares the motor commands of the cerebrum to the information coming from proprioceptors in the muscle; impulses travel from the cerebellum to both the cerebrum and muscles to coordinate movements to produce the intended action (Figure 12-11)
E. The diencephalon (Figure 12-12)
1. Located between the cerebrum and the midbrain
2. Consists of several structures located around the third ventricle: thalamus, hypothalamus, optic chiasma, pineal body, and several others
3. The thalamus
   a. Dumbbell-shaped mass of gray matter made up of many nuclei
   b. Each lateral mass forms one lateral wall of the third ventricle
   c. Intermediate mass—extends through the third ventricle and joins the two lateral masses
   d. Geniculate bodies—two of the most important groups of nuclei comprising the thalamus; located in posterior region of each lateral mass; play role in processing auditory and visual input
   e. Serves as a major relay station for sensory impulses on their way to the cerebral cortex
   f. Performs the following primary functions:
      (1) Plays two parts in mechanism responsible for sensations
         (a) Impulses produce conscious recognition of the crude, less critical sensations of pain, temperature, and touch

(b) Neurons relay all kinds of sensory impulses, except possibly olfactory, to the cerebrum
   (2) Plays part in the mechanism responsible for emotions by associating sensory impulses with feeling of pleasantness and unpleasantness
   (3) Plays part in arousal mechanism
   (4) Plays part in mechanisms that produce complex reflex movements
4. The hypothalamus
   a. Consists of several structures that lie beneath the thalamus
   b. Forms floor of the third ventricle and lower part of lateral walls
   c. Prominent structures found in the hypothalamus
      (1) Supraoptic nuclei—gray matter located just above and on either side of the optic chiasma
      (2) Paraventricular nuclei—located close to the wall of the third ventricle
      (3) Mamillary bodies—posterior part of hypothalamus, involved with olfactory sense
   d. Infundibulum—the stalk leading to the posterior lobe of the pituitary gland
   e. Small but functionally important area of the brain, performs many functions of greatest importance for survival and enjoyment
   f. Links mind and body
   g. Links nervous system to endocrine system
   h. Summary of hypothalamic functions
      (1) Regulator and coordinator of autonomic activities
      (2) Major relay station between the cerebral cortex and lower autonomic centers; crucial part of the route by which emotions can express themselves in changed bodily functions
      (3) Synthesizes hormones secreted by posterior pituitary and plays an essential role in maintaining water balance
      (4) Some neurons function as endocrine glands
      (5) Plays crucial role in arousal mechanism
      (6) Crucial part of mechanism regulating appetite
      (7) Crucial part of mechanism maintaining normal body temperature
5. Pineal body
   a. Located just above the corpora quadrigemina of the midbrain
   b. Involved in regulating the body's biological clock
   c. Produces some hormones—most notable hormone is melatonin
F. Structure of the cerebrum
   1. Cerebral cortex
      a. Largest and uppermost division of the brain; consists of right and left cerebral hemispheres; each hemisphere is divided into five lobes (Figure 12-13)
         (1) Frontal lobe
         (2) Parietal lobe
         (3) Temporal lobe
         (4) Occipital lobe
         (5) Insula (island of Riel)
      b. Cerebral cortex—outer surface made up of six layers of gray matter
      c. Gyri—convolutions; some are named: precentral gyrus, postcentral gyrus, cingulate gyrus, and hippocampal gyrus
      d. Sulci—shallow grooves
      e. Fissures—deeper grooves, divide each cerebral hemisphere into lobes; four prominent cerebral fissures
         (1) Longitudinal fissure—deepest fissure; divides cerebrum into two hemispheres
         (2) Central sulcus (fissure of Rolando)—groove between frontal and parietal lobes
         (3) Lateral fissure (fissure of Sylvius)—groove

between temporal lobe below and parietal lobes above; island of Reil lies deep in lateral fissure
         (4) Parietooccipital fissure—groove that separates occipital lobe from parietal lobes
   2. Cerebral tracts and basal ganglia
      a. Basal ganglia—islands of gray matter located deep inside the white matter of each hemisphere; include
         (1) Caudate nucleus
         (2) Lentiform nucleus—consists of putamen and globus pallidus
         (3) Amygdaloid nucleus
      b. Cerebral tracts make up cerebrum's white matter; there are three types (Figure 12-14)
         (1) Projection tracts—extensions of the sensory spinothalamic tracts and motor corticospinal tracts
         (2) Association tracts—most numerous cerebral tracts; extend from one convolution to another in the same hemisphere
         (3) Commissural tracts—extend from one convolution to a corresponding convolution in the other hemisphere; compose the corpus callosum and anterior and posterior commissures
      c. Corpus striatum—composed of caudate nucleus, internal capsule, and lentiform nucleus
G. Functions of the cerebral cortex
   1. Functional areas of the cortex—certain areas of the cerebral cortex engage in predominantly one particular function
      a. Postcentral gyrus—mainly general somatic sensory area; receives impulses from receptors activated by heat, cold, and touch stimuli
      b. Precentral gyrus—chiefly somatic motor area; impulses from neurons in this area descend over motor tracts and stimulate skeletal muscles
      c. Transverse gyrus—primary auditory area
      d. Occipital lobe—primary visual areas
   2. Sensory functions of the cortex
      a. Somatic senses—sensations of touch, pressure, temperature, proprioception, and similar perceptions that require complex sensory organs
      b. Cortex contains a "somatic sensory map" of the body
      c. Information sent to primary sensory areas is relayed to sensory association areas, as well as to other parts of the brain
      d. The sensory information is compared and evaluated, and the cortex integrates separate bits of information into whole perceptions
   3. Motor functions of the cortex
      a. For normal movements to occur, many parts of the nervous system must function
      b. Precentral gyrus—primary somatic motor area; controls individual muscles
      c. Secondary motor area—in the gyrus immediately anterior to the precentral gyrus; activates groups of muscles simultaneously
   4. Integrative functions of the cortex
      a. Consciousness
         (1) State of awareness of one's self, one's environment, and other beings
         (2) Depends on excitation of cortical neurons by impulses conducted to them by the reticular activating system
         (3) There are two current concepts about the reticular activating system
            (a) Functions as the arousal system for the cerebral cortex

(b) Its functioning is crucial for maintaining consciousness

b. Language
  (1) Ability to speak and write words and ability to understand spoken and written words
  (2) Speech centers—areas in the frontal, parietal, and temporal lobes
  (3) Left cerebral hemisphere contains speech centers in approximately 90% of the population; in the remaining 10%, contained in either the right hemisphere or both
  (4) Aphasias—lesions in speech centers

c. Emotions
  (1) Subjective experiencing and objective expression of emotions involve functioning of the limbic system
  (2) Limbic system—also known as the "emotional brain"
    (a) Most structures of limbic system lie on the medial surface of the cerebrum; they are the cingulate gyrus and hippocampus
    (b) Have primary connections with other parts of the brain, such as the thalamus, fornix, septal nuclei, amygdaloid nucleus, and hypothalamus

d. Memory
  (1) One of our major mental activities
  (2) Cortex is capable of storing and retrieving both short- and long-term memory
  (3) Temporal, parietal, and occipital lobes are among those areas responsible for short- and long-term memory
  (4) Engrams—structural traces in the cerebral cortex that comprise long-term memories
  (5) Cerebrum's limbic system plays a key role in memory

5. Specialization of cerebral hemispheres
  a. Right and left hemispheres of the cerebrum specialize in different functions; however, both sides of a normal person's brain communicate with each other to accomplish complex functions (Figure 12-16)
  b. Left hemisphere is responsible for
    (1) Language functions
    (2) Dominating control of certain hand movements
  c. Right hemisphere is responsible for
    (1) Perception of certain kinds of auditory material
    (2) Tactile perception
    (3) Perceiving and visualizing spatial relationships

6. The electroencephalogram (EEG)
  a. Records of brain electrical potentials obtained by recording brain waves
  b. Four types of brain waves based on frequency and amplitude (Figure 12-20)
    (1) Beta waves—frequency >13 Hz and relatively low voltage; "busy waves"
    (2) Alpha waves—frequency of 8 to 13 Hz and relatively low voltage; "relaxed waves"
    (3) Theta waves—frequency of 4 to 7 Hz and low voltage; "drowsy waves"
    (4) Delta waves—frequency of <4 Hz and high voltage; "deep sleep waves"

SOMATIC SENSORY PATHWAYS IN THE CENTRAL NERVOUS SYSTEM

A. For the cerebral cortex to perform its sensory functions, impulses must first be conducted to the sensory areas by sensory pathways (Figure 12-21)
B. Three main pools of sensory neurons

1. Sensory neurons I—conduct impulses from the periphery to the central nervous system
2. Sensory neurons II
  a. Conduct impulses from the cord or brain stem to the thalamus
  b. Dendrites and cell bodies are located in the gray matter of the cord and brain stem
  c. Axons ascend in ascending tracts up the cord, through the brain stem, and terminate in the thalamus where they synapse with sensory neuron III dendrites or cell bodies
3. Sensory neurons III
  a. Conduct impulses from thalamus to the postcentral gyrus of the parietal lobe
  b. Bundle of axons of sensory neurons III form the thalamocortical tracts
  c. Extend through the internal capsule to the cerebral cortex
C. Sensory pathways to the cerebral cortex are crossed
D. Two sensory pathways conduct impulses that produce sensations of touch and pressure
  1. Medial lemniscal system
    a. Consists of tracts that make up the fasciculi cuneatus and gracilis, and the medial lemniscus
    b. Axons of sensory neurons II make up medial lemniscus
    c. Functions—transmit impulses that produce discriminating touch and pressure sensations and kinesthesia
  2. Spinothalamic pathway—functions are crude touch and pressure sensations

SOMATIC MOTOR PATHWAYS IN THE CENTRAL NERVOUS SYSTEM

A. For the cerebral cortex to perform its motor functions, impulses are conducted from its motor areas to skeletal muscles by somatic motor pathways
B. Consist of motor neurons that conduct impulses from the central nervous system to skeletal muscles; some motor pathways are extremely complex and others are very simple
C. Principle of the final common path—cardinal principle about somatic motor pathways; only one final common path, the motor neuron from the anterior gray horn of the spinal cord, conducts impulses to skeletal muscles
D. Two methods used to classify somatic motor pathways
  1. Divides pathways into pyramidal and extrapyramidal tracts (Figure 12-22)
    a. Pyramidal tracts—also known as corticospinal tracts
      (1) Approximately three quarters of the fibers decussate in the medulla and extend down the cord in the crossed corticospinal tract located on the opposite side of the spinal cord in the lateral white column
      (2) Approximately one quarter of the fibers do not decussate but extend down the same side of the spinal cord as the cerebral area from which they came
    b. Extrapyramidal tracts—much more complex than pyramidal tracts
      (1) Consist of all motor tracts from the brain to the spinal cord anterior horn motor neurons except the corticospinal tracts
      (2) Within the brain, consist of numerous relays of motor neurons between motor areas of the cortex, basal ganglia, thalamus, cerebellum, and brain stem
      (3) Within the spinal cord, some important tracts are the reticulospinal tracts

(4) Conduction by extrapyramidal tracts plays a crucial part in producing large, automatic movements

(5) Conduction by extrapyramidal tracts plays an important part in emotional expressions

(6) Motor program—set of coordinated commands that control the programmed motor activity mediated by extrapyramidal pathways

CYCLE OF LIFE: CENTRAL NERVOUS SYSTEM

A. The development and degeneration of the central nervous system is the most obvious functional change over the life span

B. Development of brain and spinal cord begins in the womb

C. Lack of development in newborn is evidenced by lack of complex integrative functions
   1. Language
   2. Complex memory
   3. Comprehension of spatial relationships
   4. Complex motor skills

D. Complex functions develop by adulthood

E. Late adulthood—tissues degenerate
   1. Profound degeneration—unable to perform complex functions
   2. Milder degeneration—temporary memory lapse or difficulty with complex motor tasks

THE BIG PICTURE: THE CENTRAL NERVOUS SYSTEM AND THE WHOLE BODY

A. Central nervous system—ultimate regulator of the body; essential to survival

B. Able to integrate bits of information from all over the body, make sense of it, and make decisions

MECHANISMS OF DISEASE: DISORDERS OF THE CENTRAL NERVOUS SYSTEM

A. Neurons are destroyed by injury or disease

B. Cerebrovascular accident (stroke)—a hemorrhage from or cessation of blood flow through cerebral blood vessels
   1. Damage that occurs in a motor control area of the brain causes paralysis to appear on the side of the body opposite of the damaged side of the brain
   2. Hemiplegia—paralysis of one whole side of the body

C. Cerebral palsy—crippling childhood disease resulting from damage to brain tissue
   1. Damage is permanent and nonprogressive in motor control areas of the brain, present at birth or shortly thereafter, and remains throughout life
   2. Possible causes—prenatal infections or diseases of the mother, mechanical trauma to the head, nerve-damaging poisons, reduced oxygen to brain, other factors
   3. Spastic paralysis—paralysis characterized by involuntary contractions of affected muscles
      a. Four types
         (1) Hemiplegia—entire side paralyzed
         (2) Paraplegia—both legs paralyzed
         (3) Triplegia—both legs and one arm paralyzed
         (4) Quadriplegia—all four extremities paralyzed

D. Dementia—general term for various degenerative diseases resulting in the destruction of brain neurons that adversely affects memory, attention span, intellectual capacity, personality, and motor control
   1. Alzheimer's disease—dementia-producing lesions that develop in the cortex in middle-to-late adult years; exact cause of disease is unknown and there is no effective treatment
   2. Huntington's disease—inherited disease characterized by chorea; progresses to dementia and death
   3. Acquired immune deficiency syndrome (AIDS)—HIV virus infects neurons, causes progressive degeneration of brain, and results in dementia

E. Seizure disorders—common nervous system abnormalities characterized by seizures; sudden bursts of abnormal neuron activity
   1. Seizures are mild or severe (convulsions)
   2. Epilepsy—chronic seizure episodes of unknown cause; treatable with anticonvulsion drugs
      a. Electroencephalography (EEG)—method of diagnosing and evaluating epilepsy or other seizure disorders

REVIEW QUESTIONS

1. What term means the membranous covering of the brain and cord? What three layers compose this covering?
2. What are the large fluid-filled spaces within the brain called? How many are there? What do they contain?
3. Describe the formation and circulation of cerebrospinal fluid.
4. Describe the spinal cord's structure and general functions.
5. Explain what the term *reflex center* means.
6. List the major components of the brain stem and identify their general functions.
7. Describe the general functions of the cerebellum.
8. Describe the general functions of the thalamus.
9. Describe the general functions of the hypothalamus.
10. Describe the general functions of the cerebrum.
11. What general functions does the cerebral cortex perform?
12. Define consciousness. Name the normal states, or levels, of consciousness.
13. Explain briefly what is meant by the arousal, or alerting, mechanism.
14. Name some altered states of consciousness.
15. Compare meditation with the usual waking state.
16. Describe some of the current ideas about memory.
17. Describe the mechanism behind an electroencephalogram (EEG).
18. Identify the following kinds of brain waves according to their frequency, voltage, and the level of consciousness in which they predominate: alpha, beta, delta, theta.
19. Locate the dendrite, cell body, and axon of sensory neurons I, II, and III.
20. Explain each of the following: lower motor neuron, upper motor neuron, and final common path.
21. Compare pyramidal tract and extrapyramidal tract functions.
22. Where does a physician do a lumbar puncture? Why?
23. Explain the results of internal hydrocephalus.
24. Describe the various pain control areas.
25. What are the signs of motor pathway injury?

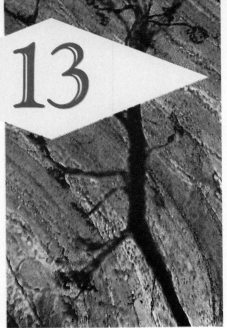

13 ▸ The Peripheral Nervous System

Nerve innervating skeletal muscle
Erlandsen/Magney: Color Atlas of Histology

OBJECTIVES

After you have completed this chapter, you should be able to:

1. Identify the cranial nerves by name and give the generalized function of each.
2. Discuss the generalized structure or branching of a typical spinal nerve.
3. Identify the location of the four major pairs of plexuses.
4. Identify the basic principles of somatic motor pathways.
5. List and discuss several of the somatic reflexes of clinical importance.
6. Identify the two major subdivisions of the autonomic nervous system.
7. Compare and contrast the structures of the parasympathetic and sympathetic pathways.
8. Identify the autonomic neurotransmitters and the fibers where they are found.
9. Discuss the function of the autonomic nervous system as a whole.
10. Compare and specify functions of the sympathetic and parasympathic divisions of the autonomic nervous system.

Cauda equina (KAU-dah e-KWI-nah) "Horse's tail"; the lower end of the spinal cord, with its attached spinal nerve roots

Cranial nerve One of the twelve pairs of nerves arising from the undersurface of the brain

Dermatome (DER-mah-tohm) Skin surface area supplied by a single spinal nerve

Ganglion (GANG-glee-on) Cluster of nerve cell bodies outside the central nervous system

Mixed nerve Nerve containing axons of sensory and motor neurons

Parasympathetic division Portion of the autonomic nervous system that arises from the brain stem and the sacral segment of the spinal cord; involved in functions such as digestion, defecation, urination ("rest-and-repair")

Plexus (PLEKS-us) Network of interwoven nerves

Ramus (RAY-mus) Branching of a spinal nerve

Reflex Action resulting from a nerve passing over a reflex arc

Spinal nerve One of the 31 pairs of nerves connected to the spinal cord

Sympathetic division Portion of the autonomic nervous system that arises from the thoracic and lumbar segments of the spinal cord; helps prepare body for immediate physical activity ("fight-or-flight")

In Chapter 12, you learned about the structure and function of the central nervous system (CNS). In this chapter we will explore the nerve pathways that lead to and from the CNS, which together comprise the peripheral nervous system (PNS). The PNS is made up of the 12 pairs of cranial nerves that emerge from the brain, the 31 pairs of spinal nerves that emerge from the spinal cord, and all of the smaller nerves that branch from these "main" nerves. Recall from earlier discussions that *afferent* fibers carry information into the CNS. Afferent fibers, part of the sensory nervous system, help us maintain homeostasis by sensing changes in our internal or external environment—providing the feedback necessary to keep the body functioning normally. Recall also that *efferent* fibers carry information away from the CNS. Efferent fibers may belong to the somatic nervous system that regulates skeletal muscles, allowing us to survive by defending ourselves, getting food, or performing other essential tasks. Efferent fibers may, on the other hand, belong to the autonomic nervous system (ANS). Autonomic regulation controls smooth and cardiac muscle and glands in ways that help us maintain homeostasis of the internal environment.

CRANIAL NERVES

Twelve pairs of **cranial nerves** arise from the undersurface of the brain (Figure 13-1), mostly from the brain stem. After leaving the cranial cavity by way of small foramina (holes) in the skull, they extend to their respective destinations. Both names and numbers identify the cranial nerves. Their names suggest either their distribution or their function. Their numbers indicate the order in which they emerge from the brain from anterior to posterior. Like all nerves, cranial nerves are made up of bundles of axons. **Mixed cranial nerves** contain axons of sensory and motor neurons. **Sensory cranial nerves** consist of sensory axons only and **motor cranial nerves** consist mainly of motor axons. Cranial nerves classified as "motor nerves" contain a small number of sensory fibers. These sensory fibers are *proprioceptive* fibers that carry information regarding tension in the muscles controlled by the motor fibers of the same motor nerve. Table 13-1 lists the name, number, and functional classification of each of the 12 pairs of cranial nerves. Details of the structure and function of each cranial nerve are described in Table 13-2 and in the paragraphs that follow.

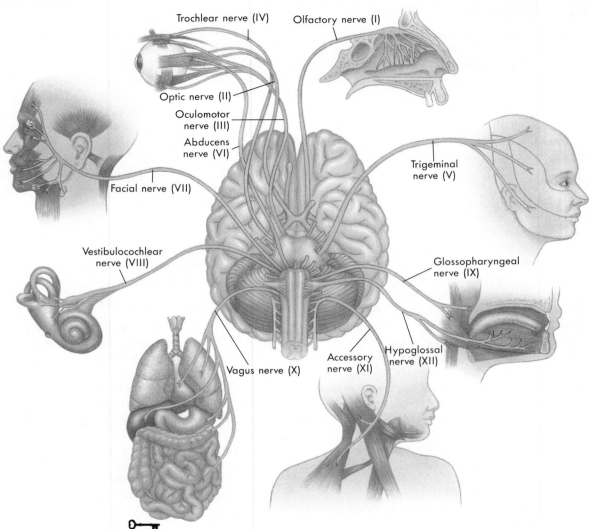

FIGURE 13-1   Cranial nerves. Ventral surface of the brain showing attachment of the cranial nerves.

TABLE 13-1   Names, numbers, and functional classifications of cranial nerves*

Name	Number	Functional Classification	Name	Number	Functional Classification
OLFACTORY	I	Sensory	**FACIAL**	VII	Mixed
OPTIC	II	Sensory	**VESTIBULOCOCHLEAR**	VIII	Sensory
OCULOMOTOR	III	Motor	**GLOSSOPHARYNGEAL**	IX	Mixed
TROCHLEAR	IV	Motor	**VAGUS**	X	Mixed
TRIGEMINAL	V	Mixed	**ACCESSORY**	XI	Motor
ABDUCENS	VI	Motor	**HYPOGLOSSAL**	XII	Motor

*The first letter of the words in the following sentence are the first letters of the name of the cranial nerves, in the correct order. Many anatomy students find that using this sentence, or one like it, helps in memorizing the names and numbers of the cranial nerves. It is "**O**n **O**ld **O**lympus' **T**iny **T**ops, **A** **F**riendly **V**iking **G**rew **V**ines **A**nd **H**ops." The functional classification of each cranial nerve can be remembered by using this sentence: "**S**ome **S**ay '**M**arry **M**oney,' **B**ut **M**y **B**rothers **S**ay '**B**ad **B**usiness, **M**arry **M**oney.'" **In this sentence, S** indicates *sensory,* **M** indicates *motor,* and **B** indicates *both* sensory and motor (mixed).

Olfactory Nerve (I)

The **olfactory nerves** are composed of axons of neurons whose dendrites and cell bodies lie in the nasal mucosa, high up along the septum and superior conchae (turbinates). Axons of these neurons form about 20 small fibers that pierce each cribriform plate and terminate in the olfactory bulbs (see Figure 14-2, p. 377). Here they synapse with the second pool of olfactory neurons, whose axons compose the olfactory tracts. The olfactory nerves carry information about the sense of smell.

Optic Nerve (II)

Axons from the innermost layer of sensory neurons of the retina compose the second pair of cranial nerves. They are called **optic nerves** because they carry visual information from the eyes to the brain. After entering the cranial cavity through the optic foramina, the two optic nerves unite to form the *optic chiasma*, in which some of the fibers of each nerve cross to the opposite side and continue in the *optic tract* of that side (see Figure 14-21, p. 394). Thus each optic nerve contains fibers only from the retina of the same side, whereas each optic tract has fibers in it from both retinas, a fact of importance in interpreting certain visual disorders. Most of the optic tract fibers terminate in the thalamus (in the portion known as the lateral geniculate nucleus). From here a new relay of fibers runs to the visual area of the occipital lobe cortex. A few optic tract fibers terminate in the superior colliculi of the midbrain, where they synapse with motor fibers to the external eye muscles (third, fourth, and sixth cranial nerves).

Oculomotor Nerve (III)

Fibers of the **oculomotor nerve** originate from cells in the oculomotor nucleus in the ventral part of the midbrain and extend to the various external eye muscles, with the exception of the superior oblique and the lateral rectus. Autonomic fibers are also present in the oculomotor nerves. They extend to the intrinsic muscles of the eye, which regulate the amount of light entering the eye and aid in focusing on near objects. Still a third group of fibers is found in the third cranial nerves, namely, sensory fibers from proprioceptors in the eye muscles.

Trochlear Nerve (IV)

Motor fibers of the **trochlear nerve** have their origin in cells in the midbrain, from which they extend to the superior oblique muscles of the eye. The name *trochlear,* which is from a Greek word that means "pulley," refers to the fact that the superior oblique muscles of the eye pass through a pulley-like ligament (see Figure 14-14 on p. 389). Afferent fibers from proprioceptors in these muscles are also contained in the trochlear nerves.

Trigeminal Nerve (V)

The fifth pair of cranial nerves is called **trigeminal nerves** because they each split into three large branches (*tri-,* "three"; *-geminal,* "pair"). The three branches of each pair are the *ophthalmic nerve, maxillary nerve,* and *mandibular nerve* (Figure 13-2). Sensory neurons in all three branches of the trigeminal nerve carry afferent impulses from the skin and mucosa of the head and from the teeth to cell bodies in the trigeminal ganglion (lodged in the petrous portion of the temporal bone). Fibers extend from the ganglion to the main sensory nucleus of the fifth cranial nerve situated in the pons. Motor fibers of the trigeminal nerve originate in the trifacial motor nucleus located in the pons just medial to the sensory nucleus. These motor fibers run to the muscles of mastication by way of the mandibular nerve.

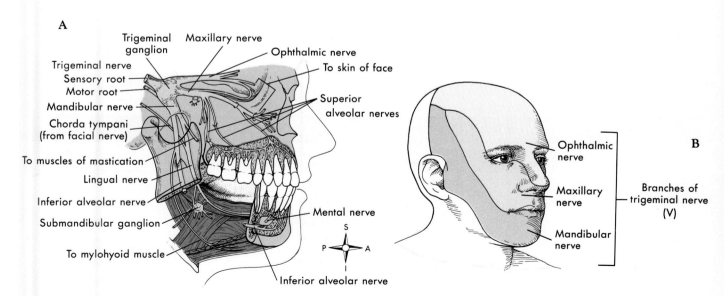

FIGURE 13-2 Trigeminal nerve (V). A, The route of the trigeminal nerve and its major branches. **B,** Sensory fibers of the trigeminal nerve form three branch nerves (ophthalmic, maxillary, and mandibular nerves), each of which conducts information from a different region of the face.

TRIGEMINAL NEURALGIA

Compression, inflammation, or degeneration of the fifth cranial nerve, the trigeminal nerve, may result in a condition called **trigeminal neuralgia** or *tic douloureux* (doo-loo-ROO). This condition is characterized by recurring episodes of intense stabbing pain, or *neuralgia*, radiating from the angle of the jaw along a branch of the trigeminal nerve.

One method of doing away with this pain is to remove the trigeminal ganglion on the posterior portion of the nerve (see Figure 13-2). This large ganglion contains the cell bodies of the nerve's afferent fibers. After such an operation the patient's face, scalp, teeth, and conjunctiva on the side treated show anesthesia. Special care, such as wearing protective goggles and irrigating the eye frequently, is therefore prescribed. The patient is instructed also to visit the dentist regularly, since one can no longer experience a toothache as a warning of diseased teeth.

Abducens Nerve (VI)

The **abducens nerve** is a motor nerve with fibers originating from a nucleus in the pons in the floor of the fourth ventricle and extending to the lateral rectus muscles of the eyes (see Figure 14-14, p. 389). The lateral rectus muscle *abducts* the eye to which it is attached, hence the name *abducens* for this nerve. The sixth cranial nerve also contains some afferent fibers from proprioceptors in the lateral rectus muscles.

Facial Nerve (VII)

The motor fibers of the **facial nerve** arise from a nucleus in the lower part of the pons, from which they extend by way of several branches to the superficial muscles of the face and scalp (Figure 13-3). Autonomic fibers of the facial nerve extend to the submaxillary and sublingual salivary glands, as well as to the lacrimal (tear) glands. Sensory fibers from the taste buds of the anterior two thirds of the tongue run in the facial nerve to cell bodies in the geniculate ganglion, a small swelling on the facial nerve, where it passes through a canal in the temporal bone. From the ganglion, fibers extend to a nucleus in the medulla.

Vestibulocochlear Nerve (VIII)

The **vestibulocochlear nerve** has two distinct divisions: the *vestibular nerve* and the *cochlear nerve* (see Figure 14-4, p. 380). Both are sensory. Fibers from the semicircular canals in the inner ear run to the vestibular ganglion (in the internal auditory meatus). Here, their cell bodies are located and from them fibers extend to the vestibular nuclei in the pons and medulla. Together, these fibers constitute the *vestibular nerve*. Some of its fibers run to the cerebellum. The vestibular nerve transmits impulses that result in sensations of equilibrium. The *cochlear nerve* consists of dendrites starting in the organ of Corti in the cochlea of the inner ear. They have their cell bodies in the spiral ganglion in the cochlea, and their axons terminate in the cochlear nuclei located between the medulla and pons. Conduction by the cochlear nerve results in sensations of hearing. Because of its role in hearing, the eighth cranial nerve is sometimes still called the *auditory* or *acoustic* nerve.

Glossopharyngeal Nerve (IX)

Both sensory and motor fibers compose the **glossopharyngeal nerve.** This nerve supplies fibers not only to the tongue and pharynx (throat), as its name implies, but also

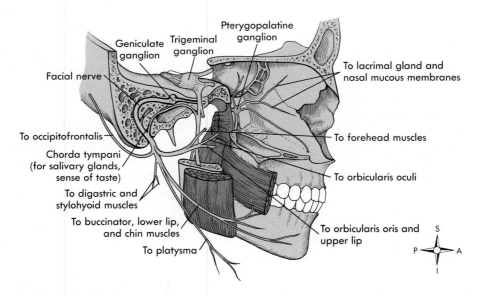

FIGURE 13-3   Facial nerve (VII). Artist's interpretation of the location of the various branches of the facial nerve.

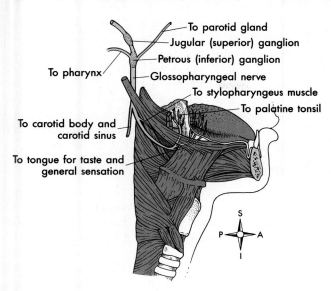

FIGURE 13-4 Glossopharyngeal nerve (IX). As its name implies (*glosso-*, "tongue"; -*pharyng-*, "throat"), the ninth cranial nerve supplies fibers to the tongue and throat. A mixed nerve, it carries both sensory and motor fibers.

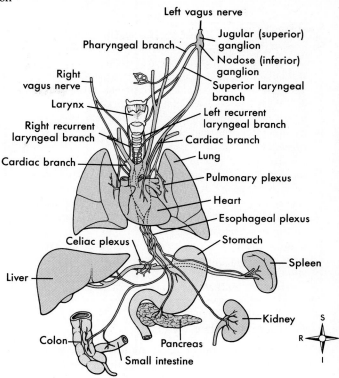

FIGURE 13-5 Vagus nerve (X). The vagus nerve is a mixed cranial nerve with many, widely distributed branches—hence the name *vagus*, which is the Latin word for "wanderer."

to other structures (Figure 13-4). One is the carotid sinus; it plays an important part in the control of blood pressure. Sensory fibers, with their receptors in the pharynx and posterior third of the tongue, have their cell bodies in the jugular (superior) and petrous (inferior) ganglia. These are located, respectively, in the jugular foramen and the petrous portion of the temporal bone. The ganglia fibers extend to a nucleus in the medulla. The motor fibers of the ninth cranial nerve originate in another medullary nucleus and run to muscles of the pharynx. Also present in this nerve are autonomic fibers that originate in a nucleus at the junction of the pons and medulla. These fibers run to the otic ganglion, from which postganglionic fibers extend to the parotid gland.

Vagus Nerve (X)

The **vagus nerve** contains both sensory and motor fibers. The name *vagus* means "wanderer" and aptly describes this nerve with many, widely distributed branches. Its sensory fibers supply the pharynx, larynx, trachea, heart, carotid body, lungs, bronchi, esophagus, stomach, small intestine, and gallbladder (Figure 13-5). Cell bodies for these sensory dendrites lie in the jugular and nodose ganglia, located, respectively, in the jugular foramen and just inferior to it on the trunk of the nerve. The sensory axons terminate in the medulla and in the pons. Somatic motor fibers of the vagus travel to the pharynx and larynx (voicebox), where they control muscles involved in swallowing. Most motor fibers of the vagus nerve are autonomic (parasympathetic) fibers. They originate in cells in the medulla and extend to various autonomic ganglia. From there, the fibers run to muscles of the pharynx, larynx, and thoracic and abdominal organs, where they control heart rate and other "visceral" activities.

HEALTH MATTERS

CRANIAL NERVE DAMAGE

Severe head injuries often damage one or more of the cranial nerves, producing symptoms analogous to the functions of the nerve affected. For example, injury of the sixth cranial nerve causes the eye to turn in because of paralysis of the abducting muscle of the eye. Injury of the eighth cranial nerve, on the other hand, produces deafness. Injury to the facial nerve results in a poker-faced expression and a drooping of the corner of the mouth from paralysis of the facial muscles.

Accessory Nerve (XI)

The **accessory nerve** is a motor nerve. Some of its fibers originate in cells in the medulla and pass by way of branches of the vagus nerve to thoracic and abdominal viscera, as well as the pharynx and larynx. Thus, this nerve can be considered an "accessory" to the vagus nerve. The rest of the fibers in the anterior gray column of the accessory nerve originate from the first five or six segments of the cervical spinal cord and extend through the spinal root of the accessory nerve to the trapezius and sternocleidomastoid muscles (Figure 13-6). Because of its spinal components, the eleventh cranial nerve was formerly called the *spinal accessory nerve.*

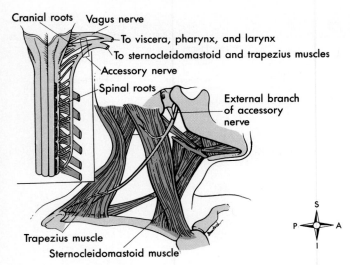

Cranial roots Vagus nerve
To viscera, pharynx, and larynx
To sternocleidomastoid and trapezius muscles
Accessory nerve
Spinal roots
External branch of accessory nerve
Trapezius muscle
Sternocleidomastoid muscle

S
P — A
I

FIGURE 13-6 Accessory nerve (XI). This cranial nerve is an "accessory" to the vagus nerve because some of its fibers join the vagus nerve before traveling on to the viscera, pharynx, and larynx. Fibers that originate in the cervical spinal segments travel through an external nerve branch to the trapezius and sternocleidomastoid muscles of the neck.

Hypoglossal Nerve (XII)

Motor fibers with cell bodies in the hypoglossal nucleus of the medulla compose the twelfth cranial nerve. They supply the muscles of the tongue (see Figure 13-1). The **hypoglossal nerve** also contains sensory fibers from proprioceptors in muscles of the tongue. The name *hypoglossal* means "under the tongue."

The main facts about the distribution and function of each of the cranial nerve pairs are summarized in Table 13-2.

QUICK CHECK

1. *What are the two efferent divisions of the peripheral nervous system called?*
2. *List the names and numbers of the twelve pairs of cranial nerves.*
3. *Distinguish between a motor nerve, sensory nerve, and mixed nerve.*

TABLE 13-2   Structure and function of the cranial nerves

Nerve	Sensory Fibers			Motor Fibers		Functions
	Receptors	Cell Bodies	Termination	Cell Bodies	Termination	
I OLFACTORY	Nasal mucosa	Nasal mucosa	Olfactory bulbs (new relay of neurons to olfactory cortex)			Sense of smell
II OPTIC	Retina	Retina	Nucleus in thalamus (lateral geniculate); some fibers terminate in superior colliculus of midbrain			Vision
III OCULOMOTOR	External eye muscles except superior oblique and lateral rectus			Midbrain (oculomotor nucleus)	External eye muscles except superior oblique and lateral rectus; autonomic fibers terminate in ciliary ganglion and then to ciliary and iris muscles	Eye movements, regulation of size of pupil, accommodation (for near vision), proprioception (muscle sense)

Continued

TABLE 13-2   Structure and function of the cranial nerves—cont'd

Nerve	Sensory Fibers			Motor Fibers		Functions
	Receptors	Cell Bodies	Termination	Cell Bodies	Termination	
IV TROCHLEAR	Superior oblique (proprioceptive)			Midbrain	Superior oblique muscle of eye	Eye movements, proprioception
V TRIGEMINAL	Skin and mucosa of head, teeth	Trigeminal ganglion	Pons (sensory nucleus)	Pons (motor nucleus)	Muscles of mastication	Sensations of head and face, chewing movements, proprioception
VI ABDUCENS	Lateral rectus (proprioceptive)			Pons	Lateral rectus muscle of eye	Abduction of eye, proprioception
VII FACIAL	Taste buds of anterior two thirds of tongue	Geniculate ganglion	Medulla (nucleus solitarius)	Pons	Superficial muscles of face and scalp; autonomic fibers to salivary and lacrimal glands	Facial expressions, secretion of saliva and tears, taste
VIII VESTIBULOCOCHLEAR						
Vestibular Branch	Semicircular canals and vestibule (utricle and saccule)	Vestibular ganglion	Pons and medula (vestibular nuclei)			Balance or equilibrium sense
Cochlear or Auditory Branch	Organ of Corti in cochlear duct	Spiral ganglion	Pons and medulla (cochlear nuclei)			Hearing

TABLE 13-2   Structure and function of the cranial nerves—cont'd

Nerve	Sensory Fibers			Motor Fibers		Functions
	Receptors	Cell Bodies	Termination	Cell Bodies	Termination	
IX GLOSSO-PHARYNGEAL	Pharynx; taste buds and other receptors of posterior one third of tongue	Jugular and petrous ganglia	Medulla (nucleus solitarius)	Medulla (nucleus ambiguus)	Muscles of pharynx	sensations of tongue, swallowing movements, secretion of saliva, aid in reflex control of blood pressure and respiration
	Carotid sinus and carotid body	Jugular and petrous ganglia	Medulla (respiratory and vasomotor centers)	Medulla at junction of pons (nucleus salivatorius)	Otic ganglion and then to parotid salivary gland	
X VAGUS	Pharynx, larynx, carotid body, and thoracic and abdominal viscera	Jugular and nodose ganglia	Medulla (nucleus solitarius), pons (nucleus of fifth cranial nerve)	Medulla (dorsal motor nucleus)	Ganglia of vagal plexus and then to muscles of pharynx, larynx, and autonomic fibers to thoracic and abdominal viscera	Sensations and movements of organs supplied; for example, slows heart, increases peristalsis, and contracts muscles for voice production
XI ACCESSORY	Trapezius and sternocleidomastoid (proprioceptive)			Medulla (dorsal motor nucleus of vagus and nucleus ambiguus)	Muscles of thoracic and abdominal viscera (autonomic) and pharynx and larynx	Shoulder movements, turning movements of head, movements of viscera, voice production, proprioception
				Anterior gray column of first five or six cervical segments of spinal cord	Trapezius and sternocleidomastoid muscle	
XII HYPOGLOSSAL	Tongue muscles (proprioceptive)			Medulla (hypoglossal nucleus)	Muscles of tongue and throat	Tongue movements, proprioception

SPINAL NERVES

Thirty-one pairs of **spinal nerves** are connected to the spinal cord. They have no special names but are merely numbered according to the level of the vertebral column at which they emerge from the spinal cavity (Figure 13-7). Although there are only 7 cervical vertebrae, there are 8 cervical nerve pairs (C1 through C8), 12 thoracic nerve pairs (T1 through T12), 5 lumbar nerve pairs (L1 through L5), 5 sacral nerve pairs (S1 through S5), and 1 coccygeal pair of spinal nerves. The first pair of cervical nerves emerge from the cord in the space above the first cervical vertebra, and the eighth cervical nerve emerges between the last cervical vertebra and the first thoracic vertebra. The rest of the cervical and all of the thoracic nerves pass out of the spinal cavity horizontally through the intervertebral foramina of their respective vertebrae.

Lumbar, sacral, and coccygeal nerve roots, on the other hand, descend from their point of origin at the lower end of the spinal cord (which terminates at the level of the first lumbar vertebra) before reaching the intervertebral foramina of their respective vertebrae, through which the nerves then emerge. This gives the lower end of the cord, with its attached spinal nerve roots, the appearance of a horse's tail. In fact, it bears the name **cauda equina,** which is the Latin equivalent for "horse's tail" (see Figure 13-7).

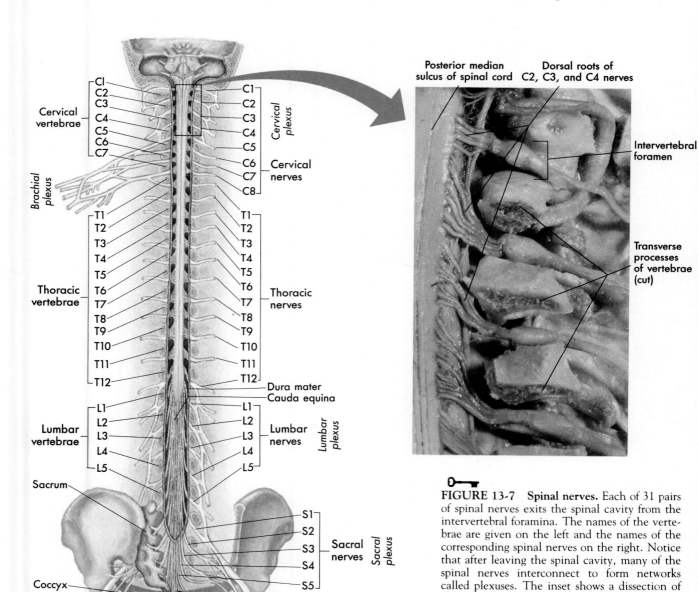

FIGURE 13-7   Spinal nerves. Each of 31 pairs of spinal nerves exits the spinal cavity from the intervertebral foramina. The names of the vertebrae are given on the left and the names of the corresponding spinal nerves on the right. Notice that after leaving the spinal cavity, many of the spinal nerves interconnect to form networks called plexuses. The inset shows a dissection of the cervical region, showing a posterior view of cervical spinal nerves exiting intervertebral foramina on the right side.

Structure of Spinal Nerves

Each spinal nerve attaches to the spinal cord by means of two short roots, a **ventral** (anterior) **root** and a **dorsal** (posterior) **root.** The dorsal root of each spinal nerve is easily recognized by a swelling called the **dorsal root ganglion,** or *spinal ganglion* (Figure 13-8). The roots and dorsal ganglia lie within the spinal cavity, as Figure 13-8 shows.

As described in the previous chapter, the ventral root includes motor neurons that carry information from the CNS and toward effectors (muscles and glands). Recall that in somatic motor pathways, a single motor fiber stretches from the anterior gray horn of the spinal cord, through the ventral root, and on through the spinal nerve toward a skeletal muscle. Autonomic fibers, which also carry motor information toward effectors, may also pass through the ventral root to become part of a spinal nerve. The dorsal root of each spinal nerve includes sensory fibers that carry information from receptors in peripheral nerves. The dorsal root ganglion contains the cell bodies of the sensory neurons. Because all spinal nerves contain both motor and sensory fibers, they are designated **mixed nerves.**

Soon after each spinal nerve emerges from the spinal cavity, it forms several large branches, each of which is called a **ramus** (plural, *rami*). As Figure 13-8 shows, each spinal nerve splits into a distinct **dorsal ramus** and **ventral ramus.** The dorsal ramus supplies somatic motor and sensory fibers to several smaller nerves. These smaller nerves, in turn, innervate the muscles and skin of the posterior surface of the head, neck, and trunk. The structure of the ventral ramus is a little more complex. Autonomic motor fibers split away from the ventral ramus,

heading toward a ganglion of the *sympathetic chain.* There, some of the autonomic fibers synapse with autonomic neurons that eventually continue on to autonomic effectors by way of *splanchnic nerves* (see Figure 13-8). However, some fibers synapse with autonomic neurons whose fibers rejoin the ventral ramus. The two thin rami formed by this splitting away, then rejoining of autonomic fibers are together called the *sympathetic rami.* Motor (autonomic and somatic) and sensory fibers of the ventral rami innervate muscles and glands in the extremities (arms and legs) and in the lateral and ventral portions of the neck and trunk.

Nerve Plexuses

The ventral rami of most spinal nerves—all but nerves T2 through T12—subdivide to form complex networks called **plexuses.** As Figure 13-7 shows, there are four major pairs of plexuses: the cervical plexus, the brachial plexus, the lumbar plexus, and sacral plexus. Table 13-3 summarizes important information about these major plexuses.

The term *plexus* is the Latin word for "braid." This is an apt name for a structure in which fibers of several different rami join together to form individual nerves. Each individual nerve that emerges from a plexus contains all the fibers that innervate a particular region of the body. In fact, the destination of each nerve serves as a basis for its name (see Table 13-3). Because spinal nerve fibers are thus rearranged according to their ultimate destination, the plexus reduces the number of nerves needed to supply each body part. And because each body region is innervated by fibers that originate in several different spinal nerves, damage to one spinal nerve does not mean a complete loss of function in any one region.

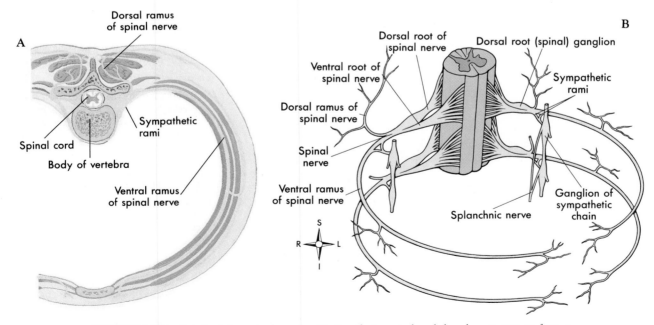

FIGURE 13-8   Rami of the spinal nerves. Notice that ventral and dorsal roots join to form a spinal nerve. The spinal nerve then splits into a dorsal *ramus* (plural, *rami*) and ventral *ramus.* The ventral ramus communicates with a chain of sympathetic (autonomic) ganglia via a pair of thin sympathetic rami. **A,** Superior view of a pair of thoracic spinal nerves. **B,** Anterior view of several pairs of thoracic spinal nerves.

TABLE 13-3   Spinal nerves and peripheral branches

Spinal Nerves	Plexuses Formed from Anterior Rami	Spinal Nerve Branches from Plexuses	Parts Supplied
CERVICAL 1 2 3 4	Cervical plexus	Lesser occipital Greater auricular Cutaneous nerve of neck Supraclavicular nerves Branches to muscles	Sensory to back of head, front of neck, and upper part of shoulder; motor to numerous neck muscles
		Phrenic nerve	Diaphragm
CERVICAL 5 6 7 8 **THORACIC (OR DORSAL)** 1	Brachial plexus	Suprascapular and dorsoscapular	Superficial muscles* of scapula
		Thoracic nerves, medial and lateral branches	Pectoralis major and minor
		Long thoracic nerve	Serratus anterior
		Thoracodorsal	Latissimus dorsi
		Subscapular	Subscapular and teres major muscles
		Axillary (circumflex)	Deltoid and teres minor muscles and skin over deltoid
2 3 4 5 6 7 8 9 10 11 12	No plexus formed; branches run directly to intercostal muscles and skin of thorax	Musculocutaneous	Muscles of front of arm (biceps brachii, coracobrachialis, and brachialis) and skin on outer side of forearm
		Ulnar	Flexor carpi ulnaris and part of flexor digitorum profundus; some of muscles of hand; sensory to medial side of hand, little finger, and medial half of fourth finger
		Median	Rest of muscles of front of forearm and hand; sensory to skin of palmar surface of thumb, index, and middle fingers
		Radial	Triceps muscle and muscles of back of forearm; sensory to skin of back of forearm and hand
		Medial cutaneous	Sensory to inner surface of arm and forearm
LUMBAR 1 2 3 4 5 **SACRAL** 1 2 3 4 5 **COCCYGEAL** 1	Lumbosacral plexus Coccygeal plexus	Iliohypogastric Sometimes fused	Sensory to anterior abdominal wall
		Ilioinguinal	Sensory to anterior abdominal wall and external genitalia; motor to muscles of abdominal wall
		Genitofemoral	Sensory to skin of external genitalia and inguinal region
		Lateral femoral cutaneous	Sensory to outer side of thigh
		Femoral	Motor to quadriceps, sartorius, and iliacus muscles; sensory to front of thigh and medial side of lower leg (saphenous nerve)
		Obturator	Motor to adductor muscles of thigh
		Tibial† (medial popliteal)	Motor to muscles of calf of leg; sensory to skin of calf of leg and sole of foot
		Common peroneal (lateral popliteal)	Motor to evertors and dorsiflexors of foot; sensory to lateral surface of leg and dorsal surface of foot
		Nerves to hamstring muscles	Motor to muscles of back of thigh
		Gluteal nerves	Motor to buttock muscles and tensor fasciae latae
		Posterior femoral cutaneous	Sensory to skin of buttocks, posterior surface of thigh, and leg
		Pudendal nerve	Motor to perineal muscles; sensory to skin of perineum

*Although nerves to muscles are considered motor, they do contain some sensory fibers that transmit proprioceptive impulses.

†Sensory fibers from the tibial and peroneal nerves unite to form the *medial cutaneous* (or sural) *nerve* that supplies the calf of the leg and the lateral surface of the foot. In the thigh the tibial and common peroneal nerves are usually enclosed in a single sheath to form the *sciatic nerve,* the largest nerve in the body with a width of approximately ¾ of an inch. About two thirds of the way down the posterior part of the thigh, it divides into its component parts. Branches of the sciatic nerve extend into the hamstring muscles.

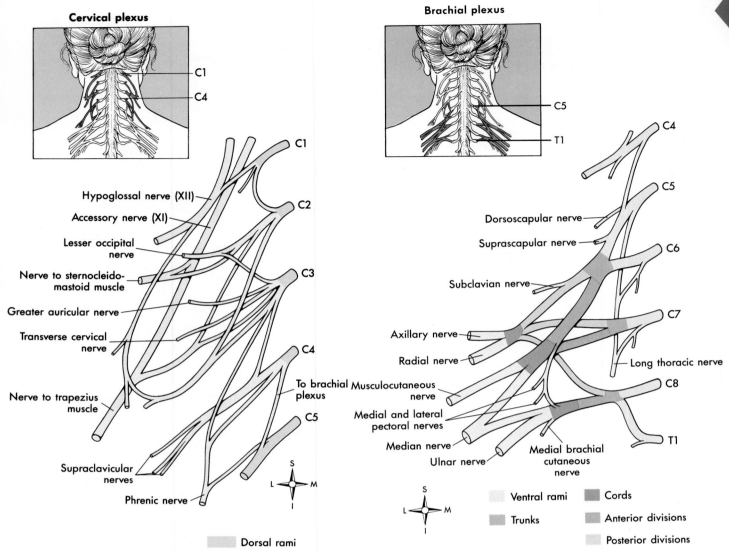

Cervical plexus

- C1
- C4

Brachial plexus

- C5
- T1

Hypoglossal nerve (XII)
Accessory nerve (XI)
Lesser occipital nerve
Nerve to sternocleido-mastoid muscle
Greater auricular nerve
Transverse cervical nerve
Nerve to trapezius muscle
Supraclavicular nerves
Phrenic nerve

C1
C2
C3
C4
C5
To brachial plexus

Dorsoscapular nerve
Suprascapular nerve
Subclavian nerve
Axillary nerve
Radial nerve
Musculocutaneous nerve
Medial and lateral pectoral nerves
Median nerve
Ulnar nerve
Medial brachial cutaneous nerve
Long thoracic nerve

C4
C5
C6
C7
C8
T1

Ventral rami · Cords · Trunks · Anterior divisions · Posterior divisions

Dorsal rami

FIGURE 13-9 Cervical plexus. Ventral rami of the first four cervical spinal nerves (C1 through C4) exchange fibers in this plexus found deep within the neck. Notice that some fibers from C5 also enter this plexus to form a portion of the phrenic nerve.

FIGURE 13-10 Brachial plexus. From the five rami, C5 through T1, the plexus forms three "trunks." Each trunk, in turn, subdivides into an anterior and posterior "division." The divisional branches then reorganize into three "cords." The cords then give rise to the individual nerves that exit this plexus.

The **cervical plexus,** shown in Figure 13-9, is found deep within the neck. Ventral rami of the first four cervical spinal nerves (C1 through C4), along with a branch of the ventral ramus of C5, exchange fibers in the cervical plexus. Individual nerves emerging from this plexus innervate the muscles and skin of the neck, upper shoulders, and part of the head. Also exiting this plexus is the **phrenic nerve,** which innervates the diaphragm. Two cranial nerves, the accessory nerve (XI) and the hypoglossal nerve (XII), receive small branches that emerge from the cervical plexus.

The **brachial plexus,** shown in Figure 13-10, is found deep within the shoulder. It passes from the ventral rami of spinal nerves C5 through T1, beneath the clavicle (collar bone), and toward the upper arm. Individual nerves that emerge from the brachial plexus innervate the lower part of the shoulder and all of the arm.

PHRENIC NERVES

The right and left **phrenic nerves,** whose fibers come from the cervical plexus, have considerable clinical interest, since they supply the diaphragm muscle. Contraction of the diaphragm permits inspiration, and relaxation of the diaphragm permits expiration. If the neck is broken in a way that severs or crushes the spinal cord above this level, nerve impulses from the brain can no longer reach the phrenic nerves, and therefore the diaphragm stops contracting. Unless artificial respiration of some kind is provided, the patient dies of respiratory paralysis as a result of the broken neck. Any disease or injury that damages the spinal cord between the third and fifth cervical segments may paralyze the phrenic nerve and, therefore, the diaphragm.

FIGURE 13-11 Lumbosacral plexus. This plexus is formed by the combination of the lumbar plexus with the sacral plexus, as shown in the inset. Notice that the ventral rami split into anterior and posterior "divisions" before reorganizing into the various individual nerves that exit this plexus.

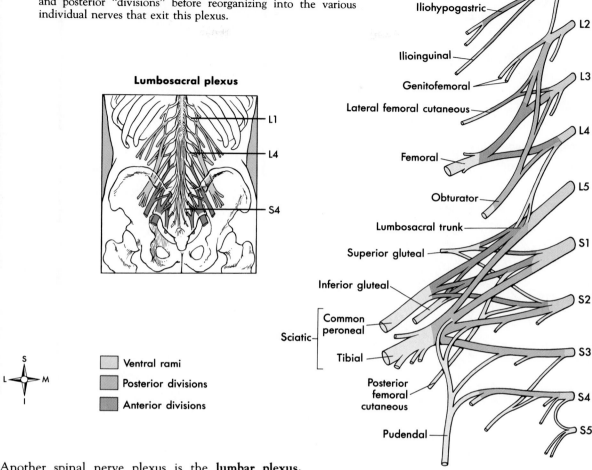

Lumbosacral plexus

Ventral rami

Posterior divisions

Anterior divisions

Another spinal nerve plexus is the **lumbar plexus**, formed by the intermingling of fibers from the first four lumbar nerves (Figure 13-11). This network of nerves is located in the lumbar region of the back in the psoas muscle. The large femoral nerve is one of several nerves emerging from the lumbar plexus. It divides into many branches supplying the thigh and leg.

Fibers from the fourth and fifth lumbar nerves (L4 and L5) and the first four sacral nerves (S1 through S4) form the **sacral plexus.** It lies in the pelvic cavity on the anterior surface of the piriformis muscle. Because of their close proximity and overlap of fibers, the lumbar and sacral plexuses are often considered together as the *lumbosacral plexus* (Figure 13-11). Among other nerves that emerge from the sacral plexus are the tibial and common peroneal nerves. In the thigh, they form the largest nerve in the body, the great sciatic nerve. It pierces the buttocks and runs down the back of the thigh. Its many branches supply nearly all the skin of the leg, the posterior thigh muscles, and the leg and foot muscles. *Sciatica* or neuralgia of the sciatic nerve is a fairly common and very painful condition.

The last sacral spinal nerve (S5), along with a few fibers from S4, join with the coccygeal nerve to form a small *coccygeal plexus*. Nerves arising from this plexus innervate the floor of the pelvic cavity and some surrounding areas.

Dermatomes

At first glance the distribution of spinal nerves does not appear to follow an ordered arrangement. But detailed mapping of the skin surface has revealed a close relationship between the spinal origin of each spinal nerve and the level of the body it innervates. Knowledge of the segmental arrangement of spinal nerves has proved useful to physicians. For instance, a neurologist can identify the site of spinal cord or nerve abnormality from the area of the body insensitive to a pinprick. Each skin surface area supplied by a single spinal nerve is called a **dermatome,** a name that means "skin section" (Figure 13-12).

QUICK CHECK

✔ 1. How many pairs of spinal nerves are there? Can you name them?

✔ 2. What is a plexus? Name the four major pairs of plexuses.

✔ 3. What is a dermatome?

HERPES ZOSTER

Herpes zoster, or **shingles,** is a unique viral infection that almost always affects the skin of a single dermatome. It is caused by the varicella zoster virus of chicken pox. About 3% of the population will suffer from shingles at some time in their lives. In most cases the disease results from reactivation of the varicella virus. The virus most likely traveled through a cutaneous nerve and remained dormant in a dorsal root ganglion for years after an episode of chicken pox. If the body's immunological protective mechanism becomes diminished in the elderly, or following stress, or in individuals undergoing radiation therapy or taking immunosuppressive drugs, the virus may reactivate. If this occurs, the virus will travel over the sensory nerve to the skin of a single dermatome. The figure shows dermatome involvement in a 13-year-old boy. The result is a painful eruption of red swollen plaques or vesicles that eventually rupture and crust before clearing in 2 to 3 weeks. In severe cases extensive inflammation, hemorrhagic blisters, and secondary bacterial infection may lead to permanent scarring. In most cases of shingles, the eruption of vesicles is preceded by 4 to 5 days of preeruptive pain, burning, and itching in the affected dermatome. Unfortunately, an attack of herpes zoster does not confer lasting immunity. Many individuals suffer three or more episodes in a lifetime.

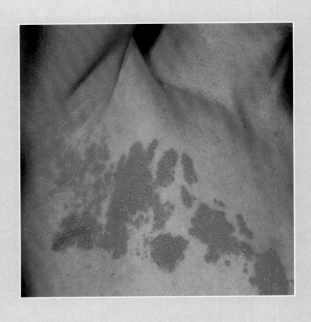

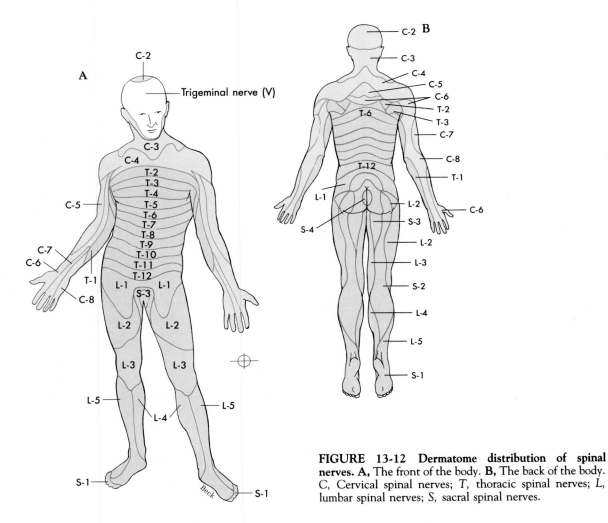

FIGURE 13-12 Dermatome distribution of spinal nerves. A, The front of the body. **B,** The back of the body. *C,* Cervical spinal nerves; *T,* thoracic spinal nerves; *L,* lumbar spinal nerves; *S,* sacral spinal nerves.

DIVISIONS OF THE PERIPHERAL NERVOUS SYSTEM

Recall from Chapter 11 that the peripheral nervous system (PNS) includes all the nervous pathways *outside* of the brain and spinal cord (see Figure 11-1, p. 282). Thus the entire PNS is comprised of the fibers present in the cranial nerves, the spinal nerves, and all of their individual branches. Although many of these nerves are *mixed nerves*—containing both sensory and motor fibers—it is often convenient to consider the PNS as having two functional divisions: the sensory division and the motor division. Details of the sensory division of the PNS will be left for Chapter 14. We will concentrate in this chapter on some essential details of the motor division of the PNS. First, we will briefly discuss the *somatic motor nervous system*, then move on to discuss the *autonomic nervous system*.

SOMATIC MOTOR NERVOUS SYSTEM
Basic Principles of Somatic Motor Pathways

The somatic motor nervous system includes all the voluntary motor pathways outside the CNS. That is, it involves the peripheral pathways to the skeletal muscles, which are the *somatic effectors*. Recall from the previous chapter that all of these pathways operate according to the principle of *final common path*. This means that all of the somatic motor pathways involve a single motor neuron whose axon stretches from the cell body in the CNS all the way to the effector innervated by that neuron. For fibers that originate in the spinal cord, this means that the axon extends from the anterior gray horn, through the ventral nerve root, and out to a skeletal muscle.

Another important principle of somatic motor pathways was mentioned in Chapter 9, when we were discussing the function of skeletal muscles. This principle states that the axon of the last somatic motor neuron—also called the *anterior horn neuron* or *lower motor neuron*—stimulates effector cells by means of the neurotransmitter **acetylcholine.** This fact will have added importance later when we compare the somatic motor nervous system with the autonomic nervous system.

Since the basic plan of the somatic motor pathways was discussed in Chapter 12, and the effect of acetylcholine on the somatic effectors (skeletal muscles) has already been outlined in Chapter 9, we will end our discussion of the somatic motor division of the PNS with a brief review of the concept of the reflex arc and how it relates to peripheral motor pathways.

Somatic Reflexes
Nature of a reflex

The action that results from a nerve impulse passing over a reflex arc is called a **reflex** (see Figure 11-6, p. 287). In other words, a reflex is a predictable response to a stimulus. It may or may not be conscious. Usually the term is used to mean only involuntary responses rather than those directly willed, that is, involving cerebral cortex activity. If the center of a reflex arc is in the brain, the response it mediates is called a **cranial reflex.** If the center of a reflex arc is in the spinal cord, the response is called a **spinal reflex.**

A reflex consists of either muscle contraction or glandular secretion. **Somatic reflexes** are contractions of skeletal muscles. Impulse conduction over somatic reflex arcs—arcs whose motor neurons are somatic motor neurons, that is, anterior horn neurons or lower motor neurons—produce somatic reflexes. **Autonomic (or visceral) reflexes** consist of contractions of smooth or cardiac muscle or secretion by glands; they are mediated by impulse conduction over autonomic reflex arcs, the motor neurons of which are autonomic neurons (discussed later). The following paragraphs describe only somatic reflexes.

Some somatic reflexes of clinical importance

Clinical interest in reflexes stems from the fact that they deviate from normal in certain diseases. Therefore the testing of reflexes is a valuable diagnostic aid. Frequently tested are the following reflexes: knee jerk, ankle jerk, Babinski reflex, corneal reflex, and abdominal reflex.

The *knee jerk*, or patellar reflex, is an extension of the lower leg in response to tapping of the patellar tendon (Figure 13-13). The tap stretches both the tendon and its muscles, the quadriceps femoris, and thereby stimulates muscle spindles (receptors) in the muscle and initiates conduction over the following two-neuron reflex arc:

1. *Sensory neurons*
   a. Dendrites—in femoral and second, third, and fourth lumbar nerves
   b. Cell bodies—second, third, and fourth lumbar ganglia
   c. Axons—in posterior roots of second, third, and fourth lumbar nerves; terminate in these segments of the spinal cord; synapse directly with lower motor neurons
2. *Reflex center*—synapses in anterior gray column between axons of sensory neurons and dendrites and cell bodies of lower motor neurons
3. *Motor neurons*
   a. Dendrites and cell bodies—in spinal cord anterior gray column
   b. Axons—in anterior roots of second, third, and fourth lumbar spinal nerves and femoral nerves; terminate in quadriceps femoris muscle

The knee jerk can be classified in various ways as follows:

♦ As a *spinal cord reflex*—because the center of the reflex arc (which transmits the impulses that activate the muscles producing the knee jerk) lies in the spinal cord gray matter
♦ As a *segmental reflex*—because impulses that mediate it enter and leave the same segment of the cord
♦ As an *ipsilateral reflex*—because the impulses that mediate it come from and go to the same side of the body

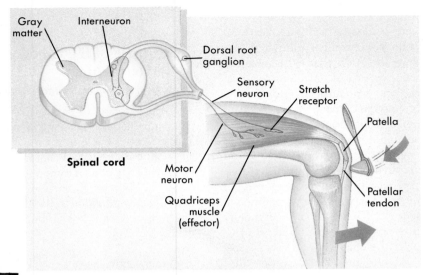

FIGURE 13-13   Patellar reflex. Neural pathway involved in the patellar (knee-jerk) reflex.

- As a *stretch reflex,* or *myotatic reflex* (from the Greek *mys,* "muscle," *tasis,* "stretching")—because of the kind of stimulation used to evoke it
- As an *extensor reflex*—because it is produced by extensors of the lower leg (muscles located on the anterior surface of the thigh, which extend the lower leg)
- As a *tendon reflex*—because tapping of a tendon is the stimulus that elicits it
- As a *deep reflex*—because of the deep location (in tendon and muscle) of the receptors stimulated to produce this reflex (as opposed to *superficial reflexes*—those elicited by stimulation of receptors located in the skin or mucosa)

When testing a patient's reflexes, a physician interprets the test results on the basis of what is known about the reflex arcs that must function to produce normal reflexes.

To illustrate, suppose that a patient has been diagnosed as having poliomyelitis. In examining him the physician finds that she cannot elicit the knee jerk when she taps the patient's patellar tendon. She knows that the poliomyelitis virus attacks anterior horn motor neurons. She also knows the information previously related about which spinal cord segments contain the reflex centers for the knee jerk. On the basis of this knowledge, therefore, she deduces that in this patient the poliomyelitis virus has damaged the second, third, and fourth lumbar segments of the spinal cord. Do you think that this patient's leg would be paralyzed, that he would be unable to move it voluntarily? What neurons would not be able to function that must function to produce voluntary contractions?

Ankle jerk, or *Achilles reflex,* is an extension (plantar flexion) of the foot in response to tapping of the Achilles tendon. Like the knee jerk, it is a tendon reflex and a deep reflex mediated by two-neuron spinal arcs. The centers for the ankle jerk lie in the first and second sacral segments of the cord.

The *Babinski reflex* is an extension of the great toe, with or without fanning of the other toes, in response to stimulation of the outer margin of the sole of the foot. Normal infants, up until they are about 1½ years old, show this Babinski reflex. By about this time, corticospinal fibers have become fully myelinated and the Babinski reflex becomes suppressed. Just why this is so is not clear. But at any rate it is, and a Babinski reflex after this age is abnormal. From then on the normal response to stimulation of the outer edge of the sole is the *plantar reflex.* It consists of a curling under of all the toes (plantar flexion) plus a slight turning in and flexion of the anterior part of the foot. A positive Babinski reflex is one of the pyramidal signs (p. 332) and is interpreted to mean destruction of pyramidal tract (corticospinal) fibers.

The *corneal reflex* is blinking in response to touching the cornea. It is mediated by reflex arcs with sensory fibers in the ophthalmic branch of the fifth cranial nerve, centers in the pons, and motor fibers in the seventh cranial nerve.

The *abdominal reflex* is drawing in of the abdominal wall in response to stroking the side of the abdomen. It is mediated by arcs with sensory and motor fibers in the ninth to twelfth thoracic spinal nerves and centers in these segments of the cord. It is classified as a superficial reflex. A decrease in this reflex or its absence occurs in lesions involving pyramidal tract upper motor neurons. It can, however, be absent without any pathological condition—in pregnancy, for example.

QUICK CHECK
1. *What are the somatic effectors?*
2. *State the principle of final common path as it applies to the somatic motor nervous system.*
3. *Explain how the knee jerk reflex can be both a stretch reflex and a spinal reflex.*

AUTONOMIC NERVOUS SYSTEM

The **autonomic nervous system (ANS)** is a subdivision of the efferent portion of the peripheral nervous system (see Figure 11-1, p. 282). It is the ANS that carries efferent neurons to the autonomic, or *visceral*, effectors—cardiac muscle, smooth muscle, and glandular epithelial tissue (Table 13-4). The major function of the ANS is to regulate heartbeat, smooth muscle contraction, and glandular secretion in ways that maintain homeostasis.

The ANS has two divisions: the **sympathetic division** and the **parasympathetic division.** The sympathetic division consists of neural pathways that are separate from the parasympathetic pathways. However, many autonomic effectors are *dually innervated.* That is, many autonomic effectors receive input from both sympathetic and parasympathetic pathways. Usually the effects of the two systems are antagonistic: one inhibits the effector, the other stimulates the effector. This allows a doubly innervated effector to be controlled with remarkable precision. It also allows a doubly innervated effector to participate in timed events, such as the sexual response, in which effectors must be stimulated then rapidly inhibited (or vice versa) in a specific timed sequence. Some autonomic effectors are singly innervated, receiving input from only the sympathetic division.

Structure of the Autonomic Nervous System
Basic plan of autonomic pathways

Each autonomic pathway, whether sympathetic or parasympathetic, is made up of autonomic nerves, ganglia, and plexuses. These structures, in turn, are made up of autonomic neurons. All autonomic neurons are efferent neurons. They conduct impulses away from the brain stem or spinal cord to autonomic effectors. Like all efferent neurons, autonomic neurons function in reflex arcs. Thus, like somatic motor regulation, autonomic regulation ultimately depends on feedback from the sensory nervous system.

A relay of two autonomic neurons conducts information from the CNS to the autonomic effectors. The first is called a *preganglionic neuron*—an awkward name, but a descriptive one. Preganglionic neurons conduct impulses from the brain stem or spinal cord to an autonomic ganglion, locating them *before* the ganglion—thus *pre*ganglionic. Within an autonomic ganglion, the preganglionic neuron synapses with a second efferent neuron. Because this second neuron conducts impulses away from the ganglion and to the effector, it is called the *postganglionic neuron.* As you can see in Figure 13-14, this plan is fundamentally different from the efferent pathways of the somatic motor nervous system. Conduction to somatic effectors requires only one efferent neuron, the somatic motor neuron that originates in the anterior gray horn of the spinal cord. Conduction to autonomic effectors, however, requires a sequence of two efferent neurons from the CNS to the effector. Essential features of the somatic motor pathways and autonomic pathways are further compared and contrasted in Table 13-5.

Structure of the sympathetic pathways

Most of the ganglia of the sympathetic division lie along either side of the anterior surface of the vertebral column (see Figure 12-3, p. 310). Both gray rami and white rami

TABLE 13-4	Visceral effector tissues and organs	
Cardiac Muscle	**Smooth Muscle**	**Glandular Epithelium**
Heart	Blood vessels	Sweat glands
	Bronchial tubes	Lacrimal glands
	Stomach	Digestive glands (salivary, gastric, pancreas, liver)
	Gallbladder	
	Intestines	Adrenal medulla
	Urinary bladder	
	Spleen	
	Eye (iris and ciliary muscles)	
	Hair follicles	

TABLE 13-5	Comparison of somatic motor and autonomic pathways	
Feature	**Somatic Motor Pathways**	**Autonomic Pathways**
Direction of information flow	Efferent	Efferent
Number of neurons between CNS and effector	One (somatic motor neuron)	Two (preganglionic and postganglionic)
Myelin sheath present	Yes	Preganglionic: yes Postganglionic: no
Location of peripheral fibers	Most cranial nerves and all spinal nerves	Most cranial nerves and all spinal nerves
Effector innervated	Skeletal muscle (voluntary)	Smooth and cardiac muscle, glands (involuntary)
Neurotransmitter	Acetylcholine	Acetylcholine or norepinephrine

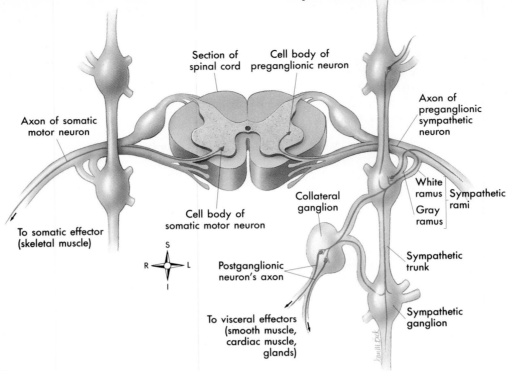

FIGURE 13-14 Autonomic conduction paths. The left side of the diagram shows that one somatic motor neuron conducts impulses all the way from the spinal cord to a somatic effector. Conduction from the spinal cord to any visceral effector, however, requires a relay of at least two autonomic motor neurons—a preganglionic and a postganglionic neuron, shown on the right side of the diagram.

are sympathetic rami. Short fibers from both rami form a trunk that connects the ganglia that lie on the same side of the vertebral column, forming a structure that resembles a chain of beads (see Figure 13-14). For this reason, they are often referred to as the "sympathetic chain ganglia."

Each chain extends from the second cervical vertebra in the neck all the way down to the level of the coccyx. There are usually 22 sympathetic chain ganglia on each side of the vertebral column: three cervical, eleven thoracic, four lumbar, and four sacral. Sympathetic preganglionic neurons have their dendrites and cell bodies in the lateral gray horns of the thoracic and lumbar segments of the spinal cord (see Figure 13-14). For this reason, the sympathetic division has also been called the *thoracolumbar division.* Axons of sympathetic preganglionic neurons leave the cord by way of the ventral roots of the thoracic and first four lumbar spinal nerves. From there, they split away from other spinal nerve fibers by means of a small branch called the *white ramus.* The white ramus gets its name from the fact that most of the sympathetic preganglionic fibers within it are myelinated axons. Note in Figure 13-14 that the sympathetic preganglionic axon extends through the white ramus to a sympathetic chain ganglion. Once inside the sympathetic chain ganglion, the preganglionic fiber may take any of three paths:

1. It can synapse with a sympathetic postganglionic neuron.

2. It can send ascending and/or descending branches through the sympathetic trunk to synapse with postganglionic neurons in other chain ganglia.

3. It can pass through one or more ganglia without synapsing.

Preganglionic neurons that pass through chain ganglia without synapsing continue on through **splanchnic nerves** to other sympathetic ganglia (see Figures 13-14 and 13-15). These **collateral ganglia,** or *prevertebral ganglia,* are pairs of sympathetic ganglia located a short distance from the spinal cord. The collateral ganglia are named for nearby blood vessels. For example, the *celiac ganglion* (also called the *solar plexus)* is a large ganglion that lies next to the celiac artery just below the diaphragm. Other examples include the *superior mesenteric ganglion* and the *inferior mesenteric ganglion,* each located close to the beginning of an artery of the same name. Some of the preganglionic fibers that enter the celiac ganglion do not synapse there but continue on to the central portion (medulla) of the *adrenal gland.* Within the adrenal medulla, they synapse with postganglionic neurons, which in turn release neurotransmitters (mostly epinephrine) into the bloodstream. These chemical messengers may reach the various sympathetic effectors, where they enhance and prolong the effects of sympathetic stimulation.

Most of the **sympathetic postganglionic neurons** have their dendrites and cell bodies in the sympathetic chain ganglia or collateral ganglia. Some postganglionic axons

return to a spinal nerve by way of a short branch called the *gray ramus,* so named because most postganglionic fibers are unmyelinated (see Figure 13-14). Once in the spinal nerve, the postganglionic fibers are distributed with other nerve fibers to the various sympathetic effectors. On the other hand, some postganglionic fibers are distributed to sympathetic effectors by way of separate autonomic nerves. The

course of postganglionic fibers through these autonomic nerves is complex, involving the redistribution of fibers in autonomic plexuses before they reach their respective destinations.

In the sympathetic division, preganglionic neurons are relatively short, and postganglionic neurons are relatively long.

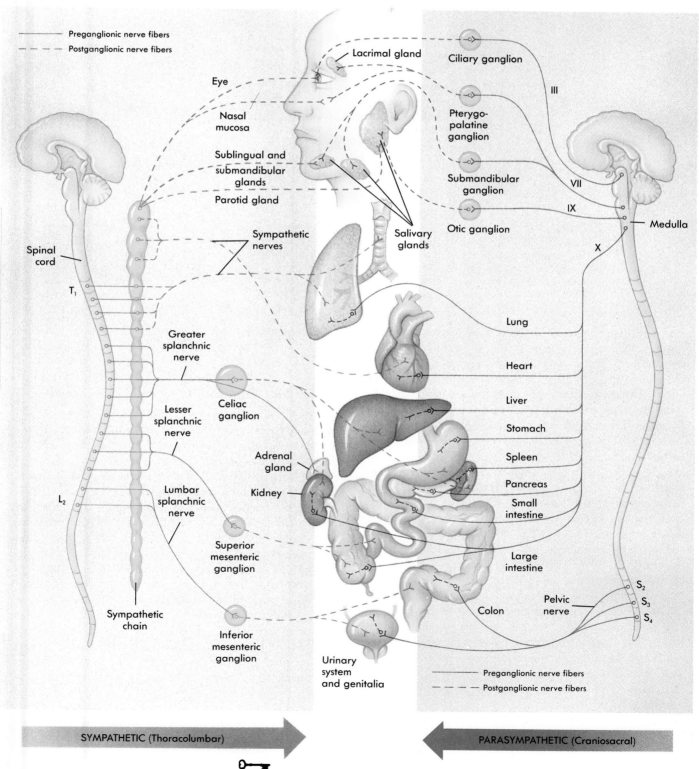

FIGURE 13-15 Major autonomic pathways.

The axon of any one sympathetic preganglionic neuron synapses with many postganglionic neurons, and these frequently terminate in widely separated organs. This anatomical fact partially explains a well-known physiological principle—sympathetic responses are usually widespread, involving many organs and not just one.

QUICK CHECK

✔ 1. *Do autonomic pathways follow the principle of final common path?*

✔ 2. *Why is the sympathetic division of the ANS also known as the thoracolumbar division?*

✔ 3. *Describe the path generally taken by an impulse along a sympathetic pathway from the CNS to an autonomic effector.*

Structure of the parasympathetic pathways

Parasympathetic preganglionic neurons have their cell bodies in nuclei in the brain stem or in the lateral gray columns of the sacral cord. For this reason, the parasym-

pathetic division has also been called the *craniosacral division*. Axons of parasympathetic preganglionic neurons are contained in cranial nerves III, VII, IX, and X and in some pelvic nerves. They extend a considerable distance before synapsing with postganglionic neurons. For example, at least 75% of all parasympathetic preganglionic fibers travel in the *vagus nerve* (X) for a distance of a foot or more before synapsing with postganglionic fibers in **terminal ganglia** near effectors in the chest and abdomen (see Figure 13-15 and Table 13-6).

Parasympathetic postganglionic neurons have their dendrites and cell bodies in parasympathetic ganglia. Unlike sympathetic ganglia that lie near the spinal column, parasympathetic ganglia lie near or embedded in autonomic effectors. For example, note the ciliary ganglion in Figure 13-15. This and the other ganglia shown near it are parasympathetic ganglia located in the skull. In a parasympathetic ganglion, preganglionic axons synapse with postganglionic neurons that send their short axons into the nearby autonomic effector. A parasympathetic preganglionic neuron, therefore, usually synapses with postgangli-

---

TABLE 13-6   Comparison of structural features of the sympathetic and parasympathetic pathways

Neurons	Sympathetic	Parasympathetic
PREGANGLIONIC NEURONS		
Dendrites and cell bodies	In lateral gray columns of thoracic and first four lumbar segments of spinal cord	In nuclei of brain stem and in lateral gray columns of sacral segments of cord
Axons	In anterior roots of spinal nerves to spinal nerves (thoracic and first four lumbar), to and through white rami to terminate in sympathetic ganglia at various levels or to extend through sympathetic ganglia, to and through splanchnic nerves to terminate in collateral ganglia	From brain stem nuclei through cranial nerve III to ciliary ganglion From nuclei in pons through cranial nerve VII to sphenopalatine or submaxillary ganglion From nuclei in medulla through cranial nerve IX to otic ganglion or through cranial nerves X and XI to cardiac and celiac ganglia, respectively
Distribution	Short fibers from CNS to ganglion	Long fibers from CNS to ganglion
Neurotransmitter	Acetylcholine	Acetylcholine
GANGLIA	Sympathetic chain ganglia (22 pairs); collateral ganglia (celiac, superior, and inferior mesenteric)	Terminal ganglia (in or near effector)
POSTGANGLIONIC NEURONS		
Dendrites and cell bodies	In sympathetic and collateral ganglia	In parasympathetic ganglia (for example, ciliary, sphenopalatine, submaxillary, otic, cardiac, celiac) located in or near visceral effector organs
Axons	In autonomic nerves and plexuses that innervate thoracic and abdominal viscera and blood vessels in these cavities In gray rami to spinal nerves, to smooth muscle of skin blood vessels and hair follicles, and to sweat glands	In short nerves to various visceral effector organs
Distribution	Long fibers from ganglion to widespread effectors	Short fibers from ganglion to single effector
Neurotransmitter	Norepinephrine (many); acetylcholine (few)	Acetylcholine

onic neurons to a single effector. For this reason, parasympathetic stimulation frequently involves response by only one organ. Sympathetic stimulation, on the other hand, usually evokes responses by numerous organs.

Autonomic neurotransmitters

Axon terminals of autonomic neurons release either of two neurotransmitters: **norepinephrine** or **acetylcholine.** Axons that release norepinephrine are known as *adrenergic fibers.* Axons that release acetylcholine are called *cholinergic fibers.* Autonomic cholinergic fibers are the axons of preganglionic sympathetic neurons and of both preganglionic and postganglionic parasympathetic neurons. This leaves the axons of postganglionic sympathetic neurons as the only autonomic adrenergic fibers, and as you can see in Figure 13-16, not all of these are adrenergic. Sympathetic postganglionic axons to sweat glands and some blood vessels are cholinergic fibers.

Norepinephrine affects visceral effectors by first binding to *adrenergic receptors* in their plasma membranes. The adrenergic receptors are of two types, one named **alpha**

receptors and the other named **beta receptors.** The binding of norepinephrine to alpha receptors in the smooth muscle of blood vessels has a stimulating effect on the muscle that causes the vessel to constrict. The binding of norepinephrine to beta receptors in smooth muscle produces opposite effects. It inhibits the muscle, causing the vessel to dilate. But the binding of norepinephrine to beta receptors in cardiac muscle has a stimulating effect that results in a faster and stronger heartbeat. Epinephrine released by the sympathetic postganglionic cells in the adrenal medulla also stimulates the adrenergic receptors, enhancing and prolonging the effects of sympathetic stimulation. Because epinephrine has a greater effect on beta receptors than norepinephrine, effectors with a large proportion of beta receptors are more sensitive to epinephrine. All of these facts point to an important principle about nervous regulation: the effect of a neurotransmitter on any postsynaptic cell is determined by the characteristics of the receptor and not by the neurotransmitter itself.

The actions of norepinephrine and epinephrine are terminated in two ways. Most of the neurotransmitter

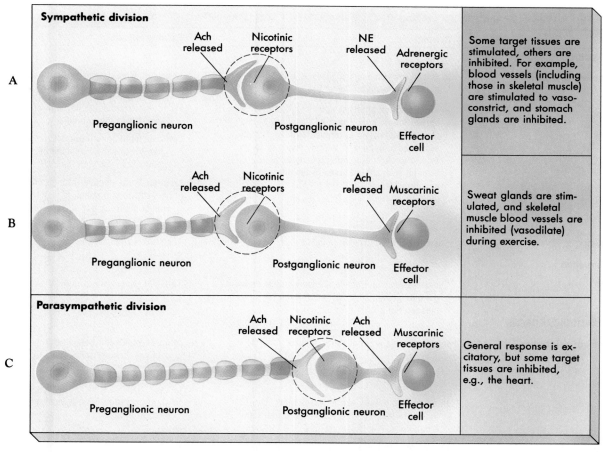

FIGURE 13-16 Neurotransmitters and receptors of the autonomic nervous system. In all pathways, preganglionic fibers are cholinergic, secreting acetylcholine *(Ach)*, which stimulates nicotinic receptors in the postganglionic neuron. Most sympathetic preganglionic fibers are adrenergic **(A)**, secreting norepinephrine *(NE)*. A few sympathetic postganglionic fibers are cholinergic, stimulating muscarinic receptors in effector cells **(B)**. All parasympathetic postganglionic fibers are cholinergic **(C)**, stimulating muscarinic receptors in effector cells.

BETA BLOCKERS

Drugs that bind to beta receptors, and thus block the binding of norepinephrine and epinephrine, are informally called *beta blockers*. Propranolol (Inderal) is one of nearly a dozen beta blockers used to treat irregular heartbeats and hypertension (high blood pressure). Beta blockers achieve their therapeutic effects by preventing the increased rate and strength of the heart's pumping action triggered by sympathetic stimulation. For patients who also have had a previous heart attack (myocardial infarction), these drugs possibly have the added benefit of preventing further attacks.

molecules are taken back up by the synaptic knobs of postganglionic neurons, where they are broken down by the enzyme *monoamine oxidase (MAO)*. The remaining neurotransmitter molecules are eventually broken down by another enzyme, *catechol-O-methyl transferase (COMT)*. Both of these mechanisms are very slow compared to the rapid deactivation of acetylcholine by acetylcholinesterase. This fact explains why adrenergic effects often linger for some time after stimulation has ceased.

Acetylcholine binds to *cholinergic receptors*. There are two types of cholinergic receptors: **nicotinic receptors** and **muscarinic receptors**. Nicotinic receptors derive their name from the fact that they were first discovered when nicotine, a drug, was shown to bind to them. Muscarinic receptors are named for the fact that their discovery came about when it was shown that muscarine, a toxin, binds to them. In the ganglia of both autonomic divisions, acetylcholine binds to nicotinic receptors in the membranes of postganglionic cells. Acetylcholine, released by all parasympathetic postganglionic cells and the few sympathetic postganglionic cells that are cholinergic, binds to muscarinic receptors in the membranes of effector cells. As mentioned previously, acetylcholine's action is quickly terminated by its being hydrolyzed by the enzyme *acetylcholinesterase*.

QUICK CHECK

✔ 1. Describe the pathway taken by an impulse traveling along a parasympathetic pathway.
✔ 2. What is the difference between a cholinergic fiber and an adrenergic fiber? Between a cholinergic receptor and an adrenergic receptor?
✔ 3. Name two major types of cholinergic receptors and two major types of adrenergic receptors.

Functions of the Autonomic Nervous System
Overview of autonomic function

The autonomic nervous system as a whole functions to regulate autonomic effectors in ways that tend to maintain or quickly restore homeostasis. Both sympathetic and parasympathetic divisions are *tonically active*, that is, they continually conduct impulses to autonomic effectors. They often exert opposite, or antagonistic, influences on them—a fact that we might call the principle of autonomic antagonism. If sympathetic impulses tend to stimulate an effector, parasympathetic impulses tend to inhibit it. Doubly innervated effectors continually receive both sympathetic and parasympathetic impulses. Summation of the two opposing influences determines the dominating or controlling effect. For example, continual sympathetic impulses to the heart tend to accelerate the heart rate while continual parasympathetic impulses tend to slow it. The actual heart rate is determined by whichever influence dominates. To find other examples of autonomic antagonism, examine Table 13-7.

The autonomic nervous system does not function autonomously as its name suggests. It is continually influenced by impulses from the so-called autonomic centers. These are clusters of neurons located at various levels in the brain whose axons conduct impulses directly or indirectly to autonomic preganglionic neurons. Autonomic centers function as a hierarchy in their control of the autonomic system (Figure 13-17). Highest ranking in the hierarchy are the autonomic centers in the cerebral cortex, for example, in the frontal lobe and limbic system (structures near medial surface of cerebrum that form a border around the corpus callosum). Neurons in these

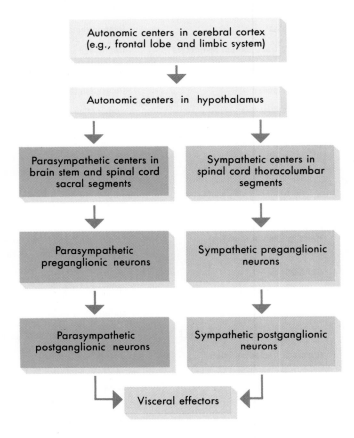

FIGURE 13-17 Central nervous system hierarchy that regulates autonomic functions.

centers send impulses to other autonomic centers in the brain, notably in the hypothalamus. Then neurons in the hypothalamus send either stimulating or inhibiting impulses to parasympathetic and sympathetic preganglionic neurons located in the lower autonomic centers of the brain stem and cord.

You may be wondering why the name *autonomic system* was ever chosen in the first place if the system is really not autonomous. Originally the term seemed appropriate. The autonomic system seemed to be self-regulating and independent of the rest of the nervous system. Common observations furnished abundant evidence of its independence from cerebral control, from direct control by the will, that is. But later, even this was found to be not entirely true. Some rare and startling exceptions were discovered, for example, a man in a brightly lighted amphitheater can change the size of his pupils from small, constricted dots (normal response to bright lights) to widely dilated circles. It is also possible to will the smooth muscle of the hairs on the arms to contract, producing gooseflesh.

TABLE 13-7 Autonomic functions

Autonomic Effector	Effect of Sympathetic Stimulation (Neurotransmitter: Norepinephrine Unless Otherwise Stated)	Effect of Parasympathetic Stimulation (Neurotransmitter: Acetylcholine)
CARDIAC MUSCLE	Increased rate and strength of contraction (beta receptors)	Decreased rate and strength of contraction
SMOOTH MUSCLE OF BLOOD VESSELS		
Skin blood vessels	Constriction (alpha receptors)	No effect
Skeletal muscle blood vessels	Dilation (beta receptors)	No effect
Coronary blood vessels	Constriction (alpha receptors) Dilation (beta receptors)	Dilation
Abdominal blood vessels	Constriction (alpha receptors)	No effect
Blood vessels of external genitals	Constriction (alpha receptors)	Dilation of blood vessels causing erection
SMOOTH MUSCLE OF HOLLOW ORGANS AND SPHINCTERS		
Bronchioles	Dilation (beta receptors)	Constriction
Digestive tract, except sphincters	Decreased peristalsis (beta receptors)	Increased peristalsis
Sphincters of digestive tract	Constriction (alpha receptors)	Relaxation
Urinary bladder	Relaxation (beta receptors)	Contraction
Urinary sphincters	Constriction (alpha receptors)	Relaxation
Reproductive ducts	Contraction (alpha receptors)	Relaxation
Eye		
Iris	Contraction of radial muscle; dilated pupil	Contraction of circular muscle; constricted pupil
Ciliary	Relaxation; accommodates for far vision	Contraction; accommodates for near vision
Hairs (pilomotor muscles)	Contraction produces goose pimples, or piloerection (alpha receptors)	No effect
GLANDS		
Sweat	Increased sweat (neurotransmitter, acetylcholine)	No effect
Lacrimal	No effect	Increased secretion of tears
Digestive (salivary, gastric, etc.)	Decreased secretion of saliva; not known for others	Increased secretion of saliva
Pancreas, including islets	Decreased secretion	Increased secretion of pancreatic juice and insulin
Liver	Increased glycogenolysis (beta receptors); increased blood sugar level	No effect
Adrenal medulla*	Increased epinephrine secretion	No effect

*Sympathetic preganglionic axons terminate in contact with secreting cells of the adrenal medulla. Thus the adrenal medulla functions, to quote someone's descriptive phrase, as a "giant sympathetic postganglionic neuron."

Biofeedback

We now know that individuals can learn to control specific autonomic effectors if two conditions are fulfilled. They must be informed that they are achieving the desired response and they must be rewarded for it. Various kinds of *biofeedback* instruments have been developed to provide these conditions. For example, a biofeedback instrument that detects slight temperature changes has been used with patients who suffer from migraine headaches. (Migraine headaches initially involve distention of blood vessels in the head.) The instrument is attached to their hands and emits a high sound each time they happen to dilate their hand blood vessels. The reward for such patients is a lessening of the migraine pain—presumably because of the shunting of blood away from the head to the hands.

Functions of the sympathetic division

Under ordinary, resting conditions the sympathetic division can act to maintain the normal functioning of doubly innervated autonomic effectors. It does this by opposing the effects of parasympathetic impulses to these structures. For example, by counteracting parasympathetic impulses that tend to slow the heart and weaken its beat, sympathetic impulses function to maintain the heartbeat's normal rate and strength. The sympathetic division also serves another important function under usual conditions. Since only sympathetic fibers innervate the smooth muscle in blood vessel walls, sympathetic impulses function to maintain the normal tone of this muscle. By so doing, the sympathetic system plays a crucial role in maintaining blood pressure under usual conditions.

The major function of the sympathetic division, however, is that it serves as an "emergency" system. When we perceive that the homeostasis of the body might be threatened—that is, when we are under physical or psychological stress—outgoing sympathetic signals increase greatly. In fact, one of the very first steps in the body's complex defense mechanism against stress is a sudden and marked increase in sympathetic activity. This brings about a group of responses that all go on at the same time. Together they make the body ready to expend maximum energy and thus to engage in the maximum muscular exertion needed to deal with the perceived threat—as, for example, in running or fighting. Walter B. Cannon coined the descriptive and now famous phrase—the **"fight-or-flight" reaction**—as his name for this group of sympathetic responses. Read Table 13-8 to find many of the "fight-or-flight" physiological changes. Some particularly important changes for maximum energy expenditure by skeletal muscles are faster, stronger heartbeat, dilated blood vessels in skeletal muscles, dilated bronchi, and increased blood sugar levels from stimulated glycogenolysis (conversion of glycogen to glucose). Also, sympathetic impulses to the medulla of each adrenal gland stimulate its secretion of epinephrine and some norepinephrine. These

TABLE 13-8	Summary of the sympathetic "fight-or-flight" reaction
Response	**Role in Promoting Energy Use by Skeletal Muscles**
Increased heart rate	Increased rate of blood flow, thus increased delivery of oxygen and glucose to skeletal muscles
Increased strength of cardiac muscle contraction	Increased rate of blood flow, thus increased delivery of oxygen and glucose to skeletal muscles
Dilation of coronary vessels of the heart	Increased delivery of oxygen and nutrients to cardiac muscle to sustain increased rate and strength of heart contractions
Dilation of blood vessels in skeletal muscles	Increased delivery of oxygen and nutrients to skeletal muscles
Constriction of blood vessels in digestive and other organs	Shunting of blood to skeletal muscles to increase oxygen and glucose delivery
Contraction of spleen and other blood reservoirs	More blood discharged into general circulation, causing increased delivery of oxygen and glucose to skeletal muscles
Dilation of respiratory airways	Increased loading of oxygen into blood
Increased rate and depth of breathing	Increased loading of oxygen into blood
Increased sweating	Increased dissipation of heat generated by skeletal muscle activity
Increased conversion of glycogen into glucose	Increased amount of glucose available to skeletal muscles

hormones reinforce and prolong effects of the norepinephrine released by sympathetic postganglionic fibers.

The fight-or-flight reaction is a normal response in times of stress. Without such a response, we might not be able to resist or retreat from something that actually threatens our well-being. However, chronic exposure to stress can lead to dysfunction of sympathetic effectors—and perhaps even to the dysfunction of the autonomic nervous system itself. Some current concepts regarding the effects of chronic stress are discussed in Chapter 21.

Functions of the parasympathetic division

The parasympathetic division is the dominant controller of most autonomic effectors most of the time. Under quiet, nonstressful conditions, more impulses reach autonomic effectors by cholinergic parasympathetic fibers than by adrenergic sympathetic fibers. If the sympathetic division

dominates during times that require "fight-or-flight," then the parasympathetic division dominates during the in-between times of "rest-and-repair." Acetylcholine, the neurotransmitter of the parasympathetic system, tends to slow the heartbeat but acts to promote digestion and elimination. For example, it stimulates digestive gland secretion. It also increases peristalsis by stimulating the smooth muscle of the digestive tract. Note other parasympathetic effects in Table 13-7.

QUICK CHECK

✔ 1. What is the principle of autonomic antagonism? Give an example.
✔ 2. Name the responses that occur in the fight-or-flight reaction. How do each of these prepare the body to expend a maximum amount of muscular energy?
✔ 3. Which division of the ANS is the dominant controller of autonomic effectors when the body is at rest?

THE BIG PICTURE
Peripheral Nervous System and the Whole Body

The peripheral nervous system is made up of all of the afferent nervous pathways coming into the CNS and all of the efferent pathways going out of the CNS. We emphasized the peripheral efferent, or motor, pathways in this chapter. We will emphasize the peripheral sensory pathways in Chapter 14. As we have stated time and again, the main role of the nervous system as a whole is to detect changes in the internal and external environment, to evaluate those changes in terms of their effect on homeostatic balance, and to regulate effectors accordingly. The peripheral motor pathways are simply those nervous pathways that lead from the integrator (CNS) to the effectors.

The somatic motor pathways lead to skeletal muscle effectors,

and the autonomic pathways lead to the cardiac muscle effectors, smooth muscle effectors, and glandular effectors. Thus all together the peripheral motor pathways serve as an information-carrying network that allows the CNS to communicate regulatory information to all of the nervous effectors in the body. Because some motor pathways communicate with endocrine glands, endocrine effectors throughout the body can also be regulated through nervous mechanisms. Every major organ of the body is thus influenced, directly or indirectly, by nervous output. In essence, the CNS is the ultimate controller of the major homeostatic functions of the body, and peripheral motor pathways are the means to exert that control.

CHAPTER SUMMARY

CRANIAL NERVES (Tables 13-1 and 13-2)

A. Overview
   1. Twelve pairs of cranial nerves arise from the brain, mostly the brain stem (Figure 13-1)
   2. Identified by name (determined by either distribution or function) and/or number (order in which they emerge, anterior to posterior)
   3. Made up of bundles of axons
      a. Mixed cranial nerve—axons of sensory and motor neurons
      b. Sensory cranial nerve—axons of sensory neurons only
      c. Motor cranial nerve—mainly axons of motor neurons and a small number of sensory fibers (proprioceptors)
B. Olfactory nerve (I)
   1. Composed of axons of neurons whose dendrites and cell bodies lie in nasal mucosa and terminate in olfactory bulbs
   2. Carries information about sense of smell
C. Optic nerve (II)
   1. Composed of axons from the innermost layer of sensory neurons of the retina
   2. Carries visual information from the eyes to the brain
D. Oculomotor nerve (III)

   1. Fibers originate from cells in the oculomotor nucleus and extend to some of the external eye muscles
   2. Autonomic fibers are also present, which extend to the intrinsic muscles of the eye to regulate amount of light entering eye and aid focusing on near objects
   3. Sensory fibers from proprioceptors in the eye muscles are also present
E. Trochlear nerve (IV)
   1. Motor fibers originate in cells of the midbrain and extend to the superior oblique muscles of the eye
   2. Also contains afferent fibers from proprioceptors in the superior oblique muscles of the eye
F. Trigeminal nerve (V)
   1. Has three branches: ophthalmic nerve, maxillary nerve, and mandibular nerve
   2. Sensory neurons carry afferent impulses from skin and mucosa of head and teeth to cell bodies in the trigeminal ganglion
   3. Motor fibers originate in trifacial motor nucleus and extend to the muscles of mastication through the mandibular nerve
G. Abducens nerve (VI)
   1. Motor nerve with fibers originating from a nucleus in the pons on the floor of the fourth ventricle and extending to the lateral rectus muscles of the eye

2. Contains afferent fibers from proprioceptors in the lateral rectus muscles

H. Facial nerve (VII)
   1. Motor fibers originate from a nucleus in lower part of pons and extend to superficial muscles of the face and scalp (Figure 13-3)
   2. Autonomic fibers extend to submaxillary and sublingual salivary glands
   3. Also contains sensory fibers from taste buds of anterior two thirds of the tongue

I. Vestibulocochlear nerve (VIII)
   1. Two distinct divisions that are both sensory: vestibular nerve and cochlear nerve
   2. Vestibular nerve fibers originate in the semicircular canals in inner ear and transmit impulses that result in sensations of equilibrium
   3. Cochlear nerve fibers originate in the organ of Corti in the cochlea of the inner ear and transmit impulses that result in sensations of hearing

J. Glossopharyngeal nerve (IX)
   1. Composed of sensory, motor, and autonomic nerve fibers
   2. Supplies fibers to tongue, pharynx, and carotid sinus (Figure 13-4)

K. Vagus nerve (X)
   1. Composed of sensory and motor fibers with many, widely distributed branches
   2. Sensory fibers supply pharynx, larynx, trachea, heart, carotid body, lungs, bronchi, esophagus, stomach, small intestine, and gallbladder (Figure 13-5)
   3. Somatic motor fibers innervate the pharynx and larynx and are mostly autonomic fibers

L. Accessory nerve (XI)
   1. Motor nerve that is an "accessory" to the vagus nerve
   2. Innervates thoracic and abdominal viscera, pharynx, larynx, trapezius, and sternocleidomastoid (Figure 13-6)

M. Hypoglossal nerve (XII)
   1. Composed of motor and sensory fibers
   2. Motor fibers innervate the muscles of the tongue
   3. Contains sensory fibers from proprioceptors in muscles of the tongue

SPINAL NERVES

A. Overview
   1. Thirty-one pairs of spinal nerves are connected to the spinal cord (Figure 13-7)
   2. No special names; are numbered by level of vertebral column at which they emerge from the spinal cavity
      a. Eight cervical nerve pairs (C1 through C8)
      b. Twelve thoracic nerve pairs (T1 through T12)
      c. Five lumbar nerve pairs (L1 through L5)
      d. Five sacral nerve pairs (S1 through S5)
      e. One coccygeal nerve pair
   3. Lumbar, sacral, and coccygeal nerve roots descend from point of origin to the lower end of the spinal cord (level of first lumbar vertebra) before reaching the intervertebral foramina of the respective vertebrae, through which the nerves emerge
   4. Cauda equina—describes the appearance of the lower end of the spinal cord and its spinal nerves as a horse's tail

B. Structure of spinal nerves
   1. Each spinal nerve attaches to spinal cord by a ventral (anterior) root and a dorsal (posterior) root
   2. Dorsal root ganglion—swelling in the dorsal root of each spinal nerve
   3. All spinal nerves are mixed nerves
   4. Ramus
      a. One of several large branches formed after each spinal nerve emerges from the spinal cavity (Figure 13-8)
      b. Dorsal ramus—supplies somatic motor and sensory fibers to smaller nerves that innervate the muscles and skin of the posterior surface of the head, neck, and trunk
      c. Ventral ramus
         (1) Structure is more complex than that of dorsal ramus
         (2) Autonomic motor fibers split from the ventral ramus and head toward a ganglion of the sympathetic chain
         (3) Some autonomic fibers synapse with neurons that continue on to autonomic effectors through splanchnic nerves; others synapse with neurons whose fibers rejoin the ventral ramus
         (4) Sympathetic rami—splitting and rejoining of autonomic fibers
         (5) Motor and sensory fibers innervate muscles and glands in the extremities and lateral and ventral portions of neck and trunk

C. Nerve plexuses
   1. Plexuses—complex networks formed by the ventral rami of most spinal nerves (**not** T2 through T12) subdividing and then joining together to form individual nerves
   2. Each individual nerve that emerges contains all the fibers that innervate a particular region of the body
   3. In plexuses, spinal nerve fibers are rearranged according to their ultimate destination, reducing the number of nerves needed to supply each body part
   4. There are four major pairs of plexuses
      a. Cervical plexus (Figure 13-9)
         (1) Located deep within the neck
         (2) Made up of ventral rami of C1 through C4 and a branch of the ventral ramus of C5
         (3) Individual nerves emerging from cervical plexus innervate the muscles and skin of the neck, upper shoulders, and part of the head
         (4) Phrenic nerve exits the cervical plexus and innervates the diaphragm
      b. Brachial plexus (Figure 13-10)
         (1) Located deep within the shoulder
         (2) Made up of ventral rami of C5 through T1
         (3) Individual nerves emerging from brachial plexus innervate the lower part of the shoulder and all of the arm
      c. Lumbar plexus (Figure 13-11)
         (1) Located in the lumbar region of the back in the psoas muscle
         (2) Formed by intermingling fibers of L1 through L4
         (3) Femoral nerve exits the lumbar plexus, divides into many branches, and supplies the thigh and leg
      d. Sacral plexus (Figure 13-11)
         (1) Located in the pelvic cavity in the anterior surface of the piriformis muscle
         (2) Formed by intermingling of fibers from L4 through S4
         (3) Tibial, common peroneal, and sciatic nerves exit the sacral plexus and supply nearly all the skin of the leg, posterior thigh muscles, and leg and foot muscles

D. Dermatomes—each skin surface area supplied by a single spinal nerve (Figure 13-12)

DIVISIONS OF THE PERIPHERAL NERVOUS SYSTEM

A. There are two functional divisions of the peripheral nervous system
   1. Sensory division
   2. Motor division
B. Motor division is divided further into the somatic motor nervous system and autonomic nervous system

SOMATIC MOTOR NERVOUS SYSTEM

A. Basic principles of somatic motor pathways
   (1) Somatic nervous system—includes all voluntary motor pathways outside the central nervous system
   (2) Somatic effectors—skeletal muscles
B. Somatic reflexes
   1. Nature of a reflex
       Reflex—action that results from a nerve impulse passing over a reflex arc; predictable response to a stimulus
       (1) Cranial reflex—center of reflex arc is in the brain
       (2) Spinal reflex—center of reflex arc is in the spinal cord
   b. Reflex consists of either muscle contraction or glandular secretion
       (1) Somatic reflex—contraction of skeletal muscles
       (2) Autonomic (visceral) reflex—either contraction of smooth or cardiac muscle or secretion by glands
   2. Some somatic reflexes of clinical importance—reflexes deviate from normal in certain diseases, and reflex testing is a valuable diagnostic aid
       a. Ankle jerk (also known as Achilles reflex)—extension of the foot in response to tapping the Achilles tendon; tendon reflex and deep reflex mediated by two-neuron spinal arcs; centers lie in first and second sacral segments of the cord
       b. Babinski reflex—extension of great toe, with or without fanning of other toes, in response to stimulation of outer margin of sole of foot; present in normal infants until approximately 1½ years of age and then becomes suppressed when corticospinal fibers become fully myelinated; in humans over 1½ years of age, a positive Babinski reflex is one of the pyramidal signs indicating destruction of corticospinal (pyramidal tract) fibers
       c. Plantar reflex—plantar flexion of all toes and a slight turning in and flexion of the anterior part of the foot in response to stimulation of the outer edge of the sole
       d. Corneal reflex—winking in response to touching the cornea; mediated by reflex arcs with sensory fibers in the ophthalmic branch of the fifth cranial nerve, centers in the pons, and motor fibers in the seventh cranial nerve
       e. Abdominal reflex—drawing in of the abdominal wall in response to stroking the side of the abdomen; superficial reflex; mediated by arcs with sensory and motor fibers in T9 through T12 and centers in these segments of the cord; decreased or absent reflex may involve lesions of pyramidal tract upper motor neurons
       f. Knee jerk (also known as patellar reflex)—extension of the lower leg in response to tapping the patellar tendon; tendon and muscles are stretched, stimulating muscle spindles and initiating conduction over a two-neuron reflex arc (Figure 13-13); may be classified in several different ways
           (1) Spinal cord reflex—center of reflex arc located in spinal cord gray matter
           (2) Segmental reflex—mediating impulses enter and leave at same cord segment
           (3) Ipsilateral reflex—mediating impulses come from and go to the same side of the body
           (4) Stretch or myotatic reflex—due to type of stimulation used to evoke reflex
           (5) Extensor reflex—produced by extensors of the lower leg
           (6) Tendon reflex—tapping tendon is stimulus that elicits reflex
           (7) Deep reflex—due to deep location of receptors stimulated to produce reflex

THE AUTONOMIC NERVOUS SYSTEM

A. Overview
   1. Subdivision of efferent portion of the peripheral nervous system
   2. Carries efferent neurons to the visceral effectors
   3. Major function—to regulate heartbeat, smooth muscle contraction, and glandular secretions to maintain homeostasis
   4. Two divisions—sympathetic division and parasympathetic division
   5. Sympathetic division consists of neural pathways that are separate from parasympathetic pathways
   6. Many autonomic effectors are dually innervated, which allows remarkably precise control of effector
B. Structure of the autonomic nervous system
   1. Basic plan of autonomic pathways (Figure 13-14)
       a. Each pathway is made up of autonomic nerves, ganglia, and plexuses, which are made of autonomic neurons
       b. All autonomic neurons are efferent neurons and function in reflex arcs
       c. Autonomic regulation ultimately depends on feedback from sensory nervous system
       d. Relay of two autonomic neurons conducts information from central nervous system to autonomic effectors
           (1) Preganglionic neuron—conducts impulses from the central nervous system to an autonomic ganglion
           (2) Postganglionic neuron—efferent neuron with which a preganglionic neuron synapses within autonomic ganglion
   2. Structure of the sympathetic pathways
       a. Sympathetic chain ganglia
           (1) Most ganglia of the sympathetic division lie along either side of the anterior surface of the vertebral column and are joined with the other ganglia located on the same side
           (2) Each chain extends from the second cervical vertebra to the level of the coccyx

(3) Usually there are 22 sympathetic chain ganglia on each side of vertebral column, three cervical, eleven thoracic, four lumbar, and four sacral
  b. Thoracolumbar division
    (1) Sympathetic preganglionic neurons with dendrite and cell bodies in lateral gray horns of the thoracic and lumbar segments of the spinal cord
    (2) Axons leave the cord by way of the ventral roots of the thoracic and first four lumbar spinal nerves and split away from other spinal nerve fibers by the white ramus to a sympathetic chain ganglion
  c. Preganglionic fiber may take one of three paths once inside the sympathetic chain ganglion
    (1) Synapse with sympathetic postganglionic neuron
    (2) Send ascending and/or descending branches through the sympathetic trunk to synapse with postganglionic neurons in other chain ganglia
    (3) Pass through one or more chain ganglia without synapsing
  d. Sympathetic postganglionic neurons
    (1) Dendrites and cell bodies are mostly in sympathetic chain ganglia or collateral ganglia
    (2) Gray ramus—short branch by which some postganglionic axons return to a spinal nerve
  e. In the sympathetic division, preganglionic neurons are relatively short, and postganglionic neurons are relatively long
  f. Axon of one sympathetic preganglionic neuron synapses with many postganglionic neurons, terminating in widely spread organs (Figure 13-15)
3. Structure of the parasympathetic pathways
  a. Parasympathetic preganglionic neurons—cell bodies are located in nuclei in the brain stem or lateral gray columns of the sacral cord; extend a considerable distance before synapsing with postganglionic neurons
  b. Parasympathetic postganglionic neurons—dendrites and cell bodies are located in parasympathetic ganglia, which are embedded in or near autonomic effectors
  c. Parasympathetic postganglionic neurons synapse with postganglionic neurons that each lead to a single effector (Figure 13-15)
4. Autonomic neurotransmitters (Figure 13-16)
  a. Axon terminal of autonomic neurons release either of two neurotransmitters, either norepinephrine or acetylcholine
  b. Adrenergic fibers—release norepinephrine; axons of postganglionic sympathetic neurons
  c. Cholinergic fibers—release acetylcholine; axons of preganglionic sympathetic neurons and of both preganglionic and postganglionic parasympathetic neurons
  d. Norepinephrine affects visceral effectors by first binding to one of two types of adrenergic receptors in plasma membranes, alpha receptors or beta receptors
    (1) Binding of norepinephrine to alpha receptors in smooth muscle of blood vessels is stimulating, causing the vessels to constrict
    (2) Binding of norepinephrine to beta receptors in smooth muscle of blood vessels is inhibitory, causing blood vessels to dilate; in cardiac muscle, has stimulating effect
  e. Epinephrine also stimulates adrenergic receptors, enhancing and prolonging effects of sympathetic stimulation
  f. Effect of a neurotransmitter on any postsynaptic cell is determined by characteristics of the receptors, not by the neurotransmitter
  g. Termination of actions of norepinephrine and epinephrine
    (1) Monoamine oxidase (MAO)—enzyme that breaks up neurotransmitter molecules taken back up by the synaptic knobs
    (2) Catechol-O-methyl transferase (COMT)—enzyme that breaks down the remaining neurotransmitter
  h. Acetylcholine binds to two types of cholinergic receptors: nicotinic receptors and muscarinic receptors
  i. Termination of action of acetylcholine is by the enzyme acetylcholinesterase
C. Functions of the autonomic nervous system
  1. Overview of autonomic function
    a. The autonomic nervous system functions to regulate visceral effectors in ways that tend to maintain or quickly restore homeostasis
    b. Sympathetic and parasympathetic divisions are tonically active, often exerting antagonistic influences on visceral effectors
    c. Doubly innervated effectors continually receive both sympathetic and parasympathetic impulses, and the summation of the two determine the controlling effect
  2. Functions of the sympathetic division
    a. Under resting conditions, the sympathetic division can act to maintain the normal functioning of doubly innervated autonomic effectors
    b. Sympathetic impulses function to maintain normal tone of the smooth muscle in blood vessel walls
    c. Major function of sympathetic division is that it serves as an "emergency" system—the "fight-or-flight" reaction (Review Table 13-8)
  3. Functions of the parasympathetic division (Review Table 13-7)
    a. Dominant controller of most autonomic effectors most of the time
    b. Acetylcholine—slows heartbeat and acts to promote digestion and elimination

THE BIG PICTURE: THE PERIPHERAL NERVOUS SYSTEM AND THE WHOLE BODY

A. The peripheral nervous system is made of all the afferent nervous pathways coming into the central nervous system and all the efferent pathways going out of the central nervous system
B. Peripheral pathways are those pathways that lead from the integrator central nervous system to the effectors
C. Peripheral motor pathways serve as an information-carrying network that allows the central nervous system to communicate regulatory information to all of the nervous effectors in the body
D. Every major organ is influenced, directly or indirectly, by peripheral nervous system output

REVIEW QUESTIONS

1. Which cranial nerves transmit impulses that result in vision? In eye movements?
2. Which cranial nerves transmit impulses that result in hearing? In taste sensations?
3. Identify the direction of the information carried by the ventral root of a spinal nerve.
4. Define mixed nerves.
5. Identify the areas innervated by the individual nerves emerging from the cervical plexus.
6. Define and explain a dermatome.
7. What pathways are found in the somatic motor nervous system?
8. Compare and contrast somatic and autonomic reflexes.
9. What is the clinical significance of Babinski's reflex?
10. Explain the various ways in which the knee jerk can be classified.
11. Explain what is meant by dual innervation in terms of the autonomic nervous system.
12. Describe an autonomic preganglionic neuron.
13. Identify the paths that the preganglionic fiber may take once it is inside the sympathetic chain ganglion.

14. Differentiate between the sympathetic preganglionic and postganglionic neurons in terms of length.
15. How do parasympathetic ganglia differ from sympathetic ganglia in terms of location?
16. Describe the actions of norepinephrine. How are these actions terminated?
17. Describe the responses caused by acetylcholine release. How are these actions terminated?
18. Explain why the name "autonomic nervous system" is misleading.
19. Describe the "fight-or-flight" reaction.
20. Which division of the autonomic nervous system is the dominant controller of most autonomic effectors most of the time?
21. Describe what happens when there is compression of the fifth cranial nerve.
22. Why do the right and left phrenic nerves have considerable clinical interest?
23. Describe the condition of shingles.
24. Give an example of a beta blocker and describe its action.

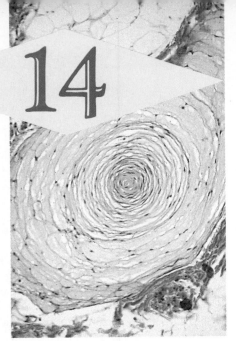

14

Sense Organs

Pacinian corpuscle
Erlandsen/Magney: Color Atlas of Histology

OBJECTIVES

After you have completed this chapter, you should be able to:

1. Compare the function of general and special sense organs.
2. Classify receptors according to the types of stimuli that activate them.
3. Identify the stimuli detected by the somatic senses.
4. Describe the receptors for pain, temperature, touch, pressure, and stretch.
5. Discuss the structure and the function of the olfactory and taste sense organs.
6. Describe the dual sensory functions of the ear.
7. List and discuss the function of the major anatomical components in the external, middle, and inner ear.
8. Identify the hearing sense organ and describe its function.
9. Discuss the physiology of hearing.
10. Identify and describe the sense organs involved in the sense of balance.
11. Identify the major anatomical structures that are visible in a horizontal section through the eyeball.
12. Describe the layers that make up the retina.
13. Compare the structure, function, and location of rods and cones in the retina.
14. Discuss the cavities and humors of the eye.
15. List and give the function of the extrinsic and intrinsic eye muscles.
16. Identify the accessory structures of the eye.
17. Discuss the four processes that focus light rays on the retina and describe the most common errors of refraction.
18. Describe the function of photopigments.

KEY TERMS

Adaptation The magnitude of the receptor potential decreases over a period of time in response to a continuous stimulus

Chemoreceptor (kee-mo-ree-SEP-tor) Receptor that responds to certain chemicals

Cochlea (KOKE-lee-ah) Spirally wound tube that contains hearing receptors in the inner ear

Crista ampullaris (KRIS-tah am-pyoo-LAIR-is) Specialized receptor located within the semicircular canals that detects head movements; important for dynamic equilibrium

Equilibrium (ee-kwi-LIB-ree-um) Refers to balance

Exteroceptors (eks-ter-o-SEP-torz) Somatic sense receptors located on the body surface

Gustatory (GUS-tah-tor-ee) Refers to taste

Labyrinth (LAB-i-rinth) Refers to the inner ear, including the cochlea, vestibule, and semicircular canals

Macula (MAK-yoo-lah) Specialized epithelium found in the utricle and saccule, associated with static equilibrium

Mechanoreceptor (mek-an-o-ree-SEP-tor) Receptor activated by mechanical stimuli that deforms the receptor

Nociceptor (no-se-SEP-tor) Pain receptor

Olfactory (ol-FAK-tor-ee) Refers to sense of smell

Photoreceptor Receptor found only in the eye, responds to light

Proprioceptor (pro-pree-o-SEP-tor) Somatic sense receptor located in muscles and joints

Refraction Bending of light rays

Retina (RET-i-nah) Innermost coat of the eyeball, contains visual receptors

Rhodopsin (ro-DOP-sin) Photopigment found in the rods

Thermoreceptor Receptor activated by changes in temperature

Visceroceptor (vis-ero-SEP-tor) Somatic sense receptor located in the internal visceral organs

The body has millions of sense organs. They fall into two main categories, general sense organs and special sense organs. Of these, by far the most numerous are the general sense organs, or receptors. Receptors function to produce the general, or somatic, senses (such as touch, temperature, and pain) and to initiate various reflexes necessary for maintaining homeostasis. Special sense organs function to produce the special senses (vision, hearing, balance, taste, and smell) and they too initiate reflexes important for homeostasis. In this chapter we begin with a description of receptors and follow with information related to the special senses.

SENSORY RECEPTORS

Sense organs called **sensory receptors** make it possible for the body to respond to stimuli caused by changes occurring in our external or internal environment. This function is crucial to survival. The abilities to see and hear, for example, may provide the necessary warning to help us avoid injury from dangers in our external environment. Internal sensations ranging from pain and pressure to hunger and thirst help us maintain homeostasis of our internal environment.

Receptor Response

The general function of receptors is to respond to stimuli by converting them to nerve impulses. As a rule, different types of receptors respond to different types of stimuli. They are specialized in that a particular type of receptor responds to a particular type of stimulus and is less able or unable to respond to others. Heat receptors, for example, do not respond to pressure or light.

When an adequate stimulus acts on a receptor, a potential develops in the receptor's membrane. It is called a **receptor potential.** The receptor potential is a graded response, graded to the strength of the stimulus (see Chapter 11, p. 292). When a receptor potential reaches a certain threshold, it triggers an action potential in the sensory neuron's axon. These impulses then travel over sensory pathways to the brain and spinal cord, where they are either interpreted as a particular **sensation** such as heat or cold or they initiate some type of reflex action such as withdrawal of a limb from a painful stimulus. A brain function called **sensory projection** pinpoints the area of the body from which the receptor potential was initiated.

Receptors exhibit a functional characteristic known as adaptation. **Adaptation** means that the magnitude of the receptor potential decreases over a period of time in

response to a continuous stimulus. As a result, the rate of impulse conduction by the sensory neuron's axon also decreases. So too does the intensity of the resulting sensation. A familiar example of adaptation is feeling the touch of your clothing when you first put it on and soon not sensing it at all. Touch receptors adapt rapidly. In contrast, the proprioceptors in our muscles, tendons, and joints adapt slowly. As long as stimulation of them continues, they continue sending impulses to the brain.

Classification of Receptors

Receptors are frequently classified into five categories based on the following types of stimuli that activate them:

1. Mechanoreceptors. Mechanoreceptors are activated by mechanical stimuli that in some way "deform" or change the position of the receptor, resulting in the generation of a receptor potential. Examples include pressure applied to the skin or to blood vessels, or caused by stretch or pressure in muscle, tendon, or lung tissue.

2. Chemoreceptors. Receptors of this type are activated by either the amount or the changing concentration of certain chemicals. Our senses of taste and smell depend on chemoreceptors. Specialized chemoreceptors in the body also "sense" the concentration of such specific chemicals as hydrogen ions (pH) and blood glucose.

3. Thermoreceptors. Thermoreceptors are activated by heat or cold.

4. Nociceptors. This type of receptor is activated by intense stimuli of any type that results in tissue damage. The cause may be a toxic chemical, intense light, sound, pressure, or heat. The sensation produced is one of *pain.*

5. Photoreceptors. This type of receptor is found only in the eye. Photoreceptors respond to light stimuli if the intensity is great enough to generate a receptor potential.

Distribution of Receptors

Receptors responsible for the **special senses** of smell, taste, vision, hearing, and equilibrium are grouped into localized areas (nasal mucosa or tongue) or into such complex organs as the eye and ear. The **general sense organs,** in contrast, consist of microscopic receptors widely distributed throughout the body in the skin, mucosa, connective tissues, muscles, tendons, joints, and viscera. Sensations produced by these receptors are often called the *somatic senses.* Their distribution, however, is not uniform in all areas. In some, it is very dense; in others, it is sparse. The skin covering the fingertips, for instance, contains many more receptors to touch than does the skin on the back. A simple procedure, the *two-point discrimination test,* demonstrates this fact. A subject reports the number of touch points felt when an investigator touches the skin simultaneously with two points of a compass. If the skin on the fingertip is touched with the compass points barely one eighth of an inch apart, the subject senses them as two points. But if the skin on the back is touched with the compass points this close together, they will be felt as only one point. Unless they are an inch or more apart, they cannot be discriminated as two points. Why this differ-

FYI   **Referred Pain**

The pain from stimulation of nociceptors in deep structures is frequently referred to surface areas. **Referred pain** is the term for this phenomenon. Pain originating in the viscera and other deep structures is generally interpreted as coming from the skin area whose sensory fibers enter the same segment of the spinal cord as the sensory fibers from the deep structure. For example, sensory fibers from the heart enter the first to fourth thoracic segments, and so do sensory fibers from the skin areas over the heart and on the inner surface of the left arm. Pain originating in the heart is referred to those skin areas, but the reason for this is not clear.

ence? Because touch receptors are so densely distributed in the fingertips that two points very close to each other stimulate two different receptors. Hence they are sensed as two points. But the situation is quite different in the skin on the back. There, touch receptors are so widely scattered that two points have to be at least an inch apart to stimulate two receptors and be felt as two points.

SOMATIC SENSES

The **somatic senses** enable us to detect sensations such as pain, temperature, pressure, and touch. In addition, the activation of specialized somatic sense receptors in muscles and tendons may provide us with a sense of body position or alert us to tension levels in these organs. Stimulation of somatic sense receptors located deep within the visceral organs of the body also alert us to pain and stretching or allow us to sense such complex feelings as hunger or thirst.

Somatic sense receptors that are located on the body surface are called **exteroceptors,** those in muscles and joints are called **proprioceptors,** and those found in the internal visceral organs are called **visceroceptors.**

PAIN AND TEMPERATURE RECEPTORS

Pain receptors, also called *nociceptors,* are widely distributed and are located on both the surface of the body (exteroceptors) and in the deep visceral organs (visceroceptors). Brain tissue is unique in that it lacks pain receptors altogether and is therefore incapable of sensing painful stimuli. Just the opposite is true of many deep visceral organs where nociceptors are present, and pain is one of the few sensations that can be evoked.

Pain receptors are essentially **free nerve endings** (Figure 14-1, A) that respond to a number of different stimuli. For example, tissue damage caused by mechanical stretching, chemicals, lack of blood flow, or extremes of temperature will all result in stimulation of free nerve endings, generation of a receptor potential, and, ultimately, the sensation of pain—often the first indication of injury or disease. Free nerve endings on the body surface are also

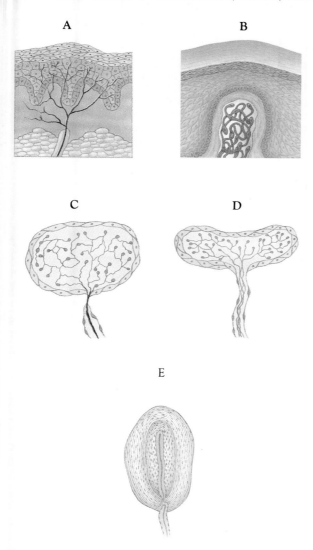

A

B

C

D

E

🔑

FIGURE 14-1 Somatic sensory receptors. A, Free nerve ending. **B,** Meissner's corpuscle. **C,** Krause's end bulb. **D,** Ruffini's corpuscle. **E,** Pacinian corpuscle. See also Figure 5-1 (p. 132), which shows a microscopic section of the skin.

involved in sensations perceived as "itching" or "tickling."

Nerve fibers that carry pain impulses from free nerve ending receptors to the brain can be divided into two types—*acute (A)* and *chronic (B) fibers.* Acute pain (A) fibers mediate sharp, intense, and generally localized pain sensations, whereas chronic (B) fibers are associated with less intense, but more persistent, pain—often described as dull or aching.

▶ TOUCH AND PRESSURE RECEPTORS

Touch and pressure receptors are exteroceptors that respond to stimuli that "deform" or change their shape or placement.

Meissner's corpuscles and structurally similar receptors, called **Krause's end bulbs,** are both involved in the sensations of touch, low frequency vibration, and two-point (texture) discrimination (Figure 14-1, *B* and *C*). These receptors are concentrated on areas of skin devoid of hair, such as the genitals and lips. **Ruffini's corpuscles** sense deep pressure and continuous touch (Figure 14-1, *D*). They are located in the dermis of the skin and are numerous in the fingers. These receptors are slow adapting, thus permitting the skin of the fingers to remain sensitive to deep pressure for long periods. The ability to grasp an object, such as the steering wheel of a car, for long periods of time and still be able to "sense" its presence between your fingers depends on these receptors. **Pacinian corpuscles** respond quickly to sensations of deep pressure, high-frequency vibration, and stretch (Figure 14-1, *E*). Although sensitive and quick to respond, these receptors adapt quickly, and the sensations they evoke seldom last for long periods. They are numerous in the deep dermis of the hands and feet and in joint capsules throughout the body.

▶ STRETCH RECEPTORS

The most important stretch receptors are associated with muscles and tendons and are classified as proprioceptors.

Two types of stretch receptors, called **muscle spindles** and **Golgi tendon receptors,** operate to provide the body with information concerning muscle length and the strength of muscle contraction. Muscle spindles are stimulated if the length of a relaxed muscle exceeds a certain limit. The result of stimulation is a **stretch reflex** that shortens a muscle or muscle group, thus aiding in the maintenance of posture or the positioning of the body or one of its extremities in a way that may be opposed by the force of gravity (see Figure 9-17, p. 239). We do this unconsciously, since these receptors do not produce a particular sensation such as pain, heat, or cold.

Golgi tendon receptors act in a way opposite to that of muscle spindles. They are stimulated by excessive muscle contraction. These receptors are located near the points of attachment of tendon to bone, and, when stimulated, they cause muscles to *relax.* This response protects muscles from tearing internally or pulling away from their tendinous points of attachment to bone because of excessive contractile forces.

The different types of somatic sense receptors, their locations, and their functions are summarized in Table 14-1.

QUICK CHECK

✔ 1. *Classify receptors into four groups based on the type of stimuli that activate them.*
✔ 2. *Distinguish between the special and the general, or somatic, senses.*
✔ 3. *Name the three types of sense receptors classified according to location in the body.*
✔ 4. *Name the receptors associated with pain, touch, pressure, and stretch responses.*

TABLE 14-1   Somatic sense receptors

Type of receptor	Main locations	Sensation or function
TOUCH AND PRESSURE		
Free nerve endings (nociceptors)	Most body tissues	Pain, temperature, itch, tickle
Meissner's corpuscles	Skin (in papillae of dermis); numerous in fingertips and lips	Two-point discrimination, fine touch, low-frequency vibration
Krause's end bulbs	Skin (dermal layer), subcutaneous tissue, mucosa of lips and eyelids, external genitals	Touch, low-frequency vibration
Ruffini's corpuscles	Skin (dermal layer) and subcutaneous tissue of fingers	Touch, deep pressure, stretch
Pacinian corpuscles	Subcutaneous, submucous, and subserous tissues, around joints, in mammary glands and external genitals of both sexes	Pressure, high-frequency vibration, stretch
STRETCH		
Muscle spindles	Skeletal muscles	Sense muscle length
Golgi tendon receptors	Near junction of tendons and muscles	Sense muscle tension

SPECIAL SENSES

The special senses are characterized by receptors grouped closely together or located in specialized organs. The senses of smell (olfaction), taste, hearing, equilibrium, and vision are considered the **special senses.**

Sense of Smell

Olfactory Receptors

The olfactory sense organ consists of epithelial support cells and specialized olfactory receptor neurons. These neurons have unique *olfactory cilia*, which touch the surface of the olfactory epithelium lining the upper surface of the nasal cavity. The olfactory receptor neurons are chemoreceptors. Receptor potentials are generated in these cells when the olfactory cilia are stimulated by gas molecules or chemicals dissolved in the mucus covering the nasal epithelium (Figure 14-2).

The olfactory epithelium is located in the most superior portion of the nasal cavity (Figure 14-2). Functionally, this is a poor location because a great deal of inspired air flows around and down the nasal passageways without contacting the olfactory receptor cells. The location of these receptors explains the necessity for sniffing, or drawing air forcefully up into the nose, to smell delicate odors.

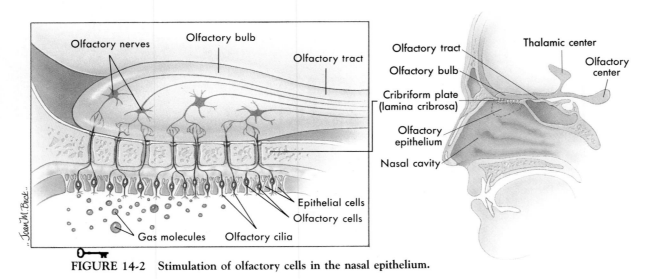

FIGURE 14-2   Stimulation of olfactory cells in the nasal epithelium.

The olfactory receptors are extremely sensitive, that is, capable of being stimulated by even very slight odors caused by just a few molecules of a particular chemical. Although humans have a sense of smell far less keen than many animals, some individuals can distinguish several thousand different odors, and most of us can easily identify at least several hundred. Most smells are complex and varying mixtures of about 30 "pure," or "primary," fragrances, or scents. Olfactory receptors are sensitive and easily fatigued—a fact that explains why odors that are at first very noticeable are not sensed at all after a short time. Examples of well-known primary scents include putrid, floral, peppermint, and musky odors. Combinations of two or more of these or other primary scents produce the wide array of odors we can identify.

Olfactory Pathway

If the level of odor-producing chemicals dissolved in the mucus surrounding the olfactory cilia reaches a threshold level, a receptor potential and, then, an action potential will be generated and passed to the olfactory nerves in the olfactory bulb. From there, the impulse passes through the olfactory tract and into the thalamic and olfactory centers of the brain for interpretation, integration, and memory storage.

The sense of smell can create powerful and long lasting memories. Memories coupled to unique sensory inputs, especially distinctive odors, often persist from early childhood to death. Dental office smells, baby smells, kitchen smells, and new car smells are examples of olfactory "triggers" that often bring back memories of events that occurred years earlier. In addition to the olfactory cortex and thalamic areas of the brain, components of the *limbic system* play a key role in coupling olfactory sense inputs to both short- and long-term memory.

SENSE OF TASTE

Taste Buds

The **taste buds** are the sense organs that respond to **gustatory,** or **taste,** stimuli. Although a few isolated taste buds are located in the lining of the mouth and throat (especially in young children), most are associated with small elevated projections on the tongue, called **papillae** (pa-PIL-e) (Figure 14-3, *B*).

Taste buds are chemoreceptors. They are stimulated by chemicals dissolved in the saliva. Each grapelike taste bud contains specialized **gustatory cells** that are surrounded by a supportive epithelial cell capsule. Tiny cilia-like structures, called **gustatory hairs,** extend from each of the gustatory cells and project into an opening called the **taste pore,** which is bathed in saliva (Figure 14-3, *C*).

The sense of taste depends on the creation of a receptor potential in gustatory cells. Only then can an action potential be generated and a nerve impulse relayed to the brain for interpretation. Generation of a receptor potential begins when specialized receptor sites on the cell membranes covering gustatory hairs bind to taste-producing chemicals in the saliva. The nature and concentration of the chemicals that bind to the receptor sites will determine how fast the receptor potential is generated.

Taste cells appear similar structurally, and all of them

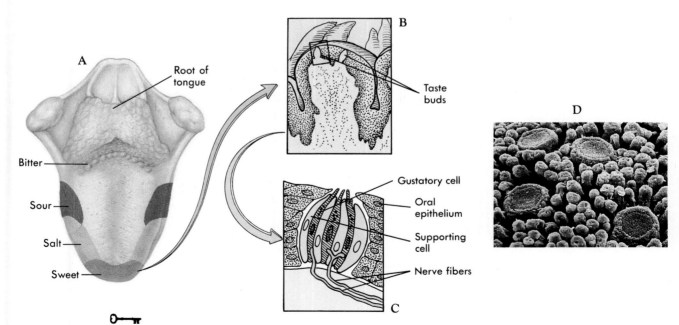

FIGURE 14-3   The tongue. A, Dorsal surface and regions sensitive to various tastes. **B,** Section through a papilla with taste buds on the side. **C,** Enlarged view of a section through a taste bud. **D,** Scanning electron micrograph of the tongue surface showing the papillae in detail.

can respond at least in some degree to most taste-producing chemicals. Functionally, however, each taste bud responds most effectively to only one of four "primary" taste sensations: sour, sweet, bitter, and salty. Our ability to detect many different flavors, or tastes, is due to combinations of the four primary sensations and to interaction with the sense of smell. The number of "primary" taste sensations (such as the pure, or primary, olfactory scents) is likely to increase with expanded research efforts. Metallic taste, for example, may be a fifth primary taste sensation rather than a mixture of other flavors.

The primary taste sensation to which a taste bud responds is closely related to its placement on the tongue. Note in Figure 14-3, A, that the tip of the tongue reacts best to sweet and, to a lesser extent, salty taste. The sides and back of the tongue respond best to sour and bitter tastes, respectively.

The exact mechanism by which a specific chemical binds to receptor sites on gustatory hairs is unknown. Chemical structure plays a part but is not the only factor involved, since substances that are very different chemically, such as artificial sweeteners and table sugar, produce a sweet taste. Some chemical compounds and specific ions, however, are definitely associated with specific tastes. Acids and other substances associated with hydrogen ions (H^+) produce sour tastes, whereas many organic substances (especially sugars) produce sweet tastes. As chemoreceptors, the taste buds, like olfactory receptors, tend to be quite sensitive but fatigue easily. Very low levels of taste-producing chemicals are required to generate a receptor potential. However, adaptation often begins within a few seconds after a taste sensation is first noticed and is generally complete in a few minutes. The adaptation times and sensitivity thresholds are different for each of the primary taste sensations. We are most sensitive to bitter tastes and can detect extremely low levels of chemicals, such as bases and alkaloids, that produce them. Since many bitter tasting compounds are poisonous, high sensitivity provides a protective response.

Neuronal Pathway for Taste

The taste sensation begins with creation of a receptor potential in the gustatory cells of a taste bud. The generation and propagation of an action potential, or nerve impulse, then transmit the sensory input to the brain.

Nervous impulses generated in the anterior two thirds of the tongue travel over the facial (VII) nerve, whereas those generated from the posterior one third are conducted by fibers of the glossopharyngeal (IX) nerve. A third cranial nerve, the vagus (X) nerve, plays a minor role in taste. It contains a few fibers that carry taste sensation from a limited number of taste buds located in the walls of the pharynx and on the epiglottis.

All three cranial nerves carry impulses into the medulla oblongata. Relays then carry the impulses into the thalamus and then into the taste, or gustatory, area of the cerebral cortex in the parietal lobe of the brain.

SENSE OF HEARING AND BALANCE: THE EAR

The ear has dual sensory functions. In addition to its role in hearing it also functions as the sense organ of balance, or equilibrium. The stimulation, or "trigger," responsible for hearing and balance involves activation of specialized mechanoreceptors called *hair cells*. Sound waves and movement are the physical forces that act on hair cells to generate receptor potentials, and then nerve impulses, which are eventually perceived in the brain as sound or balance. The ear is divided into three anatomical parts: **external ear, middle ear,** and **inner ear** (Figure 14-4).

External Ear

The external ear has two divisions: the flap, or modified trumpet, on the side of the head, called the *auricle* or *pinna*, and the tube leading from the auricle into the temporal bone, named the *external auditory meatus (ear canal)*. This canal is about 3.0 cm long and takes, in general, an inward, forward, and downward direction, although the first portion of the tube slants upward and then curves downward. Because of this curve in the auditory canal, in adults, the auricle should be pulled up and back to straighten the tube when medications are dropped into the ear. Modified sweat glands in the auditory canal secrete *cerumen* (waxlike substance), which occasionally becomes impacted and may cause pain and temporary deafness. The **tympanic membrane** (eardrum) stretches across the inner end of the auditory canal, separating it from the middle ear.

Middle Ear

The middle ear (tympanic cavity), a tiny epithelial-lined cavity hollowed out of the temporal bone, contains the three **auditory ossicles:** the *malleus, incus,* and *stapes* (see Figure 14-4). The names of these very small bones describe their shapes (hammer, anvil, and stirrup). The "handle" of the malleus is attached to the inner surface of the tympanic membrane, whereas the "head" attaches to the incus, which in turn attaches to the stapes. There are several openings into the middle ear cavity: one from the external auditory meatus, covered with the tympanic

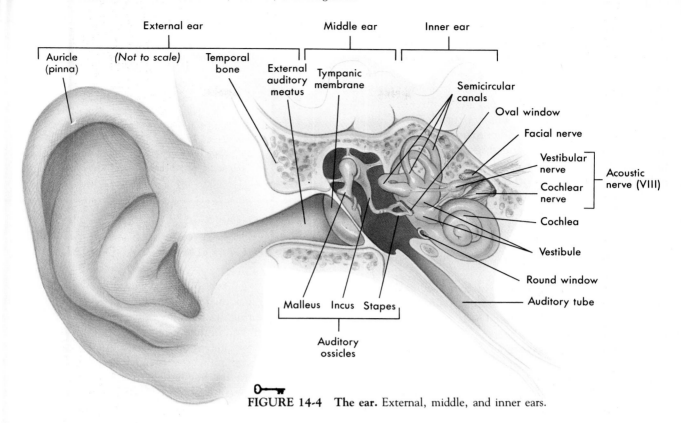

FIGURE 14-4 **The ear.** External, middle, and inner ears.

membrane; two into the internal ear, the **oval window** (into which the stapes fits), and the **round window,** which is covered by a membrane; and one into the auditory (eustachian) tube.

Posteriorly the middle ear cavity is continuous with numerous mastoid air spaces in the temporal bone. The clinical importance of these middle ear openings is that they provide routes for infection to travel. Head colds, for example, especially in children, may lead to middle ear or mastoid infections via the nasopharynx–auditory tube, middle ear–mastoid path.

The **auditory,** or **eustachian, tube** is composed partly of bone and partly of cartilage and fibrous tissue and is lined with mucosa. It extends downward, forward, and inward from the middle ear cavity to the nasopharynx, or pharyngotympanic tube, (the part of the throat behind the nose). The auditory tube serves a useful function: it makes possible equalization of pressure against inner and outer surfaces of the tympanic membrane and therefore prevents membrane rupture and the discomfort that marked pressure differences produce. The way the auditory tube equalizes tympanic membrane pressure is this: when one swallows or yawns, air spreads rapidly through the open tube. Atmospheric pressure then presses against the inner surface of the tympanic membrane. Since atmospheric pressure is continually exerted against its outer surface, the pressures are equal.

Inner Ear

The inner ear is also called the *labyrinth* because of its complicated shape. It consists of two main parts, a bony labyrinth and, inside this, a membranous labyrinth. The bony labyrinth consists of three parts: *vestibule, cochlea,* and *semicircular canals* (Figure 14-5). The membranous labyrinth consists of the *utricle* and *saccule* inside the vestibule, the *cochlear duct* inside the cochlea, and the *membranous semicircular canals* inside the bony ones. The vestibule (containing the utricle and saccule) and the semicircular canals are involved in balance; the cochlea is involved in hearing.

The term **endolymph** is used to describe the clear and potassium-rich fluid that fills the membranous labyrinth. **Perilymph,** a fluid similar to cerebrospinal fluid, surrounds the membranous labyrinth and therefore fills the space between this membranous tunnel and its contents and the bony walls that surround it (see Figure 14-5).

Cochlea and cochlear duct

The word **cochlea,** which means "snail," describes the outer appearance of this part of the bony labyrinth. When sectioned, the cochlea resembles a tube wound spirally around a cone-shaped core of bone, the *modiolus.* The modiolus houses the spiral ganglion, which consists of cell bodies of the first sensory neurons in the auditory relay. Inside the cochlea lies the membranous *cochlear duct*— the only part of the internal ear concerned with hearing. This structure is shaped like a somewhat triangular tube. It forms a shelf across the inside of the bony cochlea, dividing it into upper and lower sections all along its winding course (Figure 14-5). The upper section (above the cochlear duct) is called the *scala vestibuli,* whereas the lower section below

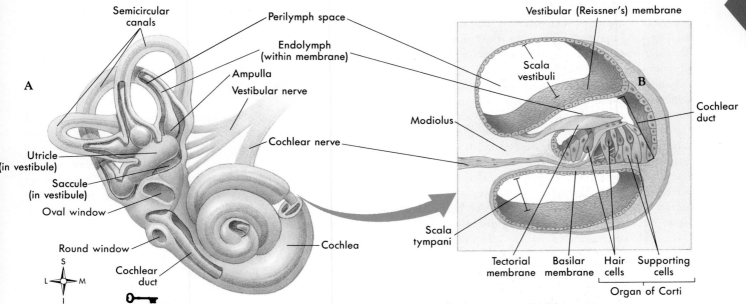

FIGURE 14-5   **The inner ear. A,** The bony labyrinth (*orange*) is the hard outer wall of the entire inner ear and includes semicircular canals, vestibule, and cochlea. Within the bony labyrinth is the membranous labyrinth (*purple*), which is surrounded by perilymph and filled with endolymph. Each ampulla in the vestibule contains a crista ampullaris that detects changes in head position and sends sensory impulses through the vestibular nerve to the brain. **B,** The inset shows a section of the membranous cochlea. Hair cells in the organ of Corti detect sound and send the information through the cochlear nerve. The vestibular and cochlear nerves join to form the eighth cranial nerve.

COCHLEAR IMPLANTS

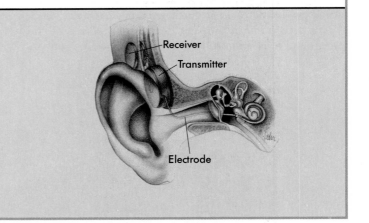

Recent advances in electronic circuitry are being used to correct some forms of nerve deafness. If the hairs on the organ of Corti are damaged, nerve deafness results—even if the vestibulocochlear nerve is healthy. A new surgically implanted device can improve this form of hearing loss by eliminating the need for the sensory hairs. As you can see in the figure, a transmitter just outside the scalp sends external sound information to a receiver under the scalp (behind the auricle). The receiver translates the information into an electrical code that is relayed down an electrode to the cochlea. The electrode, wired to the organ of Corti, stimulates the vestibulocochlear nerve endings directly. Thus even though the cochlear hair cells are damaged, sound can be perceived.

the cochlear duct is called the *scala tympani.* The roof of the cochlear duct is known as the **vestibular membrane (Reissner's membrane). Basilar membrane** is the name of the floor of the cochlear duct. It is supported by bony and fibrous projections from the wall of the cochlea. Perilymph fills the scala vestibuli and scala tympani, and endolymph fills the cochlear duct.

The hearing sense organ, named the **organ of Corti,** rests on the basilar membrane throughout the entire length of the cochlear duct. The structure of the organ of Corti consists of supporting cells, as well as of important **hair cells** that project into the endolymph and are topped by an adherent gelatinous membrane called the **tectorial membrane.** Dendrites of the sensory neurons, whose cell bodies lie in the spiral ganglion in the modiolus, have their beginnings around the bases of the hair cells of the organ of Corti. Axons of these neurons extend to form the cochlear nerve (a branch of the eighth cranial nerve) to the brain. They conduct impulses that produce the sensation of hearing.

Sense of hearing

Sound is created by vibrations that may occur in air, fluid, or solid material. When we speak, for example, the vibrating vocal cords create sound waves by producing vibrations in air passing over them.

Numerous terms are used to describe sound waves. The height, or amplitude, of a sound wave determines its perceived loudness, or **volume.** The number of sound waves that occur during a specific time unit (frequency) determines **pitch.** Our ability to hear sound waves depends in part on volume, pitch, and other acoustic properties. Sound waves must be of sufficient amplitude to initiate movement of the tympanic membrane and have a frequency that is capable of stimulating the hair cells in the organ of Corti at some point along the basilar membrane.

The basilar membrane is not the same width and thickness throughout its length. Because of this structural fact, different frequencies of sound will cause the basilar membrane to vibrate and bulge upward at different places along its length. Two bulges are shown in Figure 14-6.

High-frequency sound waves cause the narrow portion of the basilar membrane near the oval window to vibrate, whereas low frequencies vibrate the membrane near the apex of the cochlea, where it is considerably wider and thicker. This ability of sound waves of differing frequency to vibrate and cause a bulge, or upward displacement, of the basilar membrane at differing points along its length explains how specific groups of hair cells respond to specific frequencies of sound. When a particular portion of the basilar membrane bulges upward, the cilia on hair cells attached to that particular area are stimulated and, ultimately, sound of a particular pitch is perceived.

Our perception of different degrees of loudness of the same sound is determined by the amplitude, or movement, of the basilar membrane at any particular point along its length. The higher the upward bulge, the more the cilia on the attached hair cells are bent or stimulated. This causes an increase in perceived loudness. The moving wave of perilymph caused by upward displacement of the basilar membrane is soon dampened as it moves through the cochlea.

Hearing results from stimulation of the auditory area of the cerebral cortex. First, however, sound waves must be projected through air, bone, and fluid to stimulate nerve endings and set up impulse conduction over nerve fibers.

Pathway of sound waves. Sound waves in the air enter the external auditory canal with aid from the pinna. At the inner end of the canal, they strike against the tympanic membrane, setting it in vibration. Vibrations of the tympanic membrane move the malleus, whose handle attaches to the membrane. The head of the malleus attaches to the incus, and the incus attaches to the stapes. So when the malleus vibrates, it moves the incus, which moves the stapes against the oval window into which it fits so precisely. At this point, fluid conduction of sound waves begins. When the stapes moves against the oval window, pressure is exerted inward into the perilymph in the scala vestibuli of the cochlea. This starts a "ripple" in the perilymph that is transmitted through the vestibular membrane (the roof of the cochlear duct) to endolymph inside the duct and then to the organ of Corti and to the basilar membrane that supports the organ of Corti and forms the floor of the cochlear duct. From the basilar membrane the ripple is next transmitted to and through the perilymph in the scala tympani and finally expends itself against the round window—like an ocean wave as it breaks against the shore but on a much reduced scale.

The steps involved in hearing are shown in Figure 14-6.

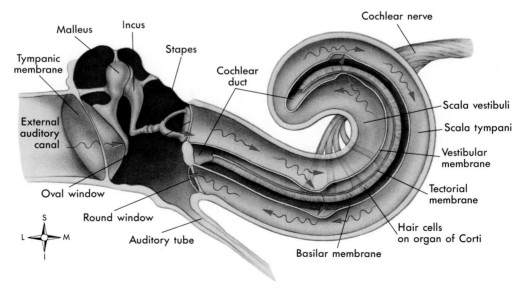

FIGURE 14-6   Effect of sound waves on cochlear structures. Sound waves strike the tympanic membrane and cause it to vibrate. This causes the membrane of the oval window to vibrate, which causes the perilymph in the bony labyrinth of the cochlea and the endolymph in the membranous labyrinth of the cochlea, or cochlear duct, to move. This movement of endolymph causes the basilar membrane to vibrate, which in turn stimulates hair cells on the organ of Corti to transmit nerve impulses along the cranial nerve. Eventually, nerve impulses reach the auditory cortex and are interpreted as sound.

Neuronal pathway of hearing. Dendrites of neurons whose cell bodies lie in the spiral ganglion and whose axons make up the cochlear nerve terminate around the bases of the hair cells of the organ of Corti, and the tectorial membrane adheres to their upper surfaces. The movement of the hair cells against the adherent tectorial membrane somehow stimulates these dendrites and initiates impulse conduction by the cochlear nerve to the brain stem. Before reaching the auditory area of the temporal lobe, impulses pass through "relay stations" in nuclei in the medulla, pons, midbrain, and thalamus.

QUICK CHECK

✔ *1. List the three anatomical divisions of the ear.*
✔ *2. Identify the three auditory ossicles.*
✔ *3. Name the divisions of both the membranous and bony labyrinth.*
✔ *4. Name the sense organ responsible for hearing.*

Vestibule and semicircular canals

The vestibule constitutes the central section of the bony labyrinth. Look again at Figure 14-5. Notice that the bony labyrinth opens into the oval and round windows from the middle ear, as well as the three semicircular canals of the internal ear. The utricle and saccule are the membranous structures within the vestibule. Both have walls of simple cuboidal epithelium and are filled with endolymph.

Three semicircular canals, each in a plane approximately at right angles to the others, are found in each temporal bone (see Figure 14-5). Within the bony semicir-

cular canals and separated from them by perilymph are the membranous semicircular canals. Each contains endolymph and connects with the utricle inside the bony vestibule. Near its junction with the utricle each canal enlarges into an *ampulla.*

Sense of balance. The sense organs involved in the sense of balance, or equilibrium, are found in the vestibule and semicircular canals. The sense organs located in the utricle and saccule function in **static equilibrium**—a function needed to sense the position of the head relative to gravity or to sense acceleration or deceleration of the body, as would occur when seated motionless in a vehicle that was increasing or decreasing in speed. The sense organs associated with the semicircular canals function in **dynamic equilibrium**—a function needed to maintain balance when the head or body itself is rotated or suddenly moved.

Static equilibrium. A small and highly specialized patchlike strip of epithelium, called the **macula** (MAK-u-lah), is found in both the utricle and saccule (Figure 14-7, A). It is specialized sensory epithelium containing receptor hair cells and supporting cells covered with a gelatinous matrix. Movements of the macula provide information related to head position or acceleration. Action potentials are generated by movement of the hair cells, which occurs when the position of the head relative to gravity changes.

Otoliths—tiny "ear stones" composed of protein and calcium carbonate—are located within the matrix of the macula (Figure 14-7, B). Now note the relative positions of the utricular and saccular maculae in Figure 14-7, A. The two maculas are oriented almost at right angles to each other: the one in the utricle is parallel to the base of the

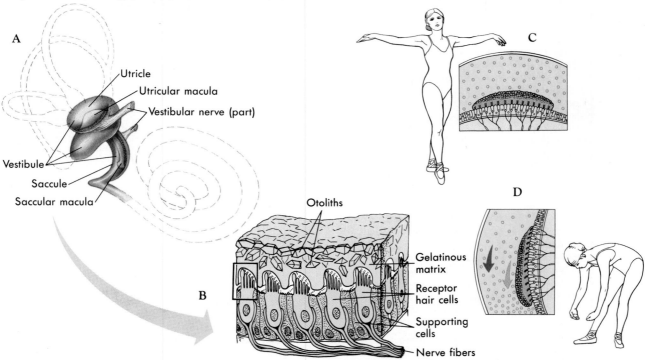

FIGURE 14-7 The macula. A, Structure of vestibule showing placement of utricular and saccular maculae. **B,** Section of macula showing otoliths. **C,** Macula stationary in upright position. **D,** Macula displaced by gravity as person bends over.

Labels in figure:
Utricle
Utricular macula
Vestibular nerve (part)
Vestibule
Saccule
Saccular macula
Otoliths
Gelatinous matrix
Receptor hair cells
Supporting cells
Nerve fibers
A   B   C   D

skull, and the one in the saccule is perpendicular. Changing the position of the head produces a change in the amount of pressure on the otolith-weighted matrix, which, in turn, stimulates the hair cells (Figure 14-7, *C* and *D*). This stimulates the adjacent receptors of the vestibular nerve. Its fibers conduct impulses to the brain that produce a sense of the position of the head and also a sensation of a change in the pull of gravity, for example, a sensation of acceleration. In addition, stimulation of the macula evokes *righting reflexes*, muscular responses to restore the body and its parts to their normal position when they have been displaced. Impulses from proprioceptors and from the eyes also activate righting reflexes. Interruption of the vestibular, visual, or proprioceptive impulses that initiate these reflexes may cause disturbances of equilibrium, nausea, vomiting, and other symptoms.

Dynamic equilibrium. Dynamic equilibrium depends on the functioning of the **crista ampullaris,** located in the ampulla of each semicircular canal. This specialized structure is a form of sensory epithelium that is similar in many ways to the maculae. Each cone-shaped crista is composed

of many hair cells, with their processes imbedded in a gelatinous cap, called the **cupula** (Figure 14-8, *A* and *B*).

The cupula is not weighted with otoliths and does not respond to the pull of gravity. It serves, instead, much like a float that moves with the flow of endolymph in the semicircular canals. Like the maculae, the semicircular canals are placed nearly at right angles to each other. This arrangement enables detection of movement in all directions. As the cupula moves, it bends the hairs imbedded in it, producing, first, a receptor and, then, an action potential that passes through the vestibular portion of the eighth cranial nerve to the medulla oblongata and, from there, to other areas of the brain and spinal cord for interpretation, integration, and response.

When a person spins (Figure 14-8, *C* to *E*), the semicircular canals move with the body, but inertia keeps the endolymph in them from moving at the same rate. The cupula therefore moves in a direction opposite to head movement until after the initial movement stops. Dynamic equilibrium is thus able to detect changes both in the direction and rate at which movement occurs.

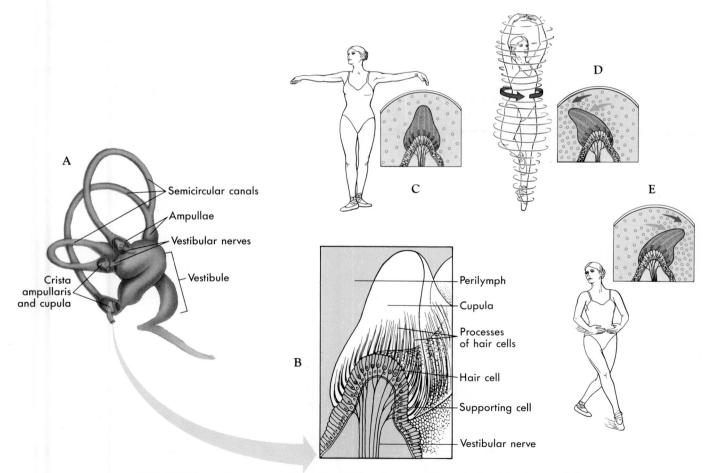

FIGURE 14-8 Structure and function of the crista ampullaris. A, Semicircular canals showing location of the crista ampullaris in ampullae. **B,** Enlargement of crista ampullaris and cupula. **C,** When a person is at rest, the crista ampullaris does not move. **D** and **E,** As a person begins to spin, the crista ampullaris is displaced by the endolymph in a direction opposite to the direction of spin.

Vision: The Eye

The eye is the body's sense organ for vision. This remarkable organ converts the energy of light into electrical nerve impulses that are interpreted by the brain as sight. We will first discuss the various structures of the eye and then move on to the ways in which these structures enable the eye to control the amount of light entering it and how the conversion to electrical stimuli actually occurs.

Structure of the Eye
Coats of the Eyeball

Approximately five sixths of the eyeball lies recessed in the orbit, protected by this bony socket. Only the small anterior surface of the eyeball is exposed. Three layers of tissues, or coats, compose the eyeball. From the outside in, they are the *sclera*, the *choroid*, and the *retina*. The sclera and the choroid coats consist of an anterior and a posterior portion. Tough, white fibrous tissue fashions the **sclera.**

The anterior portion of the sclera is called *the cornea* and lies over the colored part of the eye, the *iris* (Figure 14-9). The cornea is transparent, whereas the rest of the sclera is white and opaque, a fact that explains why the visible

CORNEAL TRANSPLANTS

Surgical removal of opaque or deteriorating corneas and replacement with donor transplants is a common medical practice. Corneal tissue is avascular; that is, the cornea is free of blood vessels. Therefore corneal tissue is seldom rejected by the body's immune system. Antibodies carried in the blood have no way to reach the transplanted tissue, and therefore long-term success following implant surgery is excellent.

anterior surface of the sclera is usually spoken of as the "white" of the eye. No blood vessels are found in the cornea or in the lens. Deep within the anterior part of the sclera at its junction with the cornea lies a ring-shaped venous sinus, the *canal of Schlemm.*

The middle, or *choroid*, coat of the eye contains a great many blood vessels and a large amount of pigment. Its anterior portion is modified into three separate structures: the ciliary body, the suspensory ligament, and the iris.

The **ciliary body** is formed by a thickening of the choroid and fits like a collar into the area between the

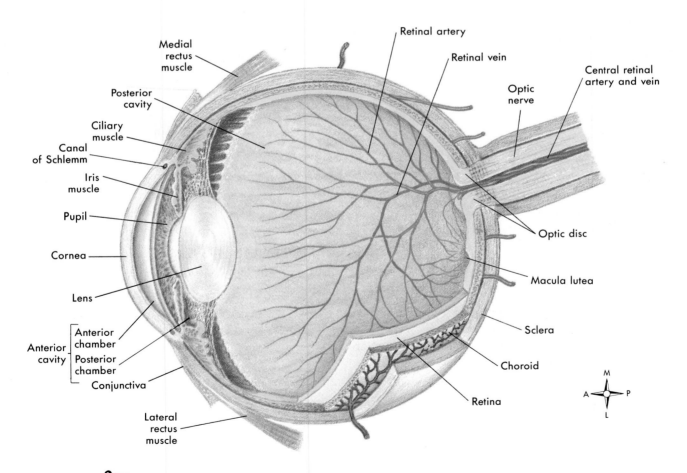

FIGURE 14-9 Horizontal section through the left eyeball. The eye is viewed from above.

anterior margin of the retina and the posterior margin of the iris (Figure 14-10). The small *ciliary muscle,* composed of both radial and circular smooth muscle fibers, lies in the anterior part of the ciliary body. Folds in the ciliary body are called *ciliary processes,* and attached to these are the *suspensory ligaments,* which blend with the elastic capsule of the *lens* and hold it suspended in place.

The **iris,** or the colored part of the eye, consists of circular and radial smooth muscle fibers arranged so as to form a doughnut-shaped structure. The hole in the middle is called the **pupil.** The iris attaches to the ciliary body.

The **retina** is the incomplete innermost coat of the eyeball—incomplete in that it has no anterior portion. Pigmented epithelial cells form the layer of the retina next to the choroid coat. Three layers of neurons make up the major portion of the retina. Named in the order in which they conduct impulses, they are *photoreceptor neurons, bipolar neurons,* and *ganglion neurons.* Identify each of these in Figure 14-11, A. The distal ends of the dendrites of the photoreceptor neurons have names that are descriptive of their shapes. Because some look like tiny rods and others look like cones, they are called **rods** and **cones,** respectively (Figure 14-11, B and C). They constitute our visual receptors, structures highly specialized for stimulation by light rays. They differ in numbers, distribution, and function. Cones are less numerous than rods and are most densely concentrated in the **fovea centralis,** a small depression in the center of a yellowish area, the **macula lutea,** found near the center of the retina (Figure 14-12).

FYI **Finding Your Blind Spot**

Demonstrate the location of the blind spot in your visual field by covering your left eye and looking at the objects below. While staring at the square, begin about 35 cm (12 in) from objects and slowly bring the figures closer to your eye. At one point, the circle will seem to disappear because its image has fallen on the blind spot.

They become less and less dense from the fovea outward. Rods, on the other hand, are absent entirely from the fovea and macula and increase in density toward the periphery of the retina. How these anatomical facts relate to rod and cone functions is revealed on p. 392.

All the axons of ganglion neurons extend back to a small circular area in the posterior part of the eyeball known as the *optic disc.* This part of the sclera contains perforations through which the fibers emerge from the eyeball as the **optic nerve** (second cranial nerve). The optic disc is also called the **blind spot** because light rays striking this area cannot be seen. Why? Because it contains no rods or cones, only nerve fibers. For an outline summary of the coats of the eyeball see Table 14-2.

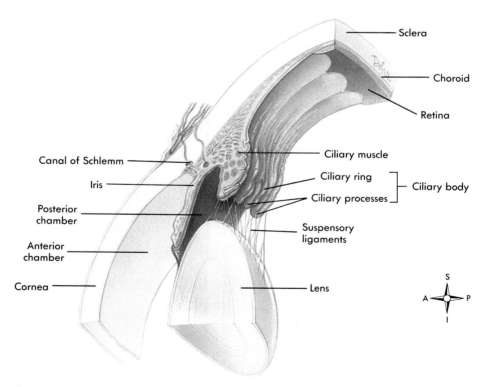

FIGURE 14-10   Lens, cornea, iris, and ciliary body. Note the suspensory ligaments that attach the lens to the ciliary body.

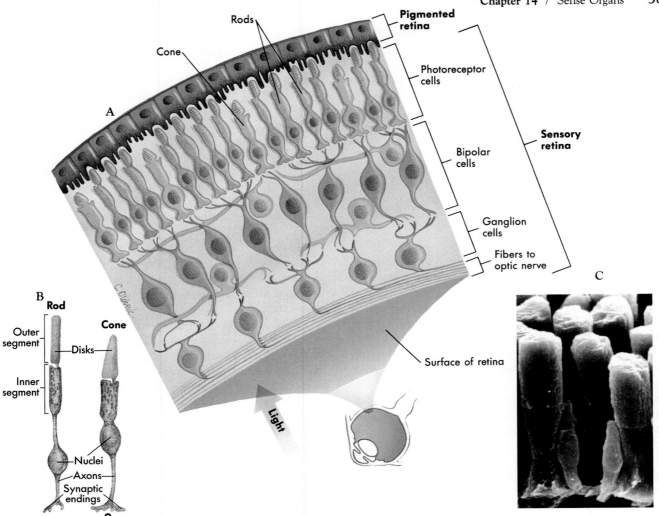

FIGURE 14-11   **Cell layers of the retina. A,** Pigmented and sensory layers of the retina. **B,** Rod and cone cells. Note their variation in the general structure of a neuron. **C,** Electron micrograph of rod and cone cells.

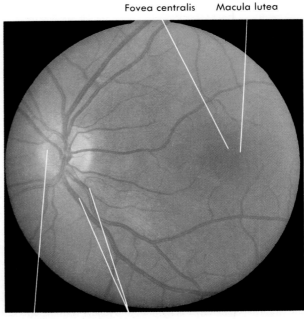

FIGURE 14-12   **Ophthalmoscopic view of the retina.** View of the retina as seen through the pupil. Note how blood vessels enter the eye through the optic disc (blind spot). Fovea centralis and macula lutea are visible.

TABLE 14-2	Coats of the eyeball		
Location	Posterior Portion	Anterior Portion	Characteristics
Outer coat (sclera)	Sclera proper	Cornea	Protective fibrous coat, cornea transparent, rest of coat white and opaque
Middle coat (choroid)	Choroid proper	Ciliary body, suspensory ligament, iris (pupil is hole in iris); lens suspended in suspensory ligament	Vascular, pigmented coat
Inner coat (retina)	Retina	No anterior portion	Nervous tissue; rods and cones (receptors for second cranial nerve) located in retina

Cavities and humors

The eyeball is not a solid sphere but contains a large interior space that is divided into two cavities, anterior and posterior.

The **anterior cavity** has two subdivisions, known as the *anterior* and *posterior chambers.* As Figure 14-9 shows, the entire anterior cavity lies in front of the lens. The posterior chamber of the anterior cavity consists of the small space directly posterior to the iris but anterior to the lens. And the anterior chamber of the anterior cavity is the space anterior to the iris but posterior to the cornea. *Aqueous humor* fills both chambers of the anterior cavity. This substance is clear and watery and often leaks out when the eye is injured.

The **posterior cavity** of the eyeball is considerably larger than the anterior, since it occupies all the space posterior to the lens, suspensory ligament, and ciliary body (see Figure 14-9). It contains *vitreous humor,* a substance with a consistency comparable to soft gelatin. This semisolid material, along with the aqueous humor, helps maintain sufficient intraocular pressure to prevent the eyeball from collapsing.

Aqueous humor forms from blood in capillaries (located mainly in the ciliary body). The ciliary body actively secretes aqueous humor into the posterior chamber, but, also, passive filtration from capillary blood contributes to aqueous humor formation. From the posterior chamber, aqueous humor moves from the area between the iris and the lens through the pupil into the anterior chamber. From here, it drains into the canal of Schlemm and moves on into small veins (Figure 14-13). Normally, aqueous humor drains out of the anterior chamber at the same rate at which it enters the posterior chamber, so the amount of aqueous humor in the eye remains relatively constant—and so, too, does intraocular pressure. But sometimes something happens to upset this balance, and intraocular pressure increases above the normal level of about 20 to 25 mm Hg pressure. The individual then has the eye disease known as *glaucoma,* which, if untreated, can lead to retinal damage and blindness. Either excess formation or, more often, decreased drainage is seen as an immediate cause of this condition, but underlying causes are unknown.

An outline summary of the cavities of the eye appears in Table 14-3.

Muscles

Eye muscles are of two types: *extrinsic* and *intrinsic.* **Extrinsic eye muscles** are skeletal muscles that attach to the outside of the eyeball and to the bones of the orbit. They move the eyeball in any desired direction and are, of course, voluntary muscles. Four of them are straight muscles, and two are oblique. Their names describe their positions on the eyeball. They are the superior, inferior, medial, and lateral rectus muscles and superior and inferior oblique muscles (Figure 14-14).

Intrinsic eye muscles are smooth, or involuntary, muscles located within the eye. Their names, as we have discussed, are the *iris* and the *ciliary muscles.* Incidentally, the eye is the only organ in the body in which both voluntary and involuntary muscles are found. The iris regulates the size of the pupil. The ciliary muscle controls the shape of the lens. As the ciliary muscle contracts, it

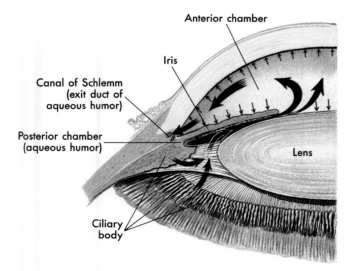

FIGURE 14-13 Formation of aqueous humor. Aqueous humor (*heavy arrows*) is believed to be formed mainly by secretion by the ciliary body into the posterior chamber. It passes into the anterior chamber through the pupil, from which it is drained away by the ring-shaped canal of Schlemm, and finally into the anterior ciliary veins. Small arrows indicate pressure of the aqueous humor.

TABLE 14-3   Cavities of the eye

Cavity	Divisions	Location	Contents
Anterior	Anterior chamber	Anterior to iris and posterior to cornea	Aqueous humor
	Posterior chamber	Posterior to iris and anterior to lens	Aqueous humor
Posterior	None	Posterior to lens	Vitreous humor

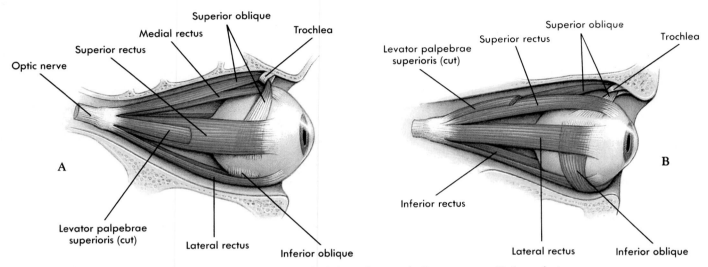

FIGURE 14-14   Extrinsic muscles of the right eye. A, Superior view. **B,** Lateral view.

releases the suspensory ligament from the backward pull usually exerted on it, and this allows the elastic lens, suspended in the ligament, to bulge, or become more convex. The role of both these muscles in vision will be discussed later in this chapter.

Accessory structures

Accessory structures of the eye include the eyebrows, eyelashes, eyelids, and the lacrimal apparatus.

Eyebrows and eyelashes. The eyebrows and eyelashes serve a cosmetic purpose and give some protection against the entrance of foreign objects into the eyes. Small glands located at the base of the lashes secrete a lubricating fluid. They frequently become infected, forming a *sty.*

Eyelids. The eyelids, or *palpebrae,* consist mainly of voluntary muscle and skin, with a border of thick connective tissue at the free edge of each lid, known as the *tarsal plate.* One can feel the tarsal plate as a ridge when turning back the eyelid to remove a foreign object. Mucous membrane, called *conjunctiva,* lines each lid (Figure 14-15). It continues over the surface of the eyeball, where it is modified to give transparency. Inflammation of the conjunctiva (conjunctivitis) is a fairly common infection. Because it produces a pinkish discoloration of the eye's surface, it is called *pinkeye* (Figure 14-16).

The opening between the eyelids bears the technical

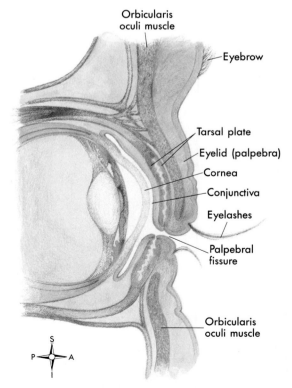

FIGURE 14-15   Accessory structures of the eye. Lateral view with eyelids closed.

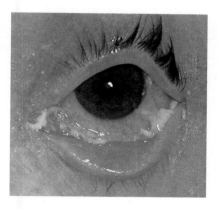

FIGURE 14-16 Acute bacterial conjunctivitis. Notice the discharge of pus characteristic of this highly contagious infection of the conjunctiva.

name of *palpebral fissure.* The height of the fissure determines the apparent size of the eyes. In other words, if the eyelids are habitually held widely opened, the eyes appear large, although there is very little difference in size between eyeballs of different adults. Eyes appear small if the upper eyelids droop. Plastic surgeons can correct this common aging change with an operation called *blepharoplasty.* The upper and lower eyelids join, forming an angle or corner known as a *canthus;* the inner canthus is the medial corner of the eye, and the outer canthus is the lateral corner.

Lacrimal apparatus. The lacrimal apparatus consists of the structures that secrete tears and drain them from the surface of the eyeball. They are the lacrimal glands, lacrimal ducts, lacrimal sacs, and nasolacrimal ducts (Figure 14-17).

The *lacrimal glands,* comparable in size and shape to a

small almond, are located in a depression of the frontal bone at the upper outer margin of each orbit. Approximately a dozen small ducts lead from each gland, draining the tears onto the conjunctiva at the upper outer corner of the eye.

The *lacrimal canals* are small channels, one above and the other below each *caruncle* (small red body at inner canthus). They empty into the lacrimal sacs. The openings into the canals are called *punctae* and can be seen as two small dots at the inner canthus of the eye. The *lacrimal sacs* are located in a groove in the lacrimal bone. The *nasolacrimal ducts* are small tubes that extend from the lacrimal sac into the inferior meatus of the nose. All the tear ducts are lined with mucous membrane, an extension of the mucosa that lines the nose. When this membrane becomes inflamed and swollen, the nasolacrimal ducts become plugged, causing the tears to overflow from the eyes instead of draining into the nose as they do normally. Hence when we have a common cold, "watering" eyes add to our discomfort.

The Process of Seeing

For vision to occur, the following conditions must be fulfilled: an image must be formed on the retina to stimulate its receptors (rods and cones), and the resulting nerve impulses must be conducted to the visual areas of the cerebral cortex for interpretation.

Formation of retinal image

Four processes focus light rays so that they form a clear image on the retina: *refraction* of the light rays, *accommodation* of the lens, *constriction* of the pupil, and *convergence* of the eyes.

Refraction of light rays. Refraction means the deflection, or bending, of light rays. It is produced by light rays passing obliquely from one transparent medium into another of different optical density; the more convex the surface of the medium, the greater its refractive power (Figure 14-18). The refracting media of the eye are the cornea, aqueous humor, lens, and vitreous humor. Light rays are bent, or refracted, at the anterior surface of the cornea as they pass from the rarer air into the denser cornea, at the anterior surface of the lens as they pass from the aqueous humor into the denser lens, and at the posterior surface of the lens as they pass from the lens into the rarer vitreous humor.

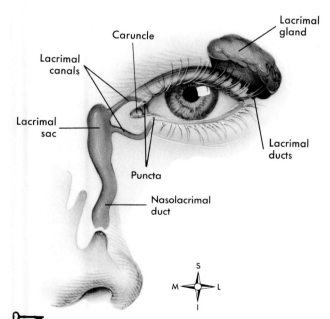

FIGURE 14-17 Lacrimal apparatus. Fluid produced by lacrimal glands (tears) streams across the eye surface, enters the canals, and then passes through the nasolacrimal duct to enter the nose.

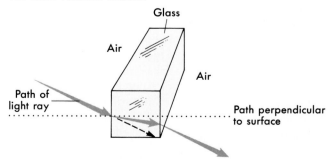

FIGURE 14-18 Refraction. Note the deflection or bending of light as it passes from one transparent medium (air) into another of different optical density (glass).

When an individual goes to an optometrist or ophthalmologist for an eye examination, the doctor performs a "refraction." In other words, by various specially designed methods, the refractory, or light-bending, power of that person's eyes is determined.

In a relaxed normal (*emmetropic*) eye the four refracting media together bend light rays sufficiently to bring to a focus on the retina the parallel rays reflected from an object 20 or more feet away. Of course a normal eye can also focus on objects located much nearer than 20 feet from the eye. This is accomplished by a mechanism known as *accommodation* (discussed next). Many eyes, however, show errors of refraction; that is, they are not able to focus the rays on the retina under the stated conditions. Some common errors of refraction that will be discussed later in the chapter are nearsightedness (myopia), farsightedness (hyperopia), and astigmatism.

Accommodation of lens. Accommodation for near vision necessitates three changes: increase in the curvature of the lens, constriction of the pupils, and convergence of the two eyes. Light rays from objects 20 or more feet away are practically parallel. The normal eye, as previously noted, refracts such rays sufficiently to focus them clearly on the retina. However, light rays from nearer objects are divergent rather than parallel. So obviously they must be bent more acutely to bring them to a focus on the retina. Accommodation of the lens or, in other words, an increase in its curvature takes place to achieve this greater refraction.

Contraction or relaxation of the ciliary muscle affects lens shape. Contraction pulls the choroid layer closer to the lens (see Figure 14-10); this, in turn, loosens the tension of the suspensory ligaments, allowing the lens to bulge. For near vision, then, the ciliary muscle is contracted and the lens is bulging, whereas for far vision the ciliary muscle is relaxed and the lens is comparatively flat (Figure 14-19). Continual use of the eyes for near work produces eyestrain because of the prolonged contraction of the ciliary muscle. Some of the strain can be avoided by looking into the distance at intervals while doing close work.

As people grow older, they tend to become farsighted because lenses lose their elasticity and therefore their ability to bulge and to accommodate for near vision. This condition is called **presbyopia.**

Constriction of pupil. The muscles of the iris play an important part in the formation of clear retinal images. Part of the accommodation mechanism consists of contraction of the circular fibers of the iris, which constricts the pupil. This prevents divergent rays from the object from entering the eye through the periphery of the cornea and lens. Such peripheral rays could not be refracted sufficiently to be brought to a focus on the retina and therefore would cause a blurred image. Constriction of the pupil for near vision is called the *near reflex* of the pupil and occurs simultaneously with accommodation of the lens in near vision. The pupil constricts also in bright light (*photopupil reflex* or *pupillary light reflex*) to protect the retina from too intense or too sudden stimulation.

Convergence of eyes. Single binocular vision (seeing only one object instead of two when both eyes are used) occurs when light rays from an object fall on corresponding points of the two retinas. The foveas and all points lying

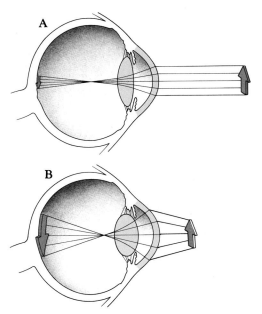

FIGURE 14-19 Accommodation of lens. A, Distant image: the lens is flattened (ciliary muscle relaxed), and the image is focused on the retina. **B,** Close image: the lens is rounded (ciliary muscle contracted), and the image is focused on the retina.

equidistant and in the same direction from the foveas are corresponding points. Whenever the eyeballs move in unison, either with the visual axes parallel (for far objects) or converging on a common point (for near objects), light rays strike corresponding points of the two retinas. Convergence is the movement of the two eyeballs inward so that their visual axes come together, or converge, at the object viewed. The nearer the object, the greater the degree of convergence necessary to maintain single vision. A simple procedure demonstrates the fact that single binocular vision results from stimulation of corresponding points on two retinas. Gently press one eyeball out of line while viewing an object. Instead of one object, you will see two. To achieve unified movement of the two eyeballs, a functional balance between the antagonistic extrinsic muscles must exist. For clear distant vision, the muscles must hold the visual axes of the two eyes parallel. For clear near vision, they must converge them. These conditions cannot be met if, for example, the internal rectus muscle of one eye should contract more forcefully than its antagonist, the external rectus muscle. That eye would then be pulled in toward the nose. The movement of its visual axis would not coordinate with that of the other eye. Light rays from an object would then fall on noncorresponding points of the two retinas, and the object would be seen double (*diplopia*). **Strabismus** (cross-eye or squint) is an exaggerated condition that cannot be overcome by neuromuscular effort. An individual with strabismus usually does not have double vision, as you might expect, because he or she learns to suppress one of the images.

The role of photopigments

Both rods and cones contain *photopigments*, or light-sensitive pigmented compounds. In the presence of light, these chemicals undergo structural changes that result in

the generation of nerve impulses, which the brain is able to interpret as sight.

Rods. The photopigment in rods is named **rhodopsin.** Rhodopsin is so highly light sensitive that even dim light causes its rapid breakdown into *opsin* (a protein) and *retinal* (a vitamin A derivative). Light causes retinal to change its shape and the opsin molecule to expand, or open. When opsin and retinal open and separate in the presence of light (a process called **bleaching**) active sites are exposed and an action potential is created in the rod cell (Figure 14-20). This signal then travels to the brain for interpretation. Objects are seen in shades of gray but not in colors. Energy is required to bring opsin back to its original shape and reattach retinal to it. Until this occurs, the photopigment is unable to respond to light.

Cones. Three types of cones are present in the retina. Each contains a different photopigment, either *erythrolabe*, *chlorolabe*, or *cyanolabe*. Each of the three primary colors (red, green, and blue) reflects light rays of a different wavelength. Each wavelength acts primarily on one type of cone, causing its particular photopigment to break down and initiate impulse conduction by the cone. Generally, light rays from red colors cause erythrolabe breakdown, green colors stimulate chlorolabe breakdown, and blue colors stimulate cyanolabe breakdown. Our perception of a range of colors is due to the combined neural input from varying numbers of the three different cone types.

Because cone photopigments are less sensitive to light than rhodopsin, brighter light is necessary for their breakdown. Cones therefore function to produce vision in bright light. In addition, cones contribute more than rods to perception of sharp images. The reason for this difference involves the way in which information generated by the stimulation of rods and cones is "processed" before it reaches the brain. Look again at Figure 14-11. Note that

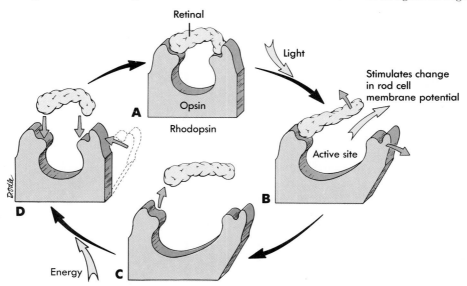

FIGURE 14-20   Rhodopsin cycle. A, Retinal is attached to opsin. **B,** Light causes retinal and opsin to change shape, causing change in rod cell membrane potential. **C,** Retinal separates from opsin (bleaching). **D,** Energy is required to bring opsin back to its original shape and to attach retinal to it.

information obtained by the bipolar cells is collated from many rods, whereas bipolar cells tend to synapse with only a single cone receptor. This property of combining input from several receptors is called *convergence*. The result of convergence is that, while the rods combine their input to make a bipolar cell fire in dimmer light, the brain is unable to determine exactly which rod was stimulated when a given bipolar or ganglion cell fires.

Convergence of impulses from cones is rare. There is almost a one-to-one relationship between cones and the ganglion cells in the retina that ultimately routes impulses to the brain. Interpretation by the brain of sharp images is therefore much better as a result of cone stimulation. The fovea contains the greatest concentration of cones and is therefore the point of clearest vision in good light. For this reason, when we want to see an object clearly in the daytime, we look directly at it to focus the image on the fovea. But in dim light or darkness, we see an object better if we look slightly to the side of it, thereby focusing the image nearer the periphery of the retina, where the more plentiful rods can collate the lesser amount of light information and generate on image.

FYI Color Blindness

Color blindness, usually an inherited condition, is caused by mistakes in producing three chemicals, called *photopigments*, in the cones. Each photopigment is sensitive to one of the three primary colors of light: green, blue, and red. In many cases, the green-sensitive photopigment is missing or deficient; other times, the red-sensitive photopigment is abnormal. (Deficiency of the blue-sensitive photopigment is rare.) Color-blind individuals see colors, but they cannot distinguish between them normally.

Figures such as those shown here are often used to screen individuals for color blindness. A person with red-green color blindness cannot see the *74* in Figure *A*, whereas a person with normal vision can. To determine which photopigment is deficient, a color-blind person may try a figure similar to *B*. Persons with a deficiency of red-sensitive photopigment can distinguish only the *2*; those deficient in green-sensitive photopigment can only see the *4*.

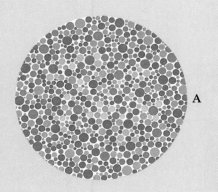

A

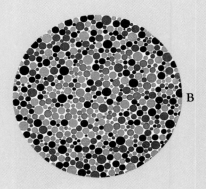

B

LASER THERAPY

Advances in laser technology have produced numerous applications in the medical field, especially in treating eye problems. For some time now, it has been common practice to use the intense light from lasers to repair detached retinas. An **opthalmologist** (eye physician) directs the laser beam at different points in the retina and makes tiny burns. Each burned area eventually forms a small, fibrous scar that holds the retina in place.

Laser therapy may soon replace procedures such as *radial keratotomy*, in which cuts made in the cornea cause it to change shape and thus correct refraction errors. Recently, researchers have been testing a new laser procedure for correcting refraction problems. Working without heat, a special laser that shoots "bursts" of light is used to microscopically shave the cornea, giving it a new shape. Although this therapy needs further testing, it promises to eliminate the need for eyeglasses or contact lenses in correcting many vision problems.

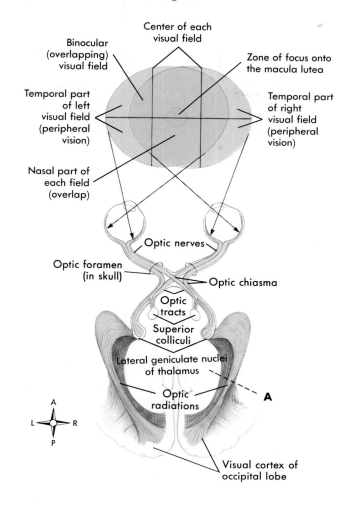

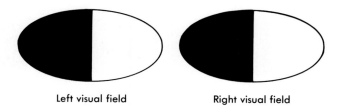

Left visual field   Right visual field

FIGURE 14-21   Visual fields and neuronal pathways of the eye. Note the structures that make up each pathway: optic nerve, optic chiasma, lateral geniculate body of thalamus, optic radiations, and visual cortex of occipital lobe. Fibers from the nasal portion of each retina cross over to opposite side at the optic chiasma and terminate in the lateral geniculate nuclei. Location of a lesion in the visual pathway determines the resulting visual defect. Damage at point A, for example, would cause blindness in the right nasal and left temporal visual fields, as the ovals beneath indicate. (Trace the visual pathway from point A back to the visual field map to see why this is so.) What would be the effect of pressure on the optic chiasma—by a pituitary tumor, for instance? (Answer: It would produce blindness in both temporal visual fields. Why? Because it destroys fibers from the nasal side of both retinas.)

Neuronal pathway of vision

Fibers that conduct impulses from the rods and cones reach the visual cortex in the occipital lobes via the optic nerves, optic chiasma, optic tracts, and optic radiations. "Relay stations" along the way include the superior colliculi and the lateral geniculate nuclei of the thalamus. Look closely at Figure 14-21. Notice that each optic nerve contains fibers from only one retina but that the optic chiasma contains fibers from the nasal portions of both retinas. Each optic tract also contains fibers from both retinas. These anatomical facts explain certain peculiar visual abnormalities that sometimes occur. Suppose a person's right optic tract were injured so that it could not conduct impulses—say, at point A in Figure 14-21. This person would be totally blind in neither eye but partially blind in both eyes. Specifically, this person would be blind in the right nasal and left temporal visual fields. Here are the reasons: the right optic tract contains fibers from the right retina's temporal area, the area that sees the right nasal visual field. In addition, the right optic tract contains fibers from the left retina's nasal area, the area that sees the left temporal visual field.

QUICK CHECK

✔ 1. *Name the layers, or coats, of the eyeball.*
✔ 2. *Identify the layers of the retina.*
✔ 3. *Name the four processes that function to focus a clear image on the retina.*
✔ 4. *Outline the steps of the rhodopsin cycle.*

Cycle of Life: *Sense Organs*

The ability of the sense organs to respond to stimuli caused by changes in the body's internal or external environment will vary during life. Ultimately, all sensory information is acquired through depolarization of sensory nerve endings. Anything that interferes with the generation of a receptor potential or its transmission to and interpretation by areas of the central nervous system will influence sensory acuity. Age, disease, structural defects, or lack of maturation all affect our ability to identify and respond to sensory input.

Structure and function response capabilities of the sense organs are related to developmental factors associated with age. For example, a newborn baby has limited sight, hearing, and tactile identification capabilities. As maturation occurs and normal development progresses, the senses become more acute. By late adulthood, presbyopia, progressive hearing loss, and a reduced sense of taste and smell are common.

Some loss of sensory capability in old age is directly related to structural change in receptor cells or other necessary sense organ structures. The lens of the eye becomes harder and less able to change shape, taste buds become less functional, and exteroceptors of all types become less responsive to stimuli because of structural deterioration.

MECHANISMS OF DISEASE

Disorders of the Sense Organs

Disorders of the Ear

Hearing problems can be divided into two basic categories: *conduction impairment* and *nerve impairment*. Conduction impairment refers to the blocking of sound waves as they are conducted through the external and middle ear to the sensory receptors of the inner ear (the conduction pathway). Nerve impairment results in insensitivity to sound because of inherited or acquired nerve damage.

The most obvious cause of conduction impairment is blockage of the external auditory canal. Waxy buildup of cerumen commonly blocks conduction of sound toward the tympanic membrane. Foreign objects, tumors, and other matter can block conduction in either the external or middle ear. An inherited bone disorder called **otosclerosis** (o-to-skle-RO-sis) impairs conduction by causing structural irregularities in the stapes. Otosclerosis usually first appears during childhood or early adulthood as **tinnitus** (tin-EYE-tus), or "ringing in the ear."

Temporary conduction impairment often results from ear infection, or **otitis**. The structure of the auditory tube, especially its connection with the nasopharynx, makes the middle ear prone to bacterial or viral *otitis media*. Otitis media often produces swelling and pus formation that blocks the conduction of sound through the middle ear. Permanent damage to structures of the middle ear occasionally occurs in severe cases.

Hearing loss due to nerve impairment is common in the elderly. Called **presbycusis** (pres-be-KYOO-sis), this progressive hearing loss associated with aging results from degeneration of nerve tissue in the ear and the vestibulocochlear nerve. A similar type of hearing loss occurs after chronic exposure to loud noises that damages receptors in the organ of Corti. Because different sound *frequencies* (tones) stimulate different regions of the organ of Corti, hearing impairment is limited to only those frequencies associated with the portion of the organ of Corti that is damaged. For example, the portion of the organ of Corti that degenerates first in presbycusis is normally

stimulated by high-frequency sounds. Thus, the inability to hear high-pitched sounds is common among the elderly.

Nerve damage can also occur in **Meniere's** (may-nee-ERZ) **disease,** a chronic inner ear disease of unknown cause. Meniere's disease is characterized by tinnitus, progressive nerve deafness, and *vertigo* (sensation of spinning).

Disorders of the Eye

Healthy vision requires three basic processes: formation of an image on the retina (refraction), stimulation of rods and cones, and conduction of nerve impulses to the brain. Malfunction of any of these processes can disrupt normal vision.

Refraction Disorders

Focusing a clear image on the retina is essential for good vision. In the normal eye, light rays enter the eye and are focused into a clear, upside-down image on the retina (Figure 14-22, A). The brain can easily right the upside-down image in our conscious perception but cannot correct an image that is not sharply focused. If our eyes are elongated the image focuses in front of the retina rather than on it. The retina receives only a fuzzy image. This condition, called **myopia** or *nearsightedness,* can be corrected by using concave contact lenses or glasses (Figure 14-22, B and D). If our eyes are shorter than normal the image focuses behind the retina, also producing a fuzzy image. This condition, called **hyperopia** or *farsightedness,* can be corrected by convex lenses (Figure 14-22, C and E).

Various other conditions can prevent the formation of a clear image on the retina. For example, the inability to focus the lens properly as we age, or *presbyopia,* has already been mentioned. Older individuals can compensate for presbyopia by using "reading glasses" when near vision is needed. An irregularity in the curvature of the cornea or lens, a condition called **astigmatism** (a-STIG-ma-tizm), can also be corrected with glasses or contact lenses that are formed with the opposite curvature. **Cataracts,** cloudy spots in the eye's lens that develop in the lens as we age, may also interfere with focusing. Cataracts are especially troublesome in dim light because weak beams of light cannot pass through the cloudy spots the way some brighter light can. This fact accounts for the trouble many older adults have with their "night vision."

Infections of the eye also have the potential to impair vision, sometimes permanently. Most eye infections begin in the conjunctiva, producing an inflammation response known as "pink-eye," or *conjunctivitis.* A variety of different pathogens can cause conjunctivitis. For example, the bacterium *Chlamydia trachomatis* that commonly infects the reproductive tract can cause a chronic infection called *chlamydial conjunctivitis,* or **trachoma.** Because chlamydia and other pathogens often inhabit the birth canal, antibiotics are routinely applied to the eyes of newborns to prevent conjunctivitis. Highly contagious *acute bacterial conjunctivitis,* characterized by drainage of a mucous pus, is most commonly caused by bacteria such as *Staphylococcus* and *Haemophilus.* Conjunctivitis may produce lesions on the inside of the eyelid that can damage the cornea and thus impair vision. Occasionally infections of the conjunctiva spread to the tissues of the eye proper and cause permanent injury—even total blindness. Besides infection, conjunctivitis may also be caused by allergies. The red, itchy, watery eyes commonly associated with allergic reactions to pollen and other substances results from an allergic inflammatory response of the conjunctiva.

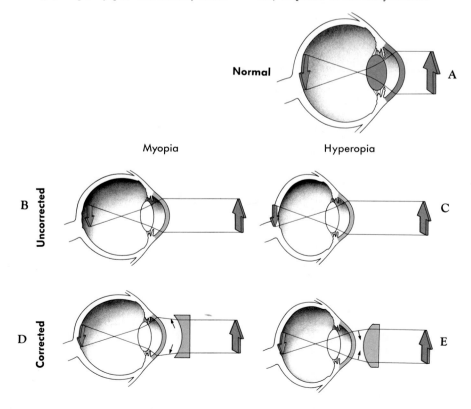

FIGURE 14-22 Refraction disorders. The upper figure (**A**) shows how light is refracted in the normal eye to form a well-focused image. The lower figures show the abnormal and corrected refraction observed in myopia (**B** and **D**) and hyperopia (**C** and **E**).

Disorders of the Retina

Damage to the retina impairs vision because even a well-focused image cannot be perceived if some or all of the light receptors do not function properly. For example, in a condition called **retinal detachment** part of the retina falls away from the tissue supporting it. This condition may result from aging, eye tumors, or from sudden blows to the head—as in a sporting injury. Common warning signs include the sudden appearance of floating spots that may decrease over a period of weeks and odd "flashes of light" that appear when the eye moves. If left untreated, the retina may detach completely and cause total blindness in the affected eye.

Diabetes mellitus, a disorder involving the hormone, insulin, may cause a condition known as **diabetic retinopathy** (ret-in-OP-ah-the). In this disorder, the diabetes causes small hemorrhages in retinal blood vessels that disrupt the oxygen supply to the photoreceptors. The eye responds by building new, but abnormal, vessels that block vision and may cause detachment of the retina. Diabetic retinopathy is one of the leading causes of blindness in the United States.

Another condition that can damage the retina is **glaucoma.** Recall that glaucoma is excessive *intraocular* (in-trah-AHK-yoo-lar) *pressure* caused by abnormal accumulation of aqueous humor. As fluid pressure against the retina increases above normal, blood flow through the retina slows. Reduced blood flow causes degeneration of the retina and thus to loss of vision. Although acute forms of glaucoma can occur, most cases of glaucoma develop slowly over a period of years. This chronic form may not produce any symptoms, especially in its early stages. For this reason, routine eye examinations typically include a screening test for glaucoma. As chronic glaucoma progresses, damage first appears at the edges of the retina—causing a gradual loss of peripheral vision. Blurred vision and headaches may also occur. As the damage becomes more extensive, "halos" are seen around bright lights. If untreated, glaucoma eventually produces total, permanent blindness.

Degeneration of the retina can cause difficulty seeing at night or in dim light. This condition, called **nyctalopia** (nik-ta-LO-pee-ah) or "night blindness," can also be caused by a deficiency of vitamin A. Recall that vitamin A is needed to make retinal, a component of rhodopsin. A deficiency of rhodopsin impairs the function of rod cells, which are needed for dim light vision.

Disorders of the Visual Pathway

Damage or degeneration in the optic nerve, the brain, or any part of the visual pathway between them, can impair vision. For example, the pressure associated with glaucoma can also damage the optic nerve. Diabetes, already cited as a cause of retina damage, can also cause degeneration of the optic nerve.

Damage to the visual pathway does not always result in total loss of sight. Depending on where the damage occurs, only a part of the visual field may be affected. For example, a certain form of neuritis (nerve inflammation), often associated with multiple sclerosis, can cause loss of only the center of the visual field—a condition called **scotoma** (sko-TO-mah).

A stroke can cause vision impairment when the resulting tissue damage occurs in one of the regions of the brain that process visual information. For example, damage to an area that processes information about colors may result in a rare condition called *acquired cortical color blindness*. This condition is characterized by difficulty in distinguishing any color—not just one or two colors as in the more common inherited forms of color blindness.

CHAPTER SUMMARY

SENSORY RECEPTORS

A. Sensory receptors make it possible for the body to respond to stimuli caused by changes occurring in our internal or external environment
B. Receptor response
1. General function—responds to stimuli by converting them to nerve impulses
2. Different types of receptors respond to different stimuli
3. Receptor potential
   a. The potential that develops when an adequate stimulus acts on a receptor; it is a graded response
   b. When a threshold is reached, an action potential in the sensory neuron's axon is triggered
   c. Impulses travel over sensory pathways to the brain and spinal cord where they are either interpreted as a particular sensation or they initiate a reflex action
   d. Sensory projection—a brain function that pinpoints the area of the body from which the receptor potential was initiated
4. Adaptation—a functional characteristic of receptors; receptor potential decreases over a period of time in response to a continuous stimulus, which leads to a decreased rate of impulse conduction and a decreased intensity of sensation
C. Classification of receptors—there are five categories based on the types of stimuli that activate them
1. Mechanoreceptors—activated by mechanical stimuli that change the position of the receptor, resulting in the generation of a receptor potential
2. Chemoreceptors—activated by the amount or the changing concentration of certain chemicals, e.g., taste and smell
3. Thermoreceptors—activated by changes in temperature
4. Nociceptors—activated by intense stimuli that result in tissue damage; the sensation produced is pain
5. Photoreceptors—found only in the eye; respond to light stimuli if the intensity is great enough to generate a receptor potential
D. Distributions of receptors
1. Receptors for special senses of smell, taste, vision, hearing, and equilibrium are grouped into localized areas or into complex organs
2. General sense organs of somatic senses are microscopic receptors widely distributed throughout the body in the skin, mucosa, connective tissue, muscles, tendons, joints, and viscera

SOMATIC SENSES

A. Somatic senses detect sensations such as pain, temperature, pressure, touch, body position, tension in muscles, hunger, thirst, etc; the three types of somatic sense receptors are:
1. Exteroceptors—somatic sense receptors located on the body surface
2. Proprioceptors—somatic sense receptors located in muscles and joints
3. Visceroceptors—somatic sense receptors located in internal visceral organs

PAIN AND TEMPERATURE RECEPTORS

A. Pain receptors (also known as nociceptors)
  1. Are widely distributed exteroceptors and visceroceptors
  2. Are free nerve endings that respond to a number of different stimuli; there are two types of nerve fibers that conduct pain impulses from free nerve endings to the brain
    a. Acute (A) fibers—mediate sharp, intense, localized pain sensations
    b. Chronic (B) fibers—associated with dull, aching pain

TOUCH AND PRESSURE RECEPTORS

A. Exteroceptors that respond to stimuli that change their shape or placement (Figure 14-1)
B. Meissner's corpuscles and Krause's end bulb—involved with sensations of touch, low frequency vibrations, and two-point discrimination; found on areas devoid of hairs
C. Ruffini's corpuscles—sense deep pressure and continuous touch; located in the dermis of the skin and are numerous in the fingers.
D. Pacinian corpuscles—respond quickly to sensations of deep pressure, high frequency vibration and stretch; found in deep dermis of the hands and feet and in joint capsules.

STRETCH RECEPTORS

A. The two most important stretch receptors are associated with muscles and tendons; classified as proprioceptors
B. Muscle spindles—provide body with information regarding muscle length; stimulated if a relaxed muscle is stretched beyond a certain limit; causes a stretch reflex that contracts the muscle
C. Golgi tendon receptors—provide body with information regarding strength of muscle contraction; stimulated by excessive muscle contraction; located near points of attachment of tendon to bone; stimulation causes muscle to relax

SPECIAL SENSES

A. Characterized by receptors grouped closely together or grouped in specialized organs; sense of smell, taste, hearing equilibrium, and vision

SENSE OF SMELL

A. Olfactory receptors
  1. Olfactory sense organs consist of epithelial support cells and specialized olfactory receptor neurons (Figure 14-2)
    a. Olfactory cilia—located on olfactory receptor neurons that touch the olfactory epithelium lining the upper surface of the nasal cavity
    b. Olfactory cells—chemoreceptors; gas molecules or chemicals dissolved in the mucus covering the nasal epithelium stimulate the olfactory cells
    c. Olfactory epithelium—located in most superior portion of the nasal cavity
    d. Olfactory receptors—extremely sensitive and easily fatigued
B. Olfactory pathway—when the level of odor-producing chemicals reaches a threshold level, the following occurs
  1. Receptor potential, and then action potential, is generated and passed to the olfactory nerves in the olfactory bulb
  2. The impulse then passes through the olfactory tract and into the thalamic and olfactory centers of the brain for interpretation, integration, and memory storage

SENSE OF TASTE

A. Taste buds—sense organs that respond to gustatory, or taste, stimuli; associated with papillae
  1. Chemoreceptors that are stimulated by chemicals dissolved in the saliva
  2. Gustatory cells—specialized cells found in taste buds; gustatory hairs extend from each gustatory cell into the taste pore
  3. Sense of taste depends on the creation of a receptor potential in gustatory cells due to taste-producing chemicals in the saliva
  4. Taste buds are similar structurally; functionally, each taste bud responds most effectively to one of four primary taste sensations: sour, sweet, bitter, and salty (Figure 14-3)
  5. Primary tastes to which a taste bud responds is related to its placement on the tongue
    a. Tip of the tongue—responds to sweet and salty
    b. Sides of the tongue—respond to sour
    c. Back of the tongue—responds to bitter
  6. Adaptation and sensitivity thresholds are different for each of the primary taste sensations
B. Neuronal pathway for taste
  1. Taste sensation begins with a receptor potential in the gustatory cells of a taste bud; generation and propagation of an action potential then transmits the sensory input to the brain
  2. Nerve impulses from the anterior two thirds of the tongue travel over the facial nerve; those from the posterior one third of the tongue travel over the glossopharyngeal nerve; vagus nerve plays a minor role in taste
  3. Nerve impulses are carried to the medulla oblongata, relayed into the thalamus, and then into the gustatory area of the cerebral cortex in the parietal lobe of the brain

SENSE OF HEARING AND BALANCE: THE EAR

A. External ear—two divisions (Figure 14-4)
  1. Auricle, or pinna—the visible portion of the ear
  2. External auditory meatus—tube leading from the auricle into the temporal bone and ending at the tympanic membrane
B. Middle ear (Figure 14-4)
  1. Tiny, epithelia-lined cavity hollowed out of the temporal bone
  2. Contains three auditory ossicles
    a. Malleus (hammer)—attached to the inner surface of the tympanic membrane
    b. Incus (anvil)—attached to the malleus and stapes
    c. Stapes (stirrup)—attached to the incus
  3. Openings into the middle ear cavity
    a. Opening from the external auditory meatus covered with tympanic membrane
    b. Oval window—opening into inner ear; stapes fits here
    c. Round window—opening into inner ear; covered by a membrane
    d. Opening into the auditory (eustachian) tube
C. Inner ear (Figure 14-5, A)
  1. Structure of the inner ear
    a. Bony labyrinth—made up of the vestibule, cochlea, and semicircular canals
    b. Membranous labyrinth—made up of utricle and saccule inside the vestibule, cochlear duct inside the cochlea, and the membranous semicircular canals inside the bony ones
    c. Vestibule and semicircular canals are involved with balance
    d. Cochlea—involved with hearing
    e. Endolymph—clear, potassium-rich fluid filling the membranous labyrinth
    f. Perilymph—similar to cerebrospinal fluid, surrounds the membranous labyrinth, filling the space between the membranous tunnel and its contents and the bony walls that surround it

2. Cochlea and cochlear duct (Figure 14-5, B)
   a. Cochlea—bony labyrinth
   b. Modiolus—cone-shaped core of bone that houses the spiral ganglion, which consists of cell bodies of the first sensory neurons in the auditory relay
   c. Cochlear duct
      (1) Lies inside the cochlea; only part of the internal ear concerned with hearing; contains endolymph
      (2) Shaped like a triangular tube
      (3) Divides the cochlea into the scala vestibuli, the upper section, and the scala tympani, the lower section; both sections filled with perilymph
      (4) Vestibular membrane—the roof of the cochlear duct
      (5) Basilar membrane—floor of the cochlear duct
      (6) Organ of Corti—rests on the basilar membrane; consists of supporting cells and hair cells
      (7) Axons of the neurons that begin around the organ of Corti, extend in the cochlear nerve to the brain to produce the sensation of hearing
3. Sense of hearing
   a. Sound is created by vibrations
   b. Ability to hear sound waves depends on volume, pitch, and other acoustic properties
   c. Sound waves must be of sufficient amplitude to move the tympanic membrane and have a frequency capable of stimulating the hair cells in the organ of Corti
   d. Basilar membrane is not the same width and thickness throughout its length; high-frequency sound waves vibrate the narrow portion near the oval window, whereas low frequencies vibrate the wider, thicker portion near the apex of the cochlea; this fact allows different hair cells to be stimulated and different pitches of sound to be perceived
   e. Perception of loudness is determined by the amplitude of movement of the basilar membrane; the greater the movement, the louder the perceived sound
   f. Hearing—results from stimulation of the auditory area of the cerebral cortex
   g. Pathway of sound waves (Figure 14-6)
      (1) Enter external auditory canal
      (2) Strike tympanic membrane, causing vibrations
      (3) Tympanic vibrations move the malleus, which in turn moves the incus and then the stapes
      (4) The stapes moves against the oval window, which begins the fluid conduction of sound waves
      (5) The perilymph in the scala vestibuli of the cochlea begins a "ripple" that is transmitted through the vestibular membrane to the endolymph inside the duct, to the basilar membrane, then to the organ of Corti
      (6) From the basilar membrane, the ripple is transmitted through the perilymph in the scala tympani and then expends itself against the round window
   h. Neuronal pathway of hearing
      (1) A movement of hair cells against the tectorial membrane stimulates the dendrites that terminate around the base of the hair cells and initiates impulse conduction by the cochlear nerve to the brain stem
      (2) Impulses pass through "relay stations" in the nuclei in the medulla, pons, midbrain, and thalamus before reaching the auditory area of the temporal lobe
4. Vestibule and semicircular canals (Figure 14-5, A)
   a. Vestibule—the central section of the bony labyrinth; the utricle and saccule are the membranous structures within the vestibule
   b. Semicircular canals—three, each at right angles to the others, are found in each temporal bone; within bony semicircular canals are membranous semicircular canals, each containing endolymph and connecting with the utricle; near this junction, each canal enlarges into an ampulla
   c. Sense of balance
      (1) Static equilibrium—ability to sense the position of the head relative to gravity or to sense acceleration or deceleration (Figure 14-7)
         (a) Movements of the macula, located in both the utricle and saccule almost at right angles to each other, provide information related to head position or acceleration
         (b) Otoliths are located within the matrix of the macula
         (c) Changing head position produces a change of pressure on the otolith-weighted matrix, which stimulates the hair cells that, in turn, stimulate the receptors of the vestibular nerve
         (d) Vestibular nerve fibers conduct impulses to the brain and produce a sensation of the position of the head and also a sensation of a change in the pull of gravity
         (e) Righting reflexes—muscular responses to restore the body and its parts to their normal position when they have been displaced; caused by stimuli of the macula and impulses from proprioceptors and from the eyes
      (2) Dynamic equilibrium—needed to maintain balance when the head or body is rotated or suddenly moved; able to detect changes both in direction and rate at which movement occurs (Figure 14-8)
         (a) Depends on the functioning of the cristae ampullaris, which are located in the ampulla of each semicircular canal
         (b) Cupula—gelatinous cap in which the hair cells of each crista are imbedded; does not respond to gravity; it moves with the flow of endolymph in the semicircular canals
         (c) Semicircular canals are almost placed at right angles to each other to detect movement in all direction
         (d) When the cupula moves, hair cells are bent, producing a receptor potential followed by an action potential; the action potential passes through the vestibular portion of the eighth cranial nerve to the medulla oblongata, where it is sent to other areas of the brain and spinal cord for interpretation, integration, and response

THE EYE

A. Structure of the eye (Figure 14-9)
   1. Coats of the eyeball—three layers of tissues compose the eyeball
      a. Sclera—outer coat
         (1) Tough, white, fibrous tissue
         (2) Cornea—the transparent anterior portion that lies over the iris; no blood vessels found in the cornea or in the lens
         (3) Canal of Schlemm—ring-shaped venous sinus found deep within the anterior portion of the sclera at its junction with the cornea
      b. Choroid—middle coat
         (1) Contains many blood vessels and a large amount of pigment

(2) Anterior portion has three different structures (Figure 14-10)
  (a) Ciliary body—thickening of choroid, fits between anterior margin of retina and posterior margin of iris; ciliary muscle lies in anterior part of ciliary body; ciliary processes—fold in the ciliary body
  (b) Suspensory ligament—attached to the ciliary processes and blends with the elastic capsule of the lens, to hold it in place
  (c) Iris—colored part of the eye; consists of circular and radial smooth muscle fibers that form in doughnut-shaped structure; attaches to the ciliary body
  c. Retina—incomplete innermost coat of the eyeball
    (1) Three layers of neurons make up the sensory retina (Figure 14-11)
      (a) Photoreceptor neurons—visual receptors, highly specialized for stimulation by light rays
        [1] Rods—absent from the fovea and macula; increased in density toward the periphery of the retina
        [2] Cones—less numerous than rods; most densely concentrated in the fovea centralis in the macula lutea
      (b) Bipolar neurons
      (c) Ganglionic neurons—all axons of these neurons extend back to the optic disc; part of the sclera, which contains perforations through which the fibers emerge from the eyeball as the optic nerve
  2. Cavities and humors
    a. Cavities—eyeball has a large interior space divided into two cavities
      (1) Anterior cavity—lies in front of the lens; has two subdivisions
        (a) Anterior chamber—space anterior to the iris and posterior to the cornea
        (b) Posterior chamber—small space posterior to the iris and anterior to the lens
      (2) Posterior cavity—larger than the anterior cavity; occupies all the space posterior to the lens, suspensory ligament, and ciliary body
    b. Humors
      (1) Aqueous humor—fills both chambers of the anterior cavity; clear, watery fluid that often leaks out when the eye is injured; formed from blood in capillaries located in the ciliary body (Figure 14-13)
      (2) Vitreous humor—fills the posterior cavity; semisolid material; helps to maintain sufficient intraocular pressure, with aqueous humor, to give the eyeball its shape
  3. Muscles—two types of eye muscles
    a. Extrinsic eye muscles (Figure 14-14)—skeletal muscles that attach to the outside of the eyeball and to the bones of the orbit; named according to their position on the eyeball; the muscles are: superior, inferior, medial, and lateral rectus muscles and superior and inferior oblique muscles
    b. Intrinsic eye muscles—smooth muscles located within the eye: iris—regulates size of pupil; ciliary muscle—controls shape of lens
  4. Accessory structures (Figure 14-15)
    a. Eyebrows and eyelashes—give some protection against foreign objects entering the eye; cosmetic purposes
    b. Eyelids—consist of voluntary muscle and skin with a tarsal plate; lined with conjunctiva, a mucous membrane; palpebral fissure—opening between the

eyelids; canthus—where the upper and lower eyelids join
    c. Lacrimal apparatus—structures that secrete tears and drain them from the surface of the eyeball (Figure 14-17)
      (1) Lacrimal glands—size and shape of a small almond; located at the upper, outer margin of each orbit; approximately a dozen small ducts lead from each gland; drain tears onto the conjunctiva
      (2) Lacrimal canals—small channels that empty into lacrimal sacs
      (3) Lacrimal sacs—located in a groove in the lacrimal bone
      (4) Nasolacrimal ducts—small tubes that extend from the lacrimal sac into the inferior meatus of the nose
B. The process of seeing
  1. Formation of retinal image
    a. Refraction of light rays—deflection, or bending, of light rays produced by light rays passing obliquely from one transparent medium into another of different optical density (Figure 14-18); cornea, aqueous humor, lens, and vitreous humor are the refracting media of the eye
    b. Accommodation of lens—increase in curvature of the lens to achieve the greater refraction needed for near vision (Figure 14-19)
    c. Constriction of pupil—muscles of iris are important to formation of a clear retinal image; pupil constriction prevents divergent rays from object from entering eye through periphery of the cornea and lens; near reflex—constriction of pupil that occurs with accommodation of the lens in near vision; photopupil reflex—pupil constricts in bright light
    d. Convergence of eyes—movement of the two eyeballs inward so that their visual axes come together at the object viewed; the closer the object, the greater the degree of convergence necessary to maintain single vision; for convergence to occur, a functional balance between antagonistic extrinsic muscles must exist
  2. The role of photopigments—light-sensitive pigmented compounds undergo structural changes that result in generation of nerve impulses, which are interpreted by the brain as sight
    a. Rods—photopigment in rods is rhodopsin; highly light sensitive; breaks down into opsin and retinal; separation of opsin and retinal in the presence of light causes an action potential in rod cells; energy is needed to reform rhodopsin (Figure 14-20)
    b. Cones—three types of cones are present in the retina with each having a different photopigment; cone pigments are less light sensitive than rhodopsin and need brighter light to breakdown
      (1) Light rays from red colors—breakdown of erythrolabe
      (2) Light rays from green colors—breakdown of chlorolabe
      (3) Light rays from blue colors—breakdown of cyanolabe
  3. Neuronal pathway of vision (Figure 14-21)
    a. Fibers that conduct impulses from the rods and cones reach the visual cortex in the occipital lobes via the optic nerves, optic chiasma, optic tracts, and optic radiations
    b. Optic nerve contains fibers from only one retina, but optic chiasma contains fibers from the nasal portion of both retinas; these anatomical facts explain peculiar visual abnormalities that sometimes occur

CYCLE OF LIFE: SENSE ORGANS

A. Sensory information is acquired through depolarization of sensory nerve endings
  1. Age, disease, structural defects, or lack of maturation affect ability to identify and respond
B. Structure and function response capabilities are related to developmental factors associated with age
C. Senses become more acute with maturation
D. Late adulthood—loss of sensory capability
  1. Structural change in receptor cells or other sense organ structures

MECHANISMS OF DISEASE: DISORDERS OF THE SENSE ORGANS

A. Disorders of the ear—two basic categories
  1. Conduction impairment—sound waves are blocked as they are conducted through the conduction pathway to the inner ear
    a. Causes of conduction impairment
      (1) Blockage of the external auditory canal by wax, foreign objects, tumors, or other matter
      (2) Otosclerosis—inherited bone disorder that causes structural irregularities in the stapes and impairs conduction; first appears as tinnitus
      (3) Otitis—ear infection causing temporary conduction impairment
      (4) The middle ear is prone to bacterial or viral otitis media, which produces swelling and pus
  2. Hearing loss from nerve impairment—common in the elderly
    a. Presbycusis—progressive hearing loss from degeneration of nerve tissue in the ear and vestibulocochlear nerve
      (1) Chronic exposure to loud noises damages the organ of Corti, leading to the inability to hear high-pitched sound
    b. Meniere's disease—chronic inner ear disease; unknown cause; characterized by tinnitus, progressive nerve deafness, and vertigo
B. Disorders of the eye
  1. Healthy vision—from three processes: refraction, stimulation of rods and cones, and conduction of nerve impulses to brain
  2. Refraction disorders
    a. Myopia (nearsightedness)—elongated eyes focus an image in front of the retina, causing it to receive a fuzzy image
    b. Hyperopia (farsightedness)—"short" eyes cause an image to focus behind the retina
    c. Presbyopia—inability to focus lense properly from aging ("old eye")
    d. Astigmatism—irregularity in the shape of the cornea
    e. Cataracts—cloudy spots developing on the lens
    f. Infections—most eye infections begin in the conjunctiva, producing conjunctivitis, an inflammation response
      (1) Conjunctivitis may be caused by infection or allergies
      (2) Permanent injury or blindness may result if infection spreads to eye tissue
  3. Disorders of the retina: images cannot be received if light receptors do not function properly
    a. Retinal detachment—part of the retina falls away from the supporting tissue; untreated, it may fall away completely, causing blindness
    b. Diabetes mellitus—diabetes causes hemorrhages in retinal blood vessels and disrupts oxygen supply to photoreceptors, causing new but abnormal vessels to form, blocking vision, and detachment of the retina (called retinopathy); it is a leading cause of blindness
    c. Glaucoma—excessive intraocular pressure caused by accumulation of aqueous humor; reduced blood flow causes degeneration of the retina and loss of vision; untreated it causes blindness
    d. Nyctalopia—degeneration of the retina causing difficulty in seeing at night or in dim light
  4. Disorders of the visual pathway
    a. Scotoma—a form of neuritis associated with multiple sclerosis causes loss of only the center of the visual field
    b. Strokes—can cause vision impairment
    c. Acquired cortical color blindness—damage to area of brain processing color; characterized by difficulty distinguishing any color

---

REVIEW QUESTIONS

1. Explain how an adequate stimulus acting on a receptor produces a particular sensation.
2. Define adaptation.
3. Describe each of the following: mechanoreceptors, chemoreceptors, thermoreceptors, nociceptors, photoreceptors.
4. Identify the various types of free nerve endings and describe the function of each.
5. Identify the pathway involved for the production of the sense of smell.
6. How are the following tastes detected: sour, sweet, bitter, salty?
7. Describe the neuronal pathway for taste.
8. Describe the main features of the middle ear.
9. Name the parts of the bony and membranous labyrinths and describe the relationship of those parts.
10. In what ear structure(s) is the hearing sense organ located? The equilibrium sense organs?
11. What is the name of the hearing sense organ? Of the equilibrium sense organs?
12. Describe the path of sound waves as they enter the ear.
13. Describe the role of the basilar membrane in hearing.
14. What is the difference between static and dynamic equilibrium? Describe the general mechanisms by which each is maintained.
15. Name the outer, middle, and inner coats of the eyeball.
16. Name two involuntary muscles in the eye. Explain their functions.
17. Define the term *refraction*. Name the refractory media of the eye.
18. Explain briefly the mechanism for accommodation for near vision.
19. Name the receptors for vision in dim light and those for vision in bright light.
20. Name the photopigments present in rods and in cones. Explain their role in vision.
21. Explain why "night blindness" may occur in marked vitamin A deficiency.
22. Describe the phenomenon of referred pain.

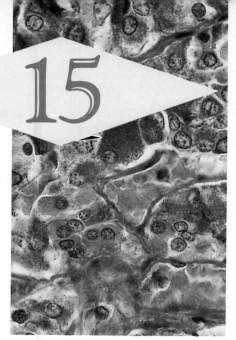

15

The Endocrine System

Pituitary gland tissue
Erlandsen/Magney: Color Atlas of Histology

CHAPTER OUTLINE

OBJECTIVES

After you have completed this chapter, you should
be able to:

1. Compare endocrine structure and function with
   nervous structure and function.
2. Identify the different ways to classify hormones.
3. Differentiate between the mechanisms of steroid
   and nonsteroid hormone action.
4. Describe endocrine reflexes.
5. Discuss the chemical nature, classification, and
   mechanism of action of prostaglandins.
6. Discuss the size, location, and anatomical
   components of the pituitary gland.
7. List the hormones of the adenohypophysis,
   describe their general functions, and identify the
   primary locations of their target cells.
8. Describe a typical negative feedback system.
9. List and identify the action of the hormones
   stored and released by the neurohypophysis.
10. Discuss the structure, location, and functions of
    the thyroid and parathyroid glands.
11. Compare and contrast the functions of the
    hormones produced by the cells of the adrenal
    cortex with those secreted by the adrenal
    medulla.
12. Describe the types of cells found in the pancreatic
    islets and their hormone secretions.
13. List and identify the functions of the pancreatic
    hormones.
14. List and identify the hormonal functions of the
    ovaries and the testes.
15. List the hormones associated with the placenta,
    the thymus, the mucous lining of the
    gastrointestinal tract, and the heart.

Adenohypophysis (ad-e-no-hye-POF-i-sis) Anterior pituitary gland; composed of regular endocrine tissue

Adrenal glands (ah-DREE-nal glands) Glands that rest on top of the kidneys; composed of the cortex and medulla

Endocrine (EN-doe-krin) Secreting into the blood or tissue fluid rather than a duct; "ductless glands"

Gonads (GO-nads) Primary sex organs, which produce reproductive cells

Hormone (HOR-mone) Substance secreted by an endocrine gland

Neurohypophysis (noo-ro-hye-POF-i-sis) Posterior pituitary gland; composed of neurosecretory tissue

Parathyroid (pair-ah-THYE-royd) Endocrine glands embedded in the posterior aspect of the thyroid gland; secrete parathyroid hormone

Prostaglandin (PG) (pross-tah-GLAN-din) Unique group of lipid molecules that have powerful, hormonelike effects

Target cell Any cell with one or more receptors for a particular hormone

Thymus (THYE-mus) Endocrine gland located in the mediastinum; vital part of the body's immune system

Thyroid (THYE-royd) Endocrine gland located in the neck whose hormones regulate cellular metabolism

Tropic hormone (TROW-pik HOR-mone) Hormone that stimulates another endocrine gland to grow and secrete its hormones

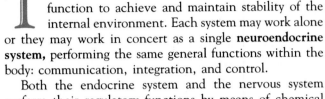

T he endocrine system and nervous system both function to achieve and maintain stability of the internal environment. Each system may work alone or they may work in concert as a single **neuroendocrine system,** performing the same general functions within the body: communication, integration, and control.

Both the endocrine system and the nervous system perform their regulatory functions by means of chemical messengers sent to specific cells. In the nervous system, neurons secrete neurotransmitter molecules to signal nearby cells that have the appropriate receptor molecules. In the endocrine system, secreting cells send **hormone** (from the Greek *hormaein,* "to excite") molecules by way of the bloodstream to signal specific **target cells** throughout the body. Tissues and organs that contain endocrine target cells are called *target tissues* and *target organs,* respectively. As with postsynaptic cells, endocrine target cells must have the appropriate receptor to be influenced by the signaling chemical. Many cells have receptors for both neurotransmitters and hormones, so they can be influenced by both types of chemicals.

While neurotransmitters are sent over very short distances across a synapse, hormones diffuse into the blood to be carried to nearly every point in the body. The nervous system can directly control only muscles and glands that are innervated with efferent fibers, whereas the endocrine system can regulate most cells in the body. The effects of neurotransmitters are rapid and short lived compared to the effects of hormones, which appear more slowly and which last longer. Table 15-1 compares endocrine structure and function with nervous structure and function.

Endocrine glands secrete their products, hormones, directly into the blood. Because they do not have ducts, they are often called "ductless glands." This characteristic distinguishes *endocrine* glands from *exocrine* glands, which secrete their products into ducts (see Chapter 4, p. 106). Many endocrine glands are made of glandular epithelium, whose cells manufacture and secrete hormones. However, a few endocrine glands are made of **neurosecretory tissue.** Neurosecretory cells are simply modified neurons that secrete chemical messengers that diffuse into the bloodstream rather than across a synapse. In such cases, the chemical messenger is called a hormone rather than a neurotransmitter. For example, when norepinephrine is released by neurons, diffuses across a synapse, and binds to an adrenergic receptor in a postsynaptic neuron, we call norepinephrine a neurotransmitter. On the other hand, we call norepinephrine a hormone when it diffuses into the

TABLE 15-1   Comparison of features of the endocrine system and nervous system

Feature	Endocrine system	Nervous system
OVERALL FUNCTION	Regulation of effectors to maintain homeostasis	Regulation of effectors to maintain homeostasis
Control by regulatory feedback loops	Yes (endocrine reflexes)	Yes (nervous reflexes)
Effector tissues	Endocrine effectors: virtually all tissues	Nervous effectors: muscle and glandular tissue only
Effector cells*	Target cells (throughout the body)	Postsynaptic cells (in muscle and glandular tissue only)
CHEMICAL MESSENGER*	Hormone	Neurotransmitter
Cells that secrete the chemical messenger	Glandular epithelial cells or neurosecretory cells (modified neurons)	Neurons
Distance traveled (and method of travel) by chemical messenger*	Long (by way of circulating blood)	Short (across a microscopic synapse)
Location of receptor in effector cell	On the plasma membrane or within the cell	On the plasma membrane
Characteristics of regulatory effects*	Slow to appear, long-lasting	Appear rapidly, short-lived

*See diagrams.

TABLE 15-2   Names and locations of major endocrine glands

Name	Location	Name	Location
Hypothalamus	Cranial cavity (brain)	Adrenal glands	Abdominal cavity (retroperitoneal)
Pituitary gland (hypophysis cerebri)	Cranial cavity	Pancreatic islets	Abdominal cavity (pancreas)
Pineal gland	Cranial cavity (brain)	Ovaries	Pelvic cavity
Thyroid gland	Neck	Testes	Scrotum
Parathyroid glands	Neck		
Thymus	Mediastinum	Placenta	Pregnant uterus

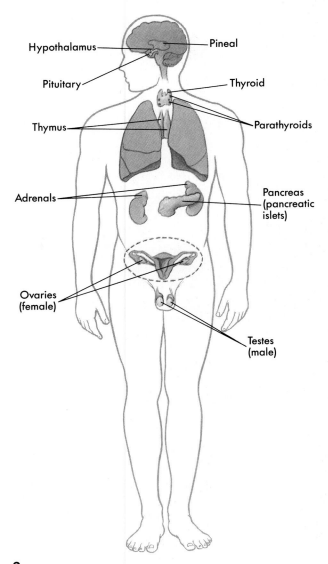

FIGURE 15-1 Locations of the major endocrine glands.

blood (because there is no postsynaptic cell present) then binds to an adrenergic receptor in a distant target cell.

Glands of the endocrine system are widely scattered throughout the body. New discoveries in *endocrinology* continue to add to the long list of hormone-secreting tissues. However, even the most newly discovered endocrine tissues and their hormones operate according to some basic physiological principles. In this chapter, we will focus our discussion primarily on the major endocrine glands. Figure 15-1 and Table 15-2 summarize the names and locations of these major endocrine glands. Once you are familiar with the basic principles of endocrinology and the major examples of glands and their hormones, you will be prepared for additional examples that you will encounter as you continue your study of the human body.

QUICK CHECK

✔ 1. *What is meant by the term* target cell?
✔ 2. *Describe how the nervous system and the endocrine system differ in the way they control effectors.*

HORMONES

Classification of Hormones

Hormone molecules can be classified in various useful ways. For example, when classified by general function, hormones can be identified as *tropic hormones* (hormones that target other endocrine glands and stimulate their growth and secretion), *sex hormones* (hormones that target reproductive tissues), *anabolic hormones* (hormones that

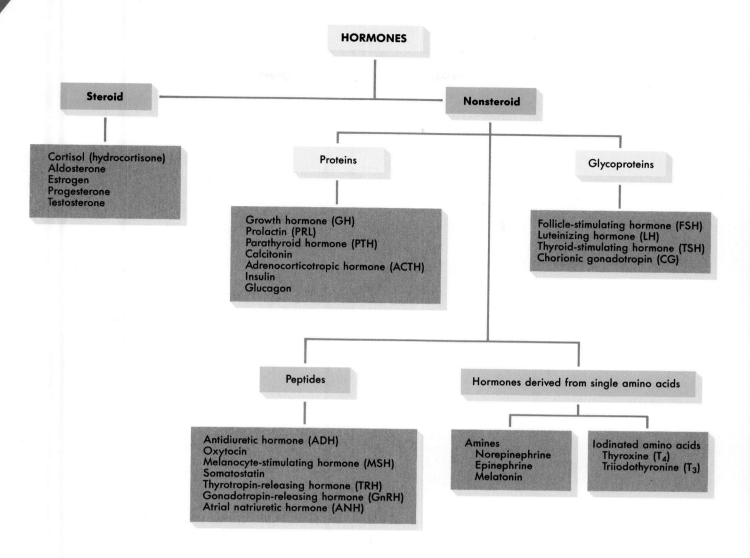

FIGURE 15-2 Chemical classification of hormones.

stimulate anabolism in their target cells), and by many other functional names. Another useful way to classify hormones is by their chemical structure. Because this method of classifying hormones is so widely used, we will briefly describe it in the following paragraphs.

Steroid hormones

All of the many hormones secreted by endocrine tissues can be classified simply as **steroid** or **nonsteroid.** Steroid hormone molecules are manufactured by endocrine cells from cholesterol, an important type of lipid in the human body (see Chapter 2, p. 55). Steroid hormones have a characteristic chemical group at the core of each molecule (see Figure 2-19). Because steroids are lipid soluble, they can easily pass through the phospholipid plasma membrane of target cells. Examples of steroid hormones include cortisol, aldosterone, estrogen, progesterone, and testosterone (Figure 15-2).

Nonsteroid hormones

Nonsteroid hormones are synthesized primarily from amino acids rather than from cholesterol (Figure 15-3). Some nonsteroid hormones are *protein hormones.* These hormones are long, folded chains of amino acids, a structure typical of protein molecules of any sort (see Chapter 2, p. 50). Included among the protein hormones are insulin, parathyroid hormone, and others listed in Figure 15-2. Protein hormones that have carbohydrate groups attached to their amino acid chains are often classified separately as *glycoprotein hormones.*

Another major category of nonsteroid hormones consists of the *peptide hormones.* Peptide hormones such as oxytocin and antidiuretic hormone (ADH) are smaller than the protein hormones. They are each made of a short chain of amino acids, as Figure 15-3, *B,* shows. Examples of peptide hormones are listed in Figure 15-2.

Yet another category of nonsteroid hormones consists of

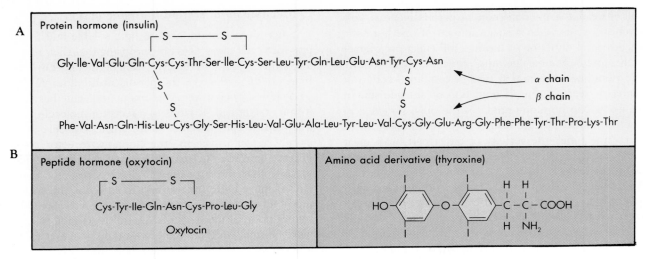

A **Protein hormone (insulin)**

Gly-Ile-Val-Glu-Gln-Cys-Cys-Thr-Ser-Ile-Cys-Ser-Leu-Tyr-Gln-Leu-Glu-Asn-Tyr-Cys-Asn

Phe-Val-Asn-Gln-His-Leu-Cys-Gly-Ser-His-Leu-Val-Glu-Ala-Leu-Tyr-Leu-Val-Cys-Gly-Glu-Arg-Gly-Phe-Phe-Tyr-Thr-Pro-Lys-Thr

α chain
β chain

B **Peptide hormone (oxytocin)**

Cys-Tyr-Ile-Gln-Asn-Cys-Pro-Leu-Gly

Oxytocin

Amino acid derivative (thyroxine)

C

FIGURE 15-3   Nonsteroid hormone structure. As these examples show, protein hormone molecules **(A)** are made of long, folded strands of amino acids. **B,** Peptide hormone molecules are smaller strands of amino acids. **C,** Amino acid derivatives are, as their name implies, each derived from a single amino acid.

the *amino-acid derivative hormones.* Each of these hormones is derived from only a single amino acid molecule. There are two major subgroups within this category. One subgroup, the *amine hormones,* is synthesized by modifying a single molecule of the amino acid, tyrosine. Amine hormones such as epinephrine and norepinephrine are produced by neurosecretory cells (where they are secreted as hormones) and by neurons (where they are secreted as neurotransmitters). Another subgroup of amino acid derivatives produced by the thyroid gland are all synthesized by adding iodine (I) atoms to a tyrosine molecule (Figure 15-3, C). Examples of hormones derived from single amino acids are listed in Figure 15-2.

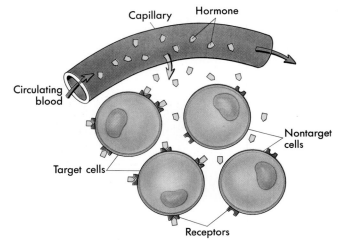

FIGURE 15-4   The target cell concept. A hormone acts only on cells that have receptors specific to that hormone, because the shape of the receptor determines which hormone can react with it. This is an example of the lock-and-key model of biochemical reactions.

QUICK CHECK

✔ 1. *How are steroid hormones able to pass through a cell's plasma membrane easily?*
✔ 2. *Name some of the different general types of nonsteroid hormones. Can you give an example of each?*

How Hormones Work
General principles of hormone action

As already stated, hormones signal a cell by binding to specific receptors on or in the cell. In a "lock-and-key" mechanism, hormones will bind only to receptor molecules that "fit" them exactly. Any cell with one or more receptors for a particular hormone is said to be a *target* of that hormone (Figure 15-4). Cells usually have many different types of receptors; therefore they are target cells of many different hormones.

Each different hormone-receptor interaction produces different regulatory changes within the target cell. These cellular changes are usually accomplished by altering the chemical reactions within the target cell. For example,

some hormone-receptor interactions initiate synthesis of new proteins. Other hormone-receptor interactions trigger the activation or inactivation of certain enzymes and thus affect the metabolic reactions regulated by those enzymes. Still other hormone-receptor interactions regulate cells by opening or closing specific ion channels in the plasma membrane. Specific mechanisms of hormone-receptor interactions are outlined in the next section.

Different hormones may work together to enhance each other's influence on a target cell. In a phenomenon called **synergism,** combinations of hormones have a greater effect on a target cell than the sum of the effects that each would have if acting alone. Combined hormone actions may

exhibit instead the phenomenon of **permissiveness.** Permissiveness occurs when a small amount of one hormone allows a second hormone to have its full effect on a target cell; the first hormone "permits" the full action of the second hormone. A common type of combined action of hormones is seen in the phenomenon of **antagonism.** In antagonism, one hormone produces the opposite effect of another hormone. Antagonism between hormones can be used to "fine tune" the activity of target cells with great accuracy, signaling the cell exactly when (and by how much) to increase or decrease a certain cellular process.

As already stated, hormones travel to their target cells by way of the circulating bloodstream. This means that all hormones travel throughout the body. Because they only affect their target cells, however, the effects of a particular hormone may be limited to specific tissues in the body. Some hormone molecules are attached to plasma proteins while they are carried along the bloodstream. Such hormones must free themselves from the plasma protein to leave the blood and combine with their receptors. Because blood carries hormones nearly everywhere in the body, even where there are no target cells, endocrine glands produce more hormone molecules than actually hit their target. Unused hormones usually are quickly excreted by the kidneys or broken down by metabolic processes.

Mechanism of steroid hormone action

Since steroid hormones are lipid soluble and thus can pass into cells easily, it is not surprising that their receptors are normally found in the cytoplasm of the target cell rather than on the surface of the plasma membrane. As Figure 15-5 shows, once a steroid hormone molecule has diffused into its target cell, it binds to a receptor molecule to form a *hormone-receptor complex.* According to a model called the *mobile-receptor hypothesis,* the hormone-receptor complex migrates into the nucleus of the target cell. Some evidence suggests that the hormone binds to a receptor that has already moved into the nucleus. In any case, the hormone-receptor complex activates a certain gene sequence to begin transcription of messenger RNA (mRNA) molecules. The newly formed mRNA molecules then move out of the nucleus into the cytoplasm, where they associate with ribosomes and begin synthesizing protein molecules. These protein molecules would not have been made if not for the arrival of the steroid hormone molecule. Steroid hormones regulate cells by regulating their production of certain critical proteins, such as enzymes that control intracellular reactions or membrane proteins that alter the permeability of a cell.

This mechanism of steroid hormone action implies several things about the effects of these hormones. For one

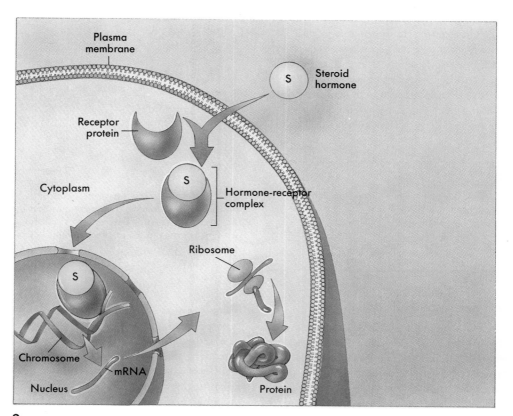

FIGURE 15-5   Steroid hormone mechanism. According to the mobile-receptor hypothesis, lipid-soluble steroid hormone molecules pass through the plasma membrane where they bind with their receptor to form a hormone-receptor complex. This mobile complex migrates into the nucleus and binds to a specific site on a DNA molecule, triggering transcription of the genetic information encoded there. The resulting mRNA molecule moves to the cytoplasm, where it associates with a ribosome, initiating synthesis of a new protein. This new protein—usually an enzyme or channel protein—produces specific effects in the target cell.

thing, the more hormone-receptor complexes formed, the more mRNA molecules are transcribed, the more new protein molecules are formed, and thus the greater the magnitude of the regulatory effect. In short, the *amount* of steroid hormone present determines the magnitude of a target cell's response. Also, because transcription and protein synthesis take some time, responses to steroid hormones are often slow—from 45 minutes to several days before the full effect is seen.

Mechanisms of nonsteroid hormone action

The second messenger mechanism. Nonsteroid hormones typically operate according to a mechanism originally called the **second messenger hypothesis.** This concept of hormone action was first proposed several decades ago by Dr. Earl W. Sutherland—a milestone in endocrinology for which he received the 1971 Nobel Prize in Medicine and Physiology. According to this concept, a nonsteroid hormone molecule acts as a "first messenger," delivering its chemical message to fixed receptors in the target cell's plasma membrane. The "message" is then passed into the cell where a "second messenger" triggers the appropriate cellular changes. This concept of nonsteroid hormone action is also called the *fixed-membrane-receptor hypothesis.*

In the example illustrated in Figure 15-6, formation of

the hormone-receptor complex causes a membrane protein, called the G protein, to bind to a nucleotide called guanosine triphosphate (GTP). This, in turn, activates another membrane protein, *adenyl cyclase.* Adenyl cyclase is an enzyme that promotes the removal of two phosphate groups from ATP molecules in the cytoplasm. The product thus formed is cyclic adenosine monophosphate *(cAMP).* The cAMP molecule acts as a "second messenger" within the cell. cAMP activates *protein kinases,* a set of enzymes that activate other types of enzymes. It is this final set of specific enzymes, which are now activated, that catalyze the cellular reactions that characterize the target cell's response. In short, the hormone "first messenger" binds to a membrane receptor, triggering formation of an intracellular "second messenger," which activates a cascade of chemical reactions that produce the target cell's response.

Since the time Sutherland first began his pioneering work, other second messenger systems have been discovered. As a matter of fact, the study of second messenger mechanisms is still a very active area of research, with new discoveries continuing to be revealed in scientific journals around the world. Although most nonsteroid hormones seem to use cAMP as the second messenger, we now know that a few hormones use compounds such as inositol triphosphate (IP_3) and cyclic GMP (guanosine monophosphate) as the second messenger. Still other hormones

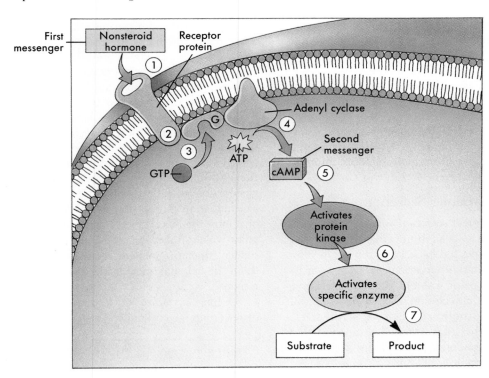

FIGURE 15-6 Example of a second-messenger mechanism. 1, A nonsteroid hormone *(first messenger)* binds to a fixed receptor in the plasma membrane of the target cell. 2, The hormone-receptor complex activates the G protein. 3, The activated G protein reacts with GTP, which in turn activates the membrane-bound enzyme adenyl cyclase. 4, Adenyl cyclase removes phosphates from ATP, converting it to cAMP *(second messenger)* 5, cAMP activates or inactivates protein kinases. 6, Protein kinases activate specific intracellular enzymes. 7, These activated enzymes then influence specific cellular reactions, thus producing the target cell's response to the hormone.

produce their effects by triggering the opening of calcium (Ca^{++}) channels in the target cell's membranes. Ca^{++} that enter the cytoplasm when the channels open bind to an intracellular molecule called *calmodulin.* The Ca^{++}-calmodulin complex thus formed acts as a second messenger, influencing the enzymes that produce the target cell's response. Recent research findings also show that in second messenger systems, the hormone-receptor complexes may be taken into the cell by means of endocytosis. Although the purpose of this may be primarily to break down the complexes and recycle the receptors, the hormone-receptor complex may continue to have physiological effects after it is taken into the cell.

The second messenger mechanism produces target cell effects that differ from steroid hormone effects in several important ways. First, the cascade of reactions produced in the second messenger mechanism greatly amplifies the effects of the hormone. Thus the effects of many nonsteroid hormones are disproportionally great when compared to the amount of hormone present. Recall that steroid hormones produce effects in proportion to the amount of hormone present. Also, the second messenger mechanism operates much more quickly than the steroid mechanism. Many nonsteroid hormones produce their full effects within seconds or minutes of initial binding to the target cell receptors—not the hours or days sometimes seen with steroid hormones.

The nuclear receptor mechanism. Not all nonsteroid hormones operate according to the second messenger model. The notable exception is the pair of thyroid hormones, thyroxine (T_4) and triiodothyronine (T_3). These small iodinated amino acids apparently enter their target cells and bind to receptors already associated with a DNA molecule within the nucleus of the target cell. Formation of a hormone-receptor complex triggers transcription of mRNA and the synthesis of new enzymes in a manner similar to the steroid mechanism. More information on these hormones is given later in this chapter.

Regulation of Hormone Secretion

The control of hormonal secretion is usually part of a negative feedback loop. Recall from Chapter 1 (Figure 1-11, p. 16) that negative feedback loops tend to reverse any deviation of the internal environment away from its stable point (the set point value). Rarely, a positive feedback loop controls the secretion of a hormone. Recall that in positive feedback control, deviation from the stable point is exaggerated rather than reversed. Responses that result from the operation of feedback loops within the endocrine system are called *endocrine reflexes,* just as responses to nervous feedback loops (reflex arcs) are called *nervous reflexes.*

For a moment, let us focus our attention on the specific mechanisms that regulate the release of hormones from endocrine cells. The simplest mechanism operates when an endocrine cell is sensitive to the physiological changes produced by its target cells (Figure 15-7). For example, parathyroid hormone (PTH) produces responses in its target cells that increase Ca^{++} concentration in the blood.

When blood Ca^{++} concentration exceeds the set point value, parathyroid cells sense it and reflexively reduce their output of PTH. Secretion by many endocrine glands is regulated by a hormone produced by another gland. For example, the pituitary gland (specifically, the anterior portion) produces thyroid-stimulating hormone (TSH), which stimulates the thyroid gland to release its hormones. The anterior pituitary responds to changes in the controlled physiological variable, as well as to changes in the blood concentration of hormones secreted by its target gland. Secretion by the anterior pituitary can, in turn, be regulated by *releasing hormones* or *inhibiting hormones* secreted by the hypothalamus. Hypothalamic secretion is responsive to changes in the controlled variable, as well as changes in blood concentration of anterior pituitary and target gland hormones. Although the target gland may be able to adjust its own output, the additional controls exerted by long feedback loops involving the anterior pituitary and hypothalamus allow more precise regulation of hormone secretion—and thus more precise regulation of the internal environment.

Another mechanism that may influence the secretion of hormones by a gland is input from the nervous system. For example, secretion by the posterior pituitary is not regulated by releasing hormones but by direct nervous input from the hypothalamus. Likewise, sympathetic nerve impulses that reach the medulla of the adrenal glands trigger the secretion of epinephrine and norepinephrine. Many other glands, including the pancreas, are also influenced to some degree by nervous input. The fact that the nervous system operates with hormonal mechanisms to produce endocrine reflexes emphasizes the close functional relationship between these two systems.

Although the operation of long feedback loops tends to minimize wide fluctuations in secretion rates, the output of several hormones typically rises and falls dramatically within a short period of time. For example, the concentration of insulin—a hormone that can correct a rise in blood glucose concentration—increases to a high level just after a meal high in carbohydrates. The level of insulin decreases only after the blood glucose concentration returns to its set point value. Likewise, threatening stimuli can cause a sudden, dramatic increase in the secretion of epinephrine from the adrenal medulla as part of the fight-or-flight response.

Specific examples of feedback control of hormone secretion are given later in this chapter.

QUICK CHECK

✔ 1. *Name some ways in which hormones can work together to regulate an endocrine effector.*

✔ 2. *Why is the concept of steroid hormone action called the mobile-receptor hypothesis?*

✔ 3. *Why is the concept of nonsteroid hormone action often called the second messenger hypothesis? Why is it also known as the fixed-membrane-receptor hypothesis?*

✔ 4. *Name some of the ways that the secretion of an endocrine cell can be controlled.*

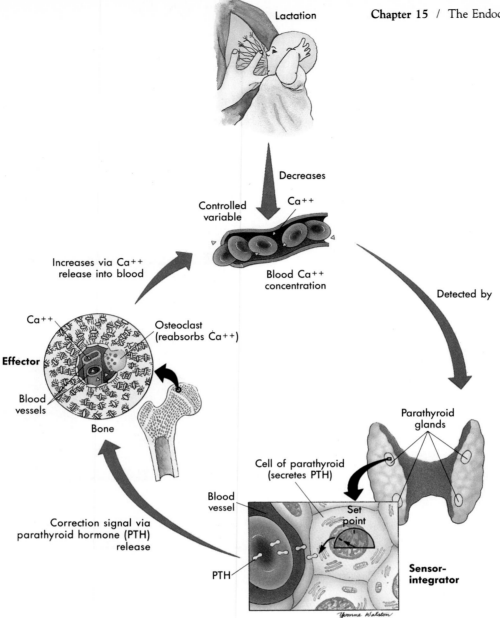

FIGURE 15-7 Endocrine feedback loop. In this example of a short feedback loop, each parathyroid gland is sensitive to changes in the physiological variable its hormone *(PTH)* controls—blood calcium (Ca^{++}) concentration. When lactation (milk production) in a pregnant woman consumes Ca^{++} and thus lowers blood Ca^{++} concentration, the parathyroids sense the change and respond by increasing their secretion of PTH (parathyroid hormone). PTH stimulates osteoclasts in bone to release more Ca^{++} from storage in bone tissue (among other effects), which increases maternal blood Ca^{++} concentration to the set point level.

HEALTH MATTERS   "TOO MUCH OR TOO LITTLE"

Diseases of the endocrine system are numerous, varied, and sometimes spectacular. Tumors or other abnormalities frequently cause the glands to secrete too much or too little of their hormones. Production of too much hormone by a diseased gland is called **hypersecretion.** If too little hormone is produced, the condition is called **hyposecretion.**

Various endocrine disorders that appear to result from hyposecretion are actually caused by a problem in the target cells. If the usual target cells of a particular hormone have damaged receptors, too few receptors, or some other abnormality, they will not respond to that hormone properly. In other words, lack of target cell response could be a sign of hyposecretion or a sign of target cell insensitivity. *Diabetes mellitus,* for example, can result from insulin hyposecretion or from the target cells' insensitivity to insulin (see essay, p. 412).

The sensitivity of a target cell to any particular hormone depends on how many receptors for that hormone it has. The more receptors, the more sensitive the target cell. Hormone receptors, like other cell components, are constantly broken down by the cell and replaced with newly synthesized receptors. This mechanism not only ensures that all cell parts are "new" and working properly, it also provides a method by which the number of receptors can be changed from time to time.

If synthesis of new receptors occurs faster than degradation of old receptors, the target cell will have more receptors and thus be more sensitive to the hormone. This phenomenon, illustrated in A, is often called *up regulation* because the number of receptors "goes up." If, on the other hand, the rate of receptor degradation exceeds the rate of receptor synthesis, the target cell's number of receptors will decrease (B). Because the number of receptors, and thus the sensitivity of the target cell, "goes down" this phenomenon is often called *down regulation.*

Endocrinologists are just now learning the mechanisms that control the process of receptor turnover in the cell and how this affects the functions of the target cell. Information uncovered so far has already led to a better understanding of important and widespread endocrine disorders such as diabetes mellitus.

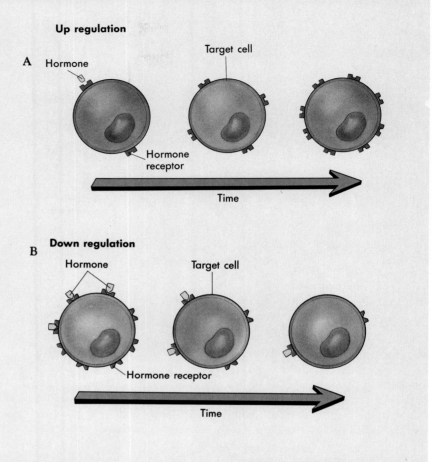

PROSTAGLANDINS

Before continuing with our discussion of endocrine glands and hormones, let us pause a moment to consider the **prostaglandins (PGs)** and related compounds. The prostaglandins are a unique group of lipid molecules that serve important and widespread integrative functions in the body but do not meet the usual definition of a hormone. The prostaglandin molecule is a 20-carbon fatty acid with a 5-carbon ring (Figure 15-8). Although they may be secreted directly into the bloodstream, they are rapidly metabolized, so that circulating levels are extremely low. The term *tissue hormone* is appropriate because the secretion is produced in a tissue and diffuses only a short distance to other cells within the same tissue. Whereas typical hormones integrate activities of widely separated organs, prostaglandins tend to integrate activities of neighboring cells.

There are at least 16 different prostaglandins, falling into nine structural classes—prostaglandins A through prostaglandins I. Prostaglandins have been isolated and identified from a variety of tissues. The first prostaglandin

FIGURE 15-8 Structure of the prostaglandin molecule. Structure of prostaglandin $F_{2\alpha}$ ($PGF_{2\alpha}$), showing the typical 20-carbon unsaturated fatty acid structure with a characteristic 5-carbon ring.

was discovered in semen, so it was attributed to the prostate gland (hence the name, prostaglandin). Later, researchers found that the seminal vesicles, not the prostate, secreted the prostaglandin that they had found. Other tissues known to secrete prostaglandins include the kidneys, lungs, iris, brain, and thymus. As a group, the prostaglandins have diverse physiological effects and are among the most varied and potent of any naturally occurring biological compounds. They are intimately involved in overall endocrine regulation by influencing adenyl cyclase–cAMP interaction within the cell's plasma membrane (see Figure 15-6).

Specific biological effects depend on the class of prostaglandin.

Intraarterial infusion of prostaglandins A (PGAs) results in an immediate fall in blood pressure accompanied by an increase in regional blood flow to several areas, including the coronary and renal systems. PGAs apparently produce this effect by causing relaxation of smooth muscle fibers in the walls of certain arteries and arterioles.

PGEs have an important role in various vascular, metabolic, and gastrointestinal functions. Vascular effects include regulation of red blood cell deformability and platelet aggregation (see Chapter 16). PGEs also have a role in systemic inflammations such as fever. Common antiinflammatory agents, such as aspirin, produce some of their effects by inhibiting PGE synthesis. PGE also regulates hydrochloric acid secretion in the stomach, helping to prevent gastric ulcers.

Prostaglandins F (PGFs) have an especially important role in the reproductive system. They cause uterine muscle contractions, so they have been used to induce labor and thus accelerate delivery of a baby. PGFs also affect intestinal motility and are required for normal peristalsis.

In addition to prostaglandins, various tissues also synthesize other fatty acid compounds that are structurally and functionally similar to prostaglandins. Examples include the *thromboxanes* and *leukotrienes*. Like prostaglandins, these compounds may also be referred to as tissue hormones because of their local, yet potent, regulatory effects.

The potential therapeutic use of prostaglandins and related compounds, which are found in almost every body tissue and are capable of regulating hormone activity on the cellular level, has been described as the most revolutionary development in medicine since the advent of antibiotics. They are likely to play increasingly important roles in the treatment of such diverse conditions as hypertension, coronary thrombosis, asthma, and ulcers.

QUICK CHECK

✔ 1. *Why are prostaglandins sometimes called tissue hormones?*

✔ 2. *Why are prostaglandins considered to be important in clinical applications?*

PITUITARY GLAND

Structure of the Pituitary Gland

The pituitary gland, or **hypophysis,** is a small but mighty structure. It measures only 1.2 to 1.5 cm (about ½ inch) across. By weight, it is even less impressive—only about 0.5 g (1/60 ounce)! And yet so crucial are the functions of the anterior lobe of the pituitary gland that years ago it was referred to as the "master gland."

The hypophysis has a well-protected location within the skull on the ventral surface of the brain (Figure 15-9). It lies in the *pituitary fossa* of the sella turcica and is covered by a portion of the dura mater called the *pituitary diaphragm.* The gland has a stemlike stalk, the **infundibulum,** which connects it to the hypothalamus of the brain.

Although the pituitary looks like one gland, it actually consists of two separate glands—the **adenohypophysis,** or anterior pituitary gland, and the **neurohypophysis,** or

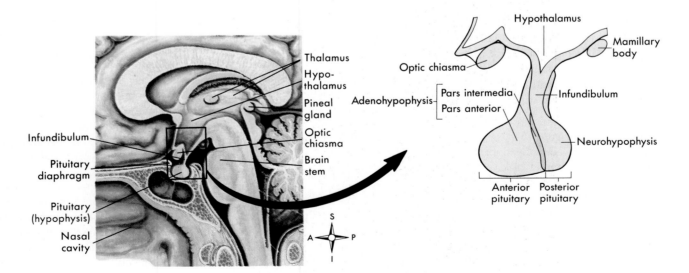

FIGURE 15-9   Location and structure of the pituitary gland (hypophysis). The pituitary gland is located within the sella turcica of the skull's sphenoid bone and is connected to the hypothalamus by a stalklike infundibulum. The infundibulum passes through a gap in the portion of the dura mater that covers the pituitary (the pituitary diaphragm). The inset shows that the pituitary is divided into an anterior portion, the adenohypophysis, and a posterior portion, the neurohypophysis. The adenohypophysis is further subdivided into the pars anterior and pars intermedia.

posterior pituitary gland. In the embryo, the adenohypophysis develops from an upward projection of the pharynx and is composed of regular endocrine tissue. The neurohypophysis, on the other hand, develops from a downward projection of the brain and is composed of neurosecretory tissue. These histological differences are incorporated into their names—*adeno* means "gland" and *neuro* means "nervous." As you may suspect, the hormones secreted by the adenohypophysis serve very different functions from those released by the neurohypophysis.

Adenohypophysis (Anterior Pituitary)

The adenohypophysis, the anterior portion of the pituitary gland, is divided into two parts—the *pars anterior* and the *pars intermedia*. The pars anterior forms the major portion of the adenohypophysis and is divided from the tiny pars intermedia by a narrow cleft and some connective tissue (Figure 15-9).

The tissue of the adenohypophysis is composed of irregular clumps of secretory cells supported by fine connective tissue fibers and surrounded by a rich vascular network. Three types of cells can be identified according to their affinity for certain types of stains:

1. **Chromophobes** (meaning "afraid of color") show little affinity for any stain. About half of all cells in the adenohypophysis fall into this category.

2. **Acidophils** (meaning "acid lover") stain easily with acid dyes and make up about 40% of the cells of the adenohypophysis. Acidophils secrete:

GH (growth hormone)
PRL (prolactin)

3. **Basophils** (meaning "base lover") stain with basic dyes and form about 10% of the adenohypophysis. Basophils secrete the following hormones:

TSH (thyroid-stimulating hormone)
ACTH (adrenocorticotropic hormone)
FSH (follicle-stimulating hormone)
LH (luteinizing hormone)
MSH (melanocyte-stimulating hormone)

All three cell types are visible in the histology photograph shown in Figure 15-10. Figure 15-11 summarizes the hormones of the adenohypophysis and shows the primary locations of their target cells.

Growth hormone

Growth hormone (GH), or **somatotropin (STH),** is thought to promote bodily growth indirectly by stimulating the liver to produce certain growth factors, which, in turn, accelerate amino acid transport into cells. Rapid entrance of amino acids from the blood into the cells allows protein anabolism within the cells to accelerate. Increased protein anabolism allows increased rate of growth. GH promotes the growth of bone, muscle, and other tissues.

In addition to stimulating protein anabolism, GH also stimulates fat metabolism. GH accelerates mobilization of lipids from storage in adipose cells and also speeds up the catabolism of those lipids after they have entered another

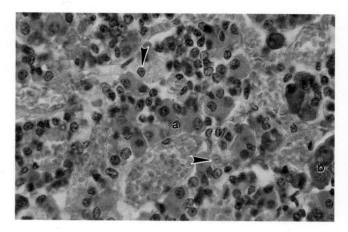

FIGURE 15-10 Histology of the adenohypophysis. In this light micrograph, nonstaining chromophobes are indicated by arrowheads. Examples of hormone-secreting cells are labeled *a* (acidophil) and *b* (basophil). (*Courtesy of Erlandsen/Magney: Color Atlas of Histology.*)

cell. In this way, GH tends to shift cell chemistry away from carbohydrate (glucose) catabolism and toward lipid catabolism as an energy source. Since less glucose is then removed from the blood by cells, the blood glucose levels tend to rise. Thus GH is said to have a hyperglycemic effect. Insulin (from the pancreas) has the opposite effect—it promotes glucose entry into cells, producing a hypoglycemic effect. Therefore GH and insulin function as antagonists. The balance between these two hormones is vital to maintaining a homeostasis of blood glucose levels.

Growth hormone affects metabolism in these ways:

♦ It promotes protein anabolism (growth, tissue repair)
♦ It promotes lipid mobilization and catabolism
♦ It indirectly inhibits glucose metabolism
♦ It indirectly increases blood glucose levels

Prolactin

Prolactin (PRL), produced by acidophils in the pars anterior, is also called *lactogenic hormone*. Both names of this hormone suggest its function in "generating" or initiating milk secretion (lactation). During pregnancy, a high level of PRL promotes the development of the breasts in anticipation of milk secretion. At the birth of an infant, PRL in the mother stimulates the mammary glands to begin milk secretion.

In addition to its stimulation of lactation, PRL has a supportive role (with luteinizing hormone) in maintaining the corpus luteum of the ovary during the last portion of the menstrual cycle (see Chapter 31). Because of this supportive role, PRL is sometimes called *luteotropic hormone* (LTH). Since PRL blood levels are high just before menstruation, it may cause temporary enlargement of the breast tissues, similar to that seen in pregnancy.

Hypersecretion of PRL may cause lactation in nonnursing women, disruption of the menstrual cycle, and can cause impotence in men. **Hyposecretion** of PRL is usually insignificant except in women who want to nurse their children. Milk production cannot be initiated or maintained without PRL.

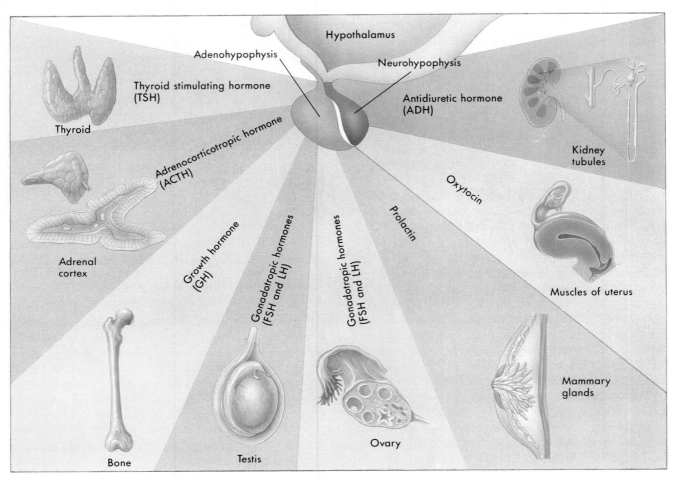

FIGURE 15-11　Pituitary hormones. Some of the major hormones of the adenohypophysis and neurohypophysis and their principal target organs.

GROWTH HORMONE ABNORMALITIES

Hypersecretion of GH during the growth years (before ossification of the epiphyseal plates) causes an abnormally rapid rate of skeletal growth. This condition is known as *gigantism* (see Figure). Hypersecretion after skeletal fusion has occurred can result in *acromegaly*, a condition in which cartilage still left in the skeleton continues to form new bone. This abnormal growth may result in a distorted appearance due to the enlargement of the hands, feet, face, jaw (causing separation of the teeth), and other body parts. Overlying soft tissue may also be affected—for instance, the skin often thickens and the pores become more pronounced.

Hyposecretion of GH during growth years may result in stunted body growth, known as *pituitary dwarfism* (see Figure at right). Formerly, patients were treated only with GH extracted from human tissues. Since the 1980s, the availability of human GH produced by genetically engineered bacteria has made the treatment obtainable for many more patients. However, concerns have been raised about possible adverse side effects associated with human GH from bacterial sources.

Gigantism and dwarfism. A pituitary giant (*far left*) and pituitary dwarf (*far right*) contrasted with two normal-sized men.

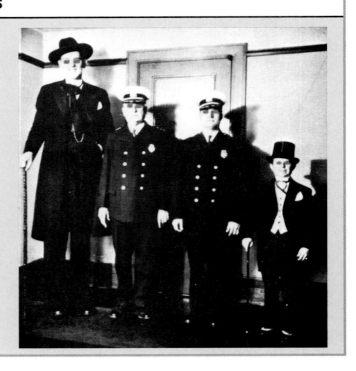

Tropic hormones

Tropic hormones are hormones that have a stimulating effect on other endocrine glands. These hormones stimulate the development of their target glands and tend to increase their secretions. Four principal tropic hormones are produced and secreted by the basophils of the pars anterior:

1. **Thyroid-stimulating hormone (TSH),** or *thyrotropin,* promotes and maintains the growth and development of its target gland—the thyroid. TSH also causes the thyroid gland to secrete its hormones.

2. **Adrenocorticotropic hormone (ACTH),** or *adrenocorticotropin,* promotes and maintains normal growth and development of the cortex of the adrenal gland. ACTH also stimulates the adrenal cortex to secrete some of its hormones.

3. **Follicle-stimulating hormone (FSH)** stimulates structures within the ovaries, primary follicles, to grow toward maturity. Each follicle contains a developing egg cell (ovum), which is released from the ovary during ovulation. FSH also stimulates the follicle cells to secrete estrogens (female sex hormones). In the male, FSH stimulates the development of the seminiferous tubules of the testes and maintains spermatogenesis (sperm production) by them.

4. **Luteinizing hormone (LH)** stimulates the formation and activity of the corpus luteum of the ovary. The corpus luteum (meaning "yellow body") is the tissue left behind when a follicle ruptures to release its egg during ovulation. The corpus luteum secretes progesterone and estrogens when stimulated by LH. LH also supports FSH in stimulating the maturation of follicles. In males, LH stimulates interstitial cells in the testes to develop and secrete testosterone (the male sex hormone).

FSH and LH are called **gonadotropins** because they stimulate the growth and maintenance of the gonads (ovaries and testes). During childhood the adenohypophysis secretes insignificant amounts of the gonadotropins. A few years before puberty, gonadotropin secretion is gradually increased. Then, suddenly, their secretion spurts, and the gonads are stimulated to develop and begin their normal functions.

Melanocyte-stimulating hormone

Basophils in the pars intermedia and pars anterior secrete **melanocyte-stimulating hormone (MSH).** This hormone gets its name from the fact that, in laboratory experiments, injection of large amounts of this hormone have stimulated melanocytes in the skin to produce more melanin and thus darken the skin. Abnormal hypersecretion of MSH can have the same effect. Whether normal amounts of MSH in the body have any significant effects on melanocytes is uncertain. The structure of the MSH molecule is very similar to that of adrenocorticotropic hormone (ACTH), a hormone that can also darken the skin when present in high concentrations. Perhaps MSH, ACTH, and other skin-darkening hormones (such as estrogen and progesterone) work together to modulate the pigmentation of normal skin. Another function that has

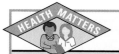

TROPIC HORMONE ABNORMALITIES

Hypersecretion of the tropic hormones may result from a pituitary tumor. The tropic hormones, produced at higher than normal levels, cause hypersecretion in their target glands. This may result in various effects throughout the body. Early hypersecretion of gonadotropins may lead to an abnormally early onset of puberty.

Hyposecretion of tropic hormones often causes their target glands to secrete less than a normal amount of their hormones. This may disrupt reproduction, kidney function, overall metabolism, and other processes.

been proposed for MSH is that it somehow helps maintain the adrenal gland's sensitivity to ACTH.

Control of secretion in the adenohypophysis

The cell bodies of neurons in certain parts of the hypothalamus synthesize chemicals that their axons secrete into the blood. These chemicals, generally called **releasing hormones,** travel through a complex of small blood vessels called the **hypophyseal portal system** (Figure 15-12). A *portal system* is an arrangement of blood vessels in which blood exiting one tissue is immediately carried to a second tissue before being returned to the heart and lungs for oxygenation and redistribution. The hypophyseal portal system carries blood from the hypothalamus directly to the adenohypophysis, where the target cells of the releasing hormones are located. The releasing hormones influence the secretion of hormones by acidophils and basophils. In this manner, the hypothalamus regulates the secretion of

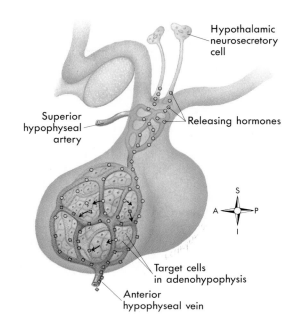

FIGURE 15-12 Hypophyseal portal system. Neurons in the hypothalamus secrete releasing hormones into veins that carry the releasing hormones directly to the vessels of the adenohypophysis, thus bypassing the normal circulatory route.

TABLE 15-3   Hormones of the hypothalamus

Hormone	Source	Target	Principal action
Growth hormone-releasing hormone (GRH)	Hypothalamus	Adenohypophysis (acidophils)	Stimulates secretion (release) of growth hormone
Growth hormone-inhibiting hormone (GIH), or somatostatin	Hypothalamus	Adenohypophysis (acidophils)	Inhibits secretion of growth hormone
Corticotropin-releasing hormone (CRH)	Hypothalamus	Adenohypophysis (basophils)	Stimulates release of adrenocorticotropic hormone (ACTH)
Thyrotropin-releasing hormone (TRH)	Hypothalamus	Adenohypophysis (basophils)	Stimulates release of thyroid-stimulating hormone (TSH)
Gonadotropin-releasing hormone (GNRH)	Hypothalamus	Adenohypophysis (basophils)	Stimulates release of gonadotropins (FSH and LH)
Prolactin-releasing hormone (PRH)	Hypothalamus	Adenohypophysis (acidophils)	Stimulates secretion of prolactin
Prolactin-inhibiting hormone (PIH)	Hypothalamus	Adenohypophysis (acidophils)	Inhibits secretion of prolactin

the adenohypophysis. You can see that the supposed "master gland" really has a master of its own—the hypothalamus.

The following is a list of some of the important hormones secreted by the hypothalamus into the hypophyseal portal system:

♦ *Growth hormone-releasing hormone (GRH)*
♦ *Growth hormone-inhibiting hormone (GIH) (also called somatostatin)*
♦ *Corticotropin-releasing hormone (CRH)*
♦ *Thyrotropin-releasing hormone (TRH)*
♦ *Gonadotropin-releasing hormone (GnRH)*
♦ *Prolactin-releasing hormone (PRH)*
♦ *Prolactin-inhibiting hormone (PIH)*

Table 15-3 lists the functions of each releasing hormone. Before consulting the table, try to deduce their functions from their names.

Through negative feedback mechanisms, the hypothalamus adjusts the secretions of the adenohypophysis, and the adenohypophysis adjusts the secretions of its target glands, which in turn adjust the activity of their target tissues. For example, Figure 15-13 shows the negative feedback control of the secretion of TSH and thyroid hormone (T_3 and T_4).

Before leaving the subject of control of pituitary secretion, we want to call attention to another concept about the hypothalamus. It functions as an important part of the body's complex machinery for coping with stress situations. For example, in severe pain or intense emotions, the cerebral cortex—especially the limbic area—sends impulses to the hypothalamus. The impulses stimulate the hypothalamus to secrete its releasing hormones into the

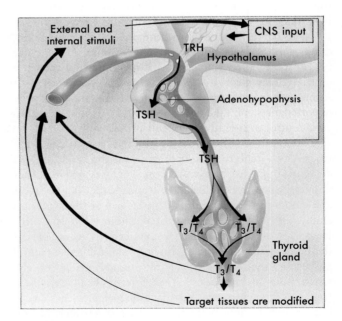

FIGURE 15-13   Negative feedback control by the hypothalamus. In this example, the secretion of thyroid hormone (T_3 and T_4) is regulated by a number of negative feedback loops. A long negative feedback loop (*thin line*) allows the CNS to influence hypothalamic secretion of thyrotropin-releasing hormone (*TRH*) by nervous feedback from the targets of T_3/T_4 (as well as from other nerve inputs). The secretion of TRH by the hypothalamus and thyroid-stimulating hormone (*TSH*) by the adenohypophysis is also influenced by shorter feedback loops (*thicker lines*), allowing great precision in the control of this system.

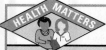

CLINICAL EVIDENCE OF FEEDBACK CONTROL

Clinical facts furnish interesting evidence about feedback control of hormone secretion by the anterior pituitary gland and its target glands. For instance, patients who have their pituitary gland removed (hypophysectomy) surgically or by radiation must be given hormone replacement therapy for the rest of their life. If not, they will develop thyroid, adrenocortical, and gonadotropic deficiencies—deficiencies, that is, of the anterior pituitary's target gland hormones. Another well-known clinical fact is that estrogen deficiency develops in women between 40 and 50 years of age. By then the ovaries seem to have tired of producing hormones and ovulating each month. They no longer respond to FSH stimulation, so estrogen deficiency develops, brings about menopause, and persists after menopause. What therefore would you deduce is true of the blood concentration of FSH after menopause? Apply the principle that a low concentration of a target gland hormone stimulates tropic hormone secretion by the anterior pituitary gland and you will understand that FSH levels rise, indeed, after menopause.

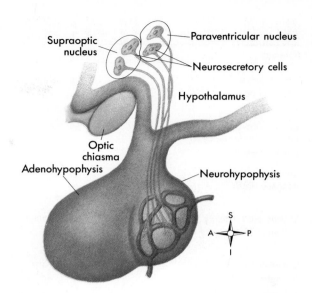

FIGURE 15-14 Relationship of the hypothalamus and neurohypophysis. Neurosecretory cells have their cell bodies in the hypothalamus and their axon terminals in the neurohypophysis. Thus hormones synthesized in the hypothalamus are actually released from the neurohypophysis.

hypophyseal portal veins. Circulating quickly to the adenohypophysis, they stimulate it to secrete more of its hormones. These, in turn, stimulate increased activity by the pituitary's target structures. In essence, what the hypothalamus does through its releasing hormones is to translate nerve impulses into hormone secretion by endocrine glands. Thus the hypothalamus links the nervous system to the endocrine system. It integrates the activities of these two great integrating systems—particularly, it seems, in times of stress. When survival is threatened, the hypothalamus can take over the adenohypophysis and thus gain control of literally every cell in the body.

The mind-body link provided by the hypothalamus has tremendous implications. It means that the cerebrum can do more than just receive sensory impulses and send out impulses to muscles and glands. It means that our thoughts and emotions—our minds—can, by way of the hypothalamus, influence the functions of all of our billions of cells. In short, the brain has two-way contact with every tissue of the body. Thus the state of the body can influence mental processes, and the state of the mind can affect the functioning of the body. Therefore both *psychosomatic* (mind influencing the body) and *somatopsychic* (body influencing the mind) relationships exist between human body systems and the brain.

QUICK CHECK

✔ 1. *What are the two main divisions of the pituitary called? How are they distinguished by location and histology?*

✔ 2. *Name three hormones produced by the adenohypophysis and give their main function.*

✔ 3. *What is a tropic hormone?*

✔ 4. *What is a releasing hormone?*

Neurohypophysis (Posterior Pituitary)

The neurohypophysis serves as a storage and release site for two hormones: **antidiuretic hormone (ADH)** and **oxytocin (OT).** The cells of the neurohypophysis, called *pituicytes,* do not themselves make these hormones. Instead, neurons whose bodies are in either the *supraoptic* or the *paraventricular nuclei* of the hypothalamus synthesize them (Figure 15-14). From the cell bodies of these neurons in the hypothalamus, the hormones pass down along axons (in the hypothalamohypophyseal tract) into the neurohypophysis. Instead of the chemical-releasing factors that triggered secretion of hormones from the adenohypophysis, release of ADH and OT into the blood is controlled by nervous stimulation.

Antidiuretic hormone

The term *antidiuresis* literally means "opposing the production of a large urine volume." And this is exactly what **antidiuretic hormone (ADH)** does—it prevents the formation of a large volume of urine. In preventing large losses of fluid through the excretion of dilute urine, ADH helps the body conserve water. In other words, ADH causes water retention by the body. ADH produces its effects by causing a portion of each tubule in the kidney to reabsorb water from the urine it has formed (see Chapter 27). When the body dehydrates, the increased osmotic pressure of the blood is detected by special *osmoreceptors* near the supraoptic nucleus. This triggers the release of ADH from the neurohypophysis. ADH causes water to be reabsorbed from the tubules of the kidney and returned to the blood. This increases the water content of the blood, restoring the osmotic pressure to its normal lower level.

Oxytocin

Oxytocin (OT) has two actions: it stimulates contraction of uterine muscles, and it causes milk ejection from the breasts of lactating women. Under the influence of OT, milk-producing *alveolar* cells release their secretion into the ducts of the breast. This is very important because milk cannot be removed by suckling unless it has first been ejected into the ducts. Throughout nursing, the mechanical and psychological stimulation of the baby's suckling action trigger the release of more oxytocin. In other words, oxytocin secretion is regulated by a *positive feedback* mechanism: the baby suckles, which increases oxytocin levels, which provides more milk, so the baby continues to suckle, which increases oxytocin levels, and so on. Oxytocin cooperates with prolactin to ensure successful nursing. Prolactin prepares the breast for milk production and stimulates cells to produce milk. The milk is not released, however, until oxytocin permits it to do so.

It is oxytocin's other action—its stimulation of uterine contractions—that gives it its name: oxytocin (literally "swift childbirth"). Oxytocin stimulates the uterus to strengthen the strong, muscular labor contractions that occur during childbirth. Oxytocin secretion is regulated here again by means of a positive feedback mechanism. Once they have begun, uterine contractions push on receptors in the pelvis, which triggers the release of more oxytocin, which again pushes on the pelvic receptors, and so on. The wavelike contractions continue to some degree after childbirth, which helps the uterus expel the placenta and then return to its unstretched shape. Commercial preparations of oxytocin have been given to stimulate

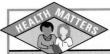

ANTIDIURETIC HORMONE (ADH) ABNORMALITIES

Hyposecretion of ADH can lead to *diabetes insipidus,* a condition in which the patient produces abnormally large amounts of urine. ADH, administered under the name vasopressin (Pitressin), can alleviate this symptom. Studies have shown that ADH may be involved in learning and memory, so investigators are looking into the possibility of administering ADH to reverse the memory loss associated with senility.

contractions after childbirth to lessen the danger of uterine hemorrhage. Important characteristics of the hormones secreted by the pituitary—both the adenohypophysis and the neurohypophysis—are summarized in Table 15-4.

PINEAL GLAND

The **pineal gland, or** *pineal body,* is a tiny (1 cm) pine cone–shaped structure located on the dorsal aspect of the brain's diencephalon region (see Figure 15-9). It is a member of two systems because it acts as a part of the nervous system (it receives visual nerve stimuli) and as a part of the endocrine system (it secretes hormones).

Although full understanding of the pineal gland is a long way off, we do know that it functions to support the body's *biological clock.* It is the biological clock that regulates our patterns of eating (hunger), sleeping, reproduction (female reproductive cycle), and behavior. One hypothesis states

TABLE 15-4 Hormones of the pituitary gland (hypophysis)

Hormone	Source	Target	Principal action
Growth hormone (GH) [somatotropin (STH)]	Adenohypophysis (acidophils)	General	Promotes growth by stimulating protein anabolism and fat mobilization
Prolactin (PRL) [lactogenic hormone]	Adenohypophysis (acidophils)	Mammary glands (alveolar secretory cells)	Promotes milk secretion
Thyroid-stimulating hormone (TSH)*	Adenohypophysis (basophils)	Thyroid gland	Stimulates development and secretion in the thyroid gland
Adrenocorticotropic hormone (ACTH)*	Adenohypophysis (basophils)	Adrenal cortex	Promotes development and secretion in the adrenal cortex
Follicle-stimulating hormone (FSH)*	Adenohypophysis (basophils)	Gonads (primary sex organs)	Female: promotes development of ovarian follicle; stimulates estrogen secretion Male: promotes development of testis; stimulates sperm production
Luteinizing hormone (LH)*	Adenohypophysis (basophils)	Gonads and mammary glands	Female: triggers ovulation; promotes development of corpus luteum Male: stimulates production of testosterone
Melanocyte-stimulating hormone (MSH)	Adenohypophysis (basophils)	Skin (melanocytes); adrenal glands	Exact function uncertain; may stimulate production of melanin pigment in skin; may maintain adrenal sensitivity
Antidiuretic hormone (ADH)	Neurohypophysis	Kidney	Promotes water retention by kidney tubules
Oxytocin (OT)	Neurohypophysis	Uterus and mammary glands	Stimulates uterine contractions; stimulates ejection of milk into mammary ducts

*Tropic hormones

that visual signals received by the pineal allow it to determine day length and lunar cycles (changing phases of the moon). Day-length information helps to keep daily and seasonal cycles "on time," whereas lunar-cycle information helps keep the menstrual cycle "on time." **Melatonin,** the principal pineal secretion, appears to inhibit LH secretion and so may be a means by which the pineal can influence the menstrual cycle.

Melatonin, whose secretion is inhibited by the presence of sunlight, may also affect a person's mood. A mental disorder, called *seasonal affective disorder (SAD),* in which a patient suffers severe depression only in winter (when day-length is shorter), has been linked to the pineal gland. Patients suffering from this "winter depression" are often advised to expose themselves to special high-intensity lights for several hours each evening during the winter months. Apparently, light stimulates the pineal gland for a longer period, which reduces the mood-altering effects of melatonin. The symptoms of depression are thus reduced or eliminated.

QUICK CHECK

✔ 1. *Where are the hormones of the neurohypophysis manufactured? From which location in the body are they released into the bloodstream?*

✔ 2. *Name the two hormones of the neurohypophysis.*

✔ 3. *How does the pineal gland adjust the body's biological clock?*

THYROID GLAND

Structure of the Thyroid Gland

Two large **lateral lobes** and a narrow connecting **isthmus** make up the **thyroid gland** (Figure 15-15). There is often a thin wormlike piece of thyroid tissue extending upward from the isthmus. The weight of the gland in the adult is variable, but it's around 30 g (1 oz). The thyroid is located in the neck, on the anterior and lateral surfaces of the trachea, just below the larynx.

Thyroid tissue is composed of tiny structural units called **follicles.** Each follicle is a small hollow sphere with a wall of simple cuboidal glandular epithelium (Figure 15-16). The interior is filled with a thick fluid called **thyroid colloid.** The colloid is produced by the cuboidal cells of the follicle wall **(follicular cells)** and contains protein-iodine complexes known as *thyroglobulins.*

Thyroid Hormone

The substance that is often called *thyroid hormone (TH)* is actually two different hormones. The most abundant TH is **tetraiodothyronine (T_4),** or **thyroxine.** The other is called **triiodothyronine (T_3).** One molecule of T_4 contains four iodine atoms, and one molecule of T_3 contains three iodine atoms. After synthesizing a preliminary form of its hormones, the thyroid gland stores considerable amounts of them before secreting them. This is unusual, since none of the other endocrine glands stores its hormones in another form for later release. Before being stored in the colloid of

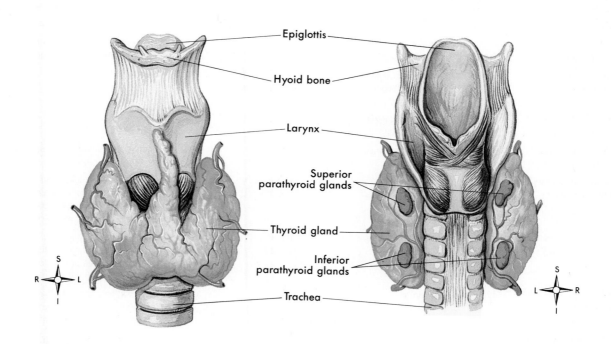

FIGURE 15-15 Thyroid and parathyroid glands. Note the relationships of the thyroid and parathyroid glands to each other, to the larynx (voice box), and to the trachea.

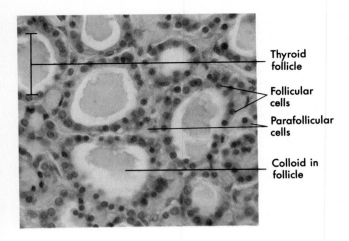

FIGURE 15-16 **Thyroid gland tissue.** Note that each of the follicles is filled with colloid. (x140.)

the follicles, T_3 and T_4 are attached to globulin molecules, forming thyroglobulin complexes. When they are to be released, T_3 and T_4 detach from the globulin and enter the blood. Once in the bloodstream, however, they attach to plasma globulins and travel as a hormone-globulin complex. When they near their target cells, T_3 and T_4 detach from the plasma globulin.

Although the thyroid gland releases about twenty times more T_4 than T_3, T_3 is considered by physiologists to be the principal thyroid hormone. Why is this? Once it enters the bloodstream, most of the T_4 secreted by the thyroid gland is converted either to T_3 or "reverse T_3" (an inactive compound). The small amount of T_4 that remains intact in the bloodstream is usually converted to T_3 once it reaches the target tissues. Add this to the fact that experiments have shown that T_3 binds more efficiently than T_4 to nuclear receptors in target cells and the evidence is overwhelming that T_3 is the principal thyroid hormone. Although T_4 may influence target cells to some extent, its major importance is as a precursor to T_3. Such hormone precursors are often called *prohormones*.

Thyroid hormone helps regulate the metabolic rate of all cells, as well as the processes of cell growth and tissue differentiation. Because thyroid hormone can potentially interact with any cell in the body, it is said to have a "general" target.

Calcitonin

Besides thyroid hormone (T_3 and T_4), the thyroid gland also produces a hormone called **calcitonin (CT).** You might wonder why some hormones of the thyroid qualify for the name "thyroid hormone," whereas calcitonin does not. The answer lies in the simple fact that for many years, we had no idea that a hormone other than thyroid hormone was produced by the thyroid gland. By the time calcitonin was discovered, and later shown to be made in the thyroid

gland, the term *thyroid hormone* was too well established to change it easily.

Produced by parafollicular cells (cells between the thyroid follicles), calcitonin influences the processing of calcium by bone cells. Calcitonin apparently controls calcium content of the blood by increasing bone formation by osteoblasts and inhibiting bone breakdown by osteoclasts. This means more calcium is removed from the blood by the osteoblasts, and less calcium is released into the blood by osteoclasts. Calcitonin, then, tends to decrease blood calcium levels and promote conservation of hard bone matrix. Parathyroid hormone, discussed later, is an antagonist to calcitonin, since it has the opposite effects. Together, calcitonin and parathyroid hormone help maintain calcium homeostasis (see Figure 15-18).

Hormones of the thyroid gland are summarized in Table 15-5.

PARATHYROID GLANDS
Structure of the Parathyroid Glands

There are usually four or five **parathyroid glands** imbedded in the posterior surface of the thyroid's lateral lobes (see Figure 15-15). They appear as tiny rounded bodies within thyroid tissue formed by compact, irregular rows of cells (Figure 15-17).

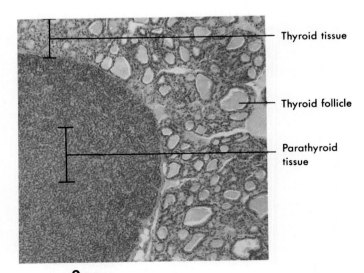

FIGURE 15-17 **Parathyroid tissue.** (x35.)

TABLE 15-5   Hormones of the thyroid and parathyroid glands

Hormone	Source	Target	Principal action
Triiodothyronine (T_3)	Thyroid gland (follicular cells)	General	Increases rate of metabolism
Tetraiodothyronine (T_4), or thyroxine	Thyroid gland (follicular cells)	General	Increases rate of metabolism (usually converted to T_3 first)
Calcitonin (CT)	Thyroid gland (parafollicular cells)	Bone tissue	Increases calcium storage in bone, lowering blood Ca^{++} levels
Parathyroid hormone (PTH) or parathormone	Parathyroid glands	Bone tissue and intestinal tract	Increases calcium removal from storage in bone and increases absorption of calcium by intestines, increasing blood Ca^{++} levels

HEALTH MATTERS

THYROID HORMONE ABNORMALITIES

Hypersecretion of thyroid hormone is a symptom of *Graves' disease,* which is thought to be an autoimmune condition (in which the immune system attacks one's own tissues). Graves' disease patients may suffer from unexplained weight loss, nervousness, increased heart rate, and *exophthalmic goiter* (protrusion of the eyeballs due in part to edema of tissue at the back of the eye socket; see Figure A).

Hyposecretion of thyroid hormone during growth years may lead to *cretinism.* Cretinism is a condition characterized by a low metabolic rate, retarded growth and sexual development, and, possibly, mental retardation. Persons with profound manifestations of this condition are said to have *deformed dwarfism* (as opposed to the proportional dwarfism caused by hyposecretion of GH). Hyposecretion later in life produces *myxedema,* a condition characterized by decreased metabolic rate, loss of mental and physical vigor, gain in weight, loss of hair, "firm" edema, and yellow dullness of the skin.

In a condition called *simple goiter,* the thyroid enlarges in an attempt to compensate for the lack of iodine in the diet (Figure B). This condition is an interesting example of how the feedback control mechanisms illustrated in Figure 15-13 operate. Since iodine is required for the synthesis of T_3 and T_4, lack of iodine in the diet results in a drop in the pro-

duction of these hormones. When the reserve (in thyroid colloid) is exhausted, feedback informs the hypothalamus and adenohypophysis of the deficiency. In response, the secretion of thyrotropin-releasing hormone (TRH) and thyroid-stimulating hormone (TSH) increases in an attempt to stimulate the thyroid to produce more thyroid hormone. Since there is no iodine available to do this, the only effect is to increase the size of the thyroid gland. This information feeds back to the hypothalamus and adenohypophysis, and both increase their secretions in response. Thus the thyroid gets larger and larger and larger—all in a futile attempt to increase thyroid hormone secretion to normal levels. This condition is still common in areas of the world where the soil and water contain little or no iodine. The use of iodized salt in the United States has dramatically reduced the incidence of simple goiter here.

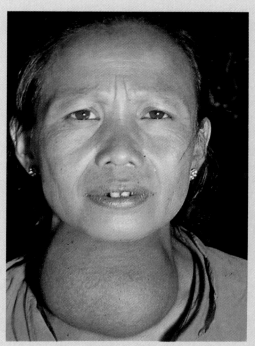

A

B

Parathyroid Hormone

The parathyroid glands secrete **parathyroid hormone (PTH),** or *parathormone* (see Table 15-5). PTH is an antagonist to calcitonin and so helps maintain calcium homeostasis. Parathyroid hormone acts on bone, kidney, and intestinal cells by increasing their input of calcium into the blood. The bone cells are especially affected, causing less new bone to be formed and more old bone to be dissolved, yielding calcium and phosphate. These minerals are then free to move into the blood, elevating blood levels of calcium and phosphate. In the kidney, however, only calcium is reabsorbed from urine into the blood. Under the influence of PTH, phosphate is secreted by kidney cells *out* of the blood and *into* the urine to be excreted. In the intestines, parathyroid hormone increases the body's absorption of calcium from food by activating *vitamin D* (cholecalciferol). Vitamin D, obtained from the diet or synthesized from cholesterol in the skin, permits Ca^{++} to be transported through intestinal cells and into the blood.

The maintenance of calcium homeostasis, achieved through the interaction of parathyroid hormone and calcitonin, is very important for healthy survival (Figure 15-18). Normal neuromuscular excitability, blood clotting, cell membrane permeability, and normal functioning of certain enzymes, all depend on the maintenance of normal levels of calcium in the blood. For example, hyposecretion of PTH can lead to hypocalcemia. Hypocalcemia increases neuromuscular irritability—sometimes so much that it produces muscle spasms and convulsions.

ADRENAL GLANDS
Structure of the Adrenal Glands

The **adrenal,** or *suprarenal,* glands are located atop the kidneys, fitting like a cap over these organs (see Figure 15-19). The outer portion of the gland is called the **adrenal cortex,** and the inner portion of the gland is called the **adrenal medulla.** Even though the adrenal cortex and

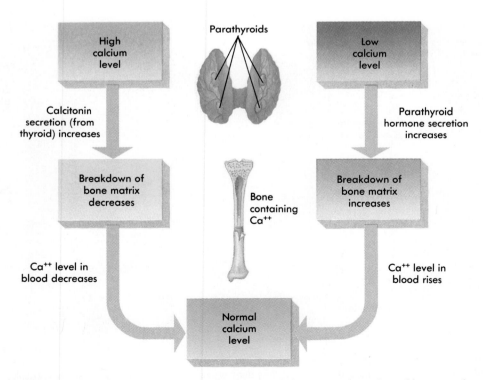

FIGURE 15-18 Regulation of blood calcium levels. Calcitonin and parathyroid hormones have antagonistic (opposite) effects on calcium concentration in the blood. Also see Figure 15-7.

TABLE 15-6   Hormones of the adrenal glands

Hormone	Source	Target	Principal action
Aldosterone	Adrenal cortex (zona glomerulosa)	Kidney	Stimulates kidney tubules to conserve sodium, which, in turn, triggers the release of ADH and the resulting conservation of water by the kidney
Cortisol (hydrocortisone)	Adrenal cortex (zona fasciculata)	General	Influences metabolism of food molecules; in large amounts, it has an antiinflammatory effect
Adrenal androgens	Adrenal cortex (zona reticularis)	Sex organs, other effectors	Exact role uncertain, but may support sexual function
Adrenal estrogens	Adrenal cortex (zona reticularis)	Sex organs	Thought to be physiologically insignificant
Epinephrine (adrenaline)	Adrenal medulla	Sympathetic effectors	Enhances and prolongs the effects of the sympathetic division of the ANS
Norepinephrine	Adrenal medulla	Sympathetic effectors	Enhances and prolongs the effects of the sympathetic division of the ANS

adrenal medulla are part of the same organ, they are structurally and functionally so different that they are often spoken of as if they were separate glands. The adrenal cortex is composed of regular endocrine tissue, but the adrenal medulla is made of neurosecretory tissue. As you might guess, each of these tissues secretes a different set of hormones (Table 15-6).

Adrenal Cortex

The adrenal cortex is composed of three distinct layers, or zones, of secreting cells (see Figure 15-19). Starting with the zone directly under the outer connective tissue capsule of the gland, they are **zona glomerulosa, zona fasciculata,** and **zona reticularis.** Cells of the outer zone secrete a class of hormones called *mineralocorticoids.* Cells of the middle

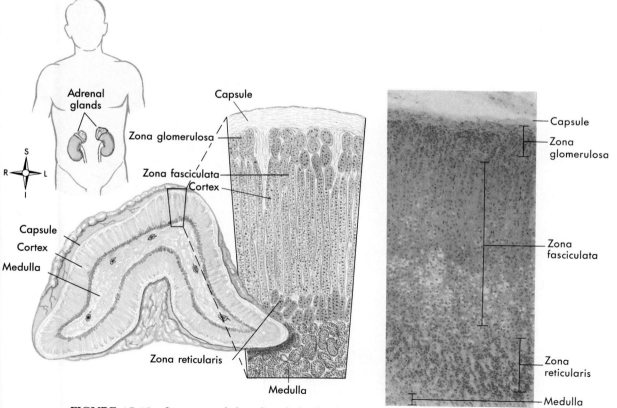

FIGURE 15-19   Structure of the adrenal gland. The zona glomerulosa of the cortex secretes aldosterone. The zona fasciculata secretes abundant amounts of glucocorticoids, chiefly cortisol. The zona reticularis secretes minute amounts of sex hormones and glucocorticoids. A portion of the medulla is visible at lower right in the photomicrograph (x35.) and at the bottom of the drawing.

zone secrete *glucocorticoids*. The inner zone secretes small amounts of *glucocorticoids* and *gonadocorticoids* (sex hormones). All of these cortical hormones are steroids, so, together, they are known as corticosteroids.

Mineralocorticoids

Mineralocorticoids, as their name suggests, have an important role in regulating how mineral salts (electrolytes) are processed in the body. In the human, **aldosterone** is the only physiologically important mineralocorticoid. Its primary function seems to be the maintenance of sodium homeostasis in the blood. Aldosterone accomplishes this by increasing sodium reabsorption in the kidneys. Sodium ions are reabsorbed from the urine back into the blood in exchange for potassium or hydrogen ions. In this way, aldosterone not only adjusts blood sodium levels but can also influence potassium and pH levels in the blood.

Because the reabsorption of sodium ions causes water to also be reabsorbed (partly by triggering the secretion of ADH), aldosterone promotes water retention by the body. Altogether, aldosterone can increase sodium and water retention and promote the loss of potassium and hydrogen ions.

Aldosterone secretion is controlled mainly by the *renin-angiotensin mechanism* and by blood potassium concentration. The renin-angiotensin mechanism (Figure 15-20) operates as indicated in this sequence of steps:

1. When the incoming blood pressure in the kidneys drops below a certain level, a piece of tissue near the vessels (the *juxtaglomerular apparatus*) secretes **renin** into the blood.
2. Renin, an enzyme, causes **angiotensinogen** (a normal constituent of blood) to be converted to **angiotensin I.**
3. Angiotensin I circulates to the lungs, where converting enzymes in the capillaries split the molecule, forming **angiotensin II.**
4. Angiotensin II circulates to the adrenal cortex, where it stimulates the secretion of aldosterone.
5. Aldosterone causes increased reabsorption of sodium, which causes increased water retention. As water is retained, the volume of blood increases. The increased volume of blood creates higher blood pressure—which then causes the renin-angiotensin mechanism to stop.

The renin-angiotensin mechanism is a negative feedback mechanism that helps maintain homeostasis of blood pressure.

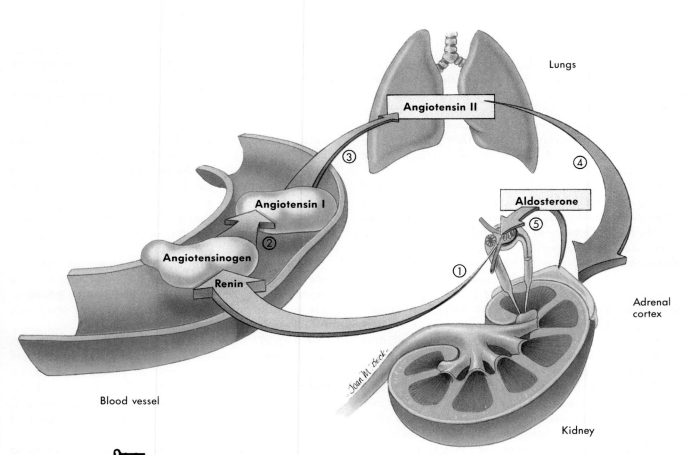

FIGURE 15-20 Renin-angiotensin mechanism for regulating aldosterone secretion. The numbers correspond to the steps outlined in the text.

ADRENAL CORTICAL HORMONES ABNORMALITIES

Hypersecretion of adrenal cortical hormones often produces a collection of symptoms called *Cushing's syndrome*. **Hypersecretion** of glucocorticoids results in a redistribution of body fat. Fat is moved from the arms and legs and is deposited in the face, shoulders, trunk, and abdomen. The fatty "moon face" characteristic of Cushing's syndrome is shown in the figure. Hypersecretion of aldosterone, *aldosteronism*, leads to increased water retention and muscle weakness due to potassium loss. Hypersecretion of androgens can result from tumors of the adrenal cortex, called *virilizing tumors*. They are so called because the increased blood level of male hormones in women can cause them to acquire male characteristics. People with Cushing's syndrome may suffer from all of the symptoms described above.

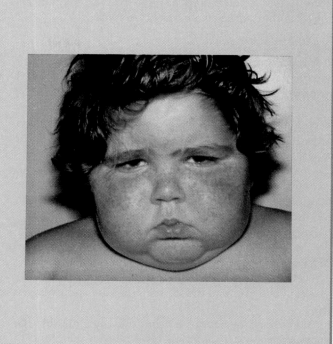

Hyposecretion of mineralocorticoids and glucocorticoids, as in *Addison's disease*, may lead to a rise in blood sodium levels, drops in blood glucose and potassium levels, dehydration, and weight loss.

Pharmacological preparations of glucocorticoids have been used for many years to temporarily relieve the symptoms of severe inflammatory conditions such as rheumatoid arthritis. More recently, over-the-counter creams and ointments containing hydrocortisone have become available for use in treating the pain, itching, swelling, and redness of skin rashes.

Glucocorticoids

The chief glucocorticoids secreted by the zona fasciculata of the adrenal cortex are **cortisol** (also called *hydrocortisone*), *cortisone*, and *corticosterone*. Of these, only cortisol is secreted in significant quantities in the human. Glucocorticoids affect every cell in the body. Although much remains to be discovered about their precise mechanisms of action, we do know enough to make some generalizations:

♦ Glucocorticoids tend to accelerate the breakdown of proteins to amino acids (except in liver cells). These "mobilized" amino acids move out of the tissue cells and into the blood. From there, they circulate to the liver cells, where they are changed to glucose in a process called *gluconeogenesis*. A prolonged high blood concentration of glucocorticoids in the blood, therefore, results in a net loss of tissue proteins ("tissue wasting") and hyperglycemia (high blood glucose). Glucocorticoids are protein-mobilizing, gluconeogenic, and hyperglycemic.

♦ Glucocorticoids tend to accelerate both mobilization of lipids from adipose cells and lipid catabolism by nearly every cell in the body. In other words, glucocorticoids tend to cause a shift from carbohydrate catabolism to lipid catabolism as an energy source. The mobilized lipids may also be used in the liver for gluconeogenesis. This effect contributes to the hyperglycemic effect already observed.

♦ Glucocorticoids are essential for maintaining a normal blood pressure. Without adequate amounts of glucocorticoids in the blood, the hormones norepinephrine and epinephrine cannot produce their vasoconstricting effect on blood vessels, and blood pressure falls. In other words, glucocorticoids exhibit *permissiveness* in that they permit norepinephrine and epinephrine to have their full effects.

♦ A high blood concentration of glucocorticoids rather quickly causes a marked decrease in the number of white blood cells, called *eosinophils*, in the blood (eosinopenia) and marked atrophy of lymphatic tissues. The thymus gland and lymph nodes are particularly affected. This, in turn, leads to a decrease in the number of lymphocytes and plasma cells in the blood. Because of the decreased number of lymphocytes and plasma cells (types of white blood cells), antibody formation decreases. Antibody formation is an important part of immunity—the body's defense against infection.

♦ Normal amounts of glucocorticoids act with epinephrine, a hormone secreted by the adrenal medulla, to bring about normal recovery from injury produced by inflammatory agents. How they act together to bring about this antiinflammatory effect is still uncertain.

♦ Glucocorticoid secretion is known to increase as part of the stress response. The advantage gained by increased secretion may be the increase in glucose available for skeletal muscles needed in fight-or-flight responses. However, prolonged stress can lead to immune dysfunction, probably as a result of prolonged exposure to high levels of glucocorticoids (see Chapter 21).

♦ Except during the stress response, glucocorticoid secretion is controlled mainly by means of a negative feedback mechanism that involves ACTH from the adenohypophysis.

Gonadocorticoids

The term *gonadocorticoid* refers to sex hormones that are released from the adrenal cortex rather than the gonads. The normal adrenal cortex secretes small amounts of both male hormones (androgens) and female hormones (estrogens). In each normal cortex, the amount of androgen produced is physiologically significant, but the amount of estrogen appears to be insignificant. Normally, there is not enough androgen produced to give women masculine characteristics, but it may contribute to female sexual characteristics such as growth of pubic hair.

Adrenal Medulla

The adrenal medulla is composed of neurosecretory tissue, that is, tissue composed of neurons specialized to secrete their products into the blood rather than across a synapse. Actually, the medullary cells are modified versions of sympathetic postganglionic fibers of the autonomic nervous system. They are innervated by sympathetic preganglionic fibers, so that when the sympathetic nervous system is activated (as in the stress response), the medullary cells secrete their hormones.

The adrenal medulla secretes two important hormones, both of which are in the class of nonsteroid hormones called *catecholamines*. **Epinephrine,** or *adrenaline*, accounts for about 80% of the medulla's secretion. The other 20% is **norepinephrine** (NE or NR). You may recall that norepinephrine is also the neurotransmitter produced by postganglionic sympathetic fibers. Sympathetic effectors such as the heart, smooth muscle, and glands have receptors for norepinephrine. Both epinephrine and norepinephrine produced by the adrenal medulla can bind to the receptors of sympathetic effectors to prolong and enhance the effects of sympathetic stimulation by the autonomic nervous system (Figure 15-21).

QUICK CHECK

✔ 1. *Distinguish between the histology of the adrenal cortex and the adrenal medulla.*
✔ 2. *Name some of the effects of cortisol in the body.*
✔ 3. *How does the function of the adrenal medulla overlap with the function of the autonomic nervous system?*

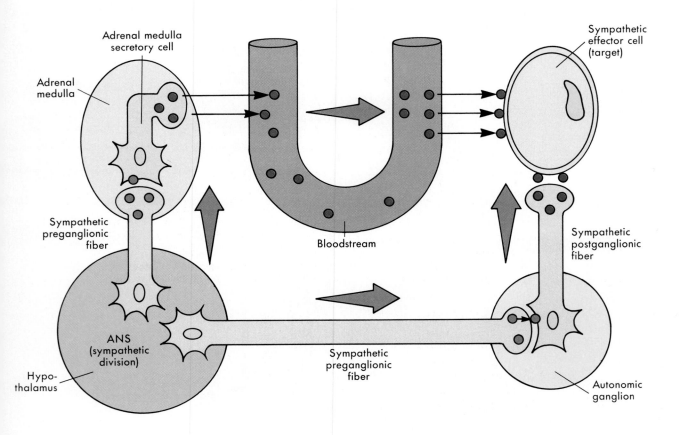

FIGURE 15-21 Combined nervous and endocrine influence on sympathetic effectors. A sympathetic center in the hypothalamus sends efferent impulses through preganglionic fibers. Some preganglionic fibers synapse with postganglionic fibers that deliver norepinephrine across a synapse with the effector cell. Other preganglionic fibers synapse with postganglionic neurosecretory cells in the adrenal medulla. These neurosecretory cells secrete epinephrine and norepinephrine into the bloodstream, where they travel to the target cells (sympathetic effectors). Compare this figure with those that appear in Table 15-1.

PANCREATIC ISLETS

Structure of the Pancreatic Islets

The pancreas is an elongated gland (12 to 15 cm long) weighing up to 100 g (Figure 15-22). The "head" of the gland lies in the C-shaped beginning of the small intestine (duodenum), with its body extending horizontally behind the stomach and its tail touching the spleen.

The tissue of the pancreas is composed of both endocrine and exocrine tissues. The endocrine portion is made up of scattered, tiny islands of cells, called **pancreatic islets** *(islets of Langerhans)* that account for only about 1% or 2% of the total mass of the pancreas. These hormone-producing islets are surrounded by cells called *acini,* which secrete a serous fluid containing digestive enzymes into ducts that lead to the small intestine (see Figure 15-22). The digestive role of the pancreas will be discussed in Chapters 24 and 25. For the moment, we will concentrate on the endocrine part of this gland, the pancreatic islets.

Each of the one to two million pancreatic islets in the pancreas contains a combination of four primary types of endocrine cells, all joined to each other by gap junctions. Each type of cell secretes a different hormone, but the gap junctions may allow for some coordination of these

functions as a single secretory unit. One type of pancreatic islet cell is the **alpha cell** (also called the *A cell*), which secretes the hormone glucagon. **Beta cells** *(B cells)* secrete the hormone, insulin; **delta cells** *(D cells)* secrete the hormone, somatostatin; and **pancreatic polypeptide cells** *(F,* or *PP,* cells) secrete pancreatic polypeptide. Current investigations are showing that other substances are secreted along with these primary endocrine hormones. For example, some beta cells are thought to secrete *amylin,* an insulin antagonist. Beta cells, which account for about three fourths of all the pancreatic islet cells, are usually found near the center of each islet, whereas cells of the other three types are more often found in the outer portion. Figure 15-23 shows how the different cell types can be distinguished by the microscope with special staining techniques.

Pancreatic Hormones

The pancreatic islets produce several hormones, the most important of which are described in Table 15-7 and in the following paragraphs.

♦ **Glucagon,** produced by alpha cells, tends to increase blood glucose levels by stimulating the conversion of glycogen to glucose in liver cells. It also stimulates

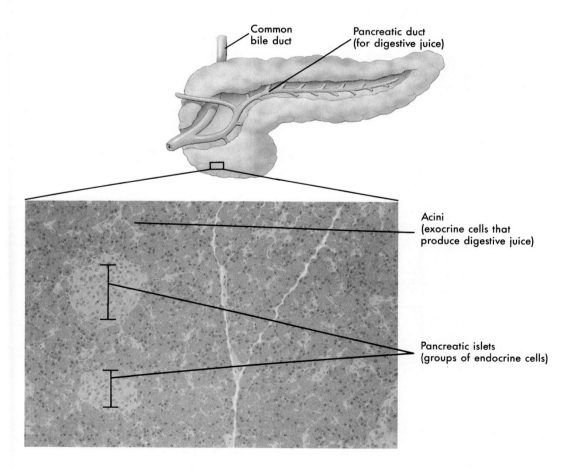

FIGURE 15-22 Pancreas. Two pancreatic islets, or hormone producing areas, evident among the pancreatic cells that produce the pancreatic digestive juice. The pancreatic islets are more abundant in the tail of the pancreas than in the body or head.

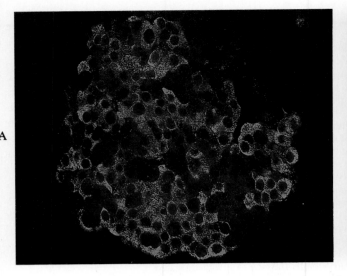
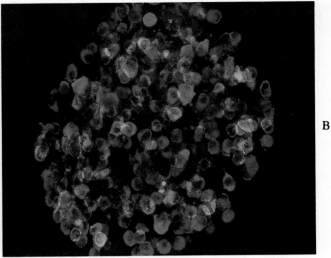

A

B

FIGURE 15-23   Cells of the pancreatic islet. A special technique of microscopy, called *immunofluorescent staining*, allows cells to be distinguished by the different molecules that they contain. **A,** Micrograph shows a central cluster of beta, or β, cells outlined by green dots. The green dots are insulin molecules stained by a special type of antibody coupled to a fluorescent dye. **B,** In this micrograph, peripheral alpha, or α, cells are stained green and delta, or δ, cells are stained red. The central beta cells are not stained in this specimen, so they appear only as black spaces in the micrograph.

TABLE 15-7	**Hormones of the pancreatic islets**		
Hormone	**Source**	**Target**	**Principal action**
Glucagon	Pancreatic islets (alpha [α] cells or A cells)	General	Promotes movement of glucose from storage and into the blood
Insulin	Pancreatic islets (beta [β] cells or B cells)	General	Promotes movement of glucose out of the blood and into cells
Somatostatin	Pancreatic islets (delta [δ] cells or D cells)	Pancreatic cells and other effectors	Can have general effects in the body, but primary role seems to be regulation of secretion of other pancreatic hormones
Pancreatic polypeptide	Pancreatic islets (pancreatic polypeptide [PP] or F cells)	Intestinal cells and other effectors	Exact function uncertain, but seems to influence absorption in the digestive tract

gluconeogenesis (transformation of fatty acids and amino acids into glucose) in liver cells. The glucose produced via the breakdown of glycogen and by gluconeogenesis is released into the bloodstream, producing a hyperglycemic effect.

♦ **Insulin,** produced by beta cells, tends to promote the movement of glucose, amino acids, and fatty acids out of the blood and into tissue cells. Hence, insulin tends to lower the blood concentrations of these food molecules and to promote their metabolism by tissue cells.

♦ **Somatostatin,** produced by delta cells, may affect many different tissues in the body, but its primary role seems to be in regulating the other endocrine cells of the pancreatic islets. Somatostatin inhibits the secretion of glucagon, insulin, and pancreatic polypeptide. It also inhibits secretion of growth hormone (somatotropin) from the anterior pituitary.

♦ **Pancreatic polypeptide** is produced by PP (or F) cells

in the periphery of pancreatic islets. Although much is yet to be learned about pancreatic polypeptide we do know that it influences the digestion and distribution of food molecules to some degree.

All four of these pancreatic hormones probably work together as a team to maintain a homeostasis of food molecules (glucose, fatty acids, and amino acids). More about their respective roles in overall nutrient metabolism is discussed in Chapter 26.

QUICK CHECK

✔ 1. *Name two of the four principal hormones secreted by the pancreatic islets.*
✔ 2. *In what way do insulin and glucagon exert antagonistic influences on the concentration of glucose in the blood?*

FOCUS ON... DIABETES MELLITUS

Diabetes mellitus is one of the most common endocrine disorders. It affects over 14 million Americans and will have a significant impact on the health of about 4% of the population at some point in their lives. Diabetes is best described as a *syndrome*—a collection of symptoms that characterize the disease. Although the symptoms of diabetes tend to vary between individuals and on the type, severity, and length of the illness, each sign or symptom is related in some way to abnormal metabolism of nutrients and the consequences that follow.

In diabetics, an inadequate amount or abnormal type of insulin may be produced. In other affected individuals, diminished numbers of insulin receptors on target cells make it impossible for glucose to enter cells even if normal insulin in adequate amounts is present. The presence of adequate amounts of normal insulin is key to the entry of glucose into cells. If there are no insulin problems and if the major target tissues such as skeletal muscle, fat, and liver have adequate insulin receptors, glucose will transfer from blood into cells. In diabetes, glucose cannot enter cells normally. The result is one of the most universal symptoms of the disease—chronic elevation in blood glucose levels, a condition called **hyperglycemia.**

Glucose is normally filtered out of the blood and then reabsorbed from the kidney tubules. However, as blood sugar levels rise in diabetes the amount of glucose filtered out of the blood exceeds the ability of the kidney tubules to resorb it. The result is a "spilling over" of sugar into the urine. This condition, called **glycosuria,** causes increased urine production **(polyuria),** since additional water is required to carry the sugar load. In effect, the excess glucose acts like an osmotic diuretic. As large quantities of water are lost in the urine, the body dehydrates. Dehydration becomes worse as high blood sugar levels increase the osmotic concentration of the blood and pull water out of cells. The resulting sense of excessive and ongoing thirst **(polydipsia)** and the tendency to drink large quantities of liquid are also classic symptoms of the disease. In addition, since cells are deprived of glucose to burn as energy, diabetics often suffer from intense and continuous hunger **(polyphagia).** Their blood sugar is high, but the cells are literally starving to death, and the body craves food. In screening for diabetes, health care professionals look for glycosuria and the "three polys": polyuria, polydipsia, and polyphagia.

The term *diabetes* is more meaningful for those who know the symptoms of the disease. Diabetes is from the Greek word meaning "pass through." Early Greek physicians noted that in diabetics large quantities of water would literally "pass through" the body as urine almost as quickly as it was consumed. In advanced cases of the disease, untreated diabetics may excrete well over 20 liters of urine a day. Almost 1500 years after diabetes was first described the name was changed to diabetes mellitus. Mellitus is from the Greek word "honey," referring to the sugary sweet nature of diabetic urine.

The untreated diabetic is unable to utilize glucose for energy, and the body is forced to burn protein and fat. This abnormal metabolic shift results in fatigue and weight loss. If glucose metabolism is severely restricted, the large quantities of fat that must be burned produce toxic quantities of acetoacedic acid and other acedic metabolites called **ketone bodies.** Buildup of these organic acids lowers blood pH, causing acidosis, and disturbs the normal acid-base balance of the body. (Mechanisms that attempt to restore homeostasis are described in Chapter 29.) Accumulation of ketone bodies in the blood is described as *diabetic ketoacidosis.* Signs and symptoms include abdominal pain, nausea, vomiting, fruity odor of the breath, possible alterations in level of consciousness, coma, and even death, if left untreated.

Types of Diabetes Mellitus

There are two major types of diabetes mellitus: **type I** and **type II.** Hereditary factors play an important role in both types.

Type I

Type I diabetes mellitus is sometimes referred to as *juvenile-onset diabetes* because it usually strikes before age 30 and most often between 11 and 13 years of age. In this form of the disease the B cells of the pancreatic islets are destroyed and there is an absolute deficiency of insulin production. Individuals with type I diabetes are required to take insulin injections daily to prevent ketosis and to control hyperglycemia. As a result, type I diabetes is also called *insulin-dependent diabetes mellitus (IDDM).* This form of the disease accounts for only about 10% of the total number of diabetic individuals.

SPORTS & FITNESS — STEROID ABUSE

Some steroid hormones are called **anabolic steroids** because they stimulate the building of large molecules (anabolism). Specifically, they stimulate the building of proteins in muscle and bone. Steroids such as testosterone and its synthetic derivatives may be abused by athletes and others who want to increase their performance. The anabolic effects of the hormones increase the mass and strength of skeletal muscles.

Unfortunately, steroid abuse has other consequences. Prolonged use of testosterone will cause a negative feedback response of the adenohypophysis: gonadotropin levels will drop in response to high blood levels of testosterone. This may lead to atrophy of the testes and, possibly, permanent sterility. Many other adverse effects, including behavioral abnormalities, are known to result from steroid abuse.

GONADS

Gonads are the primary sex organs in the male *(testes;* singular, *testis)* and in the female *(ovaries).* Each is structured differently, and each produces its own unique set of hormones.

Testes

The testes are paired organs within a sac of skin called the *scrotum,* which hangs from the groin area of the trunk (see Figure 15-1). Composed mainly of coils of sperm-producing *seminiferous tubules,* there is a scattering of endocrine *interstitial cells* found in areas between the tubules. These interstitial cells produce androgens (male sex hormones); the principal androgen is **testosterone.** Testosterone is responsible for the growth and maintenance of male sexual characteristics and for sperm production.

The cause of B cell destruction in type I diabetes is still uncertain. Current research, however, suggests that IDDM is an autoimmune disease that is probably triggered by some type of viral infection in genetically susceptible individuals. Anyone who has a parent, brother, or sister with IDDM has about a 5% to 7% chance of getting the disease. If an identical twin has IDDM the risk increases to about 50%.

Type II

Type II diabetes mellitus, or *noninsulin-dependent diabetes mellitus (NIDDM)*, is the most common form of the disease, accounting for about 90% of all cases. Because it most often occurs after age 40 it is also called *maturity onset diabetes*. In susceptible individuals who are overweight the incidence increases with age. In this form of diabetes, insulin is still produced by the B cells but generally in reduced amounts. In addition, loss of insulin receptors on the surface membranes of target cells may also reduce effectiveness of glucose uptake from the blood.

The name NIDDM is appropriate because insulin injections may not be required to control the disease. Treatment of ketosis, which requires insulin, is seldom required in type II diabetes. In these individuals, hyperglycemia will frequently respond to changes in life-style that result in eating a balanced diet, adequate exercise, and maintenance of body weight within normal limits. If life-style changes are not totally effective in lowering elevated blood glucose in NIDDM individuals, medicines, called oral hypoglycemic agents such as Micronase or Glucotrol, may be prescribed. These drugs stimulate the B cells to produce more insulin and may increase the number or effectiveness of insulin receptors on target cells.

Heredity and ethnic background are important determinants in NIDDM. Those who have a family history of the disease are particularly susceptible if overweight and sedentary. If an identical twin has NIDDM the sibling is almost certain to acquire it. Native Americans are at increased risk for type II diabetes. In addition, Hispanics and African Americans are over 50% more likely than Caucasians to develop this type of diabetes. Research published in 1992 identified a specific gene defect that was responsible for a least some forms of type II diabetes. The gene in question is involved in regulating insulin secretion.

The symptoms of type I diabetes are dramatic and, as a result, most individuals seek medical care soon after the disease occurs. The symptoms of type II diabetes, although of the same type as seen in IDDM can be much more subtle and hard to recognize. The American Diabetes Association estimates that over 7 million Americans have diabetes and are unaware of it. Unfortunately, if left untreated, over time the hyperglycemia of diabetes results in many complications affecting almost every area of the body. Reduced blood flow caused by buildup of fatty materials in blood vessels (atherosclerosis) is one of the most serious complications. It causes such diverse problems as heart attack, stroke, and reduced circulation to the extremities resulting in tingling or numbness in the feet and, in severe cases, gangrene. Retinal changes (diabetic retinopathy) may cause blindness in some diabetics who have battled the disease for decades. Kidney disease is also a common diabetic complication. Most authorities agree that careful regulation of blood sugar levels is the most important measure that diabetics can take to reduce the number of long term complications of the disease.

Treatment of Diabetes

The discovery of insulin in 1921 was one of the most important advances in medicine in this century. Problems associated with the injection of insulin harvested from animal pancreatic tissue were solved when synthetic human insulin became available.

Insulin can be introduced into the body by regular injection using a needle and syringe or by implanting miniaturized pumps to deliver the insulin as needed. Actual pancreas transplants, first performed in 1966, are becoming more numerous and successful in selected cases. Islet cell transplants are one of the newest forms of treatment. Cultured B cells from a donor are encased in plastic vials and inserted into the body of a diabetic recipient. The B cells continue to secrete insulin on demand. Because of the small pores in the covering plastic, the islet cells are safe from phagocytosis by macrophages and are therefore resistant to rejection.

Testosterone secretion is regulated principally by gonadotropin (especially LH) levels in the blood.

Ovaries

Ovaries are a set of paired glands in the pelvis (see Figure 15-1) that produce several types of sex hormones, including those described briefly in the following paragraphs:

♦ **Estrogens,** including *estradiol* and *estrone,* are steroid hormones secreted by the cells of the ovarian follicles that promote the development and maintenance of female sexual characteristics. With other hormones, they are responsible for breast development and the proper sequence of events in the female reproductive cycle (menstrual cycle). More details of its function are discussed in Chapter 31.

♦ **Progesterone** is a hormone whose name, which means "pregnancy-promoting steroid," is an indicator of its chief function. Secreted by the corpus luteum (the tissue left behind after the rupture of a follicle during ovulation), progesterone maintains the lining of the uterus necessary for successful pregnancy (gestation). This hormone, along with others, will be discussed in detail in Chapter 31.

Regulation of ovarian hormone secretion is complex, to say the least, but basically depends on the changing levels of FSH and LH (gonadotropins) from the adenohypophysis.

PLACENTA

Another reproductive tissue that also functions as an endocrine gland is the placenta. The **placenta,** the tissue that forms on the lining of the uterus as an interface between the circulatory systems of the mother and developing child, serves as a temporary endocrine gland. The

placenta produces **human chorionic gonadotropin (hCG).** This hormone is called "chorionic" because it is secreted by the chorion, a fetal tissue component of the placenta. It is called "gonadotropin" because, like the gonadotropins of the adenohypophysis, it stimulates development and secretion by ovarian tissues. Chorionic gonadotropin secretion is high during the early part of pregnancy and serves as a signal to the mother's gonads to maintain the uterine lining rather than allow it to degenerate and fall away (as in menstruation).

The discovery of hCG many years ago led to the development of early pregnancy tests. The high levels of hCG in the urine of women who are in the early part of their pregnancies can be detected through several means. The most familiar test involves the use of an over-the-counter kit, which tests for hCG in urine by means of an antigen-antibody reaction that can be easily seen.

THYMUS

The thymus is a gland in the mediastinum, just beneath the sternum (see Figure 15-1). It is large in children until puberty, when it begins to atrophy. It continues to atrophy throughout adulthood, so that by the time an individual reaches old age, the gland is but a vestige of fat and fibrous tissue.

The anatomy of the thymus is described in Chapter 19.

Although it is considered to be primarily a lymphatic organ (see Chapter 19), the hormone **thymosin** has been isolated from thymus tissue and is considered to be responsible for its endocrine activity. Thymosin actually refers to an entire family of peptides that, together, have a critical role in the development of the immune system. Specifically, thymosin is thought to stimulate production of specialized lymphocytes involved in the immune response called *T cells.* The role of T cells in the immune system is discussed in Chapter 20.

GASTRIC AND INTESTINAL MUCOSA

The mucous lining of the gastrointestinal (GI) tract, like the pancreas, contains cells that produce both endo-crine and exocrine secretions. GI hormones such as gastrin, secretin, and cholecystokinin-pancreozymin (CCK) have important regulatory roles in coordinating the secretory and motor activities involved in the digestive process. Chapter 25 describes the hormonal control of digestion in the stomach and small intestine.

HEART

The heart is another organ that has a secondary endocrine role. Although the heart's main function is to pump blood, its wall contains some hormone-producing cells. These cells produce a hormone called *atrial natriuretic hormone (ANH).* The name of this hormone reveals much about its role in the body. The term "atrial" refers to the fact that ANH is secreted by cells in an upper chamber of the heart called an *atrium.* Atrial cells increase their secretion of ANH in response to an increase in the stretch of the atrial wall caused by abnormally high blood volume or blood pressure. The term "natriuretic" refers to the fact that its principal effect is to promote the loss of sodium (Latin, *natrium*) from the body by means of the urine. When sodium is thus lost from the internal environment, water follows. Water loss results in a decrease in blood volume (and thus a decrease in blood pressure). We can then state that the primary effect of ANH is to oppose increases in blood volume or blood pressure. We can also state that ANH is an antagonist to ADH and aldosterone. Atrial natriuretic hormone is also known by several other names, including atrial natriuretic factor (ANF), atrial natriuretic peptide, and, simply, atrial peptide.

QUICK CHECK

✔ 1. What are the major hormones secreted by reproductive tissues (gonads and the placenta)?

✔ 2. Which gland produces a hormone that regulates the development of cells important to the immune system?

✔ 3. Secretin was the first substance in the body to be identified as a hormone. What structure produces secretin?

 Cycle of Life: *Endocrine system*

Endocrine regulation of body processes first begins during early development in the womb. By the time a baby is born, many of the hormones are already at work influencing the activity of target cells throughout the body. As matter of fact, new evidence suggests that it is a hormonal signal from the fetus to the mother that signals the onset of labor and delivery. Many of the basic hormones are active from birth, but most of the hormones related to reproductive function are not produced or secreted until puberty. Secretion of male reproductive hormones follows the same pattern as most nonreproductive hormones: continuous secretion from puberty until there is slight tapering off in late adulthood. The secretion of female reproductive hormones such as estrogens also declines late in life, but more suddenly and completely—often during or just at the end of middle adulthood.

THE BIG PICTURE
Endocrine System and the Whole Body

It is important to appreciate the precision of control afforded by the partnership of the two major regulatory sysems: the endocrine system and the nervous system. The neuroendocrine system is able to finely adjust the availability and processing of nutrients through a diverse array of mechanisms: growth hormone, thyroid hormone, cortisol, epinephrine, somatostatin, autonomic nervous regulation, and so on. The absorption, storage, and transport of calcium ions are kept in balance by the antagonistic actions of calcitonin and parathyroid hormone. Reproductive ability is triggered, developed, maintained, and timed by the complex interaction of the nervous system with follicle-stimulating hormone, luteinizing hormone, estrogen, progesterone, testosterone, chorionic gonadotropin, prolactin, oxytocin, and melatonin. Nearly every process in the human organism is kept in balance by the incredibly complex, but precise, interaction of all these different nervous and endocrine regulatory chemicals.

In this chapter, we have seen the many different structures and regulatory mechanisms that comprise the endocrine system. Some of the more important hormones and their characteristics are summarized in tables throughout the chapter. We have seen how the hormones interact with each other, as well as how their functions complement those of the nervous system. Although there is still much to be learned, we can see that a basic understanding of hormonal regulatory mechanisms is required to fully appreciate the nature of homeostasis in the human organism. As we continue our study of human anatomy and physiology, we will often encounter the critical integrative role played by the endocrine system.

MECHANISMS OF DISEASE

Endocrine Disorders

As we have stated throughout this chapter, endocrine disorders typically result from either elevated or depressed hormone levels (hypersecretion or hyposecretion). At first thought, this may seem very simple and straightforward. In reality, however, nothing could be further from the truth. A variety of specific mechanisms may produce hypersecretion or hyposecretion of hormones. A few of the more well-known mechanisms are briefly explained here.

Mechanisms of Hypersecretion

Excessively high blood concentration of a hormone—or any condition that mimics high hormone levels—is called *hypersecretion*. Specific types of hypersecretion are usually named by placing the prefix *hyper-* in front of the name of the source gland and the suffix *-ism* at the end. For example, hypersecretion of thyroid hormone—no matter what the specific cause—is called *hyperthyroidism*. Hyperthyroidism is not a disease itself but a condition that characterizes several different diseases (e.g., Graves' disease and toxic nodular goiter).

Any of several different mechanisms may be responsible for a particular case of hypersecretion. For example, tumors are often responsible for an abnormal proliferation of endocrine cells and the resulting increase in hormone secretion. Pituitary adenomas, for example, are benign tumors that may cause *hyperpituitarism*. As many as one in five people may have pituitary adenomas, but the majority of tumors are microscopic and asymptomatic. Larger tumors may, however, cause hyperpituitarism with a possible outcome of gigantism, or acromegaly.

Another cause of hypersecretion is a phenomenon called *autoimmunity*. In autoimmunity, the immune system attacks one's own cells instead of foreign cells such as bacteria. Autoimmune damage to endocrine cells is thought by some endocrinologists to produce the hyperthyroidism that characterizes Graves' disease.

Another possible cause of hypersecretion of a hormone is a failure of the feedback mechanisms that regulate secretion of a particular hormone. For example, a condition called *primary hyperparathyroidism* is characterized by a failure of the parathyroid gland to adjust its output to compensate for changes in blood calcium levels. Instead, the parathyroid gland seems to operate independently of the normal feedback loop and thus overproduces parathyroid hormone.

Mechanisms of Hyposecretion

Depressed blood hormone levels—or any condition that mimics low hormone levels—is termed *hyposecretion*. Specific types of hyposecretion are named in a manner similar to that in which hypersecretion disorders are named: by the addition of the *hypo-* prefix and the *-ism* suffix. For example, hyposecretion of thyroid hormone is called *hypothyroidism*.

Various different mechanisms have been shown to cause hyposecretion of hormones. For example, although most tumors cause oversecretion of a hormone, they may instead cause a gland to undersecrete its hormone(s). Tissue death, perhaps caused by a blockage or other failure of the blood supply, can also cause a gland to reduce its hormonal output. Hypopituitarism (hyposecretion by the anterior pituitary) can occur this way. Still another way in which a gland may reduce its secretion below normal levels is through abnormal opera-

tion of regulatory feedback loops. An example of this is in the case of hyposecretion of testosterone and gonadotropic hormones in males who abuse anabolic steroids. Men who take testosterone steroids increase their blood concentration of this hormone above set point levels. The body responds to this overabundance by reducing its own output of testosterone by reducing its output of gonadotropins. This may lead to sterility and other complications.

Abnormalities of immune function may also cause hyposecretion. For example, an autoimmune attack on glandular tissue sometimes has the effect of reducing hormone output. Some endocrinologists theorize that autoimmune destruction of pancreatic islet cells, perhaps in combination with viral and genetic mechanisms, is a culprit in many cases of type I (insulin-dependent) diabetes mellitus. In other cases, the gland itself is unaffected, but antibodies attack the hormone after it has been released into the blood stream. Although normal amounts of hormone may be secreted, this type of condition is still called hyposecretion because the result is a lower than normal blood concentration of hormone. Some cases of hypothyroidism are thought to result from the effects of antithyroid hormone antibodies.

Many types of hyposecretion disorders have recently been shown to be caused by insensitivity of the target cells rather than from actual hyposecretion. A few major types of abnormal responses in target cells are:

- An abnormal decrease in the number of hormone receptors
- Abnormal function of hormone receptors, resulting in failure to bind to hormones properly
- Antibodies bind to hormone receptors, thus blocking binding of hormone molecules
- Abnormal metabolic response to the hormone-receptor complex by the target cell
- Failure of the target cell to produce enough second messenger molecules

Several types of diabetes mellitus, including Type II (noninsulin dependent), are thought to be caused by target cell abnormalities that render the cells insensitive to insulin.

CASE STUDY
HYPERTHYROIDISM: GRAVES' DISEASE

Ms. R. is a 30-year-old woman who has been diagnosed as having Graves' disease. Her symptoms began 2 months ago and have been increasing in intensity since then. She has lost 25 pounds, despite an increased appetite. She has noticed that her heart seems to race and pound, even when she is at rest. She has been very irritable and has suffered from bouts of diarrhea. She states that she always feels too warm, even when others say the room is cold. The collars on her clothing feel too tight, and she has begun to experience blurred vision. Although she has always had a histoy of regular menses, she has not had a period in 2 months.

Ms. R is a thin, pale, anxious women who moves restlessly around the room. Her eyes have a bulging, staring appearance, and a mass can be seen on the anterior of her neck. Her skin is smooth, warm, and moist. She is sweating profusely, although the room temperature is 66° F. Her hair is very fine and soft. She weighs 46 kg on admission, a loss of 12 kg from her normal weight. Her vital signs are as follows: oral temperature, 99° F; heart rate, 120 at rest; respirations, 20; blood pressure, 110/50. Laboratory results are as follows: triiodothyronine (T_3), 160 ng/dl; thyroxine (T_4), 20 ug/dl; cholesterol, 10 mg/dl. She undergoes a radioactive iodine uptake test with 40% uptake in 6 hours.

1. Which of the following is the most likely cause of Ms. R's increased thyroid function?
   a. Hyperplasia of the thyroid
   b. Anterior pituitary tumor
   c. Thyroid carcinoma
   d. Autoimmune response

2. Which of the following best explains the physiologic mechanism(s) in Ms. R.'s oversection of thyroid hormones?
   a. Negative feedback loop involving the anterior pituitary, thyroid, and hypothalamus
   b. Activation of the thyroid gland by excessive circulating iodine
   c. Positive feedback loop involving the thyroid and parathyroid glands
   d. Abnormal stimulation of the thyroid gland by the adrenal glands

3. Ms. R.'s collars and neck mass suggest an enlargement of her thyroid gland, known as a goiter. In which of the following thyroid states would goiter not be found?
   a. Hypothyroidism
   b. Euthyroidism
   c. Hyperthyroidism
   d. None of the above

CHAPTER SUMMARY

INTRODUCTION

A. The endocrine and nervous systems function to achieve and maintain homeostasis

B. When the two systems work together, referred to as neuroendocrine system, they perform the same general functions: communication, integration, and control

C. In the endocrine system, secreting cells send hormone molecules via the blood to specific target cells contained in target tissues or target organs

D. Hormones—carried to almost every point in the body; can regulate most cells; effects work more slowly and last longer than those of neurotransmitters

E. Endocrine glands are "ductless glands"; many are made of glandular epithelium whose cells manufacture and secrete hormones; a few endocrine glands are made of neurosecretory tissue

F. Glands of the endocrine system are widely scattered throughout the body

HORMONES

A. Classification of hormones
   1. Classification by general function
      a. Tropic hormones—hormones that target other endocrine glands and stimulate their growth and secretion
      b. Sex hormones—hormones that target reproductive tissues
      c. Anabolic hormones—hormones that stimulate anabolism in target cells
   2. Classification by chemical structure
      a. Steroid hormones
      b. Nonsteroid hormones
   3. Steroid hormones
      a. Synthesized from cholesterol
      b. Lipid soluble and can easily pass through the phospholipid plasma membrane of target cells
      c. Examples of steroid hormones: cortisol, aldosterone, estrogen, progesterone, and testosterone
   4. Nonsteroid hormones
      a. Synthesized primarily from amino acids
      b. Protein hormones—long, folded chains of amino acids; e.g., insulin and parathyroid hormone
      c. Glycoprotein hormones—protein hormones with carbohydrate groups attached to the amino acid chain
      d. Peptide hormones—smaller than protein hormones; short chain of amino acids; e.g., oxytocin and antidiuretic hormone (ADH)
      e. Amino acid derivative hormones—each is derived from a single amino acid molecule
         (1) Amine hormones—synthesized by modifying a single molecule of tyrosine; produced by neurosecretory cells and by neurons; e.g., epinephrine and norepinephrine
         (2) Amino acid derivatives produced by the thyroid gland; synthesized by adding iodine to tyrosine

B. How hormones work
   1. General principles of hormone action
      a. Hormones signal a cell by binding to the target cell's specific receptors in a "lock-and-key" mechanism
      b. Different hormone-receptor interactions produce different regulatory changes within the target cell through chemical reactions
      c. Combined hormone actions:
         (1) Synergism—combinations of hormones acting together have a greater effect on a target cell than the sum of the effects that each would have if acting alone
         (2) Permissiveness—when a small amount of one hormone allows a second one to have its full effects on a target cell
         (3) Antagonism—one hormone produces the opposite effects of another hormone; used to "fine tune" the activity of target cells with great accuracy
      d. Endocrine glands produce more hormone molecules than actually are needed; the unused hormones are quickly secreted by the kidneys or broken down by metabolic processes
   2. Mechanism of steroid hormone action
      a. Steroid hormones are lipid soluble, and their receptors are normally found in the target cell's cytoplasm
      b. Once a steroid hormone molecule has diffused into the target cell, it binds to a receptor molecule to form a hormone-receptor complex
      c. Mobile-receptor hypothesis—the hormone-receptor complex migrates into the nucleus, and it activates a certain gene sequence to begin transcription of mRNA; newly formed mRNA molecules move into the cytoplasm, associate with ribosomes, and begin synthesizing protein molecules
      d. Steroid hormones regulate cells by regulating production of certain critical proteins
      e. The amount of steroid hormone present determines the magnitude of a target cell's response
      f. Since transcription and protein synthesis take time, responses to steroid hormones are often slow
   3. Mechanisms of nonsteroid hormone action
      a. The second messenger mechanism (also known as the fixed-membrane-receptor hypothesis)
         (1) A nonsteroid hormone molecule acts as a "first messenger" and delivers its chemical message to fixed receptors in the target cell's plasma membrane
         (2) The "message" is then passed into the cell where a "second messenger" triggers the appropriate cellular changes
         (3) Second messenger mechanism—produces target cell effects that differ from steroid hormone effects in several important ways:
            (a) The effects of the hormone are amplified by the cascade of reactions
            (b) Second messenger mechanism operates much more quickly than the steroid mechanism
      b. The nuclear receptor mechanism—small iodinated amino acids (T_4 and T_3) enter the target cell and bind to receptors associated with a DNA molecule in the nucleus; this binding triggers transcription of mRNA and synthesis of new enzymes

C. Regulation of hormone secretion
   1. Control of hormonal secretion is usually part of a negative feedback loop and are called endocrine reflexes
   2. Simplest mechanism—when an endocrine gland is sensitive to the physiological changes produced by its target cells
   3. Endocrine gland secretion may also be regulated by a hormone produced by another gland
   4. Endocrine gland secretions may be influenced by nervous system input; this fact emphasizes the close functional relationship between the two systems

PROSTAGLANDINS

A. Unique group of lipid molecules (20-carbon fatty acid with 5-carbon ring) that serves important and widespread integrative functions in the body but do not meet the usual definition of a hormone

B. Called tissue hormones, since the secretion is produced in a tissue and diffuses only a short distance to other cells within the same tissue; PGs tend to integrate activities of neighboring cells

C. Many structural classes of prostaglandins have been isolated and identified
1. Prostaglandin A (PGA)—intraarterial infusion resulting in an immediate fall in blood pressure accompanied by an increase in regional blood flow to several areas
2. Prostaglandin E (PGE)—vascular effects: regulation of red blood cell deformability and platelet aggregation; gastrointestinal effects: regulates hydrochloric acid secretion
3. Prostaglandin F (PGF)—especially important in reproductive system, causing uterine contractions; also affects intestinal motility and is required for normal peristalsis

D. Tissues known to secrete PGs—kidneys, lungs, iris, brain, thymus

E. PGs have diverse physiological effects

PITUITARY GLAND

A. Structure of the pituitary gland
1. Also known as hypophysis and called the "master gland"
2. Size: 1.2 to 1.5 cm across; weight: 0.5 g
3. Located on the ventral surface of the brain within the skull
4. Infundibulum—stemlike stalk that connects pituitary to the hypothalamus
5. Made up of two separate glands, the adenohypophysis (anterior pituitary gland) and the neurohypophysis (posterior pituitary gland)

B. Adenohypophysis (anterior pituitary)
1. Divided into two parts
   a. Pars anterior—forms the major portion of the adenohypophysis
   b. Pars intermedia
2. Tissue is composed of irregular clumps of secretory cells supported by fine connective tissue fibers and surrounded by a rich vascular network
3. Three types of cells can be identified
   a. Chromophobes—make up approximately one half of all cells in adenohypophysis
   b. Acidophils—make up approximately 40% of all cells in adenohypophysis; secrete GH and PRL
   c. Basophils—form about 10% of adenohypophysis; secrete TSH, ACTH, FSH, LH, and MSH
4. Growth hormone (GH)
   a. Also known as somatotropin (STH)
   b. Promotes growth of bone, muscle, and other tissues by accelerating amino acid transport into the cells
   c. Stimulates fat metabolism by mobilizing lipids from storage in adipose cells and speeding up catabolism of the lipids after they have entered another cell
   d. GH tends to shift cell chemistry away from glucose catabolism and toward lipid catabolism as an energy source; this leads to increased blood glucose levels
   e. GH and insulin function as antagonists and are vital to maintaining homeostasis of blood glucose levels
5. Prolactin (PRL)
   a. Produced by acidophils in the pars anterior
   b. Also known as lactogenic hormone
   c. During pregnancy, PRL promotes development of the breasts, anticipating milk secretion; after the baby is born, PRL stimulates the mother's mammary glands to produce milk
   d. PRL plays a supportive role (with luteinizing hormone) in maintaining the corpus luteum of the ovary during the final phase of the menstrual cycle; sometimes called luteotropic hormone (LTH)
6. Tropic hormones—hormones that have a stimulating effect on other endocrine glands; four principal tropic hormones are produced and secreted by the basophils of the pars anterior
   a. Thyroid-stimulating hormone (TSH), or thyrotropin—promotes and maintains the growth and development of the thyroid; also causes the thyroid to secrete its hormones
   b. Adrenocorticotropic hormone (ACTH), or adrenocorticotropin—promotes and maintains normal growth and development of the cortex of the adrenal gland; also stimulates the adrenal cortex to secrete some of its hormones
   c. Follicle-stimulating hormone (FSH)—in the female, stimulates primary graafian follicles to grow toward maturity; also stimulates the follicle cells to secrete estrogens; in the male, FSH stimulates the development of the seminiferous tubules of the testes and maintains spermatogenesis
   d. Luteinizing hormone (LH)—in the female, stimulates the formation and activity of the corpus luteum of the ovary; corpus luteum secretes progesterone and estrogens when stimulated by LH; LH also supports FSH in stimulating maturation of follicles; in the male, LH stimulates interstitial cells in the testes to develop and secrete testosterone; FSH and LH are called gonadotropins because they stimulate the growth and maintenance of the gonads
7. Melanocyte-stimulating hormone (MSH)
   a. Secreted from basophils in pars intermedia and pars anterior
   b. Works with other hormones to modulate the pigmentation of normal skin
   c. Somehow helps maintain the adrenal gland's sensitivity to ACTH
8. Control of secretion in the adenohypophysis
   a. Hypothalamus secretes releasing hormones into the blood, which are then carried to the hypophyseal portal system
   b. Hypophyseal portal system carries blood from the hypothalamus directly to the adenohypophysis where the target cells of the releasing hormones are located
   c. Releasing hormones influence the secretion of hormones by acidophils and basophils
   d. Through negative feedback, the hypothalamus adjusts the secretions of the adenohypophysis, which then adjusts the secretions of the target glands that in turn adjust the activity of their target tissues
   e. In severe pain or intense emotions, the hypothalamus translates nerve impulses into hormone secretions by endocrine glands, basically creating a mind-body link

C. Neurohypophysis (posterior pituitary)
1. Serves as storage and release site for antidiuretic hormone (ADH) and oxytocin (OT), which are synthesized in the hypothalamus
2. Release of ADH and OT into the blood is controlled by nervous stimulation
3. Antidiuretic hormone (ADH)
   a. Prevents the formation of a large volume of urine, thereby helping the body conserve water
   b. Causes a portion of each tubule in the kidney to reabsorb water from the urine it has formed
   c. With dehydration, osmoreceptors detect the decreased osmotic pressure of the blood and trigger the release of ADH
4. Oxytocin (OT)—has two actions

a. Causes milk ejection from the lactating breast; regulated by positive feedback mechanism; PRL cooperates with oxytocin
b. Stimulates contraction of uterine muscles that occurs during childbirth; regulated by positive feedback mechanism

PINEAL GLAND

A. Tiny, pine cone–shaped structure located on the dorsal aspect of the brain's diencephalon
B. Member of the nervous system, since it receives visual stimuli, and also a member of the endocrine system, since it secretes hormones
C. Pineal gland supports the body's biological clock
D. Principal pineal secretion is melatonin

THYROID GLAND

A. Structure of the thyroid gland
　1. Made up of two large lateral lobes and a narrow connecting isthmus
　2. A thin wormlike projection of thyroid tissue often extends upward from the isthmus
　3. Weight of the thyroid in an adult is approximately 30 g
　4. Located in the neck, on the anterior and lateral surfaces of the trachea, just below the larynx
　5. Composed of follicles
　　a. Small hollow spheres
　　b. Filled with thyroid colloid that contains thyroglobulins
B. Thyroid hormone
　1. Actually two different hormones
　　a. Tetraiodothyronine (T_4), or thyroxine—contains four iodine atoms; approximately 20 times more abundant than T_3; major importance is as a precursor to T_3
　　b. Triiodothyronine (T_3)—contains three iodine atoms; considered to be the principal thyroid hormone; T_3 binds efficiently to nuclear receptors in target cells
　2. Thyroid gland stores considerable amounts of a preliminary form of its hormones *prior to* secreting them
　3. Before being stored in the colloid of follicles, T_3 and T_4 are attached to globulin molecules, forming thyroglobin complexes
　4. On release, T_3 and T_4 detach from globulin and enter the bloodstream
　5. Once in the blood, T_3 and T_4 attach to plasma globulins and travel as a hormone-globulin complex
　6. T_3 and T_4 detach from plasma globulin as they near the target cells
　7. Thyroid hormone—helps regulate the metabolic rate of all cells and cell growth and tissue differentiation; it is said to have a "general" target
C. Calcitonin
　1. Produced by thyroid gland in the parafollicular cells
　2. Influences the processing of calcium by bone cells by decreasing blood calcium levels and promoting conservation of hard bone matrix
　3. Parathyroid hormone acts as antagonist to calcitonin to maintain calcium homeostasis

PARATHYROID GLANDS

A. Structure of the parathyroid glands
　1. Four or five parathyroid glands imbedded in the posterior surface of the thyroid's lateral lobes
　2. Tiny, rounded bodies within thyroid tissue formed by compact, irregular rows of cells
B. Parathyroid hormone (PTH)
　1. PTH is an antagonist to calcitonin and acts to maintain calcium homeostasis
　2. PTH acts on bone, kidney, and intestinal cells

a. Causes more bone to be dissolved, yielding calcium and phosphate, which enters the bloodstream
b. Causes phosphate to be secreted by the kidney cells into the urine to be excreted
c. Causes increased intestinal absorption of calcium by activating vitamin D

ADRENAL GLANDS

A. Structure of the adrenal glands
　1. Located on top of the kidneys, fitting like caps
　2. Made up of two portions (Table 15-6)
　　a. Adrenal cortex—composed of endocrine tissue
　　b. Adrenal medulla—composed of neurosecretory tissue
B. Adrenal cortex—all cortical hormones are steroids and known as corticosteroids
　1. Composed of three distinct layers of secreting cells
　　a. Zona glomerulosa—outermost layer, directly under the outer connective tissue capsule of the adrenal gland; secretes mineralocorticoids
　　b. Zona fasciculata—middle layer; secretes glucocorticoids
　　c. Zona reticularis—inner layer; secretes small amounts of glucocorticoids and gonadocorticoids
　2. Mineralocorticoids
　　a. Have an important role in the regulatory process of mineral salts in the body
　　b. Aldosterone
　　　(1) Only physiologically important mineralocorticoid in the human; primary function is maintenance of sodium homeostasis in the blood by increasing sodium reabsorption in the kidneys
　　　(2) Aldosterone also increases water retention and promotes the loss of potassium and hydrogen ions
　　　(3) Aldosterone secretion is controlled by the renin-angiotensin mechanism and by blood potassium concentration
　3. Glucocorticoids
　　a. Main glucocorticoids secreted by the zona fasciculata are cortisol, cortisone, and corticosterone, with cortisol the only one secreted in significant quantities
　　b. Affect every cell in the body
　　c. Are protein-mobilizing, gluconeogenic, and hyperglycemic
　　d. Tend to cause a shift from carbohydrate catabolism to lipid catabolism as an energy source
　　e. Essential for maintaining normal blood pressure by aiding norepinephrine and epinephrine to have their full effect, causing vasoconstriction
　　f. High blood concentration causes eosinopenia and marked atrophy of lymphatic tissues
　　g. Act with epinephrine to bring about normal recovery from injury produced by inflammatory agents
　　h. Secretion increases in response to stress
　　i. Except during stress response, secretion is mainly controlled by a negative feedback mechanism involving ACTH from the adenohypophysis
　4. Gonadocorticoids—sex hormones that are released from the adrenal cortex; both androgens and estrogens are secreted
C. Adrenal medulla
　1. Neurosecretory tissue—tissue composed of neurons specialized to secrete their products into the blood
　2. Adrenal medulla secretes two important hormones, epinephrine and norepinephrine; they are part of the class of nonsteroid hormones called catecholamines
　3. Both hormones bind to the receptors of sympathetic effectors to prolong and enhance the effects of sympathetic stimulation by the ANS

PANCREATIC ISLETS

A. Structure of the pancreatic islets
1. Elongated gland, weighing approximately 100 g; its head lies in the duodenum, extends horizontally behind the stomach, and, then, touches the spleen
2. Composed of endocrine and exocrine tissues
   a. Pancreatic islets (islets of Langerhans)—endocrine portion
   b. Acini—exocrine portion—secretes a serous fluid containing digestive enzymes into ducts leading to the small intestine
3. Pancreatic islets—each islet contains four primary types of endocrine glands joined by gap junctions
   a. Alpha cells (A cells)—secrete glucagon
   b. Beta cells (B cells)—secrete insulin; account for up to 75% of all pancreatic islet cells
   c. Delta cells (D cells)—secrete somatostatin
   d. Pancreatic polypeptide cells (F, or PP, cells)—secrete pancreatic polypeptides
B. Pancreatic hormones (review Table 15-7)—work as a team to maintain a homeostasis of food molecules
1. Glucagon—produced by alpha cells; tends to increase blood glucose levels; stimulates gluconeogenesis in liver cells
2. Insulin—produced by beta cells; lowers blood concentration of glucose, amino acids, and fatty acids and promotes their metabolism by tissue cells
3. Somatostatin—produced by delta cells; primary role is regulating the other endocrine cells of the pancreatic islets
4. Pancreatic polypeptide—produced by F (PP) cells; influences the digestion and distribution of food molecules to some degree

GONADS

A. Testes
1. Paired organs within the scrotum in the male
2. Composed of seminiferous tubules and a scattering of interstitial cells
3. Testosterone is produced by the interstitial cells and responsible for the growth and maintenance of male sexual characteristics
4. Testosterone secretion is mainly regulated by gonadotropin levels in the blood
B. Ovaries
1. Primary sex organs in the female
2. Set of paired glands in the pelvis that produce several types of sex hormones
   a. Estrogens—steroid hormones secreted by ovarian follicles; promote development and maintenance of female sexual characteristics
   b. Progesterone—secreted by corpus luteum; maintains the lining of the uterus necessary for successful pregnancy
   c. Ovarian hormone secretion depends on the changing levels of FSH and LH from the adenohypophysis

PLACENTA

A. Tissues that form on the lining of the uterus as a connection between the circulatory systems of the mother and developing child
B. Serves as a temporary endocrine gland that produces human chorionic gonadotropin

THYMUS

A. Gland located in the mediastinum just beneath the sternum
B. Thymus is large in children, begins to atrophy at puberty, and, by old age, the gland is a vestige of fat and fibrous tissue
C. Considered to be primarily a lymphatic organ, but the hormone thymosin has been isolated from thymus tissue
D. Thymosin—stimulates development of T cells

GASTRIC AND INTESTINAL MUCOSA

A. The mucous lining of the GI tract contains cells that produce both endocrine and exocrine secretions
B. GI hormones such as gastrin, secretin, and cholecystokinin-pancreozymin (CCK) play regulatory roles in coordinating the secretory and motor activities involved in the digestive process

HEART

A. The heart has a secondary endocrine role
B. Hormone-producing cells produce atrial natriuretic hormone (ANH)
C. ANH's primary effect is to oppose increases in blood volume or blood pressure; also an antagonist to ADH and aldosterone

THE ENDOCRINE SYSTEM AND THE WHOLE BODY

A. Nearly every process in the human organism is kept in balance by the intricate interaction of different nervous and endocrine regulatory chemicals
B. The endocrine system operates with the nervous system to finely adjust the many processes they regulate

CYCLE OF LIFE: ENDOCRINE SYSTEM

A. Endocrine regulation begins in the womb
B. Many active hormones are active from birth
1. Evidence that a hormonal signal from fetus to mother signal the onset of labor
B. Hormones related to reproduction begin at puberty
C. Secretion of male reproductive hormones—continuous production from puberty, slight decline in late adulthood
D. Secretion of female reproductive hormones declines suddenly and completely in middle adulthood

THE BIG PICTURE: THE ENDOCRINE SYSTEM AND THE WHOLE BODY

A. Neuroendocrine system adjusts nutrient supply
B. Calcitonin and parathyroid hormone balance calcium ion use
C. The nervous system and hormones regulate reproduction

MECHANISMS OF DISEASE: ENDOCRINE DISORDERS

A. Endocrine disorders result from hypersecretion or hyposecretion
B. Hypersecretion—high blood concentration of a hormone; example—hyperthyroidism is the hypersecretion of the thyroid hormone, not a disease itself but a characteristic of diseases such as Graves' disease and toxic nodular goiter
  1. Causes of hypersecretion
    a. Tumors cause abnormal proliferation of endocrine cells and result in increased hormone secretion
    b. Hyperpituitarism—caused by pituitary adenomas
    c. Autoimmunity—the immune system attacks the body's own cells instead of foreign cells, causing hypersecretion
    d. Failure of the feedback mechanisms that regulate the secretion of a hormone
C. Hyposecretion—depressed blood hormone levels; example is hypothyroidism (hyposecretion of thyroid hormone)
  1. Tumors and tissue death can cause hyposecretion
  2. Operation of regulatory feedback loops may reduce a gland's secretion
    a. Example—abuse of anabolic steroids increases levels of this hormone, and the body responds by reducing its own output
    b. Immune function abnormalities may also cause hyposecretion
    c. Many hyposecretion disorders are caused by insensitivity of target cells

REVIEW QUESTIONS

1. Compare the activity of the nervous and endocrine systems in communication, integration, and control of body functions. In what ways are the two systems alike? How do they differ?
2. Define the terms *hormone* and *target organ*.
3. Describe the characteristic chemical group found at the core of each steroid hormone.
4. Identify the major categories of nonsteroid hormones.
5. Define the terms *synergism*, *permissiveness*, and *antagonism* in relation to the endocrine system.
6. Compare and contrast the mechanisms of action of steroid and nonsteroid hormones.
7. Identify the sequence of events involved in a second messenger mechanism.
8. What is the function of calmodulin?
9. How does the size of the thyroid hormones relate to their receptor mechanism?
10. Describe a typical negative feedback loop.
11. List the classes of prostaglandins. Identify the functions of three of these classes.
12. Name the two subdivisions of the adenohypophysis.
13. Discuss and identify, by staining tendency and relative percentages, the cell types present in the anterior pituitary gland.
14. Discuss the functions of growth hormone.
15. What effect does growth hormone have on blood glucose concentration? Fat mobilization and catabolism? Protein anabolism?
16. Discuss the effect of prolactin on breast development and milk secretion. Why is the name *luteotropic hormone (LTH)* sometimes used instead of prolactin?
17. List the four tropic hormones secreted by the basophils of the anterior pituitary gland. Which of the tropic hormones are also called gonadotropins?
18. How does stress affect the functioning of the hypothalamus?
19. How does antidiuretic hormone act to alter urine volume?
20. Describe the positive feedback associated with oxytocin.
21. Discuss the synthesis and storage of thyroxine and triiodothyronine. How are they transported in the blood?
22. Discuss the functions of parathyroid hormone.
23. List the hormones produced by each "zone" of the adrenal cortex and describe the actions of these hormones.
24. Discuss the normal function of hormones produced by the adrenal medulla.
25. Identify the hormones produced by each of the cell types in the pancreatic islets.
26. Compare the physiological effects of insulin and glucagon.
27. Identify the "pregnancy promoting" hormone.
28. Where is human chorionic gonadotropin produced? What does it do?
29. Describe the role of atrial natriuretic hormone.
30. Compare and contrast up and down regulation.
31. Identify the conditions resulting from both hypersecretion and hyposecretion of growth hormone during growth years.
32. Explain how a hypophysectomy provides clinical evidence for feedback control of hormone secretion by the anterior pituitary.
33. What events result in the production of a simple goiter?
34. Describe the face of a patient with Cushing's syndrome.
35. How does exercise affect diabetes mellitus?
36. Explain what happens with steroid abuse.

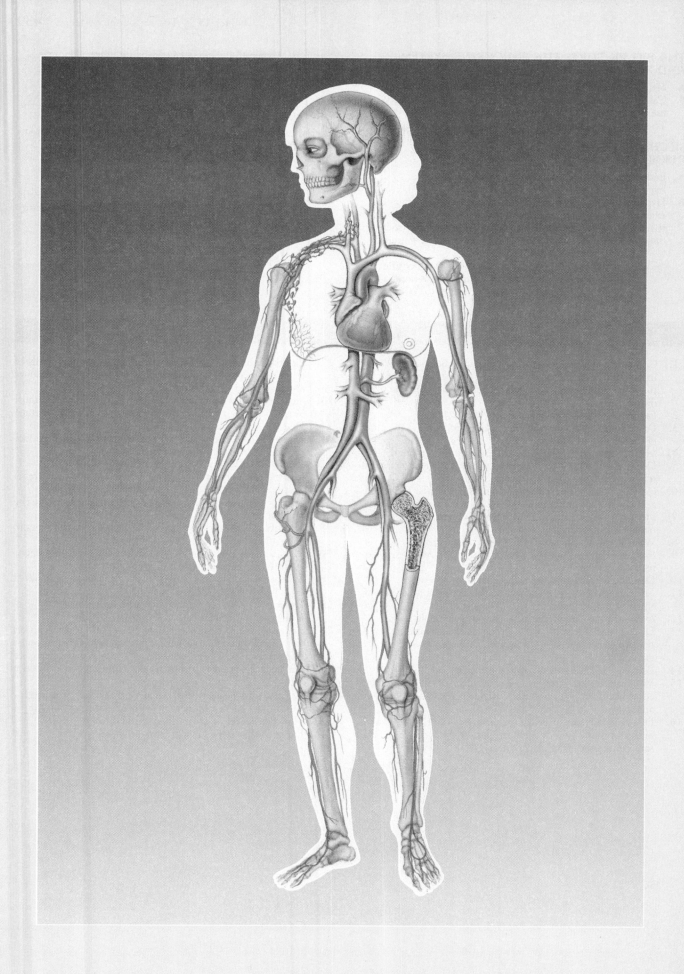

UNIT FOUR
Transportation and Defense

The chapters in Unit 4 deal with transportation, how the body defends itself, and stress. Blood (Chapter 16) is discussed as a complex fluid tissue that serves to transport respiratory gases and key nutrients to cells and carry away wastes. It fills the cardiovascular system (Chapters 17 and 18) and is moved through a closed pathway, or circuit, of vessels by the pumping action of the heart.

The elements of the lymphatic system (Chapter 19) provide an open pathway for return of fluid, proteins, and fats absorbed from the intestine to the general circulation. This system is also involved in immunity or resistance to disease and in the removal and destruction of dead red blood cells. The immune system is more fully discussed in Chapter 20. Elements of this system provide a multilayered defense mechanism, involving both phagocytic cells and specialized proteins called antibodies. Stress—and the body's frequently maladaptive response to it—is discussed in Chapter 21.

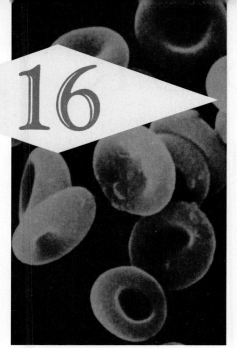

16 ▷ Blood

SEM *of red blood cells*

OBJECTIVES

After you have completed this chapter, you should be able to:

1. Describe the generalized functions of blood and explain how the packed cell volume is determined.
2. List the types of blood cells that are normally found in circulating blood and identify the most important function of each.
3. Discuss the normal appearance, size, shape, and number of erythrocytes in circulating blood.
4. Describe the structure and function of hemoglobin.
5. Describe the process of red blood cell formation (erythropoiesis) and destruction.
6. Discuss the generalized function, classification, normal appearance, size, shape, and number of leukocytes in circulating blood.
7. Compare and contrast granulocytes and agranulocytes.
8. Discuss the stages in development of granular and agranular leukocytes.
9. Discuss the appearance, size, shape, number, and function of platelets in circulating blood.
10. Discuss the important physical properties of platelets and their relationship to hemostasis.
11. Describe ABO and Rh blood grouping.
12. List the major plasma components and their generalized functions.
13. Explain the steps involved in blood coagulation and the factors that oppose and hasten clotting.

F our chapters in this unit deal with **transportation,** one of the body's vital functions. Homeostasis of the internal environment—and therefore survival itself—depends on continual transportation to and from body cells. This chapter discusses the major transportation fluid, blood. Chapters 17 and 18 consider the major transportation system, the cardiovascular system, and Chapter 19 has for its subject a supplementary transportation system, the lymphatic system.

COMPOSITION OF BLOOD

Blood is much more than the simple liquid it seems to be. Not only does it consist of a fluid but also of cells and specialized cell fragments called *platelets.* The fluid portion of blood, that is, the **plasma,** is one of the three major body fluids (interstitial and intracellular fluids are the other two). The term **formed elements** is used to designate the various kinds of blood cells and cell fragments suspended in plasma.

Blood is a complex transport medium that performs vital pickup and delivery services for the body. It picks up food and oxygen from the digestive and respiratory systems and delivers them to cells while picking up wastes from cells for delivery to excretory organs. Blood also transports hormones, enzymes, buffers, and various other biochemical substances that serve important functions. Blood serves another critical function. It is the keystone of the body's heat-regulating mechanism. Certain physical properties of blood make it especially effective in this role. Its high specific heat and conductivity enable this unique fluid to absorb large quantities of heat without an appreciable increase in its own temperature and to transfer this absorbed heat from the core of the body to its surface, where it can be more readily dissipated (see Figure 5-10).

Blood Volume

How much blood does an adult body contain? The answer, for a young adult man weighing 70 kilograms (154 pounds), is about 5 liters. However, the total blood volume varies markedly in different individuals. Age, body type, sex, and the method of measurement are all determinants.

Blood volume can be determined using either **direct** or **indirect methods** of measurement. *Direct* measurement of total blood volume can be accomplished only by complete removal of all blood from an experimental animal. In humans, *indirect* methods of measurement that employ "tagging" of red blood cells or plasma components with radioisotopes are used. The principle is simply to introduce a known amount of radioisotope into the circulation, allow the material to distribute itself uniformly in the blood, and then analyze its concentration in a representative blood sample. Accurate measurement is important in replacing blood lost because of hemorrhage or in treating other conditions, such as shock, that are influenced by abnormal blood volumes.

One of the chief variables influencing normal blood volume is the amount of body fat. Blood volume per kilogram of body weight varies inversely with the amount of

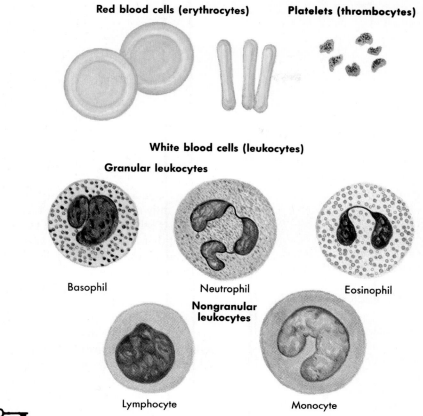

Red blood cells (erythrocytes)   **Platelets (thrombocytes)**

White blood cells (leukocytes)

Granular leukocytes

Basophil   Neutrophil   Eosinophil

Nongranular leukocytes

Lymphocyte   Monocyte

FIGURE 16-1   The formed elements of blood. Red blood cells (erythrocytes), white blood cells (leukocytes), and platelets (thrombocytes) constitute the formed elements of blood.

excess body fat. This means that the less fat there is in your body, the more blood you have per kilogram of your body weight.

FORMED ELEMENTS OF BLOOD

The formed elements of blood are illustrated in Figure 16-1. They are:

♦ Red blood cells (RBCs) (erythrocytes)
♦ White blood cells (WBCs) (leukocytes)
♦ Platelets (thrombocytes)

Plasma when separated from "whole blood" is a clear straw-colored fluid that consists of about 90% water and 10% solutes.

If a tube of whole blood, that is, plasma and formed elements, is allowed to stand or is spun in a centrifuge, separation will occur. The term **packed cell volume (PCV)**, or **hematocrit,** is used to describe the volume percent of red blood cells in whole blood. In Figure 16-2, A, a sample of normal whole blood has been separated by centrifuging so that the formed elements are forced to the bottom. The percentage of plasma is about 55% of the total sample, whereas the packed cell volume, or hematocrit, is

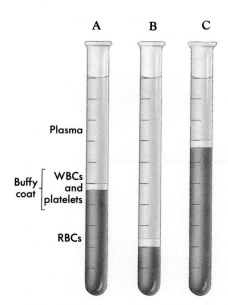

A   B   C

Plasma

Buffy coat { WBCs and platelets

RBCs

FIGURE 16-2 Hematocrit tubes showing normal blood, anemia, and polycythemia. Note the buffy coat located between the packed RBCs and the plasma. **A,** A normal percent of RBCs. **B,** Anemia (a low percent of RBCs). **C,** Polycythemia (a high percent of RBCs).

45%. A hematocrit of 45% means that in every 100 ml of whole blood there are 45 ml of red blood cells and 55 ml of fluid plasma.

Normally the average hematocrit for a man is about 45% (±7%, normal range) and for a woman about 42% (±5%). Conditions that result in decreased red blood cell numbers (Figure 16-2, B) are called **anemias** and are characterized by a reduced hematocrit value. Healthy individuals who live and work in high altitudes often have elevated red blood cell numbers and hematocrit values (Figure 16-2, C). The condition is called **physiological polycythemia** (from the Greek *polys,* "many," *kytos,* "cell," and *haima,* "blood").

White blood cells, or leukocytes, and platelets make up less than 1% of blood volume. Note in Figure 16-2 that a thin white layer of leukocytes and platelets, called the **buffy coat,** is present at the interface between the packed red cells and plasma.

QUICK CHECK

✔ *1. Name the fluid portion of whole blood.*
✔ *2. What constitutes the formed elements of blood?*
✔ *3. What factors influence blood volume?*
✔ *4. Identify the component percentages of the normal hematocrit.*

Red Blood Cells (Erythrocytes)

A normal, mature red blood cell has no nucleus and is only about 7 μm in diameter. More than 1,500 of them could be placed side by side in a 1 cm space. Before the cell reaches maturity and enters the bloodstream from the bone marrow, the nucleus is extruded, with the result that the cell caves in on both sides. As you can see in Figure 16-3, normal, mature red blood cells are shaped like tiny biconcave disks. The mature erythrocyte is also unique in that it does not contain ribosomes, mitochondria, and other organelles typical of most body cells. Instead, the primary component of each RBC is the red pigment, **hemoglobin.** It accounts for over a third of the cell volume and is critically important to its primary function.

The depression on each flat surface of the cell results in a thin center and thicker edges. This unique shape of the red blood cell gives it a very large surface area relative to its volume. The sides of the cell can actually move in and out, permitting it to undergo "deformity" or change in shape without injury as it moves through the very narrow capillary vessels. This ability to change shape is necessary for the survival of red blood cells, which are under almost constant mechanical shearing and bursting strains as they pass through the capillary system. In addition, the degree of cell deformity that is possible influences the speed of blood flow in the microcirculation.

Red blood cells are the most numerous of the formed elements in the blood. In men, red blood cell counts average about 5,500,000 per cubic millimeter (mm^3) of blood, and, in women, 4,800,000 per mm^3.

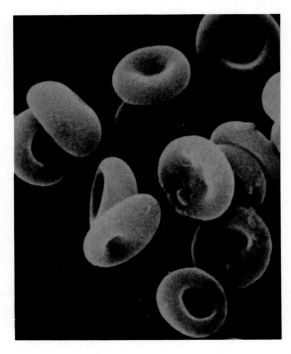

FIGURE 16-3   Erythrocytes. Color-enhanced scanning electron micrograph shows normal erythrocytes.

Function of red blood cells

Red blood cells play a critical role in the transport of both oxygen and carbon dioxide in the body. Chapter 23 will include a detailed discussion of how oxygen is transported from air in the lungs to the body cells and how carbon dioxide moves from body cells to the lungs for removal. Both of these functions are depend on hemoglobin. In addition to hemoglobin, the presence of an enzyme, called **carbonic anhydrase,** in RBCs catalyzes a reaction that joins carbon dioxide and water to form carbonic acid. Dissociation of the acid then generates **bicarbonate ions** (HCO$_3^-$), which diffuse out of the red blood cell and serve to transport CO$_2$ in the blood plasma. Bicarbonate ions also have an important role in maintaining normal blood pH levels (see Chapter 29).

Considered together, the total surface area of all the red blood cells in an adult is enormous. It provides an area larger than a football field for the exchange of respiratory gases between hemoglobin found in circulating erythrocytes and interstitial fluid that bathes the body cells. This is an excellent example of the familiar principle that structure determines function.

Hemoglobin

Packed within each red blood cell are an estimated 200 to 300 million molecules of hemoglobin, which make up about 95% of the dry weight of each cell. Each hemoglobin molecule is composed of four protein chains. Each chain, called a **globin,** is bound to a red pigment, identified in Figure 16-4, as a **heme** molecule. Each heme molecule contains one iron atom. Therefore one hemoglobin molecule contains four iron atoms. This structural fact enables

Hemoglobin

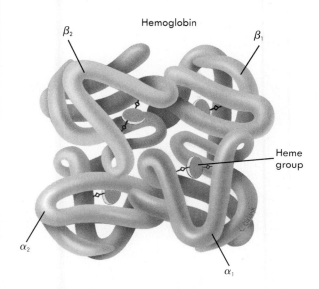

FIGURE 16-4 Hemoglobin. Four protein chains (globins), each with a heme group, form a hemoglobin molecule. Each heme contains one iron atom.

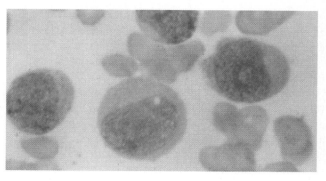

FIGURE 16-5 The hemocytoblast. The hemocytoblast gives rise to the cell lines that produce the formed elements.

one hemoglobin molecule to unite with four oxygen molecules to form oxyhemoglobin (a reversible reaction). Hemoglobin can also combine with carbon dioxide to form carbaminohemoglobin (also reversible). But in this reaction the structure of the globin part of the hemoglobin molecule, rather than of its heme part, makes the combining possible.

A man's blood usually contains more hemoglobin than a woman's. In most normal men, 100 ml of blood contains 14 to 16 g of hemoglobin. The normal hemoglobin content of a woman's blood is a little less—specifically, in the range of 12 to 14 g per 100 ml. An adult who has a hemoglobin content of less than 10 g per 100 ml of blood is diagnosed as having **anemia** (from the Greek *a-*, "not," and *haima,* "blood"). In addition, the term may be used to describe a reduction in the number or volume of functional red blood cells in a given unit of whole blood. Anemias are classified according to the size and hemoglobin content of red blood cells.

Formation of red blood cells

The entire process of red blood cell formation is called **erythropoiesis** (e-RITH-ro-poy-E-sis). In the adult, erythrocytes begin their maturation sequence in the red bone marrow from nucleated cells known as **hemocytoblasts,** or stem cells (Figure 16-5). *Stem cells* are cells that have the ability to maintain a constant population of newly differentiating cells. These stem cells divide by mitosis; some of the daughter cells remain as undifferentiated stem cells, whereas others go through several stages of development to become erythrocytes. Figure 16-6 shows the sequential cell types or stages that can be identified as transformation from the immature forms to the mature red blood cell occurs. The entire maturation process requires about 4 days and proceeds from step to step by gradual transition.

As you can see in Figure 16-6, all blood cells are derived from hemocytoblasts. In red blood cells, differentiation begins with the appearance of **proerythroblasts.** Mitotic divisions then produce **basophilic erythroblasts.** The next maturation division produces **polychromatic erythroblasts,** which produce hemoglobin. These cells subsequently lose their nuclei and become **reticulocytes.** Once released into the circulating blood, reticulocytes lose their delicate reticulum and become mature erythrocytes in about 24 to 36 hours. Note in Figure 16-6 that overall cell size decreases as the maturation sequence progresses.

FYI　**Sickle Cell Anemia**

Sickle cell anemia is a severe, sometimes fatal, hereditary disease that is characterized by an abnormal type of hemoglobin. A person who inherits only one defective gene develops a form of the disease called *sickle cell trait.* In sickle cell trait, red blood cells contain a small proportion of a type of hemoglobin that is less soluble than normal. It forms solid crystals when the blood oxygen level is low, causing distortion and fragility of the red blood cell. If two defective genes are inherited (one from each parent), then more of the defective hemoglobin is produced, and the distortion of red blood cells becomes severe. The illustration shows the characteristic shape of a red cell containing the abnormal hemoglobin.

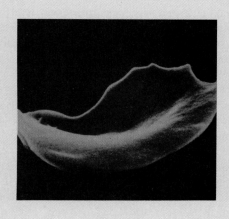

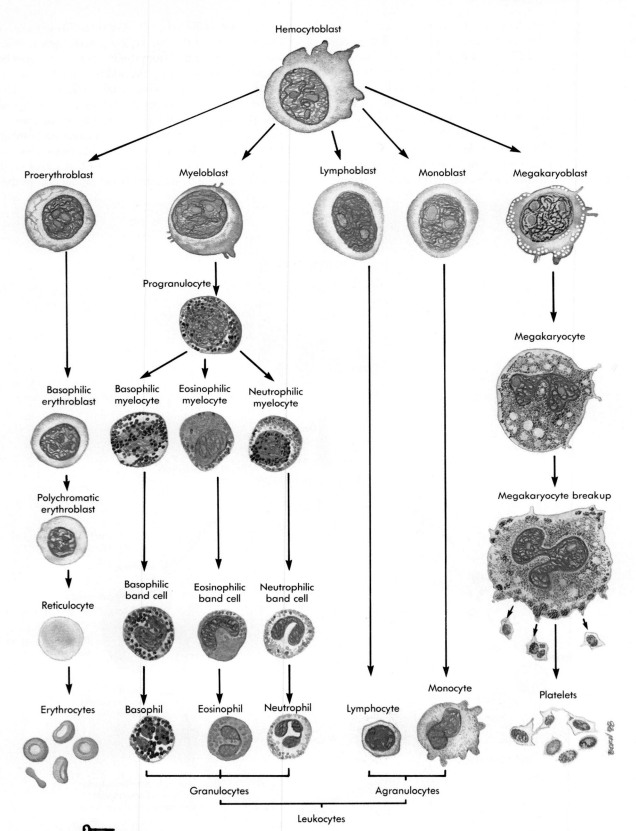

Hemocytoblast

Proerythroblast   Myeloblast   Lymphoblast   Monoblast   Megakaryoblast

Progranulocyte

Megakaryocyte

Basophilic
erythroblast

Basophilic
myelocyte

Eosinophilic
myelocyte

Neutrophilic
myelocyte

Polychromatic
erythroblast

Megakaryocyte breakup

Reticulocyte

Basophilic
band cell

Eosinophilic
band cell

Neutrophilic
band cell

Monocyte   Platelets

Erythrocytes   Basophil   Eosinophil   Neutrophil   Lymphocyte

Granulocytes   Agranulocytes

Leukocytes

FIGURE 16-6   Formation of blood cells. The hemocytoblast serves as the original stem cell from
which all formed elements of the blood are derived. Note that all five precursor cells, which
ultimately produce the different components of the formed elements, are derived from the
hemocytoblast.

Red blood cells are formed and destroyed at a breathtaking rate. Normally, every minute of every day of our adult lives, over 100 million red blood cells are formed to replace an equal number destroyed during that brief time. Since in health the number of red blood cells remains relatively constant, efficient homeostatic mechanisms must operate to balance the number of cells formed against the number destroyed. The rate of red blood cell production soon speeds up if blood oxygen levels reaching the tissues decreases. Oxygen deficiency increases RBC numbers by increasing the secretion of a glycoprotein hormone named

erythropoietin (e-RITH-ro-POY-e-tin). An inactive form of this hormone called an erythropoietinogen is released into the blood primarily from the liver on an ongoing basis. If oxygen levels decrease, the kidneys release increasing amounts of an enzyme, called **renal erythropoietic factor,** into the blood. This enzyme activates erythropoietin, which in turn stimulates bone marrow to accelerate its production of red blood cells. With increasing numbers of RBCs, oxygen delivery to tissues increases, less renal erythropoietic factor is produced, and less active erythropoietin is available to stimulate RBC production in the red

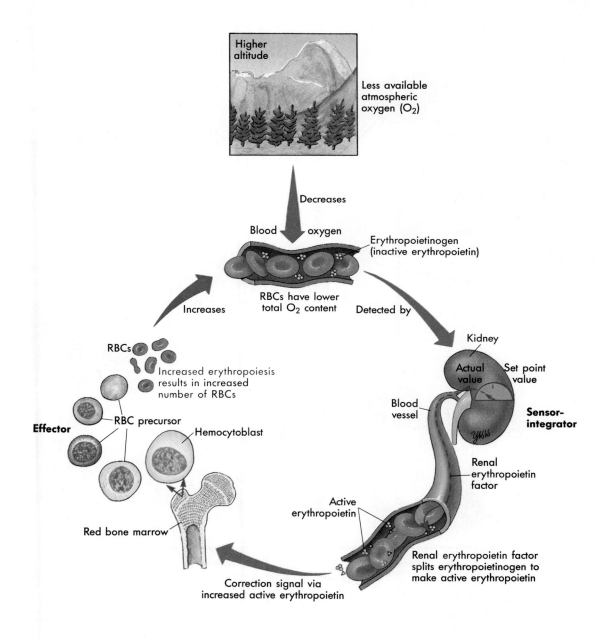

FIGURE 16-7 Erythropoiesis. In response to decreased blood oxygen, the kidneys release renal erythropoietic factor. The renal erythropoietic factor converts inactive erythropoietin from the liver into active erythropoietin, which stimulates erythrocyte production in the red bone marrow.

bone marrow. Figure 16-7 shows how erythropoietin production is controlled by a negative feedback loop that is activated by decreasing oxygen concentration in the tissues.

Destruction of red blood cells

The life spans of red blood cells circulating in the bloodstream average about 105 to 120 days. They often break apart, or fragment, in the capillaries as they age. Macrophage cells in the lining of blood vessels, particularly in the liver and spleen, phagocytose (ingest and destroy) the aged, abnormal, or fragmented red blood cells (Figure 16-8). The process results in breakdown of hemoglobin, with the release of amino acids, iron, and the pigment

bilirubin. Iron is returned to the bone marrow for use in synthesis of new hemoglobin, and bilirubin is transported to the liver, where it is excreted into the intestine as part of bile. Amino acids, released from the globin portion of the degraded hemoglobin molecule, are used by the body for energy or for synthesis of new proteins.

RETICULOCYTE COUNT

Frequently a physician needs information about the rate of erythropoiesis to help in making a diagnosis or to prescribe treatment. A **reticulocyte count** gives this information. Approximately 0.5% to 1.5% of the red blood cells in normal blood at any given time are reticulocytes. A reticulocyte count of less than 0.5% of the red blood cell count usually indicates a slowdown in the process of red blood cell formation. Conversely, a reticulocyte count higher than 1.5% usually indicates an acceleration of red blood cell formation—as occurs, for example, after treatment of anemia.

BLOOD DOPING

Reports that some Olympic athletes employ transfusions of their own blood to improve performance have surfaced repeatedly during the past 20 years. The practice—called **blood doping,** or **blood boosting**—is intended to increase oxygen delivery to muscles. A few weeks before competition, blood is drawn from the athlete and the RBCs are separated and frozen. Just before competition, the RBCs are thawed and injected. Theoretically, infused red blood cells and elevation of hemoglobin levels after transfusion should increase oxygen consumption and muscle performance during exercise. In practice, however, questions of how effective transfusions might be in affording even world-class and professional athletes substantial competitive advantages have not been resolved. Improved performance might result, but the advantage appears to be minimal.

All blood transfusions carry some risk, and unnecessary or questionably indicated transfusions are medically unacceptable.

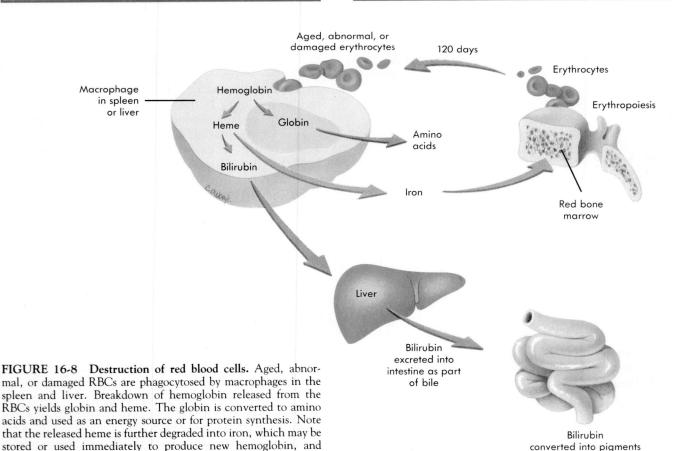

FIGURE 16-8 Destruction of red blood cells. Aged, abnormal, or damaged RBCs are phagocytosed by macrophages in the spleen and liver. Breakdown of hemoglobin released from the RBCs yields globin and heme. The globin is converted to amino acids and used as an energy source or for protein synthesis. Note that the released heme is further degraded into iron, which may be stored or used immediately to produce new hemoglobin, and bilirubin, which is ultimately excreted in the bile.

COMPLETE BLOOD COUNT

One of the most useful and frequently performed clinical blood tests is called the *complete blood count* or simply the CBC. The CBC is a collection of tests whose results, when interpreted as a whole, can yield an enormous amount of information regarding a person's health. Standard RBC, WBC, and thrombocyte counts, the differential WBC count, hematocrit, hemoglobin content, and other characteristics of the formed elements are usually included in this battery of tests. See Appendix C for tables listing some of the normal clinical values for these tests.

For the red blood cell homeostatic mechanism to succeed in maintaining a normal number of red blood cells, the bone marrow must function adequately. To do this the blood must supply it with adequate amounts of several substances with which to form the new red blood cells—vitamin B_{12}, iron, and amino acids, for example, and also copper and cobalt to serve as catalysts. In addition, the gastric mucosa must provide some unidentified intrinsic factor necessary for absorption of vitamin B_{12} (also called **extrinsic factor** because it derives from external sources in foods and is not synthesized by the body; vitamin B_{12} is also called the **antianemic principle**).

QUICK CHECK

✔ 1. *Name the red pigment found in RBCs and list the normal range (in grams/100 ml of blood) for women and men.*

✔ 2. *Trace the formation of an erythrocyte from stem cell precursor to a mature and circulating RBC.*

✔ 3. *Explain the negative feedback loop that controls erythropoiesis.*

✔ 4. *Discuss the destruction of red blood cells.*

White Blood Cells (Leukocytes)

There are five types of white blood cells (WBCs), or leukocytes, classified according to the presence or absence of granules and the staining characteristics of their cytoplasm. **Granulocytes** include the three WBCs that have large granules in their cytoplasm. They are named according to their cytoplasmic staining properties:

1. Neutrophils
2. Eosinophils
3. Basophils

There are two types of **agranulocytes** (WBCs without cytoplasmic granules):

1. Lymphocytes
2. Monocytes

As a group, the leukocytes appear brightly colored in stained preparations. In addition, they all have nuclei and are generally larger in size than RBCs. Table 16-1, p. 452, illustrates the formed elements, provides a brief description of each cell type, and lists life span and primary functions.

Granulocytes

Neutrophils. *Neutrophils* (Figure 16-9) take their name from the fact that their cytoplasmic granules stain a very light purple with neutral dyes. The granules in these cells are small and numerous and tend to give the cytoplasm a "coarse" appearance. Because their nuclei have two, three, or more lobes, neutrophils are also called **polymorphonuclear leukocytes** or, to avoid that tongue twister, simply *polys.*

Neutrophil numbers average about 65% of the total WBC count in a normal blood sample. These leukocytes are highly mobile and very active phagocytic cells that can migrate out of blood vessels and enter the tissue spaces. The process is called *diapedesis.* The cytoplasmic granules in neutrophils contain powerful lysosomes, the organelles with digestive-like enzymes that are capable of destroying bacterial cells.

Bacterial infections that produce an inflammatory response cause the release of chemicals from damaged cells that attract neutrophils and other phagocytic WBCs to the infection site. The process, called **positive chemotaxis,** helps the body concentrate phagocytic cells at focal points of infection.

Eosinophils. *Eosinophils* (Figure 16-10) contain cytoplasmic granules that are large, numerous, and stain orange with acid dyes such as eosin. Their nuclei generally have two lobes. Normally, eosinophils account for about 2% to 5% of circulating WBCs. They are numerous in body areas such as the lining of the respiratory and digestive tracts. Although eosinophils are weak phagocytes, they are capable of ingesting inflammatory chemicals and proteins associated with antigen-antibody reaction complexes. Perhaps their most important functions involve protection against infections caused by parasitic worms and allergic reactions.

Basophils. *Basophils* (Figure 16-11) have relatively large, but sparse, cytoplasmic granules that stain a dark purple with basic dyes. They are the least numerous of the WBCs, numbering only 0.5% to 1% of the total leukocyte count. Basophils are both motile and capable of diapedesis. They exhibit **S**-shaped, but indistinct, nuclei. The cytoplasmic granules of these WBCs contain *histamine* (an inflammatory chemical) and *heparin* (an anticoagulant).

Agranulocytes

Lymphocytes. *Lymphocytes* (Figure 16-12) found in the blood are the smallest of the leukocytes, averaging about 6 to 8 μm in diameter. They have large, spherical nuclei surrounded by a very limited amount of pale blue staining cytoplasm. Next to neutrophils, lymphocytes are the most numerous WBCs. They account for about 25% of all the leukocyte population. Two types of lymphocytes, called *T*

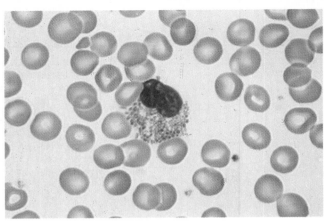

FIGURE 16-9 Neutrophil. (*Courtesy of Erlandsen/Magney: Color Atlas of Histology.*)

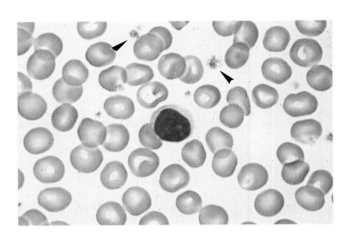

FIGURE 16-10 Eosinophil. (*Courtesy of Erlandsen/Magney: Color Atlas of Histology.*)

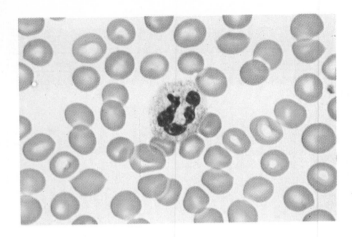

FIGURE 16-11 Basophil. (*Courtesy of Erlandsen/Magney: Color Atlas of Histology.*)

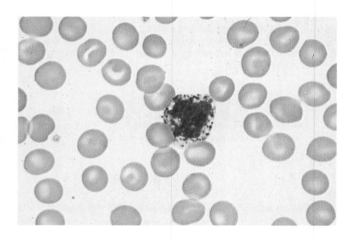

FIGURE 16-12 Lymphocyte. Arrowheads are pointing to platelets. (*Courtesy of Erlandsen/Magney: Color Atlas of Histology.*)

lymphocytes and *B lymphocytes*, have an important role in immunity. T lymphocytes function by directly attacking an infected or cancerous cell, whereas B lymphocytes produce antibodies against specific antigens. The functions of both types of lymphocytes will be fully discussed in Chapter 20.

Monocytes. *Monocytes* (Figure 16-13) are the largest of the leukocytes. They have dark kidney bean–shaped nuclei surrounded by large quantities of distinctive blue-gray cytoplasm. Monocytes are mobile and highly phagocytic cells capable of engulfing large bacterial organisms and viral infected cells.

White blood cell numbers

A cubic millimeter of normal blood usually contains about 5,000 to 9,000 leukocytes, with different percentages of each type. Because these numbers change in certain abnormal conditions, they have clinical significance. In acute appendicitis, for example, the percentage of neutro-

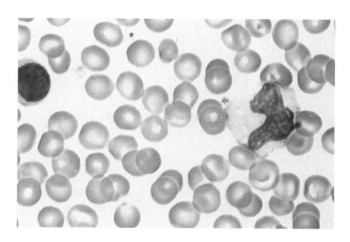

FIGURE 16-13 Monocyte. Compare the large monocyte (*right*) with the smaller lymphocyte (*left*). (*Courtesy of Erlandsen/ Magney: Color Atlas of Histology.*)

TABLE 16-1   Classes of Blood Cells

Cell Type	Description	Function	Life Span
Erythrocyte	7 μm in diameter; concave disk shape; entire cell stains pale pink; no nucleus	Transportation of respiratory gases (O_2 and CO_2)	105 to 120 days
Neutrophil	12-15 μm in diameter; spherical shape; multilobed nucleus; small, pink-purple staining cytoplasmic granules	Cellular defense—phagocytosis of small pathogenic microorganisms	Hours to 3 days
Basophil	11-14 μm in diameter; spherical shape; generally two lobed nucleus; large purple staining cytoplasmic granules	Secretes heparin (anticoagulant) and histamine (important in inflammatory response)	Hours to 3 days
Eosinophil	10-12 μm in diameter; spherical shape; generally two-lobed nucleus; large orange-red staining cytoplasmic granules	Cellular defense—phagocytosis of large pathogenic microorganisms such as protozoa and parasitic worms; releases antiinflamatory substances in allergic reactions	10 to 12 days
Lymphocyte	6-9 μm in diameter; spherical shape; round (single lobe) nucleus; small lymphocytes have scant cytoplasm	Humoral defense—secretes antibodies; involved in immune system response and regulation	Days to years
Monocyte	12–17 μm in diameter; spherical shape; nucleus generally kidney-bean or "horse-shoe" shaped with convoluted surface; ample cytoplasm often "steel blue" in color	Capable of migrating out of the blood to enter tissue spaces as a *macrophage*—an aggressive phagocytic cell capable of ingesting bacteria, cellular debris, and cancerous cells	Months
Platelet	2-5 μm in diameter; irregularly shaped fragments; cytoplasm contains very small pink staining granules	Releases clot activating substances and helps in formation of actual blood clot by forming platelet "plugs"	7 to 10 days

phils increases and so, too, does the total white blood cell count. In fact, these characteristic changes may be the deciding points for surgery.

The procedure in which the different types of leukocytes are counted and their percentage of total white blood cell count is computed is known as a **differential count.** In other words, a differential count is a percentage count of white blood cells. The different kinds of white blood cells and a normal differential count are listed in Table 16-2. A decrease in the number of white blood cells is **leukopenia.** An increase in the number of white blood cells is **leukocytosis.**

Formation of white blood cells

The hemocytoblast stem cell serves as the precursor of not only the erythrocytes but also the leukocytes and platelets in blood. Figure 16-6, p. 447, shows the maturation sequence that results in formation of the granular and the agranular leukocytes from the undifferentiated hemocytoblast stem cell.

Neutrophils, eosinophils, basophils, and a few lymphocytes and monocytes originate, as do erythrocytes, in red bone narrow (myeloid tissue). Most lymphocytes and monocytes derive from hemocytoblasts in lymphatic tissue. Although many lymphocytes are found in bone marrow,

TABLE 16-2   Differential count of white blood cells

Class	Differential count*	
	Normal range (%)	Typical value (%)
Neutrophils	65 to 75	65
Eosinophils	2 to 5	3
Basophils	½ to 1	1
Lymphocytes (large and small)	20 to 25	25
Monocytes	3 to 8	6
TOTAL	100	100

*In any differential count the sum of the percentages of the different kinds of WBCs must, of course, total 100%.

presumably most were formed in lymphatic tissues and carried to the bone marrow by the bloodstream.

Myeloid tissue (bone marrow) and lymphatic tissue together constitute the hematopoietic, or blood cell–forming, tissues of the body. Red bone marrow is myeloid tissue that is actually producing blood cells. Its red color comes from the red blood cells it contains. Yellow marrow, on the other hand, is yellow because it stores considerable fat. It is not active in the business of blood cell formation as long as it remains yellow. Sometimes, however, it becomes active and red in color when an extreme and prolonged need for red blood cell production occurs.

Platelets

To compare platelets with other blood cells in terms of appearance and size, see Figure 16-12. In circulating blood, platelets are small, nearly colorless bodies that usually appear as irregular spindles or oval disks about 2 to 4 μm in diameter.

Three important physical properties of platelets—namely, agglutination, adhesiveness, and aggregation—make attempts at classification on the basis of size or shape in dry blood smears all but impossible. As soon as blood is removed from a vessel, the platelets adhere to each other and to every surface they contact; in so doing, they assume various shapes and irregular forms.

Platelet counts in adults average about 250,000 per mm³ of blood. A range of 150,000 to 350,000 per mm³ is considered normal. Newborn infants often show reduced counts, but these rise gradually to reach normal adult values at about 3 months of age. There are no differences between the sexes in platelet count.

Functions of platelets

Platelets play an important role in both **hemostasis** (from the Greek *stasis*, "a standing") and **blood clotting** or **coagulation.** The two, although interrelated, are separate and distinct functions. Hemostasis refers to the stoppage of blood *flow* and may occur as an end result of any one of several body defense mechanisms. The role that platelets

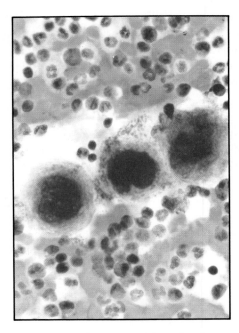

FIGURE 16-14   Megakaryocyte. Note the large number of platelets surrounding the cell.

play in the operation of the blood-clotting mechanism is discussed on pp. 457 to 458.

Within 1 to 5 seconds after injury to a blood capillary, platelets will adhere to the damaged lining of the vessel and to each other to form a hemostatic platelet plug that helps to stop the flow of blood into the tissues. At least one of the prostaglandins (PGE₂) and certain prostaglandin-like substances, called *thromboxanes* (which are both found in platelets), play roles in hemostasis and blood clotting. When released, these substances affect both local blood flow (by vasoconstriction) and platelet aggregation at the site of injury. If the injury is extensive, the blood-clotting mechanism is activated to assist in hemostasis.

Formation and life span of platelets

Platelets are formed in the red bone marrow, lungs, and, to some extent, in the spleen by fragmentation of very large (40 to 100 μm) cells known as **megakaryocytes.** These cells are characterized by large multilobular nuclei that are often bizarre in shape. Note the large number of platelets surrounding the megakaryocyte shown in Figure 16-14. Platelets have a short life span, an average of about 9 days.

BLOOD TYPES (BLOOD GROUPS)

The term *blood type* refers to the type of antigens (called *agglutinogens*) present on red blood cell membranes. (The concept of antigens and antibodies is discussed in detail in Chapter 20.) Antigens A, B, and Rh are the most important blood antigens as far as transfusions and newborn survival are concerned. Many other blood antigens have also been identified, but they are less important clinically and are seldom discussed.

The ABO System

Every person's blood belongs to one of the four ABO blood groups. Blood types are named according to the antigens present on red blood cell membranes. Here, then, are the four ABO blood types:

1. Type A—Antigen A on red blood cells
2. Type B—Antigen B on red blood cells
3. Type AB—Both antigen A and antigen B on red blood cells
4. Type O—Neither antigen A nor antigen B on red blood cells

Blood plasma may or may not contain antibodies that can react with red blood cell antigens A or B. An important principle about this is that plasma never contains antibodies against the antigens present on its own red blood cells—for obvious reasons. If it did, the antibody would react with the antigen and thereby destroy the red blood cells. But (and this is an equally important principle) plasma does contain antibodies against antigen A or antigen B if they are *not* present on its red blood cells. Applying these two principles: in type A blood, antigen A is present on its red blood cells; therefore its plasma contains no anti-A antibodies but does contain anti-B antibodies. In type B blood, antigen B is present on its red blood cells; therefore its plasma contains no anti-B antibodies but does contain anti-A antibodies (Figure 16-15).

Note in Figure 16-16, A, that type A blood donated to a type A recipient does not cause an agglutination reaction because the type B antibodies in the recipient do not combine with the type A antigens in the donated blood. However, type A blood donated to a type B recipient causes an agglutination reaction because the type A antibodies in

the recipient combine with the type A antigens in the donated blood (Figure 16-16, B). Figure 16-17 shows the results of different combinations of donor and recipient blood.

Type O blood has been referred to as *universal donor* blood, a term that implies that it can safely be given to any recipient. This, however, is not true, because the recipient's plasma may contain agglutinins other than anti-A or anti-B antibodies. For this reason the recipient's and the donor's blood— even if it is type O—should be cross-matched, that is, mixed and observed for agglutination of the donor's red blood cells.

Universal recipient (type AB) blood contains neither anti-A nor anti-B antibodies, so it cannot agglutinate type A or type B donor's red blood cells. This does not mean, however, that any type of donor blood may be safely given to an individual who has type AB blood without first cross-matching. Other agglutinins may be present in the so-called universal recipient blood and clump unidentified antigens (agglutinogens) in the donor's blood.

The Rh System

The term *Rh-positive blood* means that Rh antigen is present on its red blood cells. *Rh-negative blood,* on the other hand, is blood whose red cells have no Rh antigen present on them.

No blood normally contains anti-Rh antibodies. However, anti-Rh antibodies can appear in the blood of an Rh-negative person, provided Rh-positive red blood cells have at some time entered the bloodstream. One way this can happen is by giving an Rh-negative person a transfusion of Rh-positive blood. In a short time, the person's body makes anti-Rh antibodies, and these remain in the blood. The other way in which Rh-positive red blood cells

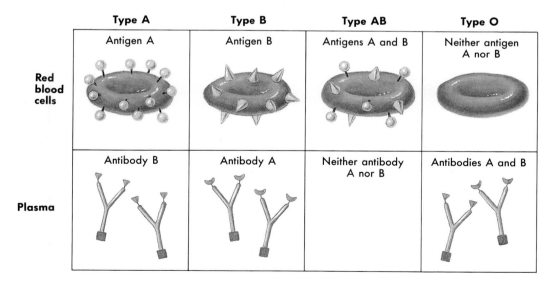

	Type A	**Type B**	**Type AB**	**Type O**
Red blood cells	Antigen A	Antigen B	Antigens A and B	Neither antigen A nor B
Plasma	Antibody B	Antibody A	Neither antibody A nor B	Antibodies A and B

FIGURE 16-15 ABO blood types. Note that antigens characteristic of each blood type are bound to the surface of RBCs. The antibodies of each blood type are found in the plasma and exhibit unique structural features that permit agglutination to occur if exposure to the appropriate antigen occurs.

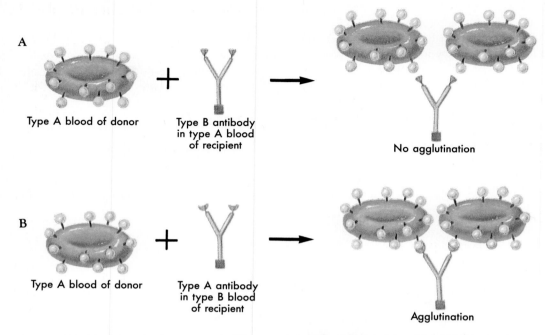

A

Type A blood of donor   +   Type B antibody in type A blood of recipient   →   No agglutination

B

Type A blood of donor   +   Type A antibody in type B blood of recipient   →   Agglutination

FIGURE 16-16   Agglutination. A, When mixing of donor and recipient blood of the same type (A) occurs, there is no agglutination because only type B antibodies are present. **B,** If type A donor blood is mixed with type B recipient blood, agglutination will occur because of the presence of type A antibodies in the type B recipient blood.

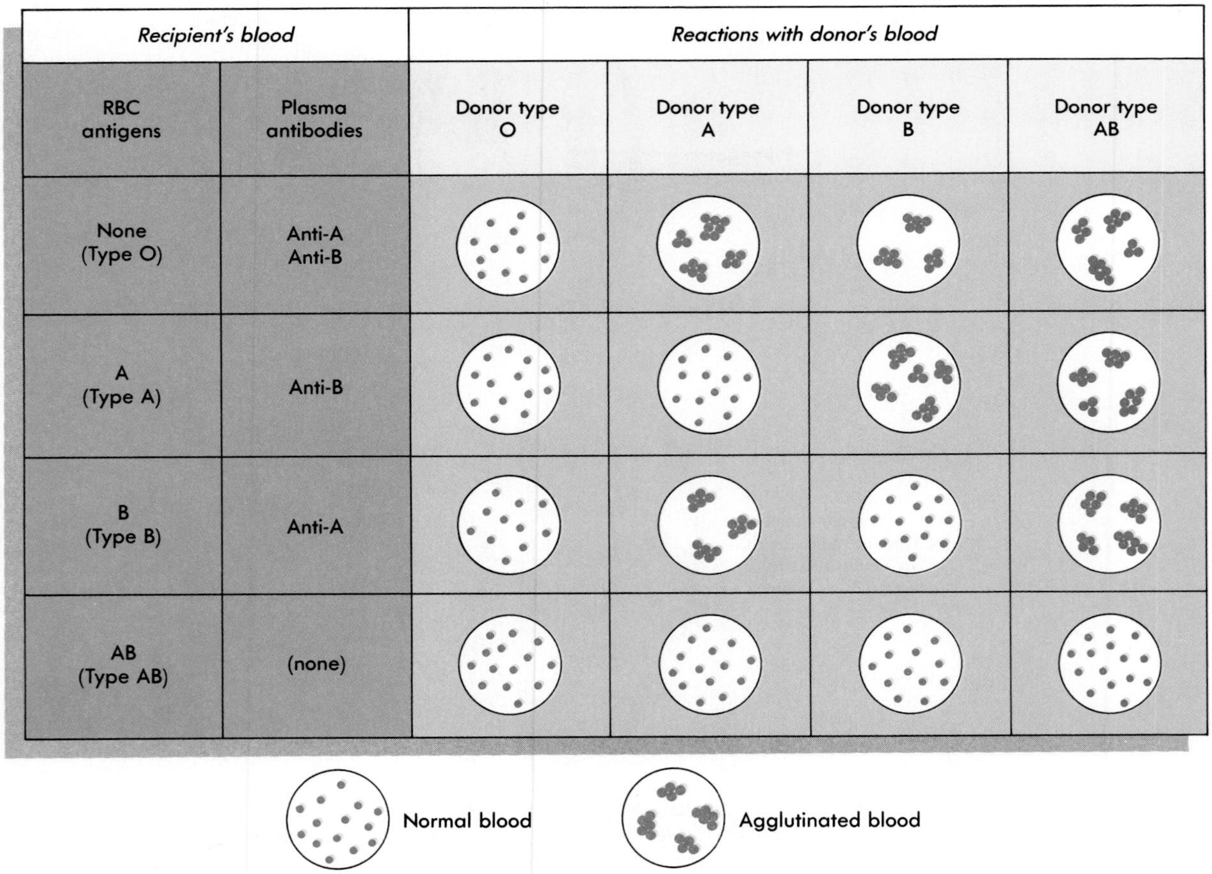

Recipient's blood		Reactions with donor's blood			
RBC antigens	Plasma antibodies	Donor type O	Donor type A	Donor type B	Donor type AB
None (Type O)	Anti-A Anti-B				
A (Type A)	Anti-B				
B (Type B)	Anti-A				
AB (Type AB)	(none)				

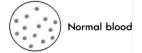 Normal blood

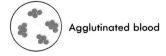 Agglutinated blood

FIGURE 16-17   Results of different combinations of donor and recipient blood. The left columns show the recipient's blood characteristics, and the top row shows the donor's blood type.

can enter the bloodstream of an Rh-negative individual can happen only to a woman during pregnancy. In this fact lies the danger for a baby born to an Rh-negative mother and an Rh-positive father. If the baby inherits the Rh-positive trait from the father, the Rh factor on the RBCs may stimulate the mother's body to form anti-Rh antibodies. Then, if she later carries another Rh-positive fetus, the fetus may develop a disease called **erythroblastosis fetalis,** caused by the mother's Rh antibodies reacting with the baby's Rh-positive cells (Figure 16-18).

Some Rh-negative mothers who carry an Rh-positive baby are treated with a protein marketed as RhoGAM. RhoGAM stops the mother's body from forming anti-Rh

antibodies and thus prevents the possibility of harm to the next Rh-positive baby.

Briefly, the only people who can ever have anti-Rh antibodies in their plasma are Rh-negative men or women who have been transfused with Rh-positive blood or Rh-negative women who have carried an Rh-positive fetus.

BLOOD PLASMA

Plasma is the liquid part of blood—whole blood minus formed elements (Figure 16-19). In the laboratory, whole blood that has not clotted is centrifuged to form plasma. This consists of a rapid whirling process that hurls the blood

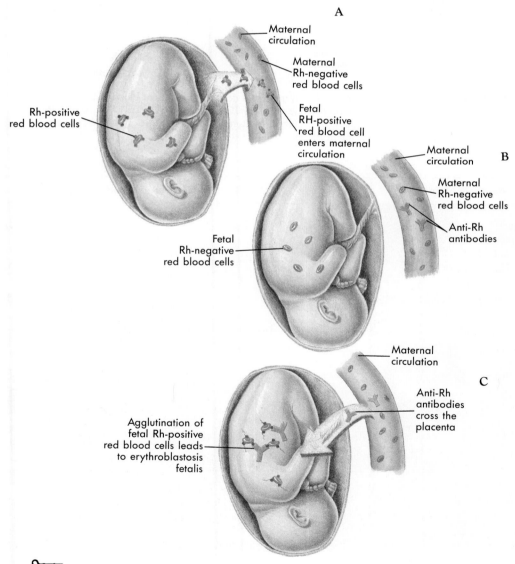

FIGURE 16-18 Erythroblastosis fetalis. A, Rh-positive blood cells enter the mother's bloodstream during delivery of an Rh-positive baby. If not treated, the mother's body will produce anti-Rh antibodies. **B,** A later pregnancy involving an Rh-negative baby is normal because there are no Rh antigens in the baby's blood. **C,** A later pregnancy involving an Rh-positive baby may result in erythroblastosis fetalis. Anti-Rh antibodies enter the baby's blood supply and cause agglutination of RBCs with the Rh antigen.

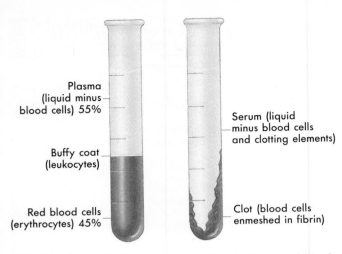

FIGURE 16-19   Difference between blood plasma and blood serum. Plasma is whole blood minus cells. Serum is whole blood minus the clotting elements. Plasma is prepared by centrifuging blood. Serum is prepared by clotting blood.

Plasma (liquid minus blood cells) 55%

Buffy coat (leukocytes)

Red blood cells (erythrocytes) 45%

Serum (liquid minus blood cells and clotting elements)

Clot (blood cells enmeshed in fibrin)

cells to the bottom of the centrifuge tube. A clear, straw-colored fluid—blood plasma—lies above the cells. Plasma consists of 90% water and 10% solutes. By far the largest quantity of these solutes is proteins; normally, they constitute about 6% to 8% of the plasma. Other solutes present in much smaller amounts in plasma are food substances (principally glucose, amino acids, and lipids), compounds formed by metabolism (e.g., urea, uric acid, creatinine, and lactic acid), respiratory gases (oxygen and carbon dioxide), and regulatory substances (hormones, enzymes, and certain other substances).

Some of the solutes present in blood plasma are true solutes, or crystalloids. Others are colloids. *Crystalloids* are solute particles less than 1 nm in diameter (ions, glucose, and other small molecules). *Colloids* are solute particles from 1 to about 100 nm in diameter (for example, proteins of all types). Blood solutes also may be classified as **electrolytes** (molecules that ionize in solution) or **nonelectrolytes**—examples: inorganic salts and proteins are electrolytes; glucose and lipids are nonelectrolytes.

The proteins in blood plasma consist of three main kinds of compounds: *albumins, globulins,* and *fibrinogen.* Measuring the amounts of these compounds indicates that 100 ml of plasma contains a total of approximately 6 to 8 g of protein. Albumins constitute about 55% of this total, globulins about 38%, and fibrinogen about 7%.

Plasma proteins are crucially important substances. Fibrinogen, for instance, and an albumin named prothrombin have key roles in the blood-clotting mechanism. Globulins function as essential components of the immunity mechanism—circulating antibodies (immune bodies) are modified gamma globulins. All plasma proteins contribute to the maintenance of normal blood viscosity, blood osmotic pressure, and blood volume. Therefore plasma proteins have an essential part in maintaining normal circulation. Synthesis of plasma proteins occurs in liver cells. They form all kinds of plasma proteins, except some

of the gamma globulins—specifically, the circulating antibodies synthesized by plasma cells.

BLOOD CLOTTING (COAGULATION)

The purpose of blood coagulation is obvious—to plug ruptured vessels so as to stop bleeding and prevent loss of a vital body fluid.

Mechanism of Blood Clotting

Because of the function of coagulation, the mechanism for producing it must be swift and sure when needed, such as when a vessel is cut or ruptured. Equally important, however, coagulation needs to be prevented from happening when it is not needed because clots can plug up vessels that must stay open if cells are to receive blood's life-sustaining cargo of oxygen.

What makes blood coagulate? Over a period of many years a host of investigators have searched for the answer to this question. They have tried to find out what events make up the coagulation mechanism and what sets it in operation. They have succeeded in gathering an abundance of relevant information, but questions about this complicated and important process still outnumber answers.

The blood-clotting mechanism presumably consists of a series of chemical reactions that take place in a definite and rapid sequence resulting in a net of fibers that traps red blood cells (Figure 16-20).

The so-called "classic theory" of coagulation was advanced in 1905 and dominated research efforts in this complex area for almost half a century. It continues as the basis of our current understanding of coagulation. This theory assumed: (1) the interaction of four plasma components in the presence of calcium ions and (2) that the interaction between the components occurred in two steps.

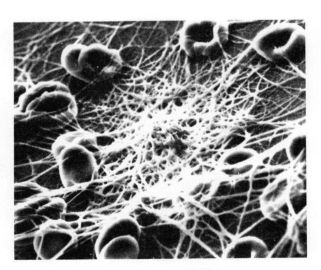

FIGURE 16-20   Blood clotting. An electron micrograph showing entrapped red blood cells in a fibrin clot.

The plasma components are as follows:

- Prothrombin
- Thrombin
- Fibrinogen
- Fibrin

Interactions between these components occurs in these steps (Figure 16-21):

Step I:

$$\text{Prothrombin} \xrightarrow{\substack{\text{Thromboplastin} \\ Ca^{++}}} \text{Thrombin}$$

Step II:

$$\text{Fibrinogen} \xrightarrow{\text{Thrombin}} \text{Fibrin}$$

It is interesting that these two basic reaction steps have been modified only by the action of a host of additional coagulation factors discovered in recent years (Table 16-3). In addition, the source of thromboplastin in step I of the current coagulation theory mechanisms is now used to divide this step into *intrinsic* and *extrinsic systems*.

The two basic coagulation steps described in the following paragraphs are shown in Figure 16-21.

Step I: In step I of the coagulation mechanism, prothrombin (a plasma protein) is converted into thrombin (a plasma enzyme) by the action of thromboplastin. In addition to thromboplastin, calcium ions are required for this reaction to occur. Thromboplastin may be released (1) from platelets or (2) from damaged tissues.

TABLE 16-3	Coagulation factors—standard nomenclature and synonyms
Factor	**Common synonym(s)**
Factor I	Fibrinogen
Factor II	Prothrombin
Factor III	Thromboplastin Thrombokinase
Factor IV	Calcium
Factor V	Proaccelerin Labile factor
Factor VI (now obsolete)	None in use
Factor VII	Serum prothrombin conversion accelerator (SPCA)
Factor VIII	Antihemophilic globulin (AHG) Antihemophilic factor (AHF)
Factor IX	Plasma thromboplastin component (PTC), Christmas factor
Factor X	Stuart factor
Factor XI	Plasma thromboplastin antecedent (PTA)
Factor XII	Hageman factor
Factor XIII	Fibrin-stabilizing factor

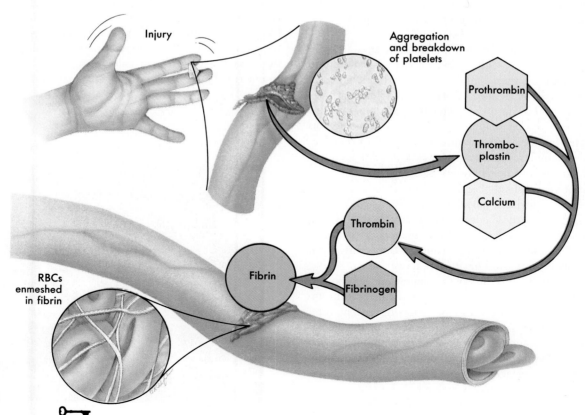

FIGURE 16-21   Blood clotting mechanism. The clotting mechanism involves release of platelet factors at the injury site, formation of thrombin, and trapping of red blood cells in fibrin to form a clot.

Intrinsic system. If thromboplastin is released from damaged platelets (*platelet thromboplastin*), step I of the coagulation process proceeds via the *intrinsic system*. In this system, plasma factors IV, V, VIII, IX, X, XI, and XII are also required to convert prothrombin to thrombin.

Extrinsic system. If thromboplastin is released from damaged tissues (*tissue thromboplastin*), step I of coagulation proceeds via the *extrinsic system*. Plasma factors IV, V, VII, and X act with tissue thromboplastin to convert prothrombin to thrombin in this system.

Step II: In step II of the coagulation mechanism, fibrinogen (a soluble plasma protein) is converted into strands of insoluble fibrin. Thrombin and plasma factors IV and XIII are required for completion of step II of the coagulation mechanism.

Unfortunately, initiation of the step I intrinsic system of coagulation often occurs when the normally very smooth endothelial lining of a blood vessel becomes "rough" because of a cut or the formation of cholesterol plaques (patchlike deposits in the vessel wall). Within a matter of seconds, clumps of platelets adhere to the injured area and disintegrate. This releases platelet thromboplastin, which then triggers the intrinsic system response in the first step of the coagulation process. As a result, prothrombin is converted to thrombin if calcium ions and the appropriate coagulation factors (IV, V, VIII, IX, X, XI, XII) are present. In step II thrombin accelerates the conversion of the soluble plasma protein fibrinogen to insoluble fibrin. Fibrin appears in blood as fine threads all tangled together. Blood cells catch in the entanglement, and, because most of the cells are red blood cells, clotted blood has a red color. The pale yellowish liquid left after a clot forms is **blood serum.** How do you think serum differs from plasma? What is plasma? To check your answers, see Figure 16-19.

Liver cells synthesize both prothrombin and fibrinogen, as they do almost all other plasma proteins. For the liver to synthesize prothrombin at a normal rate, blood must contain an adequate amount of vitamin K. Vitamin K is absorbed into the blood from the intestine. Some foods contain this vitamin, but it is also synthesized in the intestine by certain bacteria (not present for a time in newborn infants). Because vitamin K is fat soluble, its absorption requires bile. If, therefore, the bile ducts become obstructed and bile cannot enter the intestine, a vitamin K deficiency develops. The liver cannot then produce prothrombin at its normal rate, and the blood's prothrombin concentration soon falls below normal. A prothrombin deficiency gives rise to a bleeding tendency. As a preoperative safeguard, therefore, patients with obstructive jaundice are generally given some kind of vitamin K preparation.

Conditions that Oppose Clotting

Although blood clotting probably goes on continuously and concurrently with clot dissolution (fibrinolysis), several conditions operate to oppose clot formation in intact vessels. Most important by far is the perfectly smooth surface of the normal endothelial lining of blood vessels.

Platelets do not adhere to it; consequently, they do not disintegrate and release platelet factors into the blood and therefore the blood-clotting mechanism does not begin in normal vessels. As an additional deterrent to clotting, blood contains certain substances called *antithrombins*. The name suggests their function—they oppose (inactivate) thrombin. Thus antithrombins prevent thrombin from converting fibrinogen to fibrin. **Heparin,** a natural constituent of blood, acts as an antithrombin. It was first prepared from liver (hence its name), but other organs also contain heparin. Its normal concentration in blood is too low to have much effect in keeping blood fluid. However, injections of heparin are used to prevent clots from forming in vessels. *Coumarin* compounds impair the liver's utilization of vitamin K and thereby slow its synthesis of prothrombin and factors VII, IX, and X. Indirectly, therefore, coumarin compounds, such as bishydroxycoumarin, retard coagulation. Citrates keep donor blood from clotting before transfusion.

Conditions that Hasten Clotting

Two conditions particularly favor thrombus formation: a rough spot in the endothelium (blood vessel lining) and abnormally slow blood flow. Atherosclerosis, for example, is associated with an increased tendency toward thrombosis because of endothelial rough spots in the form of plaques of accumulated cholesterol-lipid material. Immobility, on the other hand, may lead to thrombosis because blood flow slows down as movements decrease. Incidentally, this fact is one of the major reasons why physicians insist that bed patients must either move or be moved frequently. Presumably, sluggish blood flow allows thromboplastin to accumulate sufficiently to reach a concentration adequate for clotting.

Once started, a clot tends to grow. Platelets enmeshed in the fibrin threads disintegrate, releasing more thromboplastin, which in turn causes more clotting, which enmeshes more platelets, and so on, in a vicious circle. Clot-retarding substances, available in recent years, have proved valuable for retarding this process.

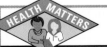

CLINICAL METHODS OF HASTENING CLOTTING

One way of treating excessive bleeding is to speed up the blood-clotting mechanism. The purpose is apparent—to increase any of the substances essential for clotting. This is accomplished in the following ways:

♦ By applying a rough surface such as gauze, by applying heat, or by gently squeezing the tissues around a cut vessel. Each of these procedures causes more platelets to disintegrate and release more platelet factors. This, in turn, accelerates the first of the clotting reactions.

♦ By applying purified thrombin (in the form of sprays or impregnated gelatin sponges that can be left in a wound). Which stage of the clotting mechanism does this accelerate?

♦ By applying fibrin foam, films, and so on.

Clot Dissolution

The physiological mechanism that dissolves clots is known as **fibrinolysis.** Evidence indicates that the two opposing processes of clot formation and fibrinolysis go on continuously. Blood contains an enzyme, fibrinolysin, that catalyzes the hydrolysis of fibrin, causing it to dissolve. Many other factors, however, presumably also take part in clot dissolution, for instance, substances that activate profibrinolysin (inactive form of fibrinolysin). Streptokinase, an enzyme from certain streptococci, can act this way and thus can cause clot dissolution and even hemorrhage.

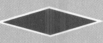

QUICK CHECK

1. List the granulocytic and agranulocytic leukocytes.
2. List the normal percentages of the different WBCs in a differential count.
3. List the four ABO blood groups and identify the antigen and antibodies (if any) associated with each.
4. Identify the two basic coagulation steps.

THE BIG PICTURE
Blood and the Whole Body

In every chapter of this book we have referred to the notion that the whole body's function is geared toward maintaining stability of the internal fluid environment—that is, *homeostasis.* The fluid that makes up the internal environment in which cells are bathed—the fluid that must be kept stable—includes the plasma of the blood. As a matter of fact, it is the blood plasma that transports substances, and even heat, around the internal environment so that all body tissues are linked together. The various tissues of the body are linked by the plasma, which flows back and forth between any two points served by blood vessels. This, of course, means that substances such as nutrients, wastes, dissolved gases, water, antibodies, and hormones can be transported between almost any two points in the body.

Blood tissue is not just plasma, however. It contains the *formed elements*—the blood cells and platelets. The red blood cells participate in the mechanisms that permit the efficient transport of the gases oxygen and carbon dioxide. White blood cells are important in the defense mechanisms of the whole body. Their presence in blood ensures that they are available to all parts of the body, all of the time, to fight cancer, resist infectious agents, and clean up injured tissues. Platelets provide mechanisms for preventing loss of the fluid that constitutes our internal environment.

All other organs and systems of the body rely on blood to perform its many functions. No organ or system can maintain proper levels of nutrients, dissolved gases, or water without direct or indirect help from the blood. On the other hand, many other systems help blood do its job. For example, the respiratory system excretes carbon dioxide from the blood and picks up oxygen. Organs of the digestive system pick up nutrients, remove some toxins, and take care of old blood cells. The endocrine system regulates the production of blood cells and the water content of the plasma. Besides removing toxic wastes such as urea, the urinary system plays a vital role in maintaining homeostasis of plasma water concentration and pH.

Of course, blood is useless unless it continually and rapidly flows around the whole body—and continues to transport, defend, and maintain balance. The next several chapters outline the structures and functions that make this possible. Chapters 17 and 18 discuss the plan of the blood circulation and how adequate blood flow is maintained. Chapter 19 discusses the role of the lymphatic system in maintaining the fluid balance of the blood. Chapters 20 and 21 deal with the defensive mechanisms of blood and other tissues. As a matter of fact, most of the remaining chapters feature the role of blood in maintaining stability of the whole body.

MECHANISMS OF DISEASE
Blood Disorders

Most blood diseases are disorders of the formed elements. Thus it is not surprising that the basic mechanism of many blood diseases is the failure of the blood-producing myeloid and lymphatic tissues to properly form blood cells. In many cases, this failure is the result of damage by drugs, toxic chemicals, or radiation. In other cases, it results from an inherited defect or even cancer.

If bone marrow failure is the suspected cause of a particular blood disorder, a sample of myeloid tissue may be drawn into a syringe. The bone marrow is obtained from inside the pelvic bone (iliac crest) or the sternum. This procedure, called *aspiration biopsy cytology (ABC),* allows examination of the tissue that may help confirm or reject a tentative diagnosis. If the bone marrow is severely damaged, the choice of a **bone marrow transplant** may be offered to the patient. In this procedure, myeloid tissue from a compatible donor is intro-

duced into the recipient intravenously. If the recipient's immune system does not reject the new tissue, which is always a danger in this type of tissue transplant, the donor cells may establish a colony of new, healthy tissue in the bone marrow. In some cases, infusion of healthy marrow follows total body irradiation. This treatment destroys the diseased marrow, permitting the new tissue to grow.

Red Blood Cell Disorders

Anemia

The term **anemia** is used to describe different disease conditions caused by an inability of the blood to carry sufficient oxygen to the body cells. Anemias can result from inadequate numbers of RBCs or a deficiency of oxygen-carrying hemoglobin. Thus anemia can occur if the hemoglobin in RBCs is inadequate, even if normal numbers of RBCs are present.

Changes in RBC number

Anemias caused by an actual change in the number of RBCs can occur if blood is lost by hemorrhage, as with accidents or bleeding ulcers or if the blood-forming tissues cannot maintain normal numbers of blood cells. Such failures occur because of cancer, radiation (x-ray) damage, and certain types of infections. If bone marrow produces an *excess* of RBCs, the result is a condition called **polycythemia** (pol-ee-sye-THEE-me-ah). The blood in individuals suffering from this condition may contain so many RBCs that it may become too thick to flow properly.

One type of anemia characterized by an abnormally low number of red blood cells is **aplastic** (a-PLAS-tik) **anemia.** Although idiopathic forms of this disease occur, most cases result from destruction of bone marrow by drugs, toxic chemicals, or radiation. Less commonly, aplastic anemia results from bone marrow destruction by cancer. Because tissues that produce other formed elements are also affected, aplastic anemia is usually accompanied by a decreased number of white blood cells and platelets. Bone marrow transplants have been successful in treating some cases of aplastic anemia.

Pernicious (per-NISH-us) **anemia** is another disorder characterized by a low number of red blood cells. Pernicious anemia sometimes results from a dietary deficiency of vitamin B_{12}. Vitamin B_{12} is used in the formation of new red blood cells in the bone marrow. In many cases, pernicious anemia results from the failure of the stomach lining to produce *intrinsic factor*—the substance that allows vitamin B_{12} to be absorbed. Pernicious anemia can be fatal if not successfully treated. One method of treatment involves intramuscular injections of vitamin B_{12}.

Folate-deficiency anemia is similar to pernicious anemia because it also causes a decrease in the RBC count due to a vitamin deficiency. In this condition, it is *folic acid* (vitamin B_9) that is deficient. Folic acid deficiencies are common among alcoholics and other malnourished individuals. Treatment for folate-deficiency anemia involves taking vitamin supplements until a balanced diet can be restored.

Of course, a significant reduction in the number of red blood cells can occur as a result of blood loss. *Acute blood-loss anemia* often occurs after hemorrhages associated with trauma, extensive surgeries, or other situations involving a sudden loss of blood. *Chronic blood-loss anemia* is also called *anemia of chronic disease* because it often results from frequent or long-lasting episodes of blood loss associated with chronic diseases such as cancer and ulcers.

Changes in hemoglobin

The amount and quality of hemoglobin within red blood cells are just as important as the number of red blood cells. In hemoglobin disorders, RBCs are sometimes classified as *hyperchromic* (abnormally high hemoglobin content) or *hypochromic* (abnormally low hemoglobin content).

Iron (Fe) is a critical component of the hemoglobin molecule, forming the central core of each heme group (see Figure 16-4). Without adequate iron in the diet, the body cannot manufacture enough hemoglobin. The result is **iron deficiency anemia**—a worldwide medical problem.

The term **hemolytic** (hee-mo-LIT-ik) **anemia** applies to any of a variety of inherited blood disorders characterized by abnormal types of hemoglobin. The term *hemolytic* means "pertaining to blood breakage" and emphasizes the fact that abnormal hemoglobin often causes red blood cells to become distorted and easily broken. An example of a hemolytic anemia is **sickle cell anemia.** Another type of hemolytic anemia is **thalassemia** (thal-as-SEE-me-ah). Like sickle cell anemia, thalassemia is an inherited disorder, with both a mild and a severe form (*Thalassemia minor* and *Thalassemia major*).

White Blood Cell Disorders

The term **leukopenia** refers to an abnormally low WBC count (under 5,000 cells/mm^3 of blood). Various disease conditions may affect the immune system and decrease the amount of circulating WBCs. Acquired immune deficiency syndrome, or AIDS, results in marked leukopenia. **Leukocytosis** refers to an abnormally high WBC count. It is a much more common problem than leukopenia and almost always accompanies bacterial infections.

The term *leukemia* applies to a group of malignant diseases characterized by transformations of stem cells that replace normal cells, leading to leukocytosis, along with anemia. The leukemic cells may eventually leave the bone marrow and infiltrate the lymph nodes, liver, spleen, central nervous system, and other parts of the body. Five percent of all cancerous deaths are due to leukemia. There are many different types of leukemias, including *acute lymphocytic leukemia (ALL)* and *acute myelogenous leukemia (AML)*.

Clotting Disorders

Unfortunately, clots sometimes form in unbroken blood vessels of the heart, brain, lungs, or other organs—a dreaded thing because clots may produce sudden death by shutting off the blood supply to a vital organ. When a clot stays in the place where it formed, it is called a **thrombus** (THROM-bus), and the condition is spoken of as **thrombosis** (throm-BO-sus). If all or part of the clot dislodges and circulates through the bloodstream, it is called an **embolus** (EM-bo-lus), and the condition is called an **embolism** (EM-bo-lizm). Physicians now have some drugs that they can use to help prevent thrombosis and embolism. Heparin, for example, can be used to prevent excessive clotting. Heparin inhibits the conversion of prothrombin to thrombin, preventing formation of a thrombus. Dicumarol, an oral *anticoagulant,* is also frequently used to prevent excessive clotting. Dicumarol blocks the stimulating effect of vitamin K on the liver, and consequently the liver cells make less prothrombin. The prothrombin content soon falls low enough to prevent abnormal clotting.

A type of X-linked inherited disorder called **hemophilia** results from a failure to form blood-clotting factor VIII, IX, or XI. These clotting factors are necessary to complete the clotting process illustrated in Figure 16-21. Thus hemophilia is characterized by a relative inability to form blood clots. Because minor blood vessel injuries are common in ordinary life, hemophilia can be a life-threatening condition.

A more common type of clotting disorder results from a decrease in the platelet count—a condition called **thrombocytopenia** (throm-bo-si-toe-PEE-nee-ah). This condition is characterized by bleeding from many small blood vessels throughout the body, most visibly in the skin and mucous membranes. Although a number of different mechanisms can result in thrombocytopenia, the usual cause is bone marrow destruction by drugs, chemicals, radiation, or cancer. Drugs may cause thrombocytopenia as a side effect. In such cases, stopping use of the drug usually solves the problem.

CASE STUDY
ACUTE LYMPHOCYTIC LEUKEMIA

M.M. is a 7-year-old girl who was brought to her physician by her mother with complaints of general fatigue, loss of appetite, and unexplained bruises and "rash" for the past 2 weeks.

M.M. was a full-term infant of an uncomplicated pregnancy and delivery. She has never been exposed to ionizing irradiation. All immunizations are current; she has had only one childhood disease, chickenpox, at age 5. M.M. has one brother, aged 4, who is in apparent good health. The family history is unremarkable with one exception: the paternal grandfather died at 61 from cancer of the colon.

M.M.'s vital signs were as follows: temperature 37°C; heart rate 92; respirations 18; blood pressure 90/60. Her height and weight are average. Her skin is pale, warm, and dry with bruises on the extremities and trunk. The "rash" was found to be minute skin hemorrhages on arms

and chest. Neither spleen nor liver was palpable; however, there were three palpable nontender lymph nodes in the submaxillary chain. The CBC with differential and platelet count results were as follows: hemoglobin 8.8g/100 ml; hematocrit 26%; red blood cell count 3.1 million/mm^3; white blood cell count 13,000; neutrophils 6500; basophils 130 mm^3; eosinophils 360 mm^3; monocytes 1170 mm^3; lymphocytes 3640; blasts 10%; platelets 50,000 mm^3.

M.M. was immediately referred to a pediatric oncologist and admitted to the children's hospital for diagnosis and treatment. A bone marrow aspiration, lumbar puncture, blood work (complete blood count with differential and platelet count, and hepatic and nephrotoxic function stuies), and chest roentgenograms, (x-ray phgotographs) were performed. A diagnosis of acute lymphocytic leukemia (ALL), was made.

---

1. A diagnosis of leukemia was suspected from the complete blood count results on the basis of:
   a. Hemoglobin of 8.8 g
   b. Total white blood cell count of 13,000
   c. Platelets of 50,000 mm^3
   d. 10% blasts

2. Which of the following factors would indicate an improved prognosis for M.M.?
   a. ALL
   b. Her age
   c. White blood cell count of 13,000 at diagnosis
   d. Absence of hepatosplenomegaly
   e. All of the above

   M.M. began chemotherapy with various chemotherapeutic drugs. Eight weeks after diagnosis, M.M.'s bone marrow and peripheral blood smear showed no evidence of leukemic cells; therefore remission was attained. With remission, M.M. moved into the maintenance phase, during which she received monthly medications, complete blood

counts, and periodic bone marrow aspirations and lumbar punctures.

Three months after the maintenance phase began, during one of her clinic visits, M.M.'s complete blood count results were as follows: hemoglobin 6.8 g; hematocrit 20.4%; red blood cell count 2.1 million/mm^3; white blood cell count totals, 340/μl; neutrophils 200 mm; basophils 3 mm^3; eosinophils 6 mm^3; monocytes 30 mm^3; lymphocytes 82 mm^3; platelets 20,000 mm^3. The physician withheld the chemotherapy dose and transfused M.M. with 1 unit packed red blood cells and 4 units of platelets.

3. The cause of M.M.'s pancytopenia (reduction of all cellular blood elements) is:
   a. Deteriorating physical status caused by the disease process
   b. A known side effect of the chemotherapy
   c. Aplastic anemia resulting from the toxic effects of the chemotherapy

CHAPTER SUMMARY

COMPOSITION OF BLOOD
A. Introduction
   1. Blood—made up of plasma and formed elements
   2. Blood—complex transport medium that performs vital pickup and delivery services for the body
   3. Blood—keystone of body's heat-regulating mechanism
B. Blood volume
   1. Young adult male has approximately 5 liters of blood
   2. Blood volume varies according to age, body type, sex, and method of measurement

FORMED ELEMENTS OF BLOOD
A. Red blood cells (erythrocytes)
   1. Description of mature red blood cells

   a. Have no nucleus and shaped like tiny biconcave disks
   b. Do not contain ribosomes, mitochondria, and other organelles typical of most body cells
   c. Primary component is hemoglobin
   d. Most numerous of the formed elements
   2. Function of red blood cells
   a. RBC's critical role in the transport of oxygen and carbon dioxide depends on hemoglobin
   b. Carbonic anhydrase—enzyme in RBCs that catalyzes a reaction that joins carbon dioxide and water to form carbonic acid
   c. Carbonic acid—dissociates and generates bicarbonate ions, which diffuse out of the RBC and

serve to transport carbon dioxide in the blood plasma

3. Hemoglobin
   a. Within each RBC are approximately 200 to 300 million molecules of hemoglobin
   b. Hemoglobin is made up of four globin chains with each attached to a heme molecule
   c. Hemoglobin is able to unite with four oxygen molecules to form oxyhemoglobin to allow RBCs to transport oxygen where it is needed
   d. A male has a greater amount of hemoglobin than a female
   e. Anemia—a decrease in number or volume of functional RBCs in a given unit of whole blood
4. Formation of red blood cells (review Figures 16-5 and 16-6)
   a. Erythropoiesis—entire process of red blood cell formation
   b. RBC formation begins in the red bone marrow as hemocytoblast stem cells and goes through several stages of development to become erythrocytes; entire maturation process requires approximately 4 days
   c. RBCs are created and destroyed at approximately 100 million per minute in an adult; homeostatic mechanisms operate to balance the number of cells formed against the number of cells destroyed
5. Destruction of red blood cells
   a. Life span of a circulating RBC averages 105 to 120 days
   b. Macrophage cells phagocytose the aged, abnormal, or fragmented RBCs
   c. Hemoglobin is broken down and amino acids, iron, and bilirubin are released

B. White blood cells (leukocytes) (review Table 16-1)
1. Granulocytes
   a. Neutrophils (review Figure 16-9)—make up approximately 65% of total WBC count in a normal blood sample; highly mobile and very active phagocytic cells; capable of diapedesis; cytoplasmic granules contain lysosomes
   b. Eosinophils (review Figure 16-10)—account for 2% to 5% of circulating WBCs; numerous in lining of respiratory and digestive tracts; weak phagocytes; capable of ingesting inflammatory chemicals and proteins associated with antigen-antibody reaction complexes; provide protection against infections caused by parasitic worms and allergic reactions
   c. Basophils (review Figure 16-11)—account for only .5% to 1% of circulating WBCs; motile and capable of diapedesis; cytoplasmic granules contain histamine and heparin
2. Agranulocytes (Figures 16-12 and 16-13)
   a. Lymphocytes—smallest of the WBCs; second most numerous WBC; account for approximately 25% of circulating WBCs; T lymphocytes and B lymphocytes play an important role in immunity; T lymphocytes directly attach an infected or cancerous cell; B lymphocytes produce antibodies against specific antigens
   b. Monocytes—largest leukocyte, mobile and highly phagocytic cells
3. White blood cell numbers—a cubic millimeter of normal blood usually contains 5,000 to 9,000 leukocytes, with different percentages for each type; WBC numbers have clinical significance, since they change with certain abnormal conditions
4. Formation of white blood cells (review Figure 16-6)
   a. Granular and agranular leukocytes mature from the undifferentiated hemocytoblast stem cell
   b. Neutrophils, eosinophils, basophils, and a few lymphocytes and monocytes originate in red bone

marrow; most lymphocytes and monocytes develop from hemocytoblasts in lymphatic tissue

C. Platelets (review Figures 16-12 and 16-14)
1. Structure
   a. In circulating blood, platelets are small, pale bodies that appear as irregular spindles or oval disks
   b. Three important properties are agglutination, adhesiveness, and aggregation
   c. Platelet counts in adults average 250,000 per mm^3 of blood; normal range is 150,000 to 350,000 per mm^3
2. Functions of platelets
   a. Important role in both hemostasis and blood coagulation
   b. Hemostasis—refers to stoppage of blood flow; however, if injury is extensive, the blood-clotting mechanism is activated to assist
3. Formation and life span of platelets (7 to 10 days)—formed in red bone marrow, lungs, and spleen by fragmentation of megakaryocytes

BLOOD TYPES (BLOOD GROUPS)

A. The ABO system (Figures 16-15 to 16-17)
1. Every person's blood belongs to one of four ABO blood groups
2. Named according to antigens present on RBC membranes
   a. Type A—antigen A on RBC
   b. Type B—antigen B on RBC
   c. Type AB—both antigen A and antigen B on RBC; known as universal recipient
   d. Type O—neither antigen A nor antigen B on RBC; known as universal donor

B. The Rh system (Figure 16-18)
1. Rh-positive blood—Rh antigen is present on the RBCs
2. Rh-negative—RBCs have no Rh antigen present
3. Anti-Rh antibodies are not normally present in blood; anti-Rh antibodies can appear in Rh-negative blood if it has come in contact with Rh-positive RBCs

BLOOD PLASMA

A. Plasma—liquid part of blood; clear, straw-colored fluid; made up of 90% water and 10% solutes (Figure 16-19)
B. Solutes—6% to 8% of plasma solutes are proteins, consisting of three main compounds
1. Albumins—key role in blood clotting
2. Globulins—essential component of the immunity mechanism
3. Fibrinogen—key role in blood clotting
C. Plasma proteins play an essential role in maintaining normal blood circulation

BLOOD CLOTTING (COAGULATION)

A. Mechanism of blood clotting—goal of coagulation is to stop bleeding and prevent loss of vital body fluid in a swift and sure method; the "classic theory" is as follows:
1. Step I—prothrombin is converted into thrombin by thromboplastin, aided by calcium ions; thromboplastin may be released from platelets or damaged tissues (Figures 16-20 and 16-21)
   a. Intrinsic system—when thromboplastin is released from damaged platelets, plasma factors IV, V, VIII, IX, X, XI, and XII are also required to convert prothrombin to thrombin
   b. Extrinsic system—when thromboplastin is released from damaged tissues, plasma factors IV, V, VII, and X are also required to convert prothrombin to thrombin
2. Step II—fibrinogen is converted into strands of insoluble fibrin, which catches blood cells to form a clot; thrombin and plasma factors IV and XIII are required to complete step II
B. Conditions that oppose clotting

1. Clot formation in intact vessels is opposed
2. Several factors oppose clotting
   a. Perfectly smooth surface of the normal endothelial lining of blood vessels does not allow platelets to adhere
   b. Antithrombins—substances in the blood that oppose or inactivate thrombin; prevent thrombin from converting fibrinogen to fibrin; e.g., heparin
C. Conditions that hasten clotting
   1. Rough spot in the endothelium
   2. Abnormally slow blood flow
D. Clot dissolution
   1. Fibrinolysis—physiological mechanism that dissolves
   2. Fibrinolysin—enzyme in the blood that catalyzes the hydrolysis of fibrin, causing it to dissolve
   3. Additional factors are presumed to aid clot dissolution; e.g., substances that activate profibrinolysin

THE BIG PICTURE: BLOOD AND THE WHOLE BODY

A. Blood plasma transports substances, including heat, around the body, linking all body tissues together
   1. Substances can be transported between almost any two points in the body
B. Blood tissue contains formed elements—blood cells and platelets
   1. Red blood cells assist in the transport of oxygen and carbon dioxide
   2. White blood cells assist in the defense mechanisms of the whole body
   3. Platelets prevent loss of the fluid that constitutes the internal environment
C. No organ or system of the body can maintain proper levels of nutrients, gases, or water without direct or indirect help from blood
   1. Other systems assist the blood
D. Blood is useless unless it continues to transport, defend, and maintain balance

MECHANISMS OF DISEASE: BLOOD DISORDERS

A. Blood diseases are disorders of the formed elements
   1. Basic mechanism of many blood diseases is failure of myeloid and lymphatic tissues to properly form blood cells
      a. This failure comes from damage by chemicals, radiation, inherited defects, or cancer
   2. Bone marrow transplants may be used to treat cases in which the bone marrow failure is the cause of a blood disorder

B. Red blood cell disorders (anemia)—various disease conditions resulting from an inability of the blood to carry sufficient oxygen to body cells; anemias result from inadequate RBCs or a deficiency of hemoglobin
   1. The number of RBCs can change if blood is lost from hemorrhage or if blood-forming tissues fail to maintain the numbers of blood cells
   2. Some types of anemias
      a. Aplastic anemia—low numbers of red blood cells from destruction of bone marrow by chemicals, drugs, or radiation
      b. Pernicious anemia—low numbers of red blood cells from lack of vitamin B_{12}
      c. Folate-deficiency anemia—low red blood cell count from lack of folic acid
      d. Acute blood loss anemia—reduction of red blood cells after hemorrhages
      e. Anemia of chronic disease (chronic blood loss anemia)—frequent or long-lasting episodes of blood loss from chronic diseases
   3. Hemoglobin disorders—RBCs are classified as hyperchromic (high hemoglobin count) or hypochromic (low hemoglobin count)
      a. Iron deficiency anemia results from insufficient iron to manufacture enough hemoglobin
      b. Hemolytic anemia—various inherited blood disorders characterized by abnormal types of hemoglobin
         1. Sickle cell anemia
         2. Thalassemia
C. White blood cell disorders
   1. Leukopenia—abnormally low WBC count
   2. Leukocytosis—abnormally high WBC count
   3. Leukemia—a group of malignant diseases characterized by the transformation of stem cells that replace normal cells, leading to leukocytosis and anemia
D. Clotting disorders—clots form in unbroken blood vessels in organs and may produce sudden death by shutting off the blood supply
   1. Thrombus—when a clot stays in the place it formed, the condition is called thrombosis
   2. Embolus—part of a clot dislodges and flows through the bloodstream, a condition called embolism
   3. Hemophilia—inherited disorder in which the blood fails to clot
   4. Thrombocytopenia—disorder resulting from a decrease in platelets; it is characterized by bleeding from many small vessels

REVIEW QUESTIONS

1. Compare different kinds of blood cells as to appearance and size, functions, formation, destruction, and life span; number per cubic millimeter of blood.
2. Define the term *hematocrit*. How is it determined?
3. What is the function of carbonic anhydrase?
4. Describe the structure of hemoglobin.
5. How does the structure of hemoglobin allow for it to combine with oxygen?
6. Discuss the steps involved in erythropoiesis.
7. What is the average life span of a circulating red blood cell?
8. How are granulocytes similar to agranulocytes? How do they differ?
9. Define the term *positive chemotaxis*.
10. List the important physical properties of platelets.
11. Explain what is meant by type AB blood. Rh-negative blood.
12. Identify the blood type referred to as universal donor blood and explain why this is the case. Do the same for universal recipient blood.
13. Which organ is responsible for the synthesis of most plasma proteins?
14. What is the normal plasma protein concentration?
15. What are some functions served by plasma proteins?
16. Describe the role of platelets in hemostasis and blood clotting.
17. What triggers blood clotting?
18. Write the two equations that show the basic chemical reactions that produce a blood clot.
19. Identify factors that oppose blood clotting. Do the same for those factors that hasten blood clotting.
20. Describe the physiological mechanism that dissolves clots.
21. Describe the hemoglobin in a person with sickle cell trait.
22. What is the purpose of blood doping?
23. Explain the condition erythroblastosis fetalis.
24. Identify three clinical methods of hastening clotting.

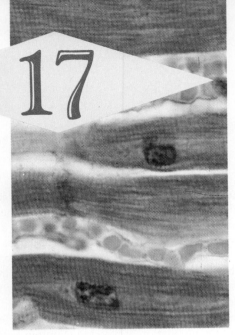

Cardiac muscle tissue
Erlandsen/Magney: Color Atlas of Histology

17 Anatomy of the Cardiovascular System

CHAPTER OUTLINE

OBJECTIVES

After you have completed this chapter, you should be able to:

1. List the primary organs of the cardiovascular system and relate each organ or group of organs to the movement and/or the direction of blood flow in the system.
2. Discuss the location, size, and position of the heart in the thoracic cavity.
3. Describe the structure of the pericardium, the function of each pericardial layer, the pericardial space, and the pericardial fluid.
4. List and discuss the three layers of the heart wall, the heart cavities, and the valves.
5. Trace blood through the heart and its coronary blood vessels.
6. List the anatomical components of the heart conduction system.
7. List, locate, and compare the primary coats or layers of tissue found in major arteries and veins.
8. Correlate structure of arteries, arterioles, veins, venules, and capillaries with their function.
9. List anatomical components of microcirculation and discuss the reservoir function of veins.
10. Trace the path of blood flow as it leaves the right side of the heart until it returns to the left side of the heart.
11. Identify the unusual anatomical characteristics of the hepatic portal circulation.
12. List and discuss the function of the six structures characteristic of the fetal circulation.
13. Discuss changes that occur in the vascular system at birth.

465

Anastomosis (ah-nas-toe-MOE-sis) Connection between blood vessels that allows collateral circulation

Arteriole (ar-TEER-ee-ole) Small artery

Artery Blood vessel that carries blood away from the heart

Atrium One of two upper chambers of the heart; "receiving chamber" of the heart

Capillary (KAP-i-lair-ee) Microscopic blood vessel carrying blood from arterioles to venules

Endocardium (en-doe-KAR-dee-um) Delicate layer of endothelial tissue lining the interior of the heart wall

Endothelium (en-doe-THEE-lee-um) Membranous tissue lining the heart and blood vessels

Epicardium (ep-i-KAR-dee-um) Visceral layer of serous pericardium, adheres to outside of heart; outer layer of heart wall

Myocardium (my-o-KAR-dee-um) Thick, contractile, middle layer of the heart wall composed of specially constructed and arranged cardiac muscle cells

Pericardium (pair-i-KAR-dee-um) Loose-fitting inextensible sac covering the heart

Pulmonary circulation Blood flow from the lungs to the heart and back to the lungs where exchange of gases occurs

Systemic circulation Blood flow from the heart through blood vessels to all parts of the body (except gas-exchange tissue of lungs) and back to the heart

Vein Large blood vessel that returns blood to the heart

Ventricle (VEN-tri-kul) One of two muscular lower chambers of the heart; "pumping chamber" of the heart

Venule (VEN-yool) Small vein

The cardiovascular system is sometimes called, simply, the *circulatory system*. It consists of the heart, which is a muscular pumping device, and a closed system of vessels called *arteries, veins,* and *capillaries.* As the name implies, blood contained in the circulatory system is pumped by the heart around a closed circle or circuit of vessels as it passes again and again through the various "circulations" of the body (discussed on p. 479).

As in the adult, survival of the developing embryo depends on the circulation of blood to maintain homeostasis and a favorable cellular environment. In response to this need, the cardiovascular system makes its appearance early in development and reaches a functional state long before any other major organ system. Incredible as it seems, the heart begins to beat regularly early in the fourth week after fertilization!

HEART

Location of the Heart

The human heart is a four-chambered muscular organ, shaped and sized roughly like a person's closed fist. It lies in the mediastinum, or middle region of the thorax, just behind the body of the sternum between the points of attachment of the second through the sixth ribs. Approximately two thirds of the heart's mass is to the left of the midline of the body and one third to the right.

Posteriorly the heart rests against the bodies of the fifth to the eighth thoracic vertebrae. Because of its placement between the sternum in front and the bodies of the thoracic vertebrae behind, it can be compressed by application of pressure to the lower portion of the body of the sternum using the heel of the hand. Rhythmic compression of the heart in this way can maintain blood flow in cases of cardiac arrest and, if combined with effective artificial respiration, the resulting procedure, called *cardiopulmonary resuscitation (CPR),* can be life saving.

The anatomical position of the heart in the thoracic cavity is shown in Figure 17-1.

The lower border of the heart, which forms a blunt point known as the *apex,* lies on the diaphragm, pointing toward the left. To count the apical beat, one must place a stethoscope directly over the apex, that is, in the space between the fifth and sixth ribs (fifth intercostal space) on a line with the midpoint of the left clavicle.

The upper border of the heart, that is, its base, lies just below the second rib. The boundaries, which, of course, indicate its size, have considerable clinical importance, since a marked increase in heart size accompanies certain

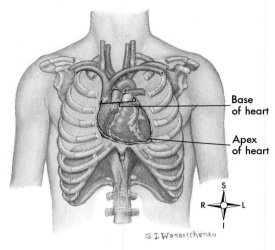

FIGURE 17-1   **Location of the heart.** The heart is located within the mediastinum, which is the middle region of the thoracic cavity.

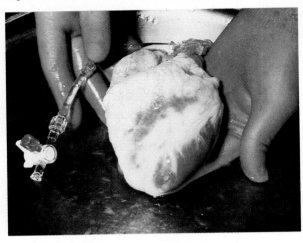

FIGURE 17-2   **Appearance of the heart.** This photograph shows a living human heart prepared for transplantation into a patient. Notice its size relative to the hands that are holding it.

types of heart disease. Therefore when diagnosing heart disorders, the physician charts the boundaries of the heart. The "normal" boundaries of the heart are, however, influenced by such factors as age, body build, and state of contraction.

Size and Shape of the Heart

At birth the heart is said to be transverse (wide) in type and appears large in proportion to the diameter of the chest cavity. In the infant, it is 1/130 of the total body weight compared to about 1/300 in the adult. Between puberty and 25 years of age the heart attains its adult shape and

weight—about 310 g is average for the male and 225 g for the female.

In the adult the shape of the heart tends to resemble that of the chest. In tall, thin individuals the heart is frequently described as elongated, whereas in short, stocky individuals it has greater width and is described as transverse. In individuals of average height and weight it is neither long nor transverse but somewhat intermediate between the two (Figure 17-2). Its approximate dimensions are length, 12 cm; width, 9 cm; and depth, 6 cm. Figure 17-3 shows details of the heart and great vessels in a posterior view and in an anterior view.

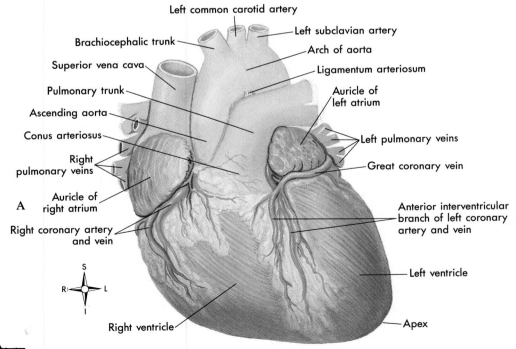

FIGURE 17-3   **The heart and great vessels. A,** Anterior view of the heart and great vessels.

Continued.

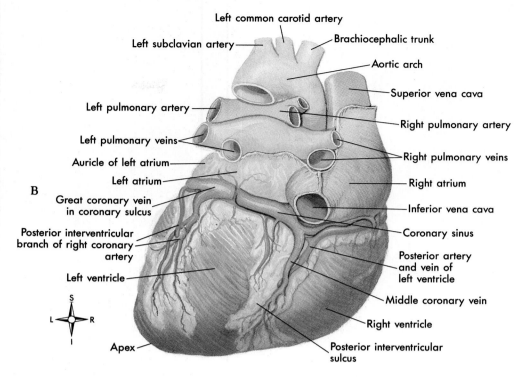

B

FIGURE 17-3 Cont'd,   **B,** Posterior view of the heart and great vessels.

Coverings of the Heart
Structure of the heart coverings

The heart has its own special covering, a loose-fitting inextensible sac called the **pericardium.** The pericardium consists of two parts: a fibrous portion and a serous portion (Figure 17-4). The sac itself is made of tough white fibrous tissue but is lined with smooth, moist serous membrane—the parietal layer of the serous pericardium. The same kind of membrane covers the entire outer surface of the heart. This covering layer is known as the visceral layer of the serous pericardium or as the **epicardium.** The fibrous sac attaches to the large blood vessels emerging from the top of the heart but not to the heart itself. Therefore it fits loosely around the heart, with a slight space between the visceral layer adhering to the heart and the parietal layer adhering to the inside of the fibrous sac. This space is called the **pericardial space.** It contains (10 to 15 ml) of lubricating fluid secreted by the serous membrane and called **pericardial fluid.**

The structure of the pericardium can be summarized in outline form as follows:

- **Fibrous pericardium**—tough, loose-fitting, and inelastic sac around the heart
- **Serous pericardium**—consisting of two layers
    - *Parietal layer*—lining inside of the fibrous pericardium
    - *Visceral layer (epicardium)*—adhering to the outside of the heart; between visceral and parietal layers is a space, the pericardial space, which contains a few drops of pericardial fluid

Function of the heart coverings

The fibrous pericardial sac with its smooth, well-lubricated lining provides protection against friction. The heart moves easily in this loose-fitting jacket with no danger of irritation from friction between the two surfaces, as long as the serous pericardium remains normal.

QUICK CHECK

✔ 1. In anatomical terms, where is the heart located?
✔ 2. Name the layers of tissue that make up the pericardium.
✔ 3. What is the function of the pericardium?

Structure of the Heart
Wall of the heart

Three distinct layers of tissue make up the heart wall (Figure 17-4) in both the atria and the ventricles.

1. The outer layer of the heart wall is called the **epicardium**—the visceral layer of the serous pericardium already described.

2. The bulk of the heart wall is the thick, contractile, middle layer of specially constructed and arranged cardiac muscle cells called the **myocardium.** The minute structure of cardiac muscle has been described in Chapters 4 and 9. Recall that cardiac muscle tissue is comprised of many branching cells that are joined into a continuous mass by *intercalated discs.* Because each intercalated disc includes many gap junctions, large areas of cardiac muscle are

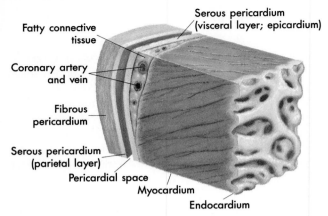

FIGURE 17-4   **Wall of the heart.** This section of the heart wall shows the fibrous pericardium, the parietal and visceral layers of the serous pericardium (with the pericardial space between them), the myocardium, and the endocardium. Notice that there is fatty connective tissue between the visceral layer of the serous pericardium (epicardium) and the myocardium. Notice also that the endocardium covers beamlike projections of myocardial muscle tissue, called trabeculae.

electrically coupled into a single functional unit called a *syncytium* (meaning "joined cells"). Because they form a syncytium, muscle cells can pass an action potential along a large area of the heart wall, stimulating contraction in each muscle fiber of the syncytium. Another advantage of the syncytium structure is that the cardiac fibers form a continuous sheet of muscle that wraps entirely around the cavities within the heart. Thus, the encircling myocardium can compress the heart cavities, and the blood within them, with great force. Recall also that cardiac muscles are *autorhythmic,* meaning that they can contract on their own in a slow, steady rhythm. Because the muscular myocardium can contract powerfully and rhythmically, the heart is an efficient pump for blood.

3. The lining of the interior of the myocardial wall is a delicate layer of endothelial tissue known as the **endocardium.** *Endothelium* is the type of membranous tissue that lines the heart and blood vessels. Endothelium resembles simple squamous epithelium, except for the fact that during embryonic development endothelium arises from different tissue than does epithelium. Notice in Figure 17-4 that the endocardium covers beamlike projections of myocardial tissue. These muscular projections are called *trabeculae.*

Chambers of the heart

The interior of the heart is divided into four cavities, or heart chambers (Figure 17-5). The two upper chambers are called *atria* (singular, *atrium*), and the two lower chambers are called *ventricles.* The left chambers are separated from the right chambers by an extension of the heart wall called the *septum.*

Atria. The two superior chambers of the heart—the **atria**—are often called the "receiving chambers" because they receive blood from vessels called *veins.* Veins are the large blood vessels that return blood from various tissues to

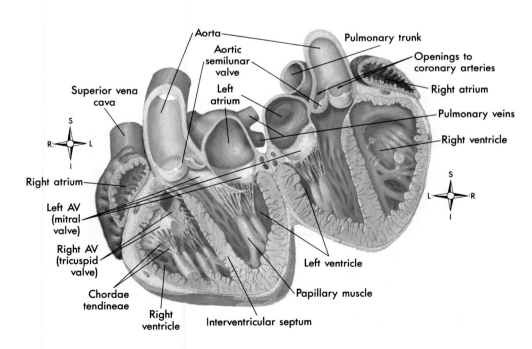

FIGURE 17-5   **Interior of the heart.** This illustration shows the heart as it would appear if it were cut along a frontal plane and opened like a book. The front portion of the heart lies to the reader's right; the back portion of the heart lies to the reader's left. The four chambers of the heart—two atria and two ventricles—are easily seen in this figure.

the heart so that the blood can be pumped out to tissues again. Figure 17-6 shows how the atria alternately relax and contract to receive blood, then push it into the lower chambers. Because the atria need not generate great pressure to move blood such a small distance, the myocardial wall of each atrium is not very thick.

If you look at Figure 17-3, A, you will notice that part of each atrium is labeled as an *auricle*. The term *auricle* (meaning "little ear") refers to the earlike flap protruding from each atrium. Thus the auricles are *part of* the atria. The terms *auricle* and *atrium* should not be used synonymously.

Ventricles

The **ventricles** are the two lower chambers of the heart. Because the ventricles receive blood from the atria and pump blood out of the heart into arteries, the ventricles are considered to be the primary "pumping chambers" of the heart. Because more force is needed to pump blood such a distance, the myocardium of each ventricle is thicker than the myocardium of either atrium. The myocardium of the left ventricle is thicker than that of the right ventricle because the left ventricle pushes blood through most vessels of the body, whereas the right ventricle pushes blood only through the vessels that serve the gas-exchange tissues of the lungs.

The pumping action of the heart chambers is summarized in Figure 17-6 and described further in Chapter 18.

Valves of the heart

The heart valves are mechanical devices that permit the flow of blood in one direction only. Four sets of valves are of importance to the normal functioning of the heart (Figures 17-6 and 17-7). Two of these, the **atrioventricular (AV) valves,** guard the openings between the atria and the ventricles (atrioventricular orifices). The atrioventricular valves are also called **cuspid valves.** The other two heart valves, the **semilunar (SL) valves,** are located where the pulmonary artery and the aorta arise from the right and left ventricles, respectively.

Atrioventricular valves. The atrioventricular valve guarding the right atrioventricular orifice consists of three flaps (cusps) of endocardium. The free edge of each flap is anchored to the *papillary muscles* of the right ventricle by several cordlike structures called **chordae tendineae.** Because the right atrioventricular valve has three flaps, it is also called the **tricuspid valve.** The valve that guards the left atrioventricular orifice is similar in structure to the right atrioventricular valve, except that it has only two flaps and is, therefore, also called the **bicuspid** or, more commonly, the **mitral valve.**

The construction of both atrioventricular valves allows blood to flow from the atria into the ventricles but prevents it from flowing back up into the atria from the ventricles. Ventricular contraction forces the blood in the ventricles hard against the valve flaps, closing the valves and thereby ensuring the movement of the blood upward into the

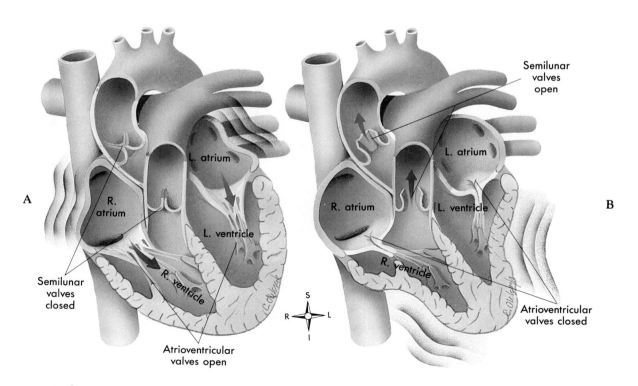

FIGURE 17-6 Chambers and valves of the heart. These illustrations depict the action of the heart chambers and valves when the atria contract (**A**) and when the ventricles contract (**B**).

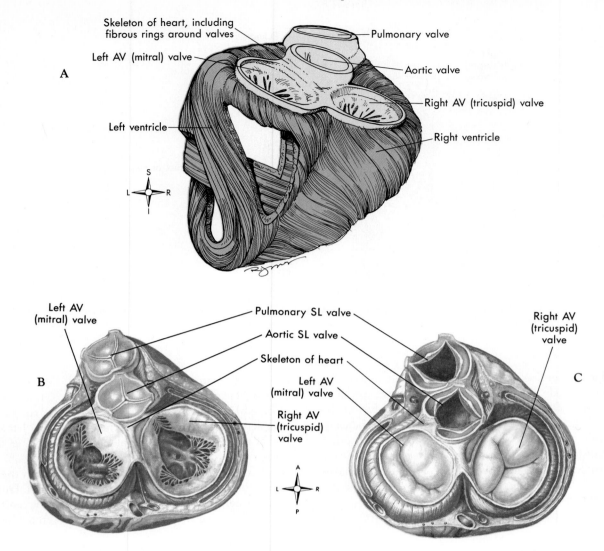

Skeleton of heart, including
fibrous rings around valves

Left AV (mitral) valve

A

Left ventricle

Pulmonary valve

Aortic valve

Right AV (tricuspid) valve

Right ventricle

Left AV
(mitral) valve

B

Pulmonary SL valve

Aortic SL valve

Skeleton of heart

Left AV
(mitral) valve

Right AV
(tricuspid)
valve

Right AV
(tricuspid)
valve

C

FIGURE 17-7 Structure of the heart valves. A, This posterior view shows part of the ventricular myocardium with the heart valves still attached. The rim of each heart valve is supported by a fibrous structure, called the skeleton of the heart, that encircles all four valves. **B,** This figure shows the heart valves as if the figure in **A** is viewed from above. Notice that the semilunar (SL) valves are closed and the atrioventricular (AV) valves are open, as when the atria are contracting (compare to Figure 17-6, A). **C,** This figure is similar to **B,** except that the semilunar valves are closed and the atrioventricular valves are open, as when the ventricles are contracting (compare to Figure 17-6, B).

pulmonary artery and aorta as the ventricles contract (see Figure 17-6).

Semilunar valves. The semilunar (SL) valves consist of half-moon shaped flaps growing out from the lining of the pulmonary artery and aorta. The semilunar valve at the entrance of the pulmonary artery (pulmonary trunk) is called the *pulmonary semilunar valve.* The semilunar valve at the entrance of the aorta is called the *aortic semilunar valve.* When these valves are closed, as in Figures 17-6, A and 17-7, B, blood fills the spaces between the flaps and the vessel wall. Each flap then looks like a tiny, filled bucket. Inflowing blood smooths the flaps against the blood vessel walls, collapsing the buckets and thereby opening the valves (see Figures 17-6, B and 17-7, C). Closure of the semilunar valves, as of the atrioventricular valves, simul-

taneously prevents backflow and ensures forward flow of blood in places where there would otherwise be considerable backflow. Whereas the atrioventricular valves prevent blood from flowing back up into the atria from the ventricles, the semilunar valves prevent it from flowing back down into the ventricles from the aorta and pulmonary artery.

Skeleton of the heart. Figure 17-7 shows the fibrous structure that is often called the *skeleton of the heart.* It is a set of connected rings that serve as a semirigid support for the heart valves (on the inside of the rings) and for the attachment of cardiac muscle of the myocardium (on the outside of the rings). The skeleton of the heart also serves as an electrical barrier between the myocardium of the atria and the myocardium of the ventricles.

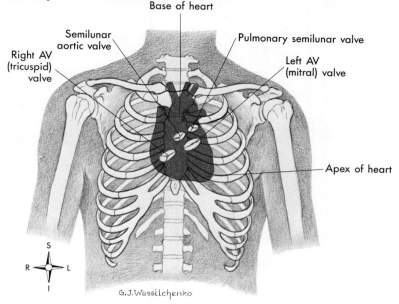

FIGURE 17-8   Relation of heart to anterior wall of thorax. Valves of the heart are projected on the anterior thoracic wall.

Surface projection. When listening to the sounds of the heart on the body surface, as with a stethoscope, one must have an idea of the relationship between the valves of the heart and the surface of the thorax. Figure 17-8 indicates the surface relationship of the four heart valves and other features of the heart. It is important to remember, however, that considerable variation within the normal range makes a precise "surface projection" outline of the heart's structure on the chest wall difficult.

Flow of blood through the heart. To understand the functional anatomy of the heart and the rest of the cardiovascular system, one should be able to trace the flow of blood through the heart. As we take you through one complete pass through the right heart, then the left side of the heart, trace the path of blood flow with your finger, using Figure 17-6.

We can trace the path of blood flow through the right side of the heart by beginning in the right atrium. From the right atrium, blood flows through the right atrioventricular (tricuspid) valve into the right ventricle. From the right

---

ECHOCARDIOGRAPHY

Echocardiography is a noninvasive technique for evaluating the internal structures and motions of the heart and great vessels. Ultrasound beams are directed into the patient's chest by a transducer. The transducer then acts as a receiver of the ultrasonic waves, or "echoes," to form images.

The images produced from the echoes are transmitted to a monitor. Echocardiography graphically demonstrates overall cardiac performance. It shows the internal dimensions of the chambers, size and motion of the intraventricular septum and posterior left ventricular wall, valve motion and anatomy, direction of blood flow, and the presence of increased pericardial fluid, blood clots, and cardiac tumors.

Three echocardiographic techniques are used in clinical practice. All use a transducer that emits ultrasonic pulses through the chest wall and receives echoes from the cardiac structures.

♦ **M-mode (motion-mode) echocardiography** produces an "ice-pick" image of a narrow area within the ultrasonic beam. It shows position and motion of cardiac structures.

♦ **Two-dimensional (2-D) echocardiography** produces a cross-sectional view and real time motion of cardiac structures. It allows the ultrasonic beam to move quickly, showing the structures and lateral movement. Together this shows the spatial relationship between the heart structures. Numerous views are possible with 2-D echocardiography.

Echocardiography is used to diagnose valvular heart disease, congenital heart disease, cardiomyopathy, congestive heart failure, pericardial disease, cardiac tumors, and intracardiac thrombi. It is also used to evaluate the function of the left ventricle after a myocardial infarction and the presence of pericardial fluid.

♦ **Doppler ultrasonography** provides continuous waves but also uses a sound, or frequency ultrasound, to record the direction of blood flow through the heart. These sound waves are reflected off red blood cells as they pass through the heart, allowing the velocity of blood to be calculated as it travels through the heart chambers.

Color Doppler mapping is a variation that converts recorded flow frequencies into different colors. These color images are then superimposed on M-mode or 2-D echocardiograms, allowing more detailed evaluation of disorders.

ventricle, blood flows through the pulmonary semilunar valve into the first portion of the pulmonary artery, the pulmonary trunk. The pulmonary trunk branches to form the left and right pulmonary arteries, which conduct blood to the gas-exchange tissues of the lung. From there, blood flows through pulmonary veins into the left atrium.

We can begin to trace the path of blood flow through the left side of the heart from the left atrium. From the left atrium, blood flows through the left atrioventricular (mitral) valve into the left ventricle. From the left ventricle, blood flows through the aortic semilunar valve into the aorta. Branches of the aorta supply all the tissues of the body except the gas-exchange tissues of the lungs. Blood leaving the head and neck tissues empties into the superior vena cava, and blood leaving the lower body empties into the inferior vena cava. Both of these large vessels conduct blood into the right atrium, bringing us back to the point where we began.

> **QUICK CHECK**
> ✔ 1. *Name the three layers of tissue that make up the wall of the heart. What is the function of each layer?*
> ✔ 2. *Name the four chambers of the heart and the valves associated with them.*
> ✔ 3. *How do atrioventricular valves differ from semilunar valves?*

Blood supply of heart tissue

Coronary arteries. Myocardial cells receive blood by way of two small vessels, the right and left coronary arteries. Since the openings into these vitally important

vessels lie behind flaps of the aortic semilunar valve, they come off of the aorta at its very beginning and are its first branches. Both right and left coronary arteries have two main branches, as shown in Figure 17-9, A.

More than a half million Americans die every year from coronary disease, and another 3.5 million or more are estimated to suffer some degree of incapacitation from this great killer. Knowledge about the distribution of coronary artery branches, therefore, has the utmost practical importance. Here, then, are some principles about the heart's own blood supply that are worth noting:

♦ Both ventricles receive their blood supply from branches of the right and left coronary arteries.

♦ Each atrium, in contrast, receives blood only from a small branch of the corresponding coronary artery.

♦ The most abundant blood supply goes to the myocardium of the left ventricle—an appropriate amount, since the left ventricle does the most work and so needs the most oxygen and nutrients delivered to it.

♦ The right coronary artery is dominant in about 50% of all hearts; the left coronary artery is dominant in about 20%; and in about 30%, neither right nor left coronary artery dominates.

Another fact about the heart's own blood supply—one of life-and-death importance—is that only a few connections, or anastomoses exist between the larger branches of the coronary arteries. An **anastomosis** consists of one or more branches from the proximal part of an artery to a more distal part of itself or of another artery. Thus anastomoses provide detours in which arterial blood can travel if the main route becomes obstructed. In short, they provide collateral circulation to a part. This explains why the

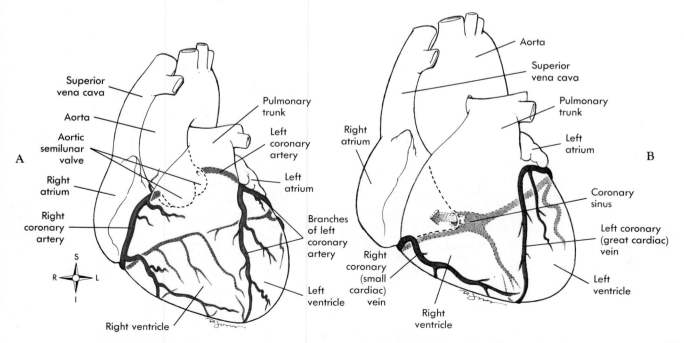

FIGURE 17-9   Coronary circulation. A, Arteries. **B,** Veins. Both illustrations are anterior views of the heart. Vessels near the anterior surface are more darkly colored than vessels of the posterior surface seen through the heart.

scarcity of anastomoses between larger coronary arteries looms so large as a threat to life. If, for example, a blood clot plugs one of the larger coronary artery branches, as it frequently does in coronary thrombosis or embolism, too little or no blood can reach some of the heart muscle cells. They become ischemic, in other words. Deprived of oxygen, metabolic function is impared and cell survival is threatened. **Myocardial infarction** (death of ischemic heart muscle cells) soon results. There is another anatomical fact, however, that brightens the picture somewhat—many anastomoses exist between very small arterial vessels in the heart, and, given time, new ones develop and provide collateral circulation to ischemic areas. In recent years, several surgical procedures have been devised to aid this process (see Figure 17-29).

Coronary veins. After blood has passed through capillary beds in the myocardium, it enters a series of coronary veins before draining into the right atrium through a common venous channel called the *coronary sinus*. Several veins that collect blood from a small area of the right ventricle do not end in the coronary sinus but instead drain directly into the right atrium. As a rule, the coronary veins (see Figure 17-9) follow a course that closely parallels that of the coronary arteries.

Conduction system of the heart

Four structures—the **sinoatrial node, atrioventricular node, atrioventricular bundle,** and **Purkinje fibers**—compose the conduction system of the heart. Each of these structures consists of cardiac muscle modified enough in structure to differ in function from ordinary cardiac muscle. The main specialty of ordinary cardiac muscle is contraction. In this, it is like all muscle, and like all muscle, ordinary cardiac muscle can also conduct impulses. But conduction alone is the specialty of the modified cardiac muscle that composes the conduction system structures.

Sinoatrial node. The sinoatrial node (SA node, or pacemaker) consists of hundreds of cells located in the right atrial wall near the opening of the superior vena cava (Figure 17-10).

Atrioventricular node. The atrioventricular node (AV node or node of Tawara), a small mass of special cardiac muscle tissue, lies in the right atrium along the lower part of the interatrial septum.

Atrioventricular bundle and Purkinje fibers. The

CARDIAC ENZYME STUDIES

When the heart muscle is damaged, enzymes contained within the muscle cells are released into the bloodstream, causing increased serum levels. Studies of cardiac enzymes (CPK, AST, LDH) are useful in confirming a myocardial infarction (MI) when they are viewed in relation to other heart function tests and the complete medical exam. The rate of release and distribution of specific enzymes over time varies after an infarction. It is this pattern of enzyme elevation that is of diagnostic importance. Enzyme determinations are also useful in following the course of a myocardial infarction and in detecting an extension of it.

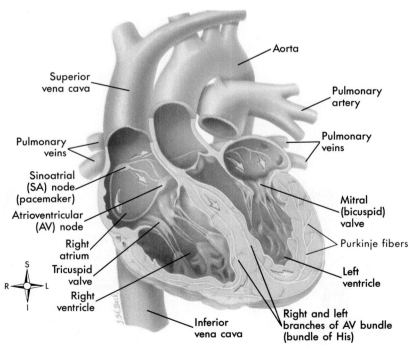

FIGURE 17-10   Conduction system of the heart. Specialized cardiac muscle cells in the wall of the heart rapidly conduct an electrical impulse throughout the myocardium. The signal is initiated by the SA node (pacemaker) and spreads to the rest of the atrial myocardium and to the AV node. The AV node then initiates a signal that is conducted through the ventricular myocardium by way of the AV bundle (of His) and Purkinje fibers.

atrioventricular bundle (AV bundle or bundle of His) is a bundle of special cardiac muscle fibers that originate in the AV node and extend by two branches down the two sides of the interventricular septum. From there, they continue as the Purkinje fibers. The latter extend out to the lateral walls of the ventricles and papillary muscles. The functioning of the conduction system of the heart will be discussed in Chapter 18.

Nerve supply of the heart

Both divisions of the autonomic nervous system send fibers to the heart. Sympathetic fibers (contained in the middle, superior, and inferior cardiac nerves) and parasympathetic fibers (in branches of the vagus nerve) combine to form **cardiac plexuses** located close to the arch of the aorta. From the cardiac plexuses, fibers accompany the right and left coronary arteries to enter the heart. Here most of the fibers terminate in the SA node, but some end in the AV node and in the atrial myocardium. Sympathetic nerves to

the heart are also called *accelerator nerves*. Vagus fibers to the heart serve as *inhibitory* or *depressor nerves*.

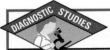

CARDIAC NUCLEAR SCANNING

Cardiac nuclear scanning is a safe method of recognizing cardiac diseases. Basically, it is used to evaluate blood flow in coronary arteries or to evaluate ventricular function. The patient is given an intravenous (IV) injection of an appropriate radioactive substance. Shortly thereafter, a gamma-ray detector is placed over the heart while the patient is lying down. The detector records the image of the heart, and a Polaroid photograph of that image is taken. The only discomfort for the patient is the initial injection.

Cardiac scanning is used for various clinical situations, particularly the following: (1) screening, evaluation of, and surveillance of adults for old and new myocardial infarctions; (2) evaluating chest pain or unusual results from other heart tests; (3) evaluation of bypass surgery or therapy.

These studies are nearly free of complications. The era of noninvasive myocardial radioscanning is just beginning; these tests will undoubtedly be a vital part of the evaluation of many patients with myocardial disease.

QUICK CHECK

✔ 1. *Briefly describe the general structure and the function of the coronary circulation. Why is an understanding of the coronary circulation so critical to understanding major types of heart disease?*
✔ 2. *What is meant by the term* conduction system of the heart?

BLOOD VESSELS
Types of Blood Vessels

There are three kinds of blood vessels: arteries, veins, and capillaries. An **artery** is a vessel that carries blood away from the heart. After birth all arteries except the pulmonary artery and its branches carry oxygenated blood. Small arteries are called **arterioles.**

A **vein,** on the other hand, is a vessel that carries blood toward the heart. All of the veins except the pulmonary veins contain deoxygenated blood. Small veins are called **venules.** Both arteries and veins are macroscopic structures; that is, they can be seen without the aid of a microscope.

Capillaries are microscopic vessels that carry blood from small arteries to small veins, that is, from arterioles to venules. They represented the "missing link" in the proof of circulation for many years—from the time William Harvey first declared that blood circulated from the heart through arteries to veins and back to the heart until the time that microscopes made it possible to find these connecting vessels between arteries and veins. Many people rejected Harvey's theory of circulation on the basis that there was no possible way for blood to get from arteries to veins. The discovery of the capillaries formed the final proof that the blood actually does circulate from the heart into arteries to arterioles, to capillaries, to venules, to veins, and back to the heart.

Structure of Blood Vessels

The walls of the larger blood vessels, the arteries and veins, have three layers (Figure 17-11 and Table 17-1).

TABLE 17-1   Structure of blood vessels

Type of vessel	Tunica intima (endothelium)	Tunica media (smooth muscle; elastic connective tissue)	Tunica adventitia (fibrous connective tissue)
ARTERIES	Smooth lining	Allows constriction and dilation of vessels; thicker than in veins; muscle innervated by autonomic fibers	Provides flexible support that resists collapse or injury; thicker than in veins; thinner than tunica media
VEINS	Smooth lining with semilunar valves to ensure one-way flow	Allows constriction and dilation of vessels; thinner than in arteries; muscle innervated by autonomic fibers	Provides flexible support that resists collapse or injury; thinner than in arteries; thicker than tunica media
CAPILLARIES	Makes up entire wall of capillary; thinness permits ease of transport across vessel wall	(Absent)	(Absent)

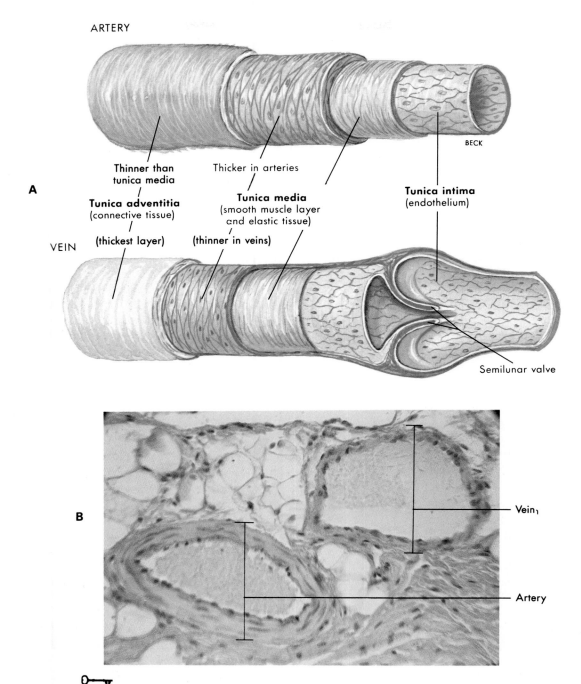

ARTERY

Thinner than
tunica media

Thicker in arteries

A

Tunica adventitia
(connective tissue)

Tunica media
(smooth muscle layer
and elastic tissue)

Tunica intima
(endothelium)

BECK

(thickest layer)

(thinner in veins)

VEIN

Semilunar valve

B

Vein₁

Artery

FIGURE 17-11   **Structure of blood vessels. A,** Drawings of a sectioned artery and vein show the three layers of large vessel walls. The outermost, *tunica adventitia,* is made of connective tissue; the middle, *tunica media,* is made of smooth muscle and elastic connective tissue; and the innermost, *tunica intima,* is made of endothelium. **B,** This light micrograph of a cross section of tissue contains both an artery *(left)* and vein *(right).* Note the prominence of the smooth muscle (tunica media) in the artery compared to the vein.

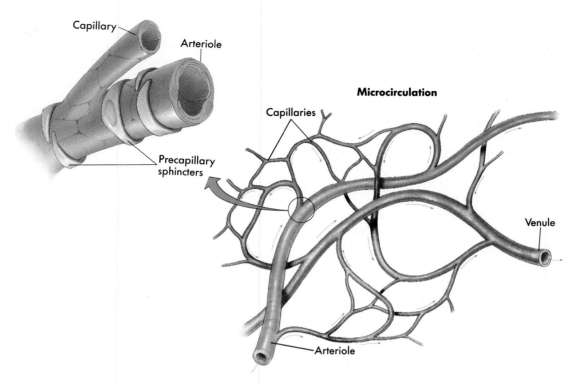

FIGURE 17-12 Microcirculation. The smaller blood vessels—arterioles, capillaries, and venules—cannot be observed without magnification. Notice that the control of blood flow through any particular region of a capillary network can be regulated by the relative contraction of precapillary sphincters in the walls of the arterioles *(see inset)*. Notice also that capillaries have a wall composed of only a single layer of flattened cells, whereas the walls of the larger vessels also have smooth muscle.

The outermost layer is called the **tunica adventitia** and is made of strong, flexible fibrous connective tissue. This layer helps hold vessels open and prevents tearing of the vessel walls during body movements. In veins, the tunica adventitia is the thickest of the three layers. In arteries, it is usually a little thinner than the middle layer, or **tunica media.** The tunica media is made of a layer of smooth muscle tissue sandwiched together with a layer of elastic connective tissue. The encircling smooth muscles, innervated by autonomic nerves and supplied with blood by tiny *vasa vasorum* ("vessels of vessels"), permit changes in blood vessel diameter. As a rule, arteries have a thicker layer of smooth muscle than do veins. The innermost layer is called the **tunica intima** and is made up of endothelium that is continuous with the endothelium that lines the heart. In arteries, the endothelium provides a completely smooth lining. In veins, however, the endothelium forms semilunar valves that help maintain the one-way flow of blood. As blood vessels decrease in diameter, the relative thickness of their walls also decreases. The smallest of the vessels, the capillaries, have only one thin coat: the endothelium. This structure is important because the thinness of the capillary wall allows for efficient exchange of materials between the blood plasma and the interstitial fluid.

Functions of Blood Vessels

The capillaries, although seemingly the most insignificant of the three kinds of blood vessels because of their diminutive size, nevertheless are the most important vessels functionally. Since the prime function of blood is to transport essential materials to and from the cells and since the actual delivery and collection of these substances take place in the capillaries, the capillaries must be regarded as the most important blood vessels. Arteries serve merely as "distributors," carrying the blood to the arterioles. Arterioles, too, serve as distributors, carrying blood from arteries to capillaries. But arterioles perform an additional function, one that is of great importance for maintaining normal blood pressure and circulation. They serve as resistance vessels. Note in Figure 17-12 that smooth muscle in the walls of arterioles forms **precapillary sphincters** near the point at which a capillary originates. These sphincters reduce the flow of blood through a network of capillaries when they contract and, thus, constrict the arterioles. Precapillary sphincters increase the flow of blood through a tissue when they relax and, thus, dilate the arterioles. The concept of flow resistance and its relationship to blood vessel diameter are discussed further in Chapter 18.

Veins function both as collectors and as reservoir vessels. They not only return blood from the capillaries to

the heart, but they also can accommodate varying amounts of blood. This reservoir function of veins, which we shall discuss later, plays an important part in maintaining normal circulation. Figure 17-13 illustrates the potential for pooling of blood between valves in one segment of a vein. Pooled blood in each valved segment is moved toward the heart by the pressure from the moving volume of blood from below. The heart acts as a "pump," keeping the blood moving through this circuit of vessels—arteries, arterioles, capillaries, venules, and veins. In short, the entire circulatory mechanism pivots around one essential, that of keeping the capillaries supplied with an amount of blood adequate to the changing needs of the cells. All the factors governing circulation operate to this one end.

Although capillaries are very tiny (on the average, only 1 mm long, or about ¹⁄₂₅ inch), their numbers are so great as to be incomprehensible. Someone has calculated that if these microscopic tubes were joined end to end, they would extend 62,000 miles, despite the fact that it takes 25 of them to reach a single inch! According to one estimate, 1 cubic inch of muscle tissue contains over 1.5 million of these important little vessels. None of the billions of cells composing the body lies very far removed from a capillary. The reason for this lavish distribution of capillaries is, of course, apparent in view of their function of keeping the cells supplied with vital materials and rid of injurious wastes.

This flow of blood through the capillary bed is referred to as the **microcirculation.** Figure 17-12 shows the anatomical components of microcirculation: arterioles, arteriole capillaries, venous capillaries, and venules.

QUICK CHECK

✔ 1. Name the three major types of blood vessels.
✔ 2. How does the structure of each major type of vessel differ from the other types?
✔ 3. How does the function of capillaries relate to the structure of their walls?

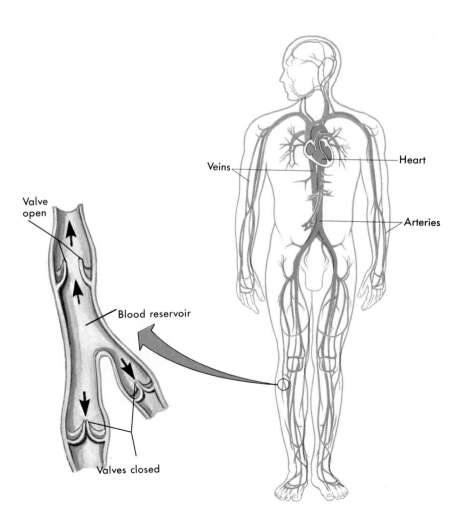

FIGURE 17-13 The reservoir function of veins. Pooled blood is moved toward the heart as valves are forced open by pressure from volume of blood from below. Closure of valves prevents backflow.

Major Blood Vessels

Circulatory Routes

The term *circulation of blood* suggests its meaning, namely, blood flow through vessels arranged to form a circuit or circular pattern. Blood flow from the heart (left ventricle) through blood vessels to all parts of the body (except the lungs) and back to the heart (to the right atrium) is spoken of as **systemic circulation.** The left ventricle pumps blood into the ascending aorta. From here it flows into arteries that carry it into the various tissues and organs of the body. Within each structure, blood moves, as indicated in Figure 17-14 (see also Figure 17-12), from arteries to arterioles to capillaries. Here the vital two-way exchange of substances occurs between the blood and cells. Blood flows next out of each organ by way of its venules and then its veins to drain eventually into the inferior or superior vena cava. These two great veins of the body return venous blood to the heart (to the right atrium) to complete the systemic circulation. But the blood does not quite come full circle back to its starting point, the left ventricle. To do this and start on its way again, it must first flow through another circuit, the **pulmonary circulation.** Observe in Figure 17-14 that venous blood moves from the right atrium to the right ventricle to the pulmonary artery to lung arterioles and capillaries (see also Figure 17-6). Here, exchange of gases between blood and air takes place, converting deoxygenated blood to oxygenated blood. This oxygenated blood then flows on through lung venules into four pulmonary veins and returns to the left atrium of the heart. From the left atrium it enters the left ventricle to be pumped again through the systemic circulation.

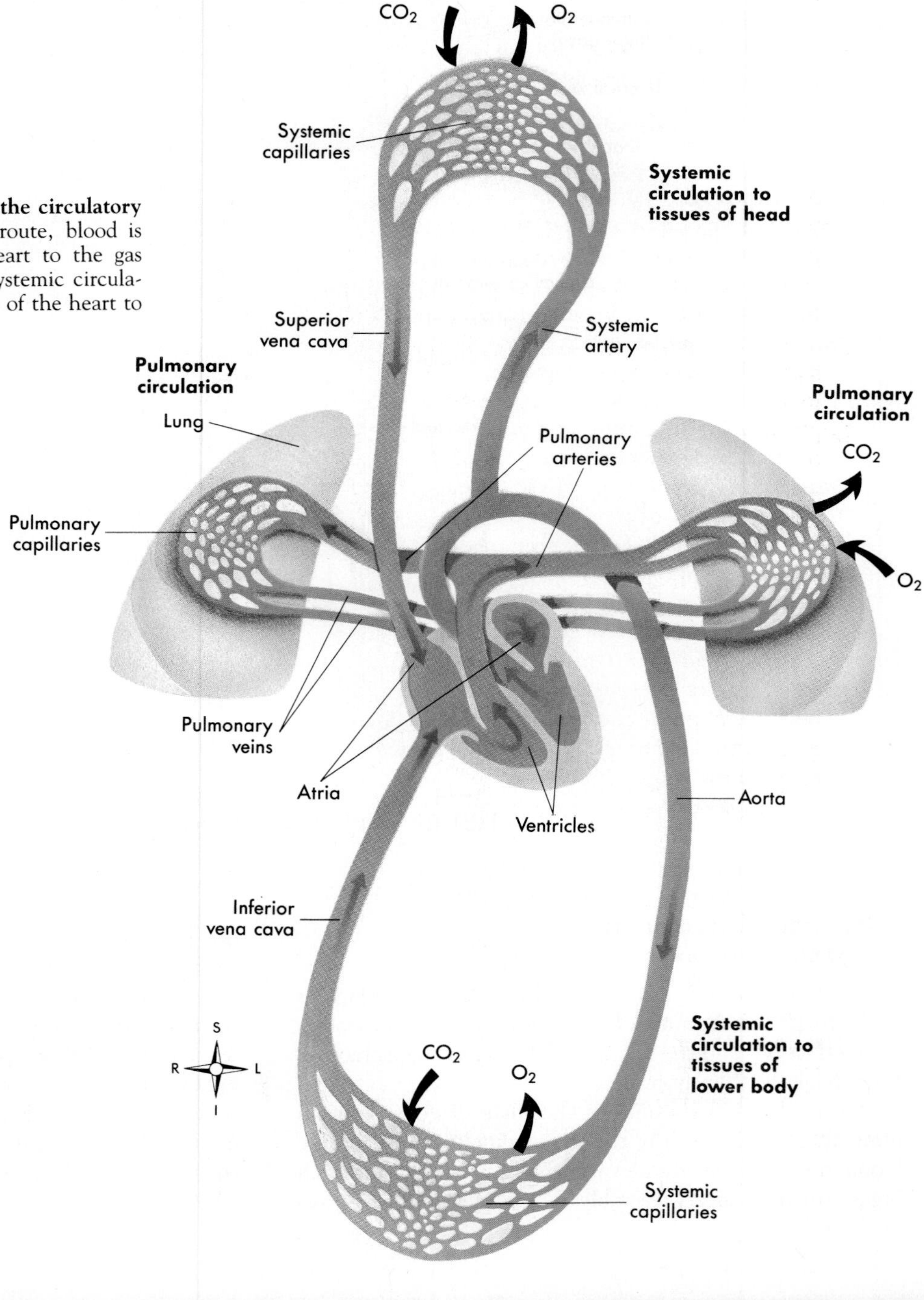

FIGURE 17-14 Blood flow through the circulatory system. In the pulmonary circulatory route, blood is pumped from the right side of the heart to the gas exchange tissues of the lungs. In the systemic circulation, blood is pumped from the left side of the heart to all other tissues of the body.

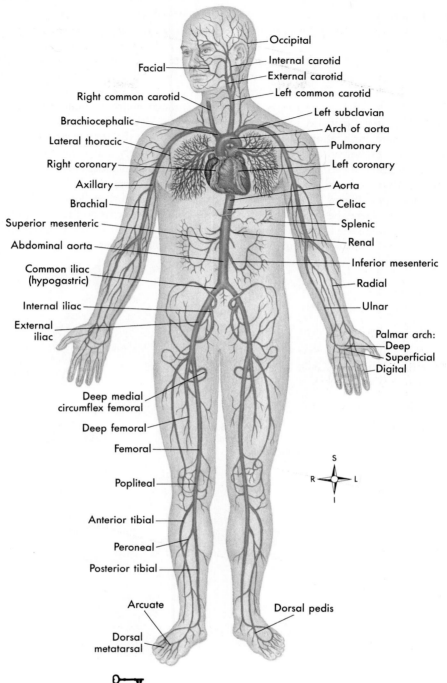

FIGURE 17-15 Principal arteries of the body.

Systemic Circulation
Systemic arteries

Locate the arteries listed in Table 17-2 (see also Figures 17-15 to 17-18). You may find it easier to learn the names of blood vessels and the relation of the vessels to each other from diagrams than from descriptions.

As you learn the names of the main arteries, keep in mind that these are only the major pipelines distributing blood from the heart to the various organs and that, in each organ, the main artery resembles a tree trunk in that it gives

off numerous branches that continue to branch and rebranch, forming ever smaller vessels (arterioles), which also branch, forming microscopic vessels, the capillaries. In other words, most arteries eventually diverge into capillaries. Arteries of this type are called **end-arteries.** Important organs or areas of the body supplied by end-arteries are subject to serious damage or death in occlusive arterial disease. As an example, permanent blindness results when the central artery of the retina, an end-artery, is occluded. Therefore occlusive arterial disease such as atherosclerosis

TABLE 17-2   Main arteries

Artery	Branches (only largest ones named)
Ascending aorta	Coronary arteries (two, to myocardium)
Aortic arch	Brachiocephalic (or innominate) artery Left subclavian Left common carotid
Brachiocephalic (Innominate)	Right subclavian Right common carotid
Subclavian (right and left)	Vertebral* Axillary (continuation of subclavian)
Axillary	Brachial (continuation of axillary)
Brachial	Radial Ulnar
Radial and ulnar	Palmar arches (superficial and deep arterial arches in hand formed by anastomosis of branches of radial and ulnar arteries; numerous branches to hand and fingers)
Common carotid (right and left)	Internal carotid (brain, eye, forehead, and nose)* External carotid (thyroid, tongue, tonsils, ear, etc.)
Descending thoracic aorta	Visceral branches to pericardium, bronchi, esophagus, mediastinum Parietal branches to chest muscles, mammary glands, and diaphragm
Descending abdominal aorta	Visceral branches: 　Celiac axis (or artery), branches into gastric, hepatic, and splenic arteries (stomach, liver, and spleen) 　Right and left suprarenal arteries (suprarenal glands) 　Superior mesenteric artery (small intestine) 　Right and left renal arteries (kidneys) 　Right and left spermatic (or ovarian) arteries (testes or ovaries) 　Inferior mesenteric artery (large intestine) Parietal branches to lower surface of diaphragm, muscles and skin of back, spinal cord, and meninges Right and left common iliac arteries—abdominal aorta terminates in these vessels in an inverted Y
Right and left common iliac	Internal iliac or hypogastric (pelvic wall and viscera) External iliac (to leg)
External iliac (right and left)	Femoral (continuation of external iliac after it leaves abdominial cavity)
Femoral	Popliteal (continuation of femoral)
Popliteal	Anterior tibial Posterior tibial
Anterior and posterior tibial	Plantar arch (arterial arch in sole of foot formed by anastomosis of terminal branches of anterior and posterior tibial arteries; small arteries lead from arch to toes)

*See text for discussion of branches of the artery.

is of great concern in clinical medicine when it affects important organs having an end-arterial blood supply.

A few arteries open into other branches of the same or other arteries. Such a communication is termed an **arterial anastomosis**. Anastomoses, we have already noted, fulfill an important protective function in that they provide detour routes for blood to travel in the event of obstruction of a main artery. The incidence of arterial anastomoses increases as distance from the heart increases, and smaller arterial branches tend to anastomose more often than larger vessels. Examples of arterial anastomoses are the palmar and plantar arches. Other examples are found around several joints, as well as in other locations.

Figure 17-6 shows the major arteries of the head and face.

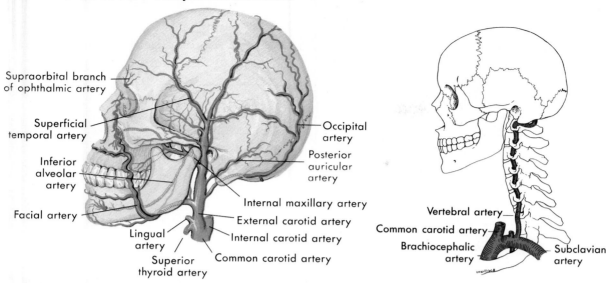

FIGURE 17-16 Major arteries of the face and head. Superficial vessels are shown in brighter colors than deep vessels.

FIGURE 17-17 Location of the vertebral artery.

FOCUS ON... HOW TO TRACE THE FLOW OF BLOOD

To list the vessels through which blood flows in reaching a designated part of the body or in returning to the heart from a part, one must remember the following:

♦ Blood always flows in this direction: from left ventricle of heart to arteries, to arterioles, to capillaries of each body part, to venules, to veins, to right atrium, right ventricle, pulmonary artery, lung capillaries, pulmonary veins, left atrium, and back to left ventricle (see the Figure).

♦ When blood is in capillaries of abdominal digestive organs, it must flow through the hepatic portal system (see Figure 17-24) before returning to the heart.
♦ Names of main arteries and veins of body.

For example, suppose glucose were instilled into the rectum. To reach the cells of the right little finger, the vessels through which it would pass after absorption from the intestinal mucosa into capillaries would be as follows: capillaries into venules of large intestine into infe-

rior mesenteric vein, splenic vein, portal vein, capillaries of liver, hepatic veins, inferior vena cava, right atrium of heart, right ventricle, pulmonary artery, pulmonary capillaries, pulmonary veins, left atrium, left ventricle, ascending aorta, aortic arch, innominate artery, right subclavian artery, right axillary artery, right brachial artery, right ulnar artery, arteries of palmar arch, arterioles, and capillaries of right little finger.

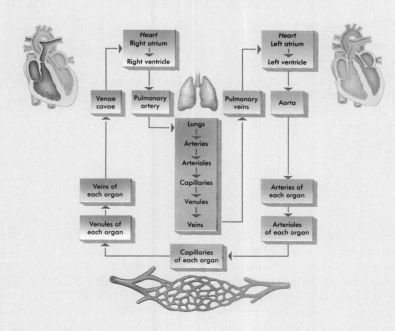

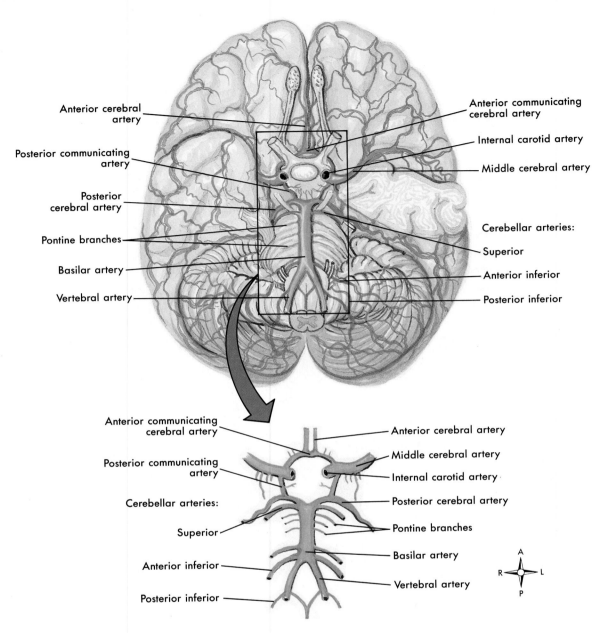

FIGURE 17-18 Arteries at the base of the brain. The arteries that compose the circle of Willis are the two anterior cerebral arteries joined to each other by the anterior communicating cerebral artery and to the posterior cerebral arteries by the posterior communicating arteries.

The right and left vertebral arteries extend from their origin as branches of the subclavian arteries up the neck (Figure 17-17), through foramina in the transverse processes of the cervical vertebrae, through the foramen magnum into the cranial cavity, and unite on the undersurface of the brain stem to form the *basilar artery,* which shortly branches into the right and left *posterior cerebral arteries* (Figure 17-18). The internal carotid arteries enter the cranial cavity in the midpart of the cranial floor where they become known as the *arterial cerebral arteries.* Small vessels, the *communicating arteries,* join the anterior and posterior cerebral arteries in such a way as to form an arterial circle (*circle of Willis*) at the base of the brain, a good example of arterial anastomosis (Figure 17-18).

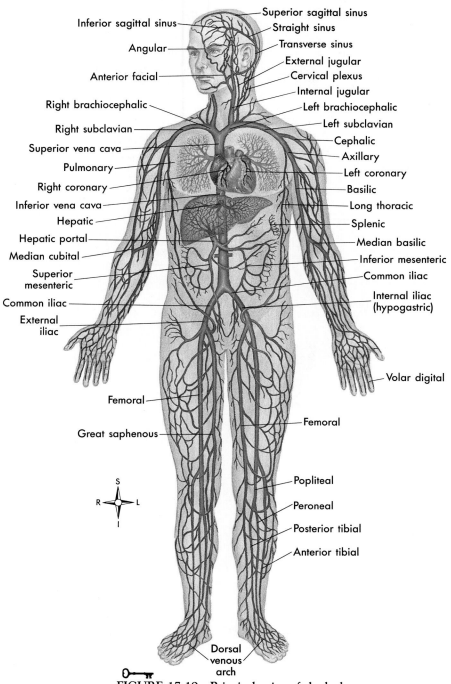

Inferior sagittal sinus

Angular

Anterior facial

Right brachiocephalic

Right subclavian

Superior vena cava

Pulmonary

Right coronary

Inferior vena cava

Hepatic

Hepatic portal

Median cubital

Superior mesenteric

Common iliac

External iliac

Superior sagittal sinus

Straight sinus

Transverse sinus

External jugular

Cervical plexus

Internal jugular

Left brachiocephalic

Left subclavian

Cephalic

Axillary

Left coronary

Basilic

Long thoracic

Splenic

Median basilic

Inferior mesenteric

Common iliac

Internal iliac (hypogastric)

Volar digital

Femoral

Great saphenous

Femoral

Popliteal

Peroneal

Posterior tibial

Anterior tibial

S
R   L
I

Dorsal venous arch

FIGURE 17-19   Principal veins of the body.

Systemic veins

The following facts should be borne in mind while learning the names of veins:

♦ Veins are the ultimate extensions of capillaries, just as capillaries are the eventual extensions of arteries. Whereas arteries branch into vessels of decreasing size to form arterioles and eventually capillaries, capillaries unite into vessels of increasing size to form venules and, eventually, veins.

♦ Many of the main arteries have corresponding veins bearing the same name and are located alongside or near the arteries (see Figure 17-13). These veins, like the arteries, lie in deep, well-protected areas, for the most part, close along the bones—example: femoral artery and femoral vein, both located along the femur bone.

♦ Veins found in the deep parts of the body are called **deep veins** in contrast to **superficial veins,** which lie near the surface. The latter are the veins that can be seen through the skin.

◆ The large veins of the cranial cavity, formed by the dura mater, are not called veins but *dural sinuses,* or, simply, **sinuses.** They should not be confused with the bony, air-filled sinuses of the skull.

◆ Veins communicate (anastomose) with each other in the same way as arteries. Such venous anastomoses provide for collateral return blood flow in cases of venous obstruction.

◆ Venous blood from the head, neck, upper extremities, and thoracic cavity, with the exception of the lungs, drains into the superior vena cava. Blood from the lower extremities and abdomen enters the inferior vena cava.

The following list identifies the major systemic veins. Locate each one as named on Figures 17-19 to 17-24.

Veins of upper extremities (Figures 17-19 and 17-21)

◆ *Deep*
Palmar (volar) arch (also superficial)
Radial (partially deep, partially superficial)
Ulnar (partially deep, partially superficial)
Brachial
Axillary (continuation of brachial)
Subclavian (continuation of axillary)

◆ *Superficial*
Veins of hand from dorsal and volar venous arches, which, together with complicated network of superficial veins of lower arm, finally pour their blood into two large veins: cephalic (thumb side) and basilic (little finger side); these two veins empty into the deep axillary vein

Veins of lower extremities (Figures 17-19 and 17-20)

◆ *Deep*
Plantar arch
Anterior tibial
Posterior tibial
Popliteal
Femoral
External iliac

◆ *Superficial*
Dorsal venous arch of foot
Great (or internal or long) saphenous
Small (or external or short) saphenous
(Great saphenous terminates in femoral vein in groin; small saphenous terminates in popliteal vein)

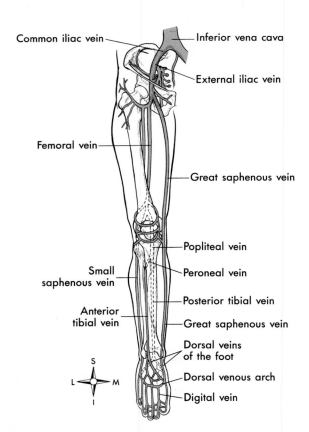

FIGURE 17-20   Major veins of the lower extremity (anterior view).

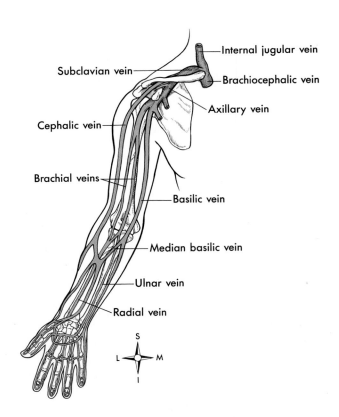

FIGURE 17-21   Major veins of the upper extremity. The median basilic (cubital) vein is commonly used for removing blood or giving intravenous infusions (anterior view).

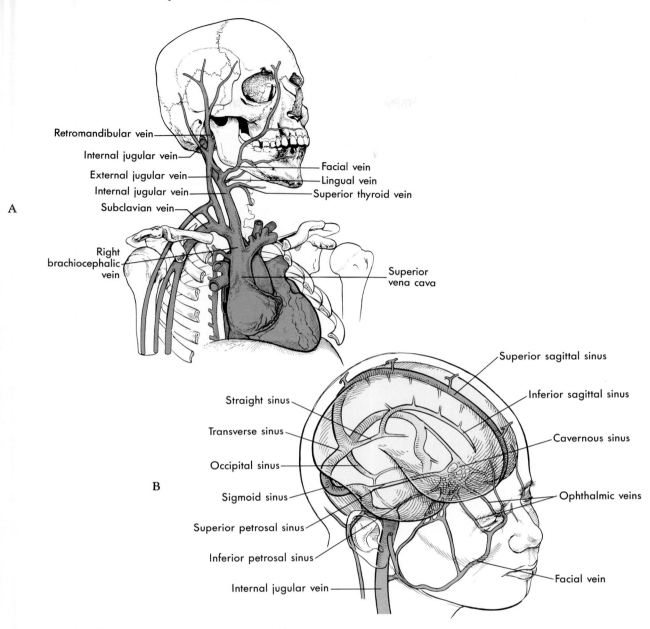

FIGURE 17-22 **Major veins of the head and neck. A,** Anterior view showing veins on the right side of the head and neck. **B,** Lateral, superior view showing the position of major veins relative to the brain.

Veins of head and neck (Figure 17-22)

- *Deep (in cranial cavity)*
  Longitudinal (or sagittal) sinus
  Inferior sagittal and straight sinus
  Numerous small sinuses
  Right and left transverse (or lateral) sinuses
  Internal jugular veins, right and left (in neck); continuations of transverse sinuses

Brachiocephalic (innominate) veins, right and left; formed by union of subclavian and internal jugulars
- *Superficial*
  External jugular veins, right and left (in neck); receive blood from small superficial veins of face, scalp, and neck; terminate in subclavian veins (small emissary veins connect veins of scalp and face with blood sinuses of cranial cavity, a fact of clinical interest as a possible avenue for infections to enter cranial cavity)

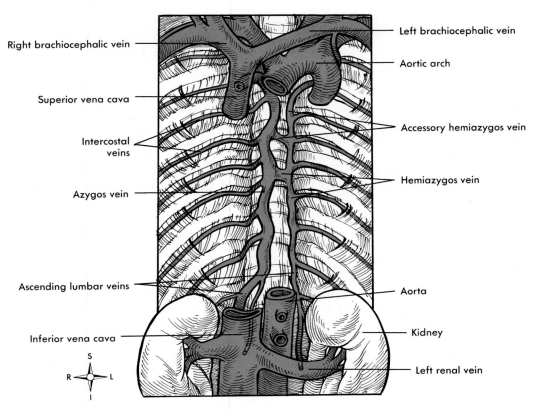

Right brachiocephalic vein

Superior vena cava

Intercostal
veins

Azygos vein

Ascending lumbar veins

Inferior vena cava

S
R — L
I

Left brachiocephalic vein

Aortic arch

Accessory hemiazygos vein

Hemiazygos vein

Aorta

Kidney

Left renal vein

FIGURE 17-23 Principal veins of the thorax. Smaller veins of the thorax drain blood into the inferior vena cava or into the azygos vein—both are shown here. The hemiazygos vein and accessory hemiazygos vein on the left drain into the azygos vein on the right.

Veins of thoracic organs (Figure 17-23)

Several small veins—such as bronchial, esophageal, pericardial—return blood from chest organs (except lungs) directly into superior vena cava or azygos vein; azygos vein lies to right of spinal column and extends from inferior vena cava (at level of first or second lumbar vertebra) through diaphragm to terminal part of superior vena cava; hemiazygos vein lies to left of spinal column, extending from lumbar level of inferior vena cava through diaphragm to terminate in azygos vein; accessory hemiazygos vein connects some of superior intercostal veins with azygos or hemiazygos vein.

Veins of abdominal organs (Figures 17-19 and 17-24)

Spermatic (or ovarian)   Suprarenal
Renal   Drain into inferior
Hepatic   vena cava

Left spermatic and left suprarenal veins usually drain into left renal vein instead of into inferior vena cava; for return of blood from abdominal digestive organs, see subsequent discussion of portal circulation.

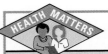

CLINICAL IMPORTANCE OF VENOUS SINUSES

Middle ear infections sometimes cause infection of the transverse sinuses with the formation of a thrombus. In such cases the internal jugular vein may be ligated (tied shut) to prevent the development of a fatal cardiac or pulmonary embolism.

Intravenous injections are perhaps most often given into the median basilic vein at the bend of the elbow. Blood that is to be used for various laboratory tests is also usually removed from this vein. In an infant, however, the longitudinal sinus is more often punctured (through the anterior fontanel) because the superficial arm veins are too tiny for the insertion of a needle.

QUICK CHECK

✔ 1. What is the difference between the pulmonary circulation and the systemic circulation?
✔ 2. What is the function of arterial anastomoses? What is the function of venous anastomoses?
✔ 3. How are systemic arteries and veins usually named?

Hepatic Portal Circulation

Veins from the spleen, stomach, pancreas, gallbladder, and intestines do not pour their blood directly into the inferior vena cava, as do the veins from other abdominal organs. Instead, they send their blood to the liver by means of the hepatic portal vein. Here the blood mingles with the arterial blood in the capillaries and is eventually drained from the liver by the hepatic veins that join the inferior vena cava. The advantages of detouring of the blood through the liver before it returns to the heart will be discussed in the chapters on the digestive system.

Figure 17-24 shows the plan of the hepatic portal system. In most individuals the hepatic portal vein is formed by the union of the splenic and superior mesenteric veins, but blood from the gastric, pancreatic, and inferior mesenteric veins drains into the splenic vein before it merges with the superior mesenteric vein.

If either hepatic portal circulation or venous return from the liver is interfered with (as often occurs in certain types of liver or heart disease), venous drainage from most of the other abdominal organs is necessarily obstructed also. The

accompanying increased capillary pressure accounts, at least in part, for the occurrence of abdominal bloating, or *ascites* ("dropsy"), under these conditions.

Fetal circulation

The basic plan of fetal circulation. Circulation in the body before birth necessarily differs from circulation after birth for one main reason—fetal blood secures oxygen and food from maternal blood instead of from fetal lungs and digestive organs. Obviously, then, there must be additional blood vessels in the fetus to carry the fetal blood into close approximation with the maternal blood and to return it to the fetal body. These structures are the two **umbilical arteries,** the **umbilical vein,** and the **ductus venosus.** Also, some structure must function as the lungs and digestive organs do postnatally, that is, a place where an interchange of gases, foods, and wastes between the fetal and maternal blood can take place. This structure is the *placenta* (Figure 17-25). The exchange of substances occurs without any actual mixing of maternal and fetal blood, since each flows in its own capillaries.

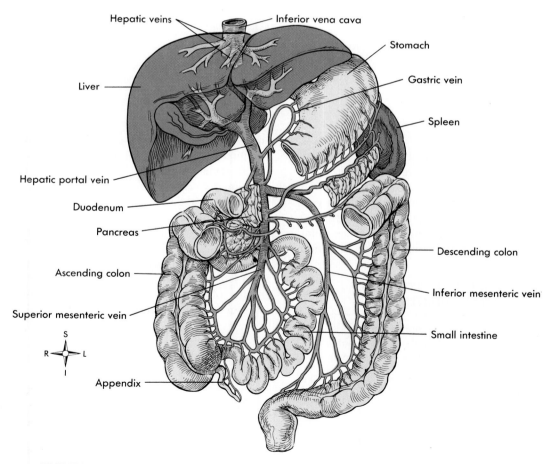

FIGURE 17-24 Hepatic portal circulation. In this unusual circulatory route, a vein is located between two capillary beds. The hepatic portal vein collects blood from capillaries in visceral structures located in the abdomen and empties it into the liver. Hepatic veins return blood to the inferior vena cava. (Organs are not drawn to scale.)

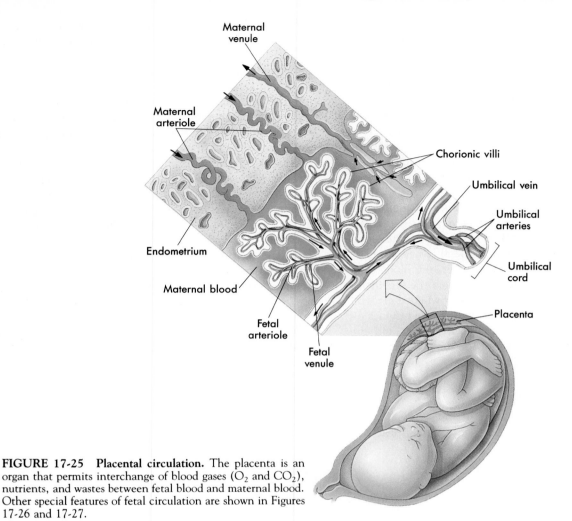

FIGURE 17-25 Placental circulation. The placenta is an organ that permits interchange of blood gases (O_2 and CO_2), nutrients, and wastes between fetal blood and maternal blood. Other special features of fetal circulation are shown in Figures 17-26 and 17-27.

HEALTH MATTERS

FETAL ALCOHOL SYNDROME

Consumption of alcohol during pregnancy can have tragic effects on a developing fetus. Educational efforts to inform pregnant women about the dangers of alcohol are now receiving national attention. Even very limited consumption of alcohol during pregnancy poses significant hazards to the developing baby because alcohol can easily cross the placental barrier and enter the fetal bloodstream.

When alcohol enters the fetal blood, the potential result, called **fetal alcohol syndrome (FAS)**, can cause tragic congenital abnormalities such as "small head" or microcephaly (my-kro-SEF-ah-lee), low birth weight, developmental disabilities such as mental retardation, and even fetal death.

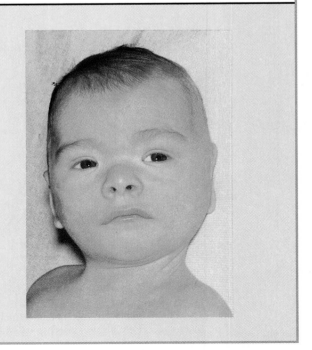

In addition to the placenta and umbilical vessels, three structures located within the fetus' own body play an important part in fetal circulation. One of them (ductus venosus) serves as a detour by which most of the blood returning from the placenta bypasses the fetal liver. The other two (foramen ovale and ductus arteriosus) provide detours by which blood bypasses the lungs. A brief description of each of the six structures necessary for fetal circulation follows (Figure 17-26).

1. The **two umbilical arteries** are extensions of the internal iliac (hypogastric) arteries and carry fetal blood to the placenta.

2. The **placenta** is a structure attached to the uterine wall. Exchange of oxygen and other substances between maternal and fetal blood takes place in the placenta, although no mixing of maternal and fetal blood occurs.

3. The **umbilical vein** returns oxygenated blood from the placenta, enters the fetal body through the umbilicus, extends up to the undersurface of the liver where it gives off two or three branches to the liver, and then continues on as the ductus venosus. Two umbilical arteries and the umbilical vein together constitute the **umbilical cord** and are shed at birth along with the placenta.

4. The **ductus venosus** is a continuation of the umbilical vein along the undersurface of the liver and drains into the inferior vena cava. Most of the blood returning from the placenta bypasses the liver. Only a relatively small amount of blood enters the liver by way of the branches from the umbilical vein into the liver.

5. The **foramen ovale** is an opening in the septum between the right and left atria. A valve at the opening of the inferior vena cava into the right atrium directs most of the blood through the foramen ovale into the left atrium so that it bypasses the fetal lungs. A small percentage of the blood leaves the right atrium for the right ventricle and pulmonary artery. But most of this blood does not flow on into the lungs. Still another detour, the ductus arteriosus, diverts it.

6. The **ductus arteriosus** is a small vessel connecting the pulmonary artery with the descending thoracic aorta. It therefore enables another portion of the blood to detour into the systemic circulation without going through the lungs.

Almost all fetal blood is a mixture of oxygenated and deoxygenated blood. Examine Figure 17-26 carefully to

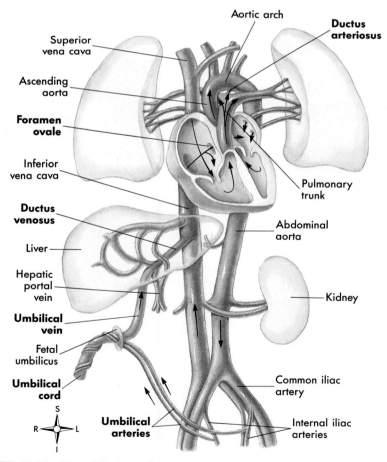

FIGURE 17-26   Plan of fetal circulation. Before birth, the human circulatory system has several special features that adapt the body to life in the womb. These features (labeled in bold type) include: two umbilical arteries, one umbilical vein, ductus venosus, foramen ovale, and ductus arteriosus. The placenta, another essential feature of the fetal circulatory plan, is shown in Figure 17-25.

determine why this is so. What happens to the oxygenated blood returned from the placenta via the umbilical vein? Note that it flows into the inferior vena cava.

Changes in circulation at birth. Since the six structures that serve fetal circulation are no longer needed after birth, several changes take place (Figure 17-27). As soon as the umbilical cord is cut, the two umbilical arteries, the placenta, and the umbilical vein obviously no longer function. The placenta is shed from the mother's body as the afterbirth with part of the umbilical vessels attached. The sections of these vessels remaining in the infant's body eventually become fibrous cords that remain throughout life (the umbilical vein becomes the round ligament of the liver). The ductus venosus, no longer needed to bypass blood around the liver, eventually becomes the ligamentum venosum of the liver. The foramen ovale normally becomes functionally closed soon after a newborn takes the first breath and full circulation through the lungs becomes established. Complete structural closure, however, usually requires 9 months or more. Eventually, the foramen ovale becomes a mere depression (fossa ovalis) in the wall of the right atrial septum. The ductus arteriosus contracts as soon as respiration is established. Eventually, it also turns into a fibrous cord, the ligamentum arteriosum. Compare the blood flow in Figures 17-26 and 17-27. Notice how separation of oxygenated and deoxygenated blood occurs *after* birth.

QUICK CHECK

✔ 1. *Compared with other routes in the systemic circulation, what is unusual about the hepatic portal circulation?*

✔ 2. *Name some of the special structural features of the fetal circulation.*

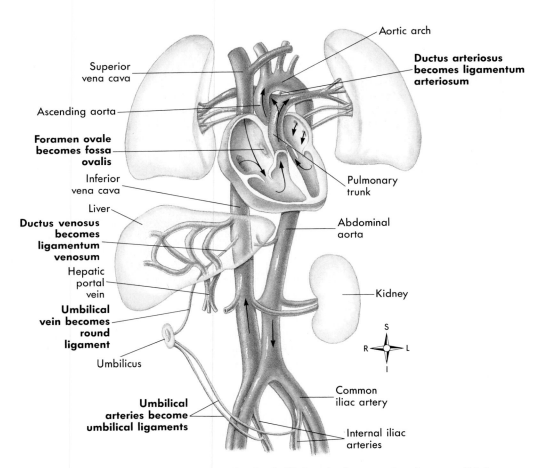

Superior vena cava

Aortic arch

Ductus arteriosus becomes ligamentum arteriosum

Ascending aorta

Foramen ovale becomes fossa ovalis

Inferior vena cava

Pulmonary trunk

Liver

Ductus venosus becomes ligamentum venosum

Abdominal aorta

Hepatic portal vein

Umbilical vein becomes round ligament

Kidney

Umbilicus

Umbilical arteries become umbilical ligaments

Common iliac artery

Internal iliac arteries

FIGURE 17-27   Changes in circulation after birth. Within the first year after the time of birth, certain changes in the circulatory plan occur to adapt the body to life outside the womb. The placenta and portions of the umbilical vessels outside the infant's body are removed or fall off at, or shortly after, the time of birth. The internal portion of the umbilical vein constricts and becomes fibrous, eventually forming the round ligament of the liver. Likewise, the internal umbilical arteries become umbilical ligaments, the ductus venosus becomes the ligamentum venosum, and the ductus arteriosus becomes the ligamentum arteriosum. The foramen ovale closes, forming a thin region of the atrial wall called the fossa ovalis.

Cycle of Life: *Cardiovascular anatomy*

As with all body structures, the heart and blood vessels undergo profound anatomical changes during early development in the womb. At birth, the switch from a placenta-dependent system causes another set of profound anatomical changes. Throughout childhood, adolescence, and adulthood, the heart and blood vessels normally maintain their basic structure and function—permitting continued survival of the individual. Perhaps the only apparent normal changes in these structures occur as a result of regular exercise. The myocardium thickens and the supply of blood vessels in skeletal muscle tissues increases in response to increased oxygen and glucose use during prolonged exercise.

As we pass through adulthood, especially later adulthood, various degenerative changes can occur in the heart and blood vessels. For example, a type of "hardening of the arteries," called atherosclerosis, can result in blockage or weakening of critical arteries—perhaps causing a myocardial infarction or stroke. The heart valves and myocardial tissues often degenerate with age, becoming hardened or fibrotic and less able to perform their functions properly. This reduces the heart's pumping efficiency and therefore threatens homeostasis of the entire internal environment.

MECHANISMS OF DISEASE

Disorders of the Cardiovascular System

Disorders of Heart Structure

Disorders Involving the Pericardium

If the pericardium becomes inflamed, a condition called **pericarditis** results. Pericarditis may be caused by various factors: trauma, viral or bacterial infection, tumors, and other factors. The pericardial edema that characterizes this condition often causes the visceral and parietal layers of the serous pericardium to rub together—causing severe chest pain. Pericardial fluid, pus, or blood (in the case of an injury) may accumulate in the space between the two pericardial layers and impair the pumping action of the heart. This is termed *pericardial effusion* and may develop into a serious compression of the heart called **cardiac tamponade.**

Pericarditis may be acute or chronic, depending on the rate, severity, and duration of symptoms. Clinical manifestations include pericardial pain that increases with respirations or coughing, a "friction rub" (a grating, scratching sound heard over the left sternal border and upper ribs) due to movement together of the swollen pericardial layers, difficulty breathing, restlessness, and an accumulation of pericardial fluid. Cardiac tamponade requires immediate pericardial drainage (pericardiocentesis). Antibiotics are usually prescribed to treat the causative organism and nonsteroidal antiinflammatory agents such as aspirin are prescribed to reduce the inflammation and thus control the symptoms.

Disorders Involving Heart Valves

Disorders of the cardiac valves can have several effects. For example, a congenital defect in valve structure can result in mild to severe pumping inefficiency. **Incompetent valves** leak, allowing some blood to flow back into the chamber from which it came. **Stenosed valves** are valves that are narrower than normal, slowing blood flow from a heart chamber.

Rheumatic heart disease results from a delayed inflammatory response to streptococcal infection that occurs most often in children. A few weeks after an improperly treated streptococcal infection, the cardiac valves and other tissues in the body may become inflamed—a condition called *rheumatic*

fever. If severe, the inflammation can result in stenosis or other deformities of the valves, chordae tendineae, or myocardium.

Mitral valve prolapse (MVP), a condition affecting the bicuspid or mitral valve, has a genetic basis in some cases but can result from rheumatic fever or other factors. A prolapsed mitral valve is one whose flaps extend back into the left atrium, causing incompetence (leaking) of the valve (Figure 17-28). Although this condition is common, occurring in up to 1 in every 20 people, most cases are asymptomatic. In severe cases, patients suffer chest pain and fatigue.

Aortic regurgitation is a condition in which blood not only ejects forward into the aorta but regurgitates back into the left ventricle because of a leaky aortic semilunar valve. This causes a volume overload on the left ventricle, with subsequent hypertrophy and dilation of the left ventricle. The left ventricle attempts to compensate for the increased load by increasing its strength of contraction—which may eventually stress the heart to the point of causing myocardial ischemia.

Damaged or defective cardiac valves can often be replaced surgically—a procedure called valvuloplasty. Artificial valves made from synthetic materials are frequently used in these valve replacement procedures.

Disorders Involving the Myocardium

One of the leading causes of deaths in the United States is **coronary artery disease (CAD).** This condition can result from many causes, all of which somehow reduce the flow of blood to the vital myocardial tissue. For example, in both coronary thrombosis and coronary embolism, a blood clot occludes, or plugs, some part of a coronary artery. Blood cannot pass through the occluded vessel and so cannot reach the heart muscle cells it normally supplies. Deprived of oxygen, these cells soon die or are damaged. In medical terms, a **myocardial infarction (MI),** or tissue death, occurs. A myocardial infarction (heart attack) is a common cause of death during middle and late adulthood. Recovery from a

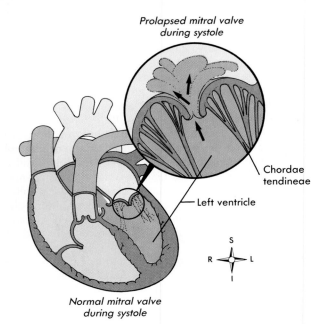

Prolapsed mitral valve during systole

Chordae tendineae

Left ventricle

Normal mitral valve during systole

FIGURE 17-28 Mitral valve prolapse. The normal mitral valve (*left*) prevents backflow of blood from the left ventricle into the left atrium during ventricular systole (contraction). The prolapsed mitral valve (*inset*) permits leakage because the valve flaps billow backward, parting slightly.

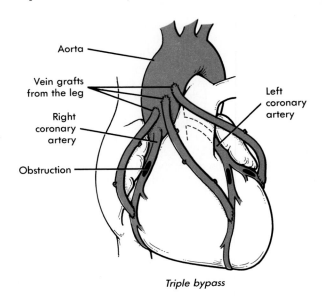

Aorta

Vein grafts from the leg

Right coronary artery

Left coronary artery

Obstruction

Triple bypass

FIGURE 17-29 Coronary bypass. In coronary bypass surgery, blood vessels are "harvested" from other parts of the body and used to construct detours around blocked coronary arteries. Artificial vessels can also be used.

myocardial infarction is possible if the amount of heart tissue damaged was small enough so that the remaining undamaged heart muscle can pump blood effectively enough to supply the needs of the rest of the heart, as well as the body.

Coronary arteries may also become blocked as a result of **atherosclerosis,** a type of "hardening of the arteries" in which lipids and other substances build up on the inside wall of blood vessels. Mechanisms of atherosclerosis are discussed later in this chapter. Coronary atherosclerosis has increased dramatically over the last few decades to become the leading cause of death in western countries. Many pathophysiologists believe this increase results from a change in life-style. They cite several important risk factors associated with coronary atherosclerosis: cigarette smoking, high-fat and high-cholesterol diets, and hypertension (high blood pressure).

The term **angina pectoris** is used to describe the severe chest pain that occurs when the myocardium is deprived of adequate oxygen. It is often a warning that the coronary arteries are no longer able to supply enough blood and oxygen to the heart muscle. **Coronary bypass surgery** is a frequent treatment for those who suffer from severely restricted coronary artery blood flow. In this procedure, veins are "harvested" from other areas of the body and used to bypass partial blockages in coronary arteries (Figure 17-29).

Heart Failure

Heart failure is the inability of the heart to pump enough blood to sustain life. Heart failure can be the result of many different heart diseases. Valve disorders can reduce the pumping efficiency of the heart enough to cause heart failure. **Cardiomyopathy,** or disease of the myocardial tissue, may reduce pumping effectiveness. A specific event such as

myocardial infarction can result in myocardial damage that causes heart failure. Arrhythmias, irregularities of heartbeat discussed in Chapter 18, can also impair the pumping effectiveness of the heart and thus cause heart failure.

Failure of the right side of the heart, or *right-sided heart failure,* accounts for about one fourth of all cases of heart failure. Right-sided heart failure often results from the progression of disease that begins in the left side of the heart. Failure of the left side of the heart results in reduced pumping of blood returning from the lungs. Blood backs up into the pulmonary circulation, then into the right side of the heart—causing an increase in pressure that the right side of the heart simply cannot overcome. Right-sided heart failure can also be caused by lung disorders that obstruct normal pulmonary blood flow and thus overload the right side of the heart—a condition called **cor pulmonale.**

Congestive heart failure (CHF), or, simply, *left-sided heart failure,* is the inability of the left ventricle to pump blood effectively. Most often, such failure results from myocardial infarction caused by coronary artery disease. It is called

HEALTH MATTERS

HEART MEDICATIONS

Although numerous drugs are used in the treatment of heart disease, the following have proven to be basic tools of the cardiologist: *anticoagulants* prevent clot formation; *beta-adrenergic blockers* block norepinephrine receptors and thus reduce the strength and rate of heart beats; *calcium channel blockers* reduce heart contractions by preventing the flow of Ca^{++} into cardiac muscle cells; *digitalis* slows and increases the strength of cardiac contractions; *nitroglycerin* dilates coronary blood vessels and thus improves O_2 supply to myocardium; and *tissue plasminogen activator (TPA)* helps dissolve clots.

congestive heart failure because it decreases pumping pressure in the systemic circulation, which in turn causes the body to retain fluids. Portions of the systemic circulation thus become congested with extra fluid. As stated previously, left-sided heart failure also causes congestion of blood in the pulmonary circulation, termed *pulmonary edema*—possibly leading to right heart failure.

Patients in danger of death because of heart failure may be candidates for heart *transplants* or heart *implants.* Heart transplants are surgical procedures in which healthy hearts from recently deceased donors replace the hearts of patients with heart disease. Unfortunately, a continuing problem with this procedure is the tendency of the body's immune system to reject the new heart as a foreign tissue. More details about the rejection of transplanted tissues are found in Chapter 19. Heart implants are artificial hearts that are made of biologically inert synthetic materials. Publicity surrounding the implantation of the Jarvik-7 artificial heart in patient Barney Clark in 1982 led many people to believe that the era of artificial hearts had arrived. However, the Jarvik-7 and most other artificial hearts require cumbersome external pumps and other devices that severely limit the recipient's mobility. Even more important, they are not efficient enough to serve as more than a temporary solution until a transplant or other medical treatment is available. Although technology in this area is improving rapidly, the promise of a practical and permanent heart replacement is still unfulfilled.

Disorders of Blood Vessels

Disorders of Arteries

As mentioned earlier in this chapter, arteries contain blood that is maintained at a relatively high pressure. This means the arterial walls must be able to withstand a great deal of force or they will burst. The arteries must also stay free of obstruction; otherwise they cannot deliver their blood to the capillary beds (and thus the tissues they serve).

A common type of vascular disease that occludes (blocks) arteries and weakens arterial walls is called **arteriosclerosis,** or *hardening of the arteries.* Arteriosclerosis is characterized by thickening of arterial walls that progresses to hardening as calcium deposits form. The thickening and calcification reduce the flow of blood to the tissues. If the blood flow slows down too much, **ischemia** results. Ischemia, or decreased blood supply to a tissue, involves the gradual death of cells and may lead to complete tissue death—a condition called **necrosis.** If a large section of tissue becomes necrotic, it may begin to decay. Necrosis that has progressed this far is called **gangrene.** Because of the tissue damage involved, arteriosclerosis is not only painful—it is life-threatening. As we have previously stated, ischemia of heart muscle can lead to *myocardial infarction.*

There are several types of arteriosclerosis, but perhaps the most well known is *atherosclerosis*—described earlier as the blockage of arteries by lipids and other matter (Figure 17-30). Eventually, the fatty deposits in the arterial walls become fibrous and perhaps calcified—resulting in sclerosis (hardening). High blood levels of triglycerides and cholesterol, which may be caused by a high-fat and high-cholesterol diet, smoking, and a genetic predisposition, are associated with atherosclerosis.

In general, arteriosclerosis develops with advanced age, diabetes, high-fat and high-cholesterol diets, hypertension (high blood pressure), and smoking. Arteriosclerosis can be treated by drugs, called *vasodilators,* that trigger the smooth muscles of the arterial walls to relax, thus causing the arteries to dilate (widen). Some cases of atherosclerosis are treated by mechanically opening the affected area of an artery, a type of procedure called **angioplasty.** In one procedure, a deflated balloon attached to a long tube, called a *catheter,* is inserted

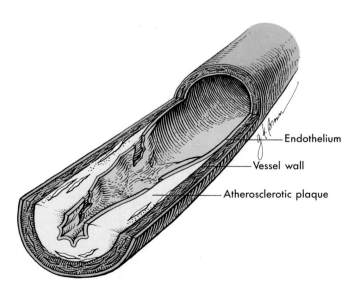

— Endothelium

— Vessel wall

— Atherosclerotic plaque

FIGURE 17-30 Partial blockage of an artery in atherosclerosis. Atherosclerotic plaque develops from the deposition of fats and other substances in the wall of the artery.

into a partially blocked artery and inflated (Figure 17-31). As the balloon inflates, the *plaque* (fatty deposits and tissue) is pushed outward, and the artery widens to allow near-normal blood flow. In a similar procedure, metal springs or mesh tubes, called *stents*, are inserted in affected arteries and hold them open. Other types of angioplasty use lasers, drills, or spinning loops of wire to clear the way for normal blood flow. Severely affected arteries can also be surgically bypassed or replaced.

Damage to arterial walls caused by arteriosclerosis or other factors may lead to the formation of an **aneurysm.** An aneurysm is a section of an artery that has become abnormally widened because of a weakening of the arterial wall. Aneurysms sometimes form a saclike extension of the arterial wall. One reason aneurysms are dangerous is because they, like atherosclerotic plaques, promote the formation of thrombi (abnormal clots). A thrombus may cause an embolism (blockage) in the heart or some other vital tissue. Another reason aneurysms are dangerous is their tendency to burst, causing severe hemorrhaging that may result in death. A brain aneurysm may lead to a *stroke,* or **cerebrovascular accident (CVA).** A stroke results from ischemia of brain tissue caused by an embolism or ruptured aneurysm. Depending on the amount of tissue affected and the place in the brain the CVA occurs, effects of a stroke may range from hardly noticeable to crippling to fatal.

Disorders of Veins

Varicose veins are enlarged veins in which blood tends to pool rather than continue on toward the heart. Varicosities, also called **varices** (singular, *varix*), commonly occur in *superficial veins* near the surface of the body. The *great saphenous vein,* the largest superficial vein of the leg (see Figure 17-20), often becomes varicose in people who stand for long periods. The force of gravity slows the return of venous blood to the heart in such cases, causing blood-engorged veins to dilate. As the veins dilate, the distance between the flaps of venous valves widens—eventually making them incompetent (leaky). Incompetence of valves causes even more pooling in affected veins—a positive-feedback phenomenon.

Hemorrhoids, or *piles,* are varicose veins in the anal canal. Excessive straining during defecation can create pressures that cause hemorrhoids.

The unusual pressures of carrying a child during pregnancy predispose expectant mothers to hemorrhoids and other varicosities.

Varicose veins can be treated by supporting the dilated veins from the outside. For instance, support stockings can reduce blood pooling in the great saphenous vein. Surgical removal of varicose veins can be performed in severe cases. Advanced cases of hemorrhoids are often treated this way. Symptoms of milder cases can be relieved by removing the pressure that caused the condition.

Several factors can cause **phlebitis,** or vein inflammation. Irritation by an intravenous catheter, for example, is a common cause of vein inflammation. **Thrombophlebitis** is acute phlebitis caused by clot (thrombus) formation. Veins are more likely sites of thrombus formation than arteries because venous blood moves more slowly and is under less pressure. Thrombophlebitis is characterized by pain and discoloration of the surrounding tissue. If a piece of a clot breaks free, it may cause an embolism when it blocks a blood vessel. **Pulmonary embolism,** for example, could result when an embolus lodges in the circulation of the lung. Pulmonary embolism can lead to death quickly if too much blood flow is blocked.

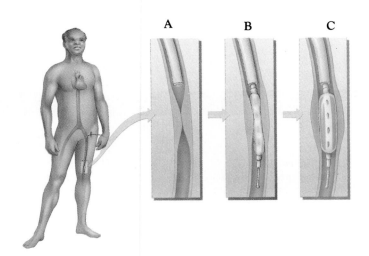

FIGURE 17-31 Balloon angioplasty. A, A catheter is inserted into the vessel until it reaches the affected region. **B,** A probe with a metal tip is pushed out the end of the catheter into the blocked region of the vessel. **C,** The balloon is inflated, pushing the walls of the vessel outward. Sometimes metal coils or tubes, called "stents," are inserted to keep the vessel open.

CASE STUDY
MYOCARDIAL INFARCTION

Mr. G. was a 49-year-old white man admitted to the critical care unit with chest pain. His health history includes the following: he has smoked two packs of cigarettes a day for 25 years; he is being treated for high blood pressure; he works as an air traffic controller; and his father died of a heart attack at the age of 51.

Mr. G. was golfing with friends when he first felt pain in his chest. At first he ignored it, but gradually the pain became a crushing sensation in his sternal area that spread down his left arm. Mr. G. then lost consciousness

and was brought to the hospital by ambulance. He was resuscitated en route to the hospital by the paramedics.

On admission to the hospital, several tests were performed. A twelve-lead electrocardiogram showed evidence of acute injury to the anterior (front) myocardium. Mr. G.'s cardiac enzymes were elevated in a pattern typical of myocardial infarction. Because of the location of the infarction, the medical team thought that it was caused by atherosclerosis in the left coronary artery.

1. Which two of Mr. G.'s alterable risk factors for atherosclerosis are most significant?
   a. Age
   b. Essential hypertension
   c. Smoking
   d. Family history of early cardiac death
   e. Male sex

2. Which of the following is the earliest pathological event leading to Mr. G.'s blocked coronary arteries?
   a. Lipid deposition in the tunica intima
   b. Ulceration of the tunica intima
   c. Cellular proliferation in fatty streaks
   d. Calcification of fibrous plaques

3. According to current knowledge, what critical event

probably produced Mr. G.'s myocardial infarction?
   a. Dysrhythmias in which synchronization of atrial and ventricular contraction is lost
   b. Complete occlusion of a coronary artery that produces myocardial ischemia for 1 hour or more
   c. Narrowing of a coronary artery lumen to less than 50% of its original size
   d. Structural changes in the cardiac valves that produce abnormal hemodynamics

4. What was the primary cause of Mr. G.'s chest pain?
   a. Myocardial ischemia
   b. Pulmonary insufficiency
   c. High cardiac output
   d. Oliguria

CHAPTER SUMMARY

HEART
A. Location of the heart
   1. Lies in the mediastinum, behind the body of the sternum between the points of attachment of ribs two through six; approximately two thirds of its mass is to the left of the midline of the body and one third to the right (Figure 17-1)
   2. Posteriorly the heart rests on the bodies of thoracic vertebrae five through eight
   3. Apex lies on the diaphragm, pointing to the left
   4. Base lies just below the second rib
   5. Boundaries of the heart are clinically important as an aid in diagnosing heart disorders
B. Size and shape of the heart (Figures 17-2 and 17-3)
   1. At birth, transverse and appears large in proportion to the diameter of the chest cavity
   2. Between puberty and 25 years of age the heart attains its adult shape and weight
   3. In adult, the shape of the heart tends to resemble that of the chest

C. Covering of the heart
   1. Structure of the heart coverings
      a. Pericardium
         (1) Fibrous pericardium—tough, loose-fitting inextensible sac
         (2) Serous pericardium—parietal layer lies inside the fibrous pericardium, and visceral layer (epicardium) adheres to the outside of the heart; pericardial space with pericardial fluid separates the two layers
   2. Function of the heart coverings—provides protection against friction
D. Structure of the heart
   1. Wall of the heart—made up of three distinct layers
      a. Epicardium—outer layer of heart wall
      b. Myocardium—thick, contractile middle layer of heart wall; compresses the heart cavities, and the blood within them, with great force
      c. Endocardium—delicate inner layer of endothelial tissue (Figure 17-4)

2. Chambers of the heart—divided into four cavities with the right and left chambers separated by the septum (Figures 17-5 and 17-6)
   a. Atria
      (1) Two superior chambers, known as "receiving chambers," since they receive blood from veins
      (2) Atria alternately contract and relax to receive blood and then push it into ventricles
      (3) Myocardial wall of each atrium is not very thick, since little pressure is needed to move blood such a small distance
      (4) Auricle—earlike flap protruding from each atrium
   b. Ventricles
      (1) Two lower chambers, known as "pumping chambers," since they push blood into the large network of vessels
      (2) Ventricular myocardium is thicker than the myocardium of the atria, since great force must be generated to pump the blood a large distance; myocardium of left ventricle is thicker than the right, since it must push blood much further
3. Valves of the heart—mechanical devices that permit the flow of blood in one direction only (Figure 17-7)
   a. Atrioventricular (AV) valves—prevent blood from flowing back into the atria from the ventricles when the ventricles contract
      (1) Tricuspid valve (right AV valve)—guards the right atrioventricular orifice; free edges of three flaps of endocardium are attached to papillary muscles by chordae tendineae
      (2) Bicuspid, or mitral, valve (left AV valve)—similar in structure to tricuspid valve except only two flaps
   b. Semilunar (SL) valves—half-moon–shaped flaps growing out from the lining of the pulmonary artery and aorta; prevents blood from flowing back into the ventricles from the aorta and pulmonary artery
      (1) Pulmonary semilunar valve—valve at entrance of the pulmonary artery
      (2) Aortic semilunar valve—valve at entrance of the aorta
   c. Skeleton of the heart
      (1) Set of connected rings that serve as a semirigid support for the heart valves and for the attachment of cardiac muscle of the myocardium
      (2) Serves as an electrical barrier between the myocardium of the atria and that of the ventricles
   d. Surface projection (review Figure 17-8)
   e. Flow of blood through heart (review Figure 17-6)
4. Blood supply of heart tissue (Figure 17-9)
   a. Coronary arteries—myocardial cells receive blood from the right and left coronary arteries
      (1) First branches to come off the aorta
      (2) Ventricles receive blood from branches of both right and left coronary arteries
      (3) Each ventricle receives blood only from a small branch of the corresponding coronary artery
      (4) Most abundant blood supply goes to the myocardium of the left ventricle
      (5) Right coronary artery is dominant in approximately 50% of all hearts and the left in about 20%, and in approximately 30%, neither coronary artery is dominant
      (6) Few anastomoses exist between the larger branches of the coronary arteries
   b. Coronary veins

      (1) As a rule, veins follow a course that closely parallels that of coronary arteries
      (2) After going through coronary veins, blood enters the coronary sinus to drain into the right atrium
      (3) Several veins drain directly into the right atrium
5. Conduction system of the heart—comprised of the sinoatrial (SA) node, atrioventricular (AV) node, AV bundle, and Purkinje fibers; made up of modified cardiac muscle (Figure 17-10)
   a. Sinoatrial node (SA node or pacemaker)—hundreds of cells in the right atrial wall near the opening of the superior vena cava
   b. Atrioventricular node (AV node or node of Tawara)—small mass of special cardiac muscle in right atrium along the lower part of interatrial septum
   c. Atrioventricular bundle (AV bundle or bundle of His) and Purkinje fibers
      (1) AV bundle originates in AV node, extends by two branches down the two sides of the interventricular septum, and continues as Purkinje fibers
      (2) Purkinje fibers extend out to the papillary muscles and lateral walls of ventricles
6. Nerve supply of the heart
   a. Cardiac plexuses—located near the arch of the aorta, made up of the combination of sympathetic and parasympathetic fibers
   b. Fibers from the cardiac plexus accompany the right and left coronary arteries to enter the heart
   c. Most fibers end in the SA node, but some end in the AV node and in the atrial myocardium
   d. Sympathetic nerves—accelerator nerves
   e. Vagus fibers—inhibitory, or depressor, nerves

BLOOD VESSELS

A. Types of blood vessels
   1. Artery—vessel that carries blood away from the heart; small artery is an arteriole
   2. Vein—vessel that carries blood toward the heart; small vein is a venule
   3. Capillary—microscopic vessel that carries blood from arterioles to venules
B. Structure of blood vessels (Figure 17-11)
   1. Arteries and veins have three layers
      a. Tunica adventitia—outermost layer; made of strong, flexible fibrous connective tissue; helps hold vessels open; prevents tearing of vessels during body movements; in veins, thickest layer; in arteries, thinner than middle layer
      b. Tunica media—middle layer; made of smooth muscle tissue sandwiched together with a layer of elastic connective tissue; permits changes in blood vessel diameter; artery tunica media thicker than that of vein
      c. Tunica intima—innermost layer; made of endothelium; in arteries, completely smooth lining; in veins, forms semilunar valves
   2. Capillaries—have only one layer, the endothelium so the capillary wall is thin enough to allow effective exchanges of material between the plasma and interstitial fluid
C. Functions of blood vessels
   1. Capillaries—most important vessels functionally because they allow the delivery and collection of substances
   2. Arteries—carry blood to arterioles
   3. Arterioles—carry blood from arteries to capillaries; also serve as resistance vessels
   4. Veins—act as collectors and as reservoir vessels

THE MAJOR BLOOD VESSELS

A. Circulatory routes (Figure 17-4)
1. Systemic circulation—blood flows from the left ventricle of the heart through blood vessels to all parts of the body (except gas-exchange tissues of lungs) and back to the right atrium
2. Pulmonary circulation—venous blood moves from right atrium to right ventricle to pulmonary artery to lung arterioles and capillaries where gases are exchanged; oxygenated blood returns to left atrium via pulmonary veins; from left atrium, blood enters the left ventricle
B. Systemic circulation
1. Systemic arteries (review Table 17-2 and Figures 17-15 to 17-18)
   a. Main arteries give off branches, which continue to rebranch, forming arterioles and then capillaries
   b. End-arteries—arteries that eventually diverge into capillaries
   c. Arterial anastomosis—arteries that open into other branches of the same or other arteries; incidence of arterial anastomoses increases as distance from the heart increases
2. Systemic veins (review Figures 17-19 to 17-23)
   a. Veins are the ultimate extensions of capillaries; unite into vessels of increasing size to form venules and then veins
   b. Large veins of the cranial cavity are called dural sinuses
   c. Veins anastomose the same as arteries
   d. Venous blood from the head, neck, upper extremities, and thoracic cavity (except lungs) drains into superior vena cava
   e. Venous blood from thoracic organs drains directly into superior vena cava or azygos vein
   f. Venous blood from the lower extremities and abdomen drains into inferior vena cava
C. Hepatic portal circulation (Figure 17-24)
1. Veins from the spleen, stomach, pancreas, gallbladder, and intestines send their blood to the liver via the hepatic portal vein
2. In the liver the venous blood mingles with arterial blood in the capillaries and is eventually drained from the liver by hepatic veins that join the inferior vena cava
D. Fetal circulation
1. The basic plan of fetal circulation—additional vessels needed to allow fetal blood to secure oxygen and nutrients from maternal blood at the placenta
   a. Two umbilical arteries—extensions of the internal iliac arteries; carry fetal blood to the placenta
   b. Placenta—attached to uterine wall; where exchange of oxygen and other substances between the separated maternal and fetal blood occurs (Figure 17-25)
   c. Umbilical vein—returns oxygenated blood from the placenta to the fetus; enters body through the umbilicus and goes to the undersurface of the liver where it gives off two or three branches and then continues as the ductus venosus
   d. Ductus venosus—continuation of the umbilical vein and drains into inferior vena cava
   e. Foramen ovale—opening in septum between the right and left atria
   f. Ductus arteriosus—small vessel connecting the pulmonary artery with the descending thoracic aorta
2. Changes in circulation at birth (Figures 17-26 and 17-27)
   a. When umbilical cord is cut, the two umbilical arteries, the placenta and umbilical vein, no longer function
   b. Umbilical vein within the baby's body becomes the round ligament of the liver
   c. Ductus venosus becomes the ligamentum venosum of the liver
   d. Foramen ovale—functionally closed shortly after a newborn's first breath and pulmonary circulation is established; structural closure takes approximately 9 months
   e. Ductus arteriosus—contracts with establishment of respiration, becomes ligamentum arteriosum

CYCLE OF LIFE: CARDIOVASCULAR ANATOMY

A. Birth—change from placenta-dependent system
B. Heart and blood vessels maintain basic structure and function from childhood through adulthood
1. Exercise thickens myocardium and increases the supply of blood vessels in skeletal muscle tissue
C. Adulthood through later adulthood—degenerative changes
1. Atherosclerosis—blockage or weakening of critical arteries.
2. Heart valves and myocardial tissue degenerate—reduces pumping efficiency

MECHANISMS OF DISEASE: DISORDERS OF THE CARDIOVASCULAR SYSTEM

DISORDERS OF HEART STRUCTURE

A. Disorders involving the pericardium
1. Pericarditis—inflammation of the pericardium, which leads to pericardial edema
2. Cardiac tamponade—fluid accumulation between the two pericardial layers, which causes a serious compression of the heart
B. Disorders involving heart valves
1. Incompetent valves—leak, allow blood to flow back into the chamber from which it came
2. Stenosed valves—narrowed valves, slow blood leaving from a heart chamber
3. Rheumatic heart disease—cardiac damage resulting from a delayed inflammatory response to streptococcal infection
4. Mitral valve prolapse (MVP)—valve flaps extend back into the left atrium, causing incompetence of the valve
5. Aortic regurgitation—when blood is ejected into the aorta and some regurgitates back into the left ventricle due to an incompetent aortic semilunar valve
C. Disorders involving the myocardium—coronary artery disease (CAD) reduces the blood flow to myocardial tissue due to various causes
1. Myocardial infarction—tissue death that occurs when a blood clot occludes some part of a coronary artery
2. Atherosclerosis—contributes to blocking of coronary arteries by lipids and other matter
3. Angina pectoris—severe chest pain due to decreased oxygen reaching the myocardium
D. Heart failure—inability of the heart to pump enough blood to sustain life; due to various causes
1. Valve disorder—can reduce the pumping efficiency of the heart
2. Cardiomyopathy—disease of the myocardial tissue
3. Arrhythmias—irregularities in heart rhythms
4. Right-side heart failure—accounts for approximately 25% of all cases of heart failure; often results from left-side heart failure; can also be caused by lung disorders that obstruct pulmonary circulation (pulmonary edema) and overload the right side of the heart; called cor pulmonale
5. Congestive heart failure (CHF)—inability of the left ventricle to pump blood effectively; most often results from myocardial infarction

DISORDERS OF BLOOD VESSELS

A. Disorders of arteries
  1. Arteriosclerosis—common vascular disease that occludes arteries and weakens arterial walls; characterized by thickening and calcification of arterial walls with a reduced blood flow occurring; can lead to ischemia and then necrosis
  2. Atherosclerosis—most well known type of arteriosclerosis, may promote the formation of thrombi
  3. Aneurysm—section of an artery becomes abnormally widened due to weakening of the arterial wall; promotes formation of thrombi, which may cause an embolism; tendency to burst, causing hemorrhaging that may be fatal

B. Disorders of veins
  1. Varicose veins—blood tends to pool rather than continue toward the heart; most commonly occur in superficial veins; hemorrhoids are varicose veins in the anal canal
  2. Phlebitis—inflammation of a vein
  3. Thrombophlebitis—acute phlebitis caused by clot formation

REVIEW QUESTIONS

1. Discuss the size, position, and location of the heart in the thoracic cavity.
2. How is compression of the heart in cardiopulmonary resuscitation effected?
3. Describe the pericardium, differentiating between the fibrous and serous portions.
4. Exactly where is pericardial fluid found? Explain its function.
5. Define the following terms: *intercalated disks, syncytium, autorhythmic.*
6. Name and locate the chambers and valves of the heart.
7. Trace the flow of blood through the heart.
8. Describe the heart's own blood supply. Explain why occlusion of a large coronary artery branch has serious consequences.
9. Identify, locate, and describe the functions of each of the following structures: SA node, AV node, AV bundle.
10. Compare arteries, veins, and capillaries as to structure and functions.
11. Differentiate between systemic, pulmonary, and portal circulations.
12. Explain why occlusion of an end-artery is more serious than occlusion of other small arteries.
13. Give the general location of the following veins: longitudinal sinus, internal jugular vein, inferior vena cava, basilic vein, coronary sinus, great saphenous vein, hepatic portal vein.
14. Identify the vessels that join to form the hepatic portal vein.
15. Describe the six structures necessary for fetal circulation.
16. Explain how the separation of oxygenated and deoxygenated blood occurs after birth.
17. Explain the purpose of echocardiography.
18. Identify the clinical significance of cardiac enzymes.
19. Why is the internal jugular vein sometimes ligated as a result of middle ear infections?
20. Briefly define the following terms: *aneurysm, atherosclerosis, phlebitis.*
21. Identify possible causes of coronary artery disease.
22. Define congestive heart failure.
23. Explain how hemorrhoids develop.

18

Physiology of the Cardiovascular System

Cross-section of vein and artery

OBJECTIVES

After you have completed this chapter, you should be able to:

1. Trace a cardiac impulse through the conduction system of the heart.
2. Discuss normal ECG deflections and intervals and their relationship to mechanical contraction.
3. Compare the results of parasympathetic and sympathetic stimulation on the heart and explain the mechanism involved in both types of autonomic control.
4. Discuss several factors that influence heart rate.
5. Discuss the major events of the cardiac cycle.
6. Discuss the physical principles that govern fluid flow and circulation.
7. Discuss how arterial blood pressure is influenced by cardiac output, stroke volume, peripheral resistance, vasomotor pressoreflex, and chemoreflex control mechanisms.
8. Explain the main determinants of peripheral resistance.
9. Identify and discuss the most important factors influencing venous return to the heart.
10. Describe the ADH mechanism in relation to total blood volume.
11. Explain how the blood pressure gradient and peripheral resistance are related to the minute volume of blood.
12. Discuss measurement of arterial blood pressure.
13. Define pulse and identify the two factors most responsible for its existence.
14. Identify those body areas where the pulse can be felt and those areas where pressure may be applied to stop arterial bleeding.

The vital role of the cardiovascular system in maintaining homeostasis depends on the continuous and controlled movement of blood through the thousands of miles of capillaries that permeate every tissue and reach every cell in the body. It is in the microscopic capillaries that blood performs its ultimate transport function. Nutrients and other essential materials pass from capillary blood into fluids surrounding the cells as waste products are removed. Blood must not only be kept moving through its closed circuit of vessels by the pumping activity of the heart, but it must be directed and delivered to those capillary beds surrounding cells that need it most. Blood flow to cells at rest is minimal. In contrast, blood is shunted to the digestive tract after a meal or to skeletal muscles during exercise. The thousands of miles of capillaries could hold far more than the body's total blood volume if it were evenly distributed. Regulation of blood pressure and flow must therefore change in response to cellular activity.

Numerous control mechanisms help to regulate and integrate the diverse functions and component parts of the cardiovascular system to supply blood to specific body areas according to need. These mechanisms ensure a constant *milieu intérieur*, that is, a constant internal environment surrounding each body cell regardless of differing demands for nutrients or production of waste products. This chapter presents information about several of the control mechanisms that regulate the pumping activity of the heart and the smooth and directed flow of blood through the complex channels of the circulation.

PHYSIOLOGY OF THE HEART

Conduction System

The anatomy of four structures that compose the conduction system of the heart—sinoatrial (SA) node, atrioventricular (AV) node, AV bundle, and Purkinje system—was discussed briefly in Chapter 17. Each of these structures consists of cardiac muscle modified enough in structure to differ in function from ordinary cardiac muscle. The specialty of ordinary cardiac muscle is contraction. In this, it is like all muscle, and like all muscle, ordinary cardiac muscle can also conduct impulses. But the conduction system structures are more highly specialized, both structurally and functionally, than ordinary cardiac muscle tissue. They are not contractile. Instead, they permit only generation or rapid conduction of an action potential through the heart.

The normal cardiac impulse that initiates mechanical contraction of the heart arises in the **SA node** (or

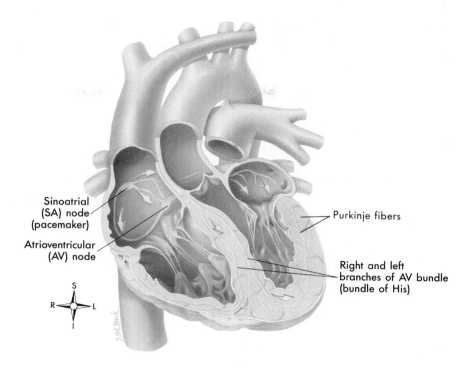

Sinoatrial
(SA) node
(pacemaker)

Atrioventricular
(AV) node

Purkinje fibers

Right and left
branches of AV bundle
(bundle of His)

FIGURE 18-1   Conduction system of the heart. Specialized cardiac muscle cells in the wall of the heart rapidly initiate or conduct an electrical impulse throughout the myocardium. The signal is initiated by the SA node (pacemaker) and spreads to the rest of atrial myocardium and to the AV node. The AV node then initiates a signal that is conducted through the ventricular myocardium by way of the AV (bundle of His) and Purkinje fibers.

pacemaker), located just below the atrial epicardium at its junction with the superior vena cava (Figure 18-1). Specialized pacemaker cells in the node possess an *intrinsic rhythm.* This means that without any stimulation by nerve impulses from the brain and cord, they themselves initiate impulses at regular intervals. Even if the heart is removed from the body and placed in a nutrient solution, completely separated from all nervous and hormonal control, it will continue to beat!

Each impulse generated at the SA node travels swiftly throughout the muscle fibers of both atria. Thus stimulated, the atria begin to contract. As the action potential enters the AV node from the right atrium, its conduction slows markedly, thus allowing for complete contraction of both atrial chambers before the impulse reaches the ventricles. After passing slowly through the AV node, conduction velocity increases as the impulse is relayed through the AV bundle (bundle of His) into the ventricles. Here, right and left branches of the bundle fibers and the Purkinje fibers in which they terminate conduct the impulses throughout the muscle of both ventricles, stimulating them to contract almost simultaneously.

Thus the SA node initiates each heartbeat and sets its pace—it is the heart's own natural pacemaker. Normally the SA node will "discharge," or "fire," at an intrinsic

rhythmical rate of 70 to 75 beats per minute. However, if for any reason the SA node loses its ability to generate an impulse, pacemaker activity will shift to another excitable component of the conduction system such as the AV node or the Purkinje fibers. Pacemakers other than the SA node are called abnormal, or **ectopic, pacemakers.** Although ectopic pacemakers fire rhythmically, their rate of discharge is generally much slower than that of the SA node. For example, a pulse of 40 to 60 beats per minute would result if the AV node were forced to assume pacemaker activity.

Electrocardiogram

Impulse conduction generates tiny electrical currents in the heart that spread through surrounding tissues to the surface of the body. This fact has great clinical importance. Why? Because from the skin, visible records of the heart's electrical activity can be made with an instrument called an *electrocardiograph.* Skilled interpretation of these records may sometimes make the difference between life and death.

The **electrocardiogram (ECG** or **EKG)** is a graphic record of the heart's electrical activity, its conduction of impulses. It is not a record of the heart's contractions but of the electrical events that precede them. To produce an electrocardiogram, electrodes of a recording voltmeter (electrocardiograph) are attached to the limbs and/or chest

FOCUS ON... ARTIFICIAL CARDIAC PACEMAKERS

Everyone has heard about **artificial pacemakers,** devices that electrically stimulate the heart at a set rhythm (continuously discharging pacemakers) or those that fire only when the heart rate decreases below a preset minimum (demand pacemakers). They do an excellent job of maintaining a steady heart rate and of keeping many individuals with damaged hearts alive for many years. Hundreds of thousands of people currently have permanently implanted cardiac pacemakers.

Several types of artificial pacemakers have been designed to deliver an electrical stimulus to the heart muscle. The stimulus passes through electrodes that are sewn directly to the epicardium on the outer surface of the heart or are inserted by a catheter into a heart chamber, such as the right ventricle, and placed in contact with the endocardium. Modern pacemakers generate a stimulus that lasts from 0.08 to 2 msec and produce a very low current output.

One common method of inserting a permanent pacemaker is by the *transvenous approach.* In this procedure a small incision is made just above the right clavicle, and the electrode is threaded into the jugular vein and then advanced to the apex of the right ventricle. Figure A shows the battery-powered stimulus generator, which is placed in a pocket beneath the skin on the right side of the chest just below the clavicle. The proximal end of the electrical lead, or catheter, is then directed through the subcutaneous tissues and attached to the power pack. Figure B shows the tip of the electrical lead in the apex of the right ventricle. Although life saving, these devices must be judged inferior to the heart's own natural pacemaker. Why? Because they cannot speed up the heartbeat when necessary (for example, to make strenuous physical activity possible), nor can they slow it down again when the need has passed. The normal SA node, influenced as it is by autonomic impulses and hormones, can produce these changes. Discharging an average of 75 times each minute, this truly remarkable bit of specialized tissue will generate well over 2 billion action potentials in an average lifetime of some 70 years!

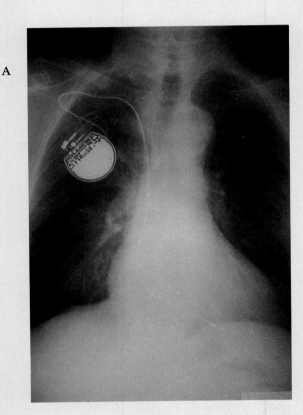

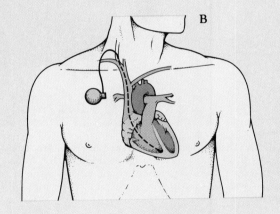

Artificial pacemaker. A, This x-ray photograph shows the stimulus generator in the subcutaneous tissue of the chest wall. **B,** Thin flexible wires extend through veins to the heart, where timed electrical impulses stimulate the myocardium.

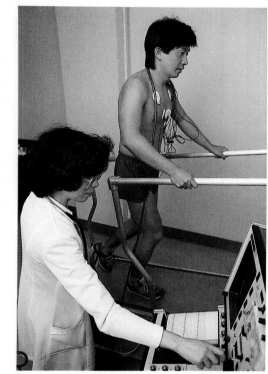

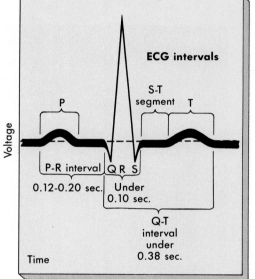

FIGURE 18-2 Electrocardiogram. A, A nurse monitors a patient's ECG as he exercises on a treadmill. **B,** Normal ECG deflections represent depolarization and repolarization of cardiac muscle tissue. **C,** Principal ECG intervals between P, QRS, and T waves.

of the subject (Figure 18-2, A). Changes in voltage, which represent changes in the heart's electrical activity, are observed as deflections of a line drawn on paper or traced on a video monitor.

Because electrocardiography is far too complex a subject to explain fully here, normal ECG *deflection waves* and the *ECG intervals* between them shall be only briefly discussed. As shown in Figure 18-2, B, the normal ECG is composed of deflection waves called the P wave, QRS complex, and T wave. (The letters do not stand for any words but were chosen as an arbitrary sequence of the alphabet.) Briefly, the P wave represents depolarization of the atria, that is, the passage of a depolarization current from the SA node through the musculature of both atria; the QRS complex represents both repolarization of the atria and depolarization of the ventricles; and the T wave reflects repolarization

of the ventricles. The principal ECG intervals between P, QRS, and T waves are shown in Figure 18-2, C. Measurement of these intervals can provide valuable information concerning the rate of conduction of an action potential through the heart. Figure 18-3 summarizes the relationship between the electrical events of the myocardium and the ECG recordings.

QUICK CHECK

✔ 1. List the principal structures of the heart's conduction system.

✔ 2. What are the three types of deflection waves seen in a typical ECG?

✔ 3. What event does each type of ECG wave represent?

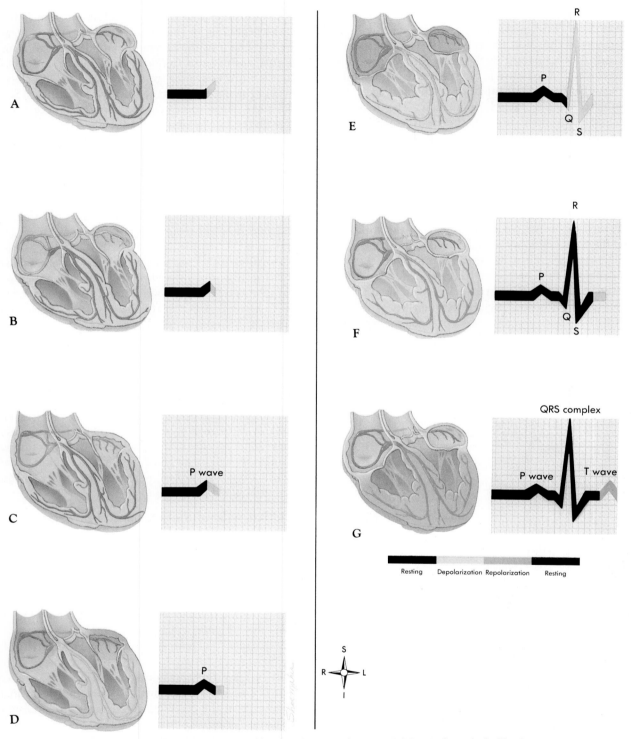

FIGURE 18-3 Events represented by the electrocardiogram (ECG). A through **C,** The P wave represents the depolarization of cardiac muscle tissue in the SA node and atrial walls. **C** and **D,** Before the QRS complex is observed, the AV node and AV bundle depolarize. **E** and **F,** The QRS complex occurs as the atrial walls repolarize and the ventricular walls depolarize. **G,** The T wave is observed as the ventricular walls repolarize. Depolarization triggers contraction in the affected muscle tissue. Thus cardiac muscle contraction occurs *after* depolarization begins.

Control of Heart Rate

Although the SA node normally initiates each heartbeat, the rate it sets is not an unalterable one. Various factors can and do change the rate of the heartbeat. One major modifier of SA node activity—and therefore of the heart rate—is the ratio of sympathetic and parasympathetic impulses conducted to the node per minute. Autonomic control of heart rate is the result of opposing influences between parasympathetic (chiefly vagus) and sympathetic (cardiac nerve) stimulation. The results of parasympathetic stimulation on the heart are inhibitory and are mediated by vagal release of acetylcholine, whereas sympathetic (stimulatory) effects result from the release of norepinephrine.

Cardiac pressoreflexes

Receptors sensitive to changes in pressure (baroreceptors) are located in two places near the heart (Figure 18-4). Called the *aortic baroreceptors* and *carotid baroreceptors*, they send afferent nerve fibers to cardiac control centers in the medulla oblongata. These stretch receptors, located in the aorta and carotid sinus, constitute a very important heart rate control mechanism because of their effect on the autonomic cardiac control centers—and therefore on

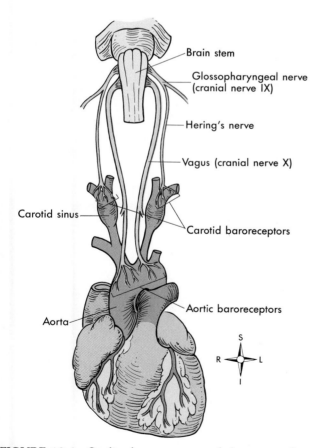

parasympathetic and sympathetic outflow. Baroreceptors operate with integrators in the cardiac control centers in feedback loops called **pressoreflexes** or *baroreflexes*.

Carotid sinus reflex. The carotid sinus is a small dilation at the beginning of the internal carotid artery just above the branching of the common carotid artery to form the internal and external carotid arteries (see Figure 18-4). The sinus lies just under the sternocleidomastoid muscle at the level of the upper margin of the thyroid cartilage. Sensory (afferent) fibers from carotid sinus baroreceptors run through the carotid sinus nerve (of Hering) and on through the glossopharyngeal (or ninth cranial) nerve to a specialized area of the medulla called the *cardiac control center*. Parasympathetic impulses from the cardiac control center reach the SA node through the vagus (or tenth cranial) nerve. Normally the heart is under the restraint of vagal inhibition, as released acetylcholine decreases the rate of SA node firing. The vagus is said to act as a "brake" on the heart.

Aortic reflex. Sensory (afferent) nerve fibers also extend from baroreceptors located in the wall of the arch of the aorta through the aortic nerve and then through the vagus (tenth cranial) nerve to terminate in the cardiac control center of the medulla (see Figure 18-4).

If blood pressure within the aorta or carotid sinus increases suddenly, it stimulates the aortic or carotid baroreceptors, as shown in Figure 18-5. Stimulation of these stretch receptors causes the cardiac control center to increase vagal inhibition, thus slowing the heart and returning blood pressure back toward normal (see Figure 18-4). A decrease in aortic or carotid blood pressure usually allows some acceleration of the heart. Details of pressoreflex activity will be included later in the chapter as part of a mechanism that tends to maintain or restore homeostasis of arterial blood pressure.

Other reflexes that influence heart rate

Reflexes involving such important factors as emotions, exercise, hormones, blood temperature, pain, and stimulation of various exteroceptors also influence heart rate. Anxiety, fear, and anger often make the heart beat faster. Grief, in contrast, tends to slow it. Emotions produce changes in the heart rate through the influence of impulses from the "higher centers" in the cerebrum via the hypothalamus. Such impulses can influence activity of the cardiac control centers.

In exercise the heart normally accelerates. The mechanism is not definitely known, but it is thought to include impulses from the cerebrum through the hypothalamus to the cardiac center. Epinephrine is the hormone most noted as a cardiac accelerator.

Increased blood temperature or stimulation of skin heat receptors tends to increase the heart rate, and decreased blood temperature or stimulation of skin cold receptors tends to slow it. Sudden, intense stimulation of pain receptors in such visceral structures as the gallbladder, ureters, or intestines can result in such slowing of the heart that fainting may result.

FIGURE 18-4 Cardiac baroreceptors and the pressoreflexes. Location of aortic and carotid baroreceptors and the sensory nerves that carry information about blood pressure back to the central nervous system.

Labels for Figure 18-4:
- Brain stem
- Glossopharyngeal nerve (cranial nerve IX)
- Hering's nerve
- Vagus (cranial nerve X)
- Carotid sinus
- Carotid baroreceptors
- Aortic baroreceptors
- Aorta

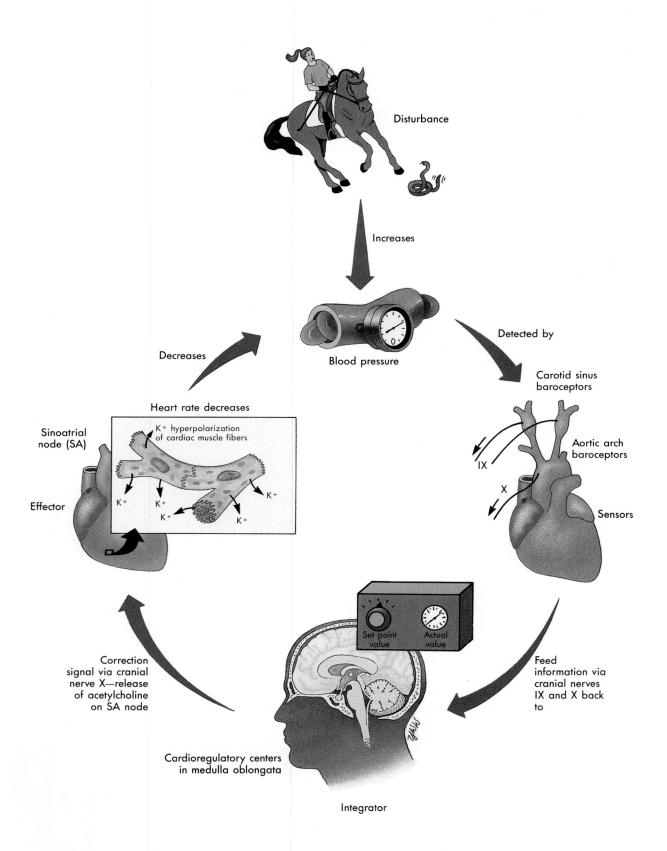

FIGURE 18-5 Aortic and carotid sinus pressoreflexes. These pressoreflexes operate in a feedback loop that maintains the homeostasis of blood pressure by decreasing the heart rate when the blood pressure surpasses the set point.

FOCUS ON... CARDIAC ARRHYTHMIA

Various conditions such as inflammation of the endocardium (endocarditis) or myocardial infarction (heart attack) can damage the heart's conduction system and thereby disturb the rhythmical beating of the heart. The term **arrhythmia** refers to an abnormality of heart rhythm.

One kind of arrhythmia is called a **heart block.** In *AV node block*, impulses are blocked from getting through to the ventricular myocardium, resulting in the ventricles contracting at a much slower rate than normal. On an ECG, there may be a large interval between the P wave and the R peak of the QRS complex (Figure A). *Complete heart block* occurs when the P waves do not match up with the QRS complexes at all—as in an ECG that shows two or more P waves for every QRS complex. A physician may treat heart block by implanting in the heart an *artificial pacemaker* (see essay on p. 503).

Bradycardia is a slow heart rhythm—below 50 beats per minute (Figure B). Slight bradycardia is normal during sleep and in conditioned athletes while they are awake (but at rest). Abnormal bradycardia can result from improper autonomic nervous control of the heart or from a damaged SA node. If the problem is severe, artificial pacemakers can be used to increase the heart rate by taking the place of the SA node.

Tachycardia is a very rapid heart rhythm—over 100 beats per minute (Figure C). Tachycardia is normal during and after exercise and during the stress response. Abnormal tachycardia can result from improper autonomic control of the heart, blood loss or shock, the action of drugs and toxins, fever, and other factors.

Sinus arrhythmia is a variation in heart rate during the breathing cycle. Typically, the rate increases during inspiration and decreases during expiration. The causes of sinus arrhythmia are not clear. This phenomenon is common in young people and usually does not require treatment.

Premature contractions, or *extrasystoles,* are contractions that occur before the next expected contraction in a series of cardiac cycles. For example, *premature atrial contractions (PACs)* may occur shortly after the ventricles contract—seen as early P waves on the ECG. Premature contractions often occur with lack of sleep, too much caffeine or nicotine, alcoholism, or heart damage. Frequent premature contractions can lead to

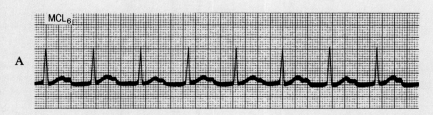

A, AV node block. Very slow ventricular contraction (25 to 45 beats/min at rest); P waves widely separated from peaks of QRS complexes.

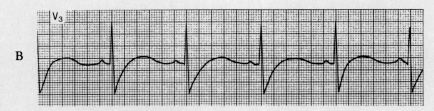

B, Bradycardia. Slow heart rhythm (less than 60 beats/min); no disruption of normal rhythm pattern.

DIAGNOSTIC STUDIES   DOPPLER ULTRASONOGRAPHY

Doppler ultrasonography uses ultrasound to record the direction of blood flow through the heart. These sound waves are reflected off red blood cells as they pass through the heart, allowing the velocity of blood to be calculated as it travels through the heart chambers.

Color Doppler mapping is a variation that converts recorded flow frequencies into different colors. These color images are then superimposed on M-mode or 2-D echocardiograms, allowing more detailed evaluation of disorders.

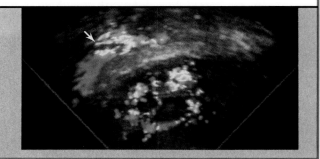

fibrillation, a condition in which cardiac muscle fibers contract out of step with each other. This event can be seen in an ECG as the absence of regular P waves or abnormal QRS and T waves. In fibrillation, the affected heart chambers do not effectively pump blood. *Atrial fibrillation* occurs commonly in mitral stenosis, rheumatic heart disease, and infarction of the atrial myocardium (Figure *D*). This condition can be treated with drugs such as digoxin (a *digitalis* preparation) or by *defibrillation*—application of electrical shock to force cardiac muscle fibers to contract in unison. *Ventricular fibrillation* is an immediately life-threatening condition in which the lack of ventricular pumping suddenly stops the flow of blood to vital tissues (Figure *E*). Unless ventricular fibrillation is corrected immediately by defibrillation or some other method, death may occur within minutes.

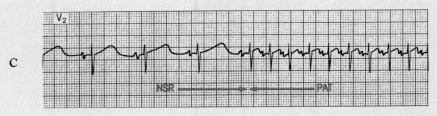

C

C, Tachycardia. Rapid heart rhythm (greater than 100 beats/min); no disruption of normal rhythm pattern.

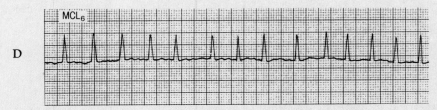

D

D, Atrial fibrillation. Irregular, rapid atrial depolarizations; P wave rapid (greater than 300/min) with irregular QRS complexes (150 to 170 beats/min)

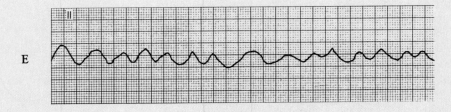

E

E, Ventricular fibrillation. Complete disruption of normal heart rhythm.

Reflexive increases in heart rate often result from an increase in sympathetic stimulation of the heart. Sympathetic impulses originate in the cardiac control center of the medulla and reach the heart via sympathetic fibers (contained in the middle, superior, and inferior cardiac nerves). Norepinephrine released as a result of sympathetic stimulation increases heart rate and strength of cardiac muscle contraction.

QUICK CHECK

✔ 1. *What is a cardiac pressoreflex?*
✔ 2. *How does the body use a pressoreflex to counteract an abnormal increase in heart rate? An abnormal decrease in heart rate?*
✔ 3. *What other factors can influence the heart rate?*

Cardiac Cycle

The term **cardiac cycle** means a complete heartbeat, or pumping cycle, consisting of contraction **(systole)** and relaxation **(diastole)** of both atria and both ventricles. The two atria contract simultaneously. Then, as the atria relax, the two ventricles contract and relax, instead of the entire heart contracting as a unit. This gives a kind of pumping action to the movements of the heart. The atria remain relaxed during part of the ventricular relaxation and then start the cycle over again. The cycle as a whole is often divided into time intervals for discussion and study. The following list relates several of the important events of the cycle to time, and Figure 18-6 is a composite chart that graphically illustrates and integrates changes in pressure gradients in the left atrium, left ventricle, and aorta with ECG and heart sound recordings. Aortic blood flow and changes in ventricular volume are also shown. Figure 18-7 summarizes the cardiac cycle only.

♦ **Atrial systole.** The contracting force of the atria completes the emptying of blood out of the atria into the ventricles. Atrioventricular (AV, or cuspid) valves are necessarily open during this phase; the ventricles are relaxed and filling with blood. The semilunar (SL) valves are closed so that blood does not reenter from the pulmonary artery or aorta. This period of the cycle begins with the P wave of the ECG. Passage of the electrical wave of depolarization is then followed almost immediately by actual contraction of the atrial musculature.

♦ **Isovolumetric ventricular contraction.** *Iso* is a combining form denoting equality or uniformity. During the brief period of isovolumetric ventricular contraction, that is, between the start of ventricular systole and the opening of the SL valves, ventricular volume remains constant, or uniform, as the pressure increases rapidly. The onset of ventricular systole coincides with the R wave of the ECG and the appearance of the first heart sound.

♦ **Ejection.** The SL valves open and blood is ejected from the heart when the pressure gradient in the ventricles exceeds the pressure in the pulmonary artery and aorta. An initial, shorter phase, called *rapid ejection,* is characterized by a marked increase in ventricular and aortic pressure and in aortic blood flow. The T wave of the ECG appears during the later, longer phase of *reduced ejection* (characterized by a less abrupt decrease in ventricular volume). It is important to note that a considerable quantity of blood, called the **residual volume,** normally remains in the ventricles at the end of the ejection period. In *heart failure* the residual volume remaining in the ventricles may greatly exceed that ejected during systole.

♦ **Isovolumetric ventricular relaxation.** Ventricular diastole, or relaxation, begins with this period of the cardiac cycle. It is the period between closure of the SL valves and opening of the AV valves. At the end of ventricular ejection the SL valves will close so that blood cannot reenter the ventricular chambers from the great vessels. The AV valves will not open until the pressure in the atrial chambers increases above that in the relaxing ventricles. The result is a dramatic fall in intraventricular

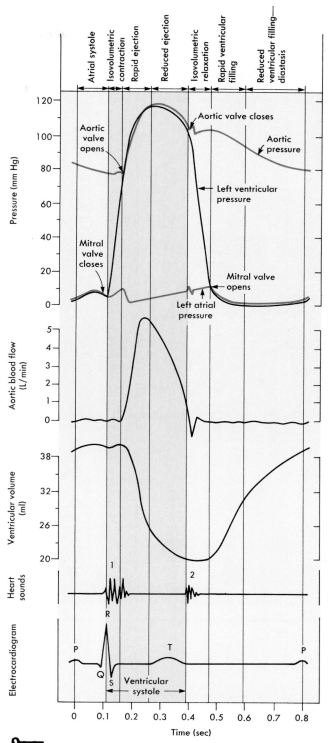

FIGURE 18-6   Composite chart of heart function. This chart is a composite of several diagrams of heart function (cardiac pumping cycle, blood pressure, blood flow, volume, heart sounds, and ECG), all adjusted to the same time scale. Although it appears daunting at first glance, you will find it a valuable reference tool as you proceed through this chapter.

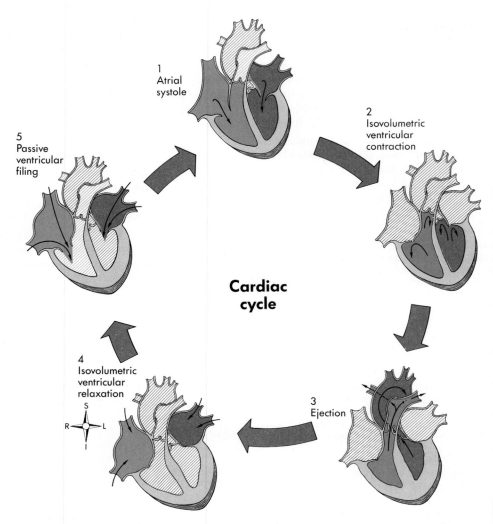

FIGURE 18-7   The cardiac cycle. The five steps of the heart's pumping cycle described in the text are shown as a series of changes in the heart wall and valves.

pressure but no change in volume. Both sets of valves are closed, and the ventricles are relaxing. The second heart sound is heard during this period of the cycle.

◆ Passive ventricular filling. Return of venous blood increases intraatrial pressure until the AV valves are forced open and blood rushes into the relaxing ventricles. The rapid influx lasts about 0.1 second and results in a dramatic increase in ventricular volume. The term *diastasis* is often used to describe a later, longer period of slow ventricular filling at the end of ventricular diastole. The abrupt inflow of blood that occurred immediately after opening of the AV valves is followed by a slow but continuous flow of venous blood into the atria and then through the open AV valves into the ventricles. Diastasis lasts about 0.2 second and is characterized by a gradual increase in ventricular pressure and volume.

Heart sounds

The heart makes certain typical sounds during each cardiac cycle that are described as sounding like "lubb-dupp" through a stethoscope. The first, or systolic, sound is believed to be caused primarily by the contraction of the ventricles and also by vibrations of the closing AV, or cuspid, valves. It is longer and lower than the second, or diastolic, sound, which is short, sharp, and thought to be caused by vibrations of the closing SL valves (see Figure 18-6).

Heart sounds have clinical significance, since they give information about the valves of the heart. Any variation from normal in the sounds indicates imperfect functioning of the valves. *Heart murmur* is one type of abnormal sound heard frequently. It may signify incomplete closing of the valves (valvular insufficiency) or stenosis (constriction, or narrowing) of them.

QUICK CHECK

✔ 1. *Using Figure 18-7, describe the major events of the cardiac cycle.*

✔ 2. *As the ventricles contract, their volume remains constant for a period of time. Can you explain why the volume does not begin to decrease immediately?*

CONTROL OF CIRCULATION

Hemodynamics

Hemodynamics is a term used to describe a collection of mechanisms that influence the active and changing—or dynamic—circulation of blood. Circulation is, of course, a vital function. It constitutes the only means by which cells can receive materials needed for their survival and can have their wastes removed. Not only is circulation necessary, but circulation of different volumes of blood per minute is also essential for healthy survival. More active cells need more blood per minute than less active cells. The reason underlying this principle is obvious. The more work cells do, the more energy they use, and the more oxygen and food they remove from the blood. Because blood circulates, it can continually bring in more oxygen and food to replace what is consumed. So the more active any part of the body is, the greater the volume of blood circulated to it per minute must be. This requires that circulation control mechanisms accomplish two functions: maintain circulation (keep blood flowing) and vary the volume and distribution of the blood circulated. The greater the activity of any part of the body, the greater the volume of blood it needs circulating through it. Therefore, as any structure increases its activity, an increased volume of blood must be distributed to it—must be shifted from the less active to the more active tissues.

To achieve these two ends, a great many factors must operate together as one smooth-running although complex machine. Incidentally, this is an important physiological principle that you have no doubt observed by now—that every body function depends on many other functions. A constellation of separate processes or mechanisms acts as a single integrated mechanism. Together, they perform one large function. For example, many mechanisms together accomplish the large function we call circulation.

Primary Principle of Circulation

Blood circulates for the same reason that any fluid flows—whether it is water in a river, in a garden hose, or in hospital tubing, or blood in vessels. A fluid flows because a pressure gradient exists between different parts of its volume (Figure 18-8). This primary fluid flow principle derives from Newton's first and second laws of motion. In essence, these laws state the following principles:

1. A fluid does not flow when the pressure is the same in all parts of it
2. A fluid flows only when its pressure is higher in one area than in another, and it flows always from its higher pressure area toward its lower pressure area

In brief, then, the primary principle about circulation is this: blood circulates from the left ventricle and returns to the right atrium of the heart because a blood pressure

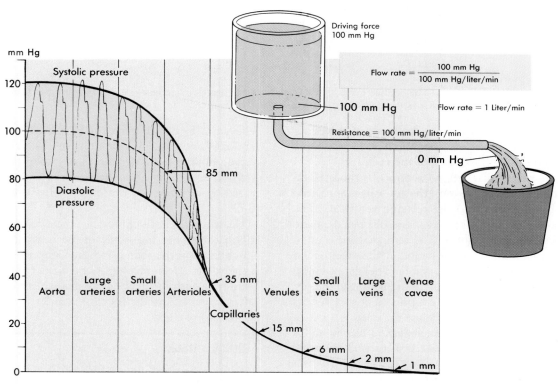

FIGURE 18-8   The primary principle of circulation. Fluid always travels from an area of high pressure to an area of low pressure. Water flows from an area of high pressure in the tank (100 mm Hg) toward the area of low pressure above the bucket (0 mm Hg). Blood tends to move from an area of high average pressure at the beginning of the aorta (100 mm Hg) toward the area of lowest pressure at the end of the venae cavae (0 mm Hg). Blood flow between any two points in the circulatory system can always be predicted by the pressure gradient.

gradient exists between these two structures. By blood pressure gradient, we mean the difference between the blood pressure in one structure and the blood pressure in another. For example, a typical normal blood pressure in the aorta, as the left ventricle contracts pumping blood into it, is 120 mm Hg; as the left ventricle relaxes, it decreases to 80 mm Hg. The mean, or average, blood pressure, therefore, in the aorta in this instance is 100 mm Hg.

Figure 18-8 shows the systolic and diastolic pressures in the arterial system and illustrates the progressive fall in pressure to 0 mm Hg by the time blood reaches the venae cavae and right atrium. The progressive fall in pressure as blood passes through the circulatory system is directly related to resistance. Resistance to blood flow in the aorta is almost zero. Although the pumping action of the heart causes fluctuations in aortic blood pressure (systolic 120 mm Hg; diastolic 80 mm Hg), the mean pressure remains almost constant, dropping perhaps only 1 or 2 mm Hg. The greatest drop in pressure (about 50 mm Hg) occurs across the arterioles because they present the greatest resistance to blood flow.

$P_1 - P_2$ is often used to stand for a pressure gradient, with P_1 the symbol for the higher pressure and P_2 the symbol for the lower pressure. For example, blood enters the arterioles at 85 mm Hg and leaves at 35 mm Hg. Which is P_1? P_2? What is the blood pressure gradient? It would cause blood to flow from the arterioles into capillaries.

Control of Arterial Blood Pressure

The primary determinant of arterial blood pressure is the volume of blood in the arteries. Arterial blood volume is directly proportional to arterial pressure. This means that an increase in arterial blood volume tends to increase arterial pressure, and, conversely, a decrease in arterial volume tends to decrease arterial pressure.

Many factors determine arterial pressure through their influence on arterial volume. Two of the most important— **cardiac output** and **peripheral resistance**—are directly proportional to blood volume (Figure 18-9).

Cardiac output

Cardiac output (CO) is determined by the volume of blood pumped out of the ventricles by each beat (**stroke volume** or *SV*) and by heart rate (HR). Because contraction of the heart is called systole, the volume of blood pumped by one contraction is known as *systolic discharge.* Stroke volume means the same thing, the amount of blood pumped by one stroke (contraction) of the ventricle.

Stroke volume. *Stroke volume and heart rate.* Stroke volume, or volume pumped per heartbeat, is one of two major factors that determines cardiac output. CO can be computed by the following simple equation:

$$SV\left(\frac{volume}{beat}\right) \times HR\left(\frac{beat}{min}\right) = CO\left(\frac{volume}{min}\right)$$

Thus the greater the stroke volume, the greater the cardiac output (but only if the heart rate remains constant). In practice, computing the CO is far from simple. It requires introducing a catheter into the right side of the heart (cardiac catheterization) and solving a computation known as *Fick's formula.*

Since the heart's rate and stroke volume determine its output, anything that changes the rate of the heartbeat or its stroke volume tends to change CO, arterial blood

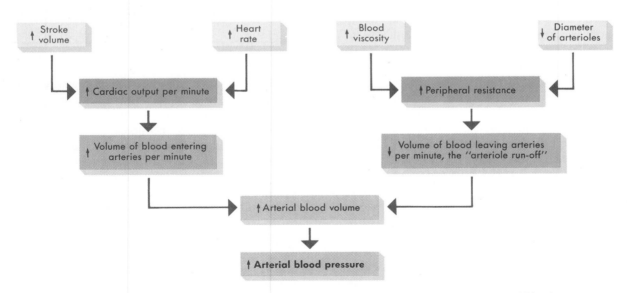

FIGURE 18-9 Relationship between arterial blood volume and blood pressure. Arterial blood pressure is directly proportional to arterial blood volume. Cardiac output (CO) and peripheral resistance (PR) are directly proportional to arterial blood volume, but for opposite reasons: CO affects blood *entering* the arteries and PR affects blood *leaving* the arteries. If cardiac minute output increases, the amount of blood entering the arteries increases and tends to increase the volume of blood in the arteries. If peripheral resistance increases, it decreases the amount of blood leaving the arteries, which tends to increase the amount of blood left in them. Thus an increase in either CO or PR results in an increase in arterial blood volume, which increases arterial blood pressure.

volume, and blood pressure in the same direction. In other words, anything that makes the heart beat faster or anything that makes it beat stronger (increases its stroke volume) tends to increase CO and therefore arterial blood volume and pressure. Conversely, anything that causes the heart to beat more slowly or more weakly tends to decrease CO, arterial volume, and blood pressure. But do not overlook the word *tends* in the preceding sentences. A change in heart rate or stroke volume does not always change the heart's output, or the amount of blood in the arteries, or the blood pressure. To see whether this is true, do the following simple arithmetic, using the simple formula for computing CO. Assume a normal rate of 72 beats per minute and a normal stroke volume of 70 ml. Next, suppose the rate drops to 60 and the stroke volume increases to 100. Does the decrease in heart rate actually cause a decrease in CO in this case? Clearly not—the CO increases. Do you think it is valid, however, to say that a slower rate *tends* to decrease the heart's output? By itself, without any change in any other factor, would not a slowing of the heartbeat cause CO volume, arterial volume, and blood pressure to fall?

Starling's law of the heart. Mechanical, neural, and chemical factors regulate the strength of the heartbeat and therefore its stroke volume. One mechanical factor that helps determine stroke volume is the length of myocardial fibers at the beginning of ventricular contraction.

Many years ago, a physiologist named Starling described a principle that later became known as **Starling's law of the heart.** In this principle he stated the factor he had observed as the main regulator of heartbeat strength in experiments performed on denervated animal hearts. Starling's law of the heart, in essence, is this: within limits, the longer, or more stretched, the heart fibers at the beginning of contraction, the stronger is their contraction. Compare this concept with the length-tension relationship in skeletal muscle described in Chapter 9 (p. 238).

The factor determining how stretched the animal hearts were at the beginning of contractions was, as you might deduce, the amount of blood in the hearts at the end of diastole. The more blood returned to the hearts per minute, the more stretched were their fibers, the stronger were their contractions, and the larger was the volume of blood they ejected with each contraction. If, however, too much blood stretched the hearts beyond a certain critical point, they seemed to lose their elasticity. They then contracted less vigorously—much as a rubber band, stretched too much, rebounds with less force (Figure 18-10).

Although it was first described in animal experiments, most physiologists agree that Starling's law of the heart operates in humans as a major regulator of stroke volume under ordinary conditions. Operation of Starling's law of the heart ensures that increased amounts of blood returned to the heart will be pumped out of it. It automatically adjusts CO to venous return under usual conditions.

Heart rate. Control mechanisms that influence heart rate have already been discussed. The pressoreflexes constitute the dominant heart rate control mechanism, although we have mentioned various other factors that also

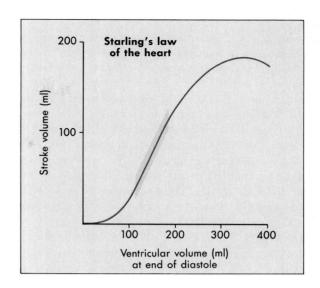

FIGURE 18-10 Starling's law of the heart. This curve represents the relationship between the stroke volume and the ventricular volume at the end of diastole. The range of values observed in a typical heart is shaded. Notice that if the ventricle has an abnormally large volume at the end of diastole (*far right portion of the curve*), the stroke volume cannot compensate.

influence heart rate. Figure 18-5 shows the relationship between homeostasis of arterial blood pressure and activation of cardiac pressoreflexes.

QUICK CHECK

✔ *1. State the primary principle of circulation.*
✔ *2. How do stroke volume and heart rate relate to cardiac output?*
✔ *3. If the amount of blood returned to the heart increases, what happens to the stroke volume? What principle explains this?*

Peripheral resistance

How resistance influences blood pressure. Peripheral resistance helps determine arterial blood pressure. Specifically, arterial blood pressure tends to vary directly with peripheral resistance. Peripheral resistance means the resistance to blood flow imposed by the force of friction between blood and the walls of its vessels. Friction develops partly because of a characteristic of blood—its *viscosity,* or stickiness—and partly from the small diameter of arterioles and capillaries. The resistance offered by arterioles, in particular, accounts for almost half of the total resistance in systemic circulation. The muscular coat with which arterioles are vested allows them to constrict or dilate and thus change the amount of resistance to blood flow. Peripheral resistance helps determine arterial pressure by controlling the rate of "arteriole runoff," the amount of blood that runs out of the arteries into the arterioles. The greater the resistance, the less the arteriole runoff, or outflow, tends to be—and therefore the more blood left in the arteries and the higher the arterial pressure tends to be.

Blood viscosity stems mainly from the red blood cells but also partly from the protein molecules present in blood. An

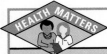

CHANGES IN BLOOD VISCOSITY

Under normal circumstances, blood viscosity changes very little. But under certain abnormal conditions, such as marked anemia or hemorrhage, a decrease in blood viscosity may be the crucial factor lowering peripheral resistance and arterial pressure, even to the point of circulatory failure.

increase in either blood protein concentration or red blood cell count tends to increase viscosity, and a decrease in either tends to decrease it.

Vasomotor control mechanism. Blood distribution patterns, as well as blood pressure, can be influenced by factors that control changes in the diameter of arterioles. Such factors might be said to constitute the **vasomotor control mechanism.** Like most physiological control mechanisms, it consists of many parts. An area in the medulla called the *vasomotor center,* or *vasoconstrictor center,* will, when stimulated, initiate an impulse outflow via sympathetic fibers that ends in the smooth muscle surrounding resistance vessels, arterioles, venules, and veins of the "blood reservoirs," causing their constriction. Thus the vasomotor control mechanism plays a role both in the maintenance of the general blood pressure and in the distribution of blood to areas of special need.

The main blood reservoirs are the venous plexuses and sinuses in the skin and abdominal organs (especially in the liver and spleen). In other words, blood reservoirs are the venous networks in most parts of the body—all but those in the skeletal muscles, heart, and brain. The term *reservoir* is apt, since these vessels serve as storage depots for blood. Blood can quickly be moved out of them and "shifted" to heart and skeletal muscles when increased activity demands (Figure 18-11). A change in either arterial blood's oxygen or carbon dioxide content sets a chemical vasomotor control mechanism in operation. A change in arterial blood pressure initiates a *vasomotor pressoreflex.*

Vasomotor pressoreflexes. A sudden increase in arterial blood pressure stimulates aortic and carotid baroreceptors—the same ones that initiate cardiac reflexes. Not only does this stimulate the cardiac control center to reduce heart rate (see Figure 18-5) but also inhibits the vasoconstrictor center. More impulses per second go out over parasympathetic fibers to the heart and to blood vessels. As a result, the heartbeat slows, and arterioles and the venules of the blood reservoirs dilate. Since sympathetic vasoconstrictor impulses predominate at normal arterial pressures, inhibition of these is considered the major mechanism of vasodilation. The nervous pathways involved in this mechanism are illustrated in Figure 18-12.

A decrease in arterial pressure causes the aortic and carotid baroreceptors to send more impulses to the medulla's vasoconstrictor centers, thereby stimulating them. These centers then send more impulses via the sympathetic fibers to stimulate vascular smooth muscle and cause vasoconstriction. This squeezes more blood out of the blood reservoirs, increasing the amount of venous blood return to the heart. Eventually, this extra blood is redistributed to more active structures such as skeletal muscles and heart because their arterioles become dilated largely from the operation of a local mechanism (discussed later). Thus the vasoconstrictor pressoreflex and the local vasodilating mechanism together serve as an important device for shifting blood from reservoirs to structures that need it more. It is an especially valuable mechanism during exercise (see Figure 18-11).

Vasomotor chemoreflexes. Chemoreceptors located in the aortic and carotid bodies are particularly sensitive to excess blood carbon dioxide (**hypercapnia**) and somewhat less sensitive to a deficiency of blood oxygen (hypoxia) and to decreased arterial blood pH. When one or more of these conditions stimulates the chemoreceptors, their fibers transmit more impulses to the medulla's vasoconstrictor

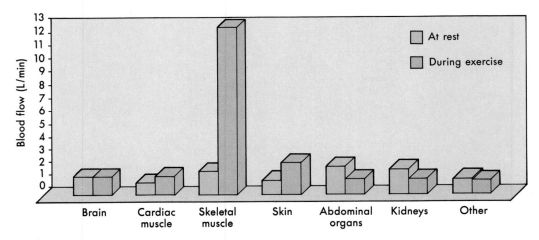

FIGURE 18-11 Changes in local blood flow during exercise. The left bar in each pair of bars shows the resting blood flow; the right bar shows the flow during exercise. During exercise, the vasomotor center of the medulla sends sympathetic signals to certain blood vessels to change diameter and thus shunt blood away from "maintenance" organs, such as the digestive organs in the abdomen, and toward the skeletal muscles. Notice that blood flow in the brain is held constant.

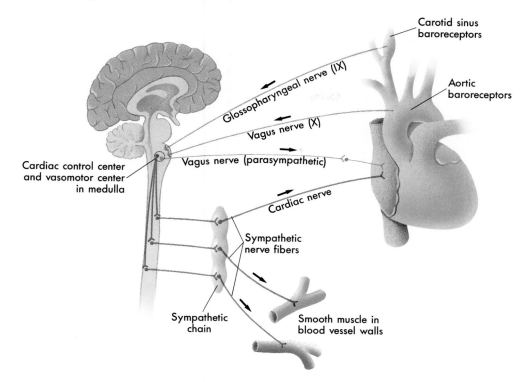

FIGURE 18-12   Vasomotor pressoreflexes. Carotid sinus and aortic baroreceptors detect changes in blood pressure and feed the information back to the cardiac control center and the vasomotor center in the medulla. In response, these control centers alter the ratio between sympathetic and parasympathetic output. If the pressure is too high, a dominance of parasympathetic impulses will reduce it by slowing heart rate, reducing stroke volume, and dilating blood "reservoir" vessels. If the pressure is too low, a dominance of sympathetic impulses will increase it by increasing heart rate and stroke volume and constricting reservoir vessels.

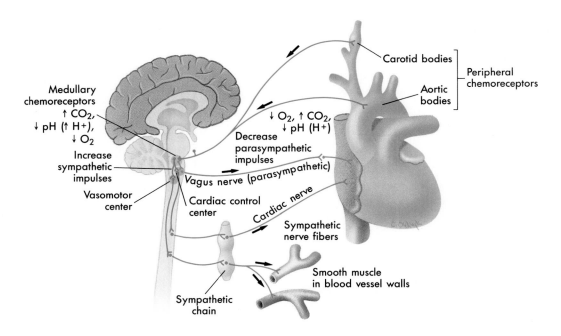

FIGURE 18-13   Vasomotor chemoreflexes. Chemoreceptors in the carotid and aortic bodies, as well as chemoreceptive neurons in the vasomotor center of the medulla itself, detect increases in carbon dioxide (CO_2), decreases in blood oxygen (O_2), and decreases in pH (which is really an increase in H^+). This information feeds back to the cardiac control center and vasomotor control center of the medulla, which in turn alter the ratio of parasympathetic and sympathetic output. When O_2 drops, CO_2 increases, and/or pH drops, a dominance of sympathetic impulses increases heart rate and stroke volume and constricts reservoir vessels, in response.

centers, and vasoconstriction of arterioles and venous reservoirs soon follows (Figure 18-13). This mechanism functions as an emergency device when hypoxia or hypercapnia endangers the stability of the internal environment.

Medullary ischemic reflex. The medullary ischemic reflex mechanism is said to exert powerful control of blood vessels during emergency situations when blood flow to the brain drops below normal. When the blood supply to the medulla becomes inadequate **(ischemic),** its neurons suffer from both oxygen deficiency and carbon dioxide excess. But, presumably, it is hypercapnia that intensely and directly stimulates the vasoconstrictor centers to bring about marked arteriole and venous constriction (Figure 18-13). If the oxygen supply to the medulla decreases below a certain level, its neurons, of course, cannot function, and the medullary ischemic reflex cannot operate.

Vasomotor control by higher brain centers. Impulses from centers in the cerebral cortex and in the hypothalamus are believed to be transmitted to the vasomotor centers in the medulla and to thereby help control vasoconstriction and dilation. Evidence supporting this view is that vasoconstriction and a rise in arterial blood pressure characteristically accompany emotions of intense fear or anger. Also, laboratory experiments on animals in which stimulation of the posterior or lateral parts of the hypothalamus leads to vasoconstriction support the belief that higher brain centers influence the vasomotor centers in the medulla.

Local control of arterioles. Several kinds of local mechanisms operate to produce vasodilation in localized areas. Although not all these mechanisms are clearly understood, they are known to function in times of increased tissue activity. For example, they probably account for the increased blood flow into skeletal muscles during exercise. They also operate in ischemic tissues, serving as a homeostatic mechanism that tends to restore normal blood flow. Norepinephrine, histamine, lactic acid,

a peptide called *endothelin,* and many other locally produced substances have been suggested as stimuli that activate the local vasodilator mechanism. Local vasodilation is also referred to as *reactive hyperemia.*

QUICK CHECK

✔ 1. *Peripheral resistance is affected by two major factors: blood viscosity and what else?*

✔ 2. *If the diameter of the arteries decreases, what effect does it have on peripheral resistance?*

✔ 3. *In general, how do vasomotor pressoreflexes affect the flow of blood?*

✔ 4. *What is a chemoreflex? How do chemoreflexes affect the flow of blood?*

Venous Return to Heart
Venous pumps

One important factor that promotes the return of venous blood to the heart is the blood-pumping action of respirations and skeletal muscle contractions. Both actions produce their facilitating effect on venous return by increasing the pressure gradient between the peripheral veins and the venae cavae (central veins).

The process of inspiration increases the pressure gradient between peripheral and central veins by decreasing central venous pressure and also by increasing peripheral venous pressure. Each time the diaphragm contracts, the thoracic cavity necessarily becomes larger and the abdominal cavity smaller. Therefore the pressures in the thoracic cavity, in the thoracic portion of the vena cava, and in the atria decrease, and those in the abdominal cavity and the abdominal veins increase. As Figure 18-14, A, shows, this change in pressure between expiration and inspiration acts as a "respiratory pump" that moves blood along the venous route. Deeper respirations intensify these effects and therefore tend to increase venous return to the heart more

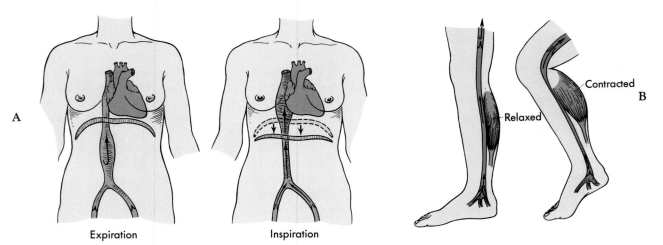

Expiration   Inspiration   Relaxed   Contracted   B   A

FIGURE 18-14   Venous pumping mechanisms. A, The respiratory pump operates by alternately decreasing thoracic pressure during inspiration (thus pulling venous blood into the central veins) and increasing pressure in the thorax during expiration (thus pushing central venous blood into the heart) . **B,** The skeletal muscle pump operates by the alternate increase and decrease in peripheral venous pressure that normally occur when the skeletal muscles are used for the activities of daily living. Both pumping mechanisms rely on the presence of semilunar valves in the veins to prevent backflow during the low-pressure points in the pumping cycle (see Figure 18-15).

than normal respirations. This is part of the reason why the principle is true that increased respirations and increased circulation tend to go hand in hand.

Skeletal muscle contractions serve as "booster pumps" for the heart. The skeletal muscle pump promotes venous return in the following way. As each skeletal muscle contracts, it squeezes the soft veins scattered through its interior, thereby "milking" the blood in them upward, or toward the heart (Figure 18-14, B). The closing of the semilunar valves present in veins prevents blood from falling back as the muscle relaxes. Their flaps catch the blood as gravity pulls backward on it (Figure 18-15). The net effect of skeletal muscle contraction plus venous valvular action, therefore, is to move venous blood toward the heart, to increase the venous return.

The value of skeletal muscle contractions in moving blood through veins is illustrated by a common experience. Who has not noticed how much more uncomfortable and tiring standing still is than walking? After several minutes of standing quietly, the feet and legs feel "full" and swollen. Blood has accumulated in the veins because the skeletal muscles are not contracting and squeezing it upward. The repeated contractions of the muscles when walking, on the other hand, keep the blood moving in the veins and prevent the discomfort of distended veins.

Changes in total blood volume

The return of venous blood to the heart can be influenced by factors that change the total volume of blood in the closed circulatory pathway. Stated simply, the more total volume of blood, the greater the volume of blood returned to the heart. What mechanisms can increase or decrease the total volume of blood? The mechanisms that change total blood volume most quickly, making them most useful in maintaining constancy of blood flow, are those that cause water to quickly move into the plasma (increasing total blood volume) or out of the plasma (decreasing total blood volume). Most of the mechanisms that accomplish such changes in plasma volume operate by altering the body's retention of water.

You have already studied the primary mechanisms for altering water retention in the body—they are the endocrine reflexes outlined in Chapter 15. One is the *ADH mechanism*. Recall that ADH (antidiuretic hormone) is released by the neurohypophysis (posterior pituitary) and acts on the kidneys in a way that reduces the amount of water lost by the body. ADH does this by increasing the amount of water that the kidneys reabsorb from urine before the urine is excreted from the body. The more ADH is secreted, the more water will be reabsorbed into the blood, and the greater the blood plasma volume will become. The ADH mechanism can be triggered by various factors such as input from baroreceptors and input from osmoreceptors (which detect the balance between water and solutes in the internal environment).

Another mechanism that changes blood plasma volume is the *renin-angiotensin mechanism* of aldosterone secretion. You may want to turn to Figure 15-20 to see that the enzyme *renin* is released when blood pressure in the kidney

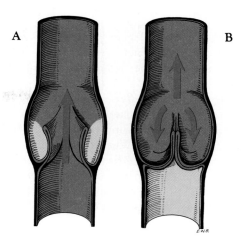

FIGURE 18-15 Semilunar valves. In veins, semilunar valves aid circulation by preventing backflow of venous blood when pressure in a local area is low. **A,** Local high blood pressure pushes the flaps of the valve to the side of the vessel, allowing easy flow. **B,** When pressure below the valve drops, blood begins to flow backward but fills the "pockets" formed by the valve flaps, pushing the flaps together and thus blocking further backward flow.

is low. Renin triggers a series of events that leads to the secretion of aldosterone, a hormone of the adrenal cortex. Aldosterone promotes sodium retention by the kidney, which in turn stimulates the osmotic flow of water from kidney tubules back into the blood plasma—but only when ADH is present to permit the movement of water. Thus low blood pressure increases secretion of aldosterone, which in turn stimulates retention of water and thus an increase in blood volume. Another effect of the renin-angiotensin mechanism is the vasoconstriction of blood vessels caused by an intermediate compound called *angiotensin II*. This complements the volume-increasing effects of the mechanism and thus also promotes an increase in overall blood flow.

Yet another mechanism that can change blood plasma volume and thus venous return of blood to the heart is the *ANH mechanism*. Recall that ANH (atrial natriuretic hormone) is secreted by specialized cells in the atrial wall in response to overstretching. Overstretching of the atrial wall, of course, occurs when venous return to the heart is abnormally high. ANH adjusts venous return back down to its set point value by promoting the loss of water from the plasma and the resulting decrease in blood volume. ANH accomplishes this feat by increasing urine sodium loss, which causes water to follow osmotically. Sodium loss also inhibits secretion of ADH. ANH may also have other complementary effects such as promoting vasodilation of blood reservoirs.

Thus various mechanisms influence blood volume and therefore venous return. These primary mechanisms are summarized in Figure 18-16. The ANH mechanism opposes ADH, renin-angiotensin, and aldosterone mechanisms to produce a balanced, precise control of blood volume. Precision of blood volume control contributes to precision in controlling venous return, which in turn contributes to precision in the overall control of blood circulation.

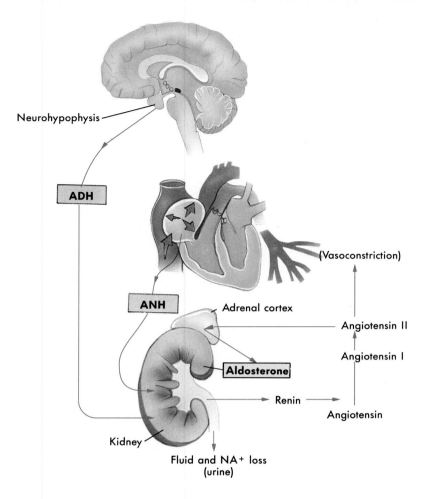

FIGURE 18-16 Three mechanisms that influence total plasma volume. The antidiuretic hormone (ADH) mechanism and renin-angiotensin and aldosterone mechanisms tend to increase water retention and thus increase total plasma volume. The atrial natriuretic hormone (ANH) mechanism antagonizes these mechanisms by promoting water loss and thus promoting a decrease in total plasma volume.

QUICK CHECK

✔ 1. *What is meant by the term* venous return?
✔ 2. *Briefly describe how the respiratory pump and skeletal muscle pump work.*
✔ 3. *What three hormonal mechanisms work together to regulate blood volume?*

Minute Volume of Blood

The volume of blood circulating through the body per minute (*minute volume*) is determined by the magnitude of both the blood pressure gradient and the peripheral resistance (Figure 18-17).

A nineteenth century physiologist and physicist, Poiseuille, (pwah-soo-EE) described the relation between these three factors—pressure gradient, resistance, and minute volume—with a mathematical equation known as *Poiseuille's law.* In general, but with certain modifications, it applies to blood circulation. We can state it in a simplified form as follows: the volume of blood circulated per minute is directly related to mean arterial pressure minus central venous pressure and inversely related to resistance:

Volume of blood circulated per minute =

$$\frac{\text{Mean arterial pressure} - \text{Central venous pressure}}{\text{Resistance}}$$

Or even more simply:

$$\text{Minute volume} = \frac{\text{Pressure gradient}}{\text{Resistance}}$$

This mathematical relationship needs qualifying with regard to the influence of peripheral resistance on circulation. For instance, according to the equation, an increase in peripheral resistance would tend to decrease blood flow. (Why? Increasing peripheral resistance increases the denominator of the fraction in the preceding equation. Increasing the denominator of any fraction necessarily does what to its value? It decreases the value of the fraction.)

Increased peripheral resistance, however, has a secondary action that opposes its primary tendency to decrease

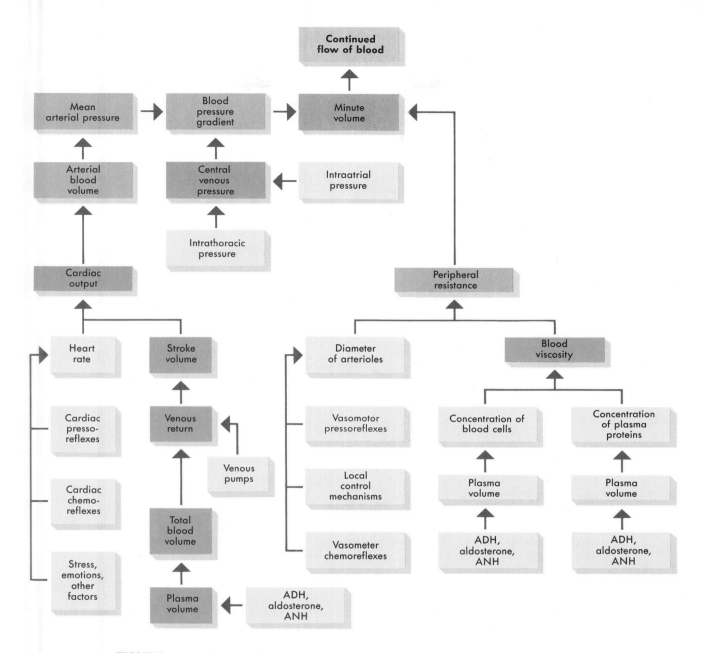

FIGURE 18-17 Factors that influence the flow of blood. The flow of blood, expressed as volume of blood flowing per minute (or *minute volume*), is determined by various factors. This chart shows only some of the major factors that influence blood flow. Notice that some factors appear more than once in the chart, indicating that they can influence blood flow in several ways.

blood flow. An increase in peripheral resistance hinders or decreases arteriole runoff. This, of course, tends to increase the volume of blood left in the arteries and so tends to increase arterial pressure. Note also that increasing arterial pressure tends to increase the value of the fraction in Poiseuille's equation. Therefore it tends to increase circulation. In short, to say unequivocally what the effect of an increased peripheral resistance will be on circulation is impossible. It depends also on arterial blood pressure—whether it increases, decreases, or stays the same when peripheral resistance increases. The clinical condition arteriosclerosis with hypertension (high blood pressure)

illustrates this point. Both peripheral resistance and arterial pressure are increased in this condition. If resistance were to increase more than arterial pressure, circulation (that is, volume of blood flow per minute) would decrease. But if arterial pressure increases proportionately to resistance, circulation remains normal.

QUICK CHECK

✔ 1. What is meant by the term *minute volume?*
✔ 2. How is *minute volume* related to *peripheral resistance?*

MEASURING BLOOD PRESSURE

Arterial Blood Pressure

Blood pressure is measured with the aid of an apparatus known as a *sphygmomanometer,* which makes it possible to measure the amount of air pressure equal to the blood pressure in an artery. The measurement is made in terms of how many millimeters (mm) high the air pressure raises a column of mercury (Hg) in a glass tube.

The sphygmomanometer usually consists of a rubber cuff attached by a rubber tube to a compressible bulb and by another tube to a column of mercury that is marked off in millimeters (Figure 18-18). The cuff is wrapped around the arm over the brachial artery, and air is pumped into the cuff by means of the bulb. In this way, air pressure is exerted against the outside of the artery. Air is added until the air pressure exceeds the blood pressure within the artery or, in other words, until it compresses the artery. At this time, no pulse can be heard through a stethoscope placed over the brachial artery at the bend of the elbow along the inner margin of the biceps muscle. By slowly releasing the air in the cuff the air pressure is decreased until it approximately equals the blood pressure within the artery. At this point the vessel opens slightly and a small spurt of blood comes through, producing sounds with a rather sharp, taplike quality. This is followed by increasingly louder sounds that suddenly change. They become more muffled, then disappear altogether. These sounds are called *Korotkoff sounds.* Nurses and physicians train themselves to hear these different sounds and simultaneously to read the column of mercury, since the first taplike sound appears when the

<table>
<tr><td>

DIAGNOSTIC STUDIES

DIASTOLIC PRESSURE

Clinically, **diastolic pressure** is considered more important than systolic pressure because it indicates the pressure, or strain, to which blood vessel walls are constantly subjected. It also reflects the condition of the peripheral vessels, since diastolic pressure rises or falls with the peripheral resistance. If, for instance, arteries are sclerosed (abnormally hardened), peripheral resistance and diastolic pressure both increase.

</td></tr>
</table>

column of mercury indicates the **systolic blood pressure.** Systolic pressure is the force with which the blood is pushing against the artery walls when the ventricles are contracting. The lowest point at which the sounds can be heard, just before they disappear, is approximately equal to the **diastolic pressure,** or the force of the blood when the ventricles are relaxed. Systolic pressure gives valuable information about the force of the left ventricular contraction, and diastolic pressure gives valuable information about the resistance of the blood vessels.

Blood in the arteries of the average adult exerts a pressure equal to that required to raise a column of mercury about 120 mm (or a column of water over 5 feet) high in a glass tube during systole of the ventricles and 80 mm high during their diastole. For the sake of brevity, this is expressed as a blood pressure of 120 over 80 (120/80). The first, or upper, figure represents systolic pressure and the second represents diastolic pressure. From the figures just given, we observe that blood pressure fluctuates considerably during each heartbeat. During ventricular systole, the force is great enough to raise the mercury column 40 mm higher than during ventricular diastole. This difference between systolic and diastolic pressure is called **pulse pressure.** It characteristically increases in arteriosclerosis, mainly because systolic pressure increases more than diastolic pressure. Pulse pressure increases even more markedly in aortic valve insufficiency because of both a rise in systolic and a fall in diastolic pressure.

Blood Pressure and Arterial vs. Venous Bleeding

Because blood exerts a comparatively high pressure in arteries and a very low pressure in veins, it gushes forth with considerable force from a cut artery but seeps in a slow, steady stream from a vein. As we have just seen, each ventricular contraction raises arterial blood pressure to the systolic level, and each ventricular relaxation lowers it to the diastolic level. As the ventricles contract, then, the blood spurts forth forcefully from the increased pressure in the artery, but as the ventricles relax, the flow ebbs to almost nothing because of the fall in pressure. In other words, blood escapes from an artery in spurts because of the alternate raising and lowering of arterial blood pressure but flows slowly and steadily from a vein because of the low, practically constant pressure. A uniform, instead of a pulsating, pressure exists in the capillaries and veins. Why? Because the arterial walls, being elastic, continue to

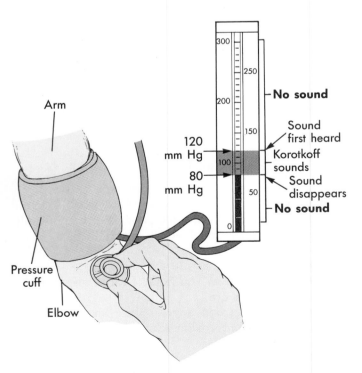

FIGURE 18-18   Sphygmomanometer. This mercury-filled pressure sensor is used in clinical and research settings to quickly and accurately measure arterial blood pressure.

squeeze the blood forward while the ventricles are in diastole. Therefore, blood enters capillaries and veins under a steady pressure (see Figure 18-8).

VELOCITY OF BLOOD

The speed with which the blood flows, that is, distance per minute, through its vessels is governed in part by the physical principle that when a liquid flows from an area of one cross-sectional size to an area of larger size, its velocity slows in the area with the larger cross section (Figure 18-19). For example, a narrow river whose bed widens flows more slowly through the wide section than through the narrow section. In terms of the blood vascular system, the total cross-sectional area of all arterioles together is greater than that of the arteries. Therefore blood flows more slowly through arterioles than through arteries. Likewise, the total cross-sectional area of all capillaries together is greater than that of all arterioles, and therefore capillary flow is slower than arteriole flow. The venule cross-sectional area, on the other hand, is smaller than the capillary cross-sectional area. Therefore the blood velocity increases in venules and again in veins, which have a still smaller cross-sectional area. In short, the most rapid blood flow takes place in arteries and the slowest in capillaries. Can you think of a valuable effect stemming from the fact that blood flows most slowly through the capillaries?

PULSE

Mechanism

Pulse is defined as the alternate expansion and recoil of an artery. Two factors are responsible for the existence of a pulse that can be felt:

1. Intermittent injections of blood from the heart into the aorta, which alternately increase and decrease the pressure in that vessel. If blood poured steadily out of the heart into the aorta, the pressure there would remain constant, and there would be no pulse.

2. The elasticity of the arterial walls, which allows them to expand with each injection of blood and then recoil. If the vessels were fashioned from rigid material such as glass, there would still be an alternate raising and lowering of pressure within them with each systole and diastole of the ventricles, but the walls could not expand and recoil, and therefore no pulse could be felt.

Pulse Wave

Each ventricular systole starts a new pulse that proceeds as a wave of expansion throughout the arteries and is known as the pulse wave. It gradually dissipates as it travels, disappearing entirely in the capillaries. The pulse felt in the radial artery at the wrist does not coincide with the contraction of the ventricles. It follows each contraction by an appreciable interval (the length of time required for the pulse wave to travel from an aorta to the radial artery). The farther from the heart the pulse is taken, therefore, the longer that interval is.

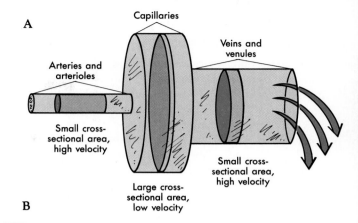

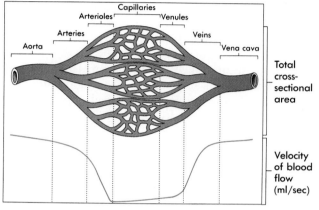

FIGURE 18-19 Relationship between cross-sectional area and velocity of blood flow. As you can see in the simple diagram **(A)** and the blood vessel chart **(B),** blood flows with great speed in the large arteries. However, branching of arterial vessels increases the total cross-sectional area of the arterioles and capillaries, reducing the flow rate. When capillaries merge into venules and venules merge into veins, the total cross-sectional area decreases, causing the flow rate to increase.

Almost everyone is aware of the diagnostic importance of the pulse. It reveals important information about the cardiovascular system, heart action, blood vessels, and circulation.

Where Pulse Can Be Felt

The pulse can be felt wherever an artery lies near the surface and over a bone or other firm background. Some of the specific locations where the **pulse point** is most easily felt are listed below and shown in Figure 18-20.

- ◆ **Radial artery**—at wrist
- ◆ **Temporal artery**—in front of ear or above and to outer side of eye
- ◆ **Common carotid artery**—along anterior edge of sternocleidomastoid muscle at level of lower margin of thyroid cartilage
- ◆ **Facial artery**—at lower margin of lower jawbone on a line with corners of mouth and in groove in mandible about one third of way forward from angle

◆ **Brachial artery**—at bend of elbow along inner margin of biceps muscle
◆ **Popliteal artery**—behind the knee
◆ **Posterior tibial artery**—behind the medial malleolus (inner "ankle bone")
◆ **Dorsalis pedis artery**—on the dorsum (upper surface) of the foot

There are six important pressure points that can be used to stop arterial bleeding:

1. **Temporal artery**—in front of ear
2. **Facial artery**—same place as pulse is taken
3. **Common carotid artery**—point where pulse is taken, with pressure back against spinal column
4. **Subclavian artery**—behind medial third of clavicle, pressing against first rib
5. **Brachial artery**—few inches above elbow on inside of arm, pressing against humerus
6. **Femoral artery**—in middle of groin, where artery passes over pelvic bone; pulse can also be felt here

In trying to stop arterial bleeding by pressure, one must always remember to apply the pressure at the pulse point, or *pressure point*, that lies between the bleeding part and the heart. Why? Because blood flows from the heart through the arteries to the part. Pressure between the heart and bleeding point therefore cuts off the source of the blood flow to that point.

Venous Pulse

A detectable pulse exists in the large veins only. It is most prominent in the veins near the heart because of changes in venous blood pressure brought about by alternate contraction and relaxation of the atria of the heart. The clinical significance of venous pulse is not as great as that of arterial pulse and thus it is less often measured.

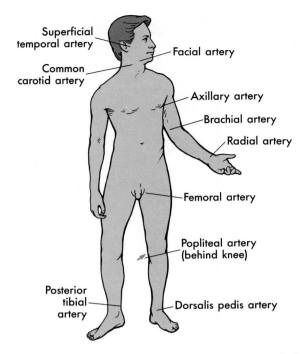

FIGURE 18-20 Pulse points. Each pulse point is named after the artery with which it is associated.

QUICK CHECK

✔ 1. *What device is used in clinical settings to measure arterial blood pressure?*
✔ 2. *Which is more important for assessing health, the systolic pressure or the diastolic pressure?*
✔ 3. *In which type of vessel is blood most likely to be flowing at a very slow rate; an artery, a capillary, or a vein?*
✔ 4. *Without being specific, where are pulse points normally located in the body?*

 Cycle of Life: *Cardiovascular physiology*

Changes in the function of the heart and blood vessels usually parallel the structural changes in these organs over the life span (see p. 492). For example, changes at the time of birth that adapt the circulatory system to life outside the womb cause changes in the blood pressure gradients that alter the flow of blood in many parts of the body. Likewise, the degenerative changes associated with aging reduce the heart's ability to maintain cardiac output and the ability of arteries to withstand high pressure.

Among the most apparent changes in the function of the cardiovascular system associated with the progression through the life cycle are changes in arterial blood pressure. In a newborn, normal arterial blood pressure is only about 90/55 mm Hg—much lower than the arterial pressure of

120/80 mm Hg in most healthy young adults. In older adults, arterial blood pressures commonly reach 150/90 mm Hg.

Another commonly observed change in cardiovascular function relates to heart rate. The heart rates of infants and children are typically more variable than in adults. Compared to adults, children often exhibit very large increases in heart rate in response to stressors such as illness, pain, tension, and exercise. Whereas a typical resting heart rate for adults is about 72 beats per minute, the resting heart rate of a newborn can range from 120 to 170 beats per minute, and the resting heart rate of a preschooler can range from 80 to 160 beats per minute. In older adults, resting heart rates range from lows of around 40 beats/min to 100 beats per minute.

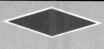

THE BIG PICTURE
Blood Flow And The Whole Body

As stated in this chapter and many times throughout this book, one of the essential concepts of homeostasis is the fact that our internal environment is a renewable fluid. If we could not maintain the chemical nature and other characteristics of our internal fluid environment, we would not survive. To maintain the constancy of the internal fluid, we must be able to shift nutrients, gases, hormones, waste products, agents of immunity, and other materials around in the body. As certain materials are depleted in one tissue and new materials enter the internal environment in another tissue, redistribution must occur. What better way than in a system of circulating fluid? This fluid not only shifts materials from place to place, it also redistributes heat and pressure. Recall from your study of the integumentary and muscular systems that shifting the flow of blood to or from warm tissues at the proper time is essential to maintaining the homeostasis of body temperature. As we will learn in a later chapter, the ability of our blood to increase or decrease blood pressure in the kidney has a great impact on that organ's vital function of filtering the internal environment. Understanding the basic mechanisms of almost any system in the body requires an understanding of the dynamics of blood flow.

What we have seen in this chapter is a wonderfully complex array of mechanisms that work together in concert with the actions of other systems to maintain the constancy of the *milieu intérieur*—the internal environment.

MECHANISMS OF DISEASE
Disorders of Cardiovascular Physiology

Hypertension

The largest number of office visits to physicians is due to a condition called **hypertension (HTN),** or high blood pressure. Over sixty million cases of HTN have been diagnosed in the United States. This condition occurs when the force of blood exerted by the arterial blood vessel exceeds a blood pressure of 140/90 mm Hg. Ninety percent of HTN cases are classified as *primary-essential,* or idiopathic, with no single known causative etiology. Another classification, *secondary HTN,* is caused by kidney disease or hormonal problems, or induced by oral contraceptives, pregnancy, or other causes.

Many risk factors have been identified in the development of HTN. Genetic factors play a large role. There is an increased susceptibility or predisposition with a family history of HTN. Males experience higher rates of HTN at an earlier age than women do, and HTN in blacks far exceeds that of whites in the United States. There is also a direct relationship between age and high blood pressure. This is because as age advances, the blood vessels become less compliant and there is a higher incidence of atherosclerotic plaque build-up. HTN can also be fatal if undetected in women taking oral contraceptives. Risk factors include high stress levels, obesity, calcium deficiencies, high levels of alcohol and caffeine intake, smoking, lack of exercise, and "type A" personalities.

There are many potential complications of untreated HTN. The risk of ischemic heart disease and heart failure, kidney failure, and stroke are some examples. As many as 400,000 people per year experience a stroke. Because HTN manifests minimal or no overt signs, it is known as the "silent killer." Headaches, dizziness, and fainting have been reported but are not always symptomatic of HTN. Regular screenings at the worksite and screening booths in malls and in hospitals often help to identify asymptomatic HTN.

Heart Failure

Heart failure is the inability of the heart to pump enough blood to sustain life. Heart failure can be the result of many different heart diseases. Valve disorders can reduce the pumping efficiency of the heart enough to cause heart failure. **Cardiomyopathy** (kar-dee-o-my-OP-ath-ee), or disease of the myocardial tissue, may reduce pumping effectiveness. A specific event such as myocardial infarction can result in myocardial damage that causes heart failure. Arrhythmias such as complete heart block or ventricular fibrillation can also impair the pumping effectiveness of the heart and thus cause heart failure.

Congestive heart failure (CHF) or, simply, *left-sided heart failure* is the inability of the left ventricle to pump blood effectively. Most often, such failure results from myocardial infarction caused by coronary artery disease. It is called *congestive heart failure* because it decreases pumping pressure in the systemic circulation, which in turn causes the body to retain fluids. Portions of the systemic circulation thus become congested with extra fluid. As previously stated, left-sided heart failure also causes congestion of blood in the pulmonary circulation (termed *pulmonary edema*)—possibly leading to right-sided heart failure.

Failure of the right side of the heart, or *right-sided heart failure,* accounts for about one fourth of all cases of heart failure. Right-sided heart failure often results from the progression of disease that begins in the left side of the heart. Failure of the left side of the heart results in reduced pumping of blood returning from the lungs. Blood backs up into the pulmonary circulation, then into the right side of the heart—causing an increase in pressure that the right side of the heart simply cannot overcome. Right-sided heart failure can also be caused by lung disorders that obstruct normal pulmonary blood flow and thus overload the right side of the heart—a condition called **cor pulmonale** (kor pul-mon-AHL-ee).

Circulatory Shock

The term **circulatory shock** refers to the failure of the circulatory system to adequately deliver oxygen to the tissues, resulting in the impairment of cell function throughout the

body. If left untreated, circulatory shock may lead to death. Circulatory failure has many causes, all of which somehow reduce the flow of blood through the blood vessels of the body. Because of the variety of causes, circulatory shock is often classified into the following types:

♦ **Cardiogenic** (kar-dee-o-JEN-ik) **shock** results from any type of heart failure, such as that after severe myocardial infarction (heart attack), heart infections, and other heart conditions. Because the heart can no longer pump blood effectively, blood flow to the tissues of the body decreases or stops.

♦ **Hypovolemic** (hye-po-vo-LEE-mik) **shock** results from the loss of blood volume in the blood vessels (*hypovolemia* means "low blood volume"). Reduced blood volume results in low blood pressure and reduced flow of blood to tissues. Hemorrhage is a common cause of blood volume loss leading to hypovolemic shock. Hypovolemia can also be caused by loss of interstitial fluid, causing a drain of blood plasma out of the vessels and into the tissue spaces. Loss of interstitial fluid is common in chronic diarrhea or vomiting, dehydration, intestinal blockage, severe or extensive burns, and other conditions.

♦ **Neurogenic** (noo-ro-JEN-ik) **shock** results from widespread dilation of blood vessels caused by an imbalance in autonomic stimulation of smooth muscle in vessel walls. You may recall from Chapter 13 that autonomic effectors

such as smooth muscle tissues are controlled by a balance of stimulation from the sympathetic and parasympathetic divisions of the autonomic nervous system. Normally, sympathetic stimulation maintains the muscle tone that keeps blood vessels at their usual diameter. If sympathetic stimulation is disrupted by an injury to the spinal cord or medulla, depressive drugs, emotional stress, or some other factor, blood vessels dilate significantly. Widespread vasodilation reduces blood pressure, thus reducing blood flow.

♦ **Anaphylactic** (an-a-fi-LAK-tik) **shock** results from an acute type of allergic reaction called *anaphylaxis*. Anaphylaxis causes the same kind of blood vessel dilation characteristic of neurogenic shock.

♦ **Septic shock** results from complications of *septicemia*, a condition in which infectious agents release toxins into the blood. The toxins often dilate blood vessels, causing shock. The situation is usually made worse by the damaging effects of the toxins on tissues combined with the increased cell activity caused by the accompanying fever. One type of septic shock is *toxic shock syndrome (TSS)*, which usually results from staphylococcal infections that begin in the vagina of menstruating women and spread to the blood.

The body has numerous mechanisms that compensate for the changes that occur during shock. However, these mechanisms may fail to compensate for changes that occur in severe cases, often resulting in death.

CASE STUDY
HYPOVOLEMIC SHOCK

Mr. C. is a healthy 20-year-old male who was injured in an industrial accident, causing him to suffer a crushed pelvis, ruptured spleen, and associated blood loss. He has an unremarkable history other than childhood diseases. He has not suffered any previous traumatic injuries and has no chronic illnesses. He has been in good health, exercises daily, is 6 feet tall, and weighs 165 pounds.

At midnight Mr. C. was working at an oil rig when a 10-ton forklift fell off its blocks onto him, pinning him at the pelvis. He was trapped for approximately 20 minutes while a crane was secured to remove the forklift. Paramedics at the scene began intravenous lactated Ringer's solution at 150 ml/hr. Vital signs were also obtained: heart rate 120; blood pressure 90/70; respirations 46. Mr. C.'s level of consciousness was reported as awake, and he complained of pelvic, back, and abdominal pain. He was pinned face down and has reduced movement of his lower extremities. His toes were mottled, pedal pulses were absent, radial pulses weak, brachial and carotid pulses palpable.

Mr. C. had tachycardia by electrocardiogram monitor and reported that his heart was "pounding in his chest." He became short of breath with conversation. He was restless and continued to complain of pain. He was pale but did not show any cyanosis. Peripheral pulses were absent with the exception of a thready brachial pulse; skin was cool and clammy; and numerous pinpoint

hemorrhages in the skin were present over his upper thorax, face, and neck.

On transport to the hospital the vital signs noted were as follows: heart rate 138, blood pressure 88/70, respirations 46, and confusion. He was diagnosed as being in hypovolemic shock (extracellular fluid volume deficit), and his intravenous fluids were increased to 300 ml/hr while blood samples were sent for type and cross-match and chemical and hematologic analysis.

Mr. C.'s condition improved after administration of 1000 ml lactated Ringer's solution and two units of red blood cells over 1 hour; heart rate 110, blood pressure 102/70, respirations 28. Chemistry results included the following: sodium 137 mEq/L, chloride 110 mEq/L, potassium 3.6 mEq/L, creatinine 1.0 mg/dl, glucose 276 mg/dl whole blood, and amylase 36 somogyi units/dl.

He was then taken to the operating room for surgical correction of a ruptured spleen; there he received six additional units of blood and was admitted to the intensive care unit with the following vital signs: heart rate 104, blood pressure 106/70, respirations 26. He was extubated and placed on 40% oxygen by face mask. He was in pelvic traction to stabilize his fractures. His continuing additional active medical problems included retroperitoneal hematoma not drained in the operating room, increased temperature, and decreased urine output.

1. Mr. C.'s extracellular fluid volume deficit occured as a result of which primary mechanism?
   a. Decreased intake of fluids and electrolytes
   b. Excessive loss of blood and fluids
   c. Shifts of fluids and electrolytes into nonaccessible areas

2. Which of Mr. C.'s signs are the result of compensatory mechanisms directed at maintaining cardiac output?
   a. Increased heart rate and oliguria
   b. Decreased blood pressure and sodium loss
   c. Respiratory acidosis and decreased heart rate
   d. All of the above

3. The mechanism *most* responsible for Mr. C.'s tachycardia is:
   a. Hypoxemia caused by atelectasis
   b. Anxiety as a result of traumatic injuries and pain
   c. Secretion of epinephrine and norepinephrine in response to decreased blood pressure
   d. Reaction to blood transfusion

4. Mr. C.'s high glucose level is the result of:
   a. Rapid infusion of lactated Ringer's solution
   b. Infusion of two units of red blood cells
   c. Decreased urinary output
   d. Compensaory mechanism in response to stress

CHAPTER SUMMARY

INTRODUCTION

A. Vital role of the cardiovascular system in maintaining homeostasis depends on the continuous and controlled movement of blood through the capillaries
B. Numerous control mechanisms help to regulate and integrate the diverse functions and component parts of the cardiovascular system to supply blood in response to specific body area needs

PHYSIOLOGY OF THE HEART

A. Conduction system (Figure 18-1)
   1. Four structures compose the conduction system of the heart
      a. Sinoatrial node (SA node)
      b. Atrioventricular node (AV node)
      c. AV bundle (bundle of His)
      d. Purkinje system
   2. Conduction system structures are more highly specialized than ordinary cardiac muscle tissue and permit only rapid conduction of an action potential through the heart
   3. SA node (pacemaker)
      a. Initiates each heartbeat and sets its pace
      b. Specialized pacemaker cells in the node possess an intrinsic rhythm
   4. Sequence of cardiac stimulation
      a. After being generated by the SA node, each impulse travels throughout the muscle fibers of both atria and the atria begin to contract
      b. As the action potential enters the AV node from the right atrium, its conduction slows to allow complete contraction of both atrial chambers before the impulse reaches the ventricles
      c. After the AV node, conduction velocity increases as the impulse is relayed through the AV bundle into the ventricles
      d. Right and left branches of the bundle fibers and Purkinje fibers conduct the impulses throughout the muscles of both ventricles, stimulating them to contract almost simultaneously

B. Electrocardiogram (ECG or EKG)
   1. Graphic record of the heart's electrical activity, its conduction of impulses; a record of the electrical events that precede the contractions of the heart
   2. To produce an ECG (Figure 18-2)
      a. Electrodes of an electrocardiograph are attached to the subject
      b. Changes in voltage are recorded that represent changes in the heart's electrical activity
   3. Normal ECG (Figure 18-3) is composed of
      a. P wave—represents depolarization of the atria
      b. QRS complex—represents repolarization of the atria and depolarization of the ventricles
      c. T wave—represents repolarization of the ventricles
      d. Measurement of the intervals between P, QRS, and T waves can provide information about the rate of conduction of an action potential through the heart
C. Control of heart rate—SA node normally initiates each heartbeat; however, various factors can and do change the rate of the heartbeat
   1. Cardiac pressoreflexes—aortic baroreceptors and carotid baroreceptors, located in the aorta and carotid sinus, are extremely important because they affect the autonomic cardiac control center, and therefore parasympathetic and sympathetic out flow, to aid in control of blood pressure (Figures 18-4 and 18-5)
      a. Carotid sinus reflex
         (1) Carotid sinus is located at the beginning of the internal carotid artery
         (2) Sensory fibers from carotid sinus baroreceptors run through the carotid sinus nerve and the glossopharyngeal nerve to the cardiac control center
         (3) Parasympathetic impulses leave the cardiac control center, travel through the vagus nerve to reach the SA node
      b. Aortic reflex—sensory fibers extend from baroreceptors located in the wall of the arch of the aorta through the aortic nerve and through the vagus nerve to terminate in the cardiac control center (Figure 18-6)

2. Other reflexes that influence heart rate—various important factors influence the heart rate; reflexive increases in heart rate often the result of increased sympathetic stimulation of the heart
   a. Anxiety, fear, and anger often increase heart rate
   b. Grief tends to decrease heart rate
   c. Emotions produce changes in heart rate through the influence of impulses from the cerebrum via the hypothalamus
   d. Exercise—heart rate normally increases
   e. Increased blood temperature or stimulation of skin heat receptors increases heart rate
   f. Decreased blood temperature or stimulation of skin cold receptors decreases heart rate

D. Cardiac cycle—a complete heartbeat consisting of contraction (systole) and relaxation (diastole) of both atria and both ventricles; the cycle is often divided into time intervals (Figures 18-6 and 18-7)
   1. Atrial systole
      a. Contraction of atria completes emptying blood out of the atria into the ventricles
      b. AV valves are open, SL valves are closed
      c. Ventricles are relaxed and filling with blood
      d. This cycle begins with the P wave of the ECG
   2. Isovolumetric ventricular contraction
      a. Occurs between the start of ventricular systole and the opening of the SL valves
      b. Ventricular volume remains constant as the pressure increases rapidly
      c. Onset of ventricular systole coincides with the R wave of the ECG and the appearance of the first heart sound
   3. Ejection
      a. SL valves open and blood is ejected from the heart when the pressure gradient in the ventricles exceeds the pressure in the pulmonary artery and aorta
      b. Rapid ejection—initial, short phase is characterized by a marked increase in ventricular and aortic pressure and in aortic blood flow
      c. Reduced ejection—characterized by a less abrupt decrease in ventricular volume, coincides with the T wave of the ECG
   4. Isovolumetric ventricular relaxation
      a. Ventricular diastole begins with this phase
      b. Occurs between closure of the SL valves and opening of the AV valves
      c. A dramatic fall in intraventricular pressure but no change in volume
      d. The second heart sound is heard during this period
   5. Passive ventricular filling
      a. Returning venous blood increases intraatrial pressure until the AV valves are forced open and blood rushes into the relaxing ventricles
      b. Influx lasts approximately 0.1 second and results in a dramatic increase in ventricular volume
      c. Diastasis—later, longer period of slow ventricular filling at the end of ventricular diastole lasting approximately 0.2 second; characterized by a gradual increase in ventricular pressure and volume
   6. Heart sounds
      a. Systolic sound—first sound, believed to be caused primarily by the contraction of the ventricles and by vibrations of the closing AV valves
      b. Diastolic sound—short, sharp sound; thought to be caused by vibrations of the closing of SL valves
      c. Heart sounds have clinical significance because they give information about the functioning of the valves of the heart

CONTROL OF CIRCULATION

A. Hemodynamics
   1. Hemodynamics—collection of mechanisms that influence the dynamic (active and changing) circulation of blood
   2. Circulation of different volumes of blood per minute is essential for healthy survival
   3. Circulation control mechanisms must accomplish two functions
      a. Maintain circulation
      b. Vary volume and distribution of the blood circulated

B. Primary principle of circulation (Figure 18-8)
   1. Blood flows because a pressure gradient exists between different parts of its bed; this is based on Newton's first and second laws of motion
   2. Blood circulates from the left ventricle to the right atrium of the heart because a blood pressure gradient exists between these two structures
   3. P_1 - P_2 is the symbol used to stand for a pressure gradient, with P_1 representing the higher pressure and P_2 the lower pressure

C. Control of arterial blood pressure—primary determinant of arterial blood pressure is the volume of blood in the arteries; a direct relationship exists between arterial blood volume and arterial pressure (Figure 18-9)
   1. Cardiac output (CO)—determined by stroke volume and heart rate
      a. Stroke volume
         (1) Stroke volume and heart rate
            (a) Stroke volume (SV)—volume pumped per heartbeat
            (b) CO (volume/min)—SV (volume/beat) HR (beats/min)
            (c) In practice, CO is computed by Fick's formula
            (d) Heart rate and stroke volume determine cardiac output, so anything that changes either also tends to change CO, arterial blood volume, and blood pressure in the same direction
         (2) Starling's law of the heart (Figure 18-10)
            (a) Within limits, the longer, or more stretched, the heart fibers at the beginning of contraction, the stronger the contraction
            (b) The amount of blood in the heart at the end of diastole determines the amount of stretch placed on the heart fibers
      b. Heart rate—pressoreflexes constitute the dominant heart rate control mechanism
   2. Peripheral resistance—resistance to blood flow imposed by the force of friction between blood and the walls of its vessels
      a. How resistance influences blood pressure
         (1) Arterial blood pressure tends to vary directly with peripheral resistance
         (2) Friction due to viscosity and small diameter of arterioles and capillaries
         (3) Muscular coat of arterioles allows them to constrict or dilate and change the amount of resistance to blood flow
         (4) Peripheral resistance helps determine arterial pressure by controlling the amount of blood that runs from the arteries to the arterioles; increased resistance, decreased arteriole run off leads to higher arterial pressure
      b. Vasomotor control mechanism—controls changes in the diameter of arterioles; plays role in maintenance of the general blood pressure and in distribution of blood to areas of special need (Figure 18-11)
         (1) Vasomotor pressoreflexes (Figure 18-12)

(a) Sudden increase in arterial blood pressure stimulates aortic and carotid baroreceptors; results in arterioles and venules of the blood reservoirs dilating

(b) Decrease in arterial blood pressure results in stimulation of vasoconstrictor centers causing vascular smooth muscle to constrict

  (2) Vasomotor chemoreflexes (Figure 18-13)—chemoreceptors located in aortic and carotid bodies are sensitive to hypercapnia, hypoxia, and decreased arterial blood pH

  (3) Medullary ischemic reflex—acts during emergency situation when there is decreased blood flow to the medulla; causes marked arteriole and venous constriction

  (4) Vasomotor control by higher brain centers— impulses from centers in cerebral cortex and hypothalamus are transmitted to vasomotor centers in medulla to help control vasoconstriction and dilation

  c. Local control of arterioles—several local mechanisms produce vasodilation in localized areas; referred to as reactive hyperemia

D. Venous return to heart

  1. Venous pumps—blood-pumping action of respirations and skeletal muscle contractions facilitate venous return by increasing pressure gradient between peripheral veins and venae cavae (Figure 18-14)

    a. Respirations—inspiration increases the pressure gradient between peripheral and central veins by decreasing central venous pressure and also by increasing peripheral venous pressure

    b. Skeletal muscle contractions—promote venous return by squeezing veins through a contracting muscle and milking the blood toward the heart

  2. Changes in total blood volume—mechanisms that change total blood volume most quickly are those that cause water to quickly move into or out of the plasma (Figure 18-16)

    a. ADH mechanism—decreases the amount of water lost by the body by increasing the amount of water that kidneys reabsorb from urine before the urine is excreted from the body; triggered by input from baroreceptors and osmoreceptors

    b. Renin-angiotensin mechanism

      (1) Renin—released when blood pressure in kidney is low; leads to increased secretion of aldosterone, which stimulates retention of sodium, causing increased retention of water and an increase in blood volume

      (2) Angiotensin II—intermediate compound that causes vasoconstriction, which complements the volume-increasing effects of renin and promotes an increase in overall blood flow

    c. ANH mechanism—adjusts venous return from an abnormally high level by promoting the loss of water from plasma, causing a decrease in blood volume; increases urine sodium loss which causes water to follow osmotically

E. Minute volume of blood (Figure 18-17)

  1. Minute volume—determined by the magnitude of the blood pressure gradient and peripheral resistance

  2. Poiseuille's law—Minute volume = Pressure gradient ÷ Resistance

MEASURING BLOOD PRESSURE

A. Arterial blood pressure

  1. Measured with the aid of a sphygmomanometer and stethoscope; listen for Korotkoff sounds as the pressure in the cuff is gradually decreased (Figure 18-18)

  2. Systolic blood pressure—force of the blood pushing against the artery walls while ventricles are contracting

  3. Diastolic blood pressure—force of the blood pushing against the artery walls when ventricles are relaxed

  4. Pulse pressure—difference between systolic and diastolic blood pressure

B. Relation to arterial and venous bleeding

  1. Arterial bleeding—blood escapes from artery in spurts due to alternating increase and decrease of arterial blood pressure

  2. Venous bleeding—blood flows slowly and steadily due to low, practically constant pressure

VELOCITY OF BLOOD

A. Velocity of blood is governed by the physical principle that when a liquid flows from an area of one cross-sectional size to an area of larger size, its velocity decreases in the area with the larger cross section (Figure 18-19)

B. Blood flows more slowly through arterioles than arteries because total cross-sectional area of arterioles is greater than that of arteries and capillary blood flow is slower than arteriole blood flow

C. Venule cross-sectional area is smaller than capillary cross-sectional area, causing blood velocity to increase in venules and then veins with a still smaller cross-sectional area

PULSE

A. Mechanism

  1. Pulse—alternate expansion and recoil of an artery; clinical significance is that it reveals important information regarding the cardiovascular system, blood vessels, and circulation

  2. Existence of pulse is due to two factors

    a. Alternating increase and decrease of pressure in the vessel

    b. Elasticity of arterial walls allows walls to expand with increased pressure and recoil with decreased pressure

B. Pulse wave

  1. Each pulse that starts with ventricular contraction and proceeds as a wave of expansion throughout the arteries

  2. Gradually dissipates as it travels, disappearing in the capillaries

C. Where pulse can be felt—wherever an artery lies near the surface and over a bone or other firm background (Figure 18-20)

D. Venous pulse—detectable pulse exists only in large veins; most prominent near the heart; not of clinical importance

CYCLE OF LIFE: CARDIOVASCULAR PHYSIOLOGY

A. Functional changes of heart and blood vessels parallels their structural changes

B. Arterial blood pressure—increases with age

C. Heart rates of children more variable than in adults

THE BIG PICTURE: BLOOD FLOW AND THE WHOLE BODY

A. Blood flow shifts materials from place to place and redistributes heat and pressure

B. Vital to maintaining homeostasis of internal environment

MECHANISMS OF DISEASE: DISORDERS OF CARDIOVASCULAR PHYSIOLOGY

A. Disorders of cardiovascular physiology

  1. Hypertension (HTN)—high blood pressure occurs when the force of blood exerted by the arterial blood vessel exceeds a pressure of 140/90 mm Hg

    a. Primary-essential class—no known etiology

    b. Secondary HTN class—caused by kidney disease, hormonal problems, oral contraceptives, pregnancy

    c. Risk factors include inherited factors, race, age,

maleness, stress, obesity, calcium deficiencies, alcohol, caffeine, smoking, lack of exercise, and type "A" personalities
  d. HTN manifests minimal symptoms, ("silent killer")
2. Heart failure—the inability of the heart to pump enough blood to sustain life
  a. Cardiomyopathy—disease of the myocardial tissue may reduce pumping
  b. Congestive heart failure—inability of the left ventricle to pump properly (also called left-sided heart failure)
  c. Right-sided heart failure—often results from disease that began on the left side or can be caused by cor pulmonale

B. Circulatory shock—failure of the circulatory system to deliver oxygen to tissue
  1. Cardiogenic shock results from any type of heart failure
  2. Hypovolemic shock results from low blood volume in the vessels
  3. Neurogenic shock results from widespread dilation of blood vessels from an imbalance in stimulation of muscles in vessel walls
  4. Anaphylactic shock results from an acute allergic reaction called anaphylaxis
  5. Septic shock results from septicemia; infectious agents release toxins into the blood

REVIEW QUESTIONS

1. Identify, locate, and describe the function of each of the following structures: SA node, AV node, AV bundle, and Purkinje fibers.
2. Compare the intrinsic rhythm of the SA node with other components of the heart's conduction system. What is an ectopic pacemaker?
3. What does an electrocardiogram measure and record? List the normal ECG deflection waves and intervals. What do the various ECG waves represent?
4. Discuss and compare the effects of sympathetic and parasympathetic stimulation on heart rate. What effect would vagal stimulation have on heart rate?
5. Explain the mechanism of action of the cardiac pressoreflexes on heart rate.
6. List and give the effect of several factors such as grief or pain on heart rate.
7. What is meant by the term *cardiac cycle?*
8. List the "periods" of the cardiac cycle and briefly describe the events that occur in each.
9. What is meant by the term *residual volume* as it applies to the heart?
10. Describe and explain the origin of the heart sounds.
11. State in your own words the basic principle of fluid flow.
12. What blood vessels present the greatest resistance to blood flow?
13. What is the primary determinant of arterial blood pressure?
14. List the two most important factors that indirectly determine arterial pressure by their influence on arterial volume.
15. How is cardiac output determined?

16. What is Starling's law of the heart?
17. What mechanisms control peripheral resistance? Cite an example of the operation of one or more parts of this mechanism to increase resistance; to decrease it.
18. What is arteriole runoff? What is the relationship of arteriole runoff to peripheral resistance?
19. What are the components of the vasomotor control mechanism?
20. Explain the mechanism of action of the medullary ischemic reflex.
21. How are respirations and skeletal muscle contractions related to venous return?
22. Explain how antidiuretic hormone can change the total blood volume.
23. What is the effect of low blood pressure in relation to aldosterone and antidiuretic hormone secretion?
24. State in your own words Poiseuille's law. Give an example of increased circulation to illustrate application of this law. Give an example of decreased circulation to illustrate application of this law.
25. Describe the measurement of arterial blood pressure.
26. Identify the eight locations where the pulse point is most easily felt. List the six pressure points at which pressure can be applied to stop arterial bleeding distal to that point.
27. Describe the various types of cardiac arrhythmias.
28. What two factors determine blood viscosity? What does viscosity mean? Give an example of a condition in which blood viscosity decreases. Explain its effect on circulation.

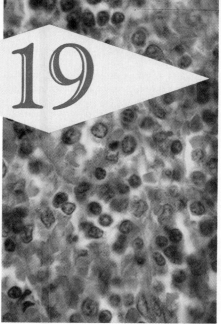

Spleen tissue
Erlandsen/Magney: Color Atlas of Histology

OBJECTIVES

After you have completed this chapter, you should be able to:

1. Describe the generalized functions of the lymphatic system and list the primary lymphatic structures.
2. Compare the chemical structure of lymph and interstitial fluid.
3. Discuss the formation, distribution, and general body plan of lymphatic drainage through the right lymphatic duct and the thoracic duct.
4. Compare the structure of lymphatic vessels and veins.
5. Discuss the specialized function of the lymphatic system in absorption of fats and other nutrients from the small intestine.
6. Discuss the "lymphatic pump" and other lymphokinetic actions that result in central movement, or flow, of lymph.
7. Describe and correlate the structure of lymph nodes with their function as biological filters.
8. Give the location of the major groups, or clusters, of lymph nodes in the body and identify their two primary functions.
9. Discuss the lymphatic drainage of the breast.
10. Locate the thymus in the body and discuss its gross and microscopic anatomy.
11. Discuss the functions of the thymus that result in its designation as a primary central organ of the lymphatic system.
12. Discuss the location, structure, and functions of the spleen.

KEY TERMS

Chyle (kile) Milky lymph found in lacteals after digestion

Lacteal (LAK-tee-al) Lymphatics in the villi of the small intestine

Lymph node Biological filter found in lymphatics

Lymph Clear, watery fluid found in lymphatic vessels; similar to blood plasma in composition; returns excess fluid and protein molecules to the blood

Lymphatic vessels Vessels that return lymph to the circulatory system

Lymphokinetics (lim-fo-ke-NE-tiks) Activities that result in the central flow of lymph

Spleen Lymphoid organ located in the upper left quadrant; its functions are defense, hematopoiesis, blood cell destruction and reservoir for blood

Thymus A primary central organ of the lymphatic system, plays vital role in immunity

Tonsils Masses of lymphoid tissue located in the back of the throat; protect against bacteria

OVERVIEW OF THE LYMPHATIC SYSTEM
Importance of the Lymphatic System

The lymphatic system serves various functions in the body. The two most important functions of this system are maintenance of fluid balance in the internal environment and immunity. Although both of these important functions will be discussed in this chapter, details of immunity will be discussed more fully in Chapter 20.

The importance of the lymphatic system in maintaining a balance of fluid in the internal environment is best explained by the diagram shown in Figure 19-1. As this figure shows, plasma filters into interstitial spaces from blood flowing through capillaries. Much of this interstitial fluid is absorbed by tissue cells or reabsorbed by the blood before it flows out of the tissue. However, a small percentage of the interstitial fluid remains behind. If this continued over even a brief period of time, the increased interstitial fluid would cause massive edema (swelling) of the tissue. This edema could cause tissue destruction or perhaps even death. This problem is avoided by the presence of lymphatic vessels that act as "drains" to collect

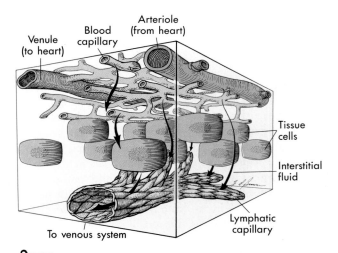

FIGURE 19-1 Role of the lymphatic system in fluid balance. Fluid from plasma flowing through the capillaries moves into interstitial spaces. Although much of this interstitial fluid is either absorbed by tissue cells or reabsorbed by capillaries, some of the fluid tends to accumulate in the interstitial spaces. As this fluid builds up, it tends to drain into lymphatic vessels that eventually return the fluid to the venous blood.

531

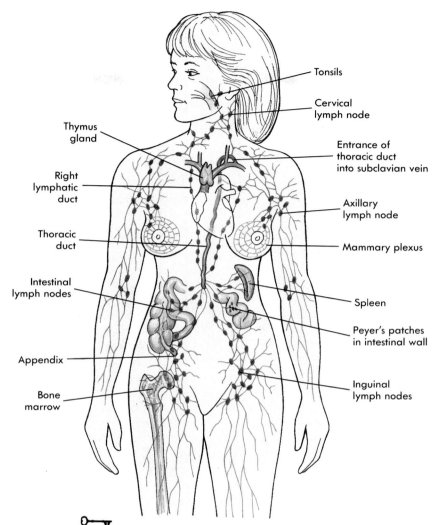

FIGURE 19-2   Principal organs of the lymphatic system.

the excess tissue fluid and return it to the venous blood just before it reaches the heart.

The lymphatic system is actually a specialized component of the circulatory system, since it consists of a moving fluid (lymph) derived from the blood and tissue fluid and a group of vessels (lymphatics) that returns the lymph to the blood. In general, the lymphatic vessels that drain the peripheral areas of the body parallel the venous return. In addition to lymph and the lymphatic vessels, the system includes lymph nodes located along the paths of the collecting vessels, isolated nodules of lymphatic tissue such as Peyer's patches in the intestinal wall, and specialized lymphatic organs such as the tonsils, thymus, and spleen (Figure 19-2).

Although it serves a unique transport function by returning tissue fluid, proteins, fats, and other substances to the general circulation, lymph flow differs from the true "circulation" of blood seen in the cardiovascular system. The lymphatic vessels do not, like vessels in the blood vascular system, form a closed ring, or circuit, but instead begin blindly in the intercellular spaces of the soft tissues of the body (Figure 19-1).

LYMPH AND INTERSTITIAL FLUID

Lymph is the clear, watery appearing fluid found in the lymphatic vessels. Interstitial fluid, which fills the spaces between the cells, is not the simple fluid it seems to be. Recent studies show that it is a complex and organized material. In some tissues, it is part of a semifluid ground substance. In others, it is the bound water in a gelatinous ground substance. Interstitial fluid and blood plasma together constitute the extracellular fluid, or in the words of Claude Bernard, the "internal environment of the body"—the fluid environment of cells in contrast to the atmosphere, or external environment, of the body.

Both lymph and interstitial fluid closely resemble blood plasma in composition. The main difference is that they contain a lower percentage of proteins than does plasma. Lymph is isotonic and almost identical in chemical composition to interstitial fluid when comparisons are made between the two fluids taken from the same area of the body. However, the average concentration of protein (4 g/100 ml) in lymph taken from the thoracic duct (Figure 19-2) is about twice that found in most interstitial fluid

LOSS OF LYMPHATIC FLUID

Lymph does not clot. Therefore if damage to the main lymphatic trunks in the thorax should occur as a result of penetrating injury, the flow of lymph must be stopped surgically or death ensues. It is impossible to maintain adequate serum protein levels by dietary means if significant loss of lymph continues over time. As lymph is lost, rapid emaciation occurs, with a progressive and eventually fatal decrease in total blood fat and protein levels.

samples. The elevated protein level of thoracic duct lymph (a mixture of lymph from all areas of the body) results from protein-rich lymph flowing into the duct from the liver and small intestine. A little over one half of the 2,500 to 2,800 ml total daily lymph flowing through the thoracic duct is derived from these two organs.

LYMPHATIC VESSELS

Distribution of Lymphatic Vessels

Lymphatic vessels originate as microscopic blind-end vessels called **lymphatic capillaries.** (Those originating in the villi of the small intestine are called **lacteals;** see Chapter 25.) The wall of the lymphatic capillary consists of a single layer of flattened endothelial cells. Each blindly ending capillary is attached, or fixed, to surrounding cells by tiny connective tissue filaments. Networks of lymphatic capillaries, which branch and anastomose extensively, are located in the intercellular spaces and are widely distributed throughout the body. As a rule, lymphatic and blood capillary networks lie side by side but are always independent of each other.

As twigs of a tree join to form branches, branches join to form larger branches, and large branches join to form the tree trunk, so do lymphatic capillaries merge, forming slightly larger lymphatics that join other lymphatics to form still larger vessels, which merge to form the main lymphatic trunks: the **right lymphatic ducts** and the **thoracic duct** (see Figure 19-2). Lymph from the entire body, except the upper right quadrant (Figure 19-3), drains eventually into the thoracic duct, which drains into the left subclavian vein at the point where it joins the left internal jugular vein. Lymph from the upper right quadrant of the body empties into the right lymphatic duct (or, more commonly, into three collecting ducts) and then into the right subclavian vein. Since most of the lymph of the body returns to the bloodstream by way of the thoracic duct, this vessel is considerably larger than the other main lymph channels, the right lymphatic ducts, but is much smaller than the large veins, which it resembles in structure. It has an average diameter of about 5 mm and a length of about 40 cm. It originates as a dilated structure, the **cisterna chyli,** in the lumbar region of the abdominal cavity and ascends a curving pathway to the root of the neck, where it joins the subclavian vein as just described (see Figure 19-2).

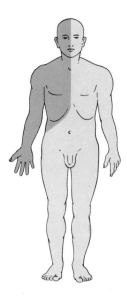

FIGURE 19-3 Lymph drainage. The right lymphatic ducts drain lymph from the parts of the body indicated by the green-colored area. Lymph from all the rest of the body (blue-colored area) enters the general circulation by way of the thoracic duct.

Structure of Lymphatic Vessels

Lymphatics resemble veins in structure with these exceptions:
- Lymphatics have thinner walls
- Lymphatics contain more valves
- Lymphatics contain lymph nodes located at certain intervals along their course

The lymphatic capillary wall is formed by a single layer of large but very thin and flat endothelial cells (Figure 19-4). Very small intercellular openings (clefts) exist between adjacent endothelial cells in the lymphatic capillary, allowing fluid and small particles to enter easily.

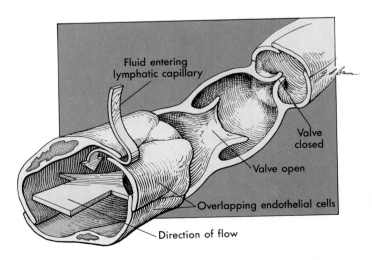

FIGURE 19-4 Structure of a typical lymphatic capillary. Notice that interstitial fluid enters through clefts between overlapping endothelial cells that form the wall of the vessel. Semilunar valves ensure one-way flow of lymph out of the tissue.

As lymph flows from the thin-walled capillaries into vessels with a larger diameter (0.2 to 0.3 mm), the walls become thicker and exhibit the three coats, or layers, typical of arteries and veins (see Table 17-1, p. 475). Interlacing elastic fibers and several strata of circular smooth muscle bundles are found in both the tunica media and the tunica adventitia of the large lymphatic vessel wall. Boundaries between layers, or coats, are less distinct in the thinner lymphatic vessel walls than in arteries or veins.

Semilunar valves are extremely numerous in lymphatics of all sizes and give the vessels a somewhat varicose and beaded appearance. Valves are present every few millimeters in large lymphatics and are even more numerous in the smaller vessels. Formed from folds of the tunica intima, each valve projects into the vessel lumen in a slightly expanded area circled by bundles of smooth muscle fibers.

Experimental evidence suggests that most lymph vessels have the capacity for repair or regeneration when damaged. Formation of new lymphatic vessels occurs by extension of solid cellular cores, or sprouts, formed by mitotic division of endothelial cells in existing vessels, which later become "canalized."

Functions of Lymphatic Vessels

The lymphatics play a critical role in numerous interrelated homeostatic mechanisms. The high degree of permeability of the lymphatic capillary wall permits large molecular weight substances and even particulate matter, which cannot be absorbed into a blood capillary, to be removed from the interstitial spaces. Proteins that accumulate in the tissue spaces can return to blood only via lymphatics. This fact has great clinical importance. For instance, if anything blocks lymphatic return, blood protein concentration and blood osmotic pressure soon fall below normal, and fluid imbalance and death will result (discussed in Chapter 28).

Lacteals (lymphatics in the villi of the small intestine) serve an important function in the absorption of fats and other nutrients. The milky lymph found in lacteals after digestion contains 1% to 2% fat and is called **chyle.** Interstitial fluid has a much lower lipid content than chyle (see Chapter 25).

CIRCULATION OF LYMPH

Water and solutes continually filter out of capillary blood into the interstitial fluid (see Figure 19-1). To balance this outflow, fluid continually reenters blood from the interstitial fluid. Newer evidence has disproved the old idea that healthy capillaries do not "leak" proteins. In truth, each day about 50% of the total blood proteins leak out of the capillaries into the tissue fluid and return to the blood by way of the lymphatic vessels. For more details about fluid exchange between blood and interstitial fluid, see Chapter 28. From lymphatic capillaries, lymph flows through progressively larger lymphatic vessels to eventually reenter blood at the junction of the internal jugular and subclavian veins (see Figure 19-2).

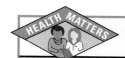

HEALTH MATTERS

LYMPHATIC DRAINAGE AND ARTIFICIAL LIMBS

An understanding of the anatomy of lymphatic drainage of the skin is critically important in surgical amputation of an extremity. A majority of the lymphatics draining the skin are located in plexuslike networks lying on the deep fascia. To prevent stasis of lymph in the stump following amputation, the surgeon retains the deep fascia and its lymphatic vessels with the skin flaps that are used to cover the cut end of the extremity. This procedure results in minimal edema and swelling—a matter of obvious importance to the patient and artificial-limb fitter.

The Lymphatic Pump

Although there is no muscular pumping organ connected with the lymphatic vessels to force lymph onward as the heart forces blood, still lymph moves slowly and steadily along in its vessels. Lymph flows through the thoracic duct and reenters the general circulation at the rate of about 125 ml per hour. This occurs despite the fact that most of the flow is against gravity, or "uphill." It moves through the system in the right direction because of the large number of valves that permit fluid flow only in the central direction. What mechanisms establish the pressure gradient required by the basic law of fluid flow? Two of the same mechanisms that contribute to the blood pressure gradient in veins also establish a lymph pressure gradient. These are breathing movements and skeletal muscle contractions (see Figure 18-14, p. 517).

Activities that result in central movement, or flow, of lymph are called **lymphokinetic** actions (from the Greek *kinetos,* "movable"). X-ray films taken after radiopaque material is injected into the lymphatics (*lymphography*) show that lymph pours into the central veins most rapidly at the peak of inspiration.

The mechanism of inspiration, resulting from the descent of the diaphragm, causes intraabdominal pressure to increase as intrathoracic pressure decreases. This simultaneously causes pressure to increase in the abdominal portion of the thoracic duct and to decrease in the thoracic portion. In other words, the process of inspiring establishes a pressure gradient in the thoracic duct that causes lymph to flow upward through it.

Research studies have shown that thoracic duct lymph is literally "pumped" into the venous system during the inspiration phase of pulmonary ventilation. The rate of flow, or ejection, of lymph into the venous circulation is proportional to the depth of inspiration. The total volume of lymph that enters the central veins during a given time period depends on both the depth of the inspiration phase and the overall breathing rate.

Contracting skeletal muscles also exerts pressure on the lymphatics to push the lymph forward. During exercise, lymph flow may increase as much as 10- to 15-fold. In addition, segmental contraction of the walls of the lym-

phatics themselves results in lymph being pumped from one valved segment to the next.

Other pressure-generating factors that can compress the lymphatics also contribute to the effectiveness of the "lymphatic pump." Examples of such lymphokinetic factors include arterial pulsations, postural changes, and passive compression (massage) of the body soft tissues.

The fact that changes in the rate and depth of pulmonary ventilation, coupled with other lymphokinetic activities, can effectively regulate levels of oxygen and lymph (a complex nutritive) in circulating blood may have great clinical significance in athletes and patients with cardiovascular or pulmonary diseases. In addition to metabolic interaction and regulation, this mechanism of "biocarburetion," or controlled mixing of blood oxygen and lymph, may also have immunological significance.

QUICK CHECK

✔ 1. *What is the overall function of the lymphatic system?*
✔ 2. *What is the origin of lymph?*
✔ 3. *Compare lymphatic vessels to blood vessels.*

LYMPH NODES

Structure of Lymph Nodes

Lymph nodes, or glands, as some people call them, are oval-shaped or bean-shaped structures. Some are as small as a pinhead and others are as large as a lima bean. Each

lymph node (from 1 mm to over 20 mm in diameter) is enclosed by a fibrous capsule. Note in Figure 19-5 that lymph moves into a node via several afferent lymphatic vessels and emerges by one efferent vessel. Think of a lymph node as a biological filter placed in the channel of several afferent lymph vessels. Once lymph enters the node, it "percolates" slowly through the spaces known as sinuses before draining into the single efferent exit vessel. One-way valves in both the afferent and efferent vessels keep lymph flowing in one direction.

Fibrous septa, or *trabeculae*, extend from the covering capsule toward the center of the node. **Cortical nodules** found within sinuses along the periphery, or outer region, of the node are separated from each other by these connective tissue trabeculae. Each cortical nodule is composed of packed lymphocytes that surround a less dense area called a *germinal center* (Figures 19-5 and 19-6).

Figure 19-6 is a low-power (×35.) light micrograph of a portion of a typical lymph node. When an infection is present, germinal centers form and the node begins to release lymphocytes. Lymphocytes begin their final stages of maturation within the less dense germinal center of the nodule and then are pushed to the more densely packed outer layers as they mature to become antibody-producing plasma cells. The center, or medulla, of a lymph node is composed of sinuses and cords (see Figure 19-5). Both the cortical and medullary sinuses are lined with specialized reticuloendothelial cells (fixed macrophages) capable of phagocytosis.

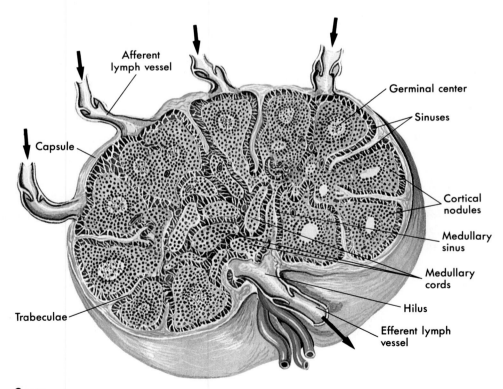

FIGURE 19-5 Structure of a lymph node. Several afferent valved lymphatics bring lymph to the node. A single efferent lymphatic leaves the node at the hilus. Note that the artery and vein also enter and leave at the hilus. Arrows show the direction of lymph flow.

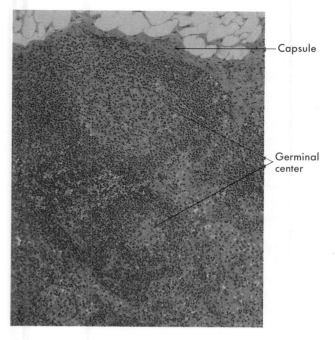

FIGURE 19-6 **Photomicrograph of lymph node cortex.** Note the presence of numerous germinal centers. (×35.)

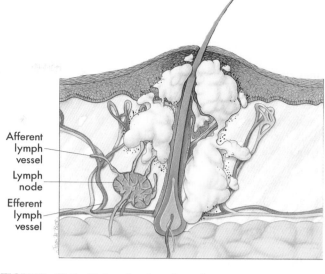

FIGURE 19-7 **Role of a lymph node in a skin infection.** Yellow areas represent dead and dying cells (pus). Black dots around the yellow areas represent bacteria. Leukocytes phagocytose many bacteria in tissue spaces. Others may enter the lymph nodes by way of afferent lymphatics. The nodes filter out those bacteria; here reticuloendothelial cells usually destroy them all by phagocytosis. (Lymph node is shown smaller than actual size.)

Locations of Lymph Nodes

With the exception of comparatively few single nodes, most of the lymph nodes occur in groups, or clusters, in certain areas. The group locations of greatest clinical importance are as follows:

♦ **Submental and submaxillary groups** in the floor of the mouth—lymph from the nose, lips, and teeth drains through these nodes.

♦ **Superficial cervical lymph nodes** in the neck along the sternocleidomastoid muscle—these nodes drain lymph (which has already passed through other nodes) from the head and neck (see Figure 19-2).

♦ **Superficial cubital,** or **supratrochlear, lymph nodes** located just above the bend of the elbow—lymph from the forearm passes through these nodes (see Figure 19-2).

♦ **Axillary lymph nodes** (20 to 30 large nodes clustered deep within the underarm and upper chest regions)—lymph from the arm and upper part of the thoracic wall, including the breast, drains through these nodes.

♦ **Inguinal lymph nodes** in the groin—lymph from the leg and external genitals drains through these nodes.

Functions of Lymph Nodes

Lymph nodes perform two distinct functions: defense and hematopoiesis.

Defense functions: filtration and phagocytosis

The structure of the sinus channels within lymph nodes slows the lymph flow through them. This gives the reticuloendothelial cells that line the channels time to remove the microorganisms and other injurious particles—

soot, for example—from the lymph and phagocytose them (Figure 19-7). Sometimes, however, such hordes of microorganisms enter the nodes that the phagocytes cannot destroy enough of them to prevent their injuring the node. An infection of the node, *adenitis*, then results. Also, because cancer cells often break away from a malignant tumor and enter lymphatics, they travel to the lymph nodes, where they may set up new growths. This may leave too few channels for lymph to return to the blood. For example, if tumors block axillary node channels, fluid accumulates in the interstitial spaces of the arm, causing the arm to become markedly swollen.

Hematopoiesis

The lymphatic tissue of lymph nodes serves as the site of the final stages of maturation for some types of lymphocytes and monocytes that have migrated from the bone marrow.

LYMPHATIC DRAINAGE OF THE BREAST

Cancer of the breast is one of the most common forms of malignancy in women. Unfortunately, cancerous cells from a single "primary" tumor in the breast often spread to other areas of the body through the lymphatic system. An understanding of the lymphatic drainage of the breast is therefore of particular importance in the diagnosis and treatment of this type of malignancy. Refer to Figure 19-8 as you study the lymphatic drainage of the breast.

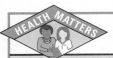

LYMPHEDEMA AFTER BREAST SURGERY

Surgical procedures called *mastectomies*, in which some or all of the breast tissues are removed, are sometimes done to treat breast cancer. Because cancer cells can spread so easily through the extensive network of lymphatic vessels associated with the breast (see Figure 19-8), the lymphatic vessels and their nodes are sometimes also removed. Occasionally, such procedures interfere with the normal flow of lymph fluid from the arm. When this happens, tissue fluid may accumulate in the arm—resulting in swelling, or **lymphedema.**

Distribution of Lymphatics in the Breast

The breast—mammary gland and surrounding tissues—is drained by two sets of lymphatic vessels:

1. Lymphatics that originate in and drain the skin over the breast with the exception of the areola and nipple
2. Lymphatics that originate in and drain the substance of the breast itself, as well as the skin of the areola and nipple

Superficial vessels that drain lymph from the skin and surface areas of the breast converge to form a diffuse *cutaneous lymphatic plexus.* Communication between the cutaneous plexus and large lymphatics that drain the secretory tissue and ducts of the breast occurs in the *subareolar plexus (plexus of Sappey)* located under the areola surrounding the nipple.

Lymph Nodes Associated with the Breast

Over 85% of the lymph from the breast enters the lymph nodes of the axillary region (Figure 19-8). Most of the

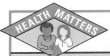

LYMPHATIC ANASTOMOSES AND BREAST CANCER

Anastomoses (connections) also occur between superficial lymphatics from both breasts across the middle line. Such communication can result in the spread of cancerous cells to previously healthy tissue in the other breast.

Both superficial and deep lymphatic vessels also communicate with lymphatics in the fascia of the pectoralis major muscle. Removal of a wide area of deep fascia is therefore required in surgical treatment of advanced or diffuse breast malignancy (radical mastectomy). In addition, cancer cells from a breast tumor sometimes reach the abdominal cavity because of lymphatic communication through the upper part of the linea alba.

remainder enters lymph nodes along the lateral edges of the sternum.

Several very large nodes in the axillary region are in actual physical contact with an extension of breast tissue called the *axillary tail of Spence.* Because of the physical contact between these nodes and breast tissue, cancerous cells may spread by both lymphatic extension and contiguity of tissue. Other nodes in the axilla will enlarge and swell after being "seeded" with malignant cells as lymph from a cancerous breast flows through them.

QUICK CHECK

✔ 1. *Describe the overall structure of a typical lymph node.*
✔ 2. *Where are lymph nodes usually found?*
✔ 3. *What functions are carried out by lymph nodes?*
✔ 4. *How do the lymphatic structures of the breast relate to breast cancer?*

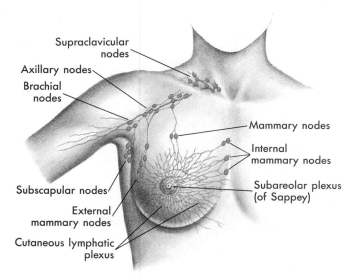

FIGURE 19-8 Lymphatic drainage of the breast. Note the extensive network of lymphatic vessels and nodes that receive lymph from the breast.

TONSILS

Masses of lymphoid tissue, called **tonsils,** are located in a protective ring under the mucous membranes in the mouth and back of the throat (Figure 19-9). They help protect against bacteria that may invade tissues in the area around the openings between the nasal and oral cavities. The **palatine tonsils** are located on each side of the throat. The **pharyngeal tonsils,** known as *adenoids* when they become swollen, are near the posterior opening of the nasal cavity. A third type of tonsil, the **lingual tonsils,** is near the base of the tongue. The tonsils serve as the first line of defense from the exterior and as such are subject to chronic infection, or **tonsillitis.** They are sometimes removed surgically if antibiotic therapy is not successful or if swelling impairs breathing. This procedure, called *tonsillectomy,* has become controversial because of the critical immunological role played by the lymphoid tissue.

THYMUS
Location and Appearance of the Thymus

Intensive study and experimentation have identified the thymus as a primary organ of the lymphatic system. It is an unpaired organ consisting of two pyramidal-shaped lobes with delicate and finely lobulated surfaces. The thymus is located in the mediastinum, extending up into the neck as far as the lower edge of the thyroid gland and inferiorly as far as the fourth costal cartilage (Figure 19-10, A). Its size relative to the rest of the body is largest in a child about 2 years old. Its absolute size is largest at puberty, when its weight ranges between 35 and 40 g. From then on, it gradually atrophies until, in advanced old age, it may be largely replaced by fat, weigh less than 10 g, and be barely

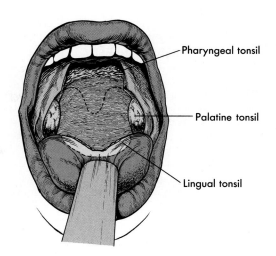

FIGURE 19-9   Location of the tonsils. Small segments of the roof and floor of the mouth have been removed to show the protective ring of tonsils (lymphoid tissue) around the internal openings of the nose and throat.

recognizable. The process of shrinkage of an organ in this manner is called **involution.** The thymus is pinkish gray in color early in childhood but, with advancing age, becomes yellowish as lymphatic tissue is replaced by fat.

Structure of the Thymus

The pyramid-shaped lobes of the thymus are subdivided into small (1 to 2 mm) lobules by connective tissue septa that extend inward from a fibrous covering capsule. Each lobule is composed of a dense cellular cortex and an inner, less dense medulla (Figure 19-10, B). Both cortex and medulla are composed of lymphocytes in an epithelial

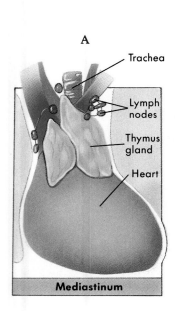
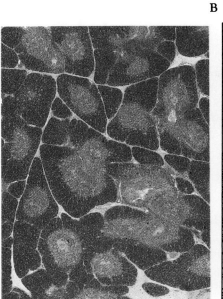
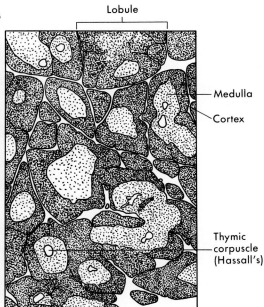

FIGURE 19-10   Thymus. A, Location of the thymus within the mediastinum. **B,** Microscopic structure of the thymus showing several lobules, each with a cortex and a medulla.

framework quite different from the supporting connective tissue seen in other lymphoid organs.

In stained histological sections of thymus, medullary tissue can be identified by the presence of rather large (30 to 150 μm) laminated spherical structures called **thymic corpuscles,** or *Hassall's corpuscles.* Composed of concentric layers of keratinized epithelial cells, thymic corpuscles have a unique whorl-like appearance.

Function of the Thymus

One of the body's best-kept secrets has been the function of the thymus. Before 1961 there were no significant clues as to its role. Then a young Briton, Dr. Jacques F.A.P. Miller, removed the thymus glands from newborn mice. His findings proved startling and crucial. Almost like a chain reaction, further investigations followed and led to at least a partial uncovering of the thymus' long-held secret. It is now clear that this small structure plays a critical part in the body's defenses against infections—in its vital immunity mechanism (see Chapter 20).

The thymus performs two important functions. First, it serves as the final site of lymphocyte development before birth. (The fetal bone marrow forms immature lymphocytes, which then move to the thymus.) Many lymphocytes leave the thymus and circulate to the spleen, lymph nodes, and other lymphatic tissues. Second, soon after birth the thymus begins secreting a group of hormones collectively called *thymosin,* that enables lymphocytes to develop into mature *T cells.* Since T cells attack foreign or abnormal cells and also serve as regulators of immune function, the thymus functions as part of the immune mechanism. It probably completes its essential work early in childhood.

SPLEEN

Location of the Spleen

The spleen is located in the left hypochondrium directly below the diaphragm, above the left kidney and descending colon, and behind the fundus of the stomach (see Figure 19-2).

Structure of the Spleen

As Figure 19-11 shows, the spleen is roughly ovoid in shape. Its size varies in different individuals and in the same individual at different times. For example, it hypertrophies during infectious diseases and atrophies in old age.

Like other lymphoid organs, the spleen is surrounded by a fibrous capsule with inward extensions that roughly divide the organ into compartments. One such compartment is shown in Figure 19-11, *B.* Arteries leading into each compartment are surrounded by dense masses (nodules) of developing lymphocytes. Because of its whitish appearance, this tissue is called *white pulp.* Near the outer regions of each compartment is tissue called *red pulp,* made up of a network of fine reticular fibers submerged in blood that comes from the nearby arterioles. After passing through the reticular meshwork, blood collects in venous sinuses and then returns to the heart through veins.

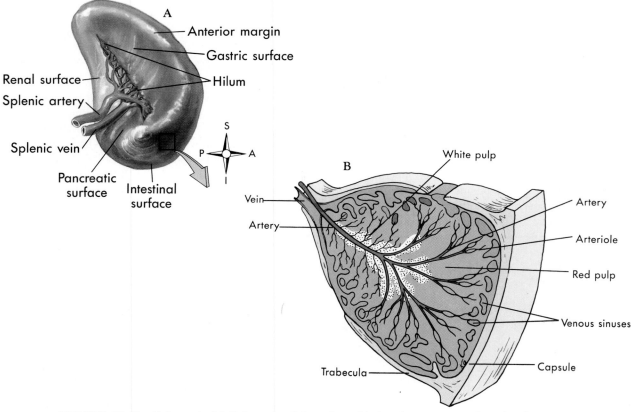

FIGURE 19-11   Spleen. A, Medial aspect of the spleen. Notice the concave surface that fits against the stomach within the abdominopelvic cavity. **B,** Section showing the internal organization of the spleen.

Functions of the Spleen

The spleen has many and sundry functions, including defense, hematopoiesis, and red blood cell and platelet destruction; it also serves as a reservoir for blood.

♦ **Defense.** As blood passes through the sinusoids of the spleen, reticuloendothelial cells (macrophages) lining these venous spaces remove microorganisms from the blood and destroy them by phagocytosis. Therefore the spleen plays a part in the body's defense against microorganisms.

♦ **Hematopoiesis.** Nongranular leukocytes, that is, monocytes and lymphocytes, complete their development to become activated in the spleen. Before birth, red blood cells are also formed in the spleen, but, after birth, the spleen is said to form red blood cells only in extreme hemolytic anemia.

♦ **Red blood cell and platelet destruction.** Macrophages lining the spleen's sinusoids remove worn-out red blood cells and imperfect platelets from the blood and destroy them by phagocytosis. They also break apart the hemoglobin molecules from the destroyed red blood cells and salvage their iron and globin content by returning them to the bloodstream for storage in bone marrow and liver.

♦ **Blood reservoir.** The pulp of the spleen and its venous sinuses store considerable blood. Its normal volume of about 350 ml is said to decrease about 200 ml in less than a minute's time following sympathetic stimulation that produces marked constriction of its smooth-muscle capsule. This "self-transfusion" occurs, for example, as a response to the stress imposed by hemorrhage.

SPLENOMEGALY

Splenomegaly, or abnormal spleen enlargement, is observed in various disorders. For example, infectious conditions such as scarlet fever, syphilis, and typhoid fever may be characterized by splenomegaly. Spleen enlargement sometimes accompanies hypertension. Splenomegaly also accompanies some forms of hemolytic anemia in which red blood cells appear to be broken apart at an abnormally fast rate. Surgical removal of the spleen often prevents death in such cases.

Although the spleen's functions make it a most useful organ, it is not a vital one. Dr. Charles Austin Doan in 1933 took the daring step of performing the first splenectomy. He removed the spleen from a 4-year-old girl who was dying of hemolytic anemia. Presumably, he justified his radical treatment on the basis of what was then merely conjecture, that is, that the spleen destroys red blood cells. The child recovered, and Dr. Doan's operation proved to be a landmark. It created a great upsurge of interest in the spleen and led to many investigations of it.

QUICK CHECK

✔ 1. Where are each of these lymphatic organs located in the body: tonsils, thymus, spleen?
✔ 2. How does the endocrine function of the thymus relate to its lymphatic function?
✔ 3. List four general functions of the spleen.

Cycle of Life: *Lymphatic system*

Many of the structural features of the lymphatic system exhibit dramatic changes as a person progresses through a life span. Most of the organs containing masses of developing lymphocytes appear before birth and continue growing through most of childhood until just before puberty. After puberty, these lymphoid organs typically begin to slowly atrophy until they reach much smaller size by late adulthood. These organs—including the thymus, lymph nodes, tonsils, and other lymphoid structures—shrink in size and become fatty or fibrous. The notable exception to this principle is the spleen, which develops early in life and remains intact until very late adulthood. Despite the fact that lymphocyte-producing lymphoid tissues decline after puberty, the overall function of the immune system is maintained until late adulthood. During the late adult years, deficiency of the immune system permits a greater risk of infections and cancer, and hypersensitivity of the immune system may make autoimmune conditions more likely to occur.

THE BIG PICTURE
The Lymphatic System and the Whole Body

One way to imagine the role of the lymphatic system in the "society of cells" that comprises the human body is as a sort of waste-water system. Like waste-water systems used in the cities of human society, the lymphatic system drains away excess, or "runoff," water from large areas. After collecting the body's runoff, or lymph, the lymphatic system conducts it through a network of lymphatic vessels, or "drain pipes," to miniature "treatment facilities" called lymph nodes. Contaminants are there removed from lymph, just as contaminants are removed in a waste-water treatment plant. The "clean" fluid is then returned to the bloodstream much as clean waste water is returned to a nearby river or lake. Like municipal waste-water systems, the lymphatic

system not only prevents dangerous fluid buildups, or "floods," but also prevents the spread of disease.

All systems of the body benefit from the fluid-balancing and immune functions of the lymphatic system. Some parts of the body such as the digestive and respiratory tracts make special use of the defensive capacities of lymphatic organs such as Peyer's patches and tonsils. Likewise, body structures such as the breasts and limbs make more use of the fluid-draining capacities of the lymphatic system than do other regions of the body. Overall, however, the entire body benefits from the fluid balance and freedom from disease conferred by the proper functioning of this important body system.

MECHANISMS OF DISEASE

Disorders of the Lymphatic System

Disorders Associated with Lymphatic Vessels

Lymphedema is an abnormal condition in which swelling of tissues in the extremities occurs because of an obstruction of the lymphatics and accumulation of lymph. The most common type of lymphedema is *congenital lymphedema* (lymphedema praecox), more often seen in women between the ages of 15 and 25 years of age. The obstruction in lymphedema can be in both the lymphatic vessels and lymph nodes themselves. Initially, the swelling, or edema, in the extremity will be soft, but as the condition progresses, it becomes firm, painful, and unresponsive to treatment. Frequent infections, involving high fever and chills, may occur with chronic lymphedema. Diuretics (agents that cause water loss) to reduce the swelling have been shown to be effective, along with strict bedrest and elevation of the involved extremities. If the edema is severe and unresponsive to these measures, or infection has occurred, or the person's mobility is severely compromised, surgical removal of the involved subcutaneous tissue and fascia may be required. Other procedures involving surgically "shunting" superficial lymphatic drainage into the deep lymphatic system have been tried.

Lymphedema may be caused by small parasitic worms called *filaria* that infest the lymph vessels. This condition is rare in the United States and is more often seen in the tropics. The flow of lymph is blocked, causing edema in the affected extremities that, in severe cases, become so swollen that they resemble an elephant's limbs. For this reason, the condition is referred to as *elephantiasis*—literally "condition of being like an elephant." Chronic swelling, thickening of the subcutaneous tissue, and frequent bouts of infections are common in this condition.

Lymphangitis, an acute inflammation of the lymphatic vessels, stems from invasion of an infectious organism. This condition is characterized by thin, red streaks extending from an infected region up the arm or leg. The lymph nodes also become enlarged, tender, and reddened. Necrosis, or tissue

death, along with development of an abscess (collection of fluid) can occur, leading to a condition known as *suppurative lymphadentitis.* The lymph nodes commonly involved are in the groin, axilla, and cervical regions. The infectious agents that cause lymphangitis may eventually spread into the bloodstream, causing *septicemia* (blood poisoning) and possible death from septic shock, but this is rare if the proper antibiotic therapy is initiated early enough.

Disorders Associated with Lymph Nodes and Other Lymphatic Organs

Lymphoma is a term that refers to a tumor of the cells of lymphoid tissue. Lymphomas are often malignant but, in rare cases, can be benign. They usually originate in isolated lymph nodes but can involve lymphoid tissue in the liver, spleen, and gastrointestinal tract. Widespread involvement is common because the disease spreads from node to node through the many anastomoses of the lymphatic vessels throughout the body. The exact cause of these neoplasms remains unknown.

Two principal categories of lymphomas are *Hodgkin's* and *non-Hodgkin's* lymphoma. Hodgkin's lymphoma is a malignancy with an uncertain etiology. Some pathophysiologists believe that it originates as a viral-induced tumor of T cells, although there is currently no evidence to support this. This condition usually begins as painless, nontender, enlarged lymph nodes in the neck or axilla. Soon, lymph nodes in other regions enlarge in the same manner. If they involve the trachea or esophagus, pressure results in difficulty breathing or swallowing.

Lymphedema caused by blockage of lymph nodes may cause enlargement of the extremities. Occasionally the disease may obstruct flow into or out of the liver, leading to liver enlargement and failure. Anemia, leukocytosis, fever, and weight loss occur as the condition progresses. Hodgkin's lymphoma is potentially curable with radiation therapy, provided it has not spread beyond the lymphatic system.

Chemotherapy is used in addition to radiation therapy in more advanced cases. Infection, from both the disease and the treatments, is a common complication.

Non-Hodgkin's lymphoma is the name given to a malignancy of lymphoid tissue other than Hodgkin's lymphoma. Again, the etiology is uncertain but has been hypothesized to be caused by a virus. Patients with immunodeficiencies such as AIDS often develop this condition. Manifestations are similar to Hodgkin's lymphoma, but there is usually a more generalized involvement of lymph nodes. The central nervous system is also often involved. Radiation and chemotherapy are treatments of choice.

The tonsils, composed of lymphatic tissue, which serve as the first line of defense from the exterior, are also subject to acute or chronic infection, known as **tonsillitis.** Fever, sore throat, and difficulty swallowing are common signs and symptoms. Enlarged pharyngeal tonsils (*adenoids*) may cause nasal obstruction. The infection may extend to the middle ear by way of the eustachian tubes, causing *acute otitis media* (middle ear infection) and possible deafness if left untreated. Antibiotics are usually initiated after diagnosis of tonsillitis. If these are unsuccessful, and swelling has endangered the airway and breathing, a *tonsillectomy*, or surgical removal of the tonsils, may be performed.

CHAPTER SUMMARY

OVERVIEW OF THE LYMPHATIC SYSTEM

A. Importance of the lymphatic system
1. Two most important functions—maintain fluid balance in the internal environment and immunity
2. Lymph vessels act as "drains" to collect excess tissue fluid and return it to the venous blood just before it returns to the heart
3. Lymphatic system—specialized component of the circulatory system; made up of lymph, lymphatic vessels, lymph nodes, isolated nodules of lymphatic tissue, tonsils, thymus, and spleen
4. Transports tissue fluid, proteins, fats, and other substances to the general circulation
5. Lymphatic vessels begin blindly in the intercellular spaces of the soft tissues; do not form a closed circuit

B. Lymph and interstitial fluid
1. Lymph
   a. Clear, watery appearing fluid found in the lymphatic vessels; closely resembles blood plasma in composition but has a lower percentage of protein; isotonic
   b. Elevated protein level in thoracic duct lymph due to protein-rich lymph from the liver and small intestine
2. Interstitial fluid
   a. Complex, organized fluid that fills the spaces between the cells; resembles blood plasma in composition with a lower percentage of protein
   b. Along with blood plasma, constitutes the extracellular fluid

LYMPHATIC VESSELS

A. Distribution of lymphatic vessels
1. Lymphatic capillaries—microscopic blind-end vessels where lymphatic vessels originate; wall consists of a single layer of flattened endothelial cells; networks branch and anastomose freely
2. Lymphatic capillaries merge to form larger lymphatics and eventually form the main lymphatic trunks, the right lymphatic ducts, and the thoracic duct
3. Lymph from upper right quadrant empties into right lymphatic duct and then into right subclavian vein
4. Lymph from rest of the body empties into the thoracic duct, which then drains into the left subclavian vein; thoracic duct originates as the cisterna chyli

B. Structure of lymphatic vessels
1. Similar to veins except lymphatic vessels have thinner walls, more valves, and contain lymph nodes
2. Lymphatic capillary wall is formed by a single layer of thin, flat endothelial cells
3. As the diameter of lymphatic vessels increases from capillary size, the walls become thicker and have three layers
4. Semilunar valves are present every few millimeters in large lymphatics and even more frequently in smaller lymphatics

C. Functions of the lymphatic vessels
1. Remove large molecular weight substances and even particulate matter from interstitial spaces
2. Lacteals absorb fats and other nutrients from the small intestine

CIRCULATION OF LYMPH

A. The lymphatic pump
1. Lymph moves through the system in the right direction due to the large number of valves
2. Breathing movements and skeletal muscle contractions establish a lymph pressure gradient, as they do with venous blood
3. Lymphokinetic actions—activities that result in a central flow of lymph

LYMPH NODES

A. Structure of lymph nodes
1. Lymph nodes are oval-shaped structures enclosed by a fibrous capsule
2. Nodes are similar to a biological filter
3. Once lymph enters a node, it moves slowly through sinuses to drain into the efferent exit vessel
4. Trabeculae extend from the covering capsule toward the center of the node
5. Cortical and medullary sinuses are lined with specialized reticuloendothelial cells capable of phagocytosis

B. Locations of lymph nodes
1. Most lymph nodes occur in groups
2. Location of groups with greatest clinical importance are submental and submaxillary groups, and superficial cervical, superficial cubital, axillary, and inguinal lymph nodes

C. Functions of lymph nodes—perform two distinct functions
 1. Defense functions: filtration and phagocytosis—reticuloendothelial cells remove microorganisms and other injurious particles from lymph and phagocytose them; if overwhelmed, lymph nodes can become infected or damaged
 2. Hematopoiesis—lymphatic tissue is the site for the final stages of maturation of some lymphocytes and monocytes

LYMPHATIC DRAINAGE OF THE BREAST
A. Distribution of lymphatics in the breast
 1. Drained by two sets of lymphatic vessels
  a. Lymphatics that drain the skin over the breast with the exception of the areola and nipple
  b. Lymphatics that drain the substance of the breast, as well as the skin of the areola and nipple
 2. Superficial vessels converge to form a diffuse, cutaneous lymphatic plexus
 3. Subareolar plexus—located under the areola surrounding the nipple; where communication between the cutaneous plexus and large lymphatics that drain the secretory tissue and ducts of the breast occurs
B. Lymph nodes associated with the breast
 1. Over 85% of the lymph from the breast enters the lymph nodes of the axillary region
 2. Remainder of lymph enters lymph nodes along the lateral edges of the sternum

TONSILS
A. Located under the mucous membranes in the mouth and back of the throat
 1. Palatine tonsils—located on each side of the throat
 2. Pharyngeal tonsils—located near the posterior opening of the nasal cavity
 3. Lingual tonsils—located near the base of the tongue
B. Protect against bacteria that may invade tissues around the openings between the nasal and oral cavities

THYMUS
A. Location and appearance of the thymus
 1. Primary central organ of lymphatic system
 2. Single, unpaired organ located in the mediastinum, extending upward to the lower edge of the thyroid and inferiorly as far as the fourth costal cartilage
 3. Thymus is pinkish gray in childhood; with advancing age, becomes yellowish as lymphatic tissue is replaced by fat
B. Structure of the thymus
 1. Pyramid-shaped lobes are subdivided into small lobules
 2. Each lobule is composed of a dense cellular cortex and an inner, less dense, medulla
 3. Medullary tissue can be identified by presence of thymic corpuscles
C. Function of the thymus
 1. Plays vital role in immunity mechanism
 2. Source of lymphocytes before birth
 3. Shortly after birth, thymus secretes thymosin, which enables lymphocytes to develop into T cells

SPLEEN
A. Location of the spleen—in the left hypochondrium, directly below the diaphragm, above the left kidney and descending colon, and behind the fundus of the stomach

B. Structure of the spleen
 1. Ovoid in shape
 2. Surrounded by fibrous capsule with inward extensions that divide the organ into compartments
 3. White pulp—dense masses of developing lymphocytes
 4. Red pulp—near outer regions, made up of a network of fine reticular fibers submerged in blood that comes from nearby arterioles
C. Functions of the spleen
 1. Defense—macrophages lining the sinusoids of the spleen remove microorganisms from the blood and phagocytose them
 2. Hematopoiesis—monocytes and lymphocytes complete their development in the spleen
 3. Red blood cell and platelet destruction—macrophages remove worn-out RBCs and imperfect platelets and destroy them by phagocytosis; also salvage iron and globin from destroyed RBCs
 4. Blood reservoir—pulp of spleen and its sinuses store blood

CYCLE OF LIFE: LYMPHATIC SYSTEM
A. Dramatic changes throughout life
B. Organs with lymphocytes appear before birth and grow until puberty
C. Postpuberty
 1. Organs atrophy through late adulthood
  a. Shrink in size
  b. Become fatty or fibrous
 2. Spleen—develops early, remains intact
D. Overall function maintained until late adulthood
 1. Later adulthood
  a. Deficiency permits risk of infection and cancer
  b. Hypersensitivity—likelihood of autoimmune conditions

THE BIG PICTURE: THE LYMPHATIC SYSTEM AND THE WHOLE BODY
A. Lymphatic system drains away excess water from large areas
B. Lymph is conducted through lymphatic vessels to nodes, where contaminants are removed
C. Lymphatic system benefits the whole body by maintaining fluid balance and freedom from disease

MECHANISMS OF DISEASE: DISORDERS OF THE LYMPHATIC SYSTEM
A. Disorders of the lymphatic system
 1. Disorders associated with lymphatic vessels
  a. Lymphedema—swelling of lymphatic tissues in the extremities occurring due to obstruction of lymphatics and accumulation of lymph
  b. Lymphaginitis—acute inflammation of lymphatic vessels due to an infectious organism
 2. Disorders associated with lymph nodes and other lymphatic organs
  a. Lymphoma—tumor of cells of lymphoid tissue; usually malignant
   (1) Hodgkin's lymphoma—malignancy with an uncertain etiology
   (2) Non-Hodgkin's lymphoma—malignancy of lymphoid tissues other than Hodgkins lymphoma
  b. Tonsillitis—acute or chronic infection of the tonsils

REVIEW QUESTIONS

1. List the anatomical components of the lymphatic system.
2. Why is the term *circulation* more appropriate in describing the movement of blood than of lymph?
3. How do interstitial fluid and lymph differ from blood plasma?
4. How do lymphatic vessels originate?
5. Briefly describe the anatomy of the lymphatic capillary wall.
6. Lymph from what body areas enters the general circulation by way of the thoracic duct? The right lymphatic ducts?
7. What is the cisterna chyli?
8. Where does lymph enter the blood vascular system?
9. In general, lymphatics resemble veins in structure. List three exceptions to this general rule.
10. What are the specialized lymphatics that originate in the villi of the small intestine called?
11. What is chyle? Where is it formed?
12. How is lymph formed?
13. Discuss the "lymphatic pump."
14. Discuss the importance of valves in the lymphatic system.
15. List several important groups, or clusters, of lymph nodes.
16. Discuss how lymph nodes function in body defense and hematopoiesis.
17. If cancer cells from breast cancer were to enter the lymphatics of the breast, where do you think they might lodge and start new growths? Explain, using your knowledge of the anatomy of the lymphatic and circulatory systems.
18. Locate the thymus and describe its appearance and size at birth, at maturity, and in old age.
19. Discuss the function of the thymus.
20. Describe the location and functions of the spleen.
21. What happens when there is a loss of lymphatic fluid?
22. Explain why lymphedema may occur after breast surgery.
23. Discuss the importance of lymphatic anastomoses in the spread of breast cancer.

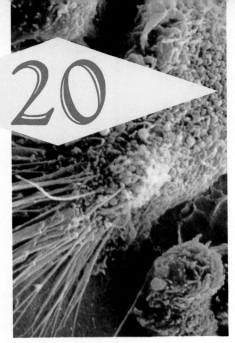

Macrophage ingesting cells.

20 Immune System

CHAPTER OUTLINE

OBJECTIVES

After you have completed this chapter, you should be able to:

1. Differentiate between the two major categories of immune mechanisms.
2. Describe species resistance as it relates to the human species.
3. Explain the inflammatory response and phagocytosis as it relates to inflammation.
4. Compare and contrast specific and nonspecific mechanisms of immunity.
5. Discuss the formation of lymphocytes and identify the two major classes of these specialized cells.
6. Compare the chemistry and functional activity of antigens and antibodies.
7. Discuss the development and activation of B cells.
8. Discuss the relationship between chemical structure and functional activity of antibodies or immunoglobulins.
9. List and compare the five classes of immunoglobulins.
10. Discuss the role of complement in the immune system.
11. Briefly discuss the two basic tenets of the clonal selection theory.
12. Describe the development, activation, and functions of T cells.
13. Explain the role of the thymus in cell-mediated immunity.
14. Discuss the origin and function of lymphokines and lymphotoxin.
15. Describe the different classifications of acquired immunity.

nemies of many kinds and in great numbers assault the body during a lifetime. Among the most threatening are hordes of microorganisms. We live our lives in a virtual sea of protozoans, bacteria, and viruses. So ever-present and potentially lethal are these small but formidable foes that no newborn could live through infancy, much less survive to adulthood or old age, without effective defenses against them. However, there are also enemies within. Inside the body, abnormal cells appear on an irregular but continual basis. If allowed to survive, these abnormal cells would reproduce to form a tumor. A tumor by itself can be life-threatening and there is always the possibility that it could become cancerous and spread (metastasize) to many other locations within the body. Without an internal "security force" to deal with such abnormal cells when they first appear, we would live very short lives. This chapter presents a brief overview of the system that provides defenses against both external and internal enemies—the **immune system.**

ORGANIZATION OF THE IMMUNE SYSTEM

Like any security force, the components and mechanisms of the immune system are organized in an efficient—almost military—manner. They are not just ready on a moment's notice, they are *continually* patrolling the body for foreign or internal enemies and shoring up the various lines of defense for a possible attack. Before we begin studying the specifics of immunity, we should spend a moment to map out the overall defensive strategy of the immune system.

In human society, any good security force employs numerous different strategies to guard their territory and take action if necessary. It is not surprising, then, to discover that the body's "society of cells" also employs a system that uses many different kinds of mechanisms to ensure the integrity and survival of the internal environment. All of these defense mechanisms can be categorized into one of two major categories of immune mechanisms: **nonspecific immunity** and **specific immunity.**

Nonspecific immunity includes mechanisms that resist a variety of threatening agents or conditions. The term

nonspecific implies that these immune mechanisms do not act on only one or two specific invaders, but rather provide a more general defense by simply acting against anything recognized as *not self*. Specific immunity, on the other hand, involves mechanisms that recognize *specific* threatening agents and respond by targeting their activity against these agents—and these agents only. Specific immune mechanisms often take some time to recognize their targets and react with sufficient force to overcome the threat, at least on their first exposure to a specific kind of threatening agent. Nonspecific mechanisms, however, have the advantage of being able to meet an enemy as soon as it presents itself. As we discuss examples of each type of immunity, you will come to understand how each type works and appreciate the distinction between them. You will also come to appreciate the value in having two complementary strategies for defending the body.

As in any body system, the work of the immune system is done by cells or substances made by cells. The primary types of cells involved in nonspecific immunity are these: neutrophils, monocytes, macrophages, and so-called natural killer (NK) cells. The primary types of cells involved in specific immunity are two types of lymphocytes called T cells and B cells. Awesome indeed is the army of cells and molecules that make up the immune system. Over one trillion lymphocytes, for example, and 100 million trillion (10^{20}) plasma proteins (antibodies) are a few of the many agents that help your body resist damage and disease.

QUICK CHECK

✔ 1. What is the difference between specific and nonspecific *immunity*?

✔ 2. What is the difference between self and not self?

NONSPECIFIC IMMUNITY

The general, nonspecific defensive mechanisms of the body are many and varied (Table 20-1). Only the major types of nonspecific immune mechanisms are listed here; many other examples appear in other chapters throughout this book. As a matter of fact, you will probably recognize examples here that you have encountered already in previous chapters.

Species Resistance

Species resistance refers to a phenomenon in which the genetic characteristics common to a particular kind of organism, or *species*, provide defense against certain *pathogens* (disease-causing agents). The human species (*Homo sapiens*), for example, is resistant to many life-threatening infections and infestations that often spread easily among plants and other animals. For example, humans do not have to worry about getting Dutch elm disease or becoming infected with canine viral distemper to which young dogs are susceptible. Usually, species resistance in humans results from the fact that our internal environment is not suitable for certain pathogens.

TABLE 20-1	Mechanisms of nonspecific defense
Mechanism	**Description**
SPECIES RESISTANCE	Genetic characteristics of the human species protect the body from certain pathogens
MECHANICAL AND CHEMICAL BARRIERS	Physical impediments to the entry of foreign cells or substances
Skin and mucosa	Forms a continuous wall that separates the internal environment from the external environment, preventing the entry of pathogens
Secretions	Secretions such as sebum, mucus, and enzymes chemically inhibit the activity of pathogens
INFLAMMATION	The inflammatory response isolates the pathogen and stimulates the speedy arrival of large numbers of immune cells
PHAGOCYTOSIS	Ingestion and destruction of pathogens by phagocytic cells
Neutrophils	Granular leukocytes that are usually the first phagocytic cell to arrive at the scene of an inflammatory response
Macrophages	Monocytes that have enlarged to become giant phagocytic cells capable of consuming many pathogens; often called by other, more specific, names when found in specific tissues of the body
NATURAL KILLER (NK) CELLS	Group of lymphocytes that kill many different types of cancer cells and virus-infected cells
INTERFERON	Chemical produced by cells after they become infected by a virus; inhibits the spread or further development of a viral infection
COMPLEMENT	Group of plasma proteins (inactive enzymes) that produce a cascade of chemical reactions that ultimately causes lysis (rupture) of a foreign cell; the complement cascade can be triggered by either specific or nonspecific immune mechanisms

Mechanical and Chemical Barriers

The internal environment of the human body is protected by a continuous mechanical barrier formed by the cutaneous membrane (skin) and mucous membranes (see Figure 4-27, p. 119). Often called the *first line of defense*, these membranes provide several layers of densely packed cells and other materials—forming a sort of "castle wall" that protects the internal environment from invasion by foreign cells (Figure 20-1).

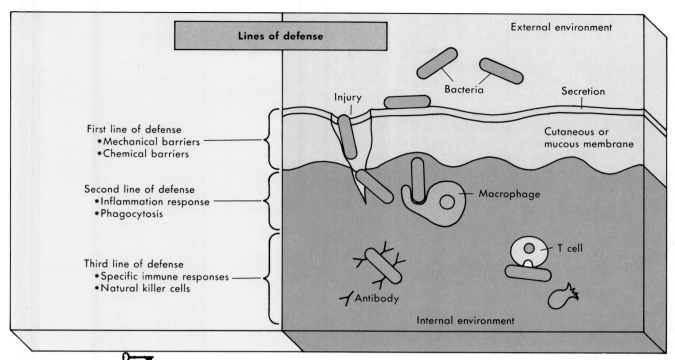

FIGURE 20-1   Lines of defense. Immune function, that is, defense of the internal environment against foreign cells, includes three layers of protection. The first line of defense is a set of barriers between the internal and external environment, the second line of defense involves the nonspecific inflammatory response (including phagocytosis), and the third line of defense includes the specific immune responses and the nonspecific defense offered by NK cells. Of course, tumor cells that arise in the body are already past the first two lines of defense and must be attacked by the third line of defense. This diagram is a simplification of the complex function of the immune system; in reality, there is a great deal of crossover of mechanisms between these "lines of defense."

Besides forming a protective wall, the skin and mucous membranes operate various additional immune mechanisms. For example, substances such as *sebum* (contains pathogen-inhibiting agents), *mucus* (pathogens may stick and be swept away), enzymes (may hydrolyze pathogens), and hydrochloric acid in gastric mucosa (may destroy pathogens) may also be present to act as nonspecific defense mechanisms. These chemical barriers act as a sort of "moat" around the castle wall formed by the membranes.

Inflammation
The inflammatory response

If bacteria or other invaders should break through the chemical and mechanical barriers formed by the membranes and their secretions, then the body has a *second line of defense:* the inflammatory response (see Figure 20-1). The **inflammatory response** has already been discussed in some detail (see Chapter 4, p. 116). Recall that in this nonspecific defensive mechanism, tissue damage elicits a host of responses that counteract the injury and promote a return to normal. An example of how local inflammation works is illustrated in Figure 20-2. In the example, bacteria

cause tissue damage that, in turn, triggers the release of mediators from cells such as the *mast cell* found in connective tissues. These inflammation mediators include *histamine, kinins, prostaglandins,* and related compounds. Many of these mediators are chemotactic factors, that is, substances that attract white blood cells to the area in a process called *chemotaxis.* Additionally, many of these factors produce the characteristic signs of inflammation: heat, redness, pain, and swelling. These signs result from increased blood flow and vascular permeability in the affected region, which help phagocytic white blood cells reach the general area and then enter the affected tissue. Besides local inflammation, systemic inflammation may occur when the inflammatory response occurs on a body-wide basis.

Phagocytosis

A major component of the body's second line of defense is the mechanism of **phagocytosis**—the ingestion and destruction of microorganisms or other small particles. There are many types of *phagocytes,* that is, cells capable of

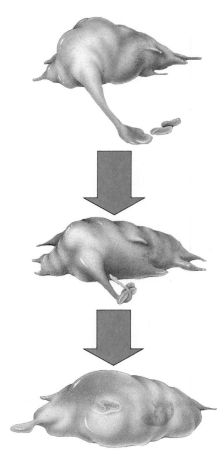

FIGURE 20-3 Phagocytosis of bacteria. This drawing is based on electron micrograph studies of phagocytosis by large phagocytes called macrophages. The plasma membrane extends (as a pseudopod) toward the bacterial cells, then envelopes them. Once trapped within a phagosome inside the macrophage, they can be destroyed by enzymes and other chemicals stored in lysosomes.

phagocytosis, in the body. As Figure 20-3 shows, when phagocytes approach a microbe, they extend footlike projections (pseudopods) toward it. Soon the pseudopods encircle the organism and form a complete sac, called a *phagosome* around it (Figure 20-3). The phagosome then moves into the interior of the cell where a lysosome fuses with it. The contents of the lysosome, chiefly digestive enzymes and hydrogen peroxide, drain into the phagosome and destroy the microbes in it. Because phagocytosis defends us against various kinds of agents, it is classified as a nonspecific defense.

The most numerous type of phagocyte is the *neutrophil,* a granular, neutral-staining type of white blood cell (WBC). Neutrophils attracted by chemotactic factors are usually the first type of phagocyte to arrive at the site of an injury during an inflammatory response. In a process called **diapedesis,** they migrate out of the bloodstream and around (or even through) the tissue cells to the injury site. They

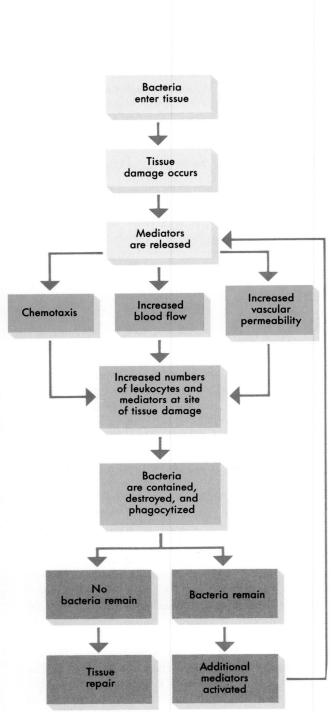

FIGURE 20-2 Inflammatory response. Tissue damage caused by bacteria triggers a series of events that produces the inflammatory response and promotes phagocytosis at the site of injury. These responses tend to inhibit or destroy the bacteria.

have a very short life span, and thus dead neutrophils tend to "pile up" at the site—forming most of the white substance called **pus** (see Figure 19-7). Another common type of phagocyte is the **macrophage** (meaning "large eater"). Macrophages are phagocytic monocytes (a non-granular WBC) that have grown to several times their original size after migrating out of the bloodstream. Macrophages are present in many areas of the body, even on the outside surface of some mucous membranes (for example, in the respiratory tract). Macrophages are often known by more specific names that designate their location. For instance, they are known as *histiocytes* in connective tissue, as *microglia* in the nervous system, and as *Kupffer cells* in the liver.

Natural Killer Cells

Besides phagocytes, the body has another important set of cells that provides nonspecific defense of the body. These are the **natural killer (NK) cells.** NK cells are a group of lymphocytes that kill many types of tumor cells and cells infected by different kinds of viruses. Because they have such a broad action, they are usually included among the nonspecific immune strategies. NK cells use several different methods for killing cells, most of which involve lysing (breaking apart) cells by damaging their plasma membranes.

Interferon

Several types of cells, if invaded by viruses, respond rapidly by synthesizing the protein **interferon** and releasing some of it into the circulation. Interferon comes in several varieties. As the name suggests, interferon proteins interfere with the ability of viruses to cause disease. They do this by preventing viruses from multiplying in cells. Leukocyte interferon, fibroblast interferon, and immune interferon are the three major types of interferon proteins. All three have now been produced by using gene-splicing techniques. Studies exploring antiviral and anticancer activities of interferons are currently underway.

 Interferon

Although interferon has not proven to be the cancer cure researchers may have once hoped, some useful therapies have emerged from early research into clinical uses of this protein. One type of leukocyte interferon, a protein called interferon alpha-2b, is now commercially available. Having both anticancer and antiviral properties, it is sometimes used with other therapies in the treatment of AIDS-related Kaposi's sarcoma (see p. 136), hairy-cell leukemia, and genital warts. Interferon alpha-2b is manufactured by splicing human genes that encode the amino acid sequence for this protein into the DNA of a common type of bacteria called *Escherichia coli.* These easily cultured bacteria, which normally would not make interferon, are thus converted into miniature "interferon factories."

Complement

Complement is the name given to each of a group of about 20 inactive enzymes in the plasma. Complement molecules are activated in a cascade of chemical reactions triggered by either specific or nonspecific mechanisms. Ultimately, the complement cascade causes lysis (rupture) of the foreign cell that triggered it. The function of complement is discussed further late in this chapter.

QUICK CHECK
✔ 1. Why are the skin and mucous membranes together called the body's first line of defense?
✔ 2. Name some of the events of the inflammatory response. How does each help protect the body?
✔ 3. What is the role of macrophages in the defense of the body?

OVERVIEW OF SPECIFIC IMMUNITY

Unlike the nonspecific mechanisms of immunity, the various types of specific immune mechanisms attack *specific* agents that the body recognizes as *not self*. Specific immunity, part of the body's *third line of defense*, is orchestrated by two different classes of a type of white blood cell called the lymphocyte. Originally, lymphocytes are formed in the red bone marrow of the fetus. They, like all blood cells, derive from primitive cells known as hematopoietic stem cells (see Chapter 16). The stem cells destined to become lymphocytes follow two developmental paths and differentiate into two major classes of lymphocytes—*B lymphocytes* and *T lymphocytes*, or simply, **B cells** and **T cells** (Figure 20-4). B cells do not attack pathogens themselves but, instead, produce molecules called *antibodies* that attack the pathogens or direct other cells, such as phagocytes, to attack them. B-cell mechanisms are therefore often classified as **antibody-mediated immunity.** Because T cells attack pathogens more directly, T-cell immune mechanisms are classified as **cell-mediated immunity.**

The densest populations of lymphocytes occur in the bone marrow, thymus gland, lymph nodes, and spleen. From these structures, lymphocytes pour into the blood and then distribute themselves throughout the tissues of the body. After wandering through the tissue spaces, they eventually find their way into lymphatic capillaries. Lymph flow transports the lymphocytes through a succession of lymph nodes and lymphatic vessels and empties them by way of the thoracic and right lymphatic ducts into the subclavian veins. Thus returned to the blood, the lymphocytes embark on still another long journey—through blood, tissue spaces, lymph, and back to blood. The survival value of the continued recirculation of lymphocytes and of their widespread distribution throughout body tissues seems apparent. It provides these major cells of the immune system ample opportunity to perform their functions of searching out, recognizing, and destroying foreign invaders.

Before continuing our discussion of specific immunity, we shall define the following terms to be used:

♦ **Antigens**—macromolecules (very large molecules) that induce the immune system to make certain responses. Most antigens are foreign proteins. Some, however, are polysaccharides, and some are nucleic acids. Many antigens that enter the body are macromolecules located in the walls or outer membranes of microorganisms or the outer coats of viruses.

Of course, antigens on the surfaces of some tumor cells are not really from outside the body, but are "foreign" in the sense that they are recognized as not belonging. The membrane molecules that identify all the normal cells of the body are called *self-antigens*. Foreign and tumor-cell antigens can be called *non-self-antigens*.

♦ **Antigenic determinants**—variously shaped, small regions on the surface of an antigen molecule; a less cumbersome name is *epitopes*. In a protein molecule, for instance, an epitope consists of a sequence of only about 10 amino acids that are part of a much longer, folded chain of amino acids. The sequence of the amino acids in an epitope determines its shape. Since the sequence differs in different kinds of antigens, each kind of antigen usually has specific and uniquely shaped epitopes.

♦ **Antibodies**—plasma proteins of the class called *immunoglobulins*. Unlike most antigens, all antibodies are native molecules, that is, they are normally present in the body.

♦ **Combining sites**—two small concave regions on the surface of an antibody molecule. Like epitopes, combining sites have specific and unique shapes. An antibody's combining sites are so shaped that an antigen's epitope that has a complementary shape can fit into the combining site and thereby bind the antigen to the antibody to form an **antigen-antibody complex.** Because combining sites receive and bind antigens, they are also called *antigen receptors* and *antigen-binding sites*.

♦ **Clone**—family of cells, all of which have descended from one cell.

♦ **Complement**—a group of proteins that, when activated, work together to destroy foreign cells.

QUICK CHECK

✔ 1. What is an antigen? What is the difference between a self-antigen and a non-self-antigen?

✔ 2. What is meant by the term clone?

B Cells and Antibody-Mediated Immunity

Development and Activation of B Cells

The development of the lymphocytes called B cells occurs in two stages (Figure 20-4). In chickens the first stage of B cell development occurs in the bursa of Fabricius—hence the name B cells. Since humans do not have a bursa of Fabricius, another organ must serve as the site for the first stage of B cell development. The bursa-equivalent tissue in humans is the yolk sac, then the red marrow or the fetal liver. At any rate, by the time a human infant is a few months old, its pre-B cells have completed the first stage of their development. They are then known as inactive B cells.

Inactive B cells synthesize antibody molecules but secrete few if any of them. Instead, they insert on the surface of their plasma membranes perhaps 100,000 antibody molecules. The combining sites of these surface antibody molecules stand ready to serve as receptors for a specific antigen if it comes by.

After being released from the bone marrow, inactive B cells circulate to the lymph nodes, spleen, and other lymphoid structures.

The second major stage of B cell development occurs when the inactive B cells become activated. Activation of a B cell must be initiated by an encounter between an

FIGURE 20-4 Development of B cells and T cells. Both types of lymphocytes originate from stem cells in the red bone marrow. Pre-B cells that are formed by dividing stem cells develop in the "bursa-equivalent" tissues in the yolk sac, fetal liver, and bone marrow. Likewise, pre-T cells migrate to the thymus, where they continue developing. Once they are formed, B cells and T cells circulate to the lymph nodes and spleen.

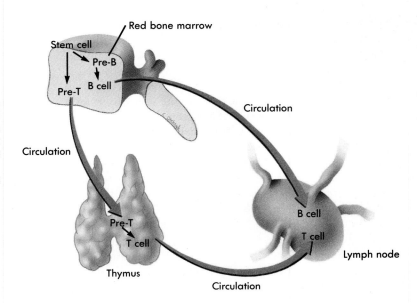

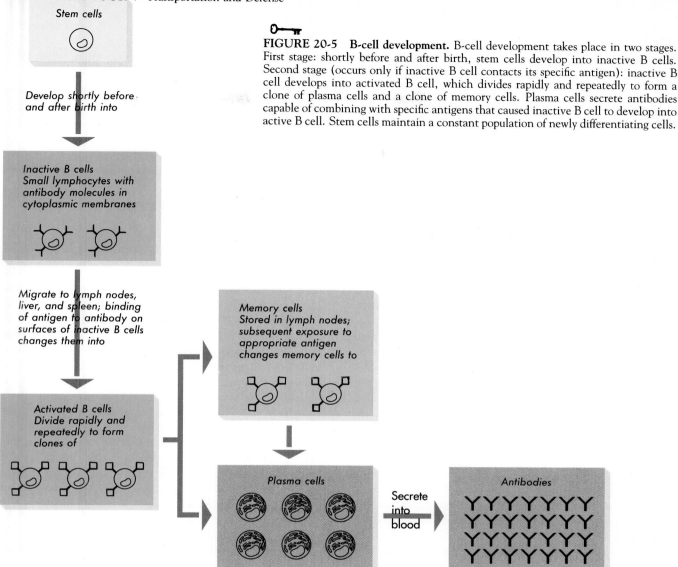

FIGURE 20-5 **B-cell development.** B-cell development takes place in two stages. First stage: shortly before and after birth, stem cells develop into inactive B cells. Second stage (occurs only if inactive B cell contacts its specific antigen): inactive B cell develops into activated B cell, which divides rapidly and repeatedly to form a clone of plasma cells and a clone of memory cells. Plasma cells secrete antibodies capable of combining with specific antigens that caused inactive B cell to develop into active B cell. Stem cells maintain a constant population of newly differentiating cells.

inactive B cell and its specific antigen, that is, one whose epitopes fit the combining sites of the B cell's surface antibodies (Figure 20-5).

The antigen binds to these antibodies on the B cell's surface. Antigen-antibody binding activates the B cell, triggering a rapid series of mitotic divisions. By dividing repeatedly, a single B cell produces a clone, or family of identical B cells. Some of them become differentiated to form plasma cells. Others do not differentiate completely but remain in the lymphatic tissue as the so-called memory B cells. Plasma cells synthesize and secrete copious amounts of antibody. A single plasma cell, according to one estimate, secretes 2,000 antibody molecules per second during the few days that it lives. All the cells in a clone of plasma cells secrete identical antibodies because they have all descended from the same B cell. Memory B cells do not themselves secrete antibodies, but if they are later exposed to the antigen that triggered their formation, memory B cells then become plasma cells, and the plasma cells secrete antibodies that can combine with the initiating antigen (see Figure 20-5). Thus the ultimate function of B cells is to serve as ancestors of antibody secreting plasma cells.

Antibodies (Immunoglobulins)
Structure of antibody molecules

Antibodies are proteins of the family called *immunoglobulins*. Like all proteins, they are very large molecules and are composed of long chains of amino acids (polypeptides). Each immunoglobulin molecule consists of four polypeptide chains—two heavy chains and two light chains. Each polypeptide chain is intricately folded to form globular regions that are joined together in such a way that the immunoglobulin molecule as a whole is Y-shaped. Look now at Figure 20-6, A. The twisted strands of red spheres (amino acids) in the diagram represent the light chains, and the two twisted strands of blue spheres represent the heavy chains. Each heavy chain consists of 446 amino acids. Heavy chains therefore are about twice as long and weigh about twice as much as light chains.

The regions with colored bars seen in Figure 20-6, *B*, represent *variable regions*, that is, regions in which the sequence of amino acids varies in different antibody molecules. Note the relative positions of the variable regions of the light and heavy chains; they lie directly

opposite each other. Because amino acid sequence determines conformation or shape, and because different sequences of amino acids occur in the variable regions of different antibodies, the shapes of the sites between the variable regions also differ. At the end of each "arm" of the Y-shaped antibody molecule, the unique shapes of the variable regions form a cleft that serves as the antibody's combining sites, or antigen-binding sites. Each one of us is thought normally to have millions of different kinds of antibody molecules in our bodies. And each one of these, almost unbelievably, has its own uniquely shaped combining sites. It is this structural feature that enables antibodies to recognize and combine with specific antigens, both of which are crucial first steps in the body's defense against invading microbes and other foreign cells.

In addition to its variable region, each light chain in an antibody molecule also has a constant region. The constant region consists of 106 amino acids whose sequence is identical in all antibody molecules. Each heavy chain of an antibody molecule consists of three constant regions in addition to its one variable region. Identify the constant and variable regions of the light and heavy chains in Figure 20-6. Note the location of two *complement-binding sites* on the antibody molecule (one on each heavy chain).

In summary, an immunoglobulin, or antibody molecule, consists of two heavy and two light polypeptide chains. Each light chain consists of one variable region and one constant region. Each heavy chain consists of one variable region and three constant regions. Disulfide bonds join the two heavy chains to each other; they also bind each heavy chain to its adjacent light chain. An antibody has two antigen-binding sites—one at the top of each pair of variable regions—and two complement-binding sites located as shown in Figure 20-6.

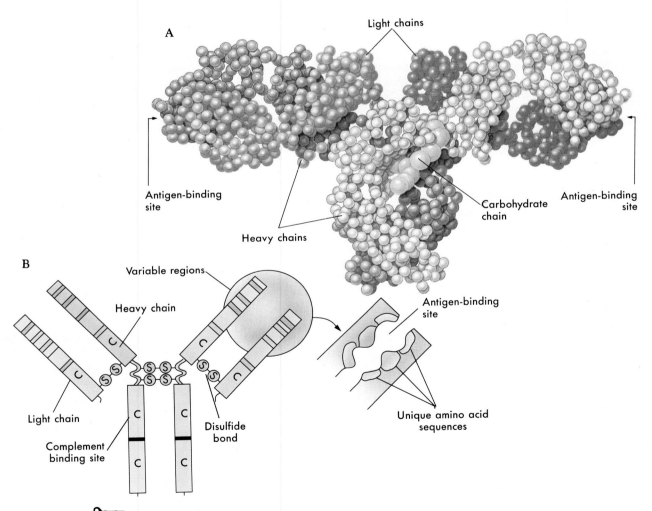

FIGURE 20-6 Structure of the antibody molecule. A, In this molecular model of a typical antibody molecule, the light chains are represented by strands of red spheres (each represents an individual amino acid). Heavy chains are represented by strands of blue spheres. Notice that the heavy chains can complex with a carbohydrate chain. **B,** This simplified diagram shows the variable regions, highlighted by colored bars, that represent amino acid sequences that are unique to that molecule. Constant regions of the heavy and light chains are marked "C." The inset shows that the variable regions at the end of each arm of the molecule form a cleft that serves as an antigen-binding site.

IMMUNIZATION

Active immunity can be established artificially by using a technique called *vaccination*. The original vaccine was a live *cowpox* virus that was injected into healthy people to cause a mild cowpox infection. The term vaccine literally means "cow substance." Because the cowpox virus is similar to the deadly *smallpox* virus, vaccinated individuals developed antibodies that imparted immunity against both cowpox and smallpox viruses.

Modern vaccines work on a similar principle; substances that trigger the formation of antibodies against specific pathogens are introduced orally or by injection. Some of these vaccines are killed pathogens or live, attenuated (weakened) pathogens. Such pathogens still have their specific antigens intact, so they can trigger formation of the proper antibodies, but they are no longer *virulent* (able to cause disease). Although rare, these vaccines sometimes backfire and actually cause an infection. Many of the newer vaccines avoid this potential problem by using only the part of the pathogen that contains antigens. Because the disease-causing portion is missing, such vaccines cannot cause infection.

The amount of antibodies in a person's blood produced in response to vaccination or an actual infection is called the *antibody titer*. As you can see in the graph, the initial injection of vaccine triggers a rise in the antibody titer that gradually diminishes. Often, a *booster shot*, or second injection, is given to keep the antibody titer high or to raise it to a level

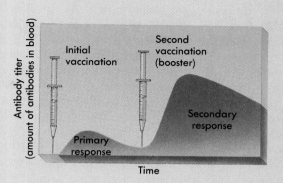

that is more likely to prevent infection. The secondary response is more intense than the primary response because memory B cells are ready to produce a large number of antibodies at a moment's notice. A later accidental exposure to the pathogen will trigger an even more intense response—thus preventing infection.

Toxoids are similar to vaccines but use an altered form of a bacterial toxin to stimulate production of antibodies. Injection of toxoids imparts protection against toxins, whereas administration of vaccines imparts protection against pathogenic organisms and viruses.

Diversity of antibodies

Every normal baby is born with an enormous number of different clones of B cells populating its bone marrow, lymph nodes, and spleen. All the cells of each clone are committed to synthesizing a specific antibody with a sequence of amino acids in its variable regions that is different from the sequence synthesized by any other of the innumerable clones of B cells.

How does this astounding diversity originate? One suggested answer is called the *somatic recombination hypothesis*. According to this explanation, our chromosomes do not contain whole genes for producing the heavy and light polypeptide chains that make up each antibody molecule. Instead, the genetic code is a set of separate sequences that are assembled into whole genes as a B cell develops. Because the separate sequences can be assembled in an astounding number of different combinations to form each whole gene, and since several different polypeptides are needed to make one antibody, any particular B cell is not likely to synthesize *exactly* the same antibody as any other B cell. Thus somatic recombination is a sort of "genetic lottery" that produces millions of unique genes by combining different gene segments and millions of unique antibodies by combining different polypeptides.

Antibody diversity may also be influenced by occasional mutations in the gene segments used to form the genes needed to produce antibodies. Evidence from several different studies show that random genetic mutations—

slight changes in the master DNA code—result in slight differences in the variable regions of antibodies.

If these hypotheses about antibody diversity are correct, it is possible to produce B cells that produce antibodies against self-antigens. It is thought that although most such B cells are eliminated early in their development, before they produce antibodies that attack a person's own cells, all humans have some "anti-self" B cells in their bodies.

Classes of antibodies

There are five classes of antibodies, identified by letter names as immunoglobulins M, G, A, E, and D (Figure 20-7). *IgM* (abbreviation for immunoglobulin M) is the antibody that immature B cells synthesize and insert into their plasma membranes. It is also the predominant class of antibody produced after initial contact with an antigen. The most abundant circulating antibody, the one that normally makes up about 75% of all the antibodies in the blood, is *IgG*. It is the predominant antibody of the secondary antibody response, that is, following subsequent contacts with a given antigen. *IgA* is the major class of antibody present in the mucous membranes of the body, in saliva, and in tears. *IgE*, although minor in amount, can produce major harmful effects such as those associated with allergies. *IgD* is present in the blood in very small amounts, and its precise function is as yet unknown. As Figure 20-7 shows, some immunoglobulin molecules are formed by the joining of several basic antibody units.

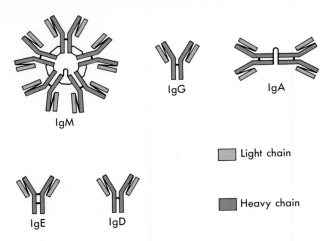

FIGURE 20-7 Classes of antibodies. Antibodies are classified into five major groups: immunoglobulin M (IgM), immunoglobulin G (IgG), immunoglobulin A (IgA), immunoglobulin E (IgE), and immunoglobulin D (IgD). Notice that each IgM molecule is comprised of five Y-shaped basic antibody units, IgA of two basic antibody units, and the others of a single basic antibody unit.

Functions of antibodies

The function of antibody molecules—some 100 million trillion of them—is to produce **antibody-mediated immunity.** This type of immunity is also called *humoral immunity* because it occurs within plasma, which is one of the humors, or fluids, of the body.

Antigen-antibody reactions. Antibodies fight disease first by recognizing substances that are foreign or abnormal. In other words, they distinguish non-self-antigens from self-antigens. Recognition occurs when an antigen's epitopes (small regions on its surface) fit into and bind to an antibody molecule's antigen-binding sites. The binding of the antigen to antibody forms an antigen-antibody complex that may produce one or more effects. For example, it transforms antigens that are toxins (chemicals poisonous to cells) into harmless substances. It agglutinates antigens that

are molecules on the surface of microorganisms. In other words, it makes them stick together in clumps, and this, in turn, makes it possible for macrophages and other phagocytes to dispose of them more rapidly by ingesting and digesting large numbers of them at one time. The binding of antigen to antibodies frequently produces still another effect—it alters the shape of the antibody molecule, not very much, but enough to expose the molecule's previously hidden complement-binding sites. This seems a trivial enough change, but it is not so. It initiates an astonishing series of reactions that culminate in the destruction of microorganisms and other foreign cells.

Complement. Complement is a component of blood plasma that consists of about 20 protein compounds. They are inactive enzymes that become activated in a definite sequence to catalyze a series of intricately linked reactions. The binding of an antibody to an antigen located on the surface of a cell alters the shape of the antibody molecule in a way that exposes its complement-binding sites (see Figure 20-6, *B*). By binding to these sites, complement protein 1 becomes activated and touches off the catalytic activity of the next complement protein in the series. A rapid sequence, or cascade, of activity by the next protein, then the next, and the next, follows until the entire series of enzymes has functioned. The end result of this rapid-fire activity challenges the imagination. Molecules formed by these reactions assemble themselves on the enemy cell's surface in such a way as to form a doughnut-shaped structure—complete with a hole in the middle (Figure 20-8). In effect, the complement has drilled a hole through the foreign cell's surface membrane! Ions and water rush into the cell; it swells and bursts. **Cytolysis** is the technical name for this process. Briefly, then, complement functions to kill foreign cells by cytolysis. In addition, various complement proteins serve other functions. Some, for example, cause vasodilation in the invaded area and some enhance phagocytosis.

The complement cascade can also be initiated by nonspecific immune mechanisms. Complement protein 3

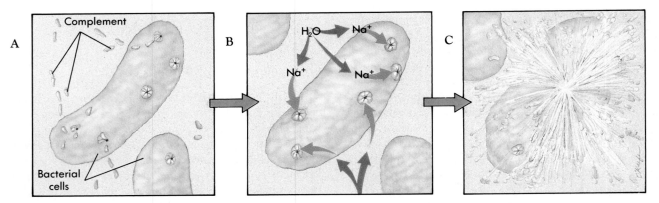

FIGURE 20-8 Complement fixation. A, Complement molecules activated by antibodies form doughnut-shaped complexes in a bacterium's plasma membrane. **B,** Holes in the complement complex allow sodium (Na⁺) and then water (H₂O) to diffuse into the bacterium. **C,** After enough water has entered, the swollen bacterium bursts.

MONOCLONAL ANTIBODIES

Techniques that have permitted biologists to produce large quantities of pure and very specific antibodies have resulted in dramatic advances in medicine. As a new medical technology, the development of **monoclonal antibodies** has been compared in importance with advances in recombinant DNA, or genetic engineering.

Monoclonal antibodies are specific antibodies produced or derived from a population or culture of identical, or **monoclonal,** cells. In the past, antibodies produced by the immune system against a specific antigen had to be "harvested" from serum containing a large number of other antibodies. The total amount of a specific antibody that could be recovered was very limited, so the cost of recovery was high. Monoclonal antibody techniques are based on the ability of immune system cells to produce individual antibodies that bind to and react with very specific antigens. We know, for example, that if the body is exposed to the varicella virus of chickenpox, B cells will produce an antibody that will react specifically with that virus and no other. With monoclonal antibody techniques, B lymphocytes that are activated after the injection of a specific antigen are "harvested" and then "fused" with other cells that have been transformed to grow and divide indefinitely in a tissue culture medium. These fused, or hybrid, cells, called *hybridomas,* continue to produce the same antibody produced by the original lymphocyte. The result is a rapidly growing population of identical, or monoclonal, cells that produce large quantities of a very specific antibody. Monoclonal antibodies have now been produced against an array of

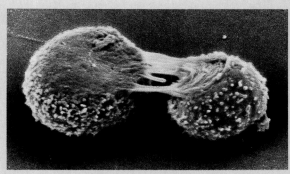

Hybridoma dividing.

different antigens, including disease-producing organisms and various types of cancer cells.

The availability of very pure antibodies against specific disease-producing agents has been used in the commercial preparation of diagnostic tests that can be used to identify viruses, bacteria, and even specific cancer cells in the blood or other body fluids. The use of monoclonal antibodies serves as the basis for specific treatment of many human diseases.

---

can become activated without any stimulation by an antigen. Protein 3 is normally inactivated by enzymes, but it can produce the full complement effect if it binds to bacteria or viruses in the presence of a compound called *properdin.* Thus lysis of various foreign cells and viruses by complement can occur even when antibodies are not involved. This method of activating the complement cascade is often called the "alternate pathway," to distinguish it from the "classical pathway" involving antibodies.

Clonal selection theory. The *clonal selection theory,* which deals with antigen destruction, was first proposed in 1959 by Sir Macfarlane Burnet. It has two basic tenets. First, it holds that the body contains an enormous number of diverse clones of cells, each committed by certain of its genes to synthesize a different antibody. Second, the clonal selection theory postulates that when an antigen enters the body, it selects the clone whose cells are committed to synthesizing its specific antibody and stimulates these cells to proliferate and to thereby produce more antibody. We now know that the clones selected by antigens consist of lymphocytes. We also know how antigens select lymphocytes—by the shape of antigen receptors on the lymphocyte's plasma membrane. An antigen recognizes receptors that fit its epitopes and combines with them. By thus selecting the precise clone committed to making its specific antibody, each antigen provokes its own destruction.

QUICK CHECK

✔ 1. How do B cells get their name?
✔ 2. How do B cells help defend the body against pathogens?
✔ 3. How does the structure of an antibody relate to its function?
✔ 4. Describe the mechanism by which complement destroys foreign cells.

T CELLS AND CELL-MEDIATED IMMUNITY
Development of T Cells

T cells, by definition, are lymphocytes that have made a detour through the thymus gland before migrating to the lymph nodes and spleen (see Figure 20-4). During their residence in the thymus, pre-T cells develop into **thymocytes,** cells that proliferate as rapidly as any in the body. Thymocytes divide up to three times a day, and, as a result, their numbers increase enormously in a relatively short time. They stream out of the thymus into the blood and find their way to a new home in areas of the lymph nodes and spleen called *T-dependent zones.* From this time on, they are known as T cells.

Activation and Functions of T Cells

Each T cell, like each B cell, displays antigen receptors on its surface membrane. They are not immunoglobulins as are B cell receptor molecules but are compounds similar to them. When an antigen (preprocessed and presented by macrophages) encounters a T cell whose surface receptors fit the antigen's epitopes, the antigen binds to the T cell's receptors. This activates or sensitizes the T cell, causing it to divide repeatedly to form a clone of identical *sensitized T cells*. The process of T cell development and activation is summarized in Figure 20-9.

The sensitized T cells then travel to the site where the antigens originally entered the body. There, in the inflamed tissue, the sensitized T cells bind to antigens of the same kind that led to their formation. However, T cells will bind to their specific antigen only if the antigen is presented by a macrophage. The antigen-bound sensitized T cells then release chemical messengers into the inflamed tissues (Figure 20-10). These chemical messengers, or factors, have been called **lymphokines.** Because some

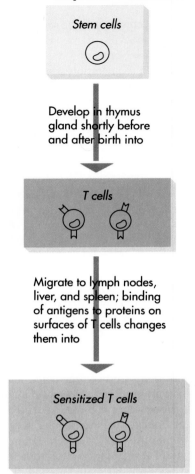

FIGURE 20-9 T-cell development. The first stage occurs in the thymus gland shortly before and after birth. Stem cells maintain a constant population of newly differentiating cells as they are needed. The second stage occurs only if a T cell is presented an antigen, which combines with certain proteins on the T cell's surface.

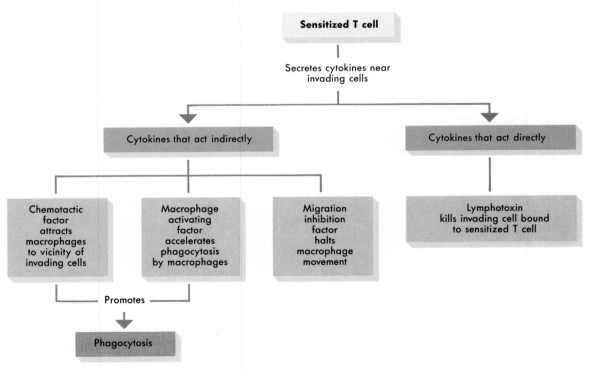

FIGURE 20-10 T-cell function. Sensitized T cells produce cell-mediated immunity by releasing various compounds in the vicinity of invading cells. Some act directly, and some act indirectly to kill invading cells.

lymphokines have been shown to be secreted by other types of cells besides lymphocytes, however, they are now more often called **cytokines.** Names of some individual cytokines are chemotactic factor, migration inhibition factor, macrophage activating factor, and lymphotoxin. *Chemotactic factor* attracts macrophages, causing hundreds of them to migrate into the vicinity of the antigen-bound, sensitized T cell. *Migration inhibition factor* halts macrophage migration. *Macrophage activating factor* prods the assembled macrophages to destroy antigens by phagocytosing them at a rapid rate. **Lymphotoxin** is a powerful poison that acts more directly, quickly killing any cell it attacks. Figure 20-11 shows how a type of lymphotoxin called perforin kills cells by lysing them. Compare this mechanism with the complement mechanism illustrated in Figure 20-8.

Sensitized T cells that release lymphotoxin are called *killer T cells*, or *cytotoxic T cells*. Two other types of T cells, helper T cells and suppressor T cells, regulate the function of B cells. *Helper T cells* help B cells differentiate into antibody-secreting plasma cells by secreting cytokines such as *interleukin-2* and *interleukin-4* (B cell differentiating factor). *Suppressor T cells* act to suppress B cell differentiation into plasma cells. The antagonistic action allows the immune system to finely tune its antibody-mediated response. Helper T cells and suppressor T cells also regulate other T cells, which refines the system's cell-mediated immune system.

Summarizing briefly, the function of T cells is to produce **cell-mediated immunity.** They search out, recognize, and bind to appropriate antigens located on the surfaces of cells. This kills the cells—the ultimate function of killer T cells. Usually these are not the body's own normal cells but are cells that have been invaded by viruses, that have become malignant, or that have been transplanted into the body. Killer T cells therefore function to defend us from viral diseases and cancer, but they also bring about rejection of

HEALTH MATTERS

IMMUNITY AND CANCER

One of the many functions of the immune system is to constantly guard against the development of cancer. Cell mutations occur frequently in the normal body, and many of the mutated cells formed are cancer cells. Cancer cells, you may recall, are cells capable of forming tumors in many different parts of the body—unless they are destroyed before this can happen. Abnormal antigens on cancer cells, called *tumor-specific antigens,* are present in the plasma membranes of some cancer cells in addition to self-antigens called *histocompatibility antigens.* Lymphocytes, in their continual wanderings through body tissues, are almost sure to come in contact with newly formed cancer cells. Hopefully, the lymphocytes recognize these cells by their abnormal antigens and quickly initiate reactions that kill the cancer cells. It often happens, however, that newly formed cancer cells escape destruction by cells of the immune system. This could happen because there are too many cancer cells to overcome, because the cancer cells somehow "fool" the immune system into not attacking, because the immune system is impaired, or because of a combination of mechanisms. The relationship between cancer and the immune system continues to be an area of intense research by scientists looking for effective cancer treatments.

transplanted tissues or organs. T cells also serve as overall regulators of specific immune mechanisms.

QUICK CHECK

✔ 1. How do T cells get their name?
✔ 2. What causes a T cell to become sensitized?
✔ 3. How do cytotoxic T cells destroy pathogens?

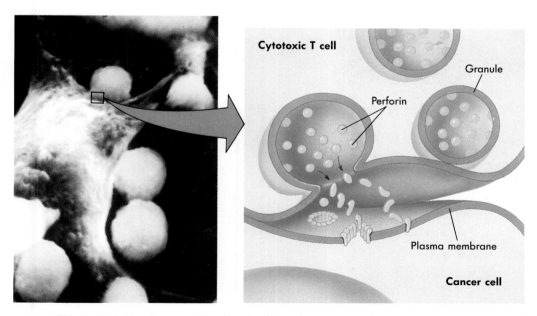

FIGURE 20-11 Cytotoxic T cells. The blue spheres seen in this scanning electron microscope view are cytotoxic T cells attacking a much larger cancer cell. T cells are a significant part of our defense against cancer and other abnormal or foreign cells. The inset shows how the lymphotoxin *perforin* acts to kill cells by puncturing holes in their plasma membranes.

TYPES OF SPECIFIC IMMUNITY

Both B cell immunity and T cell immunity, the two major types of specific immunity, can be further classified according to the manner in which they develop. For instance, specific immunity can be described as either **inherited immunity** or **acquired immunity** (Table 20-2). Inherited immunity, also called *inborn immunity*, occurs when specific or nonspecific immune mechanisms are put in place by genetic mechanisms during the early stages of human development in the womb. For example, our inborn resistance to certain diseases that affect animals that are biologically similar to us is often the result of the presence of enough antibodies or T cells to prevent infection.

Acquired immunity is resistance that develops after we are born. Acquired immunity may be further classified as either *natural immunity* or *artificial immunity*, depending on how the body is exposed to the antigen. Natural exposure is not deliberate and occurs in the course of everyday living. We are naturally exposed to many disease-causing agents on a regular basis. Artificial, or deliberate, exposure to potentially harmful antigens is called *immunization*.

Natural and artificial immunity may be "active" or "passive." Active immunity occurs when an individual's own immune system responds to a harmful agent, regardless of whether that agent was naturally or artificially encountered. Passive immunity results when immunity to a disease that has developed in another individual or animal is transferred to an individual who was not previously immune. For example, antibodies in a mother's milk confer passive immunity to her nursing infant. Active immunity generally lasts longer than passive immunity. Passive immunity, although temporary, provides immediate protection. Table 20-2 lists the various forms of specific immunity and gives examples of each.

QUICK CHECK

✔ 1. What is the difference between inherited and acquired immunity?

✔ 2. What is the difference between natural and artificial immunity?

✔ 3. What is the difference between active and passive immunity?

TABLE 20-2   Types of specific immunity

Type	Description or Example
INHERITED IMMUNITY	Immunity to certain diseases develops before birth; also called inborn immunity
ACQUIRED IMMUNITY	
Natural immunity	Exposure to the causative agent is not deliberate
Active (exposure)	A child develops measles and acquires an immunity to a subsequent infection
Passive (exposure)	A fetus receives protection from the mother through the placenta, or an infant receives protection via the mother's milk
Artificial immunity	Exposure to the causative agent is deliberate
Active (exposure)	Injection of the causative agent, such as a vaccination against polio, confers immunity
Passive (exposure)	Injection of protective material (antibodies) that was developed by another individual's immune system

---

THE BIG PICTURE
The Immune System and the Whole Body

The "big picture" of the immune system's role in maintaining the relative constancy of the internal environment is probably easier to "see" than any other system. After all, its agents—antibodies, lymphocytes, and other substances and cells—are everywhere in the body. They even stand guard on the outside surface of the body. Without the defensive activity of the immune system, our internal constancy would be ruined by cancer, infections, and even minor injuries.

In describing the various mechanisms of the immune system, we have used the analogy of a militaristic-style security force. As useful as this analogy might be, it may mislead us into believing that the immune system is a completely independent group of defensive agents. Nothing could be further from the truth. First of all, recent evidence has shown scientists that the immune system is regulated to some degree by the nervous and endocrine systems. These systems, as you already know, are in turn influenced by feedback from all parts of the body. Secondly, the agents of the immune system are not a separate, distinct group of cells and substances. They include blood cells, skin cells, mucosal cells, brain cells, and liver cells, as well as many other types of cells and their secretions. Thus the immune system is more like a self-defense force made up of ordinary citizens that work shoulder-to-shoulder with military specialists.

MECHANISMS OF DISEASE

Disorders of the Immune System

There are two basic mechanisms for disorders of immunity. The immune defenses can either overreact to antigens or fail to react to an antigen to produce disease. We will briefly describe a few examples of each of these mechanisms.

Hypersensitivity of the Immune System

Hypersensitivity is a type of inappropriate or excessive response of the immune system. There are three major types of immune hypersensitivity: allergy, autoimmunity, and isoimmunity. Each of these three types is discussed in the following sections.

Allergy

The term **allergy** is used to describe hypersensitivity of the immune system to relatively harmless environmental antigens. Antigens that trigger an allergic response are often called **allergens** (AL-ler-jenz). One in six Americans has a genetic predisposition to an allergy of some kind.

Immediate allergic responses involve antigen-antibody reactions. Before such a reaction occurs, a susceptible person must be exposed repeatedly to an allergen—triggering the production of antibodies. After a person is thus *sensitized,* exposure to an allergen causes antigen-antibody reactions that trigger the release of histamine, kinins, and other inflammatory substances. These responses usually cause typical allergy symptoms such as runny nose, conjunctivitis, and *urticaria* (hives). In some cases, however, these substances may cause constriction of the airways, relaxation of blood vessels, and irregular heart rhythms that can progress to a life-threatening condition called *anaphylactic shock* (see Chapter 18, p. 525). Drugs called *antihistamines* are sometimes used to relieve the symptoms of this type of allergy.

Delayed allergic responses, on the other hand, involve cell-mediated immunity. In *contact dermatitis,* for example, T cells trigger events that lead to local skin inflammation a few hours or days after initial exposure to an antigen. Exposure to poison ivy, soaps, and certain cosmetics may cause contact dermatitis in this manner. Hypersensitive individuals may use *hypoallergenic* products (products without common allergens) to avoid such allergic reactions.

Autoimmunity

Autoimmunity is an inappropriate and excessive response to self-antigens. Disorders that result from autoimmune responses are called **autoimmune diseases.** Examples of autoimmune diseases are given in Table 20-3. Self-antigens are molecules that are native to a person's body and that are used by the immune system to identify components of "self." In autoimmunity, the immune system inappropriately attacks these antigens.

A common autoimmune disease is **systemic lupus erythematosus (SLE),** or, simply, *lupus.* Lupus is a chronic inflammatory disease that affects many tissues in the body: joints, blood vessels, kidney, nervous system, and skin. The name lupus erythematosus refers to the red rash that often develops on the face of those afflicted with SLE. The "systemic" part of the name comes from the fact that the disease affects many systems throughout the body. The systemic nature of SLE results from the production of antibodies against many different self-antigens.

Isoimmunity

Isoimmunity is excessive reaction of the immune system to antigens from a different individual of the same species. Isoimmunity is important in two situations: pregnancy and tissue transplants.

During pregnancy, antigens from the fetus may enter the mother's blood supply and sensitize her immune system. Antibodies that are formed as a result of this sensitization may enter the fetal circulation and cause an inappropriate immune reaction. One example, erythroblastosis fetalis, was discussed in Chapter 16. Other pathological conditions may also be caused by damage to developing fetal tissues resulting from attack by the mother's immune system. Examples include congenital heart defects, Graves' disease, and myasthenia gravis.

Tissue or organ **transplants** are medical procedures in which tissue from a donor is surgically *grafted* into the body. For example, skin grafts are often performed to repair damage caused by burns. Donated whole blood tissue is often transfused into a recipient after massive hemorrhaging. A kidney is sometimes removed from a living donor and grafted into a person suffering from kidney failure. Unfortunately, the immune system sometimes reacts against foreign antigens in the grafted tissue, causing what is often called a *rejection syndrome.* The antigens commonly involved in transplant rejection are called **human lymphocyte antigens (HLAs).**

Rejection of grafted tissues can occur in either of two ways. One is called *host versus graft rejection* because the recipient's immune system recognizes foreign HLAs and attacks it, destroying the donated tissue. The other is *graft versus host rejection* because the donated tissue (for example, bone marrow) attacks the recipient's HLAs, destroying tissue throughout the recipient's body. Graft versus host rejection may lead to death.

There are two ways to prevent rejection syndrome. One strategy is called *tissue typing* in which HLAs and other antigens of a potential donor and recipient are identified. If they match, tissue rejection is unlikely to occur. Another strategy is the use of **immunosuppressive drugs** in the recipient. Immunosuppressive drugs such as *cyclosporine* and *prednisone* suppress the immune system's ability to attack the foreign antigens in the donated tissue.

Deficiency of the Immune System

Immune deficiency, or *immunodeficiency,* is the failure of immune system mechanisms in defending against pathogens. Immune system failure usually results from disruption of lymphocyte (B cell or T cell) function. The chief characteristic of immune deficiency is the development of unusual or recurring severe infections or cancer. Although immune deficiency by itself does not cause death, the resulting infections or cancer can.

There are two broad categories of immune deficiencies, based on the mechanism of lymphocyte dysfunction: *congenital* and *acquired.* Each of these types is outlined in the following discussion.

TABLE 20-3   Examples of autoimmune diseases

Disease	Possible self-antigen	Description
Addison's Disease	Surface antigens on adrenal cells	Hyposecretion of adrenal hormones, resulting in weakness, reduced blood sugar, nausea, loss of appetite, and weight loss
Cardiomyopathy	Cardiac muscle	Disease of cardiac muscle (i.e., the myocardium), resulting in loss of pumping efficiency (heart failure)
Diabetes Mellitus (Insulin-Dependent)	Pancreatic islet cells, insulin, insulin receptors	Hyposecretion of insulin by the pancreas, resulting in extremely elevated blood glucose levels (in turn causing a host of metabolic problems, even death if untreated)
Glomerulonephritis	Blood antigens that form immune complexes that deposit in kidney	Disease of the filtration apparatus of the kidney (renal corpuscle), resulting in fluid and electrolyte imbalance and possibly total kidney failure and death
Graves' Disease (Type of Hyperthyroidism)	TSH receptors on thyroid cells	Hypersecretion of thyroid hormone and resulting increase in metabolic rate.
Hemolytic Anemia	Surface antigens on RBCs	Condition of low RBC count in the blood due to excessive destruction of mature RBCs (hemolysis)
Multiple Sclerosis	Antigens in myelin sheaths of nervous tissue	Progressive degeneration of myelin sheaths, resulting in widespread impairment of nerve function (especially muscle control)
Myasthenia Gravis	Antigens at neuromuscular junction	Muscle disorder characterized by progressive weakness and chronic fatigue
Myxedema	Antigens in thyroid cells	Hyposecretion of thyroid hormone in adulthood, causing decreased metabolic rate and characterized by reduced mental and physical vigor, weight gain, hair loss, and edema
Pernicious Anemia	Antigens on parietal cells, intrinsic factor	Abnormally low RBC count resulting from the inability to absorb vitamin B_{12}, a substance critical to RBC production
Reproductive Infertility	Antigens on sperm or tissue surrounding ovum (egg)	Inability to produce offspring (in this case, due to destruction of gametes)
Rheumatic Fever	Cardiac cell membranes (cross-reaction with group A streptococcal antigen)	Rheumatic heart disease; inflammatory cardiac damage (especially to the endocardium/valves)
Rheumatoid Arthritis	Collagen	Inflammatory joint disease characterized by synovial inflammation that spreads to other fibrous tissues
Systemic Lupus Erythematosus	Numerous	Chronic inflammatory disease with widespead effects and characterized by arthritis, a red rash on the face, and other signs
Ulcerative Colitis	Mucus cells of colon	Chronic inflammatory disease of the colon characterized by watery diarrhea containing blood, mucus, and pus

Congenital immune deficiency

Congenital immune deficiency, which is rare, results from improper lymphocyte development before birth. Depending on which stage of the development of stem cells, B cells, or T cells the defect occurs, different diseases can result. For example, improper B cell development can cause insufficiency or absence of antibodies in the blood. If stem cells are disrupted, a condition called **severe combined immune deficiency (SCID)** results. In most forms of SCID, both humoral immunity and cell-mediated immunity are defective. Temporary immunity can be imparted to children with SCID by injecting them with a preparation of antibodies (gamma globulin). Bone marrow transplants, which replace the defective stem cells with healthy donor cells, have proven effective in treating some cases of SCID.

Acquired immune deficiency

Acquired immune deficiency develops after birth (and is not related to genetic defects). Many factors can contribute to acquired immune deficiency: nutritional deficiencies, immunosuppressive drugs or other medical treatments, trauma, stress, and viral infection.

One of the best known examples of acquired immune deficiency is **acquired immune deficiency syndrome (AIDS).** AIDS was first recognized as a new disease by the Centers for Disease Control in 1981. This syndrome is caused by the *human immunodeficiency virus,* or *HIV.* HIV, a retrovirus, contains RNA that produces its own DNA inside infected cells. The viral DNA often becomes part of the cell's DNA. When the viral DNA is activated, it directs the cell to synthesize viral RNA and viral proteins—producing new

retroviruses. The HIV virus thus "steals" raw materials from the cell. When this occurs in T cells, the cell is destroyed and immunity is impaired. As the T cell dies, it releases new retroviruses that can spread the HIV infection. The mechanism of HIV infection is outlined in Figure 20-12.

Although HIV can invade several types of human cells, including brain cells, it has its most obvious effects in helper T cells. When helper T cell function is impaired, infectious organisms and cancer cells can grow and spread much more easily than normal. Unusual conditions such as *pneumocystosis* (a protozoal condition) and *Kaposi's sarcoma* (a type of skin cancer) may also appear. Since their immune system is deficient, AIDS patients usually die from one of these infections or cancers.

After they are infected with HIV, T cells may not show signs of AIDS for years. AIDS does not appear until the viral DNA is activated and begins to destroy many T cells.

There are several strategies for controlling AIDS and related conditions. Many agencies are trying to slow the spread of AIDS by educating people about how to avoid contact with the HIV retrovirus. HIV is spread by direct contact of body fluids, so preventing such contact reduces HIV transmission. Sexual relations, blood transfusions, and intravenous use of contaminated needles are the usual modes of HIV transmission. Several research teams are working on vaccines to impart immunity to people who are not yet infected by HIV. Like many viruses, such as those that cause the common cold, HIV changes rapidly enough to make development of a vaccine difficult.

A way to inhibit symptoms of the disease is by means of chemicals such as *azidothymidine (AZT)* that block HIV's ability to reproduce within infected cells. At least 80 such compounds are currently being evaluated for use in halting the progress of HIV infections.

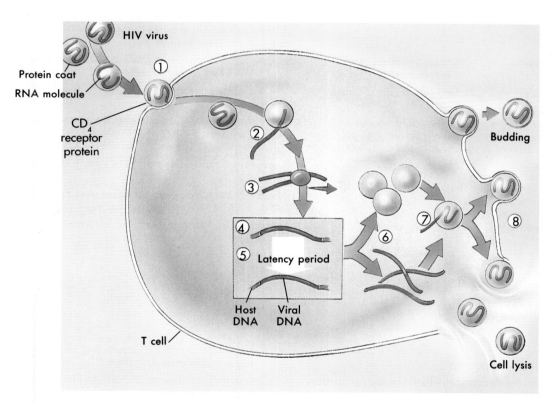

FIGURE 20-12 Mechanisms of HIV infection. The HIV is an RNA-containing virus (*retrovirus*) that appears to infect T cells via the mechanism described in the following steps:

1. HIV attaches to specific receptors on the surface of a T cell.
2. Viral RNA is released into cytoplasm.
3. Viral RNA is used to synthesize a DNA molecule ("reverse transcription").
4. The new "viral" DNA is spliced into the cell's chromosomal DNA.
5. After a latency period (usually several years), the viral DNA initiates transcription of RNA molecules.
6. The new RNA molecules direct synthesis of both HIV RNA and HIV proteins.
7. Whole HIV particles are assembled by the cell.
8. New HIV viruses are released from the cell by lysis (breakage of the plasma membrane) or by budding (exocytosis). They are now free to infect more cells.

CASE STUDY
ACQUIRED IMMUNE DEFICIENCY SYNDROME (AIDS)

D.L., a 32-year-old Caucasian male, was admitted to the hospital with symptoms of a depressed immune system, including oral and groin candidiasis (thrush) and a herpes virus infection. This is his fourth admission in less than 2 years.

D.L. is a homosexual and has engaged in anal and oral intercourse since the age of 20. In October 1988 he presented with oral and groin candidiasis and perianal lesion, which on culture revealed a herpes infection. His past sexual behavior and current viral infections warranted HIV antibody testing (enzyme-linked immunosorbent assay [ELISA] and western blot), which yielded positive findings.

D.L. has had four inpatient and three outpatient admissions between November 1988 and December 1989. In December 1988, he went to the clinic complaining of diarrhea, a loss of appetite resulting in a 10-pound weight loss over a 2-week period, a nonproductive cough, some shortness of breath, and a perianal

lesion. He tested HIV-positive for the second time. Symptoms were severe enough to recommend hospital admission. Admitting chest roentgenogram (x-ray) revealed diffuse infiltrates compatible with *Pneumocystis carinii* pneumonia (PCP). This diagnosis was confirmed with other lab tests. D.L. was treated with antibiotics and anti-AIDS drugs. As is true of many persons with AIDS, D.L. was admitted to the hospital on numerous occasions with *Pneumocystis carinii* pneumonia, anemia, gastrointestinal problems, and Kaposi's sarcoma of the skin.

In December of 1989, he came back to the hospital in a state of extreme mental confusion. He also had muscle wasting of his extremities, including hands and feet, and was unable to fully control the muscles of his feet and legs. This resulted in a condition known as foot drop. His condition continued to deteriorate until his death a few weeks later.

---

1. AIDS manifests as reduced resistance to opportunistic infections and malignancies because of:
   a. An acquired deficiency of isoimmunity
   b. An exaggerated immune response to counteract a T cell deficiency
   c. Impaired functioning of one or more components of the immune/inflammatory response
   d. Deficiency of antigen (HIV) recognition

2. The T_4 helper cell to T_8 suppressor cell ratio is normally about $2:1$. This ratio in D.L. was severely reduced because:
   a. Bone marrow depression reduces hemoglobin and hematocrit levels
   b. The HIV virus infects and kills the host T_4 cells, decreasing their numbers
   c. The T_8 suppressor cells are stimulated by the HIV virus, and their numbers increase
   d. γ-Interferon is reduced, thereby reduce cytotoxic T_8 suppressor cell activity

3. PCP (a protozoan), normally nonpathogenic (non-disease-causing), caused severe respiratory problems for D.L. because of:
   a. Bacterial invasion of PCP into his lungs
   b. Pleural effusion (leakage) caused by the increase of the PCP
   c. Filling of the alveoli with eosinophilic exudate and PCP
   d. Other superimposed infections masked by his medication

4. D.L.'s confusion and distal peripheral neuropathy, in light of negative serological findings, was most likely caused by:
   a. Primary or secondary central nervous system lymphoma
   b. Toxoplasma, a protozoon infecting the central nervous system
   c. Disturbance in normal brain patterns from drug therapy
   d. Presence of HIV virus in peripheral nerves and the central nervous system

CHAPTER SUMMARY

INTRODUCTION

A. The immune system protects against assaults on the body
   1. External assaults include microorganisms—protozoans, bacteria, and viruses
   2. Internal assaults—abnormal cells reproduce and form tumors that may become cancerous and spread

ORGANIZATION OF THE IMMUNE SYSTEM

A. The immune system is continually at work
B. Two major categories of immune mechanisms—nonspecific immunity and specific immunity
C. Nonspecific immunity provides a general defense against anything that is not "self"
D. Specific immunity acts against specific threatening agents
E. Primary cells for nonspecific immunity—neutrophils, monocytes, macrophages, and natural killer cells
F. Primary types of cells for specific immunity—lymphocytes called T cells and B cells

NONSPECIFIC IMMUNITY (Table 20-1)

A. Species resistance—genetic characteristics of an organism or species defends against pathogens
B. Mechanical and chemical barriers—first line of defense (Figure 20-1)
   1. The internal environment of the body is protected by a barrier formed by the skin and the mucous membranes
   2. Skin and mucous membranes provide additional immune mechanisms—sebum, mucus, enzymes, and hydrochloric acid in the stomach
C. Inflammation—second line of defense (Figure 20-2)
   1. Inflammatory response—tissue damage elicits responses to counteract injury and promote normalcy
      a. Inflammation mediators include histamine, kinins, prostaglandins, and related compounds
      b. Chemotactic factors—substances that attract white blood cells to the area in a process called chemotaxis
      c. Characteristic signs of inflammation—heat, redness, pain, and swelling
      d. Systemic inflammation—occurs from a body-wide inflammatory response
   2. Phagocytosis—ingestion and destruction of micro-organisms or other small particles by phagocytes (Figure 20-3)
      a. Neutrophil—most numerous phagocyte; usually first to arrive at site of injury, migrates out of bloodstream during diapedesis, forms pus
      b. Macrophage—phagocytic monocytes grow larger after migrating from bloodstream; examples are histiocytes in connective tissue, microglia in nervous system, and Kupffer cells in liver
D. Natural killer cells—lymphocytes that kill tumor cells and cells infected by viruses
   1. Method of killing cells—lysing cells by damaging plasma membranes
E. Interferon—protein synthesized and released into circulation by certain cells if invaded by viruses
F. Complement—group of enzymes that lyse cells when activated by either specific or nonspecific mechanisms

OVERVIEW OF SPECIFIC IMMUNITY

A. Specific immunity is part of the third line of defense consisting of lymphocytes—two different classes of a type of white blood cell
B. Two classes of lymphocytes (Figure 20-4)—B lymphocytes (B cells) and T lymphocytes (T cells)
C. Lymphocytes are densest where they develop—in bone marrow, thymus gland, lymph nodes, and spleen
D. Lymphocytes flow through the bloodstream, become distributed in tissues, and return to the bloodstream in a continuous recirculation
E. B-cell mechanisms—antibody-mediated immunity; produce antibodies that attack pathogens
F. T cells attack pathogens move directly—classified as cell-mediated immunity

B CELLS AND ANTIBODY-MEDIATED IMMUNITY

A. B cells develop in two stages
   1. Pre-B cells develope by a few months of age
   2. The second stage occurs in lymph nodes and spleen—activation of a B cell when it binds a specific antigen
   3. B cells serve as ancestors to antibody-secreting plasma cells
B. Antibodies—proteins (immunoglobins) secreted by activated B cell (Figure 20-5)
C. An antibody molecule consists of two heavy and two light polypeptide chains; each molecule has two antigen-binding sites and two complement-binding sites (Figure 20-6)
D. Babies are born with different clones of B cells in bone marrow, lymph nodes, and spleen; cells of the clone synthesize a specific antibody with a sequence of amino acids in its variable region that is different from the sequence synthesized by other clones
E. Five classes of antibodies (Figure 20-7)—immunoglobulins M, G, A, E, and D
   1. IgM—antibody inactive B cells synthesize and insert into their own plasma membranes; it is the predominant class produced after initial contact with an antigen
   2. IgG—makes up 75% of antibodies in the blood; predominant antibody of the secondary antibody response
   3. IgA—major class of antibody in the mucous membranes, in saliva and tears
   4. IgE—small amount; produces harmful effects such as allergies
   5. IgD—small amount in blood; precise function unknown
F. Antibody molecules produce antibody-mediated immunity (humoral immunity)—within plasma
G. Antibodies resist disease first by recognizing foreign or abnormal substances
   1. Epitopes bind to an antibody molecule's antigen-binding sites, which forms an antigen-antibody complex that may produce several effects
H. Complement—a component of blood plasma consisting of a number of protein compounds; serves to kill foreign cells by cytolysis (Figure 20-8), causes vasodilation, enhances phagocytosis, and other functions
I. Complement activity can be initiated by nonspecific immune mechanisms (the alternate pathway)
   1. Complement protein 3—activated without antigen stimulation—produces full complement effect by binding to bacteria or viruses in presence of properdin
J. Clonal selection theory
   1. The body contains many diverse clones of cells, each committed by its genes to synthesize a different antibody
   2. When an antigen enters the body, it selects the clone whose cells are synthesizing its antibody and stimulates them to proliferate and create more antibody
   3. The clones selected by antigens consist of lymphocytes and are selected by the shape of antigen receptors on the lymphocyte's plasma membrane

T CELLS AND CELL-MEDIATED IMMUNITY

A. T cells—lymphocytes that go through the thymus gland before migrating to the lymph nodes and spleen

1. Pre-T cells develop into thymocytes while in the thymus
2. Thymocytes stream into the blood and are carried to the T-dependent zones in the spleen and the lymph nodes
B. T cells display antigen receptors on their surface membranes; the T cell is activated when an antigen (presented by a macrophage) binds to its receptors, causing it to divide repeatedly to form a clone of identical sensitized T cells (Figure 20-9)
   1. Sensitized T cells go to the site where the antigen entered, bind to antigens, and release cytokines (lymphokines)
C. Killer T cells—T cells release lymphotoxin to kill cells
D. Helper T cells—regulate the function of B cells
E. Suppressor T cells—suppress B cell differentiation into plasma cells
F. T cells function to produce cell-mediated immunity (Figures 20-10 and 20-11) and regulate specific immunity in general

TYPES OF SPECIFIC IMMUNITY (Table 20-2)

A. Inherited immunity (inborn immunity)—genetic mechanisms put specific or nonspecific immune mechanisms in place during development in the womb
B. Acquired immunity; resistance developed after birth; two types
   1. Natural immunity results fom nondeliberate exposure to antigens
   2. Artificial immunity results from deliberate exposure to antigens, called immunization
C. Natural and artificial immunity may be active or passive
   1. Active immunity—when the immune system responds to a harmful agent regardless of whether it was natural or artificial; lasts longer than passive
   2. Passive immunity—immunity developed in another individual is transferred to an individual who was not previously immune; it is temporary but provides immediate protection

THE BIG PICTURE: THE IMMUNE SYSTEM AND THE WHOLE BODY

A. Immune system regulated to some degree by nervous and endocrine systems
B. Agents of the immune system include blood cells, skin cells, mucosal cells, brain cells, liver cells, and other types of cells and their secretions

MECHANISMS OF DISEASE: DISORDERS OF THE IMMUNE SYSTEM

A. Disorders of the immune system—two basic mechanisms— defenses overreact or fail to react to antigens

B. Hypersensitivity of the immune system—excessive response
   1. Allergy—hypersensitivity to harmless environmental antigens
      a. Immediate allergic responses—a susceptible person becomes sensitized to antigens and the body produces antibodies, then exposure to allergen causes antigen-antibody reactions that trigger release of histamine, kinins, and other inflammatory substances
      b. Delayed allergic responses—involve cell-mediated immunity
   2. Autoimmunity—excessive response to self-antigen diseases are called autoimmune diseases
   3. Isoimmunity—excessive reaction to antigens from a different individual of the same species
      a. Pregnancy—antigens from fetus may enter mother's blood supply and sensitize her immune system, causing antibodies formed by this to enter fetal circulation and cause inappropriate immune reactions
      b. Tissue or organ transplants—the immune system may react to foreign antigens in the grafted tissue; the most common antigens are called human lymphocyte antigens (HLAs)
         (1) Two types of rejection: host versus graft rejection and graft versus host rejection
         (2) Two ways to prevent rejection: tissue typing and immunosuppressive drugs
C. Immune deficiency (immunodeficiency)—failure of immune system mechanisms to defend against pathogens usually resulting from lymphocyte cell function; the chief characteristic is the development of unusual or recurring infections or cancer, which may lead to death
   1. Congenital immune deficiency—results from improper lymphocyte development from birth
   2. Acquired immune deficiency—develops after birth from factors such as nutritional deficiencies, immunosuppressive drugs, other medical treatments, trauma, stress, and viral infection
      a. Acquired immune deficiency syndrome (AIDS)— caused by human immunodeficiency virus (HIV), which invades cells—especially cells of the immune system—inhibiting immune response (Figure 20-12)
         (1) Opportunistic infections, such as pneumocystosis or Kaposi's sarcoma—occur and lead to death
         (2) HIV is spread through direct contact of body fluids—via sexual relations, blood transfusions, and intravenous use of contaminated needles
         (3) Azidothymidine (AZT) is a drug used to inhibit HIV's ability to reproduce within infected cells

REVIEW QUESTIONS

1. Define the term *nonspecific immunity*.
2. List several of the mechanisms of nonspecific defense and give a brief description of each one.
3. Identify the body's first line of defense.
4. Describe chemotactic factors released from the mast cell.
5. Discuss the steps involved in phagtocytosis.
6. What is the function of interferon?
7. Define the following terms: *antigens, antibodies, antigenic determinants, combining sites, clone.*
8. What two terms are synonyms for combining sites?
9. When an antigen-antibody complex is formed, what region on the antigen molecule fits into what region on the antibody molecule?
10. Activated B cells develop into clones of what two kinds of cells?
11. What cells synthesize and secrete copious amounts of antibodies?

12. Explain the function of memory cells.
13. Antibodies belong to what class of compounds? Diagram and describe the structure of an antibody molecule.
14. What is complement and how does it function?
15. Explain the two basic tenets of Burnet's clonal selection theory.
16. Describe several functions performed by T cells after they have been activated or sensitized.
17. What are lymphokines? Lymphotoxins?
18. Differentiate between the classifications of natural and artificial immunity.
19. Describe the process behind the functioning of modern vaccines.
20. What are monoclonal antibodies and how do they function?
21. How is the immune system related to the development of cancer?

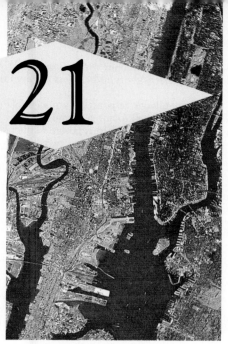

21 ▶ Stress

New York City by Satellite in 1985

OBJECTIVES

After you have completed this chapter, you should be able to:

1. Define the term *stress* and give examples of agents, or stressors, that produce stress.
2. Explain the three differing stages of the general adaptation syndrome.
3. Discuss the mechanisms of stress and Selye's hypothesis about the activation of the stress response.
4. Compare the alarm reaction responses resulting from hypertrophy of the adrenal cortex and from increased sympathetic activity.
5. Outline and discuss current concepts of the stress syndrome.
6. Explain the relationship between corticoids and resistance to stress.
7. Discuss psychological stress.

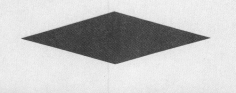
Stress affects people of all ages and in all walks of life. Children at play, students preparing for an exam, or workers on the job are all subject to stress. Although some stress can result in beneficial outcomes, excessive and long-term stress is often responsible for disastrous consequences on the health and quality of life for many people.

People experiencing severe stress are often overwhelmed by tension, anger, fear, and frustration. As a result, adrenalin levels rise, blood pressure and heart rate increase, and breathing patterns change. Blood levels of such nutrients as glucose and fatty acids deviate from their normal set points and the immune system becomes less effective. People who have difficulty in dealing with stress over time suffer from a number of "stress-related illnesses." Every individual responds somewhat differently to stress, making diagnosis and treatment difficult. There is often a decrease in efficiency of study habits or job-related problem-solving skills, susceptibility to infections increases, and complaints of stomach pains, heart palpitations, fatigue, and muscle aches are common. Sleep disorders and depression frequently accompany stress, and affected individuals often have difficulty in staying with tasks.

Contemporary medicine now recognizes excessive and long-term stress as a critically important and widespread cause of disease. It causes disruption in homeostasis of numerous physiological control systems in the body and must be controlled to ensure good health.

SELYE'S CONCEPT OF STRESS

In 1935 Hans Selye of McGill University in Montreal made an accidental discovery that launched him on a lifelong career and led him to conceive the idea of stress. This chapter will tell the story briefly of how Selye developed his stress concept and will describe the mechanism of stress that he postulated. It will then present some current ideas about stress.

Development of the Stress Concept

Selye made his accidental discovery in 1935 when he was trying to learn whether there might be another sex hormone besides those already known. He had injected rats with various extracts made from ovaries and placenta, expecting to find that different changes had occurred in animals injected with different hormonal preparations. But to his surprise and puzzlement, he found the same three changes in all of the animals. The cortex of their adrenal glands were enlarged but their lymphatic organs—thymus glands, spleens, and lymph nodes—were atrophied, and bleeding ulcers of the stomach and duodenum had developed in every animal. Then he tried injecting many other substances, for example, extracts from pituitary glands, kidneys, and spleens, and even a poison, formaldehyde. Every time he found the same three changes: enlarged adrenals, shrunken lymphatic organs, and bleeding gastrointestinal ulcers. They seemed to be a syndrome, he thought. A syndrome, according to the classical definition, is a set of signs and symptoms that occur together and that are characteristic of one particular disease. The three changes, or "stress triad," Selye had observed occurred together, but they seemed to be characteristic not of any one particular kind of injury but of any and all kinds of harmful stimuli. More experiments using various chemicals and injurious agents confirmed for him that the three changes truly were a syndrome of injury. His first publication on the subject was a short paper entitled "A syndrome produced by diverse nocuous agents"; it appeared in the July, 1936, issue of the British journal, *Nature*. Years later, in 1956, he published his monumental technical treatise, *The Stress of Life*.

Definitions

Stress, according to Selye's usage of the word, is a state, or condition, of the body produced by "diverse nocuous agents" and manifested by a syndrome of changes. He named the agents that produce stress **stressors** and coined a name—**general adaptation syndrome**—for the syndrome or group of changes that make the presence of stress in the body known.

Stressors

A stressor is any agent or stimulus that produces stress. Precise classification of stimuli as stressors or nonstressors is not possible. We can, however, make the following generalizations about the character of stressors.

1. Stressors are extreme stimuli—too much or too little of almost anything. In contrast, almost anything in moderation or mild stimuli are nonstressors. Thus coolness, warmth, and soft sounds are nonstressors, whereas extreme cold, extreme heat, and extremely loud sounds almost always act as stressors. Not only extreme excesses but also extreme deficiencies may act as stressors. One example of this kind of stressor is an extreme lack of social contact stimuli. Solitary confinement in a prison, space travel, social isolation because of blindness or deafness, or, in some cases, old age have all been identified as stressors. But the opposite extreme, an excess of social contact stimuli, for example (caused by overcrowding), also acts as a stressor (Figure 21-1, A to D).

2. Stressors very often are injurious, unpleasant, or painful stimuli—but not always. "A painful blow and a passionate kiss," Selye wrote, "can be equally stressful."

3. Anything that an individual perceives as a threat, whether real or imagined, arouses fear or anxiety. These emotions act as stressors. So, too, does the emotion of grief.

4. Stressors differ in different individuals and in one individual at different times. A stimulus that is a stressor for you may not be a stressor for me. A stimulus that is a stressor for you today may not be a stressor for you tomorrow and might not have been a stressor for you yesterday. Many factors—including one's physical and mental health, heredity, past experiences, coping habits, and even one's diet—determine which stimuli are stressors for each individual.

QUICK CHECK

✔ 1. Define the terms stress syndrome and stress.
✔ 2. Identify the three changes Hans Selye called the "stress triad."
✔ 3. List four characteristics of stressors.

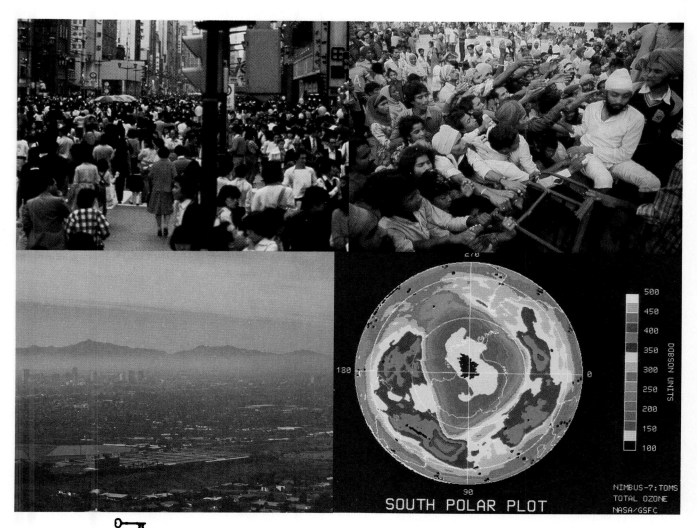

FIGURE 21-1   Causes of stress. A, Overcrowding (Asia). **B,** Food shortages (refugee camp, New Delhi, India). **C,** Air pollution (Los Angeles). **D,** Environmental dangers (Antarctic ozone hole).

General Adaptation Syndrome
Manifestations

Stress, like health or any other state or condition, is an intangible phenomenon. It cannot itself be seen, heard, tasted, smelled, felt, or measured directly. How, then, can we know that stress exists? It can be inferred to exist when certain visible, tangible, measurable responses occur. Selye, for example, inferred that the animals on which he experimented were in a state of stress when he found the syndrome of the three changes previously noted—hypertrophied adrenals, atrophied lymphatic organs, and bleeding gastrointestinal ulcers. Because this syndrome indicated the presence of stress and consisted of three changes, he called them the "stress triad." Eventually he found that many other changes also took place as a result of stress. The entire group of changes or responses he named the **general adaptation syndrome (GAS).** For each word in this name he gave his reasons. By the word "general" he wanted to suggest that the syndrome was "produced only by agents which have a general effect upon large portions of the body." The word "adaptation" was meant to imply that the syndrome of changes made it possible for the body to adapt, to cope successfully with stress. Selye thought that these responses seemed to protect the animals from serious damage by extreme stimuli and to promote their healthy survival. He looked on the general adaptation syndrome as a crucial part of the body's complex defense mechanism.

Stages

Changes that make up the general adaptation syndrome do not all take place simultaneously but over a period of time in three stages. Selye named these stages: (1) the alarm reaction, (2) the stage of resistance or adaptation, and (3) the stage of exhaustion. A different syndrome of changes, he noted, characterized each stage. Among the responses characteristic of the **alarm reaction,** for example, were the stress triad already described—hypertrophied adrenal cortex, atrophied lymphatic organs (thymus, spleen, lymph nodes), and bleeding gastric and duodenal ulcers. In addition, the adrenal cortex increased its secretion of glucocorticoids, the number of lymphocytes decreased markedly, and so, too, did the number of eosinophils. Also, the sympathetic nervous system and the adrenal medulla greatly increased their activity. Each of these changes, in turn, produced other widespread changes. Figure 21-2 indicates some of the changes stemming from adrenal cortical hypertrophy. Figure 21-3 shows responses produced by increased sympathetic activity and increased secretion by the adrenal medulla of its hormone, epinephrine (adrenalin).

Quite different responses characterize the **stage of resistance** of the general adaptation syndrome. For instance, the adrenal cortex and medulla both return to their normal rates of secreting hormones. The changes that had taken place in the alarm stage as a result of increased

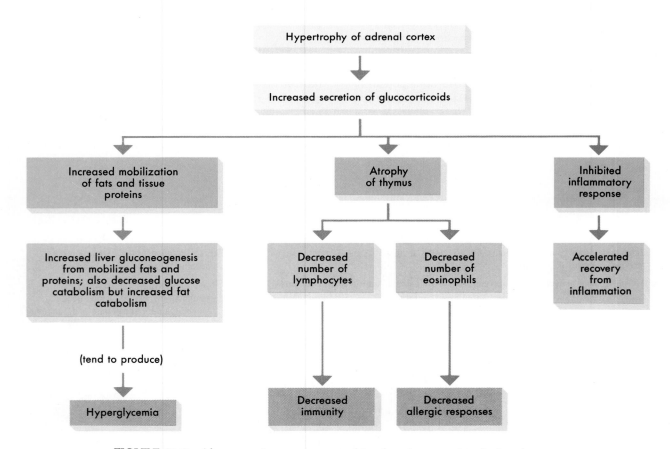

FIGURE 21-2 Alarm reaction responses resulting from hypertrophy of adrenal cortex.

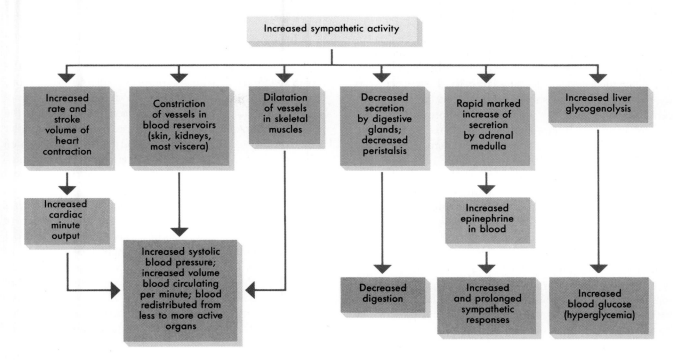

FIGURE 21-3 Alarm reaction responses resulting from increased sympathetic activity. Note that these are the responses commonly referred to as the "fight-or-flight" reaction.

corticoid secretion disappear during the stage of resistance. All of us go through the first and second stages of the stress syndrome many times in our lifetimes. Stressors of one kind or another act on most of us every day. They may upset or alarm us, but we soon resist them successfully. In short, we adapt; we cope.

The **stage of exhaustion** develops only when stress is extremely severe or when it continues over long periods of time. Otherwise when stress is mild and of short duration, it ends with a successful stage of resistance and adaptation to the stressor. If stress continues to the stage of exhaustion, corticoid secretion and adaptation eventually de-

crease markedly. The body can no longer cope successfully with the stressor and death may ensue. For a brief summary of the changes characteristic of the three stages of the general adaptation syndrome, see Table 21-1.

Mechanism of Stress

Stressors produce a state of stress. A state of stress, in turn, inaugurates a series of responses that Selye called the general adaptation syndrome. More simply, a state of stress turns on the stress response mechanism. It activates the organs that produce the responses that make up the general adaptation syndrome. But just how stress—a state of the

TABLE 21-1 The three stages of the general adaptation syndrome		
Alarm	**Resistance**	**Exhaustion**
Increased secretion of glucocorticoids and resultant changes (Figure 21-2)	Glucocorticoid secretion returns to normal	Increased glucocorticoid secretion but eventually marked decreased secretion
Increased activity of sympathetic nervous system	Sympathetic activity returns to normal	Stress triad (hypertrophied adrenals, atrophied thymus and lymph nodes, bleeding ulcers in stomach and duodenum)
Increased norepinephrine secretion by adrenal medulla	Norepinephrine secretion returns to normal	
"Fight-or-flight" syndrome of changes (Figure 21-3)	"Fight-or-flight" syndrome disappears	
Low resistance to stressors	High resistance (adaptation) to stressor	Loss of resistance to stressor; may lead to death

Stress and Aging

It has been postulated that stress and stress-related syndromes tend to augment age-related changes in the body. For example, researchers have shown that stress promotes anatomical and biochemical changes occurring at the neuromuscular junction to reduce available transmitter chemicals, such as acetylcholine, and decrease communication between nerve and muscle cells. The result is chronic fatigue and reduced physical strength.

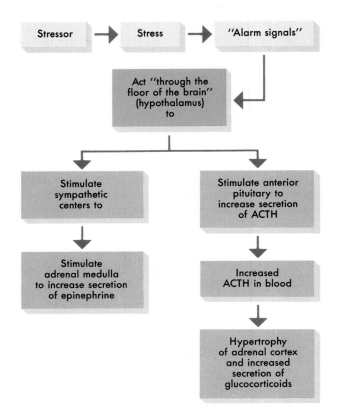

FIGURE 21-4 Selye's hypothesis about activation of the stress mechanism.

body—does this is not clear. Selye could only guess about it and his terms were vague. For instance, he postulated that, by some unknown "alarm signals," stress "acted through the floor of the brain" (presumably the hypothalamus) to stimulate the sympathetic nervous system and the pituitary gland. In Figure 21-4 you can see Selye's hypothesis in diagram form.

QUICK CHECK

✔ 1. *What is the general adaptation syndrome?*
✔ 2. *Identify the three stages of the general adaptation syndrome. How do they differ?*
✔ 3. *Discuss the types of responses produced in the body by increased sympathetic activity.*

Stress and Disease

Stress, as we have observed several times, produces different results in different individuals and in the same individual at different times. In one person, a certain amount of stress may induce responses that maintain or even enhance health. But in another person the same amount of stress appears to cause sickness. Whether stress is "good" or "bad" for you seems to depend more on your own body's responses to it than on the severity of the stressors inducing it. You may recall that Selye emphasized the adaptive nature of stress responses. He coined the term *general adaptation syndrome* because he believed that stress responses usually enable the body to adapt successfully to the many stressors that assail it. He held that the state of stress activates physiological mechanisms to meet the challenge imposed by stressors. But a challenge issued does not necessarily mean a challenge successfully met. Selye proposed that sometimes the body's adaptive mechanisms fail to meet the challenge issued by stressors and that, when they fail, disease results—diseases of adaptation, he called them.

Around the middle of this century, one of the problems studied was the relationship of blood glucocorticoid concentration to disease. If stress is adaptive, the investigators reasoned, and helps the body combat the effects of many kinds of stressors (for example, infection, injury, burns, and the like), then possibly various diseases might be treated by adding to the body's natural output of glucocorticoids. So often has it been used and so numerous have been the

Type "A" Behavior and Stress

High stress, hard-driving, and competitive "type A" individuals have been found to be at greater risk of elevated systolic blood pressure and coronary disease than individuals with the more relaxed and less impatient "type B" personality. Recent studies show that type A behavior patterns can be reduced by psychological guidance. Counseling of type A individuals helps them reduce stress and may cut in half their chances of suffering a heart attack because of stress.

articles written about cortisone that today it would be almost impossible to find an adult who has never heard of this hormone.

SOME CURRENT CONCEPTS ABOUT STRESS

Definitions

Some of today's physiologists use the terms *stressors* and *stress* as Selye did—that is, they define stressors as stimuli that produce stress, a state (or condition) of the body. In contrast, many physiologists now use the word "stress" to mean something much more specific than a state of the

body. **Stress,** according to their operational definition, is any stimulus that directly or indirectly stimulates neurons of the hypothalamus to release corticotropin-releasing hormone (CRH). CRH acts as a trigger that initiates many diverse changes in the body. Together, these changes constitute a syndrome now commonly called the **stress syndrome,** or, simply, the **stress response.**

Stress Syndrome

Look now at Figure 21-5. It summarizes some current major ideas about the syndrome of stress responses. Beginning at the top of the diagram, note that the initiator of the stress syndrome is stress—any factor that stimulates

the hypothalamus to release CRH. Most often, stress consists of injurious or extreme stimuli. These may act directly on the hypothalamus to stimulate it. Instead, or in addition, they may act indirectly on the hypothalamus. An example of stress that stimulates the hypothalamus directly is hypoglycemia. A lower than normal concentration of glucose in the blood circulating to the hypothalamus stimulates it to release CRH. Indirect stimulation of the hypothalamus occurs in this way: stress stimulates the limbic lobe—the so-called emotional brain—and other parts of the cerebral cortex, and these regions then send stimulating impulses to the hypothalamus, causing it to release CRH. In addition to releasing CRH, note in Figure

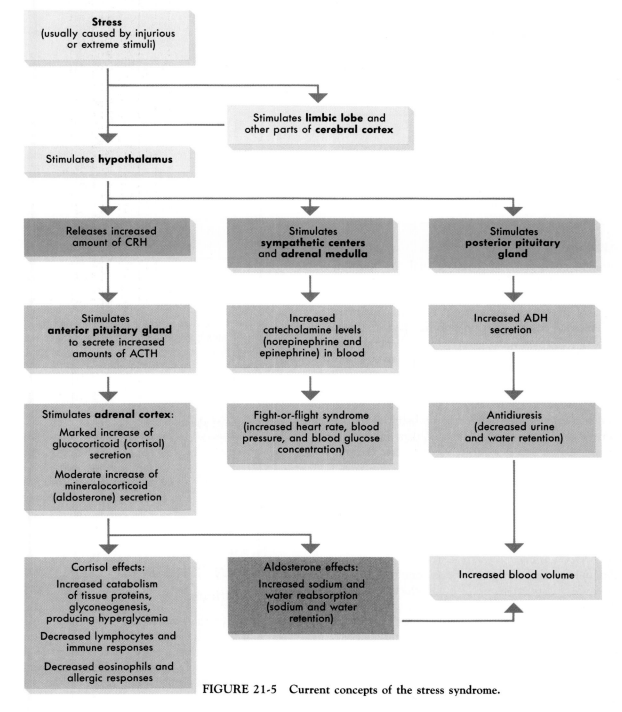

FIGURE 21-5 Current concepts of the stress syndrome.

21-5 that the stress-stimulated hypothalamus sends stimulating impulses to sympathetic centers and to the posterior pituitary gland.

CRH stimulates the anterior pituitary gland to secrete increased amounts of ACTH. ACTH stimulates the adrenal cortex to secrete greatly increased amounts of cortisol and more moderately increased amounts of aldosterone. These two hormones induce various stress responses. Some important ones worth remembering are listed in Figure 21-5 under cortisol effects and aldosterone effects.

Stimulation of sympathetic centers by impulses from the stress-stimulated hypothalamus leads to many stress responses, those known collectively as the **fight-or-flight reaction.** They include such important changes as an increase in the rate and strength of heartbeat, a rise in blood pressure, and hyperglycemia. Other sympathetic stress responses are pallor and coolness of the skin, sweaty palms, and dry mouth.

Water retention and an increase in blood volume are common stress responses. They stem, as you can see in Figure 21-5, from increased ADH and increased aldosterone secretion.

Indicators of Stress

Whether or not an individual's body is responding to stress stimuli can be determined by certain measurements and observations. Examples are the following: an increase in the rate and force of heartbeat, a rise in systolic blood pressure, an increase in blood and urine concentration of epinephrine and norepinephrine, sweating of the palms of the hands, and dilatation of pupils. The heart rate has been shown to increase in response to such varied stress stimuli as anesthesia and annoying sounds. Even anticipation by patients in a coronary care unit of their transferral to a less closely supervised convalescent unit has been identified as a stress stimulus that causes the heart rate to speed up. A decrease in the number of eosinophils and lymphocytes in the blood indicates that the individual is responding to stress stimuli. Soldiers stressed by prolonged marching, for example, have been found to have fewer circulating eosinophils than normal. This same stress indicator has been observed in college oarsmen when they were anticipating performing in an exhibition and in heart patients when they were anticipating the various types of procedures required to treat their illness.

The amount of urinary adrenocorticoids is often used as a measure of stress. It has been found to increase in depressed persons who feel hopeless and doomed, in test pilots, and in college students taking examinations or attending exciting movies. In contrast, urinary corticoids were found to drop markedly in persons watching unexciting nature study films.

The level of adrenocorticoids in the blood plasma of disturbed patients having acute psychotic episodes has been found to be 70% higher than that in normal individuals or in calm patients. Another study showed that the plasma corticoid levels of chronically depressed patients were significantly lower than those of acutely anxious patients. Smoking and exposure to nicotine have also been shown to be stressors that caused a marked rise in plasma adrenocorticoids—by as much as 77% in human beings and in experimental animals.

Corticoids and Resistance to Stress

Selye thought that the increase in corticoids that occurred in his stressed animals enabled them to adapt to and to resist stress. Today many physiologists doubt this. No one questions that adrenocortical hormones increase during stress. That fact has been clearly established. But what many do question is how essential this increase is for resisting stress. No one has proved by an unequivocal experiment that a higher than normal blood level of corticoids increases an animal's ability to adapt to stress. Some clinical evidence, however, seems to indicate that it does increase a human's coping ability. For instance, patients who have been taking cortisol for some time are known to require increased doses of this hormone to successfully resist stresses such as surgery or severe injury.

Psychological Stress

Stress as defined by Selye is physiological stress, that is, a state of the body. Psychological stress, in contrast, might be defined as a state of the mind. It is caused by psychological stressors and manifested by a syndrome. A **psychological stressor** is anything that an individual perceives as a threat—a threat to survival or to self-image. Moreover, the threat does not need to be real—it needs only to be real to the individual, who must see it as a threat, although in truth, it may not be. Psychological stressors produce a syndrome of subjective and objective responses. Dominant among the subjective reactions is a feeling of anxiety. Other emotional reactions such as anger, hate, depression, fear, and guilt are also common subjective responses to psychological stressors. Some characteristic objective responses are restlessness, fidgeting, criticizing, quarreling, lying, and crying. Another objective indicator of psychological stress is that the concentration of lactate in the blood increases.

Does psychological stress relate to physiological stress? The answer is clearly "yes." Physiological stress usually is accompanied by some degree of psychological stress. And, conversely, in most people, psychological stress produces some physiological stress responses. Ancient peoples intuitively recognized this fact. For example, it is said that in ancient times when the Chinese suspected a person of lying, they made him chew rice powder and then spit it out. If the powder came out dry, not moistened by saliva, they judged the suspect guilty. They seemed to know that lying makes a person "nervous"—and that nervousness makes a person's mouth dry. We "moderns" also know these facts but we describe them with more technical language. Lying, we might say, induces psychological stress, and psychological stress acts in some way to cause the physiological stress response of decreased salivation.

Within recent years, a scientific discipline called *psycho-*

physiology has come into being. Psychophysiologists, using accepted research methods and sophisticated instruments—including polygraphs designed especially for this type of research—have investigated various physiological responses made by individuals subjected to psychological stressors. Their findings amply confirm the principle that psychological stressors often produce physiological stress responses. They have found, however, that identical psychological stressors do not necessarily induce identical physiological responses in different individuals. For example, stress may result in heart rate and blood pressure changes in one individual and changes in breathing patterns in another. Another of their interesting discoveries is that some organ systems become less responsive after they have been stimulated a number of times.

In summary, here are some principles to remember about psychological stress.

♦ Physiological stress almost always is accompanied by some degree of psychological stress.
♦ In most people, psychological stress leads to some physiological stress responses. Many of these are measurable autonomic responses, for example, accelerated heart rate and increased systolic blood pressure.
♦ Identical psychological stressors do not always induce identical physiological responses in different individuals.
♦ In any one individual, certain autonomic responses are better indicators of psychological stress than others.

QUICK CHECK

✔ 1. *What is meant by the phrase "diseases of adaptation?"*
✔ 2. *What is the "fight-or-flight" reaction?*
✔ 3. *List four indicators of stress.*
✔ 4. *What is the difference between physiological and psychological stress?*

THE BIG PICTURE
Stress and the Whole Body

Although our understanding of physiological and psychological stress is still evolving, we are certain of this: stress affects the entire body. Stress responses involve a large number of physiological mechanisms, many of which are suspected to occur but are as yet unproven.

So far, we understand that stress responses involve nearly every system of the body. The nervous system detects and integrates the factors, or stressors, that trigger the stress responses. Physiological stress responses result from signals sent from the nervous system directly—or by way of the endocrine system. Many of these "stress signals" have only recently been discovered. Many different kinds of neurotransmitters, hormones, and perhaps other regulatory chemicals influence the function of the skeletal muscles, the digestive system, the urinary system, the reproductive system, the respiratory system, the cardiovascular system, the integumentary system—perhaps every system, organ, and tissue in the body.

Because the regulatory agents associated with stress influence the function of blood cells, stress can have a profound effect on the function of the immune system. Although the fact that stress inhibits immune function has been known for quite some time, many of the exact mechanisms that accomplish this have only recently been discovered. Biologists now better appreciate the link between the mind and the immune system and thus are better able to explain how stress causes disease—and even death. In fact, there is an emerging field within human biology devoted to studying the mind-immunity link—the field of neuroimmunology. Some researchers in this new field have found themselves overlapping such diverse fields as endocrinology, psychology, and hematology.

Considering the effect that stress can have on the entire internal environment—the whole body—and the high levels of stress that characterize the modern cultures in which many of us live and work, advancements in stress research hold the promise of improving the length and quality of our lives.

CHAPTER SUMMARY

SELYE'S CONCEPT OF STRESS

A. Development of the concept
  1. Through many experiments, Selye exposed animals to noxious agents and found that they all responded with the same syndrome of changes, or "stress triad."

B. Definitions
  1. Stress—a state, or condition, of the body produced by "diverse nocuous agents" and manifested by a syndrome of changes
  2. Stressors—agents that produce stress
  3. General adaptation syndrome—group of changes that manifest the presence of stress

C. Stressors
  1. Stressors are extreme stimuli—too much or too little of almost anything
  2. Stressors are very often injurious or painful stimuli
  3. Anything an individual perceives as a threat is a stressor for that individual
  4. Stressors are different for different individuals and for one individual at different times

D. The general adaptation syndrome
   1. Manifestations—stress triad (hypertrophied adrenals, atrophied thymus and lymph nodes, and bleeding ulcers); many other changes
   2. Stages—three successive phases, namely, alarm reaction, stage of adaptation or resistance, and stage of exhaustion, each characterized by a different syndrome of changes (Table 21-1)
E. Mechanism of stress
   1. Consists of group (syndrome) of responses to internal condition of stress; stress responses nonspecific in that same syndrome of responses occurs regardless of kind of extreme change that produced stress
   2. Stimulus that produces stress and thereby activates stress mechanism is nonspecific in that it can be any kind of extreme change in environment
   3. Stress responses, Selye thought, were adaptive, tend to enable body to adapt to and survive extreme change; Selye referred to this syndrome of stress responses as general adaptation syndrome
   4. Numerous factors influence stress responses—individual's physical and mental condition, age, sex, socioeconomic status, heredity, and previous experience with similar stressors
   5. Stress most often is coped with successfully—namely, results in adaptation, healthy survival, and increased resistance; sometimes, however, stress produces exhaustion and death
   6. See Figure 21-4, then 21-2 and 21-3, for a summary of Selye's stress mechanism
F. Stress and disease
   1. Selye held that stress could result in disease instead of adaptation

SOME CURRENT CONCEPTS ABOUT STRESS

A. Definitions
   1. Stress—any stimulus that directly or indirectly stimulates hypothalamus to release CRH
   2. Stress syndrome—also called the stress response; many diverse changes initiated by stress
B. The stress syndrome—(see Figure 21-5)
C. Indicators of stress
   1. Changes caused by increased sympathetic activity; for example, faster, stronger heartbeat; higher blood pressure; sweaty palms; dilated pupils
   2. Changes resulting from increased corticoids—eosinopenia, lymphocytopenia, increased adrenocorticoids in blood and urine
D. Corticoids and resistance to stress
   1. Still controversial; not proved that increased corticoids increase an animal's or man's ability to resist stress but is proved that corticoid levels in blood increase in stress
E. Psychological stress
   1. Psychological stressors—anything that an individual perceives as a threat to survival or self-image
   2. Psychological stress—a mental state characterized by a syndrome of subjective and objective responses; dominant subjective response is anxiety; some characteristic objective responses are restlessness, quarrelsomeness, lying, crying
   3. Relation to physiological stress (see summary of principles on p. 574)

THE BIG PICTURE: STRESS AND THE WHOLE BODY

A. Stress affects the entire body
B. Nervous system detects and integrates stressors that trigger stress responses
C. Stress affects immune system
   1. Neuroimmunology studies mind-immunity link

REVIEW QUESTIONS

1. Describe the experimental results that led Selye to conceive his idea of stress.
2. Define the terms *stress*, *stressor*, and *general adaptation syndrome*.
3. Make a few generalizations about the kinds of stimuli that constitute stressors.
4. What three stages make up the general adaptation syndrome? What changes characterize each stage?
5. What changes constituted the "stress triad"?
6. An increase in what three hormones brought about the changes that he named the general adaptation syndrome?
7. What function, according to Selye, does the general adaptation syndrome serve?
8. What relation, if any, exists between stress and disease?
9. Stress, according to Selye, is a state, or condition, of the body. According to a current operational definition, what is stress?
10. What part of the brain, according to current ideas, plays the key role in initiating stress syndrome responses?

11. What parts of the nervous system other than that named in question 10 are involved in inducing the stress syndrome?
12. Briefly, what role, if any, do each of the following hormones play when the body is subjected to stress: ACTH, ADH, aldosterone, cortisol, CRH, epinephrine, and norepinephrine?
13. Describe some changes you might observe that would indicate an individual was being subjected to stress.
14. What has been proved about corticoids in relation to stress?
15. What is the controversial issue about corticoids in relation to stress?
16. Define psychological stressor and psychological stress.
17. Cite several examples of psychological stressors.
18. Are psychological stressors the same for all individuals?
19. Give some examples of subjective indicators of psychological stress.
20. Give some examples of objective responses that are part of the syndrome of psychological stress.
21. State three of four principles about the relationship between psychological stress and physiological stress.

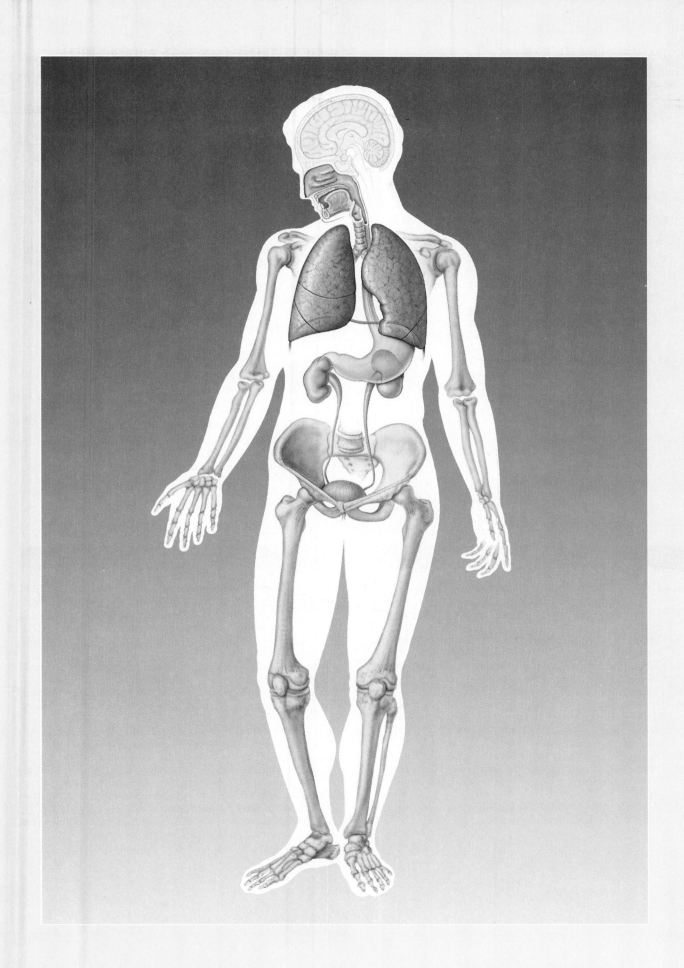

UNIT FIVE

Respiration, Nutrition, and Excretion

The chapters of Unit 5 deal with respiration, digestion, the processing of nutrients, and the excretion of wastes by the urinary system. The unit ends with chapters discussing both fluid and electrolyte and acid-base balance.

Ultimately, all homeostatic mechanisms function to maintain constancy at the cellular level. Delivery of oxygen and the elimination of carbon dioxide and other wastes resulting from the metabolism of nutrients must be regulated within narrow limits so that cellular function remains normal. Maintaining the constancy of fluid and electrolyte and acid-base balance at the cellular level is required for the myriad metabolic reactions that are required for survival. The anatomic structures and functional control mechanisms discussed in this unit all relate, in the last analysis, to cellular homeostasis.

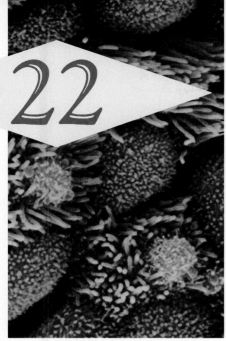

22

Anatomy of the Respiratory System

SEM of epithelium in bronchus

CHAPTER OUTLINE

OBJECTIVES

After you have completed this chapter, you should be able to:

1. List and locate the organs of the respiratory system.
2. List the generalized functions of the respiratory system.
3. Describe and correlate the anatomy of the nose with its specialized functions.
4. Locate the paranasal sinuses in the skull and describe how they drain into the nose.
5. List the anatomical divisions of the pharynx and name the openings into and between its divisions.
6. Identify and locate the tonsils.
7. Discuss the location, structure, and specialized functions of the larynx.
8. Describe the structure and function of the trachea, bronchi, bronchioles, and alveoli.
9. Identify the lobes of the lungs and the bronchopulmonary segments.
10. Discuss the gross surface anatomy and generalized functions of the lungs.
11. Discuss the structure and function of the thorax and mediastinum in respiration.

OVERVIEW OF THE RESPIRATORY SYSTEM

Functions of the Respiratory System

The respiratory system functions as an air distributor and a gas exchanger so that oxygen may be supplied to and carbon dioxide be removed from the body's cells. Since most of our billions of cells lie too far from air to exchange gases directly with it, air must first exchange gases with blood, blood must circulate, and, finally, blood and cells must exchange gases. These events require the functioning of two systems, namely, the respiratory system and the circulatory system. All parts of the respiratory system—except its microscopic-sized sacs called *alveoli*—function as air distributors. Only the alveoli and the tiny passageways that open into them serve as gas exchangers.

In addition to air distribution and gas exchange, the respiratory system effectively filters, warms, and humidifies the air we breathe. Respiratory organs also influence sound production, including speech used in communicating oral language. Specialized epithelium in the respiratory tract make the sense of smell (olfaction) possible. The respiratory system also assists in the regulation, or homeostasis, of pH in the body.

Structural Plan of the Respiratory System

For purposes of study, the respiratory system may be divided into upper and lower tracts, or divisions. The organs of the upper respiratory tract are located outside of the thorax, or chest cavity, whereas those in the lower tract, or division, are located almost entirely within it (Figure 22-1).

The **upper respiratory tract** is composed of the nose, nasopharynx, oropharynx, laryngopharynx, and larynx. The **lower respiratory tract,** or division, consists of the trachea, all segments of the bronchial tree, and the lungs. Functionally, the respiratory system also includes a number of accessory structures, including the oral cavity, rib cage, and diaphragm. Together these structures constitute the lifeline, the air supply line of the body. This chapter describes the functional anatomy of these organs. The physiology of the respiratory system as a whole will be discussed in Chapter 23.

Cells require a constant supply of oxygen for the vital energy-conversion process carried out within each cell's mitochondria—a process called cellular respiration (Chapter 26). Cellular respiration produces carbon dioxide (CO_2) as a waste product that must be removed before it accumulates to dangerously high levels.

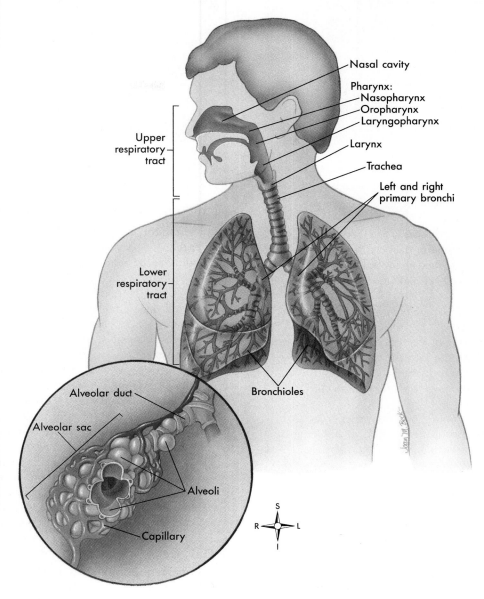

FIGURE 22-1 Structural plan of the respiratory system. The inset shows the alveolar sacs where the interchange of oxygen and carbon dioxide takes place through the walls of the grapelike alveoli. Capillaries surround the alveoli.

Upper Respiratory Tract

Nose

Structure of the nose

The nose consists of an external and an internal portion. The external portion, that is, the part that protrudes from the face, consists of a bony and cartilaginous framework overlaid by skin containing many sebaceous glands. The two nasal bones meet above where they are surrounded by the frontal bone to form the root of the nose. The nose is surrounded by the maxilla laterally and inferiorly at its base. The flaring cartilaginous expansion forming and supporting the outer side of each oval nostril opening is called the *ala*.

The internal nose, or nasal cavity, lies over the roof of the mouth where the palatine bones, which form the floor of the nose and the roof of the mouth, separate the nasal cavities from the mouth cavity. Sometimes the palatine bones fail to unite completely, producing a condition known as **cleft palate.** When this abnormality exists, the mouth is only partially separated from the nasal cavity, and difficulties arise in swallowing and speaking.

The roof of the nose is separated from the cranial cavity by a portion of the ethmoid bone called the **cribriform plate** (Figures 22-2 and 22-3). The cribriform plate is perforated by many small openings that permit branches of

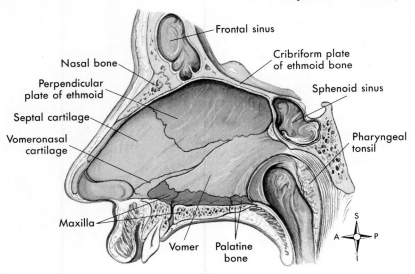

FIGURE 22-2   Nasal septum. The nasal septum consists of the perpendicular plate of the ethmoid bone, the vomer, and the septal and vomeronasal cartilages.

the olfactory nerve responsible for the special sense of smell to enter the cranial cavity and reach the brain.

Separation of the nasal and cranial cavities by a thin, perforated plate of bone presents real hazards. If the cribriform plate is damaged as a result of trauma to the nose, it is possible for potentially infectious material to pass directly from the nasal cavity into the cranial fossa and surround the brain.

The hollow nasal cavity is separated by a midline partition, the **septum** (see Figure 22-2), into a right and a left cavity. Note in Figure 22-2 that the nasal septum is made up of four main structures: the perpendicular plate of the ethmoid bone above, the vomer bone, and the

vomeronasal and septal nasal cartilages below. In the adult the nasal septum is frequently deviated to one side or the other, interfering with respiration and with drainage of the nose and sinuses. The nasal septum has a rich blood supply. Nosebleeds, or *epistaxis*, often occur as a result of septal contusions caused by a direct blow to the nose.

Each nasal cavity is divided into three passageways (superior, middle, and inferior meati) by the projection of the **turbinates** or **conchae,** from the lateral walls of the internal portion of the nose (see Figure 22-3). The superior and middle turbinates are processes of the ethmoid bone, whereas the inferior turbinates are separate bones.

The external openings into the nasal cavities (nostrils)

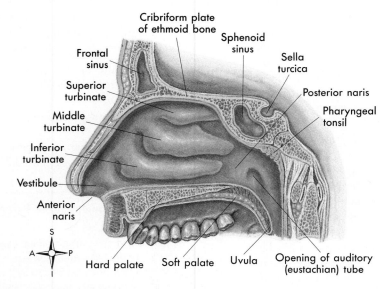

FIGURE 22-3   Nasal cavity. In this midsagittal section through the nose and nasal cavity, the nasal septum has been removed to reveal the turbinates (nasal conchae) of the lateral wall of the nasal cavity.

have the technical name of *anterior nares* (singular, naris). They open into an area covered by skin that is reflected from the wings (ala) of the nose. This area, called the **vestibule,** is located just inside the nasal cavity below the inferior meatus. Coarse hairs called **vibrissae,** sebaceous glands, and numerous sweat glands are found in the skin of the vestibule. Once air has passed over the skin of the vestibule, it enters the **respiratory portion** of each nasal passage. This area extends from the inferior meatus to the small funnel-shaped orifices of the *posterior nares*. The posterior nares are openings that allow air to pass from the nasal cavity into the next major segment of the upper respiratory tract—the pharynx.

If one were to "trace" the movement of air through the nose into the pharynx, it would pass through several structures on the way. The sequence is as follows:

1. Anterior nares (nostrils)
2. Vestibule
3. Inferior, middle, and superior meati, simultaneously
4. Posterior nares

Nasal mucosa. Once air has passed over the skin of the vestibule and enters the respiratory portion of the nasal passage, it passes over the highly specialized **respiratory mucosa.** This mucous membrane has a pseudostratified ciliated columnar epithelium rich in goblet cells (Figure 22-4). The respiratory mucosa possesses a rich blood supply, especially over the inferior turbinate, and is bright pink or red in color. Near the roof of the nasal cavity and over the superior turbinate and opposing portion of the septum, the mucosa turns pale and has a yellowish tint. In this area it is referred to as the **olfactory epithelium.** This specialized membrane contains many olfactory nerve cells and has a rich lymphatic plexus. Ciliated mucous membrane lines the rest of the respiratory tract down as far as the smaller bronchioles.

Paranasal sinuses. The four pairs of paranasal sinuses are air-containing spaces that open, or drain, into the nasal cavity and take their names from the skull bones in which they are located (see Chapter 7). These paranasal sinuses are the frontal, maxillary, ethmoid, and sphenoid (Figure 22-5). Like the nasal cavity, each paranasal sinus is lined by respiratory mucosa. The mucous secretions produced in the sinuses are continually being swept into the nose by the ciliated surface of the respiratory membrane.

The right and left frontal sinuses are located just above the corresponding orbit, whereas the maxillary, the largest

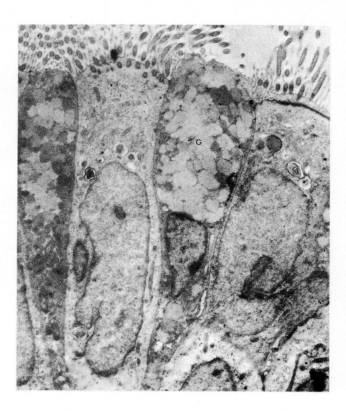

FIGURE 22-4 Electron micrograph of respiratory mucosa. This epithelium is typically ciliated and exhibits numerous goblet cells, G, filled with mucus. (x7,600.)

of the sinuses, extends into the maxilla on either side of the nose. The sphenoid sinuses lie in the body of the sphenoid bone on either side of the midline in very close proximity to the optic nerves and pituitary gland.

Note in Figure 22-5 that the ethmoid sinuses are not single large cavities but a collection of small air cells divided into anterior, middle, and posterior groups that open independently into the upper part of the nasal cavity. The paranasal sinuses drain as follows:

♦ Into the middle meatus (passageway below middle turbinate)—frontal, maxillary, anterior, and middle ethmoidal sinuses
♦ Into the superior meatus—posterior ethmoidal sinuses
♦ Into the space above the superior turbinates (sphenoethmoidal recess)—sphenoidal sinuses

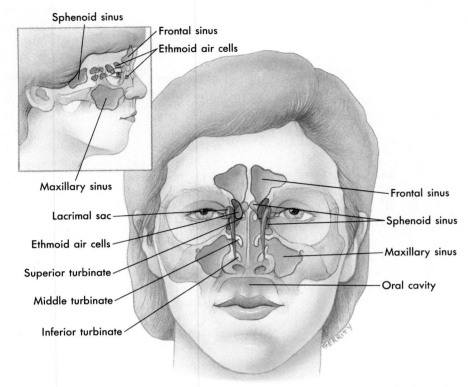

FIGURE 22-5   The paranasal sinuses. The anterior view shows the anatomical relationship of the paranasal sinuses to each other and to the nasal cavity. The inset is a lateral view of the position of the sinuses.

Functions of the nose

The nose serves as a passageway for air going to and from the lungs. However, if the nasal passages are obstructed, it is possible for air to bypass the nose and enter the respiratory tract directly through the mouth. Air that enters the system through the nasal cavity is filtered of impurities, warmed, moistened, and chemically examined for substances that might prove irritating to the delicate lining of the respiratory tract. The vibrissae, or hairs, in the vestibule serve as an initial "filter" to screen particulate matter from air that is entering the system. The turbinates, or conchae, then serve as baffles to provide a large mucus-covered surface area over which air must pass before reaching the pharynx. The respiratory membrane produces copious quantities of mucus and possesses a rich blood supply, especially over the inferior turbinates, which permits rapid warming and moistening of the dry inspired air. Mucous secretions provide the final "trap" for removal of remaining particulate matter from air as it moves through the nasal passages. Fluid from the lacrimal ducts (see Figure 22-5) and additional mucus produced in the paranasal sinuses also help to trap particulate matter and moisten air passing through the nose. In addition, the hollow sinuses act to lighten the bones of the skull and serve as resonating chambers for speech. Deflection of air by the middle and superior concha over the olfactory epithelium makes the special sense of smell possible.

QUICK CHECK

✔ 1. *What are the overall functions of the respiratory system? Which other body system is involved in accomplishing these functions?*

✔ 2. *What are the principal organs of the upper respiratory tract. What are the principal organs of the lower respiratory tract?*

✔ 3. *What are the paranasal sinuses? What is their anatomical relationship to the nose?*

Pharynx
Structure of the pharynx

Another name for the pharynx is the throat. It is a tubelike structure about 12.5 cm (5 inches) long that extends from the base of the skull to the esophagus and lies just anterior to the cervical vertebrae. It is made of muscle, is lined with mucous membrane, and has three anatomical divisions: the **nasopharynx,** located behind the nose and extending from the posterior nares to the level of the soft palate; the **oropharynx,** located behind the mouth from the soft palate above to the level of the hyoid bone below; and the **laryngopharynx,** which extends from the hyoid bone to its termination in the esophagus. Figure 22-6 shows the divisions of the pharynx.

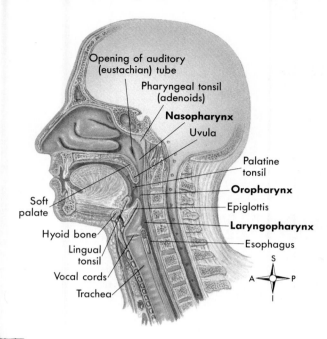

FIGURE 22-6 **Pharynx.** This midsagittal section shows the three divisions of the pharynx (nasopharynx, oropharynx, and laryngopharynx) and nearby structures.

Seven openings are found in the pharynx (see Figures 22-3 and 22-6).

♦ Right and left auditory (eustachian) tubes opening into the nasopharynx
♦ Two posterior nares into the nasopharynx
♦ The opening from the mouth, known as the *fauces,* into the oropharynx
♦ The opening into the larynx from the laryngopharynx
♦ The opening into the esophagus from the laryngopharynx

The **pharyngeal tonsils** are located in the nasopharynx on its posterior wall opposite the posterior nares. The pharyngeal tonsils are called *adenoids* when they are enlarged. Although the cavity of the nasopharynx differs from the oral and laryngeal divisions in that it does not collapse, it may become obstructed. If these tonsils enlarge to become adenoids, they may fill the space behind the posterior nares and make it difficult or impossible for air to travel from the nose into the throat. When this happens, the individual keeps the mouth open to breathe and is described as having an "adenoidy" appearance.

Two pairs of organs are found in the oropharynx: the **palatine tonsils,** located behind and below the pillars of the fauces, and the **lingual tonsils,** located at the base of the tongue. The palatine tonsils are the ones most commonly removed by a tonsillectomy. Only rarely are the lingual tonsils also removed.

Functions of the pharynx

The pharynx serves as a common pathway for the respiratory and digestive tracts, since both air and food must pass through this structure before reaching the appropriate tubes. It also affects phonation (speech production). For example, only by the pharynx changing its shape can the different vowel sounds be formed.

Larynx
Location of the larynx

The **larynx,** or voice box, lies between the root of the tongue and the upper end of the trachea just below and in front of the lowest part of the pharynx (see Figure 22-1). It might be described as a vestibule opening into the trachea from the pharynx. It normally extends between the third, fourth, fifth, and sixth cervical vertebrae but is often somewhat higher in females and during childhood. The lateral lobes of the thyroid gland and the carotid artery in its covering sheath touch the sides of the larynx.

Structure of the larynx

The triangular-shaped larynx consists largely of cartilages that are attached to one another and to surrounding structures by muscles or by fibrous and elastic tissue components (Figure 22-7). It is lined by a ciliated mucous membrane. The cavity of the larynx extends from its triangular-shaped inlet at the epiglottis to the circular outlet at the lower border of the cricoid cartilage where it is continuous with the lumen of the trachea (Figure 22-8). The mucous membrane lining the larynx forms two pairs of folds that jut inward into its cavity and divide it into three compartments, or divisions. The upper pair is called the **vestibular,** or **false, vocal folds** for the rather obvious reason that they play no part in vocalization. The lower pair serves as the **true vocal cords.** The slitlike space between the true vocal cords, called the *rima glottidis,* is the narrowest part of the larynx. The true vocal cords and the space between them (rima glottidis) are together designated as the **glottis.** Endoscopic views of the vocal cords and related structures are shown in Figure 22-9.

The division, or compartment, of the laryngeal cavity above the false, or vestibular, vocal folds is called the **vestibule.** The very short middle portion of the cavity between the false and true vocal cords is the ventricular division, or **ventricle.** The lower compartment extending from the true vocal cords to the outlet is referred to as the *infraglottic larynx* (see Figure 22-8).

Cartilages of the larynx

Nine cartilages form the framework of the larynx. The three largest of these—the thyroid cartilage, the epiglottis, and the cricoid cartilage—are single structures. There are three pairs of smaller accessory cartilages, namely, the arytenoid, corniculate, and cuneiform cartilages.

Single laryngeal cartilages

♦ The **thyroid cartilage** (Adam's apple) is the largest cartilage of the larynx and is the one that gives the characteristic triangular shape to its anterior wall. It is usually larger in men than in women and has less of a fat pad lying over it—two reasons why a man's thyroid cartilage protrudes more than a woman's.

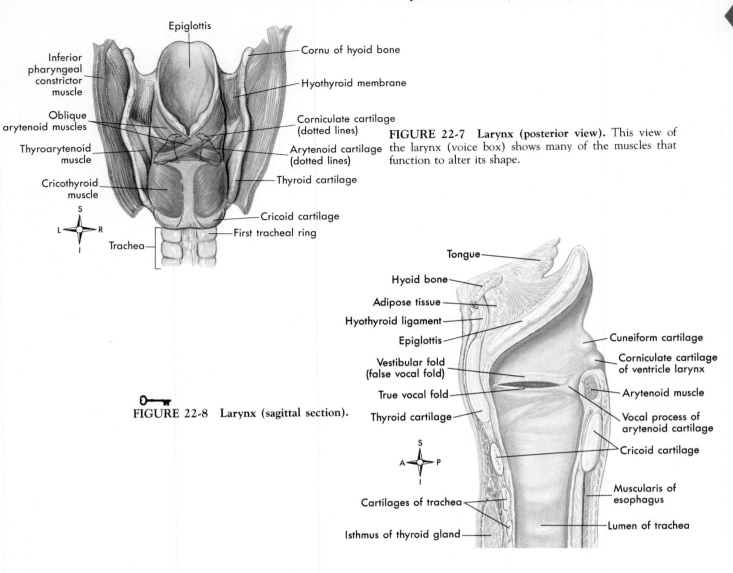

Epiglottis

Inferior pharyngeal constrictor muscle

Oblique arytenoid muscles

Thyroarytenoid muscle

Cricothyroid muscle

Trachea

Cornu of hyoid bone

Hyothyroid membrane

Corniculate cartilage (dotted lines)

Arytenoid cartilage (dotted lines)

Thyroid cartilage

Cricoid cartilage

First tracheal ring

FIGURE 22-7 Larynx (posterior view). This view of the larynx (voice box) shows many of the muscles that function to alter its shape.

FIGURE 22-8 Larynx (sagittal section).

Tongue

Hyoid bone

Adipose tissue

Hyothyroid ligament

Epiglottis

Vestibular fold (false vocal fold)

True vocal fold

Thyroid cartilage

Cartilages of trachea

Isthmus of thyroid gland

Cuneiform cartilage

Corniculate cartilage of ventricle larynx

Arytenoid muscle

Vocal process of arytenoid cartilage

Cricoid cartilage

Muscularis of esophagus

Lumen of trachea

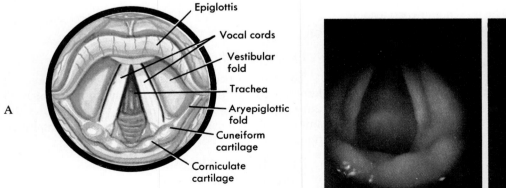

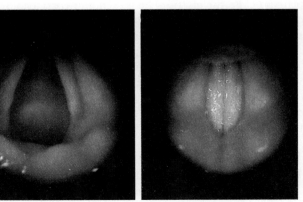

Epiglottis

Vocal cords

Vestibular fold

Trachea

Aryepiglottic fold

Cuneiform cartilage

Corniculate cartilage

A

B

Vocal cords open

Vocal cords closed

FIGURE 22-9 Vocal cords. A, This sketch shows how the vocal cords within the larynx can be viewed with an optic device called an endoscope. **B,** Photographs taken with an endoscope show the vocal cords in the open position *(left)* and the closed position *(right).*

♦ The **epiglottis** is a small leaf-shaped cartilage that projects upward behind the tongue and hyoid bone. It is attached below to the thyroid cartilage, but its free superior border can move up and down during swallowing to prevent food or liquids from entering the trachea (see Figures 22-7 and 22-8).

♦ The **cricoid,** or signet ring, **cartilage,** so called because its shape resembles a signet ring (turned so the signet forms part of the posterior wall of the larynx), is the most inferiorly placed of the nine cartilages.

Paired laryngeal cartilages

♦ The pyramid-shaped **arytenoid** cartilages are the most important of the paired laryngeal cartilages. The base of each cartilage articulates with the superior border of the cricoid cartilage (see Figure 22-7). The anterior angles of these cartilages serve as points of attachment for the vocal cords.

♦ The nodular **corniculate** cartilages are small and conical in shape. Note in Figure 22-7 that they rest on the apex of each arytenoid cartilage.

♦ The two small **cuneiform** cartilages are rod-shaped structures located near the base of the epiglottis. They are closely related to the arytenoid cartilages.

Muscles of the larynx

Muscles of the larynx are often divided into intrinsic and extrinsic groups. Intrinsic muscles have both their origin and insertion on the larynx. They are important in controlling vocal cord length and tension and in regulating the shape of the laryngeal inlet. Extrinsic muscles insert on the larynx but have their origin on some other structure—such as the hyoid bone. Therefore contraction of the extrinsic muscles actually moves or displaces the larynx as a whole. Muscles in both groups play important roles in respiration, vocalization, and swallowing. During swallowing, for example, contraction of the intrinsic aryepiglottic muscles (those that connect the arytenoid cartilages with the epiglottis) prevents "swallowing down the wrong throat" by squeezing the laryngeal inlet shut.

Two other pairs of intrinsic laryngeal muscles function to open and close the glottis. The posterior cricoarytenoid muscles (between cricoid and arytenoid cartilages) open the glottis by adducting the true vocal cords. The lateral cricoarytenoid muscles close the glottis by adducting the true vocal cords. These events are crucial to both respiration and voice production. Certain other intrinsic muscles of the larynx function to influence the pitch of the voice by either lengthening and tensing or shortening and relaxing the vocal cords.

Functions of the larynx

The larynx functions in respiration, since it constitutes part of the vital airway to the lungs. This unique passageway, like the other components of the upper respiratory tract, is lined with a ciliated mucous membrane that helps in the removal of dust particles and in the warming and humidification of inspired air. In addition, it protects the airway against the entrance of solids or liquids

during swallowing. It also serves as the organ of voice production—hence its popular name, the voice box. Air being expired through the glottis, narrowed by partial adduction of the true vocal cords, causes them to vibrate. Their vibration produces the voice. Several other structures besides the larynx contribute to the sound of the voice by acting as sounding boards or resonating chambers. Thus the size and shape of the nose, mouth, pharynx, and bony sinuses help determine the quality of the voice.

QUICK CHECK

✔ 1. *What are the three main divisions of the pharynx?*
✔ 2. *Can you describe where the tonsils are located?*
✔ 3. *Distinguish between true and false vocal cords.*

LOWER RESPIRATORY TRACT

Trachea
Structure of the trachea

The trachea, or windpipe, is a tube about 11 cm (4.5 inches) long that extends from the larynx in the neck to the primary bronchi in the thoracic cavity (Figure 22-10). Its diameter measures about 2.5 cm (1 inch). Smooth muscle, in which are imbedded **C**-shaped rings of cartilage at regular intervals, fashions the wall of the trachea (Figure 22-11). The cartilaginous rings are incomplete on the posterior surface (see Figure 22-11). They give firmness to the wall, tending to prevent it from collapsing and shutting off the vital airway.

Note in Figure 22-12 that the trachea is lined with the type of pseudostratified ciliated columnar epithelium typical of the respiratory tract as a whole.

Function of the trachea

The trachea performs a simple but vital function—it furnishes part of the open passageway through which air can reach the lungs from the outside. Obstruction of this airway for even a few minutes causes death from asphyxiation.

Bronchi and Alveoli
Structure of the bronchi

The trachea divides at its lower end into two **primary bronchi,** of which the right bronchus is slightly larger and more vertical than the left. This anatomical fact explains why aspirated foreign objects frequently lodge in the right bronchus. In structure the bronchi resemble the trachea. Their walls contain incomplete cartilaginous rings before

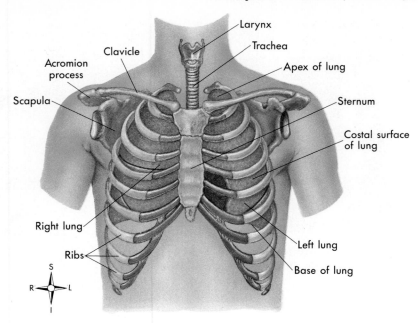

FIGURE 22-10   Lungs and trachea in relation to the rib cage and clavicles. Note that the apex of each lung projects above the clavicle.

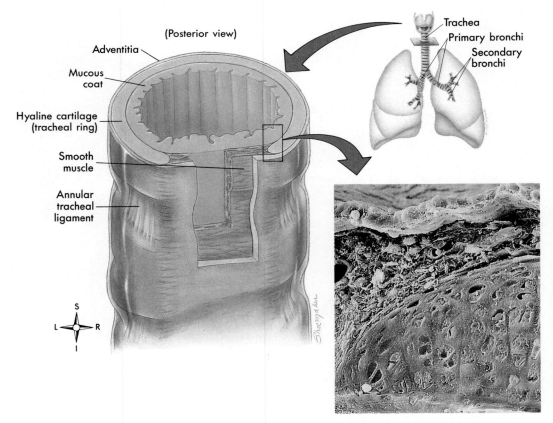

FIGURE 22-11   Cross section of the trachea. The inset at the top shows where the section was cut. The scanning electron micrograph shows details of the mucous coat, the tip of a cartilage ring, and the adventitia that form the wall of the trachea. (x300.)

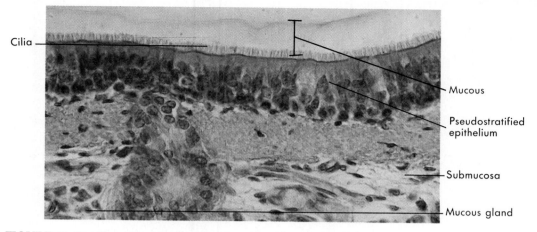

Cilia

Mucous

Pseudostratified epithelium

Submucosa

Mucous gland

FIGURE 22-12 **Transverse section of trachea.** Note the mucosa of ciliated epithelium. Hyaline cartilage occurs below the glandular submucosa and is not visible in this section. (x70.)

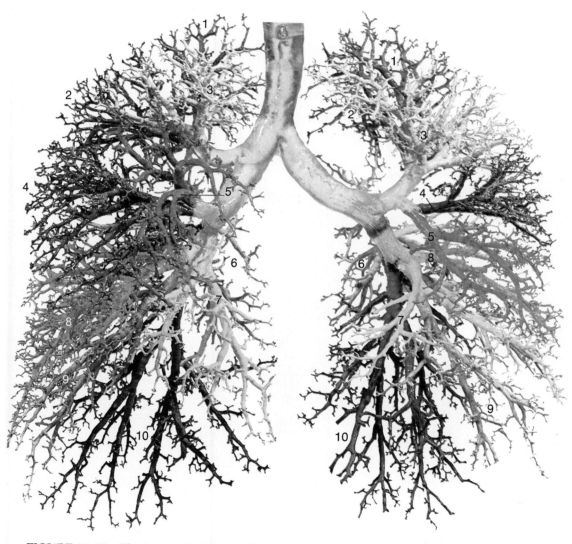

FIGURE 22-13 **Plastic cast of air spaces of the lungs.** The cast was prepared by pouring liquid plastic into the airways of a human lung—a different color for each bronchopulmonary segment supplied by its own tertiary bronchus. After the plastic hardened, the soft tissue was removed, leaving the branched form of the lower respiratory tract that is pictured here (compare this photograph to Figure 22-18).

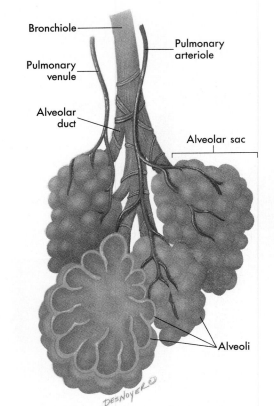

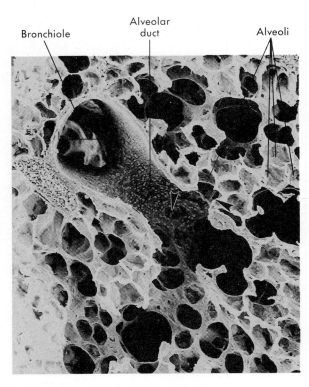

FIGURE 22-14   Alveoli. A, Bronchioles subdivide to form tiny tubes called *alveolar ducts,* which end in clusters of alveoli called *alveolar sacs.* **B,** Scanning electron micrograph of a bronchiole, alveolar ducts, and surrounding alveoli. Arrowheads indicate opening of alveoli into the alveolar duct.

the bronchi enter the lungs, but they become complete within the lungs. Ciliated mucosa lines the bronchi, as it does the trachea.

Each primary bronchus enters the lung on its respective side and immediately divides into smaller branches called **secondary bronchi.** The secondary bronchi continue to branch, forming tertiary bronchi and small **bronchioles.** The trachea and the two primary bronchi and their many branches resemble an inverted tree trunk with its branches and are therefore spoken of as the *bronchial tree* (Figure 22-13). The bronchioles subdivide into smaller and smaller tubes, eventually terminating in microscopic branches that divide into **alveolar ducts,** which terminate in several alveolar sacs, the walls of which consist of numerous **alveoli** (Figure 22-14; see also Figure 22-1). The structure of an alveolar duct with its branching alveolar sacs can be likened to a bunch of grapes—the stem represents the alveolar duct, each cluster of grapes represents an alveolar sac, and each grape represents an alveolus. Some 300 million alveoli are estimated to be present in our two lungs.

The structure of the secondary and tertiary bronchi and bronchioles shows some modification of the primary bronchial structure. The cartilaginous rings become irregular and disappear entirely in the smaller bronchioles. By the time the branches of the bronchial tree have dwindled

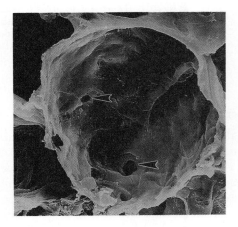

FIGURE 22-15   Alveolus. This scanning electron micrograph of a lung section shows an alveolus with arrowheads pointing to pores that allow air to flow into surrounding alveoli.

sufficiently to form the alveolar ducts and sacs and the alveoli, only the internal surface layer of cells remains. In other words, the walls of these microscopic structures consist of a single layer of simple, squamous epithelial tissue (Figures 22-15 and 22-16). As we shall see, this structural fact makes possible the performance of their functions.

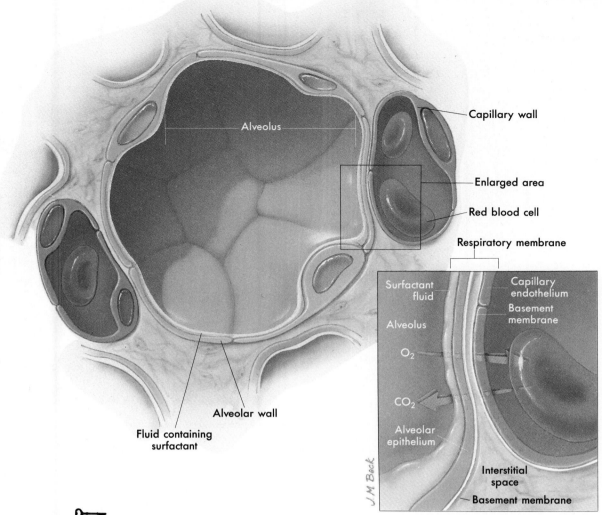

FIGURE 22-16 The gas-exchange structures of the lung. Each alveolus is continually ventilated with fresh air. The inset shows a magnified view of the respiratory membrane composed of the alveolar wall (fluid coating, epithelial cells, and basement membrane), interstitial fluid, and the wall of a pulmonary capillary (basement membrane and endothelial cells). The gases, CO_2 (carbon dioxide) and O_2 (oxygen), diffuse across the respiratory membrane.

Structure of the alveoli

The alveoli are the primary gas exchange structures of the respiratory tract (see Figures 22-15 and 22-16). Alveoli are very effective in the exchange of carbon dioxide (CO_2) and oxygen (O_2) because each alveolus is extremely thin walled, each alveolus lies in contact with blood capillaries, and there are millions of alveoli in each lung. The barrier across which gases are exchanged between alveolar air and the blood is called the **respiratory membrane** (see Figure 22-16, *inset*). The respiratory membrane consists of the alveolar epithelium, the capillary endothelium, and their joined basement membranes. The surface of the respiratory membrane inside each alveolus is coated with a fluid containing **surfactant.** Surfactant helps reduce surface tension—the force of attraction between water molecules—of the fluid. Thus it helps prevent each alveolus

from collapsing and "sticking shut" as air moves in and out during respiration.

Functions of the bronchi and alveoli

The tubes composing the bronchial tree perform the same function as the trachea—that of distributing air to the lung's interior. The alveoli, enveloped as they are by networks of capillaries, accomplish the lung's main and vital function, that of gas exchange between air and blood. Someone has observed that "the lung passages all serve the alveoli" just as "the circulatory system serves the capillaries."

Recall that in addition to serving as air distribution passageways or gas exchange surfaces, the anatomical components of the respiratory tract and lungs cleanse, warm, and humidify inspired air. Air entering the nose is

KEEPING THE TRACHEA OPEN

Often a tube is placed in the trachea **(endotracheal intubation)** before patients leave the operating room, especially if they have been given a muscle relaxant. The purpose of the tube is to ensure an open airway (Figure A). Another procedure done frequently in today's modern hospitals is a **tracheostomy,** that is, the cutting of an opening into the trachea (Figure B). A surgeon may do this so that a suction device can be used to remove secretions from the bronchial tree or so that a machine such as the intermittent positive pressure breathing (IPPB) machine can be used to improve ventilation of the lungs.

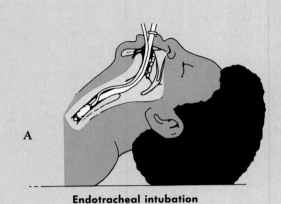

A

Endotracheal intubation

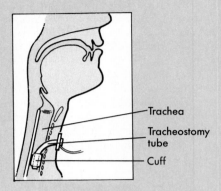

Trachea

Tracheostomy tube

Cuff

B

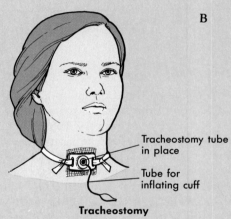

Tracheostomy tube in place

Tube for inflating cuff

Tracheostomy

generally contaminated with one or more common irritants; examples include insects, dust, pollen, and bacterial organisms. A remarkably effective air purification mechanism removes almost every form of contaminant before inspired air reaches the alveoli or terminal air sacs in the lungs.

The layer of protective mucus that covers a large portion of the membrane that lines the respiratory tree serves as the most important air purification mechanism. Over 125 ml of respiratory mucus is produced daily. It forms a continuous sheet, called a *mucus blanket,* that covers the lining of the air distribution tubes in the respiratory tree. This layer of cleansing mucus moves upward to the pharynx from the lower portions of the bronchial tree on millions of hairlike cilia that cover the epithelial cells in the respiratory mucosa

(see Figure 22-12). The microscopic cilia that cover epithelial cells in the respiratory mucosa beat or move in only one direction. The result is movement of mucus toward the pharynx. Cigarette smoke paralyzes these cilia and results in accumulations of mucus and the typical smoker's cough, an effort to clear the secretions.

QUICK CHECK

✔ 1. *How are the trachea and primary bronchi held open so that they do not collapse during inspiration?*
✔ 2. *What is meant by the term* bronchial tree?
✔ 3. *What characteristics of the alveoli enable them to efficiently exchange gases with the blood?*

RESPIRATORY DISTRESS

Respiratory distress results from the body's relative inability to inflate the alveoli of the lungs normally. **Respiratory distress syndrome** is a condition most often caused by absence or impairment of the surfactant in the fluid that lines the alveoli.

Infant respiratory distress syndrome or **IRDS** is a very serious, life-threatening condition that often affects prematurely born infants of less than 37 weeks' gestation or those who weigh less than 2.2 kg (5 lbs) at birth. IRDS is the leading cause of death among premature infants in the United States, claiming over 5000 premature babies each year. The disease, characterized by a lack of surfactant in the alveolar air sacs, affects 50,000 babies annually.

Surfactant is manufactured by specialized cells in the walls of the alveoli. Surfactant reduces the surface tension of the fluid on the free surface of the alveolar walls and permits easy movement of air into and out of the lungs. The ability of the body to manufacture this important substance is not fully developed until shortly before birth—normally about 40 weeks after conception.

In newborn infants who are unable to manufacture surfactant, many air sacs collapse during expiration because of the increased surface tension. The effort required to reinflate these collapsed alveoli is much greater than that needed to reinflate normal alveoli with adequate surfactant. The baby soon develops labored breathing, and symptoms of respiratory distress appear shortly after birth.

In the past, treatment of IRDS was limited to keeping the alveoli open so that delivery and exchange of oxygen and carbon dioxide could occur. To accomplish this, a tube is inserted into the respiratory tract and oxygen-rich air delivered under sufficient pressure to keep the alveoli from collapsing at the end of expiration. A newer treatment delivers air under pressure and applies prepared surfactant directly into the baby's airways by means of a tube.

Adult respiratory distress syndrome (ARDS) is caused by impairment or removal of surfactant in the alveoli. For example, accidental inhalation of foreign substances such as water, vomit, smoke, or chemical fumes can cause ARDS. Edema of the alveolar tissue can impair surfactant and reduce the alveoli's ability to stretch, causing respiratory distress.

Lungs
Structure of the lungs

The lungs are cone-shaped organs, large enough to fill the pleural portion of the thoracic cavity completely (see Figure 22-10). They extend from the diaphragm to a point slightly above the clavicles and lie against the ribs both anteriorly and posteriorly. The medial surface of each lung is roughly concave to allow room for the mediastinal structures and for the heart, but the concavity is greater on the left than on the right because of the position of the heart. The primary bronchi and pulmonary blood vessels (bound together by connective tissue to form what is known as the **root** of the lung) enter each lung through a slit on its medial surface called the **hilum.**

The broad inferior surface of the lung, which rests on the diaphragm, constitutes the **base,** whereas the pointed upper margin is the **apex** (Figure 22-17). Each apex projects above a clavicle. The **costal surface** of each lung lies against the ribs and is rounded to match the contours of the thoracic cavity.

Each lung is divided into lobes by fissures. The left lung is partially divided into two lobes (superior and inferior) and the right lung into three lobes (superior, middle, and inferior). Note in Figure 22-18 that an **oblique fissure** is present in both lungs. In the right lung a **horizontal fissure** is also present that separates the superior from the middle lobe. After the primary bronchi enter the lungs, they branch into *secondary,* or *lobar,* bronchi that enter each lobe. Thus, in the right lung, three secondary bronchi are formed, which enter the superior, middle, and inferior lobes. Each secondary bronchus is named for the lung lobe that it enters; for example, the superior secondary bronchus enters the superior lobe. The left primary bronchus divides

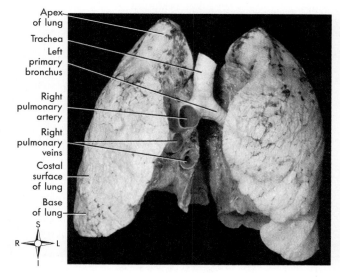

Apex of lung
Trachea
Left primary bronchus
Right pulmonary artery
Right pulmonary veins
Costal surface of lung
Base of lung

FIGURE 22-17 Trachea, bronchi, and lungs. The lower respiratory tract has been dissected from a cadaver and its organs separated to show them clearly.

into two secondary bronchi entering the superior and inferior lobes of that lung.

The lobes of the lung can be further subdivided into functional units called **bronchopulmonary segments.** Each bronchopulmonary segment is served by a tertiary bronchus. There are 10 segments in the right lung and 8 in the left (see Figure 22-18). The interior of each bronchopulmonary segment consists of the almost innumerable tubes of dwindling diameters that make up the bronchial tree and serve as air distributors. The smallest tubes terminate in the smallest but functionally most important structures of the lung—the alveoli, or "gas exchangers."

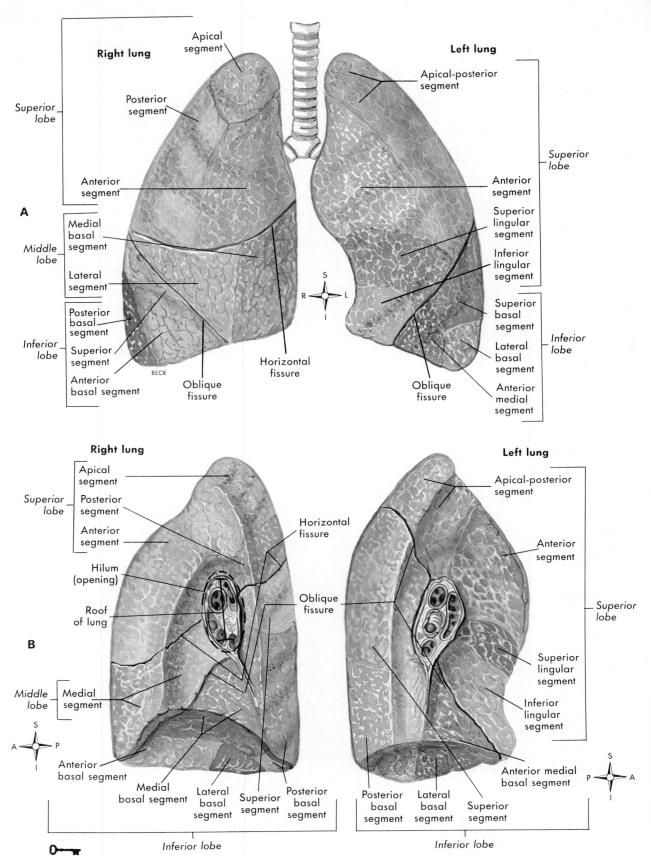

FIGURE 22-18   Lobes and segments of the lungs. A, Anterior view of the left and right lungs, bronchi, and trachea. **B,** Medial views of the right and left lungs.

Visceral pleura covers the outer surfaces of the lungs and adheres to them much as the skin of an apple adheres to the apple.

Functions of the lungs

The lungs perform two functions—air distribution and gas exchange. Air distribution to the alveoli is the function of the tubes of the bronchial tree. Gas exchange between air and blood is the joint function of the alveoli and the networks of blood capillaries that envelop them. These two structures—one part of the respiratory system and the other part of the circulatory system—together serve as highly efficient gas exchangers. Why? Because they provide an enormous surface area, the **respiratory membrane,** where the very thin-walled alveoli and equally thin-walled pulmonary capillaries come in contact (see Figures 22-15 and 22-16). This makes possible extremely rapid diffusion of gases between alveolar air and pulmonary capillary blood. Someone has estimated that if the lungs' 300 million or so alveoli could be opened up flat, they would form a surface about the size of a tennis court, that is, about 85 square meters, or well over 40 times the surface area of the entire body! No wonder such large amounts of oxygen can be so quickly loaded into the blood while large amounts of carbon dioxide are rapidly unloaded from it.

Thorax
Structure of the thoracic cavity

As described in Chapter 1, the thoracic cavity has three divisions, separated from each other by partitions of pleura. The parts of the cavity occupied by the lungs are the pleural divisions. The space between the lungs occupied mainly by the esophagus, trachea, large blood vessels, and heart is the mediastinum (Figure 22-19).

The parietal layer of the pleura lines the entire thoracic cavity. It adheres to the internal surface of the ribs and the superior surface of the diaphragm, and it partitions off the mediastinum. A separate pleural sac thus encases each lung. Since the outer surface of each lung is covered by the visceral layer of the pleura, the visceral pleura lies against the parietal pleura, separated only by a potential space (pleural space) that contains just enough pleural fluid for lubrication (Figure 22-19). Thus, when the lungs inflate with air, the smooth, moist visceral pleura coheres to the smooth, moist parietal pleura. Friction is thereby avoided, and respirations are painless. In *pleurisy* (pleuritis), on the other hand, the pleura is inflamed and respirations become painful.

Functions of the thoracic cavity

The thorax plays a major role in respirations. Because of the elliptical shape of the ribs and the angle of their attachment to the spine, the thorax becomes larger when the chest is raised and smaller when it is lowered. Lifting up the chest raises the ribs so that they no longer slant downward from the spine, and because of their elliptical shape this enlarges both depth (from front to back) and width of the thorax. (If this does not sound convincing to you, examine a skeleton to see why it is so.) An even greater change in thoracic volume occurs when the diaphragm contracts and relaxes. When the diaphragm contracts it flattens out and thus pulls the floor of the

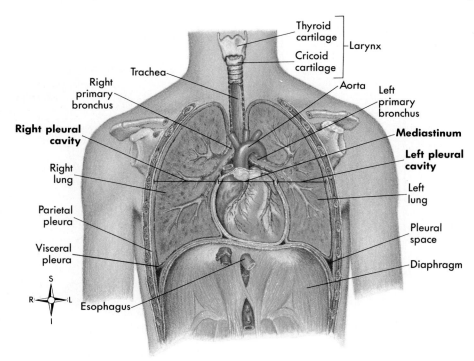

FIGURE 22-19   The thoracic cavity. The thoracic cavity, or "chest cavity," is divided into three subdivisions (left and right pleural cavities and mediastinum) by a partition formed by a serous membrane called the *pleura.*

thoracic cavity downward—enlarging the volume of the thorax. When the diaphragm relaxes, it returns to its resting, dome-like shape—thus reducing the volume of the thoracic cavity. It is these changes in thorax size that bring about inspiration and expiration (discussed on pp. 602 in Chapter 23).

> **QUICK CHECK**
>
> ✔ 1. What is meant by the term lobe of the lung? What is a bronchopulmonary segment?
> ✔ 2. How does the structure of the diaphragm enable it to participate in breathing movements?

 ## Cycle of Life: *Respiratory system*

Respiration involves the exchange of O_2 and CO_2 between the organism and its environment. The exchange must occur between air in the lungs and the blood and, then, between the blood and every body cell. In addition to the structural components of the body through which the respiratory gases must pass, hemoglobin plays a vital role in the respiration process. Each component of the system may be affected by developmental defect, by age-related structural change, or by loss of function during the life cycle.

Premature birth can cause potentially fatal respiratory problems. A very low birth weight baby may have inadequate blood flow to the lungs, an inability to ventilate properly, and inadequate quantities of surfactant. Other diseases that cause serious respiratory problems are also associated with specific age groups. Examples include cystic fibrosis and asthma in children and certain types of obstructive pulmonary disease and emphysema in older adults. Pneumothorax occurs more frequently in young adult females.

Numerous age-related changes affect vital capacity, make ventilation difficult, or reduce the oxygen or carbon dioxide carrying capacity of the blood. For example, in older adulthood the ribs and sternum tend to become more fixed and less able to expand during inspiration, the respiratory muscles are less effective, and hemoglobin levels are often reduced. The result is a general reduction in respiratory efficiency in old age.

 # MECHANISMS OF DISEASE

Disorders Associated with Respiratory Anatomy

Disorders of the Upper Respiratory Tract

Inflammation and infection

Any infection localized in the mucosa of the upper respiratory tract (nose, pharynx, and larynx) can be called an **upper respiratory infection (URI)** and is often named for the specific structure involved. **Rhinitis** (from the Greek *rhinos*, "nose") is an inflammation of the mucosa of the nasal cavity. It is commonly caused by a viral infection, as in the common cold (caused by rhinoviruses) or flu (caused by influenza viruses). Rhinitis can also be caused by nasal irritants or an allergic reaction to airborne allergens. Allergic rhinitis, or "hayfever," occurs in sensitive people in a seasonal pattern depending on the allergens involved (for example, pollen). Excessive mucus production that results from the inflammatory response involved in rhinitis can cause fluid to drip down the pharynx and into the esophagus and lower respiratory tract. This dripping may cause sore throat, coughing, and upset stomach. Irritation of the nasal mucosa itself often triggers the sneeze reflex. Elimination of the causative factor, rest, and use of antihistamines and decongestants usually relieves these symptoms.

Pharyngitis is inflammation or infection of the pharynx. Commonly referred to as a "sore throat," it is often due to viral invasion. Bacterial infection by *Streptococcus* bacteria is termed "strep throat." The common complaint is a sore throat, but redness and difficulty swallowing often accompany it. Throat lozenges, rest, and fluid intake are encouraged, and antibiotics are prescribed for severe infections.

Laryngitis, or inflammation of the mucous lining of the larynx, is characterized by edema of the vocal cords, resulting in hoarseness or loss of voice. Besides infections, inhalation of toxic or irritating fumes (i.e., smoking), endotracheal intubation, vocal abuse (i.e., public speaking), and alcohol ingestion can precipitate laryngitis. In children under 5 years of age, it may cause difficulty in breathing, a condition often called *croup*. Conservative treatment, including limiting speech, is usually effective.

Because the upper respiratory mucosa is continuous with the mucous lining of the sinuses, eustachian tube, middle ear, and lower respiratory tract, URIs have the unfortunate tendency to spread. It is not unusual to see a common cold progress to *sinusitis* (sinus infection) or *otitis media* (middle ear infection).

Anatomical disorders

Nasal obstruction can be caused by displacement of the nasal septum from the midline of the nasal cavity, called a **deviated septum.** Most individuals have a small amount of septal cartilage protrusion into one nasal passage; however, some are born with a congenital defect that results in various degrees of blockage to one or both sides of the nasal cavity. Damage from injury or infection may also cause a deviated septum. If breathing is impaired, surgical intervention is required to correct the deformity.

Trauma to the nose can occur because the nose projects some distance from the front of the head. Usually, however, common bumps and other injuries cause little, if any, serious damage. **Epistaxis,** or nosebleed, can be caused by violent sneezing or nose blowing, chronic infection or inflammation (as in rhinitis), hypertension, or a strong bump or blow to the nose. Immediate direct pressure such as ice packs will slow or stop the bleeding.

Disorders of the Lower Respiratory Tract

A range of conditions can interfere with the functions of the lower respiratory tract functions of gas exchange and ventilation. Some of these disorders, such as restrictive and obstructive conditions, will be discussed in Chapter 23. For now, we will concentrate on infections and lung cancer.

Lower respiratory infection

Acute bronchitis is a common condition characterized by acute inflammation of the tracheobronchial tree, most commonly caused by infection. Part of, or preceded by, an acute URI, it is most prevalent in winter. Predisposing factors include chilling, fatigue, malnutrition, and exposure to air pollutants. The protective functions of the bronchial epithelium are disturbed and excessive fluid accumulates in the bronchi. Acute bronchitis often begins with a nonproductive cough, but malaise, slight fever, back and muscle pain, and a sore throat occur if a URI is present. Rest is indicated until the fever subsides and cough suppressants may be used if the cough is troublesome.

An inflammation of the lower respiratory tract that involves the airways of the lungs proper is called **pneumonia**. Exposure to an infectious organism through inhalation, aspiration (breathing of a fluid or solid), or spread through the blood supply causes the alveoli and bronchi to become plugged with thick fluid (exudate). Pneumonia is characterized by a high fever, severe chills, headache, cough, and chest pain. A test examining the arterial blood may reveal *hypoxia*, or deficiency of oxygen in the blood. The white blood cell count is often elevated. Involvement of an entire lung lobe is called lobar pneumonia; involvement of small parts of a lobe only is referred to as lobular pneumonia. The majority of cases result from infection by *Streptococcus* bacteria but can also be caused by viruses or fungi. Therapeutic management involves antimicrobial drugs to control the infection and supportive therapy of the respiratory compromise or failure. Removal of tracheobronchial secretions may be vital to maintain airway integrity.

Tuberculosis (TB) is a chronic bacillus infection caused by *Mycobacterium tuberculosis*. It is a highly contagious disease, transmitted by airborne mechanisms (that is, inhalation of

---

DIAGNOSTIC STUDIES   PULMONARY RADIOLOGY

Chest x-ray examinations account for more than half of all radiographs taken in the United States each year. They can be extremely useful in the diagnosis of pathological conditions and in the study of normal respiratory anatomy and physiology.

Terminology

Most chest radiographs are taken during full inspiration while the individual is standing erect with the anterior surface of the chest against the film holder. The x-ray machine is about 2 meters (7 feet) behind the subject so that the x-ray beam enters the body from behind (posterior) and exits through the front (anterior). This type of exposure is appropriately called a PA (posterior to anterior) radiograph

and is the most common type of chest x-ray examination. If the subject is asked to turn completely around so that the x-ray beam enters the chest from in front (anterior) and exits from the back (posterior), the resulting radiograph is termed an AP (anterior to posterior) film. Figure A is a normal PA chest x-ray film of a young adult. Figure B is an AP radiograph of the same individual. Note that in the PA film the clavicles are superimposed over the upper lungs with the medial aspect of each bone lower than the distal, or lateral, end. Note also that structural details of the cervical and thoracic vertebrae are more clearly visible in the PA film. In an AP chest film the clavicles are higher and the heart appears larger and is more clearly defined.

A
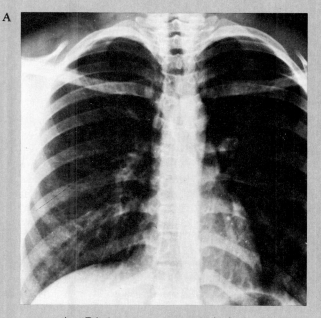

A, PA (posterior to anterior) view.

B
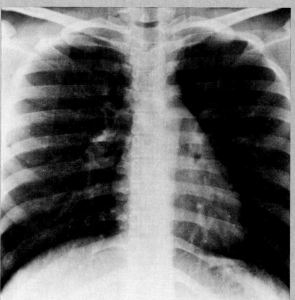

B, AP (anterior to posterior) view.

infectious droplets). Inflammatory lesions called "tubercles" form around colonies of TB bacilli in the lung, producing the characteristic symptoms of nonproductive cough, fatigue, chest pain, weight loss, and fever. As TB progresses, lung hemorrhage and dyspnea (labored breathing) may develop. If large areas of the lung are infected and tissue is destroyed, scar tissue may develop, causing reduced lung volume and restrictive lung disease. TB can invade other tissues or organs such as the lymphatic system, genitourinary system, or bone tissue. Because of the advancement of modern antimicrobial agents the incidence of TB in the United States dropped dramatically in the last half century. However, various factors have allowed TB to emerge once again as a major health threat in the United States. While the incidence of TB has not yet reached epidemic proportions, health authorities in many large cities are working hard to prevent a major health crisis.

Lung Cancer

Lung cancer is a malignancy of pulmonary tissue that not only destroys the vital gas exchange tissues of the lungs but like other cancers may also invade other parts of the body (metastasis). Lung cancer most often develops in damaged or diseased lungs. The most common predisposing condition associated with lung cancer is cigarette smoking (accounting for about 75% of lung cancer cases). Other factors thought to cause lung cancer include exposure to "second-hand" cigarette smoke, asbestos, chromium, coal products, petroleum products, rust, and ionizing radiation (as in radon gas).

Lung cancer may be arrested if detected early in routine chest x-ray films or other diagnostic procedures. Depending on the exact type of malignancy involved and the extent of lung involvement, several strategies are available for treatment. Chemotherapy can cause a cure or remission in selected cases, as can radiation therapies. Surgery is the most effective treatment known, but only half of the persons diagnosed as having lung cancer are good candidates for surgery because the damage is too extensive. In a *lobectomy* only the affected lobe of a lung is removed. *Pneumonectomy* is the surgical removal of an entire lung.

Specialized Radiographs

X-ray examinations that provide specialized information such as *bronchograms* and *arteriograms* are particularly useful in diagnosis of disease. Figure C is a bronchogram of the right lung. The bronchi of the right lower lobe show "crowding," indicating collapse and compression of lung tissue. Figure D is an arteriogram of the right lung showing the presence of an embolus, or moving blood clot (*black arrow*), at the root of the right pulmonary artery. The white arrowheads outline the normal lumen of the artery distal to the clot.

C
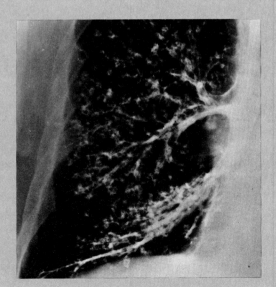

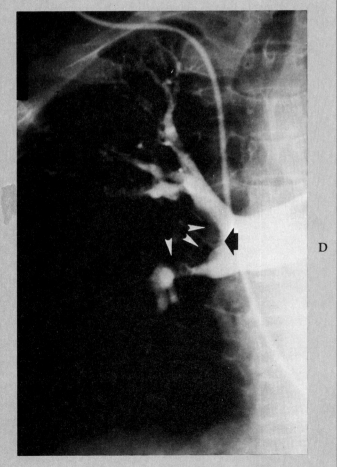
D

C, Lung bronchogram.

D, Pulmonary arteriogram.

CHAPTER SUMMARY

OVERVIEW OF THE RESPIRATORY SYSTEM

A. The respiratory system functions as an air distributor and gas exchanger—supplying oxygen and removing carbon dioxide from cells (Figure 22-1)
 1. Alveoli—sacs that serve as gas exchangers; all other parts of respiratory system serve as air distributors
 2. The respiratory system also warms, filters, and humidifies air
 3. Respiratory organs influence speech, homeostasis of body pH, and olfaction

B. The respiratory system is divided into two divisions
 1. Upper respiratory tract—the organs are located outside of the thorax and consist of the nose, nasopharnyx, oropharnyx, laryngopharnyx, and larnyx
 2. Lower respiratory tract—the organs are located within the thorax and consist of the trachea, the bronchial tree, and the lungs
 3. Accessory structures include the oral cavity, rib cage, and diaphragm

UPPER RESPIRATORY TRACT

A. The nose
 1. The external portion of the nose consists of a bony and cartilaginous frame covered by skin containing sebaceous glands—the two nasal bones meet and are surrounded by the frontal bone to form the root; the nose is surrounded by the maxilla (Figure 22-2)
 2. The internal nose (nasal cavity) lies over the roof of the mouth, separated by the palatine bones
  a. Cleft palate—the palatine bones fail to unite completely and only partially separate the nose and the mouth, producing difficulty in swallowing
  b. Cribriform plate—separates the roof of the nose from the cranial cavity
  c. Septum—separates the nasal cavity into a right and left cavity; it consists of four structures: the perpendicular plate of the ethmoid bone, the vomer bone, the vomeronasal cartilages, and the septal nasal cartilage
 3. Each nasal cavity is divided into three passageways: superior, middle, and inferior meati (Figure 22-3)
 4. Anterior nares—external openings to the nasal cavities, open into the vestibule
 5. The sequence of air through the nose into the pharynx—anterior nares to vestibule to all three meati simultaneously to posterior nares
 6. Nasal mucosa—a mucous membrane that air passes over; it contains a rich blood supply
  a. Olfactory epithelium—specialized membrane containing many olfactory nerve cells and a rich lymphatic plexus
 7. Paranasal sinuses—four pairs of air-containing spaces that open or drain into nasal cavity and each is lined with respiratory mucosa (Figure 22-5)
 8. The nose is a passageway for air traveling to and from the lungs—it filters the air, aids speech, and makes possible the sense of smell

B. Pharynx (throat)
 1. The pharynx is a tubelike structure extending from the base of the skull to the esophagus; it is made of muscle and divided into three parts (Figure 22-6)
  a. Nasopharynx
  b. Oropharynx
  c. Laryngopharynx
 2. The pharyngeal tonsils—located in the nasopharynx, called adenoids when they become enlarged

 3. The oropharynx contains two pair of organs—the palatine tonsils (most commonly removed) and the lingual tonsils (rarely removed)
 4. The pharynx is the pathway for the respiratory and digestive tracts

C. Larynx (Figures 22-7 and 22-8)
 1. The larynx is located between the root of the tongue and the upper end of the trachea
 2. The larynx consists of cartilages attached to each other by muscle and is lined by a ciliated mucous membrane, which forms two pairs of folds (Figure 22-9)
  a. The vestibular (false) vocal folds
  b. The true vocal cords
 3. The framework of the larynx is formed by nine cartilages
  a. Single laryngeal cartilages—the three largest cartilages: the thyroid cartilage, the epiglottis, and the cricoid cartilages
  b. The paired laryngeal cartilages—three pairs of smaller cartilages: the arytenoid, the corniculate, and the cuneiform cartilages
 4. The larynx functions as part of the airway to the lungs and produces the voice

LOWER RESPIRATORY TRACT

A. The trachea (windpipe) extends from the larynx to the primary bronchi (Figure 22-11)
 1. The trachea furnishes part of the open airway to the lungs—obstruction causes death

B. Bronchi and alveoli
 1. The lower end of the trachea divides into two primary bronchi, one on the right and one on the left, which enters the lung and divides into secondary bronchi that branch into bronchioles, which eventually divide into alveolar ducts (Figure 22-13)
 2. The alveoli are the primary gas exchange structures
  a. The respiratory membrane—the barrier between which gases are exchanged by the alveolar air and the blood (Figure 22-16)
  b. The respiratory membrane consists of the alveolar epithelium, the capillary endothelium, and their joined basement membranes
  c. Surfactant—a component of the fluid coating the respiratory membrane that reduces surface tension
 3. The bronchi and alveoli distribute air to the lung's interior

C. Lungs
 1. The lungs are cone-shaped organs extending from the diaphragm to above the clavicles (Figure 22-18)
  a. The hilum—slit on lung's medial surface where the primary bronchi and pulmonary blood vessels enter
  b. The base—the inferior surface of the lung that rests on the diaphragm
  c. The costal surface—lies against the ribs
  d. The left lung is divided into two lobes—superior and inferior
  e. The right lung is divided into three lobes—superior, middle, and inferior
  f. The lobes are further divided into functional units—bronchopulmonary segments
   (1) Ten segments in the right lung
   (2) Eight segments in the left lung
 2. The lungs have two functions—air distribution and gas exchange

D. The thorax (Figure 22-19)
 1. The thoracic cavity has three divisions divided by pleura

a. Pleural divisions—the part occupied by the lungs
b. Mediastinum—part occupied by the esophagus, trachea, large blood vessels, and heart
2. The thorax functions to bring about inspiration and expiration

CYCLE OF LIFE: RESPIRATORY SYSTEM

A. Respiration may be affected by developmental defects, age-related structural changes, or loss of function throughout the life cycle
B. Age-related changes affect vital capacity, make ventilation difficult, or reduce the oxygen or carbon dioxide carrying capacity of blood
C. Respiratory efficiency is reduced in old age as a result of changes in ribs, respiratory muscles, and hemoglobin levels

MECHANISMS OF DISEASE: DISORDERS ASSOCIATED WITH RESPIRATORY ANATOMY

A. Disorders of the upper respiratory tract
1. Upper respiratory infection (URI)—any infection localized in the upper respiratory tract; often named for the structure involved
   a. Examples:
      (1) Rhinitis—inflammation of the mucosa of the nasal cavity
      (2) Pharyngitis—inflammation of the pharynx
      (3) Laryngitis—inflammation of the mucous lining of the larynx
2. URIs may spread and become more serious because the mucosa is continuous with the mucous lining of the sinuses, eustachian tube, middle ear, and lower respiratory tract
3. Anatomical disorders may lead to problems

a. Deviated septum—displacement of the nasal septum from the midline of the nasal cavity may cause nasal obstruction
b. Epistaxis—nosebleeds caused by sneezing, nose-blowing, chronic infection, hypertension, or a blow to the nose
B. Disorders of the lower respiratory tract
1. Acute bronchitis—characterized by inflammation of the tracheobronchial tree; caused by infection; the protective functions of the bronchial epithelium are disturbed and fluid accumulates in the bronchi
2. Pneumonia—an inflammation that involves the airways of the lungs; it is characterized by fever, chills, headaches, and chest pain
   a. Lobar pneumonia—entire lung lobe is involved
   b. Lobular pneumonia—small parts of a lobe involved
3. Tuberculosis—a highly contagious infection transmitted by airborne mechanisms; tubercles form around TB bacilli in the lung and produce symptoms such as nonproductive cough, fatigue, chest pain, weight loss, and fever—as it progresses, lung hemorrhage and dyspnea develop; scar tissue may develop and cause reduced lung volume and restrictive lung disease; TB may also spread to other tissues or organs; TB dropped in the U.S. as a result of modern antibiotics but has recently reemerged as a major health threat
C. Lung cancer—a malignancy of pulmonary tissue that destroys the gas exchange tissues of the lung and may invade other parts of the body; it develops in damaged lungs and is caused by cigarette smoking and exposure to other air pollutants
1. Treatment may involve chemotherapy or surgery

REVIEW QUESTIONS

1. Identify the major anatomical structures of the nose.
2. Discuss the structure and function of the respiratory mucosa.
3. How are the turbinates arranged in the nose? What are they?
4. Trace the movement of air through the nose into the pharynx.
5. Describe the draining of the paranasal sinuses.
6. What organs are found in the nasopharynx?
7. What tubes open into the nasopharynx?
8. List the anatomical subdivisions of the pharynx.
9. The pharynx is common to what two systems?
10. List the divisions of the larynx.
11. What is the voice box? Of what is it composed? What is the Adam's apple?
12. What is the epiglottis? What is its function?
13. What are the true vocal cords? What name is given to the opening between the cords?
14. Which of the paired laryngeal cartilages are most important? Why?
15. Describe the structure and function of the trachea.
16. Discuss the component parts of the bronchial tree.
17. Make a diagram showing the termination of a bronchiole in an alveolar duct with alveoli.
18. Identify the location of surfactant and explain its function.
19. How many lobes are in the right lung? The left? What are the bronchopulmonary segments?
20. What is the difference between air distribution and gas exchange in the respiratory system? What organs serve as air distributors? As gas exchangers?
21. Explain how the fluid in the pleural space eliminates friction during respirations.
22. Describe the changes in thorax size during respiration.
23. List the organs that are included in the upper respiratory tract. Do the same for the lower respiratory tract.
24. Discuss the importance of the "danger area of the face."
25. What is the purpose of an endotracheal intubation?
26. Describe infant and adult respiratory distress syndromes.

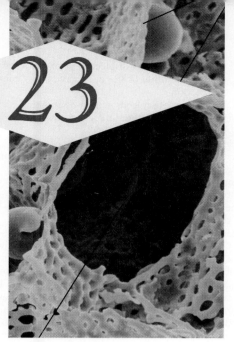

23 Physiology of the Respiratory System

SEM *of alveolus and capillaries*

CHAPTER OUTLINE

OBJECTIVES

After you have completed this chapter, you should be able to:

1. List and briefly discuss the regulated and integrated processes that ensure tissues of an adequate oxygen supply and prompt removal of carbon dioxide.
2. Define pulmonary ventilation and outline the mechanism of normal, quiet inspiration and expiration.
3. List by names and explain the volume of air exchanged in pulmonary ventilation.
4. Define the following: tidal volume, expiratory reserve volume, inspiratory reserve volume, residual volume, minimal volume, inspiratory capacity, functional residual capacity, total lung capacity.
5. Demonstrate the principles of partial pressures (Dalton's law) in explaining movement of respiratory gases between alveolar air and blood moving through pulmonary capillaries.
6. Discuss the major factors that determine the volume of oxygen entering lung capillary blood.
7. Explain how blood transports oxygen and carbon dioxide.
8. Interpret changes in an oxygen-hemoglobin dissociation curve at various blood pH levels.
9. Discuss gas exchange in tissue capillaries between arterial blood and cells.
10. Explain the reciprocal interaction of oxygen and carbon dioxide on blood gas transport (Bohr vs. Haldane effect).
11. Discuss the primary factors that influence the respiratory control center and thereby control respirations.

RESPIRATORY PHYSIOLOGY

In Chapter 22 the anatomy of the respiratory system was presented as a basis for understanding the physiological principles that regulate air distribution and gas exchange. This chapter deals with **respiratory physiology**—a complex series of interacting and coordinated processes that play a critical role in maintaining the stability, or constancy, of our internal environment. The proper functioning of the respiratory system ensures the tissues of an adequate oxygen supply and prompt removal of carbon dioxide. This process is complicated by the fact that control mechanisms must permit maintenance of homeostasis throughout a wide range of ever-changing environmental conditions and body demands. Adequate and efficient regulation of gas exchange between body cells and circulating blood under changing conditions is the essence of respiratory physiology. This complex function would not be possible without integration between numerous physiological control systems, including acid-base, water, and electrolyte balance, circulation, and metabolism.

Functionally, the respiratory system is composed of an integrated set of regulated processes that include pulmonary ventilation (breathing), gas exchange in the lungs and tissues, transport of gases by the blood, and overall regulation of respiration.

PULMONARY VENTILATION

Pulmonary ventilation is a technical term for what most of us call breathing. One phase of it, inspiration, moves air into the lungs and the other phase, expiration, moves air out of the lungs.

Mechanism of Pulmonary Ventilation

Air moves in and out of the lungs for the same basic reason that any fluid, that is, a liquid or a gas, moves from one place to another—briefly, because its pressure in one place is different from that in the other place. Or stated differently, the existence of a pressure gradient (a pressure difference) causes fluids to move. A fluid always moves down its pressure gradient. This means that a fluid moves from the area where its pressure is higher to the area where its pressure is lower. Under standard conditions, air in the atmosphere exerts a pressure of 760 mm Hg. Air in the alveoli at the end of one expiration and before the beginning of another inspiration also exerts a pressure of 760 mm Hg. This fact explains why, at that moment, air is neither entering nor leaving the lungs. The mechanism that produces pulmonary ventilation is one that establishes a gas pressure gradient between the atmosphere and the alveolar air.

When atmospheric pressure is greater than pressure within the lung, air flows down this gas pressure gradient. Then air moves from the atmosphere into the lungs. In other words, inspiration occurs. When pressure in the lungs becomes greater than atmospheric pressure, air again moves down a gas pressure gradient. But now, this means that it moves in the opposite direction. This time, air moves out of the lungs into the atmosphere. The pulmonary ventilation mechanism, therefore, must somehow establish these two gas pressure gradients—one in which intraalveolar pressure (pressure within the alveoli of the lungs) is lower than atmospheric pressure to produce inspiration and one in which it is higher than atmospheric pressure to produce expiration.

These pressure gradients are established by changes in the size of the thoracic cavity, which, in turn, are produced by contraction and relaxation of respiratory muscles. An understanding of *Boyle's law* is important for understanding the pressure changes that occur in the lungs and thorax during the breathing cycle. It is a familiar principle, stating that the volume of a gas varies inversely with pressure at a

Atmospheric pressure = 760 mm Hg

Atmospheric pressure = 760 mm Hg

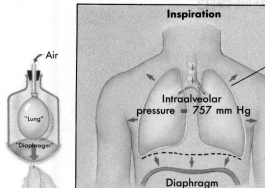

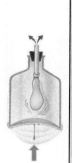

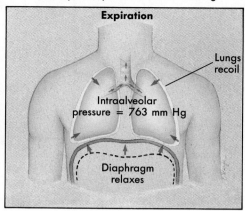

FIGURE 23-1   Mechanics of ventilation. During *inspiration,* the diaphragm contracts, increasing the volume of the thoracic cavity. This increase in volume results in a decrease in pressure, which causes air to rush into the lungs. During *expiration,* the diaphragm returns to an upward position, reducing the volume in the thoracic cavity. Air pressure increases then, forcing air out of the lungs. Insets show the classic model in which a jar represents the rib cage, a rubber sheet represents the diaphragm, and a balloon represents the lungs.

constant temperature. Application: expansion of the thorax (increase in volume) results in a decreased intrapleural (intrathoracic) pressure. This leads to a decreased intraalveolar pressure that causes air to move from the outside into the lungs. The mechanics of ventilation are summarized in Figure 23-1.

Inspiration

Contraction of the diaphragm alone, or of the diaphragm and the external intercostal muscles, produces quiet inspiration. As the diaphragm contracts, it descends, and this makes the thoracic cavity longer. Contraction of the external intercostal muscles pulls the anterior end of each rib up and out. This also elevates the attached sternum and enlarges the thorax from front to back and from side to side. In addition, contraction of the sternocleidomastoid, pectoralis minor, and serratus anterior muscles can aid in elevation of the sternum and rib cage during forceful inspiration. As the size of the thorax increases, the intrapleural (intrathoracic) and intraalveolar pressures decrease (Boyle's law) and inspiration occurs. At the end of an expiration and before the beginning of the next inspiration, intrathoracic pressure is about 4 mm Hg less than atmospheric pressure (frequently written−4 mm Hg). During quiet inspiration, intrathoracic pressure decreases further to−6 mm Hg. As the thorax enlarges, it pulls the lungs along with it because of cohesion between the moist pleura covering the lungs and the moist pleura lining the thorax. Thus the lungs expand and the pressure in their tubes and alveoli necessarily decreases. Intraalveolar pressure decreases from an atmospheric level to a subatmospheric level—typically to about−3 mm Hg. The moment intraalveolar pressure becomes less than atmospheric pressure, a pressure gradient exists between the

atmosphere and the interior of the lungs. Air then necessarily moves into the lungs. For a summary of the mechanism of inspiration just described, see Figure 23-2.

Expiration

Quiet expiration is ordinarily a passive process that begins when those pressure changes, or gradients, that resulted in inspiration are reversed. The inspiratory muscles relax, causing a decrease in the size of the thorax and an increase in intrapleural (intrathoracic) pressure from about−6 mm Hg to a preinspiration level of−4 mm Hg. It is important to understand that this pressure between the parietal and visceral pleura is always negative, that is, less than atmospheric pressure. The negative intrapleural (intrathoracic) pressure is required to overcome the so-called collapse tendency of the lungs caused by surface tension of the fluid lining the alveoli and the stretch of elastic fibers that are constantly attempting to recoil.

As alveolar pressure increases from about −3 mm Hg to +3 or +4 mm Hg, a positive-pressure gradient is established from alveoli to atmosphere, and expiration occurs as air flows outward through the respiratory passageways. In forced expiration, contraction of the abdominal and internal intercostal muscles can increase intraalveolar pressure to over 100 mm Hg. Figure 23-3 summarizes the mechanism of expiration just described.

QUICK CHECK

✔ 1. What is meant by the term pulmonary ventilation?
✔ 2. What effect does enlargement of the thoracic cavity have on the air pressure inside the lungs?
✔ 3. Which requires more expenditure of energy during normal, quiet breathing—inspiration or expiration?

FIGURE 23-2   Mechanism of inspiration.

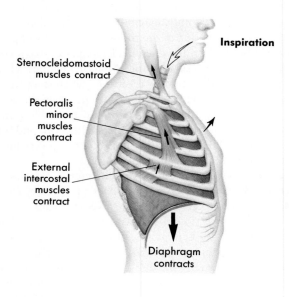

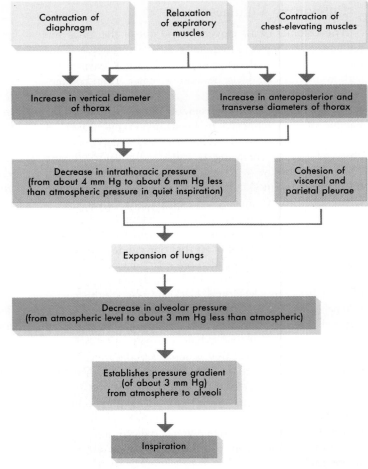

Contraction of diaphragm

Relaxation of expiratory muscles

Contraction of chest-elevating muscles

Increase in vertical diameter of thorax

Increase in anteroposterior and transverse diameters of thorax

Decrease in intrathoracic pressure (from about 4 mm Hg to about 6 mm Hg less than atmospheric pressure in quiet inspiration)

Cohesion of visceral and parietal pleurae

Expansion of lungs

Decrease in alveolar pressure (from atmospheric level to about 3 mm Hg less than atmospheric)

Establishes pressure gradient (of about 3 mm Hg) from atmosphere to alveoli

Inspiration

Sternocleidomastoid muscles contract

Pectoralis minor muscles contract

External intercostal muscles contract

Diaphragm contracts

Inspiration

FIGURE 23-3   Mechanism of expiration.

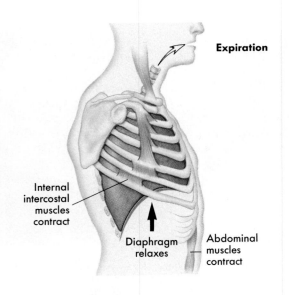

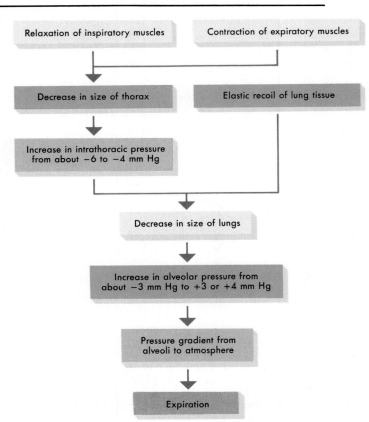

Relaxation of inspiratory muscles

Contraction of expiratory muscles

Decrease in size of thorax

Elastic recoil of lung tissue

Increase in intrathoracic pressure from about −6 to −4 mm Hg

Decrease in size of lungs

Increase in alveolar pressure from about −3 mm Hg to +3 or +4 mm Hg

Pressure gradient from alveoli to atmosphere

Expiration

Internal intercostal muscles contract

Diaphragm relaxes

Abdominal muscles contract

Expiration

Pulmonary Volumes

The volumes of air moved in and out of the lungs and remaining in them are matters of great importance. They must be normal so that a normal exchange of oxygen and carbon dioxide can occur between alveolar air and pulmonary capillary blood.

An apparatus called a **spirometer** is used to measure the volume of air exchanged in breathing (Figure 23-4). A graphic recording of the changing pulmonary volumes observed during breathing is called a *spirogram* (Figure 23-5). The volume of air exhaled normally after a typical inspiration is termed *tidal volume* (TV). As you can see in Figure 23-5, the normal volume of tidal air for an adult is approximately 500 ml. After an individual has expired tidal air, he/she can force still more air out of the lungs. The largest additional volume of air that one can forcibly expire after expiring tidal air is called the *expiratory reserve volume* (ERV). An adult, as Figure 23-5 shows, normally has an ERV of between 1,000 and 1,200 ml. *Inspiratory reserve volume (IRV)* is the amount of air that can be forcibly inspired over and above a normal inspiration. It is measured by having the individual exhale normally after a forced inspiration. The normal IRV is about 3,300 ml. No matter how forcefully one exhales, one cannot squeeze all the air out of the lungs. Some of it remains trapped in the alveoli. This amount of air that cannot be forcibly expired is known as *residual volume* (RV) and amounts to about 1,200 ml. Between breaths, an exchange of oxygen and carbon dioxide occurs between the trapped residual air in the alveoli and the blood. This process helps to "level off" the amounts—or restore the set point values—of oxygen and carbon dioxide in the blood during the breathing cycle. In pneumothorax the RV is eliminated when the lung collapses. Even after the RV is forced out, the collapsed lung has a porous, spongy texture and will float in water because of trapped air called the *minimal volume*, which is about 40% of the RV.

Notice in Figure 23-5 that *vital capacity* (VC) is the sum of IRV + TV + ERV. A pulmonary "capacity" is the sum of two or more pulmonary "volumes." The vital capacity represents the largest volume of air an individual can move in and out of the lungs. It is determined by measuring the largest possible expiration after the largest possible inspiration. How large a vital capacity a person has depends on many factors—the size of the thoracic cavity, posture, and various other factors. In general, a larger person has a larger vital capacity than a smaller one. An individual has a larger vital capacity when standing erect than when stooped over or lying down. The volume of blood in the lungs also affects the vital capacity. If the lungs contain more blood than normal, alveolar air space is encroached on and vital capacity accordingly decreases. This becomes a very important factor in congestive heart disease.

Excess fluid in the pleural or abdominal cavities also decreases vital capacity. So, too, does the disease, *emphysema*. In the latter condition, alveolar walls become stretched—that is, lose their elasticity—and are unable to recoil normally for expiration. This leads to an increased RV. In severe emphysema, the RV may increase so much that the chest occupies the inspiratory position even at rest. Excessive muscular effort is therefore necessary for inspiration, and because of the loss of elasticity of lung tissue, greater effort is required, too, for expiration.

In diagnosing lung disorders a physician may need to know the inspiratory capacity and the functional residual capacity of the patient's lungs. *Inspiratory capacity (IC)* is the maximal amount of air an individual can inspire after a normal expiration. From Figure 23-5, you can deduce that IC = TV + IRV. With the volumes given in the figure, how many milliliters is the IC? *Functional residual capacity*

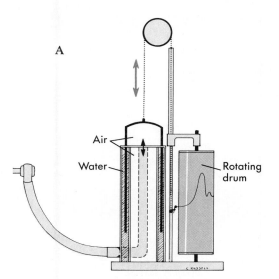

A

Air
Water
Rotating drum

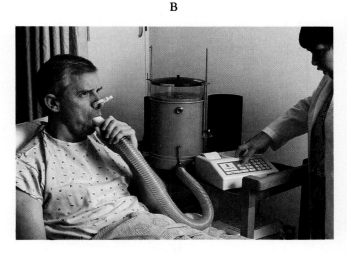

B

FIGURE 23-4 Spirometer. Spirometers are devices that measure the volume of gas that the lungs inhale and exhale, usually as a function of time. **A,** Diagram of a classic spirometer design, showing how the volume of air exhaled and inhaled is recorded as a rising and falling line. **B,** A simple spirometer attached to a computerized recording device. This apparatus is used frequently for routine assessment of ventilation.

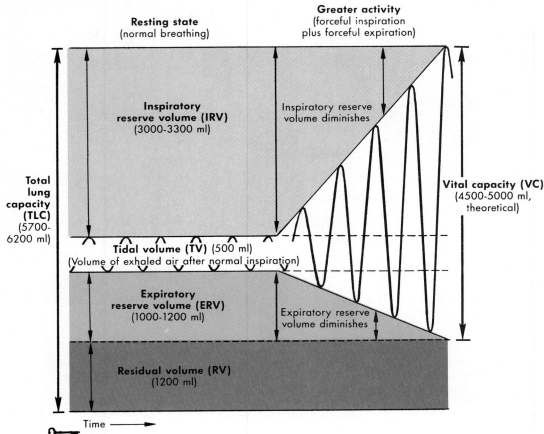

FIGURE 23-5 Spirogram. During normal, quiet respirations the atmosphere and lungs exchange about 500 ml of air (TV). With a forcible inspiration, about 3,300 ml more air can be inhaled (IRV). After a normal inspiration and normal expiration, approximately 1,000 ml more air can be forcibly expired (ERV). Vital capacity is the amount of air that can be forcibly expired after a maximal inspiration and indicates, therefore, the largest amount of air that can enter and leave the lungs during respiration. Residual volume is the air that remains trapped in the alveoli.

(FRC) is the amount of air left in the lungs at the end of a normal expiration. Therefore, as Figure 23-5 implies, FRC = ERV + RV. With the volumes given, the functional residual capacity is 2,200 to 2,400 ml. The total volume of air a lung can hold is called the *total lung capacity (TLC)*. It is, as Figure 23-5 indicates, the sum of all four lung volumes.

The term **alveolar ventilation** means the volume of inspired air that actually reaches, "ventilates," the alveoli. Only this volume of air takes part in the exchange of gases between air and blood. (Alveolar air exchanges some of its oxygen for some of the blood's carbon dioxide.) With every breath we take, part of the entering air necessarily fills our air passageways—nose, pharynx, larynx, trachea, and bronchi. This portion of air does not descend into any alveoli and therefore cannot take part in gas exchange. In this sense, it is "dead air." Appropriately, the larger air passageways it occupies are said to constitute the **anatomical dead space.** One rule of thumb estimates the volume of air in the anatomical dead space as the same number of milliliters as the individual's weight in pounds. Another generalization says that the anatomical dead space approximates 30% of the TV. TV minus dead space volume equals

alveolar ventilation volume. Suppose you have a normal TV of 500 ml and that 30% of this, or 150 ml, fills the anatomical dead space. The amount of air reaching your alveoli—your alveolar ventilation volume—is then 350 ml per breath, or 70% of your TV. Emphysema and certain other abnormal conditions, in effect, increase the amount of dead space air. Consequently, alveolar ventilation decreases and this, in turn, decreases the amount of oxygen that can enter blood and the amount of carbon dioxide that can leave it. Inadequate air-blood gas exchange, therefore, is the inevitable result of inadequate alveolar ventilation. Stated differently, the alveoli must be adequately ventilated for an adequate gas exchange to take place in the lungs.

QUICK CHECK

✔ 1. *What is the difference between a pulmonary volume and a pulmonary capacity?*

✔ 2. *The volume of air that is expired after a normal inspiration during normal, quiet breathing goes by what name?*

✔ 3. *What is meant by the term* vital capacity?

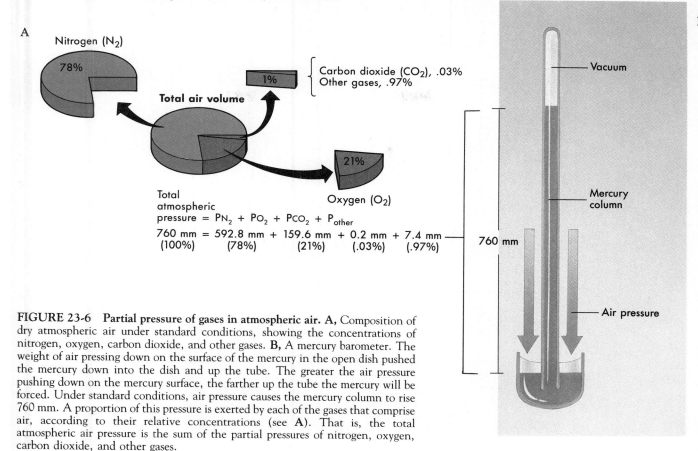

FIGURE 23-6 Partial pressure of gases in atmospheric air. A, Composition of dry atmospheric air under standard conditions, showing the concentrations of nitrogen, oxygen, carbon dioxide, and other gases. **B,** A mercury barometer. The weight of air pressing down on the surface of the mercury in the open dish pushed the mercury down into the dish and up the tube. The greater the air pressure pushing down on the mercury surface, the farther up the tube the mercury will be forced. Under standard conditions, air pressure causes the mercury column to rise 760 mm. A proportion of this pressure is exerted by each of the gases that comprise air, according to their relative concentrations (see **A**). That is, the total atmospheric air pressure is the sum of the partial pressures of nitrogen, oxygen, carbon dioxide, and other gases.

PULMONARY GAS EXCHANGE

Partial Pressure

Before discussing the exchange of gases across the respiratory membranes, we need to understand the **law of partial pressures** (*Dalton's law*). The term *partial pressure* means the pressure exerted by any one gas in a mixture of gases or in a liquid. According to the law of partial pressures, the partial pressure of a gas in a mixture of gases is directly related to the concentration of that gas in the mixture and to the total pressure of the mixture. Figure 23-6 shows how each gas in atmospheric air contributes to the total atmospheric pressure. The partial pressure of each gas is directly related to its concentration in the total mixture. Suppose we apply this principle to compute the partial pressure of oxygen in the atmosphere. The concentration of oxygen in the atmosphere is about 21%, and the total pressure of the atmosphere is 760 mm Hg under standard conditions. Therefore:

$$\text{Atmospheric } P_{O_2} = 21\% \times 760 = 159.6 \text{ mm Hg}$$

The symbol used to designate partial pressure is the capital letter P preceding the chemical symbol for the gas. Examples: alveolar air P_{O_2} is about 100 mm Hg; arterial blood P_{O_2} is also about 100 mm Hg; venous blood P_{O_2} is about 37 mm Hg. The word **tension** is often used as a synonym for the term *partial pressure*—oxygen tension means the same thing as P_{O_2}.

The partial pressure of a gas in a liquid is directly determined by the amount of that gas dissolved in the liquid, which in turn is determined by the partial pressure of the gas in the environment of the liquid. Gas molecules diffuse into a liquid from its environment and dissolve in the liquid until the partial pressure of the gas in solution becomes equal to its partial pressure in the environment of the liquid. Alveolar air constitutes the environment of blood moving through pulmonary capillaries. Standing between the blood and the air are only the very thin alveolar and capillary membranes, and both of these are highly permeable to oxygen and carbon dioxide. By the time blood leaves the pulmonary capillaries as arterial blood, diffusion and approximate equilibration of oxygen and carbon dioxide across the membranes have occurred. Arterial blood P_{O_2} and P_{CO_2} therefore usually equal or very nearly equal alveolar P_{O_2} and P_{CO_2} (Table 23-1).

Exchange of Gases in the Lungs

Exchange of gases in the lungs takes place between alveolar air and blood flowing through lung capillaries. Gases move in both directions through the respiratory membrane (see Figure 22-16). Oxygen enters blood from the alveolar air because the P_{O_2} of alveolar air is greater than the P_{O_2} of incoming blood. Another way of saying this is that oxygen diffuses "down" its pressure gradient. Simultaneously, carbon dioxide molecules exit from the blood by diffusing down the carbon dioxide pressure gradient out into the alveolar air. The P_{CO_2} of venous blood is much higher than the P_{CO_2} of alveolar air. This

PNEUMOTHORAX

Air in the pleural space may accumulate when the visceral pleura ruptures and air from the lung rushes out or when atmospheric air rushes in through a wound in the chest wall and parietal pleura. In either case, the lung collapses and normal respiration is impaired. Air in the thoracic cavity is a condition known as **pneumothorax.** To apply some of the information you have learned about the respiratory mechanism, let us suppose that a surgeon makes an incision through the chest wall into the pleural space, as is done in one of the dramatic, modern open-chest operations. What change, if any, can you deduce takes place in respirations? Compare your deductions with those in the next paragraph.

Intrathoracic pressure, of course, immediately increases from its normal subatmospheric level to the atmospheric level.

More pressure than normal is therefore exerted on the outer surface of the lung and causes its collapse. It could even collapse the other lung. Why? Because the mediastinum is a mobile rather than a rigid partition between the two pleural sacs. This anatomical fact allows the increased pressure in the side of the chest that is open to push the heart and other mediastinal structures over toward the intact side, where they exert pressure on the other lung. Pneumothorax can also result from disruption of the visceral pleura and the resulting flow of pulmonary air into the pleural space.

Pneumothorax results in many respiratory and circulatory changes. They are of great importance in determining medical and nursing care but lie beyond the scope of this book.

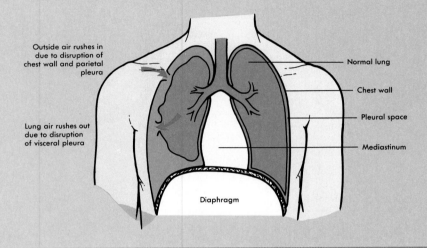

Outside air rushes in due to disruption of chest wall and parietal pleura

Lung air rushes out due to disruption of visceral pleura

Normal lung

Chest wall

Pleural space

Mediastinum

Diaphragm

TABLE 23-1	Oxygen and carbon dioxide pressure gradients			
	Atmo-sphere	Alveolar air	Systemic arterial blood	Systemic venous blood
P_{O_2}	160*	100	100	40
P_{CO_2}	0.2	40	40	46

*Figures indicate approximate mm Hg pressure under usual conditions.

two-way exchange of gases between alveolar air and pulmonary blood converts deoxygenated blood to oxygenated blood (Figure 23-7).

The amount of oxygen that diffuses into blood each minute depends on several factors, notably these four:

1. The oxygen pressure gradient between alveolar air and incoming pulmonary blood (alveolar P_{O_2} blood— P_{O_2})
2. The total functional surface area of the respiratory membrane

3. The respiratory minute volume (respiratory rate per minute times volume of air inspired per respiration)
4. Alveolar ventilation (discussed on p. 605)

All four of these factors bear a direct relation to oxygen diffusion. Anything that decreases alveolar P_{O_2}, for instance, tends to decrease the alveolar-blood oxygen pressure gradient and therefore tends to decrease the amount of oxygen entering the blood. Application: Alveolar air P_{O_2} decreases as altitude increases, and therefore less oxygen enters the blood at high altitudes. At a certain high altitude, alveolar air P_{O_2} equals the P_{O_2} of blood entering the pulmonary capillaries. How would this affect oxygen diffusion into blood?

Anything that decreases the total functional surface area of the respiratory membrane also tends to decrease oxygen diffusion into the blood (functional surface area is meant as that which is freely permeable to oxygen). Application: in emphysema the total functional area decreases and is one of the factors responsible for poor blood oxygenation in this condition.

Anything that decreases the respiratory minute volume

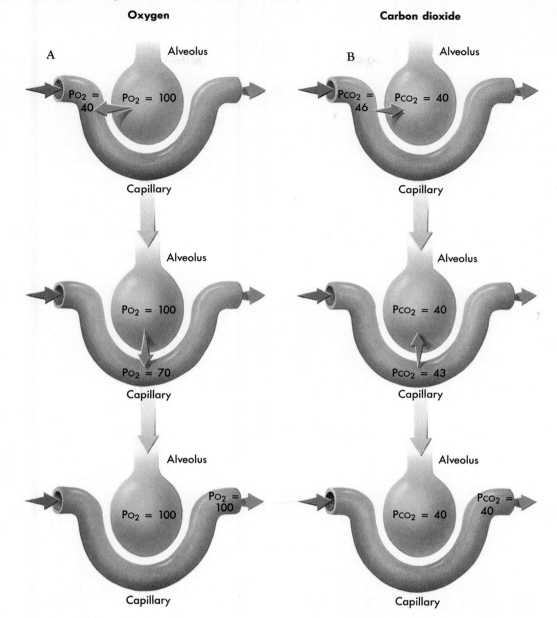

FIGURE 23-7 Pulmonary gas exchange. A, As blood enters a pulmonary capillary, O_2 diffuses down its pressure gradient (into the blood). O_2 continues diffusing into the blood until equilibration has occurred (or until the blood leaves the capillary). **B,** As blood enters a pulmonary capillary, CO_2 diffuses down its pressure gradient (out of the blood). As with O_2, CO_2 continues diffusing as long as there is a pressure gradient.

also tends to decrease blood oxygenation. Application: morphine slows respirations and therefore decreases the respiratory minute volume (volume of air inspired per minute) and tends to lessen the amount of oxygen entering the blood.

Several times we have stated the principle that structure determines function. You may find it interesting to note the application of this principle to gas exchange in the lungs. Several structural facts facilitate oxygen diffusion from the alveolar air into the blood in lung capillaries:

♦ The walls of the alveoli and of the capillaries together form a very thin barrier for the gases to cross

(estimated at not more than 0.004 mm thick—see Figure 22-16).

♦ Alveolar and capillary surfaces both are extremely large.

♦ Lung capillaries accommodate a large amount of blood at one time. The lung capillaries of a small individual—one who has a body surface area of 1.5 square meters—contain about 90 ml of blood at one time under resting conditions.

♦ Blood is distributed through the capillaries in a layer so thin (equal only to the diameter of one red blood cell) that each red blood cell comes close to alveolar air.

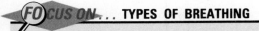

FOCUS ON... TYPES OF BREATHING

The alternate movement of air into and out of the lungs that we call breathing can occur in distinctive patterns that can be recognized and designated by name (see table below).

Eupnea is the term used to describe normal quiet breathing. During eupnea, the need for oxygen and carbon dioxide exchange is being met and the individual is usually not conscious of the breathing pattern. Ventilation occurs spontaneously at the rate of 12 to 17 breaths per minute.

Hyperpnea means increased breathing that is regulated to meet an increased demand by the body for oxygen. During hyperpnea, there is always an increase in pulmonary ventilation. The hyperpnea caused by exercise may meet the need for increased oxygen by an increase in tidal volume alone or by both an increase in tidal volume and breathing frequency.

Hyperventilation is characterized by an increase in pulmonary ventilation in excess of the need for oxygen. It sometimes results from a conscious voluntary effort preceding exertion or from psychogenic factors (hysterical hyperventilation). **Hypoventilation** is a decrease in pulmonary ventilation that results in elevated blood levels of carbon dioxide.

Dyspnea refers to labored or difficult breathing and is often associated with hypoventilation. A person suffering from dyspnea is aware, or conscious, of the breathing pattern and is generally uncomfortable and in distress. **Orthopnea** refers to dyspnea while lying down. It is relieved by sitting or standing up. This condition is common in patients with heart disease.

Several terms are used to describe the cessation of breathing. **Apnea** refers to the temporary cessation of breathing at the end of a normal expiration. It may occur during sleep or when swallowing. **Apneusis** is the cessation of breathing in the inspiratory position. Failure to resume breathing following a period of apnea, or apneusis, is called **respiratory arrest.**

Cheyne-Stokes respiration is a periodic type of abnormal breathing often seen in terminally ill or brain-damaged patients. It is characterized by cycles of gradually increasing tidal volume for several breaths followed by several breaths with gradually decreasing tidal volume. These cycles repeat in a type of crescendo-decrescendo pattern.

Biot's breathing is characterized by repeated sequences of deep gasps and apnea. This type of abnormal breathing pattern is seen in individuals suffering from increased intracranial pressure.

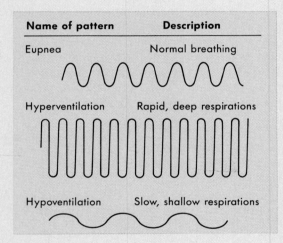

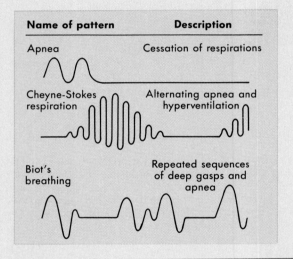

QUICK CHECK

✔ 1. *How does the partial pressure of a gas relate to its concentration?*

✔ 2. *What determines the direction in which oxygen will diffuse across the respiratory membrane?*

✔ 3. *List two of the four major factors that influence how much oxygen diffuses into pulmonary blood per minute.*

HOW BLOOD TRANSPORTS GASES

Blood transports oxygen and carbon dioxide as solutes and as parts of molecules of certain chemical compounds. Immediately on entering the blood, both oxygen and carbon dioxide dissolve in the plasma, but, because fluids can hold only small amounts of gas in solution, most of the oxygen and carbon dioxide rapidly form a chemical union with some other blood constituent. In this way, comparatively large volumes of the gases can be transported.

Transport of Oxygen

Since oxygenated blood has a PO_2 of 100 mm Hg, it contains only about 0.3 ml of dissolved O_2 per 100 ml of blood. Many times that amount, however, combines with the hemoglobin in 100 ml of blood to form **oxyhemoglobin.** Since each gram of hemoglobin can unite with 1.34 ml of oxygen, the exact amount of oxygen in blood depends mainly on the amount of hemoglobin present. Normally, 100 ml of blood contains about 15 g of hemoglobin. If 100% of it combines with oxygen, then 100 ml of blood will contain 15×1.34, or 20.1 ml, oxygen in the form of oxyhemoglobin. Figure 23-8 shows how hemoglobin increases the oxygen-carrying capacity of blood.

Perhaps a more common way of expressing blood oxygen content is in terms of volume percent. Normal arterial blood, with P_{O_2} of 100 mm Hg, contains about 20 vol% O_2 (meaning 20 ml of oxygen in each 100 ml of blood).

Blood that contains more hemoglobin can, of course, transport more oxygen, and that which contains less

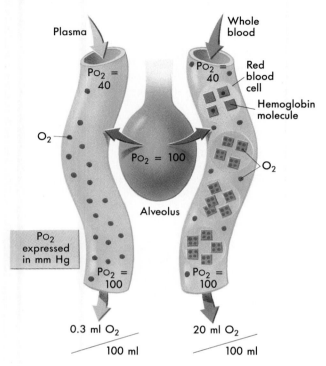

FIGURE 23-8 **Oxygen-carrying capacity of blood.** If blood consisted only of plasma, the maximum O_2 that could be transported is only about 0.3 ml of O_2 per 100 ml of blood. Because the red blood cells contain hemoglobin molecules, which act as "oxygen sponges," the blood can actually carry up to 20 ml of dissolved O_2 per 100 ml of blood.

hemoglobin can transport less oxygen. Hence hemoglobin-deficiency anemia decreases oxygen transport and may produce marked cellular hypoxia (inadequate oxygen supply).

To combine with hemoglobin, oxygen must, of course, diffuse from plasma into the red blood cells where millions of hemoglobin molecules are located. Several factors influence the rate at which hemoglobin combines with oxygen in lung capillaries. For instance, as the following equation and the *oxygen-hemoglobin dissociation curve* (Figure 23-9) show, an increasing blood P_{O_2} accelerates hemoglobin association with oxygen:

$$Hb + O_2 \xrightarrow{\text{Increasing } P_{O_2}} HbO_2$$

Decreasing P_{O_2}, on the other hand, accelerates oxygen dissociation from oxyhemoglobin, that is, the reverse of the preceding equation. Oxygen associates with hemoglobin rapidly—so rapidly, in fact, that about 97% of the blood's hemoglobin has united with oxygen by the time the blood leaves the lung capillaries to return to the heart. In other words, the average oxygen saturation of hemoglobin in oxygenated blood is about 97%.

Summing up, we can say that oxygen travels in two forms: as dissolved O_2 in the plasma and associated with hemoglobin (oxyhemoglobin). Of these two forms of transport, oxyhemoglobin carries the vast majority of the total oxygen transported by the blood.

Transport of Carbon Dioxide

Carbon dioxide is carried in the blood in several ways, the most important of which are described briefly in the following paragraphs.

A small amount of CO_2 dissolves in plasma and is transported as a solute (dissolved CO_2 produces the P_{CO_2} of blood).

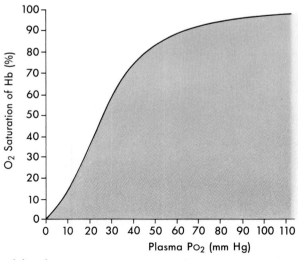

FIGURE 23-9 **Hemoglobin dissociation curve.** This graph represents the relationship between P_{O_2} and O_2 saturation of hemoglobin (Hb-O_2 affinity). Notice that at high plasma P_{O_2} values, hemoglobin (Hb) is fully loaded with O_2. At low plasma P_{O_2} values, Hb is only partially loaded with O_2.

Somewhat less than one quarter of blood carbon dioxide unites with the NH$_2$ (amine) groups of hemoglobin and certain other proteins to form carbamino compounds. Most of these are formed and transported in the red blood cells, since hemoglobin is the main protein to combine with carbon dioxide. The compound formed has a tongue-twisting name—**carbaminohemoglobin.** Carbon dioxide association with hemoglobin is accelerated by an increased P$_{CO_2}$ and is slowed by a decrease in P$_{CO_2}$ (Figure 23-10).

More than two thirds of the CO$_2$ is carried in the plasma as bicarbonate ions (HCO$_3^-$). When CO$_2$ dissolves in water (as in blood plasma), some of the CO$_2$ molecules associate with H$_2$O to form carbonic acid (H$_2$CO$_3$). Once formed, some of the H$_2$CO$_3$ molecules dissociate to form H$^+$ and bicarbonate (HCO$_3^-$) ions. This process, which is catalyzed by enzymes present in red blood cells, is summarized by this chemical equation:

$$CO_2 + H_2O \leftrightarrows H_2CO_3 \leftrightarrows H^+ + HCO_3^-$$

According to basic principles of chemistry, as more CO$_2$ is added to the plasma, more will be converted to bicarbonate ions. Because enzymes in the blood are continually "pulling" the equilibrium toward the bicarbonate side, CO$_2$ molecules will continually be removed from the solution (Figure 23-11). This allows even more CO$_2$ to dissolve in the plasma—thus increasing the CO$_2$-carrying capacity of the blood.

> ### QUICK CHECK
> ✔ 1. *Most oxygen carried by the blood is transported in what form?*
> ✔ 2. *Most carbon dioxide carried by the blood is transported in what form?*
> ✔ 3. *What is oxyhemoglobin? What is carbaminohemoglobin?*

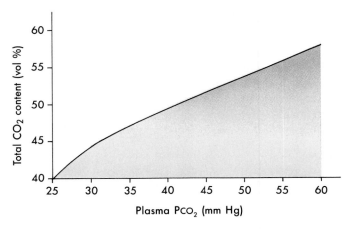

FIGURE 23-10 Carbon dioxide dissociation curve. The relationship between P$_{CO_2}$ and total CO$_2$ content (vol%) is graphed as a nearly straight line. Notice that the CO$_2$-carrying capacity of blood increases as the plasma P$_{CO_2}$ increases.

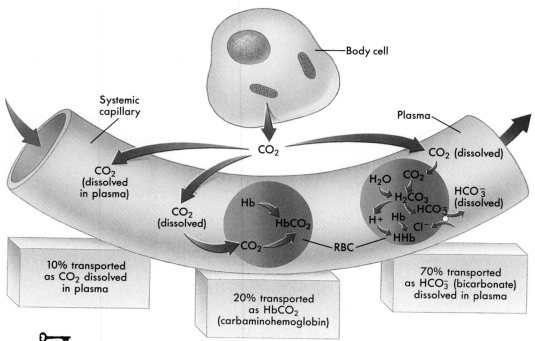

FIGURE 23-11   Carbon dioxide transport in the blood. As the illustration shows, CO$_2$ dissolves in the plasma. Some of the dissolved CO$_2$ enters red blood cells (RBCs) and combines with hemoglobin (Hb) to form carbaminohemoglobin (HbCO$_2$). Some of the CO$_2$ entering RBCs combines with H$_2$O to form carbonic acid (H$_2$CO$_3$), a process facilitated by an enzyme (carbonic anhydrase) present inside each cell. Carbonic acid then dissociates to form H$^+$ and bicarbonate (HCO$_3^-$). The H$^+$ combines with Hb, while the HCO$_3^-$ diffuses down its concentration gradient into the plasma. As HCO$_3^-$ leaves each RBC, Cl$^-$ enters and prevents an imbalance in charge—a phenomenon called the chloride shift, which will be discussed in Chapter 29.

Systemic Gas Exchange

Exchange of gases in tissues takes place between arterial blood flowing through tissue capillaries and cells (Figure 23-12). It occurs because of the principle already noted—that gases move down a gas pressure gradient. More specifically, in the tissue capillaries, oxygen diffuses out of arterial blood because the oxygen pressure gradient favors its outward diffusion (see Figure 23-12). Arterial blood Po_2 is about 100 mm Hg, interstitial fluid Po_2 is considerably lower, and intracellular fluid Po_2 is still lower. Although interstitial fluid and intracellular fluid Po_2 are not definitely established, they are thought to vary considerably—perhaps from around 60 mm Hg down to about 1 mm Hg. As activity increases in any structure, its cells necessarily use oxygen more rapidly. This decreases intracellular and interstitial Po_2, which in turn tends to increase the oxygen pressure gradient between blood and tissues and to accel-erate oxygen diffusion out of the tissue capillaries. In this way, the rate of oxygen use by cells automatically tends to regulate the rate of oxygen delivery to cells. As dissolved oxygen diffuses out of arterial blood, blood Po_2 decreases, and this accelerates oxyhemoglobin dissociation to release more oxygen into the plasma for diffusion out to cells, as indicated in Figure 23-13 and the following equation:

$$Hb + O_2 \xleftarrow{\text{Decreasing } Po_2} HbO_2$$

Because of oxygen release to tissues from tissue capillary blood, Po_2, oxygen saturation, and total oxygen content are less in venous blood than in arterial blood, as shown in Table 23-2.

Carbon dioxide exchange between tissues and blood takes place in the opposite direction from oxygen exchange. Catabolism produces large amounts of CO_2 inside cells. Therefore intracellular and interstitial Pco_2 are

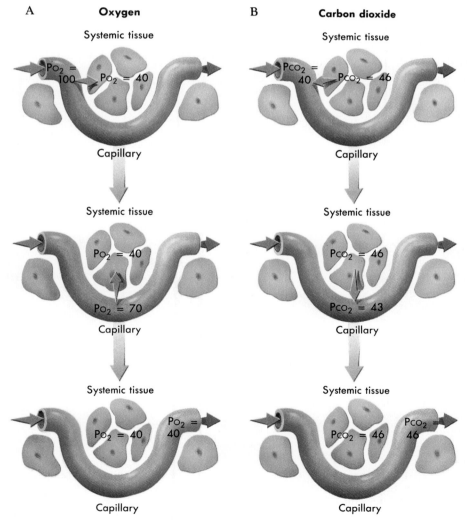

A **Oxygen**

Systemic tissue

$Po_2 = 100$ $Po_2 = 40$

Capillary

Systemic tissue

$Po_2 = 40$

$Po_2 = 70$

Capillary

Systemic tissue

$Po_2 = 40$ $Po_2 = 40$

Capillary

B **Carbon dioxide**

Systemic tissue

$Pco_2 = 40$ $Pco_2 = 46$

Capillary

Systemic tissue

$Pco_2 = 46$

$Pco_2 = 43$

Capillary

Systemic tissue

$Pco_2 = 46$ $Pco_2 = 46$

Capillary

FIGURE 23-12 Systemic gas exchange. A, As blood enters a systemic capillary, O_2 diffuses down its pressure gradient (out of the blood). O_2 continues diffusing out of the blood until equilibration has occurred (or until the blood leaves the capillary). **B,** As blood enters a pulmonary capillary, CO_2 diffuses down its pressure gradient (into the blood). As with O_2, CO_2 continues diffusing as long as there is a pressure gradient.

higher than arterial blood P_{CO_2}. This means that the CO_2 pressure gradient causes diffusion of CO_2 from the tissues into the blood flowing along through tissue capillaries (see Figure 23-12). Consequently, the P_{CO_2} of blood increases in tissue capillaries from its arterial level of about 40 mm Hg to its venous level of about 46 mm Hg. This increasing

P_{CO_2} and decreasing P_{O_2} together produce two effects—they favor oxygen dissociation from oxyhemoglobin and carbon dioxide association with hemoglobin to form carbaminohemoglobin. This reciprocal interrelationship between oxygen and carbon dioxide transport mechanisms is contrasted in Figure 23-14. Note that increased P_{CO_2} decreases the affinity between hemoglobin and oxygen—this is called a "right shift," or the Bohr effect. The Haldane effect refers to the increased CO_2 loading caused by a decrease in P_{O_2}.

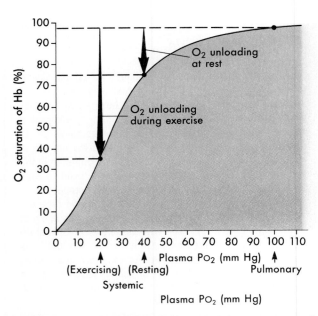

FIGURE 23-13 O_2 unloading at rest and during exercise. At rest, fully saturated Hb unloads almost 25% of its O_2 load when it reaches the low P_{O_2} (40 mm Hg) environment in systemic tissues. During exercise, the tissue P_{O_2} is even lower (20 mm Hg)—thus causing fully saturated Hb to unload about 70% of its O_2 load. As you can see, a slight drop in tissue P_{O_2} causes a large increase in O_2 unloading.

QUICK CHECK

✔ 1. What factors can cause the attraction between hemoglobin and oxygen to decrease?

✔ 2. What factors can cause an increase in the amount of carbon dioxide loaded into the systemic blood?

TABLE 23-2	Blood oxygen	
	Systemic venous blood	**Systemic arterial blood**
P_{O_2}	40 mm Hg	100 mm Hg
Oxygen saturation	75%	97%
Oxygen content	15 ml O_2 per 100 ml blood	20 ml O_2 per 100 ml blood*

*Oxygen use by tissues = difference between oxygen contents of arterial and venous blood (20-15) = 5 ml O_2 per 100 ml blood circulated per minute.

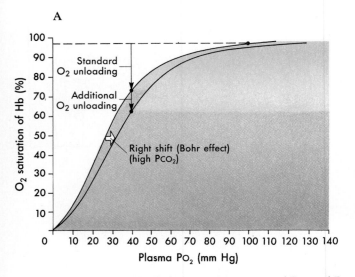

A

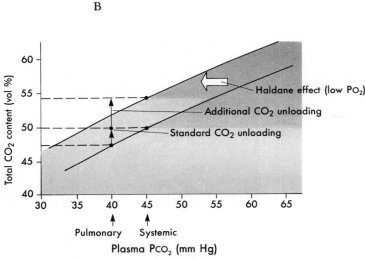

B

FIGURE 23-14 Interaction of P_{O_2} and P_{CO_2} on gas transport by the blood. A, The increased P_{CO_2} in systemic tissues decreases the affinity between hemoglobin and O_2, shown as a right shift of the oxygen-hemoglobin dissociation curve. This phenomenon is known as the *Bohr effect.* **B,** At the same time, the decreased P_{O_2} commonly observed in systemic tissues increases the CO_2 content of the blood, shown as a left shift of the CO_2 dissociation curve. This phenomenon is known as the *Haldane effect.*

REGULATION OF BREATHING
Respiratory control centers

Various mechanisms operate to maintain relative constancy of the blood P_{O_2} and P_{CO_2}. This homeostasis of blood gases is maintained primarily by means of changes in ventilation—the rate and depth of breathing. The main integrators that control the nerves that affect the inspiratory and expiratory muscles are located within the brain stem and are together simply called the *respiratory centers* (Figure 23-15).

The basic rhythm of the respiratory cycle of inspiration and expiration seems to be generated by the **medullary rhythmicity area.** This area of the medulla consists of two interconnected control centers: the *inspiratory center* and the *expiratory center.* As their names imply, output from the inspiratory center stimulates inspiration, and output from the expiratory center stimulates expiration. Because normal, quiet breathing involves stimulation of inspiratory muscles (mainly the diaphragm) alternating with relaxation of the same muscles, the inspiratory area is thought to act as the primary respiratory pacemaker. The expiratory area seems to be active only when expiratory muscles are needed during forced expiration.

A current hypothesis suggests that the basic breathing rhythm can be altered by different inputs to the medullary rhythmicity area. For example, input from the **apneustic center** in the pons stimulates the inspiratory center to increase the length and depth of inspiration; breathing characterized by abnormally long, deep inspirations is sometimes called "apneustic breathing." The **pneumotaxic center,** also in the pons, normally inhibits both the apneustic center and the inspiratory center. This prevents overinflation of the lungs and thus permits a normal rhythm of breathing.

Diving Reflex

A protective physiological response called the **diving reflex** is responsible for the astonishing recovery of apparent drowning victims who have been submerged for over 40 minutes. Survivors are most often preadolescent children who have been immersed in water below 20° C. Apparently, the colder the water, the better the chance of survival. Victims initially appear dead when pulled from the water. Breathing has stopped; they have fixed, dilated pupils; they are cyanotic; and the pulse has stopped.

Studies have shown that when the head and face are immersed in ice-cold water there is immediate shunting of blood to the core body areas with peripheral vasoconstriction and slowing of the heart (bradycardia). Metabolism is slowed, and tissue requirements for oxygen and nutrients decrease. The diving reflex is a protective response of the body to cold water immersion and is a function of such physiological and environmental parameters as water temperature, age, lung volume, and posture.

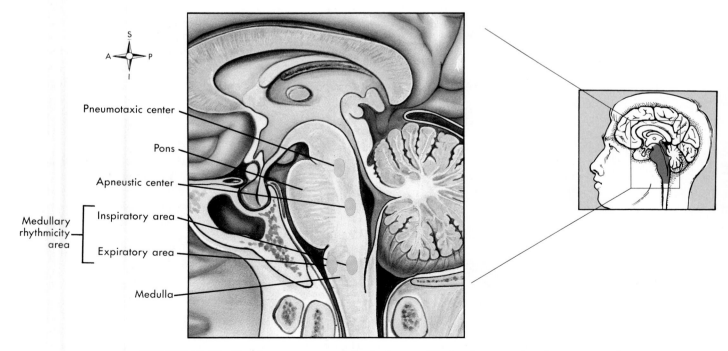

FIGURE 23-15 Respiratory centers of the brainstem. The inspiratory and expiratory areas of the medulla comprise the medullary rhythmicity area. The pneumotaxic center and apneustic center of the pons influence the basic respiratory rhythm by means of neural inputs to the medullary rhythmicity area.

Factors that Influence Breathing

Feedback information to the medullary rhythmicity area comes from sensors throughout the nervous system, as well as from other control centers. For example, changes in the P_{CO_2}, P_{O_2}, and pH of systemic arterial blood all influence the medullary rhythmicity area. The P_{CO_2} acts on chemoreceptors located in the medulla. Chemoreceptors, in this case, are cells that are sensitive to changes in arterial blood's CO_2 and hydrogen ion concentrations (pH). The normal range for arterial P_{CO_2} is about 38 to 40 mm Hg. When it increases even slightly above this, it has a stimulating effect, mainly on *central chemoreceptors* (postulated to be present in the medulla). Large but tolerable increases in arterial P_{CO_2} stimulate *peripheral chemoreceptors* in the carotid bodies and aorta. Stimulation of chemoreceptors by increased arterial P_{CO_2} results in faster breathing, with a greater volume of air moving in and out of the lungs per minute. Decreased arterial P_{CO_2} produces opposite effects—it inhibits central and peripheral chemoreceptors, which leads to inhibition of the medullary rhythmicity area and slower respirations. In fact, breathing stops entirely for a few moments (apnea) when arterial P_{CO_2} drops moderately—to about 35 mm Hg, for example.

A decrease in arterial blood pH (increase in acid), within certain limits, has a stimulating effect on chemoreceptors located in the carotid and aortic bodies.

The role of **arterial blood P_{O_2}** in controlling respirations is not entirely clear. Presumably, it has little influence as long as it stays above a certain level. But neurons of the respiratory centers, like all body cells, require adequate amounts of oxygen to function optimally. Consequently, if they become hypoxic, they become depressed and send fewer impulses to respiratory muscles. Respirations then decrease or fail entirely. This principle has important clinical significance. For example, the respiratory centers cannot respond to stimulation by an increasing blood CO_2 if, at the same time, blood P_{O_2} falls below a critical level—a fact that may become life or death important during anesthesia.

However, a decrease in arterial blood P_{O_2} below 70 mm Hg but not so low as the critical level stimulates chemoreceptors in the carotid and aortic bodies and causes reflex stimulation of the inspiratory center. This constitutes an emergency respiratory control mechanism. It does not help regulate respirations under usual conditions when arterial blood P_{O_2} remains considerably higher than 70 mm Hg—the level necessary to stimulate the chemoreceptors.

Arterial blood pressure helps control breathing through the respiratory pressoreflex mechanism. A sudden rise in arterial pressure, by acting on aortic and carotid baroreceptors, results in reflex slowing of respirations. A sudden drop in arterial pressure brings about a reflex increase in rate and depth of respirations. The pressoreflex mechanism is probably not of great importance in the control of respirations. It is, however, of major importance in the control of circulation.

The **Hering-Breuer reflexes** also help control respirations, particularly their depth and rhythmicity. They are

believed to regulate the normal depth of respirations (extent of lung expansion), and therefore the volume of tidal air, in the following way. Presumably, when the tidal volume of air has been inspired, the lungs are expanded enough to stimulate stretch receptors located within them. The stretch receptors then send inhibitory impulses to the inspiratory center, relaxation of inspiratory muscles occurs, and expiration follows the Hering-Breuer expiratory reflex. Then, when the tidal volume of air has been expired, the lungs are sufficiently deflated to inhibit the lung stretch receptors and allow inspiration to start again—the Hering-Breuer inspiratory reflex.

The **cerebral cortex** also influences breathing. Impulses to the respiratory center from the motor area of the cerebrum may either increase or decrease the rate and strength of respirations. In other words, an individual may voluntarily speed up or slow down the breathing rate. This voluntary control of respirations, however, has certain limitations. For example, one may will to stop breathing and do so for a few minutes, but holding the breath results in an increase in the CO_2 content of the blood, since it is not being removed by respirations. Carbon dioxide is a powerful respiratory stimulant. So when arterial blood P_{CO_2} increases to a certain level, it stimulates the inspiratory center (directly and reflexively) to send motor impulses to the respiratory muscles, and breathing is resumed, even though the individual may still will contrarily. This knowledge that the CO_2 content of the blood is a more

powerful regulator of respirations than cerebral impulses is of practical value when dealing with children who hold their breath to force the granting of their wishes. The best treatment is to ignore such behavior, knowing that respirations will start again as soon as the amount of CO_2 in arterial blood increases to a certain level.

Miscellaneous factors also may influence breathing. Among these are blood temperature and sensory impulses from skin thermal receptors and from superficial or deep pain receptors:

♦ *Sudden painful stimulation* produces a reflex apnea, but continued painful stimuli cause faster and deeper respirations.

♦ *Sudden cold stimuli* applied to the skin cause reflex apnea.

♦ *Afferent impulses* initiated by stretching the anal sphincter produce reflex acceleration and deepening of respirations. Use has sometimes been made of this mechanism as an emergency measure to stimulate respirations during surgery.

♦ *Stimulation of the pharynx or larynx* by irritating chemicals or by touch causes a temporary apnea. This is the choking reflex, a valuable protective device. It operates, for example, to prevent aspiration of food or liquids during swallowing.

The major factors that influence breathing are summarized in Figure 23-16.

QUICK CHECK

✔ 1. *Where are the chief regulatory centers of the respiratory function located?*

✔ 2. *Name several factors that can influence the breathing rate of an individual; tell whether each triggers an increase or decrease in breathing rate.*

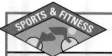

CONTROL OF RESPIRATIONS DURING EXERCISE

Respirations increase abruptly at the beginning of exercise and decrease even more markedly as it ends. This much is known. The mechanism that accomplishes this, however, is not known. It is not identical to the one that produces moderate increases in breathing. Numerous studies have shown that arterial blood P_{CO_2}, P_{O_2}, and pH do not change enough during exercise to produce the degree of hyperpnea (faster, deeper respirations) observed. Presumably, many chemical and nervous factors and temperature changes operate as a complex, but still unknown, mechanism for regulating respirations during exercise.

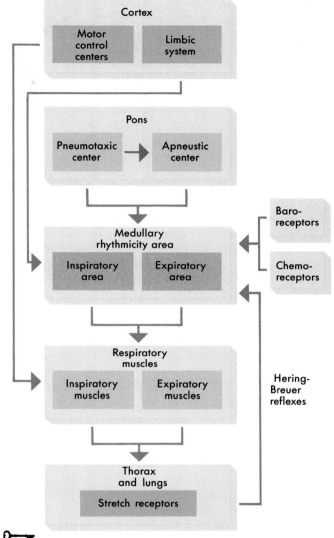

FIGURE 23-16 Scheme of respiratory regulation. This diagram summarizes the fact that the actions of the respiratory muscles are governed by control centers in the medulla, with feedback or other input from other regions of the nervous system.

HEALTH MATTERS

HEIMLICH MANEUVER

The Heimlich maneuver is an effective and often lifesaving technique that can be used to open a windpipe that is suddenly obstructed. The maneuver (see figures) uses air already present in the lungs to expel the object obstructing the trachea. Most accidental airway obstructions result from pieces of food aspirated during a meal; the condition is sometimes referred to as a "cafe coronary." Other objects such as chewing gum or balloons are frequently the cause of obstructions in children. Individuals trained in emergency procedures must be able to tell the difference between airway obstruction and other conditions such as heart attacks that produce similar symptoms. The key question they must ask the person who appears to be choking is, "Can you talk?" A person with an obstructed airway will not be able to speak, even while conscious. The Heimlich maneuver, if the victim is standing, consists of the rescuer's grasping the victim with both arms around the victim's waist just below the rib cage and above the navel. The rescuer makes a fist with one hand, grasps it with the other, and then delivers an upward thrust against the diaphragm just below the xiphoid process of the sternum. Air trapped in the lungs is compressed, forcing the object that is choking the victim out of the airway.

Technique if victim can be lifted (A)

1. Rescuer stands behind the victim and wraps both arms around the victim's chest slightly below the rib cage and above the navel. Victim is allowed to fall forward with head, arms, and chest over the rescuer's arms.
2. Rescuer makes a fist with one hand and grasps it with the other hand, pressing thumb side of fist against victim's abdomen just below the end of the xiphoid process and above the navel.
3. The hands only are used to deliver the upward subdiaphragmatic thrusts. It is performed with sharp flexion of the elbows, in an upward rather than inward direction, and is usually repeated four times. It is very important *not* to compress the rib cage or actually press on the sternum during the Heimlich maneuver.

Technique if victim has collapsed or cannot be lifted (B)

1. Rescuer places victim on floor face up.
2. Facing victim, rescuer straddles the hips.
3. Rescuer places one hand on top of the other, with the bottom hand on the victim's abdomen slightly above the navel and below the rib cage.
4. Rescuer performs a forceful upward thrust with the heel of the bottom hand, repeating several times if necessary.

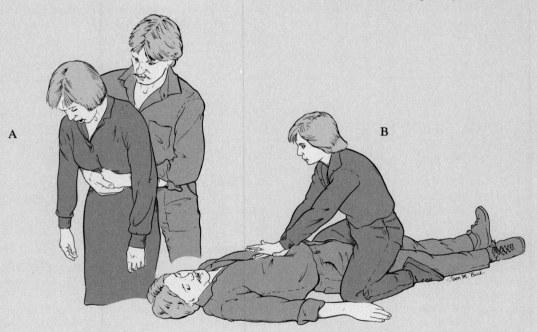

The Heimlich maneuver for dislodging aspirated foods or other foreign objects from the airway.
A, Technique if victim is standing. **B,** Technique if victim is collapsed.

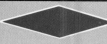

THE BIG PICTURE
The Respiratory System And The Whole Body

The homeostatic balance of the entire body, and thus the survival of each and every cell, depends on the proper functioning of the respiratory system. Because the mitochondria in each cell require oxygen for their energy conversions, and because each cell produces toxic carbon dioxide as a waste product of the very same energy conversions, the internal environment must continually get new oxygen and dump out carbon dioxide. If each cell were immediately adjacent to the external environment—atmospheric air—this would require no special system. However, because almost every one of the 100 trillion cells that make up the body are far away from the outside air, there must be a way of satisfying this condition—this is where the respiratory system comes in. By the process of ventilation, fresh external air continually flows less than a hair's breadth away from the circulating fluid of the body—the blood. By means of diffusion, oxygen enters the internal environment and carbon dioxide leaves. The efficiency of this process is enhanced by the presence of "oxygen sponges," called hemoglobin molecules, that immediately take O_2 molecules out of solution in the plasma so that more O_2 can rapidly diffuse into the blood. The blood, the circulating fluid tissue of the cardiovascular system, carries the blood gases throughout the body—picking up gases where there is an excess and unloading them where there is a deficiency. In this manner, each cell of the body is continually bathed in a fluid environment that offers a constant supply of oxygen and a bottomless dump for carbon dioxide.

Specific mechanisms involved in respiratory function show the interdependence between body systems observed throughout our study of the human body. For example, without blood and the maintenance of blood flow by the cardiovascular system, blood gases could not be transported between gas-exchange tissues of the lungs and the various systemic tissues of the body. Without regulation by the nervous system, ventilation could not be adjusted to compensate for changes in the oxygen or carbon dioxide content of the internal environment. Without the skeletal muscles of the thorax, the airways could not maintain the flow of fresh air that is so vital to respiratory function. The skeleton itself provides a firm outer housing for the lungs and an arrangement of bones that facilitates the expansion and recoil of the thorax, which is needed to accomplish inspiration and expiration. Without the immune system, pathogens from the external environment could easily colonize the respiratory tract and possibly cause a fatal infection.

Even more subtle interactions between the respiratory system and other systems can be found. For example, the language function of the nervous system is limited without the speaking ability provided by the larynx and other structures of the respiratory tract. The homeostasis of pH, which is regulated by a variety of systems, is influenced by the respiratory system's ability to adjust the body's CO_2 levels (and thus the levels of carbonic acid).

MECHANISMS OF DISEASE
Disorders Associated With Respiratory Function

Many things can interfere with functions of gas exchange and ventilation, causing respiratory failure. A few of the more important disorders are briefly described.

Restrictive Pulmonary Disorders

Restrictive pulmonary disorders involve restriction of the alveoli, leading to decreased lung inflation. The hallmark of these disorders, regardless of their cause, is decreased lung volumes and capacities such as inspiratory reserve volume and vital capacity. Factors that restrict breathing can originate either within the lung or outside of it. Causes of restrictive lung disorders include alveolar fibrosis (scarring) secondary to occupational exposure to asbestos, toxic fumes, coal dust, or other contaminants; immunological diseases, as in rheumatoid lung; obesity; and metabolic disorders, such as uremia. Restriction of breathing can also be caused by pain that accompanies pleurisy (inflammation of the pleurae) or mechanical injuries (such as a fractured or bruised rib). Patients with restrictive lung disease classically experience *dyspnea* (labored breathing) and do not tolerate increased activity,

which reduces their ability to work or perform normal daily activities. Therapy involves eliminating the cause of the restriction, ensuring adequate gas exchange, and improving exercise tolerance.

Obstructive Pulmonary Disorders

Different conditions may cause obstruction of the airways. Exposure to cigarette smoke and other common air pollutants can trigger a reflexive constriction of bronchial airways. Obstructive disorders may obstruct both inspiration and expiration, whereas restrictive disorders mainly restrict inspiration.

Chronic obstructive pulmonary disease (COPD) is a broad term used to describe conditions of progressive irreversible obstruction of expiratory air flow. People with COPD have difficulty breathing, mainly emptying their lungs, and develop visibly hyperinflated chests. They have a productive cough and intolerance of activity. The major disorders observed in people with COPD are bronchitis, emphysema, and asthma (Figure 23-17).

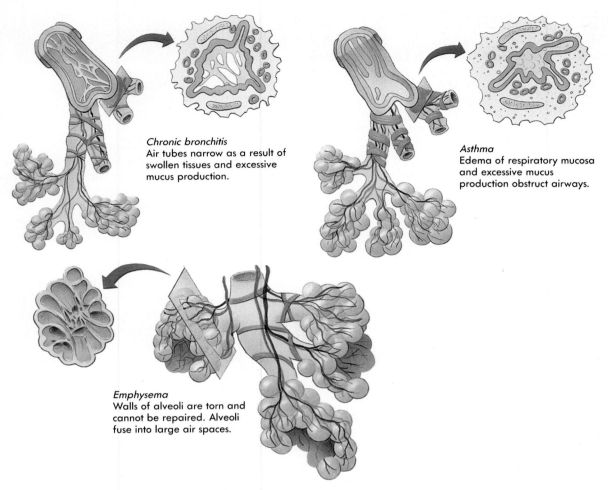

Chronic bronchitis
Air tubes narrow as a result of swollen tissues and excessive mucus production.

Asthma
Edema of respiratory mucosa and excessive mucus production obstruct airways.

Emphysema
Walls of alveoli are torn and cannot be repaired. Alveoli fuse into large air spaces.

FIGURE 23-17 **Disorders commonly observed in chronic obstructive pulmonary disease (COPD).**

Bronchitis

In **bronchitis,** the person produces excessive tracheobronchial secretions that obstruct air flow, and there is enlargement of the bronchial mucous glands. Risk factors include cigarette smoking (accounting for 80% to 90% of the risk of developing COPD), a normal decline in pulmonary function due to age, and environmental exposures to dust and chemicals. With impairment of the alveoli and loss of capillary beds, there is inefficient gas exchange, which in turn produces hypoxia.

Asthma

Asthma is an obstructive lung disorder that is characterized by recurring spasms of the smooth muscles in the walls of the bronchial air passages. These contractions narrow airways, making breathing difficult. Inflammation (edema and excessive mucus production) that usually accompanies the spasms further obstructs the airways. Initial onset of asthma can occur in children or adults. Acute episodes of asthma—so-called "asthma attacks"—can be triggered by stress, heavy exercise, infection, or exposure to allergens or other irritants such as

dusts, vapors, or fumes. Many patients with asthma have a family history of allergies. Dyspnea is the major symptom of asthma, but hyperventilation, headaches, numbness, and nausea can occur.

Acute respiratory failure can occur when any of the disorders that produce COPD become intense. Heart failure due to the pulmonary disease and vascular resistance that develops with COPD is another possible outcome. Although there is no cure, limiting symptoms can improve quality of life. Bronchodilators and corticosteroids have been used to relieve some of the airway obstruction involved in COPD.

Emphysema

In **emphysema,** there is enlargement of air spaces distal to the terminal bronchioles due to damage of lung connective tissue. As the alveoli enlarge, their walls rupture, fuse into large irregular spaces, and gas exchange units are destroyed. Although the etiology is not fully understood, this condition is believed to be caused by proteolytic enzymes that destroy lung tissue. Emphysema victims often develop hypoxia.

CASE STUDY
EMPHYSEMA

Mr. T. is a 65-year-old man who is the vice president of a prestigious law firm. As a result of the stressful nature of his job, Mr. T. has a habit of smoking two packs of cigarettes a day and has done so for 35 years. During the last few years Mr. T. has experienced slight shortness of breath and a mild cough with activity and on arising in the morning. Recently, Mr. T. has noticed that he has difficulty climbing the stairs at work, which not only produces fatigue but often requires him to stop at intervals so that he may catch his breath. He has also noticed that besides being short of breath with exertion, he tends to experience dyspnea (labored breathing) at rest. He has had an approximate weight loss of 10 pounds within the last 2 months. Mr. T. is brought into the emergency room by his wife for evaluation because of his increasing dyspnea and because he must sleep sitting up with the aid of several pillows.

On admission to the emergency room, Mr. T. is a thin, frail-looking man in acute respiratory distress. He is restless and breathing rapidly. He is sitting on the side of the bed, leaning on an over-the-bed table. His vital signs are as follows: heart rate, 120; respirations, 30; blood pressure, 140/80. A chest roentgenogram (x ray) is taken and arterial blood gases are drawn. Mr. T.'s arterial blood gases indicate the following: Po_2, 39 mm Hg; Pco_2, 52 mm Hg; pH, 7.32; bicarbonate, 36 mEq/L. He is now placed on 2 L of nasal oxygen and is sent to the medical intensive care unit for further evaluation and treatment.

The chest roentgenogram reveals a flat, low diaphragm and hyperinflation of the lungs. The lung fields are relatively clear but appear translucent, and no gross cardiac enlargement is seen. Auscultation (listening) of the lungs reveals decreased breath sounds with expiratory wheezes. Mr. T.'s chest has an increased anteroposterior diameter, and the accessory muscles are used for ventilation. Pulmonary function tests reveal decreased tidal volume, decreased vital capacity, increased total lung capacity, and prolonged forced expiratory volume. Arterial blood gases are redrawn. The results are as follows: Po_2, 49 mm Hg; Pco_2, 50 mm Hg; pH, 7.25; bicarbonate, 32 mEq/L.

1. Which of the following is the most likely cause of Mr. T.'s dyspnea?
   a. Increased lung compliance
   b. Decreased elastic recoil
   c. Decreased lung compliance
   d. Increased elastic recoil

2. The causes of Mr. T.'s barrel chest include:
   a. Excessive secretions
   b. Severe hypoxemia
   c. Hyperinflation of the lungs
   d. Thickening of bronchial mucosa

3. Retention of carbon dioxide in individuals with chronic obstructive pulmonary disease (COPD), such as Mr. T., is caused by what mechanism?
   a. Hyperventilation
   b. Dilation of the bronchial tree
   c. Hypoventilation
   d. Hypoxemia

4. Which of the following best explains Mr. T.'s weight loss?
   a. Cigarette smoking decreases his sense of taste, thus decreasing his appetite
   b. COPD increases the body's need for calories
   c. Cigarette smoking increases the basal metabolism rate, leading to weight loss
   d. COPD prevents the body from using nutrients

CHAPTER SUMMARY

RESPIRATORY PHYSIOLOGY

A. The respiratory system includes pulmonary ventilation, gas exchange in the lungs and tissues, transport of gases by the blood, and regulation of respiration

PULMONARY VENTILATION

A. Pulmonary ventilation (breathing)
   1. Inspiration—moves air into the lungs
   2. Expiration—moves air out of the lungs
B. Mechanism of pulmonary ventilation

1. The pulmonary ventilation mechanism must establish two gas pressure gradients
   a. One in which the pressure within the alveoli of the lungs is lower than atmospheric pressure to produce inspiration
   b. One in which the pressure in the alveoli of the lungs is higher than atmospheric pressure to produce expiration
2. Pressure gradients are established by changes in the size of the thoracic cavity that are produced by contraction and relaxation of muscles (Figure 23-1)

3. Boyle's law—the volume of gas varies inversely with pressure at a constant temperature
   a. Expansion of the thorax results in decreased intrapleural pressure, leading to a decreased alveolar pressure causing air to move into the lungs
4. Inspiration—contraction of the diaphragm produces inspiration—as it contracts, it makes the thoracic cavity larger (Figure 23-2)
5. Expiration—a passive process that begins when the inspiratory muscles are relaxed, decreasing the size of the thorax and increasing intrapleural pressure from about -6 mm Hg to a preinspiration level of -4 mm Hg (Figure 22-3)
   a. The pressure between parietal and visceral pleura is always less than atmospheric pressure
C. Pulmonary volumes—the amount of air moved in and out and remaining is important in order that a normal exchange of oxygen and carbon dioxide can take place (Figure 23-5)
   1. Spirometer—instrument used to measure volume of air (Figure 23-4)
   2. Tidal volume—amount of air exhaled after normal inspiration
   3. Expiratory reserve volume—largest volume of additional air that can be forcibly exhaled (between 1.0 and 1.2 L is normal ERV)
   4. Inspiratory reserve volume—amount of air that can be forcibly inhaled after normal inspiration (normal IRV is 3.3 liters)
   5. Residual volume—amount of air that cannot be forcibly exhaled (1.2 liters)
   6. Pulmonary capacity—the sum of two or more pulmonary volumes
   7. Vital capacity—the sum of IRV + TV + ERV
   8. Minimal volume—the amount of air remaining after RV
   9. A person's vital capacity depends on many factors, including the size of the thoracic cavity and posture
   10. Functional residual capacity—the amount of air at the end of a normal respiration
   11. Total lung capacity—the sum of all four lung volumes—the total amount of air a lung can hold
   12. Alveolar ventilation—volume of inspired air that reaches the alveoli
   13. Anatomical dead space—air in passageways that do not participate in gas exchange
   14. Alveoli must be properly ventilated for adequate gas exchange

PULMONARY GAS EXCHANGE

A. Partial pressure of gases—pressure exerted by a gas in a mixture of gases or a liquid (Figure 23-6)
   1. Law of partial pressures (Dalton's law)—the partial pressure of a gas in a mixture of gases is directly related to the concentration of that gas in the mixture and to the total pressure of the mixture
   2. Arterial blood P_{O_2} and P_{CO_2} equal alveolar P_{O_2} and P_{CO_2}
B. The exchange of gases in the lungs takes place between alveolar air and blood flowing through lung capillaries (Figure 23-7)
   1. Four factors determine the amount of oxygen that diffuses into blood
      a. The oxygen pressure gradient between alveolar air and blood
      b. The total functional surface area of the respiratory membrane
      c. The respiratory minute volume
      d. Alveolar ventilation
   2. Structural facts that facilitate oxygen diffusion from the alveolar air to the blood

a. The fact that the walls of the alveoli and capillaries form only a very thin barrier for gases to cross
b. The fact that alveolar and capillary surfaces are large
c. The fact that the blood is distributed through the capillaries in a thin layer so each red blood cell comes close to alveolar air

HOW BLOOD TRANSPORTS GASES

A. Oxygen and carbon dioxide are transported as solutes and as parts of molecules of certain chemical compounds
B. Transport of oxygen—oxygenated blood contains about 0.3 ml of dissolved O_2 per 100 ml of blood
   1. Hemoglobin increases the oxygen-carrying capacity of blood (Figure 23-8)
   2. Oxygen travels in two forms: as dissolved O_2 in plasma and associated with hemoglobin (oxyhemoglobin)
      a. Oxyhemoglobin carries the majority of the total oxygen transported by blood
C. Transport of carbon dioxide (Figure 23-11)
   1. A small amount of CO_2 dissolves in plasma and is transported as a solute (10%)
   2. Less than one fourth of blood carbon dioxide combines with NH_2 (amine) groups of hemoglobin and other proteins to form carbaminohemoglobin (20%)
   3. More than two thirds of the carbon dioxide is carried in plasma as bicarbonate ions (70%)

SYSTEMIC GAS EXCHANGE

A. The exchange of gases in tissues takes place between arterial blood flowing through tissue capillaries and cells (Figure 23-12)
   1. Oxygen diffuses out of arterial blood because the oxygen pressure gradient favors its outward diffusion
   2. As dissolved oxygen diffuses out of arterial blood, blood P_{O_2} decreases, which accelerates oxyhemoglobin dissociation to release more oxygen to plasma for diffusion to cells
B. Carbon dioxide exchange between tissues and blood takes place in the opposite direction from oxygen exchange
   1. Bohr effect—increased P_{CO_2} decreases the affinity between oxygen and hemoglobin
   2. Haldane effect—increased carbon dioxide loading caused by a decrease in P_{O_2}

REGULATION OF BREATHING

A. Respiratory control centers—the main integrators that control the nerves that affect the inspiratory and expiratory muscles are located in the brainstem (Figure 23-15)
   1. Medullary rhythmicity center—generates the basic rhythm of the respiratory cycle
      a. This area consists of two interconnected control centers
         (1) The inspiratory center stimulates inspiration
         (2) The expiratory center stimulates expiration
   2. The basic breathing rhythm can be altered by different inputs to the medullary rhythmicity center (Figure 23-16)
      a. Input from the apneustic center in the pons stimulates the inspiratory center to increase the length and depth of inspiration
      b. The pneumotaxic center—in the pons—inhibits the apneustic center and inspiratory center to prevent overinflation of the lungs
B. Factors that influence breathing—sensors from the nervous system provide feedback to the medullary rhythmicity center
   1. Changes in the P_{O_2}, P_{CO_2}, and pH of arterial blood influence the medullary rhythmicity area
      a. P_{CO_2} acts on chemoreceptors in the medulla—if it increases the result is faster breathing; if it decreases, the result is slower breathing

b. A decrease in blood pH stimulates chemoreceptors in the cartoid and aortic bodies

c. Arterial blood P_{O_2} presumably has little influence if it stays above a certain level

2. Arterial blood pressure controls breathing through the respiratory pressoreflex mechanism

3. Hering-Breuer reflexes help control respirations by regulating depth of respirations and the volume of tidal air

4. The cerebral cortex influences breathing by increasing or decreasing the rate and strength of respirations

THE BIG PICTURE: THE RESPIRATORY SYSTEM AND THE WHOLE BODY

A. The internal system must continually get new oxygen and rid itself of carbon dioxide because each cell requires oxygen and produces carbon dioxide as a result of energy conversion

B. Specific mechanisms involved in respiratory function:

1. Blood gases need blood and the cardiovascular system to be transported between gas exchange tissues of the lungs and various systemic tissues of the body

2. Regulation by the nervous system adjusts ventilation to compensate for changes in oxygen or carbon dioxide in the internal environment

3. The skeletal muscles of the thorax aid the airways in maintaining the flow of fresh air

4. The skeleton houses the lungs, and the arrangement of bones facilitates the expansion and recoil of the thorax

5. The immune system prevents pathogens from colonizing the respiratory tract and causing infection

MECHANISMS OF DISEASE: DISORDERS ASSOCIATED WITH RESPIRATORY FUNCTION

A. Restrictive pulmonary disorders restrict the alveoli, which leads to decreased lung inflation—these disorders have decreased lung volumes and capacities

1. Causes of restrictive lung disorders include

a. Alveolar fibrosis secondary to occupational exposure to contaminants such as asbestos, coal dust, or toxic fumes

b. Immunological diseases

c. Obesity

d. Metabolic disorders

2. Therapy involves removing the cause of the restriction ensuring adequate gas exchange, and improving exercise tolerance

B. Obstructive pulmonary disorders

1. Obstructive pulmonary disorders are caused by numerous different conditions and may obstruct inspiration and expiration—unlike restrictive disorders, which only obstruct inspiration

2. Chronic obstructive pulmonary disease (Figure 23-17)—a broad term used to describe conditions of progressive irreversible obstruction of expiratory air flow

a. COPD produces difficult breathing, visibly inflated chests, a productive cough, and intolerance of activity

b. Major COPD disorders

(1). Bronchitis—the body produces excessive tracheobronchial secretions that obstruct air flow, and the bronchial mucous glands become enlarged; risk factors include smoking, aging, and exposure to environmental contaminants

(2). Emphysema—air spaces distal to the terminal bronchioles enlarge due to damage of lung connective tissues. Victims of this disease often develop hypoxia

(3). Asthma—this disorder is characterized by recurring spasms of the smooth muscles in the walls of the bronchial passages, which narrows airways and makes breathing difficult

(1) Major symptoms—dyspnea, hyperventilation, headaches, and nausea

4. Acute respiratory failure can occur if the COPD disorders become intense

REVIEW QUESTIONS

1. Define *respiratory physiology.*

2. Respiratory control mechanisms must function throughout a range of changing environmental conditions and body demands. Is this more difficult than steady-state homeostasis? Why?

3. Name at least three physiological control systems that work closely with the respiratory system in maintenance of homeostasis.

4. Define *pulmonary ventilation.* Are the lungs active or passive during this process? Explain.

5. What is the main inspiratory muscle?

6. How is inspiration accomplished? Expiration?

7. Identify the separate volumes that make up the total lung capacity.

8. Normally, about what percentage of the tidal volume fills the anatomical dead space?

9. Normally, about what percentage of the tidal volume is useful air, that is, ventilates the alveoli?

10. Discuss the reciprocal interaction of oxygen and carbon dioxide on blood-gas transport. What is meant by the term *Bohr effect? Haldane effect?*

11. One gram of hemoglobin combines with how many milliliters of oxygen?

12. Suppose your blood has a hemoglobin content of 15 g per 100 ml and an oxygen saturation of 97%. How many milliliters of oxygen would 100 ml of your arterial blood contain?

13. Explain the law of partial pressures (Dalton's law).

14. What factors influence the amount of oxygen that diffuses into the blood from the alveoli?

15. Make a generalization about the effect of a moderate increase in the amount of blood carbon dioxide on circulation and respiration. What advantage can you see in this effect?

16. Make a generalization about the effect of a moderate decrease in oxygen on circulation and respiration.

17. Identify the major factors that influence breathing.

18. If an opening is made into the pleural cavity from the exterior, what happens? Why?

19. Orthopnea is a symptom of what type of disease?

20. Dyspnea is often associated with what type of breathing?

21. Define the diving reflex and explain its physiological importance.

22. Describe four nonrespiratory breathing reflexes.

23. Describe the changes in respirations during a period of exercise.

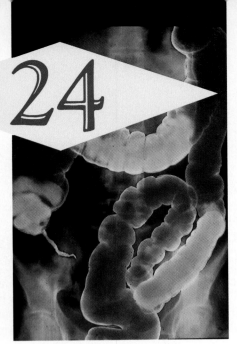

X-ray of large intestine

24 ▷ Anatomy of the Digestive System

OBJECTIVES

After you have completed this chapter, you should be able to:

1. Discuss the generalized function of the digestive system.
2. List, in sequence, each of the component parts or segments of the alimentary canal from mouth to anus and identify the accessory organs of digestion that are located within or open into the gastrointestinal tract.
3. List and describe the four layers of the wall of the GI tract.
4. Discuss the major modifications of the layers of the digestive tract.
5. List and describe the structures of the mouth.
6. Identify and compare the structure and secretions of the salivary glands.
7. Discuss the structural components of a typical tooth and identify by name and number the deciduous and permanent teeth.
8. Define the term *deglutition* and identify the structural divisions of the pharynx.
9. Discuss the size, position, divisions, curves, sphincters, layers, and glands of the stomach.
10. Compare the structure and the functional activity of chief cells, parietal cells, and mucus-producing cells of the stomach.
11. Discuss the size, position, divisions, and wall of the small and large intestines.
12. Locate and discuss the significance of the vermiform appendix.
13. Discuss the peritoneum and its reflections.
14. Discuss the structure and functions of the liver and gallbladder.
15. Explain the relationship between cell types and function in the pancreas.

Alimentary canal Digestive tract as a whole; gastrointestinal tract

Bile Substance that reduces large fat globules into smaller droplets of fat that are more easily broken down

Colon Middle portion of the large intestine joining cecum and rectum; consists of ascending, transverse, descending, and sigmoid segments

Deglutition (deg-loo-TISH-un) Swallowing

Duodenum (doo-o-DEE-num) First subdivision of the small intestine; where most chemical digestion occurs

Esophagus (e-SOF-ah-gus) Tube extending from the pharynx to the stomach

Gastrointestinal tract Digestive tract; alimentary canal

Hepatic (he-PAT-ik) Refers to the liver

Mastication Chewing

Mucosa (myoo-KO-sah) Mucous membrane

Peritoneum (pair-i-toe-NEE-um) Serous membrane that lines the abdominopelvic cavity and covers its organs

Villus (VIL-us) Fingerlike fold covering the plicae of the small intestine

OVERVIEW OF THE DIGESTIVE SYSTEM
Role of the Digestive System

This chapter deals with the anatomy of the digestive system. The organs of the digestive system together perform a vital function—that of preparing food for absorption and for use by the millions of body cells. Most food when eaten is in a form that cannot reach the cells (because it cannot pass through the intestinal mucosa into the bloodstream), nor could it be used by the cells even if it could reach them. It must therefore be modified as to both chemical composition and physical state so that nutrients can be absorbed and used by the body cells. The digestive tract and accessory organs comprise the system by which these complex changes in ingested food materials can occur. Part of the digestive system, the large intestine, serves also as an organ of elimination. Ingested food material that cannot be put into an absorbable form becomes waste material (feces) that is ultimately eliminated from the body.

The process of altering the chemical and physical composition of food so that it can be absorbed and used by the body cells (known as **digestion**) is the function of the digestive system. The process of digestion depends on both endocrine and exocrine secretions and the controlled movement of ingested food materials through the tract so that absorption can occur. The physiology of the digestive system is discussed in Chapter 25.

Organization of the Digestive System
Organs of digestion

The main organs of the digestive system (Figure 24-1) form a tube all the way through the ventral cavities of the body. It is open at both ends. This tube is usually referred to as the **alimentary canal** (tract) or, more commonly, the **gastrointestinal (GI) tract.** It is important to realize that ingested food material passing through the lumen of the gastrointestinal tract is actually outside the internal environment of the body, even though the tube itself is inside the ventral body cavity. Table 24-1 lists the main organs of the digestive system, that is, the segments of the alimentary canal, and the accessory organs located in the main digestive organs or opening into them.

Wall of the GI tract

Layers. The GI tract is essentially a tube with walls fashioned of four layers of tissues: a mucous lining, a submucous coat of connective tissue in which are imbedded

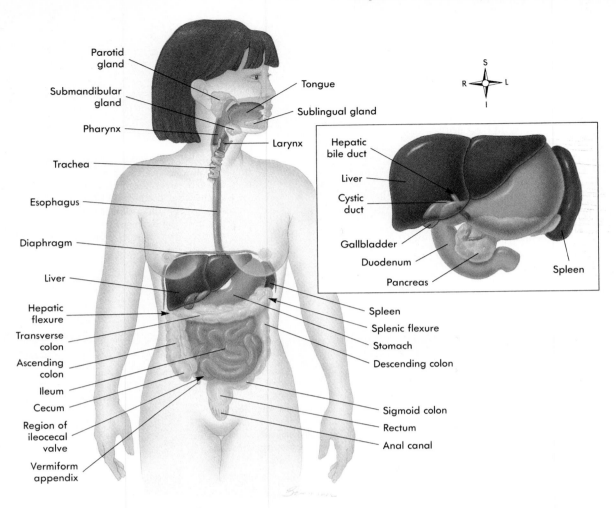

FIGURE 24-1   Location of digestive organs.

TABLE 24-1	Organs of the digestive system		

SEGMENTS OF THE GASTROINTESTINAL TRACT	Mouth	**ACCESSORY ORGANS**	Salivary glands
			  Parotid
			  Submandibular
			  Sublingual

	Oropharynx		
	Esophagus		
	Stomach		Tongue

	Small intestine
	  Duodenum
	  Jejunum
	  Ileum
			Teeth

			Liver
			Gallbladder

	Large intestine
	  Cecum
	  Colon
	    Ascending colon
	    Transverse colon
	    Descending colon
	    Sigmoid colon
			Pancreas

			Vermiform appendix
	Rectum		
	Anal canal		

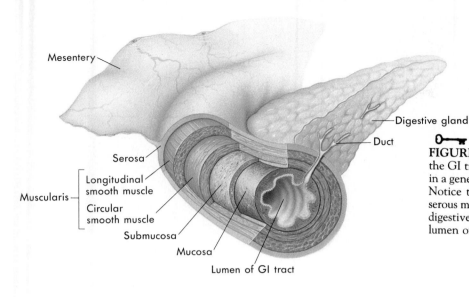

FIGURE 24-2   Wall of the GI tract. The wall of the GI tract is made of up to four layers, shown here in a generalized diagram of a segment of the GI tract. Notice that the serosa is continuous with a fold of serous membrane called a *mesentery.* Notice also that digestive glands may empty their products into the lumen of the GI tract by way of ducts.

TABLE 24-2	Modifications of layers of the digestive tract wall		
Organ	**Mucosa**	**Muscularis**	**Serosa**
Esophagus	Stratified squamous epithelium resists abrasion	Two layers—inner one of circular fibers and outer one of longitudinal fibers; striated muscle in upper part and smooth muscle in lower part of esophagus and in rest of tract	Outer layer fibrous (adventitia); serous around part of esophagus in thoracic cavity
Stomach	Arranged in flexible longitudinal folds, called *rugae;* allow for distention; contains gastric pits with microscopic gastric glands	Has three layers instead of usual two—circular, longitudinal, and oblique fibers; two sphincters—cardiac at entrance of stomach and pyloric at its exit, formed by circular fibers	Outer layer, visceral peritoneum; hangs in double fold from lower edge of stomach over intestines, forming apronlike structure; greater omentum; lesser omentum connects stomach to liver
Small intestine	Contains permanent circular folds, plicae circulares Microscopic fingerlike projections, villi with brush border Crypts (of Lieberkühn) Microscopic duodenal (Brunner's) mucous glands Clusters of lymph nodules (Peyer's patches) Numerous single lymph nodes, called solitary nodes	Two layers—inner one of circular fibers and outer one of longitudinal fibers	Outer layer, visceral peritoneum, continuous with mesentery
Large intestine	Solitary lymph nodes Intestinal mucous glands Anal columns form in anal region	Outer longitudinal layer condensed to form three tapelike strips (taeniae coli); small sacs (haustra) give rest of wall of large intestine puckered appearance; internal anal sphincter formed by circular smooth fibers; external; anal sphincter formed by striated fibers	Outer layer, visceral peritoneum, continuous with mesocolon

the main blood vessels of the tract, a muscular layer, and a fibroserous layer (Figure 24-2).

The innermost layer of the GI wall—the layer facing the lumen, or open space, of the tube—is called the *mucosa* because it is made of mucous membrane. Supporting the mucosa is a layer of connective tissue called the *submucosa*. The submucosa contains blood vessels and nerves that supply and control the digestive tract. A thick layer of muscle tissue wraps around the submucosa. This layer, called the *muscularis*, often contains several sheets of muscle cells. Muscle tissue that encircles the GI tube is called circular muscle, and muscle tissue with fibers that are parallel to the GI tube is called longitudinal muscle. The outermost layer of the GI wall, the *serosa*, is made of serous membrane. The serosa is actually the visceral layer of the *peritoneum*—the serous membrane that lines the abdominopelvic cavity and covers its organs. The fold of serous membrane that connects the parietal and visceral portions is called a *mesentery*. Notice in Figure 24-2 that exocrine glands empty their secretions into the lumen of the GI tract through ducts.

Modifications of layers. Although the same four tissue layers form the various organs of the GI tract, their structures vary in different organs. Variations in the epithelial layer of the mucosa, for example, range from stratified layers of squamous cells that provide protection from abrasion in the upper esophagus to the simple columnar epithelium, designed for absorption and secretion, that is found throughout most of the tract. Some of these modifications are listed in Table 24-2 and should be referred to when each of these organs is studied in detail.

QUICK CHECK

✔ *1. What is another name for the digestive tract?*
✔ *2. Can you name the four layers of the digestive tract wall?*

◆ MOUTH

Structure of the Oral Cavity

The **mouth** is also called the *oral cavity*. The following structures form the oral cavity (*buccal cavity*): the lips surrounding the orifice of the mouth and forming the anterior boundary of the oral cavity, the cheeks (side walls), the tongue and its muscles (floor), and the hard and soft palates (roof) (Figure 24-3).

Lips

The **lips** are covered externally by skin and internally by mucous membrane that continues into the oral cavity and

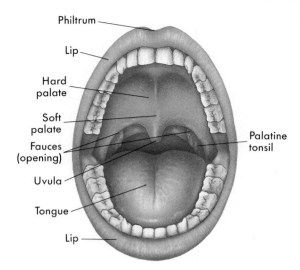

FIGURE 24-3   The oral cavity.

lines the mouth. The junction between skin and mucous membrane is highly sensitive and easily irritated. The upper lip is marked near the midline by a shallow vertical groove called the *philtrum*, which ends at the junction between skin and mucous membrane in a slight prominence called the *tubercle*. When the lips are closed, the line of contact is called the *oral fissure*.

Cheeks

The **cheeks** form the lateral boundaries of the oral cavity. They are continuous with the lips in front and are lined by mucous membrane that is reflected onto the gingiva, or gums, and the soft palate. The cheeks are formed in large part by the buccinator muscle, which is sandwiched with a considerable amount of adipose, or fat, tissue between the outer skin and mucous membrane lining. Numerous small mucus-secreting glands are placed between the mucous membrane and the buccinator muscle; their ducts open opposite the last molar teeth.

Hard and soft palates

The **hard palate** consists of portions of four bones: two maxillae and two palatines (see Figure 7-5). The **soft palate,** which forms a partition between the mouth and nasopharynx (see Figure 22-3), is fashioned of muscle arranged in the shape of an arch. The opening in the arch leads from the mouth into the oropharynx and is named the *fauces.* Suspended from the midpoint of the posterior border of the arch is a small cone-shaped process, the *uvula.*

Tongue

The **tongue** is a solid mass of skeletal muscle components (intrinsic muscles) covered by a mucous membrane. The intrinsic muscles of the tongue have, by definition, both their origin and their insertion in the tongue itself. Intrinsic muscles have their fibers oriented in all directions, thus providing a basis for extreme maneuverability. Changes in the size and shape of the tongue caused by intrinsic muscle contraction assist in placement of food materials between the teeth during **mastication** (chewing). Extrinsic tongue muscles are those that insert into the tongue but have their origin on some other structure, such as a skull bone. Examples of extrinsic tongue muscles are the genioglossus, which protrudes the tongue, and the hyoglossus, which depresses it (see Figure 24-4, C).

Contraction of the extrinsic muscles is important during **deglutition,** or swallowing, and speech.

Note in Figure 24-4 that the tongue has a blunt *root,* a *tip,* and a central *body.* The upper, or dorsal, surface of the tongue is normally moist, pink, and covered by rough elevations, called *papillae.* The three types of papillae—vallate, fungiform, and filiform—are all located on the sides or upper surface (dorsum) of the tongue. Note in Figure 24-4, *A,* that the large vallate papillae form an inverted V-shaped row extending from a median pit named the *foramen cecum* on the posterior part of the tongue. There are from 10 to 14 of these large, mushroomlike papillae. You can readily distinguish them if you look at your own tongue. Figure 24-5 shows two micrographs of vallate papillae. Dissolved substances to be tasted enter a

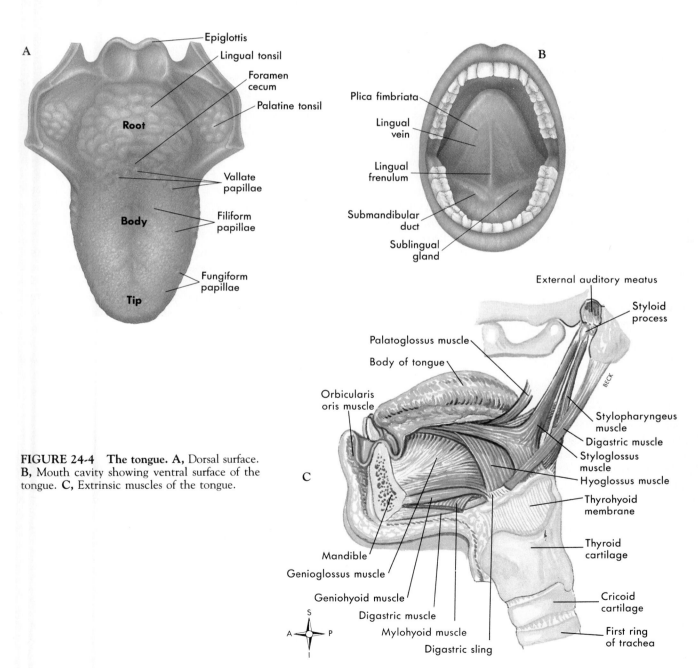

FIGURE 24-4   The tongue. A, Dorsal surface. **B,** Mouth cavity showing ventral surface of the tongue. **C,** Extrinsic muscles of the tongue.

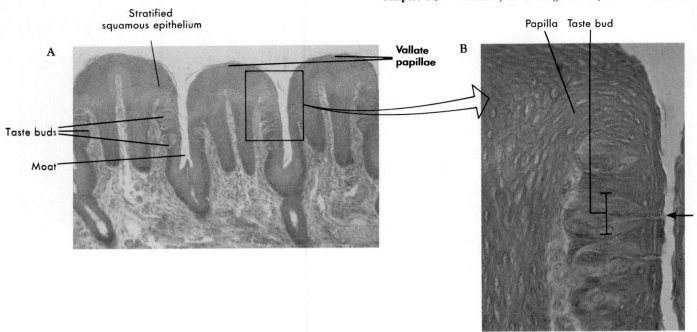

FIGURE 24-5 **Vallate papillae on surface of tongue. A,** Taste buds are located on lateral surfaces of papillae. Several taste buds can be seen opening into the moat from the sides of the papillae. (×35.) **B,** Taste buds. Enlargement of photomicrograph of taste buds in **A.** Arrow points to pore in outer surface of taste bud. (×140.)

moatlike depression surrounding the papillae to contact taste buds located on their lateral surface. Taste buds are also located on the sides of the fungiform papillae, which are found chiefly on the sides and tip of the tongue. The numerous filiform papillae are filamentous and threadlike in appearance. They have a whitish appearance and are distributed over the anterior two thirds of the tongue. Filiform papillae do not contain taste buds.

The *lingual frenulum* (Figure 24-4, *B*) is a fold of mucous membrane in the midline of the undersurface of the tongue that helps to anchor the tongue to the floor of the mouth. If the frenulum is too short for freedom of tongue movements—a congenital condition called *ankyloglossia*—the individual is said to be tongue-tied, and speech is faulty.

Folds of mucous membrane called the *plica fimbriata* (see Figure 24-4, *B*) extend toward the apex of the tongue on either side of the lingual frenulum. The floor of the mouth and undersurface of the tongue are richly supplied with blood vessels. The deep lingual vein can be seen (see Figure 24-4, *B*) shining through the mucous membrane between the lingual frenulum and plica fimbriata. In this region many vessels are extremely superficial and are covered only by a very thin layer of mucosa. Soluble drugs are absorbed into the circulation rapidly if placed under the tongue.

Salivary Glands

Three pairs of compound tubuloalveolar glands (Figures 24-1 and 24-6)—the parotids, submandibulars, and sublinguals—secrete a major amount (about 1 liter) of the saliva produced each day. The small glands (buccal glands) that

occur in the mucosa lining the cheeks and mouth contribute less than 5% of the total salivary volume. Buccal gland secretion is important, however, to the hygiene and comfort of the mouth tissues. The salivary glands are typical of the accessory glands associated with the digestive system. They are located outside of the alimentary canal and convey their exocrine secretions by ducts from the glands into the lumen of the tract (Figure 24-6, *A*). The functions of saliva in the digestive process will be discussed in Chapter 25.

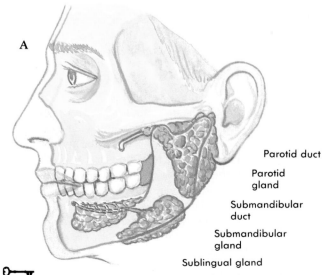

FIGURE 24-6 **Salivary glands. A,** Location of the salivary glands.

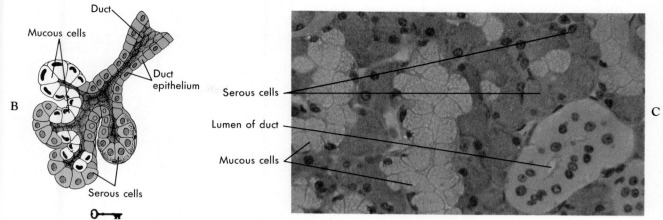

FIGURE 24-6 cont'd, B and C, Detail of submandibular salivary gland. This mixed- or compound-type gland produces mucus from the mucous cells and enzymatic secretion from the serous cells. Duct cross sections are also visible. (×140.)

HEALTH MATTERS

MUMPS

Mumps is an acute viral disease characterized by swelling of the parotid salivary glands. Most of us think of mumps as a childhood disease because it most often affects children between the ages of 5 and 15 years of age. However, it can occur in adults—often producing a more severe infection. The mumps infection can affect other tissues in addition to the parotid gland, including the joints, pancreas, myocardium, and kidneys. In about 25% of infected men, mumps causes inflammation of the testes, or *orchitis*. Orchitis resulting from mumps only very rarely causes enough damage to render a man sterile.

Parotid glands

The pyramidal-shaped **parotids** are the largest of the paired salivary glands (see Figure 24-6, A). They are located between the skin and underlying masseter muscle in front of and below the external ear. The parotids produce a watery, or serous, type of saliva containing enzymes but not mucus. The parotid (Stensen's) ducts are about 5 cm long. They penetrate the buccinator muscle on each side and open into the mouth opposite the upper second molars.

Submandibular glands

Submandibular glands (see Figure 26-6, A) are called mixed or compound glands because they contain both serous (enzyme) and mucus-producing elements (Figure 24-6, B and C). These glands are located just below the mandibular angle. You can feel the gland by placing your index finger on the posterior part of the floor of the mouth and your thumb medial to and just in front of the angle of the mandible. The gland is irregular in form and about the size of a walnut. The ducts of the submandibular glands (Wharton's ducts) open into the mouth on either side of the lingual frenulum.

Sublingual glands

Sublingual glands are the smallest of the salivary glands (see Figure 24-6, A). They lie in front of the submandibular

glands, under the mucous membrane covering the floor of the mouth. Each sublingual gland is drained by 8 to 20 ducts (ducts of Rivinus) that open into the floor of the mouth. Unlike the other salivary glands, the sublingual glands produce only a mucous type of saliva.

Teeth

The teeth are the organs of **mastication,** or chewing. They are designed to cut, tear, and grind ingested food so it can be mixed with saliva and swallowed. During the process of mastication, food is ground into small bits. This increases the surface area that can be acted on by the digestive enzymes.

Typical tooth

A typical tooth (Figure 24-7) can be divided into three main parts: crown, neck, and root. The **crown** is the exposed portion of a tooth. It is covered by enamel—the hardest and chemically most stable tissue in the body. Enamel consists of approximately 97% calcified (inorganic) material and only 3% organic material and water. It is ideally suited to withstand the very abrasive process of mastication. The **neck** of a tooth is the narrow portion shown in Figure 24-7 that is surrounded by the gingivae, or gums. It joins the crown of the tooth to the root. It is the **root** that fits into the socket of the alveolar process of either upper or lower jaw. The root of a tooth may be a single peglike structure or have two or three separate conical projections. The root is not rigidly anchored to the alveolar process by cement but is suspended in the socket by the fibrous *periodontal membrane* (Figure 24-7).

In addition to enamel, the outer shell of each tooth is composed of two additional dental tissues—*dentin* and *cementum* (Figure 24-7). Dentin makes up the greatest proportion of the tooth shell. It is covered by enamel in the crown and by cementum in the neck and root area. The dentin contains a **pulp cavity** consisting of connective tissue, blood and lymphatic vessels, and sensory nerves.

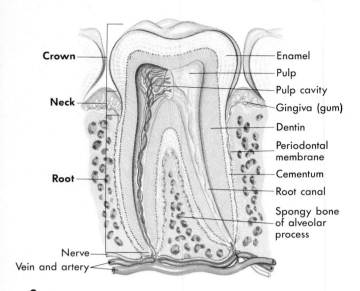

Crown
Neck
Root
Nerve
Vein and artery

Enamel
Pulp
Pulp cavity
Gingiva (gum)
Dentin
Periodontal membrane
Cementum
Root canal
Spongy bone of alveolar process

FIGURE 24-7 Typical tooth. A molar tooth sectioned to show its bony socket and details of its three main parts: crown, neck, and root. Enamel (over the crown) and cementum (over the neck and root) surround the dentin layer. The pulp contains nerves and blood vessels.

more a month until all 20 have appeared. There is, however, great individual variation in the age at which teeth erupt. Deciduous teeth are shed generally between the ages of 6 and 13 years. The third molars (wisdom teeth) are the last to appear, erupting usually sometime after 17 years of age.

TABLE 24-3 Dentition		
	Number per jaw	
Name of tooth	**Deciduous set**	**Permanent set**
Central incisors	2	2
Lateral incisors	2	2
Canines (cuspids)	2	2
Premolars (bicuspids)	0	4
First molars (tricuspids)	2	2
Second molars	2	2
Third molars (wisdom teeth)	0	2
TOTAL (per jaw)	10	16
TOTAL (per set)	20	32

Types of teeth

Twenty **deciduous teeth,** or so-called baby teeth, appear early in life and are later replaced by 32 **permanent teeth** (Figure 24-8). The names and numbers of teeth in both sets are given in Table 24-3.

The first deciduous tooth usually erupts at about 6 months of age. The remainder follow at the rate of 1 or

QUICK CHECK

✔ 1. *What are the boundaries of the oral cavity?*
✔ 2. *Describe the location of the taste buds in the mouth.*
✔ 3. *What are the names of the three types of salivary glands? How do their names describe their locations?*
✔ 4. *What are the three main parts of a typical tooth?*

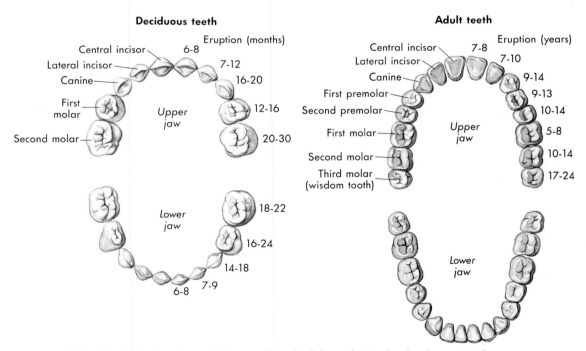

Deciduous teeth

Central incisor — 6-8   Eruption (months)
Lateral incisor —   7-12
Canine —   16-20
First molar   12-16
Second molar —   20-30
Upper jaw

Lower jaw   18-22
                          16-24
                          14-18
6-8   7-9

Adult teeth

Central incisor — 7-8   Eruption (years)
Lateral incisor —   7-10
Canine —   9-14
First premolar —   9-13
Second premolar —   10-14
First molar —   5-8
Second molar —   10-14
Third molar (wisdom tooth) — 17-24
Upper jaw

Lower jaw

FIGURE 24-8 Deciduous (baby) teeth and adult teeth. In the deciduous set, there are no premolars and only two pairs of molars in each jaw. Generally the lower teeth erupt before the corresponding upper teeth.

TOOTH DECAY

Tooth decay, or dental **caries** (KAIR-ees), is a common disease throughout the world. It is a disease of the enamel, dentin, and cementum of teeth that results in the formation of a permanent defect called a *cavity*. Most people living in the United States, Canada, and Europe are significantly affected by the disease. Decay occurs on tooth surfaces where food debris, acid-secreting bacteria, and plaque accumulate.

If the disease is untreated, tooth decay results in infection, loss of teeth, and inflammation of the soft tissues in the mouth. Bacteria may also invade the paranasal sinuses or extend to the surface of the face and neck, causing serious complications.

Gingivitis (jin-ji-VYE-tis) is the general term for inflam-
mation or infection of the gums. Most cases of gingivitis result from poor oral hygiene—inadequate brushing and no flossing. Gingivitis may also be a complication of other conditions such as diabetes mellitus, vitamin deficiency, or pregnancy.

Periodontitis (pair-ee-o-don-TIE-tis) is the inflammation of the periodontal membrane, or *periodontal ligament,* that anchors the tooth to the bone of the jaw. Periodontitis is often a complication of advanced or untreated gingivitis and may spread to the surrounding bony tissue. Destruction of periodontal membrane and bone results in loosening and eventually complete loss of teeth. Periodontitis is the leading cause of tooth loss among adults.

PHARYNX

The act of swallowing, or **deglutition,** moves a rounded mass of food, called a *bolus*, from the mouth to the stomach. As the food bolus passes from the mouth, it enters the oropharynx by passing through a constricted opening called the *fauces*. The oropharynx is the second division of the pharynx (see Figure 22-6). During respiration, air passes through all three pharyngeal divisions. However, only the terminal portions of the pharynx serve the digestive system. Once a bolus has passed through the pharynx, it enters the digestive tube proper—the portion of the digestive tract that serves only the digestive system. The anatomy of the pharynx is discussed in more detail on p. 584.

ESOPHAGUS

The esophagus, a collapsible tube about 25 cm (10 inches) long, extends from the pharynx to the stomach, piercing the diaphragm in its descent from the thoracic to the abdominal cavity. It lies posterior to the trachea and heart.

The esophagus is the first segment of the digestive tube proper, and the four layers that form the wall of the GI tract organs can be identified (Figure 24-9). The esophagus is normally flattened and thus the lumen is practically nonexistent in the resting state. The stratified squamous epithelium of the esophageal mucosa seen in Figure 24-9 provides a thick, abrasion-resistant lining that protects the esophagus from injury. The inner circular and outer longitudinal layers of the muscular layer are striated (voluntary) in the upper third, mixed (striated and smooth) in the middle third, and smooth (involuntary) in the lower third of the tube.

STOMACH

Size and Position of the Stomach

Just below the diaphragm, the digestive tube dilates into an elongated pouchlike structure, the stomach (Figure
24-10), the size of which varies according to several factors, notably the amount of distention. For some time after a meal, the stomach is enlarged because of distention of its walls, but, as food leaves, the walls partially collapse, leaving the organ about the size of a large sausage. In adults, the stomach usually holds a volume of up to 1.0 to 1.5 L.

The stomach lies in the upper part of the abdominal cavity under the liver and diaphragm, with approximately five sixths of its mass to the left of the median line (see Figure 24-1). In other words, it is described as lying in the *epigastrium* and left *hypochondrium* (Figure 1-15). Its position, however, alters frequently. For example, it is pushed downward with each inspiration and upward with each expiration. When it is greatly distended from an unusually large meal, its size interferes with the descent of the diaphragm on inspiration, producing the familiar feeling of dyspnea that accompanies overeating. In this state, the stomach also pushes upward against the heart and may give rise to the sensation that the heart is being crowded.

Divisions of the stomach

The **fundus, body,** and **pylorus** are the three divisions of the stomach. The fundus is the enlarged portion to the left and above the opening of the esophagus into the stomach. The body is the central part of the stomach, and the pylorus is its lower portion (Figure 24-10).

Curves of the stomach

The curve formed by the upper right surface of the stomach is known as the *lesser curvature;* the curve formed by the lower left surface is known as the *greater curvature* (see Figure 24-10).

Sphincter Muscles

Sphincter muscles guard both stomach openings. A sphincter muscle consists of circular fibers so arranged that there is an opening in the center of them (like the hole in a doughnut) when they are relaxed and no opening when they are contracted.

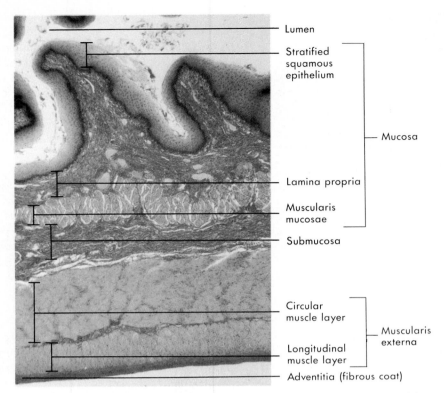

FIGURE 24-9 Wall of the esophagus. This low-power micrograph of a section of the esophageal wall shows the four layers that form the GI wall. Notice that the mucosa contains a thin layer of smooth muscle, called the *muscularis mucosae,* or "muscle of the mucosa." Notice also that in the portion of the esophagus shown, the outer layer is fibrous rather than serous (see Table 24-2).

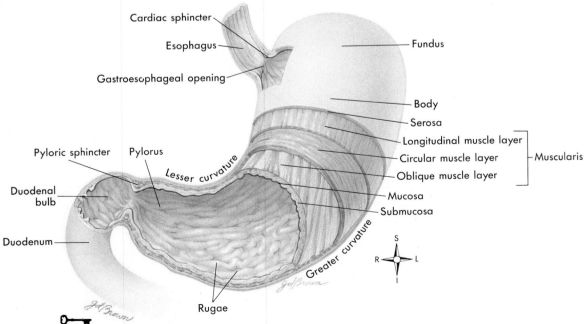

FIGURE 24-10 Stomach. A portion of the anterior wall has been cut away to reveal the muscle layers of the stomach wall. Notice that the mucosa lining the stomach forms folds called *rugae.*

DISORDERS OF THE PYLORIC SPHINCTER

The pyloric sphincter is of clinical importance because **pylorospasm** is a fairly common condition in infants. The pyloric fibers do not relax normally to allow food to leave the stomach, and the infant vomits food instead of digesting and absorbing it. The condition is relieved by the administration of a drug that relaxes smooth muscles. Another abnormality of the pyloric sphincter is **pyloric stenosis,** an obstructive narrowing of its opening.

The **cardiac sphincter** controls the opening of the esophagus into the stomach, and the **pyloric sphincter** controls the opening from the pyloric portion of the stomach into the first part of the small intestine (duodenum).

Stomach Wall

Each of the four layers of the stomach wall suits the function of this organ, as summarized in Table 24-2, p. 626, and shown in Figures 24-10 and 24-11. Of particular interest are the modifications to the stomach mucosa and muscularis, both of which are briefly described below.

Gastric mucosa

The epithelial lining of the stomach is thrown into folds, called *rugae,* and marked by depressions called *gastric pits.* Numerous coiled tubular-type glands, *gastric glands,* are found below the level of the pits, particularly in the fundus and body of the stomach. Figure 24-11 illustrates the anatomical relationship of the gastric pits and gastric glands. The glands secrete most of the gastric juice, a mucous fluid containing digestive enzymes and hydrochloric acid (HCl). Figure 24-12, *A,* is a low-power micrograph of the mucosal lining in the body of the stomach, showing numerous gastric pits and a uniform underlying layer of coiled gastric glands. The mucosal lining is easily differentiated from the deeper submucosal layer in this section. Figure 24-12, *B,* shows an enlarged view of gastric pits and gastric glands isolated from the submucosa and surrounding tissues.

In addition to the mucus-producing cells that cover the entire surface of the stomach and line the pits, the gastric glands contain two major secretory cells—**chief cells** and **parietal cells** (Figure 24-11). Chief cells (zymogenic cells) secrete the enzymes of gastric juice. Parietal cells secrete hydrochloric acid and are also thought to produce the important substance known as *intrinsic factor.* Intrinsic factor binds to vitamin B_{12} molecules to protect them from digestive juices until they reach the small intestine. Once they reach the small intestine, B_{12} molecules can be absorbed into the internal environment only if they are bound to intrinsic factor.

Gastric muscle

The thick layer of muscle in the stomach wall—the **muscularis**—is made of three distinct sublayers of smooth muscle tissue. As Figure 24-10 shows, there is the usual layer of longitudinal muscles and circular muscles, as well as an additional, underlying oblique layer. The crisscrossing pattern of smooth muscle fibers formed by this arrangement gives the stomach wall the ability to contract strongly at many angles—making the mixing action of this organ very efficient.

Functions of the Stomach

The stomach carries on the following functions:

♦ It serves as a reservoir, storing food until it can be partially digested and moved farther along the gastrointestinal tract.

♦ It secretes *gastric juice,* containing acid and enzymes, to aid in the digestion of food.

♦ Through contractions of its muscular coat, it churns the food, breaking it into small particles and mixing them well with the gastric juice. In time, it moves the gastric contents into the duodenum.

♦ It secretes *intrinsic factor.*

♦ It carries on a limited amount of absorption—of some water, alcohol, and certain drugs.

♦ It produces the hormone **gastrin,** which helps regulate digestive functions.

♦ It helps protect the body by destroying pathogenic bacteria swallowed with food or with mucus from the respiratory tract.

The digestive functions of the stomach will be discussed further in Chapter 25.

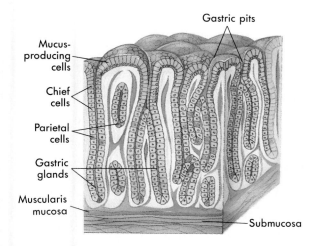

FIGURE 24-11   Gastric pits and gastric glands. Gastric pits are depressions in the epithelial lining of the stomach. At the bottom of each pit is one or more tubular *gastric glands.* Chief cells produce the enzymes of gastric juice, and parietal cells produce stomach acid.

Labels on figure:
Gastric pits
Mucus-producing cells
Chief cells
Parietal cells
Gastric glands
Muscularis mucosa
Submucosa

QUICK CHECK
✔ 1. *What is the primary digestive function of the pharynx?*
✔ 2. *Describe the location of the esophagus.*
✔ 3. *What are the three main divisions of the stomach?*
✔ 4. *What are gastric pits?*

A

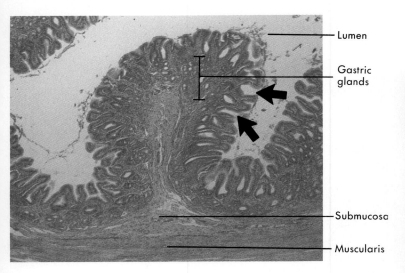

— Lumen

— Gastric glands

— Submucosa

— Muscularis

B

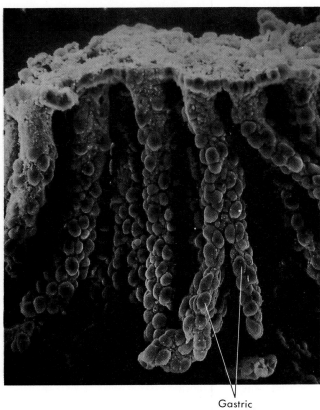

Gastric glands

FIGURE 24-12 Gastric mucosa. A, Low-power light micrograph showing that folds of gastric mucosa (*rugae*) have numerous gastric pits (*thick arrows*) and underlying gastric glands. (×14.) **B,** Scanning electron micrograph showing epithelium that has been isolated from the gastric mucosa. Again, notice the gastric pits (*thick arrows*) that have gastric glands at their bases. The outer surfaces of the parietal cells are seen as prominent dome-shaped bulges. (×500.) (*Courtesy of Erlandsen/Magney: Color Atlas of Histology.*)

DIAGNOSTIC STUDIES

UPPER GASTROINTESTINAL (UGI) X-RAY STUDY

The upper GI study consists of a series of x-rays of the lower esophagus, stomach, and duodenum, usually using barium sulfate as the contrast medium. The test is used to detect ulcerations, tumors, inflammations, or anatomical malpositions such as *hiatal hernia* (rupture of the diaphragm that allows the stomach to protrude). Obstruction of the upper GI tract is also easily detected.

In this test the patient is asked to drink a flavored drink containing barium sulfate. As the contrast medium travels through the system, the lower esophagus, gastric wall, pyloric channel, and duodenum are each evaluated for defects. Benign peptic ulcer is a common pathological condition affecting these areas. Tumors, cysts, or enlarged organs near the stomach can also be identified by an anatomical distortion of the outline of the upper GI tract. Can you identify the outline of the stomach in the figure?

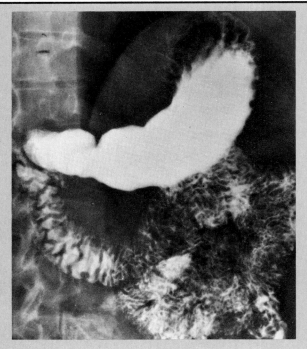

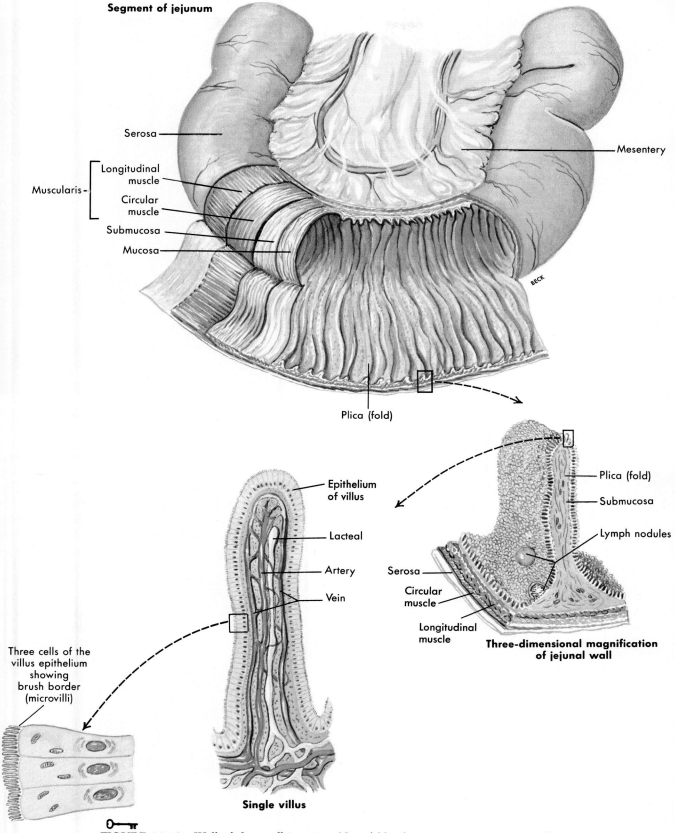

Segment of jejunum

Serosa

Mesentery

Muscularis
— Longitudinal muscle
— Circular muscle

Submucosa

Mucosa

Plica (fold)

Epithelium of villus

Lacteal

Artery

Vein

Plica (fold)

Submucosa

Lymph nodules

Serosa

Circular muscle

Longitudinal muscle

Three-dimensional magnification of jejunal wall

Three cells of the villus epithelium showing brush border (microvilli)

Single villus

FIGURE 24-13   Wall of the small intestine. Note folds of mucosa are covered with villi and each villus is covered with epithelium, which increases the surface area for absorption of food.

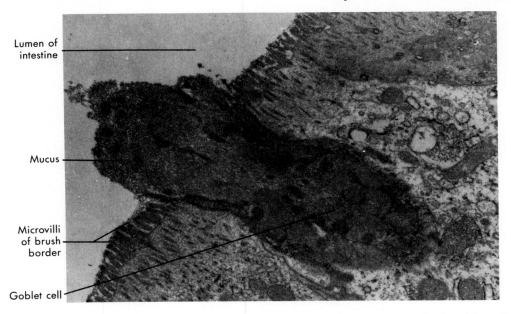

Lumen of intestine

Mucus

Microvilli of brush border

Goblet cell

FIGURE 24-14   Intestinal goblet cell. This transmission electron micrograph of a goblet cell in a villus of the jejunum shows the cell discharging mucus. Note the microvilli of the brush border on the adjacent epithelial cell. It is estimated that there are over 200 million microvilli per mm^2 on intestinal mucosa. (×22,500.)

SMALL INTESTINE

Size and Position of the Small Intestine

The small intestine is a tube measuring about 2.5 cm (1 inch) in diameter and 6 m (20 feet) in length. Its coiled loops fill most of the abdominal cavity (see Figure 24-1).

Divisions of the Small Intestine

The small intestine consists of three divisions: the duodenum, the jejunum, and the ileum. The **duodenum** is the uppermost division and is the part to which the pyloric end of the stomach attaches. It is about 25 cm (10 inches) long and is shaped roughly like the letter **C**. The name duodenum, meaning "12 fingerbreadths," refers to the short length of this intestinal division. The duodenum becomes **jejunum** at the point where the tube turns abruptly forward and downward. The jejunal portion continues for approximately the next 2.5 m (8 feet), where it becomes the **ileum,** but without any clear line of demarcation between the two divisions. The ileum is about 3.5 m (12 feet) long.

 Fractal Geometry of the Body

Biologists have just begun applying the principles of the new field of **fractal geometry** to human anatomy. Specialists in fractal geometry often study surfaces with a seemingly infinite area, such as the lining of the small intestine. Fractal surfaces have bumps that have bumps that have bumps, and so on. The fractal-like nature of the intestinal lining is represented in Figure 24-13. The plicae (folds) have villi, the villi have microvilli, and even the microvilli have bumps that cannot be seen in the figure. Thus, the absorptive surface area of the small intestine is almost limitless.

Wall of the Small Intestine

Notice in Figure 24-13 that the intestinal lining has circular *plicae* (folds) that have many tiny projections called **villi.** Villi are important modifications of the mucosal layer of the small intestine. Millions of these projections, each about 1 mm in height, give the intestinal mucosa a velvety appearance. Each villus contains an arteriole, venule, and lymph vessel (lacteal) (Figure 24-13). Epithelial cells on the surface of villi can be seen by microscopy (Figure 24-14) to have a surface resembling a fine brush. This so-called *brush border* is formed by about 1,700 ultrafine *microvilli* per cell. Intestinal digestive enzymes are produced in these brush border cells toward the top of the villi. The presence of villi and microvilli increases the surface area of the small intestine hundreds of times, making this organ the main site of digestion and absorption. Mucus-secreting goblet cells are found in large numbers on villi and in crypts (Figure 24-15). Intestinal crypts serve as a site of rapid mitotic cell division. As new cells are produced, older cells are pushed up and out of each crypt, eventually moving to the distal end of a villus, where they are shed. Thus the intestinal mucosa is continually renewed. At the base of each crypt, secretory cells produce an enzyme that is thought to inhibit bacterial growth in the small intestine. See Table 24-2 and Figure 24-13 for more information on the layers of the small intestine.

QUICK CHECK

✔ 1. What are the three main divisions of the small intestine?

✔ 2. What are intestinal villi? What is their function?

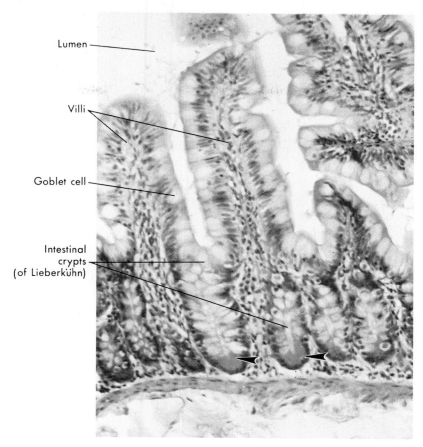

Lumen

Villi

Goblet cell

Intestinal
crypts
(of Lieberkühn)

FIGURE 24-15 Intestinal villi and crypts. This micrograph shows a section of the ileum wall with several villi and the intestinal crypts (of Lieberkühn) between them. Cells at the base of each crypt secrete bactericidal enzymes (*arrowheads*). (*Courtesy of Erlandsen/Magney: Color Atlas of Histology.*)

LARGE INTESTINE

Size of the Large Intestine

The lower part of the alimentary canal bears the name *large intestine* because its diameter is noticeably larger than that of the small intestine. Its length, however, is much less, being about 1.5 to 1.8 m (5 or 6 feet). Its average diameter is approximately 6 cm (2½ inches), but this decreases toward the lower end of the tube.

Divisions of the Large Intestine

The large intestine is divided into the cecum, colon, and rectum (Figure 24-16).

Cecum

The first 5 to 8 cm (2 or 3 inches) of the large intestine are named the cecum. It is a blind pouch located in the lower right quadrant of the abdomen (see Figure 24-1).

Colon

The colon is divided into the following portions: ascending, transverse, descending, and sigmoid (Figure 24-16).

♦ The **ascending colon** lies in a vertical position, on the right side of the abdomen, extending up to the lower border of the liver. The ileum joins the large intestine at the junction of the cecum and ascending colon, the place of attachment resembling the letter **T** in formation (Figure 24-16). The *ileocecal valve* permits material to pass from the ileum into the large intestine but not in the reverse direction.

♦ The **transverse colon** passes horizontally across the abdomen, below the liver, stomach, and spleen. Note that this part of the colon is above the small intestine (see Figure 24-1). The transverse colon extends from the *hepatic flexure* to the *splenic flexure*, the two points at which the colon bends on itself to form 90-degree angles.

♦ The **descending colon** lies in the vertical position, on the left side of the abdomen, extending from a point below the stomach and spleen to the level of the iliac crest.

♦ The **sigmoid colon** is that portion of the large intestine that courses downward below the iliac crest. It is called sigmoid (meaning "S-shaped") because it describes an S-shaped curve. The lower part of the curve, which joins the rectum, bends toward the left, the anatomical reason for placing a patient on the left side when giving an enema. In this position, gravity aids the flow of the enema fluid from the rectum into the sigmoid flexure.

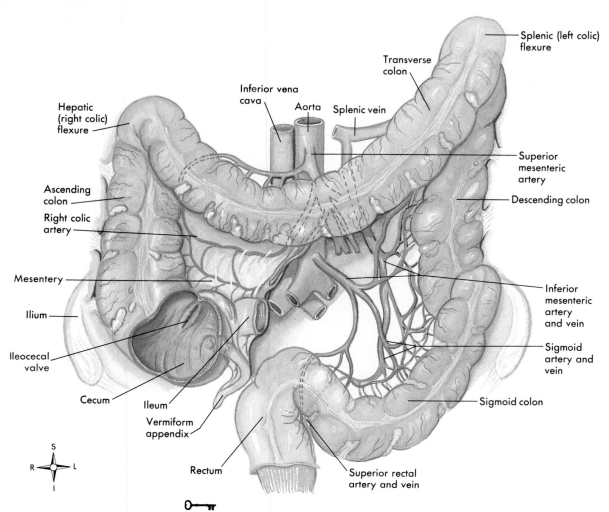

FIGURE 24-16   Divisions of the large intestine.

BARIUM ENEMA STUDY

The barium enema (BE) study, or lower GI series, consists of a series of x-ray films of the colon that are used to detect and locate polyps, tumors, and diverticula (abnormal "pouches" in the lining of the intestine). Abnormalities in organ position can also be detected.

The test begins with the rectal instillation (enema) of approximately 500 to 1500 ml of fluid containing barium sulfate. The patient is placed in various positions, and the progress of the barium's flow through the intestine is followed on a fluoroscope. Small polyps and early changes in ulcerative colitis are more easily detected with an *air-contrast barium enema study*. In this study, after the bowel is outlined with a thin coat of barium, air is added to enhance the contrast and outline of small lesions. After the x-ray films are taken, the patient is allowed to expel the barium.

The figure shows a colorized x-ray film of a barium enema produced by a special technique that produces a clear image of the large intestine and its position relative to the skeleton. Can you identify the divisions of the large intestine in this figure?

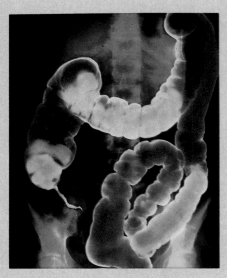

The large intestine.

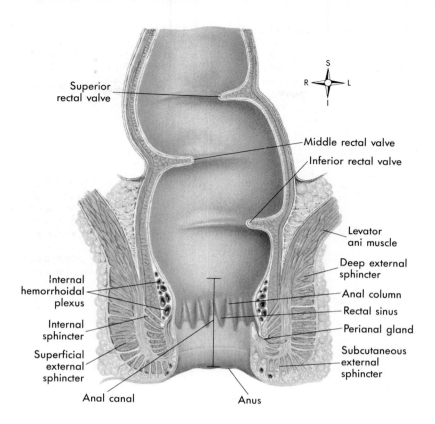

FIGURE 24-17 The rectum and anus.

Rectum

The last 7 or 8 inches of the intestinal tube are called the **rectum** (Figure 24-17). The terminal inch of the rectum is called the **anal canal.** Its mucous lining is arranged in numerous vertical folds known as *anal columns,* each of which contains an artery and a vein. **Hemorrhoids** (or piles) are enlargements of the veins in the anal canal. The opening of the canal to the exterior is guarded by two sphincter muscles—an internal one of smooth muscle and an external one of striated muscle. The opening itself is called the *anus.* The general direction of the rectum is shown in Figure 24-20. Note that the anus is directed slightly posteriorly and is therefore at almost a right angle to the rectum.

Wall of the Large Intestine

Table 24-2 (see p. 626) summarizes the modifications of GI wall seen in the large intestine. One of the most notable of these is the presence of intestinal mucous glands, which produce the lubricating mucus that coats the feces as they are formed (see Figure 24-18). Another notable feature of the wall of the colon is the uneven distribution of fibers in the muscle layer. The longitudinal muscles are grouped into tapelike strips called *taeniae coli,* and the circular muscles are grouped into rings that produce pouchlike *haustra* between them (see Figure 24-16). In the rectum, rings of circular muscle form the rectal valves seen in Figure 24-17.

VERMIFORM APPENDIX

The vermiform appendix (from the Latin *vermis,* meaning "worm," and *forma,* meaning "shape") is, as the name implies, a wormlike tubular organ. It averages 8 to 10 cm in length and is most often found just behind the cecum or over the pelvic rim. The lumen of the appendix communicates with the cecum about 3 cm below the ileocecal valve, making it an accessory organ of the digestive system (see Figure 24-16). Its functions are not certain, but some biologists believe that the appendix serves as a sort of "breeding ground" for intestinal bacteria. Often called simply the *intestinal flora,* a community of various bacterial populations normally inhabits the colon. The predominance of nonpathogenic bacteria under normal conditions is thought to help prevent disease. Some of the nonpathogenic bacteria are also thought to aid in the digestion or absorption of essential nutrients.

PERITONEUM

Now that you have completed the route through the digestive tube, consider for a moment the membrane covering most of these organs and holding them loosely in place. The peritoneum is a large, continuous sheet of serous membrane. It lines the walls of the entire abdominal cavity (parietal layer) and also forms the serous outer coat of the organs (visceral layer). In several places the peritoneum

forms reflections, or extensions, that bind abdominal organs together (Figures 24-19 and 24-20). The **mesentery** is a fan-shaped projection of the parietal peritoneum from the lumbar region of the posterior abdominal wall. The attached posterior border of this great fan is just 15 to 20 cm long, yet the loose outer edge enclosing the jejunum and ileum is 6 m long. The mesentery allows free movement of each coil of the intestine and helps prevent strangulation of the long tube. A similar but less extensive fold of peritoneum, called the **transverse mesocolon,** attaches the transverse colon to the posterior abdominal wall. The **greater omentum** is a continuation of the serosa of the greater curvature of the stomach and the first part of the duodenum to the transverse colon. Spotty deposits of fat accumulate in the omentum and give it the appearance of a lace apron hanging down loosely over the intestines. In cases of localized abdominal inflammation such as appendicitis the greater omentum envelops the inflamed area, walling it off from the rest of the abdomen. The **lesser omentum** attaches from the liver to the lesser curvature of the stomach and the first part of the duodenum. The falciform ligament extends from the liver to the anterior abdominal wall. Examine the relations of peritoneal extensions in Figure 24-20.

QUICK CHECK

✔ 1. *What are the four main divisions of the colon?*
✔ 2. *What are haustra?*
✔ 3. *Where is the vermiform appendix located?*
✔ 4. *Why is the greater omentum sometimes called the* lace apron?

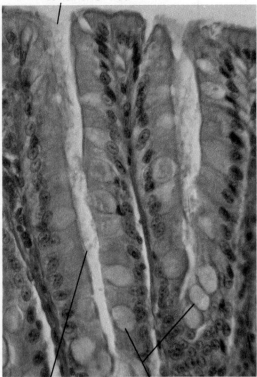

FIGURE 24-18   **Wall of the colon.** Note the straight nature of the intestinal glands. Many of the columnar epithelial cells are mucus-producing goblet cells. (×140.)

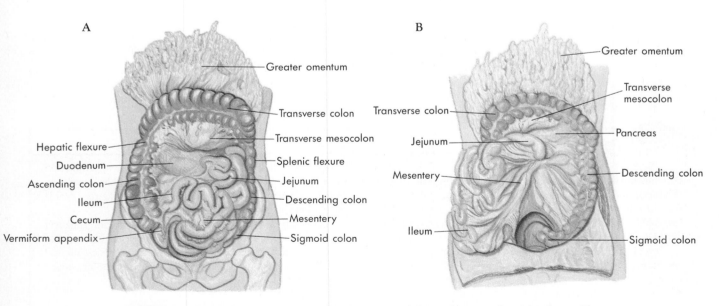

FIGURE 24-19   **Projections of the peritoneum. A,** Abdominal viscera from the front. The transverse colon and the greater omentum are elevated to reveal the flexures of the colon and the loops of the small intestine. **B,** The transverse colon and greater omentum are raised and the small intestine is pulled to the side to show the transverse mesocolon and mesentery.

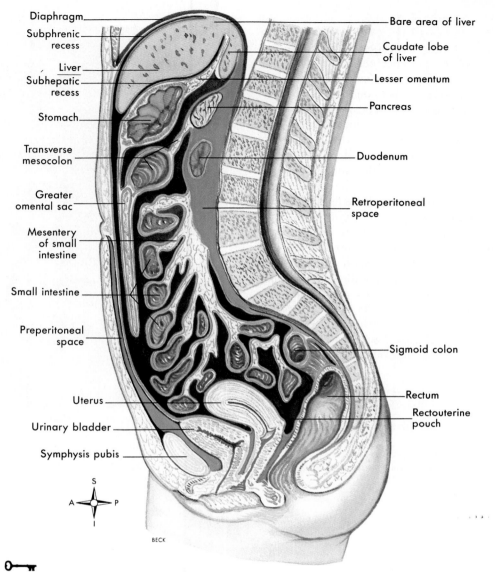

FIGURE 24-20 Peritoneum. Sagittal view of the abdomen showing the peritoneum and its reflections. *Intraperitoneal* spaces are shown in black and *extraperitoneal* spaces in green. The portion of the extraperitoneal space along the posterior wall of the abdomen is often called the *retroperitoneal space*.

APPENDICITIS

If the mucous lining of the appendix becomes inflamed, the resulting condition is the well-known affliction, **appendicitis.** As you can see in Figure 24-16, the appendix is very close to the rectal wall. For patients with suspected appendicitis, a physician often evaluates the appendix by a digital rectal examination.

The opening between the lumen of the appendix and the cecum is quite large in children and young adults—a fact of clinical significance because food or fecal material trapped in the appendix will irritate and inflame its mucous lining,

causing appendicitis. The opening between the appendix and the cecum is often completely obliterated in elderly persons, which explains the low incidence of appendicitis in this population.

If infectious material becomes trapped in an inflamed appendix, the appendix may rupture and release the material into the abdominal cavity. Infection of the peritoneum and other abdominal organs may result—with sometimes tragic consequences.

LIVER

Location and Size of the Liver

The liver is the largest gland in the body. It weighs about 1.5 kg (3 to 4 pounds), lies immediately under the diaphragm, and occupies most of the right hypochondrium and part of the epigastrium (see Figure 24-1).

Liver Lobes and Lobules

The liver consists of two lobes separated by the falciform ligament (Figure 24-21). The **left lobe** forms about one sixth of the liver, whereas the **right lobe** makes up the remainder. The right lobe has three parts designated as the *right lobe proper,* the *caudate lobe* (a small oblong area on the posterior surface), and the *quadrate lobe* (a four-sided section on the undersurface). Each lobe is divided into numerous lobules by small blood vessels and by fibrous strands that form a supporting framework (the capsule of Glisson) for them. The capsule of Glisson is an extension of the heavy connective tissue capsule that envelops the entire liver.

The **hepatic lobules** (Figure 24-22), the anatomical units of the liver, are tiny hexagonal or pentagonal cylinders about 2 mm high and 1 mm in diameter. A small branch of the hepatic vein extends through the center of each lobule. Around this central (intralobular) vein, in plates or irregular walls radiating outward, are arranged the hepatic cells. On the outer corners of each lobule, several sets of tiny tubes—branches of the hepatic artery, of the portal vein (interlobular veins), and of the hepatic duct (interlobular bile ducts)—are arranged. From these, irregular branches (sinusoids) of the interlobular veins extend between the radiating plates of hepatic cells to join the central vein. Minute bile canaliculi are formed by the spaces around each cell that collect bile secreted by the hepatic cells.

Consider the function of the hepatic lobule while carefully examining Figures 24-22 and 24-23. Blood enters a lobule from branches of the hepatic artery and portal vein. Arterial blood oxygenates the hepatic cells, whereas blood from the portal system passes through the liver for "inspection." Sinusoids in the lobule have many reticuloendothelial cells (mainly *Kupffer cells*) along their lining. These phagocytic cells can remove bacteria, worn RBCs,

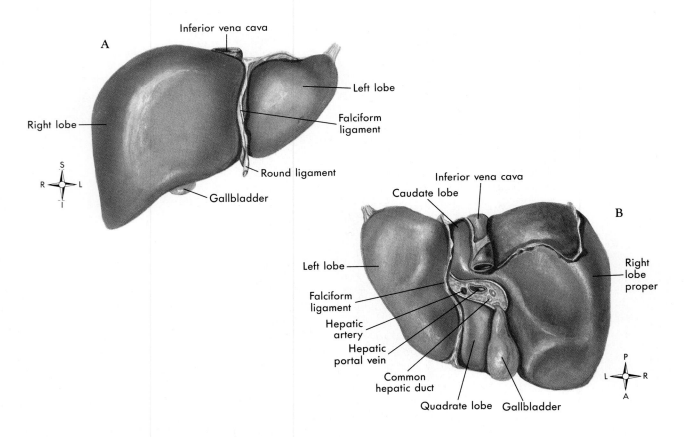

FIGURE 24-21 Gross structure of the liver. A, Anterior view. **B,** Inferior view.

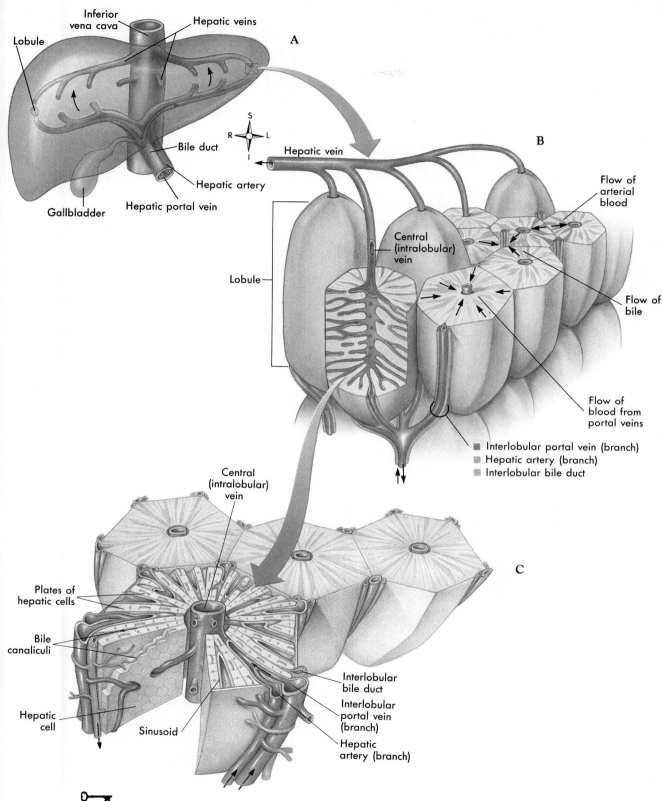

FIGURE 24-22 Microscopic structure of the liver. A, This diagram shows the location of liver lobules relative to the overall circulatory scheme of the liver. **B** and **C,** Enlarged views of several lobules show how blood from the hepatic portal veins and hepatic arteries flows through sinuses and thus past plates of hepatic cells toward a central vein in each lobule. Hepatic cells form bile, which flows through bile canaliculi toward hepatic ducts that eventually drain the bile from the liver.

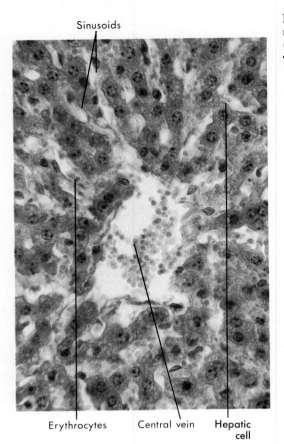

Sinusoids

Erythrocytes   Central vein   Hepatic cell

FIGURE 24-23   Liver tissue. Note the continuity permitting passage of red blood cells from sinusoids to central vein. The glandular hepatocytes (hepatic cells) form plates between the sinusoids. (Verhoeff's stain, ×140.)

and other particles from the bloodstream. Ingested vitamins and other nutrients to be stored or metabolized by liver cells enter the hepatic cells that form radiating walls of the lobule. Dissolved toxins in the blood are also absorbed into hepatic cells, where they are detoxified (rendered harmless). Blood continues along sinusoids to a vein at the center of the lobule. Such central, intralobular veins eventually lead to the main hepatic veins that drain into the inferior vena cava. Bile formed by hepatic cells passes through canaliculi to the periphery of the lobule to join small bile ducts.

Bile Ducts

The small bile ducts within the liver join to form two larger ducts that emerge from the undersurface of the organ as the right and left hepatic ducts. These immediately join to form one **hepatic duct.** The hepatic duct merges with the *cystic duct* from the gallbladder, forming the *common bile duct* (Figure 24-24), which opens into the duodenum in a small raised area called the major duodenal papilla. This papilla is located 7 to 10 cm below the pyloric opening from the stomach.

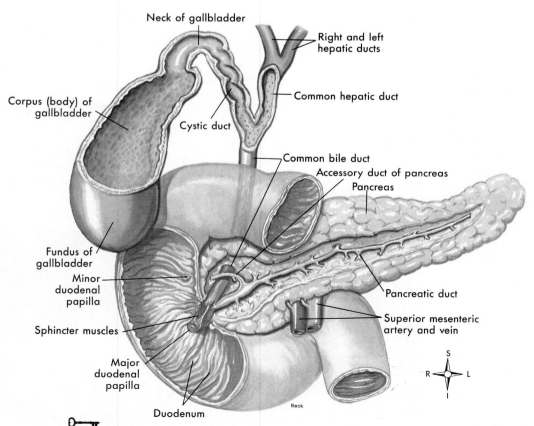

Neck of gallbladder

Right and left hepatic ducts

Corpus (body) of gallbladder

Common hepatic duct

Cystic duct

Common bile duct

Accessory duct of pancreas

Pancreas

Fundus of gallbladder

Minor duodenal papilla

Pancreatic duct

Sphincter muscles

Superior mesenteric artery and vein

Major duodenal papilla

Duodenum

FIGURE 24-24   Common bile duct and its tributaries. Obstruction of either the hepatic or the common bile duct by stone or spasm prevents bile from being ejected into the duodenum.

Functions of the Liver

The liver is one of the most vital organs of the body. Here, in brief, are its main functions:

♦ Liver cells detoxify various substances.

♦ Liver cells secrete about a pint of bile a day.

♦ Liver cells carry on numerous important steps in the metabolism of all three kinds of foods—proteins, fats, and carbohydrates.

♦ Liver cells store several substances—iron, for example, and vitamins A, B_{12}, and D.

♦ The liver produces important plasma proteins and serves as a site of hematopoiesis (blood cell production) during fetal development.

Detoxification by liver cells

Numerous poisonous substances enter the blood from the intestines. They circulate to the liver where, through a series of chemical reactions, they may be changed to nontoxic compounds. Ingested substances—alcohol, acetaminophen, and various other drugs, for example—and toxic substances formed in the intestines can be detoxified in the liver.

Bile secretion by liver

The main components of bile are bile salts, bile pigments, and cholesterol. Bile salts (formed in the liver from cholesterol) are the most essential part of bile. They aid in absorption of fats and then are themselves absorbed into the ileum. Eighty percent of bile salts are recycled in the liver to again become part of bile. Bile also serves as a pathway for elimination of certain breakdown products of red blood cells. The pigments bilirubin (reddish yellow) and biliverdin (greenish yellow), derived from hemoglobin, give bile its characteristic color. Because it secretes bile into ducts, the liver qualifies as an exocrine gland.

Liver metabolism

Although all liver functions are important for healthy survival, some of its metabolic processes are crucial for survival itself. A fairly detailed description of the role of the liver in metabolism will be given in Chapter 26.

GALLBLADDER

Size and Location of the Gallbladder

The gallbladder is a pear-shaped sac from 7 to 10 cm (3 to 4 inches) long and 3 cm broad at its widest point (see Figure 24-24). It can hold 30 to 50 ml of bile. It lies on the undersurface of the liver and is attached there by areolar connective tissue.

Structure of the Gallbladder

Serous, muscular, and mucous layers compose the wall of the gallbladder. The mucosal lining is arranged in folds called rugae, similar in structure to those of the stomach.

Functions of the Gallbladder

The gallbladder stores bile that enters it by way of the hepatic and cystic ducts. During this time, the gallbladder concentrates bile five- to ten-fold. Then later, when digestion occurs in the stomach and intestines, the gallbladder contracts, ejecting the concentrated bile into the duodenum.

HEALTH MATTERS — DISORDERS OF THE BILE TRACT

Inflammation of the lining of the gallbladder is called *cholecystitis*. **Cholecystectomy** is the surgical removal of the gallbladder, often necessitated by *cholelithiasis*, or stones in this organ (see figure). *Jaundice*, a yellow discoloration of the skin and mucosa, results when obstruction of the hepatic, or common, bile duct occurs. Bile is thereby denied its normal exit from the body in the feces. Instead, it is absorbed into the blood. An excess of bile pigments in the blood gives it a yellow hue. Feces, deprived of their normal amount of bile pigments, become a grayish (so-called clay) color.

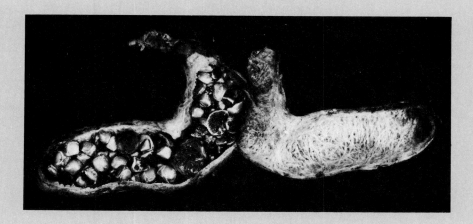

Cholelithiasis. This gallbladder, which has been removed from a patient and cut open, contains numerous calculi, or gallstones.

PANCREAS

Size and Location of the Pancreas

The pancreas is a grayish-pink–colored gland about 12 to 15 cm (6 to 9 inches) long, weighing about 60 g. It resembles a fish with its head and neck in the C-shaped curve of the duodenum, its body extending horizontally behind the stomach, and its tail touching the spleen (see Figures 24-1 and 24-25). According to an old anatomical witticism, the "romance of the abdomen" is the pancreas lying "in the arms of the duodenum." Locate the pancreas in the abdominal section pictured in the mini-atlas, p. 876.

Structure of the Pancreas

The pancreas is composed of two different types of glandular tissue, one exocrine and one endocrine. Most of the tissue is exocrine, with a compound acinar arrangement. The word *acinar* means that the cells are in a grapelike formation and that they release their secretions into a microscopic duct within each unit (see Figure 24-25, B). The word *compound* indicates that the ducts have

branches. These tiny ducts unite to form larger ducts that eventually join the main pancreatic duct, which extends throughout the length of the gland from its tail to its head. It empties into the duodenum at the same point as the common bile duct, that is, at the major duodenal papilla. An accessory duct is frequently found extending from the head of the pancreas into the duodenum, about 2 cm above the major papilla (see Figure 24-24).

Imbedded between the exocrine units of the pancreas, like so many little islands, lie clusters of endocrine cells called **pancreatic islets** (Figure 24-25). Although there are about a million of these tiny islands, they constitute only about 2% of the total mass of the pancreas. Special staining techniques have revealed that several kinds of cells— mainly alpha cells and beta cells—make up the islets. They are secreting cells, but their secretion passes into blood capillaries rather than into ducts. Thus the pancreas is a dual gland—an exocrine, or duct, gland because of the acinar units and an endocrine, or ductless, gland because of the pancreatic islets.

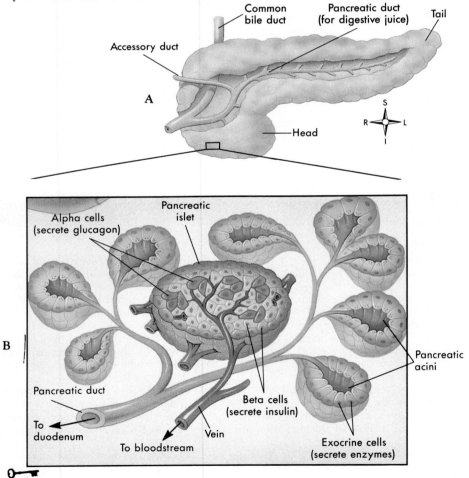

FIGURE 24-25 Pancreas. A, Pancreas dissected to show main and accessory ducts. The main duct may join the common bile duct, as shown here, to enter the duodenum by a single opening at the major duodenal papilla (see Figure 24-24), or the two ducts may have separate openings. The accessory pancreatic duct is usually present and has a separate opening into the duodenum. **B,** Exocrine glandular cells (around small pancreatic ducts) and endocrine glandular cells of pancreatic islets (adjacent to blood capillaries). Exocrine pancreatic cells secrete pancreatic juice, alpha endocrine cells secrete glucagon, and beta cells secrete insulin.

Functions of the Pancreas

♦ The acinar units of the pancreas secrete the digestive enzymes found in pancreatic juice. Hence the pancreas plays an important part in digestion (Chapter 25).

♦ Beta cells of the pancreas secrete **insulin,** a hormone that exerts a major control over carbohydrate metabolism (see Figure 24-14, p. 684).

♦ Alpha cells secrete *glucagon.* It is interesting to note that glucagon, which is produced so closely to insulin, has a directly opposite effect on carbohydrate metabolism.

QUICK CHECK

✔ 1. Where is the liver located?
✔ 2. Name three of the many functions of the liver.
✔ 3. Trace the route of bile from the gallbladder to the duodenum.
✔ 4. What is the function of the acinar units of the pancreas?

Cycle of Life: *Digestive system*

The digestive process involves both mechanical and chemical breakdown of ingested food, followed by absorption of nutrients through the intestinal mucosa. Significant changes in both the structure and the function of the digestive system are age related. Such changes result in numerous diseases or pathological conditions and may occur in any segment of the intestinal tract from mouth to anus. In addition, life cycle changes also involve accessory organs of digestion such as the teeth, salivary glands, liver, gallbladder, and pancreas.

Because of the immaturity of intestinal mucosa in young infants, some types of intact proteins can pass through the epithelial cells that line the tract. The result may be an early allergic response caused by the protein triggering the baby's immune system. Lactose intolerance is another age-related example of a common digestive system problem. Intestinal lactase, needed for digestion of lactose, or milk sugar, is almost always present at the time of birth. Levels may rapidly diminish in some babies, however, and such individuals soon become unable to digest lactose.

Inflammation of the parotid salivary gland (mumps) is a common disease of children, whereas appendicitis occurs more frequently in adolescents. Incidence of appendicitis then decreases with age because the size of the opening between appendix and intestinal lumen decreases. Gallbladder disease and ulcers are primarily problems of middle age. In more elderly individuals, a decrease in volume of digestive fluids coupled with a slowing of peristalsis and reduced physical activity often results in constipation and diverticulosis.

CHAPTER SUMMARY

OVERVIEW OF THE DIGESTIVE SYSTEM

A. Role of the digestive system
  1. Prepares food for absorption and utilization by all the cells of the body
  2. Food material not absorbed becomes feces that is eliminated
  3. Digestion depends on both endocrine and exocrine secretions and the controlled movement of ingested food materials through the gastrointestinal (GI) tract
B. Organization of the digestive system (Figure 24-1)
  1. Organs of digestion
    a. Main organs of the digestive system form the GI tract that extends through the abdominopelvic cavity
    b. Ingested food material passing through the lumen of the GI tract is outside the internal environment of the body
  2. Wall of the GI tract (Figure 24-2)
    a. Layers—GI tract is made of four layers of tissues: mucosa, submucosa, muscularis, and serosa
    b. Modifications of layers—layers of the GI tract have various modifications to enable it to perform a variety of functions

MOUTH

A. Structure of the oral cavity (buccal cavity) (Figure 24-3)
  1. Lips—covered externally by skin and internally by mucous membrane; junction between skin and mucous membrane is highly sensitive; when lips are closed, line of contact is oral fissure
  2. Cheeks—lateral boundaries of the oral cavity, continuous with lips and lined by mucous membrane; formed in large part by buccinator muscle covered by adipose tissue; contain mucus-secreting glands
  3. Hard and soft palates
    a. Hard palate consists of portions of four bones, two maxillae and two palatines
    b. Soft palate forms partition between the mouth and nasopharynx and is made of muscle arranged in an arch
    c. Suspended from midpoint of the posterior border of the arch is the uvula
  4. Tongue—solid mass of skeletal muscle covered by a mucous membrane; extremely maneuverable (Figure 24-4)
    a. Important for mastication and deglutition
    b. Has three parts: root, tip, and body

c. Papillae located on dorsal surface of tongue

d. Lingual frenulum anchors the tongue to the floor of the mouth

B. Salivary glands—three pairs of compound tubuloalveolar glands (Figure 24-6) secrete approximately 1 liter of saliva each day; buccal glands contribute less than 5% of total salivary volume but provide for hygiene and comfort of oral tissues

1. Parotid glands—largest of the paired salivary glands; produce watery saliva containing enzymes

2. Submandibular glands—compound glands that contain enzyme and mucus-producing elements

3. Sublingual glands—smallest of the salivary glands; produce a mucous type of saliva

C. Teeth—organs of mastication

1. Typical tooth (Figure 24-7)

a. Crown—exposed portion of a tooth, covered by enamel; ideally suited to withstand abrasion during mastication

b. Neck—narrow portion that joins the crown to the root; surrounded by the gingivae

c. Root fits into the socket of the alveolar process and is suspended by fibrous periodontal membrane

d. Outer shell contains two additional tissues: dentin and cementum

(1) Dentin makes up the greatest portion of the tooth shell; at crown, covered by enamel, and at neck and root, by cementum

(2) Pulp cavity—located in dentin, contains connective tissue, blood, and lymphatic vessels and sensory nerves

2. Types of teeth (Figure 24-8)

a. Deciduous teeth—20 baby teeth, which appear early in life

b. Permanent teeth—32 teeth, which replace the deciduous teeth

PHARYNX

A. Tube through which a bolus passes when moved from the mouth to the esophagus by the process of deglutition

ESOPHAGUS

A. Tube that extends from the pharynx to the stomach

B. First segment of digestive tube

STOMACH (Figure 24-10)

A. Size and position of the stomach

1. Size varies according to factors such as gender and amount of distention

a. When no food is in the stomach, it is about the size of a large sausage

b. In adults, capacity ranges from 1.0 to 1.5 L

2. Stomach location: upper part of abdominal cavity under the liver and diaphragm

B. Divisions of the stomach

1. Fundus—enlarged portion to the left and above the opening of the esophagus into the stomach

2. Body—central portion of the stomach

3. Pylorus—lower part of the stomach

C. Curves of the stomach

1. Lesser curvature—upper right curve of the stomach

2. Greater curvature—lower left curve of the stomach

D. Sphincter muscles—circular fibers arranged so that there is an opening in the center when relaxed and no opening when contracted

1. Cardiac sphincter controls the opening of the esophagus into the stomach

2. Pyloric sphincter controls the outlet of the pyloric portion of the stomach into the duodenum

E. Stomach wall

1. Gastric mucosa

a. Epithelial lining has rugae marked by gastric pits (Figure 24-11)

b. Gastric glands—found below level of the pits; secrete most of gastric juice

c. Chief cells—secretory cells found in the gastric glands; secrete the enzymes of gastric juice

d. Parietal cells—secretory cells found in the gastric glands; secrete hydrochloric acid; thought to produce intrinsic factor needed for vitamin B_{12} absorption

2. Gastric muscularis—thick layer of muscle with three distinct sublayers of smooth muscle tissue arranged in a crisscrossing pattern; this pattern allows the stomach to contract strongly at many angles

F. Functions of the stomach

1. Reservoir for food until it is partially digested and moved further along the GI tract

2. Secretes gastric juice to aid in digestion of food

3. Breaks food into small particles and mixes them with gastric juice

4. Secretes intrinsic factor

5. Limited absorption

6. Produces gastrin

7. Helps protect body from pathogenic bacteria swallowed with food

SMALL INTESTINE

A. Size and position of the small intestine—tube approximately 2.5 cm in diameter and 6 m in length; coiled loops fill most of the abdominal cavity

B. Divisions of the small intestine

1. Duodenum—uppermost division; approximately 25 cm long, shaped roughly like the letter **C**

2. Jejunum—approximately 2.5 m long

3. Ileum—approximately 3.5 m long

C. Wall of the small intestine (Figure 24-13)

1. Intestinal lining has plicae with villi

2. Villi—important modifications of the mucosal layer

a. Each villus contains an arteriole, venule, and lacteal

b. Covered by a brush border made up of 1,700 ultrafine microvilli per cell

c. Villi and microvilli increase the surface area of the small intestine hundreds of times

LARGE INTESTINE

A. Size of the large intestine—average diameter, 6 cm; length, approximately 1.5 to 1.8 m

B. Divisions of the large intestine (Figure 24-16)

1. Cecum—first 5 to 8 cm of the large intestine, blind pouch located in lower right quadrant of the abdomen

2. Colon

a. Ascending colon—vertical position on right side of abdomen; ileocecal valve prevents material passing from the large intestine into the ileum

b. Transverse colon passes horizontally across the abdomen, above the small intestine; extends from the hepatic flexure to the splenic flexure

c. Descending colon—vertical position on left side of the abdomen

d. Sigmoid colon joins descending colon to rectum

e. Rectum—last 7 or 8 inches of the intestinal tube; terminal inch is the anal canal with the opening called the anus (Figure 24-17)

C. Wall of the large intestine
  1. Intestinal mucous glands produce lubricating mucus that coats feces as they are formed
  2. Uneven distribution of fibers in the muscle coat

VERMIFORM APPENDIX

A. Accessory organ of digestive system; 8 to 10 cm in length; communicates with cecum

PERITONEUM

A. Large, continuous sheet of serous membrane (Figure 24-19)
B. Made up of parietal and visceral layers
C. Mesentery—projection of the parietal peritoneum; allows free movement of each coil of the intestine and helps prevent strangulation of the long tube (Figure 24-20)
D. Transverse mesocolon—extension of peritoneum that supports the transverse colon

LIVER

A. Location and size of the liver (Figure 24-21)—largest gland in the body, weighs approximately 1.5 kg; lies under the diaphragm; occupies most of the right hypochondrium and part of the epigastrium
B. Liver lobes and lobules—two lobes separated by the falciform ligament
  1. Left lobe—forms about one sixth of the liver
  2. Right lobe—forms about five sixths of the liver; divides into right lobe proper, caudate lobe, and quadrate lobe
  3. Hepatic lobules—anatomical units of the liver; small branch of hepatic vein extends through the center of each lobule (Figure 24-22)
C. Bile ducts (Figure 24-24)
  1. Small bile ducts form right and left hepatic ducts
  2. Right and left hepatic ducts immediately join to form one hepatic duct
  3. Hepatic duct merges with cystic duct to form the common bile duct, which opens into the duodenum
D. Functions of the liver
  1. Detoxification by liver cells—ingested toxic substances and toxic substances formed in the intestines may be changed to nontoxic substances
  2. Bile secretion by liver—bile salts are formed in the liver from cholesterol and are the most essential part of bile; liver cells secrete approximately 1 pint of bile per day
  3. Liver metabolism carries out numerous important steps in the metabolizing of proteins, fats, and carbohydrates
  4. Storage of substances such as iron and some vitamins
  5. Production of important plasma proteins

GALLBLADDER

A. Size and location of the gallbladder—pear-shaped sac from 7 to 10 cm long and 3 cm wide at its broadest point; holds 30 to 50 ml of bile; lies on undersurface of liver
B. Structure of the gallbladder—serous, muscular, and mucous layers compose the gallbladder wall; mucosal lining has rugae
C. Functions of the gallbladder
  1. Storage of bile
  2. Concentration of bile five- to ten-fold
  3. Ejection of the concentrated bile into duodenum

PANCREAS

A. Size and location of the pancreas—grayish-pink–colored gland; 12 to 15 cm long; weighs approximately 60 g; runs from the duodenum, behind the stomach, to the spleen
B. Structure of the pancreas (Figure 24-25)—composed of endocrine and exocrine glandular tissue
  1. Exocrine portion makes up majority of pancreas; has a compound acinar arrangement; tiny ducts unite to form the main pancreatic duct, which empties into the duodenum
  2. Endocrine portion—embedded between exocrine units; called pancreatic islets; constitute only 2% of the total mass of the pancreas; made up of alpha cells and beta cells; pass secretions into capillaries
C. Functions of the pancreas
  1. Acinar units secrete digestive enzymes
  2. Beta cells secrete insulin
  3. Alpha cells secrete glucagon

CYCLE OF LIFE: DIGESTIVE SYSTEM

A. Changes in digestive function and structure are age related
  1. Result in diseases or pathological conditions
  2. May occur in any segment of intestinal tract
  3. Changes involve accessory organs: teeth, salivary glands, liver, gallbladder, and pancreas
B. Infants—immature intestinal mucosa; intact proteins can pass through epithelial cells lining the tract and trigger allergic response
C. Lactose intolerance affects infants who lack the enzyme lactase
D. Young age—mumps common in children; appendicitis more common in adolescents and then decreases with advancing age
E. Middle age—ulcers and gallbladder disease common
F. Old age—decreased digestive fluids, slowing of peristalsis, and reduced physical activity lead to constipation and diverticulosis

REVIEW QUESTIONS

1. List the component parts or segments of the GI tract and the accessory organs of digestion.
2. Name and describe the four tissue layers that form the wall of GI tract organs.
3. Identify the structures that form the mouth.
4. Define the following terms associated with the mouth and pharynx: *philtrum, oral fissure, hard* and *soft palates, fauces, uvula, foramen cecum, lingual frenulum.*
5. What is ankyloglossia?
6. How do the intrinsic and extrinsic muscles of the tongue differ?
7. Identify the types of tongue papillae. What is the relationship between papillae and taste buds?
8. List and give the location of the paired salivary glands. Identify by name the ducts that drain the saliva from these glands into the mouth.
9. What type of saliva is produced by the parotid glands? What is meant by the term *mixed* or *compound* salivary gland?
10. Describe a typical tooth. Name the specific types of teeth.
11. Discuss deciduous and permanent teeth.
12. What is meant by the term *deglutition?*
13. How does the muscular layer of the esophagus differ from the muscle layer typical of the GI tract as a whole?
14. List the divisions of the stomach. What is the difference between gastric pits and gastric glands?
15. Identify the two major cell types of the gastric glands. What cell type produces HCl? Gastric enzymes?
16. Describe the seven functions of the stomach.
17. List the divisions of the small intestine from proximal to distal.
18. Compare the rugae of the stomach with the circular plicae of the small intestine.
19. In what area of the gastrointestinal tract do you find villi? Haustra? Taenia coli?
20. List the divisions of the large intestine.
21. What is believed to be the function of the vermiform appendix?
22. Discuss the peritoneum and its reflections.
23. Discuss the anatomy of a typical liver lobule.
24. Identify the ducts of the liver and gallbladder.
25. Explain the functions of the gallbladder.
26. Differentiate between endocrine and exocrine functions of the pancreas.
27. Identify the condition that may result in men infected with mumps.
28. What is the difference between dental caries and periodontitis?
29. What is pyloric stenosis?
30. Describe an upper gastrointestinal x-ray study and a barium enema study.
31. Why is appendicitis more common in children and young adults than in the elderly?
32. Define cholelithiasis.

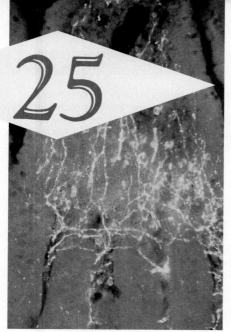

25 ▷ Physiology of the Digestive System

Nerve fibers in villus of small intestine
Erlandsen/Magney: Color Atlas of Histology

CHAPTER OUTLINE

OBJECTIVES

After you have completed this chapter, you should be able to:

1. Describe the primary mechanisms of the digestive system.
2. Define and compare mechanical and chemical digestion.
3. Discuss the function of mastication.
4. List and explain the three main steps or stages of deglutition.
5. Differentiate between peristalsis and segmentation.
6. Explain the process of emptying the stomach.
7. Define the different processes involved in mechanical digestion and identify the organ(s) that accomplish(es) each process.
8. Define chemical digestion.
9. Define the term *enzyme* and classify enzymes according to the type of chemical reactions catalyzed.
10. List and discuss six important enzyme properties.
11. List the most important digestive juices and enzymes, the food product each digests, and the resulting products.
12. Compare and contrast protein, fat, and carbohydrate digestion.
13. Discuss the control of salivary, gastric, pancreatic, biliary, and intestinal exocrine secretions.
14. Identify and discuss the absorption of nutrients resulting from the digestive process and the structures into which they are absorbed.
15. Define the terms *micelles, chylomicrons, vasoactive intestinal peptide, gastric inhibitory peptide,* and *hydrolysis.*
16. Discuss elimination and defecation.

KEY TERMS

Absorption Passage of substances through GI mucosa into the blood or lymph

Chylomicron (ki-lo-MYE-kron) Microscopic droplet of fat in blood

Chyme (kime) Partially digested food mixture leaving the stomach and traveling through the intestines

Defecation (de-fi-KAY-shun) Discharge of feces

Elimination Excretion of materials not absorbed

Gastrin Hormone that stimulates secretion of gastric juice

Hydrolysis Chemical process of adding a molecule of water to a compound that then splits into simpler compounds

Ingestion Taking in, or consuming, foods by mouth

Micelle (mye-SELL) Sphere formed by bile salts that surround lipids to make them temporarily water soluble

Motility Movement

Peristalsis (pair-i-STAL-sis) Rhythmic contractions of the muscle of a hollow organ that propel the contents

Secretin (se-KREE-tin) Hormone that helps adjust production of pancreatic fluid; first hormone discovered

Segmentation Mixing movement

◢ OVERVIEW OF DIGESTIVE FUNCTION

Now that we are familiar with the structural organization of the digestive system (Chapter 24), we are ready to understand the physiological organization of this system. The primary function of the digestive system is to bring essential nutrients into the internal environment so that they are available to each cell of the body. To accomplish this function, the digestive system uses various mechanisms (Table 25-1). For example, complex foods must first be taken in—a process called **ingestion.** Then, complex nutrients are broken down into simpler nutrients in the process that gives this system its name: **digestion.** To physically break large chunks of food into smaller bits and to move it along the tract, movement (or **motility**) of the GI wall is required. Chemical digestion—that is, breakdown of large molecules into small molecules—requires **secretion** of digestive enzymes into the lumen of the gastrointestinal (GI) tract. After being digested, nutrients are ready for the process of **absorption,** or movement through the GI mucosa into the internal environment. The material that is not absorbed must then be excreted to make room for more material—a process known as **elimination.**

After we have explored the various mechanisms of the digestive process in this chapter we will be ready for Chapter 26, which discusses the fate of nutrients once they have been absorbed.

◢ DIGESTION

Once food is ingested (taken into the mouth), the process of digestion begins immediately. Digestion is the overall name for all the processes that physically and mechanically break complex foods into simpler nutrients that can be easily absorbed. We will begin our discussion with a brief overview of **mechanical digestion,** then later move on to a discussion of **chemical digestion.**

Mechanical Digestion

Mechanical digestion consists of all movement (motility) of the digestive tract that brings about the following:

- ◆ Change in the physical state of ingested food from comparatively large solid pieces into minute particles, thereby facilitating chemical digestion
- ◆ Churning of the contents of the GI lumen in such a way that they become well mixed with the digestive juices and all parts of them come in contact with the surface of the intestinal mucosa, thereby facilitating absorption
- ◆ Propelling the food forward along the digestive tract, finally eliminating the digestive wastes from the body

Mastication

Mechanical digestion begins in the mouth when the particle size of ingested food material is reduced by chewing movements, or *mastication.* The tongue, cheeks, and lips play an important role in keeping food material between the cutting or grinding surfaces of the teeth when chewing. In addition to reducing particle size, chewing movements serve to mix food with saliva in preparation for swallowing.

TABLE 25-1	Primary mechanisms of the digestive system
Mechanism	**Description**
Ingestion	Process of taking food into the mouth, starting it on its journey through the digestive tract
Digestion	A group of processes that break complex nutrients into simpler ones, thus facilitating their absorption; *mechanical digestion* physically breaks large chunks into small bits; *chemical digestion* breaks molecules apart
Motility	Movement by the muscular components of the digestive tube, including processes of mechanical digestion; examples include *peristalsis* and *segmentation*
Secretion	Release of digestive juices (containing enzymes, acids, bases, mucus, bile, or other products that facilitate digestion); some digestive organs also secrete endocrine hormones that regulate digestion or metabolism of nutrients
Absorption	Movement of digested nutrients through the GI mucosa and into the internal environment
Elimination	Excretion of the residues of the digestive process (feces) from the rectum, through the anus; defecation

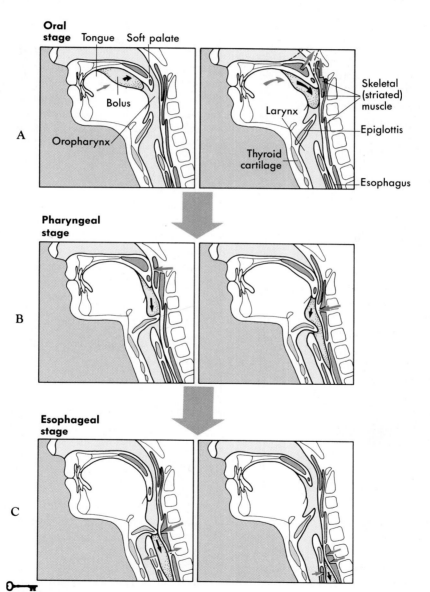

FIGURE 25-1 **Deglutition. A,** Oral stage. During this stage of deglutition (swallowing), a bolus of food is voluntarily formed on the tongue and pushed against the palate and into the oropharynx. Notice that the soft palate acts as a valve that prevents food from entering the nasopharynx. **B,** Pharyngeal stage. Once the bolus is in the oropharynx, involuntary reflexes push the bolus down toward the esophagus. Notice that upward movement of the larynx and downward movement of the bolus close the epiglottis and thus prevent food from entering the lower respiratory tract. **C,** Esophageal stage. Involuntary reflexes of skeletal (striated) and smooth muscle in the wall of the esophagus move the bolus through the esophagus toward the stomach.

Deglutition

The process of swallowing, or *deglutition*, may be divided into formation and movement of a food bolus from mouth to stomach in three main steps or stages (Figure 25-1):

1. Oral stage (mouth to oropharynx)
2. Pharyngeal stage (oropharynx to esophagus)
3. Esophageal stage (esophagus to stomach)

The first step, which is voluntary and under control of the cerebral cortex, involves the formation of a food bolus to be swallowed on a depression or groove in the middle of the tongue. During the oral stage, the bolus is pressed against the palate by the tongue and then moved back into the oropharynx. The pharyngeal and esophageal stages, both involuntary, consist of movement of food from the pharynx into the esophagus and, finally, into the stomach.

To propel food from the pharynx into the esophagus, three openings must be blocked: mouth, nasopharynx, and larynx. Continued elevation of the tongue seals off the mouth. The soft palate, including the uvula, is elevated and tensed, causing the nasopharynx to be closed off. Food is denied entrance into the larynx by muscle action that causes the epiglottis to block this opening. The mechanism involves raising of the larynx, a process easily noted by palpation of the thyroid cartilage during swallowing. As a result, the bolus slips over the back of the epiglottis to enter the laryngopharynx. A combination of gravity and contractions of pharynx and esophagus compresses the bolus into and through the esophageal tube. These steps are involuntary and under control of the *deglutition center* in the medulla. The presence of a bolus stimulates sensory receptors in the mouth and pharynx, thus initiating reflex pharyngeal contractions. Consequently anesthesia of sensory nerves from the mucosa of the mouth and pharynx by a drug such as procaine makes swallowing difficult or impossible.

Swallowing is a complex process requiring the coordination of many muscles and other structures in the head and neck. Not only does the process occur smoothly, but it must also take place rapidly, since respiration is inhibited for the one to three seconds required for food to clear the pharynx during each swallowing.

Peristalsis and segmentation

After food enters the lower portion of the esophagus, smooth muscle tissue in the wall of the GI tract takes on primary responsibility for its movement. Motility produced by smooth muscle is of two main types: peristalsis and segmentation.

Peristalsis is often described as a wavelike ripple of the muscle layer of a hollow organ. Figure 25-2 shows a step-by-step diagram that shows how peristalsis occurs. A bolus stretches the GI wall, triggering a reflex contraction of circular muscle that pushes the bolus forward. This, in turn, triggers a reflex contraction in that location, pushing the bolus even farther. This continues as long as the stretch reflex is activated by the presence of food. Peristalsis is a progressive kind of motility, that is, a type of motion that

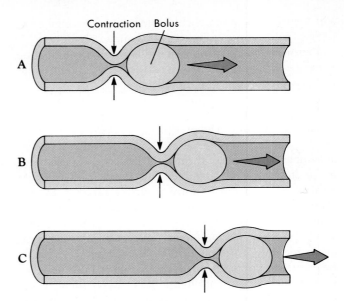

FIGURE 25-2 Peristalsis. Peristalsis is a progressive type of movement, propelling material from point to point along the GI tract. **A,** A ring of contraction occurs where the GI wall is stretched, pushing the bolus forward. **B,** The moving bolus triggers a ring of contraction in the next region, which pushes the bolus even farther along. **C,** The ring of contraction moves like a wave along the GI tract, pushing the bolus forward.

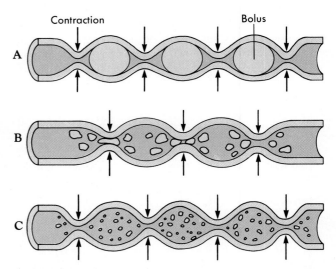

FIGURE 25-3 Segmentation. Segmentation is a back-and-forth action that breaks apart chunks of food and mixes in digestive juices. **A,** Ring-like regions of contraction occur at intervals along the GI tract. **B,** Previously contracted regions relax and adjacent regions now contract, effectively "chopping" the contents of each segment into smaller chunks. **C,** Locations of the contracted regions continue to alternate back and forth, chopping and mixing the contents of the GI lumen.

produces forward movement of ingested material along the GI tract.

Segmentation can be simply described as mixing movement. Segmentation occurs when digestive reflexes cause a forward-and-backward movement within a single region, or segment, of the GI tract (Figure 25-3). Such movement helps mechanically break down food particles, mixes food

and digestive juices thoroughly, and brings digested food in contact with intestinal mucosa to facilitate absorption.

Peristalsis and segmentation can occur in an alternating sequence. When this occurs, food is churned and mixed as it slowly progresses along the GI tract.

Regulation of motility

Gastric motility. The process of emptying the stomach takes about 2 to 6 hours after a meal, depending on the amount and contents of the meal. During its "storage time" in the stomach food is churned with gastric juices to form a thick, milky material known as **chyme,** which is ejected about every 20 seconds into the duodenum. Since the volume of the stomach is large and that of the duodenum is small, gastric emptying must be regulated to prevent overburdening of the duodenum. Such control occurs by two principal mechanisms, one hormonal and one nervous. Fats and other nutrients in the duodenum stimulate the intestinal mucosa to release a hormone called **gastric inhibitory peptide (GIP)** into the bloodstream. When it reaches the stomach wall via the circulation, GIP has an inhibitory effect on gastric muscle, decreasing its peristalsis and thus slowing passage of food into the duodenum. A hormone called *enterogastrone* may be involved in this endocrine reflex, but its existence has never been proven. Nervous control results from receptors in the duodenal mucosa that are sensitive to the presence of acid and to distention. Sensory and motor fibers in the vagus nerve then cause a reflex inhibition of gastric peristalsis. This nervous mechanism is known as the **enterogastric reflex.**

Intestinal motility. Intestinal motility includes both peristaltic contractions and segmentation. Segmentation in the duodenum and upper jejunum mixes the incoming chyme with digestive juices from the pancreas, liver, and intestinal mucosa. This mixing action also allows the products of digestion to contact the intestinal mucosa, where they can be absorbed into the internal environment. Peristalsis continues as the chyme nears the end of the jejunum—moving the food through the rest of the small intestine and into the large intestine. After leaving the stomach, it normally takes about 5 hours for chyme to pass all the way through the small intestine.

Several mechanisms are involved in the control of intestinal motility. Peristalsis is regulated in part by the intrinsic stretch reflexes already described. It is also thought to be stimulated by the hormone **cholecystokinin-pancreozymin (CCK),** which is secreted by endocrine cells of the intestinal mucosa when chyme is present.

A list of definitions of the different processes involved in mechanical digestion, with the organs that accomplish them, are given in Table 25-2.

QUICK CHECK

✔ 1. What is meant by the term motility?
✔ 2. Is deglutition a voluntary or involuntary process?
✔ 3. What is the purpose of peristalsis?
✔ 4. What triggers the enterogastric reflex to inhibit gastric emptying?

TABLE 25-2	Processes of mechanical digestion	
Organ	**Mechanical Process**	**Nature of Process**
Mouth (teeth and tongue)	Mastication	Chewing movements—reduce size of food particles and mix them with saliva
	Deglutition	Swallowing—movement of food from mouth to stomach
Pharynx	Deglutition	
Esophagus	Deglutition	
	Peristalsis	Rippling movements that squeeze food downward in tract; constricted ring forms first in one section, the next, etc., causing waves of contraction to spread along entire canal
Stomach	Churning	Forward and backward movement of gastric contents, mixing food with gastric juices to form chyme
	Peristalsis	Waves starting in body of stomach about three times per minute and sweeping toward closed pyloric sphincter; at intervals, strong peristaltic waves press chyme past sphincter into duodenum
Small intestine	Segmentation (mixing contractions)	Forward and backward movement within segment of intestine; purpose, to mix food and digestive juices thoroughly and to bring all digested food in contact with intestinal mucosa to facilitate absorption; purpose of peristalsis, on the other hand, to propel intestinal contents along digestive tract
	Peristalsis	
Large intestine		
Colon	Segmentation	Churning movements within haustral sacs
	Peristalsis	
Descending colon	Mass peristalsis	Entire contents moved into sigmoid colon and rectum; occurs three or four times a day, usually after a meal
Rectum	Defecation	Emptying of rectum, so-called bowel movement

Chemical Digestion

Chemical digestion consists of all the changes in chemical composition that foods undergo in their travel through the digestive tract. These changes result from the hydrolysis of foods. **Hydrolysis** is a chemical process in which a compound unites with water and then splits into simpler compounds (see Chapter 2, p. 57). Numerous enzymes in the various digestive juices catalyze the hydrolysis of foods.

Enzymes

Enzymes and their actions were briefly discussed in Chapter 2 (see p. 52). Although our interest here primarily concerns *digestive enzymes*, this is a good opportunity for a comprehensive discussion of enzymes.

Enzymes are usually defined simply as "organic catalysts," that is, they are organic compounds, and they accelerate chemical reactions without appearing in the final products of the reaction. Enzymes are vital substances. Without them, the chemical reactions necessary for life could not take place. So important are they that someone has even defined life as the "orderly functioning of hundreds of enzymes."

Chemical structure of enzymes. Enzymes are proteins. Frequently their molecules contain a nonprotein part called the *prosthetic group* of the enzyme molecule (if this group readily detaches from the rest of the molecule, it is called the *coenzyme*). Some prosthetic groups contain inorganic ions (e.g., Ca^{++}, Mg^{++}, Mn^{++}). Many of them contain vitamins. In fact, every vitamin of known function constitutes part of a prosthetic group of some enzyme.

Classification and naming of enzymes. Two systems used for naming enzymes are as follows: the suffix *-ase* is used with the root name of the substance whose chemical reaction is catalyzed (the substrate chemical, that is) or with the word that describes the kind of chemical reaction catalyzed. Thus, according to the first method, sucrase is an enzyme that catalyzes a chemical reaction in which sucrose takes part. According to the second method, sucrase might also be called a hydrolase because it catalyzes the hydrolysis of sucrose. Enzymes investigated before these methods of nomenclature were adopted still are called by older names, such as *pepsin* and *trypsin*.

Classified according to the kind of chemical reactions catalyzed, enzymes fall into several groups:

♦ **Oxidation-reduction enzymes.** These are known as oxidases, hydrogenases, and dehydrogenases. Energy release for muscular contraction and all physiological work depends on these enzymes.

♦ **Hydrolyzing enzymes,** or hydrolases. Digestive enzymes belong to this group. These are named after the substrate acted on, for example, lipase, sucrase, and maltase.

♦ **Phosphorylating enzymes.** These add or remove phosphate groups and are known as phosphorylases or phosphatases.

♦ **Enzymes that add or remove carbon dioxide.** These are known as carboxylases or decarboxylases.

♦ **Enzymes that rearrange atoms within a molecule.** These are known as mutases or isomerases.

♦ **Hydrases.** These add water to a molecule without splitting it, as hydrolases do.

Enzymes are also classified as intracellular or extracellular, depending on whether they act within cells or outside them in the surrounding medium. Most enzymes act intracellularly in the body; an important exception is the digestive enzymes. All digestive enzymes are classified as hydrolases because they catalyze the hydrolysis of food molecules.

Properties of enzymes. In general, the chemical properties of enzymes are the same as those of all proteins, since enzymes are proteins themselves. For example, they form colloidal solutes in water and are precipitated or coagulated by various agents such as high temperatures and salts of heavy metals. Hence, these agents inactivate enzymes. Other important enzyme properties are as follows:

♦ Most enzymes are *specific in their action,* that is, they act only on a specific substrate. This is attributed to a "key-in-a-lock" kind of action, the configuration of the enzyme molecule fitting the configuration of some part of the substrate molecule (Figure 25-4).

♦ Enzymes *function optimally at a specific pH* and become inactive if this deviates beyond narrow limits (Figure 25-5).

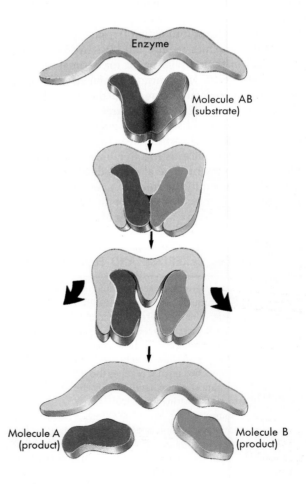

Enzyme

Molecule AB
(substrate)

Molecule A
(product)

Molecule B
(product)

FIGURE 25-4 Model of enzyme action. Enzymes are functional proteins whose molecular shape allows them to catalyze chemical reactions. Molecule *AB* is acted on by a digestive enzyme, yielding simpler molecules *A* and *B*.

This effect is produced by the fact that changes in H^+ concentration influence the chemical attractions that hold all protein molecules—including enzymes—in their complex, multidimensional shapes. In short, changing the pH changes the shape of an enzyme molecule—possibly rendering it inactive.

♦ *Various physical and chemical agents inactivate or inhibit enzyme action, by changing the shape of enzyme molecules.* For example, such agents include x-rays and ionizing radiation (this presumably accounts for some of the ill effects of excessive radiation), certain antibiotic drugs, or unfavorable pH or extreme temperatures (Figure 25-5).

♦ *Most enzymes catalyze a chemical reaction in both directions,* the direction and rate of the reaction being governed by the law of mass action. An accumulation of a product slows the reaction and tends to reverse it. A practical application of this fact is the slowing of digestion when absorption is interfered with and the products of digestion accumulate.

♦ *Enzymes are continually being destroyed in the body* and therefore have to be continually synthesized, even though they are not used up in the reactions they catalyze.

♦ *Many enzymes are synthesized* as inactive **proenzymes.** Substances that convert proenzymes to active enzymes are often called *kinases,* for example, enterokinase changes inactive trypsinogen into active trypsin.

Different enzymes require different hydrogen ion concentrations in their environment for optimal functioning. This is because H^+ concentration influences the shape of each enzyme molecule. Amylase, the main enzyme in saliva, functions best in the neutral to slightly acid pH range characteristic of saliva. It is gradually inactivated by the marked acidity of gastric juice. In contrast, pepsin, an enzyme in gastric juice, is inactive unless sufficient hydrochloric acid is present. Therefore, in diseases characterized by gastric hypoacidity (pernicious anemia, for example), dilute hydrochloric acid is given orally before meals.

Although we eat six main types of chemical substances

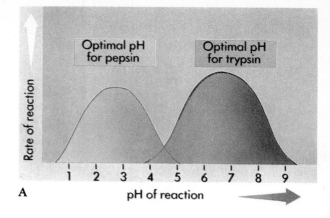

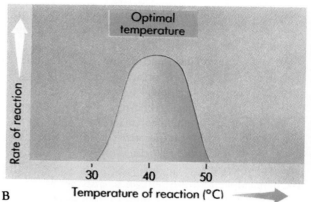

FIGURE 25-5   Effects of pH and temperature on enzyme function. The rate of reactions catalyzed can be affected by chemical or physical properties of the surrounding medium. **A,** Enymes catalyze chemical reactions with greatest efficiency within a narrow range of pH. For example, *pepsin* (a protein-digesting enzyme in gastric juice) operates within a low pH range, whereas *trypsin* (a protein-digesting enzyme in pancreatic juice) operates within a higher pH range. **B,** Most enzymes in the human body work best within a narrow range of temperatures near 40° C.

(carbohydrates, proteins, fats, vitamins, mineral salts, and water), only the first three have to be chemically digested to be absorbed.

Carbohydrate digestion

Carbohydrates are saccharide compounds. This means that their molecules contain one or more saccharide groups ($C_6H_{10}O_5$). Polysaccharides, notably starches and glycogen, contain many of these groups. Disaccharides (sucrose, lactose, and maltose) contain two of them, and monosaccharides (glucose, fructose, and galactose) contain only one. Polysaccharides are hydrolyzed to disaccharides by enzymes known as *amylases,* found in saliva and pancreatic juice (salivary amylase was formerly called ptyalin). The enzymes that catalyze the final steps of carbohydrate digestion are *sucrase, lactase,* and *maltase* (Figure 25-6). These enzymes are located in the cell membrane of epithelial cells covering villi and, therefore, lining the intestinal lumen. The substrates (disaccharides) bind onto the enzymes at the surface of the brush border, giving the name **"contact digestion"** to the process. The resulting end products of digestion, mainly glucose, are conveniently located at the site of absorption (and are not floating around somewhere in the lumen).

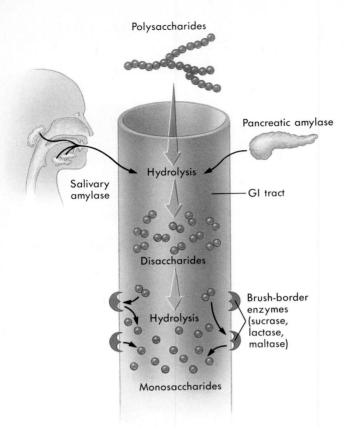

FIGURE 25-6   **Carbohydrate digestion.** Amylase in saliva and pancreatic juice hydrolyze polysaccharides into disaccharides. Brush-border disaccharidases in lining of the small intestine then promote hydrolysis of the disaccharides into monosaccharides.

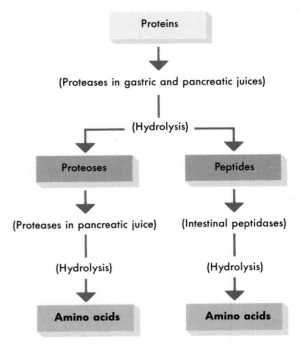

FIGURE 25-7   **Protein digestion.** Gastric juice protease (pepsin) and pancreatic juice protease (trypsin and chymotrypsin) hydrolyze proteins to proteoses and peptides. Protein digestion is then completed by pancreatic proteases, which hydrolyze proteoses to amino acids, and intestinal peptidases, which hydrolyze peptides to amino acids.

Protein digestion

Protein compounds have very large molecules made up of folded or twisted chains of amino acids, often hundreds in number. Enzymes called proteases catalyze the hydrolysis of proteins into intermediate compounds, for example, proteoses and peptides, and, finally, into amino acids (Figure 25-7). The main proteases are pepsin in gastric juice, trypsin in pancreatic juice, and peptidases of the intestinal brush border.

Fat digestion

Because fats are insoluble in water, they must be emulsified, that is, dispersed as very small droplets, before they can be digested. Bile emulsifies fats in the small intestine. This mechanical process facilitates chemical digestion of fats by providing a greater contact area between fat molecules and pancreatic *lipase,* the main fat-digesting enzyme (Figure 25-8). Triglycerides, important dietary fats, are broken down by lipase to yield fatty acids, monoglycerides, and glycerol molecules.

For a summary of the actions of each digestive juice, see Table 25-3.

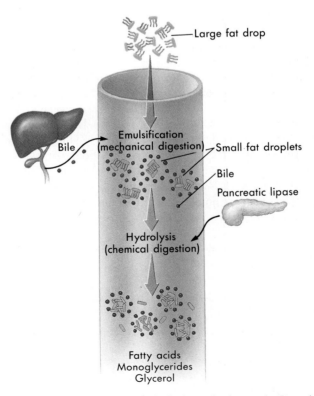

FIGURE 25-8   Fat digestion (hydrolysis) by lipase, facilitated first by emulsion of fats by bile.

TABLE 25-3   Chemical digestion

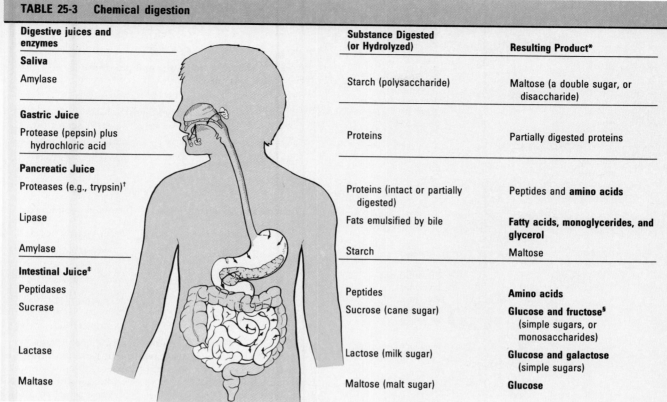

Digestive juices and enzymes	Substance Digested (or Hydrolyzed)	Resulting Product*
Saliva		
Amylase	Starch (polysaccharide)	Maltose (a double sugar, or disaccharide)
Gastric Juice		
Protease (pepsin) plus hydrochloric acid	Proteins	Partially digested proteins
Pancreatic Juice		
Proteases (e.g., trypsin)†	Proteins (intact or partially digested)	Peptides and **amino acids**
Lipase	Fats emulsified by bile	**Fatty acids, monoglycerides, and glycerol**
Amylase	Starch	Maltose
Intestinal Juice‡		
Peptidases	Peptides	**Amino acids**
Sucrase	Sucrose (cane sugar)	**Glucose and fructose§** (simple sugars, or monosaccharides)
Lactase	Lactose (milk sugar)	**Glucose and galactose** (simple sugars)
Maltase	Maltose (malt sugar)	**Glucose**

*Substances in boldface type are end products of digestion (that is, completely digested foods ready for absorption).
†Secreted in inactive form (trypsinogen); activated by enterokinase, an enzyme in the intestinal juice.
‡Brush-border enzymes.
§Glucose is also called *dextrose;* fructose is also called *levulose.*

Residues of digestion

Certain components of food resist digestion and are eliminated from the intestines in the **feces.** Including among these *residues of digestion* are cellulose (a carbohydrate, also known as "dietary fiber") and undigested connective tissue from meat (mostly collagen). These substances remain undigested because humans lack the enzymes required to hydrolyze them. The residues of digestion also include undigested fats. Some fat molecules remain undigested because they have combined with dietary minerals such as calcium and magnesium, which renders the fats undigestible. In addition to these wastes, feces consist of bacteria, pigments, water, and mucus.

QUICK CHECK

✔ 1. *What type of reaction do all digestive enzymes catalyze?*

✔ 2. *List some factors that alter the shape of an enzyme, thus altering its function.*

✔ 3. *Name the final digestive products of each of these food molecules: protein, carbohydrate, triglyceride.*

◤ CONTROL OF DIGESTIVE GLAND SECRETION

Exocrine digestive glands secrete when food is present in the digestive tract or when it is seen, smelled, or imagined. Complicated nervous and hormonal reflex mechanisms control the flow of digestive juices in such a way that they appear in proper amounts when and for as long as needed.

Salivary Secretion

As far as is known, only reflex mechanisms control the secretion of saliva. Chemical, mechanical, olfactory, and visual stimuli initiate afferent impulses to centers in the brain stem that send out efferent impulses to salivary glands, stimulating them. Chemical and mechanical stimuli come from the presence of food in the mouth. Olfactory and visual stimuli come, of course, from the smell and sight of food.

Gastric Secretion

Stimulation of gastric juice secretion occurs in three phases controlled by reflex and chemical mechanisms. Because stimuli that activate these mechanisms arise in the head, stomach, and intestines, the three phases are known as the cephalic, gastric, and intestinal phases, respectively. As you read the description of each phase, glance at the diagrams shown in Figure 25-9.

The **cephalic phase** is also spoken of as the "psychic phase" because psychic (mental) factors activate the mechanism. For example, the sight, smell, taste, or even thought of food that is pleasing to an individual activate control centers in the medulla oblongata from which parasympathetic fibers of the vagus nerve conduct efferent impulses to the gastric glands. Vagal nerve impulses also stimulates production of **gastrin,** a hormone secreted by the gastric mucosa. Gastrin stimulates gastric secretion, thus prolonging and enhancing the response.

During the **gastric phase** of gastric secretion, the following chemical control mechanism dominates. Products of protein digestion in foods that have reached the pyloric portion of the stomach stimulate its mucosa to release *gastrin* into the blood in stomach capillaries. When it circulates to the gastric glands, gastrin greatly accelerates their secretion of gastric juice, which has a high pepsinogen and hydrochloric acid content (Table 25-4). Hence this seems to be a mechanism for ensuring that when food is in the stomach, there will be enough enzymes there to digest it. Gastrin release is also stimulated by distention of the stomach (caused by the presence of food), which activates local and parasympathetic reflexes in the pylorus.

The **intestinal phase** of gastric juice secretion is less clearly understood than the other two phases. Various different mechanisms seem to adjust gastric juice secretion as chyme passes to and through the intestinal tract. Experiments show that gastric secretions are inhibited when chyme containing fats, carbohydrates, and acid (low pH) are present in the duodenum. This probably occurs by means of endocrine reflexes that involve the hormones **gastric inhibitory peptide (GIP), secretin, CCK,** and perhaps several others. These hormones are secreted by endocrine cells in the mucosa of the duodenum. Gastric secretion may also be inhibited by the parasympathetic *enterogastric reflex.* We have discussed how this reflex inhibits gastric motility as food begins to fill the duodenum; now we see that it may inhibit gastric secretion as well.

In summary, we see that the rate of gastric secretion can be adjusted by nervous and endocrine reflex mechanisms in ways that improve the efficiency of the system. Anticipation of swallowing food causes the stomach to prepare itself by increasing its secretion of enzymes and acid. Thus food enters a stomach already partially filled with gastric juice. The rate of gastric secretion can then be adjusted according to the amount of food present and whether it contains proteins (the only food that can be chemically digested by gastric juice). Gastric secretion—and thus chemical digestion in the stomach—can be slowed when the duodenum becomes full. This prevents the stomach from finishing its task before the small intestine is ready to receive the chyme.

Pancreatic Secretion

Several hormones released by intestinal mucosa are known to stimulate pancreatic secretion. One of these, *secretin,* evokes production of pancreatic fluid low in enzyme content but high in bicarbonate (HCO_3^-). This

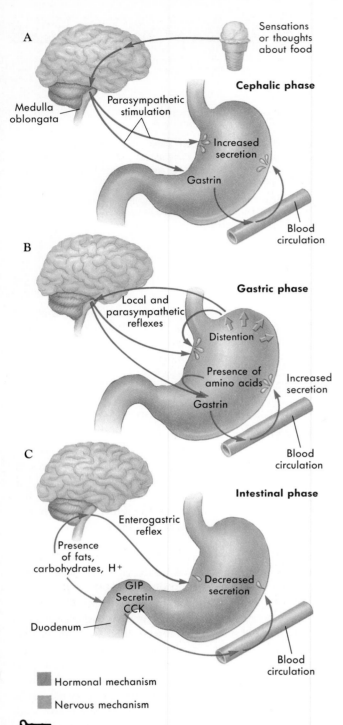

FIGURE 25-9   Phases of gastric secretion. A, Cephalic phase. Sensations of thoughts about food are relayed to the brainstem, where parasympathetic signals to the gastric mucosa are initiated. This directly stimulates gastric juice secretion and also stimulates the release of gastrin, which prolongs and enhances the effect. **B,** Gastric phase. The presence of food, specifically the distention it causes, triggers local and parasympathetic nervous reflexes that increase secretion of gastric juice and gastrin (which further amplifies gastric juice secretion). Products of protein digestion can also trigger the gastrin mechanism. **C,** Intestinal phase. As food moves into the duodenum, the presence of fats, carbohydrates, and acid stimulate hormonal and nervous reflexes that inhibit stomach activity.

TABLE 25-4   Actions of some digestive hormones summarized

Hormone	Source	Action
Gastrin	Formed by gastric mucosa in presence of partially digested proteins, when stimulated by the vagus nerve, or when the stomach is stretched	Stimulates secretion of gastric juice rich in pepsin and hydrochloric acid
Gastric inhibitory peptide (GIP)	Formed by intestinal mucosa in presence of fats and perhaps other nutrients	Inhibits gastric secretion and motility
Secretin	Formed by intestinal mucosa in presence of acid, partially digested proteins, and fats	Inhibits gastric secretion; stimulates secretion of pancreatic juice low in enzymes and high in alkalinity (bicarbonate); stimulates ejection of bile by the gallbladder
Cholecystokinin-pancreozymin (CCK)	Formed by intestinal mucosa in presence of fats, partially digested proteins, and acids	Stimulates ejection of bile from gallbladder and secretion of pancreatic juice high in enzymes; opposes the action of gastrin, reducing the pH of gastric juice

alkaline fluid acts to neutralize the acid (chyme) entering the duodenum. As you might expect, the presence of acid in the duodenum serves as the most potent stimulator of secretin. (Additional control involving the same hormone is shown by the fact that fats in the duodenum also elicit secretin, which then influences the gallbladder to increase its ejection of the fat emulsifier bile.)

The other intestinal hormone, known as **cholecystokinin-pancreozymin (CCK),** was originally thought to be two separate substances. It has now been identified as one chemical with several important functions: (1) it causes the pancreas to increase exocrine secretions high in enzyme content, (2) it opposes the influence of gastrin on gastric parietal cells, thus inhibiting HCl secretion by the stomach, and (3) it also stimulates contraction of the gallbladder so that bile can pass into the duodenum.

Secretion of Bile

Bile is secreted continually by the liver and is stored in the gallbladder until it is needed by the duodenum. The hormones secretin and CCK, as described, stimulate ejection of bile from the gallbladder (see Table 25-4).

Intestinal Secretion

Relatively little is known about the regulation of intestinal exocrine secretions. Some evidence suggests that the intestinal mucosa, stimulated by hydrochloric acid and food products, releases hormones into the blood including *vasoactive intestinal peptide (VIP),* which brings about increased production of intestinal juice. Intestinal secretions appear to contain bicarbonate, which, along with pancreatic bicarbonate, neutralizes acid from the stomach. Bicarbonate secretion is probably regulated by a reflex sensitive to changes in pH of the chyme. Presumably, neural mechanisms also help control the secretion of intestinal juice.

QUICK CHECK

✔ 1. Can you name the three phases of gastric secretion?
✔ 2. What is the function of gastric inhibitory peptide (GIP)?

ABSORPTION

Process of Absorption

Absorption is the passage of substances (notably digested foods, water, salts, and vitamins) through the intestinal mucosa into the blood or lymph. As stated earlier, most absorption occurs in the small intestine, where the large surface area provided by the intestinal villi and microvilli (Figure 25-10) facilitates this process.

Mechanisms of Absorption

Absorption of some substances, such as water, is simple and straightforward: simple diffusion, or osmosis. However, some substances depend on more complex mechanisms in order to be absorbed. Sodium is a good example. Epithelial cells that form the outer wall of the villus (see Figure 25-10) constantly pump sodium from the GI lumen into the internal environment through a complex process called *secondary active transport.* As Figure 25-11, A, shows, active transport carriers on the basal side, or "back side," of the cell continually pump Na^+ out of the cell. This mechanism maintains a low Na^+ concentration inside the cell. Thus it is likely that Na^+ in the GI lumen will diffuse into the low-Na^+ cell. As Na^+ diffuses in through passive carriers in the cell's luminal surface, or "lumen side," it is removed by active transport pumps in the cell's basal membrane. In short, Na^+ moves out of the GI lumen only because it is being pumped from the other side of the intestinal cells.

Another good example of a complex transport process is that involving glucose. Although considered an "end product of digestion," glucose is a relatively large molecule and cannot pass freely through the brush border membrane of an intestinal mucosa cell. In addition to physical size, the lipid nature of the cell membrane (Figure 3-2, p. 66) presents another barrier to glucose absorption. Only lipid-soluble (hydrophobic) molecules the size of glucose can pass freely (passively) through the lipid cell barrier. Since glucose is too large physically and is hydrophilic (water-soluble) in nature, it must be transported across the membrane by a carrier to enter the cell. In a process called *sodium cotransport* or coupled transport, carriers that bind both sodium ions and glucose molecules passively transport these molecules *together* out of the GI lumen (Figure 25-11,

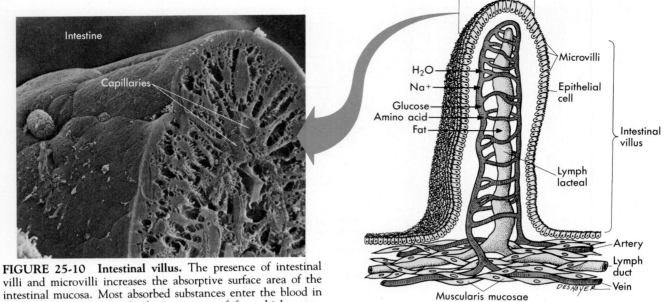

FIGURE 25-10   Intestinal villus. The presence of intestinal villi and microvilli increases the absorptive surface area of the intestinal mucosa. Most absorbed substances enter the blood in intestinal capillaries, with the exception of fat, which enters lymph by way of the intestinal lacteals.

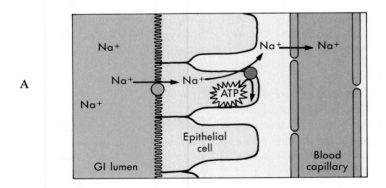

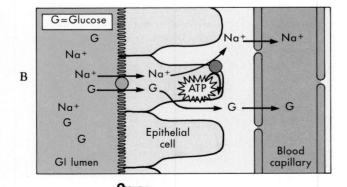

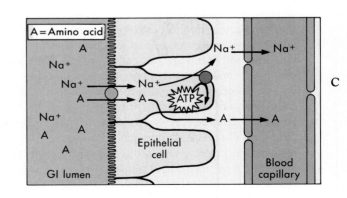

FIGURE 25-11   Absorption of sodium, glucose, and amino acids. Absorption of sodium **(A)**, glucose **(B)**, and amino acids **(C)** are all forms of *secondary active transport* because each involves two carriers, one of which is active. The active carrier on the basal side of the epithelial cell maintains a sodium gradient, which facilitates passive transport of sodium, and perhaps another molecule, out of the GI lumen via a passive carrier on the luminal side of the cell.

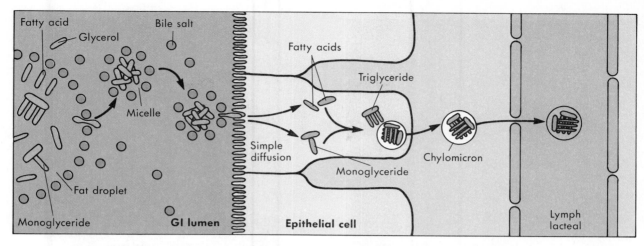

FIGURE 25-12 **Absorption of fats.** Fats such as triglycerides are chemically digested within emulsified fat droplets, yielding fatty acids, monoglycerides, and glycerol (*left*). Fatty acids and other lipid-soluble compounds (such as cholesterol) leave the fat droplets in small spheres coated with bile salts (micelles). When a micelle reaches the plasma membrane of an absorptive cell, individual fat-soluble molecules diffuse directly into the cytoplasm. The endoplasmic reticulum of the cell resynthesizes fatty acids and monoglycerides into triglycerides. A Golgi body within the cell packages the fats into vesicles called chylomicrons, which then exit the absorptive cell by exocytosis and enter a lymphatic lacteal.

B). However, this is another case of secondary active transport because this movement does not occur without the Na^+ concentration gradient maintained by the active transport of Na^+ out of the cell's basal membrane. Amino acids and several other compounds are thought to also be absorbed by such a secondary active transport mechanism (Figure 25-11, C).

Other mechanisms of transporting glucose and amino acids have also been proposed. One hypothesis suggests that these compounds are transported by passive carriers on both the luminal and the basal surfaces of the absorptive cells. Another hypothesis suggests that the brush border enzymes also act as carriers. It should also be noted that some short polypeptides can diffuse into absorptive cells where they are hydrolyzed into amino acids that can move into the blood.

Fatty acids and monoglycerides (products of fat digestion) and cholesterol are transported with the aid of bile salts from the watery intestinal lumen to absorbing cells on villi. Bile salts form spheres called **micelles** that surround the lipid, making it temporarily water-soluble (Figure 25-12). As micelles approach the brush border of absorbing cells, lipids are released to pass through the cell membrane (since its lipid bilayer is receptive to lipids) by simple diffusion. Once inside the cell, fatty acids are rapidly reunited with monoglycerides to form triglycerides (neutral fats). The final step in lipid transport by the intestine is the formation of **chylomicrons,** which are composed mainly of neutral fats and some cholesterol covered by a delicate protein envelope. This important envelope allows fats to be transported through lymph and into the bloodstream (Table 25-5).

Vitamins A, D, E, and K, known as the "fat-soluble vitamins," also depend on bile salts for their absorption. Some water-soluble vitamins, such as certain of the B group, are small enough to be absorbed by simple diffusion; however, most require carrier-mediated transport. Many drugs (sedatives, analgesics, antibiotics) appear to be absorbed by simple diffusion, since many are lipid soluble.

Note that after absorption, food does not pass directly

DIAGNOSTIC STUDIES | **FECAL FAT TEST**

Impaired fat absorption (malabsorption), prevalent in numerous diseases, produces large, greasy, and foul-smelling stools, or *steatorrhea.*

The fecal fat test measures the fat content in the stool. The total output of fecal fat per 24 hours in a 3- to 5-day stool collection provides the most reliable measurement. Each stool specimen throughout the period is collected in a clean, dry container and is sent immediately to the lab. The 3- to 5-day collection period is necessary to eliminate daily variations in the amount of fecal fat.

A standard fat content diet is begun 2 or 3 days before collection begins and continues until collection is done. Usually 100 g of fat per day is suggested for adults. In children and infants who cannot eat 100 g of fat per day a fat retention coefficient is determined by using the formula:

$$\frac{\text{Ingested fat} - \text{Fecal fat}}{\text{Ingested fat}} \times 100\%$$

If this fat retention coefficient is lower than 95%, the patient may have steatorrhea.

Analysis of fecal fat is useful in monitoring malabsorption in cystic fibrosis or in any condition characterized by malabsorption, maldigestion, or increased fecal fat.

TABLE 25-5   Food absorption

Form Absorbed	Structures into which Absorbed	Circulation
Protein—as amino acids	Blood in intestinal capillaries	Portal vein, liver, hepatic vein, inferior vena cava to heart, etc.
Perhaps minute quantities of some short-chain polypeptides and whole proteins absorbed, for example, some antibodies		
Carbohydrates—as simple sugars	Same as amino acids	Same as amino acids
Fats		
Glycerol and monoglycerides	Lymph in intestinal lacteals	During absorption, that is, while in epithelial cells of intestinal mucosa, glycerol, and fatty acids recombine to form micro-
Fatty acids combine with bile salts to form water-soluble substance	Lymph in intestinal lacteals	scopic packages of fats (chylomicrons); lymphatics carry them by way of thoracic duct to left subclavian vein, superior vena
Some finely emulsified, undigested fats absorbed	Small fraction enters intestinal blood capillaries	cava, heart, etc.; some fats transported by blood in form of phospholipids or cholesterol esters

into the general circulation. Instead, it first travels by way of the hepatic portal system to the liver (see Figure 17-24, p. 448). Following absorption, blood entering the liver via the portal vein contains greater concentrations of glucose, amino acids, and fats than blood leaving the liver via the hepatic vein for the systemic circulation. Clearly, the excess of these food substances over and above the normal blood levels has remained behind in the liver.

What the liver does with them is part of the story of nutrition and metabolism, our topic for discussion in the next chapter.

ELIMINATION

The process of elimination is simply the expulsion of the residues of digestion—*feces*—from the digestive tract. Formation of feces is the primary function of the colon. The act of expelling feces is called **defecation.** Defecation is a reflex brought about by stimulation of receptors in the rectal mucosa. Normally the rectum is empty until mass peristalsis moves fecal matter out of the colon into the rectum. This distends the rectum and produces the desire to defecate. Also, it stimulates colonic peristalsis and initiates reflex relaxation of the internal sphincters of the anus (see Figure 24-17). Voluntary straining efforts and relaxation of the external anal sphincter may then follow as a result of the desire to defecate. Together, these several responses bring about defecation. Note that this is a reflex partly under voluntary control. If one voluntarily inhibits it, the rectal receptors soon become depressed and the urge to defecate usually does not recur until hours later, when mass peristalsis again takes place.

Constipation occurs when the contents of the lower colon and rectum move at a rate that is slower than normal. Extra water is absorbed from the fecal mass, producing a hardened, or constipated, stool. **Diarrhea** may occur as a result of increased motility of the small intestine. Chyme moves through the small intestine too quickly, reducing the amount of absorption of water and electrolytes. Diarrhea

may also result from bacterial toxins that damage the water-reabsorption mechanisms of the intestinal mucosa. The large volume of material arriving in the large intestine exceeds the limited capacity of the colon for absorption, so a watery stool results. Prolonged diarrhea can be particularly serious, even fatal, in infants because they have a minimal reserve of water and electrolytes.

QUICK CHECK

✔ 1. *Explain the term* secondary active transport.
✔ 2. *Describe how fatty acids are absorbed by cells of the GI mucosa.*
✔ 3. *What triggers the defecation reflex?*

HEALTH MATTERS   ORAL REHYDRATION THERAPY

Simple diarrhea can result in life-threatening dehydration if significant water loss occurs during a short period of time. Diarrhea is one of the leading causes of infant mortality in developing countries. For example, many countries have recently experienced outbreaks of **cholera,** a diarrheal condition caused by the bacterium, *Vibrio cholerae.* Cholera bacteria in the GI tract poison certain transport mechanisms in absorptive cells, preventing them from reabsorbing water from the lumen of the GI tract. In cholera and similar conditions, diarrhea may become so severe that it causes death within a day or two. New attempts to educate people about the dangers of diarrhea and to provide them with a simple, easily prepared remedy have saved hundreds of thousands of lives in recent years. This treatment is called **oral rehydration therapy (ORT).** It involves liberal doses of an easily prepared solution containing sugar and salt. This salt-sugar solution utilizes undamaged transport mechanisms and thus replaces water, nutrients, and electrolytes lost in diarrheal fluid. Because the replacement fluid can be prepared from readily available and inexpensive ingredients, it is particularly valuable in the treatment of infant diarrhea in third-world countries.

THE BIG PICTURE
Digestion and the Whole Body

The process of digestion, like any other vital function, provides a means of survival for the entire body and also requires the function of other systems. The digestive system's primary contribution to overall homeostasis is its ability to maintain a constancy of nutrient concentration in the internal environment. It accomplishes this by breaking large, complex nutrients into smaller, simpler nutrients so they can be absorbed. The digestive system also provides the means of absorption—the cellular mechanisms that operate in the absorptive cells of the intestinal mucosa. The digestive system also provides some secondary, less vital functions. For example, the teeth and tongue aid the nervous system and respiratory system in producing spoken language. Also, acid in the stomach assists the immune system by destroying potentially harmful bacteria. Some of the various vital and nonvital roles played by the different organs that make up the digestive system are summarized in Figure 25-13.

To accomplish its functions, the digestive system requires functional contributions by other systems of the body. Regulation of digestive motility and secretion require the active participation of both the nervous system and the endocrine system. The oxygen needed for digestive activity requires the proper functioning of both the respiratory system and the circulatory system. The body's framework (integumentary and skeletal systems) are required to support and protect the digestive organs. The skeletal muscles must function if ingestion, mastication, deglutition, and defecation are to occur normally. As you can see, the digestive system cannot operate alone—nor can any system or organ, for that matter. The body is truly an integrated system, not a collection of independent components.

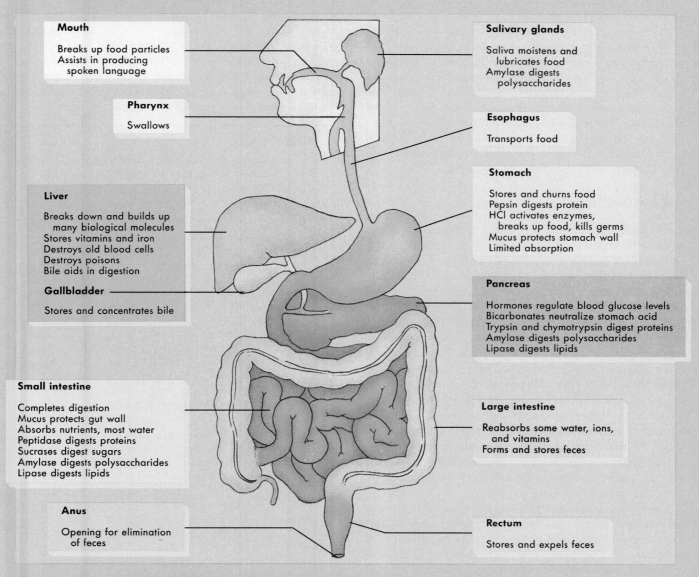

FIGURE 25-13 Summary of digestive functions of digestive organs.

MECHANISMS OF DISEASE

Disorders of the Digestive System

Disorders of the GI Tract

Gastroenterology (gas-tro-en-ter-AHL-o-jee) is the study of the stomach (*gastro-*) and intestines (*entero-*) and their diseases. The gastrointestinal tract is the potential site of numerous diseases and conditions, some of which are briefly described in this section. Many of these disorders, particularly those that primarily affect the stomach or duodenum, are characterized by one or more of these signs and symptoms:

- **Gastroenteritis**—stomach inflammation (gastritis) and intestinal inflammation (enteritis)
- **Anorexia**—chronic loss of appetite
- **Nausea**—unpleasant feeling that often leads to vomiting
- **Emesis**—vomiting (Figure 25-14)
- **Diarrhea**—elimination of liquid feces, perhaps accompanied by abdominal cramps
- **Constipation**—decreased motility of colon, resulting in difficulty in defecation

An **ulcer** is an open wound or sore in an area of the digestive system that is acted on by acid gastric juice. The two most common sites for ulcers are the stomach (gastric ulcers) and the upper part of the small intestine, or duodenum (duodenal ulcers). Although most people think of ulcers as occurring in the stomach, most are duodenal. Ulcers cause disintegration, loss, and death of tissue as they erode the layers in the wall of the stomach or duodenum. Left untreated, ulcers cause persistent pain and may perforate the wall of the digestive tube, causing massive hemorrhage and widespread inflammation of the abdominal cavity and its contents. Usually perforation does not occur, but small, repeated hemorrhages over long periods cause anemia. There is an old saying in medicine, "No acid, no ulcer." Most experts now agree that too much gastric acid secretion (that is, prolonged hyperacidity) is one of the most important factors in ulcer formation. If the protective layer of mucus is insufficient or if there is inadequate dilution and buffering of acid gastric juices by swallowed food and the alkaline juices of the small

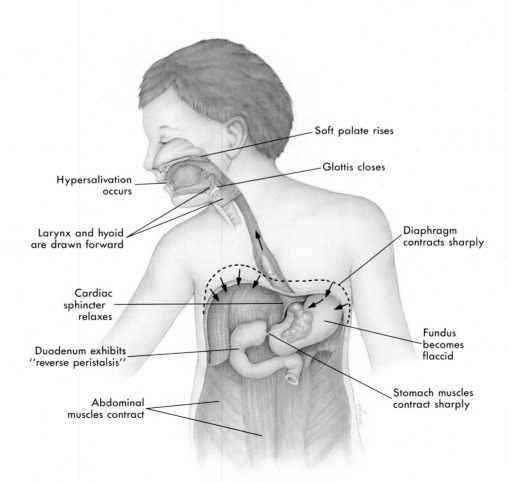

Soft palate rises

Glottis closes

Hypersalivation occurs

Larynx and hyoid are drawn forward

Diaphragm contracts sharply

Cardiac sphincter relaxes

Duodenum exhibits "reverse peristalsis"

Fundus becomes flaccid

Stomach muscles contract sharply

Abdominal muscles contract

FIGURE 25-14   Emesis. This illustration shows the key events of the vomiting reflex—a common sign of gastrointestinal disorders.

intestine, ulcers may form. Hyperacidity is influenced by nervous system factors and by anxiety, other emotional states, and stress.

The drugs ranitidine (Zantac), cimetidine (Tagamet), and other medications that reduce hydrochloric acid formation in the stomach are widely prescribed in the treatment of ulcers. In addition to excess acid, a bacterium called *Campylobacter (Helicobacter) pylori*, found in many ulcer patients, may also be a cause of both ulcers and chronic indigestion. This bacterium, which was first discovered in 1982 and linked to gastrointestinal inflammation in 1987, now joins hyperacidity as a potential cause of ulcers.

Stomach cancer has been linked to excessive alcohol consumption, use of chewing tobacco, and eating smoked or heavily preserved food. Most stomach cancers, usually *adenocarcinomas*, have already metastasized before they are found because patients treat themselves for the early warning signs of heartburn, belching, and nausea. Later warning signs of stomach cancer include chronic indigestion, vomiting, anorexia, stomach pain, and blood in the feces. Surgical removal of the malignant tumors has been the most successful method of treating stomach cancer.

Malabsorption syndrome is a general term referring to a group of symptoms resulting from the failure of the small intestine to absorb nutrients properly. These symptoms include anorexia, abdominal bloating, cramps, anemia, and fatigue. Numerous underlying conditions can cause malabsorption syndrome. For example, certain enzyme deficiencies can result in an absorption failure because there are no digested nutrients to absorb. Cystic fibrosis and other genetic conditions can also cause malabsorption syndrome.

Diverticulitis is an inflammation of abnormal saclike outpouchings of the intestinal wall called *diverticula*. Diverticula often develop in adults over 50 years of age who eat low-fiber foods. Diverticulitis is characterized by constipation.

Colitis refers to any inflammatory condition of the large intestine. Symptoms of colitis include diarrhea and abdominal cramps or constipation. Some forms of colitis may also produce bleeding and intestinal ulcers. Colitis may be a result of emotional stress, as in *irritable bowel syndrome*. It may also result from an autoimmune disease, as in *ulcerative colitis*. Another type of autoimmune colitis is *Crohn's disease*, which often also affects the small intestine. If more conservative treatments fail, colitis may be corrected by surgical removal of the affected portions of the colon.

Colorectal cancer is a malignancy, usually *adenocarcinoma*, of the colon or rectum. Colorectal cancer occurs most frequently after the age of 50, and a low-fiber, high-fat diet and genetic predisposition are known risk factors. Early warning signs of this common type of cancer include changes in bowel habits, fecal blood, rectal bleeding, abdominal pain, unexplained anemia or weight loss, and fatigue.

Disorders of the Liver and Pancreas

Hepatitis is a general term referring to inflammation of the liver. Hepatitis is characterized by jaundice (yellowish discoloration of body tissues), liver enlargement, anorexia, abdominal discomfort, gray-white feces, and dark urine. Various different conditions can produce hepatitis. Alcohol, drugs, or other toxins may cause hepatitis. It may also be a complication of bacterial or viral infection or parasite infestation. *Hepatitis A*, for example, results from infection by the hepatitis A virus. Contaminated food is often a source of infection. Hepatitis A occurs commonly in young people and ranges in severity from mild to life-threatening. Another viral hepatitis, *hepatitis B*, is usually more severe. It is also called *serum hepatitis* because it is often transmitted by contaminated blood serum.

Hepatitis, chronic alcohol abuse, malnutrition, or infection may lead to a degenerative liver condition known as **cirrhosis.** The liver's ability to regenerate damaged tissue is well known, but it has its limits. For example, when the toxic effects of alcohol accumulate faster than the liver can regenerate itself, damaged tissue is replaced with fibrous or fatty tissue instead of normal tissue. *Cirrhosis* is the name given to such degeneration.

Besides the endocrine disorders discussed in Chapter 15 such as diabetes mellitus, the pancreas may be involved in numerous other diseases. For example, **pancreatitis,** or inflammation of the pancreas, can be caused by a variety of factors. *Acute pancreatitis* usually results from blockage of the common bile duct. The blockage causes pancreatic enzymes to "back up" into the pancreas and digest it. This is a very serious condition; about half of all cases prove to be fatal. Another condition that blocks the flow of pancreatic enzymes is *cystic fibrosis (CF)*. You may recall from Chapter 3 that this inherited disorder disrupts cell transport and causes exocrine glands to produce excessively thick secretions. Thick pancreatic secretions may build up and block pancreatic ducts, disrupting the flow of pancreatic enzymes and damaging the pancreas.

Another serious pancreatic disorder is **pancreatic cancer.** Usually a form of *adenocarcinoma*, pancreatic cancer claims the lives of nearly all its victims within 5 years after diagnosis.

CASE STUDY
PEPTIC ULCER DISEASE

Ms. K. is a 44-year-old white female who was admitted for diagnostic work-up of persistent abdominal pain. She is the mother of three young children, is employed in a management position with an advertising agency, and works long hours. Her husband died this past year.

Ms. K. has a history of smoking 1 pack/day of cigarettes. Her eating habits are irregular, and she frequently takes aspirin for mild headaches. Having been diagnosed with rheumatoid arthritis 2 years ago, Ms. K. takes 10 mg/day of prednisone. Family history revealed that her father has complained of "gastritis-like" symptoms for the last 10 years, although he has never tested for this problem. The remaining history was unremarkable except for incidences of hypertension in a paternal uncle and a maternal grandmother.

Ms. K. is 5 feet 3 inches and currently weighs 100 pounds, having lost 10 pounds during the last 6 weeks.

Blood pressure on admission was 150/88. For the last 3 months she has been experiencing mild intermittent epigastric pain. The week before admission the pain became more continuous, becoming intense 1 to 2 hours after eating. Eating occasionally relieves the pain temporarily. She also experienced slight nausea throughout the last week with one episode of vomiting blood the evening before admission.

On admission, Ms. K. was kept on "nothing by mouth" status. A nasogastric tube was inserted and connected to intermittent suction, and intravenous fluid replacement begun. Ms. K. experienced a massive upper gastrointestinal bleed 2 days after admission. In 24 hours she had a recurrence of bleeding bright red blood through the nasogastric tube and her rectum. Gastroscopy indicated gastric and pyloric ulcers.

She was taken to the operating room for surgical treatment of her ulcers.

---

1. In Ms. K.'s case, what contributing factors are relevant to her developing a peptic ulcer disease?
   a. Smoking
   b. Ingesting antiinflammatory agents
   c. Hereditary factors
   d. All of the above

2. Before admission, Ms. K. experienced a great deal of abdominal discomfort about 1 to 2 hours after eating. The underlying physiological reason for this is:
   a. Edematous duodenum resulting from the mechanical pressure placed on it by the semidigested meal

   b. Increased back-diffusion of ingested acid into the mucosa
   c. Increased gastrin concentration in the antral mucosa caused by the ingestion of proteins

3. Which of the following are the most frequent complications of peptic ulcer disease?
   a. Gastric outlet obstruction
   b. Hemorrhage
   c. Anemia
   d. Perforation

CHAPTER SUMMARY

OVERVIEW OF DIGESTIVE FUNCTION

A. Primary function of digestive system—to bring essential nutrients into the internal environment so that they are available to each cell of the body
B. Mechanisms used to accomplish primary function of digestive system
   1. Ingestion—food is taken in
   2. Digestion—breakdown of complex nutrients into simple nutrients
   3. Motility of GI wall—physically breaks down large chunks of food material and moves food along the tract
   4. Secretion of digestive enzymes allows chemical digestion
   5. Absorption—movement of nutrients through the GI mucosa into the internal environment
   6. Elimination—excretion of material that is not absorbed

DIGESTION

A. Mechanical digestion—movements of the digestive tract
   1. Change ingested food from large particles into minute particles, facilitating chemical digestion
   2. Churn contents of the GI lumen to mix with digestive juices and come in contact with the surface of the intestinal mucosa, facilitating absorption
   3. Propel food along the alimentary tract, eliminating digestive waste from the body
   4. Mastication—chewing movements
      a. Reduces size of food particles
      b. Mixes food with saliva in preparation for swallowing
   5. Deglutition—process of swallowing; complex process requiring coordinated and rapid movements (Figure 25-1)
      a. Oral stage (mouth to oropharynx)—voluntarily

controlled; formation of a food bolus in the middle of the tongue; tongue presses bolus against the palate and food is then moved into the oropharynx
   b. Pharyngeal stage (oropharynx to esophagus)— involuntary movement; to propel bolus from the pharynx to the esophagus, the mouth, nasopharynx, and larynx must be blocked; combination of contractions and gravity move bolus into esophagus
   c. Esophageal stage (esophagus to stomach)— involuntary movement; contractions and gravity move bolus through esophagus and into stomach
6. Peristalsis and segmentation—two main types of motility produced by the smooth muscle of the GI tract; can occur together, in an alternating fashion (Figures 25-2 and 25-3)
   a. Peristalsis—wavelike ripple of the muscle layer of a hollow organ; progressive motility that produces forward movement of matter along the GI tract
   b. Segmentation—mixing movement; digestive reflexes cause a forward-and-backward movement with a single segment of the GI tract; helps break down food particles, mixes food and digestive juices, and brings digested food in contact with intestinal mucosa to facilitate absorption
7. Regulation of motility
   a. Gastric motility—emptying the stomach takes approximately 2 to 6 hours; while in the stomach, food is churned and mixed with gastric juices to form chyme; chyme is ejected about every 20 seconds into the duodenum; gastric emptying is controlled by hormonal and nervous mechanisms
      (1) Hormonal mechanism—fats in duodenum stimulate the release of gastric inhibitory peptide, which acts to decrease peristalsis of gastric muscle and slows passage of chyme into duodenum
      (2) Nervous mechanism—enterogastric reflex; receptors in the duodenal mucosa are sensitive to presence of acid and to distention; impulses over sensory and motor fibers in the vagus nerve cause a reflex inhibition of gastric peristalsis
   b. Intestinal motility includes peristalsis and segmentation
      (1) Segmentation in duodenum and upper jejunum mixes chyme with digestive juices from the pancreas, liver, and intestinal mucosa
      (2) Rate of peristalsis picks up as chyme approaches end of jejunum, moving it through the rest of the small intestine into the large intestine; after leaving stomach, normally takes approximately 5 hours for chyme to pass all the way through the small intestine
      (3) Peristalsis—regulated in part by intrinsic stretch reflexes; stimulated by cholecystokinin-pancreozymin (CCK)
B. Chemical digestion—changes in chemical composition of food as it travels through the digestive tract; these changes are the result of hydrolysis
1. Enzymes—"organic catalysts"
   a. Chemical structure of enzymes—enzymes are proteins; may contain a prosthetic group
   b. Classification and naming of enzymes
      (1) Oxidation-reduction enzymes—known as oxidases, hydrogenases, and dehydrogenases; energy release depends on these enzymes
      (2) Hydrolyzing enzymes—hydrolases; digestive enzymes belong to this group
      (3) Phosphorylating enzymes—phosphorylases or phosphatases; add or remove phosphate groups
      (4) Enzymes that add or remove carbon dioxide— carboxylases or decarboxylases

      (5) Enzymes that rearrange atoms within a molecule—mutases or isomerases
      (6) Hydrases add water to a molecule without splitting substrate
   c. Properties of enzymes are those of proteins; other important enzyme properties (Figure 25-4)
      (1) Specific in their action
      (2) Function optimally at a specific pH
      (3) A variety of physical and chemical agents inactivate or inhibit enzymes action by changing the shape of enzyme molecules (Figure 25-5)
      (4) Most enzymes catalyze a chemical reaction in both directions
      (5) Enzymes are continually being destroyed in the body and must continually be synthesized
      (6) Many enzymes are synthesized as inactive proenzymes
2. Carbohydrate digestion
   a. Carbohydrates are saccharide compounds
   b. Polysaccharides are hydrolyzed by amylases to form disaccharides
   c. Final steps of carbohydrate digestion are catalyzed by sucrase, lactase, and maltase, which are found in the cell membrane of epithelial cells covering the villi that line the intestinal lumen
3. Protein digestion (Figure 25-7)
   a. Protein compounds are made up of twisted chains of amino acids
   b. Proteases catalyze hydrolysis of proteins into intermediate compounds and, finally, into amino acids
   c. Main proteases: pepsin in gastric juice, trypsin in pancreatic juice, peptidases in intestinal brush border
4. Fat digestion (Figure 25-8)
   a. Fats must be emulsified by bile in small intestine before being digested
   b. Pancreatic lipase is the main fat-digesting enzyme
5. Residues of digestion—some compounds of food resist digestion and are eliminated as feces

CONTROL OF DIGESTIVE GLAND SECRETION

A. Salivary secretion
1. Only reflex mechanisms control the secretion of saliva
2. Chemical and mechanical stimuli come from the presence of food in the mouth
3. Olfactory and visual stimuli come from the smell and sight of food
B. Gastric secretion—three phases (Figure 25-9)
1. Cephalic phase—"psychic phase," since mental factors activate the mechanism; parasympathetic fibers in branches of the vagus nerve conduct stimulating efferent impulses to the glands; stimulate production of gastrin
2. Gastric phase—when products of protein digestion reach the pyloric portion of the stomach, they stimulate release of gastrin; gastrin accelerates secretion of gastric juice, ensuring enough enzymes present to digest food
3. Intestinal phase—various mechanisms seem to adjust gastric secretion as chyme passes to and through the intestinal tract; endocrine reflexes involving gastric inhibitory peptide, secretin, and CCK inhibit gastric secretions
C. Pancreatic secretion stimulated by several hormones released by intestinal mucosa
1. Secretin evokes production of pancreatic fluid low in enzyme content but high in bicarbonate
2. Cholecystokinin-pancreozymin (CCK)—several functions
   a. Causes increased exocrine secretion from the pancreas
   b. Opposes gastrin, thus inhibiting gastric HCl secretion
   c. Stimulates contraction of the gallbladder so that bile is ejected into the duodenum

D. Secretion of bile—bile secreted continually by the liver; secretin and CCK stimulate ejection of bile from the gallbladder

E. Intestinal secretion—little is known how intestinal secretion is regulated; suggested that the intestinal mucosa is stimulated to release hormones that increase the production of intestinal juice

ABSORPTION

A. Process of absorption
 1. Passage of substances through the intestinal mucosa into the blood or lymph
 2. Most absorption occurs in the small intestine

B. Mechanisms of absorption
 1. For some substances such as water, absorption occurs by simple diffusion or osmosis
 2. Other substances are absorbed through more complex mechanisms (Figure 25-11)
    a. Secondary active transport—how sodium is transported
    b. Sodium cotransport (coupled transport)—how glucose is transported
    c. Fatty acids, monoglycerides, and cholesterol are transported with the aid of bile salts from the lumen to absorbing cells of the villi (Figure 25-12)
 3. After food is absorbed, it travels to the liver via the portal system

ELIMINATION

A. Elimination—the expulsion of feces from the digestive tract; act of expelling feces is called defecation

B. Defecation occurs as a result of a reflex brought about by stimulation of receptors in the rectal mucosa that is produced when the rectum is distended

C. Constipation—contents of the lower colon and rectum move at a slower than normal rate; extra water is absorbed from the feces, resulting in a hardened stool

D. Diarrhea—result of increased motility of the small intestine, causing decreased absorption of water and electrolytes and a watery stool

THE BIG PICTURE: DIGESTION AND THE WHOLE BODY

A. Primary contribution of the digestive system to overall homeostasis is to provide a constant nutrient concentration in the internal environment (Figure 25-13)

B. Secondary roles of digestive system
 1. Absorption of nutrients
 2. Teeth and tongue, along with respiratory system and nervous system, important in producing spoken language
 3. Gastric acids aid the immune system by destroying potentially harmful bacteria

C. To accomplish its functions, digestive system needs other systems to contribute
 1. Regulation of digestive motility and secretion requires the nervous system and endocrine system
 2. Oxygen for digestive activity needs the proper functioning of respiratory and circulatory systems
 3. Integumentary and skeletal systems support and protect the digestive organs
 4. Muscular system is needed for ingestion, mastication, deglutition, and defecation to occur normally

MECHANISMS OF DISEASE: DISORDERS OF THE DIGESTIVE SYSTEM

A. Disorders of the GI tract
 1. Specific disorders
    a. Gastroenteritis—stomach and intestinal inflammation
    b. Anorexia—chronic loss of appetite
    c. Nausea—unpleasant feeling that often leads to vomiting
    d. Emesis—vomiting (Figure 25-14)
    e. Diarrhea—elimination of liquid feces
    f. Constipation—decreased motility of colon
 2. Ulcer—open wound or sore in an area of the digestive system acted on by acid gastric juice; two common types are gastric ulcers and duodenal ulcers
 3. Stomach cancer—linked to excessive alcohol consumption, use of chewing tobacco, and eating smoked or heavily preserved food; usually metastasized before being discovered
 4. Malabsorption syndrome—general term for a group of symptoms resulting from failure of the small intestine to properly absorb nutrients; symptoms include anorexia, abdominal bloating, cramps, anemia, and fatigue
 5. Diverticulitis—inflammation of diverticula of the intestinal wall; characterized by constipation
 6. Colitis—any inflammatory condition of the large intestine; may produce bleeding and intestinal ulcers
 7. Colorectal cancer—malignancy of the colon or rectum; early warning signs: change in bowel habits, fecal blood, rectal bleeding, abdominal pain, unexplained anemia or weight loss, and fatigue

B. Disorders of the liver and pancreas
 1. Hepatitis—general term for inflammation of the liver; characterized by jaundice, liver enlargement, anorexia, abdominal discomfort, gray-white feces, and dark urine
 2. Cirrhosis—degenerative liver condition; damaged liver tissue is replaced with fibrous or fatty tissue
 3. Pancreatitis—inflammation of the pancreas
 4. Pancreatic cancer—usually adenocarcinoma; usually fatal within 5 years after diagnosis

REVIEW QUESTIONS

1. Define and give the purpose of digestion. What is the difference between mechanical and chemical digestion?
2. List four of the mechanical processes that occur during digestion.
3. List the three steps, or stages, in deglutition.
4. Discuss the function of the deglutition center in the medulla.
5. What is the difference between peristalsis and segmentation contractions?
6. How does gastric inhibitory peptide influence emptying of the stomach? What is the enterogastric reflex?
7. Discuss the functions of gastric juice.
8. What is chyme?
9. Differentiate trypsin from pepsin; proteins from proteoses; polysaccharides from monosaccharides.
10. Describe the classification of enzymes.
11. Discuss three important enzyme properties.
12. What juices digest proteins? Carbohydrates? Fats?
13. What is meant by the term *cephalic phase* of gastric secretion? *Gastric phase? Intestinal phase?*
14. What digestive functions does the pancreas perform?
15. Describe the absorption of glucose from the lumen of the small intestine.
16. What vitamins depend on bile salts for their absorption?
17. Compare carbohydrate, fat, and protein absorption.
18. Name each hormone that controls ejection of bile, stimulation of gastric enzymes, inhibition of gastric emptying, secretion of alkaline fluid from pancreas, secretion of pancreatic enzymes.
19. Discuss responses that collectively result in defecation.
20. Describe the treatment called oral rehydration therapy.

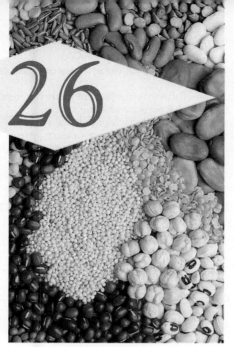

Legumes—a dietary source of protein

26 ▶ Nutrition and Metabolism

CHAPTER OUTLINE

OBJECTIVES

After you have completed this chapter, you should be able to:

1. Define the terms *nutrition* and *metabolism*.
2. Outline the processes of anabolism and catabolism and discuss the role of ATP adenosine triphosphate/adenosine diphosphate (ATP/ADP) system and its role in metabolism.
3. Discuss the various dietary sources of carbohydrates.
4. Discuss the steps involved in glycolysis.
5. Explain the role of the electron transport system and oxidative phosphorylation in the citric acid cycle.
6. Explain what is meant by the terms *anaerobic respiration* and *aerobic respiration*.
7. Discuss glycogenesis and glycogenolysis.
8. Discuss glucose homeostasis.
9. Discuss the dietary sources of lipids.
10. Identify the major lipid constituents in blood and discuss their mechanisms of transport.
11. Discuss the metabolism of lipids.
12. Outline hormonal control of fat metabolism.
13. Discuss protein anabolism and catabolism.
14. Discuss the two kinds of protein or nitrogen imbalance.
15. Discuss the importance of vitamins and minerals.
16. Define the term *metabolic rate* and discuss how it can be expressed.
17. Discuss the major factors that influence the basal metabolic rate (BMR).
18. Discuss the relationship between energy intake, output and balance, and body weight.
19. Discuss the mechanisms for regulating food intake.

OVERVIEW OF NUTRITION AND METABOLISM

Nutrition and *metabolism* are words that are often used together—but what do they mean? **Nutrition** is a term that refers to the food (nutrients) that we eat. Proper nutrition requires a balance of the three basic food types, *carbohydrates*, *fats*, and *proteins*, plus essential *vitamins* and *minerals*. Malnutrition is a deficiency or imbalance in the consumption of food, vitamins, and minerals.

A good phrase to remember in connection with the word **metabolism** is "use of foods" because basically this is what metabolism is—the use the body makes of foods after they have been digested, absorbed, and circulated to cells. It uses them in two ways: as an energy source and as building blocks for making complex chemical compounds. Before they can be used in these two ways, foods have to be *assimilated.* Assimilation occurs when nutrient molecules enter cells and undergo many chemical changes. Metabolism is a complex process made up of two major processes, **catabolism** and **anabolism.** Each of these, in turn, consists of a series of enzyme-catalyzed chemical reactions known as metabolic pathways.

Catabolism breaks food molecules down into smaller molecular compounds and, in so doing, releases energy from them. Anabolism does the opposite. It builds nutrient molecules up into larger molecular compounds and, in so doing, uses energy. Catabolism is a decomposition process. Anabolism is a synthesis process. Both catabolism and anabolism take place inside cells. Both processes go on continually and concurrently.

Catabolism releases energy in two forms—heat and chemical energy. The amount of heat generated is relatively large—so large, in fact, that it would hardboil cells if it were released in one large burst. Fortunately, this does not happen. Catabolism releases heat in frequent, small bursts. Heat is practically useless as an energy source for cells in that they cannot use it to do their work. However, this heat is important in maintaining the homeostasis of body temperature. In contrast, chemical energy released by catabolism is more obviously useful. It cannot, however, be used directly for biological reactions. First it must be transferred to high-energy bonds ($\sim$) of adenosine triphosphate (ATP) molecules. The term **high-energy bonds** indicates that they break —and thus give up their energy—more easily than other types of chemical bonds.

ATP is one of the most important compounds in the world. Why? Because it supplies energy directly to the

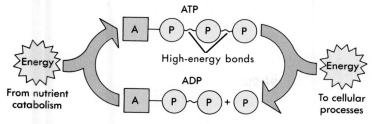

FIGURE 26-1   The role of ATP in metabolism. ATP temporarily stores energy in its last high-energy phosphate bond. When that bond is broken, energy is released to do cellular work. The ADP and phosphate groups that result can be resynthesized into ATP capturing additional energy from nutrient catabolism. This cycle is called the ATP/ADP system.

energy-using reactions of all cells in all kinds of living organisms from one-celled plants to trillion-celled humans. ATP functions as the universal biological currency. It pays the energy bills for all cells and is as important in the world of cells as money is in the world of contemporary society.

Look now at Figure 26-1. The structural formula at the top of the diagram shows three phosphate groups attached to the rest of the ATP molecule, two of them by high-energy bonds. The breaking of the last one of these bonds yields a phosphate group (P), adenosine diphosphate (ADP), and energy, which, as the diagram indicates, is used for anabolism and other cell work. The diagram also shows that P and ADP then use energy released by catabolism to recombine and form ATP. This cycle is called the *ATP/ADP system.*

Metabolism is not identical in all cells. It differs mainly with regard to rate and the kind of products synthesized by anabolism. More active cells have a higher metabolic rate than less active cells. Anabolism in different kinds of cells produces different compounds. In liver cells, for example, anabolism synthesizes various blood protein compounds. Not so in beta cells of the pancreas. Anabolism here produces a different compound—insulin.

The bulk of this chapter discusses concepts related to the many and varied metabolic pathways of the human body. It is important to note that our discussion, and the diagrams that accompany the discussion, have been simplified to facilitate understanding of the basic concepts.

CARBOHYDRATES
Dietary Sources of Carbohydrates

Carbohydrates are found in most of the foods that we eat. *Complex carbohydrates*—polysaccharides such as starches found in vegetables, grains, and other plant tissues and *glycogen* found in meat and other animal tissues—are broken down into simpler carbohydrates before they are absorbed. **Cellulose,** a major component of most plant tissues, is an important exception to this principle. Since humans do not make enzymes that chemically digest this complex carbohydrate, it passes through our system without being broken down. Also called *dietary fiber* or "roughage," cellulose and other indigestible polysaccharides keep chyme thick enough for the digestive system to push it easily. They

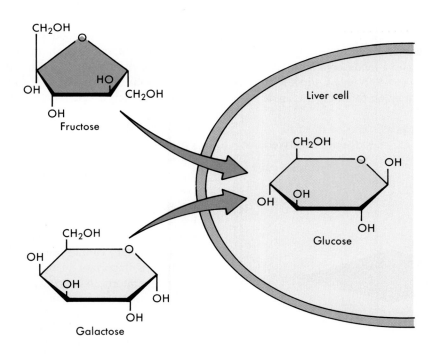

FIGURE 26-2   Conversion of monosaccharides. Monosaccharides *fructose* and *galactose* are usually converted to *glucose* by liver cells. Although simplified in this diagram, conversion to glucose requires several steps. Glucose is the carbohydrate used universally by all cells in the body.

also help mix chyme, much like the ball inside a can of spray paint. Most biologists believe that a high-fiber diet reduces the risk of many forms of cancer, including colorectal cancer.

Disaccharides such as those found in refined sugar must also be chemically digested before they can be absorbed. Monosaccharides in fruits and some "diet foods" are already in an absorbable form, so they can move directly into the internal environment without initially being processed. The monosaccharide *glucose* is the carbohydrate that is most useful to the typical human cell. As Figure 26-2 shows, other important monosaccharides, *fructose* and *galactose,* are usually converted by liver cells into glucose for use by other cells of the body.

QUICK CHECK

✔ 1. *Name the two types of metabolism and distinguish between them.*

✔ 2. *Why must energy in nutrient molecules be transferred to ATP?*

Carbohydrate Metabolism

The body metabolizes carbohydrates by both catabolic and anabolic processes. Since most human cells use carbohydrates—mainly glucose—as their first or preferred energy fuel, they catabolize most of the carbohydrate absorbed and anabolize a relatively small portion of it. When the amount of glucose entering cells is inadequate for their energy needs, they catabolize fats, and then proteins.

Glucose transport and phosphorylation

Carbohydrate metabolism starts with the movement of glucose through cell membranes. Immediately on reaching the interior of a cell, glucose reacts with ATP to form glucose-6-phosphate. This step, named **glucose phosphorylation,** prepares glucose for further metabolic reactions. In most cells of the body, it is an irreversible reaction. However, in a few cells—namely, those of the intestinal mucosa, liver, and kidney tubules—glucose phosphorylation is reversible. These cells contain phosphatase, an enzyme that splits phosphate off from glucose-6-phosphate. This reverse glucose phosphorylation reaction forms glucose, which then moves out of the cells into the blood.

FOCUS ON... TRANSFERRING CHEMICAL ENERGY

The ability to transfer energy from molecule to molecule is, as you might imagine, essential to life. We have already discussed the critical role played by the nucleotide adenosine triphosphate (ATP) in transferring energy within living cells. ATP can accept energy from catabolic reactions and transfer that energy to energy-requiring catabolic reactions (see Figure 26-1). Although we say that ATP is an "energy storage molecule," do not suppose that the energy is stored for very long periods. In fact, an ATP molecule exists for only a brief period before its last phosphate group is broken off and its energy is transferred to another molecule in some metabolic pathway. Long-term storage of energy can be accomplished only by nutrient molecules such as glucose, glycogen, and triglycerides.

In addition to ATP, various other energy-transfer molecules are essential to human life. When atoms in a molecule absorb energy, some of their electrons may move outward to a higher energy level (shell). Electrons often become so energized that they leave the atom completely. As this occurs, pairs of "high-energy" electrons can be picked up and transferred to another molecule by an electron carrier such as **flavine adenine dinucleotide (FAD)** or **nicotinamide adenine dinucleotide (NAD).** The figure shows how NAD^+ picks up a pair of energized electrons to become NADH. It should be noted here that electrons always travel with a proton (H^+) in the metabolic pathways described in this chapter. The electrons do not stay with the electron carrier for long, however. They are immediately transferred to molecules in another metabolic pathway, as the figure shows. In the cell, pairs of electrons (and their energy) can thus be transferred from pathway to pathway by NAD and FAD.

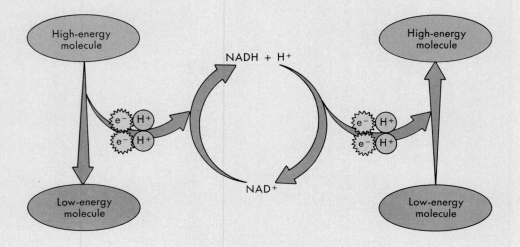

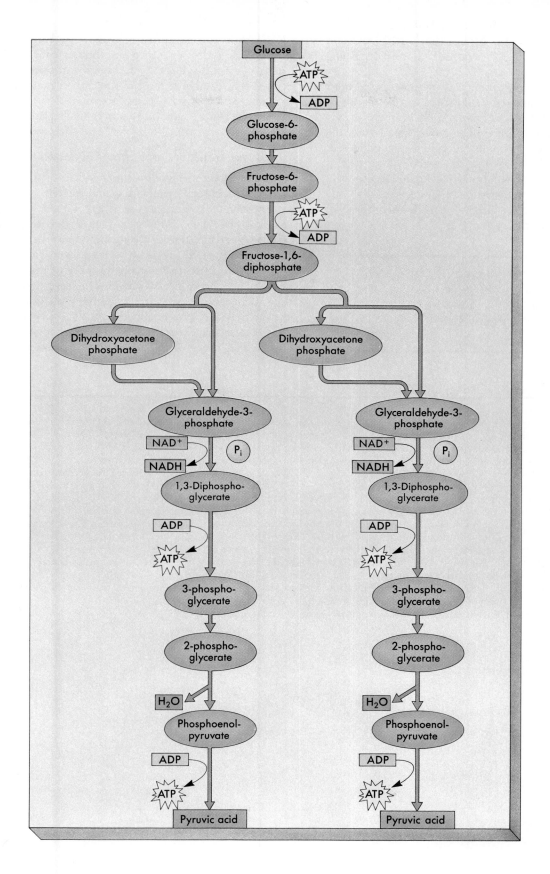

FIGURE 26-3   Glycolysis. The series of enzyme-catalyzed reactions that make up the portion of the catabolic pathway for carbohydrates called *glycolysis.*

(Glucose-6-phosphate cannot pass through cell membranes.) Depending on their energy needs of the moment, cells either catabolize (break apart) or anabolize (bind together) glucose-6-phosphate.

Glycolysis

Glycolysis is the first process of carbohydrate catabolism. It breaks apart one glucose molecule to form two pyruvic acid molecules. (A glucose molecule contains six carbon atoms, and a pyruvic acid molecule contains three carbon atoms. See Figure 26-4.) Glycolysis consists, as Figure 26-3 shows, of a series of chemical reactions. A specific enzyme catalyzes each of these reactions. Probably the most important facts for you to remember about glycolysis are the following:

♦ Glycolysis occurs in the cytoplasm of all human cells.

♦ Glycolysis is an **anaerobic** process, that is, it does not use oxygen. It is the only process that provides cells with energy when their oxygen supply is inadequate or even absent.

♦ Glycolysis breaks the chemical bonds in glucose molecules and thereby releases about 5% of the energy stored in them. Much of the released energy appears as heat, but some of it is transferred to the high-energy bonds of ATP molecules. For every molecule of glucose undergoing glycolysis, a net of two molecules of ATP is formed. About 8 kilocalories (kcal) of energy (under normal physiological conditions) is stored in the high-energy bonds

that bind phosphate to ADP to form 1 mole (6.02×10^{23} molecules) of ATP.

♦ Glycolysis is an essential process because it prepares glucose for the second step in catabolism, namely, the *citric acid cycle*. Glucose itself cannot enter the cycle but must first be converted to pyruvic acid.

Citric acid cycle

Essentially, the **citric acid cycle** converts two pyruvic acid molecules to six carbon dioxide and six water molecules. But many chemical reactions intervene. Figure 26-4 shows that one glucose molecule is changed by

FIGURE 26-4 Catabolism of glucose. Glycolysis splits one molecule of glucose (six carbon atoms) into two molecules of pyruvic acid (three carbon atoms each). The citric acid cycle converts each pyruvic acid molecule into three carbon dioxide molecules (one carbon atom each).

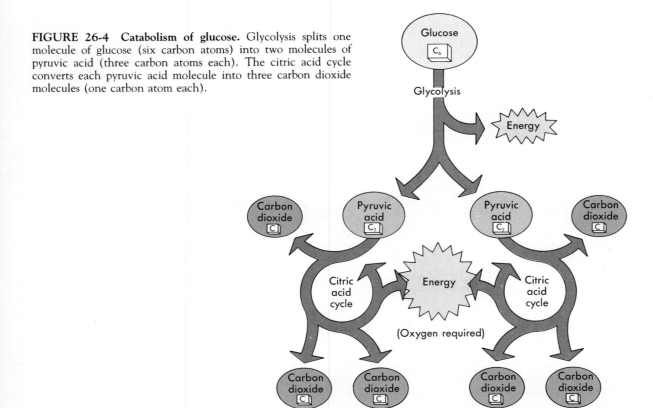

glycolysis to two pyruvic acid molecules, which, by means of the citric acid cycle, yield six carbon dioxide molecules. Figure 26-5 shows the details of the citric acid cycle.

Glycolysis takes place in the cytoplasm of cells, whereas the citric acid cycle occurs in their mitochondria. Before it can enter the citric acid cycle, each pyruvic acid molecule combines with coenzyme A, as shown in Figure 26-5, after splitting off CO_2 and a pair of high-energy electrons (with their accompanying protons, H^+) from pyruvic acid, thus forming acetyl-CoA. Coenzyme A then detaches from acetyl CoA, leaving a two-carbon acetyl group, which enters the citric acid cycle by combining with oxaloacetic acid to form citric acid. This is what gives the citric acid cycle its name. The cycle is also called the **tricarboxylic acid (TCA) cycle** because citric acid is also called *tricarboxylic acid*. For many years, this cycle was called the **Krebs cycle**,

after Sir Hans Krebs, whose brilliant work in discovering this metabolic pathway earned him the 1935 Nobel Prize.

You probably do not need to memorize the names of the intermediate products formed during the citric acid cycle but notice that all of them are acids. Observe, too, that for each pyruvic acid molecule entering this pathway, three CO_2 molecules are formed and that certain reactions yield high-energy electrons. The next paragraph describes how these high-energy electrons are used.

Electron transport system and oxidative phosphorylation

High-energy electrons removed during the citric acid cycle enter a chain of carrier molecules, which is embedded in the inner membrane of mitochondria and is known as the **electron transport system.** Figure 26-6 indicates that

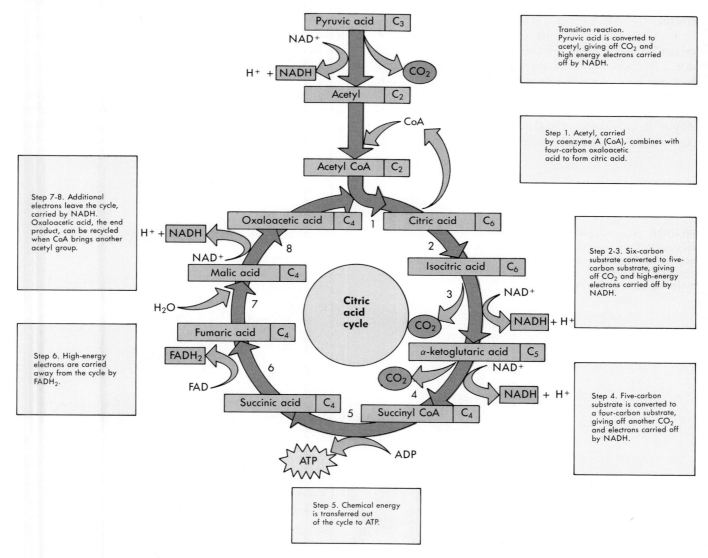

FIGURE 26-5 Citric acid cycle. Each pyruvic acid molecule is prepared to enter the citric acid cycle by the transition reaction, which yields a pair of high-energy electrons and a CO_2 molecule. The acetyl group that is thus formed is picked up by coenzyme A (CoA) and led into the citric acid cycle proper, which is described here as a recurring series of 8 steps.

high-energy electrons are carried to the electron transport system by NAD. The electrons quickly move down the chain, from cytochrome to cytochrome, to their final acceptor, oxygen. At this time, a proton (H^+) also joins the oxygen, forming water. The most important fact about electron transport is this: as electrons move down the carrier chain, they release small bursts of energy used to pump protons into the compartment between the inner and outer membranes of the mitochondrion. The diffusion of protons back into the matrix in the inner compartment drives the process of oxidative phosphorylation. **Oxidative phosphorylation** refers to the joining of a phosphate group to ADP to form ATP—a reaction whose importance can scarcely be overemphasized.

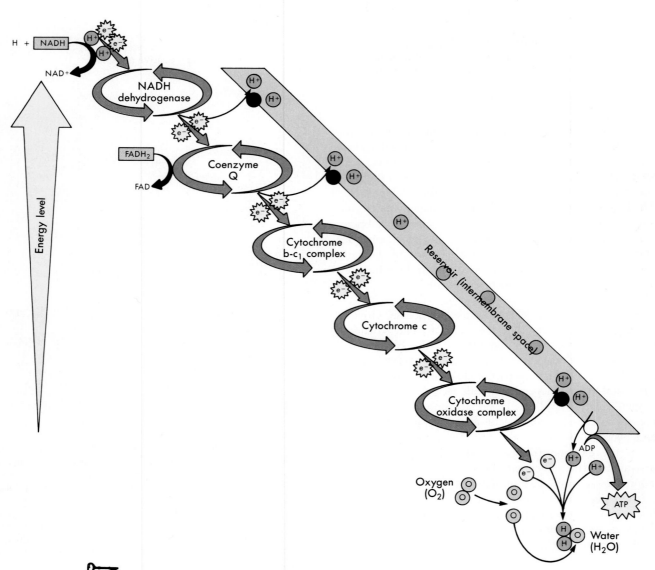

FIGURE 26-6   Electron transport system. Pairs of high-energy electrons and their accompanying protons (H^+) are transferred to the components (cytochromes) of the electron transport system by NAD and FAD. They then jump from cytochrome to cytochrome, losing energy along the way. The energy is used to pump protons (H^+) into the compartment between the inner and outer mitochondrial membranes. The diffusion of protons back into the inner compartment drives the phosphorylation of ADP to form ATP. The protons are joined with oxygen and low-energy electrons at the end of the cytochrome chain to form water molecules.

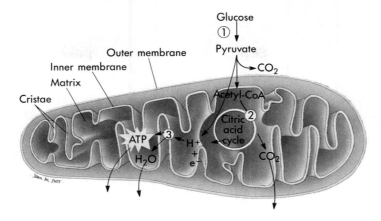

FIGURE 26-7 **Cell machinery for glucose catabolism. 1,** Glycolysis occurs in cytoplasm. **2,** Citric acid cycle mostly in mitochondrial matrix. **3,** Electron transport and oxidative phosphorylation on inner membrane of mitochondria.

The arrangement of catabolic "machinery" within the cell, as currently viewed, is shown in Figure 26-7. Glycolytic enzymes in the cytoplasm produce pyruvic acid, which diffuses into mitochondria. The enzymes of the citric acid cycle have been localized mostly in the soluble matrix. The high-energy electrons and their accompanying protons are then carried to the cristae of the inner membrane, where the electron transport carriers and mechanism for phosphorylation are located. Since so many of the cell's energy-releasing enzymes are located within the mitochondria, these tiny structures are aptly described as the "power plants" of the cell.

The breakdown of ATP molecules, of course, provides virtually all the energy that does cellular work. Therefore the process that produces some 90% of the ATP formed during carbohydrate catabolism, namely, oxidative phosphorylation, is the crucial part of catabolism (Figure 26-8). This vital process depends on cells receiving an adequate oxygen supply. Why? Briefly, because only when oxygen is present in cells to serve as the final acceptor of electrons and hydrogen ions can electrons then continue moving down the electron transport chain. If oxygen becomes unavailable, the movement of electrons and hydrogen ions stops. Cessation of ATP formation by oxidative phosphorylation necessarily follows. All too soon, cells have an inadequate energy supply—a lethal condition if it persists for more than a few minutes.

We can summarize the long series of chemical reactions in glucose catabolism with one short equation:

$$C_6H_{12}O_6 + 6\ O_2 \rightarrow 6\ CO_2 + 6\ H_2O + 36 \text{ (or 38) ATP} + \text{Heat}$$

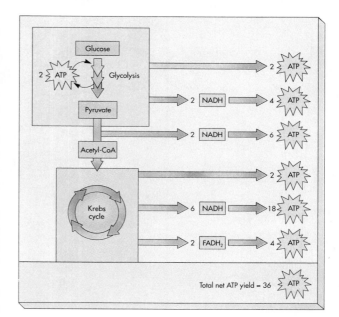

FIGURE 26-8 **Energy extracted from glucose.** Energy release from the breakdown of glucose is released mostly as heat, but some of it is transferred to a usable form—the high-energy bonds of ATP. In most human cells, one glucose molecule produces enough usable chemical energy to synthesize or "charge up" 36 ATP molecules. Some cells, such as heart and liver cells, shuttle electrons more efficiently and may be able to synthesize up to 38 ATP molecules. This represents an energy conversion efficiency of 38% to 44%, much better than the 20% to 25% typical of most machines.

QUICK CHECK

✔ 1. *What is glycolysis? How much energy is transferred to ATP through this process?*

✔ 2. *What happens to a nutrient molecule as it proceeds through the citric acid cycle?*

✔ 3. *What is the purpose of the electron transport system?*

Anaerobic pathway

What if the amount of oxygen required to maintain adequate ATP levels is not available? You may recall from Chapter 9 that this often occurs in skeletal muscle cells, especially during strenuous exercise. You may also recall from Chapter 9 that another pathway for the catabolism of glucose exists. It is sometimes called the anaerobic pathway

(or anaerobic respiration) because it transfers energy to ATP using only glycolysis—a process that does not require oxygen. Although reliance on glycolysis for energy cannot continue indefinitely, it can gain a few extra minutes before cellular functions must stop.

As Figure 26-9 shows, there are two main pathways that glucose or its derivatives can take. One is the pathway that ends with oxidative phosphorylation of ATP. This pathway, described in the previous sections, is called the aerobic pathway, or aerobic respiration, because it requires the presence of oxygen. If oxygen is not available to maintain adequate ATP levels, then the cell will rely solely on glycolysis to produce ATP. Even though this process does not extract the maximum amount of energy from a glucose molecule, it is the only ATP-producing process that can operate under anaerobic conditions. Since the pyruvic acid molecule produced by glycolysis cannot enter the citric acid cycle, it is converted to *lactic acid* rather than

acetyl-CoA. Lactic acid cannot enter the citric acid cycle.

Once oxygen becomes available again, then some of the lactic acid is converted back to pyruvic acid by the cell (see Figure 26-9). Notice that such reconversion requires energy from ATP. Thus reconversion can only occur after oxidative phosphorylation has resumed and produced enough ATP to allow reconversion to occur. Once it is converted to pyruvic acid, the molecule can follow the aerobic pathway and become completely catabolized. Much of the lactic acid produced during anaerobic respiration diffuses into the blood and is removed by the liver. Inside liver cells, ATP produced by oxidative respiration is used to convert the lactic acid back into glucose. Because anaerobic respiration requires the later use of ATP molecules produced via oxidative respiration, we can say that an **oxygen debt** is incurred. The oxygen debt is repaid when oxygen becomes available to form the extra ATP needed to convert lactic acid back into pyruvic acid or glucose.

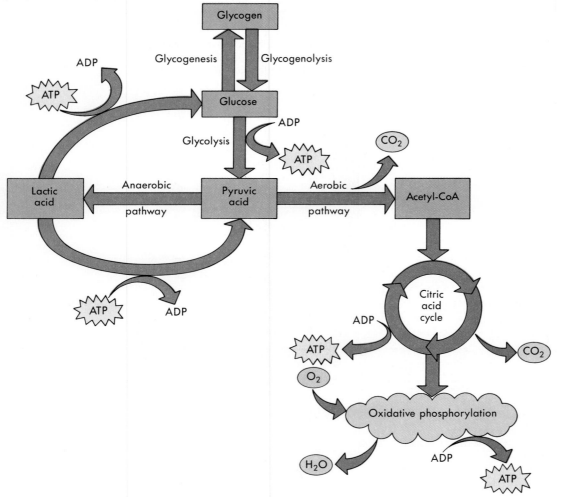

FIGURE 26-9   Summary of glucose metabolism. Glucose is catabolized to pyruvic acid in the process of glycolysis. If oxygen is available, pyruvic acid is converted to acetyl-CoA and then enters the citric acid cycle and transfers energy to the maximum number of ATP molecules via oxidative phosphorylation. If oxygen is not available, pyruvic acid is converted to lactic acid, incurring an oxygen debt. The oxygen debt is later repaid when ATP produced via oxidative phosphorylation is used to convert lactic acid back into pyruvic acid or all the way back to glucose. If there is an excess of glucose, the cell may convert it to glycogen (glycogenesis). Later, individual glucose molecules can be removed from the glycogen chain by the process of glycogenolysis. Although NAD and FAD play important roles in these pathways, they have been left out of this diagram for the sake of simplicity.

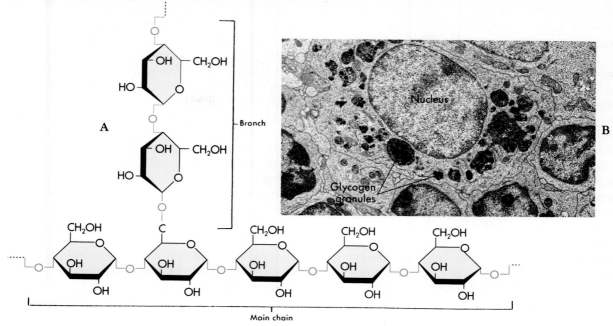

FIGURE 26-10   Glycogen. A, A portion of the highly-branched strand of glucose subunits called *glycogen.* **B,** Transmission electron micrograph of glycogen granules in a liver cell. Glycogen (a polysaccharide) is the form in which human cells store glucose (a monosaccharide).

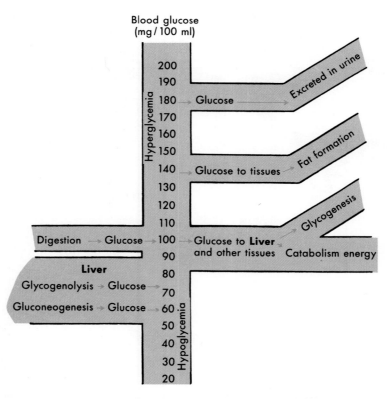

FIGURE 26-11   Homeostasis of blood glucose level. When blood glucose level starts to decrease toward lower normal, liver cells increase the rate at which they convert glycogen, amino acids, and fats to glucose (glycogenolysis and gluconeogenesis) and release it into blood. But when blood glucose level increases, liver cells increase the rate at which they remove glucose molecules from blood and convert them to glycogen for storage (glycogenesis). At still higher levels, glucose leaves blood for tissue cells to be anabolized into adipose tissue and, at still higher levels, is excreted in the urine.

Glycogenesis

Imagine what happens in a cell if glucose catabolism is proceeding at maximum rate. What do you do if you see a traffic jam ahead with no possible way to get through it? Probably take an alternate route, if available. Similarly, if glycolytic pathways are "saturated" because of high levels of glucose entering the cell, a "traffic jam" of glucose-6-phosphate will result. Unable to enter glycolysis, glucose-6-phosphate will begin an alternate route, that is, it will enter the anabolic pathway of glycogen formation. The process of glycogen formation, called **glycogenesis** (see Figure 26-9), is a series of chemical reactions in which glucose molecules are joined together to form a structure made of a branched strand of stored glucose "beads" (Figure 26-10).

The process of glycogenesis is part of a homeostatic mechanism that operates when the blood glucose level increases above the midpoint of its normal range (80 to 100 mg/100 ml of blood), as indicated in Figure 26-11. Example: Soon after a meal, while glucose is being absorbed rapidly, blood is quickly shunted to the liver via the portal system. Here a great many glucose molecules leave the blood for storage as glycogen. As a result of glycogenesis, the blood glucose level decreases, ordinarily enough to reestablish its normal level.

Glycogenolysis

Glycogen molecules do not remain in the cell permanently but are eventually broken apart (hydrolyzed). This process of "splitting glycogen" is called **glycogenolysis** (Figure 26-12; see Figure 26-9). It is, in essence, a reversal of glycogenesis. What are the products of glycogenolysis? The answer depends on the cell. Although all cells presumably have the enzymes to break glycogen to glucose-6-phosphate, only a few cell types (liver, kidney, intestinal mucosa) have the enzyme phosphatase, which allows free

FYI   **Glycogenesis**

Muscle cells, like liver cells, have a high rate of glycogenesis. Certain cells, however, cannot carry out the process at all. Brain cells, for instance, cannot form or store glycogen. They must depend on the circulation of blood to deliver glucose to them.

glucose to form and possibly leave the cell. So the term *glycogenolysis* means different things in different cells. In muscles, glucose-6-phosphate is the product, which then undergoes glycolysis. But liver glycogenolysis (see Figure 26-12) results in free glucose that can leave the cell and increase the blood glucose level. Accordingly, liver glycogenolysis acts as a part of the homeostatic mechanism to maintain the blood glucose level. Example: A few hours after a meal, when the blood glucose level decreases (see Figure 26-11), liver glycogenolysis accelerates. However glycogenolysis alone can probably maintain homeostasis of blood glucose concentration for only a few hours, since the body can store only small amounts of glycogen.

Gluconeogenesis

Literally **gluconeogenesis** means the formation of "new" glucose—"new" in the sense that it is made from proteins or, less frequently, from the glycerol of fats—not from carbohydrates. The process occurs chiefly in the liver. It consists of many complex chemical reactions. The new glucose produced from proteins or fats by gluconeogenesis (Figure 26-13) diffuses out of liver cells into the blood. Gluconeogenesis, therefore, can add glucose to the blood when needed. So, too, can the process of liver glycogenolysis. Obviously, then, the liver is a most important organ for maintaining blood glucose homeostasis.

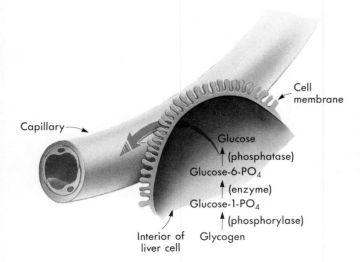

FIGURE 26-12   **Glycogenolysis in a liver cell.**

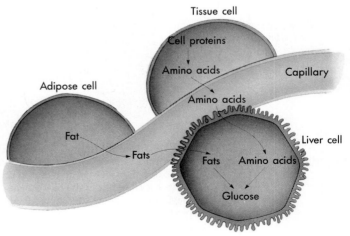

FIGURE 26-13   Gluconeogenesis. Liver cells can form glucose from mobilized tissue proteins and fats.

Control of glucose metabolism

The complex mechanism that normally maintains homeostasis of blood glucose concentration consists of hormonal and neural devices. At least five endocrine glands—pancreatic islets, anterior pituitary gland, adrenal cortex, adrenal medulla, and thyroid gland—and at least eight hormones secreted by those glands function as key parts of the glucose homeostatic mechanism.

Beta cells of the pancreatic islets secrete the most famous sugar-regulating hormone of all—**insulin.** Insulin decreases blood glucose level (Figure 26-14). Although the exact details of its mechanism of action are still being worked out, insulin is known to accelerate glucose transport through cell membranes. It also increases the activity of the enzyme glucokinase. Glucokinase catalyzes glucose phosphorylation, the reaction that must occur before either glycogenesis or glucose catabolism can take place. By applying these facts, you can deduce the main methods by which insulin decreases blood glucose level: increased glycogenesis and increased catabolism of glucose. Figure 26-14 mentions these effects of insulin. By studying this figure, you can also deduce some of the prominent

metabolic defects resulting from insulin deficiency such as in diabetes mellitus. Slow glycogenesis and low glycogen storage, decreased glucose catabolism, and increased blood glucose all result from insulin deficiency.

Alpha cells of the pancreatic islets secrete the sugar-regulating hormone, **glucagon.** Whereas insulin tends to decrease the blood glucose level, glucagon tends to increase it. Glucagon increases the activity of the enzyme phosphorylase (see Figure 26-12), and this necessarily accelerates liver glycogenolysis and releases more of its product, glucose, into the blood.

Epinephrine is a hormone secreted in large amounts by the adrenal medulla in times of emotional or physical stress. Like glucagon, epinephrine increases phosphorylase activity. This makes glycogenolysis occur at a faster rate. Epinephrine accelerates both liver and muscle glycogenolysis, whereas glucagon accelerates only liver glycogenolysis. Both hormones increase the blood glucose level. Epinephrine is the only hormone whose release into the systemic circulation (and therefore its effects on metabolism) are directly under the control of the nervous system.

Adrenocorticotropic hormone (ACTH) and *glucocorticoids*

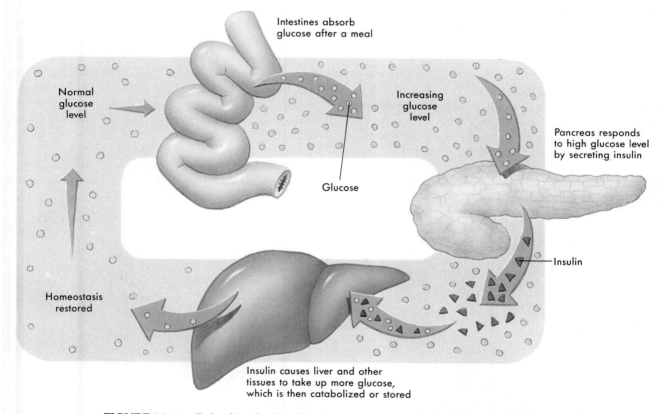

Intestines absorb glucose after a meal

Normal glucose level

Increasing glucose level

Pancreas responds to high glucose level by secreting insulin

Glucose

Insulin

Homeostasis restored

Insulin causes liver and other tissues to take up more glucose, which is then catabolized or stored

FIGURE 26-14   Role of insulin. Insulin operates in a negative feedback loop that prevents blood glucose concentration from increasing too far above the normal, or set point, level. Insulin promotes uptake of glucose by all cells of the body, enabling them to catabolize and/or store it. Thus excess glucose is removed from the bloodstream. If the glucose level should fall below the set point level (see Figure 26-11), then the hormone glucagon promotes the release of glucose from storage into the bloodstream.

(e.g., cortisone) are two more hormones that increase blood glucose concentration. ACTH stimulates the adrenal cortex to increase its secretion of glucocorticoids. Glucocorticoids accelerate gluconeogenesis. They do this by mobilizing proteins, that is, the breakdown, or hydrolysis, of tissue proteins to amino acids. More amino acids enter the circulation and are carried to the liver. Liver cells step up their production of "new" glucose from the mobilized amino acids. More glucose streams out of liver cells into the blood and adds to the blood glucose level.

Growth hormone (GH), made by the anterior pituitary, also increases blood glucose level, but by a different mechanism. Growth hormone causes a shift from carbohydrate to fat catabolism. It does this by limiting the storage of fat in fat depots. Instead, more fats are mobilized and catabolized. In this way, growth hormone "spares" carbohydrates from catabolism, and the level of glucose in the blood is increased.

Thyroid-stimulating hormone (TSH) from the anterior pituitary gland and its target secretion, *thyroid hormone* (T_3 and T_4), have complex effects on metabolism. Some of these raise, and some lower, the glucose level. One of the effects of thyroid hormone is to accelerate catabolism, and since glucose is the body's "preferred fuel," the result may be a decrease in blood glucose level.

A look at a summary of hormone control shown in Figure 26-15 indicates that most hormones cause the blood

  Abnormal blood glucose concentration

The term **hyperglycemia,** which literally means "condition of too much sugar in the blood," is used any time the blood glucose concentration becomes higher than the normal set point level. Hyperglycemia is most frequently associated with untreated diabetes mellitus but can occur in newborns when too much intravenous glucose is given or in other situations.

Hypoglycemia occurs when the blood glucose concentration dips below the normal set point level. Hypoglycemia can occur in various conditions, including starvation, hypersecretion of insulin by the pancreatic islets, or injection of too much insulin. Symptoms of hypoglycemia include weakness, hunger, headache, blurry vision, anxiety, and personality changes—perhaps leading to coma and death if untreated.

level of glucose to rise. These hormones are called *hyperglycemic.* The one notable exception is insulin, which is *hypoglycemic* (tends to decrease blood glucose).

In looking back over the cells' accomplishments in catabolism, we are struck by a sense of wonder, well-expressed by Wayne Becker[*]: "Respiratory metabolism can be regarded as a marvel of design and engineering. No

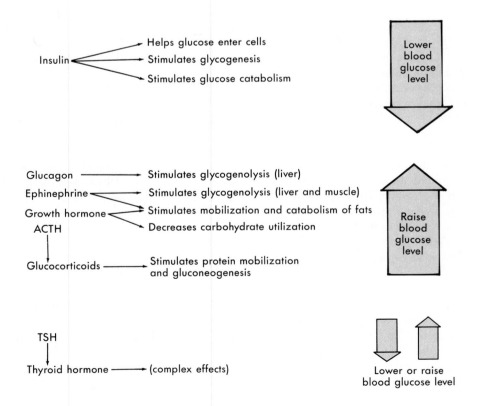

FIGURE 26-15   Hormonal control of blood glucose level. Insulin lowers blood glucose level and is therefore hypoglycemic. Most hormones raise blood glucose level and are called hyperglycemic, or anti-insulin, hormones.

transistors, no mechanical parts, no noise, no pollution—and all done in units of organization that require an electron microscope to visualize. Yet the process goes on routinely and continuously in almost every living cell with a degree of integration, efficiency, fidelity, and control that we can scarcely understand well enough to appreciate, let alone aspire to reproduce in our test tubes."

QUICK CHECK

✔ 1. *Why might a cell switch to anaerobic respiration as its major source of usable energy?*

✔ 2. *What is meant by the term* oxygen debt?

✔ 3. *Distinguish between glycogenesis and glycogenolysis. Under what circumstances might each occur?*

✔ 4. *List three of the hormones that affect glucose metabolism.*

LIPIDS

Dietary Sources of Lipids

Recall from Chapter 2 that **lipids** are a class of organic compounds that includes fats, oils, and related substances. The most common lipids in the diet are **triglycerides,** which are comprised of a *glycerol* subunit to which are attached three *fatty acids.* Other important dietary lipids include *phospholipids* and *cholesterol.* Dietary fats are often classified as either **saturated** or **unsaturated.** Saturated fats contain fatty acid chains in which there are no double bonds—that is, all available bonds of its hydrocarbon chain are filled (saturated) with hydrogen atoms (see Figure 2-16, p. 53). Saturated fats are usually solid at room temperature. Unsaturated fats contain fatty acid chains, of which there are some double bonds, meaning that not all sites for hydrogen are filled. Unsaturated fats are usually liquid at room temperature.

Triglycerides are found in nearly every food that we eat. However, the amount of triglycerides in each type of food varies considerably, as does the proportion of saturated to unsaturated types. Phospholipids are also found in nearly all foods because they form the cellular membranes in and around each cell of all living organisms. Cholesterol, however, is found only in foods of animal origin. Cholesterol concentration also varies. For example, it is particularly high in liver and the yolks of eggs.

Transport of Lipids

Lipids are transported in blood as chylomicrons, lipoproteins, and free fatty acids. **Chylomicrons** are small fat droplets present in blood following fat absorption. Fatty acids and monoglyceride products of fat digestion, combine during absorption to again form fats (triglycerides, or triacylglycerols). These triglycerides plus small amounts of cholesterol and phospholipids compose the chylomicrons.

During fat absorption, the so-called *absorptive state,* blood may contain so many of these fat droplets that it appears turbid or even yellowish in color. But during the *postabsorptive state*—usually within about 4 hours after a meal—few, if any, chylomicrons remain in the blood. Their contents have moved mostly into adipose tissue cells.

In the postabsorptive state, when chylomicrons are virtually absent from the circulation, some 95% of the lipids in blood are transported in the form of lipoproteins. **Lipoproteins** are produced mainly in the liver, and, as their name suggests, they consist of lipids (triglycerides, cholesterol, and phospholipids) and protein. At all times, blood contains three types of lipoproteins, namely, very-low-density lipoproteins, low-density lipoproteins, and high-density lipoproteins. Usually, they are designated by their abbreviations: VLDL, LDL, and HDL. Diets high in saturated fats and cholesterol tend to produce an increase in blood LDL concentration, which, in turn, is associated with a high incidence of coronary artery disease and atherosclerosis. A high blood HDL concentration, in contrast, is associated with a low incidence of heart disease. One might therefore think of the LDLs as the "bad lipoproteins" and the HDLs as the "good lipoproteins." Considerable evidence indicates that exercise tends to elevate HDL concentration. This may partially account for the beneficial effects of exercise.

Fatty acids, on entering the blood from adipose tissue or other cells, combine with albumin to form the so-called **free fatty acids (FFAs).** Fatty acids are transported from cells of one tissue to those of another in the form of free fatty acids. Whenever the rate of fat catabolism increases—as it does in starvation or diabetes—the free fatty acid content of blood increases markedly.

Lipid Metabolism
Lipid Catabolism

Lipid catabolism, like carbohydrate catabolism, consists of several processes. Each of these, in turn, consists of a series of chemical reactions. Triglycerides are first hydrolyzed to yield fatty acids and glycerol. Glycerol is then converted to glyceraldehyde-3-phosphate, which enters the glycolysis pathway (see Figure 26-3). Fatty acids, as Figure 26-16 shows, are broken down by a process called *beta-oxidation* into two-carbon pieces, the familiar acetyl-CoA. These are then catabolized via the citric acid cycle. The final process of lipid catabolism therefore consists of the same reactions as carbohydrate catabolism. Catabolism of lipids, however, yields considerably more energy than catabolism of carbohydrates. Whereas catabolism of 1 g of carbohydrates yields only 4.1 kcal of heat, catabolism of 1 g of fat yields 9 kcal. It is not surprising, then, that lipids are the preferred energy source for muscle tissue.

When fat catabolism occurs at an accelerated rate, as in diabetes mellitus (when glucose cannot enter cells) or fasting, excessive numbers of acetyl-CoA units are formed. Liver cells then temporarily condense acetyl-CoA units together to form four-carbon acetoacetic acid. Acetoacetic acid is classified as a *ketone body* and can be converted to

*From Becker WM: Energy and the living cell, Philadelphia, 1977, JB Lippincott Co., pp. 125-126.

FO)CUS ON... LIPOPROTEINS

As stated in the text, high blood concentrations of LDLs (low-density lipoproteins) are associated with a high risk of atherosclerosis. Atherosclerosis is a form of "hardening of the arteries" that occurs when lipids accumulate in cells lining the blood vessels and promote the development of a plaque that eventually impedes blood flow and may trigger clot formation. Atherosclerosis may also weaken the wall of a blood vessel to the point that it ruptures. In any case, a person with atherosclerosis of the coronary arteries risks a heart attack when blood flow to cardiac muscle is impaired. If vessels in the brain are affected, there is risk of a *cerebrovascular accident (CVA)*, or "stroke." Figure A is a simplified version of a concept of LDL's function first proposed by Nobel laureates Michael Brown and Joseph Goldstein, at the University of Texas. According to their model, LDL delivers cholesterol to cells for use in synthesizing steroid hormones

and stabilizing the plasma membrane. Most, if not all, cells have many LDL receptors embedded in the outer surface of their plasma membranes. These receptors attract cholesterol-bearing LDL. Once the LDL molecule binds to the receptor, specific mechanisms operate to release the cholesterol it carries into the cell. Excess cholesterol is stored in droplets near the center of the cell. It seems that, in some individuals at least, cells have so many LDL receptors that they accumulate too much cholesterol. This has been proposed as a cause for the lipid accumulation characteristic of atherosclerosis.

High blood concentrations of HDLs (high-density lipoproteins) have been associated with a low risk of developing atherosclerosis and its many possible complications. Although the exact details of how this works have yet to be worked out or confirmed, some scientists have made some progress. Jack Oram, a cell biologist

working at the University of Washington, has proposed the mechanism illustrated in Figure B. According to his model, HDL molecules are attracted to HDL receptors imbedded in the plasma membranes. Once they bind to their receptors, the cell is stimulated to release some of its cholesterol from storage. The released cholesterol migrates to the plasma membrane where it may attach to the HDL molecule and be whisked away to the liver for excretion in bile.

Apparently, high blood LDL levels (over 180 mg LDL per 100 ml of blood) signify that a large amount of cholesterol is being delivered to cells. High blood HDL levels (over 40 mg/100 HDL) apparently indicate that a large amount of cholesterol is being removed from cells and delivered to the liver for excretion from the body. Currently, researchers are using this information to develop treatments that may prevent—or even cure—atherosclerosis and the disorders it causes.

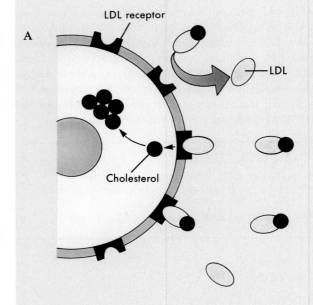

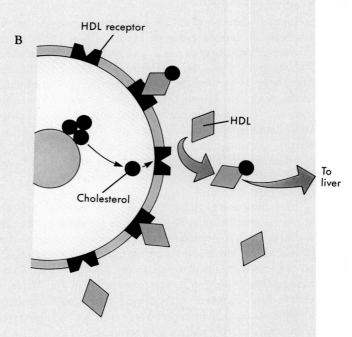

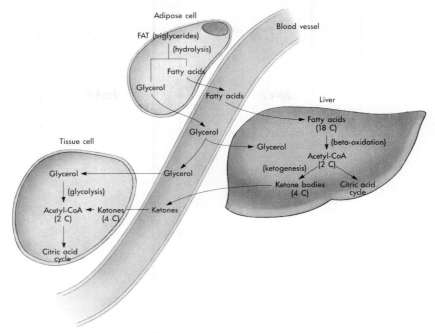

FIGURE 26-16 Fat mobilization and catabolism. Notice the role of the liver as the chief site of ketogenesis. Numbers of carbon atoms are in parentheses.

two other types of ketone bodies, namely, acetone and betahydroxybutyric acid—hence the name **ketogenesis** for this process. Liver cells oxidize a small portion of the ketone bodies for their own energy needs, but most of them are transported by the blood to other tissue cells for the change back to acetyl-CoA and oxidation via the citric acid cycle (see Figure 26-16).

Lipid anabolism

Lipid anabolism, also called **lipogenesis,** consists of the synthesis of various types of lipids, notably *triglycerides, cholesterol, phospholipids,* and *prostaglandins.* Triglycerides and structural lipids (e.g., phospholipids) are synthesized from fatty acids and glycerol or from excess glucose or amino acids. So it is possible to "get fat" from foods other than fat. Triglycerides are stored mainly in adipose tissue cells. These fat depots constitute the body's largest reserve energy source—often too large, unfortunately. Almost limitless pounds of fat can be stored. In contrast, only a few hundred grams of carbohydrates can be stored as liver and muscle glycogen.

Most fatty acids can be synthesized by the body. Certain of the unsaturated fatty acids must be provided by the diet and so are called **essential fatty acids.** Some of the essential fatty acids serve as a source within the body for synthesis of an important group of lipids called *prostaglandins.* These hormonelike compounds, first discovered in the 1930s in semen, have in recent years gained increasing recognition for their occurrence in various tissues, with a wide spectrum of biological activities (see Chapter 15).

Control of lipid metabolism

Lipid metabolism is controlled mainly by the following hormones:

- ◆ Insulin
- ◆ Growth hormone
- ◆ ACTH
- ◆ Glucocorticoids

You probably recall from our discussion of these hormones in connection with carbohydrate metabolism that they regulate fat metabolism in such a way that the rate of fat catabolism is inversely related to the rate of carbohydrate catabolism. If some condition such as diabetes mellitus causes carbohydrate catabolism to decrease below energy needs, increased secretion of growth hormone, ACTH, and glucocorticoids soon follows. These hormones, in turn, bring about an increase in fat catabolism. But, when carbohydrate catabolism equals energy needs, fats are not mobilized out of storage and catabolized. Instead, they are spared and stored in adipose tissue. "Carbohydrates have a 'fat-sparing' effect," so says an old physiological maxim. Or stating this truth more descriptively: "Carbohydrates have a 'fat-storing' effect."

QUICK CHECK

✔ 1. *In what forms are lipids transported to cells?*
✔ 2. *How can glycerol and fatty acids enter the citric acid cycle?*
✔ 3. *Which fatty acids cannot be made by the body?*

PROTEINS

Sources of Proteins

Recall from Chapter 2 that proteins are very large molecules composed of chemical subunits called **amino acids** (see Figure 2-15, p. 51). Proteins are assembled from a pool of 20 different kinds of amino acids. If any one type of amino acid is deficient, several vital proteins cannot be

synthesized—a serious health threat. One way your body maintains a constant supply of amino acids is by synthesizing them from other compounds already present in the body. Only about half of the required 20 types of amino acid can be made by the body, however. The remaining types of amino acids must be supplied in the diet. Nutritionists often refer to the amino acids that must be in the diet as essential, or nondispensible, amino acids. Table 26-1 lists amino acids according to whether they are considered essential in the diet or nonessential (dispensible) in the diet (synthesized by the body).

Both amino acids and whole proteins are obtained in the diet from various sources. Muscle meat and other animal tissues particularly high in proteins contain the essential amino acids. Food from a single plant or other nonanimal source usually does not contain all the essential amino acids. Therefore it is important to include meat in the diet or a mixture of different vegetables that provide all the amino acids needed by the body. Plant tissues that are particularly high in protein content include cereal grains, nuts, and legumes such as peas and beans.

Protein Metabolism

In protein metabolism, anabolism is primary, and catabolism is secondary. In carbohydrate and fat metabolism, the opposite is true—catabolism is primary, and anabolism is secondary. Proteins are primarily tissue-building foods. Carbohydrates and fats are primarily energy supplying foods.

Protein anabolism

Protein anabolism is the process by which proteins are synthesized by the ribosomes of all cells. The specific mechanisms of protein anabolism were discussed in Chapter 2 (see pp. 48 to 50).

Every cell synthesizes its own structural proteins and its own enzymes. In addition, many cells, such as liver and glandular cells, synthesize special proteins for export. For

TABLE 26-1   Amino acids

Essential (nondispensible)	Nonessential (dispensible)
Histidine*	Alanine
Isoleucine	Arginine
Leucine	Asparagine
Lysine	Aspartic acid
Methionine	Cysteine
Phenylalanine	Glutamic acid
Threonine	Glutamine
Tryptophan	Glycine
Tyrosine†	Proline
Valine	Serine

*Essential in infants and, perhaps, adult males.
†Can be synthesized from phenylalanine.

example, liver cells manufacture the plasma proteins found in blood (Figure 26-17). The cell's genes determine the specific proteins synthesized. Protein anabolism is truly "big business" in the body. Consider, for instance, that protein anabolism constitutes the major process of growth, reproduction, tissue repair, and the replacement of cells destroyed by daily wear and tear. Red blood cell replacement alone runs into millions of cells per second!

Protein catabolism

The first step in protein catabolism takes place in liver cells. Called **deamination,** it consists of the splitting off of an amino (NH_2) group from an amino acid molecule to form a molecule of ammonia and one of keto acid (e.g., alpha-ketoglutaric acid). Most of the ammonia is converted by liver cells to *urea* and later excreted in the urine. The keto acid may be oxidized via the citric acid cycle (see Figure 26-5) or may be converted to glucose via gluconeogenesis (Figure 26-18) or to fat (lipogenesis). Both protein catabolism and anabolism go on continually. Only their rates differ from time to time. With a protein-deficient diet, for example, protein catabolism exceeds protein anabolism. Various hormones, as we shall see, also influence the rates of protein catabolism and anabolism.

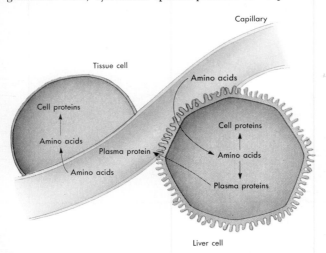

FIGURE 26-17 Protein anabolism. Proteins are synthesized from amino acids transported to each cell by the blood. Notice that the liver cell—like a few other specialized cells in the human body—makes some proteins for export.

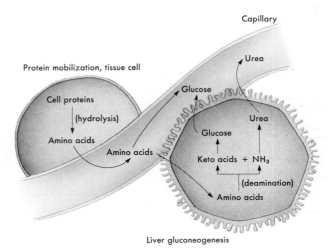

FIGURE 26-18 Protein mobilization and catabolism. Glucocorticoids tend to accelerate these processes, therefore are classed as protein catabolic hormones.

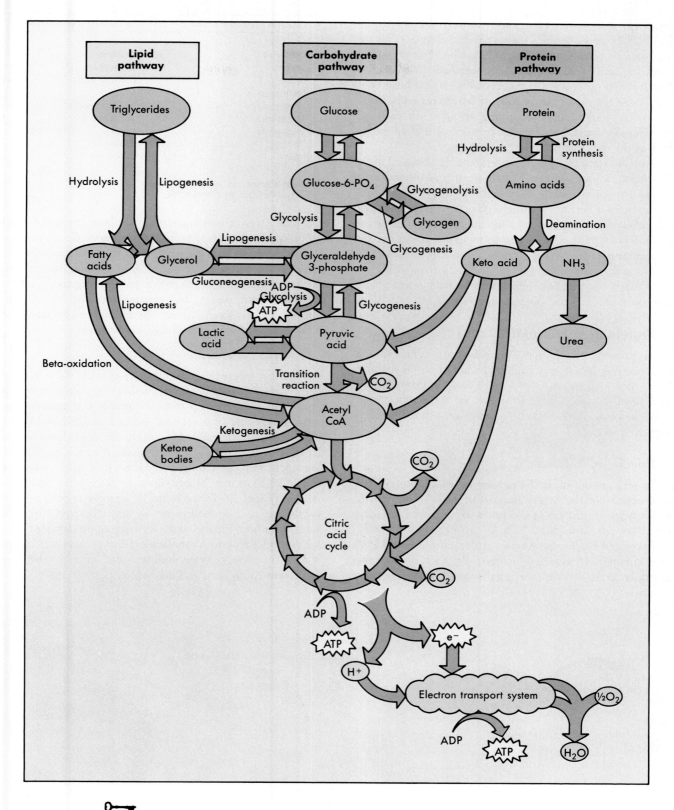

FIGURE 26-19   Summary of metabolism. Notice the central role played by the citric acid cycle and electron transport system. Notice also how different molecules can be converted to forms that may enter other pathways.

TABLE 26-2   Metabolism

Nutrient	Anabolism	Catabolism
Carbohydrates	Temporary excess changed into glycogen by liver cells in presence of insulin; stored in liver and skeletal muscles until needed and then changed back to glucose	Oxidized, in presence of insulin, to yield energy (4.1 kcal per g) and wastes (carbon dioxide and water)
	True excess beyond body's energy requirements converted into adipose tissue; stored in various fat depots of body	$C_6H_{12}O_6 + 6\ O_2 \rightarrow$ Energy $+ 6\ CO_2 + 6\ H_2O$
Fats	Built into adipose tissue; stored in fat depots of body	Fatty acids $\downarrow$ (beta-oxidation) Acetyl-CoA $\leftrightarrows$ Ketones $\downarrow$ (tissues; citric acid cycle) Energy (9.3 kcal per g) $+\ CO_2 + H_2O$ — Glycerol $\downarrow$ (glycolysis) Acetyl-CoA
Proteins	Synthesized into tissue proteins, blood proteins, enzymes, hormones, etc.	Deaminated by liver, forming ammonia (which is converted to urea) and keto acids (which are either oxidized or changed to glucose or fat)

Protein balance and nitrogen balance

Usually a state of **protein balance** exists in the normal healthy adult body, that is, the rate of protein anabolism equals or balances the rate of protein catabolism. When the body is in protein balance, it is also in a state of **nitrogen balance.** For then the amount of nitrogen taken into the body (in protein foods) equals the amount of nitrogen in protein catabolic waste products excreted in the urine, feces, and sweat.

It is important to realize that there are two kinds of protein, or nitrogen, imbalance. When protein catabolism exceeds protein anabolism, the amount of nitrogen in the urine exceeds the amount of nitrogen in the protein foods ingested. The individual is then said to be in a state of **negative nitrogen balance,** or in a state of "tissue-wasting"—because more of the tissue proteins are being catabolized than are being replaced by protein synthesis. Protein-poor diets, starvation, and wasting illnesses, for example, produce a negative nitrogen balance. A **positive nitrogen balance** (nitrogen intake in foods greater than nitrogen output in urine) indicates that protein anabolism is occurring at a faster rate than protein catabolism. A state of positive nitrogen balance therefore characterizes any condition, in which large amounts of tissue are being synthesized, such as during growth, pregnancy, and convalescence from an emaciating illness.

Control of protein metabolism

Protein metabolism, like that of carbohydrates and fats, is controlled largely by hormones rather than by the nervous system. Growth hormone and the male hormone testosterone both have a stimulating effect on protein synthesis or anabolism. For this reason, they are referred to as *anabolic* hormones. The protein *catabolic* hormones of greatest consequence are glucocorticoids. They speed up tissue protein mobilization, that is, the hydrolysis of cell proteins to amino acids, their entry into the blood, and their subsequent catabolism (see Figure 26-18). ACTH functions indirectly as a protein catabolic hormone because

of its stimulating effect on glucocorticoid secretion.

Thyroid hormone is necessary for and tends to promote protein anabolism and therefore growth when plenty of carbohydrates and fats are available for energy production. On the other hand, under different conditions, for example, when the amount of thyroid hormone is excessive or when the energy foods are deficient, this hormone may then promote protein mobilization and catabolism.

Some of the facts about metabolism set forth in the preceding sections are summarized in Table 26-2 and Figure 26-19.

QUICK CHECK

✔ 1. *What is meant by the term* essential amino acid?
✔ 2. *What happens when an amino acid is deaminated? What is the purpose of this process?*
✔ 3. *What is meant by the term* nitrogen balance?

VITAMINS AND MINERALS

One glance at the label of any packaged food product reveals the importance we place on vitamins and minerals. We know that carbohydrates, fats, and proteins are used by our bodies to build important molecules and to provide energy. So why do we need vitamins and minerals?

Vitamins

Vitamins are organic molecules needed in small quantities for normal metabolism throughout the body. Vitamin molecules attach to enzymes or coenzymes and help them work properly. **Coenzymes** are organic, nonprotein catalysts that often act as "molecule carriers." Many enzymes or coenzymes are totally useless without the appropriate vitamins to attach to them and thus give them the shape that allows them to function properly. For example, coenzyme A (CoA), an important carrier molecule associated with the citric acid cycle, has *pantothenic acid* (vitamin B_5) as one of its major components.

TABLE 26-3   Major vitamins

Vitamin	Dietary Source	Functions	Symptoms of Deficiency
Vitamin A	Green and yellow vegetables, dairy products, and liver	Maintains epithelial tissue and produces visual pigments	Night blindness and flaking skin
B-complex vitamins			
B_1 (thiamine)	Grains, meat, and legumes	Helps enzymes in the critic acid cycle	Nerve problems (beriberi), heart muscle weakness, and edema
B_2 (riboflavin)	Green vegetables, organ meats, eggs, and dairy products	Aids enzymes in the citric acid cycle	Inflammation of skin and eyes
B_3 (niacin)	Meat and grains	Helps enzymes in the citric acid cycle	Pellagra (scaly dermatitis and mental disturbances) and nervous disorders
B_5 (pantothenic acid)	Organ meat, eggs, and liver	Aids enzymes that connect fat and carbohydrate metabolism	Loss of coordination (rare)
B_6 (pyridoxine)	Vegetables, meats, and grains	Helps enzymes that catabolize amino acids	Convulsions, irritability, and anemia
B_{12} (cyanocobalamin)	Meat and dairy products	Involved in blood production and other processes	Pernicious anemia
Biotin	Vegetables, meat, and eggs	Helps enzymes in amino acid catabolism and fat and glycogen synthesis	Mental and muscle problems (rare)
Folic acid	Vegetables	Aids enzymes in amino acid catabolism and blood production	Digestive disorders and anemia
Vitamin C (ascorbic acid)	Fruits and green vegetables	Helps in manufacture of collagen fibers	Scurvy and degeneration of skin, bone, and blood vessels
Vitamin D (calciferol)	Dairy products and fish liver oil	Aids in calcium absorption	Rickets and skeletal deformity
Vitamin E (tocopherol)	Green vegetable and seeds	Protects cell membranes from being catabolized	Muscle and reproductive disorders (rare)

Most vitamins cannot be made by the body, so we must eat them in our food. The body can store fat-soluble vitamins—A, D, E, and K—in the liver for later use. Because the body cannot store significant amounts of water-soluble vitamins such as B vitamins and vitamin C, they must be continually supplied in the diet. Table 26-3 lists some of the more well-known vitamins, their sources, functions, and symptoms of deficiency.

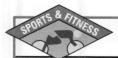

VITAMIN SUPPLEMENTS FOR ATHLETES

Because a deficiency of vitamins (*avitaminosis*) can cause poor athletic performance, many athletes regularly consume vitamin supplements. However, research suggests that vitamin supplementation has little or no effect on a person's athletic performance. A reasonably well-balanced diet supplies more than enough vitamins for even the elite athlete. The use of vitamin supplements therefore has fueled controversy among exercise experts. Opponents of vitamin supplements cite the cost and the possibility of liver damage associated with some forms of *hypervitaminosis*, whereas supporters cite the benefit of protecting against vitamin deficiency.

Minerals

Minerals are just as important as vitamins. Minerals are inorganic elements or salts found naturally in the earth. Like vitamins, mineral ions can attach to enzymes or other organic molecules and help them work. Minerals also function in various other vital chemical reactions. For example, sodium, calcium, and other minerals are required for nerve conduction and for contraction in muscle fibers. Without these minerals, the brain, heart, and respiratory tract would cease to function. Calcium, phosphorus, and magnesium are required to build the strong structural components of the skeleton. Information about some of the more important minerals is summarized in Table 26-4.

Like vitamins, minerals are beneficial only when taken in the proper amounts. Many of the minerals listed in Table 26-4 are required in trace amounts. Any intake of such minerals beyond the recommended trace amount may become toxic—perhaps even life threatening.

QUICK CHECK

✔ 1. *What is a vitamin?*
✔ 2. *List two functions of minerals in the body.*

TABLE 26-4   Major minerals

Mineral	Dietary Source	Functions	Symptoms of Deficiency
Calcium (Ca)	Dairy products, legumes, and vegetables	Helps blood clotting, bone formation, and nerve and muscle function	Bone degeneration and nerve and muscle malfunction
Chlorine (Cl)	Salty foods	Aids in stomach acid production and acid-base balance	Acid-base imbalance
Cobalt (Co)	Meat	Helps vitamin B_{12} in blood cell production	Pernicious anemia
Copper (Cu)	Seafood, organ meats, and legumes	Involved in extracting energy from the citric acid cycle and in blood production	Fatigue and anemia
Iodine (I)	Seafood and iodized salt	Required for thyroid hormone synthesis	Goiter (thyroid enlargement) and decrease of metabolic rate
Iron (Fe)	Meat, eggs, vegetables, and legumes	Involved in extracting energy from the citric acid cycle and in blood production	Fatigue and anemia
Magnesium (Mg)	Vegetables and grains	Helps many enzymes	Nerve disorders, blood vessel dilation, and heart rhythm problems
Manganese (Mn)	Vegetables, legumes, and grains	Helps many enzymes	Muscle and nerve disorders
Phosphorus (P)	Dairy products and meat	Aids in bone formation and is used to make ATP, DNA, RNA, and phospholipids	Bone degeneration and metabolic problems
Potassium (K)	Seafood, milk, fruit, and meats	Helps muscle and nerve function	Muscle weakness, heart problems, and nerve problems
Sodium (Na)	Salty foods	Aids in muscle and nerve function and fluid balance	Weakness and digestive upset
Zinc (Zn)	Many foods	Helps many enzymes	Metabolic problems

METABOLIC RATES

The term **metabolic rate** means the amount of energy released in the body in a given time by catabolism. It represents energy expended for accomplishing various kinds of work. In short, metabolic rate actually means catabolic rate, or rate of energy release.

Metabolic rates are expressed in either of two ways: (1) in terms of the number of kilocalories of heat energy expended per hour or per day or (2) as normal or as a definite percentage above or below normal.

Basal Metabolic Rate

The **basal metabolic rate (BMR)** is the body's rate of energy expenditure under "basal conditions," namely, when the individual:

♦ Is awake but resting, that is, lying down and, as far as possible, not moving a muscle
♦ Is in the postabsorptive state (12 to 18 hours after the last meal)
♦ Is in a comfortably warm environment (the so-called thermoneutral zone, a temperature range at which metabolism is independent of ambient temperature)

Note that the BMR is not the minimum metabolic rate. It does not indicate the smallest amount of energy that must be expended to sustain life. It does, however, indicate the smallest amount of energy expenditure that can sustain life and also maintain the waking state and a normal body temperature in a comfortably warm environment.

Factors influencing basal metabolic rate

The BMR is not identical for all individuals because of the influence of various factors (Figure 26-20), some of which are described in the following paragraphs.

Size. In computing the BMR, size is usually indicated by the amount of the body's surface area. It is computed from the individual's height and weight. A large individual has the same BMR as a small person per square meter of body surface, if other conditions are equal. However, because a large individual has more square meters of surface area, the BMR is greater than that of a small individual. For example, the BMR for a man in his twenties is about 40 kcal per square meter of body surface per hour (Table 26-5). However, a large man with a body surface area of 1.9 square meters would have a BMR of 76 kcal per hour, whereas a smaller man with a surface area of perhaps 1.6 square

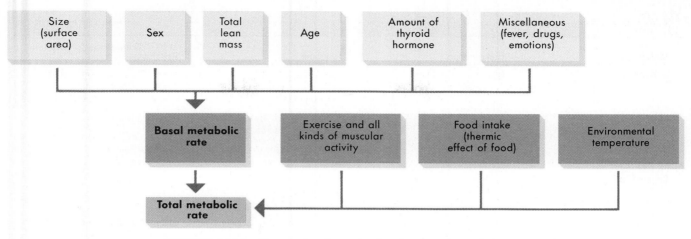

FIGURE 26-20   Factors that determine basal and total metabolic rates.

TABLE 26-5	Basal metabolism	
	Kilocalories per hour per square meter body surface	
Age (yr)	**Male**	**Female**
10-12	51.5	50.0
12-14	50.0	46.5
14-16	46.0	43.0
16-18	43.0	40.0
18-20	41.0	38.0
20-30	39.5	37.0
30-40	39.5	36.5
40-50	38.5	36.0
50-60	37.5	35.0
60-70	36.5	34.0

meters would have a BMR of only 64 kcal per hour. The average surface area for American adults is 1.6 square meters for women and 1.8 square meters for men.

Sex. Men oxidize their food approximately 5% to 7% faster than women. Therefore their BMRs are about 5% to 7% higher for a given size and age. A man 5 feet, 6 inches tall, weighing 140 pounds, for example, has a 5% to 7% higher BMR than a woman of the same height, weight, and age. This gender difference in BMR probably results from the difference in the proportion of body fat determined by sex hormones. Women tend to have a higher percentage of body fat (and thus a lower total lean mass) than men. Fat tissue is less metabolically active than lean tissues such as muscle. Differences in total lean mass not related to gender also affect the BMR.

Age. That the fires of youth burn more brightly than those of age is a physiological and a psychological fact. In general, the younger the individual, the higher the BMR for a given size and sex (see Table 26-5). Exception: the BMR is slightly lower at birth than it is a few years later.

That is to say, the rate increases slightly during the first 3 to 6 years, then starts to decrease and continues to do so throughout life.

Thyroid hormone. Thyroid hormone (T_3 and T_4) stimulates basal metabolism. Without a normal amount of this hormone in the blood, a normal BMR cannot be maintained. When an excess of thyroid hormone is secreted, foods are catabolized faster, much as wood in a fireplace is burned faster when the draft is open. Deficient thyroid secretion, on the other hand, slows the rate of metabolism.

Body temperature. Fever increases the BMR. For every degree Celsius increase in body temperature, metabolism increases about 13%. A decrease in body temperature (hypothermia) has the opposite effect. Metabolism decreases, and, because it does, cells use less oxygen than they normally do. This knowledge has been applied clinically by using hypothermia in certain situations—for example, in open-heart surgery. Because circulation is reduced or interrupted during this procedure, oxygen supply necessarily decreases. Cells can tolerate this decreased oxygen supply reasonably well if their oxygen need has also decreased. Induced hypothermia decreases their rate of metabolism and thereby decreases their use of oxygen.

Drugs. Certain drugs, such as caffeine, amphetamine, and dinitrophenol, increase the BMR.

Other factors. Other factors, such as *emotions*, *pregnancy*, and *lactation* (milk production), also influence basal metabolism. All of these factors increase the BMR.

Total Metabolic Rate

Total metabolic rate is the amount of energy used or expended by the body in a given time. It is expressed in kilocalories per hour or per day. Most of the factors that determine the total metabolic rate are shown in Figure 26-20. Of these, the main direct determinants are the following:

Factor 1. The basal metabolic rate, that is, the energy used to do the work of maintaining life under the basal conditions previously described. Basal metabolic rate usu-

FOCUS ON... HOW BMR IS DETERMINED

BMR can be determined by a method called **indirect calorimetry.** The rationale underlying this method is the fact that BMR (expressed as the number of kilocalories of heat produced per unit of time) can be calculated from the amount of oxygen consumed in a given time. BMR can then be expressed as normal, or as a definite percent above or below normal, by dividing the actual kilocalorie rate by the known average kilocalorie rate for normal individuals of the same size, sex, and age. Statistical tables, based on research, list these normal BMRs. (If BMR were calculated to be 10% above normal, for example, it would be reported as + 10.) Are you curious to know the average BMR for a person of your size, sex, and age? If so, take the following steps:

1. Start with your weight in kilograms and your height in centimeters. (Convert pounds to kilograms by dividing pounds by 2.2. Convert inches to approximate centimeters by multiplying inches by 2.5.) For example, 110 pounds = 50 kg; 5 feet, 3 inches = 158 cm.

2. Convert your weight and height to square meters, using the chart shown here. For example, weight 50 kg and height 158 cm = about 1.5 square meters of body surface area.

3. Find your age and sex in Table 26-5 and then multiply the number of kilocalories per square meter per hour given there by your square meters of surface area and then by 24. For example, average BMR per day for a 25-year-old female, weight 110 pounds and height 5 feet, 3 inches = 1,332 kcal (37 × 1.5 × 24).

A quick rule of thumb for estimating a young woman's BMR in kcal/hr is to multiply her weight in pounds by 12.

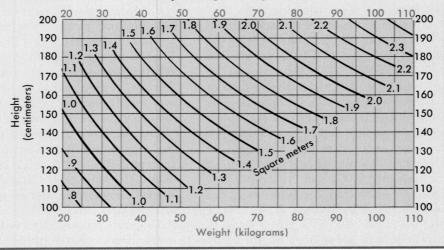

ally constitutes about 55% to 60% of the total metabolic rate.

Factor 2. The energy used to do all kinds of skeletal muscle work—from the simplest activities such as feeding oneself or sitting up in bed to the most strenuous kind of physical labor or exercise.

Factor 3. The *thermic effect,* or *specific dynamic action* (SDA), of foods. The metabolic rate increases for several hours after a meal, apparently because of the energy needed for metabolizing foods. Carbohydrates and fats have a thermic effect of about 5%. Proteins have a much higher thermic effect, about 30%. This means that, for 100 kcal of protein, 30 kcal are used for such processes as deamination and oxidation of the protein, leaving just 70 kcal available for other cell work. For this reason, proteins are "worth" fewer calories and are popular as diet foods.

Energy Balance and Body Weight

When we say that the body maintains a state of energy balance, we mean that its energy input equals its energy output. Energy input per day equals the total calories (kilocalories) in the food ingested per day. Energy output equals the total metabolic rate expressed in kilocalories. You may be wondering what energy intake, output, and balance have to do with body weight. "Everything" would be a fairly good one-word answer. Or, to be somewhat more explicit, the following basic principles describe the relationships between these factors:

♦ Body weight remains constant (except for possible variations in water content) when the body maintains energy balance—when the total calories in the food ingested equals the total metabolic rate. Example: If you have a total metabolic rate of 2,000 kcal per day and if the food you eat per day yields 2,000 kcal, your body will be maintaining energy balance and your weight will stay constant.

♦ Body weight increases when energy input exceeds energy output—when the total calories of food intake per day is greater than the total calories of the metabolic rate. A small amount of the excess energy input is used to

synthesize glycogen for storage in the liver and muscles. The rest is used for synthesizing fat and storing it in adipose tissue. If you were to eat 3,000 kcal each day for a week and if your total metabolic rate were 2,000 kcal per day, you would gain weight. How much you would gain you can discover by doing a little simple arithmetic:

$$
\begin{aligned}
\text{Total energy input for week} &= 21,000 \text{ kcal} \\
\text{Total energy output for week} &= \underline{14,000 \text{ kcal}} \\
\text{Excess energy input for week} &= 7,000 \text{ kcal}
\end{aligned}
$$

Approximately 3,500 kcal are used to synthesize 1 pound of adipose tissue. Hence, at the end of this 1 week of "overeating"—of eating 7,000 kcal over and above your total metabolic rate—you would have gained about 2 pounds.

♦ Body weight decreases when energy input is less than energy output—when the total number of calories in the food eaten is less than the total metabolic rate. Suppose you were to eat only 1,000 kcal a day for a week and that you have a total metabolic rate of 2,000 kcal per day. By the end of the week, your body would have used a total of 14,000 kcal of energy for maintaining life and doing its many kinds of work. All 14,000 kcal of this actual energy expenditure had to come from catabolism of foods, since this is the body's only source of energy. Catabolism of ingested food supplied 7,000 kcal, and catabolism of stored food supplied the remaining 7,000 kcal. That week your body would not have maintained energy balance, nor would it have maintained weight balance. It would have incurred an energy deficit paid out of the energy stored in approximately 2 pounds of body fat. In short, you would have lost about 2 pounds.

Foods are stored primarily as glycogen and fats. As you will recall, cells catabolize them preferentially in this same order: carbohydrates, then fats. If there is no food intake, almost all of the glycogen is estimated to be used up in a matter of 1 or 2 days. Then, with no more carbohydrate to act as a fat sparer, fat is catabolized. How long it takes to deplete all of this reserve food depends, of course, on how much adipose tissue the individual has when starting the starvation diet. Finally, with no more fat available, tissue proteins are catabolized (see Figure 26-19). Because significant amounts of protein are not "stored" for use in catabolism, this means that important structural and functional proteins are quickly depleted. Thus death soon ensues.

MECHANISMS FOR REGULATING FOOD INTAKE

Mechanisms for regulating food intake are still not clearly established. That the hypothalamus plays a part in these mechanisms, however, seems certain. Numerous studies seem to indicate that a cluster of neurons in the lateral hypothalamus function as an **appetite center**—meaning that impulses from them bring about increased appetite. Other data suggest that a group of neurons in the ventral medial nucleus of the hypothalamus functions as a **satiety center**—meaning that impulses from these neurons decrease appetite so that we feel sated, or satisfied. What acts directly on these centers to stimulate or depress them is still a matter of theory rather than fact. One theory (the "*thermostat theory*") holds that it is the temperature of the blood circulating to the hypothalamus that influences the centers. A moderate decrease in blood temperature stimulates the appetite center (and inhibits the satiety center). Result: the individual has an appetite, wants to eat, and probably does. An increase in blood temperature produces the opposite effect, a depressed appetite (**anorexia**). One well-known instance of this is the loss of appetite in persons who have a fever.

Another theory (the "*glucostat theory*") says that it is the blood glucose concentration and rate of glucose use that influences the hypothalamic feeding centers. A low blood glucose concentration or low glucose use stimulates the appetite center, whereas a high blood glucose concentration inhibits it. Unquestionably, many factors operate together as a complex mechanism for regulating food intake. Some of these factors are indicated in Figure 26-21.

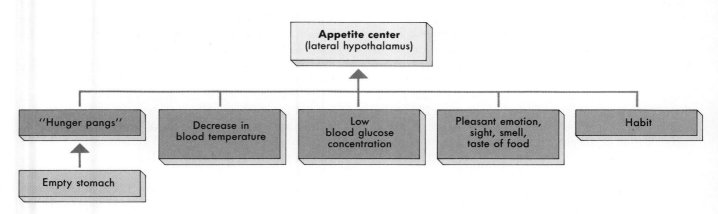

FIGURE 26-21   Factors that affect food intake. These factors directly or indirectly stimulate the appetite center (and inhibit the satiety center) in the hypothalamus and thereby tend to increase the amount of food eaten.

QUICK CHECK

✔ 1. *Give one of the two ways in which metabolic rates can be expressed.*

✔ 2. *Name three of the factors that influence basal metabolic rate.*

✔ 3. *Distinguish between* basal metabolic rate *and* total metabolic rate.

✔ 4. *In which division of the brain would you find the control centers for regulating food intake?*

Cycle of Life: *Nutrition and Metabolism*

The importance of proper nutrition to an individual's well-being begins at the very moment of conception and continues until death. In the womb, various nutrients must be obtained from the mother's blood in sufficient quantity to ensure normal growth and development. One critical nutrient during fetal development, infancy, and childhood is protein. Sufficient proteins, containing all the essential amino acids, are required to permit normal development of the nervous system, muscle tissues, and other vital structures. Another critical nutrient during the early years of life is the mineral, calcium. Large quantities of calcium are needed by a growing body to maintain normal development of the skeleton and other tissues. In the womb, a steady supply of calcium in the mother's blood is maintained by increased levels of the parathyroid hormone (PTH). Recall from Chapter 15 that PTH increases blood calcium levels by removing it from storage in the bones. Unless a pregnant woman consumes enough calcium to replace this calcium, she may suffer from the bone-softening effects of calcium

deficiency. If proteins, calcium, or other necessary nutrients are in short supply anytime before the beginning of adulthood, the consequences may be permanent. For example, bone deformities resulting from a lack of calcium during childhood could become permanent if not corrected or compensated for before the skeleton ossifies completely.

In late adulthood, the number of food calories needed declines because the metabolic rate declines. This metabolic decline is thought to result largely from age-related changes in the balance of metabolic hormones such as thyroid hormones (T_3 and T_4). Even though the number of required food calories declines, the overall balance of nutrients consumed must be maintained to maintain proper metabolic function. Some nutrients, such as calcium, may be needed in greater quantity in the older adult to compensate for (or avoid) age-related bone loss or other conditions.

THE BIG PICTURE
Nutrition, Metabolism, and the Whole Body

Of all the topics we have discussed so far, the topic of nutrition and metabolism has the most easily seen role in the "big picture" of human body function. Each cell in the body cannot remain alive without maintaining the operation of its metabolic pathways. Anabolic pathways are required to build the various structural and functional components of the cells. Catabolic pathways are required to convert energy to a usable form. Catabolic pathways are also needed to degrade large molecules into small subunits that can be used in anabolic pathways. Of course, the basic nutrient molecules—carbohydrates, fats, and proteins of the correct type—must be available to each cell to carry out these metabolic processes. Besides the basic nutrient molecules, cells also require small amounts of specific vitamins and minerals needed to produce the structural and functional components necessary for cellular metabolism.

Various body systems operate to make sure that essential nutrients reach the cells as needed to maintain metabolism in a manner that preserves relative constancy of the internal environment. For example, the nervous, skeletal, and muscular systems help us obtain complex foods from our external environment. The digestive system reduces complex nutrients to simpler, more

usable, nutrients—then provides the mechanisms that allows us to absorb them into the internal environment. The circulatory system—both the cardiovascular and the lymphatic circulations—transports the absorbed nutrients to the individual cells for immediate use or to the liver or other organs for temporary storage. The endocrine system regulates the balance between immediate use and storage. The respiratory system, working with the cardiovascular system, provides the oxygen needed for oxidative phosphorylation—that is, using the citric acid cycle and electron transport system to transfer energy to ATP. These two systems also provide a mechanism for removing waste CO_2 generated by catabolism of nutrient molecules. Likewise, the urinary system provides a mechanism for removing waste urea generated by protein catabolism. Even the integumentary system becomes involved, by producing vitamin D in the presence of sunlight.

Metabolism, with all the physiological mechanisms that support it, could be described as the essential process of life. It is, after all, the sum total of all the biochemical processes that distinguish a living organism from a nonliving object.

MECHANISMS OF DISEASE
Metabolic and Nutritional Disorders

Disorders characterized by a disruption or imbalance of normal metabolism can be caused by several different factors. For example, **inborn errors of metabolism** are a group of genetic conditions involving a deficiency or absence of a particular enzyme. Specific enzymes are required by cells to carry out each step of every metabolic reaction. Although an abnormal genetic code may affect the production of only a single enzyme, the resulting abnormal metabolism may have widespread effects. Specific diseases resulting from inborn errors of metabolism, such as *phenylketonuria,* are discussed in Chapter 33.

A number of metabolic disorders are complications of other conditions. For example, you may recall from Chapter 15 that both hyperthyroidism and hypothyroidism have profound effects on the basal metabolic rate (BMR). Diabetes mellitus affects metabolism throughout the body when an insulin deficiency limits the amount of glucose available for use by the cells.

Some metabolic disorders result from normal mechanisms in the body that maintain homeostasis. For example, the body has several mechanisms that maintain a relatively constant level of glucose in the blood—glucose required by cells for life-sustaining catabolism. As mentioned earlier in this chapter, during starvation or in certain eating disorders, these mechanisms are taken to the extreme as they attempt to maintain blood glucose homeostasis. A few of the more well-known eating and nutrition disorders are briefly described in the following.

Eating Disorders

Eating disorders have been a part of the medical literature for many years, but there has been growing interest and concern as the numbers of reported cases have increased dramatically. The two most common eating disorders are called **anorexia nervosa** and **bulimia.** Neither illness is completely understood, and successful treatment is often varied and sometimes controversial.

Anorexia nervosa

Anorexia nervosa is primarily a disease of young adults. Most individuals affected are female (90% to 95%) from 12 to 25 years of age. As many as 4% of college-age students suffer from this condition to some degree. These individuals have a disturbed body image, an intense fear of obesity, and diet with a vengeance. They almost always develop unusual eating rituals and scrupulously monitor and restrict their food intake. Anorexic individuals literally starve themselves and, as a result, develop serious medical complications. The illness is characterized by a 20% to 25% loss of body mass, accompanied by slowed or impaired intellectual functioning. People affected are usually involved in excessive exercise and pursue thinness, regardless of their health. In women, menstruation ceases (amenorrhea), and the BMR is decreased as a result of starvation. These individuals suffer from many skin abnormalities and an assortment of psychological, cardiovascular, and hormonal problems. They are at increased risk of sudden death from complications directly related to excessive weight loss and nutritional deficiency. Treatment is directed at resolution of both medical and psychological problems. In addition to

psychotherapy and weight stabilization, pharmacological treatment with antidepressants has been used to improve mood and self-image.

Bulimia

Bulimia is an illness characterized by an eating and vomiting, or purging, cycle. It is sometimes referred to as **binge-purge syndrome.** The disease, which has only been classified as a distinct disorder since 1980, is said to affect about 1% of college-age students. Most people who have bulimia are relatively young, single, white females. The mean age is 25, but the age of patients appears to be increasing. People suffering from bulimia have an uncontrollable urge for food that leads to massive overeating (binging) that is followed by repeated forced vomiting and laxative abuse (purging). Loss of gastric and intestinal contents often results in serious fluid and electrolyte imbalance. The result is often the development of neurological problems such as convulsions, tetany, and seizures. Vomiting may also cause aspiration pneumonia, erosion of tooth enamel, trauma of the mouth and esophagus, and infection of the salivary glands. The longer the disease is allowed to continue without treatment, the greater the increase in mortality from medical complications. Many people who have bulimia suffer from major depression and have concomitant social problems such as alcohol abuse. About a fourth of bulimics are chemically dependent, and many have been victims of sexual abuse. They are especially prone to self-mutilation and suicide attempts. Nutritional counseling, psychotherapy, and treatment with antidepressants help bulimic patients cope with stress and break the binge/purge cycle.

Obesity

Obesity is not an eating disorder itself but may be a symptom of chronic overeating behavior. Like anorexia nervosa and bulimia, eating disorders characterized by chronic overeating usually have an underlying emotional cause. Obesity may also result from metabolic disorders. Obesity is defined as an abnormal increase in the proportion of fat in the body. Most of the excess fat is stored in the subcutaneous tissue and around the viscera. Obesity is a risk factor in various life-threatening diseases, including many forms of cancer, diabetes, and heart disease.

Nutritional Disorders
Protein-calorie malnutrition

Protein-calorie malnutrition (PCM) is an abnormal condition resulting from a deficiency of calories in general and protein in particular. PCM is likely to result from reduced intake of food but may also be caused by increased nutrient loss or increased use of nutrients by the body. Mild cases occur frequently in illness; as many as one in five patients admitted to the hospital are significantly malnourished. More severe cases of PCM are likely to occur in parts of the world where food, especially protein-rich food, is relatively unavailable. There are two forms of advanced PCM: *marasmus* and *kwashiorkor* (kwah-shee-OR-kor). Marasmus results from an overall lack of calories and proteins, such as when sufficient quantities of food are not available. Marasmus is characterized

by progressive wasting of muscle and subcutaneous tissue accompanied by fluid and electrolyte imbalances. Kwashiorkor results from a protein deficiency in the presence of sufficient calories, as when a child is weaned from milk to low-protein foods. Kwashiorkor also causes wasting of tissues, but unlike marasmus, it also causes pronounced ascites (abdominal bloating) and flaking dermatitis. The ascites results from a deficiency of plasma proteins, which changes the osmotic balance of the blood and thus promotes osmosis of water from the blood into the peritoneal space.

Vitamin disorders

Vitamin deficiency, or **avitaminosis** (a-vye-tah-min-OS-is), can lead to severe metabolic problems. For example, *avitaminosis* C (vitamin C deficiency) can lead to **scurvy.** Scurvy results from the inability of the body to manufacture and maintain collagen fibers. As you may have gathered from your studies thus far, collagen fibers compose the connective

tissues that hold most of the body together. In scurvy, the body literally falls apart in the same way that a neglected house eventually falls apart. Other details about scurvy and other types of avitaminosis are given in Table 26-3.

Some forms of **hypervitaminosis**— or vitamin excess— can be just as serious as a deficiency of vitamins. For example, chronic *hypervitaminosis* A can occur if large amounts of vitamin A— over 10 times the U.S. Recommended Daily Allowance (RDA)— are consumed daily over a period of 3 months or more. This condition first manifests itself with dry skin, hair loss, anorexia (appetite loss), and vomiting. However, it may progress to severe headaches and mental disturbances, liver enlargement, and occasionally cirrhosis. Acute hypervitaminosis A, characterized by vomiting, abdominal pain, and headache, can occur if a massive overdose is ingested. Excesses of the fat-soluble vitamins (A, D, E, and K) are generally more serious than excesses of the water-soluble vitamins (B complex and C).

CHAPTER SUMMARY

OVERVIEW OF NUTRITION AND METABOLISM

A. Nutrition refers to the food we eat— involves three basic nutrient types: carbohydrates, fats, and proteins plus vitamins and minerals
  1. Malnutrition— a deficiency in the consumption of food, vitamins, and minerals
B. Metabolism— the use of nutrients— a process made up of two major processes
  1. Catabolism breaks food down into smaller molecular compounds and releases two forms of energy— heat and chemical energy
  2. Anabolism— a synthesis process
  3. Both processes take place inside of cells continuously and concurrently
  4. Chemical energy released by catabolism must be transferred to high-energy bonds of ATP, which supply energy directly to the energy-using reactions of all cells (Figure 26-1)

CARBOHYDRATES

A. Dietary sources of carbohydrates
  1. Complex carbohydrates
    a. Polysaccharides— starches; are found in vegetables and grains; glycogen is found in meat
    b. Cellulose— a component of most plant tissue; passes through the system without being broken down
    c. Disaccharides— found in refined sugar; must be broken down before they can be absorbed
    d. Monosaccharides— found in fruits; move directly into the internal environment without being processed directly
      (1) Glucose— carbohydrate most useful to the human cell
B. Carbohydrate metabolism— human cells catabolize most of the carbohydrate absorbed and anabolize a small portion of it
  1. Glucose transport and phosphorylation— glucose reacts with ATP to form glucose-6-phosphate; this step prepares glucose for further metabolic reactions
    a. This step is irreversible except in the intestinal mucosa, liver, and kidney tubules
  2. Glycolysis— the first process of carbohydrate

catabolism; consists of a series of chemical reactions (Figure 26-3)
    a. Glycolysis occurs in the cytoplasm of all human cells
    b. An anaerobic process— the only process that provides cells with energy under conditions of inadequate oxygen
    c. It breaks down chemical bonds in glucose molecules and releases about 5% of the energy stored in them
    d. It prepares glucose for the second step in catabolism— the citric acid cycle
  3. Citric acid cycle converts two pyruvic acid molecules to six carbon dioxide and six water molecules (Figure 26-5)
    a. It is also called the tricarboxylic acid (TCA) cycle because citric acid is also called tricarboxylic acid
    b. It was once called Krebs cycle after Sir Hans Krebs who discovered this process
  4. Electron transport system (Figure 26-6)— high-energy electrons removed during the citric acid cycle enter a chain of molecules that are embedded in the inner membrane of the mitochondria; as electrons move down the chain, they release small bursts of energy in order to pump protons between the inner and outer membrane of the mitochondrion
  5. Oxidative phosphorylation— the joining of a phosphate group to ADP to form ATP
  6. The anaerobic pathway— a pathway for the catabolism of glucose; transfers energy to ATP using only glycolysis; ultimately ends with the oxidative phosphorylation of ATP (paying the "oxygen debt")
  7. Glycogenesis— a series of chemical reactions in which glucose molecules are joined to form a strand of glucose beads; a process that operates when the blood glucose level increases above the midpoint of its normal range
  8. Glycogenolysis (Figure 26-12)— the reversal of glycogenesis, it means different things in different cells
  9. Gluconeogenesis (Figure 26-13)— the formation of new glucose, which occurs chiefly in the liver
  10. Control of glucose metabolism— hormonal and neural devices maintain homeostasis of blood glucose concentration (Figure 26-15)

a. Insulin- secreted by beta cells to decrease blood glucose level (Figure 26-14)
b. Glucogen increases the blood glucose level by increasing the activity of the enzyme phosphorylase
c. Epinephrine—hormone secreted in times of stress; increases phosphorylase activity
d. Adrenocorticotropic hormone stimulates the adrenal cortex to increase its secretion of glucocorticoids
e. Glucocorticoids accelerate gluconeogenesis
f. Growth hormone increases blood glucose level by shifting from carbohydrate to fat catabolism
g. Thyroid-stimulating hormone has complex effects on metabolism
11. Hormones that cause the blood glucose level to rise are called hyperglycemic
12. Insulin is hypoglycemic because it causes the blood glucose level to decrease

LIPIDS

A. Dietary sources of lipids
1. Triglycerides—the most common lipids—composed of a glycerol subunit that is attached to three fatty acids
2. Phospholipids—an important lipid found in all foods
3. Cholesterol—an important lipid found only in animal foods
4. Dietary fats
a. Saturated fats contain fatty acid chains in which there are no double bonds
b. Unsaturated fats contain fatty acid chains in which there are some double bonds
B. Transport of lipids—they are transported in blood as chylomicrons, lipoproteins, and fatty acids
1. In the absorptive state, many chylomicrons are present in the blood
2. Postabsorptive state—95% of lipids are in the form of lipoproteins
a. Lipoproteins consist of lipids and protein and are formed in the liver
(1) Blood contains three types of lipoproteins: very low density, low density and high density
b. Fatty acids are transported from the cells of one tissue to the cells of another in the form of free fatty acids
C. Lipid metabolism
1. Lipid catabolism—triglycerides are hydrolyzed to yield fatty acids and glycerol; glycerol is converted to glyceraldehyde-3-phosphate, which enters the glycolysis pathway; fatty acids are broken down by beta-oxidation and are then catabolized through the citric acid cycle (Figure 26-16)
2. Lipid anabolism consists of the synthesis of triglycerides, cholesterol, phospholipids, and prostaglandins
3. Control of lipid metabolism is through the following hormones
a. Insulin
b. Growth hormone
c. ACTH
d. Glucocorticoids

PROTEINS

A. Sources of proteins
1. Proteins are assembled from a pool of 20 different amino acids
2. The body synthesizes amino acids from other compounds in the body
3. Only about half of the necessary types of amino acids can be produced by the body; the rest are supplied through diet—they are found in both meat and vegetables

B. Protein metabolism—anabolism is primary and catabolism is secondary
1. Protein anabolism—the process by which proteins are synthesized by the ribosomes of the cells (Figure 26-17)
2. Protein catabolism—deamination takes place in the liver cells and forms an ammonia molecule, which is converted to urea and excreted in urine, and a keto acid molecule, which is oxidized or converted to glucose or fat (Figure 26-18)
3. Protein balance-the rate of protein anabolism balances the rate of protein catabolism
4. Nitrogen balance—the amount of nitrogen taken in equates the nitrogen in protein catabolic waste
5. Two kinds of protein or nitrogen imbalance
a. Negative nitrogen balance—protein catabolism exceeds protein anabolism; more tissue proteins are catabolized than are replaced by protein synthesis
b. Positive nitrogen balance—protein anabolism exceeds protein catabolism
6. Control of protein metabolism—achieved by hormones

VITAMINS AND MINERALS

A. Vitamins (Table 26-3)—organic molecules necessary for normal metabolism; they attach to enzymes and help them work
1. The body does not make most of the necessary vitamins; they must be obtained through diet
a. The body stores fat-soluble vitamins and does not store water soluble vitamins
B. Minerals (Table 26-4)—inorganic elements or salts found in the earth; they attach to enzymes and help them work and function in chemical reactions
1. Large amounts of some minerals may be toxic

METABOLIC RATES

A. Metabolic rate means the amount of energy released by catabolism
B. Metabolic rates are expressed in two ways
1. The number of kilocalories of heat energy expended per hour or per day
2. As normal or as a percentage above or below normal
C. Basal metabolic rate—the rate of energy expended under basal conditions
1. Factors: size, sex, age, thyroid hormone, body temperature, drugs, other factors
D. Total metabolic rate (Figure 26-20)—the amount of energy used in a given time
1. Main determinates
a. The basal metabolic rate
b. The energy used to do skeletal muscle work
c. The thermic effect of foods
E. Energy balance and weight—the body maintains a state of energy balance
1. The body maintains a weight when the total calories in the food ingested equals the total metabolic rate
2. Body weight increases when energy input exceeds energy output
3. Body weight decreases when energy output exceeds energy input

MECHANISMS FOR REGULATING FOOD INTAKE (FIGURE 26-21)

A. The hypothalamus plays a part in food intake
B. Appetite center—cluster of neurons in the lateral hypothalamus that if stimulated brings about increased appetite
C. Satiety center—a group of neurons in the ventral medial nucleus of the hypothalamus that if stimulated brings about decreased appetite

CYCLE OF LIFE: NUTRITION/METABOLISM

A. Fetus—obtains nutrients from mother's blood
B. Critical nutrients from fetal stage through childhood
  1. Proteins required for development of nervous system, muscle tissues, and other vital structures
  2. Calcium required for development of skeleton and other tissues
  3. Lack of necessary nutrients may result in permanent structural and functional problems
C. Late adulthood—metabolic rate declines; increased nutrients may be necessary to alleviate age-related conditions

THE BIG PICTURE: NUTRITION, METABOLISM, AND THE WHOLE BODY

A. Every cell in the body needs the maintenance of the metabolic pathways in order to stay alive
B. Anabolic pathways build the various structural and functional components of the cells
C. Catabolic pathways convert energy to a usable form and degrade large molecules into subunits used in anabolic pathways
D. Cells require appropriate amounts of vitamins and minerals to produce structural and functional components necessary for cellular metabolism
E. Other body mechanisms operate to ensure that nutrients reach the cells

MECHANISMS OF DISEASE: METABOLIC AND NUTRITIONAL DISORDERS

A. Inborn errors of metabolism—group of genetic conditions involving a deficiency of a particular enzyme
B. Eating disorders
  1. Anorexia—disease that primarily affects young adults (individuals have disturbed body image, fear of obesity, and diet to excess, thus literally starving themselves); characterized by a loss of 20% to 25% total body mass with impaired intellectual functioning; treatment involves both medical and psychological intervention
  2. Bulimia—illness characterized by an eating and purging cycle; treatment involves both nutritional and psychological counseling and, sometimes, antidepressants
  3. Obesity—a symptom of chronic overeating, it is defined as an abnormal increase in the fat proportion of the body; can also result from metabolic disorders
C. Nutritional disorders
  1. Protein-calorie malnutrition—a condition that results from a deficiency of calories in general and protein in particular; it is likely to occur in parts of the world where protein-rich food is unavailable; two of the types are:
    a. Marasmus results from an overall lack of calories and proteins and is characterized by progressive wasting of muscle and subcutaneous tissue and fluid and electrolyte imbalances
    b. Kwashiorkor results from a protein deficiency in the presence of sufficient calories; it is characterized by wasting tissue, ascites, and flaking dermatitis
  2. Vitamin disorders
    a. Vitamin deficiency (avitaminosis) can lead to severe metabolic problems such as scurvy
    b. Vitamin excess (hypervitaminosis) is also serious; excesses of fat-soluble vitamins (A, D, E, and K) are generally more serious than that of water-soluble vitamins (B complex and C)

REVIEW QUESTIONS

1. What is metabolism? Nutrition?
2. What two processes make up the process of metabolism?
3. Does the body digest dietary fiber? Why or why not?
4. Describe briefly glycolysis, the first process of carbohydrate catabolism.
5. Where does glycolysis occur?
6. Describe briefly the process of carbohydrate catabolism known as the citric acid cycle.
7. Explain why mitochondria are called "power plants" of cells.
8. Differentiate between anaerobic and aerobic respiration.
9. Describe the process of "splitting glycogen."
10. Explain the processes and hormones involved in maintaining homeostasis of blood glucose concentration.
11. How are dietary fats classified?
12. Explain how lipids are transported in blood.
13. Describe lipid catabolism briefly.
14. List the hormones involved in the control of lipid metabolism.
15. What are the essential amino acids?
16. Describe protein catabolism briefly.
17. Compare the functions proteins, carbohydrates, and fats serve in the body.
18. Why are vitamins important to the body?
19. What does the term *metabolic rate* mean?
20. Differentiate between basal and total metabolic rates.
21. List the various factors that influence basal metabolic rate.
22. Identify a quick way to estimate a young woman's basal metabolic rate.
23. What happens to body weight when energy input exceeds energy output? Why?
24. Describe various factors that influence the amount of food a person eats.
25. Define the term *calorie*.
26. High blood concentrations of low-density lipoproteins may lead to what disease state? Why?

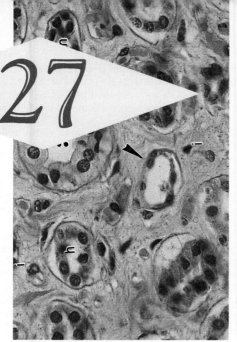

Tissue of medulla portion of kidney
Erlandsen/Magney: Color Atlas of Histology

27 The Urinary System

OBJECTIVES

**After you have completed this chapter, you should
be able to:**

1. List the major organs of the urinary system.
2. Discuss and compare the structure and the
   functions of the ureters, urinary bladder, and
   urethra.
3. Locate or position the kidneys in the abdominal
   cavity and identify the gross internal structures
   visible in a coronal section.
4. Name the parts of a nephron and describe the
   role of each component in the formation of urine.
5. Describe the renal blood supply and trace blood
   flow through the specialized vessels of the kidney.
6. Discuss how the kidneys form urine and trace
   urine from its point of formation to the exterior of
   the body.
7. Discuss filtration, reabsorption, and secretion in
   relation to the formation of urine by the kidneys.
8. Discuss the countercurrent mechanisms for
   concentrating or diluting urine.
9. Explain how urine volume is regulated under
   normal conditions.
10. Describe the physical characteristics of normal
    urine.

OVERVIEW OF THE URINARY SYSTEM

The principal organs of the urinary system are the **kidneys,** which process blood and form **urine** as a waste to be excreted (removed from the body). The excreted urine travels from the kidney to the outside of the body via accessory organs: **ureters, urinary bladder,** and **urethra.**

We often think of the urinary system primarily as a "urine producer," which it certainly is. However, a better image of the system is that of "blood plasma balancer." Each kidney processes incoming blood plasma in ways that allow it to leave the kidney in better condition. The water content is adjusted so that the body does not have too much or too little water to maintain constancy of the internal environment. Likewise, the blood content of important ions such as sodium and potassium is adjusted to match set point levels. Even the pH of the blood can be altered to match the set point level. In these ways, the urinary system regulates the content of blood plasma so that the homeostasis of the entire internal fluid environment can be maintained within normal limits.

Excretion

The urinary system's chief function is to regulate the volume and composition of body fluids and excrete unwanted materials, but it is not the only system in the body that is able to excrete unneeded substances.

The table compares the excretory functions of several systems. Although all of these systems contribute to the body's effort to remove wastes, only the urinary system can finely adjust the water and electrolyte balance to the degree required for normal homeostasis of body fluids.

System	Organ	Excretion
Urinary	Kidney	Nitrogen compounds Toxins Water Electrolytes
Integumentary	Skin— sweat glands	Nitrogen compounds Water Electrolytes
Respiratory	Lung	Carbon dioxide Water
Digestive	Intestine	Digestive wastes Bile pigments Salts of heavy metals

ANATOMY OF THE URINARY SYSTEM

Gross Structure

Kidney

The kidneys resemble lima beans in shape, that is, roughly oval with a medial indentation (Figure 27-1, A). An average-sized kidney measures approximately 11 cm by 7 cm by 3 cm. The left kidney is often slightly larger than the right. The kidneys lie in a *retroperitoneal* position, meaning posterior to the parietal peritoneum, against the posterior wall of the abdomen (Figure 27-1, B). They are located on either side of the vertebral column and extend from the level of the last thoracic vertebra (T12) to the third lumbar vertebra (L3). Usually the right kidney is a little lower than the left, presumably because the liver takes up some of the space above the right kidney. A heavy cushion of fat normally encases each kidney and holds it in position. Very thin individuals (those with little body fat) may suffer from *ptosis* (dropping) of one or both of these organs. Connective tissue, the renal fasciae, anchors the kidneys to surrounding structures and also helps maintain their normal positions.

The medial surface of each kidney has a concave notch called the **hilum.** Structures enter or leave the kidney

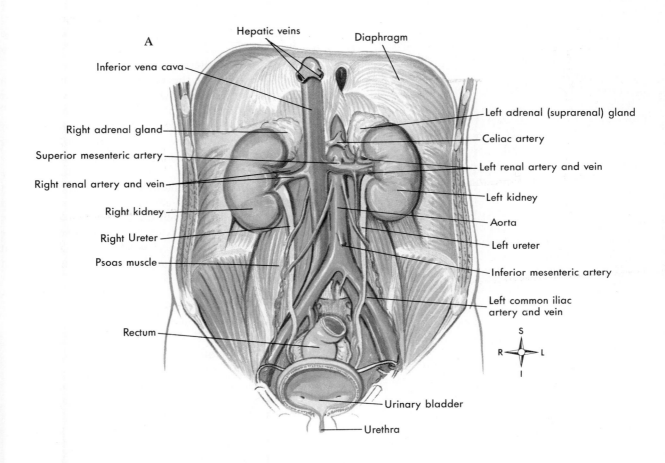

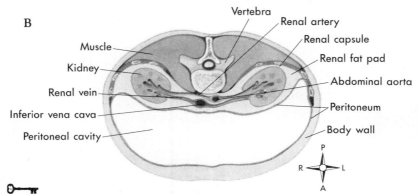

FIGURE 27-1   Location of urinary system organs. A, Anterior view of urinary organs with the peritoneum and visceral organs removed. **B,** Horizontal (transverse) section of the abdomen showing the retroperitoneal position of the kidneys.

through this notch. A tough, white, fibrous capsule encases each kidney (Figure 27-2).

The coronal section of the right kidney shown in Figure 27-2 portrays the major internal structures of the kidney. Identify the **cortex,** or outer region, and the **medulla,** or inner region. A dozen or so distinct triangular wedges, the **renal pyramids,** make up much of the medullary tissue. The *base* of each pyramid faces outward, and the narrow *papilla* of each faces toward the hilum. Notice that the cortical tissue dips into the medulla between the pyramids, forming areas known as **renal columns.**

Each renal papilla (point of a pyramid) juts into a cuplike structure called a **calyx.** The calyces are considered the beginnings of the "plumbing system" of the urinary system, for it is here that urine leaving the renal papilla is collected for transport out of the body. The calyces join together to form a large collection reservoir called the **renal pelvis.** The pelvis of the kidney narrows as it exits the hilum to become the ureter.

Ureter

The **ureter** is an approximately 28 cm long tube composed of three layers of tissue: a mucous lining, a muscular middle layer, and a fibrous outer layer. The muscular layer is composed of smooth muscle, which propels the urine by peristalsis. The ureter of each kidney conducts urine inferiorly from the kidney to the **urinary bladder** below. The entry to the bladder is guarded by a valvelike narrow region that prevents backflow from the bladder toward the kidneys.

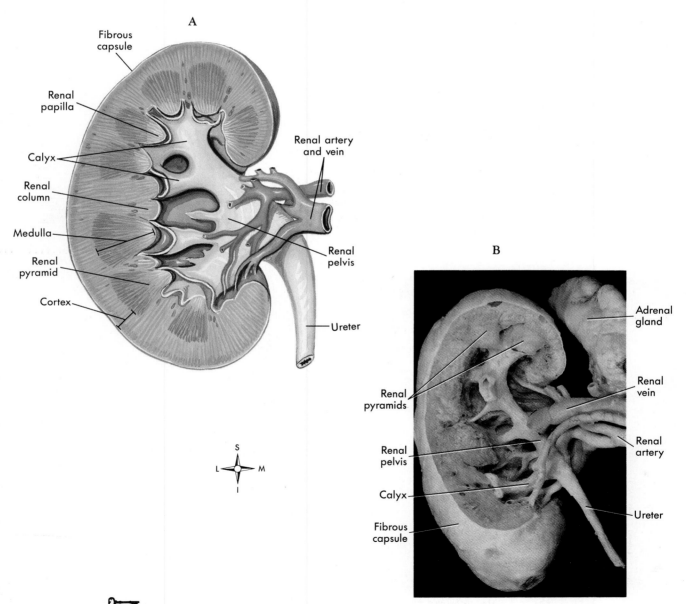

FIGURE 27-2   Internal structure of the kidney. A, Coronal section of the right kidney in an artist's rendering and in a photo, **B,** of a preserved human kidney.

Urinary bladder

The urinary bladder is a collapsible bag that is located directly behind the symphysis pubis (Figure 27-3, *B*). It lies below the parietal peritoneum, which covers only its superior surface.

The wall of the bladder is made mostly of smooth muscle tissue. Often called the **detrusor muscle,** the muscle layer is formed by a network of crisscrossing bundles of smooth muscle fibers. The bundles run in all directions: circular, oblique, and lengthwise. The bladder is lined with mucous transitional epithelium that forms folds called *rugae* (Figure 27-3, *A*). Because of the folds and the extensibility of transitional epithelium, the bladder can distend considerably. There are three openings in the floor of the bladder—two from the ureters and one into the **urethra.** The ureter openings lie at the posterior corners of the triangle-shaped floor (the *trigone*), and the urethral opening at the anterior, lower corner.

The bladder performs two major functions:

1. It serves as a reservoir for urine before it leaves the body.
2. Aided by the urethra, it expels urine from the body.

The mechanism for voiding urine begins with the voluntary relaxation of the external sphincter muscle of the bladder. In rapid succession, different regions of the detrusor muscle in the bladder wall contract reflexively. This forces the urine out of the bladder and through the urethra. Parasympathetic fibers transmit the impulses that cause contractions of the bladder and relaxation of the internal sphincter. Voluntary contraction of the external sphincter to stop voiding is learned. Voluntary control of *micturition* (voiding, or urination) is possible only if the nerves supplying the bladder and urethra, the projection tracts of the CNS, and the motor areas of the brain are all intact and functioning properly. Injury to any of these parts

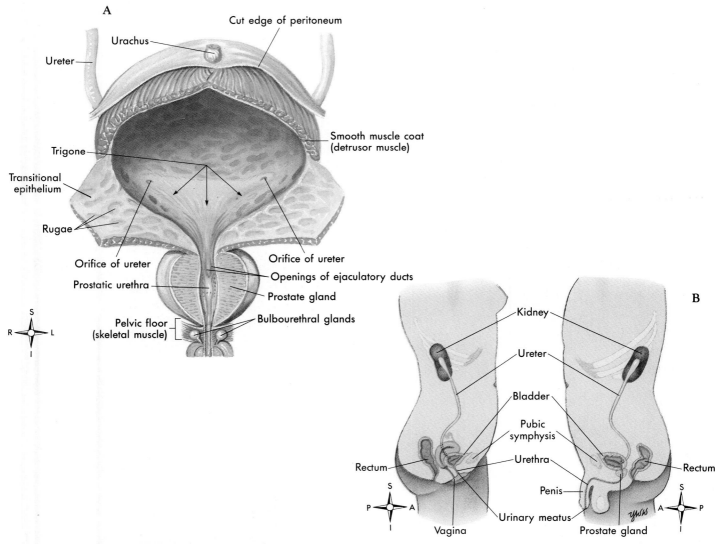

FIGURE 27-3   Structure and location of the urinary bladder. A, Frontal view of a dissected urinary bladder (male) in a fully distended position. **B,** Sagittal section of the female urinary system (*left*) and male urinary system (*right*), each showing a partially distended bladder.

URINATION DIFFICULTIES

Occasionally, an individual is unable to void even though the bladder contains an excessive amount of urine. This condition is known as *retention*. It often follows pelvic operations and childbirth. *Catheterization* (introduction of a sterile flexible tube through the urethra into the bladder) is used to relieve the discomfort accompanying retention. A more serious complication, which is also characterized by the inability to void, is called *suppression*. In this condition, the patient cannot void because the kidneys are not excreting any urine, and therefore the bladder is empty. Catheterization, of course, gives no relief for this condition.

of the nervous system, by a cerebral hemorrhage or spinal cord injury, for example, results in involuntary emptying of the bladder at intervals. Involuntary micturition is called *incontinence*. In the average bladder, 250 ml of urine will cause a moderately distended sensation and therefore the desire to void.

Urethra

The urethra is a small tube lined with mucous membrane leading from the floor of the bladder (*trigone*) to the exterior of the body. In the female, it lies directly behind the symphysis pubis and anterior to the vagina (Figure 27-3, B). It extends down and forward from the bladder for a distance of about 3 cm. The male urethra, on the other hand, extends along a winding path for about 20 cm (Figure 27-3, B). The male urethra passes through the center of the *prostate gland* just after leaving the bladder. Within the prostate, it is joined by two *ejaculatory ducts*. After leaving the prostate, the urethra extends down, forward, then up to

enter the base of the penis. It then travels through the center of the penis and ends as a *urinary meatus* at the tip of the penis.

Because the male urethra is joined by the ejaculatory ducts, it serves as a pathway for *semen* (fluid containing sperm) as it is ejaculated out of the body through the penis. Thus we can say that the male urethra is a part of two different systems: the urinary system (when it is used to void urine) and the reproductive system (when it is used to ejaculate semen). Urine is prevented from mixing with semen during ejaculation by a reflex closure of sphincter muscles guarding the bladder's opening. The female urethral tract is separate from the lower reproductive tract (vagina), which lies just behind the urethra (see Figure 27-3, B).

URETHRAL STRICTURE

Urethral stricture is narrowing or blockage of the urethra, which usually results in retention of urine. A common cause of urethral stricture is chronic infection of the urethra. As one after another infection occurs, the resulting damage to tissues may lead to the formation of fibrous scar tissue. These fibrous masses may eventually block the urethra.

QUICK CHECK

1. *Name the accessory organs of the urinary system.*
2. *What is the general function of the urinary system?*
3. *Distinguish between the renal cortex and the renal medulla.*
4. *What are the two major functions of the bladder?*

URODYNAMIC TESTING

Urodynamics is a term that literally means "pertaining to the active force of urine." Urodynamic testing, then, measures the force of urine flow in the urinary tract. Various specific methods of urodynamic testing have been developed, each measuring the ability of the urinary system to conduct, store, and void urine normally. One type of urodynamic test, called *uroflowmetry*, measures the rate of urine flow as the subject urinates into a funnel containing a flow meter. Uroflowmetry is useful in screening for obstructions in the urethra or weakness in the detrusor muscle. *Urinary electromyography* (EMG) can also be used in urodynamic testing. Urinary electromyography is a variation of the same voltage-recording technique used in electrocardiography (ECG). Urinary EMGs graphically record the electrical activity of the muscles used in voiding urine. Several urodynamic tests make use of special radiological imaging techniques. For example, the figure shows a special type of radiographic (x-ray) image in which the urinary tract has been filled with a contrast material that absorbs x rays. Such a radiograph may show where obstructions or muscle abnormalities are located.

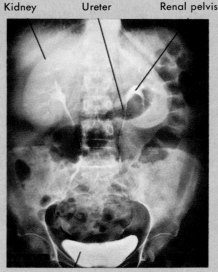

Kidney   Ureter   Renal pelvis

Urinary bladder

Microscopic Structure

Microscopic functional units, named **nephrons** and numbering about 1.25 million per kidney, make up the bulk of the kidney. The shape of the nephron is unusual, unmistakable, and uniquely suited to its function of blood plasma processing and urine formation (Figure 27-4). It resembles a tiny funnel with a long, winding stem.

As Figure 27-4 shows, each nephron contains these structures, in the order in which fluid flows through them:

♦ Bowman's capsule
♦ Proximal convoluted tubule
♦ Loop of Henle
♦ Distal convoluted tubule
♦ Collecting duct

As you read the brief description of each of these nephron components, refer often to Figure 27-4, B, which shows a schematic diagram of the nephron.

Bowman's capsule

The **Bowman's capsule** is the cup-shaped mouth of a nephron. It is formed by two layers of epithelial cells with a space between them. The parietal (outer) wall is composed of simple squamous epithelium. The visceral (inner) wall, however, is quite different. It is composed of special epithelial cells called *podocytes* (meaning "cells with feet"). The scanning electron micrograph in Figure 27-5 reveals the odd shapes of the podocytes. Notice that the primary branches extending from the cell bodies divide into a network of branches that terminate in little "feet" called *pedicels*. The pedicels are packed so closely together that only narrow slits of space lie between them. These spaces are called *filtration slits*. The slits are not merely open spaces, however. There is a mesh of fine connective tissue fibers called the *slit diaphragm* that prevents the slits from enlarging under pressure while still maintaining the permeability of the slit.

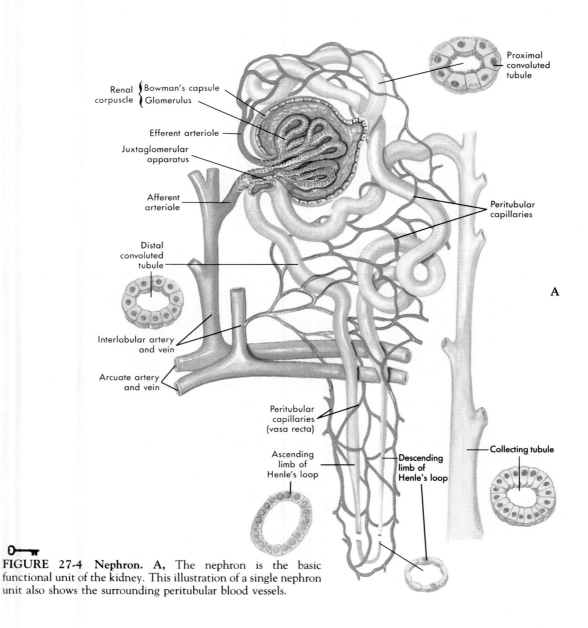

FIGURE 27-4 Nephron. A, The nephron is the basic functional unit of the kidney. This illustration of a single nephron unit also shows the surrounding peritubular blood vessels.

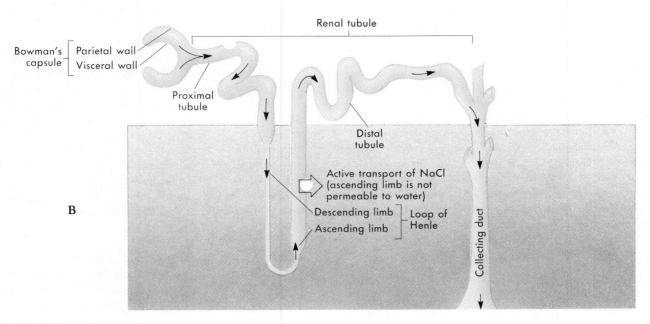

FIGURE 27-4, cont'd
B, Schematic diagram showing the essential components of a nephron. Arrows indicate direction of flow of fluid within the tubule.

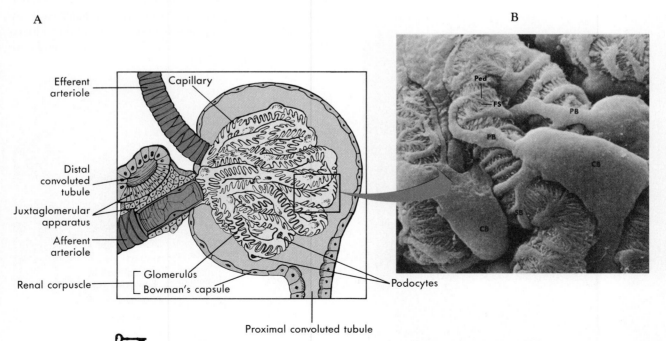

FIGURE 27-5   Renal corpuscle. The renal corpuscle is made up of Bowman's capsule and the glomerulus within it. **A,** Diagram showing renal corpuscle opened to show the visceral wall of Bowman's capsule that covers vessels of the glomerulus. **B,** Scanning electron micrograph shows podocytes forming filtration slits *(FS)*. Extending from each cell body *(CB)* are primary branches *(PB)* that divide to form secondary branches *(SB)*, which continue branching until small pedicels *(Ped)* are formed. The filtration slits are gaps between adjacent pedicels.

Look again at Figure 27-4, A. Pay particular attention to the structure fitted neatly into the Bowman's capsule. It is a network of fine capillaries that has a special name—**glomerulus.** The glomerulus is probably the body's most famous capillary network and is surely one of its most important ones for survival. A glomerulus and its Bowman's capsule together are called a **renal corpuscle** (see Figures 27-4 and 27-5). Renal corpuscles lie in the cortex of the kidney. Like all capillaries, glomeruli have thin, membranous walls that are composed of a single layer of endothelial cells. Many pores, or *fenestrations* (meaning "windows"), are present in the glomerular endothelium (Figure 27-6).

Between a glomerulus and its Bowman's capsule lies a *basement membrane* (basal lamina). It consists of a thin layer of fine fibrils imbedded in a matrix of glycoprotein. The visceral layer of Bowman's capsule contacts the basement membrane by means of countless pedicels ("feet" of the podocytes). The glomerular endothelium, the basement membrane, and the visceral layer of Bowman's capsule constitute the **glomerular-capsular membrane,** a structure well suited to its function of filtration (Figure 27-7).

Proximal tubule

The **proximal tubule** is the second part of the nephron but the first part of the *renal tubule.* As its name suggests, the proximal tubule is the segment proximal, or nearest, to the Bowman's capsule. Because it follows a winding, convoluted course it is also called the *proximal convoluted tubule.* Its wall consists of one layer of epithelial cells. These cells have a brush border facing the lumen of the tubule (Figure 27-4, A). Thousands of microvilli form the brush border and greatly increase its luminal surface area—a structural fact of importance to its function, as we shall see. Like renal corpuscles, proximal tubules are located in the cortex of the kidney (see Figure 27-8).

Loop of Henle

The **loop of Henle** is the segment of renal tubule just beyond the proximal tubule. It consists of a descending limb, a sharp turn, and an ascending limb. Note in Figure 27-8 that the thin segment of the descending limb loops around to become the thick ascending limb. Figure 27-8 shows a nephron in which the loop of Henle dips deep into the medulla. A nephron with a loop of Henle that dips far into the medulla is called a *juxtamedullary nephron* (meaning "nephron near the medulla"). Nephrons with loops of Henle that do not dip into the medulla, instead remaining within the cortex, are called *cortical nephrons.*

Distal tubule

The **distal tubule,** or *distal convoluted tubule,* is a convoluted portion of the tubule beyond (distal to) the loop of Henle. Notice in Figure 27-8 that it has a rather thick wall much like that of the thick ascending limb of the loop of Henle.

Collecting duct

The **collecting duct** is a straight tubule joined by the distal tubules of several nephrons. Collecting ducts join larger ducts, and all the larger collecting ducts of one renal pyramid converge to form one tube that opens at a renal papilla into one of the small calyces. Bowman's capsules and both convoluted tubules lie in the cortex of the kidney, whereas the loops of Henle and collecting ducts extend into the medulla (Figure 27-8).

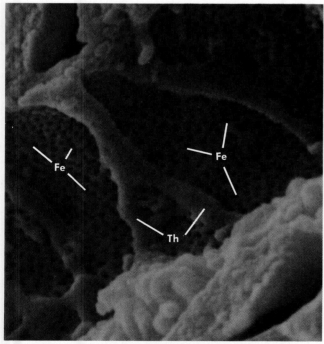

FIGURE 27-6 Fenestrations in the wall of the glomerulus. Scanning electron micrograph showing numerous fenestrations *(Fe)* in endothelium of a glomerular capillary. Thickened branches *(Th)* of the endothelial cells contain no fenestrations. (x17,525.)

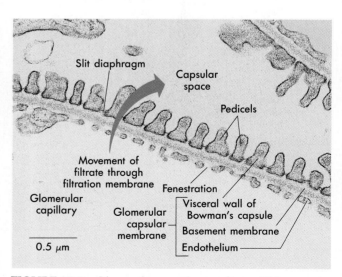

FIGURE 27-7 Glomerular-capsular membrane. This transmission electron micrograph shows a cross sectional view of filtration membrane that is formed when footlike extensions (pedicels) of cells forming the visceral wall of Bowman's capsule share a basement membrane with the fenestrated endothelial cells that form the wall of glomerular capillaries.

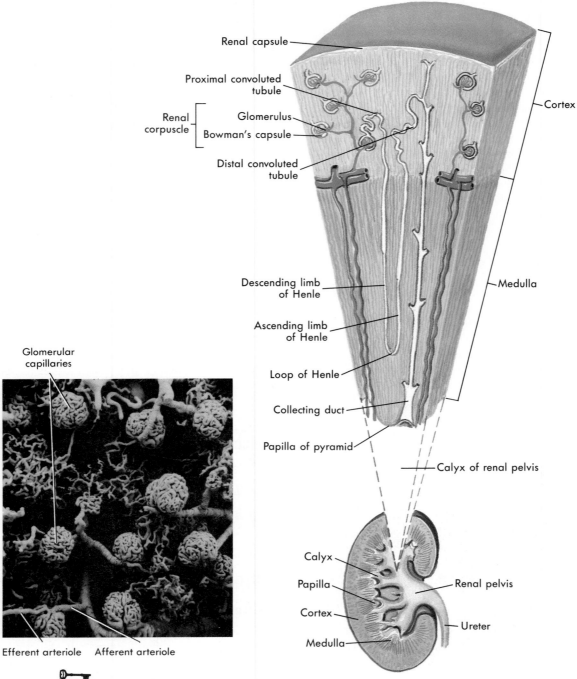

Renal capsule

Proximal convoluted tubule

Renal corpuscle

Glomerulus

Bowman's capsule

Distal convoluted tubule

Cortex

Descending limb of Henle

Ascending limb of Henle

Loop of Henle

Collecting duct

Papilla of pyramid

Medulla

Calyx of renal pelvis

Glomerular capillaries

Calyx

Papilla

Cortex

Medulla

Renal pelvis

Ureter

Efferent arteriole   Afferent arteriole

FIGURE 27-8 Location of the nephron. Magnified wedge shows that renal corpuscles (Bowman's capsules with invaginated glomeruli) and both proximal and distal convoluted tubules are located in cortex of kidney. Medulla contains loops of Henle and collecting tubules. The inset shows a scanning electron micrograph of several glomeruli and their associated blood vessels. Note that the blood vessel that brings blood to the glomerulus (afferent arteriole) has a larger diameter than the blood vessel that drains blood from it (efferent arteriole).

Blood Vessels of the Kidneys

The kidneys are highly vascular organs (Figure 27-9). Every minute about 1,200 ml of blood flows through them. Stated another way, approximately one fifth of all the blood pumped by the heart per minute goes to the kidneys. From this fact one might guess, and correctly so, that the kidneys process the blood in some way before returning it to the general circulation. A large branch of the abdominal aorta—the *renal artery*—brings blood into each kidney. Between the pyramids of the kidney's medulla, the renal artery branches to form *interlobar arteries* that extend out toward the cortex, then arch over the bases of the pyramids to form the *arcuate arteries*. From the arcuate arteries, *interlobular arteries* penetrate the cortex.

Branches of the interlobular arteries are the afferent arterioles shown in Figure 27-4. As you can see in that figure, afferent arterioles branch into the capillary networks called glomeruli. Recall that the usual direction of blood flow is:

Arteries
↓
Arterioles
↓
Capillaries
↓
Venules
↓
Veins

This is not entirely true for the kidney, however. Here, blood leaving the glomerular capillaries flows into efferent arterioles, not into venules. From the efferent arterioles, blood moves into the **peritubular capillaries** (meaning "capillaries around the tubules"). One portion of the peritubular circulation, the *vasa recta*, loops down into the medulla and back to supply the loop of Henle and collecting ducts. The blood then returns to veins via venules. The flow of blood through the kidney tissue follows this pattern:

Interlobular artery
↓
Afferent arteriole
↓
Glomerulus
↓
Efferent arteriole
↓
Peritubular capillaries (including vasa recta)
↓
Venules
↓
Interlobular vein

The *juxtaglomerular apparatus* (meaning "structure near the glomerulus") is found at the point where the afferent arteriole brushes past the distal tubule (see Figures 27-4 and 27-5). This structure is important in maintaining homeostasis of blood flow because it reflexively secretes *renin* when blood pressure in the afferent arteriole drops. Recall from Chapter 18 that renin triggers a mechanism that produces *angiotensin*, a substance that causes vasoconstriction and the resulting increase in blood pressure (see Figure 27-19).

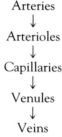
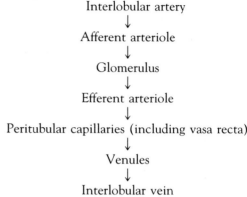

FIGURE 27-9 Circulation of blood through the kidney. This illustration shows the major arteries and veins of the renal circulation. (Refer to Figure 27-4 for detail of circulation to the individual nephron.)

Interlobular arteries
Interlobular veins
Arcuate arteries and veins
Interlobar arteries and veins
Renal artery
Renal vein
Ureter
Renal pyramid

QUICK CHECK

✔ 1. *Name the segments of the nephron in the order in which fluid flows through them.*
✔ 2. *What characteristics of the glomerular-capsular membrane permit filtration?*
✔ 3. *What is the name of the capillary network that surrounds the nephron?*

FUNCTION OF THE KIDNEY

Overview of Kidney Function

The chief functions of the kidney are to process blood plasma and excrete urine, life-preserving functions because homeostasis depends on it. For example, the kidneys are the most important organs in the body for maintaining fluid-electrolyte and acid-base balances. They do this by varying the amount of water and electrolytes leaving the blood in the urine so that they equal the amounts of these substances entering the blood from various other avenues. Nitrogenous wastes from protein metabolism, notably *urea*, leave the blood by way of the kidneys.

Here are just a few of the blood constituents that cannot be held within their normal concentration ranges if the kidneys fail:

- ◆ Sodium
- ◆ Potassium
- ◆ Chloride
- ◆ Nitrogenous wastes (especially urea)

In short, kidney failure means homeostatic failure and, if not relieved, inevitable death.

In addition to processing blood plasma and forming urine, the kidneys also perform other important functions. They influence the rate of secretion of the hormones ADH (antidiuretic hormone) and aldosterone and synthesize the hormone *erythropoietin*, the active form of vitamin D, and certain prostaglandins.

As you already know, the basic functional unit of the kidney is the nephron. It has two main parts—renal corpuscle and renal tubule—which form urine by means of three processes:

1. **Filtration,** or the movement of water and solutes from the plasma in the glomerulus, across the glomerular-capsular membrane, and into the capsular space of the Bowman's capsule.

2. **Reabsorption,** or the movement of molecules out of the tubule and into the peritubular blood.

3. **Secretion,** or the movement of molecules out of the peritubular blood and into the tubule for excretion.

These three mechanisms are used in concert to process blood plasma and form urine. First, a hydrostatic pressure gradient drives the filtration of much of the plasma into the nephron (Figure 27-10). Since the filtrate contains materials that the body must conserve (save), the walls of the tubules start reabsorbing those materials back into the blood. As the filtrate (urine) begins to leave the nephron,

the kidney may secrete a few "last minute" items into the urine for excretion. In short, the kidney does not selectively filter out only harmful or excess material. It first filters out much of the plasma, then reabsorbs what should not be "thrown out" before the filtrate reaches the end of the tubule and becomes urine. This mechanism allows very fine adjustments to blood homeostasis, as we shall see. Figure 27-11 shows the amounts of some important molecules that are filtered, then reabsorbed, by the nephron.

Filtration

Filtration, the first step in blood processing, is a physical process that occurs in the kidneys' 2.5 million renal corpuscles (see Figures 27-5 and 27-7). As blood flows through glomerular capillaries, water and small solutes filter out of the blood into the Bowman's capsules. The only blood constituents that do not move out are the blood solids (cells) and most plasma proteins. The filtration takes place through the glomerular-capsular membrane.

Filtration from glomeruli into Bowman's capsules occurs for the same reason filtration from other capillaries into interstitial fluid occurs—because of the existence of a pressure gradient. The main factor establishing the pressure gradient between the blood in the glomeruli and the filtrate in the Bowman's capsule is the hydrostatic pressure of

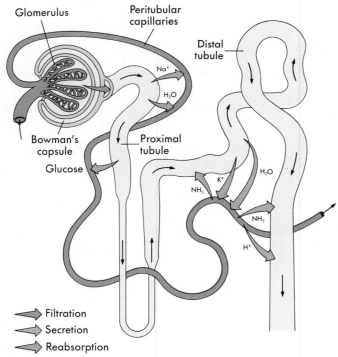

FIGURE 27-10 Mechanism of urine formation. Diagram shows the mechanisms of urine formation and where they occur in the nephron: filtration, reabsorption, and secretion.

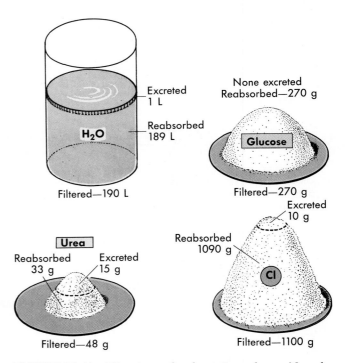

FIGURE 27-11 Filtration and reabsorption volumes. Note the enormous volume of water that is filtered out of glomerular blood per day—190 L, or many times the total volume of blood in the body. Only a small proportion of this, however, is excreted into the urine. More than 99% of it (188 L) is reabsorbed into tubular blood. One other substance shown is also filtered and reabsorbed in large amounts. Which one? Another is normally entirely reabsorbed. Which one? Somewhat more than half of the filtered amount of which substance is reabsorbed?

glomerular blood (Figure 27-12). It tends to cause filtration out of the glomerular blood plasma into Bowman's capsules. However, exerting force in the opposite direction are the osmotic pressure of glomerular blood plasma and the hydrostatic pressure of the capsular filtrate. The net or effective filtration pressure (EFP) therefore equals glomerular hydrostatic pressure minus the sum of glomerular osmotic pressure plus capsular hydrostatic pressure (Figure 27-12). For example, assume the following pressures:

- Glomerular hydrostatic pressure = 60 mm Hg
- Glomerular osmotic pressure = 32 mm Hg
- Capsular hydrostatic pressure = 18 mm Hg
- Capsular osmotic pressure = negligent (±0 mm Hg)

The EFP (effective filtration pressure) using these figures equals (60 + 0) − (32 + 18), or 10 mm Hg. An effective filtration pressure of 1 mm Hg, according to some investigators, produces a glomerular filtration rate of 12.5 ml per minute (including both kidneys). With an EFP of 10 mm Hg, the glomerular filtration rate would be 125.0 ml per minute, a normal rate.

Filtration occurs more rapidly out of glomeruli than out of other tissue capillaries. One reason for this is a structural difference between the endothelium of glomeruli and that of tissue capillaries. Glomerular endothelium has many more pores (fenestrations) in it, so it is more permeable than tissue capillary endothelium. Another reason for more rapid glomerular filtration than tissue capillary filtration is that glomerular hydrostatic pressure is higher than tissue capillary pressure. The reason for this, briefly, is that the efferent arteriole has a smaller diameter than the afferent arteriole. Therefore it offers more resistance to blood flow out of the glomerulus than venules offer to blood flow out of tissue capillaries.

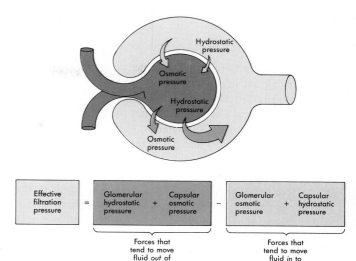

FIGURE 27-12 Forces affecting glomerular filtration. Effective filtration pressure (EFP) is determined by comparing the forces that push fluid into the capillary to those that push it out of the capillary.

The glomerular filtration rate (GFR) can be altered by changes in the diameters of the afferent and efferent arterioles or by changes in the systemic blood pressure. For example, stress may lead to intense sympathetic stimulation of the arterioles with greater constriction of the afferent than of the efferent arteriole. Consequently, glomerular hydrostatic pressure falls. In severe stress, it may even drop to a level so low that the EFP falls to zero. No glomerular

---

FOCUS ON... **CHANGES IN GLOMERULAR FILTRATION RATE**

In some types of kidney disease, the permeability of the glomerular endothelium increases sufficiently to allow plasma proteins to filter out into the capsule. Osmotic pressure then develops in the capsular filtrate. To determine how this would affect the effective filtration pressure in the glomeruli, first calculate the normal EFP using the figures that the table lists as "Normal" in the formula given in Figure 27-12. After you have obtained your answer, apply the formula to the figures given under "Kidney disease" in the table. You should have found an EFP of 10 mm Hg with the normal figures and an EFP of 15 mm Hg with the kidney disease figures. A change in the glomerular effective filtration pressure produces a similar change in the glomerular filtration rate. Therefore, a loss of plasma protein into the urine increases not only the EFP but also the glomerular filtration rate.

Intense exercise causes temporary proteinuria in many individuals. Some exercise physiologists believed that intense athletic activities cause kidney damage, but subsequent research has ruled out that explanation. One current hypothesis is that hormonal changes during strenuous exercise increase the permeability of the nephron's filtration membrane, allowing more plasma proteins to enter the filtrate. Some degree of postexercise proteinuria is usually considered normal.

Normal and abnormal pressures in the renal corpuscle

	Hydrostatic Pressure	Osmotic Pressure
NORMAL		
Glomerular blood	60 mm Hg	32 mm Hg
Capsular filtrate	18 mm Hg	0 mm Hg
KIDNEY DISEASE		
Glomerular blood	60 mm Hg	32 mm Hg
Capsular filtrate	18 mm Hg	5 mm Hg

filtration then occurs. The kidneys "shut down," or, in technical language, renal suppression occurs.

Glomerular hydrostatic pressure and filtration are directly related to systemic blood pressure. That is, a decrease in blood pressure tends to produce a decrease in both the glomerular pressure and the glomerular filtration rate. The converse is also true. However, when arterial pressure increases, a smaller increase in glomerular pressure follows because the afferent arterioles constrict. This decreases blood flow into the glomeruli and prevents a marked rise in glomerular pressure or glomerular filtration. For instance, when the mean arterial blood pressure doubles, glomerular filtration reportedly increases only 15% to 20%.

QUICK CHECK

✔ 1. *What are the three basic processes a nephron uses to form urine?*

✔ 2. *What is the GFR? Why is a high GFR important to kidney function?*

✔ 3. *How does blood pressure affect filtration in the kidney?*

Reabsorption

Reabsorption, the second step in urine formation, takes place by means of both passive and active transport mechanisms from all parts of the renal tubules. A major portion of water and electrolytes and (normally) all nutrients are, however, reabsorbed from proximal tubules. The rest of the renal tubule reabsorbs comparatively little of the filtrate. Researchers are still investigating the exact mechanisms of reabsorption in the various segments of the nephron. We have summarized only the essential principles of some of the current concepts in the following paragraphs.

Reabsorption in the proximal tubule

As already stated, most of the filtrate that enters the renal tubule from the Bowman's capsule does not get very far. More than two thirds of it is reabsorbed before it reaches the end of the proximal tubule.

Proximal tubules are thought to reabsorb sodium and other major ions in this manner: sodium ions (Na^+) are actively transported out of the lumen of the tubule and into the peritubular blood by the mechanism summarized in Figure 27-13. The microvilli on the luminal surface of each epithelial cell in the tubule wall form a brush border that increases the absorptive surface area of the entire inner face of the tubule. As sodium ions accumulate in the interstitial fluid, the interstitial fluid becomes temporarily positive with respect to the tubule fluid. This electrical gradient (difference in net charge) drives the diffusion of negative ions from the filtrate, into the interstitial fluid, and, eventually, into the peritubular blood. In other words, the attraction between negative and positive ions is used to drive the passive transport of chloride (Cl^-), phosphate ($PO_4^{\equiv}$) and other negative ions out of the tubule.

As the concentration of ions in the peritubular blood increases, the blood becomes momentarily hypertonic to the tubule fluid. Through the process of osmosis, water dif-

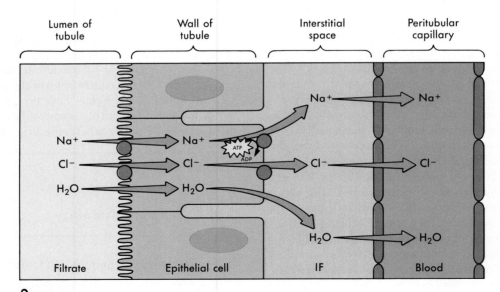

FIGURE 27-13   Mechanisms of tubular reabsorption. Sodium ions *(Na⁺)* are pumped from tubule cell to interstitial fluid *(IF)*, increasing interstitial Na^+ concentration to a level that drives diffusion of Na^+ into blood. As Na^+ is pumped out of the cell, more Na^+ passively diffuses in from filtrate to maintain an equilibrium of concentration. Enough Na^+ moves out of the tubule and into blood that an electrical gradient is established (blood is positive relative to filtrate). Electrical attraction between oppositely charged particles drives diffusion of negative ions in filtrate, such as chloride (Cl^-), into blood. As ion concentration in blood increases, osmosis of water from the tubule occurs. Thus active transport of sodium creates a situation that promotes passive transport of negative ions and water.

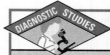

GLUCOSE IN THE URINE

Occasionally, the maximum cotransport capacity is greatly reduced, and glucose appears in the urine (*glycosuria*), even though the blood sugar level may be normal. This condition is known as *renal diabetes* or *renal glycosuria*. It is a congenital defect.

Of course, the most common cause of glycosuria is *diabetes mellitus* (see Chapter 15). In this condition, insulin deficiency or target cell dysfunction causes glucose to accumulate in the blood, causing hyperglycemia. The high glucose content of the filtrate formed in the renal corpuscle exceeds the maximum capacity of the cotransport mechanism. Therefore glycosuria results.

BLOOD INDICATORS OF RENAL DYSFUNCTION

Elevated urea levels in blood, as measured in a BUN (blood urea nitrogen) test, was one of the earliest clinical measurements of kidney dysfunction. Elevated BUN levels indicate the failure of the kidney to *clear* urea and therefore other substances as well. (*Renal clearance* is the volume of plasma from which a substance is removed from the blood by the kidneys per minute.)

Blood levels of creatinine are also used to test renal function. Creatinine levels in the blood seldom change significantly because they are determined by skeletal muscle mass—which seldom changes much. Therefore a reduced clearance rate of serum creatinine is considered to be a reliable indicator of depressed renal function.

fuses rapidly from the tubule fluid and into the peritubular blood, making the two fluids isotonic. In short, ion transport out of the proximal tubules causes water osmosis out of them. This is obligatory water reabsorption—obligatory because it is demanded by the principle of osmosis.

Proximal tubules reabsorb nutrients from the tubule fluid, notably glucose and amino acids, into peritubular blood by a special type of active transport mechanism called **sodium cotransport.** Recall from Chapter 25 that, in this mechanism, a carrier molecule in the cell membrane first binds to both sodium and glucose (see Figure 25-11, p. 663). The carrier then passively transports both substances through the brush border of a proximal tubule cell into the cell's interior (facilitated diffusion). Sodium moves into the cell because of a concentration gradient maintained by the active transport of sodium out the other side of the cell. Glucose actually moves up its concentration gradient, but no energy is required because it is "riding the coat-tails" of sodium. Once inside the cell, the substances dissociate from the carrier molecule and diffuse to the far side of the cell. The substances move out of the epithelial cell by different mechanisms. Sodium is transported actively and glucose is transported passively. Normally, all of the glucose that has filtered out of the glomeruli returns to the blood by this sodium cotransport mechanism. Therefore very little glucose is lost in the urine. If, however, the blood glucose level exceeds a threshold amount (about 150 mg/100 ml), then not all of the glucose can be reabsorbed. The excess glucose remains in the urine. The maximum capacity for moving glucose molecules back into the blood is determined by the number of cotransport carriers available.

Urea is a nitrogen-containing waste formed as a result of protein catabolism (see Chapter 26). Actually, toxic ammonia is formed first but much of it is quickly transformed into the less toxic urea. Urea in the tubule fluid remains in the proximal tubule as sodium, chloride, and water are reabsorbed into the blood. Once these materials are gone, a tubule fluid high in urea is left. Because the urea concentration in the tubule is now greater than in the peritubular blood, urea passively diffuses into the blood. About half the urea present in the tubule fluid leaves the proximal tubule this way.

Reabsorption in the proximal tubules can be summarized in this manner:

1. Sodium is actively transported out of the tubule fluid and into the blood.

2. Glucose and amino acids "hitch a ride" with sodium and passively move out of the tubule fluid by means of the sodium cotransport mechanism.

3. Chloride ions passively move into the blood plasma because of an imbalance in electrical charges (positive sodium ions have already moved out, making the plasma positive and the tubule fluid negative).

4. The movement of sodium and chloride out of the tubule fluid into the plasma creates an osmotic imbalance (the blood is hypertonic to the filtrate), so water is obliged by the principle of osmosis to passively move into the blood.

5. About half the urea present in the tubule fluid passively moves out of the tubule, leaving half the urea to move on to the loop of Henle.

6. The total content of the filtrate has been reduced greatly by the time it is ready to leave the proximal tubule. Most of the water and solutes have been recovered by the blood, leaving only a small volume of fluid to continue to the next portion of the tubule, the loop of Henle.

QUICK CHECK

✔ 1. How are NaCl and water reabsorbed in the proximal tubule?
✔ 2. What is sodium cotransport?

Reabsorption in the loop of Henle

In juxtamedullary nephrons—those low in the cortex, near the medulla—the loop of Henle and its vasa recta participate in a very unique type of process called a **counter-current mechanism.** A counter-current structure is any set of parallel passages in which the contents flow in opposite directions (Figure 27-14). The loop of Henle is a counter-current structure because the contents of the ascending limb travel in an opposite direction to the flow of urine in the descending limb. The vasa recta also have a

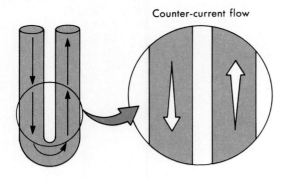

Counter-current flow

FIGURE 27-14 Concept of counter-current flow. Counter-current flow simply refers to flow in opposite directions, as the inset shows. Tubule filtrate in the loop of Henle flows in a counter-current manner, as does blood flowing with the vasa recta of the peritubular capillary network.

counter-current structure because arterial blood flows down into the medulla and venous blood flows up toward the cortex. The kidney's counter-current mechanism functions to keep the solute concentration of the medulla extremely high. We will discuss the reason why this is important later. For now, we will briefly discuss how the counter-current mechanism achieves this goal.

Before we can understand the kidney's counter-current mechanisms, we must appreciate the histology of the loop of Henle. The descending limb is formed by a much thinner wall than the thick part of the ascending limb (see Figure 27-4). Even more important, the permeability and transport abilities of the two walls are very different. The thin-walled descending limb allows water and urea to diffuse freely into or out of the tubule, depending on their concentration gradients. The thick-walled ascending limb, however, limits diffusion of most molecules (including water, sodium, chloride, and urea) while actively transporting selected molecules out of the tubule and into the interstitial fluid.

Given the characteristics of each limb, we can see how the system illustrated in Figure 27-15 can develop in the loop of Henle. Look at the ascending limb. You will see that this limb is actively pumping sodium and chloride out of the tubule fluid and into the interstitial fluid. This probably occurs by the same mechanism that operates in the proximal tubule. Normally, sodium and chloride ions would simply diffuse right back into the tubule fluid to achieve an equilibrium. The ascending limb prevents the diffusion of these ions, so they are "trapped" in the interstitial area. Under normal circumstances, water would move from the tubule fluid to the interstitial fluid to achieve an osmotic balance. However, the wall of the ascending limb is relatively impermeable to water. In short, salt ions are pumped out of the ascending limb, water is prevented from following osmotically, and thus the tubule fluid develops a low solute concentration (low osmotic pressure) and the interstitial fluid develops a high solute concentration (high osmotic pressure).

The ion pumps in the ascending limb can maintain an

osmotic difference of 200 mOsm across the wall of the tubule. Notice in Figure 27-15 that the movement of salt out of the tubule at any horizontal level creates a difference of 200 mOsm between the tubule fluid and the interstitial fluid. Because the salt is continually added to the interstitial fluid, the interstitial fluid becomes very concentrated (up to 1200 mOsm in our model). This high solute concentration of the interstitial fluid of the medulla is created and maintained by the constant pumping of salt by the ascending limb. For this reason, this process is often called a *counter-current multiplier mechanism.*

You will notice that the tubule fluid in the descending limb equilibrates easily with the interstitial fluid. Because the interstitial fluid has a high solute concentration (created by the ion pumps in the ascending limb), the fluid in the descending limb loses water osmotically. Thus the solute concentration of the tubule fluid becomes increasingly higher. Urea, a solute that is also highly concentrated in the renal medulla, diffuses into the tubule fluid in the descending limb—increasing the solute concentration of the tubule fluid even more. However, as the fluid "rounds

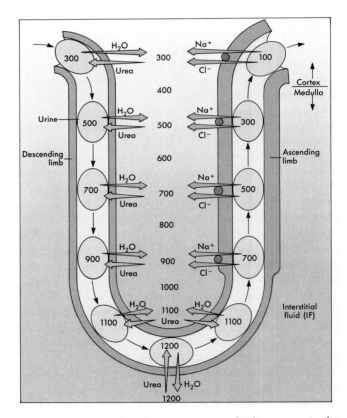

FIGURE 27-15 The counter-current multiplier system in the loop of Henle. Na^+ and Cl^- is pumped from the ascending limb and moved into interstitial fluid *(IF)* to maintain a high osmolality there. Because salt content of the medullary IF increases, this is called a "multiplier" mechanism. Because ion pumping also lowers the tubule fluid's osmolality by 200 mOsm, fluid leaving the loop of Henle is only 100 mOsm (hypotonic), compared to 300 mOsm (isotonic) when it entered the loop. Numbers in the diagram are expressed in mOsm.

Osmolality

Kidney physiologists prefer to express relative solute concentrations in terms of osmotic pressure. **Osmolality** is the osmotic pressure of a solution expressed as the number of *osmols* of pressure per kilogram of water. One osmol is equivalent to the osmotic pressure produced when 1 g of nonionizing substance is dissolved in 1 kg of water. Here, we are using the smaller unit *milliosmol (mOsm)* to make the numbers easier to handle. Most body fluids have an osmolality of about 300 mOsm.

The advantage of using a solution's osmolality as an expression of its solute concentration is that it takes into account *all types* of solute particles present. In the human kidney, sodium ions, chloride ions, and urea are all present in varying amounts—and all contribute to the osmotic pressure of the fluid in which they are found.

See Appendix B for more information on how the osmotic pressure of a solution is determined.

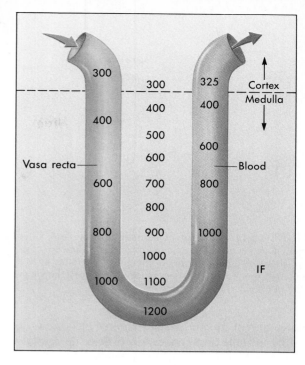

FIGURE 27-16   The counter-current exchange mechanism in the vasa recta. Because the vasa recta forms a counter-current loop, blood leaving the capillary bed has only a slightly higher solute content than when it entered. Thus high osmolality of medullary tissue fluid is maintained. If peritubular blood instead traveled straight through the tissue, all excess solute in the medulla would be removed, and the osmolality of medullary interstitial fluid would be equivalent to that of the cortex. Numbers in the diagram are expressed in mOsm.

the bend" and begins moving into the thick portion of the ascending limb, its Na^+ and Cl^- are removed and it becomes increasingly lower in solute concentration.

When tubule fluid enters the loop of Henle it is about 300 mOsm (isotonic to most body fluids). When it leaves the loop of Henle, it is about 100 mOsm (hypotonic to most body fluids). Because water is reabsorbed from the fluid in the descending limb, there is a net reduction of tubule fluid volume. And because urea entered the tubule fluid in the descending limb, there is a net increase in urea concentration in the tubule fluid.

You might think that the blood of the vasa recta (a portion of the peritubular capillary network) would remove the excess solute from the medulla's interstitial fluid as it flows through the tissue. Usually it would, but the vasa recta has its own counter-current mechanism. Figure 27-16 shows how the looping of the vasa recta, down into the medulla, then back up to the cortex, prevents it from accumulating too much solute. Consider also that the blood flow through the vasa recta is sluggish; it cannot remove anything very efficiently. Just enough solute is removed to prevent the medulla from crystallizing completely due to a high solute concentration. Thus the tissues of the medulla have the benefits of a blood supply without much loss of its high solute concentration.

The primary functions of the loop of Henle are summarized here.

1. The loop of Henle reabsorbs water from the tubule fluid (and picks up urea from the interstitial fluid) in its descending limb. It reabsorbs sodium and chloride from the tubule fluid in the ascending limb.

2. By reabsorbing salt from its ascending limb, it makes the tubule fluid dilute (hypotonic).

3. Reabsorption of salt in the ascending limb also creates and maintains a high osmotic pressure, or high solute concentration, of the medulla's interstitial fluid.

QUICK CHECK

✔ 1. *What is a counter-current mechanism?*
✔ 2. *How does the function of the descending limb of the loop of Henle differ from the function of the thick ascending limb?*
✔ 3. *What is the purpose of the counter-current multiplier mechanism of the loop of Henle?*

Reabsorption in the distal tubules and collecting ducts

The distal tubule is similar to the proximal tubule in that it also reabsorbs some sodium by active transport but in much smaller amounts. Left to themselves, the cells that form the distal tubule's walls are relatively impermeable to water. This means that sodium can be removed, but water cannot follow osmotically, so the solute concentration of the tubule fluid continues to decrease. Recall that the tubule fluid is already hypotonic to most body fluids at this point due to the counter-current system in the loop of Henle.

Unless influenced from outside the kidney, the cells that form the wall of the collecting duct also prevent water from leaving the filtrate by osmosis. Even though the collecting duct conducts the tubule fluid through the hypertonic medullary region, equilibration does not occur.

Given no other circumstances, the kidney will produce

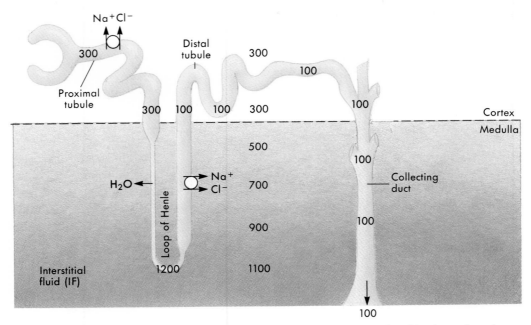

FIGURE 27-17 Production of hypotonic urine. Hypotonic urine is produced by the nephron by the mechanism shown here. Isotonic (300 mOsm) tubule fluid that enters the loop of Henle becomes hypotonic (100 mOsm) by the time it enters the distal tubule. The tubule fluid remains hypotonic as it is conducted out of the kidney because the walls of the distal tubule and collecting duct are impermeable to H_2O, Na^+, and Cl^-. (Values are expressed in mOsm.)

and excrete only very dilute (hypotonic) urine (Figure 27-17). This would be catastrophic because the body would soon dehydrate. A regulatory mechanism centered outside the kidney normally prevents excessive loss of water. This mechanism, illustrated in Figure 27-18, involves *antidiuretic hormone* (ADH), a hormone secreted by the neurohypophysis (posterior pituitary). ADH targets cells of the distal and collecting tubules and causes them to become more permeable to water. When this happens, water is permitted to flow osmotically out of the tubule and into the interstitial fluid, toward an equilibrium. The more ADH present, the more water is allowed out of the tubule, and the closer the tubule fluid's solute concentration matches that of the surrounding tissue fluid. In this way, the tubule fluid's osmotic pressure could go as high as 1200 mOsm, since the medulla's interstitial fluid can be that high. The solute concentration of the urine excreted depends in large part on the amount of ADH present.

Notice in Figure 27-18 that reabsorption of urea also occurs in the collecting duct when water is reabsorbed under the influence of ADH. As water is reabsorbed from the fluid descending through the collecting duct, the urea concentration of the fluid rises. Because the urea concentration is higher inside the lower part of the collecting duct than it is in the surrounding interstitial fluid, urea diffuses

out of the lower collecting duct. The addition of urea to the medullary interstitial fluid assists in maintaining a high solute concentration in the medulla. Less than half the urea that leaves the collecting duct is removed by the vasa recta. The dashed line in Figure 27-18 shows that much of the urea in the medullary interstitial fluid diffuses into the descending limb of the loop of Henle. Thus urea participates in a sort of counter-current multiplier mechanism that, with the counter-current mechanisms of the loop of Henle and vasa recta, maintains the high osmotic pressure needed to form concentrated urine and thus avoid dehydration.

Tubular Secretion

In addition to reabsorption, tubule cells also secrete certain substances. Tubular secretion means the movement of substances out of the blood and into the tubular fluid. Recall that the descending limb of the loop of Henle removes urea by means of diffusion. Distal and collecting tubules secrete potassium, hydrogen, and ammonium ions. They actively transport potassium ions (K^+) or hydrogen ions (H^+) out of the blood into the tubule fluid in exchange for sodium ions (Na^+) that diffuse back into the blood (H^+ transport will be discussed further in Chapter 29). Potassium secretion increases when blood aldosterone concen-

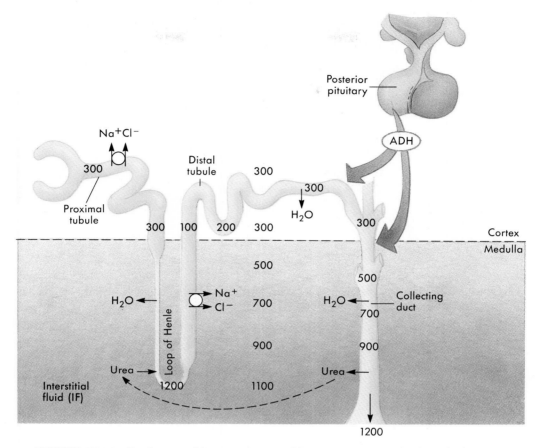

FIGURE 27-18 **Production of hypertonic urine.** Hypertonic urine can be formed when ADH is present. ADH, a posterior pituitary hormone, increases the water permeability of the distal tubule and collecting duct. Thus hypotonic (100 mOsm) tubule fluid leaving the loop of Henle can equilibrate first with the isotonic (300 mOsm) interstitial fluid (*IF*) of the cortex, then with the increasingly hypertonic (400-1200 mOsm) IF of the medulla. As H$_2$O leaves the collecting duct by osmosis, the filtrate becomes more concentrated with the solutes left behind. The concentration gradient causes urea to diffuse into the IF, where some of it is eventually picked up by tubule fluid in the descending limb of the loop of Henle (*dashed line*). This counter-current movement of urea helps maintain high solute concentration in the medulla. (Values are expressed in mOsm.)

tration increases. *Aldosterone,* a hormone of the adrenal cortex, targets distal and collecting tubule cells and causes them to increase activity of sodium-potassium pumps that move sodium out of the tubule and potassium into the tubule. Hydrogen ion secretion increases when blood hydrogen ion concentration increases. Ammonium ions are secreted into the tubule fluid by diffusing out of the tubule cells where they are synthesized. Tubule cells also secret certain drugs, for example, penicillin and paraaminohippuric acid (PAH).

Table 27-1 summarizes the functions of the different parts of the nephron in forming urine.

Regulation of Urine Volume

ADH has a central role in the regulation of urine volume. Control of the solute concentration of urine

translates into control of urine volume. If no water is reabsorbed by the distal and collecting tubules, then the urine volume is relatively high—and water loss from the body is high. As water is reabsorbed under the influence of ADH, the total volume of urine is reduced by the amount of water removed from the tubules. Thus ADH reduces water loss by the body.

Another hormone that tends to decrease urine volume—and thus conserves water—is aldosterone, a secretion of the adrenal cortex. It increases distal and collecting tubule absorption of sodium, which in turn causes an osmotic imbalance that drives the reabsorption of water from the tubule. Because water reabsorption in the distal and collecting tubule portions requires ADH, the aldosterone mechanism must work in concert with the ADH mechanism if homeostasis of fluid content in the body is to

TABLE 27-1   Summary of nephron function

Part of Nephron	Function	Substance Moved
Renal corpuscle	Filtration (passive)	Water Smaller solute particles (ions, glucose, etc.)
Proximal tubule	Reabsorption (active)	Active transport: Na^+ Cotransport: glucose and amino acids
	Reabsorption (passive)	Diffusion: Cl^-, $PO_4^=$, urea, other solutes Osmosis: water
Loop of Henle		
Descending limb	Reabsorption (passive)	Osmosis: water
	Secretion (passive)	Diffusion: urea
Ascending limb	Reabsorption (active)	Active transport: Na^+
	Reabsorption (passive)	Diffusion: Cl^-
Distal tubule	Reabsorption (active)	Active transport: Na^+
	Reabsorption (passive)	Diffusion: Cl^-, other anions Osmosis: water (only in presence of ADH)
	Secretion (passive)	Diffusion: ammonia
	Secretion (active)	Active transport: K^+, H^+, some drugs
Collecting duct	Reabsorption (active)	Active transport: Na^+
	Reabsorption (passive)	Diffusion: urea Osmosis: water (only in presence of ADH)
	Secretion (passive)	Diffusion: ammonia
	Secretion (active)	Active transport: K^+, H^+, some drugs

be maintained. The cooperative roles of ADH and aldosterone in regulating urine volume—and thus regulating fluid balance in the whole body—are summarized in Figure 27-19.

You may recall from Chapters 15 and 18 that another hormone, specifically **atrial natriuretic hormone (ANH),** also influences water reabsorption in the kidney. ANH is secreted by specialized muscle fibers in the atrial wall of the heart. Its name implies its function: ANH promotes natriuresis (loss of Na^+ via the urine). ANH indirectly acts as an antagonist of aldosterone, promoting the secretion of sodium into the kidney tubules rather than sodium reabsorption. Thus ANH reduces the plasma and interstitial fluid Na^+ concentration, which, in turn, reduces the reabsorption of water by having the opposite effect as aldosterone. ANH also inhibits secretion of aldosterone and opposes the aldosterone-ADH mechanism to reabsorb less water and thus produce more urine. In short, ANH inhibits the ADH mechanism—thus inhibiting water conservation by the internal environment and increasing urine volume.

Urine volume also relates to the total amount of solutes other than sodium excreted in the urine. Generally, the more solutes, the more urine. Probably the best known example of this occurs in untreated diabetes mellitus. The

HEALTH MATTERS

HYPERTENSION

The relationship of high blood pressure, *hypertension,* and renal function is still being debated. Clinical observation and animal experiments have established that destruction of a large proportion of kidney tissue usually results in the development of hypertension. This happens frequently in patients who have severe renal arteriosclerosis. Many experiments have been performed and various theories devised to explain the mechanism responsible for "renal hypertension." The initiating factor is believed to be ischemia of the kidneys (reduced blood flow through them). When this occurs, the cells of the juxtaglomerular apparatus secrete renin. Renin is a proteolytic enzyme that hydrolyzes one of the plasma proteins to produce angiotensin. One effect of angiotensin is to increase blood pressure by constricting arterioles (see Figure 18-16 and 27-19 for a review of the renin-angiotensin reactions).

symptom that often brings an undiagnosed diabetic to a physician is the voiding of abnormally large amounts of urine. Excess glucose "spills over" into urine, increasing solute concentration of the urine (and decreasing solute concentration of the plasma), which in turn leads to diuresis.

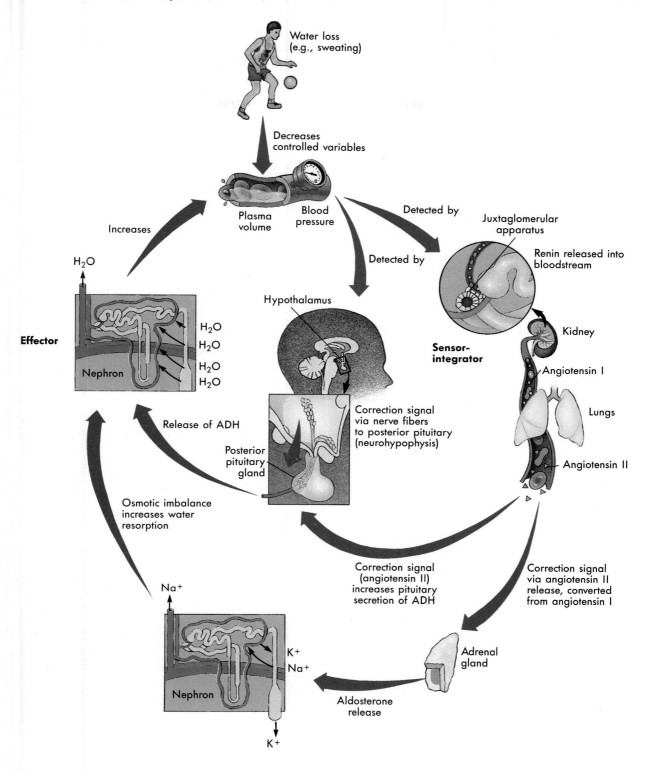

FIGURE 27-19 Cooperative roles of ADH and aldosterone in regulating urine and plasma volume. The drop in blood pressure that accompanies loss of fluid from the internal environment triggers the hypothalamus to rapidly release ADH from the posterior pituitary gland. ADH increases water reabsorption by the kidney by increasing water permeability of the distal tubules and collecting ducts. The drop in blood pressure is also detected by each nephron's juxtaglomerular apparatus, which responds by secreting renin. Recall from Chapter 15 that renin triggers the formation of angiotensin II, which stimulates release of aldosterone from the adrenal cortex. Aldosterone then slowly boosts water reabsorption by the kidneys by increasing reabsorption of Na^+. Because angiotensin II also stimulates secretion of ADH, it serves as an additional link between the ADH and aldosterone mechanisms.

Urine volume is not normally altered by changes in the glomerular filtration rate, which remains remarkably constant. The glomerular filtration rate usually does not change, even in the presence of changes in renal blood flow and in glomerular hydrostatic pressure. Under certain pathological conditions, however, the glomerular filtration rate may change enough to alter urine volume.

Urine Composition

The physical characteristics of normal urine are listed in Table 27-2. Notice that both normal and abnormal characteristics are listed.

Urine is approximately 95% water, in which are dissolved several kinds of substances; the most important are discussed here:

♦ **Nitrogenous wastes** from protein catabolism such as urea (the most abundant solute in urine), uric acid, ammonia, and creatinine.
♦ **Electrolytes**—mainly the following ions: sodium, potassium, ammonium, chloride, bicarbonate, phosphate, and sulfate. The amounts and kinds of minerals vary with diet and other factors.
♦ **Toxins**—during disease bacterial poisons leave the body in the urine. One reason for "forcing fluids" on patients suffering with infectious diseases is to dilute the toxins that might damage the kidney cells if they were eliminated in a concentrated form.
♦ **Pigments**—especially *urochromes*—yellowish pigments derived from the products of the breakdown of old red blood cells in the liver and elsewhere. Various foods and drugs may contain, or be converted to, pigments that are cleared from plasma by the kidneys and are therefore found in the urine.

♦ **Hormones**—high hormone levels sometimes result in significant amounts of hormone in the filtrate (and therefore in urine).
♦ **Abnormal constituents,** such as blood, glucose, albumin (a plasma protein), casts (chunks of material, such as mucus, that harden inside the urinary passages and then are washed out in the urine), or calculi (small stones).

QUICK CHECK
✔ 1. Does ADH promote water loss from the internal environment or water conservation by the internal environment?
✔ 2. How does aldosterone influence secretion in the kidney tubules?
✔ 3. How does aldosterone cause the body to conserve water?
✔ 4. What gives urine its characteristic yellowish color?

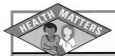

CLINICAL TERMS ASSOCIATED WITH URINE ABNORMALITIES

Glycosuria or **glucosuria** Sugar (glucose) in the urine
Hematuria Blood in the urine
Pyuria Pus in the urine
Dysuria Painful urination
Polyuria Unusually large amounts of urine
Oliguria Scant urine
Anuria Absence of urine

TABLE 27-2 Characteristics of urine

Normal Characteristics	Abnormal Characteristics
COLOR	
Transparent yellow, amber, or straw color	Abnormal colors or cloudiness, which may indicate presence of blood, bile, bacteria, drugs, food pigments, or high-solute concentration
COMPOUNDS	
Mineral ions (for example, Na^+, Cl^-, K^+)	Acetone
Nitrogenous wastes: ammonia, creatinine, urea, uric acid	Albumin
Suspended solids (sediment)*: bacteria, blood cells, casts (solid matter)	Bile
Urine pigments	Glucose
ODOR	
Slight odor	Acetone odor, which is common in diabetes mellitus
pH	
4.6-8.0 (freshly voided urine is generally acidic)	High in alkalosis; low in acidosis
Specific Gravity	
1.001-1.035	High specific gravity can cause precipitation of solutes and formation of kidney stones

*Occasional trace amounts.

GOUT

Gout is a condition characterized by excessive levels of uric acid in the blood. The body produces uric acid from metabolism of ingested purines in the diet (especially in glandular meats) or from purine turnover in our own bodies. Blood uric acid levels may become elevated because of increased dietary intake, excessive production in the body, or defective excretion by the kidneys. Because uric acid is not as soluble as many waste products, it tends to be deposited in the joints and tissues of the body if blood levels become elevated. The figure shows the results of uric acid deposits in the soft tissues and joints of the hand in a patient with gout. Deposit in the kidneys produces uric acid stones or calculi.

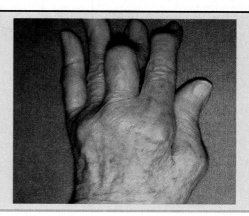

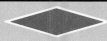

Cycle of Life: *Urinary system*

The kidney plays a critical role in homeostasis by regulating the levels of many substances in the blood. Primary renal functions include filtration, reabsorption, and secretion. All are interrelated by complex control systems involving both central nervous system activity and hormonal secretions. Over 1 million nephron units in each kidney serve as the structural framework permitting normal function to occur.

Normally, life cycle changes in kidney structure and function occur only within rather narrow limits. Significant structural changes, such as dramatic decreases in the numbers of nephron units, almost always indicate serious disease or result from trauma such as crush injuries. Functionally, the kidney is able to operate normally throughout life under a wide array of conditions. If, however, the kidneys cannot cope with extreme conditions, such as water deprivation or disease, death from the buildup of toxins in the blood will occur.

Initially, kidney function in the newborn is less efficient than in the older child and adult. As a result, the urine is less concentrated because the regulatory mechanisms required to retain water are not fully operative. Incontinence, or the inability to control urination, is normal in very young children. Reflex emptying occurs when the bladder fills, but normal sphincter activity keeps urine in the bladder until filling occurs. In contrast, many older adults have problems with incontinence because of loss of sphincter tone, or control.

Renal clearance is the ability of the kidneys to clear, or cleanse, the blood of a certain substance in a given unit of time, generally 1 minute. This value for certain substances tends to decrease with advanced age, indicating deterioration of kidney function. Changes in the porosity of the filtration membrane also occurs in the elderly. Loss of functional nephron units is yet another consequence of aging. It contributes to the gradual decline in renal function in this age group.

THE BIG PICTURE
The Urinary System and the Whole Body

As our study of the urinary system has shown us, the homeostasis of water and electrolytes in body fluids depends largely on the proper functioning of the kidneys. Each nephron within the kidney processes blood plasma in a way that adjusts its content to maintain a relative constancy of the internal environment of the body. Without renal processing, blood plasma characteristics would soon move out of their set point range. On the other hand, without the blood pressure generated by cardiovascular mechanisms, the kidney could not filter blood plasma and thus could not process blood plasma. Thus the urinary system and the cardiovascular system are interdependent.

Regulation of urinary function, we have seen, is often centered outside the kidney—mainly in the form of endocrine hormone action. Urinary function is also regulated to some extent by nerve reflexes. Thus both the endocrine system and nervous system must operate properly to ensure efficient kidney function.

The urinary system also interacts with many other body systems and tissues. For example, the kidneys clear the blood plasma of nitrogenous wastes and excess metabolic acids produced by the chemical activity of nearly every cell in the body. The kidneys can also clear some toxins and other compounds that enter the blood via the digestive tract, skin, or respiratory tract.

In the next chapter, entitled *Fluid and Electrolyte Balance*, we will apply some of what we know about urinary function to a broader study of water and ion homeostasis within the human body's internal environment. After that, we will move on to Chapter 29, in which we discuss the role of the urinary system and other body systems in maintaining a relatively constant pH in the body's internal environment.

MECHANISMS OF DISEASE

Renal and Urinary Disorders

You may have experienced the discomfort and pain of a bladder infection or know someone who has. Bladder infection is the most common urinary disorder, but it usually is not serious if promptly treated. There are numerous renal and urinary disorders that are very serious, however. Any disorder that significantly reduces the effectiveness of the kidneys is immediately life-threatening. In this section, we will discuss some life-threatening kidney diseases, as well as a few of the less serious, but more common, disorders.

Obstructive Disorders

Obstructive urinary disorders are abnormalities that interfere with normal urine flow anywhere in the urinary tract. The severity of obstructive disorders depends on where the interference occurs and to what degree the flow of urine is impaired. Obstruction of urine flow usually results in "backing up" of the urine, perhaps all the way to the kidney itself. When urine backs up into the kidney, causing swelling of the renal pelvis and calyces, the condition is called **hydronephrosis.** A few of the more important obstructive conditions are summarized here.

Renal Calculi

Renal calculi, or *kidney stones,* are crystallized mineral chunks that develop in the renal pelvis or calyces. Many calculi develop as calcium and other minerals crystallize on the renal papillae, then break off into the urine. *Staghorn calculi* are large, branched stones that form in the pelvis and branched calyces.

If the stones are small enough, they will simply pass through the ureters and eventually be voided with the urine. Larger stones may obstruct the ureters, causing an intense pain called **renal colic** as rhythmic muscle contractions of the ureter attempt to dislodge it. Hydronephrosis may occur if the stone does not move from its obstructing position.

Neurogenic Bladder

Disruption of the nerve input to the bladder results in loss of normal control of voiding. Such paralysis or abnormal activity of the bladder is termed **neurogenic bladder.** Neurogenic bladder is characterized by involuntary retention of urine, subsequent distention (bulging) of the bladder, and perhaps a burning sensation or fever with chills.

Tumors and Other Obstructions

Tumors of the urinary system typically obstruct urine flow, possibly causing hydronephrosis in one or both kidneys. Most kidney tumors are malignant neoplasms called *renal cell carcinomas.* They usually occur only in one kidney. Bladder cancer occurs about as frequently as renal cancer (each accounts for about two in every hundred cancer cases).

Various other conditions can obstruct the normal flow of urine. For example, a person with a low proportion of body fat may lack the pad of fat that normally surrounds the kidneys. One or both kidneys may then drop, a condition called **renal ptosis** (TO-sis). In renal ptosis, the ureters that drain urine out of the kidney may kink and thus obstruct the normal flow of urine. Urinary passages may also be abnormally narrowed through scarring, inflammation, or external pressure—a condition known as a **stricture.**

Urinary Tract Infections

Most *urinary tract infections (UTIs)* are caused by bacteria, most often gram-negative types. UTIs can involve the urethra, bladder, ureter, and/or kidneys. Common types of urinary tract infections are summarized below.

Urethritis is inflammation of the urethra that commonly results from bacterial infection, often *gonorrhea.* Nongonococcal urethritis is usually caused by a *chlamydia* infection. Males suffer from urethritis more often than females.

Cystitis is a term that refers to any inflammation of the bladder. Cystitis commonly occurs as a result of infection but can also accompany calculi, tumors, or other conditions. Bacteria usually enter the bladder through the urethra. Cystitis occurs more frequently in women than in men because the female urethra is shorter and closer to the anus (a source of bacteria) than in the male. Bladder infections are characterized by pelvic pain, an urge to urinate frequently, and hematuria.

Nephritis is a general term referring to kidney disease, especially inflammatory conditions. **Pyelonephritis** is literally "pelvis nephritis" and refers to inflammation of the renal pelvis and connective tissues of the kidney. As with cystitis, pyelonephritis is usually caused by bacterial infection but can also result from viral infection, mycosis, calculi, tumors, pregnancy, and other conditions.

Glomerular Disorders

Glomerular disorders, collectively called **glomerulonephritis,** result from damage to the glomerular-capsular membrane. This damage can be caused by immune mechanisms, heredity, and other factors. Without successful treatment, glomerular disorders can progress to kidney failure.

Nephrotic syndrome is a collection of signs and symptoms that accompany various glomerular disorders. This syndrome is characterized by:

- **Proteinuria**—presence of proteins (especially *albumin*) in the urine. Protein, normally absent from urine, filters through damaged glomerular-capsular membranes and is not reabsorbed by the kidney tubules.
- **Hypoalbuminemia**—low albumin concentration in the blood, resulting from the loss of albumin from the blood through holes in the damaged glomeruli. Albumin is the most abundant plasma protein. Since it normally cannot leave the blood vessels, it usually remains as "permanent" solute in the plasma. This keeps plasma water concentration low and thus prevents osmosis of large amounts of water out of the blood and into tissue spaces. In hypoalbuminemia, this function is lost and fluid leaks out of the blood vessels and into tissue spaces, causing widespread edema.
- **Edema**—general tissue swelling caused by accumulation of fluids in the tissue spaces. The edema associated with nephrotic syndrome is caused by the loss of plasma protein (albumin) and the resulting osmosis of fluid out of the blood.

Acute glomerulonephritis is the most common form of kidney disease. It may be caused by a delayed immune response to streptococcal infection—the same mechanism that causes valve damage in rheumatic heart disease (see Chapter 17). For

this reason, it is sometimes called *postinfectious glomerulonephritis*. If antibiotic treatment is not successful, it may progress to a chronic form of glomerulonephritis.

Chronic glomerulonephritis is the general name for various noninfectious glomerular disorders that are characterized by progressive kidney damage leading to renal failure. Immune mechanisms are believed to be the major causes of chronic glomerulonephritis. One immune mechanism involves antigen-antibody complexes that form in the blood when antibodies bind with foreign antigens (or possible self-antigens). These antigen-antibody complexes lodge in the glomerular-capsular membrane and trigger an inflammation response. Less commonly, formation of antibodies that directly attack the glomerular-capsular membrane causes chronic glomerulonephritis.

Kidney Failure

Kidney failure, or **renal failure,** is simply the failure of the kidney to properly process blood plasma and form urine. Renal failure can be classified as acute or chronic.

Acute renal failure is an abrupt reduction in kidney function that is characterized by oliguria and a sharp rise in nitrogenous compounds in the blood. The concentration of nitrogenous wastes in the blood is often assessed by the *BUN (blood urea nitrogen)* test—a high BUN result indicates failure of the kidneys to remove urea from the blood. Acute renal failure can be caused by various factors that alter blood pressure or otherwise affect glomerular filtration. For example, hemorrhage, severe burns, acute glomerulonephritis or pyelonephritis, and obstruction of the lower urinary tract may each progress to kidney failure. If the underlying cause of renal failure is attended to, recovery is usually rapid and complete.

Chronic renal failure is a slow, progressive condition resulting from the gradual loss of nephrons. There are dozens of diseases that may result in the gradual loss of nephron function, including infections, glomerulonephritis, tumors, systemic autoimmune disorders, and obstructive disorders. As kidney function is lost, the glomerular filtration rate (GFR) decreases, causing the blood urea nitrogen (BUN) levels to climb (Figure 27-20). Chronic renal failure can be described as progressing through the three stages given here:

Stage 1. During the first stage, some nephrons are lost but the remaining healthy nephrons compensate by enlarging and taking over the function of the lost nephrons. As Figure 27-20 shows, BUN is kept within normal limits even though up to 75% of the nephrons are lost (as indicated by a 75% drop in GFR). This stage is often asymptomatic and may last for years, depending on the underlying cause.

Stage 2. The second stage is often called *renal insufficiency*. It is during this stage that the kidney can no longer adapt to the loss of nephrons. The remaining healthy nephrons cannot handle the urea load, and the BUN levels climb dramatically (see Figure 27-20). Because the kidney's ability to concentrate urine is impaired, polyuria and dehydration may occur.

Stage 3. The final stage of chronic renal failure is called *uremia*, or *uremic syndrome*. Uremia literally means "high blood urea" and is characterized by a very high BUN value caused by loss of kidney function (see Figure 27-20). During this stage, low GFR causes low urine production and oliguria. Because fluids are retained by the body rather than eliminated by the kidneys, edema and hypertension often occur. The uremic syndrome includes a long list of other symptoms caused directly or indirectly by the loss of kidney function. Unless an artificial kidney is used or a new kidney is transplanted, the progressive loss of kidney function will eventually cause death.

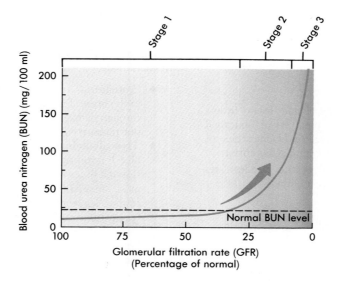

FIGURE 27-20 The three stages of chronic renal failure. Stage 1: As nephrons are lost (indicated by decreasing *GFR*), remaining healthy nephrons compensate—keeping BUN values within the normal range. Stage 2: As more than 75% of kidney function is lost, BUN levels begin to climb. Stage 3: Uremia (elevated BUN) results from massive loss of kidney function.

FOCUS ON... ARTIFICIAL KIDNEY

The artificial kidney is a mechanical device that uses the principle of dialysis to remove or separate waste products from the blood. In the event of kidney failure, the process, appropriately called **hemodialysis** (Greek *haima*—"blood" and *lysis*—"separate"), is a reprieve from death for the patient. During a hemodialysis treatment, a semipermeable membrane is used to separate large (nondiffusible) particles such as blood cells from small (diffusible) ones such as urea and other wastes. Figure A shows blood from the radial artery passing through a porous (semipermeable) cellophane tube that is housed in a tanklike container. The tube is surrounded by a bath, or dialysis solution, containing varying concentrations of electrolytes and other chemicals. The pores in the membrane are small and allow only very small molecules, such as urea, to escape into the surrounding fluid. Larger molecules and blood cells cannot escape and are returned through the tube to reenter the patient via a wrist or leg vein. By constantly replacing the bath solution in the dialysis tank with freshly mixed solution, levels of waste materials can be kept at low levels. As a result, wastes such as urea in the blood rapidly pass into the surrounding wash solution. For a patient with complete kidney failure, two or three hemodialysis treatments a week are required. New dialysis methods are now being developed, and dramatic advances in treatment are expected in the next few years.

Another technique used in the treatment of renal failure is called **continuous ambulatory peritoneal dialysis (CAPD).** In this procedure, 1 to 3 L of sterile dialysis fluid is introduced directly into the peritoneal cavity through an opening in the abdominal wall (Figure B). Peritoneal membranes in the abdominal cavity transfer waste products from the blood into the dialysis fluid, which is then drained back into a plastic container after about 2 hours. This technique is less expensive than hemodialysis and does not require the use of complex equipment.

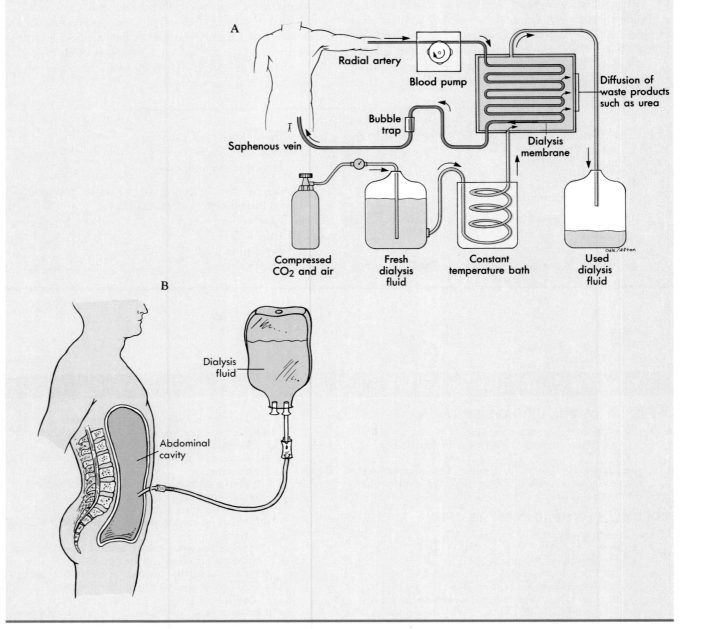

A

Radial artery

Blood pump

Diffusion of waste products such as urea

Saphenous vein

Bubble trap

Dialysis membrane

Compressed CO$_2$ and air

Fresh dialysis fluid

Constant temperature bath

Used dialysis fluid

B

Dialysis fluid

Abdominal cavity

CASE STUDY
KIDNEY STONES (UROLITHIASIS)

Mr. G. is a 42-year-old farmer from Texas. For the past week he has noted "twinges" in his right lower back. At 3:00 a.m. today, he awoke with excruciating pain in his back, unresponsive to a hot water bottle or aspirin. He spent 4 hours pacing the house with intermittent bouts of vomiting. Finally his wife persuaded him to go to the emergency room 2 hours away.

While pacing in the emergency room examining room, Mr. G. reported good health in the past with no prior medical illnesses or surgical procedures. He takes no medications, does not use tobacco or alcohol, and is very active on the farm, working 12 to 14 hours a day. He denied any recent or remote trauma.

Family history revealed that he was raised in a rural Texas hamlet. His family was healthy, although his father and grandfather occasionally "passed blood in their water." No cause for this problem was ever sought.

Physical examination revealed a healthy-appearing Caucasian male, appearing somewhat older than stated age, pacing and wincing in acute discomfort. Vital signs were as follows: temperature 97° F; heart rate 140, respirations 28, blood pressure 138/78.

Findings included cool and moist skin, moderate to severe right flank pain, and abdominal pain without rebound tenderness. Mr. G. vomited three times during the examination. Urinalysis revealed 1+ protein, 3+ occult blood, trace of glucose. High-power field microscopic examination revealed 20 to 30 red blood cells, 8 to 10 white blood cells, 1 to 2 bacteria, 6 to 10 calcium oxalate crystals. No other abnormal findings were noted.

A brief nutritional assessment was obtained from Mrs. G. while her husband was having blood drawn. She indicated that her husband eats milk, ice cream, and cheese several times daily. She serves mashed potatoes with cream sauce three times a week, and he eats rhubarb, garlic, asparagus, spinach, fruit juices, chocolate milk drink (Ovaltine), beef, pork, and chicken several times a week.

1. The most likely cause of Mr. G.'s discomfort is:
   a. Acute pyelonephritis
   b. Cystitis
   c. Urolithiasis
   d. Nephrotic syndrome

2. Which is the least likely risk factor for urolithiasis in Mr. G.?
   a. Age, sex, race
   b. Region of the country where he lives
   c. Family history of calculi
   d. Diet
   e. Hyperparathyroidism

3. Which of the following is the most potentially damaging side effect of renal calculi?
   a. Obstruction
   b. Pain
   c. Hematuria
   d. Crystalluria

CHAPTER SUMMARY

OVERVIEW OF THE URINARY SYSTEM

A. Kidneys—principal organs of the urinary system; accessory organs are ureters, urinary bladder, and urethra (Figure 27-1)
B. Urinary system—regulates the content of blood plasma to maintain the homeostasis of the internal fluid environment within normal limits

ANATOMY OF THE URINARY SYSTEM

A. Gross structure (Figure 27-2)
   1. Kidney
      a. Shape, size, and location
         (1) Roughly oval with a medial indentation
         (2) Approximately 11 cm × 7 cm × 3 cm
         (3) Left kidney often larger than right; right kidney is a little lower

(4) Located in a retroperitoneal position
(5) Lie on either side of the vertebral column between T12 and L3
(6) Renal fasciae anchors the kidneys to surrounding structures
(7) Heavy cushion of fat surrounds each kidney
      b. Internal structures of the kidney
         (1) Cortex and medulla
         (2) Renal pyramids comprise much of the medullary tissue
         (3) Renal columns—where cortical tissue dips into the medulla between the pyramids
         (4) Calyx—cuplike structure at each renal papilla to collect urine; join together to form renal pelvis
         (5) Renal pelvis narrows as it exits the kidney to become the ureter

2. Ureter—tube running from each kidney to urinary bladder; composed of three layers: mucous lining, muscular middle layer, and fibrous outer layer
3. Urinary bladder (Figure 27-3)
   a. Structure—collapsible bag located behind the symphysis pubis made mostly of smooth muscle tissue; lining forms rugae; can distend considerably
   b. Functions
      (1) Reservoir for urine prior to its leaving the body
      (2) Aided by the urethra, it expels urine from the body
   c. Mechanism for voiding
      (1) Voluntary relaxation of external sphincter muscle
      (2) Regions of the detrusor muscle contract reflexively
      (3) Urine is forced out of the bladder and through the urethra
4. Urethra
   a. Small mucous membrane lined tube, extending from the trigone to the exterior of the body
   b. In female, lies posterior to symphysis pubis and anterior to the vagina; approximately 3 cm long
   c. In male, after leaving the bladder, passes through prostate gland where it is joined by two ejaculatory ducts; from prostate, it extends to the base of the penis, then through the center of the penis, and ends as the urinary meatus; approximately 20 cm long; male urethra is part of the urinary system, as well as part of the reproductive system

B. Microscopic structure—nephrons, the microscopic functional units, make up the bulk of the kidney; each nephron is made up of various structures (Figure 27-4)
   1. Bowman's capsule—cup-shaped mouth of nephron
      a. Formed by parietal and visceral walls with a space between them
      b. Pedicels in visceral layer are packed closely together to form filtration slits; slit diaphragm prevents filtration slits from enlarging under pressure
      c. Glomerulus—network of fine capillaries in Bowman's capsule; together called renal corpuscle; located in cortex of kidney (Figure 27-5)
      d. Basement membrane lies between glomerulus and Bowman's capsule
      e. Glomerular-capsular membrane—formed by glomerular endothelium, basement membrane, and visceral layer of Bowman's capsule; function is filtration
   2. Proximal tubule—first part of renal tubule nearest to Bowman's capsule; follows a winding, convoluted course; also known as proximal convoluted tubule
   3. Loop of Henle (Figure 27-8)
      a. Renal tubule segment just beyond proximal tubule
      b. Consists of a thin descending limb, a sharp turn, and a thick ascending limb
      c. Juxtamedullary nephron—a nephron with a loop of Henle that dips into the medulla
      d. Cortical nephron—a nephron with a loop of Henle that does not dip into the medulla but remains within the cortex
   4. Distal tubule—convoluted tubule beyond the loop of Henle; also known as distal convoluted tubule
   5. Collecting duct
      a. Straight tubule joined by the distal tubules of several nephrons
      b. Joins larger ducts; larger collecting ducts of one renal pyramid converge to form one tube that opens at a renal papilla into a calyx

C. Blood vessels of the kidneys—kidneys are highly vascular
   1. Renal artery—large branch of abdominal aorta; brings blood into each kidney
   2. Interlobular arteries—between the pyramids of the medulla, the renal artery branches; interlobular arteries extend toward the cortex, arch over the bases of the pyramids, and form arcuate arteries; from arcuate arteries, interlobular arteries penetrate the cortex
   3. Pattern of blood flow through kidneys—interlobular artery → afferent arteriole → glomerulus → efferent arteriole → peritubular capillaries (including vasa recta) → venules → interlobular vein (Figure 27-9)
   4. Juxtaglomerular apparatus—located where the afferent arteriole brushes past the distal tubule; important to maintenance of blood flow homeostasis by reflexively secreting renin when blood pressure in the afferent arteriole drops

FUNCTION OF THE KIDNEY

A. Overview of kidney function
   1. Chief functions of the kidney are to process blood and form urine
   2. Basic functional unit of the kidney is nephron; forms urine through three processes (Figure 27-10)
      a. Filtration—movement of water and solutes from the plasma in the glomerulus into the capsular space of the Bowman's capsule
      b. Reabsorption—movement of molecules out of the tubule and into the peritubular blood
      c. Secretion—movement of molecules out of the peritubular blood and into the tubule for excretion

B. Filtration—first step in blood processing that occurs in the renal corpuscles
   1. From blood in the glomerular capillaries, water and solutes filter into Bowman's capsule; takes place through the glomerular-capsular membrane
   2. Filtration occurs due to existence of a pressure gradient
   3. Glomerular capillary filtration occurs rapidly due to the increased number of fenestrations
   4. Glomerular hydrostatic pressure and filtration are directly related to systemic blood pressure

C. Reabsorption—second step in urine formation; occurs due to passive and active transport mechanisms from all parts of the renal tubules; major portion of reabsorption occurs in proximal tubules (Figure 27-11)
   1. Reabsorption in the proximal tubule—most water and solutes are recovered by the blood, leaving only a small volume of tubule fluid to move on to the loop of Henle
      a. Sodium—actively transported out of tubule fluid and into blood (Figure 27-13)
      b. Glucose and amino acids—passively transported out of tubule fluid by means of the sodium cotransport mechanism
      c. Chloride, phosphate, and bicarbonate ions passively move into blood due to an imbalance of electrical charge
      d. Water—movement of sodium and chloride into blood causes an osmotic imbalance, moving water passively into blood
      e. Urea—approximately one half of urea passively moves out of tubule with the remaining urea moving on to the loop of Henle
   2. Reabsorption in the loop of Henle (Figure 27-15)
      a. Water is reabsorbed from the tubule fluid, and urea is picked up from the interstitial fluid in the descending limb
      b. Sodium and chloride is reabsorbed from the filtrate in the ascending limb, where the reabsorption of salt makes the tubule fluid dilute and creates and

maintains a high osmotic pressure of the medulla's interstitial fluid

C. Reabsorption in the distal tubules and collecting ducts
  1. Distal tubule reabsorbs sodium by active transport but in smaller amounts than in the proximal tubule
  2. ADH is secreted by the posterior pituitary and targets the cells of distal tubules and collecting ducts to make them more permeable to water
  3. With the reabsorption of water in the collecting duct, the urea concentration of the tubule fluid increases, causing urea to diffuse out of the collecting duct into the medullary interstitial fluid
  4. Urea participates in a counter-current multiplier mechanism that, along with counter-current mechanisms of the loop of Henle and vasa recta, maintains the high osmotic pressure needed to form concentrated urine and avoid dehydration

D. Tubular secretion
  1. Tubular secretion—the movement of substances out of the blood and into tubular fluid
  2. Descending limb of the loop of Henle secretes urea via diffusion
  3. Distal and collecting tubules secrete potassium, hydrogen, and ammonium ions
  4. Aldosterone—hormone that targets the cells of the distal and collecting tubule cells, causes increased activity of the sodium-potassium pumps
  5. Secretion of hydrogen ions increases with increased blood hydrogen ion concentration

E. Regulation of urine volume (Figure 27-19)
  1. ADH influences water reabsorption; as water is reabsorbed, the total volume of urine is reduced by the amount of water removed by the tubules; ADH reduces water loss
  2. Aldosterone, secreted by adrenal cortex, increases distal tubule absorption of sodium, raising the sodium concentration of blood and thus promoting reabsorption of water
  3. Atrial natriuretic hormone (ANH), secreted by specialized atrial muscle fibers, promotes loss of sodium via urine; opposes aldosterone, causing the kidneys to reabsorb less water and thereby produce more urine
  4. Urine volume—also related to total amount of solutes other than sodium excreted in the urine; generally, the more solutes, the more urine

F. Urine composition—approximately 95% water with several substances dissolved in it; the most important are
  1. Nitrogenous wastes—result of protein metabolism; e.g., urea, uric acid, ammonia, and creatinine
  2. Electrolytes—mainly the following ions: sodium, potassium, ammonium, chloride, bicarbonate, phosphate, and sulfate; amounts and kinds of minerals vary with diet and other factors
  3. Toxins—during disease, bacterial poisons leave the body in the urine
  4. Pigments—especially urochromes
  5. Hormones—high hormone levels may spill into the filtrate
  6. Abnormal constituents—such as blood, glucose, albumin, casts, or calculi

CYCLE OF LIFE: URINARY SYSTEM

A. Changes in kidney structure and function occur within narrow limits
  1. Structure—changes indicate disease or result from trauma
  2. Function—operates under various conditions, but inability to cope with extreme conditions will result in death from the buildup of toxins in the blood

B. Infants—kidney function less efficient; urine less concentrated
C. Young children—incontinence; reflex emptying when bladder fills, normal sphincter activity keeps urine in bladder until it fills
D. Older adults
  1. Incontinence from loss of sphincter control
  2. Decreased renal clearance
  3. Loss of functional nephron units; gradual decline in renal function

THE BIG PICTURE: THE URINARY SYSTEM AND THE WHOLE BODY

1. Homeostasis of water and electrolytes in body fluids relies on proper functioning of the kidneys; nephrons process blood to adjust its content to maintain a relatively constant internal environment
2. Urinary and cardiovascular systems are interdependent
3. Endocrine and nervous systems must operate properly to ensure efficient kidney function

MECHANISMS OF DISEASE: RENAL AND URINARY DISORDERS

A. Obstructive disorders—abnormalities that interfere with normal urine flow anywhere in the urinary tract; if urine backs up into kidney, hydronephrosis occurs
  1. Renal calculi—also called kidney stones
    a. Crystallized mineral chunks that develop in the renal pelvis or calyces
    b. Staghorn calculi—large branched stones that form in renal pelvis and branched calyces
    c. If small, calculi will be voided with urine; large stones may obstruct the ureters, causing renal colic
  2. Neurogenic bladder
    a. Disruption of nerve input to the bladder results in loss of normal control of voiding
    b. Characterized by involuntary retention of urine, distention of the bladder, and, perhaps, a burning sensation or fever with chills
  3. Tumors and other obstructions
    a. Typically obstruct urine flow
    b. Renal cell carcinomas—most common kidney tumor; malignant neoplasms
    c. Renal ptosis—"drooping" kidneys; ureters may kink and obstruct normal flow of urine

B. Urinary tract infections (UTIs) involve urethra, bladder, ureter, and/or kidneys; caused by bacteria
  1. Urethritis—inflammation of the urethra; occurs more commonly in males
  2. Cystitis—inflammation of the bladder; more common in females; characterized by pelvic pain, frequent urge to urinate, and hematuria
  3. Nephritis—general term referring to kidney disease, especially inflammatory conditions; pyelonephritis—inflammation of the renal pelvis and connective tissues of the kidney

C. Glomerular disorders—collectively called glomerulonephritis; without successful treatment, kidney failure can result
  1. Nephrotic syndrome—collection of signs and symptoms that accompany various glomerular disorders; characterized by
    a. Proteinuria—presence of protein in urine
    b. Hypoalbuminemia—low albumin concentration in the blood, resulting from the loss of blood albumin through holes in the damaged glomeruli; causes widespread edema
    c. Edema—general tissue swelling caused by accumulation of fluids in tissue spaces

2. Acute glomerulonephritis—most common form of kidney disease, due to streptococcal infection; can progress to chronic glomerulonephritis
3. Chronic glomerulonephritis—general name for various noninfectious glomerular disorders; characterized by progressive kidney damage leading to renal failure
D. Kidney failure—renal failure; failure of kidney to properly process blood and form urine
   1. Acute renal failure
      a. Abrupt reduction in kidney function; characterized by oliguria and sharp increase in nitrogenous compounds in the blood
      b. Assessed by blood urea nitrogen (BUN) test
      c. Caused by various factors—e.g., hemorrhage, severe burns, acute glomerulonephritis, obstruction of lower urinary tract
      d. Recovery is usually rapid and complete if underlying cause is identified and successfully treated
   2. Chronic renal failure (Figure 27-20)

   a. Slow, progressive condition resulting from the gradual loss of nephrons; with decreased kidney function, glomerular filtration rate (GFR) decreased, causing BUN levels to increase
   b. Progresses through three stages
      (1) Stage 1—some nephrons are lost, but remaining nephrons compensate; BUN level stays within normal limits; GFR decreased; often asymptomatic and may last for years
      (2) Stage 2—renal insufficiency—can no longer adapt to loss of nephrons; BUN level increases dramatically; polyuria and dehydration may occur
      (3) Stage 3—uremia or uremic syndrome—very high BUN value caused by loss of kidney function; low GFR causes low urine production and oliguria; edema and hypertension often occur; unless an artificial kidney is used or a new kidney is transplanted, progressive loss of kidney function will lead to death

REVIEW QUESTIONS

1. List the principal and accessory organs of the urinary system.
2. Name, locate, and give the main function(s) of each organ of the urinary system.
3. Identify the beginnings of the "plumbing system" of the urinary system.
4. How does the mechanism for voiding urine start?
5. Define the terms *micturition* and *incontinence*.
6. The male urethra is part of two different systems. Identify them.
7. Describe the microscopic structure of the kidney.
8. Describe the mechanism of urine formation, relating each step to the part of the nephron that performs it.
9. Diagram the flow of blood through the kidney tissue.
10. Define the terms *filtration*, *reabsorption*, and *secretion*.
11. How is effective filtration pressure calculated?
12. In the proximal tubules, why and how do chloride ions move into the blood?
13. Describe the solute concentration of the interstitial fluid of the medulla.
14. What happens to sodium and chloride in the ascending limb of the loop of Henle?
15. How does antidiuretic hormone alter the solute concentration of the urine?
16. What happens to potassium secretion when blood aldosterone concentration increases?
17. Identify two drugs secreted by tubule cells.
18. Identify and describe the function of the antagonist of aldosterone that is secreted by muscle fibers in the atrial wall of the heart.
19. What is the normal pH range for freshly voided urine?
20. Identify three body systems in addition to the urinary system that also excrete unneeded substances.
21. Define retention.
22. What is the most common cause of glycosuria?
23. What do elevated BUN levels indicate?
24. Define the term *osmolality*.

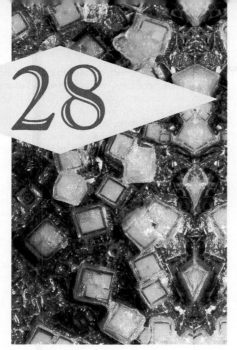

28

Fluid and Electrolyte Balance

Sodium chloride crystals

CHAPTER OUTLINE

OBJECTIVES

After you have completed this chapter, you should be able to:

1. Define the phrase *fluid and electrolyte balance.*
2. Discuss total body water content in terms of body weight, sex, and age.
3. List, describe, and compare the body fluid compartments and their subdivisions.
4. Discuss avenues by which water enters and leaves the body.
5. Discuss the objectives of fluid and electrolyte therapy in the practice of medicine.
6. Explain the mechanisms that maintain homeostasis of the body fluid compartments and of total body fluid volume.
7. Contrast the chemical composition of the three fluid compartments.
8. Discuss the measuring of electrolyte concentration and reactivity.
9. Explain the regulation of water and electrolyte levels in plasma and interstitial fluid.
10. Discuss edema and the mechanisms of edema formation.
11. Explain how water and electrolyte levels are regulated in intracellular fluid.
12. Discuss the regulation of sodium and potassium levels in body fluids.
13. Discuss dehydration.

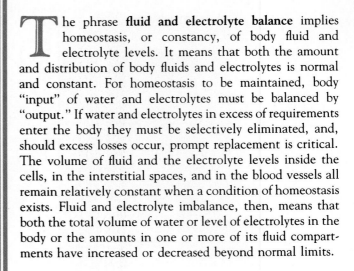

The phrase **fluid and electrolyte balance** implies homeostasis, or constancy, of body fluid and electrolyte levels. It means that both the amount and distribution of body fluids and electrolytes is normal and constant. For homeostasis to be maintained, body "input" of water and electrolytes must be balanced by "output." If water and electrolytes in excess of requirements enter the body they must be selectively eliminated, and, should excess losses occur, prompt replacement is critical. The volume of fluid and the electrolyte levels inside the cells, in the interstitial spaces, and in the blood vessels all remain relatively constant when a condition of homeostasis exists. Fluid and electrolyte imbalance, then, means that both the total volume of water or level of electrolytes in the body or the amounts in one or more of its fluid compartments have increased or decreased beyond normal limits.

INTERRELATIONSHIP OF FLUID AND ELECTROLYTE BALANCE

Several of the basic physical properties of matter discussed in Chapter 2 will help explain the mechanisms of fluid and electrolyte balance. The concept of chemical bonding is a good example. The type of chemical bonds between molecules of certain chemical compounds such as sodium chloride (NaCl) permits breakup, or dissociation, into separate particles (Na^+ and Cl^-). Recall that such compounds are known as **electrolytes.** The dissociated particles of an electrolyte are called **ions** and carry an electrical charge. Organic substances such as glucose, however, have a type of bond that does not permit the compound to break up, or **dissociate,** in solution. Such compounds are known as **nonelectrolytes.**

Many electrolytes and their dissociated ions are of critical importance in fluid balance. Fluid balance and electrolyte balance are so interdependent that if one deviates from normal, so does the other. A discussion of one therefore necessitates a discussion of the other.

TOTAL BODY WATER

The fluid or water content of the human body will range from 40% to 60% of its total weight. Normal values for fluid volume, however, vary considerably—mainly according to the fat content of the body. Fat people have a lower water content per kilogram of body weight than slender people. Women have a relatively lower water content than men—but mainly because a woman's body contains a higher percentage of fat. Total fluid volume and fluid distribution also vary with age. In infants, total body water constitutes about 75% of body weight. This percentage will decrease rapidly during the first 10 years of life. As an adult individual grows older the amount of body water continues to decrease so that in the aged individual fluid makes up a smaller percentage of body weight. In young adults the percent of body weight represented by water will average about 57% for males and 47% for females.

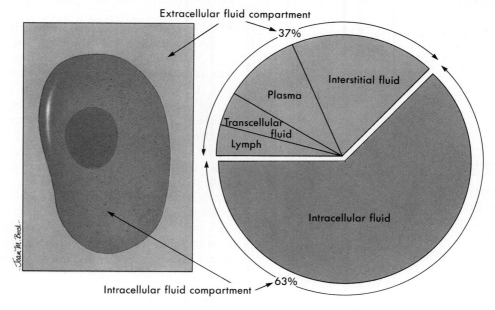

FIGURE 28-1   **Distribution of total body water.**

Body Fluid Compartments

Functionally, the total body water can be subdivided into two major **fluid compartments** called the **extracellular** and the **intracellular fluid compartments. Extracellular fluid (ECF)** consists mainly of the *plasma* found in the blood vessels and the *interstitial fluid* that surrounds the cells. In addition, the lymph and so-called *transcellular fluid*—such as cerebrospinal fluid and the specialized joint fluids—are also considered as extracellular fluid. The distribution of body water by compartment is shown in Figure 28-1. **Intracellular fluid (ICF)** refers to the water inside the cells.

Extracellular fluid, as you already know, constitutes the internal environment of the body. It therefore serves the dual vital functions of providing a relatively constant environment for cells and of transporting substances to and from them. Intracellular fluid, on the other hand, because it is a solvent, functions to facilitate intracellular chemical reactions that maintain life. Compared according to volume, intracellular fluid is the largest, plasma the smallest, and the interstitial fluid in between. Figure 28-2 gives the typical normal fluid volumes in a young adult female. Note that intracellular fluid constitutes 33% of body weight, interstitial fluid 10%, and blood plasma 4%. Or, expressed differently, for every kilogram of its weight,

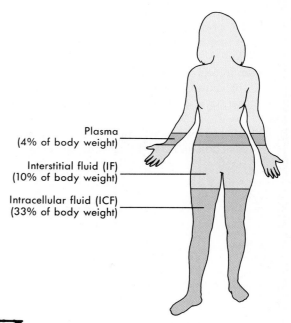

FIGURE 28-2   **Relative volumes of three body fluids.** Values represent fluid distribution in a young adult female.

the body contains about 330 ml of intracellular fluid, 100 ml of interstitial fluid, and 40 ml of plasma. With the exception of plasma, fluid volumes are proportionately larger in infants and children than in adults. Table 28-1 lists the volumes of body fluid compartments as a percentage of body weight.

TABLE 28-1   Volumes of body fluid compartments*

Body Fluid	Infant	Adult Male	Adult Female
Extracellular fluid			
Plasma	4	4	4
Interstitial fluid	26	15	10
Intracellular fluid	45	38	33
TOTAL	75	57	47

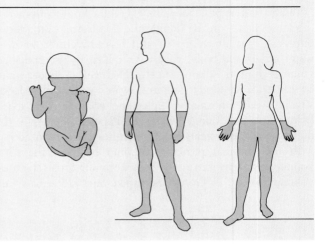

*Percentage of body weight

FOCUS ON... FLUID AND ELECTROLYTE THERAPY

The term **parenteral therapy** is used to describe the administration of nutrients, special fluids, and/or electrolytes by injection. The term implies that whatever is administered enters the body by injection and not through the alimentary canal. Examples of parenteral routes include **intravenous** (into veins) and **subcutaneous** (under the skin). Significant quantities of nutrient or electrolyte solutions that are injected subcutaneously must be isotonic with plasma or cellular damage will occur. Such solutions may be administered intravenously, however, regardless of tonicity, if correct rates of administration are used. The intravenous route is the preferred route for all fluid and electrolyte solutions. It permits the body to adjust its fluid compartments in the same way that it does following the ordinary intake of water and food. The ideal route for the absorption of nutrients and fluids is, of course, the digestive tract. However, if for any reason that route cannot be used, parenteral administration of these substances is required to sustain life.

Parenteral solutions are generally given to accomplish one or more of three primary objectives:

1. To meet current maintenance needs for nutrients, fluids, and electrolytes
2. To replace past losses
3. To replace concurrent losses (additional losses that are in excess of maintenance needs)

Although many different types and combinations of nutrients and electrolytes in solution are available to meet almost every medical need, 85% to 95% of all individuals needing fluid therapy are treated with one or more of the seven basic solutions listed below:

1. Carbohydrate in water
2. Carbohydrate in various strengths of saline
3. Normal saline (0.9% NaCl)
4. Potassium solutions
5. Ringer's solution
6. Lactate solutions
7. Ammonium chloride solutions

Carbohydrate and water solutions not only supply water for body needs but also provide calories required for energy. Dextrose (glucose) and fructose (levulose) are the common parenteral carbohydrates. Perhaps the most frequently used parenteral solution is 5% dextrose in water (D5W).

Various carbohydrate and saline solutions are also available for parenteral use. Such solutions are of primary value in individuals who have a chloride deficit, as well as ongoing fluid and caloric needs. Patients who are vomiting or undergoing gastric suction that results in the loss of chloride in hydrochloric acid need these solutions. Prolonged and heavy sweating and diarrhea also produce chloride deficits.

Current trends in parenteral therapy have reduced the frequency of normal saline use administered independently of other electrolytes or carbohydrates. In recent years, normal saline as a general purpose electrolyte has been replaced by the use of Ringer's solution, which provides more of the essential electrolytes in physiological proportions. Ringer's solution is often described as normal saline modified by the addition of calcium and potassium in amounts approximating those found in plasma. Normal saline is still useful and widely used in cases where chloride loss is equal to or greater than the loss of sodium, however.

Potassium, lactate, and ammonium chloride solutions are specialty fluids used in the treatment of such conditions as acid-base imbalances or, in the case of potassium, administered during the healing phase of severe burns or in cases of actual potassium deficiency. In acidosis, lactate is rapidly converted by the liver to bicarbonate ions, and administration of ammonium chloride is useful in treating alkalosis. Acid-base imbalances and their treatment will be discussed in Chapter 29.

AVENUES BY WHICH WATER ENTERS AND LEAVES THE BODY

Water enters the body, as everyone knows, from the digestive tract—in the liquids one drinks and in the foods one eats. But, in addition, and less universally known, water enters the body, that is, is added to its total fluid volume, from its billions of cells. Each cell produces water by catabolizing foods, and this water enters the bloodstream. Water normally leaves the body by four exits: kidneys (urine), lungs (water in expired air), skin (by diffusion and by sweat), and intestines (feces). In accord with the cardinal principle of fluid balance, the total volume of water entering the body normally equals the total volume leaving. In short, fluid intake normally equals fluid output. Figure 28-3 illustrates the portals of water entry and exit, and Table 28-2 gives their normal volumes. These, however, can vary considerably and still be considered normal.

QUICK CHECK

✔ 1. *Name the type of chemical compound that breaks up, or dissociates, in solution to form ions.*

✔ 2. *Plasma and interstitial fluid are subdivisions of what major body fluid compartment?*

✔ 3. *List the volumes of body fluid compartments in a young adult female as a percentage of body weight.*

✔ 4. *List the major "portals" of water entry and exit from the body.*

TABLE 28-2	Typical normal values for each portal of water entry and exit (with wide variations)		
Intake		**Output**	
Water in foods	700 ml	Lungs (water in expired air)	350 ml
Ingested liquids	1,500 ml	Skin	
Water formed by catabolism	200 ml	By diffusion	350 ml
		By sweat	100 ml
		Kidneys (urine)	1,400 ml
		Intestines (in feces)	200 ml
TOTALS	2,400 ml		2,400 ml

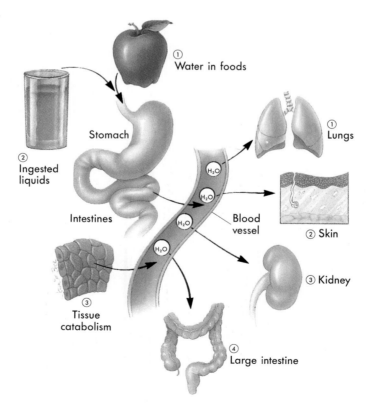

FIGURE 28-3   Sources of fluid intake and output.

SOME GENERAL PRINCIPLES ABOUT FLUID BALANCE

The cardinal principle about fluid balance is this: fluid balance can be maintained only if intake equals output. Obviously, if more water or less leaves the body than enters it, imbalance will result. Total fluid volume will increase or decrease but cannot remain constant under those conditions.

Devices for varying output so that it equals intake constitute the most crucial mechanisms for maintaining fluid balance, but mechanisms for adjusting intake to output also operate. Figure 28-4 summarizes the **aldosterone mechanism** for decreasing fluid output (urine volume) to compensate for decreased intake. Cells in the outer zone of the adrenal cortex that secrete aldosterone are also influenced by highly specialized cells (**juxtaglomerular cells**) in the kidney. If blood pressure or the level of blood Na$^+$ ions decreases, additional aldosterone will be secreted. When stimulated, specialized juxtaglomerular cells in the kidney secrete *renin*, which, in turn, acts on angiotensinogen in the bloodstream to form angiotensin I, which is eventually converted in lung tissue to angiotensin II (see Figure 15-20, p. 425). Angiotensin II increases aldosterone

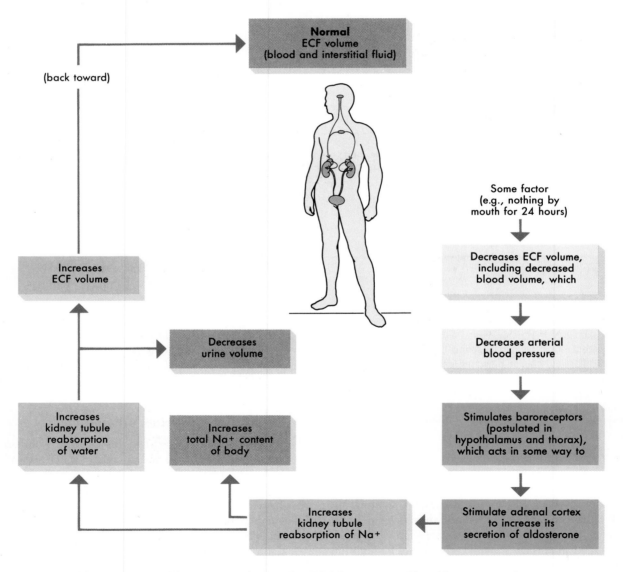

FIGURE 28-4 Aldosterone mechanism for ECF homeostasis. The aldosterone mechanism tends to restore normal extracellular fluid (ECF) volume when it decreases below normal. Excess aldosterone, however, leads to excess extracellular fluid volume, that is, excess blood volume (hypervolemia) and excess interstitial fluid volume (edema), and, also, to an excess of the total Na$^+$ content of the body. The renin-angiotensin mechanism influencing aldosterone secretion is shown in Figure 15-20.

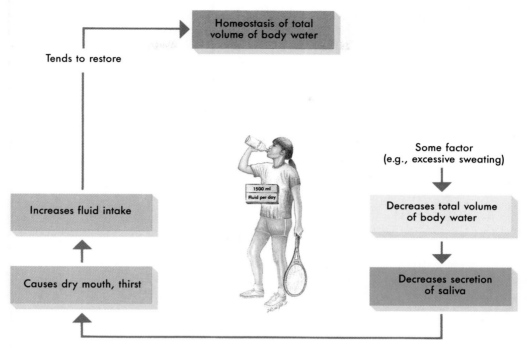

FIGURE 28-5   **Homeostasis of the total volume of body water.** A basic mechanism for adjusting intake to compensate for excess output of body fluid is diagrammed.

 Thirst Center

The sensation of thirst is apparently regulated, at least in part, by highly specialized cells in the **thirst center** of the hypothalamus of the brain. These cells, called **osmoreceptors,** detect the increase in solute concentration in the extracellular fluid caused by water loss. Signals generated by the osmoreceptors are sent directly to the cerebrum where they are interpreted, like the sensation of a dry mouth, as thirst.

secretion and also acts on the brain, stimulating the sensation of thirst. Thirst is associated with any factor, such as blood loss or hemorrhage, that decreases total volume of body water. Simple dehydration caused by sweating also results in reduced saliva secretion and thirst. Details of the **renin-angiotensin mechanism** are discussed in Chapter 15. Figure 28-5 diagrams a postulated mechanism for adjusting intake to compensate for excess output.

Mechanisms for controlling water movement between the fluid compartments of the body constitute the most rapid-acting fluid balance devices. They serve first of all to maintain normal blood volume at the expense of interstitial fluid volume.

MECHANISMS THAT MAINTAIN HOMEOSTASIS OF TOTAL FLUID VOLUME

Under normal conditions, homeostasis of the total volume of water in the body is maintained or restored primarily by devices that adjust output (urine volume) to intake and secondarily by mechanisms that adjust fluid intake.

Regulation of Urine Volume

Two factors together determine urine volume: the glomerular filtration rate and the rate of water reabsorption by the renal tubules. The glomerular filtration rate, except under abnormal conditions, remains fairly constant —hence it does not normally cause urine volume to fluctuate. The rate of tubular reabsorption of water, on the other hand, fluctuates considerably. The rate of tubular reabsorption therefore, rather than the glomerular filtration rate, normally adjusts urine volume to fluid intake. The amount of antidiuretic hormone (ADH) and of aldosterone secreted regulates the amount of water reabsorbed by the kidney tubules (discussed on p. 720; see also Figure 28-4). In other words, urine volume is regulated chiefly by hormones secreted by the posterior lobe of the pituitary

gland (ADH) and by the adrenal cortex (aldosterone). The regulation of aldosterone secretion by renin-angiotensin has also been discussed.

Although changes in the volume of fluid loss via the skin, the lungs, and the intestines also affect the fluid intake-output ratio, these volumes are not automatically adjusted to intake volume, as is the volume of urine.

Factors that Alter Fluid Loss Under Abnormal Conditions

The rate of respiration and the volume of sweat secreted may greatly alter fluid output under certain abnormal conditions. For example, a patient who hyperventilates for an extended time loses an excessive amount of water via the expired air. If, as frequently happens, the individual also takes in less water by mouth than normal, the fluid output then exceeds intake and a fluid imbalance, namely, dehydration (that is, a decrease in total body water) develops. Other abnormal conditions such as vomiting, diarrhea, or intestinal drainage also cause fluid and electrolyte output to exceed intake and so produce fluid and electrolyte imbalances.

Regulation of Fluid Intake

Physiologists disagree about the details of the mechanism for controlling intake so that it increases when output increases and decreases when output decreases. In general, it operates in this way: when dehydration starts to develop, salivary secretion decreases, producing a "dry mouth feeling" and the sensation of thirst. The individual then drinks water, thereby increasing fluid intake to offset increased output, and this tends to restore fluid balance (Figure 28-5). If, however, an individual takes nothing by mouth for several days, fluid balance cannot be maintained despite every effort of homeostatic mechanisms to compensate for the zero intake. Obviously, under this condition, the only way balance could be maintained would be for fluid output to also decrease to zero. But this cannot occur. Some output is obligatory. Why? Because as long as respirations continue, some water leaves the body by way of the expired air. Also, as long as life continues, an irreducible minimum of water diffuses through the skin.

QUICK CHECK

1. *How does aldosterone secretion restore normal extracellular fluid (ECF) volume when it decreases below normal?*
2. *Identify the two substances that are most important in regulating the amount of water reabsorbed by the kidney tubules.*
3. *Name the two most important factors that alter fluid loss under abnormal conditions.*

CHEMICAL CONTENT, DISTRIBUTION, AND MEASUREMENT OF ELECTROLYTES IN BODY FLUIDS

We have defined an electrolyte as a compound that will break up or dissociate into charged particles called ions when placed in solution. Sodium chloride, when dissolved in water, provides a positively charged sodium ion (Na^+) and a negatively charged chloride ion (Cl^-).

If two electrodes charged with a weak current are placed in an electrolyte solution, the ions will move, or migrate, in opposite directions according to their charge. Positive ions such as Na^+ will be attracted to the negative electrode (cathode) and are called **cations.** Negative ions such as Cl^- will migrate to the positive electrode (anode) and are called **anions.** Various anions and cations serve critical nutrient or regulatory roles in the body. Important cations include sodium (Na^+), calcium (Ca^{++}), potassium (K^+), and magnesium (Mg^{++}). Important anions include chloride (Cl^-), bicarbonate (HCO_3^-), phosphate ($HPO_4^=$), and many proteins.

The importance of electrolytes in controlling the movement of water between the body fluid compartments will be discussed in this chapter. Their role in maintaining acid-base balance will be examined in Chapter 29.

Extracellular vs. Intracellular Fluids

Compared chemically, plasma and interstitial fluid (the two extracellular fluids) are almost identical. Intracellular fluid, on the other hand, shows striking differences from either of the two extracellular fluids. Let us examine first the chemical structure of plasma and interstitial fluid as shown in Figure 28-6, p. 740, and Table 28-3.

TABLE 28-3   Electrolyte composition of blood plasma

Cations	Anions
142 mEq Na^+	102 mEq Cl^-
4 mEq K^+	26 HCO_3^-
5 mEq Ca^{++}	17 protein
	6 other
2 mEq Mg^{++}	2 $HPO_4^=$
TOTAL 153 mEq/L plasma	153 mEq/L plasma

Perhaps the first difference between the two extracellular fluids that you notice (Figure 28-6) is that blood contains a slightly larger total of electrolytes (ions) than do interstitial fluids. If you compare the two fluids, ion for ion, you will discover the most important difference between blood plasma and interstitial fluid. Look at the anions (negative ions) in these two extracellular fluids. Note that blood contains an appreciable amount of protein anions. Interstitial fluid, in contrast, contains hardly any protein anions. This is the only functionally important difference between blood and interstitial fluid. It exists because the normal capillary membrane is practically impermeable to proteins. Hence almost all protein anions remain behind in the blood instead of filtering out into the interstitial fluid. Because proteins remain in the blood, certain other differences also exist between blood and interstitial fluid—

notably, blood contains more sodium ions and fewer chloride ions than does interstitial fluid.

Extracellular fluids and intracellular fluid are more unlike than alike chemically. Chemical difference predominates between the extracellular and intracellular fluids. Chemical similarity predominates between the two extracellular fluids. Study Figure 28-6 and make some generalizations about the main chemical differences between the extracellular and intracellular fluids. For example: what is the most abundant cation in the extracellular fluids? In the intracellular fluid? What is the most abundant anion in the extracellular fluids? In the intracellular fluid? What about the relative concentrations of protein anions in extracellular fluids and intracellular fluid?

The only reason we called attention to the chemical structure of the three body fluids is that here, as elsewhere,

*Key to symbols:

Na^+	Sodium
K^+	Potassium
Mg^{++}	Magnesium
Ca^{++}	Calcium
Cl^-	Chloride
$SO_4^=$	Sulfate
HCO_3^-	Bicarbonate
$HPO_4^=$	Phosphate
$H \cdot HCO_3$	Carbonic acid

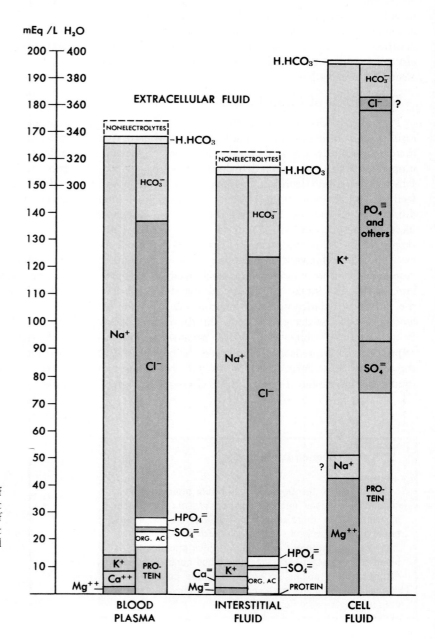

FIGURE 28-6 **Chief chemical constituents of three fluid compartments.** The column of figures at the left (200, 190, 180, etc.) indicates amounts of cations or of anions, whereas the figures on the right (400, 380, 360, etc.) indicate the *sum* of cations and anions.

structure determines function. In this instance the chemical structure of the three fluids helps control water and electrolyte movement between them. Or, phrased differently, the chemical structure of body fluids, if normal, functions to maintain homeostasis of fluid distribution and, if abnormal, results in fluid imbalance. **Hypervolemia** (excess blood volume) is a case in point. **Edema** (discussed in detail on p. 743), too, frequently stems from changes in the chemical structure of body fluids.

Before discussing mechanisms that control water and electrolyte movement between blood, interstitial fluid, and intracellular fluid, it is important to understand the units used for measuring electrolytes.

Measuring Electrolyte Reactivity

Once the important electrolytes and their constituent ions in the body fluid compartments had been established, physiologists needed to measure changes in their levels to understand the mechanisms of fluid balance. To have meaning, measurement units used to report electrolyte levels must be related to actual physiological activity. In the past, only the weight of an electrolyte in a given amount of solution—its *concentration*—was measured. The number of milligrams per 100 ml of solution (mg%) was one of the most frequently used units of measurement. However, simply reporting the concentration of an important electrolyte such as sodium or calcium in milligrams per 100 ml of blood (mg%) gives no direct information about its chemical combining power or physiological activity in body fluids. The importance of valence and electrovalent or ionic bonding in chemical reactions was discussed in Chapter 2. The reactivity or combining power of an electrolyte depends not just on the number of molecular particles present but also on the total number of ionic charges (valence). Univalent ions such as sodium (Na^+) carry only a single charge, but the divalent calcium ion (Ca^{++}) carries two units of electrical charge.

The need for a unit of measurement more related to activity has resulted in increasing use of a more meaningful measurement yardstick—the **milliequivalent.** Milliequivalents measure the number of ionic charges or electrovalent bonds in a solution and therefore serve as an accurate measure of the chemical (physiological) combining power, or reactivity, of a particular electrolyte solution. The number of milliequivalents of an ion in a liter of solution (mEq/L) can be calculated from its weight in 100 ml (mg%) using a convenient conversion formula.

Conversion of milligrams per 100 ml (mg%) to milliequivalents per liter (mEq/L):

$$mEq/L = \frac{mg/100\ ml \times 10 \times Valence}{Atomic\ weight}$$

Example: Convert 15.6 mg% K^+ to mEq/L

Atomic weight of K^+ = 39
Valence of K^+ = 1

$$mEq/L = \frac{15.6 \times 10 \times 1}{39} = \frac{156}{39} = 4$$

Therefore 15.6 mg/100 ml K^+ = 4 mEq/L.

QUICK CHECK

✔ 1. *List three important cations and anions that serve critical nutrient or regulatory roles in the body.*
✔ 2. *Name the most abundant chemical constituent in blood plasma, interstitial fluid, and intracellular fluid.*
✔ 3. *Identify the units used to describe electrolyte concentration and electrolyte reactivity.*

REGULATION OF WATER AND ELECTROLYTE LEVELS IN PLASMA AND INTERSTITIAL FLUID (ECF)

Over 70 years ago, English physiologist Ernest Starling advanced a hypothesis about the nature of the mechanism that controls water movement between plasma and interstitial fluid—that is, across the capillary membrane. This hypothesis has since become one of the major premises of physiology and is often spoken of as Starling's **law of the capillaries.** According to this law, the control mechanism for water exchange between plasma and interstitial fluid consists of four pressures: **blood hydrostatic** and **colloid osmotic pressures** on one side of the capillary membrane and **interstitial fluid hydrostatic** and **colloid osmotic pressures** on the other side.

We are ready now to try to answer the following question: how does the chemical structure of body fluids control water movement between them and thereby control fluid distribution in the body?

According to the physical laws governing filtration and osmosis, **blood hydrostatic pressure (BHP)** tends to force fluid out of capillaries into interstitial fluid (IF), but **blood colloid osmotic pressure (BCOP)** tends to draw it back into them. **Interstitial fluid hydrostatic pressure (IFHP),** in contrast, tends to force fluid out of the interstitial fluid into the capillaries, and **interstitial fluid colloid osmotic pressure (IFCOP)** tends to draw it back out of capillaries. In short, two of these pressures constitute vectors in one direction and two in the opposite direction. This process is similar in many ways to the mechanism responsible for formation of glomerular filtrate studied in the last chapter. The movement of fluids and electrolytes between plasma and interstitial fluid caused by hydrostatic and colloid osmotic pressure is illustrated in Figure 28-7.

The difference between the two sets of opposing forces obviously represents the net or effective filtration pressure—in other words, the effective force tending to produce the net fluid movement between blood and interstitial fluid. In general terms, therefore, we may state Starling's law of the capillaries this way: the rate and direction of fluid exchange between capillaries and interstitial fluid is determined by the hydrostatic and colloid

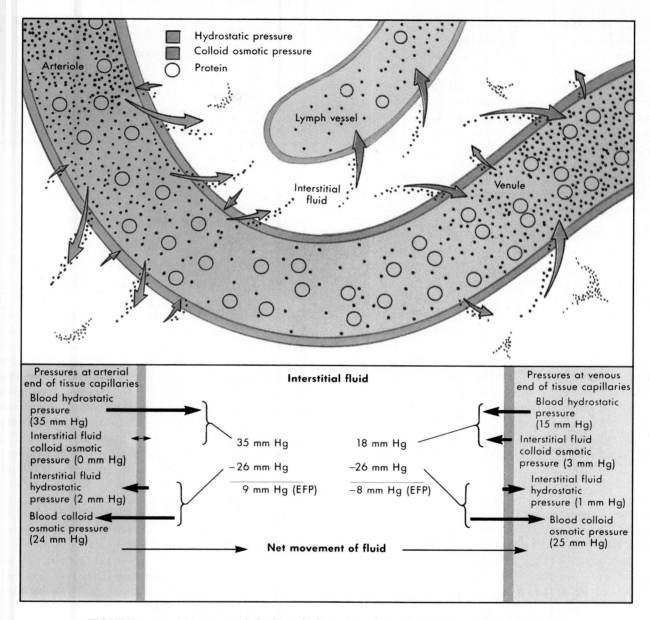

FIGURE 28-7   Movement of fluids and electrolytes between plasma and interstitial fluid caused by hydrostatic and colloid osmotic pressure. See text for discussion.

osmotic pressures of the two fluids. Or, we may state it more specifically as a formula:

$$(BHP + IFCOP) - (IFHP + BCOP) = EFP^*$$

Note that the factors enclosed in the first set of parentheses tend to move fluid out of capillaries and that those in the second set oppose this movement—they tend to move fluid into the capillaries.

---

*BHP, blood hydrostatic pressure; *IFCOP*, interstitial fluid colloid osmotic pressure; *IFHP*, interstitial fluid hydrostatic pressure; *BCOP*, blood colloid osmotic pressure; *EFP*, effective filtration pressure between blood and interstitial fluid.

To illustrate operation of Starling's law (Figure 28-7), let us consider how it controls water exchange at the arterial ends of tissue capillaries. The lower left-hand portion of Figure 28-7 gives typical normal pressures. Using these figures in Starling's law of the capillaries we get (35 + 0) – (2 + 24) = 9 mm Hg net pressure (EFP), causing water to filter out of blood at arterial ends of capillaries into interstitial fluid.

The same law operates at the venous end of capillaries (see the lower right-hand portion of Figure 28-7). Again, apply Starling's law of the capillaries. What is the net effective pressure at the venous ends of capillaries? In which direction does it cause water to move? Assuming

that the figures given are normal, do you agree that theoretically "the same amount of water returns to the blood at the venous ends of the capillaries as left it from the arterial ends"?

On the basis of our discussion thus far, we can formulate some principles about the transfer of water between blood and interstitial fluid.

1. No net transfer of water occurs between blood and interstitial fluid as long as the effective filtration pressure (EFP) equals 0, that is, when

$$(BHP + IFCOP) = (IFHP + BCOP)$$

2. A net transfer of water, a "fluid shift," occurs between blood and interstitial fluid whenever the EFP does not equal 0, that is, when

$$\begin{array}{c} (BHP + IFCOP) \\ \text{does not equal} \\ (IFHP + BCOP) \end{array}$$

3. Since (BHP + IFCOP) is a force that tends to move water out of capillary blood, fluid shifts out of blood into interstitial fluid whenever

$$\begin{array}{c} (BHP + IFCOP) \\ \text{is greater than} \\ (IFHP + BCOP) \end{array}$$

4. Since (IFHP + BCOP) is a force that tends to move water out of interstitial fluid into capillary blood, fluid shifts out of interstitial fluid into blood whenever

$$\begin{array}{c} (IFHP + BCOP) \\ \text{is greater than} \\ (BHP + IFCOP) \end{array}$$

Or, stated the other way around, fluid shifts out of interstitial fluid into blood whenever

$$\begin{array}{c} (BHP + IFCOP) \\ \text{is less than} \\ (IFHP + BCOP) \end{array}$$

Edema

Edema may be defined as the presence of abnormally large amounts of fluid in the intercellular tissue spaces of the body. The condition is a classic example of fluid imbalance and may be caused by disturbances in any of the factors that govern the interchange between blood plasma and interstitial fluid compartments. Examples include:

♦ Retention of electrolytes (especially Na^+) in the extracellular fluid as a result of increased aldosterone secretion or following serious renal disease such as acute glomerulonephritis.

♦ An increase in capillary blood pressure. Normally fluid is drawn from the tissue spaces into the venous end of a tissue capillary because of the low venous hydrostatic pressure and the high water-pulling force of the plasma proteins (Figure 28-7). This balance is upset by anything that will increase the capillary hydrostatic pressure. The generalized venous congestion of heart failure is the most common cause of widespread edema. In patients with this condition, blood cannot flow freely through the capillary beds, and therefore the pressure will increase until venous return of blood improves.

♦ A decrease in the concentration of plasma proteins normally retained in the blood (Figures 28-8 and 28-9). This may occur as a result of increased capillary permeability caused by infection, burns, or shock.

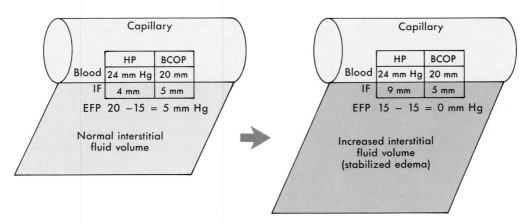

FIGURE 28-8 Edema formation. The mechanism of edema formation can be initiated by a decrease in blood protein concentration and, therefore, in blood colloid osmotic pressure. In the diagram on the left, blood osmotic pressure has just decreased to 20 from the normal 25 mm Hg. This increases the effective filtration pressure (EFP) to 5 mm Hg from a normal of 0 (see Starling's formula, p. 742). The EFP of 5 mm Hg causes fluid to shift out of blood into interstitial fluid (IF) until the EFP again equals 0—in this case, when the interstitial fluid volume has increased enough to raise interstitial fluid hydrostatic pressure to 9 mm Hg, as shown in the diagram on the right. At this point a new equilibrium is established, and equal amounts of water once more are exchanged between the blood and interstitial fluid. Thus the increased interstitial fluid volume—that is, the edema—becomes stabilized.

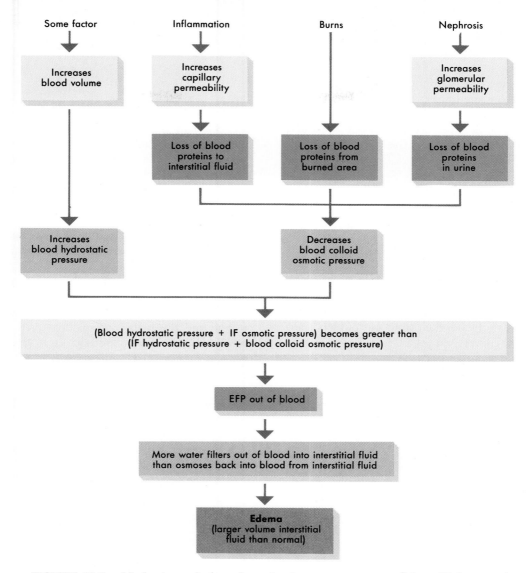

FIGURE 28-9   Mechanisms of edema formation in some common conditions. *IF,* Interstitial fluid; *EFP,* effective or net filtration pressure (see also Figure 28-4).

REGULATION OF WATER AND ELECTROLYTE LEVELS IN INTRACELLULAR FLUID (ICF)

It is the plasma membrane that separates the intracellular and extracellular fluid compartments. We know that chemical difference predominates between these two fluids, and it is the plasma membrane that plays a critical role in the regulation of intracellular fluid composition.

The mechanism that regulates water movement through cell membranes is similar to the one that regulates water movement through capillary membranes. In other words, interstitial fluid and intracellular fluid hydrostatic and colloid osmotic pressures regulate water transfer between these two fluids. But because the colloid osmotic pressures of interstitial and intracellular fluids vary more than their hydrostatic pressures, their colloid osmotic pressures serve

as the chief regulators of water transfer across cell membranes. Their colloid osmotic pressures, in turn, are directly related to the electrolyte concentration gradients—notably sodium and potassium—maintained across cell membranes. As Figure 28-6 shows, most of the body sodium is outside the cells. A concentration of 138 to 143 mEq/L makes sodium the chief electrolyte by far in interstitial fluid. The intracellular fluid's main electrolyte is potassium salt. Therefore a change in the sodium or the potassium concentrations of either of these fluids causes the exchange of fluid between them to become unbalanced.

Pores in the selectively permeable cell membrane retain large molecules, such as proteins, inside the cell but permit many smaller ions such as sodium and potassium to either diffuse through or be selectively transported across the membrane. The electrical charge difference that is created by unequal concentration of electrolytes on either side of

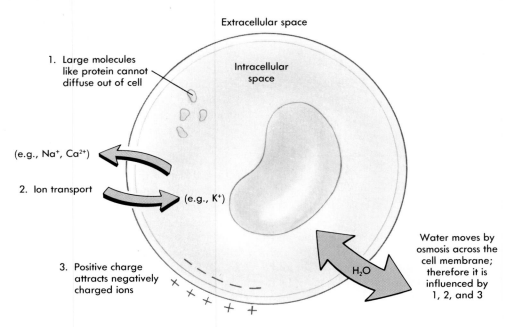

1. Large molecules like protein cannot diffuse out of cell

(e.g., Na⁺, Ca²⁺)

2. Ion transport

(e.g., K⁺)

3. Positive charge attracts negatively charged ions

Extracellular space

Intracellular space

Water moves by osmosis across the cell membrane; therefore it is influenced by 1, 2, and 3

H_2O

FIGURE 28-10   Mechanisms regulating movement of water and solutes between ECF and ICF spaces. Osmotic pressure is influenced by large protein molecules in the ICF, which cannot diffuse through the small pores of the cell membrane. In addition, electrolyte transport and diffusion and the charge difference across the cell membrane also influence water movement by osmosis.

the cell membrane also influences the composition of intracellular fluid. Mechanisms that regulate the movement of solutes and water between ECF and ICF spaces are summarized in Figure 28-10.

Any change in the solute concentration of extracellular fluid will have a direct effect on water movement across the cell membrane in one direction or another. If for any reason dehydration occurs, the concentration of solutes in the extracellular fluid will increase, and osmosis will cause water to move from the intracellular space into the extracellular space. In severe dehydration, the increasing concentration of intracellular fluid caused by water loss to the extracellular space results in abnormal metabolism or cellular death. Increased movement of water into the cell is caused by decreased concentration of solutes in the extracellular fluids.

A decrease in interstitial fluid sodium concentration immediately decreases interstitial fluid colloid osmotic pressure, making it hypotonic to intracellular fluid colloid osmotic pressure. In other words, a decrease in interstitial fluid sodium concentration establishes a colloid osmotic pressure gradient between interstitial and intracellular fluids. This causes net osmosis to occur out of interstitial fluid into cells. In short, interstitial fluid and intracellular fluid electrolyte concentrations are the main determinants of their colloid osmotic pressures; their colloid osmotic pressures regulate the amount and direction of water transfer between the two fluids, and this regulates their volumes. Hence fluid balance depends on electrolyte balance. Conversely, electrolyte balance depends on fluid balance. An imbalance in one produces an imbalance in the other (Figure 28-11).

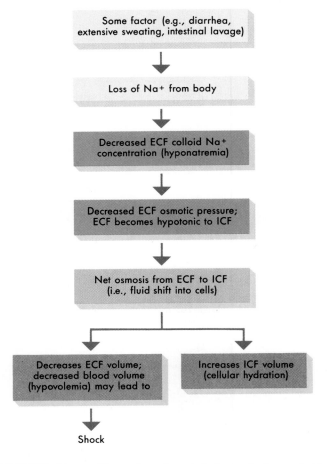

Some factor (e.g., diarrhea, extensive sweating, intestinal lavage)

↓

Loss of Na⁺ from body

↓

Decreased ECF colloid Na⁺ concentration (hyponatremia)

↓

Decreased ECF osmotic pressure; ECF becomes hypotonic to ICF

↓

Net osmosis from ECF to ICF (i.e., fluid shift into cells)

↓

Decreases ECF volume; decreased blood volume (hypovolemia) may lead to

Increases ICF volume (cellular hydration)

↓

Shock

FIGURE 28-11   How electrolyte imbalance leads to fluid imbalances. The schematic uses the example of sodium deficit and resulting hypovolemia (cellular hydration). *ECF,* Extracellular fluid; *ICF,* intracellular fluid.

REGULATION OF SODIUM AND POTASSIUM LEVELS IN BODY FLUIDS

Normal sodium concentration in interstitial fluid and potassium concentration in intracellular fluid depend on many factors but especially on the amount of ADH and aldosterone secreted. As shown in Figure 28-12, ADH regulates extracellular fluid electrolyte concentration and colloid osmotic pressure by regulating the amount of water reabsorbed into blood by renal tubules. Aldosterone, on the other hand, regulates extracellular fluid volume by regulating the amount of sodium reabsorbed into blood by renal tubules (see Figure 28-4).

If for any reason conservation of body sodium is required, the normal kidney is capable of excreting an essentially sodium-free urine and is therefore considered the chief regulator of sodium levels in body fluids. Sodium lost in sweat can become appreciable with elevated environmental temperatures or fever. However the thirst that results may lead to replacement of water but not the lost sodium and, as a result of the increased fluid intake, the remaining sodium pool may be diluted even more. Sweat loss of sodium is not therefore considered a normal means of regulation.

In addition to the well-regulated movement of sodium into and out of the body and between the three primary fluid compartments, there is a continuous movement or circulation of this important electrolyte between a number of internal secretions. Over 8 *liters* of various internal secretions such as saliva, gastric and intestinal secretions, bile, and pancreatic fluid are produced every day (Figure 28-13). The total daily secretion of sodium into these alimentary tract fluids alone will average between 1,200 and 1,400 mEq. A 70 kg (154 lb) adult has a total body sodium pool of only 2,800 to 3,000 mEq. Precise regulatory and conservation mechanisms for sodium are required for survival.

Chloride is the most important extracellular anion and is almost always linked to sodium. Generally ingested together, they provide in large part for the isotonicity of extracellular fluid. Chloride ions are generally excreted in

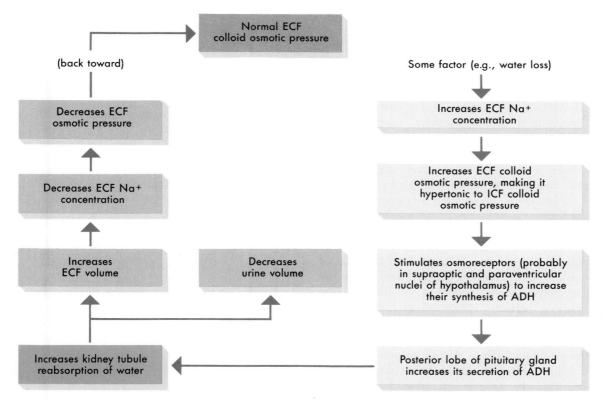

FIGURE 28-12   Antidiuretic hormone (ADH) mechanism for ECF homeostasis. Antidiuretic hormone (ADH) mechanism helps maintain homeostasis of extracellular fluid (ECF) colloid osmotic pressure by regulating its volume and thereby its electrolyte concentration, that is, mainly ECF Na$^+$ concentration. *ECF,* Extracellular fluid; *ICF,* intracellular fluid.

the urine as a potassium salt, and therefore chloride deficiency—**hypochloremia**—is often found in cases of potassium loss.

Total body potassium content in the average-sized adult is approximately 4,000 mEq. Because the majority of body potassium is intracellular, serum determinations, which normally fall between 4.0 and 5.0 mEq/L, may not be the best index to reflect imbalances. The body may lose one

third to one half of its intracellular potassium reserves before the loss is reflected in lowered serum potassium levels.

Potassium deficit, or **hypopotassemia,** occurs whenever there is cell breakdown, as in starvation, burns, trauma, or dehydration. As individual cells disintegrate, potassium enters the extracellular fluid and is rapidly excreted because it is not reabsorbed efficiently by the kidney.

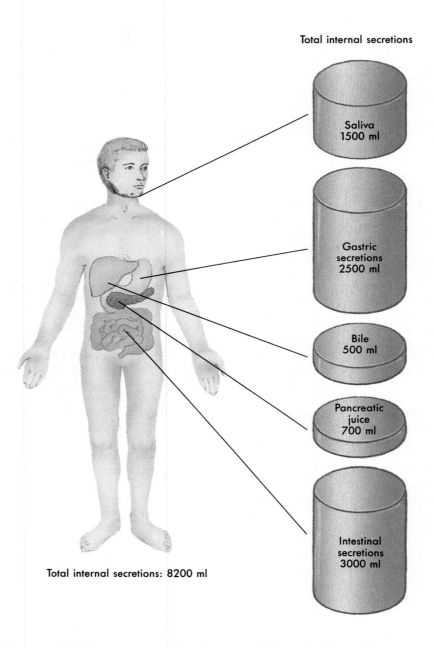

FIGURE 28-13   Sodium-containing internal secretions. The total volume of these secretions may reach 8,000 or more milliliters in a 24-hour period.

MECHANISMS OF DISEASE

Fluid and Electrolyte Disorders

Total body water constitutes 45% to 65% of body weight in adults and is regulated by homeostatic mechanisms of the neuroendocrine system, heart, kidneys, and blood vessels. Normal water losses, known as insensible water losses, occur through expired air from the lungs and from the skin, constituting about 0.4 to 0.5 ml/h/kg body weight. Abnormally excessive water losses constitute a volume deficit and can lead to a state of dehydration, or **hypovolemia,** where there is inadequate fluid volume in the extracellular compartment (see essay, p. 749). If left untreated, it can result in hypovolemic shock. Many causes for dehydration exist. The most common cause of dehydration includes fluid losses from the gastrointestinal tract due to vomiting or diarrhea. Another cause occurs when an individual fails to take in sufficient oral fluid because of depression, nausea, or oral trauma. *Diaphoresis,* or excessive perspiration, may also cause dehydration, with rapid respirations leading to water vapor losses. Any disorder of the kidneys that increases urine excretion can lead to dehydration, such as nephritis. Fluid can also shift into a space outside the normal fluid compartments during disease states that include ascites, burns, pancreatitis, or traumatic injuries. Regardless of the cause, fluid volume deficits cause low blood pressure and cardiac output, electrolyte disturbances, or acid-base abnormalities. Symptoms include dizziness, light-headedness, weakness, poor skin turgor, and tachycardia. Restoration of the fluid losses is the main goal of therapy. If the deficit is mild, volume can be replaced orally. If the dehydration is severe, fluid volume is replaced via the intravenous route. Examining the person's weight, skin turgor, blood pressure, and urine output will provide important data on fluid volume stability.

Fluid volume excess, or **hypervolemia,** is an expansion of fluid volume in the body. This can occur if the kidneys retain a large amount of sodium and water, as in congestive heart failure, nephrotic syndrome, renal failure, and liver failure. Manifestations include weight gain (most consistent sign), edema, dyspnea, tachycardia, and pulmonary congestion. Approaches to this disorder are to treat the original pathological process (i.e., renal failure), monitor the person's weight closely, and use diuretics cautiously to remove the excess fluid.

Water intoxication may result from over administration of water or hypotonic solutions to persons unable to dilute and excrete urine normally. This may occur with kidney insufficiency or patients with abnormal "thirst" mechanisms due to neurological disorders. Water content is elevated and serum sodium levels are diluted. Development of subtle mental changes such as confusion and lethargy occur. If intoxication is severe, stupor, seizures, and coma may result. Correction of the neurological impairment along with water restriction can reverse the symptoms.

Disturbances in electrolytes can occur in fluid volume abnormalities and many different disease states. **Hyponatremia** is a condition of decreased serum sodium concentration below the normal range (less than 136 mEq/L) and is usually due to an excess of water relative to solute. It may also be due to excessive losses of sodium. Causes of hyponatremia include skin losses via profuse perspiration, overzealous use of salt-wasting diuretics, adrenal insufficiency, renal or liver failure, low salt intake, or excessive water intake (diluting sodium content). Signs and symptoms include muscle cramps, nausea and vomiting, postural blood pressure changes, poor skin turgor, fatigue, and difficulty breathing. Cerebral swelling can occur in severe cases, causing confusion, hemiparesis (motor weakness on one side of the body), seizures, and coma. Management includes neurological assessment and the administration of sodium orally or intravenously. Water restriction can also suffice.

Hypernatremia is elevation of the serum sodium concentration above 145 mEq/L. It is usually indicative of a body water deficit relative to sodium, but can also result from grossly elevated sodium intake. Causes include lack of fluid intake, diarrhea, diabetes insipidus, loss of water via the respiratory tract, heart disease or congestive heart failure, renal failure, or ingestion of salt in abnormal amounts. Signs and symptoms are similar to those of dehydration and include thirst, disorientation, lethargy, and seizures. The neurological symptoms are thought to be due to cellular dehydration. Replacement with a hypotonic solution will help lower the sodium slowly, thereby reducing the risk of cerebral edema.

A common type of electrolyte imbalance is **hypokalemia,** a condition in which potassium is lost from the body, resulting in a serum potassium level below 3.5 mEq/L. Causes include potassium-wasting diuretics, increased urine output with loss of potassium, and vomiting or gastric suctioning without potassium replacement. Hypokalemia can be life threatening and includes manifestations of anorexia, muscle weakness, decreased reflexes, low blood pressure, and cardiac arrhythmias. Potassium can be replaced through the diet, with potassium-rich foods, or intravenously with caution.

The opposite of hypokalemia is **hyperkalemia,** or a serum potassium level above 5.5 mEq/L. This can be even more dangerous than hypokalemia because the myocardial muscle can be profoundly affected. The common cause of hyperkalemia is kidney disease but other factors such as vomiting, diarrhea, potassium-conserving diuretics, extensive tissue damage as in burns or trauma victims, severe infections, and Cushing's syndrome are also cited. Notable changes can be seen on the electrocardiogram, such as peaked T waves. Hyperkalemia can induce ventricular dysrhythmias, leading to possible cardiac arrest. Because potassium is a part of neuromuscular functions, the person may experience extremity muscle weakness or failure of the respiratory muscles. Intermittent diarrhea, nausea, and intestinal colic are also manifested. Dietary restriction of potassium is sufficient in mild cases but emergent intravenous administration of calcium gluconate may be required to correct cardiac symptoms. Also, correction of the underlying condition (i.e., trauma) and dialysis to remove the excess potassium can be instituted to correct severe hyperkalemia.

FOCUS ON... DEHYDRATION

The term **dehydration** is used to describe the condition that results from excessive loss of body water. Water deprivation or loss triggers a complex series of protective responses designed to maintain homeostasis of both water and electrolyte levels. Unfortunately, the term *dehydration* is incomplete. It does not, by definition, include the loss of electrolytes. To understand the control mechanisms that ensure fluid and electrolyte balance or properly interpret the clinical signs and symptoms of dehydration in disease states, it is important to realize that in any process of dehydration, water loss is always accompanied by loss of electrolytes. If water intake is reduced to the point of dehydration, there must be removal of a corresponding quantity of electrolytes to maintain the normal ionic content of body fluids. The same is true in the case of electrolyte loss when an accompanying loss of water must occur to maintain homeostasis of both fluid and electrolyte levels. Understanding the close interrelationships of water and electrolyte loss in dehydration provides the rationale for effective treatment. Water alone is inadequate; treatment of dehydration also requires appropriate electrolyte replacement therapy.

As discussed in Chapter 5, maintaining a constant core body temperature in a hot environment is an important function of the skin. As sweat evaporates, excess body heat can be eliminated. In hot weather or during extended periods of strenuous physical activity, the volume of water lost because of sweat production can reach 10 liters a day. If water intake is inadequate, signs of dehydration will appear very rapidly. As body water levels decrease, the initial defense mechanisms are directed toward maintaining an adequate blood volume.

In addition to water, sweat contains significant quantities of sodium and chloride. However, the relative loss of water in sweat is greater than the loss of electrolytes. Therefore as water is shifted from the interstitial fluid compartment to the plasma to compensate for fluid loss, the kidneys excrete the excess electrolytes to preserve normal ionic concentrations in the two compartments.

The chemical composition and actual volume of fluids lost from the body will also affect the type and effectiveness of defense mechanisms that occur. For example, fluids lost through vomiting or diarrhea will have differing ratios of fluid to electrolytes than sweat, and the actual electrolyte composition and concentration will also be different. As a result, the type of electrolyte excretion or retention by the kidneys that will be needed to maintain ionic balance in the fluid compartments will also change.

There is a lag in the volume-electrolyte adjustment mechanism triggered by dehydration. Shifts in fluid occur more quickly between compartments than the adjustment in electrolyte levels. However, if water and electrolyte losses are limited and the interval between loss and replacement is short, the symptoms of dehydration will be mild and transitory.

In severe and prolonged water deprivation or loss, the initial shift of interstitial fluid to plasma will be followed by movement of water from the intracellular compartment as well. Over time the extra- and intracellular fluid losses are about equal.

Extracellular (interstitial) water is more "expendable" and quickly assessable as a fluid source to maintain blood volume in the early stages of body fluid loss. It is said to serve as the "first line of defense" against dehydration. As extracellular fluid is depleted, intracellular water must be used to prolong survival time. Ultimately, the volume of the extracellular fluid can be reduced by almost 60% and intracellular fluids by 30% before death occurs.

CHAPTER SUMMARY

INTERRELATIONSHIP OF FLUID AND ELECTROLYTE BALANCE

A. Fluid and electrolyte balance—implies homeostasis
B. Electrolytes have chemical bonds that allow dissociation into ions, which carry an electrical charge; of critical importance in fluid balance
C. Fluid balance and electrolyte balance are interdependent

TOTAL BODY WATER

A. Fluid content of human body ranges from 40% to 60% of its total weight
B. Fluid content varies according to age, gender, and weight

BODY FLUID COMPARTMENTS

A. Two major fluid compartments (Figure 28-1)
B. Extracellular fluid (ECF) constitutes the internal environment of the body
   1. Consists mainly of plasma and interstitial fluid
   2. Lymph, cerebrospinal fluid, and specialized joint fluids are considered extracellular
   3. Functions of ECF provide a relatively constant environment for cells and transport substances to and from the cells
C. Intracellular fluid (ICF)—water inside the cells
   1. Functions to facilitate intracellular chemical reactions that maintain life
   2. By volume, ICF is the largest body fluid compartment

AVENUES BY WHICH WATER ENTERS AND LEAVES THE BODY

A. Water enters the body via the digestive tract; water is also added to the total fluid volume from each cell as it catabolizes food, and the resulting water enters the bloodstream (Figure 28-3)
B. Water leaves the body via four exits (Figure 28-3)
   1. As urine through the kidney
   2. As water in expired air through the lungs
   3. As sweat through the skin
   4. As feces from the intestine

SOME GENERAL PRINCIPLES ABOUT FLUID BALANCE

A. Cardinal principle of fluid balance is: fluid balance can be maintained only if intake equals output
B. Mechanisms are available to adjust output and intake to maintain fluid balance, e.g., aldosterone mechanism (Figure 28-4), renin-angiotensin mechanism
C. Most rapid fluid balance devices are mechanisms for controlling water movement between fluid compartments of the body; will maintain normal blood volume at the expense of interstitial fluid volume

MECHANISMS THAT MAINTAIN HOMEOSTASIS OF TOTAL FLUID VOLUME

A. Under normal conditions, homeostasis of total volume of water is maintained or restored primarily by adjusting urine

volume and secondarily by fluid intake (Figure 28-5)

B. Regulation of urine volume—two factors determine urine volume
  1. Glomerular filtration rate, except under abnormal conditions, remains fairly constant
  2. Rate of tubular reabsorption of water fluctuates considerably; normally adjusts urine volume to fluid intake; influenced by amount of antidiuretic hormone and of aldosterone

C. Factors that alter fluid loss under abnormal conditions—rate of respiration and volume of sweat secreted may alter fluid output under certain abnormal conditions; vomiting, diarrhea, or intestinal drainage can produce fluid and electrolyte imbalances

D. Regulation of fluid intake—when dehydration begins to develop, salivary secretion decreases, producing the sensation of thirst; increased fluid intake to offset increased output tends to restore fluid balance

CHEMICAL CONTENT, DISTRIBUTION, AND MEASUREMENT OF ELECTROLYTES IN BODY FLUID

A. Extracellular vs. intracellular fluids
  1. Plasma and interstitial fluid (ECFs) are almost identical in chemical make-up, with intracellular fluid showing striking differences (Figure 28-6)
  2. Extracellular fluids
     a. Difference between blood and interstitial fluid—blood contains a slightly larger total of ions than interstitial fluid
     b. Functionally important difference between blood and interstitial fluid is the number of protein anions; blood has an appreciable amount, whereas interstitial fluid has hardly any; since the capillary membrane is practically impermeable to proteins, almost all protein anions remain in the blood
  3. Intracellular fluids—ICF and ECF are more dissimilar than similar
  4. The chemical structure of plasma, interstitial fluid, and intracellular fluid helps control water and electrolyte movement between them

B. Measuring electrolyte concentration—done to help understand the mechanisms of fluid balance; converted to milliequivalents to provide information on actual physiological activity

C. Measuring electrolyte reactivity—milliequivalent—measures the number of ionic charges or electrocovalent bonds in a solution; accurately measures the physiological combining power of an electrolyte solution

D. Regulation of water and electrolyte levels in plasma and interstitial fluid (ECF)
  1. Law of capillaries—the control mechanism for water exchange between plasma and interstitial fluid consists of four pressures: blood hydrostatic and colloid osmotic pressures on one side of the capillary membrane and interstitial fluid hydrostatic and colloid osmotic pressures on the other side; two of the pressures for a vector in one direction and the other two in the opposite direction
     a. Blood hydrostatic pressure (BHP) forces fluid out of capillaries into interstitial fluid (IF)
     b. Blood colloid osmotic pressure (BCOP) draws fluid from IF into capillaries
     c. Interstitial fluid hydrostatic pressure (IFHP) forces fluid out of IF into capillaries
     d. Interstitial fluid colloid osmotic pressure (IFCOP) draws fluid from capillaries to interstitial fluid
  2. The rate and direction of fluid exchange between capillaries and interstitial fluid are determined by the hydrostatic and colloid osmotic pressures of the two fluids (Figure 28-7)
  3. Some principles about transfer of water between blood and interstitial fluid
     a. No net transfer of water occurs as long as (BHP + IFCOP) = (IFHP + BCOP)
     b. A net transfer of fluid occurs when (BHP + IFCOP) $\neq$ (IFHP + BCOP)
     c. Fluid shifts out of blood into interstitial fluid whenever (BHP + IFCOP) > (IFHP + BCOP)
     d. Fluid shifts out of interstitial fluid into blood whenever (BHP + IFCOP) < (IFHP + BCOP)

EDEMA

A. Edema—presence of abnormally large amounts of fluid in the intercellular tissue spaces of the body

B. Classic example of fluid imbalance; may be due to
  1. Retention of electrolytes in the extracellular fluid
  2. Increase in capillary blood pressure
  3. Decrease in the concentration of plasma proteins normally retained in the blood (Figure 28-8)

REGULATION OF WATER AND ELECTROLYTE LEVELS IN INTRACELLULAR FLUID (ICF)

A. Plasma membrane plays critical role in regulating intracellular fluid composition

B. IF and ICF hydrostatic and colloid pressures regulate water transfer between ECF and ICF; colloid osmotic pressures are chief regulators of water transfer across cell membranes, and these are directly related to the electrolyte concentration gradients maintained across cell membranes (Figure 28-10)

REGULATION OF SODIUM AND POTASSIUM LEVELS IN BODY FLUIDS

A. Normal sodium concentration in IF and potassium concentration in ICF depends on various factors, but especially on amount of ADH and aldosterone secreted
  1. ADH regulates ECF electrolyte concentration and colloid osmotic pressure by regulating amount of water reabsorbed into blood by renal tubules
  2. Aldosterone regulates ECF volume by regulating the amount of sodium reabsorbed into blood by renal tubules

B. When conservation of body sodium is required, the kidneys excrete an essentially sodium-free urine; kidneys are considered the chief regulator of sodium levels

C. Chloride—most important extracellular anion and is almost always linked to sodium; chloride ions are generally excreted in urine as potassium salts, thus hypochloremia is often associated with cases of potassium loss

D. Hypopotassemia occurs where there is cell breakdown; as cells disintegrate, potassium enters ECF and is rapidly excreted because it is not reabsorbed efficiently by the kidney

MECHANISMS OF DISEASE: FLUID AND ELECTROLYTE DISORDERS

A. Dehydration—inadequate fluid volume in the extracellular compartment; if left untreated, can result in hypovolemic shock
  1. Various causes: vomiting, diarrhea, insufficient fluid intake, diaphoresis with tachypnea, disorders of the kidney (which increases urine excretion), and fluid shifts due to certain disease states
  2. Regardless of cause, dehydration causes low blood pressure and cardiac output, electrolyte disturbances, or acid-base abnormalities
  3. Main goal of therapy is restoration of fluid

B. Hypervolemia—an expansion of fluid volume in the body; fluid volume excess
  1. Can occur when kidneys retain large amounts of sodium and water
  2. Goal is to treat original cause, monitor weight, and use diuretics
C. Water intoxication results from over administration of water or hypotonic solutions to individuals unable to dilute and excrete urine normally; water content elevated and serum sodium levels decreased
  1. Due to kidney insufficiency or patients with abnormal thirst mechanism due to neurological disorders
  2. Treat underlying cause
D. Hyponatremia—decreased serum sodium concentration
  1. Usually due to an excess of water relative to solute; may be due to excessive losses of sodium
  2. Treatment—neurological assessment, administration of sodium, water restriction
E. Hypernatremia—elevated serum sodium concentration
  1. Due to decreased fluid intake, diarrhea, diabetes insipidus, loss of water via respiratory tract, heart disease or congestive heart failure, renal failure, or ingestion of salt in abnormal amounts
  2. Treatment—replacement with a hypotonic solution to lower sodium slowly and reduce risk of cerebral edema
F. Hypokalemia—one of the common types of electrolyte imbalances; low serum potassium level
  1. Due to potassium-wasting diuretics, increased urine output with loss of potassium and vomiting or gastric suctioning without potassium replacement; can be life threatening
G. Hyperkalemia—elevated serum potassium levels; can profoundly affect myocardium
  1. Due to kidney disease most commonly; can be due to vomiting, diarrhea, potassium-conserving diuretics, extensive tissue damage, severe infections and Cushing's syndrome
  2. Treatment—mild cases, dietary restriction of potassium; more severe, administration of calcium gluconate, address underlying cause

REVIEW QUESTIONS

1. Define fluid and electrolyte balance.
2. Discuss the changes in total body water content from an infant to an adult.
3. How does total body water content differ in men and women?
4. List the compartments of extracellular fluid.
5. What are the four exits by which water normally leaves the body?
6. What are the objectives for giving parenteral solutions?
7. What is the cardinal principle about fluid balance?
8. Describe the aldosterone mechanism.
9. How is urine volume regulated?
10. Define the terms *cation* and *anion*.
11. Are plasma and interstitial fluid chemically similar or different? Explain.
12. Describe the electrolyte composition of blood plasma.
13. Compare the chemical composition of extracellular fluids and interstitial fluid.
14. Define the term *milliequivalent*. How is it used to measure electrolyte reactivity?
15. Explain Starling's law of the capillaries.
16. What are the four pressures involved in Starling's law?
17. When effective filtration pressure equals 0, what is the net transfer of water between blood and interstitial fluid?
18. When does a "fluid shift" occur between blood and interstitial fluid?
19. Identify various mechanisms that may lead to edema.
20. What role does the plasma membrane play in the regulation of intracellular fluid composition?
21. How does hyponatremia lead to fluid imbalances?
22. How does the antidiuretic hormone mechanism maintain homeostasis of extracellular fluid colloid osmotic pressure?
23. In your own words, define dehydration.
24. How do the cells in the thirst center of the hypothalamus work?

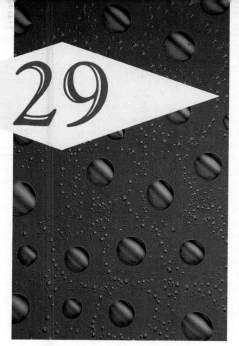

29

Acid-Base Balance

Water—formed when acids and bases combine

OBJECTIVES

After you have completed this chapter, you should be able to:

1. Define acid-base balance.
2. Discuss the concept of pH.
3. List four acids that contribute hydrogen ions to body fluids and identify the source of each.
4. Give examples of acid- and base-forming elements and identify dietary sources for each.
5. Identify and contrast chemical and physiological buffers.
6. Contrast strong and weak acids and bases.
7. Compare the buffering of a strong acid and base with a weak acid and base.
8. Explain how the chloride shift makes it possible for carbon dioxide to be buffered in red blood cells and then carried as bicarbonate in the plasma.
9. Contrast the respiratory and urinary mechanisms of pH control.
10. Compare the effects of hypoventilation and hyperventilation on blood pH.
11. Discuss the function of the distal renal tubule in acidification of urine.

Acid Any substance that increases the H⁺ ion concentration of a solution

Acidosis Excessive proportion of acid in the blood; metabolic acidosis = bicarbonate deficit; respiratory acidosis = carbonic acid excess

Alkaline Any substance that decreases the amount of H⁺ in a solution; same as base

Alkalosis Relative increase in alkaline content of body fluids; metabolic alkalosis = bicarbonate excess; respiratory alkalosis = carbonic acid deficit

Buffer Substance that prevents marked changes in the pH of a solution

Chloride shift Process that allows carbon dioxide to be buffered in the RBC and then carried as bicarbonate in the plasma

pH Symbol indicating hydrogen ion concentration

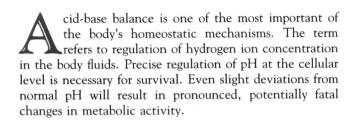

A cid-base balance is one of the most important of the body's homeostatic mechanisms. The term refers to regulation of hydrogen ion concentration in the body fluids. Precise regulation of pH at the cellular level is necessary for survival. Even slight deviations from normal pH will result in pronounced, potentially fatal changes in metabolic activity.

MECHANISMS THAT CONTROL pH OF BODY FLUIDS

Meaning of Term *pH*

Recall from Chapter 2 that water and all water solutions contain hydrogen ions (H⁺) and hydroxide ions (OH⁻). *pH* is a symbol used to mean the hydrogen ion (H⁺) concentration of a solution (see Figure 29-1). (Actually, pH stands for the negative logarithm of the hydrogen ion concentration, see FYI box, p. 46). pH indicates the degree of **acidity** or **alkalinity** of a solution. As the concentration of hydrogen ions increases, the pH goes down and the solution becomes more acid; a decrease in hydrogen ion concentration makes the solution more alkaline and the pH goes up. A pH of 7 indicates neutrality (equal amounts of H⁺ and OH⁻), a pH of less than 7 indicates acidity (more H⁺ than OH⁻), and a pH greater than 7 indicates alkalinity (more OH⁻ than H⁺). With a pH of about 1.6, gastric juice is the most acid substance in the body. Saliva has a pH of 7.7, on the alkaline side. The overall pH range is often expressed numerically on a logarithmic scale of 1 to 14. Keep in mind that a change of 1 pH unit on this type of scale represents a 10-fold difference in actual concentration of hydrogen ions.

Arterial and venous blood are slightly alkaline because both have a pH slightly higher than 7.0. The slight increase in acidity of venous blood (pH 7.36) compared to arterial blood (pH 7.41) results primarily from carbon dioxide entering venous blood as a waste product of cellular metabolism. The lungs remove the equivalent of over 30 *liters* of 1-normal carbonic acid each day from the venous blood by elimination of carbon dioxide, and yet 1 liter of venous blood contains only about 1/100,000,000 g more hydrogen ions than 1 liter of arterial blood. What incredible constancy! The pH homeostatic mechanism does indeed control effectively—astonishingly so.

Sources of pH-Influencing Elements

Both acids and bases continually enter the blood from absorbed foods and from the metabolism of nutrients at the cellular level. Therefore some kind of mechanism for

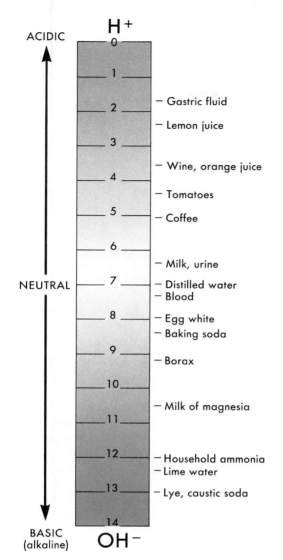

ACIDIC

NEUTRAL

BASIC
(alkaline) **OH⁻**

— Gastric fluid
— Lemon juice

— Wine, orange juice
— Tomatoes
— Coffee

— Milk, urine
— Distilled water
— Blood
— Egg white
— Baking soda
— Borax

— Milk of magnesia

— Household ammonia
— Lime water
— Lye, caustic soda

FIGURE 29-1 The pH range. Note that as concentration of H⁺ increases, the solution becomes increasingly acidic and the pH value decreases. As OH⁻ concentration increases, the pH value also increases and the solution becomes more and more basic, or alkaline.

neutralizing or eliminating these substances is necessary if blood pH is to remain constant. Although both acid and basic components are important, the homeostasis of body pH largely depends on the control of hydrogen ion concentration in the extracellular fluid. Hydrogen ions are continually entering the body fluids from (1) **carbonic,** (2) **lactic,** (3) **sulfuric,** and (4) **phosphoric acids** and (5) **acidic ketone bodies.**

Carbonic and lactic acids are produced by the aerobic and anaerobic metabolism of glucose, respectively. Sulfuric acid is produced when sulfur-containing amino acids are oxidized, and phosphoric acid accumulates when certain phosphoproteins and nucleoproteins are broken down for energy purposes. Acidic ketone bodies, which include **acetone, acetoacetic acid,** and **beta-hydroxybutyric acid,**

accumulate during the incomplete breakdown of fats. Each of these acids contributes H⁺ ions in varying amounts to the extracellular fluid and influences acid-base balance. Toxic accumulation of acidic ketone bodies is a common complication of untreated diabetes mellitus.

Minerals that remain after food has been metabolized are said to be either **acid** or **base** forming, depending on whether they contribute to formation of an acid or basic medium when in solution. Acid-forming elements include **chlorine, sulfur,** and **phosphorus**—all are abundant in high-protein foods such as meat, fish, poultry, and eggs. These foods are often designated as **acid-forming foods.** After metabolism is complete, most mixed diets contain a surplus of acid-forming mineral elements that must be continually buffered to maintain acid-base balance. Extremely high-protein diets that produce a predominantly *acid mineral residue* when metabolized may tax the body's ability to remain in acid-base balance if consumed over prolonged periods.

Mineral elements that are alkaline, or basic, in solution include **potassium, calcium, sodium,** and **magnesium.** All of these elements are found in fruits and vegetables, which nutritionists often label as **base-forming foods.** The predominantly **basic residue** that results after metabolism of a strict vegetarian diet may also tax the ability of the body to maintain acid-base balance because of a high influx of alkaline components into the extracellular fluid.

Foods containing acids that cannot be metabolized, such as rhubarb (oxalic) or cranberries (benzoic), are said to be direct acid-forming foods, whereas antacids such as sodium bicarbonate or calcium carbonate are examples of direct base-forming substances.

QUICK CHECK

✔ 1. *Define the term pH.*

✔ 2. *Is a solution with a pH above 7 acid or alkaline?*

✔ 3. *Identify three acid-forming and three base-forming elements.*

✔ 4. *Does CO₂ entering venous blood increase or decrease the pH level?*

TABLE 29-1   pH control systems

Type	Response Time	Example
Chemical buffer systems	Rapid	Bicarbonate buffer system
		Phosphate buffer system
		Protein buffer system
Physiological buffer systems	Delayed	Respiratory response system
		Renal response system

Types of pH Control Mechanisms

The two major types of control systems listed in Table 29-1—*chemical* and *physiological*—operate to maintain the constancy of body pH.

In the discussion that follows, buffer action is defined and the specific types of **chemical** and **physiological buffer systems** are discussed. The rapid-acting **chemical buffers** immediately combine with any added acid or alkali that enters the body fluids and thus prevent drastic changes in hydrogen ion concentration and pH. As explained later, all buffers act to prevent swings in pH even if hydrogen ion concentrations change. If the immediate action of chemical buffers cannot stabilize pH, the **physiological buffers** serve as a secondary defense against harmful shifts in pH of body fluids. pH shifts that are not halted by the immediate effects of chemical buffering cause the respiratory system to respond in 1 to 2 minutes, and changes in the rate and depth of breathing will occur. For reasons explained later, such changes in carbon dioxide levels will alter hydrogen ion concentration and help stabilize pH. If respiratory mechanisms are unable to stop the pH shift, a more powerful but slower acting renal physiological buffer system involving the excretion of either an acid or alkaline urine will be initiated. Collectively, these devices—buffers, respirations, and kidney excretion of acids and bases—might be said to constitute the pH homeostatic mechanism.

Effectiveness of pH Control Mechanisms— Range of pH

The most eloquent evidence of the effectiveness of the pH control mechanism is the extremely narrow range of blood pH—normally 7.36 to 7.41.

BUFFER MECHANISM FOR CONTROLLING pH OF BODY FLUIDS
Buffers Defined

In terms of action a buffer is a substance that prevents marked changes in the pH of a solution when an acid or a base is added to it. Let us suppose that a small amount of a strong acid, hydrochloric acid, is added to a solution that contains a buffer (e.g., blood) and that its pH decreases from 7.41 to 7.27. If the same amount of hydrochloric acid were added to pure water containing no buffers, its pH would decrease much more markedly, from 7 to perhaps 3.4. In both instances, pH decreased on addition of the acid but much less so with buffers present than without them.

In terms of chemical composition, buffers consist of two kinds of substances and are therefore often referred to as **buffer pairs.**

Buffer Pairs Present in Body Fluids

Most of the body fluid buffer pairs consist of a weak acid and a salt of that acid. The main buffer pairs present in body fluids are as follows:

$$\text{Bicarbonate pairs } \frac{NaHCO_3}{H_2CO_3}, \frac{KHCO_3}{H_2CO_3}, \text{ etc.}$$

$$\text{Plasma protein pair } \frac{Na \cdot Proteinate}{Proteins \text{ (weak acids)}}$$

$$\text{Hemoglobin pairs } \frac{K \cdot Hb}{Hb} \text{ and } \frac{K \cdot HbO_2}{HbO_2}$$

$$(\text{Hb and } HbO_2 \text{ are weak acids})$$

$$\text{Phosphate buffer pair } \frac{Na_2HPO_4 \text{ (basic phosphate)}}{NaH_2PO_4 \text{ (acid phosphate)}}$$

Action of Buffers to Prevent Marked Changes in pH of Body Fluids

Buffers react with a relatively strong acid (or base) to replace it with a relatively weak acid (or base). That is, an acid that highly dissociates to yield many hydrogen ions is replaced by one that dissociates less highly to yield fewer hydrogen ions. Thus by the buffer reaction, instead of the strong acid remaining in the solution and contributing many hydrogen ions to drastically lower the pH of the solution, a weaker acid takes its place, contributes fewer additional hydrogen ions to the solution, and thereby lowers its pH only slightly. Because blood contains buffer pairs, its pH fluctuates much less widely than it would without them. In other words, blood buffers constitute one of the devices for preventing marked changes in blood pH.

Let us consider, as a specific example of buffer action, how the sodium bicarbonate ($NaHCO_3$)–carbonic acid (H_2CO_3) system works in the presence of a strong acid or base.

Addition of a strong acid, such as hydrochloric acid (HCl), to the sodium bicarbonate–carbonic acid buffer system would initiate the reaction shown in Figure 29-2. Note how this reaction between HCl and the base bicarbonate ($NaHCO_3$) applies the principle of buffering. As a result of the buffering action of $NaHCO_3$, the weak acid, $H \cdot HCO_3$, replaces the very strong acid, HCl, and therefore the hydrogen ion concentration of the blood increases much less than it would have if HCl were not buffered.

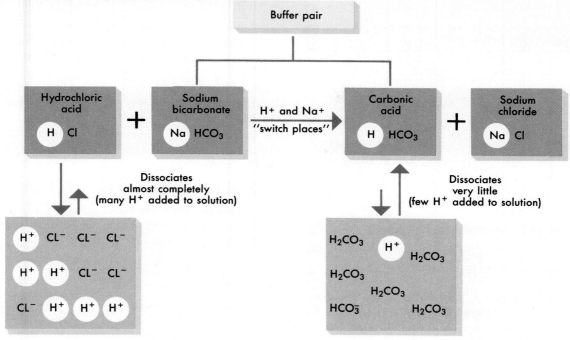

FIGURE 29-2 **Buffering action of sodium bicarbonate.** Buffering of acid HCl by $NaHCO_3$. As a result of the buffer action, the strong acid (HCl) is replaced by a weaker acid (H · HCO_3). Note that HCl, as a strong acid, "dissociates" almost completely and releases more H^+ than H_2CO_3. Buffering decreases the number of H^+ in the system.

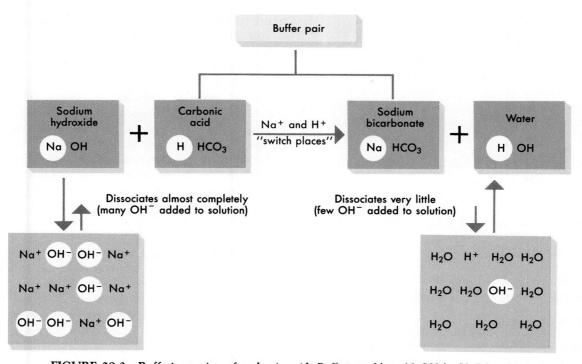

FIGURE 29-3 **Buffering action of carbonic acid.** Buffering of base NaOH by H_2CO_3. As a result of buffer action, the strong base (NaOH) is replaced by $NaHCO_3$ and H_2O. As a strong base, NaOH "dissociates" almost completely and releases large quantities of OH^-. Dissociation of H_2O is minimal. Buffering decreases the number of OH^- in the system.

If, on the other hand, a strong base such as sodium hydroxide (NaOH) is added to the same buffer system, the reaction shown in Figure 29-3 would take place. The hydrogen ion of H · HCO₃, the weak acid of the buffer pair, combines with hydroxyl ion (OH⁻) of the strong base NaOH to form water. Note what this accomplishes. It decreases the number of hydroxyl ions added to the solution, and this, in turn, prevents the drastic rise in pH that would occur in the absence of buffering.

The principles of buffer action illustrated by the reaction of HCl and NaOH with the sodium bicarbonate buffer pair can be applied equally to the plasma protein, hemoglobin, and phosphate buffer systems.

Carbon dioxide and other acid waste products are continuously being formed as a result of cellular metabolism. The formation of carbonic acid from carbon dioxide and water requires the enzyme carbonic anhydrase, which is found in the red blood cells. Carbonic acid is buffered primarily by the potassium salt of hemoglobin inside the red blood cell, as shown in Figure 29-4.

It is interesting to note that the KHCO₃ formed by the buffering of carbonic acid dissociates in the red blood cell, and the bicarbonate ion diffuses down its concentration gradient into the blood plasma. Due to the movement of these negatively charged ions out of the red blood cell, chloride ions move *into* the cell from the plasma to maintain the electrical balance on both sides of the RBC membrane. The process of exchanging a bicarbonate ion formed in the red blood cell with a chloride ion from the plasma is called the **chloride shift**. This process makes it possible for carbon dioxide to be buffered in the red blood cell and then carried as bicarbonate in the plasma. Figure 29-5 summarizes the reactions of the chloride shift.

Nonvolatile, or fixed, acids, such as hydrochloric acid,

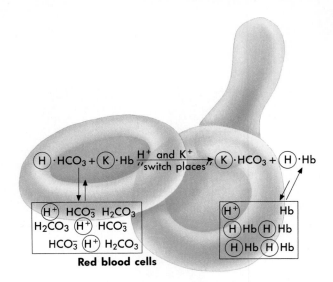

FIGURE 29-4 Buffering of the volatile carbonic acid (H · HCO₃) inside red blood cell by potassium salt of hemoglobin. Note that each molecule of carbonic acid is replaced by a molecule of acid hemoglobin. Since hemoglobin is a weaker acid than carbonic acid, fewer of these hemoglobin molecules dissociate to form hydrogen ions. Hence fewer hydrogen ions are added to red blood cell intracellular fluid than would be added by unbuffered carbonic acid. Also, since some of the carbonic acid in the red blood cell has come from plasma, fewer hydrogen ions remain in blood than would have if there were no buffering of carbonic acid.

lactic acid, and ketone bodies, are buffered mainly by sodium bicarbonate (Figure 29-6).

Normal blood pH and acid-base balance depend on a base bicarbonate to carbonic acid buffer pair ratio of 20/1 in the extracellular fluid. Actually, in a state of acid-base

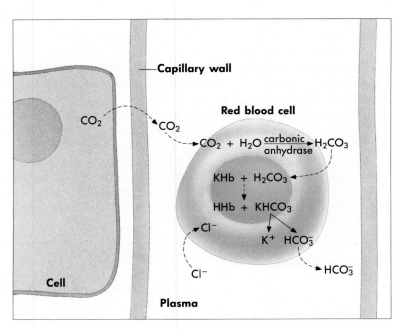

FIGURE 29-5   Chloride shift. Concentration of chloride ions (Cl⁻) in RBCs increases as bicarbonate ions (HCO₃⁻) diffuse out of the cell. Bicarbonate ions form as a result of the buffering of carbonic acid by the potassium salt of hemoglobin.

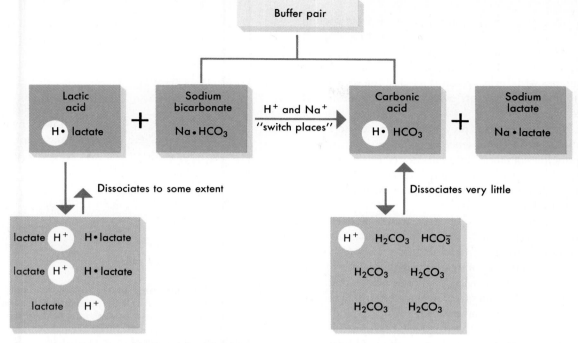

FIGURE 29-6 Lactic acid buffered by sodium bicarbonate. Lactic acid (H· lactate) and other "fixed" acids are buffered by $NaHCO_3$ in the blood. Carbonic acid (H· HCO_3 or H_2CO_3, a weaker acid than lactic acid) replaces lactic acid. As a result, fewer H^+ are added to blood than would be if lactic acid were not buffered.

balance a liter of plasma contains 27 mEq of $NaHCO_3$ as base bicarbonate (BB)—ordinary baking soda—and 1.3 mEq of carbonic acid (CA):

$$\frac{27 \text{ mEq NaHCO}_3}{1.3 \text{ mEq H}_2\text{CO}_3} = \frac{BB}{CA} = \frac{20}{1} = pH\ 7.4$$

The **ratio** of base to acid is critical. If the ratio is maintained, acid-base balance (pH) will remain near normal despite changes in the absolute amounts of either component of the buffer pair. For example, a BB/CA ratio of 40/2 or 10/0.5 would result in a compensated state of acid-base balance. However, an increase in the ratio causes an increase in pH **(uncompensated alkalosis),** and a decrease in the ratio causes a decrease in pH **(uncompensated acidosis).** The ability of the body to regulate the amount of either component of the bicarbonate buffer pair—to maintain the correct ratio for acid-base balance—makes this system one of the most important for controlling pH of body fluids.

The relationship between the hydrogen ion concentration of body fluids and the ratio of base bicarbonate to carbonic acid has been expressed as a mathematical formula called the **Henderson-Hasselbalch equation.** This equation is useful in clinical medicine to predict the blood pH changes that will occur if the sodium bicarbonate–carbonic acid buffer system ratio is altered by drugs or disease.

Evaluation of Role of Buffers in pH Control

Buffering alone cannot maintain homeostasis of pH. As we have seen, hydrogen ions are added continually to capillary blood despite buffering. If even a few more hydrogen ions were added every time blood circulated and no way were provided for eliminating them, blood hydrogen ion concentration would necessarily increase and thereby decrease blood pH. "Acid blood," in other words, would soon develop. Respiratory and urinary devices must therefore function concurrently with buffers in order to remove from the blood and from the body the hydrogen ions continually being added to blood. Only then can the body maintain constancy of pH over time.

FYI **Calculation of Henderson-Hasselbalch Equation**

Henderson-Hasselbalch equation for the bicarbonate buffer system is calculated as follows:

$$pH = 6.1 + \frac{\log[\text{HCO}_3^-]}{[\text{CO}_2]}$$ (molar concentrations of bicarbonate ion and of dissolved carbon dioxide)

QUICK CHECK

✔ 1. *Define the term* buffer.
✔ 2. *Identify the two major types of buffer systems in the body. Which buffer system is the most rapid acting?*
✔ 3. *Use equations to explain the buffering of HCl by sodium bicarbonate and NaOH by carbonic acid.*
✔ 4. *Define the terms* Henderson-Hasselbalch equation *and* chloride shift.

RESPIRATORY MECHANISM OF pH CONTROL

Explanation of Mechanism

Respirations play a vital part in controlling pH. With every expiration, carbon dioxide and water leave the body in the expired air. The carbon dioxide has come from the venous blood—has diffused out of it as it moves through the lung capillaries. Therefore less carbon dioxide remains in the arterial blood leaving the lung capillaries. The lower P_{CO_2} in arterial blood reduces the amount of carbonic acid and the number of hydrogen ions that can be formed in red blood cells by the following reactions:

$$CO_2 + H_2O \xrightarrow{\text{(carbonic anhydrase)}} H_2CO_3$$

$$H_2CO_3 \rightarrow H^+ + HCO_3^-$$

Arterial blood therefore has a lower hydrogen ion concentration and a higher pH than venous blood. A typical average pH for venous blood is 7.36, and 7.41 is a typical average pH for arterial blood.

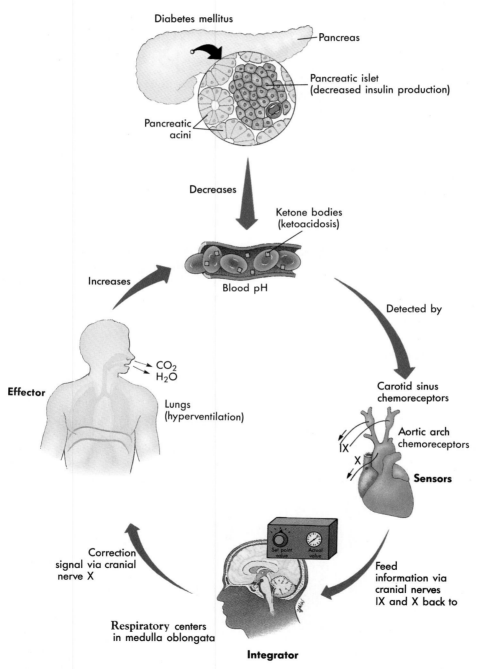

FIGURE 29-7 **Respiratory mechanism of pH control.** A rise in arterial blood CO_2 content or a drop in its pH (below about 7.38) stimulates respiratory center neurons. Hyperventilation results. Less CO_2 and therefore less carbonic acid and fewer hydrogen ions remain in the blood so that blood pH increases, often reaching the normal level.

Adjustment of Respirations to pH of Arterial Blood

For respirations to serve as a mechanism of pH control, there must be some mechanism for increasing or decreasing respirations as needed to maintain or restore normal pH. Suppose that blood pH has decreased; that is, the hydrogen ion concentration has increased. Respirations then need to increase to eliminate more carbon dioxide from the body and thereby leave less carbonic acid and fewer hydrogen ions in the blood.

One mechanism for adjusting respirations to arterial blood carbon dioxide content or pH operates in this way: neurons of the respiratory center are sensitive to changes in arterial blood carbon dioxide content and to changes in its pH. If the amount of carbon dioxide in arterial blood increases beyond a certain level, or if arterial blood pH decreases below about 7.38, the respiratory center is stimulated and respirations accordingly increase in rate and depth. This, in turn, eliminates more carbon dioxide, reduces carbonic acid and hydrogen ions, and increases pH back toward the normal level (Figure 29-7). The carotid chemoreflexes are also devices by which respirations adjust to blood pH and, in turn, adjust pH.

Some Principles Relating Respirations and pH of Body Fluids

♦ A decrease in blood pH below normal (acidosis) tends to stimulate increased respirations (hyperventila-

tion), which tends to increase pH back toward normal. In other words, acidosis causes hyperventilation, which in turn acts as a compensating mechanism for the acidosis.

♦ Prolonged hyperventilation may increase blood pH enough to produce alkalosis.

♦ An increase in blood pH above normal (or alkalosis) causes hypoventilation, which serves as a compensating mechanism for the alkalosis by decreasing blood pH back toward normal.

♦ Prolonged hypoventilation may decrease blood pH enough to produce acidosis.

URINARY MECHANISM OF pH CONTROL

General Principles about Mechanism

Because the kidneys can excrete varying amounts of acid and base, they, like the lungs, play a vital role in pH control. Kidney tubules, by excreting many or few hydrogen ions in exchange for reabsorbing many or few sodium ions, control urine pH and thereby help control blood pH. If, for example, blood pH decreases below normal, kidney tubules secrete more hydrogen ions from blood to urine and, in exchange for each hydrogen ion, reabsorb a sodium ion from the urine back into the blood. This, of course, decreases urine pH. But simultaneously—and of far more importance—it increases blood pH back toward normal. This urinary mechanism of pH control is a device for excreting varying amounts of hydrogen ions from the body to match the amounts entering the blood. It constitutes a much more effective device for adjusting hydrogen output to hydrogen input than does the body's only other

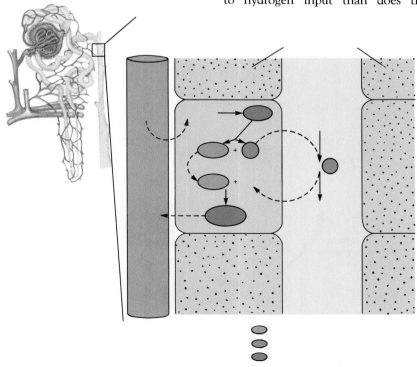

FIGURE 29-8   Acidification of urine and conservation of base by distal renal tubule excretion of H⁺. See text for discussion of mechanism.

mechanism for expelling hydrogen ions, namely, the respiratory mechanisms previously described. But abnormalities of any one of the three pH control mechanisms soon throw the body into a state of acid-base imbalance. Only when all three parts of this complex mechanism—buffering, respirations, and urine secretion—function adequately can acid-base balance be maintained.

Let us turn our attention now to mechanisms that adjust urine pH to counteract changes in blood pH.

Mechanisms that Control Urine pH

A decrease in blood pH accelerates the renal tubule ion-exchange mechanisms that both acidify urine and conserve blood's base; thereby tending to increase blood pH back to normal. The following paragraphs describe these mechanisms.

♦ Distal and collecting tubules secrete hydrogen ions into the urine in exchange for basic ions, which they reabsorb. Refer to Figure 29-8 as you read the rest of this paragraph. Note that carbon dioxide diffuses from tubule capillaries into distal tubule cells, where the enzyme carbonic anhydrase accelerates the combining of carbon dioxide with water to form carbonic acid. The carbonic acid dissociates into hydrogen ions and bicarbonate ions. The hydrogen ions then diffuse into the tubular urine, where they displace basic ions (most often sodium) from a basic salt of a weak acid and thereby change the basic salt to an acid salt or to a weak acid that is eliminated in the urine. While this is happening, the displaced sodium or other basic ion diffuses into a tubule cell. Here, it combines with the bicarbonate ion left over from the carbonic acid dissociation to form sodium bicarbonate. The sodium bicarbonate then diffuses—is reabsorbed—into the blood. Consider the various results of this mechanism. Sodium bicarbonate (or other base bicarbonate) is conserved for the body. Instead of all the basic salts that filter out of glomerular blood leaving the body in the urine, considerable amounts are recovered into peritubular capillary blood. In addition, extra hydrogen ions are added to the urine and thereby eliminated from the body. Both the reabsorption of base bicarbonate into blood and the excretion of hydrogen ions into urine tend to increase the ratio of the bicarbonate buffer pair $B \cdot HCO_3/H \cdot HCO_3$ (BB/CA) present in blood. This automatically increases blood pH. In short, kidney tubule base bicarbonate reabsorption and hydrogen ion excretion both tend to alkalinize blood by acidifying urine.

Renal tubules can excrete hydrogen or potassium in exchange for the sodium they reabsorb. Therefore, in general, the more hydrogen ions they excrete, the fewer potassium ions they can excrete. In acidosis, tubule excretion of hydrogen ions increases markedly and potassium ion excretion decreases—an important factor because it may lead to **hyperkalemia** (excessive blood potassium), a condition that can cause heart block and death.

♦ Distal and collecting tubule cells excrete ammonia into the tubular urine. As Figure 29-9 shows, the ammonia combines with hydrogen to form an ammonium ion. The

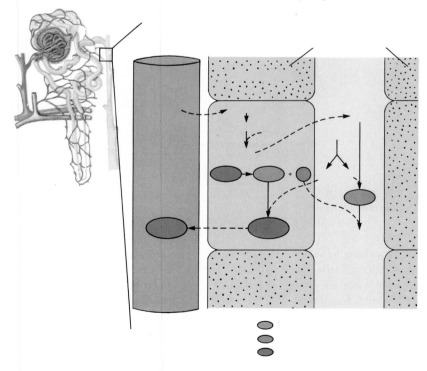

FIGURE 29-9   Acidification of urine by tubule excretion of ammonia (NH_3). An amino acid (glutamine) moves into tubule cell and loses an amino group (NH_2) to form ammonia, which is secreted into urine. In exchange, the tubule cell reabsorbs a basic salt (mainly $NaHCO_3$) into blood from urine.

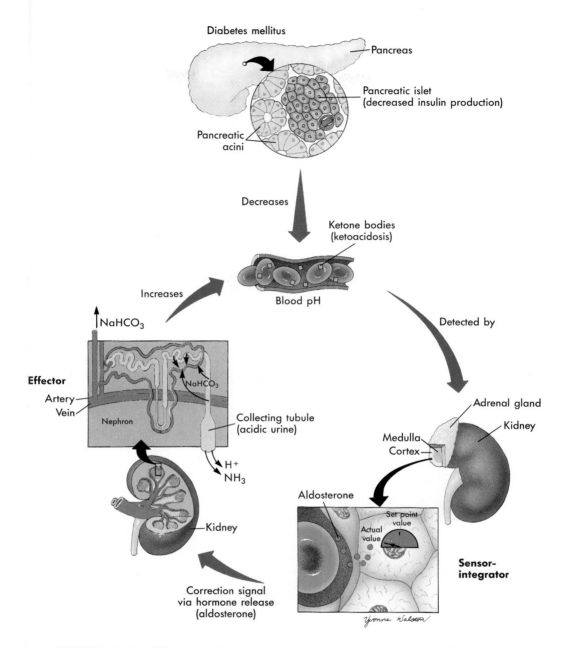

FIGURE 29-10 Scheme to show main parts of urinary mechanism for maintaining homeostasis of blood pH.

ammonium ion displaces sodium or some other basic ion from a salt of a fixed (nonvolatile) acid to form an ammonium salt. The basic ion then diffuses back into a tubule cell and combines with bicarbonate ion to form a basic salt, which in turn diffuses into tubular blood. Thus, like the renal tubules' excretion of hydrogen ions, their excretion of ammonia and its combining with hydrogen to form ammonium ions also tends to increase the blood bicarbonate buffer pair ratio and therefore tends to increase blood pH. Quantitatively, however, ammonium ion excretion is more important than hydrogen ion excretion.

Renal tubule excretion of hydrogen and ammonia is

controlled at least in part by the blood pH level. As indicated in Figure 29-10, a decrease in blood pH accelerates tubule excretion of both hydrogen and ammonia. An increase in blood pH produces the opposite effects.

QUICK CHECK

✔ 1. *What is the function of carbonic anhydrase in buffer action?*

✔ 2. *How does respiratory rate affect blood pH levels?*

✔ 3. *List two ways in which acidification of urine occurs.*

MECHANISMS OF DISEASE

Acid-base Imbalances

All of the buffer pairs present in body fluids play an important role in acid-base balance. However, only in the bicarbonate system can the body regulate quickly and precisely the levels of both chemical components in the buffer pair. Carbonic acid levels can be regulated by the respiratory system and bicarbonate ion by the kidneys. Recall that a 20/1 ratio of base bicarbonate to carbonic acid (BB/CA) will, according to the Henderson-Hasselbalch equation, maintain acid-base balance and normal blood pH. Therefore, from a clinical standpoint, disturbances in acid-base balance can be considered dependent on the relative quantities of carbonic acid and base bicarbonate present in the extracellular fluid. Two types of disturbances, metabolic and respiratory, can alter the proper ratio of these components. Metabolic disturbances affect the bicarbonate element, and respiratory disturbances affect the carbonic acid element of the buffer pair.

Metabolic and respiratory acidosis, for example, are separate and very different types of acid-base imbalances. Both are treated by the intravenous infusion of solutions containing **sodium lactate.** The infused lactate ions are metabolized by liver cells and converted to bicarbonate ions. This therapy helps replace depleted bicarbonate reserves required to restore acid-base balance in metabolic acidosis. In respiratory acidosis the additional bicarbonate ions function to offset elevated carbonic acid levels.

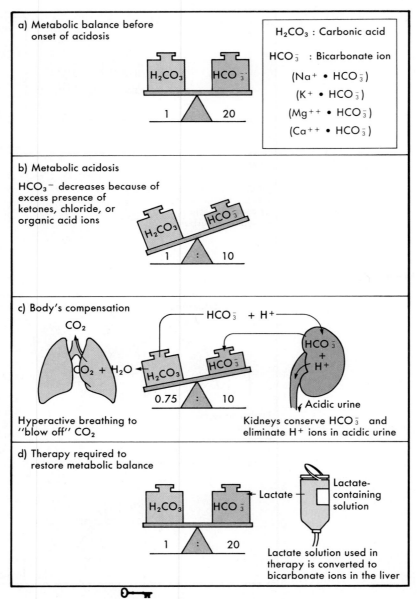

FIGURE 29-11   Metabolic acidosis.

Metabolic Disturbances
Metabolic Acidosis (Bicarbonate Deficit)

During the course of certain diseases such as untreated diabetes mellitus, or during starvation, abnormally large amounts of acids enter the blood. The ratio of BB/CA is altered as the base bicarbonate component of the buffer pair reacts with the acids. The result may be a new ratio near 10/1. The decreasing ratio will lower the blood pH, and the respiratory center will be stimulated (Figure 29-11). The resulting hyperventilation will result in a "blow-off" of carbon dioxide, with a decrease in carbonic acid. This compensatory action of the respiratory system, coupled with excretion of H^+ and NH_3 in exchange for reabsorbed Na^+ by the kidneys, may be sufficient to adjust the *ratio* of BB/CA, and therefore blood pH, to normal. (The compensated BB/CA ratio may approach 10/0.5.) If, despite these compensating homeostatic devices, the ratio and pH cannot be corrected, uncompensated metabolic acidosis develops.

Increased blood hydrogen ion concentration, that is, decreased blood pH, as we have noted, stimulates the respiratory center. For this reason, hyperventilation is an outstanding clinical sign of acidosis. Increases in hydrogen ion concentration above a certain level depress the central nervous system and therefore produce such symptoms as disorientation and coma. In a terminal illness, death from acidosis is likely to follow coma, whereas death from alkalosis generally follows tetany and convulsions.

Metabolic Alkalosis (Bicarbonate Excess)

Patients suffering from chronic stomach problems such as hyperacidity sometimes ingest large quantities of alkali—often plain baking soda, or sodium bicarbonate—for extended periods of time. Such improper use of antacids or excessive vomiting can produce metabolic alkalosis. Initially the condition results in an increase in the BB/CA ratio to perhaps 40/1 (Figure 29-12). Compensatory mechanisms are aimed at

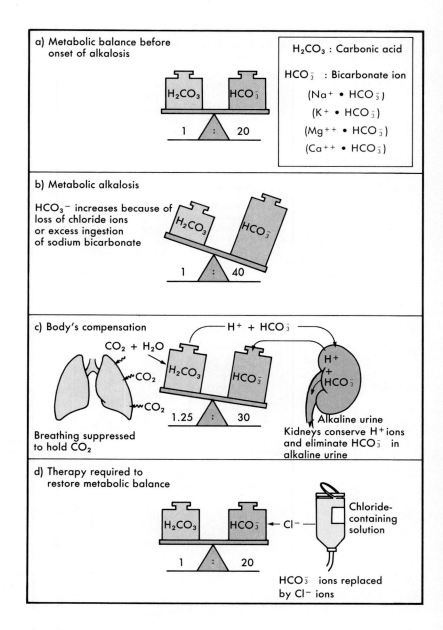

FIGURE 29-12   Metabolic alkalosis.

increasing carbonic acid and decreasing the bicarbonate load. With breathing suppressed and the kidneys excreting bicarbonate ions, a compensated ratio of 30/1.25 might result. Such a ratio would restore acid-base balance and blood pH to normal. In uncompensated metabolic alkalosis the ratio, and therefore the pH, remain increased.

Respiratory Disturbances

Respiratory Acidosis (Carbonic Acid Excess)

Clinical conditions such as pneumonia or emphysema tend to cause retention of carbon dioxide in the blood. Also, drug abuse or overdose, such as barbiturate poisoning, will suppress breathing and result in respiratory acidosis (Figure 29-13). The carbonic acid component of the bicarbonate buffer pair increases above normal in respiratory acidosis. Body compen-

---

SPORTS & FITNESS

BICARBONATE LOADING

The buildup of lactic acid in the blood, released as a waste product from working muscles, has been blamed for the soreness and fatigue that sometimes accompanies strenuous exercise. Some athletes adopt a technique called **bicarbonate loading**, ingesting large amounts of sodium bicarbonate ($NaHCO_3$) to counteract the effects of lactic acid buildup. Their theory is that fatigue is avoided because the $NaHCO_3$, a base, buffers the lactic acid. Unfortunately, the excess bicarbonate intake and diarrhea that often results can trigger fluid and electrolyte imbalances. Long-term $NaHCO_3$ abuse can lead to metabolic alkalosis and its disastrous effects.

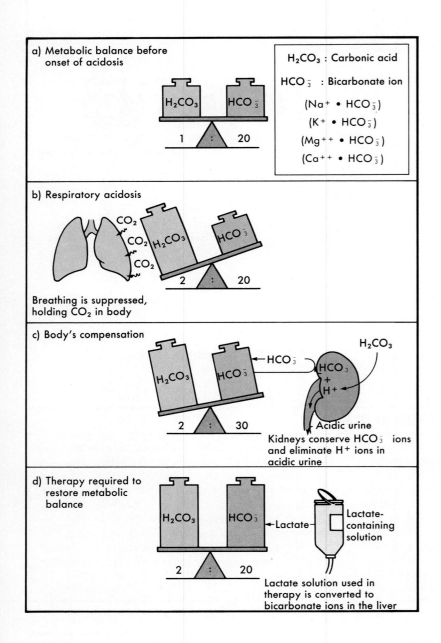

a) Metabolic balance before onset of acidosis

H_2CO_3 : Carbonic acid

HCO_3^- : Bicarbonate ion

$(Na^+ \cdot HCO_3^-)$

$(K^+ \cdot HCO_3^-)$

$(Mg^{++} \cdot HCO_3^-)$

$(Ca^{++} \cdot HCO_3^-)$

H_2CO_3 HCO_3^-

1 : 20

b) Respiratory acidosis

CO_2 CO_2 CO_2

H_2CO_3 HCO_3^-

2 : 20

Breathing is suppressed, holding CO_2 in body

c) Body's compensation

H_2CO_3 HCO_3^- HCO_3^- H_2CO_3

$HCO_3^- + H^+$

2 : 30

Acidic urine

Kidneys conserve HCO_3^- ions and eliminate H^+ ions in acidic urine

d) Therapy required to restore metabolic balance

H_2CO_3 HCO_3^- Lactate   Lactate-containing solution

2 : 20

Lactate solution used in therapy is converted to bicarbonate ions in the liver

FIGURE 29-13   Respiratory acidosis.

sation, if successful, increases the bicarbonate fraction so that a new BB/CA ratio (perhaps 20/2) will return blood pH to normal or near normal levels.

Respiratory Alkalosis (Carbonic Acid Deficit)

Hyperventilation caused by fever or mental disease (hysteria) can result in excessive loss of carbonic acid and lead to respiratory alkalosis (Figure 29-14) with a bicarbonate buffer pair ratio of 20/0.5. Compensatory mechanisms may adjust the ratio to 10/0.5 and return blood pH to near normal.

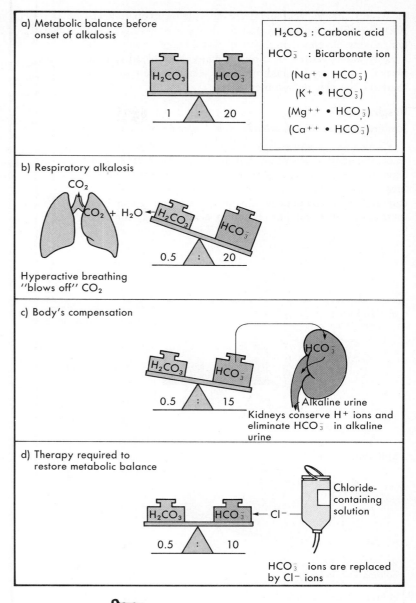

a) Metabolic balance before onset of alkalosis

H_2CO_3 : Carbonic acid

HCO_3^- : Bicarbonate ion

$(Na^+ \cdot HCO_3^-)$
$(K^+ \cdot HCO_3^-)$
$(Mg^{++} \cdot HCO_3^-)$
$(Ca^{++} \cdot HCO_3^-)$

H_2CO_3 HCO_3^-

1 : 20

b) Respiratory alkalosis

CO_2

$CO_2 + H_2O$ H_2CO_3 HCO_3^-

0.5 : 20

Hyperactive breathing "blows off" CO_2

c) Body's compensation

H_2CO_3 HCO_3^- HCO_3^-

0.5 : 15

Alkaline urine
Kidneys conserve H^+ ions and eliminate HCO_3^- in alkaline urine

d) Therapy required to restore metabolic balance

H_2CO_3 HCO_3^- Cl^-

Chloride-containing solution

0.5 : 10

HCO_3^- ions are replaced by Cl^- ions

FIGURE 29-14 **Respiratory alkalosis.**

METABOLIC ALKALOSIS CAUSED BY VOMITING

Vomiting, sometimes referred to as **emesis,** is the forcible emptying or expulsion of gastric and occasionally intestinal contents through the mouth. It occurs as a result of many stimuli, including foul odors or tastes, irritation of the stomach or intestinal mucosa, and some vomitive or *emetic* drugs such as *ipecac*. A "vomiting center" in the brain regulates the many coordinated (but primarily involuntary) steps involved (see Figure 25-14, p. 667). Severe vomiting such as **pernicious vomiting** of pregnancy or the repeated vomiting associated with pyloric obstruction in infants can be life threatening. One of the most frequent and serious complica-

tions of vomiting is metabolic alkalosis. The bicarbonate excess of metabolic alkalosis results because of the massive loss of chloride from the stomach as hydrochloric acid. It is the loss of chloride that causes a compensatory increase of bicarbonate in the extracellular fluid. The result is **metabolic alkalosis.** Therapy includes intravenous administration of chloride-containing solutions such as **normal saline** (0.9% NaCl in water). The chloride ions of the solution replace bicarbonate ions and thus help relieve the bicarbonate excess responsible for the imbalance.

CASE STUDY
INSULIN-DEPENDENT DIABETES MELLITUS (IDDM)

David is a 10-year-old boy who has been admitted to the hospital with type I insulin-dependent diabetes mellitus. He had recently experienced several symptoms, such as weight loss, constant hunger, and increased urination. In the past few days, however, he felt increasingly sick and had begun vomiting this morning. He became progressively less responsive, and when he arrived at the hospital, he was confused and nearly unconscious, with rapid, deep respirations.

His lab results showed the following: blood glucose, 450 mg/dl; positive urine and blood ketones; arterial blood pH, 7.25; P_{CO_2}, 40 mm Hg (millimeters of mercury); and HCO_3^- (bicarbonate), 16 mEq/L (milliequivalent per liter). An hour later, his blood pH was 7.35; HCO_3^-, was 18 mEq/L; and P_{CO_2}, was 30 mm Hg.

He was treated with intravenous fluids containing electrolytes and insulin and began improving rapidly.

1. Which of the following best explains David's symptoms of increased thirst and increased urination?
   a. Breakdown of fatty acids in the liver
   b. Hyperosmolar body fluids from high blood glucose
   c. Loss of potassium in the urine
   d. Gluconeogenesis in the liver

2. David's initial arterial blood gas values represent which of following conditions?
   a. Normal
   b. Uncompensated metabolic alkalosis

   c. Uncompensated metabolic acidosis
   d. Compensated metabolic acidosis

3. The body has many ways of responding to alterations in blood pH, and David's respiratory pattern is one of these. What is the effect on blood pH of his rapid, deep respirations?
   a. Increases pH
   b. Decreases pH
   c. Maintains pH

CHAPTER SUMMARY

INTRODUCTION
A. Acid-base balance is one of the most important of the body's homeostatic mechanisms
B. Acid-base balance refers to regulation of hydrogen ion concentration in body fluids
C. Precise regulation of pH at the cellular level is necessary for survival
D. Slight pH changes have dramatic effects on cellular metabolism

MECHANISMS THAT CONTROL pH OF BODY FLUIDS
A. Meaning of pH—negative logarithm of hydrogen ion concentration of solution (Figure 29-1)
B. Sources of pH-influencing elements
   1. Carbonic acid—formed by aerobic glucose metabolism
   2. Lactic acid—formed by anaerobic glucose metabolism
   3. Sulfuric acid—formed by oxidation of sulfur-containing amino acids
   4. Phosphoric acid—formed in breakdown of phosphoproteins and nucleoproteins
   5. Acidic ketone bodies—formed in breakdown of fats
      a. Acetone
      b. Acetoacetic acid
      c. Beta-hydroxybutyric acid
C. Acid-forming potential of foods—determined by chloride, sulfur, and phosphorus elements
D. Types of pH control mechanisms
   1. Chemical—rapid action buffers

   a. Bicarbonate buffer system
   b. Phosphate buffer system
   c. Protein buffer system
   2. Physiological—delayed action buffers
      a. Respiratory response
      b. Renal response
   3. Summary of pH control mechanisms
      a. Buffers
      b. Respirations
      c. Kidney excretion of acids and bases
E. Effectiveness of pH control mechanisms; range of pH—extremely effective, normally maintain pH within very narrow range of 7.36 to 7.41

BUFFER MECHANISM FOR CONTROLLING pH OF BODY FLUIDS
A. Buffers defined
   1. Substances that prevent marked change in pH of solution when acid or base added to it
   2. Consist of weak acid (or its acid salt) and basic salt of that acid
B. Buffer pairs present in body fluids—mainly carbonic acid, proteins, hemoglobin, acid phosphate, and sodium and potassium salts of these weak acids
C. Action of buffers to prevent marked changes in pH of body fluids
   1. Nonvolatile acids, such as hydrochloric acid, lactic acid, and ketone bodies, buffered mainly by sodium bicarbonate

2. Volatile acids, chiefly carbonic acid, buffered mainly by potassium salts of hemoglobin and oxyhemoglobin (Figure 29-4)
3. The chloride shift makes it possible for carbonic acid to be buffered in the red blood cell and then carried as bicarbonate in the plasma (Figure 29-5)
4. Bases buffered mainly by carbonic acid (when homeostasis of pH at 7.4 exists)

$$\text{Ratio} \frac{B \cdot HCO_3}{H_2CO_3} = \frac{20}{1}$$

5. The Henderson-Hasselbalch equation is a mathematical formula that explains the relationship between hydrogen ion concentration of body fluids and the ratio of base bicarbonate to carbonic acid
D. Evaluation of role of buffers in pH control—cannot maintain normal pH without adequate functioning of respiratory and urinary pH control mechanisms

RESPIRATORY MECHANISM OF pH CONTROL

A. Adjustment of respirations to pH of arterial blood
B. Explanation of mechanism
1. Amount of blood carbon dioxide directly relates to amount of carbonic acid and therefore to concentration of H^+
2. With increased respirations, less carbon dioxide remains in blood, hence less carbonic acid and fewer H^+; with decreased respirations, more carbon dioxide remains in blood, hence more carbonic acid and more H^+
C. Some principles relating respirations and pH of body fluids
1. Acidosis → hyperventilation
↓
increases elimination of CO_2
↓
decreases blood CO_2
↓
decreases blood H_2CO_3
↓
decreases blood H^+, that is, increases blood pH
↓
tends to correct acidosis, that is, to restore normal pH
2. Prolonged hyperventilation, by decreasing blood H^+ excessively, may produce alkalosis
3. Alkalosis causes hypoventilation, which tends to correct alkalosis by increasing blood CO_2 and therefore blood H_2CO_3 and H^+
4. Prolonged hypoventilation, by eliminating too little CO_2 causes increase in blood H_2CO_3 and consequently in blood H^+, thereby may produce acidosis

URINARY MECHANISM OF pH CONTROL

A. General principles about mechanism—plays vital role in acid-base balance because kidneys can eliminate more H^+ from body while reabsorbing more base when pH tends toward acid side and eliminate fewer H^+ while reabsorbing less base when pH tends toward alkaline side
B. Mechanisms that control urine pH
1. Secretion of H^+ into urine—when blood CO_2, H_2CO_3, and H^+ increase above normal, distal tubules secrete more H^+ into urine to displace basic ion (mainly sodium) from a urine salt and then reabsorb sodium into blood in exchange for the H^+ excreted
2. Secretion of NH_3—when blood hydrogen ion concentration increases, distal tubules secrete more NH_3, which combines with H^+ of urine to form ammonium ion, which displaces basic ion (mainly sodium) from a salt; basic ion then reabsorbed back into blood in exchange for ammonium ion excreted

MECHANISMS OF DISEASE: ACID-BASE IMBALANCES

Metabolic disturbances affect the bicarbonate and respiratory disturbances the carbonic acid element of the bicarbonate buffer pair
A. Metabolic disturbances
1. Metabolic acidosis (bicarbonate deficit) results in decreased alkaline reserve (mainly $NaHCO_3$), but ratio BB/CA maintained at normal 20/1 by proportionately decreasing blood carbonic acid by hyperventilation coupled with excretion of H^+ and NH_3 in exchange for reabsorbed Na^+ by the kidneys (Figure 29-11)
2. Metabolic alkalosis (bicarbonate excess)—initially the condition results in an increase in BB/CA; compensatory mechanisms are aimed at increasing carbonic acid and decreasing the bicarbonate load (Figure 29-12)
B. Respiratory disturbances
1. Respiratory acidosis (carbonic acid excess)—the carbonic acid component of the bicarbonate buffer pair increases above normal in respiratory acidosis; compensatory mechanisms are aimed at increasing the bicarbonate fraction to return blood pH to normal or near normal levels (Figure 29-13)
2. Respiratory alkalosis (carbonic acid deficit)—characterized by excessive loss of carbonic acid caused by hyperventilation (Figure 29-14)

REVIEW QUESTIONS

1. Explain, in your own words, what pH means.
2. How are carbonic and lactic acids produced?
3. Are fruits and vegetables acid-forming or base-forming foods?
4. Identify several acid-forming elements.
5. What is a physiological buffer?
6. What is the normal range of blood pH?
7. Describe the buffering action of sodium bicarbonate.
8. Identify the main buffer pairs in body fluids.
9. Define and explain the purpose of the chloride shift.
10. How is the Henderson-Hasselbalch equation used?
11. What effect does a decrease in blood pH have on respiration?
12. How is the distal renal tubule involved in the acidification of urine and the conservation of base?
13. How can sodium lactate be useful in the treatment of both metabolic and respiratory acidosis?
14. What type of acid-base imbalance occurs with prolonged vomiting? Explain.
15. Discuss hyperventilation in relation to the development of an acid-base imbalance.

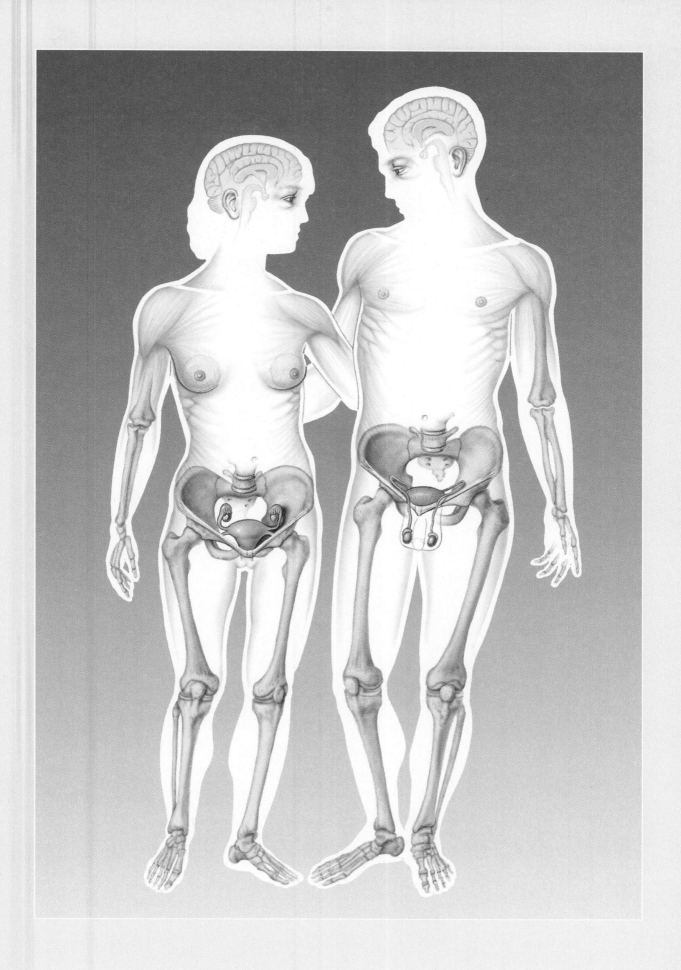

UNIT SIX

Reproduction and Development

The chapters of Unit 6 deal with human reproduction, growth, development, genetics, and heredity. The anatomic structures and complex control mechanisms characteristic of the male and the female reproductive systems are intended to ensure survival of the human species. These systems in men and women are adapted structurally and functionally for the specific sequence of events that permit development of sperm or ova, followed by fertilization, normal development, and then birth of a baby. Chapter 32 details the developmental changes that occur from fertilization to death. Chapter 33 discusses the scientific study of genetics and heredity.

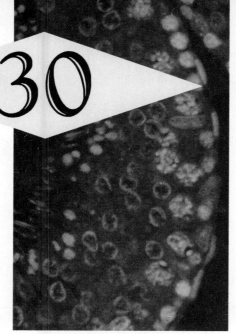

30 ▶ Male Reproductive System

Seminiferous tubules
Erlandsen/Magney: Color Atlas of Histology

CHAPTER OUTLINE

OBJECTIVES

**After you have completed this chapter, you should
be able to:**

1. Explain how the normal functioning of the
   reproductive system differs from the end result of
   normal function measured in any other organ
   system of the body.
2. List the essential and accessory organs of the male
   reproductive system and give the generalized
   function of each.
3. Describe the gross and microscopic anatomy of
   the testes.
4. Discuss the primary functions of testosterone and
   identify the cell type responsible for its secretion.
5. Describe the structure of a mature spermatozoon.
6. Trace the passage of an individual sperm cell from
   its point of formation, in sequence, through the
   genital ducts to the exterior of the body.
7. Describe the structure, location, and function(s)
   of the following: seminal vesicles, prostate gland,
   bulbourethral glands.
8. Identify the components and functions of the
   male external genitals.
9. Discuss the composition and course of seminal
   fluid.
10. Discuss male fertility.
11. List and discuss the male functions in
    reproduction.

Androgen (AN-dro-jen) Male sex hormone

Bulbourethral (bul-bo-yoo-REE-thral) Pertaining to the bulb of the urethra

Epididymis (ep-i-DID-i-mis) Tightly coiled tube where sperm mature and develop the ability to swim; lies along the top and behind the testes

Gamete (GAM-eet) Sex cell

Genital Pertaining to genitalia

Genitalia External reproductive organs

Gonads (GO-nads) Sex glands in which reproductive cells are formed

Penis Male organ of copulation and urinary excretion

Prostate Gland located below the bladder; secretion helps activate sperm and helps them maintain motility

Scrotum Pouchlike sac containing the testes

Semen Male reproductive fluid

Seminal vesicle Paired glands that contribute a fluid rich in fructose to seminal fluid

Sertoli cells (ser-TOE-lee sells) Elongated cells to which spermatids become attached

Spermatogenesis (sper-mah-toe-JEN-e-sis) Production of sperm cells

Spermatozoa (sper-mah-tah-ZOE-ah) Mature sperm cells

Testes (TES-teez) Male gonads

Testosterone (tes-TOS-te-rone) Male sex hormone produced by interstitial cells in the testes

---

The importance of normal reproductive system function is notably different from the end result of "normal function" as measured in any other organ system of the body. The proper functioning of the reproductive system and of its enormously complex control mechanisms ensures survival not of the individual but of the species. In both sexes, organs of the reproductive system are adapted for the specific sequence of functions that are concerned primarily with propagation of the species. Production of hormones that permits development of the secondary sex characteristics occurs as a result of normal reproductive system activity. In humans, sexual maturity and the ability to reproduce occur at puberty. The male reproductive system consists of those organs whose functions are to produce, transfer, and ultimately introduce mature sperm into the female reproductive tract, where fertilization can occur. It is the testes that secrete androgens, or male sex hormones, notably testosterone.

MALE REPRODUCTIVE ORGANS

Organs of the reproductive system (Figure 30-1) may be classified as either **essential organs** for the production of gametes (sex cells) or **accessory organs** that play some type of supportive role in the reproductive process.

In both sexes the essential organs of reproduction that produce the gametes, or sex cells (sperm or ova), are called **gonads.** The gonads of the male are the testes.

The accessory organs of reproduction in the male include genital ducts, glands, and supporting structures.

Genital ducts serve to convey sperm to the outside of the body. The ducts are a pair of epididymides (singular: epididymis), the paired vasa deferentia (singular: vas deferens), a pair of ejaculatory ducts, and the urethra.

Accessory glands in the reproductive system produce secretions that serve to nourish, transport, and mature sperm. The glands are a pair of seminal vesicles, one prostate, and a pair of bulbourethral (Cowper's) glands.

The supporting structures are the scrotum, the penis, and a pair of spermatic cords.

Perineum

The perineum in the male (Figure 30-1, *B*) is a roughly diamond-shaped area between the thighs. It extends from the symphysis pubis anteriorly to the coccyx posteriorly. Its most lateral boundary on either side is the ischial tuberosity. A line drawn between the two ischial tuberosities divides the area into a larger **urogenital triangle,** which contains the external genitals (penis and scrotum), and the **anal triangle,** which surrounds the anus.

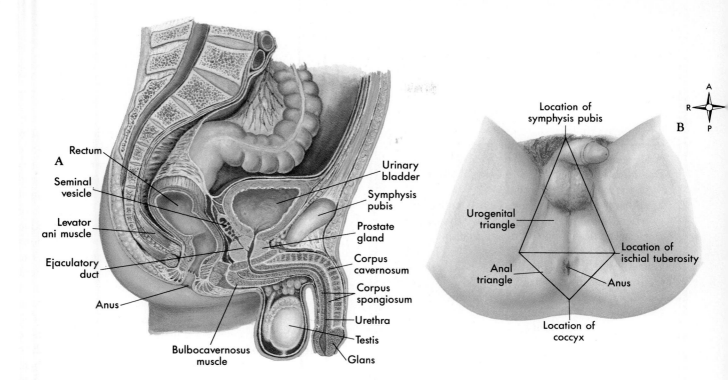

FIGURE 30-1 **Male reproductive organs. A,** Sagittal section of pelvis showing placement of male reproductive organs. **B,** Male perineum showing outline of the urogenital and anal triangles.

QUICK CHECK

✔ 1. How is the normal function of the reproductive system different from the end result of "normal function" in other organ systems?

✔ 2. Identify the essential and accessory organs of reproduction.

✔ 3. Describe the location, shape, and subdivisions of the perineum.

TESTES

Structure and Location

The testes are small ovoid glands that are somewhat flattened from side to side, measure about 4 or 5 cm in length, and weigh 10 to 15 g each. The left testis is generally located about 1 cm lower in the scrotal sac than the right. Both testes are suspended in the pouch by attachment to scrotal tissue and by the spermatic cords (Figure 30-2). Note in Figure 30-2, A, that testicular blood vessels, collectively called the *vas afferens*, reach the testes by passing through the spermatic cord. A dense, white fibrous capsule called the **tunica albuginea** encases each testis and then enters the gland, sending out partitions (septa) that radiate through its interior, dividing it into 200 or more cone-shaped lobules.

Each lobule of the testis contains specialized **interstitial cells** (of Leydig) and one to three tiny, coiled **seminiferous tubules,** which, if unraveled, would measure about 75 cm in length. The tubules from each lobule come together to form a plexus called the *rete testis*. A series of sperm ducts called *efferent ductules* then drain the rete testis and pierce the tunica albuginea to enter the head of the epididymis (Figure 30-2).

Microscopic Anatomy of the Testis

Figure 30-3 is a low-power (×70) light micrograph of testicular tissue showing a number of cut seminiferous tubules and numerous interstitial cells (Leydig cells) in the surrounding connective tissue septa. In this figure, maturing sperm appear as dense nuclei with their tails projecting into the lumen of the tubule. The wall of each seminiferous tubule may contain five or more layers of cells. At puberty, when sexual maturity begins, spermatogenic cells in diverse stages of development appear, and the hormone-producing interstitial cells become more prominent in the surrounding septa. Actively dividing cells at the periphery of each tubule in Figure 30-3 can be identified by their chromosomes. Irregular elongated **Sertoli cells** are supportive cells that provide mechanical support and protection for the developing germ cells. They also play an active role in eventual release of mature spermatozoa into the lumen of the seminiferous tubule. An isolated Sertoli cell is shown in Figure 30-4.

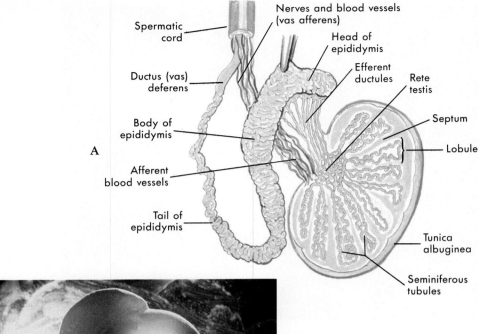

A

Spermatic cord

Ductus (vas) deferens

Body of epididymis

Afferent blood vessels

Tail of epididymis

Nerves and blood vessels (vas afferens)

Head of epididymis

Efferent ductules

Rete testis

Septum

Lobule

Tunica albuginea

Seminiferous tubules

B

FIGURE 30-2 Tubules of the testis and epididymis. A, Illustration showing epididymis lifted free of testis. The ducts and tubules are exaggerated in size. **B,** Transilluminated photograph, the testicle is the darker sphere in the center.

FIGURE 30-3 Testis. Low-power view showing several seminiferous tubules surrounded by septa containing interstitial (Leydig) cells. (×70.)

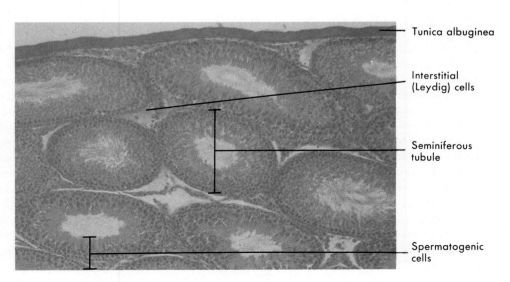

Tunica albuginea

Interstitial (Leydig) cells

Seminiferous tubule

Spermatogenic cells

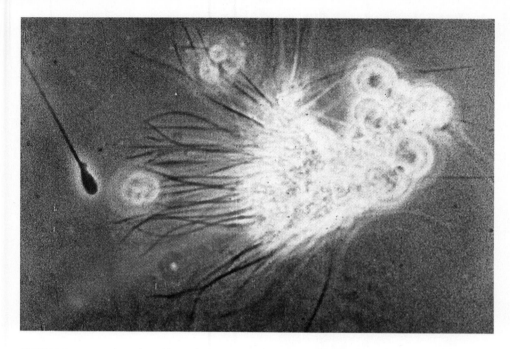

FIGURE 30-4 Sertoli cell. Light micrograph of an isolated Sertoli cell. Note the many developing sperm cells (spermatids) that are imbedded in the cell membrane. A single mature sperm cell is seen in the lumen (*left*) of the seminiferous tubule.

Testes Functions

The testes perform two primary functions: spermatogenesis and secretion of hormones.

1. **Spermatogenesis,** the production of spermatozoa (sperm), the male gametes, or reproductive cells. The seminiferous tubules produce the sperm. Details of spermatogenesis will be discussed in Chapter 32.
2. **Secretion of hormones,** chiefly testosterone (androgen or masculinizing hormone) by interstitial cells (Leydig cells). Testosterone serves the following general functions:
   a. It promotes "maleness," or development and maintenance of male secondary sex characteristics, accessory organs such as the prostate, seminal vesicles, and adult male sexual behavior.
   b. It helps regulate metabolism and is sometimes referred to as "the anabolic hormone" because of its marked stimulating effect on protein anabolism. By stimulating protein anabolism, testosterone promotes growth of skeletal muscles (responsible for greater male muscular development and strength) and growth of bone. Testosterone also promotes closure of the epiphyses (see Chapter 6, p. 160).

Early sexual maturation leads to early epiphyseal closure. The converse also holds true: late sexual maturation, delayed epiphyseal closure, and tallness tend to go together.

   c. It plays a part in fluid and electrolyte metabolism. Testosterone has a mild stimulating effect on kidney tubule reabsorption of sodium and water; it also promotes kidney tubule excretion of potassium.
   d. It inhibits anterior pituitary secretion of gonadotropins, namely, FSH and LH.

The anterior pituitary gland controls the testes by means of its gonadotropic hormones—specifically, FSH and LH, just mentioned. FSH stimulates the seminiferous tubules to produce sperm more rapidly. In the male, LH stimulates interstitial cells to increase their secretion of testosterone. Soon the blood concentration of testosterone reaches a high level that inhibits anterior pituitary secretion of FSH and LH. Thus a negative feedback mechanism operates between the anterior pituitary gland and the testes. A high blood concentration of gonadotropins stimulates testosterone secretion. But a high blood concentration of testosterone inhibits (has a negative effect on) gonadotropin secretion (Figure 30-5).

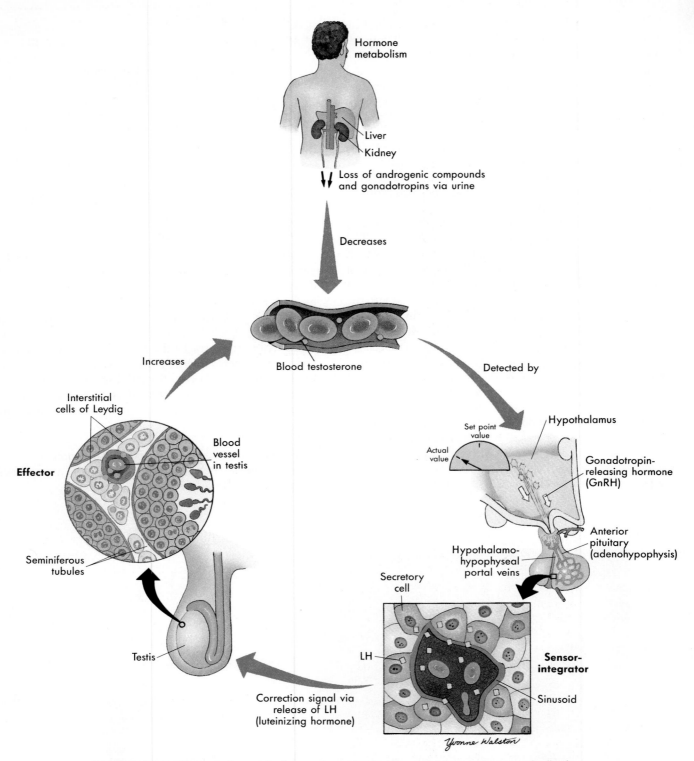

FIGURE 30-5 Hormonal control of testosterone levels. Diagram shows the negative feedback mechanism that controls anterior pituitary gland secretion of LH and interstitial cell secretion of testosterone.

Structure of Spermatozoa

The elongated tail-bearing spermatozoa seen in the seminiferous tubules (Figure 30-6, A) appear fully formed. We know, however, that they undergo a process of "ripening," or maturation, as they pass through the genital ducts before ejaculation. Although anatomically complete and highly motile when ejaculated, sperm must still undergo a complex process called **capacitation** before they are actually capable of fertilizing an ovum (female gamete). Normally, capacitation occurs in sperm only after they have been introduced into the vagina of the female.

Figure 30-6, B, shows the characteristic parts of a spermatozoon: head, middle piece, and elongated, lashlike tail. The head of a spermatozoon is, in essence, a highly compact package of genetic chromatin material covered by a specialized *acrosome* and *acrosomal (head) cap*. The acrosome contains hydrolytic (splitting) enzymes, which are released during capacitation. During the process of capacitation, these specialized acrosomal enzymes first break down cervical mucus, allowing sperm to pass into the uterus and uterine tubes. If an ovum is present in the female reproductive tract when semen is introduced, continued release of capacitation enzymes will assist the sperm cells to digest and penetrate the outer covering of the egg and initiate fertilization. This is the primary reason a high sperm count is essential for male fertility.

The cylindrical middle piece is characterized by a helical arrangement of mitochondria arranged end to end around a central core. It is this mitochondrial sheath that provides energy for sperm locomotion. The tail is divided into a principal piece and short end-piece, both typical in appearance of all flagella capable of motility.

QUICK CHECK

✔ 1. Describe the location, the size, and the shape of the testes.
✔ 2. List the two primary functions of the testes and identify the cell type or structure involved in each function.
✔ 3. List the general functions of testosterone.
✔ 4. Identify the structural components of a spermatozoon and give the function of each.

REPRODUCTIVE (GENITAL) DUCTS

Epididymis

Structure and Location

Each epididymis consists of a single, tightly coiled tube enclosed in a fibrous casing. The tube has a very small diameter (just barely macroscopic) but measures approximately 6 meters (20 feet) in length. It lies along the top and behind the testis (see Figure 30-2). The comma-shaped epididymis is divided into a blunt superior **head** (which is connected to the testis by the efferent ductules), a central **body,** and a tapered inferior portion that is continuous with the vas deferens and called the **tail.** If the epididymis is cut or sectioned and a slide prepared as in Figure 30-7, the compact and highly coiled nature of the tubule is apparent.

Functions

The epididymis serves the following functions:
♦ It serves as one of the ducts through which sperm pass in their journey from the testis to the exterior.

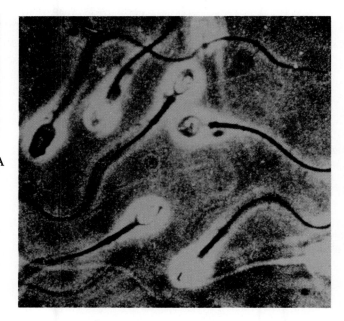
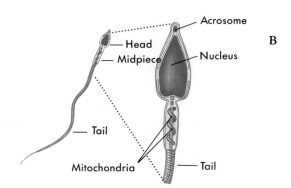

FIGURE 30-6 Human sperm. A, Micrograph showing the heads and portions of the long slender tails of several sperm. **B,** Illustration showing components of a mature sperm cell and an enlargement of a sperm head.

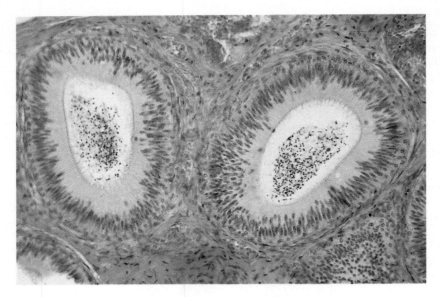

FIGURE 30-7 Epididymis section. Two cross sections of this extensively coiled tubule are visible. Note the presence of spermatozoa within the lumina of the tubules. (Courtesy of Erlandsen/Magney: Color Atlas of Histology.)

◆ It contributes to the maturation of sperm, which spend from 1 to 3 weeks in this segment of the duct system.

◆ It secretes a small part of the seminal fluid (semen).

Vas Deferens (Ductus Deferens)
Structure and Location

The vas deferens, like the epididymis, is a tube. In fact, the duct of the vas is an extension of the tail of the epididymis. The vas deferens has thick muscular walls (Figure 30-8) and can be palpated in the scrotal sac as a smooth movable cord. Note in Figure 30-8 that the muscular layer of the vas has three layers: a thick intermediate circular layer of muscle fibers and inner and outer longitudinal layers. The muscular layers of the vas help in propelling sperm through the duct system. The vas deferens from each testis ascends from the scrotum and

Vasectomy

Severing of the vas deferens—that is, a **vasectomy**, usually done through incision in the scrotum—makes a man sterile. Why? Because it interrupts the route to the exterior from the epididymis. To leave the body, sperm must journey in succession through the epididymis, vas deferens, ejaculatory duct, and urethra.

passes through the inguinal canal as part of the spermatic cord—enclosed by fibrous connective tissue with blood vessels, nerves, and lymphatics—into the abdominal cavity. Here it extends over the top and down the posterior surface of the bladder where an enlarged and tortuous

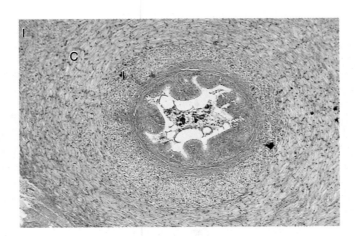

FIGURE 30-8 Transverse section of vas deferens. Mucosa protrudes into the lumen in several low folds. Note the thick muscular coat surrounding the mucosa. (Courtesy of Erlandsen/Magney: Color Atlas of Histology.)

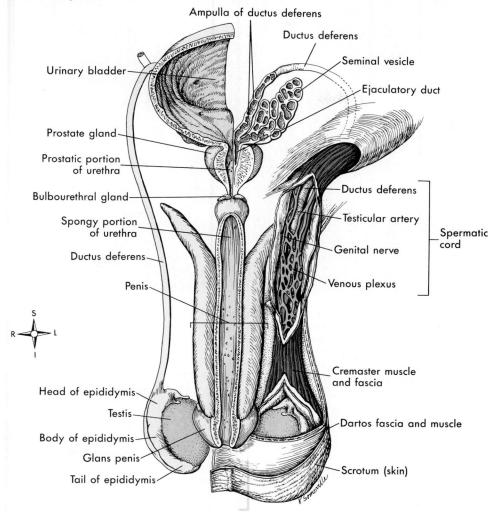

FIGURE 30-9 The male reproductive system. Illustration shows the testes, epididymis, ductus deferens, and glands of the male reproductive system in an isolation/dissection format.

portion called the *ampulla* joins the duct from the seminal vesicle to form the ejaculatory duct (Figures 30-1 and 30-9).

Function

The vas deferens serves as one of the male genital ducts connecting the epididymis with the ejaculatory duct. Sperm remain in the vas deferens for varying periods of time depending on the degree of sexual activity and frequency of ejaculation. Storage time may exceed 1 month with no loss of fertility.

Ejaculatory Duct

The two ejaculatory ducts are short tubes that pass through the prostate gland to terminate in the urethra. As Figure 30-9 shows, they are formed by the union of the vas deferens with the ducts from the seminal vesicles.

Urethra

The urethra in males serves a dual function, which involves both the reproductive system and the urinary system. Refer back to Chapter 27, p. 707, for a discussion of this duct.

◢ ACCESSORY REPRODUCTIVE GLANDS

Seminal Vesicles
Structure and Location

The seminal vesicles are convoluted pouches that lie along the lower part of the posterior surface of the bladder, directly in front of the rectum (Figure 30-1). The highly branched and convoluted nature of the secretory epithelium filling the lumen of the seminal vesicles is apparent in Figure 30-10.

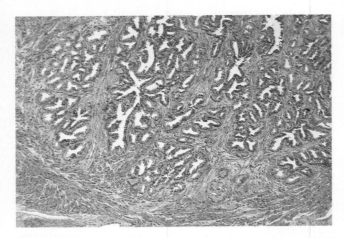

FIGURE 30-10 Seminal vesicle. Note the highly branched and convoluted nature of the secretory epithelium.

Function

The seminal vesicles secrete an alkaline, viscous liquid component of the semen rich in fructose. As a simple sugar, fructose serves as an energy source for sperm motility after ejaculation. The thick, yellowish secretion also contains prostaglandins, substances postulated to influence cyclic AMP formation (p. 409). Normal secretory activity of the seminal vesicles depends on adequate levels of testosterone.

Prostate Gland
Structure and Location

The prostate is a compound tubuloalveolar gland that lies just below the bladder and is shaped like a doughnut. The fact that the urethra passes through the small hole in the center of the prostate is a matter of considerable clinical significance. Many older men suffer from enlargement of this gland. As it enlarges, it squeezes the urethra, frequently closing it so completely that urination becomes impossible. Urinary retention results. Surgical removal of the gland (prostatectomy) is required as a cure for this condition when other less radical methods of treatment fail.

Function

The prostate secretes a thin alkaline substance that constitutes about 30% of the seminal fluid volume. Its alkalinity helps protect the sperm from acid present in the male urethra and female vagina and thereby increases sperm motility. Acid depresses or, if strong enough, kills sperm. Sperm motility is greatest in neutral or slightly alkaline media.

Bulbourethral Glands
Structure and Location

The two bulbourethral, or Cowper's, glands resemble peas in size and shape. You can see the location of these

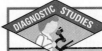

DETECTING PROSTATE CANCER

Many of the 32,000 men who die each year from prostate cancer—the most common cancer in American men—could be saved if the cancer was detected early enough for effective treatment. Several screening tests are now available for early detection of prostate cancer. For example, physicians can sometimes detect prostate cancer early by palpating the prostate through the wall of the rectum. Unfortunately, by the time prostate cancer can be palpated, it has usually spread to other organs. Many cancer experts believe prostate cancer screening can be more effective if rectal examinations are performed along with a newer type of screening test called the *PSA test.* This test is a type of blood analysis that screens for *prostate specific antigen (PSA),* a substance often found in the blood of men with prostate cancer. Because of the prevalence of this disease, adult men are encouraged to have regular prostate examinations and to report *any* urinary or sexual difficulty to their physicians.

compound tubuloalveolar glands in Figure 30-9. A duct approximately 2.5 cm (1 inch) long connects them with the penile portion of the urethra.

Function

Like the prostate gland and seminal vesicles, the bulbourethral glands secrete an alkaline fluid that is important for counteracting the acid present in both the male urethra and the female vagina. Mucus produced in these glands serves to lubricate the urethra.

QUICK CHECK

✔ 1. List, in sequence, the reproductive ducts sperm pass through from formation to ejaculation.
✔ 2. What is the structural relationship between the prostate gland and the urethra?
✔ 3. Compare the volume, viscosity, pH, and composition of the secretions produced by the accessory reproductive glands.

▶ SUPPORTING STRUCTURES
Scrotum

The **scrotum** is a skin-covered pouch suspended from the perineal region. Internally, it is divided into two sacs by a septum, each sac containing a testis, epididymis, and lower part of a spermatic cord.

The *dartos* fascia and muscle are located just below the skin of the scrotum (see Figure 30-9). Contraction of the dartos muscle fibers causes slight elevation of the testes and wrinkling of the scrotal pouch. In addition, contraction of the *cremaster muscle,* also seen in Figure 30-9, will cause

significant elevation of the testes. As a result of contraction, the testes are pulled upward against the perineum. Sexual arousal and cold temperature provide the stimulus for contraction of both the dartos and cremaster muscles.

By elevating the scrotum closer to the warmth of the perineal wall, the temperature of the testicles, which are located outside the abdominal cavity, can be maintained at a more constant level in a cold environment. The temperature required for sperm formation is about 3° C below normal body temperature. This is the "functional" reason that justifies placement of the testicles outside of the body cavity where they are exposed and subject to traumatic injury. In a warm environment the scrotum is longer and its skin loose and wrinkle free, permitting the testes to descend. In the cold, the scrotum elevates and becomes heavily wrinkled, effectively pulling the testes upward toward the body wall.

Penis (External)
Structure

Three cylindrical masses of erectile, or cavernous, tissue, enclosed in separate fibrous coverings and held together by a covering of skin, compose the penis (Figure 30-11). The two larger and uppermost of these cylinders are named the **corpora cavernosa,** whereas the smaller, lower one, which contains the urethra, is called the **corpus spongiosum.**

The distal part of the corpus spongiosum overlaps the terminal end of the two corpora cavernosa to form a slightly bulging structure, the **glans penis,** over which the skin is folded doubly to form a more or less loose-fitting, retractable casing known as the **prepuce,** or foreskin. Surgical removal of the foreskin is called *circumcision.* It is performed most frequently for religious or cultural reasons. The opening of the urethra at the tip of the glans is called the *external urinary meatus.*

Functions

The penis contains the urethra, the terminal duct for both urinary and reproductive tracts. During sexual arousal, the erectile tissue of the penis fills with blood, causing the organ to become rigid and enlarge in both diameter and length. The result, called an *erection,* permits the penis to serve as a penetrating copulatory organ during sexual intercourse. The scrotum and penis together constitute the **external genitals** of the male.

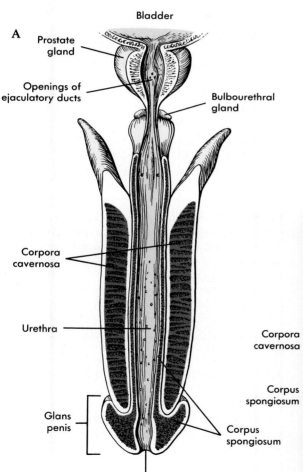

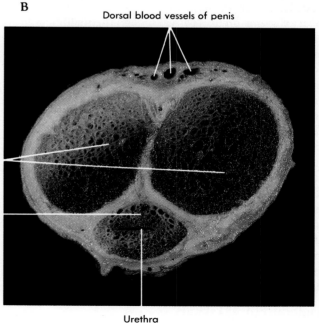

FIGURE 30-11   The penis. A, In this sagittal section of the penis viewed from above, the urethra is exposed throughout its length and can be seen exiting from the bladder and passing through the prostate gland before entering the penis to end at the external urethral orifice. **B,** Photograph of a cross section of the shaft of the penis showing the three columns of erectile, or cavernous, tissue. Note the urethra within the substance of the corpus spongiosum.

Spermatic Cords (Internal)

The **spermatic cords** are cylindrical casings of white, fibrous tissue located in the inguinal canals between the scrotum and the abdominal cavity. They enclose the vas deferens, blood vessels, lymphatics, and nerves (see Figure 30-9).

COMPOSITION AND COURSE OF SEMINAL FLUID

The following structures secrete the substances that, together, make up the seminal fluid or **semen:**

♦ Testes and epididymis—their secretions constitute less than 5% of the seminal fluid volume.

♦ Seminal vesicles—their secretions are reported to contribute about 60% of the seminal fluid volume.

♦ Prostate gland—its secretions constitute about 30% of the seminal fluid volume.

♦ Bulbourethral glands—their secretions are said to constitute less than 5% of the seminal fluid volume.

Besides contributing slightly to the fluid part of semen, the testes also add hundreds of millions of sperm. In traversing the distance from their place of origin to the exterior, the sperm must pass from the testis through the epididymis, vas deferens, ejaculatory duct, and urethra. Note that sperm originate in the testes, glands located outside the body (that is, not within a body cavity), travel inside, and finally are expelled outside.

MALE FERTILITY

Male fertility relates to many factors—most of all to the number of sperm ejaculated but also to their size, shape, and motility. Fertile sperm have a uniform size and shape. They are highly motile. Although only one sperm fertilizes an ovum, millions of sperm seem to be necessary for fertilization to occur. According to one estimate, when the sperm count falls below about 50 million per milliliter of semen, functional sterility results.

One hypothesis suggested to explain this puzzling fact is this: semen that contains an adequate number of sperm also contains enough hyaluronidase and other hydrolytic enzymes to liquefy the intercellular substance between the cells that encase each ovum. Without this, a single sperm cannot penetrate the layers of cells around the ovum and hence cannot fertilize it. Infertility may also be caused by production of antibodies some men make against their own sperm. This type of sperm destruction or inactivation called

 Neural Control of the Male Sex Act

Recall that all body functions but one have for their ultimate goal survival of the individual. Only the function of reproduction serves a different, a longer range, and, no doubt in nature's scheme, a more important purpose—survival of the human species. Male functions in reproduction consist of the production of male sex cells (spermatogenesis, discussed in Chapter 3) and introduction of these cells into the female body (coitus, copulation, or sexual intercourse). For coitus to take place, erection of the penis must first occur, and for sperm to enter the female body, both the sex cells and secretions from the accessory glands must be introduced into the urethra (emission) and semen ejaculated from the penis.

Erection is a parasympathetic reflex initiated mainly by certain tactile, visual, and mental stimuli. It consists of dilation of the arteries and arterioles of the penis, which in turn floods and distends spaces in its erectile tissue and compresses its veins. Therefore more blood enters the penis through the dilated arteries than leaves it through the constricted veins. Hence it becomes larger and rigid, or, in other words, erection occurs.

Emission is the reflex movement of sex cells, or spermatozoa, and secretions from the genital ducts and accessory glands into the prostatic urethra. Once emission has occurred, ejaculation will follow.

Ejaculation of semen is also a reflex response. It is the usual outcome of the same stimuli that initiate erection. Ejaculation and various other responses—notably accelerated heart rate, increased blood pressure, hyperventilation, dilated skin blood vessels, and intense sexual excitement—characterize the male **orgasm,** or sexual climax.

"immune infertility" is caused by an antigen-antibody reaction. In immune infertility, a sperm surface protein called fertilization antigen (FA-1) triggers antibody production, which results in infertility.

QUICK CHECK

✔ 1. *What two structures constitute the external genitals of the male?*

✔ 2. *What is the function of the dartos and cremaster muscles? How does this function influence fertility?*

✔ 3. *Identify by name the three cylindrical masses of erectile tissue in the penis.*

✔ 4. *What factors influence male fertility?*

Cycle of Life: *Male reproductive system*

The reproductive system is unlike all other systems of the male body with regard to normal changes that occur throughout the life span. All other systems perform their functions from the time they develop *in utero* until advanced old age, when degeneration may cause loss of function and, perhaps, death. The male reproductive system, however, does not begin to perform its functions until puberty—usually during the early teenage years. Of course, the biological advantage of this "late start" is that a person does not have the biological, psychological, or social maturity to become a parent before that time.

Initial development of the male reproductive organs begins before birth, when the reproductive tract differentiates into the male form rather than the female form. A couple of months before birth, the immature testes descend behind the parietal peritoneum into the scrotum, guided by the fibrous gubernaculum. It is not uncommon for them to be late in completing the trip, perhaps not arriving in the scrotum until several weeks after birth. The testes and other reproductive organs remain in an immature form—and thus remain incapable of providing reproductive function—until puberty, when high levels of reproductive hormones stimulate the final stages of their development. From puberty until advanced old age the male reproductive system continues to operate efficiently enough to permit successful reproduction. A gradual decline in hormone production during late adulthood may decrease sexual desire and fertility to some degree, but a man can usually father a child until the time of death.

MECHANISMS OF DISEASE

Disorders of the Male Reproductive System

Several disorders of the male reproductive system cause **infertility.** Infertility is an abnormally low ability to reproduce. If there is a complete inability to reproduce, the condition is called **sterility.** Infertility or sterility involve an abnormally reduced capacity to deliver healthy sperm to the female reproductive tract. Reduced reproductive capacity may result from factors such as a decrease in the testes' production of sperm, structural abnormalities in the sperm, or obstruction of the reproductive ducts.

Disorders of the Testes

Disruption of the sperm-producing function of the seminiferous tubules can result in decreased sperm production, a condition called **oligospermia** (ol-i-go-SPER-mee-ah). If the *sperm count* is too low, infertility may result. A large number of sperm is needed to ensure that many sperm will reach the ovum and dissolve its coating—allowing a single sperm to unite with the ovum. Oligospermia can result from factors such as infection, fever, radiation, malnutrition, and high temperature in the testes. In some cases, oligospermia is temporary—as in some acute infections. Oligospermia is a leading cause of infertility. Of course, total absence of sperm production results in sterility.

Early in fetal life the testes are located in the abdominal cavity near the kidneys but normally descend into the scrotum about 2 months before birth. Occasionally a baby is born with undescended testes, a condition called **cryptorchidism** (krip-TOR-ki-dizm), which is readily observed by palpation of the scrotum at delivery. The word *cryptorchidism* is from the Greek words *kryptikos* ("hidden") and *orchis* ("testis"). Failure of the testes to descend may be caused by hormonal imbalances in the developing fetus or by a physical deficiency or obstruction. Regardless of cause, in the cryptorchid infant the testes remain "hidden" in the abdominal cavity. Because the higher temperature inside the body cavity inhibits spermatogenesis,

measures must be taken to bring the testes down into the scrotum to prevent permanent sterility. Early treatment of this condition by surgery or by injection of testosterone, which stimulates the testes to descend, may result in normal testicular and sexual development.

Most testicular cancers arise from the sperm-producing cells of the seminiferous tubules. Malignancies of the testes are most common among men 25 to 35 years old. Besides age, this type of cancer is associated with genetic predisposition, trauma or infection of the testis, and cryptorchidism. Treatment of testicular cancer is most effective when the diagnosis is made early in the development of the tumor. Many physicians encourage male patients to perform regular self-examination of their testes, especially if they are in a high-risk group.

Disorders of the Prostate

A noncancerous condition called **benign** (bee-NINE) **prostatic hypertrophy** (hye-PER-tro-fee) is a common problem in older men. The condition is characterized by an enlargement or hypertrophy of the prostate gland. As discussed earlier in this chapter, the fact that the urethra passes through the center of the prostate after exiting from the bladder is a matter of considerable clinical significance in this condition. As the prostate enlarges, it squeezes the urethra, frequently closing it so completely that urination becomes very difficult or even impossible. In such cases, surgical removal of a part or all of the gland, a procedure called **prostatectomy** (pros-ta-TEK-toe-me), may become necessary.

Disorders of the Penis and Scrotum

The penis is subject to numerous sexually transmitted infections, as well as structural abnormalities. One such structural abnormality is **phimosis** (fi-MO-sis), a condition in which the foreskin fits so tightly over the glans that it cannot retract. The usual treatment for this condition is circumci-

sion—a procedure in which the foreskin is cut along the base of the glans and removed. Severe phimosis can obstruct the flow of urine, possibly causing the death of an infant born with this condition. Milder phimosis can result in accumulation of dirt and organic matter under the foreskin, possibly causing severe infections.

Failure to achieve an erection of the penis is called **impotence** (IM-po-tens). Although impotence does not affect sperm production, it may cause infertility because normal intercourse may not be possible. Anxiety and psychological stress are often cited as causes of impotence. Impotence may also result from an abnormality in the erectile tissues of the penis or a failure of the nerve reflexes that control erection. Drugs and alcohol can cause temporary impotence by interfering with the nerves and blood vessels involved in producing an erection.

Swelling of the scrotum can be caused by various conditions. One of the most common causes of scrotal swelling is an accumulation of fluid called **hydrocele** (HYE-dro-seel). Hydroceles may be congenital, resulting from structural abnormalities present at birth. In adults, hydrocele often occurs when fluid produced by the serous membrane lining the scrotum is not absorbed properly. The cause of adult hydrocele is not always known, but in some cases, it can be linked to trauma or infection.

Swelling of the scrotum may also occur when the intestines push through the weak area of the abdominal wall that separates the abdominopelvic cavity from the scrotum. This condition is a form of **inguinal** (IN-gwi-nal) **hernia** (see Figure 9-22). If the intestines protrude into the scrotum, the digestive tract may become obstructed—resulting in death. Inguinal hernia often occurs while lifting heavy objects because of the high internal pressure generated by the contraction of abdominal muscles. Inguinal hernia may also be congenital. Small inguinal hernias may be treated with external supports that prevent organs from protruding into the scrotum; more serious hernias must be repaired surgically.

CHAPTER SUMMARY

INTRODUCTION

A. Proper functioning of the reproductive system ensures the survival of the species
B. Male reproductive system consists of organs whose functions are to produce, transfer, and introduce mature sperm into the female reproductive tract where fertilization can occur

MALE REPRODUCTIVE ORGANS

A. Classified as essential organs for production of gametes or accessory organs that support the reproductive process (Figure 30-1, A)
  1. Essential organs—gonads of the male; testes
  2. Accessory organs of reproduction
    a. Genital ducts convey sperm to outside of body; pair of epididymides, paired vasa deferentia, pair of ejaculatory ducts, and the urethra
    b. Accessory glands produce secretions that nourish, transport, and mature sperm; pair of seminal vesicles, the prostate, and pair of bulbourethral glands
    c. Supporting structures—scrotum, penis, and pair of spermatic cords
B. Perineum—in males, roughly diamond-shaped area between thighs; extends anteriorly from symphysis pubis to coccyx posteriorly; lateral boundary is the ischial tuberosity on either side; divided into the urogenital triangle and the anal triangle (Figure 30-1, B)

TESTES

A. Structure and location
  1. Several lobules composed of seminiferous tubules and interstitial cells (of Leydig), separated by septa, encased in fibrous capsule called the tunica albuginea (Figure 30-2)
  2. Seminiferous tubules in testis open into a plexus called rete testis, which is drained by a series of efferent ductules that emerge from the top of the organ and enter the head of epididymis
  3. Located in scrotum, one testis in each of two scrotal compartments
B. Microscopic anatomy

C. Functions
  1. Spermatogenesis—formation of mature male gametes (spermatozoa) by seminiferous tubules
  2. Secretion of hormone (testosterone) by interstitial cells
D. Structure of spermatozoa (Figure 30-6)—consists of a head (covered by acrosome), neck, midpiece, and tail; tail is divided into a principal piece and a short end-piece

REPRODUCTIVE (GENITAL) DUCTS

A. Epididymis
  1. Structure and location
    a. Single tightly coiled tube enclosed in fibrous casing (Figure 30-7)
    b. Lies along top and side of each testis
    c. Anatomical divisions include head, body, and tail
  2. Functions
    a. Duct for seminal fluid
    b. Also secretes part of seminal fluid
    c. Sperm become capable of motility while they are passing through the epididymis
B. Vas deferens (ductus deferens)
  1. Structure and location
    a. Tube, extension of epididymis
    b. Extends through inguinal canal, into abdominal cavity, over top and down posterior surface of bladder
    c. Enlarged terminal portion called ampulla—joins duct of seminal vesicle
  2. Function
    a. One of excretory ducts for seminal fluid
    b. Connects epididymis with ejaculatory duct
C. Ejaculatory duct
  1. Formed by union of vas deferens with duct from seminal vesicle
  2. Passes through prostate gland, terminating in urethra
D. Urethra (see p. 707)

ACCESSORY REPRODUCTIVE GLANDS

A. Seminal vesicles
  1. Structure and location—convoluted pouches on posterior surface of bladder

2. Function—secrete the viscous nutrient-rich part of seminal fluid (60%)
B. Prostate gland
   1. Structure and location
      a. Doughnut-shaped
      b. Encircles urethra just below bladder
   2. Function—adds alkaline secretion to seminal fluid (30%)
C. Bulbourethral gland
   1. Structure and location
      a. Small, pea-shaped structures with 1-inch long ducts leading into urethra
      b. Lie below prostate gland
   2. Function—secrete alkaline fluid that is part of semen (5%)

SUPPORTING STRUCTURES

A. Scrotum
   1. Skin-covered pouch suspended from perineal region
   2. Divided into two compartments
   3. Contains testis, epididymis, and lower part of a spermatic cord
   4. Dartos and cremaster muscles elevate the scrotal pouch
B. Penis (Figure 30-11)
   1. Structure—composed of three cylindrical masses of erectile tissue, one of which contains urethra
   2. Functions—penis contains the urethra, the terminal duct for both urinary and reproductive tracts; during sexual arousal, penis becomes erect, serving as a penetrating copulatory organ during sexual intercourse
C. Spermatic cords (internal)
   1. Fibrous cylinders located in inguinal canals
   2. Enclose seminal ducts, blood vessels, lymphatics, and nerves

COMPOSITION AND COURSE OF SEMINAL FLUID

A. Consists of secretions from testes, epididymides, seminal vesicles, prostate, and bulbourethral glands
B. Each ml contains millions of sperm

C. Passes from testes through epididymis, vas deferens, ejaculatory duct, and urethra

MALE FERTILITY

A. Relates to many factors—number of sperm; size, shape, and motility
B. Infertility may be caused by antibodies some men make against their own sperm

CYCLE OF LIFE: MALE REPRODUCTIVE SYSTEM

A. Reproductive functions begin at time of puberty
B. Development of organs begins before birth; immature testes descend into scrotum before or shortly after birth
C. Puberty—high levels of hormones stimulate final stages of development
D. System operates to permit reproduction until advanced old age
E. Late adulthood—gradual decline in hormone production may decrease sexual appetite and fertility

MECHANISMS OF DISEASE: DISORDERS OF THE MALE REPRODUCTIVE SYSTEM

A. Infertility—reduced ability to reproduce
B. Sterility—complete incapacity to deliver healthy sperm to the female reproductive tract
C. Oligospermia—decreased sperm production; leading cause of male infertility
D. Cryptorchidism—undescended testes
E. Testicular cancer—second most common malignancy among 25- to 35-year-old males
F. Benign prostatic hypertrophy (BPH)—enlarged prostate
G. Erectile dysfunction—impotence, failure of penis to achieve an erection
H. Hydrocele—accumulation of fluid in the scrotum
I. Inguinal hernia—a weakening of the abdominal wall allowing intestines to protrude

REVIEW QUESTIONS

1. How does the function of reproduction differ from all other body functions?
2. Name the accessory glands of the male reproductive system.
3. List the genital ducts in the male.
4. List the supporting structures of the male reproductive system.
5. What is tunica albuginea? How does it aid in dividing the testis into lobules?
6. What is the relationship between the rete testis, seminiferous tubules, and efferent ductules?
7. What are the two primary functions of the testes?
8. What are the general functions of testosterone?
9. Discuss the structure of a mature spermatozoon.
10. What is meant by the term *capacitation?*
11. List the three functions of the epididymis.
12. List the anatomical divisions of the epididymis.
13. Discuss the formation of the ejaculatory ducts.
14. What is the relationship of the prostate gland to the urethra?
15. Discuss the type of secretion typical of the prostate gland and seminal vesicles.
16. What and where are the bulbourethral (Cowper's) glands?
17. Describe the structure, location, and function(s) of the scrotum.
18. Name the three cylindrical masses of erectile, or cavernous, tissue in the penis.
19. What and where is the glans penis? The prepuce, or foreskin?
20. What is the spermatic cord? From what does it extend, and what does it contain?
21. Of what is the seminal fluid composed? Trace its course from its formation in the gonads to the exterior.
22. Identify and define the male functions in reproduction.

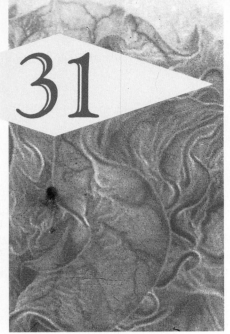

31 Female Reproductive System

Oocyte traveling through uterine tube
Erlandsen/Magney: Color Atlas of Histology

CHAPTER OUTLINE

OBJECTIVES

After you have completed this chapter, you should be able to:

1. List the essential and accessory sex organs of the female reproductive system and give the generalized functions of each.
2. Discuss the structure of the uterus, including details of its wall layers, size, shape, cavities, blood supply, and ligaments.
3. Locate the uterus in the pelvic cavity and compare its position with the abnormal position of retroflexion.
4. Identify the functions of the uterus.
5. Discuss the location, structure, divisions, and functions of the uterine tubes.
6. Describe the structure of the ovaries and explain the steps in development of mature ova from ovarian follicles.
7. Identify the two functions of the ovaries.
8. Discuss the location, structure, and primary functions of the vagina.
9. Identify the structures that together constitute the female external genitals.
10. Explain the clinical importance of the perineum during childbirth.
11. Describe the structure of the breasts and the mechanism controlling lactation.
12. Identify the phases of the endometrial, or menstrual, cycle.
13. Explain the hormonal control of cyclical changes that occur in the ovaries and in the uterus.
14. Discuss the importance of the female reproductive cycles.
15. Compare and contrast menarche and menopause.

787

I n the previous chapter, we discussed the structure and function of the male reproductive system. In this chapter, we will discuss the structure and function of the female reproductive system. As you study this chapter, keep in mind that *both systems* must function properly if successful reproduction of offspring is to occur.

OVERVIEW OF THE FEMALE REPRODUCTIVE SYSTEM

Function of the Female Reproductive System

The physiological importance of the female reproductive system is best understood in terms of its final outcome: production of offspring and continued existence of the genetic code. The female reproductive system produces gametes which may unite with a male gamete to form the first cell of the offspring. The function of conception emphasizes the similarity between the male and female reproductive systems. However, this is where the similarity ends. Unlike the male system, the female reproductive system also provides protection and nutrition to the developing offspring for up to several years after conception, as we shall see.

Structural Plan of the Female Reproductive System

So many organs make up the female reproductive system that we need to look first at the structural plan of the system as a whole. Reproductive organs can be classified as **essential** or **accessory,** depending on how directly they are involved in producing offspring.

Essential organs

The essential organs of reproduction in women, the *gonads,* are the paired **ovaries.** The female gametes, or **ova,** are produced by the ovaries.

Accessory organs

The accessory organs of reproduction in women consist of the following structures:

♦ A series of ducts or modified duct structures that extend from near the ovaries to the exterior. This group of organs includes the *uterine tubes, uterus,* and *vagina.* Along with the ovaries, these organs are sometimes collectively called the "internal genitals."

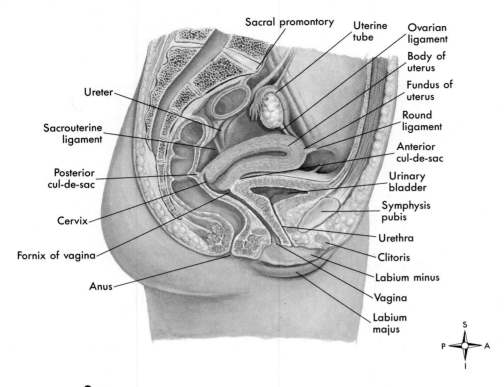

FIGURE 31-1   Female reproductive organs. Sagittal section of pelvis shows location of female reproductive organs.

♦ The *vulva*, or external reproductive organs. These organs are often called the "external genitals."

♦ Additional sex glands, including the *mammary glands*, which have an important reproductive function only in women.

Many of the essential and accessory organs of the female reproductive system can be seen in Figures 31-1 and 31-2. Refer to these illustrations often as you read about each structure in the pages that follow.

UTERUS

Structure of the Uterus
Size and shape of the uterus

In a woman who has never been pregnant, the uterus is pear shaped and measures approximately 7.5 cm (3 inches) in length, 5 cm (2 inches) in width at its widest part, and 3 cm (1 inch) in thickness. Note in Figure 31-2 that the uterus has two main parts: a wide, upper portion, the **body**, and a lower, narrow "neck," the **cervix.** Did you notice

that the body of the uterus rounds into a bulging prominence above the level at which the uterine tubes enter? This bulging upper component of the body is called the **fundus.**

Wall of the uterus

Three layers compose the walls of the uterus: the inner endometrium, a middle myometrium, and an outer incomplete layer of parietal peritoneum.

1. The lining of mucous membrane, called the **endometrium,** is composed of three layers of tissues: a compact surface layer of partially ciliated, simple columnar epithelium called the *stratum compactum,* a spongy middle, or intermediate, layer of loose connective tissue—the *stratum spongiosum*—and a dense inner layer termed the *stratum basale* that attaches the endometrium to the underlying myometrium. During menstruation and following delivery of a baby, the compact and spongy layers slough off. The endometrium varies in thickness from 0.5 mm just after the menstrual flow to about 5 mm near the end of the endometrial cycle. The endometrium has a rich supply of blood capillaries, as well as numerous exocrine glands that secrete mucus and other substances onto the endometrial surface.

2. A thick, middle layer (the **myometrium**) consists of three layers of smooth muscle fibers that extend in all directions, longitudinally, transversely, and obliquely, and give the uterus great strength. The bundles of smooth

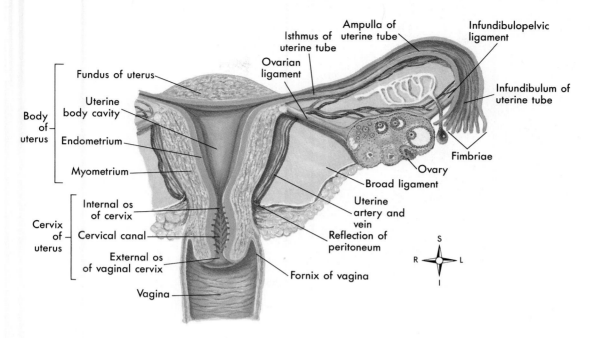

FIGURE 31-2   Female pelvic organs in a frontal section. The entire uterus is shown, with the upper portion of the vagina and the left uterine tube and ovary.

muscle fibers interlace with elastic and connective tissue components and generally blend into the endometrial lining with no sharp line of demarcation between the two layers. The myometrium is thickest in the fundus and thinnest in the cervix—a good example of the principle of structural adaptation to function. To expel a fetus, that is, move it down and out of the uterus, the fundus must contract more forcibly than the lower part of the uterine wall, and the cervix must be stretched or dilated.

3. An external layer of serous membrane, the **parietal peritoneum,** is incomplete, since it covers none of the cervix and only part of the body (all except the lower one fourth of its anterior surface). The fact that the entire uterus is not covered with peritoneum has clinical significance because it makes it possible to perform operations on this organ without the risk of infection that attends cutting into the peritoneum.

Cavities of the uterus

The cavities of the uterus are small because of the thickness of its walls (Figure 31-2). The body cavity is flat and triangular. Its apex is directed downward and constitutes the **internal os,** which opens into the **cervical canal.** The cervical canal is constricted on its lower end also, forming the **external os,** which opens into the vagina. The uterine tubes open into the body cavity at its upper, outer angles.

Blood supply of the uterus

The uterus receives a generous supply of blood from uterine arteries, branches of the internal iliac arteries (Figure 31-2). In addition, blood from the ovarian and vaginal arteries reaches the uterus by anastomosis with the

uterine vessels. Tortuous arterial vessels enter the layers of the uterine wall as arterioles and then break up into capillaries between the endometrial glands.

Uterine, ovarian, and vaginal veins return venous blood from the uterus to the internal iliac veins.

Location of the Uterus

Figure 31-3 shows the location of the uterus in the pelvic cavity between the urinary bladder in front and the rectum behind. Age, pregnancy, and distention of related pelvic viscera such as the bladder will alter the position of the uterus.

Between birth and puberty the uterus descends gradually from the lower abdomen into the true pelvis. At menopause the uterus begins a process of involution that results in a decrease in size and a position deep in the pelvis. Some variation among women in uterine placement within the pelvis is common.

Position of the Uterus

Normally the uterus is flexed between the body and cervix, with the body lying over the superior surface of the bladder, pointing forward and slightly upward (see Figure 31-1). The cervix points downward and backward from the point of flexion, joining the vagina at approximately a right angle. Several ligaments hold the uterus in place but allow its body considerable movement, a characteristic that often leads to malpositions of the organ. Fibers from several muscles that form the pelvic floor (see Figure 10-10 on p. 261) converge to form a node called the **perineal body** (see Figure 31-9), which also serves an important role in support of the uterus.

The uterus may lie in any one of several abnormal

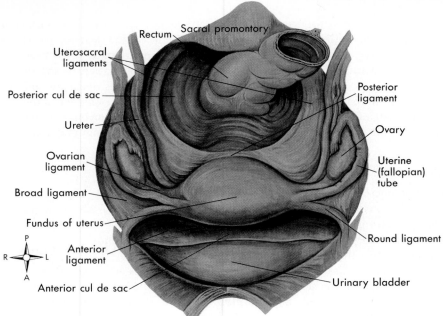

FIGURE 31-3   Female pelvic organs. Internal view shows the organs of the female pelvis as seen from above and in front.

positions. A common one is *retroflexion,* or backward tilting, of the entire organ. Retroflexion may allow the uterus to prolapse, or descend, into the vaginal canal, as shown in Figure 31-4, *B.*

Eight *uterine ligaments* (three pairs, two single ones) hold the uterus in its normal position by anchoring it in the pelvic cavity. These ligaments include the broad (paired), uterosacral (paired), posterior (single), anterior (single), and round (paired) ligaments. Six of these so-called ligaments are actually extensions of the parietal peritoneum in different directions. The round ligaments are fibromuscular cords. Many of the following structures can be identified in Figure 31-3.

♦ The *two* **broad ligaments** are double folds of parietal peritoneum that form a kind of partition across the pelvic cavity. The uterus is suspended between these two folds.

♦ The *two* **uterosacral ligaments** are foldlike extensions of the peritoneum from the posterior surface of the uterus to the sacrum, one on each side of the rectum.

♦ The **posterior ligament** is a fold of peritoneum extending from the posterior surface of the uterus to the rectum. This ligament forms a deep pouch known as the **posterior cul-de-sac (of Douglas),** or rectouterine pouch, between the uterus and rectum. Since this is the lowest point in the pelvic cavity, pus collects here in pelvic inflammations. To secure drainage, an incision may be made at the top of the posterior wall of the vagina (posterior colpotomy).

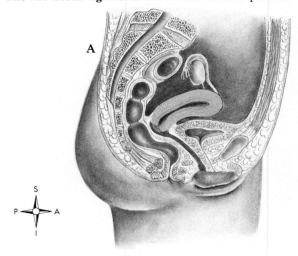
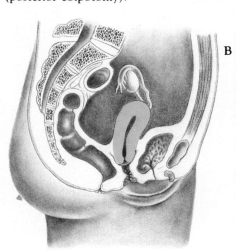

FIGURE 31-4   Normal and abnormal positions of the uterus. A, Normal position of the uterus, tilted over the urinary bladder. **B,** Prolapsed uterus. Prolapse results from weakening of the supporting ligaments of the uterus, which permits retroflexion (backward tilt) and descent of the uterus into the vagina. In severe cases, the uterus may descend outside the exterior opening of the vagina.

♦ The **anterior ligament** is the fold of peritoneum formed by the extension of the peritoneum on the anterior surface of the uterus to the posterior surface of the bladder. This fold forms an **anterior cul-de-sac,** which is less deep than the posterior cul-de-sac.

♦ The *two* **round ligaments** are fibromuscular cords extending from the upper, outer angles of the uterus through the inguinal canals and terminating in the labia majora.

Functions of the Uterus

The uterus, or womb, has many functions important to successful reproductive function. The uterus serves as part of the female reproductive tract, permitting sperm from the male to ascend toward the uterine tubes. If fusion of gametes (*fertilization,* or *conception*) occurs, the developing offspring implants itself in the endometrial lining of the uterus and continues its development during the term of pregnancy (*gestation*). The tiny endometrial glands are specialized to produce nutrient secretions—sometimes called "uterine milk"—to sustain the developing offspring until a *placenta* can be produced. The placenta is a unique organ that permits the exchange of materials between the offspring's blood and the maternal blood. The rich network of endometrial capillaries promotes efficiency of this exchange function. Rhythmic contractions of the myometrium are inhibited during gestation but are allowed to occur as the time of delivery approaches. Myometrial contractions are the "labor contractions" that help push the offspring out of the mother's body.

If conception or the implantation of the offspring does not occur successfully, then the outer layers of the endometrium are shed during *menstruation.* Menstruation is a regular event of the female reproductive cycle that permits the endometrium to renew itself in anticipation of conception and implantation during the next cycle. The myometrial contractions seem to aid menstruation by promoting the complete sloughing of the outer endometrial layers. Fatigue of the myometrial muscle tissues may contribute to the abdominal cramping sometimes associated with menstruation.

UTERINE TUBES

The **uterine tubes** are also sometimes called *fallopian tubes* or *oviducts.*

Location of the Uterine Tubes

The uterine tubes are about 10 cm (4 inches) long and are attached to the uterus at its upper outer angles (see Figures 31-1 to 31-3). They lie in the upper free margin of the broad ligaments and extend upward and outward toward the sides of the pelvis and then curve downward and backward.

Structure of the Uterine Tubes
Wall of the uterine tubes

The same three layers (mucous, smooth muscle, and serous) of the uterus compose the tubes (Figure 31-5). The mucosal lining of the tubes, however, is directly continuous with the peritoneum lining the pelvic cavity. This fact has great clinical significance because the tubal mucosa is also continuous with that of the uterus and vagina and therefore often becomes infected by gonococci or other organisms

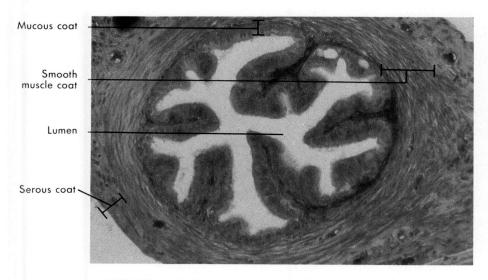

Mucous coat

Smooth muscle coat

Lumen

Serous coat

FIGURE 31-5 Transverse. section of uterine tube. This micrograph shows a section of the uterine (fallopian) tube in the isthmus region. Notice the highly folded nature of the epithelial, mucous lining of the tube. The folding is even more exaggerated in the ampulla region.

introduced into the vagina. Inflammation of the tubes **(salpingitis)** may readily spread to become inflammation of the peritoneum **(peritonitis),** a serious condition. In the male, there is no such direct route by which microorganisms can reach the peritoneum from the exterior.

Divisions of the uterine tubes

Each uterine tube consists of three divisions (see Figure 31-2).

1. A medial third that extends from the upper outer angle of the uterus called the **isthmus.**

2. An intermediate dilated portion called the **ampulla** that follows a winding path over the ovary.

3. A funnel-shaped terminal component called the **infundibulum** that opens directly into the peritoneal cavity. The open outer margin of the infundibulum resembles a fringe in its irregular outline. The fringelike projections are known as **fimbriae** (see Figure 31-2).

Histology of the uterine tubes

Figure 31-5 is a low-power micrograph that illustrates the mucosal lining of the oviduct cut in cross section. These shapes are typical of the appearance of the mucosal lining throughout most of the duct. Note the extensive

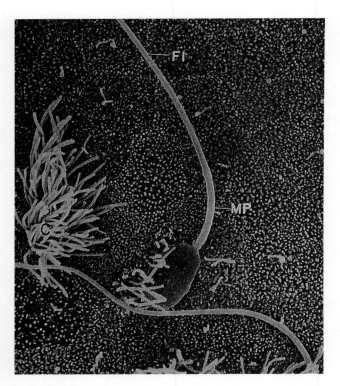

FIGURE 31-6 Luminal surface of the uterine tube. This scanning electron micrograph shows the surface of the uterine tube wall that faces the lumen, as well as some spermatozoa that are present (*Fl,* flagellum; *MP,* midpiece). The smaller, raised projections are microvilli, and the larger projections, found in clumps, are cilia (*C*).

folds of mucosa that project as shelves into the lumen of the tube. An area of smooth muscle (muscularis layer) can be seen surrounding the mucosa in this section. Cilia, which are important in maintaining currents within the tube to move the ovum toward the uterus, can be seen projecting from the luminal surface in Figure 31-6.

Function of the Uterine Tubes

The uterine tubes are extensions of the uterus that communicate loosely with the ovaries. This arrangement allows an ovum released from the surface of the ovary to be collected by the fimbriae and swept along the uterine tube toward the body of the uterus by ciliary action. The uterine tubes serve as more than mere transport channels, however. The uterine tube is also the site of fertilization. A few of the sperm deposited in the vagina during sexual intercourse move up the uterine tube, where they meet the ovum traveling toward them. It is generally here, within the uterine tube, that the primary function of human sexual reproduction occurs: recombination of the genetic information from both parents in the first cell of the offspring.

QUICK CHECK

✔ 1. Name the three principal layers of the uterine wall.
✔ 2. Describe the anatomical position of the uterus. How is it held in place?
✔ 3. List the chief functions of the uterus.
✔ 4. What are the functions of the uterine (fallopian) tubes?

OVARIES

Location of the Ovaries

The female gonads, or ovaries, are homologous (similar in origin) to the testes in the male. They are nodular glands that after puberty present a puckered, uneven surface, resemble large almonds in size and shape, and are located one on each side of the uterus, below and behind the uterine tubes. Each ovary weighs about 3 g and is attached to the posterior surface of the broad ligament by the *mesovarian* ligament. The ovarian ligament anchors it to the uterus. The distal portion of the uterine tube curves about the ovary in such a way that the fimbriae cup over the ovary but do not actually attach to it (see Figure 31-2). Here, then, is a gland whose duct is detached from it, a fact that makes possible pregnancy in the pelvic cavity instead of in the uterus as is normal. Development of the fetus in a location other than the uterus is referred to as an **ectopic pregnancy** (from the Greek *ektopos,* "displaced").

Microscopic Structure of the Ovaries

The surface of the ovary is covered by a layer of small epithelial cells called the **germinal epithelium.** The term

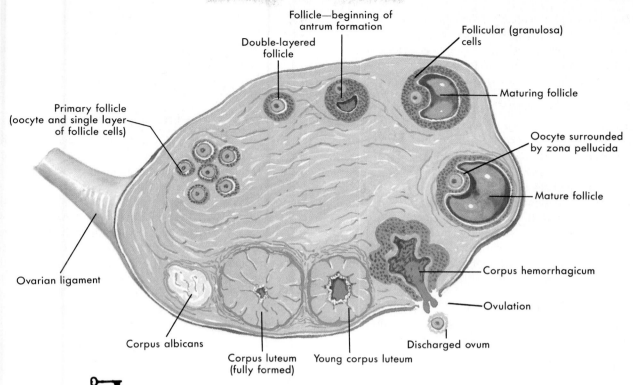

FIGURE 31-7 Stages of ovarian follicle development. Artist's rendition shows the successive stages of ovarian follicle and oocyte development. Begin with the first stage (primary follicle) and follow around clockwise to the final stage (corpus albicans). Remember, however, that all the stages shown occur over time to a *single* follicle, and the presence of all these stages at a single point in time is an artificial construct for learning purposes only.

germinal is misleading because the epithelial cells of this layer do not give rise to ova. Deep to the surface layer of epithelial cells, thousands of microscopic structures known as **ovarian follicles** are imbedded in a connective tissue matrix. The follicles contain the immature female sex cells, or *oocytes,* and after puberty are present in varying stages of development. The primary follicles consist of an oocyte surrounded by hormone-secreting follicular, or granulosa, cells. Refer to Figure 31-7 and trace the development of a female sex cell from its most primitive state through ovulation. Figure 31-8 shows oocytes at different stages of development surrounded by several layers of hormone-secreting follicular (granulosa) cells. The developing antrum, or cavity, and shell-like zona pellucida surrounding the oocyte are both apparent in Figure 31-8. Dividing cells are common in the follicular, or granulosa, portion of the rapidly maturing follicle. After the oocyte is released from the ovary, it is known as an *ovum.*

The process of egg formation, *oogenesis,* is discussed in greater detail in the next chapter.

Functions of the Ovaries

Recall that the ovaries are considered to be the *essential* organs of the female reproductive system. This means that it is the ovaries that produce female gametes, or ova. The process that culminates in the release of an ovum is called *oogenesis,* a term that literally means "egg production." The release of an ovum at the end of oogenesis is an event called *ovulation.* As Figure 31-7 shows, ovulation involves the rupture of an ovarian follicle and the subsequent release of fluid and an ovum. The ovum, surrounded by a coat of follicular cells, moves into the uterine tube where it may draw a sperm cell into it and thus become the first cell of an offspring.

The ovaries are also endocrine organs, secreting the female sex hormones. Estrogens (chiefly estradiol and estrone) and progesterone are secreted by cells of ovarian tissues. These hormones help regulate reproductive function in the female—making the ovaries even more essential to female reproductive function.

More details of oogenesis and fertilization will be discussed in Chapter 32. Further discussion of hormonal regulation of reproductive functions, as well as associated changes within the ovaries, appears later in this chapter.

QUICK CHECK

✔ 1. Briefly describe the location of the ovaries.
✔ 2. What are ovarian follicles?
✔ 3. List the two major functions of the ovaries.

A

Zona pellucida   Follicular (granulosa) cells

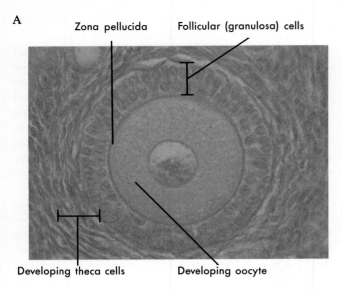

Developing theca cells   Developing oocyte

B

Antrum   Developing oocyte

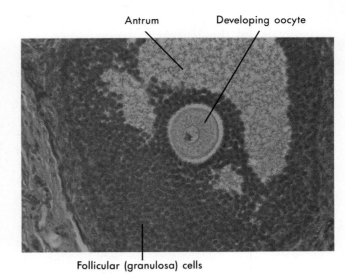

Follicular (granulosa) cells

FIGURE 31-8   Ovarian follicles. Female gametes mature within follicles in the outer region of an ovary. Follicles in both early stages of development **(A)** and late stages of development **(B)** exhibit a developing oocyte (immature ovum) surrounded by hormone-secreting follicular (granulosa) cells. Notice that the more mature ovarian follicle in **B** has a fluid-filled cavity called the antrum.

VAGINA

Location of the Vagina

The vagina is a tubular organ situated between the rectum, which lies posterior to it, and the urethra and bladder, which lie anterior to it. It extends upward and backward from its external orifice in the vestibule between the labia minor of the vulva to the cervix (see Figures 31-1, 31-2 and 31-9).

Structure of the Vagina

The vagina is a collapsible tube about 7 or 8 cm long that is capable of great distention. It is composed mainly of smooth muscle and is lined with mucous membrane arranged in rugae. It should be noted that the anterior wall of the vagina is shorter than the posterior wall because of the way the cervix protrudes into the uppermost portion of the tube (see Figure 31-1). In some cases—especially in young girls—a fold of mucous membrane, the **hymen,** forms a border around the external opening of the vagina, partially closing the orifice. Occasionally, this structure completely covers the vaginal outlet, a condition referred to as **imperforate hymen.** Perforation must be performed at puberty before the menstrual flow can escape.

Functions of the Vagina

The vagina is a portion of the female reproductive tract that has several important functions. During sexual intercourse, the lining of the vagina stimulates the glans penis, which in turn triggers the ejaculation of semen. Thus the vagina also serves as a receptacle for semen, which often pools in the corner, or *fornix,* of the vagina where it meets the cervix of the uterus. Sperm within the semen may move

further into the female reproductive tract by "climbing" along fibrous strands of mucus in the cervical canal.

The vagina also serves as the lower portion of the birth canal. At the time of delivery, the offspring is pushed from the body of the uterus, through the cervical canal, and finally through the vagina and out of the mother's body. The placenta, or "afterbirth," is also expelled through the vagina.

Another important function of the vagina is transport of tissue shed from the lining of the uterus during menstruation.

 The "G Spot"

In 1950, Dr. Ernest Gräfenberg first described what he called an "erotic zone" about the size of a dime on the anterior wall of the vagina midway between the symphysis pubis and cervix. The area was later named the "Gräfenberg spot," or "G spot."

The G spot is said to consist of glandular tissue that secretes a prostatic-like fluid into the urethra during sexual stimulation. This fluid may constitute the female "ejaculate" reported to occur in some women during orgasm. The existence of G spot tissue in significant numbers of women continues to be the subject of heated controversy in the medical community. Although existence of the G spot is supported by some physicians and professionals in the area of human sexuality, most clinicians remain skeptical because of conflicting and inconsistent data.

VULVA

Structure of the Vulva

Figure 31-9 shows the structures that, together, constitute the female external genitals (reproductive organs), or **vulva:** mons pubis, labia majora, labia minora, clitoris, urinary meatus (urethral orifice), vaginal orifice, and Bartholin's, or the greater vestibular, glands.

The **mons pubis** is a skin-covered pad of fat over the symphysis pubis. Coarse pubic hairs appear on this structure at puberty and persist throughout life.

The **labia majora** (Latin, "large lips") are covered with pigmented skin and hair on the outer surface and are smooth and free from hair on the inner surface. Each labium majus is composed mainly of fat and connective tissue with numerous sweat and sebaceous glands on the inner surface. The labia majora are homologous to the scrotum in the male.

The **labia minora** (Latin, "small lips") are located medial to the labia majora. Each labium minus is covered with hairless skin. The two labia minora come together anteriorly in the midline. The area between the labia minora is the **vestibule.** The labia minora are homologous to the corpus spongiosum of the penis in the male.

The **clitoris** is a small organ composed of erectile tissue, located just behind the junction of the labia minora and homologous to the corpora cavernosa and glans of the penis. The *prepuce*, or foreskin, covers the clitoris, as it does the glans penis in the male.

The **urinary meatus** (urethral orifice) is the small opening of the urethra, situated between the clitoris and the vaginal orifice.

The **vaginal orifice** is an opening that is larger than the urinary meatus. It is located posterior to the meatus.

The **greater vestibular glands** are two bean-shaped glands, one on each side of the vaginal orifice. Each gland opens by means of a single, long duct into the space between the hymen and the labium minus. These glands, which are also called *Bartholin's glands,* are of clinical importance because they can be infected (bartholinitis or Bartholin's abscess), particularly by gonococci. They are homologous to the bulbourethral glands in the male, and they secrete a lubricating fluid. Opening into the vestibule near the urinary meatus by way of two small ducts is a group of tiny mucous glands, the **lesser vestibular glands.** Also called *Skene's glands,* they have clinical interest because gonococci that lodge there are difficult to eradicate.

Functions of the Vulva

The various components of the external genitals of the female operate alone or separately to accomplish several functions important to successful reproduction. The protective features of the mons pubis and labia help prevent injury to the delicate tissues of the clitoris and vestibule. The vestibular glands produce lubricating fluids that reduce friction between the penis and the vagina during sexual intercourse. The clitoris becomes erect during sexual stimulation and, like the male glans, possesses a large number of sensory receptors that feed back information to the sexual response areas of the brain. Of course, the vaginal orifice serves as the boundary between the internal and external female genitals.

PERINEUM

The perineum is the skin-covered muscular region between the vaginal orifice and the anus (Figure 31-9). This area has great clinical importance because of the danger of its being torn during childbirth. Such tears are often deep, have irregular edges, and extend all the way through the perineum, the muscular perineal body, and even through the anal sphincter, resulting in involuntary

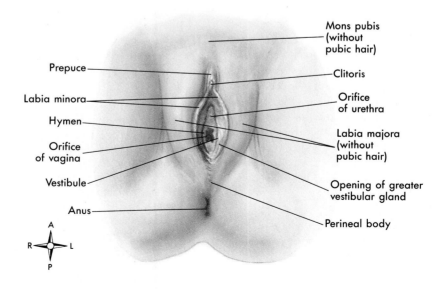

FIGURE 31-9 **External genitals of the female.**

seepage from the rectum until the laceration is repaired. In addition, injuries to the perineal body can result in partial uterine or vaginal prolapse if this important support structure is weakened. To avoid these possibilities in a woman prone to such injuries, a surgical incision known as an **episiotomy** may be made in the perineum, particularly at the birth of a first baby. In current medical practice, episiotomy procedures are decreasing in frequency and are no longer performed on a routine basis preceding vaginal delivery of a baby.

QUICK CHECK

✔ *1. List three functions of the vagina.*
✔ *2. What is another name for the external genitals of the female?*
✔ *3. How are the clitoris of the female and the glans penis of the male similar in structure and function?*

BREASTS
Location and Size of the Breasts

The breasts lie over the pectoral muscles and are attached to them by a layer of connective tissue (fascia) (Figure 31-10). Estrogens and progesterone, two types of ovarian hormones, control breast development during puberty. Estrogens stimulate growth of the ducts of the mammary glands, whereas progesterone stimulates development of the actual secreting cells. Breast size is determined more by the amount of fat around the glandular tissue than by the amount of glandular tissue itself. Hence the size of the breast is not related to its functional ability.

Structure of the Breasts

Each breast consists of several lobes separated by septa (walls) of connective tissue. Each lobe consists of several lobules, which, in turn, are composed of connective tissue in which are imbedded the secreting cells (alveoli) of the gland, arranged in grapelike clusters around the minute ducts. The ducts from the various lobules unite, forming a single lactiferous (milk-carrying) duct for each lobe, or between 15 and 20 in each breast. These main ducts converge toward the nipple, like the spokes of a wheel. They enlarge slightly before reaching the nipple into small lactiferous sinuses (Figure 31-10, A). Each of these main ducts terminates in a tiny opening on the surface of the nipple. Adipose tissue is deposited around the surface of the gland, just under the skin, and between the lobes.

The nipples are bordered by a circular pigmented area, the **areola.** It contains numerous sebaceous glands that appear as small nodules under the skin. Sebum produced by these glands helps reduce irritating dryness of the areolar skin associated with nursing. In Caucasians (other than those with very dark complexions) the areola and nipple change color from delicate pink to brown early in pregnancy—a fact of value in diagnosing a first pregnancy. The color decreases after lactation has ceased but never entirely returns to the original hue. In darker-skinned women, no noticeable color change in the areola or nipple heralds the first pregnancy.

Knowledge of the lymphatic drainage of the breast is important in clinical medicine because cancerous cells from malignant breast tumors often spread to other areas of the body through the lymphatics. Lymphatic drainage of the breast is presented in Chapter 19 (see Figure 19-8).

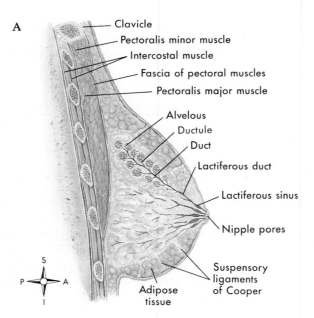

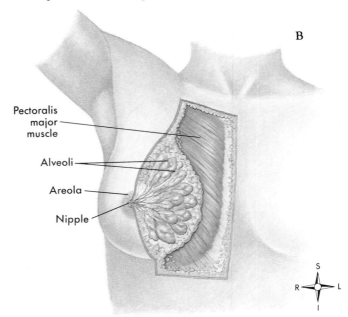

FIGURE 31-10 The female breast. A, Sagittal section of a lactating breast. Notice how the glandular structures are anchored to the overlying skin and to the pectoral muscles by the suspensory ligaments of Cooper. Each lobule of glandular tissue is drained by a lactiferous duct that eventually opens through the nipple. **B,** Anterior view of a lactating breast. Overlying skin and connective tissue has been removed from the medial side to show the internal structure of the breast and underlying skeletal muscle. In nonlactating breasts the glandular tissue is much less prominent, with adipose tissue comprising most of each breast.

Function of the Breasts

The function of the mammary glands is lactation, that is, the secretion of milk for the nourishment of newborn infants.

Mechanism controlling lactation

Very briefly, lactation is controlled as follows:

♦ The ovarian hormones, estrogens and progesterone, act on the breasts to make them structurally ready to secrete milk. Estrogens promote development of the ducts of the breasts. Progesterone acts on the estrogen-primed breasts to promote completion of the development of the ducts and development of the alveoli, the secreting cells of the breasts. This is an example of hormonal *permissiveness;* estrogen permits progesterone to have its full effect. A high blood concentration of estrogens during pregnancy also inhibits anterior pituitary secretion of prolactin.

♦ Shedding of the placenta following delivery of the baby cuts off a major source of estrogens. The resulting rapid drop in the blood concentration of estrogens stimulates anterior pituitary secretion of prolactin. Also, the suckling movements of a nursing baby stimulate both anterior pituitary secretion of prolactin and posterior pituitary secretion of oxytocin.

♦ Prolactin stimulates lactation, that is, stimulates alveoli of the mammary glands to secrete milk. Milk secretion starts about the third or fourth day after delivery of a baby, supplanting a thin, yellowish secretion called *colostrum.* With repeated stimulation by the suckling

infant, plus various favorable mental and physical conditions, lactation may continue for extended periods.

♦ Oxytocin stimulates the alveoli of the breasts to eject milk into the ducts, thereby making it accessible for the infant to remove by suckling.

The mechanism is summarized in Figure 31-11 (see also Figure 15-7, p. 411).

The importance of lactation

The process of lactation plays an important role in the ultimate success of the reproductive system. The biological goal of human reproduction does not lie solely in delivering a healthy infant—the infant must also survive until reproductive age. If a child does not survive to reproduce, then the genetic code cannot be passed on to successive generations and the ultimate goal of reproduction has not been met. Humans and other mammals help ensure survival of offspring for up to several years by producing nutrient-rich milk. Nursing from the mother's breast provides several advantages for human offspring:

♦ Milk is a rich source of proteins, fat, calcium, vitamins, and other nutrients in proportions needed by a young, developing body.

♦ Human milk provides passive immunity to the offspring in the form of maternal antibodies present in the milk.

♦ Nursing appears to enhance the emotional bond between mother and child. Such bonding may foster healthy psychological development in the child and strengthen family relationships that contribute to successful human development.

QUICK CHECK

✔ 1. Briefly describe the network of ducts and secreting cells that form the mammary glands.
✔ 2. List the hormones that prepare the breast structurally for lactation.
✔ 3. Which hormone causes milk to be ejected into the lactiferous ducts?

◀ FEMALE REPRODUCTIVE CYCLES

Recurring Cycles

Many changes recur periodically in the female during the years between the onset of the menses (*menarche*) and their cessation (*menopause,* or *climacteric*). Most obvious, of course, is menstruation—the outward sign of changes in the endometrium. Most women also note periodic changes in their breasts. But these are only two of many changes that occur over and over again at fairly uniform intervals during the approximately three decades of female reproductive maturity.

First we shall describe the major cyclical changes, and then we shall discuss the mechanisms that produce them.

Ovarian cycle

Before a female child is born, precursor cells in her ovarian tissue begin a type of cell division characterized as

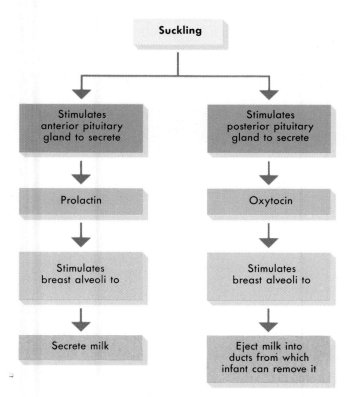

FIGURE 31-11 Lactation. This diagram summarizes the mechanism that controls secretion and ejection of milk.

meiosis. Meiotic cell division differs from mitotic cell division in that it reduces the number of chromosomes in the daughter cells by half (recall Chapter 3, pp. 91-92). By the time the child is born, her ovaries contain many primary follicles, each containing an oocyte that has temporarily suspended the meiotic process before it is complete. Once each month, on about the first day of menstruation, the oocytes within several primary follicles resume meiosis. At the same time, the follicular cells surrounding them proliferate and start to secrete estrogens (and tiny amounts of progesterone). Usually, only one of these developing follicles matures and migrates to the surface of the ovary. Just before ovulation, the meiosis within the oocyte of the mature follicle halts again. It is this cell, which has not quite completed meiosis, that is expelled from the ruptured wall of the mature follicle during ovulation. Meiosis is completed only when, and if, the head of a sperm cell is later drawn into the ovum during the process of fertilization.

When does ovulation occur? This is a question of great practical importance and one that in the past was given many answers. Today it is known that ovulation usually occurs 14 days before the next menstrual period begins. (Only in a 28-day menstrual cycle is this also 14 days after the beginning of the preceding menstrual cycle, as explained subsequently.)

Immediately after ovulation, cells of the ruptured follicle enlarge and, because of the appearance of lipoid substances in them, become transformed into a golden-colored body, the **corpus luteum.** The corpus luteum grows for 7 or 8 days. During this time, it secretes progesterone in increasing amounts. Then, provided fertilization of the ovum has not taken place, the size of the corpus luteum and the amount of its secretions gradually diminish. In time, the last components of each nonfunctional corpus luteum are reduced to a white scar called the **corpus albicans,** which moves into the central portion of the ovary and eventually disappears (see Figure 31-7).

Endometrial or menstrual cycle

During menstruation, bits of the compact and spongy layers of the endometrium slough off, leaving denuded bleeding areas. Following menstruation, the cells of these layers proliferate, causing the endometrium to reach a thickness of 2 or 3 mm by the time of ovulation. During this period, endometrial glands and arterioles grow longer and more coiled—two factors that also contribute to the thickening of the endometrium. After ovulation, the endometrium grows still thicker, reaching a maximum of

about 4 to 6 mm. Most of this increase, however, is believed to be caused by swelling produced by fluid retention rather than by further proliferation of endometrial cells. The increasingly coiled endometrial glands start to secrete during the time between ovulation and the next menses. Then, the day before menstruation starts again, the tightly coiled arterioles constrict, producing endometrial ischemia. This leads to death of the tissue, sloughing, and once again, menstrual bleeding.

The menstrual cycle is customarily divided into phases, named for major events occurring in them: menses, postmenstrual phase, ovulation, and premenstrual phase.

1. The **menses,** or **menstrual period,** occurs on days 1 to 5 of a new cycle. There is some individual variation, however.

2. The **postmenstrual phase** occurs between the end of the menses and ovulation. Therefore it is the **preovulatory phase,** as well as the postmenstrual phase. In a 28-day cycle, it usually includes cycle days 6 to 13 or 14. However, the length of this phase varies more than the others. It lasts longer in long cycles and ends sooner in short cycles. This phase is also called the **estrogenic,** or **follicular, phase** because of the high blood estrogen level resulting from secretion by the developing follicle. **Proliferative phase** is still another name for it because proliferation of endometrial cells occurs at this time.

3. Ovulation, that is, rupture of the mature follicle with expulsion of its ovum into the pelvic cavity (Figure 31-12), occurs frequently on cycle day 14 in a 28-day cycle. However, it occurs on different days in different-length cycles, depending on the length of the preovulatory phase. For example, in a 32-day cycle the preovulatory phase would probably last until cycle day 18, and ovulation would

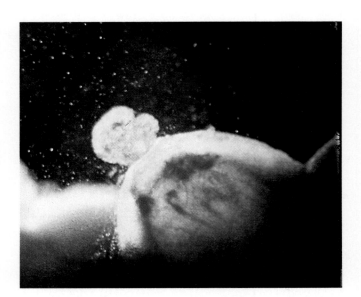

FIGURE 31-12 Ovulation. The rupture of a mature follicle on the surface of an ovary results in the release of an ovum into the pelvic cavity. This process of *ovulation* often occurs on day 14 in a 28-day menstrual cycle, but its exact timing depends on the length of the postmenstrual (preovulatory) phase. Notice in this photograph that the ovum released during ovulation is surrounded by a mass of cells.

 Mittelschmerz

A few women experience pain within a few hours after ovulation. This is referred to as **mittelschmerz**—German for "middle pain." It has been ascribed to irritation of the peritoneum by hemorrhage from the ruptured follicle on about day 14 of the menstrual cycle.

Simply knowing the length of previous cycles cannot ensure with any degree of accuracy how many days the preovulatory phase will last in the next or some future cycle. This physiological fact probably accounts for most of the unreliability of the calendar rhythm methods of fertility planning and its replacement by other more sophisticated "natural" methods that are not based on a knowledge of previous cycle lengths to predict the day of ovulation. Instead, such natural methods base their judgments about fertility at any point in a woman's cycle on other changes. Examples: measurement of basal body temperature and recognition of cyclic changes in the amount and consistency of cervical mucus, both of which occur in response to changes in circulating hormones that control ovulation. The time of ovulation can also be determined by over-the-counter urine tests that detect the high levels of luteinizing hormone (LH) associated with ovulation.

then occur on cycle day 19 instead of 14. In short, because the majority of women show some month-to-month variation in the length of their cycles, the day of ovulation in a current or future cycle cannot be predicted with accuracy based on the length of previous cycles.

4. The **premenstrual phase,** or **postovulatory phase,** occurs between ovulation and the onset of the menses. This phase is also called the **luteal phase,** or more simply, the **secretory phase,** because the corpus luteum secretes only during this time. It is also called the **progesterone phase** because the corpus luteum secretes mainly this hormone. The length of the premenstrual phase is fairly constant, lasting usually 14 days—or cycle days 15 to 28 in a 28-day cycle. Differences in length of the total menstrual cycle therefore exist mainly because of differences in duration of the postmenstrual rather than of the premenstrual phase.

Myometrial cycle

The myometrium contracts mildly but with increasing frequency during the 2 weeks preceding ovulation. Contractions decrease or disappear between ovulation and the next menses, thereby lessening the probability of expulsion

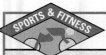

AMENORRHEA IN FEMALE ATHLETES

Failure to have a menstrual period is called **amenorrhea.** Amenorrhea occurs in some female athletes, probably resulting from a body fat composition that is too low to sustain normal reproductive function. Although it keeps the hematocrit (RBC level) higher than during menstruation, it is not considered a desirable condition. Besides infertility, amenorrhea may cause other problems. For example, the low blood levels of estrogen associated with long-term amenorrhea may cause osteoporosis (loss of bone mass).

of a fertilized ovum that may have implanted in the endometrium.

Gonadotropic cycle

The adenohypophysis (anterior pituitary gland) secretes two hormones called gonadotropins that influence female reproductive cycles. Their names are **follicle-stimulating hormone (FSH)** and **luteinizing hormone (LH).** The amount of each gonadotropin secreted varies with a rhythmic regularity that can be related, as we shall see, to the rhythmic ovarian and uterine changes just described.

QUICK CHECK

✔ 1. *What is the function of the corpus luteum? How is it formed?*

✔ 2. *What is the difference between the* ovarian cycle *and the* menstrual (endometrial) cycle?

✔ 3. *Why is the postmenstrual phase of the menstrual cycle sometimes called the* proliferative phase?

Control of Female Reproductive Cycles

Physiologists agree that hormones play a major role in producing the cyclic changes characteristic in women during the reproductive years. The development of a method called *radioimmunoassay* has made it possible to measure blood levels of gonadotropins. By correlating these with the monthly ovarian and uterine changes, investigators have worked out the main features of the control mechanism.

A brief description follows of the mechanisms that produce cyclical changes in the ovaries and uterus and in the amounts of gonadotropins secreted.

Control of cyclical changes in the ovaries

Cyclical changes in the ovaries result from cyclical changes in the amounts of gonadotropins secreted by the anterior pituitary gland. An increasing FSH blood level has two effects: it stimulates one or more primary follicles and their oocytes to start growing, and it stimulates the follicular cells to secrete estrogens. (Developing follicles also secrete very small amounts of progesterone.) Because of the influence of FSH on follicle secretion, the level of estrogens in blood increases gradually for a few days during the postmenstrual phase. Then suddenly, on about the twelfth cycle day, it leaps upward to a maximum peak. Scarcely 12 hours after this "estrogen surge," an "LH surge" occurs and presumably triggers ovulation a day or two later. This hormone surge is the basis of the over-the-counter "ovulation test" (see box on ovulation). The control of cyclical ovarian changes by the gonadotropins FSH and LH is summarized in Figure 31-13. As Figure 31-13 shows, LH brings about the following changes:

1. Completion of growth of the follicle and oocyte with increasing secretion of estrogens before ovulation. LH and FSH act as synergists to produce these effects.

2. Rupturing of the mature follicle with expulsion of its

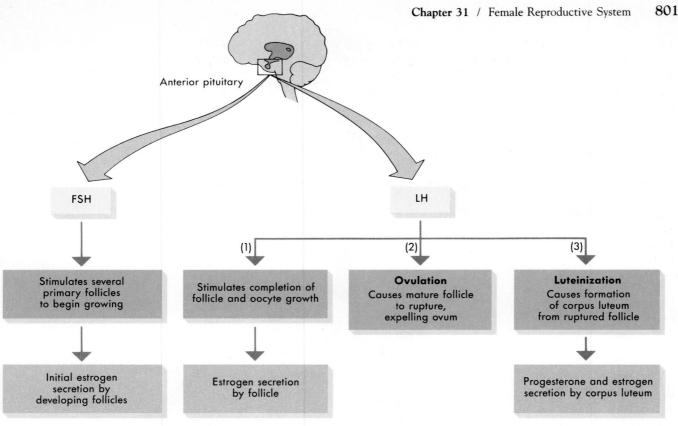

FIGURE 31-13 The primary effects of gonadotropins on the ovaries. Follicle-stimulating hormone (*FSH*) gets its name from the fact that it triggers development of primary ovarian follicles and stimulates follicular cells to secrete estrogens. Luteinizing hormone (*LH*) has several effects on ovaries: (1) LH acts as synergist to FSH to enhance its effects on follicular development and secretion; (2) LH presumably triggers ovulation—hence it is called "the ovulating hormone"; and (3) there is a luteinizing effect of LH (for which the hormone was named); recent evidence shows that FSH is also necessary for luteinization.

ovum (ovulation). Because of this function, LH is sometimes also called "the ovulating hormone."

3. Formation of a golden body, the corpus luteum, in the ruptured follicle (process called **luteinization**). The name *luteinizing hormone* refers, obviously, to this LH function—a function to which, experiments have shown, FSH also contributes. The corpus luteum functions as a temporary endocrine gland. It secretes only during the luteal (postovulatory, or premenstrual) phase of the menstrual cycle. Its hormones are progestins (the important one of which is progesterone) and also estrogens. The blood level of progesterone rises rapidly after the "LH surge" described earlier. It remains at a high level for about a week, then it decreases to a very low level approximately 3 days before menstruation begins again. This low blood level of progesterone persists during both the menstrual and the postmenstrual phases. Its sources? Not from the corpus luteum, which secretes only during the luteal phase, but from the developing follicles and the adrenal cortex. Blood's estrogen content increases during the luteal phase but to a lower level than develops before ovulation.

If pregnancy does not occur, lack of sufficient LH and FSH causes the corpus luteum to regress in about 14 days. The corpus luteum is then replaced by the corpus albicans. Look again at Figure 31-7, which shows the cyclical changes in the ovarian follicles.

Control of cyclical changes in the uterus

Cyclical changes in the uterus are brought about by changing blood concentrations of estrogens and progestrone. As blood estrogens increase during the postmenstrual phase of the menstrual cycle, they produce the following main changes in the uterus.

♦ Proliferation of endometrial cells, producing a thickening of the endometrium
♦ Growth of endometrial glands and of the spiral arteries of the endometrium
♦ Increase in the water content of the endometrium
♦ Increased myometrial contractions

Increasing blood progesterone concentration during the premenstrual phase of the menstrual cycle produces prog-

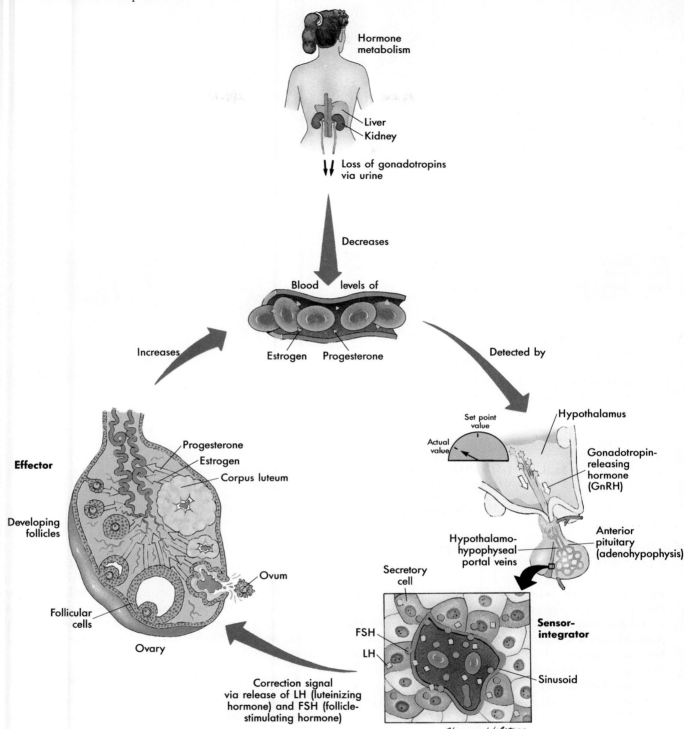

Hormone metabolism

Liver
Kidney

Loss of gonadotropins via urine

Decreases

Blood   levels of

Estrogen   Progesterone

Increases

Detected by

Set point value

Actual value

Hypothalamus

Gonadotropin-releasing hormone (GnRH)

Hypothalamo-hypophyseal portal veins

Anterior pituitary (adenohypophysis)

Secretory cell

Effector

Progesterone
Estrogen
Corpus luteum

Developing follicles

Ovum

Follicular cells

Ovary

Sensor-integrator

FSH

LH

Sinusoid

Yvonne Walston

Correction signal via release of LH (luteinizing hormone) and FSH (follicle-stimulating hormone)

FIGURE 31-14 Control of FSH and estrogen secretion. A negative feedback mechanism controls anterior pituitary secretion of follicle-stimulating hormone (FSH) and ovarian secretion of estrogens. A high blood level of FSH stimulates estrogen secretion, whereas the resulting high estrogen level inhibits FSH secretion. How does this compare with the LH testosterone feedback mechanism in the male? (See Figure 30-5, p. 777 if you want to check your answer.) According to the diagram, what effect does a high blood concentration of estrogens have on anterior pituitary secretion of FSH? of LH?

Fertility Signs

Rhythmic changes associated with female reproductive cycles take place in the ovaries, in the amount and consistency of the cervical mucus produced during each cycle, in the myometrium, in the vagina, in gonadotropin secretion, in body temperature, and even in mood or "emotional tone." Many of these recurring events that a woman may recognize on almost a monthly schedule during her reproductive years are called **"fertility signs"** and are manifestations of those body changes required for successful reproductive function.

 The Pill

Establishment of the relationship between sex hormone levels and ovulation led eventually to the development of "the Pill" for preventing pregnancy. Contraceptive pills contain synthetic progesterone-like compounds, sometimes combined with synthetic estrogens. By sustaining a high blood concentration of these substances, contraceptive pills prevent the monthly development of a follicle. With no ovum to be expelled, ovulation does not occur, and therefore pregnancy cannot occur. The next menses, however, does take place if the progesterone and estrogen dosage is stopped in time to allow their blood level to decrease as they normally do near the end of the cycle to bring on menstruation.

For this reason, the Pill can be used to regulate the menstrual cycle, as well as prevent pregnancy.

Actually the effects of contraceptive pills—indeed, of estrogens and progesterone—are much more complex than our explanation indicates. They have widespread effects on the body quite independent of their action on the reproductive and endocrine systems and still are not completely understood. Side effects that may limit or prohibit use of the Pill often result because of the related activity of these complex hormones on systems other than the reproductive system.

estational changes in the uterus—that is, changes favorable for pregnancy—specifically:

♦ Secretion by endometrial glands, thereby preparing the endometrium for implantation of a fertilized ovum
♦ Increase in the water content of the endometrium
♦ Decreased myometrial contractions

As mentioned earlier, low levels of FSH and LH cause regression of the corpus luteum if pregnancy does not occur. This in turn causes a drop in estrogen and progesterone levels, with the result that their maintenance of a thick, vascular endometrium ceases. Thus a drop in estrogen and progesterone levels at the end of the premenstrual phase triggers the endometrial sloughing that characterizes the menstrual phase.

Control of cyclical changes in gonadotropin secretion

Both negative and positive feedback mechanisms help control anterior pituitary secretion of the gonadotropins FSH and LH. These mechanisms involve the ovaries' secretion of both estrogens and progesterone and the hypothalamus' secretion of releasing hormones. Figure 31-14 describes a negative feedback mechanism that controls gonadotropin secretion. Examine it carefully. Note particularly the effects of a sustained high blood concentration of estrogens and progesterone on anterior pituitary gland secretion and the effect of a low blood concentration of FSH on follicular development: essentially, follicles do not mature and ovulation does not occur.

Several observations and animal experiments strongly suggest that sustained high blood levels of estrogens and progesterone inhibit pituitary secretion of FSH and LH. These ovarian hormones appear to inhibit certain neurons of the hypothalamus (part of the central nervous system) from secreting FSH-releasing and LH-releasing hormones into the hypophyseal portal vessels (see Figure 15-12, p. 416). Without the stimulating effects of these releasing hormones, the pituitary's secretion of FSH and LH decreases.

A positive feedback mechanism has also been postulated to control LH secretion. The sudden and marked increase

in blood's estrogen content that occurs late in the follicular phase of the menstrual cycle is thought to stimulate the hypothalamus to secrete LH-releasing hormone (LH-RH) into the hypophyseal portal vessels. As its name suggests, this hormone stimulates the release of LH by the anterior pituitary, which in turn accounts for the "LH surge" that triggers ovulation. The fact that a part of the brain—the hypothalamus—secretes FSH- and LH-releasing hormones has interesting implications. This may be the crucial part of the pathway by which changes in a woman's environment or in her emotional state can alter her menstrual cycle. That this occurs is a matter of common observation. Stress, for example—such as intense fear of either becoming or not becoming pregnant—often delays menstruation.

Importance of Female Reproductive Cycles

There are several important functional roles played by the female reproductive cycles. As Figure 31-15 shows, the changes associated with the different cycles are all closely interrelated. The primary role of the ovarian cycle, for example, is to produce an ovum at regular enough intervals to make reproductive success likely. The ovarian cycle's secondary role is to regulate the endometrial (menstrual) cycle by means of the sex hormones estrogen and progesterone. The role of the endometrial cycle, in turn, is to ensure that the lining of the uterus is suitable for the implantation of an embryo should fertilization of the ovum occur. The constant renewal of the endometrium makes successful implantation more likely. The cyclic mechanisms of female reproductive function, and the fact that an

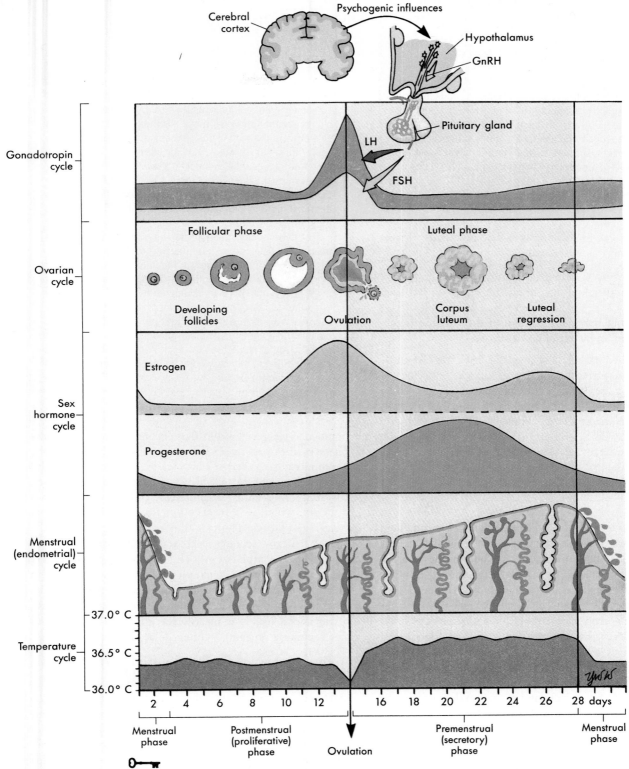

FIGURE 31-15 Female reproductive cycles. This diagram illustrates the interrelationships among the cerebral, hypothalamic, pituitary, ovarian, and uterine functions throughout a standard 28-day menstrual cycle. The variations in basal body temperature are also illustrated.

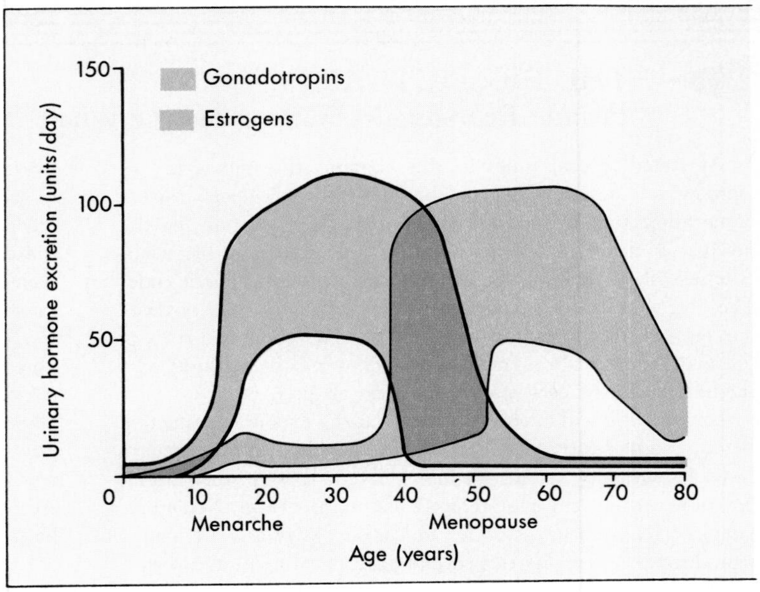

FIGURE 31-16 Gonadotropin and estrogen levels over the life span. This graph shows changes in hormone levels as reflected in urinary excretion rates from birth to advanced old age. Note that an increase in the gonadotropin level at the time of menarche sustains a high but variable level of estrogens until the time of menopause, when ovarian follicular cells cease to respond to gonadotropins. A negative feedback mechanism tries to increase estrogen levels to their former high levels by increasing the gonadotropin levels—a strategy that always fails.

ovum will unite with a sperm only during the first 24 hours or so after ovulation, result in the fact that a woman is fertile only a few days out of each monthly cycle. This limited fertility increases the likelihood that conception will occur only when the woman's body is at its reproductive peak.

Menarche and Menopause

The menstrual flow first occurs **(menarche)** at puberty, at about the age of 13 years, although there is individual variation according to race, nutrition, health, and heredity. Normally, it recurs about every 28 days for about three decades, except during pregnancy, and then ceases **(menopause, or climacteric).** The average age at which menstruation ceases is reported to have increased markedly—from about age 40 a few decades ago to between ages 45 and 50 more recently.

Figure 31-16 shows how the changes just described relate to changes in hormone levels over the life span. Relatively low concentrations of gonadotropins (FSH and LH) sustain a peak of estrogen secretion from menarche to menopause. After menopause, estrogen concentration decreases dramatically—which causes a negative feedback response that increases the gonadotropin levels. Because the follicular cells are no longer sensitive to gonadotropins after menopause, the increased gonadotropin level has no effect on estrogen secretion.

QUICK CHECK
✔ 1. A surge in FSH and LH is associated with what major event of the ovarian cycle?
✔ 2. How does an increase in estrogen level affect the uterine lining?
✔ 3. What is menopause? What causes menopause to occur?

Cycle of Life: *Female reproductive system*

As mentioned in the previous chapter, the reproductive system is unlike any other body system with regard to the normal changes that occur during the life span. Unlike other systems, the female reproductive system does not begin to perform its functions until the teenage years (puberty), and unlike the male reproductive system, the female reproductive system ceases its principal functions in middle adulthood.

The female organs begin their initial stages of development in the womb. As a matter of fact, the first stage of meiotic development of all the ova that will ever be produced by a woman is completed by the time she is born. However, full development of the reproductive organs—and the gametes within the ovaries—does not resume until puberty. At puberty, reproductive hormones stimulate the organs of the reproductive tract to become functional and produce a mature ovum one at a time. Reproductive function then continues in a cyclic fashion until **menopause.** Menopause is an event that is usually marked by the passage of at least 1 full year without menstruation. After that time, a woman may continue to enjoy normal sexual activity, but she cannot produce more offspring.

THE BIG PICTURE
Female Reproductive System and the Whole Body

As stated several times in this chapter, the importance of reproductive function lies in the fact that it imparts virtual immortality to our genes. This is important not only to the survival of the human species but to the survival of life itself. After all, life as we know it could not exist without a genetic code. The "big picture" of human reproduction requires two systems, one reproductive system in each parent. The combined roles of the male and female reproductive systems will be explored as a single topic in the early part of the next chapter.

For now, we will take a closer look at the female reproductive system and its relationships with other systems within a woman's body. As with any system, the female reproductive system cannot function without the maintenance functions of the circulatory, immune, respiratory, digestive, and urinary systems. The female reproductive system shares a special anatomical relationship with the urinary system. These two systems develop in close proximity to each other and thus share a common structure: the vulva. A special anatomical relationship with the skeletal muscular system is evident in the structure known as the perineum. Of course the skeletal and muscular systems both support and protect the internal organs of the female reproductive system. An even more special relationship with the integumentary system should be noted. The breasts, containing the milk-producing mammary glands, are actually modifications of the skin. Although structurally the breasts can be thought of as belonging to the integumentary system, functionally they are best considered as a part of the reproductive system. Nervous and endocrine regulation of female reproductive function has been outlined in this chapter and will be explored further in the next chapter.

MECHANISMS OF DISEASE

Disorders of the Female Reproductive System

Hormonal and Menstrual Disorders

Dysmenorrhea (dis-men-o-REE-ah), or painful menstruation, can have several causes. *Primary dysmenorrhea* occurs in the absence of an associated pelvic disease, such as an infection or tumor. Primary dysmenorrhea is thought to be caused by an abnormally increased concentration of certain prostaglandins produced by the uterine lining. High concentrations of prostaglandin E_2 (PGE_2) and prostaglandin $F_{2\alpha}$ ($PGF_{2\alpha}$) cause painful spasms of uterine muscle. For this reason, prostaglandin-inhibiting drugs such as *ibuprofen* are sometimes used to relieve the symptoms of primary dysmenorrhea. Oral contraceptives, which are hormone preparations that inhibit contractions of the uterine lining, are sometimes prescribed for this condition. *Secondary dysmenorrhea* can result from several conditions, including inflammatory conditions and cervical stenosis (narrow cervix). Treatment of secondary dysmenorrhea involves treating the underlying disorder.

Amenorrhea (a-men-o-REE-ah) is the absence of normal menstruation. *Primary amenorrhea* is the failure of menstrual cycles to begin and may be caused by various factors, such as hormone imbalances, genetic disorders, brain lesions, or structural deformities of the reproductive organs. *Secondary amenorrhea* occurs when a woman who has previously menstruated slows to three or fewer cycles per year. Secondary amenorrhea may be a symptom of weight loss, pregnancy, lactation, menopause, or disease of the reproductive organs. Treatment of amenorrhea involves treating the underlying disorder or condition.

Dysfunctional uterine bleeding (DUB) is irregular or excessive uterine bleeding that results from a hormonal imbalance rather than from an infection or other disease condition. In DUB, hormonal imbalances may cause excessive growth (*hyperplasia*) of the endometrium—or abnormally frequent menstrual flows. Excessive uterine bleeding from any cause can result in life-threatening anemia because of the chronic loss of blood.

Premenstrual syndrome (PMS) is a condition that involves a collection of symptoms that regularly occur in many women during the premenstrual phase of their reproductive cycles. Symptoms include irritability, fatigue, nervousness, depression, and other problems that are often distressing enough to affect personal relationships. Because the cause of PMS is still unclear, current treatments focus on relieving the symptoms.

Infection and Inflammation

Infections of the female reproductive tract are often classified as *exogenous* or *endogenous*. Exogenous infections result from pathogenic organisms transmitted from another person, such as **sexually transmitted diseases (STDs)**. Endogenous infections result from pathogens that normally inhabit the intestines, vulva, or vagina. You may recall from Chapter 1 that many areas of the body are normally inhabited by pathogenic microbes but that they cause infection only when there is a change in conditions or they are moved to a new area.

Pelvic inflammatory disease (PID) is an acute inflammatory condition caused by infection. PID can involve the uterus, uterine (fallopian) tubes, or ovaries. Uterine tube inflammation is termed **salpingitis** (sal-pin-JYE-tis), which simply means "tube inflammation." Inflammation of the ovaries is termed **oophritis** (o-o-FRY-tis). PID can be caused by several different pathogens, which usually spread upward from the vagina. Pelvic inflammatory disease may be accompanied by fever and pain or may have no symptoms at all. PID

can lead to serious complications, including infertility resulting from obstruction or other damage to the reproductive tract. The infection may also spread to other tissues, including the blood—where it may cause septic shock and death (see Chapter 18).

Vaginitis (vaj-in-EYE-tis) is inflammation or infection of the vaginal lining. Vaginitis most often results from STDs or from a "yeast infection." So-called yeast infections are usually opportunistic infections of the fungus *Candida albicans*, producing *candidiasis* (kan-did-EYE-as-is). Candidiasis infections are characterized by a whitish discharge—a symptom known as **leukorrhea** (loo-ko-REE-ah).

Tumors and Related Conditions

Myoma (my-O-ma), or **fibromyoma** (fi-BRO-my-o-ma), is a benign tumor of smooth muscle and fibrous connective tissue that accounts for many uterine tumors. Also called *fibroids*, these tumors are usually small and often produce no symptoms. If they interfere with pregnancy or become large enough to cause a health risk, they may be removed surgically.

Ovarian cysts are benign enlargements on one or both ovaries. An ovarian cyst is a fluid-filled sac that develops from a follicle that fails to rupture completely or from a corpus luteum that fails to degenerate. Ovarian cysts rarely become dangerous and often disappear within a few months of their appearance.

Another benign condition that commonly affects the female reproductive tract is **endometriosis** (en-doe-me-tree-O-sis). Endometriosis is the presence of functioning endometrial tissue outside the uterus. The displaced endometrial tissue can occur in many different places throughout the body but is most often found in or on pelvic and abdominal organs. The tissue reacts to ovarian hormones in the same way as the normal endometrium—exhibiting a cycle of growth and sloughing off.

Malignancies of reproductive and related organs, especially the breasts, account for the majority of cancer cases among women. Some studies show that 1 in 10 women eventually get breast cancer, often a form of adenocarcinoma. Breast cancer often metastasizes to the ovaries, where it produces ovarian cancer. Of course, cancer can begin in the ovaries independent of breast cancer. Usually a type of adenocarcinoma, ovarian cancer is difficult to detect early—often it is not easily apparent until it has grown into a large mass.

Cancer in the uterus can affect the body of the uterus or the cervix. Cancers of the uterine body most often involve the endometrium and mostly affect women beyond childbearing years. Cervical cancer occurs most frequently in women between the ages of 30 to 50. Cervical cancer is often diagnosed early, through screening tests such as the **Papanicolaou** (pap-a-nik-o-LAH-oo) **test,** or *Pap smear.* In this test, cells swabbed from the cervix are smeared on a glass slide and examined microscopically to determine whether any abnormalities exist. The American Cancer Society recommends two Pap smears 1 year apart beginning at age 21. If these two Pap smears are negative (that is, revealing no abnormalities), subsequent Pap smears should occur every 1 to 3 years thereafter. Because early or promiscuous intercourse is a risk factor for cervical cancer, sexually active young women should have their first Pap smear much earlier—and have follow-ups more often. Because screening tests and other early detection methods have been so successful, the death rates for uterine cancers have dropped dramatically over the last few decades.

Sexually Transmitted Diseases

Sexually transmitted diseases (STDs), or *venereal diseases,* are infections caused by communicable pathogens such as viruses, bacteria, fungi, and protozoans. The factor that links all these diseases and gives this disease category its name is the fact that they can all be transmitted by sexual contact. The term *sexual contact* refers to sexual intercourse in addition to any contact between the genitals of one person and the body of another person. Diseases classified as STDs can be transmitted sexually but do not have to be. For example, *acquired immune deficiency syndrome (AIDS)* is a viral condition that can be spread through sexual contact but is also spread by transfusion of infected blood and use of contaminated medical instruments such as intravenous needles and syringes. Candidiasis, or yeast infection, is a common opportunistic infection, but it can also be transmitted through sexual contact. Sexually transmitted diseases are the most common of all communicable diseases. Table 31-1 summarizes a few of the principal STDs.

TABLE 31-1   Examples of sexually transmitted diseases (STDs)

Disease	Pathogen	Description
Acquired immune deficiency syndrome (AIDS)	*Virus:* Human immunodeficiency virus (HIV)	HIV is transmitted by direct contact of body fluids, often during sexual contact. After a latent period that sometimes lasts several years, HIV infection produces the condition known as AIDS. AIDS is characterized by damage to lymphocytes (T cells), resulting in immune system impairment. Death results from secondary infections or tumors.
Candidiasis	*Fungus: Candida albicans*	Yeast infection characterized by a white discharge (leukorrhea), peeling of skin, and bleeding. Although it can occur as an ordinary opportunistic infection, it may be transmitted sexually.
Chancroid	*Bacterium: Haemophilus ducreyi*	Highly contagious STD, characterized by papules on the skin of the genitals that eventually ulcerate. About 90% of cases are reported by men.
Genital herpes	*Virus:* Herpes simplex virus (HSV)	HSV causes blisters on the skin of the genitals. The blisters may disappear temporarily but may reappear occasionally, especially as a result of stress.
Genital warts	*Virus:* Human papilloma virus (HPV-6, HPV-7)	Genital warts are nipplelike neoplasms of skin covering the genitals.

Continued

TABLE 31-1   Examples of sexually transmitted diseases (STDs)—(cont'd)

Disease	Pathogen	Description
Giardiasis	*Protozoan: Giardia lamblia*	Intestinal infection; may be spread by sexual contact. Symptoms range from mild diarrhea to malabsorption syndrome, with about half of all cases being asymptomatic.
Gonorrhea	*Bacterium: Neisseria gonorrhoeae*	Gonorrhea primarily involves the genital and urinary tracts but can affect throat, conjunctiva, or lower intestines. It may progress to PID.
Hepatitis	*Virus:* Hepatitis B virus (HBV)	This acute-onset liver inflammation may develop into a severe chronic disease, perhaps ending in death.
Lymphogranuloma venereum (LGV)	*Bacterium: Chlamydia trachomatis*	Chronic STD, characterized by genital ulcers, swollen nodes, headache, fever, and muscle pain. *C. trachomatis* infection may cause other syndromes, including conjunctivitis, urogenital infections, and systemic infections. *C. trachomatis* infections constitute the most common STD in the U.S.
Scabies	*Animal: Sarcoptes scabiei*	Scabies is caused by infestation by the *itch mite,* which burrows into the skin to lay eggs. About 2 to 4 months after initial contact, a hypersensitivity reaction occurs, causing a rash along each burrow that itches intensely. Secondary bacterial infection is possible.
Syphilis	*Bacterium: Treponema pallidum*	Although transmitted sexually, syphilis can affect any system. *Primary syphilis* is characterized by a chancre sore on an exposed area of the skin. If untreated, *secondary syphilis* may appear 2 months after the chancre disappears. The secondary stage occurs when the spirochete has spread through the body, presenting a variety of symptoms, and is still highly contagious—even through kissing. *Tertiary syphilis* may appear years later, possibly resulting in death.
Trichomoniasis	*Protozoan: Trichomonas vaginalis*	Urogenital infection, asymptomatic in most women and nearly all men. Vaginitis may occur, characterized by itching or burning and a foul-smelling discharge.

CASE STUDY
BREAST CANCER

Mrs. C. is a 47-year-old Caucasian female who has been admitted to the general surgical floor with a lump in her right breast. Mrs. C. has generally enjoyed good health up to this admission. She neither smokes nor drinks, and she follows a daily exercise regimen. Approximately 2 months ago, Mrs. C.'s husband noticed a small lump in her right breast. She gave this finding little attention, assuming that the lump was like the many others that she tends to experience around her menses. The lump, however, failed to follow the general pattern of resolution and Mrs. C. became concerned when it seemed to grow bigger. Mrs. C. is married and the mother of two children, one 8 and the other 6. Mrs. C. took birth control pills for 5 years after the birth of her second child. Last year she chose to discontinue use of these and turned to an alternative method of birth control. Mrs. C. is the only child born to her parents late in their life. Her father is alive and well, but her mother died of breast cancer 5 years ago. A further family history reveals that there is a strong history of both heart disease and cancer on both sides of Mrs. C.'s family.

On examination a 2- to 3-cm mass was felt in the upper quadrant of her right breast. This mass felt firm, was fixed to the chest wall, and was tender to the touch. The remaining breast skin was normal in appearance with no discoloration or retraction of the skin. The nipple was neither inverted nor draining. One node, approximately the size of a pea, was palpated under the right axilla. Mammography confirmed the presence of a 3-cm mass in the upper right quadrant of the right breast. Mrs. C. was taken to surgery 3 days later, the node was identified as malignant, and a modified radical mastectomy was performed.

1. Mrs. C. is considered to be at increased risk for developing breast cancer. Which of the following factors is most positively related to this high-risk profile?
   a. History of breast cancer in family members
   b. History of cystic breast disease
   c. Early onset of menarche
   d. Trauma related to the birth of her children

2. Which of the following best explains the existence of an enlarged right axillary lymph node in Mrs. C.?
   a. The lymph node is the result of an inflammatory reaction that normally occurs with the onset of her current menses
   b. The existence of the node is the result of an increased strain on the lymphatic system as a result of cellular degeneration
   c. The lymph node exists to provide nutrients to the rapidly growing cancer cells

   d. The lymph node is the result of cancer cells' spreading to different tissues within the body

3. On the third postoperative day, Mrs. C.'s right arm became increasingly swollen and painful. What is the best explanation for the lymphedema presently being experienced?
   a. An electrolyte imbalance is creating an increase in the hydrostatic pressure in the lymphatic system
   b. A postoperative infection has produced the beginnings of an immune response from the T lymphocytes
   c. The lymphatic channels are congested with fat particles that are being displaced from the trauma of surgery
   d. The remaining lymph nodes are inadequate to handle the lymph flow, creating an increase in the hydrostatic pressure

CHAPTER SUMMARY

OVERVIEW OF THE FEMALE REPRODUCTIVE SYSTEM
A. Function of the female reproductive system
   1. The function of the female reproductive system is to produce offspring and thereby ensure continuity of the genetic code
   2. It produces eggs, or female gametes, which each may unite with a male gamete to form the first cell of an offspring
   3. It also can provide nutrition and protection to the offspring for up to several years after conception
B. Structural plan of the female reproduction system
   1. Reproductive organs are classified as essential or accessory (Figure 31-1)
      a. Essential organs—gonads are the paired ovaries; gametes are ova produced by the ovaries—the ovaries are also internal genitals
      b. Accessory organs
         (1) Internal genitals—uterine tubes, uterus, and vagina—ducts or duct structures that extend from the ovaries to the exterior
         (2) External genitals—the vulva
         (3) Additional sex glands such as the mammary glands

UTERUS
A. Structure of the uterus (Figure 31-2)
   1. Size and shape of the uterus
      a. The uterus is pear shaped and has two main parts—the cervix and the body
   2. The wall of the uterus is composed of three layers—the inner endometrium, the middle myometrium, and the outer incomplete layer of parietal peritoneum

   3. Cavities of the uterus—the cavities are small because of the thickness of the uterine walls
      a. The body cavity's apex constitutes the internal os and opens into the cervical canal, which is constricted at its lower end and forms the external os that opens into the vagina
   4. The blood to the uterus is supplied by uterine arteries
B. Location of the uterus
   1. The uterus is located in the pelvic cavity between the urinary bladder and the rectum (Figure 31-3)
   2. The position of the uterus (Figure 31-4) is altered by age, pregnancy, and distension of related pelvic viscera
   3. The uterus descends, between birth and puberty, from the lower abdomen to the true pelvis
   4. The uterus begins to decrease in size at menopause
C. Functions of the uterus
   1. The uterus is part of the reproductive tract and permits sperm to ascend toward the uterine tubes
   2. If conception occurs, the offspring develops in the uterus
      a. The embryo is supplied with nutrients by endometrial glands until the production of the placenta
      b. The placenta is an organ that permits the exchange of materials between the mother's blood and the fetal blood but keeps the two circulations separate
      c. Myometrial contractions occur during labor and help push the offspring out of the mother's body
   3. If conception does not occur, outer layers of endometrium are shed during menstruation
      a. Menstruation is a cyclical event that allows the endometrium to renew itself

UTERINE TUBES

A. Uterine tubes are also called fallopian tubes or oviducts.
B. Uterine tubes are attached to the uterus at its upper outer angles and extend upward and outward toward the sides of the pelvis
C. Structure of the uterine tubes
   1. Uterine tubes consist of mucous, smooth muscle, and serous lining (Figure 31-5)
   2. Mucosal lining is directly continuous with the peritoneum lining the pelvic cavity
      a. Tubal mucosa is continuous with that of the vagina and uterus, which means it may become infected with organisms introduced into the vagina
   3. Each uterine tube has three divisions: isthmus, ampulla, and infundibulum
D. Function of the uterine tubes
   1. Uterine tubes serve as transport channels for ova and as the site of fertilization

OVARIES

A. Location of the ovaries
   1. The ovaries are nodular glands located on each side of the uterus, below and behind the uterine tubes (Figure 31-2)
   2. Ectopic pregnancy—development of the fetus in a place other than the uterus
B. Microscopic structure of the ovaries
   1. The surface of the ovaries is covered by the germinal epithelium
   2. Ovarian follicles contain the developing female sex cells (Figure 31-7)
   3. Ovum—an oocyte released from the ovary
C. Functions of the ovaries
   1. Ovaries produce ova—the female gametes
   2. Oogenesis—process that results in formation of a mature egg
   3. The ovaries are endocrine organs that secrete the female sex hormones (estrogens and progesterone)

VAGINA

A. The vagina is a tubular organ located between the rectum, urethra, and bladder
B. Structure of the vagina
   1. The vagina is a collapsible tube capable of distension, composed of smooth muscle, and lined with mucous membrane arranged in rugae
   2. The anterior wall is shorter than the posterior wall because the cervix protrudes into its uppermost portion
   3. Hymen—a mucous membrane that typically forms a border around the vagina in young premenstrual girls
C. Functions of the vagina
   1. The lining of the vagina stimulates the penis during sexual intercourse and acts as a receptacle for semen
   2. The vagina is the lower portion of the birth canal
   3. The vagina is a transport for tissue shed during menstruation

VULVA

A. The vulva consists of the female external genitals: mons pubis, labia majora, labia minora, clitoris, urinary meatus, vaginal orifice, and greater vestibular glands (Figure 31-9)
B. Functions of the vulva
   1. The mons pubis and labia protect the clitoris and vestibule
   2. Vestibular glands produce lubrication to reduce friction during intercourse
   3. The clitoris contains sensory receptors that send information to the sexual response area of the brain
   4. The vaginal orifice is the boundary between the internal and external genitals

PERINEUM

A. The perineum is the skin-covered region between the vaginal orifice and the rectum
B. This area may be torn during childbirth

BREASTS

A. Location and size
   1. The breasts lie over the pectoral muscles
   2. Estrogens and progesterone control breast development
   3. Breast size is determined by the amount of fat around glandular tissue (Figure 31-10)
B. Function of the breasts
   1. The function of mammary glands is lactation
   2. Mechanism of lactation (Figure 31-11)
      a. The ovarian hormones make the breasts structurally ready to produce milk
      b. Shedding of the placenta results in a decrease of estrogens and thus stimulates prolactin
      c. Prolactin stimulates lactation
   3. Lactation can provide nutrient-rich milk to offspring for up to several years from birth; some advantages are
      a. Nutrients
      b. Passive immunity from antibodies present in the milk
      c. Emotional bonding between mother and child

FEMALE REPRODUCTIVE CYCLES

A. The female reproductive system has many cyclical changes that start with the beginning of menses
   1. Ovarian cycle—ovaries from birth contain oocytes in primary follicles in which the meiotic process has been suspended. At the beginning of menstruation each month, several of the oocytes resume meiosis. Meiosis will stop again just before the cell is released during ovulation (Figure 31-12)
   2. Menstrual cycle (endometrial cycle) is divided into four phases
      a. Menses
      b. Postmenstrual phase
      c. Ovulation
      d. Premenstrual phase
   3. Myometrial phase
   4. Gonadotropic cycle
B. Control of female reproductive cycles
   1. Hormones control cyclical changes
   2. Cyclical changes in the ovaries result from changes in the gonadotropins secreted by the pituitary gland (Figures 31-13 and 31-14)
   3. Cyclical changes in the uterus are caused by changes in estrogens and progesterone (Figure 31-15)
   4. Low levels of FSH and LH cause regression of the corpus luteum if pregnancy does not occur. This causes a decrease in estrogen and progesterone, which triggers endometrial sloughing of the menstrual phase
   5. Control of cyclical changes in gonadotropin secretion is caused by positive and negative feedback mechanisms and involves estrogens, progesterone, and the hypothalamus' secretion of releasing hormones
C. Importance of the female reproductive cycles
   1. The ovarian cycle's primary function is to produce ovum at regular intervals
      a. It s secondary function is to regulate the endometrial cycle through estrogen and progesterone
   2. The function of the endometrial cycle is to make the uterus suitable for implantation of a new offspring
   3. The cyclical nature of the reproductive system and the fact that fertilization will occur within 24 hours after ovulation mean that a woman is only fertile a few days of each month

D. Menstrual flow begins at puberty, and the menstrual cycle continues for about three decades

CYCLE OF LIFE: FEMALE REPRODUCTIVE SYSTEM

A. Reproductive functions begin at time of puberty (first menses)
B. Principal cyclic functions cease in middle adulthood
C. Initial stages of development begin in fetus—first stage of meiotic development of all ova to be produced is completed by time of birth
D. Puberty—hormones stimulate the organs to become functional and produce mature ova
E. Function continues in a cycle until menopause
F. Menopause—marked by 1 full year without menstruation
G. Postmenopause—normal sexual activity, no procreation

THE BIG PICTURE: THE FEMALE REPRODUCTIVE SYSTEM AND THE WHOLE BODY

A. The female reproductive system shares a special relationship with
  1. The urinary system because of their close proximity and because they share the vulva
  2. The skeletal muscles in the perineum
  3. The integumentary system because breasts are actually modifications of the skin

MECHANISMS OF DISEASE: DISORDERS OF THE FEMALE REPRODUCTIVE SYSTEM

A. Hormonal and menstrual disorders
  1. Dysmenorrhea—painful menstruation
    a. Primary dysmenorrhea occurs in the absence of an associated pelvic disease and is possibly caused by an increased concentration of prostaglandins; prostaglandin-inhibiting drugs are used to relieve symptoms
    b. Secondary dysmenorrhea—caused by various conditions, including inflammatory conditions and cervical stenosis; treatment involves the underlying cause

  2. Amenorrhea—absence of normal menstruation
    a. Primary—failure to begin menstrual cycles caused by numerous factors
    b. Secondary—previously regular menstrual cycles slow to only a few times a year
    c. Treatment of both types involves the underlying disorder
  3. Dysfunctional uterine bleeding—excessive or irregular uterine bleeding caused by a hormonal imbalance
  4. Premenstrual syndrome—disturbing and unpleasant emotional and physical symptoms occurring regularly in the premenstrual phase
B. Infection and inflammation
  1. Infections of the female reproductive system are classified as:
    a. Exogenous—infections resulting from pathogenic organisms transmitted from another person
    b. Endogenous—infections resulting from pathogens that normally inhabit the intestines, vulva, or vagina
  2. Pelvic inflammatory disease—an inflammatory condition caused by infection that can involve the uterus, uterine tubes, or ovaries
  3. Vaginitis—an inflammation or infection of the vaginal lining, often from STDs or yeast infections
C. Tumors and related conditions
  1. Myoma or fibromyoma—a benign tumor of smooth muscle and fibrous connective tissue that makes up uterine tumors; it may be surgically removed
  2. Ovarian cysts are benign enlargements on one or both ovaries
  3. Endometriosis—the presence of functioning endometrial tissue outside the uterus
  4. Breast cancer—a form of adenocarcinoma, often metastasizes in the ovaries where it causes ovarian cancer
  5. Uterine cancer can affect the uterus or the cervix
D. Sexually transmitted diseases (venereal disease)—infections caused by communicable pathogens that can be transmitted by sexual contact (Table 31-1)

REVIEW QUESTIONS

1. Identify the essential and accessory organs in the female reproductive system.
2. Describe the three layers that compose the wall of the uterus.
3. Identify the vessels that supply blood to the uterus.
4. List the eight ligaments that hold the uterus in a normal position.
5. How does the uterus serve as part of the female reproductive tract?
6. What and where are the uterine tubes? Approximately how long are they? With what are they lined? Their lining is continuous on their distal ends with what? With what on their proximal ends? Why is an infection of the lower part of the female reproductive tract likely to develop into a serious condition?
7. Trace the development of a female sex cell from its most primitive state through ovulation.
8. What hormones are secreted by the cells in ovarian tissue?
9. Identify all the vaginal functions.

10. List all the structures that make up the female external genitals.
11. Define the term *episiotomy*.
12. How do estrogens and progesterone act to control lactation?
13. Identify the advantages that nursing from the mother's breast provides human offspring.
14. List the phases of the menstrual cycle.
15. Discuss the level of luteinizing hormone at the time of ovulation.
16. What method makes it possible to measure blood levels of gonadotropins?
17. Summarize the control of cyclical ovarian changes brought on by FSH and LH.
18. Compare and contrast the actions of estrogens and progesterone on the uterus.
19. Explain how contraceptive pills prevent pregnancy.
20. Describe the hormonal changes during menopause.
21. Define the term *mittelschmerz*.
22. Why does amenorrhea occur in some female athletes?

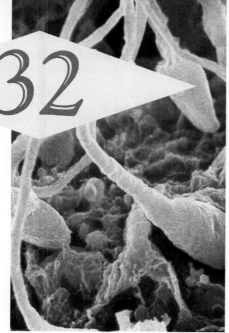

Fertilization

32

Growth and Development

OBJECTIVES

After you have completed this chapter, you should be able to:

1. Explain the meaning of developmental biology.
2. Describe the process of meiosis and how it differs from mitosis.
3. Compare and contrast spermatogenesis and oogenesis.
4. Outline the steps involved in fertilization and implantation.
5. Describe the early stages in the development of the human embryo.
6. Identify the developmental and structural features of the placenta.
7. Outline the developmental changes that occur during the first trimester of pregnancy.
8. Explain the meanings of histogenesis and organogenesis.
9. Identify and describe the stages of labor.
10. Explain two different processes that can result in twinning, or double birth.
11. Give a brief description of the four most common postnatal periods, including the major growth and developmental changes.
12. Describe, by body system, the effects of aging.

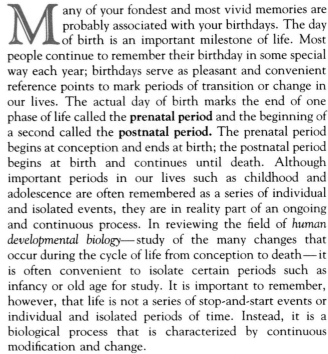

Many of your fondest and most vivid memories are probably associated with your birthdays. The day of birth is an important milestone of life. Most people continue to remember their birthday in some special way each year; birthdays serve as pleasant and convenient reference points to mark periods of transition or change in our lives. The actual day of birth marks the end of one phase of life called the **prenatal period** and the beginning of a second called the **postnatal period.** The prenatal period begins at conception and ends at birth; the postnatal period begins at birth and continues until death. Although important periods in our lives such as childhood and adolescence are often remembered as a series of individual and isolated events, they are in reality part of an ongoing and continuous process. In reviewing the field of *human developmental biology*—study of the many changes that occur during the cycle of life from conception to death—it is often convenient to isolate certain periods such as infancy or old age for study. It is important to remember, however, that life is not a series of stop-and-start events or individual and isolated periods of time. Instead, it is a biological process that is characterized by continuous modification and change.

This chapter discusses some of the basic concepts of important events and changes that occur in the ongoing development of the individual from conception to death. Study of development during the prenatal period is followed by a review of changes occurring during infancy and adulthood, and, finally, some of the more important changes that occur in the individual organ systems of the body as a result of aging are discussed.

BEGINNING OF HUMAN LIFE

Production of Sex Cells

Before a human life is first formed, some preliminary processes must occur. Of utmost importance is the production of mature *gametes,* or sex cells, by each parent. Spermatozoa, gametes of the male parent, are produced by a process called **spermatogenesis.** Ova, gametes of the female parent, are produced by a process called **oogenesis.**

Meiosis

Both types of gamete production require a special form of cell division characterized by **meiosis.** Recall from Chapter 3 that meiosis is the orderly arrangement and distribution of chromosomes that, unlike *mitosis,* reduces the number of chromosomes in each daughter cell to half the number present in the parent cell. The necessity for

813

Meiosis I (first division)

Meiosis II (second division)

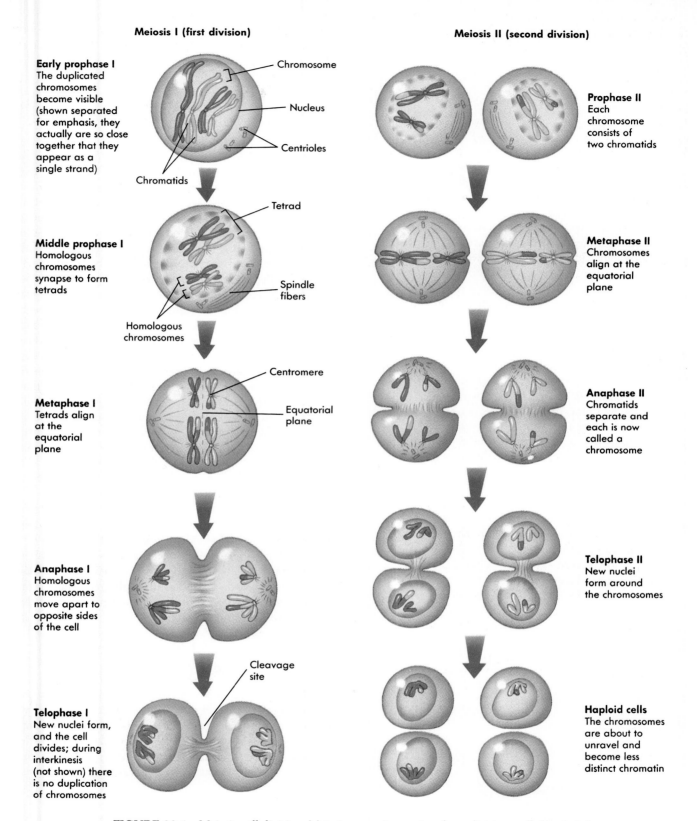

Early prophase I
The duplicated chromosomes become visible (shown separated for emphasis, they actually are so close together that they appear as a single strand)

Chromosome

Nucleus

Centrioles

Chromatids

Prophase II
Each chromosome consists of two chromatids

Middle prophase I
Homologous chromosomes synapse to form tetrads

Tetrad

Spindle fibers

Homologous chromosomes

Metaphase II
Chromosomes align at the equatorial plane

Metaphase I
Tetrads align at the equatorial plane

Centromere

Equatorial plane

Anaphase II
Chromatids separate and each is now called a chromosome

Anaphase I
Homologous chromosomes move apart to opposite sides of the cell

Telophase II
New nuclei form around the chromosomes

Cleavage site

Telophase I
New nuclei form, and the cell divides; during interkinesis (not shown) there is no duplication of chromosomes

Haploid cells
The chromosomes are about to unravel and become less distinct chromatin

FIGURE 32-1   Meiotic cell division. Meiosis occurs in a series of two divisions called meiosis I and meiosis II. Notice that four daughter cells, each with the haploid number of chrosomes, is produced from each each parent cell that enters meiotic cell division. For simplicity's sake, only four chromosomes are shown in the parent cell instead of the usual 46.

chromosome reduction as a preliminary to union of the sex cells is explained by the fact that the cells of each species of living organisms contain a specific number of chromosomes. Human cells, for example, contain 23 pairs or a total of 46 chromosomes. This total of 46 chromosomes per body cell is known as the **diploid** number of chromosomes. Diploid comes from the Greek *diploos,* meaning "twofold." If the male and female cells united without first halfing their respective chromosomes, the resulting cell would contain twice as many chromosomes as is normal for human beings. Mature ova and sperm therefore contain only 23 chromosomes, or half as many, as other human cells. This total of 23 chromosomes per sex cell is known as the **haploid** number of chromosomes (from the Greek *haploos,* meaning "single").

Meiotic division consists of two cell divisions that take place one after the other in succession. They are referred to as meiotic division I and meiotic division II, and in both, an interphase, prophase, metaphase, anaphase, and telophase occur (Figure 32-1). In the interphase that precedes prophase I (of meiotic division I) the same events occur as take place in the interphase of mitotic division. Specifically, each DNA molecule replicates itself and thereby becomes a pair of chromatids, attached to each other only at the centromere. The term *chromosome* applies to any condensed chromatin with its own centromere. For simplicity's sake, in both Figures 32-1 and 32-2, only four of the 46 chromosomes are shown. Notice that early in meiosis I homologous pairs of chromosomes are moved together to form groupings called *tetrads.* It is the tetrads that split during anaphase I, not chromosomes as in mitosis. It is also in meiosis I that the phenomenon of "crossing over" occurs. During crossing over a chromatid segment of each chromosome crosses over and becomes part of the adjacent chromosome in the pair (see Figure 33-3). This is a highly significant event and will be discussed in some detail in the next chapter. Since each chromatid

segment consists of specific genes, the crossing over of chromatids reshuffles the genes—that is, it transfers some of them from one chromosome to another. This exchange of genetic material can add almost infinite variety to the ultimate genetic makeup of an individual.

Metaphase I follows the last stage of prophase I, and as in mitosis, the chromosomes align themselves along the equator of the spindle fibers as Figure 32-1 shows. But in anaphase the two chromatids that make up each chromosome do not separate from each other as they do in mitosis to form two new chromosomes out of each original one. In anaphase I, only one of each pair of chromosomes move to each pole of the parent cell. Therefore, as you can see in Figures 32-1 and 32-2, when the parent cell divides to form two cells, each daughter cell contains two chromosomes or half as many as the parent cell had, although remember, of course, that each chromosome still consists of two sister chromatids joined at the centromere. Thus the daughter cells formed by meiotic division I contain a haploid number of chromosomes, or half as many as the diploid number in the parent cell.

As you can see in Figure 32-1, meiotic division II is essentially the same as mitotic division. In both spermatogenesis and oogenesis the second meiotic division reproduces each of the two cells formed by meiotic division I and so forms four cells, each with the haploid number of chromosomes.

Spermatogenesis

Spermatogenesis is the process by which the primitive sex cells, or **spermatogonia,** in the seminiferous tubules of a newborn baby boy become transformed into mature sperm, or *spermatozoa.* Spermatogenesis begins at about the time of puberty and usually continues throughout a man's life. Figure 32-2, A, diagrams some of the major steps of spermatogenesis. Meiotic division I reproduces one primary spermatocyte to form two secondary spermatocytes, each

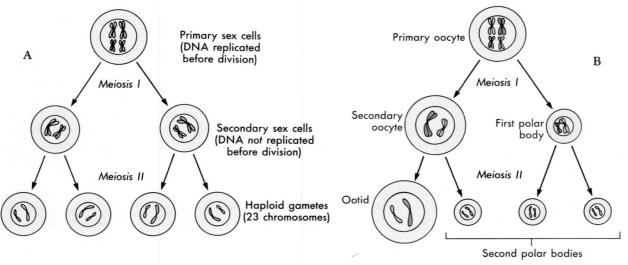

FIGURE 32-2 Gamete production. A, Spermatogenesis. A primary spermatocyte undergoes meiotic division to produce four haploid daughter spermatids. **B,** Oogenesis. A primary oocyte undergoes meiotic division to produce a single ovum and three small polar bodies.

Arrest of Oocyte Development

The unique processes that characterize oogenesis have functional advantages that may not be apparent at first glance. For instance, the arrest of oocyte development during prophase I is an excellent example of how the body uses an opportunity to its advantage. Why halt meiosis as it is just beginning? By arresting meiosis at this stage, the oocyte has four copies of the genetic code available (two copies of each chromosome, each of which contain two identical chromatids). With four copies of the DNA code available, the oocyte can quickly synthesize the enormous number of RNA molecules needed for the early stages of human development—the period just after fertilization and before development of the placenta. As soon as enough tRNA, rRNA, and mRNA molecules to regulate the cell through the early developmental stages are made, DNA activity slows and the oocyte waits for the hormonal signal that will stimulate completion of oogenesis. When the hormonal signal, in the form of gonadotropins, arrives, meiosis resumes. Before the cell is released during ovulation, however, the previously transcribed RNA molecules will have directed the synthesis of enough protein to cause a 500-fold increase in the size of the oocyte! The released ovum finally completes meiosis after the sperm DNA has entered it. The fertilized ovum—or zygote, as it is now called—has enough cytoplasmic material to sustain it through some of the most critical phases of early human development.

with a haploid number of chromosomes (23). Meiotic division II then reproduces each of the two secondary spermatocytes to form a total of four spermatids. Spermatids then differentiate to form heads and tails, thereby becoming mature *spermatozoa*. Thus spermatogenesis forms four spermatozoa (singular, *spermatozoon*)—each with only 23 chromosomes—from one primary spermatocyte that had 23 pairs or 46 chromosomes.

Oogenesis

Oogenesis is the process by which primitive female sex cells, or **oogonia,** become mature ova. During the fetal period, oogonia reproduce in the ovaries by mitosis to form primary oocytes—about a half million of them by the time a baby girl is born. Most of the primary oocytes develop to prophase I of meiosis before birth. There they stay until puberty. Then, once each cycle, a few primary oocytes resume meiosis as their surrounding follicles begin to mature and migrate toward the surface of the ovary. Usually only one oocyte matures enough to reach the surface of the ovary, where ovulation occurs and meiosis again halts—this time at metaphase II. Meiosis in the released ovum resumes only when, and if, the head of a sperm cell enters the ovum.

Note that during oogenesis, the cytoplasm is not equally divided among the daughter cells, thus only one mature ovum is produced from each primary oocyte, plus three *polar bodies* which disintegrate (Figure 32-2, B). How many

mature sperm cells are formed from each primary spermatocyte in spermatogenesis? This difference may be accounted for by the fact that for reproductive success, an ovum must have a huge store of cytoplasm with all of its component organelles, nutrients, and regulatory molecules. In other words, nearly all the cytoplasm is conserved by the one daughter oocyte that survives.

Ovulation and Insemination

The second preliminary step necessary for conception of a new individual consists in bringing the sperm and ovum into proximity with each other so that the union of the two can take place. Two processes are involved in the accomplishment of this step.

1. Ovulation or expulsion of the mature ovum from the mature ovarian follicle into the pelvic cavity, from which it enters one of the uterine tubes.

2. Insemination or expulsion of the seminal fluid from the male urethra into the female vagina. Several million sperm enter the female reproductive tract with each ejaculation of semen. By lashing movements of their flagella-like tails, assisted by various processes in the female reproductive tract, the sperm make their way into the external os of the cervix, through the cervical canal and uterine cavity, and into the uterine tubes.

Fertilization

After ovulation the discharged ovum first enters the abdominal cavity and then finds its way into the uterine (fallopian) tubes, where conception, or **fertilization,** takes place (Figure 32-3). Sperm cells "swim" up the uterine tubes toward the ovum. Look at the relationship of the ovary, the uterine tube, and the uterus in Figure 32-4. Recall from Chapter 31 that each uterine tube extends outward from the uterus for about 10 cm. It then ends in the abdominal cavity near the ovary, as you can see in Figure 32-4, in an opening surrounded by fringelike processes, the fimbriae. Sperm cells that are deposited in the vagina must enter and "swim" through the uterus and then move out of the uterine cavity and through the uterine tube to meet the ovum. Fertilization most often occurs in the outer one third of the oviduct, as shown in Figure 32-4. The process of sperm movement has been shown to be assisted by mechanisms within the female reproductive tract. For example, mucus strands in the cervical canal guide the sperm on their way into the uterus. The ovum also takes an active role in the process of fertilization. Experiments show that the ovum actually attracts and "traps" sperm with special receptor molecules on the surface of the ovum. As soon as the head and neck of one spermatozoon enter the ovum (the tail drops off), complex mechanisms in the egg are activated to ensure that no more sperm enter. The 23 chromosomes from the sperm combine with the 23 chromosomes already in the ovum to restore the diploid number of 46 chromosomes.

Inasmuch as the ovum lives only a short time (probably only 24 hours) after leaving the ruptured follicle, the fertilization "window" occurs around the time of ovulation. Since sperm may live up to a few days after entering the

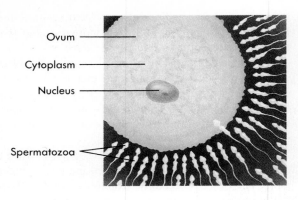
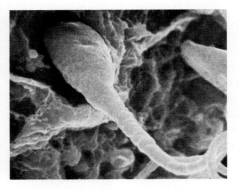

Ovum
Cytoplasm
Nucleus
Spermatozoa

FIGURE 32-3 Fertilization. Fertilization is a specific biological event. It occurs when the male and female sex cells fuse. After union between a sperm cell and the ovum has occurred, the cycle of life begins. The scanning electron micrograph shows spermatozoa attached to the surface of an ovum. Only one will enter the ovum.

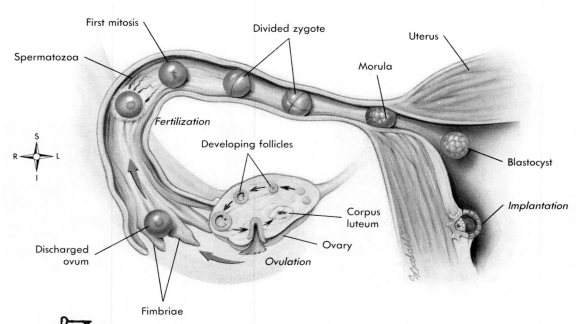

Uterine (Fallopian) tube
First mitosis
Divided zygote
Uterus
Spermatozoa
Morula
Fertilization
Developing follicles
Blastocyst
S
R — L
I
Corpus luteum
Implantation
Discharged ovum
Ovary
Ovulation
Fimbriae

FIGURE 32-4 Fertilization and implantation. At ovulation, an ovum is released from the ovary and begins its journey through the uterine tube. While in the tube, the ovum unites with a sperm to form the single-celled zygote. After a few days of rapid mitotic division, a ball of cells called a *morula* is formed. After the morula develops into a hollow ball called a *blastocyst,* implantation occurs.

female tract, sexual intercourse any time from about 3 days before ovulation to 1 day after ovulation may result in fertilization.

The fertilized ovum, or **zygote** (ZYE-gote), is genetically complete; it represents a new single-celled individual. Time, nourishment, and a proper prenatal environment are all that is needed for expression of characteristics such as sex, hair, and skin color that were determined at the time of fertilization.

QUICK CHECK

✔ 1. *What is the function of meiotic division?*
✔ 2. *How does meiosis differ from mitosis?*
✔ 3. *Where in the female reproductive tract does fertilization usually occur?*
✔ 4. *What is the technical name for the fertilized ovum?*

PRENATAL PERIOD

The **prenatal stage** of development begins at the time of conception, or fertilization (that is, at the moment the female ovum and the male sperm cell unite). The period of prenatal development continues until the birth of the child about 39 weeks later. The science of the development of the individual before birth is called **embryology** (em-bree-OL-o-jee). It is a story of biological marvels, describing the means by which a new human life is begun and the steps by which a single microscopic cell is transformed into a complex human being.

Cleavage and Implantation

As you can see in Figure 32-4, once the zygote is formed it immediately begins to cleave, or divide, and in about 3 days a solid mass of cells called a **morula** (MOR-yoo-lah) is formed. The cells of the morula begin to form an inner cavity as they continue to divide, and by the time the developing embryo reaches the uterus, it is a hollow ball of cells called a **blastocyst** (BLAS-toe-sist). In about 10 days from the time of fertilization the blastocyst is completely implanted in the uterine lining—before nutrients from the mother are available to nourish it.

The rapid cell division taking place up to the blastocyst stage occurs with no significant increase in total mass compared to the zygote (Figure 32-5). One of the specializations of the ovum is its incredible store of nutrients that support this embryonic development until implantation has occurred.

Note in Figure 32-6 that the blastocyst consists of an outer layer of cells and an **inner cell mass.** The outer wall of the blastocyst is called the **trophoblast** (Figure 32-6). As the blastocyst develops further, the inner cell mass forms a structure with two cavities called the **yolk sac** and **amniotic** (am-ne-OT-ik) **cavity** (Figures 32-7 and 32-8). The yolk sac is most important in animals such as birds that depend heavily on yolk as a nutrient for the developing embryo. In these animals the yolk sac digests the yolk and provides the resulting nutrients to the embryo. Because the uterine lining provides nutrients to the developing embryo in humans, the function of the yolk sac is not a nutritive one. Instead, it has other functions, including production of

---

FYI   In Vitro Fertilization

The world's first "test tube" baby was born in Oldham, England, on July 25, 1978. This historic and highly publicized event represented a major medical breakthrough and set the stage for ongoing and sophisticated research in the area of reproductive physiology that continues in many fertility clinics and laboratories around the world.

Nine months before the historic birth, baby Louise Brown was conceived in vitro (in VEE-tro). The Latin term *in vitro* means, literally, "within a glass" and refers to the glass laboratory container in which a mature ovum from the mother was fertilized by the father's sperm. The ovum was obtained from the mother using a specialized optical viewing tube called a **laparoscope** (LAP-ah-ro-skope) in a procedure sometimes called "belly button surgery." After 2½ days' growth in a temperature-controlled environment, the fertilized ovum (at the 8-cell stage) was returned by the physicians to the mother's uterus. Implantation was successful, and the subsequent pregnancy and birth were normal in every respect. (The figure shows Louise Brown at age 10.)

Current in vitro fertilization procedures also use a laparoscope. The newer and smaller fiberoptic instruments can be inserted through a very small incision in the woman's abdomen or through the bladder wall after being introduced through the urethra in a specially designed catheter. Once in the abdominal cavity the device allows the physician to see the ovary and also to puncture a mature follicle. The mature ovum that is released is sucked up and then transferred to and maintained in a specialized medium until it is fully mature and capable of being fertilized.

Most women who seek in vitro fertilization as a treatment for infertility have uterine tubes that either do not

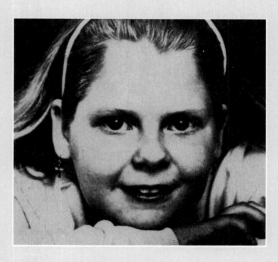

permit movement of sperm to reach and fertilize the ovum or that do not allow a developing zygote to reach the uterus after fertilization has occurred. Treatment begins with hormone treatment to induce simultaneous development and maturation of several ova, which are then collected and fertilized in vitro with spermatozoa that have been treated to bring about their full maturation.

After the developing zygotes have reached the 8- or 16-cell stage, they are then transferred to the mother's uterus. Under the best conditions, only an estimated 50% of fertilized ova successfully implant. In the most successful fertility clinics in the United States, 20% to 30% of in vitro fertilization cases result in implantation and a pregnancy that progresses to term.

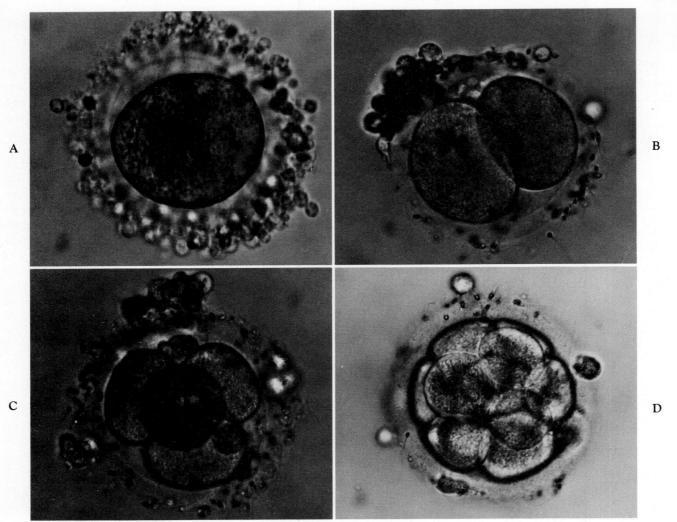

FIGURE 32-5   Early stages of human development. A, Fertilized ovum, or zygote. **B** to **D,** Early cell divisions produce more and more cells. The solid mass of cells shown in **D** forms the morula—an early stage in embryonic development.

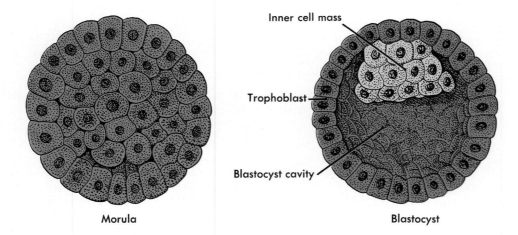

Inner cell mass

Trophoblast

Blastocyst cavity

Morula

Blastocyst

FIGURE 32-6   Early stages in the development of the human embryo. The morula consists of an almost solid spherical mass of cells. The embryo reaches this stage about 3 days after fertilization. The blastocyst (hollowing) stage develops later, prior to implantation in the uterine lining.

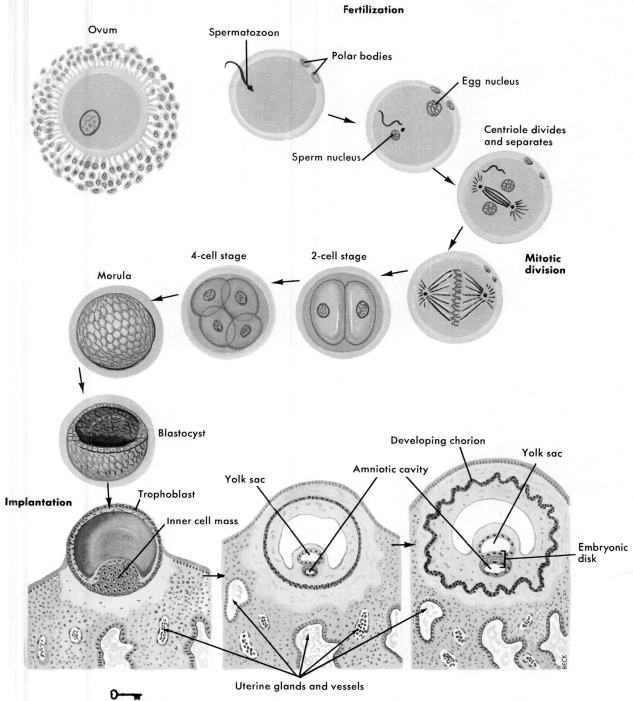

Fertilization

Ovum

Spermatozoon

Polar bodies

Egg nucleus

Sperm nucleus

Centriole divides and separates

Mitotic division

4-cell stage

2-cell stage

Morula

Blastocyst

Implantation

Trophoblast

Inner cell mass

Yolk sac

Amniotic cavity

Developing chorion

Yolk sac

Embryonic disk

Uterine glands and vessels

BECK

FIGURE 32-7   Fertilization to implantation and development of the yolk sac. Rapid growth of uterine glands and vessels covers the developing blastocyst at the time of implantation.

blood cells. It is the inner cell mass that eventually forms the tissues of the offspring's body. The trophoblast, on the other hand, forms the support structures described in the following paragraphs.

The amniotic cavity becomes a fluid-filled, shock-absorbing sac, sometimes called the "bag of waters," in which the embryo floats during development. The **chorion** (KO-ree-on), shown in Figures 32-7 to 32-9, develops from

the trophoblast to become an important fetal membrane in the **placenta** (plah-SEN-tah). The *chorionic villi* shown in Figures 32-8 and 32-9 connect the blood vessels of the chorion to the placenta. The placenta (Figure 32-9) anchors the developing fetus to the uterus and provides a "bridge" for the exchange of nutrients and waste products between mother and baby.

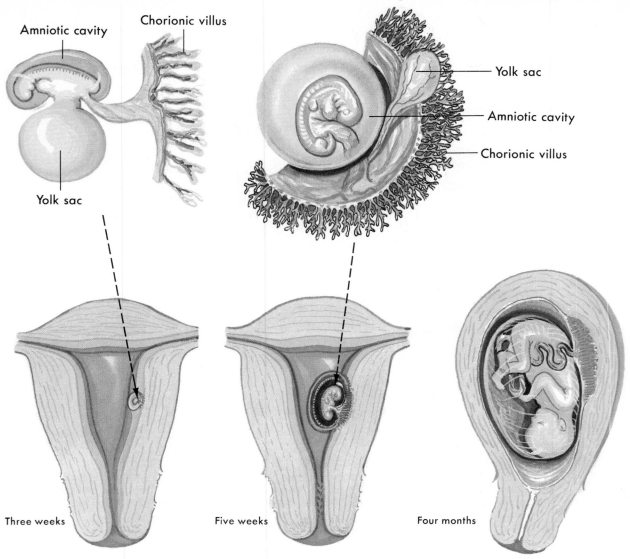

Amniotic cavity

Chorionic villus

Yolk sac

Yolk sac

Amniotic cavity

Chorionic villus

Three weeks

Five weeks

Four months

FIGURE 32-8 Development of the chorion and amnion. Development of the chorion and amniotic cavity to 4 months of gestation.

FYI   Developmental biology

Developmental biology is the name given to the branch of life science that studies the process of change over the life cycle. This process of change is called *development*. It is important to realize that in developmental biology, the terms *growth* and *development* do not mean the same thing. Growth is simply an increase in body mass. Development, on the other hand, refers to the complex series of changes that occur at various times of life. Early stages of development—particularly the prenatal stages—are characterized by rapid growth, whereas later stages of development are characterized by little, if any, growth of body tissues.

In this chapter, we briefly discuss various subtopics within the field of human developmental biology. For example, prenatal development is studied by a branch of developmental biology called **embryology.** The biological changes observed during late adulthood are studied by a branch of developmental biology called **gerontology.**

Basic concepts of embryology, gerontology, and other

subdisciplines of developmental biology seem to have taken on a greater practical importance during the past decade than ever before. One reason is the explosion in knowledge of developmental processes and our ability to treat the abnormalities that we can now find. Procedures such as fetal surgery, electrocardiography, and ultrasound permit physicians to diagnose and treat the fetus much like any other patient. Recent discoveries about the processes of aging—as well as the rapidly growing population of aged individuals—have spawned new methods of recognizing and treating physical and psychological problems in the elderly. Another reason developmental biology has taken on great practical importance is that it is a field that serves to unify human biology into a framework that integrates anatomy, physiology, cell biology, molecular biology, medicine, and other disciplines. Thus developmental biology gives us an excellent view of the "big picture" of the human body.

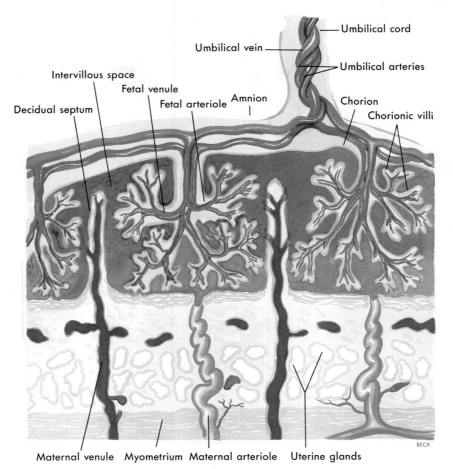

FIGURE 32-9   Structural features of the placenta. The close placement of the fetal blood supply and the maternal blood supply permits diffusion of nutrients and other substances. It also forms a thin barrier to prevent diffusion of most harmful substances. No mixing of fetal and maternal blood occurs.

Placenta

The placenta is a unique and highly specialized structure that has a temporary but very important series of functions during pregnancy. It is composed of tissues from mother and child and functions not only as structural "anchor" and nutritive bridge, but also as an excretory, respiratory, and endocrine organ.

Placental tissue normally separates the maternal and fetal blood supplies so that no intermixing occurs. The very thin layer of placental tissue that separates maternal and fetal blood also serves as an effective "barrier" that can protect the developing baby from many harmful substances that may enter the mother's bloodstream. Unfortunately, toxic substances such as alcohol and some infectious organisms may penetrate this protective placental barrier and injure the developing baby. The virus responsible for German measles (rubella), for example, can easily pass through the placenta and cause tragic developmental defects in the fetus.

Placental tissue also has important endocrine functions. As Figure 32-10 shows, placental tissue secretes large amounts of **human chorionic gonadotropin (HCG)** early in

pregnancy. HCG secretion peaks about 8 or 9 weeks after fertilization then drops to a continuous low level by about week 16. The function of HCG, as its name implies, is to act as a gonadotropin and stimulate the corpus luteum to continue its secretion of estrogen and progesterone. Recall from Chapter 31 (Figure 31-15, p. 804) that reduced levels of the anterior pituitary gonadotropins (FSH and LH) after ovulation normally cause a corresponding reduction in luteal secretion of the estrogen and progesterone needed to sustain the uterine lining. The drop in estrogen and progesterone secretion results from the fact that the FSH and LH needed to maintain the corpus luteum are now in short supply. To prevent menstruation and to allow successful implantation and development of the offspring, the cells of the trophoblast and, later, the placenta secrete enough HCG to maintain the corpus luteum and thus keep luteal estrogen and progesterone levels high. As the placenta develops, it begins to secrete its own estrogen and progesterone. As Figure 32-10 shows, as more estrogen and progesterone are secreted from the placenta, a corresponding decrease in HCG secretion produces a drop in luteal secretion of these hormones. After about 3 months, the

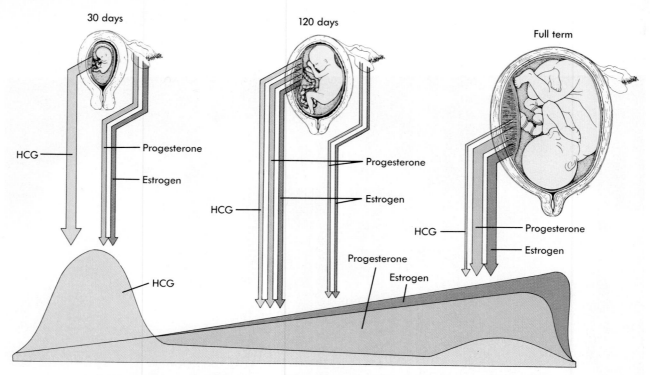

30 days

120 days

Full term

HCG

Progesterone

Estrogen

HCG

Progesterone

Estrogen

HCG

Progesterone

Estrogen

HCG

Progesterone

Estrogen

FIGURE 32-10 Hormone levels during pregnancy. This diagram shows the changes that occur in the blood concentration of human chorionic gonadotropin (HCG), estrogen, and progesterone during gestation. Note that high HCG levels produced by placental tissue early in pregnancy maintain estrogen and progesterone secretion by the corpus luteum. This prevents menstruation and promotes maintenance of the uterine lining. As the placenta takes over the job of secreting estrogen and progesterone, HCG levels drop, and the corpus luteum subsequently ceases secreting these hormones.

corpus luteum has degenerated and the placenta has completely taken over the job of secreting the estrogen and progesterone needed to sustain the pregnancy. Over-the-counter "early pregnancy" tests detect the presence of the HCG that is excreted in the urine during the first couple of months of a pregnancy.

QUICK CHECK

✔ *1. What is a morula? What is a blastocyst?*
✔ *2. What structures are derived from the trophoblast (outer wall) of the blastocyst?*
✔ *3. What placental hormone maintains the corpus luteum during the early weeks of pregnancy?*

Periods of Development

The length of pregnancy (about 39 weeks)—called the *gestation period*—is divided into three 3-month segments called *trimesters.* Several terms are used to describe development during these periods known as the first, second, and third trimesters of pregnancy.

During the first trimester, or 3 months, of pregnancy, numerous terms are used. Zygote is used to describe the

ovum just after fertilization by a sperm cell. After about 3 days of constant cell division, the solid mass of cells, identified earlier as the morula, enters the uterus. Continued development transforms the morula into the hollow blastocyst, which then implants into the uterine wall (see Figure 32-7).

The embryonic phase of development extends from fertilization until the end of week 8 of gestation. During this period in the first trimester, the term *embryo* is used to describe the developing individual. The fetal phase is used to indicate the development extending from week 8 to 39. During this period, the term *embryo* is replaced by the term *fetus.*

By day 35 of gestation (Figure 32-11, A), the heart is beating and, although the embryo is only 8 mm (about ⅜ inch) long, the eyes and so-called limb buds, which ultimately form the arms and legs, are clearly visible. Figure 32-11, C, shows the stage of development at the end of the first trimester of gestation, when the offspring becomes known as a fetus. Body size is about 7 to 8 cm (3.2 inches) long. The facial features of the fetus are apparent, the limbs are complete, and sex can be identified. By month 4 (Figure 32-11, D) all organ systems are formed and functioning. Growth of the embryo to 4 months is summarized in the photographs in Figure 32-11.

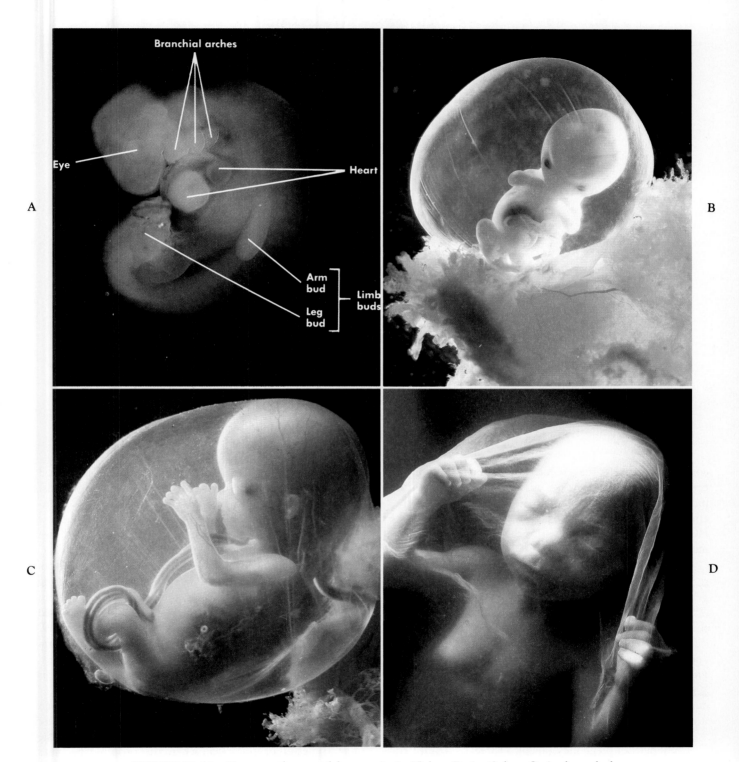

Branchial arches

Eye

Heart

Arm bud

Leg bud

Limb buds

A

B

C

D

FIGURE 32-11   Human embryos and fetuses. A, At 35 days. **B,** At 49 days. **C,** At the end of the first trimester. **D,** At 4 months.

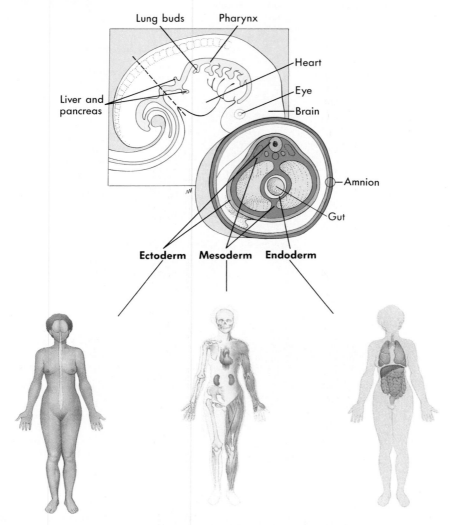

FIGURE 32-12   The primary germ layers. Illustration shows the primary germ layers and the body systems into which they develop.

Formation of the Primary Germ Layers

Early in the first trimester of pregnancy, three layers of specialized cells develop that embryologists call the **primary germ layers.** Cells of the *embryonic disk* seen in Figure 32-7 differentiate into distinct types that form each of these three primary germ layers. Cells in each continue to differentiate and thus give rise to the various specific organs and systems of the body such as the skin, nervous tissue, muscles, or digestive organs (Figure 32-12). Each primary germ layer is called, respectively, **endoderm** (EN-doe-derm), or inside layer; **ectoderm** (EK-toe-derm), or outside layer; and **mesoderm** (MEZ-o-derm), or middle layer.

Endoderm

The inner germ layer, or *endoderm,* forms the linings of various tracts, as well as several glands. For example, the lining of the respiratory tract and GI tract, including some of the accessory structures such as tonsils, are derived from the endoderm. The linings of the pancreatic ducts, hepatic ducts, and urinary tract also have an endodermal origin.

The glandular epithelium of the thymus, thyroid, and parathyroid glands is also derived from the endoderm.

Ectoderm

The outer germ layer, or *ectoderm,* forms many of the structures around the periphery of the body. For example, the epidermis of the skin, enamel of the teeth, and cornea and lens of the eye are derived from the ectoderm. Other structures near the body's outer boundary, such as the muscles and bones of the face, the framework of the outer ear, and the nasal cavity, are also derived from the ectoderm. Besides these peripheral structures, various components of the nervous system—including the brain and the spinal cord—also have an ectodermal origin.

Mesoderm

The middle germ layer, or *mesoderm,* forms most of the organs and other structures between those formed by the endoderm and ectoderm. For example, the dermis of the skin, most of the skeletal muscles and bones (except some

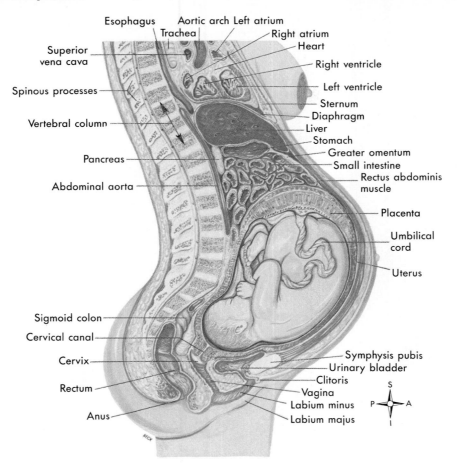

FIGURE 32-13   Full-term pregnancy.

in the head), many of the glands of the body, kidneys, gonads, and components of the circulatory system are derived from the mesoderm. Look carefully at Figure 32-12 to discern the logical pattern exhibited by germ layer development and differentiation.

Histogenesis and Organogenesis

The process by which the primary germ layers develop into many different kinds of tissues is called **histogenesis** (his-toe-JEN-e-sis). The way that these tissues arrange themselves into organs is called **organogenesis** (or-gah-no-JEN-e-sis). The fascinating story of histogenesis and organogenesis in human development is long and complicated; its telling belongs to the science of embryology. But for the beginning student of anatomy and physiology, it seems sufficient to appreciate that life begins when two sex cells unite to form a single-celled zygote and that the new human body evolves by a series of processes consisting of cell differentiation, multiplication, growth, and rearrangement, all of which take place in a definite, orderly sequence. Development of structure and function go hand in hand, and from 4 months of gestation, when every organ system is in place and functioning, until term (about 280 days), development of the fetus is mainly a matter of growth.

Figure 32-13 shows the normal intrauterine position of a fetus just before birth in a full-term pregnancy. The large size of the pregnant uterus toward the end of pregnancy affects the normal function of the mother's body greatly. For example, you might be able to tell from Figure 32-13 that a woman's center of gravity is shifted forward. This can make walking and other movements of the body difficult—or even hazardous—because the sensory and motor control systems often do not compensate completely for this shift. The pregnant uterus presses on the rectum, sometimes adversely affecting intestinal motility and thus may cause constipation and/or hemorrhoids. Pressure on the bladder reduces its urine-storing capacity, which results in frequent urination. Upward pressure pushes the abdominal organs against the diaphragm, making deep breathing difficult and sometimes causing the stomach to protrude into the thoracic cavity—a condition called *hiatal hernia.*

QUICK CHECK

✔ 1. *What is a trimester?*
✔ 2. *What is the difference between an* embryo *and a* fetus?
✔ 3. *Name the three primary germ layers.*
✔ 4. *What is histogenesis?*

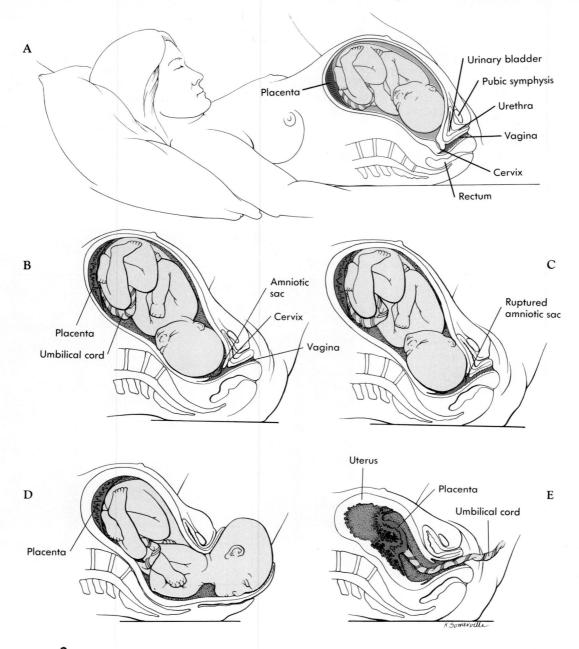

FIGURE 32-14   Parturition. A, The relation of the fetus to the mother. **B,** The fetus moves into the opening of the birth canal, and the cervix begins to dilate. **C,** Dilation of the cervix is complete. **D,** The fetus is expelled from the uterus. **E,** The placenta is expelled.

◣ BIRTH, OR PARTURITION

Birth, or **parturition**, is the point of transition between the prenatal and postnatal periods of life. As the fetus signals the end of pregnancy, the uterus becomes "irritable" and, ultimately, muscular contractions begin and cause the cervix to dilate (open), thus permitting the fetus to move from the uterus through the vagina or "birth canal" to the exterior. The process normally begins with the fetus taking a head-down position against the cervix (Figure 32-14, A). When contractions occur, the amniotic sac, or "bag of waters," ruptures, and labor begins.

Stages of Labor

Labor is the term used to describe the process that results in the birth of the baby. It is divided into three stages (Figure 32-14, B to E):

1. Stage one—period from onset of uterine contractions until dilation of the cervix is complete
2. Stage two—period from the time of maximal cervical dilation until the baby exits through the vagina
3. Stage three—process of expulsion of the placenta through the vagina

The time required for normal vaginal birth varies widely and may be influenced by many variables, including

ANTENATAL DIAGNOSIS AND TREATMENT

DIAGNOSTIC STUDIES

Advances in **antenatal** (from the Latin *ante*, "before," *natus*, "birth") **medicine** now permit extensive diagnosis and treatment of disease in the fetus much like any other patient. This new dimension in medicine began with techniques by which Rh$^+$ babies could be given transfusions before birth.

Current procedures using images provided by ultrasound equipment (see Figures A and B) allow physicians to prepare for and perform, before the birth of a baby, corrective surgical procedures such as bladder repair. These procedures also allow physicians to monitor the progress of other types of treatment on a developing fetus. Figure A shows placement of the ultrasound transducer on the abdominal wall. The resulting image (Figure B), called an *ultrasonogram*, shows a 6-week embryo. The image plane is showing the head and trunk. The hollow cerebral ventricles are visible in the image.

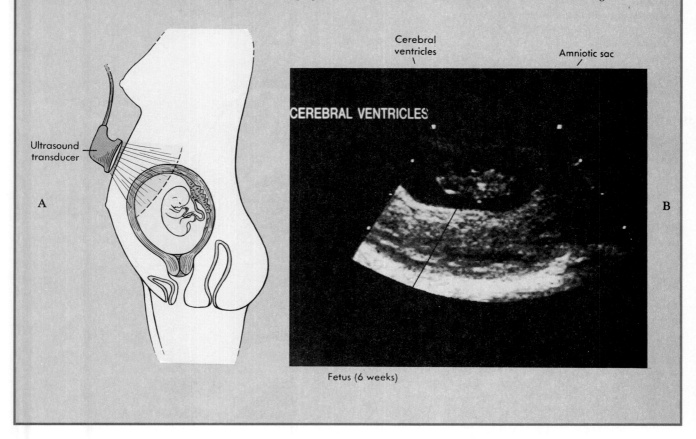

Cerebral ventricles

Amniotic sac

Ultrasound transducer

CEREBRAL VENTRICLES

A

B

Fetus (6 weeks)

whether the woman has previously had a child. In most cases, stage one of labor lasts from 6 to 24 hours, and stage two lasts from a few minutes to an hour. Delivery of the placenta (stage three) normally occurs within 15 minutes after the birth of the baby.

If abnormal conditions of the mother or fetus (or both) make normal vaginal delivery hazardous or impossible, physicians may suggest a **cesarean** (se-ZAIR-ee-an) **section.** Often called simply a *C-section,* it is a surgical procedure in which the newborn is delivered through an incision in the abdomen and uterine wall.

Multiple Births

The term *multiple birth* refers to the birth of two or more infants from the same pregnancy. The birth of twins is more common than the birth of triplets, quadruplets, or quintuplets. Multiple-birth babies are often born prematurely, so they are at a greater than normal risk of complications in

infancy. However, premature infants that have modern medical care available have a much lower risk of complications than those without such care.

Twinning, or double births, can result from two different processes:

1. **Identical twins** result from the splitting of embryonic tissue from the same zygote early in development. One way this happens is that, during the blastocyst stage of development, the inner cell mass divides into two masses. Each inner cell mass thus formed develops into a separate individual. As Figure 32-15, A, shows, identical twins usually share the same placenta but have separate umbilical cords. This is not surprising because in this type of twinning there is a single, shared trophoblast. Because they develop from the same fertilized egg, identical twins have the same genetic code. Despite this, identical twins are not absolutely identical in terms of structure and function. Different environmental factors and personal experiences lead to

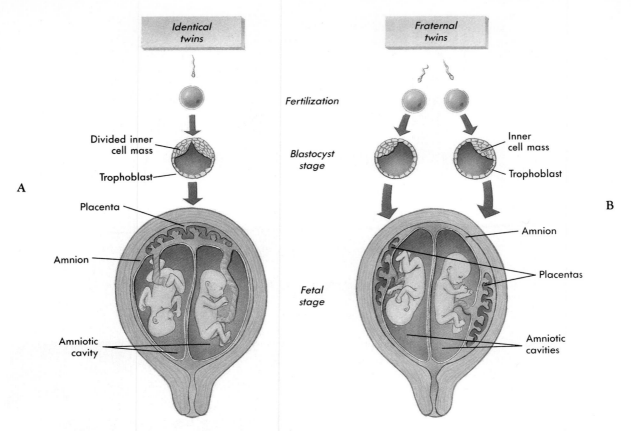

FIGURE 32-15   Multiple births. A, Identical twins develop when embryonic tissue from a single zygote splits to form two individuals. Notice that, because the trophoblast is shared, the placenta and the part of the amnion separating the amniotic cavities are shared by the twins. **B,** Fraternal twins develop when two ova are fertilized at the same time, producing two separate zygotes. Notice that each fraternal twin has its own placenta and amnion.

individuality even in genetically identical twins.

2. **Fraternal twins** result from the fertilization of two different ova by two different spermatozoa (Figure 32-15, B). Fraternal twinning requires the production of more than one mature ovum during a single menstrual cycle, a trait that is often inherited. Multiple ovulation may also occur in response to certain fertility drugs, especially the gonadotropin preparations. Fraternal twins are no more closely related genetically than any other brother-sister relationship. Because two separate fertilizations must occur, it is even possible for fraternal twins to have different biological fathers. Triplets, quadruplets, and other multiple births may be identical, fraternal, or any combination.

QUICK CHECK
✔ 1. *Briefly describe the three stages of labor.*
✔ 2. *How does identical twinning differ from fraternal twinning?*

POSTNATAL PERIOD

The **postnatal period** begins at birth and lasts until death. It is often divided into major periods for study but

people need to understand and appreciate the fact that growth and development are continuous processes that occur throughout the life cycle. Gradual changes in the physical appearance of the body as a whole and in the relative proportions of the head, trunk, and limbs are quite noticeable between birth and adolescence. Note in Figure 32-16 the obvious changes in the size of bones and in the proportionate sizes between different bones and body areas. The head, for example, becomes proportionately smaller. Whereas the infant head is approximately one fourth the total height of the body, the adult head is only about one eighth the total height. The facial bones also show several changes between infancy and adulthood. In an infant the face is one eighth of the skull surface, but in an adult the face is half of the skull surface. Another change in proportion involves the trunk and lower extremities. The legs become proportionately longer and the trunk proportionately shorter. In addition, the thoracic and abdominal contours change from round to elliptical.

Such changes are good examples of the ever-changing and ongoing nature of growth and development. It is unfortunate that many of the changes that occur in the later years of life do not result in an increased function. These degenerative changes are certainly important, how-

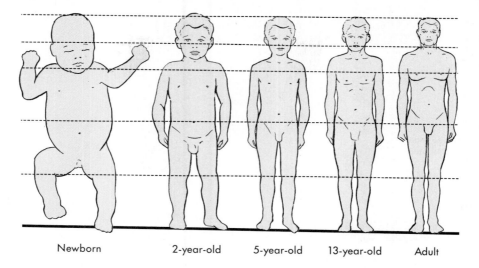

| Newborn | 2-year-old | 5-year-old | 13-year-old | Adult |

FIGURE 32-16 **Changes in the proportions of body parts from birth to maturity.** Note the dramatic differences in head size.

ever, and will be discussed later in this chapter (see pp. 832 and 833). The following are the most common postnatal periods: (1) **infancy,** (2) **childhood,** (3) **adolescence** and **adulthood,** and (4) **older adulthood.**

Infancy

The period of infancy begins abruptly at birth and lasts about 18 months. The first 4 weeks of infancy are often referred to as the **neonatal period** (Figure 32-17). Dramatic changes occur at a rapid rate during this short but critical period. **Neonatology** (nee-o-nay-TOL-o-jee) is the medical and nursing specialty concerned with the diagnosis and treatment of disorders of the newborn. Advances in this area have resulted in dramatically reduced infant mortality.

Many of the changes that occur in the cardiovascular and respiratory systems at birth are necessary for survival (see Figure 17-25, p. 489, and Figure 17-26, p. 490). Whereas the fetus totally depended on the mother for life support, the newborn infant, to survive, must become totally self-supporting in terms of blood circulation and respiration immediately after birth. A baby's first breath is deep and forceful. The stimulus to breathe results primarily from the increasing amounts of carbon dioxide (CO_2) that accumulate in the blood after the umbilical cord is cut shortly after delivery.

Many developmental changes occur between the end of the neonatal period and 18 months of age. Birth weight generally doubles during the first 4 months and then triples by 1 year. The baby also increases in length by 50% by the twelfth month. The "baby fat" that accumulated under the skin during the first year begins to decrease, and the plump infant becomes leaner.

Early in infancy the baby has only one spinal curvature (Figure 32-18). The lumbar curvature appears between 12 and 18 months, and the once-helpless infant becomes a toddler who can stand (Figure 32-19). One of the most striking changes to occur during infancy is the rapid

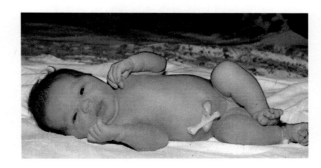

FIGURE 32-17 **The neonate infant.** The umbilical cord has been cut.

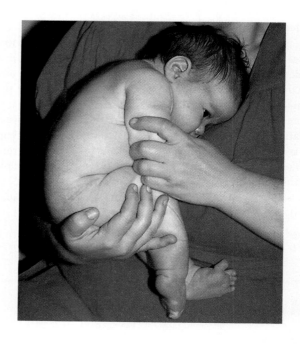

FIGURE 32-18 **Normal curvature of the infant's spine.**

development of the nervous and muscular systems. This permits the infant to follow a moving object with the eyes (2 months); lift the head and raise the chest (3 months); sit when well supported (4 months); crawl (10 months); stand alone (12 months); and run, although a bit stiffly (18 months).

Childhood

Childhood extends from the end of infancy to sexual maturity, or puberty—12 to 14 years in girls and 14 to 16 years in boys. Overall, growth during early childhood continues at a rather rapid pace, but month-to-month gains become less consistent. By the age of 6 the child appears more like a preadolescent than an infant or toddler. The child becomes less chubby, the potbelly becomes flatter, and the face loses its babyish look. The nervous and muscular systems continue to develop rapidly during the middle years of childhood; by 10 years of age the child has developed numerous motor and coordination skills.

The *deciduous teeth*, which began to appear at about 6 months of age, are lost during childhood, beginning at about 6 years of age. The *permanent teeth*, with the possible

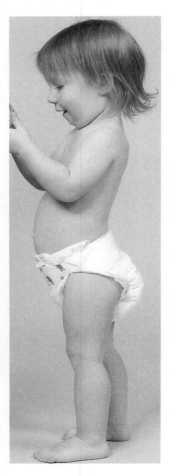

FIGURE 32-19   Normal lumbar curvature of a toddler's spine.

exception of the third molars (wisdom teeth), have all erupted by age 14.

Adolescence and Adulthood

The average age range of **adolescence** varies but generally the teenage years (13 to 19) are used. The period is marked by rapid and intense physical growth, which ultimately results in sexual maturity. Many of the developmental changes that occur during this period are controlled by the secretion of sex hormones and are classified as **secondary sex characteristics.** Breast development is often the first sign of approaching puberty in girls, beginning about age 10. Most girls begin to menstruate at 12 to 13 years of age. In boys the first sign of puberty is often enlargement of the testicles, which begins between 10 and 13 years of age. Both sexes show a spurt in height during adolescence. In girls the spurt in height begins between the ages of 10 and 12 and is nearly complete by 14 or 15. In boys the period of rapid growth begins between 12 and 13 and is generally complete by 16.

Many developmental changes that begin early in childhood are not completed until the early or middle years of **adulthood.** Examples include the maturation of bone, resulting in the full closure of the growth plates, and changes in the size and placement of other body components such as the sinuses. Many body traits do not become apparent for years after birth. Normal balding patterns, for example, are determined at the time of fertilization by heredity but do not appear until maturity. As a rule, adulthood is characterized by the maintenance of existing body tissues. With the passage of years the ongoing effort of maintenance and repair of body tissues becomes more and more difficult. As a result, degeneration begins. This is part of the process of aging, and it culminates in death.

Older Adulthood and Senescence

Most body systems are in peak condition and function at a high level of efficiency during the early years of adulthood. As a person grows older, a gradual but certain decline takes place in the functioning of every major organ system in the body. The study of aging is called *gerontology.* Unfortunately, the mechanisms and causes of aging are not well understood.

Some gerontologists believe that an important aging mechanism is the limit on cell reproduction. Laboratory experiments have shown that many types of human cells cannot reproduce more than 50 times—thus limiting the maximum life span. Cells die continually, no matter what a person's age, but in older adulthood, many dead cells are not replaced—causing degeneration of tissues. Perhaps the cells are not replaced because the surrounding cells have reached their limit of reproduction. Perhaps differences in each individual's aging process result from differences in the reproductive capacity of cells. This mechanism seems to operate in individuals with **progeria** (pro-JAIR-ee-ah), a rare, inherited condition in which a person appears to age rapidly.

Various factors that affect the rates of cell death and cell

reproduction have been cited as causes of aging. Some gerontologists believe that nutrition, injury, disease, and other environmental factors affect the aging process. A few have even proposed that aging results from cellular changes caused by slow-acting "aging" viruses found in all living cells. Other gerontologists have proposed that aging is caused by "aging" genes—genes in which aging is "preprogrammed." Yet another proposed cause of aging is autoimmunity. You may recall from Chapter 20 that autoimmunity occurs when the immune system attacks a person's own tissues.

Although the causes and basic mechanisms of aging are yet to be understood, at least many of the signs of aging are obvious. The remainder of this chapter deals with some of the more common degenerative changes that frequently characterize **senescence** (se-NES-ens), or older adulthood.

QUICK CHECK

✔ 1. *Name the four major phases of the postnatal period.*

✔ 2. *When does the neonatal period of human development occur?*

✔ 3. *What signs characterize the adolescent period of human development?*

✔ 4. *Briefly describe one of the proposed mechanisms of the aging process.*

EFFECTS OF AGING

Skeletal System

In older adulthood, bones undergo changes in texture, degree of calcification, and shape. Instead of clean-cut margins, older bones develop indistinct and shaggy-appear-ing margins with spurs—a process called *lipping*. This type of degenerative change restricts movement because of the piling up of bone tissue around the joints. With advancing age, changes in calcification may result in reduction of bone size and in bones that are porous and subject to fracture. The lower cervical and thoracic vertebrae are the site of frequent fractures. The result is curvature of the spine and the shortened stature so typical of late adulthood. Degenerative joint diseases such as **osteoarthritis** (OS-tee-o-ar-THRYE-tis) are also common in elderly adults.

Integumentary System (Skin)

With advancing age the skin becomes dry, thin, and inelastic. It "sags" on the body because of increased wrinkling and skinfolds. Pigmentation changes and the thinning or loss of hair are also common problems associated with aging (Figure 32-20).

Urinary System

The number of nephron units in the kidney decreases by almost 50% between the ages of 30 and 75. Also, because less blood flows through the kidneys as an individual ages, there is a reduction in overall function and excretory capacity or the ability to produce urine. In the bladder, significant age-related problems often occur because of diminished muscle tone. Muscle atrophy (wasting) in the bladder wall results in decreased capacity and inability to empty, or void, completely.

Respiratory System

In older adulthood the costal cartilages that connect the ribs to the sternum become hardened or calcified. This makes it difficult for the rib cage to expand and contract as it normally does during inspiration and expiration. In time

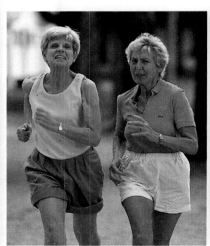
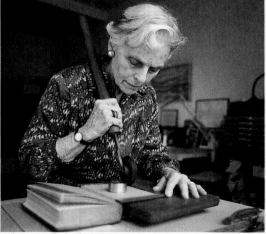
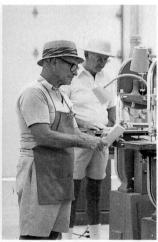

FIGURE 32-20 Some physical changes associated with maturity and aging. These include graying, thinning, or loss of hair *(left)*; skin pigmentation changes *(center)*; and barrel chest *(right)*. Although physical capacity may decline with age, older adults can lead active productive lives.

the ribs gradually become "fixed" to the sternum, and chest movements become difficult. When this occurs the rib cage remains in a more expanded position, respiratory efficiency decreases, and a condition called "barrel chest" results (Figure 32-20). With advancing years a generalized atrophy, or wasting, of muscle tissue takes place as the contractile muscle cells are replaced by connective tissue. This loss of muscle cells decreases the strength of the muscles associated with inspiration and expiration.

Cardiovascular System

Degenerative heart and blood vessel disease is one of the most common and serious effects of aging. Fatty deposits build up in blood vessel walls and narrow the passageway for the movement of blood, much as the buildup of scale in a water pipe decreases flow and pressure. The resulting condition, called **atherosclerosis** (ath-er-o-skle-RO-sis), often leads to eventual blockage of the coronary arteries and a "heart attack." If fatty accumulations or other substances in blood vessels calcify, actual hardening of the arteries, or **arteriosclerosis** (ar-te-ree-o-skle-RO-sis), occurs. Rupture of a hardened vessel in the brain (stroke) is a frequent cause of serious disability or death in the older adult. **Hypertension,** or high blood pressure, is also more common.

Special Senses

The sense organs, as a group, all show a gradual decline in performance and capacity as a person ages. Most people are farsighted by age 65 because eye lenses become hardened and lose elasticity; the lenses cannot become curved to accommodate for near vision. This hardening of the lens is called **presbyopia** (pres-bee-O-pee-ah), which means "old eye." Many individuals first notice the change at about 40 or 45 years of age, when it becomes difficult to do close-up work or read without holding printed material at arm's length. This explains the increased need, with advancing age, for bifocals (glasses that incorporate two lenses) to automatically accommodate for near and distant vision. Loss of transparency of the lens or its covering capsule is another common age-related eye change. If the lens actually becomes cloudy and significantly impairs vision, it is called a **cataract** and must be removed surgically. The incidence of **glaucoma** (glaw-KO-mah), the most serious age-related eye disorder, increases with age. Glaucoma causes an increase in the pressure within the eyeball and, unless treated, often results in blindness.

In many elderly people a very significant loss of hair cells in the organ of Corti (inner ear) causes a serious decline in the ability to hear certain frequencies. In addition, the eardrum and attached ossicles become more fixed and less able to transmit mechanical sound waves. Some degree of hearing impairment is universally present in the older adult.

EXERCISE AND AGING

A sound exercise program throughout life can reduce some of the common effects of aging. The improved cardiovascular condition associated with aerobic training can prevent or reduce the effects of several age-related circulatory problems. The loss of bone mass that occurs in older adulthood is less troublesome if one enters this period with a higher than average bone density developed through exercising. Finally, the relatively high ratio of body fat to muscle often seen in the elderly can be avoided entirely by a life-long fitness program.

The sense of taste is also decreased. This loss of appetite may be caused, at least in part, by the replacement of taste buds with connective tissue cells. Only about 40% of the taste buds present at age 30 remain fully functional in an individual at age 75.

Reproductive Systems

Although most men and women remain sexually active throughout their later years, mechanisms of sexual response may change, and fertility declines. In men, erection may be more difficult to achieve and maintain, and urgency for sex may decline. In women, lubrication of the vagina may decrease. Although men can continue to produce gametes as they age, women experience a cessation of reproductive cycling between the ages of 45 and 60—**menopause.** Menopause results from a decrease in the cyclic production of the primary sex hormones, especially estrogen, with advancing age. The decrease in estrogen accounts for the common symptoms of menopause: cessation of menstrual cycles, hot flashes, and thinning of the vaginal wall. The exact mechanism of hot flashes is not clearly understood, but it is related to the hormonal changes that occur during menopause. Rarely serious, hot flashes usually subside over time. Estrogen therapy may be used to relieve menopause symptoms in some cases.

The decrease in estrogen levels associated with menopause may also contribute to *osteoporosis*. This condition, characterized by loss of bone mass (see Chapter 6), is often treated with estrogen therapy when it occurs in postmenopausal women.

QUICK CHECK

✔ 1. *What kinds of changes occur in the skeleton as one ages?*

✔ 2. *List some age-related changes in the cardiovascular system.*

✔ 3. *How can age affect a person's vision?*

MECHANISMS OF DISEASE

Disorders of Pregnancy and Early Development

Implantation Disorders

A pregnancy has the best chance of a successful outcome, the birth of a healthy baby, if the blastocyst is implanted properly in the uterine wall. However, proper implantation does not always occur. Many offspring are lost before implantation occurs, often for unknown reasons. As mentioned in this and the previous chapter, implantation outside the uterus results in an ectopic pregnancy. If the blastocyst implants in a region of endometriosis or normal peritoneal membrane, the pregnancy may be successful if there is room for the developing fetus to grow. Ectopic pregnancies that do succeed must be delivered by C-section rather than by normal vaginal birth. If an ectopic pregnancy occurs in a uterine tube, which cannot stretch to accommodate the developing offspring, the tube may rupture and cause life-threatening hemorrhaging. So-called **tubal pregnancies** are the most common type of ectopic pregnancy.

Occasionally, the blastocyst implants in the uterine wall near the cervix. This in itself may present no problem, but if the placenta grows too close to the cervical opening a condition called **placenta previa** (PRE-vee-ah) results. The normal dilation and softening of the cervix that occurs in the third trimester often causes painless bleeding as the placenta near the cervix separates from the uterine wall. The massive blood loss that may result can be life-threatening for both mother and offspring (Figure 32-21, A).

Separation of the placenta from the uterine wall can occur even when implantation occurs in the upper part of the uterus. When this occurs in a pregnancy of 20 weeks or more, the condition is called **abruptio placentae** (ab-RUP-chee-o pla-SEN-tay). Complete separation of the placenta causes the immediate death of the fetus. The severe hemorrhaging that often results, sometimes hidden in the uterus, may cause circulatory shock and death of the mother within minutes. A cesarean section and perhaps also a hysterectomy must be performed immediately to prevent blood loss and death (Figure 32-21, B).

Preeclampsia

Preeclampsia (pre-ee-KLAMP-see-ah), also called *toxemia of pregnancy*, is a serious disorder that occurs in about 1 in every 20 pregnancies. This disorder is characterized by the onset of acute hypertension after the twenty-fourth week, accompanied by proteinuria and edema. The causes of preeclampsia are largely unknown, despite intense research efforts. Preeclampsia can result in complications such as abruptio placentae, stroke, hemorrhage, fetal malnutrition, and low birth weight. This condition can progress to *eclampsia*, a life-threatening form of toxemia that causes severe convulsions, coma, kidney failure, and perhaps death of the fetus and mother.

Fetal Death

A *miscarriage* is the loss of an embryo or fetus before the twentieth week (or a fetus weighing less than 500 g or 1.1 lb). Technically known as a **spontaneous abortion,** the most common cause of such a loss is a structural or functional defect in the developing offspring. Abnormalities of the mother, such

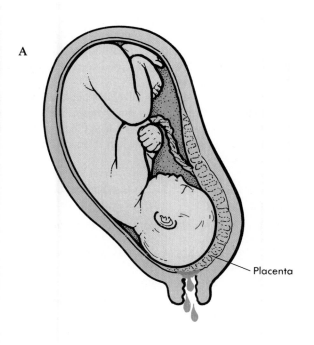

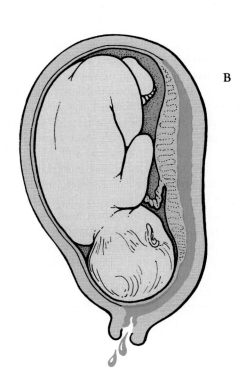

A

Placenta

B

FIGURE 32-21   Placental disorders. A, Placenta previa. **B,** Abruptio placentae.

as hypertension, uterine abnormalities, and hormonal imbalances, can also cause spontaneous abortions. After 20 weeks, delivery of a lifeless infant is termed a **stillbirth.**

Birth Defects

Birth defects, also called **congenital abnormalities,** include any structural or functional abnormality present at birth. Congenital defects may be inherited or may be acquired during gestation or delivery. Inherited defects will be discussed in the next chapter. Acquired defects result from agents called **teratogens** (TAIR-et-o-jenz) that disrupt normal histogenesis and organogenesis. Some teratogens are chemicals such as alcohol, antibiotics, and other drugs. Microbes, such as those that cause rubella (a viral infection), can also cross the placental barrier and disrupt normal embryonic development. Radiation and other physical factors can also cause birth defects. Some teratogens are mutagens because they do their damage by changing the genetic code in cells of the developing embryo.

Postpartum Disorders

Puerperal (poo-ER-per-al) **fever,** or *childbed fever,* is a syndrome of postpartum mothers characterized by bacterial infection that progresses to septicemia (blood infection) and possibly death. Until the 1930s, puerperal fever was the leading cause of maternal death—claiming the lives of more than 20% of postpartum women. Modern antiseptic techniques prevent most postpartum infections now. Puerperal infections that do occur are usually treated successfully by an immediate and intensive program of antibiotic therapy.

After a child is born, it needs the nourishment of milk to survive. However, several disorders of *lactation* (milk production) may occur to prevent a mother from nursing her infant. For example, anemia, malnutrition, emotional stress, and structural abnormalities of the breast can all interfere with normal lactation. **Mastitis** (mas-TIE-tis), or breast inflammation, often caused by infection, can result in lactation problems or production of milk contaminated with pathogenic organisms. In many cultures, the availability of other nursing mothers or breast-milk substitutes allows proper nourishment of the infant, even when lactation problems develop. Most breast-milk substitutes are formulations of milk from another mammal such as the cow. Infants who lack the enzyme *lactase* may not be able to digest the lactose present in human or animal milk, resulting in a condition called **lactose intolerance.** Infants with lactose intolerance are sometimes given a lactose-free milk substitute made from soy or other plant products.

CHAPTER SUMMARY

INTRODUCTION

A. Prenatal period—period beginning with conception and ending at birth
B. Postnatal period—period beginning with birth and continuing until death
C. Human developmental biology—study of changes occurring during the cycles of life from conception to death

BEGINNING OF HUMAN LIFE

A. Production of sex cells—spermatozoa are produced by spermatogenesis; ova are produced by oogenesis
  1. Meiosis (Figure 32-1)
    a. Special form of cell division that reduces the number of chromosomes in each daughter cell to one-half of those in the parent cell
    b. Mature ova and sperm contain only 23 chromosomes, half as many as other human cells
    c. Meiotic division—two cell divisions that occur one after another in succession
      (1) Meiotic division I and meiotic division II
      (2) Both divisions made up of an interphase, prophase, metaphase, anaphase, and telophase
    d. During prophase I of meiosis "cross over" occurs where genetic material is "shuffled"
    e. Daughter cells formed by meiotic division I contain a haploid number of chromosomes
    f. Meiotic division II—essentially the same as mitotic division; reproduces each of the two cells formed by meiotic division I and forms four cells, each with the haploid number of chromosomes
  2. Spermatogenesis (Figure 32-2, A)—process by which primitive male sex cells become transformed into mature sperm; begins at approximately puberty and continues throughout a man's life

    a. Meiotic division I—one primary spermatocyte forms two secondary spermatocytes, each with 23 chromosomes
    b. Meiotic division II—each of the two secondary spermatocytes form a total of four spermatids
  3. Oogenesis (Figure 32-2, B)—process by which primitive female sex cells become transformed into mature ova
    a. Mitosis—oogonia reproduce to form primary oocytes; most primary oocytes begin meiosis and develop to prophase I before birth; there they stay until puberty
    b. Once during each menstrual cycle, a few primary oocytes resume meiosis and migrate toward the surface of the ovary; usually only one oocyte matures enough for ovulation, and meiosis again halts at metaphase II
    c. Meiosis resumes only if the head of a sperm cell enters the ovum
B. Ovulation and insemination
  1. Ovulation—expulsion of the mature ovum from the mature ovarian follicle, into the pelvic cavity, and then the uterine tube
  2. Insemination—expulsion of seminal fluid from the male into the female vagina; sperm travel through the cervix and uterus and into the uterine tubes
C. Fertilization—also known as conception
  1. Most often occurs in the outer one third of the oviduct
  2. Ovum attracts and "traps" sperm with special receptor molecules on its surface
  3. When one spermatozoon enters the ovum, the ovum stops collecting sperm on its surface
  4. 23 chromosomes from the sperm head and 23 chromosomes in the ovum comprise a total of 46 chromosomes
  5. Zygote—fertilized ovum; genetically complete

PRENATAL PERIOD

A. Begins with conception and continues until the birth of a child

B. Cleavage and implantation (Figure 32-4)—once zygote is formed, it immediately begins to divide
  1. Morula—solid mass of cells formed from zygote; takes approximately 3 days; continues to divide (Figure 32-5)
  2. Blastocyst—by the time the developing embryo reaches the uterus, it has formed a hollow ball of cells, which implants into the uterine lining (Figure 32-6)
  3. Approximately 10 days pass from fertilization until implantation in the uterine lining; ovum has a store of nutrients that support this embryonic development until implantation has occurred
  4. Blastocyst has an outer layer of cells and an inner cell mass
    a. Trophoblast—outer wall of the blastocyst
    b. Inner cell mass—as blastocyst develops, yolk sac and amniotic cavity are formed (Figure 32-7)
      (1) In humans, yolk sac's functions are largely non-nutritive
      (2) Amniotic cavity becomes a fluid-filled, shock-absorbing sac (bag of waters) in which the embryo floats during development (Figure 32-8)
    c. Chorion develops from trophoblast to become an important fetal membrane in the placenta
  5. Placenta (Figure 32-9)
    a. Anchors fetus to the uterus and provides a "bridge" for the exchange of nutrients and waste products between mother and baby
    b. Also serves as an excretory, respiratory, and endocrine organ
    c. Placental tissue normally separates maternal and fetal blood supplies
    d. Has important endocrine functions—secretes large amounts of HCG, which stimulate the corpus luteum to continue its secretion of estrogen and progesterone (Figure 32-10)

C. Periods of development
  1. Gestation period—approximately 39 weeks; divided into three 3-month segments called trimesters
  2. Embryonic phase extends from fertilization until the end of week 8 of gestation
  3. Fetal phase—weeks 8 to 39 (Figure 32-11)

D. Formation of the primary germ layers
  1. Three layers of specialized cells develop early in the first trimester of pregnancy
  2. Cells of embryonic disk differentiate and form each of the three primary germ layers
  3. Each primary germ layer gives rise to specific organs and systems of the body (Figure 32-12)
  4. There are three primary germ layers
    a. Endoderm—inside layer
    b. Ectoderm—outside layer
    c. Mesoderm—middle layer

E. Histogenesis and organogenesis
  1. Histogenesis—process by which primary germ layers develop into different kinds of tissues
  2. Organogenesis—how tissues arrange themselves into organs

BIRTH, OR PARTURITION

A. Transition between prenatal and postnatal periods of life

B. Stages of labor (Figure 32-14)
  1. Stage one—period from onset of uterine contractions until cervical dilation is complete
  2. Stage two—period from maximal cervical dilation until the baby exits through the vagina
  3. Stage three—process of expulsion of the placenta through the vagina

C. Multiple births—birth of two or more infants from the same pregnancy; twins are most common (Figure 32-15)
  1. Identical twins result from the splitting of embryonic tissue from the same zygote early in development
  2. Fraternal twins result from the fertilization of two different ova by two different spermatozoa

POSTNATAL PERIOD

A. Begins at birth and continues until death; commonly divided into a number of periods

B. Infancy begins at birth and lasts until approximately 18 months (Figure 32-17)
  1. Neonatal period—first 4 weeks of infancy; dramatic changes occur at a rapid rate
  2. Changes allow the infant to become totally self-supporting, especially respiratory and cardiovascular systems

C. Childhood extends from end of infancy to sexual maturity, or puberty
  1. Early childhood—growth continues at a rapid pace but slows down
  2. By age 6, child looks more like a preadolescent than an infant or toddler
  3. Nervous and muscular system develop rapidly during the middle years of childhood
  4. Deciduous teeth are lost during childhood, beginning at approximately 6 years of age
  5. Permanent teeth have erupted by age 14, except for the third molars (wisdom teeth)

D. Adolescence and adulthood
  1. Adolescence is considered to be the teenage years (from 13 to 19); marked by rapid and intense physical growth, resulting in sexual maturity
  2. Adulthood—characterized by maintenance of existing body tissues

E. Older adulthood and senescence
  1. As a person grows older, a gradual decline occurs in every major organ system in the body
  2. Gerontologists theorize an important aging mechanism is the limit on cell reproduction

EFFECTS OF AGING—COMMON DEGENERATIVE CHANGES THAT FREQUENTLY CHARACTERIZE SENESCENCE

A. Skeletal system
  1. Bones change in texture, degree of calcification, and shape
  2. Lipping occurs, which can limit range of motion
  3. Decreased bone size and density lead to increased risk of fracture

B. Integumentary system (skin)
  1. Skin becomes dry, thin, and inelastic
  2. Pigmentation changes and thinning hair are common problems associated with aging

C. Urinary system
  1. Number of nephron units in the kidney decrease by almost 50% between the ages of 30 and 75
  2. Decreased blood flow through kidneys reduces overall function and excretory capacity
  3. Diminished muscle tone in bladder results in decreased capacity and inability to empty, or void, completely

D. Respiratory system
  1. Costal cartilages become calcified
  2. Respiratory efficiency decreases
  3. Decreased strength of respiratory muscles

E. Cardiovascular system

1. Degenerative heart and blood vessel disease—one of the most common and serious effects of aging
2. Atherosclerosis—build-up of fatty deposits on blood vessel walls narrows the passageway for blood
3. Arteriosclerosis—"hardening" of the arteries
4. Hypertension—high blood pressure

F. Special senses
1. Sense organs—gradual decline in performance and capacity with aging
2. Presbyopia—far-sightedness due to hardening of lens
3. Cataract—cloudy lens, which impairs vision
4. Glaucoma—increased pressure within the eyeball; if left untreated, often results in blindness
5. Decreased hearing
6. Decreased taste

G. Reproductive systems
1. Mechanism of sexual response may change
2. Fertility decreases
3. In females, menopause occurs between ages of 45 and 60

MECHANISMS OF DISEASE: DISORDERS OF PREGNANCY AND EARLY DEVELOPMENT

A. Implantation disorders
1. Ectopic pregnancy—implantation outside the uterus; tubal pregnancy is most common type
2. Placenta previa—blastocyst implants in uterine wall near the cervix, and placenta blocks the cervical opening (Figure 32-21, A)

3. Abruptio placenta—separation of placenta from uterine wall occurring after 20th week of pregnancy (Figure 32-21, B)

B. Preeclampsia—toxemia
1. Acute hypertension occurring after the 24th week of pregnancy
2. Accompanied by proteinuria and edema; can result in many complications
3. Can progress to eclampsia—life threatening

C. Fetal death
1. Miscarriage—loss of an embryo or fetus before the 20th week; also known as spontaneous abortion
2. Stillbirth—delivery of a lifeless infant after 20 weeks

D. Birth defects
1. Structural or functional abnormalities at birth; also known as congenital abnormalities
2. May be inherited or acquired during gestation or delivery
3. Acquired defects result form teratogens that disrupt normal histogenesis and organogenesis

E. Postpartum disorders
1. Puerperal fever—childbed fever; syndrome characterized by bacterial infection that progresses to septicemia and possibly death
2. Mastitis—breast inflammation in mother
3. Lactose intolerance—inability to digest lactose present in human or animal milk due to lack of lactase

REVIEW QUESTIONS

1. Define the terms *developmental biology, growth,* and *development.*
2. How is the process of meiosis different from mitosis?
3. Identify the number of chromosomes in mature ova and sperm. Why is this number necessary?
4. Outline the major steps in spermatogenesis. Do the same for oogenesis.
5. Identify the two processes necessary to bring the sperm and ovum into proximity with each other.
6. During fertilization, how does the ovum attract or "trap" sperm?
7. At what developmental stage does implantation occur?
8. Describe the structural changes between a morula and a blastocyst.
9. How does the placenta develop?
10. What functions does the placenta provide?
11. Outline the hormonal levels of human chorionic gonadotropin (HCG), estrogen, and progesterone.
12. During what period is the term *embryo* replaced by the term *fetus?*
13. List the various structures derived from each of the three primary germ layers.
14. How do the terms *histogenesis* and *organogenesis* differ?
15. Explain the three stages of labor.
16. When would a physician perform a cesarean section?
17. What is the difference between identical and fraternal twins?
18. From birth to maturity, how does the size of the head compare to the rest of the body?
19. What are the time spans of the following postnatal periods: infancy, childhood, adolescence?
20. During what postnatal period do the secondary sex characteristics develop? What initiates this development?
21. What structural changes may result from the changes in bone calcification due to aging?
22. Define the terms *atherosclerosis, arteriosclerosis,* and *hypertension.*
23. Identify and describe the most serious age-related eye disorder.
24. Explain the cause(s) of menopause.
25. What is the purpose of a laparoscope in the process of *in vitro* fertilization?
26. How can a sound exercise program reduce some of the common effects of aging?

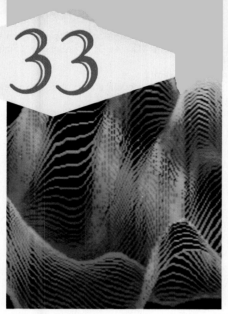

33 Genetics and Heredity

Molecule of DNA

OBJECTIVES

After you have completed this chapter, you should be able to:

1. Describe the mechanism of gene function.
2. Outline the steps in meiosis.
3. Differentiate between the following sets of terms: *dominant/recessive, genotype/phenotype, homozygous/heterozygous.*
4. Explain the term *codominance* and give an example.
5. Explain how certain traits demonstrate sex-linked inheritance.
6. Identify how genetic mutations may occur.
7. Differentiate between genetic diseases resulting from an abnormality in a single gene and those resulting from chromosome breakage or from the abnormal presence or absence of entire chromosomes.
8. Describe the following single-gene diseases: cystic fibrosis, phenylketonuria (PKU), Tay-Sachs disease.
9. Describe the following chromosomal diseases: Down's syndrome; Klinefelter's syndrome; Turner's syndrome.
10. Identify the various hypotheses relating to the genetic basis of cancer.
11. Define and describe a pedigree, the Punnett square, and the karyotype.
12. Identify various treatments of genetic diseases.

I t seems that today we are hearing more and more about the importance of **genetics,** the scientific study of inheritance, to all fields of human biology—especially anatomy, physiology, and medicine. Popular news magazines are running story after story on the revolution in treating fatal inherited disorders by using something called *gene therapy*. Health and science columns in newspapers keep us informed of the latest discoveries of disease-causing genes. Television programs outline the progress of the largest coordinated biological quest that anyone can remember: mapping the entire human genetic code. Clearly, one cannot be informed about human biology today without some knowledge of basic genetics and heredity. In this chapter, we will briefly review the essential concepts of genetics and explain how heredity affects every structure and function in the body.

THE SCIENCE OF GENETICS

History shows that humans have been aware of patterns of inheritance—or *heredity*—for thousands of years, but it was not until the 1860s that the scientific study of these patterns—genetics—was born. At that time, a monk living in Brno, Moravia was the first to discover the basic mechanism by which traits are transmitted from parents to offspring. That man, Gregor Mendel, proved that independent units (which we now call genes) are responsible for the inheritance of biological traits.

The science of genetics developed from Mendel's quest to explain how normal biological characteristics are inherited. As time went by and more genetic studies were done, it became clear that certain diseases also have a genetic basis. As you may recall from Chapter 1 (p. 28), some diseases are inherited directly. For example, the group of blood-clotting disorders called *hemophilia* is inherited by children from parents who have the genetic code for hemophilia. Directly inherited diseases are often called "hereditary diseases." Other diseases are only partly determined by genetics—that is, they involve genetic risk factors (Chapter 1, p. 30). For example, certain forms of skin cancer are thought to have a genetic basis. A person who inherits the genetic code associated with skin cancer will develop the disease only if the skin is also heavily exposed to the ultraviolet radiation in sunlight.

CHROMOSOMES AND GENES
Mechanism of Gene Function

Mendel proposed that the genetic code is transmitted to offspring in discrete, independent units that we now call **genes** (jeens). Recall from Chapters 2 and 3 that each gene

839

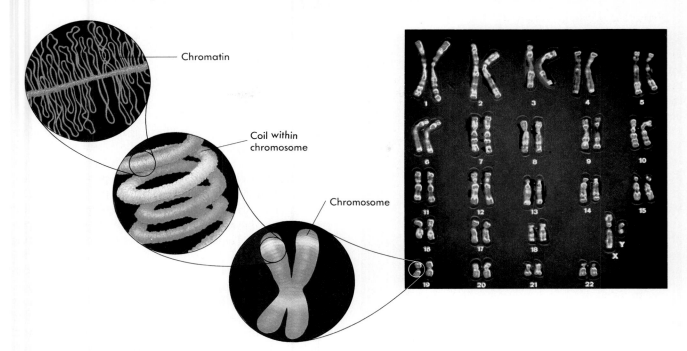

FIGURE 33-1 **Human chromosomes.** Each of the 46 human chromosomes, arranged here in numbered pairs for easy reference, is a coiled mass of chromatin (DNA).

is a sequence of nucleotide bases in the deoxyribonucleic acid (DNA) molecule. Each gene contains a genetic code that the cell transcribes to a messenger RNA (mRNA) molecule. Each mRNA molecule associates with a ribosome where the code is translated to form a specific polypeptide molecule. Many of the protein molecules formed from these polypeptides are *enzymes,* functional proteins that regulate the various metabolic pathways of the body. Because enzymes regulate the biochemistry of the body, they regulate the entire structure and function of the body. Thus genes determine the structure and function of the human body by producing a set of specific regulatory enzymes and other proteins.

As described in Chapters 2 and 3, genes are simply segments of a DNA molecule. While the genetic codes of its genes are being actively transcribed, the DNA is in a threadlike form called *chromatin.* During cell division, each replicated strand of chromatin coils upon itself to form a compact mass called a *chromosome* (Figure 33-1). Thus each DNA molecule can be called either a chromatin strand or a chromosome. Throughout this chapter we will use the term "chromosome" for DNA and the term "gene" for each distinct code within a DNA molecule.

Distribution of Chromosomes to Offspring
Meiosis

Each cell of the human body contains 46 chromosomes. The only exceptions to this principle are the **gametes** —male *spermatozoa* and female *ova.* Recall from Chapter 32 that a special form of nuclear division called **meiosis** (see Figure 32-1, p. 814) produces gametes with only 23 chromosomes—exactly half the usual number. This num-

 Mapping the Human Genome

The entire collection of genetic material in the nucleus of each cell of the human body is called the **genome** (JEE-nome). The typical human genome consists of 46 individual chromosomes. An intense, coordinated effort is currently underway to map all the gene locations in the human genome and read the different genetic codes possible at each location. Many scientists predict that all 100,000 or so human genes will be identified within the next few years. Already the location of about 2,000 genes that may cause disease have been identified. For example, the gene that causes one form of hypothyroidism (Chapter 15) is located in a specific region of chromosome 8. Much of the work is being done under a government-sponsored program called the Human Genome Project. The first director of this project was James Watson, one of a team of researchers that originally discovered the structure of DNA in the early 1950s.

ber is called the *haploid number.* When a sperm (with its 23 chromosomes) unites with an ovum (with its 23 chromosomes) at conception, a *zygote* with 46 chromosomes is formed. Thus the zygote has the same number of chromosomes (46, the *diploid number*) as each body cell in the parents.

As the photograph in Figure 33-1 shows, the 46 human chromosomes can be arranged in 23 pairs according to size. One pair called the **sex chromosomes** may not match, but the remaining 22 pairs of **autosomes** (AW-toe-sohms) always appear to be nearly identical to each other.

Principle of independent assortment

Because half of an offspring's chromosomes are from the mother and half are from the father, a unique blend of inherited traits is formed. According to principles first discovered by Mendel, each chromosome assorts itself independently during meiosis. This **principle of independent assortment** states that as sperm are formed, chromosome pairs separate and the maternal and paternal chromosomes get mixed up and redistribute themselves independently of the other chromosome pairs (Figure 33-2). Thus each sperm is likely to have a *different* set of 23 chromosomes. Since the ova are formed in the same manner, each ovum is likely to be genetically different from the ovum that preceded it. Independent assortment of chromosomes assures that each offspring from a single set of parents is very likely to be genetically unique—a phenomenon known as *genetic variation*.

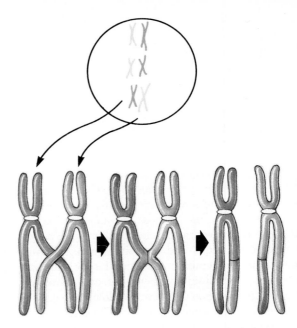

FIGURE 33-3 Crossing over. Genes (or linked groups of genes) from one chromosome are exchanged with matching genes in the other chromosome of a pair during meiosis.

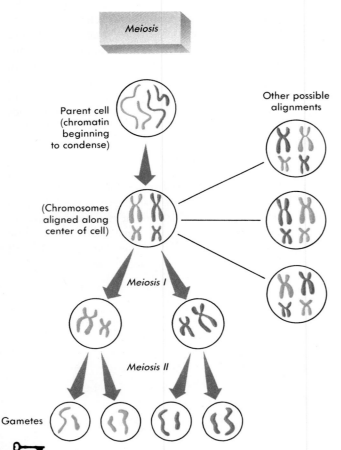

FIGURE 33-2 Meiosis and the principle of independent assortment. In meiosis, a series of two divisions results in the production of gametes with half the number of chromosomes of the original parent cell. In this figure, the original cell has four chromosomes and the gametes each have two chromosomes. During the first division of meiosis, pairs of similar chromosomes line up along the cell's equator for even distribution to daughter cells. Because different pairs assort independently of each other, any of four (2^2) different combinations of chromosomes may occur. Because human cells have 23 pairs of chromosomes, over 8 million (2^{23}) different combinations are possible.

The principle of independent assortment also applies to individual genes or groups of genes. During one phase of meiosis, pairs of matching chromosomes line up along the equator of the cell and exchange genes with one another. This process is called **crossing over** because genes from a particular location cross over to the same location on the matching chromosome (Figure 33-3). Sometimes an entire group of genes stay together and cross over as a single unit—a phenomenon called **gene linkage**. Crossing over introduces additional opportunities for genetic variation among the offspring of a single set of parents.

QUICK CHECK

✔ 1. How do genes produce biological traits?
✔ 2. Who might be considered the founder of the scientific study of genetics?
✔ 3. What is the difference between an autosome and a sex chromosome?
✔ 4. List some mechanisms that increase genetic variation among human offspring.

GENE EXPRESSION
Hereditary Traits
Dominance

Mendel discovered that the genetic units we now call genes may be expressed differently among individual offspring. After rigorous experimentation with pea plants, he discovered that each inherited trait is controlled by two sets of similar genes, one from each parent. We now know that each autosome in a pair matches its partner in the type

of genes it contains. In other words, if one autosome has a gene for hair color, its partner will also have a gene for hair color—in the same location on the autosome. Although both genes specify hair color, they may not specify the *same* hair color. Mendel discovered that some genes are **dominant** and some are **recessive**. A dominant gene is one whose effects are seen and which are capable of masking the effects of a recessive gene for the same trait.

Consider the example of **albinism** (AL-bin-izm), a total lack of melanin pigment in the skin and eyes. Because they lack dark pigmentation, people with this condition have difficulty seeing and protecting themselves from burns in direct sunlight. The genes that cause albinism are recessive; the genes that cause normal melanin production are dominant. By convention, dominant genes are represented by upper case letters and recessive genes by lower case letters. One can represent the gene for albinism as *a* and the gene for normal skin pigmentation as *A*. A person with the gene combination *AA* has two dominant genes—and so will exhibit a normal skin color. The code *AA* is called a **genotype.** A person with a genotype of two identical forms of a trait is said to be **homozygous** (from *homo*, "same," and *zygo*, "joined") for that trait. The manner in which a genotype is expressed is called the **phenotype.** Thus in a person who is homozygous dominant (*AA*) for skin color will have a normal phenotype (that is, normal skin pigmentation). Someone with the gene combination *Aa* will also have normal skin color because the normal gene *A* is dominant over the recessive albinism gene *a*. A person with genotype *Aa* is said to be **heterozygous** (from *hetero*, "different," and *zygo*, "joined") and will express the normal phenotype. Only a person with the homozygous recessive genotype of *aa* will have the abnormal phenotype, albinism, since there is no dominant gene to mask the effects of the two recessive genes.

In the example of albinism, a person with the heterozygous genotype of *Aa* is said to be a genetic **carrier** of albinism. This means that the person can transmit the albinism gene, *a*, to offspring. Thus two normal parents, each having the heterozygous genotype *Aa*, can produce both normal children and children who have albinism (Figure 33-4).

Codominance

What happens if two different dominant genes occur together? Suppose there is a gene A^1 for light skin and a gene A^2 for dark skin. In a form of dominance called **codominance,** they will simply have equal effects, and a person with the heterozygous genotype A^1A^2 will exhibit a phenotype of skin color that is something between light and dark. Recall from Chapter 16 (p. 461) that the genes for sickle-cell anemia behave this way. A person with two sickle-cell genes will have *sickle cell anemia*, whereas a person with one normal gene and one sickle cell gene will have a milder form of the disease called *sickle cell trait*.

The case of sickle cell inheritance is a good example of how the mechanism of codominance works. The hemoglobin molecules within all red blood cells (RBCs) are quaternary proteins that are each comprised of four polypeptide chains—two alpha chains and two beta chains (see Figure 16-4, p. 446). The sickle cell gene is actually an abnormal version of the gene that contains the code for the beta chains of the hemoglobin molecule. Any beta chain that is produced by this code has one (of 146) amino acids replaced by the wrong amino acid—making it different enough to drastically alter its function. The RBCs of a person with one sickle cell gene and one normal beta-chain gene contain hemoglobin in which about half the total number of beta chains are abnormal and about half are normal. The RBCs of a person with two sickle cell genes contain hemoglobin in which all of the beta chains are abnormal. Thus in sickle cell trait only some of the hemoglobin molecules are defective but in sickle cell anemia *all* of the hemoglobin molecules are defective.

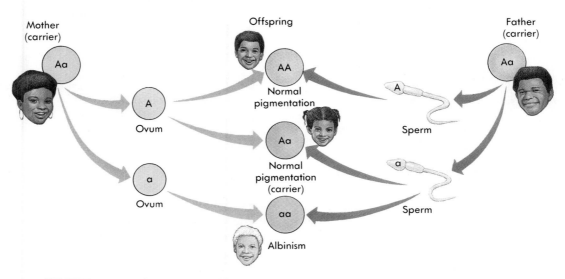

FIGURE 33-4   Inheritance of albinism. Albinism is a recessive trait, producing abnormalities only in those with two recessive genes (*a*). Presence of the dominant gene (*A*) prevents albinism.

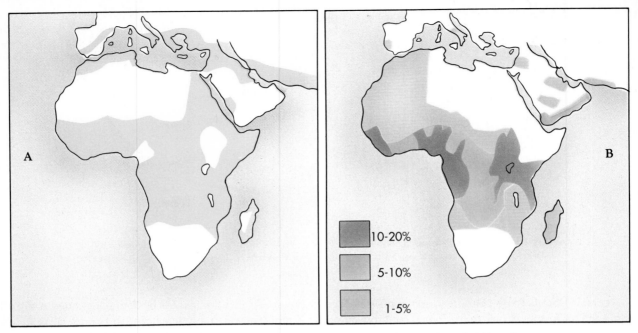

Distribution of *falciparum* malaria   Frequency of sickle cell gene

FIGURE 33-5   Relationship between the frequency of sickle cell trait and the distribution of malaria. A, The distribution of the most deadly form of malaria in Africa correlates closely with **(B)** the frequency of occurrence of the sickle cell gene. The sickle cell trait provides resistance to malaria, and thus, heterozygous individuals are more likely than homozygous individuals to survive and reproduce—spreading the abnormal gene further in the population.

The frequency of occurrence of the abnormal sickle cell gene is an example of an interesting epidemiological phenomenon. Because sickle cell trait provides resistance to the parasite that causes the most deadly form of malaria (*falciparum malaria*), sickle cell disorders persist in areas of the world in which malaria is still prevalent (Figure 33-5). Malaria is a sometimes fatal condition caused by blood-cell parasites (*Plasmodium* species) and is characterized by fever, anemia, swollen spleen, and possible relapse months or years later. The unique distribution of falciparum malaria results from the fact that people without sickle cell trait more often die of this condition before producing offspring than those with the malaria-resistant sickle cell trait. Thus the "bad" sickle cell gene is more likely to be transmitted to the next generation than the "good" genes for normal hemoglobin. The sickle cell/malaria relationship points to an important concept in medical genetics: "disease" genes often provide some biological advantage for a human population in certain circumstances. It is only when circumstances change that the gene is seen to do more harm than good. Genes for many other hereditary diseases (for example, *thalassemia* and *Tay-Sachs disease*) are now known to impart protection against pathogenic conditions in heterozygous individuals.

Sex-Linked Traits

Recall from our earlier discussion that besides the 22 pairs of autosomes, there is one pair of sex chromosomes. Notice in the lower right portion of the photograph in

 Timing and Sex Determination

Research has shown that X-bearing sperm swim more slowly than do Y-bearing sperm. An interesting hypothesis stems from this fact. If insemination occurs on the day of ovulation, Y-bearing sperm, being faster than X-bearing sperm, should reach the ovum first. Therefore a Y-bearing sperm would be more apt to fertilize the ovum. And since Y-bearing sperm produce males, there should be a greater probability of having a baby boy when insemination occurs on the day of ovulation. Statistical evidence supports this view.

Figure 33-1 that the chromosomes of this pair do not have matching structures. The larger sex chromosome is called the X *chromosome,* and the smaller one is called the Y *chromosome.* The X chromosome is sometimes called the "female chromosome" because it includes genes that determine female sexual characteristics. If a person has only X chromosomes, she is genetically a female. The Y chromosome is often called the "male chromosome" because anyone possessing a Y chromosome is genetically a male. Thus all normal females have the sex chromosome combination XX, and all normal males have the combination XY. Because men produce both X-bearing and Y-bearing sperm, any two parents can produce either male or female children (see Figure 33-6).

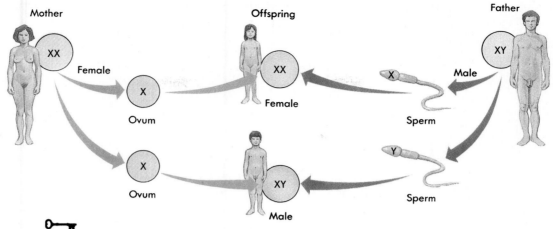

FIGURE 33-6 **Sex determination.** The presence of the Y chromosome specifies maleness. In the absence of a Y chromosome, an individual develops into a female.

The large X chromosome contains many genes besides those needed for female sexual traits. Genes for producing certain clotting factors, photopigments in the retina of the eye, and many other proteins are also found on the X chromosome. The tiny Y chromosome, on the other hand, contains few genes other than those that determine male sexual characteristics. Thus both males and females need at least one normal X chromosome—otherwise genes for clotting factors and other essential proteins would be missing. Nonsexual traits thus carried on sex chromosomes are called **sex-linked traits.** Most sex-linked traits are called *X-linked traits* because they are determined by genes in the large X chromosome.

Dominant X-linked traits appear in each person, as one would expect for any dominant trait. In females, recessive X-linked genes are masked by dominant genes in the other X chromosome. Only females with two recessive X-linked genes can exhibit the recessive trait. Since males inherit only one X chromosome (from the mother), the presence of only one recessive X-linked gene is enough to produce the recessive trait. In short, in males there are no matching genes in the Y chromosome to mask recessive genes in the X chromosome. For this reason, X-linked recessive traits appear much more frequently in males than in females.

An example of a recessive X-linked condition is *red-green color blindness,* which involves a deficiency of photopigments in the retina (see Chapter 14, p. 393). In this condition, male children of a parent who carries the recessive abnormal gene on an X chromosome may be color-blind (Figure 33-7). A female can inherit this form of

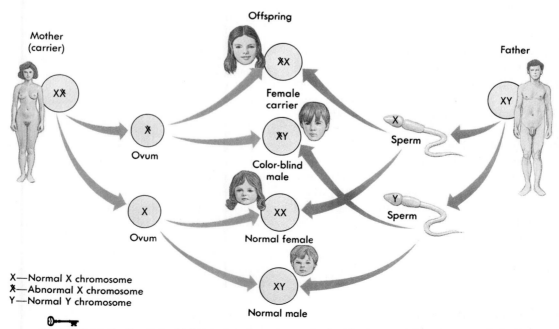

X—Normal X chromosome
✗—Abnormal X chromosome
Y—Normal Y chromosome

FIGURE 33-7 **Sex-linked inheritance.** Some forms of color blindness involve recessive X-linked genes. In this case, a female carrier of the abnormal gene can produce male children who are color blind.

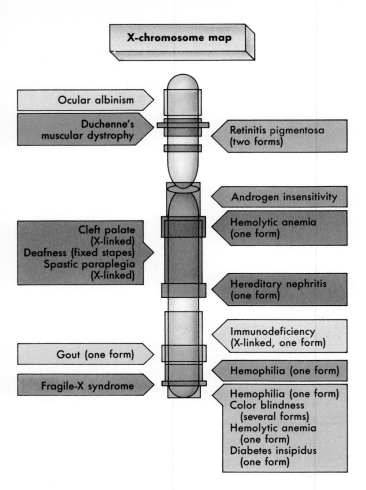

X-chromosome map

Ocular albinism

Duchenne's muscular dystrophy

Retinitis pigmentosa (two forms)

Androgen insensitivity

Hemolytic anemia (one form)

Cleft palate (X-linked)
Deafness (fixed stapes)
Spastic paraplegia (X-linked)

Hereditary nephritis (one form)

Immunodeficiency (X-linked, one form)

Gout (one form)

Hemophilia (one form)

Fragile-X syndrome

Hemophilia (one form)
Color blindness (several forms)
Hemolytic anemia (one form)
Diabetes insipidus (one form)

FIGURE 33-8   Map of the X chromosome. This diagram shows only a few of the many regions of the X chromosome that have been identified to contain specific bits of genetic information.

color blindness only if her father is color-blind *and* her mother is either color-blind (homozygous recessive) or a color-blindness carrier (heterozygous). The X chromosome has been studied in great detail, and general locations for genes causing at least 59 distinct X-linked diseases have been identified (Figure 33-8).

Genetic Mutations

The term *mutation* simply means "change." A **genetic mutation** is a change in the genetic code. Mutations may occur spontaneously without the influence of factors outside the DNA itself. However, most genetic mutations are believed to be caused by **mutagens**—agents that cause mutations. Genetic mutagens include chemicals, some forms of radiation, and even viruses. Some mutations involve a change in the genetic code within a single gene, perhaps a slight rearrangement of the nucleotide sequence. Other mutations involve damage to a portion of a chromosome or a whole chromosome. For example, a portion of a chromosome may completely break away.

Beneficial mutations allow organisms to adapt to their environments. Because such mutant genes benefit survival, they tend to spread throughout a population over the course of several generations. Harmful mutations inhibit survival and therefore are not likely to spread through the population. Most harmful mutations kill the organism in which they occur or at least prevent successful reproduction—and so are never passed to offspring. If a harmful mutation is only mildly harmful, it may persist in a population over many generations. Consider also the case of the mutations that cause sickle-cell anemia, thalassemia, and Tay-Sachs disease—the heterozygous genotype produces a phenotype that resists a specific disease, and the homozygous genotype produces a phenotype characterized by a different disease condition.

QUICK CHECK

✔ 1. What is a dominant *genetic trait*? A recessive trait?
✔ 2. What is codominance?
✔ 3. How can a mutant gene benefit a human population?
✔ 4. What is X-linked inheritance?

MEDICAL GENETICS
Mechanisms of Genetic Disease
Nuclear inheritance

Genetic diseases are diseases produced by an abnormality in the genetic code. Many genetic diseases are caused by individual mutant genes in the nuclear DNA that is passed from one generation to the next—making them *single-gene* diseases. In single-gene diseases, the mutant gene may make an abnormal product that causes disease or it may fail to make a product required for normal function. As discussed previously, some disease conditions result from the combined effects of inheritance and environmental factors. Since they are not solely caused by genetic mechanisms, such conditions are not genetic diseases in the usual sense of the word—they are instead said to involve a *genetic predisposition.*

Some genetic diseases do not result from an abnormality in a single gene. Instead, these diseases result from chromosome breakage or from the abnormal presence or absence of entire chromosomes. For example, a condition called **trisomy** (TRY-so-mee) may occur where there is a triplet of autosomes rather than a pair. Trisomy results from a mistake in meiosis called **nondisjunction** when a pair of

 Congenital Disorders

A congenital disorder is any pathological condition present at birth. As explained in the previous chapter, congenital disorders may have a genetic cause. For example, one form of a facial deformity known as *cleft palate* is an X-linked inherited condition (Figure 33-8). However, some congenital disorders are not inherited. For example, *fetal alcohol syndrome* is a group of congenital deformities that result from exposing a developing fetus to alcohol consumed by the mother (see Chapter 17, p. 489). Thus not all congenital disorders are inherited disorders.

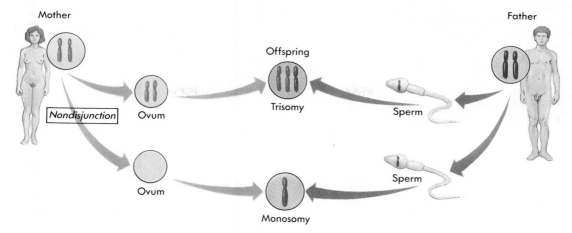

FIGURE 33-9 Effects of nondisjunction. Nondisjunction, failure of a chromosome pair to separate during gamete production, may result in trisomy or monosomy in the offspring.

chromosomes fails to separate. This produces a gamete with two autosomes that are "stuck together" instead of the usual one. When this abnormal gamete joins with a normal gamete to form a zygote, the zygote has a triplet of autosomes (Figure 33-9). Trisomy of any autosome pair is usually fatal. However, if trisomy occurs in autosome pair 13, 15, 18, 21, or 22, a person may survive for a time—but with profound developmental defects. **Monosomy** (MAHN-o-so-mee), the presence of only one autosome instead of a pair, may also result from conception involving a gamete produced by nondisjunction (Figure 33-9). Like trisomy, monosomy may produce life-threatening abnormalities. Since most trisomic and monosomic individuals die before they can reproduce, these conditions are not usually passed from generation to generation. Trisomy and monosomy are congenital conditions that are sometimes referred to as *chromosomal* genetic diseases.

Mitochondrial inheritance

Mitochondria are tiny, bacteria-like organelles present in every cell of the body (see Figure 3-6, p. 71). Like a bacterium, each mitochondrion has its own DNA molecule, sometimes called **mitochondrial DNA (mDNA).** Inheritance of mDNA occurs only through one's mother because the sperm does not contribute any mitochondria to the ovum during fertilization. Because mDNA contains the only genetic code for several important enzymes, it has the potential for carrying mutations that produce disease. Mitochondrial inheritance is now known to transmit genes for several degenerative nerve and muscle disorders. One such disease is *Leber's hereditary optic neuropathy.* In this disease, young adults begin losing their eyesight as the optic nerve degenerates—resulting in total blindness by age 30. Some medical researchers believe that *Parkinson's disease,* which is a nervous disorder that produces characteristic muscle tremors, may also be inherited through mitochondrial DNA.

QUICK CHECK

✔ 1. *How are single-gene diseases different from chromosomal conditions?*

✔ 2. *What is nondisjunction? How can it cause trisomy?*

✔ 3. *What is mitochondrial inheritance?*

Single-Gene Diseases

There are many examples of single-gene diseases. Only a few of the more important single-gene diseases are discussed here and summarized in Table 33-1.

Cystic fibrosis (CF), briefly mentioned in Chapter 3 (p. 93), is caused by recessive genes in chromosome pair 7. The primary effect of the recessive genes is impairment of chloride ion transport across cell membranes. Disruption of chloride transport causes exocrine cells to secrete thick mucus and concentrated sweat. The thickened mucus is especially troublesome in the GI and respiratory tracts, where it can cause obstruction that leads to death. This condition is often treated by the continuous use of drugs and other therapies that relieve the symptoms. CF occurs almost exclusively among caucasians.

Phenylketonuria (feen-il-kee-toe-NOO-ree-ah) **(PKU)** is caused by recessive genes that fail to produce the enzyme *phenylalanine hydroxylase.* This enzyme is needed to convert the amino acid phenylalanine into another amino acid, tyrosine. Thus phenylalanine absorbed from ingested food accumulates in the body—resulting in the abnormal presence of phenylketone in the urine (hence the name phenylketonuria). A high concentration of phenylalanine destroys brain tissue; babies born with this condition are at risk of progressive mental retardation and, perhaps, death. Many PKU victims are identified at birth by state-mandated screening tests. Once identified, PKU victims are put on diets low in phenylalanine—thus avoiding a toxic accumulation of this amino acid. You may be familiar with the printed warning for phenylketonurics commonly

TABLE 33-1   Examples of genetic conditions

Chromosome Location	Disease	Description
SINGLE-GENE INHERITANCE (NUCLEAR DNA)		
Dominant		
7, 17	Osteogenesis imperfecta	Group of connective tissue disorders is characterized by imperfect skeletal development that produces brittle bones
17	Multiple neurofibromatosis	Disorder is characterized by multiple, sometimes disfiguring benign tumors of the Schwann cells (neuroglia) that surround nerve fibers
5	Hypercholesterolemia	High blood cholesterol may lead to atherosclerosis and other cardiovascular problems
4	Huntington's disease (HD)	Degenerative brain disorder is characterized by chorea (purposeless movements) progressing to severe dementia and death generally by age 55
Codominant		
11	Sickle cell anemia Sickle cell trait	Blood disorder in which abnormal hemoglobin causes red blood cells (RBCs) to deform into a sickle shape; sickle cell anemia is the severe form, and sickle cell trait the milder form
11, 16	Thalassemia	Group of inherited hemoglobin disorders is characterized by production of hypochromic, abnormal RBCs
Recessive (autosomal)		
7	Cystic fibrosis (CF)	Condition is characterized by excessive secretion of thick mucus and concentrated sweat, often causing obstruction of gastrointestinal or respiratory ducts
15	Tay-Sachs disease	Fatal condition in which abnormal lipids accumulate in the brain and cause tissue damage leading to death by age 4
12	Phenylketonuria (PKU)	Excess of phenylketones in the urine is caused by accumulation of phenylalanine in the tissues; it may cause brain injury and death if phenylalanine (amino acid) intake is not managed properly
11	Albinism (total)	Lack of the dark brown pigment *melanin* in the skin and eyes results in vision problems and susceptibility to sunburn and skin cancer
20	Severe combined immune deficiency (SCID)	Failure of the lymphocytes to develop properly causes failure of the immune system's defense of the body; it is usually caused by adenosine deaminase (ADA) deficiency
Recessive (X-linked)		
23 (X)	Hemophilia	Group of blood clotting disorders is caused by failure to form clotting factors VIII, IX, or XI
23 (X)	Duchenne's muscular dystrophy (DMD)	Muscle disorder is characterized by progressive atrophy of skeletal muscle without nerve involvement
23 (X)	Red-green color blindness	Inability to distinguish red and green light results from a deficiency of photopigments in the cone cells of the retina
23 (X)	Fragile X syndrome	Mental retardation results from breakage of X chromosome in males

Continued

TABLE 33-1   Examples of genetic conditions—cont'd

Chromosome Location	Disease	Description
Recessive (X-linked)—cont'd		
23 (X)	Ocular albinism	Form of albinism in which the pigmented layers of the eyeball lack melanin results in hypersensitivity to light and other problems
23 (X)	Androgen insensitivity	Inherited insensitivity to androgens (steroid sex hormones associated with maleness) results in reduced effects of these hormones
23 (X)	Cleft palate (X-linked form)	One form of a congenital deformity in which the skull fails to develop properly, it is characterized by a gap in the palate (plate separating mouth from nasal cavity)
23 (X)	Retinitis pigmentosa	Condition causes blindness, characterized by clumps of melanin in retina of eyes
SINGLE-GENE INHERITANCE (MITOCHONDRIAL DNA)		
mDNA	Leber's hereditary optic neuropathy	Optic nerve degeneration in young adults results in total blindness by age 30
mDNA	Parkinson's disease (?)	Nervous disorder is characterized by involuntary trembling and muscle rigidity
CHROMOSOMAL ABNORMALITIES		
Trisomy		
21	Down's syndrome	Condition is characterized by mental retardation and multiple structural defects
23	Klinefelter's syndrome	Condition is caused by the presence of two or more X chromosomes in a male (XXY); it is characterized by long legs, enlarged breasts, low intelligence, small testes, sterility, and chronic pulmonary disease
Monosomy		
23	Turner's syndrome	Condition is caused by monosomy of the X chromosome (XO); it is characterized by immaturity of sex organs (causing sterility), webbed neck, cardiovascular defects, and learning disorders

seen on products that contain aspartame (Nutrasweet) or other substances made from phenylalanine.

Tay-Sachs (tay saks) **disease** is a recessive condition involving failure to make an essential lipid-processing enzyme. Abnormal lipids accumulate in the brain tissue of Tay-Sachs victims causing severe retardation and death by age 4. There is currently no specific therapy for this condition. Tay-Sachs disease is most prevalent among certain Jewish populations. Some epidemiologists believe that this ethnic distribution is related to the hypothesis that heterozygous carriers of the Tay-Sachs gene have a higher than normal resistance to tuberculosis (TB)—a potentially fatal disease that once killed millions in the crowded Jewish ghettos of many large cities. Residents of these TB-infested areas who carried the Tay-Sachs gene apparently survived longer—and reproduced more frequently—than noncarriers.

Osteogenesis (os-tee-o-JEN-es-is) **imperfecta** is a dom-inant genetic disorder of connective tissues. Its name, which means "imperfect bone formation," describes its chief characteristic. The bones of people with osteogenesis imperfecta are often so brittle that the slightest trauma can result in serious fractures. There are many different forms of the disease. In its most severe form, it results in fractures of the fetal skeleton in utero—often progressing to death shortly after birth. In a form seen in infancy, this disease is characterized by short, deformed limbs, a thin, enlarged skull, and easily fractured bones (Figure 33-10). In a less severe form, symptoms appear when a child begins to walk and become milder until after puberty—when the symptoms usually disappear.

Multiple neurofibromatosis is a dominant inherited disorder previously discussed in Chapter 11 (p. 303). This disorder, sometimes called *Elephant Man's disease*, is characterized by multiple, sometimes disfiguring, benign tumors of the glial cells that surround nerve fibers.

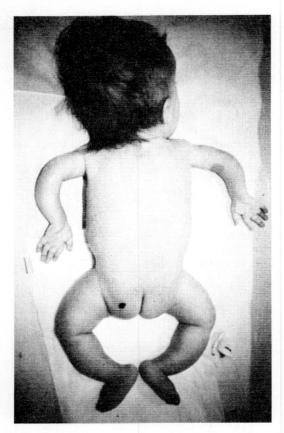

FIGURE 33-10 Osteogenesis imperfecta. The infantile form of this inherited disease is characterized by imperfect bone development that results in the appearance of this child: curved, brittle bones in the limbs and a thin, enlarged skull.

Other important inherited disorders include Duchenne's muscular dystrophy (DMD), hypercholesterolemia, sickle cell anemia, albinism, certain forms of hemophilia, and Huntington's disease. These and other conditions are summarized in Table 33-1.

Chromosomal Diseases

As described earlier, some genetic disorders are not inherited in the usual sense but result from nondisjunction during formation of the gametes. As Figure 33-9 shows, nondisjunction results in gametes that produce either trisomy or monosomy in the cells of offspring. A few of these chromosomal disorders are summarized here and in Table 33-1.

The most well-known chromosomal disorder is *trisomy 21*, which produces a group of symptoms called **Down's syndrome.** As Figure 32-11, A, shows, in this condition there is a triplet of chromosomes 21 rather than the usual pair. In the general population, trisomy 21 occurs in only 1 of every 600 or so live births. After age 35, however, a mother's chances of producing a trisomic child increase dramatically—to as high as 1 in 80 births by age 40. One recent hypothesis that explains this phenomenon states that as a woman ages, her reproductive system becomes less likely to spontaneously abort abnormal blastocysts that have implanted in the endometrium. Thus nondisjunction may occur equally among young and middle-aged women.

Down's syndrome results from trisomy 21 and rarely from other genetic abnormalities (which can be inherited). This syndrome is characterized by mental retardation (ranging from mild to severe) and multiple defects that include

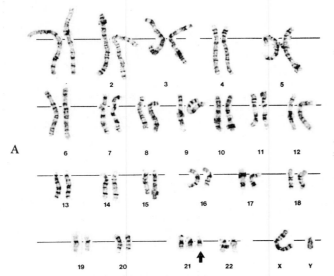

A

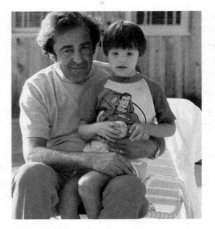

B

FIGURE 33-11 Down's syndrome. A, Down's syndrome is usually associated with trisomy of chromosome 21. **B,** A child with Down's syndrome sitting on his father's knee. Notice the distinctive anatomical features: "oriental" folds around the eyes, flattened nose, round face, and small hands with short fingers.

FOCUS ON... MATERNAL AND PATERNAL GENES

According to principles based on Mendel's work, inheritance of a dominant gene for a disease causes that disease regardless of whether the gene originally came from the mother or from the father. However, many geneticists now believe that maternal genes can be temporarily *imprinted* or chemically marked differently than paternal genes. Inherited maternal imprints are changed to the paternal form when males produce spermatozoa. Like-

wise, all paternal imprints inherited by a woman are changed to maternal imprints when she produces ova. The concept of maternal and paternal imprinting is summarized in the figure.

Imprinting may explain why some dominant inherited disorders have different forms in different people—it depends on whether the abnormal gene was inherited from the mother or from the father. For example, 90% of Huntington's dis-

ease (HD) victims inherit the dominant abnormal gene from their fathers. These HD victims show symptoms rather late in adulthood. The remaining 10% of those who have this disease inherited the abnormal gene from their mothers—and experience symptoms early in childhood. Thus the form of Huntington's disease one experiences depends on whether the abnormal gene is from the mother or from the father.

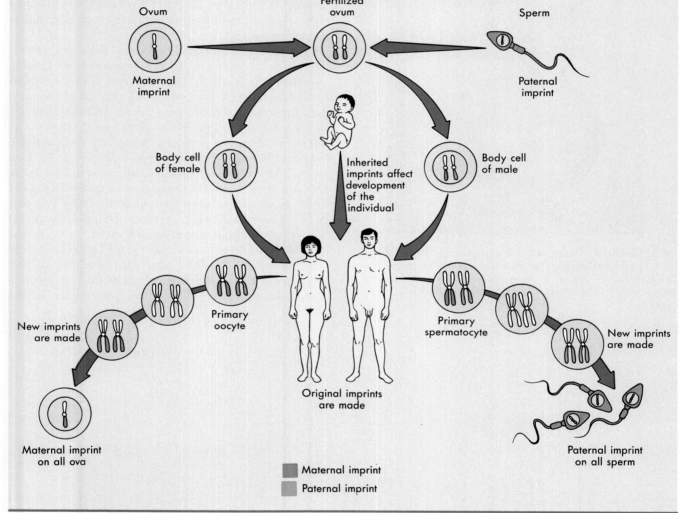

distinctive facial appearance (Figure 33-11, *B*), enlarged tongue, short hands and feet with stubby digits, congenital heart disease, and susceptibility to acute leukemia. People with Down's syndrome have a shorter than average life expectancy but can survive to old age.

Klinefelter's syndrome is another genetic disorder resulting from nondisjunction of chromosomes. This disorder occurs in males with a Y chromosome and at least two X chromosomes, typically the XXY genotype. Characteristics of Klinefelter's syndrome include long legs, enlarged

breasts, low intelligence, small testes, sterility, and chronic pulmonary disease (Figure 33-12).

Turner's syndrome, sometimes called *XO syndrome,* occurs in females with a single sex chromosome, X. Like the conditions described above, it results from nondisjunction during gamete formation. Turner's syndrome is characterized by failure of the ovaries and other sex organs to mature (causing sterility), cardiovascular defects, dwarfism or short stature, a webbed neck, and possible learning disorders (Figure 33-13). Symptoms of Turner's syndrome

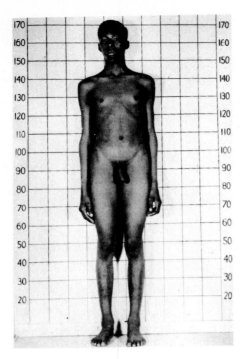

FIGURE 33-12 Klinefelter's syndrome. This young man exhibits many of the characteristics of Klinefelter's syndrome: small testes (but normal sized genitals), some development of the breasts, sparse body hair, and long limbs. This syndrome results from the presence of two or more X chromosomes with a Y chromosome (genotypes XXY or XXXY, for example).

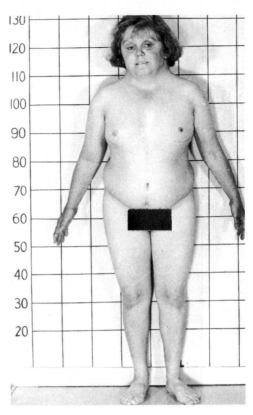

FIGURE 33-13 Turner's syndrome. This woman exhibits many of the characteristics of Turner's syndrome, including short stature, webbed neck, and sexual immaturity. Turner's syndrome results from monosomy of sex chromosomes (genotype XO).

can be reduced by hormone therapy using estrogens and growth hormone. Cardiovascular defects may be repaired surgically.

Genetic Basis of Cancer

Recall from Chapter 4 that some forms of cancer are thought to be caused, at least in part, by abnormal genes called **oncogenes.** One hypothesis states that most normal cells contain such cancer-causing genes. However, it is uncertain how these genes become activated and produce cancer. Perhaps oncogenes can transform a cell into a cancer cell only when certain environmental conditions occur.

Another hypothesis states that normal cells contain another class of genes, **tumor suppressor genes.** According to this hypothesis, such genes regulate cell division so that it proceeds normally. When a tumor suppressor gene is nonfunctional because of a genetic mutation, it then allows cells to divide abnormally—possibly producing cancer.

Yet another possible genetic basis for cancer relates to the genes that govern the cell's ability to repair damaged DNA. For example, a rare genetic disorder called *xeroderma pigmentosum* (see Chapter 5, p. 136) is characterized by the inability of skin cells to repair genetic damage caused by the ultraviolet (UV) radiation in sunlight. Individuals with this condition nearly always develop skin cancer when exposed to direct sunlight. In this condition, the genetic abnormality does not cause skin cancer directly but inhibits the cell's cancer-preventing mechanisms.

Cancer researchers are now working intensely to determine the exact role various genes play in the development of cancer. The more we understand the genetic basis of cancer, the more likely it is that we will find effective treatments—or even cures.

QUICK CHECK

✔ 1. *How does avoidance of phenylalanine in the diet reduce the problems associated with phenylketonuria (PKU)?*
✔ 2. *Briefly describe the mechanism of Tay-Sachs disease.*
✔ 3. *What is trisomy 21?*
✔ 4. *How might the genetic code be involved in the development of cancer?*

PREVENTION AND TREATMENT OF GENETIC DISEASES

Genetic Counseling

The term *genetic counseling* refers to professional consultations with families regarding genetic diseases. Trained genetic counselors may help a family determine the risk of producing children with genetic diseases. Parents with a high risk of producing children with genetic disorders may decide to avoid having children of their own.

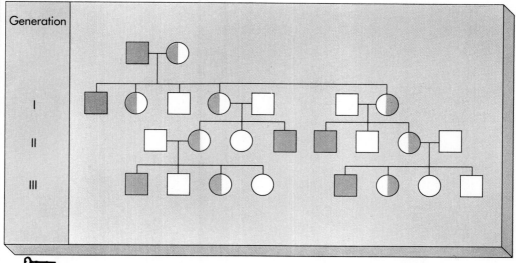

FIGURE 33-14 Pedigree. Pedigrees chart the genetic history of family lines. Squares represent males, and circles represent females. Fully shaded symbols indicate affected individuals, partly shaded symbols indicate carriers, and unshaded symbols indicate unaffected noncarriers. Roman numerals indicate the order of generations. This pedigree reveals the presence of an X-linked recessive trait.

Genetic counselors may also help evaluate whether any offspring already have a genetic disorder and offer advice on treatment or care. A growing list of tools is available to genetic counselors, some of which are described in the following.

Pedigree

A **pedigree** is a chart that illustrates genetic relationships in a family over several generations (Figure 33-14). Using medical records and family histories, genetic counselors assemble the chart beginning with the client and moving backward through as many generations as are known. Squares represent males; circles represent females. Fully shaded symbols represent affected individuals, and unshaded symbols represent normal individuals. Partially shaded symbols represent carriers of a recessive trait. A horizontal line between symbols designates a sexual relationship that produced offspring.

The pedigree is useful in determining the possibility of producing offspring with certain genetic disorders. It also may tell a person whether he or she might have a genetic disorder that appears late in life, such as Huntington's disease. In either case, a family can prepare emotionally, financially, and medically before a crisis occurs.

Punnett square

The **Punnett** (PUN-et) **square,** named after the English geneticist Reginald Punnett, is a grid used to determine the mathematical *probability* of inheriting genetic traits. As Figure 33-15, A, shows, genes in the mother's gametes are represented along the horizontal axis of the grid and genes in the father's gametes along the vertical axis. The ratio of different gene combinations in the offspring predict their probability of occurrence in the next generation. Thus offspring produced by two carriers of PKU (a recessive disorder) have a one in four (25%) chance of inheriting this recessive condition (Figure 33-15 A). The same grid shows that there is a two in four (50%) chance that a child produced will be a PKU carrier. Figure 33-15, B, however, shows that offspring between a carrier and a noncarrier cannot inherit PKU. What is the chance of an individual offspring being a PKU carrier in this case? The grid in Figure 33-15, C, shows the probability of producing an affected offspring when a PKU victim and a PKU carrier have children. Figure 33-15, D, shows the genetic probability when a PKU victim and a noncarrier produce children.

Karyotype

Disorders that involve trisomy (extra chromosomes), monosomy (missing chromosomes), and broken chromosomes can be detected after a **karyotype** (KAIR-ee-o-type) is produced. The first step in producing a karyotype is getting a sample of cells from the individual to be tested. This can be done by scraping cells from the lining of the cheek or from a blood sample containing white blood cells (WBCs). Fetal tissue can be collected by **amniocentesis** (am-nee-o-sen-TEE-sis), a procedure in which fetal cells floating in the amniotic fluid are collected with a syringe (Figure 33-16). **Chorionic villus sampling** is a newer procedure in which cells from chorionic villi that surround a young embryo (see Chapter 32, p. 821) are collected through the opening of the cervix. Collected cells are grown in a special culture medium and allowed to reproduce. Cells in metaphase (when the chromosomes are most distinct) are stained and photographed using a

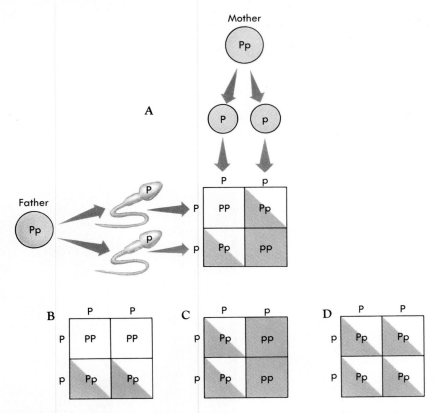

FIGURE 33-15 **Punnett square.** The Punnett square is a grid used to determine relative probabilities of producing offspring with specific gene combinations. Phenylketonuria (PKU) is a recessive disorder caused by the gene *p*. P is the normal gene. **A,** Possible results of cross between two PKU carriers. Because one in four of the offspring represented in the grid have PKU, a genetic counselor would predict a 25% chance that this couple will produce a PKU baby at each birth. **B,** Cross between a PKU carrier and a normal noncarrier. **C,** Cross between a PKU victim and a PKU carrier. **D,** Cross between a PKU victim and a normal noncarrier.

microscope. The chromosomes are cut out of the photo and pasted on a chart in pairs according to size, as in Figures 33-1 and 33-11, A. Genetic counselors then examine the karyotype, looking for chromosome abnormalities. What chromosome abnormality is visible in Figure 33-11, A? Is this a male or female karyotype?

QUICK CHECK

✔ 1. *What is genetic counseling?*
✔ 2. *How are pedigrees used by genetic counselors?*
✔ 3. *How is a Punnett square used to predict mathematical probabilities of inheriting specific genes?*
✔ 4. *How is a karyotype prepared? What is its purpose?*

FIGURE 33-16 **Amniocentesis.** In amniocentesis, a syringe is used to collect amniotic fluid. Ultrasound imaging is used to guide the tip of the syringe needle to prevent damage to the placenta and fetus. Fetal cells in the collected amniotic fluid can then be chemically tested or used to produce a karyotype of the developing baby.

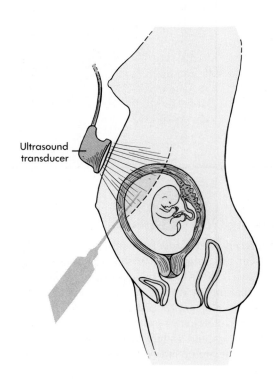

FO CUS ON... DNA ANALYSIS

As a result of the intense efforts underway to map the entire human genome, new techniques have been developed to analyze the genetic makeup of individuals. Automated machines can now chemically analyze chromosomes and "read" their sequence of nucleotide bases—the genetic code. One method by which this is done is called **electrophoresis** (el-ek-tro-fo-REE-sis), which means "electric separation" (Figure A). In electrophoresis, DNA fragments are chemically processed, then placed in a thick fluid or gel. An electric field in the gel causes the DNA fragments to separate into groups according to their relative sizes. The resulting pattern (Figure B) represents the sequence of codons in the DNA fragment. This process is also the basis for so-called DNA *fingerprinting*. Like a fingerprint pattern, each person's DNA sequence is unique. Once the exact sequences for specific diseases have been discovered, genetic counselors will be able to provide more details about the genetic makeup of their clients.

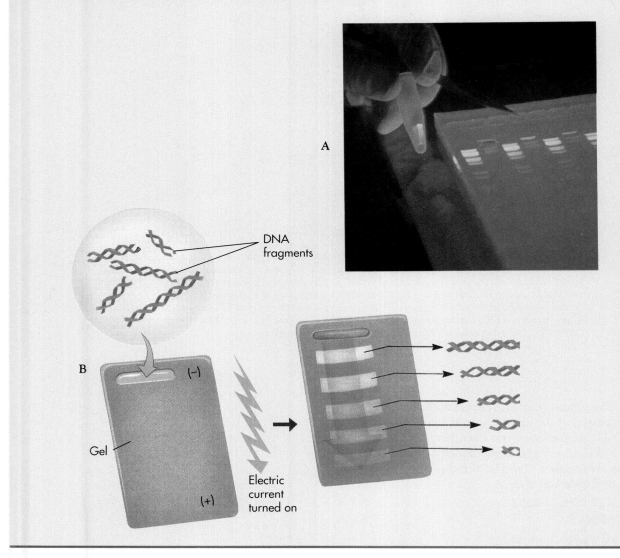

A

DNA fragments

B

Gel

(−)

(+)

Electric current turned on

Treating Genetic Diseases

Until recently, the only hope of treating genetic diseases was to treat the symptoms. In some diseases such as PKU this works well. If PKU victims simply avoid large amounts of phenylalanine in their diets, especially during critical stages of development, severe complications can be avoided. In Klinefelter's syndrome and Turner's syndrome, hormone therapy and surgery can alleviate some symptoms. However, there are no effective treatments for the majority of genetic disorders. Fortunately, medical science now offers us some hope of treating genetic disorders through **gene therapy.**

In a therapy called **gene replacement,** genes that specify production of abnormal, disease-causing proteins are replaced by normal or "therapeutic" genes. To get the therapeutic genes to cells that need them, researchers have proposed using genetically altered viruses as carriers. Recall from Chapter 1 and Chapter 4 that viruses are easily

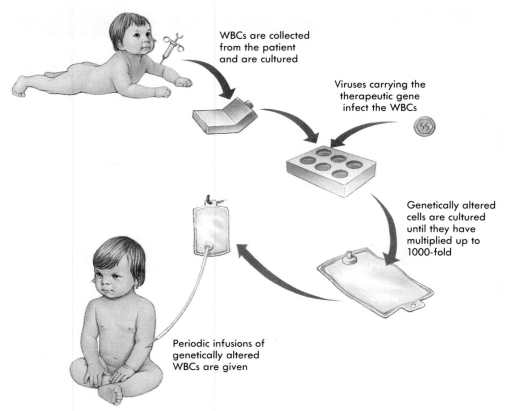

WBCs are collected
from the patient
and are cultured

Viruses carrying the
therapeutic gene
infect the WBCs

Genetically altered
cells are cultured
until they have
multiplied up to
1000-fold

Periodic infusions of
genetically altered
WBCs are given

FIGURE 33-17   Gene therapy. This method of gene augmentation therapy was used to treat children stricken with a form of severe combined immune deficiency (SCID). White blood cells taken from the patient were infected with the viruses carrying the therapeutic gene. The altered cells were reproduced and injected into the blood stream, reducing the immunity inhibiting effects of SCID.

capable of inserting new genes into the human genome. If the therapeutic genes behave as expected, a cure may result. Thus the goal of gene replacement therapy is to genetically alter existing body cells in the hope of eliminating the cause of a genetic disease.

In a therapy called **gene augmentation,** normal genes are introduced with the hope that they will augment (add to) the production of the needed protein. In one form of gene augmentation, virus-altered cells are injected into the blood or implanted under the skin of a patient to produce increased amounts of the missing protein. Gene augmentation attempts to add genetically altered cells to the body, rather than change existing body cells, as in gene replacement therapy.

The use of genetic therapy began in 1990 with a group of young children having **adenosine deaminase deficiency.** In this rare recessive disorder, the gene for producing the enzyme *adenosine deaminase (ADA)* is missing from both autosomes in pair 20. Deficiency of ADA results in *severe combined immune deficiency (SCID)*, making its victims highly susceptible to infection (see Chapter 20, p. 561). As Figure 33-17 shows, white blood cells (WBCs) from each patient were collected and infected with viruses carrying

therapeutic genes. After reproducing a thousand-fold, the genetically altered WBCs were then injected into the patient. The first attempts at gene therapy in humans have met with some success but many technical problems are yet to be overcome before it can be used widely.

A gene therapy for cystic fibrosis (CF) is also being developed. Researchers propose introducing therapeutic genes in cold viruses that are sprayed into the respiratory tract of CF victims. The viruses will insert genes for normal chloride channels in the lung tissues, thus reducing the abnormally thick fluid secretions that characterize this disease. This treatment is not expected to be a one-time cure, but it should raise the average life expectancy of CF patients well beyond the current 27 years. It is too early to say for sure, but there may soon come a time when many genetic diseases are treated—or even cured—with gene therapy.

QUICK CHECK

✔ *1. How are most genetic disorders treated today?*
✔ *2. How does gene replacement therapy work?*

SCIENCE OF GENETICS

A. Genetics—scientific study of inheritance; developed to explain how normal biological characteristics are inherited
B. Directly inherited diseases are often called hereditary diseases

CHROMOSOMES AND GENES

A. Mechanism of gene function
  1. Genetic code is transmitted via genes
  2. Each gene is a sequence of nucleotide bases in the DNA molecule, which the cell transcribes to an mRNA molecule
  3. Each mRNA molecule associates with a ribosome, which translates the code to form a specific polypeptide molecule
  4. Genes determine the structure and function of the human body by producing a set of specific regulatory enzymes and other proteins
B. Distribution of chromosomes to offspring
  1. Meiosis (Figure 33-2)
    a. Produces gametes with the haploid number of chromosomes
    b. When a sperm unites with an ovum at conception, it forms a zygote with 46 chromosomes
  2. Principle of independent assortment
    a. As sperm and ovum are formed, chromosome pairs separate and the maternal and paternal chromosomes get mixed up and redistributed independently of the other chromosome pairs, resulting in each sperm having a different set of 23 chromosomes
    b. Genetic variation—independent assortment of chromosomes assures that each offspring from a single set of parents is genetically unique
    c. Applies to individual genes or groups of genes
    d. Crossing over—during one phase of meiosis, pairs of matching chromosomes line up along the equator and exchange genes with one another (Figure 33-3)
    e. Gene linkage—sometimes a whole group of genes stays together and crosses over as a single unit

GENE EXPRESSION

A. Hereditary traits
  1. Dominance
    a. Each inherited trait is controlled by two sets of similar genes, one from each parent
    b. Each autosome in a pair matches its partner in the type of gene it contains
    c. Different types of genes (Figure 33-4)
      (1) Dominant gene—effects are seen; capable of masking the effects of a recessive gene for the same trait
      (2) Recessive gene—effects are masked by the effects of a dominant gene for the same trait
    d. Genotype—combination of genes within the cells of an individual
      (1) Homozygous—genotype with two identical forms of a trait
      (2) Heterozygous—genotype with two different forms of a trait
    e. Phenotype—manner in which genotype is expressed; how an individual looks due to genotype
    f. Carrier—person who possesses the gene for a recessive trait but does not exhibit the trait
  2. Codominance—when two different dominant genes occur together, each will have an equal effect
B. Sex-linked traits (Figure 33-7)

  1. X chromosome—"female chromosome"; larger than Y chromosome; includes genes that determine female sexual characteristics, as well as nonsexual characteristics (Figure 33-8)
  2. Y chromosomes—"male chromosome"; smaller than X chromosome; contains few genes other than male sexual characteristics
  3. Sex-linked traits—nonsexual traits carried on sex chromosomes; also known as X-linked traits
C. Genetic mutations
  1. Change in the genetic code
  2. Can occur without outside influence
  3. Mutagens—agents that cause most genetic mutations

MEDICAL GENETICS

A. Mechanisms of genetic disorders
  1. Nuclear inheritance
    a. Single-gene diseases—diseases caused by individual mutant genes in nuclear DNA that pass from one generation to the next
    b. Genetic predisposition—disease occurring due to combined effects of inheritance and environmental factors
    c. Chromosomal genetic diseases—congenital conditions such as trisomy and monosomy that produce life-threatening abnormalities; trisomic and monosomic individuals die before they can reproduce (Figure 33-9)
  2. Mitochondrial inheritance
    a. Mitochondrial DNA (mDNA)—each mitochondrion has its own DNA molecule
    b. Inheritance of mDNA occurs through one's mother, since sperm does not contribute mitochondria to the ovum during fertilization
    c. mDNA contains the only genetic code for several important enzymes
B. Single-gene diseases
  1. Cystic fibrosis
    a. Results due to recessive genes in chromosome pair 7
    b. Impairment of chloride ion transport across cell membranes causes exocrine cells to secrete thick mucus and concentrated sweat; thickened mucus may obstruct respiratory and gastrointestinal tracts, leading to death
    c. Treatment—use of drugs and other therapies
  2. Phenylketonuria (PKU)
    a. Results from recessive genes that fail to produce phenylalanine hydroxylase
    b. Phenylalanine cannot be converted into tyrosine and thus accumulates
    c. High concentrations of phenylalanine destroy brain tissue
    d. Treatment—diets low in phenylalanine
  3. Tay-Sachs disease
    a. Recessive condition involving failure to make an essential lipid-processing enzyme
    b. Abnormal lipids accumulate in the brain, causing severe retardation and death by age 4
    c. No specific therapy available
  4. Osteogenesis imperfecta (Figure 33-10)
    a. Dominant genetic disorder of connective tissues resulting in imperfect bone formation
    b. Bones are very brittle
  5. Multiple neurofibromatosis
    a. Dominant inherited disorder
    b. Characterized by multiple benign tumors of glial cells that surround nerve fibers

C. Chromosomal diseases—genetic disorders resulting from nondisjunction during formation of the gametes; produce either trisomy or monosomy
  1. Trisomy 21—triplet of chromosomes 21 rather than a pair; characterized by Down's syndrome's mental retardation and multiple defects (Figure 33-11)
  2. Klinefelter's syndrome occurs in males with a Y chromosome and at least two X chromosomes; characteristics include long legs, enlarged breasts, low intelligence, small testes, sterility, and chronic pulmonary disease (Figure 33-12)
  3. Turner's syndrome— XO syndrome occurs in females with a single X chromosome; characterized by failure of ovaries and other organs to mature, sterility, cardiovascular defects, dwarfism, webbed neck, and learning disorders; symptoms can be reduced by hormone therapy (Figure 33-13)

D. Genetic basis of cancer
  1. Oncogenes—abnormal genes thought to cause some forms of cancer
  2. Tumor suppressor genes regulate cell division so it proceeds normally; when nonfunctional due to a genetic mutation, it allows cells to divide abnormally
  3. Also, genetic abnormalities may inhibit the cell's cancer-preventing mechanisms

PREVENTION AND TREATMENT OF GENETIC DISEASES

A. Genetic counseling—professional consultations with families regarding genetic diseases
  1. Pedigree—chart that illustrates genetic relationships in a family over several generations; helpful in determining the possibility of producing offspring with certain genetic disorders (Figure 33-14)
  2. Punnett square—grid used to determine the mathematical probability of inheriting genetic traits (Figure 33-15)
  3. Karyotype—ordered arrangement of photographs of chromosomes from a single cell; used in genetic counseling to identify chromosomal disorders

B. Treating genetic diseases
  1. Gene replacement—abnormal, disease-causing proteins are replaced by "therapeutic" genes; goal is to genetically alter existing body cells in the hope of eliminating the cause of a genetic disease (Figure 33-17)
  2. Gene augmentation—normal genes are introduced to augment the production of the needed protein

REVIEW QUESTIONS

1. Who was the first person to discover the basic mechanism by which traits are transmitted from parents to offspring?
2. How do enzymes regulate the entire structure and function of the body?
3. How does meiosis differ from mitosis?
4. What is meant by the principle of independent assortment?
5. Describe albinism in relation to dominance, recessiveness, and genotype.
6. Explain the difference between a genotype and a phenotype.
7. Define codominance and give an example of a condition that demonstrates codominance.
8. Compare the male chromosome and the female chromosome.
9. If a certain trait is identified as X-linked recessive, describe the genotype of a female expressing the given trait.
10. Identify several genetic mutagens.
11. How can mutations be beneficial to a species?
12. What role do environmental factors play in relation to certain genetic diseases?
13. Explain the "mistake" in meiosis that results in the condition called trisomy.
14. Describe the genetic inheritance of cystic fibrosis, phenylketonuria (PKU), Tay-Sachs disease.
15. Identify the chromosomal disorder that involves trisomy 21.
16. Differentiate between oncogenes and tumor suppressor genes.
17. Explain the medical significance of a pedigree. Do the same for a karyotype.
18. How are gene replacement and gene augmentation therapies used to treat genetic diseases?
19. Define the term *genome*.
20. When is a disorder classified as congenital?

Mini-atlas of Human Anatomy

Dissection and direct observation of the human body is perhaps the best way to study the form, placement, and relationships that exist between anatomic structures. This mini-atlas is intended as a study guide for students who are unable to participate in a cadaver dissection laboratory.

Actual dissection photographs show the highly integrated nature of the body organ systems. Regional dissection of body areas provides an opportunity to visualize the skeletal, muscular, vascular, and nervous system components as they relate to surface landmarks. In selected instances, medical imaging provides additional information in the form of MRI or x-ray visuals. Several plates show "cast" preparations of anatomic structures. In these specialized teaching aids, structures such as blood vessels and components of the respiratory tree or biliary tract branches are filled with colored latex, which remains after the surrounding tissue is removed. The result is a model of the filled structures showing anatomic detail that cannot be visualized by any other means.

Anatomy is truly a study of self. Use your own body as a teaching model. Begin with surface anatomy and then visualize internal structures using written descriptions and illustrations provided throughout the text and in the mini-atlas.

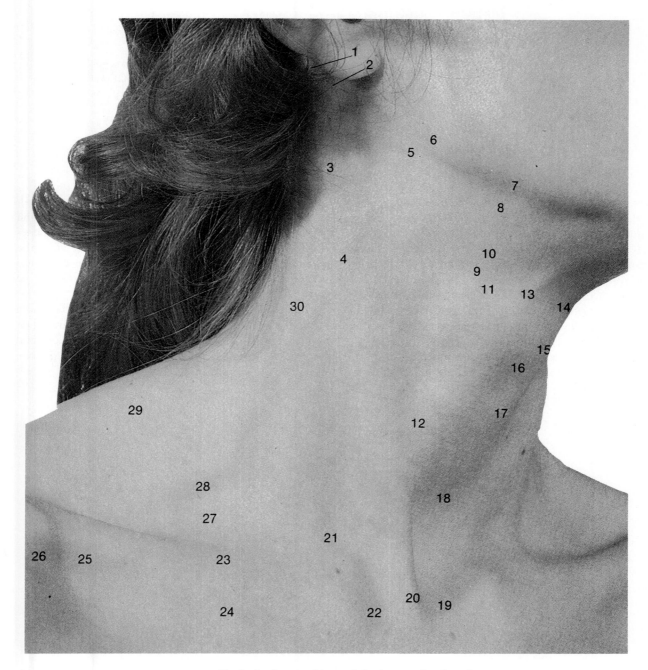

Neck. Surface markings of the front and right side.

1	Mastoid process	13	Anterior jugular vein	23	Clavicle
2	Tip of transverse process of atlas	14	Body of hyoid bone	24	Pectoralis major muscle
3	Sternocleidomastoid muscle	15	Laryngeal prominence (Adam's apple)	25	Infraclavicular fossa and cephalic vein
4	External jugular vein	16	Vocal fold	26	Deltoid muscle
5	Lowest part of parotid salivary gland	17	Arch of cricoid cartilage	27	Inferior belly of omohyoid
6	Angle of mandible	18	Isthmus of thyroid gland	28	Upper trunk of brachial plexus
7	Anterior border of masseter muscle and facial artery	19	Jugular notch and trachea	29	Accessory nerve passing under anterior border of trapezius muscle
8	Submandibular salivary gland	20	Sternal head ⎤ of sternocleido-	30	Accessory nerve emerging from sternocleidomastoid muscle
9	Tip of greater horn of hyoid bone	21	Clavicular head ⎦ mastoid muscle		
10	Hypoglossal nerve	22	Sternoclavicular joint and union of internal jugular and subclavian veins to form brachiocephalic vein		
11	Internal laryngeal nerve				
12	Site for palpation of common carotid artery				

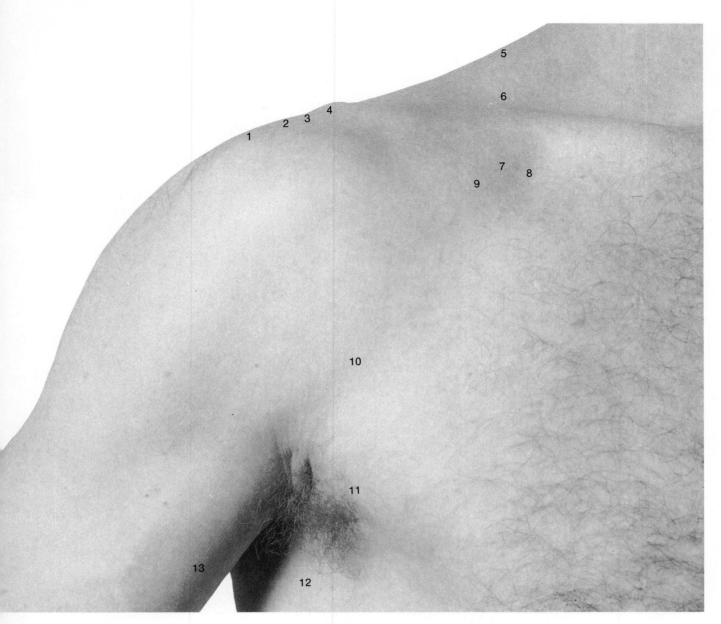

Right shoulder, from the front.

1	Deltoid muscle overlying greater tubercle of humerus	7	Infraclavicular fossa
2	Acromion	8	Upper margin of pectoralis major muscle
3	Acromioclavicular joint	9	Anterior margin of deltoid muscle
4	Acromial end of clavicle	10	Deltopectoral groove and cephalic vein
5	Trapezius muscle	11	Lower margin of pectoralis major muscle
6	Supraclavicular fossa	12	Serratus anterior muscle
		13	Biceps brachii muscle

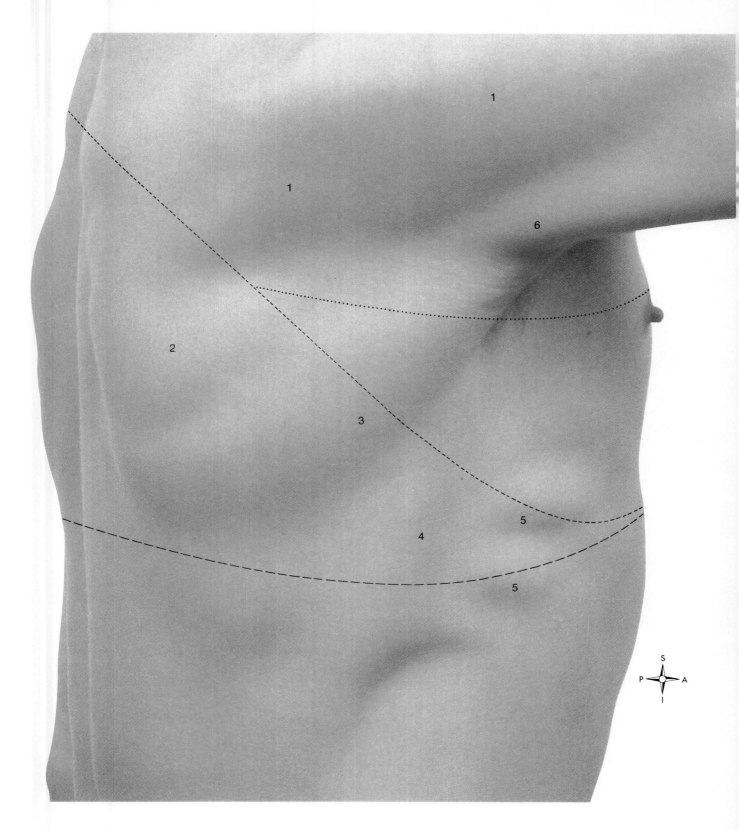

Right side of the thorax, from the right.

1   Deltoid muscle
2   Infraspinatus muscle
3   Teres major muscle
4   Latissimus dorsi muscle
5   Serratus anterior muscle
6   Long head of triceps brachii muscle

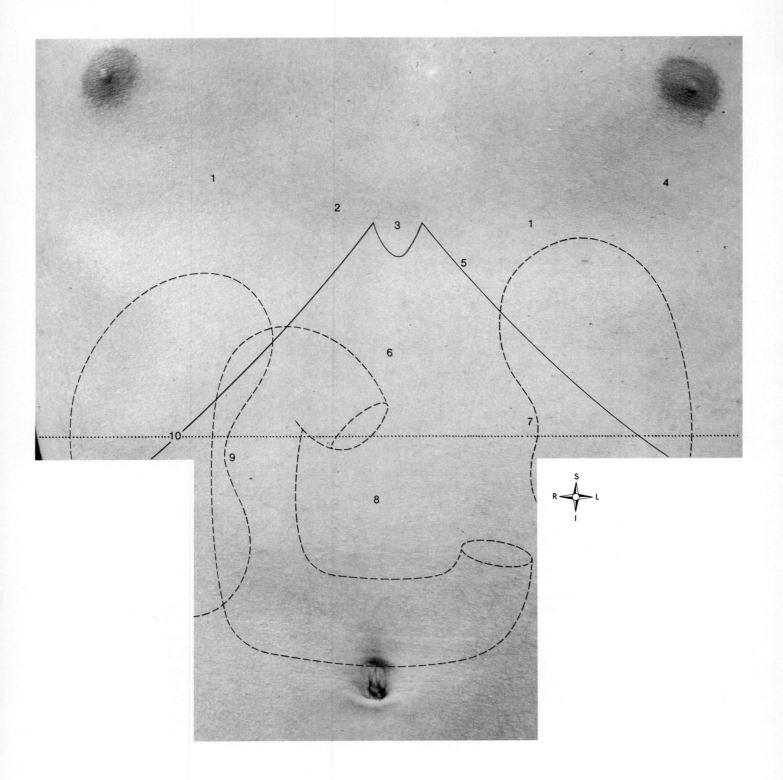

Anterior abdominal wall above the umbilicus. Surface markings.

1 Dome of diaphragm and upper
   margin of liver
2 Foramen for inferior vena cava in
   diaphragm
3 Xiphisternal joint
4 Apex of heart in fifth intercostal space
5 Esophageal ⎱ opening in
6 Aortic   ⎰ diaphragm

7 Hilum of left kidney
8 Head of pancreas and level of
   second lumbar vertebra
9 Hilum of right kidney
10 Fundus of gallbladder, and junction
   of ninth costal cartilage and lateral
   border of rectus (muscle) sheath

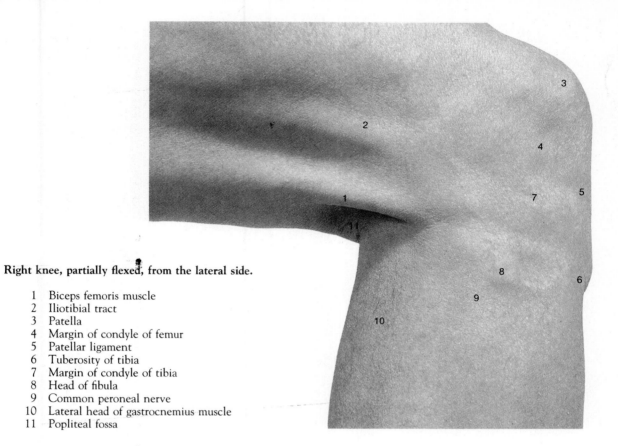

Right knee, partially flexed, from the lateral side.

1 Biceps femoris muscle
2 Iliotibial tract
3 Patella
4 Margin of condyle of femur
5 Patellar ligament
6 Tuberosity of tibia
7 Margin of condyle of tibia
8 Head of fibula
9 Common peroneal nerve
10 Lateral head of gastrocnemius muscle
11 Popliteal fossa

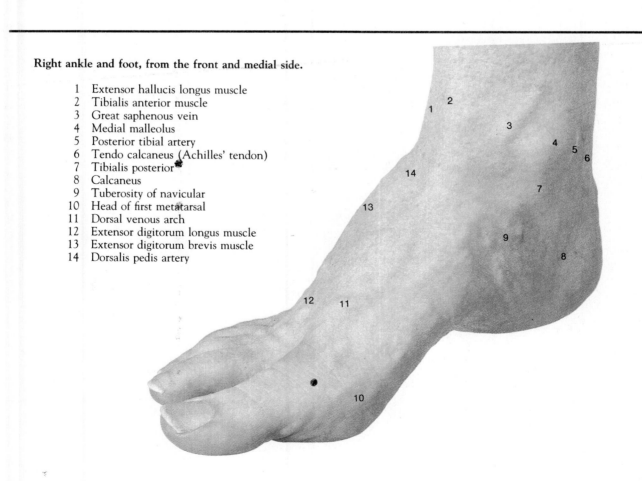

Right ankle and foot, from the front and medial side.

1 Extensor hallucis longus muscle
2 Tibialis anterior muscle
3 Great saphenous vein
4 Medial malleolus
5 Posterior tibial artery
6 Tendo calcaneus (Achilles' tendon)
7 Tibialis posterior
8 Calcaneus
9 Tuberosity of navicular
10 Head of first metatarsal
11 Dorsal venous arch
12 Extensor digitorum longus muscle
13 Extensor digitorum brevis muscle
14 Dorsalis pedis artery

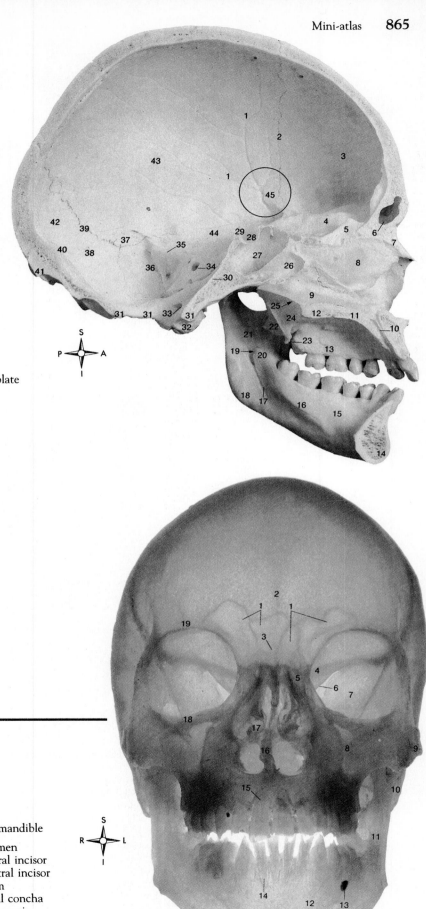

Left half of the skull. Sagittal section.

1   Grooves for middle meningeal vessels
2   Coronal suture
3   Squamous part of frontal bone
4   Orbital part of frontal bone
5   Crista galli of ethmoid bone
6   Frontal sinus
7   Nasal bone
8   Perpendicular plate of ethmoid bone
9   Vomer
10 Incisive canal
11 Palatine process of maxilla
12 Horizontal plate of palatine bone
13 Alveolar process of maxilla
14 Mental protuberance
15 Body of mandible
16 Mylohyoid line
17 Groove for mylohyoid nerve
18 Angle of mandible
19 Mandibular foramen
20 Lingula
21 Ramus of mandible
22 Lateral pterygoid plate
23 Pterygoid hamulus of medial pterygoid plate
24 Medial pterygoid plate
25 Posterior nasal aperture (choana)
26 Right sphenoidal sinus
27 Left sphenoidal sinus
28 Pituitary fossa (sella turcica)
29 Dorsum sellae
30 Clivus
31 Margin of foramen magnum
32 Occipital condyle
33 Hypoglossal canal
34 Internal acoustic meatus in petrous part of temporal bone
35 Groove for superior petrosal sinus
36 Groove for sigmoid sinus
37 Mastoid (posterior inferior) angle of parietal bone
38 Groove for transverse sinus
39 Lambdoidal suture
40 Internal occipital protuberance
41 External occipital protuberance
42 Occipital bone
43 Parietal bone
44 Squamous part of temporal bone
45 Pterion (encircled)

Skull, from the front. Cleared specimen, illuminated from behind.

1   Frontal sinus
2   Frontal crest
3   Crista galli
4   Lesser wing of sphenoid
5   Ethmoidal sinus
6   Superior orbital fissure
7   Greater wing of sphenoid
8   Maxillary sinus
9   Zygomatic arch
10 Mastoid process
11 Ramus } of mandible
12 Body
13 Mental foramen
14 Root of lateral incisor
15 Root of central incisor
16 Nasal septum
17 Inferior nasal concha
18 Infraorbital margin
19 Supraorbital margin

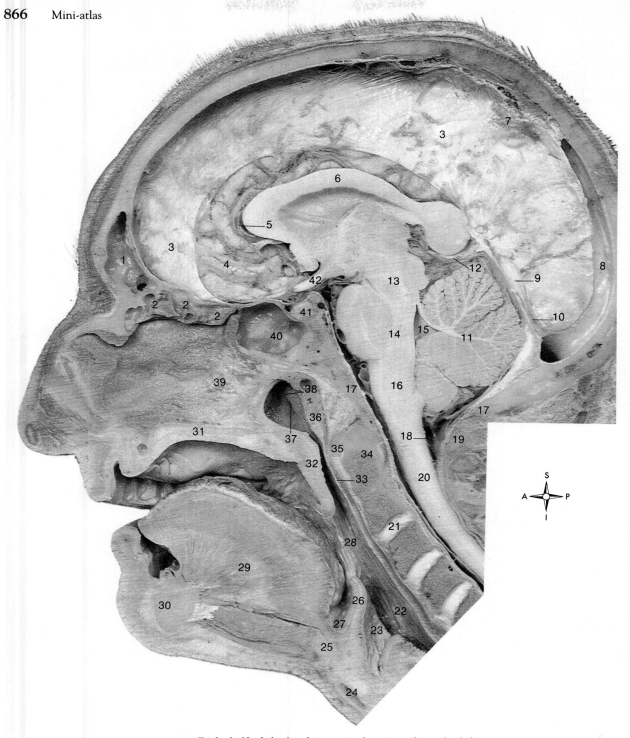

Right half of the head, in sagittal section, from the left.

1	Left frontal sinus	15	Fourth ventricle	28	Oropharynx
2	Left ethmoidal air cells	16	Medulla oblongata	29	Tongue
3	Falx cerebri	17	Margin of foramen magnum	30	Mandible
4	Medial surface of right cerebral	18	Cerebellomedullary cistern (cisterna	31	Hard palate
	hemisphere		magna)	32	Soft palate
5	Anterior cerebral artery	19	Posterior arch of atlas	33	Nasopharynx
6	Corpus callosum	20	Spinal cord	34	Dens of axis
7	Arachnoid granulations	21	Intervertebral disc between axis and	35	Anterior arch of atlas
8	Superior sagittal sinus		third cervical vertebra	36	Pharyngeal tonsil
9	Tentorium cerebelli	22	Laryngopharynx	37	Opening of auditory tube
10	Straight sinus	23	Inlet of larynx	38	Choana (posterior nasal aperture)
11	Cerebellum	24	Thyroid cartilage	39	Nasal septum
12	Great cerebral vein	25	Hyoid bone	40	Sphenoidal sinus
13	Midbrain	26	Epiglottis	41	Pituitary gland
14	Pons	27	Vallecula	42	Optic chiasma

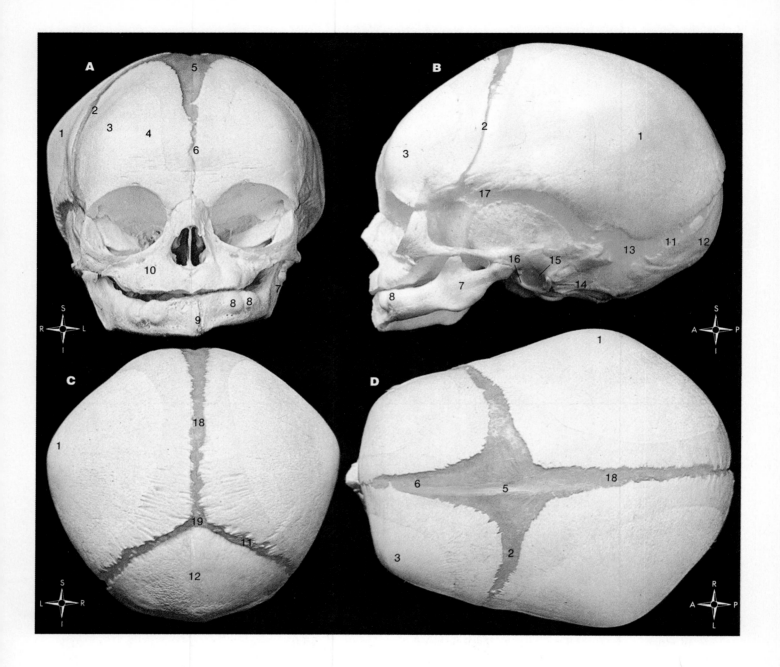

Skull of a full-term fetus. A, From the front; B, from the left and slightly below; C, from behind; D, from above.

1 Parietal tuberosity
2 Coronal suture
3 Frontal tuberosity
4 Half of frontal bone
5 Frontal (anterior) fontanel
6 Frontal suture
7 Ramus of mandible
8 Elevations over deciduous teeth in body of mandible
9 Symphysis menti

10 Maxilla
11 Lambdoid suture
12 Occipital bone
13 Mastoid (posterolateral) fontanel
14 Stylomastoid foramen
15 External acoustic meatus
16 Tympanic ring
17 Sphenoidal (anterolateral) fontanel
18 Sagittal suture
19 Occipital (posterior) fontanel

Brain, from below.

1	Orbital sulcus
2	Olfactory bulb
3	Olfactory tract
4	Gyrus rectus
5	Anterior perforated substance
6	Optic nerve
7	Optic chiasma
8	Optic tract
9	Pituitary stalk (infundibulum)
10	Tuber cinereum and median eminence
11	Mamillary body
12	Posterior perforated substance
13	Anterior cerebral artery
14	Middle cerebral artery
15	Internal carotid artery
16	Posterior communicating artery
17	Posterior cerebral artery
18	Oculomotor nerve
19	Superior cerebellar artery
20	Trochlear nerve
21	Crus of cerebral peduncle
22	Uncus
23	Parahippocampal gyrus
24	Collateral sulcus
25	Pons
26	Trigeminal nerve
27	Abducens nerve
28	Facial nerve
29	Vestibulocochlear nerve
30	Flocculus of cerebellum
31	Choroid plexus from lateral recess of fourth ventricle
32	Roots of glossopharyngeal, vagus and accessory nerves
33	Spinal part of accessory nerve
34	Rootlets of hypoglossal nerve (superficial to marker)
35	Vertebral artery
36	Medulla oblongata
37	Tonsil of cerebellum
38	Posterior inferior cerebellar artery
39	Pyramid ⎫ of medulla
40	Olive ⎭ oblongata
41	Anterior inferior cerebellar artery
42	Basilar artery

Right half of the brain, in a midline sagittal section, from the left.

1	Anterior cerebral artery
2	Rostrum ⎫
3	Genu ⎬ of corpus callosum
4	Body ⎭
5	Cingulate gyrus
6	Precentral gyrus
7	Central sulcus
8	Postcentral gyrus
9	Parietooccipital sulcus
10	Calcarine sulcus
11	Lingual gyrus
12	Cerebellum
13	Medulla oblongata
14	Median aperture of fourth ventricle
15	Fourth ventricle
16	Pons
17	Basilar artery
18	Tegmentum ⎫
19	Aqueduct ⎬ of midbrain
20	Inferior colliculus ⎪
21	Superior colliculus ⎭
22	Posterior commissure
23	Pineal body
24	Suprapineal recess
25	Great cerebral vein
26	Splenium of corpus callosum
27	Fornix
28	Cut edge of septum pellucidum
29	Body of lateral ventricle
30	Thalamus
31	Interthalamic connection
32	Hypothalamic sulcus
33	Hypothalamus
34	Posterior perforated substance
35	Mamillary body
36	Tuber cinereum and median eminence
37	Infundibular recess (base of pituitary stalk)
38	Optic chiasma
39	Supraoptic recess
40	Lamina terminalis
41	Anterior commissure
42	Anterior column of fornix
43	Interventricular foramen and choroid plexus

MRI scan of head and neck

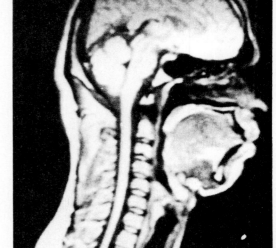

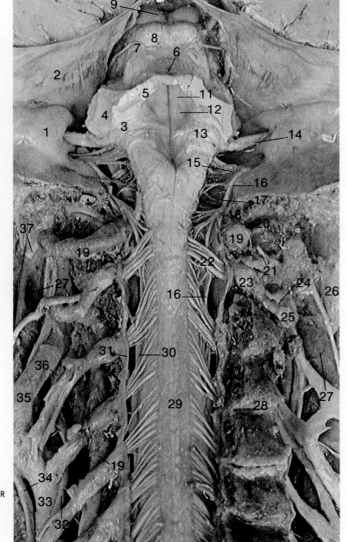

Brainstem and upper part of the spinal cord, from behind.

1 Petrous part of temporal bone
2 Tentorium cerebelli
3 Inferior ⎫
4 Middle   ⎬ cerebellar
5 Superior ⎭ peduncle
6 Superior medullary velum
7 Trochlear nerve
8 Inferior ⎫
9 Superior ⎬ colliculus
10 Straight sinus
11 Medial eminence
12 Facial colliculus
13 Medullary striae
14 Facial and vestibulocochlear nerves and internal acoustic meatus
15 Glossopharyngeal, vagus, and accessory nerves and jugular foramen
16 Spinal part of accessory nerve
17 Rootlets of hypoglossal nerve and hypoglossal canal
18 Margin of foramen magnum
19 Vertebral artery
20 Lateral mass of atlas
21 Ventral ramus of first cervical nerve
22 Dorsal rootlets ⎫ of
23 Dorsal root ganglion ⎬ second
24 Ventral ramus ⎪ cervical
25 Dorsal ramus ⎭ nerve
26 Posterior belly of digastric muscle
27 Internal jugular vein
28 Zygapophyseal joint
29 Spinal cord
30 Denticulate ligament
31 Dura mater
32 Sympathetic trunk
33 Common carotid artery
34 Vagus nerve
35 Internal carotid artery
36 Superior cervical sympathetic ganglion
37 Hypoglossal nerve

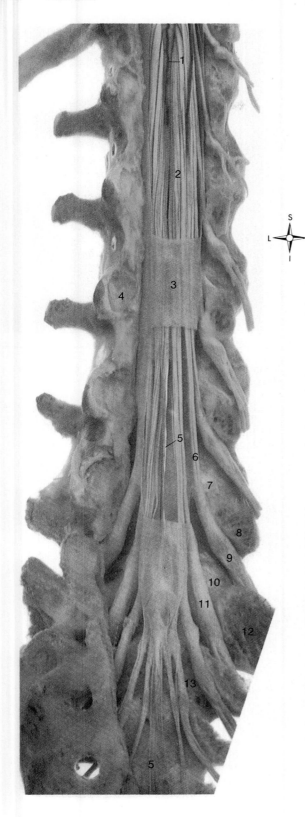

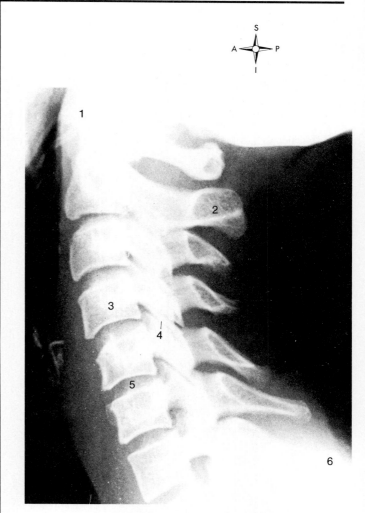

**Vertebral column, lumbar and
sacral regions, from behind.**

1 Conus medullaris of spinal cord
2 Cauda equina
3 Dura mater
4 Superior articular process of third lumbar vertebra
5 Filum terminale
6 Roots of fifth lumbar nerve
7 Fourth lumbar intervertebral disk
8 Pedicle of fifth lumbar vertebra
9 Dorsal root ganglion of fifth lumbar nerve
10 Fifth lumbar (lumbosacral) intervertebral disk
11 Dural sheath of first sacral nerve roots
12 Lateral part of sacrum
13 Second sacral vertebra

**Radiograph of vertebrae,
cervical region, lateral view.**

1 Anterior arch of atlas
2 Spinous process of axis
3 Body of fourth cervical vertebra
4 Zygapophyseal (facet) joint
5 Intervertebral disk space
6 Spinous process of seventh cervical vertebra

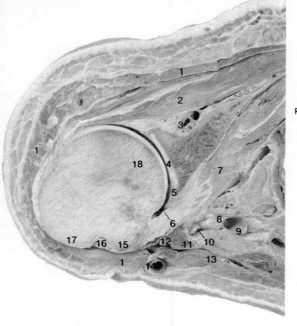

Right shoulder joint, horizontal section.

1	Deltoid	10	Musculocutaneous nerve
2	Infraspinatus	11	Coracobrachialis
3	Suprascapular nerve and vessels	12	Short head of biceps
		13	Pectoralis major
4	Glenoid cavity	14	Cephalic vein
5	Glenoid labrum	15	Lesser tubercle
6	Capsule	16	Tendon of long head of biceps in intertubercular groove
7	Subscapularis		
8	Cords of brachial plexus	17	Greater tubercle
9	Axillary artery	18	Head of humerus

Left shoulder joint, opened from behind.

1	Capsule	7	Middle glenohumeral ligament
2	Head of humerus	8	Opening into subscapularis bursa
3	Long head of biceps	9	Superior glenohumeral ligament
4	Glenoid labrum	10	Supraspinatus
5	Glenoid cavity	11	Acromion
6	Inferior glenohumeral ligament		

Right brachial plexus and branches.

1 Lateral cord
2 Posterior cord
3 Medial cord
4 Pectoralis minor and lateral pectoral nerve
5 Musculocutaneous nerve
6 Axillary nerve
7 Lateral root of median nerve
8 Radial nerve
9 Medial root of median nerve
10 Upper subscapular nerves
11 Thoracodorsal nerve
12 Lower subscapular nerve
13 Medial cutaneous nerve of arm
14 Ulnar nerve
15 Medial cutaneous nerve of forearm
16 Intercostobrachial nerve
17 Subscapularis
18 Teres major
19 Latissimus dorsi
20 Long head of triceps
21 Lateral head of triceps
22 Medial head of triceps
23 Radial nerve branches to triceps
24 Median nerve
25 Coracobrachialis
26 Biceps
27 Deltoid

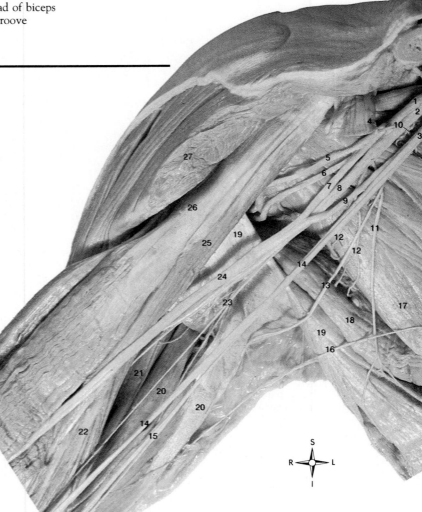

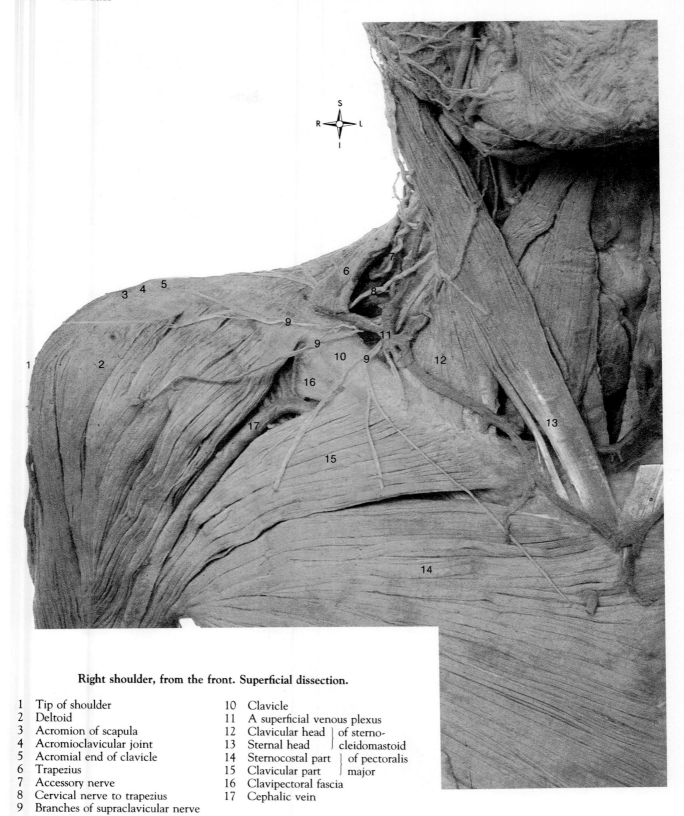

Right shoulder, from the front. Superficial dissection.

1	Tip of shoulder	10	Clavicle
2	Deltoid	11	A superficial venous plexus
3	Acromion of scapula	12	Clavicular head ⎫ of sterno-
4	Acromioclavicular joint	13	Sternal head ⎬ cleidomastoid
5	Acromial end of clavicle	14	Sternocostal part ⎫ of pectoralis
6	Trapezius	15	Clavicular part ⎬ major
7	Accessory nerve	16	Clavipectoral fascia
8	Cervical nerve to trapezius	17	Cephalic vein
9	Branches of supraclavicular nerve		

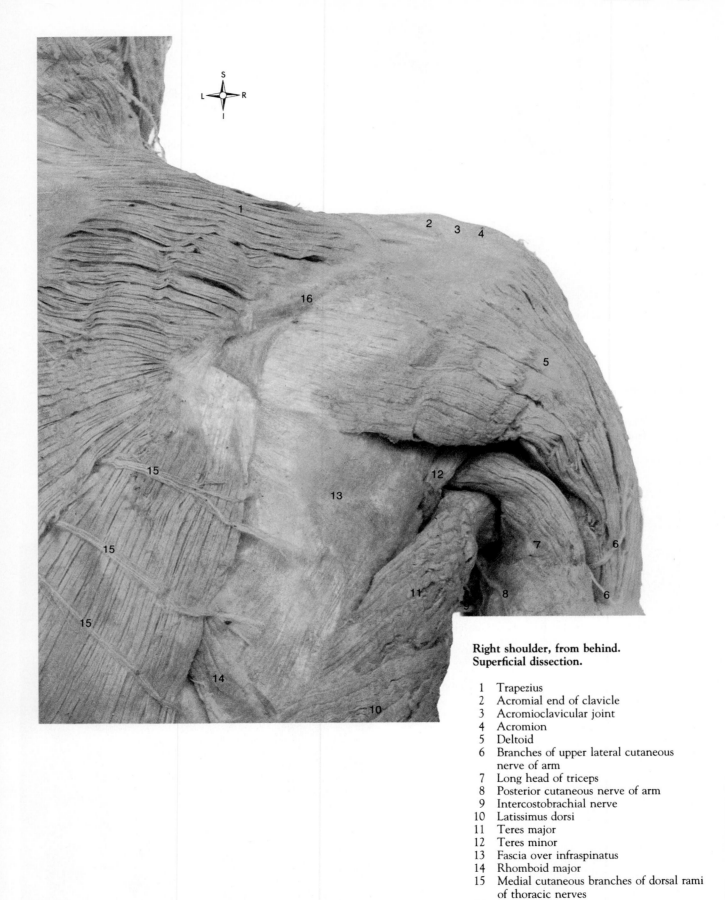

Right shoulder, from behind.
Superficial dissection.

1 Trapezius
2 Acromial end of clavicle
3 Acromioclavicular joint
4 Acromion
5 Deltoid
6 Branches of upper lateral cutaneous
   nerve of arm
7 Long head of triceps
8 Posterior cutaneous nerve of arm
9 Intercostobrachial nerve
10 Latissimus dorsi
11 Teres major
12 Teres minor
13 Fascia over infraspinatus
14 Rhomboid major
15 Medial cutaneous branches of dorsal rami
   of thoracic nerves
16 Spine of scapula

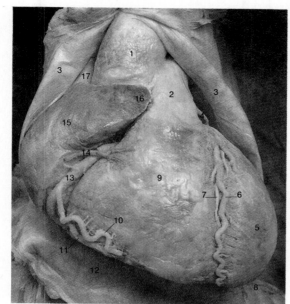

Heart and pericardium, from the front.

1   Ascending aorta
2   Pulmonary trunk
3   Serous pericardium overlying fibrous pericardium (turned laterally)
4   Auricle of left atrium
5   Left ventricle
6   Anterior interventricular branch of left coronary artery
7   Great cardiac vein
8   Diaphragm
9   Right ventricle
10   Marginal branch of right coronary artery
11   Small cardiac vein
12   Pericardium fused with tendon of diaphragm
13   Right coronary artery
14   Anterior cardiac vein
15   Right atrium
16   Auricle of right atrium
17   Superior vena cava

Thoracic inlet and mediastinum, from the front.

1   Arch of cricoid cartilage
2   Isthmus   ⎫ of thyroid
3   Lateral lobe   ⎭ gland
4   Trachea
5   Inferior thyroid veins
6   Left common carotid artery
7   Left vagus nerve
8   Internal jugular vein
9   Subclavian vein
10   Thoracic duct
11   Internal thoracic vein
12   Internal thoracic artery
13   Phrenic nerve
14   Parietal pleura (cut edge) over lung
15   Left brachiocephalic vein
16   A thymic artery
17   Thymic veins
18   Thymus
19   Superior vena cava
20   Right brachiocephalic vein
21   First rib
22   Brachiocephalic trunk
23   Right common carotid artery
24   Right subclavian artery
25   Right recurrent laryngeal nerve
26   Right vagus nerve
27   Unusual cervical tributary of 20
28   Vertebral vein
29   Thyrocervical trunk
30   Suprascapular artery
31   Scalenus anterior
32   Upper trunk of brachial plexus
33   Superficial cervical artery
34   Ascending cervical artery
35   Inferior thyroid artery
36   Sympathetic trunk

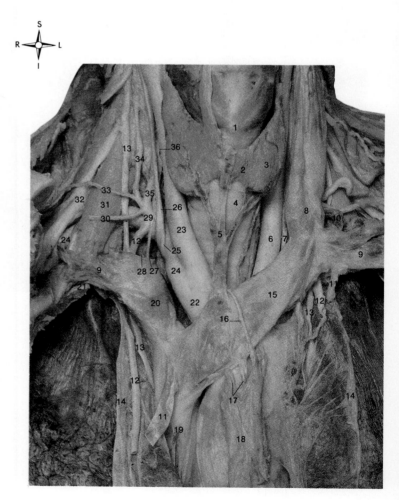

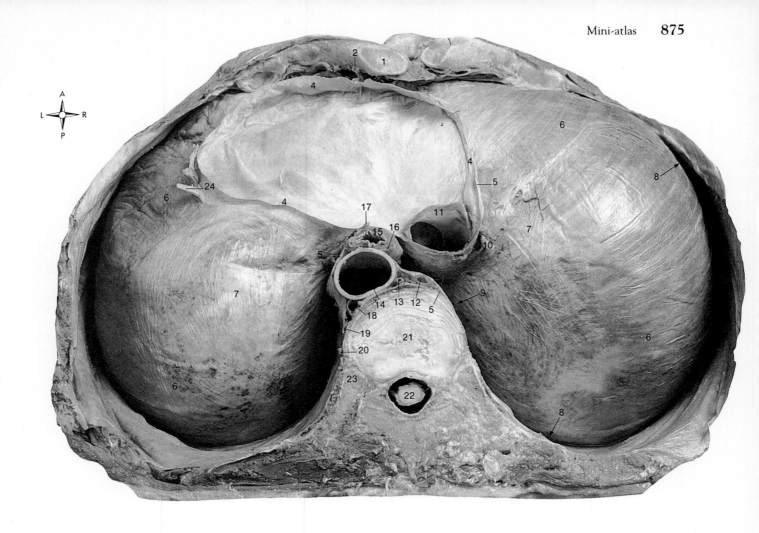

Diaphragm, from above.

1   Seventh left costal cartilage
2   Left internal thoracic artery
3   Left musculophrenic artery
4   Fibrous pericardium (cut edge)
5   Pleura (cut edge)
6   Muscle of diaphragm
7   Tendon of diaphragm
8   Costodiaphragmatic recess
9   Costomediastinal recess
10 Right phrenic nerve
11 Inferior vena cava
12 Azygos vein
13 Thoracic duct
14 Thoracic aorta
15 Esophagus
16 Posterior vagal trunk
17 Anterior vagal trunk
18 Hemiazygos vein
19 Left greater splanchnic nerve
20 Left sympathetic trunk
21 Intervertebral disc
22 Spinal cord
23 Head of left ninth rib
24 Left phrenic nerve

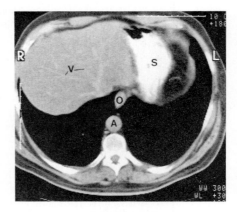

CT scan of the lower thorax.

V   Hepatic vein tributaries
S   Stomach
O   Esophagus
A   Thoracic aorta

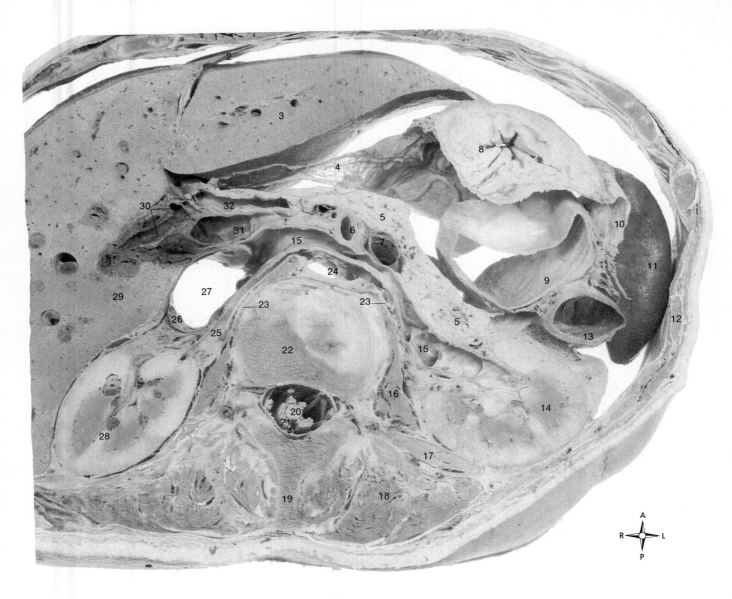

Upper abdominal viscera, in transverse section.

1   Right rectus abdominis
2   Falciform ligament
3   Left lobe of liver
4   Lesser omentum
5   Pancreas
6   Superior mesenteric artery
7   Splenic vein
8   Stomach
9   Transverse colon
10 Greater omentum
11 Spleen
12 Tenth rib
13 Descending colon
14 Left kidney
15 Left renal vein
16 Psoas major
17 Quadratus lumborum
18 Erector spinae
19 Spine of first lumbar vertebra
20 Conus medullaris of spinal cord
21 Nerve roots of cauda equina
22 Body of first lumbar vertebra
23 Sympathetic trunk
24 Abdominal aorta
25 Right renal artery
26 Right renal vein
27 Inferior vena cava
28 Right kidney
29 Right lobe of liver
30 Hepatic ducts
31 Portal vein
32 Hepatic artery

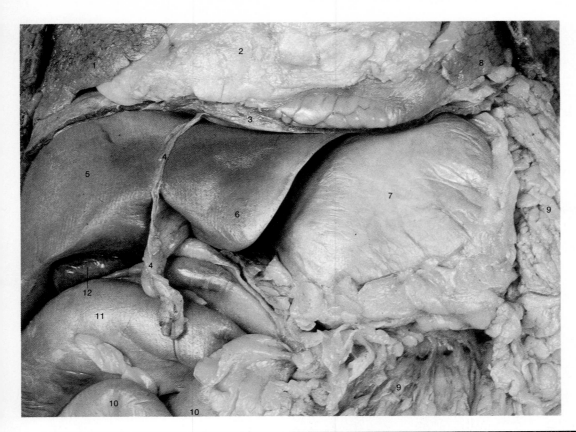

Upper abdominal viscera, from the front.

1 Inferior lobe of right lung
2 Pericardial fat
3 Diaphragm
4 Falciform ligament
5 Right lobe of liver
6 Left lobe of liver
7 Stomach
8 Inferior lobe of left lung
9 Greater omentum
10 Small intestine
11 Transverse colon
12 Gallbladder

**Stomach, with vessels
and vagus nerves,
from the front.**

1 Right lobe of liver
2 Fissure of ligamentum venosum
3 Caudate lobe of liver
4 Lesser omentum (cut edge)
5 Left gastric artery
6 Left gastric vein
7 Posterior vagal trunk
8 Esophageal branches of left gastric vessels
9 Anterior vagal trunk
10 Esophagus
11 Esophageal opening in diaphragm
12 Fundus ⎫
13 Body ⎬ of stomach
14 Greater curvature ⎭
15 Greater omentum
16 Lower end of spleen
17 Branches of left gastroepiploic vessels
18 Right gastroepiploic vessels and branches
19 Pyloric part of stomach
20 Pylorus
21 Superior (first) part of duodenum
22 Right gastric artery
23 Lesser curvature

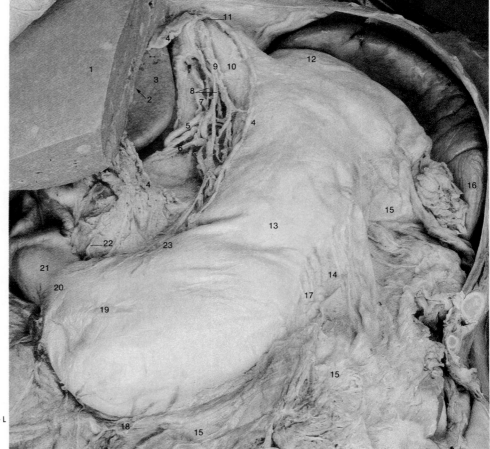

Liver, from above and behind.

1 Left triangular ligament
2 Diaphragm
3 Left lobe
4 Gastric impression
5 Esophageal groove
6 Lesser omentum in fissure for ligamentum venosum
7 Inferior vena cava
8 Caudate lobe
9 Diaphragm on part of bare area
10 Bare area
11 Inferior layer of coronary ligament
12 Right triangular ligament
13 Renal impression
14 Right lobe
15 Colic impression
16 Duodenal impression
17 Suprarenal impression
18 Caudate process
19 Right free margin of lesser omentum in porta hepatis
20 Portal vein
21 Hepatic artery
22 Common hepatic duct
23 Gallbladder
24 Quadrate lobe
25 Ligamentum teres and falciform ligament in fissure for ligamentum teres
26 Omental tuberosity

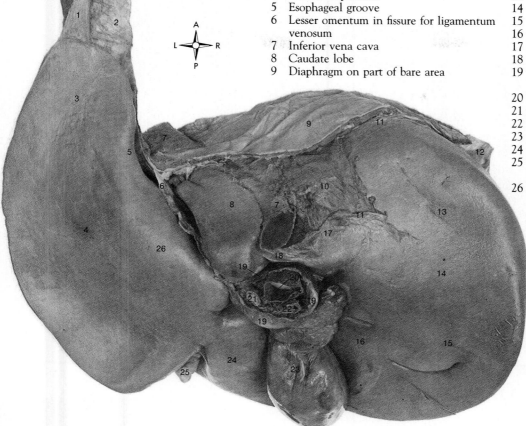

Small intestine.
Coil of typical jejunum,
coil of typical ileum.

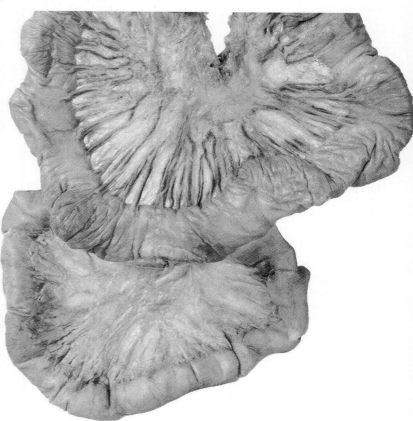

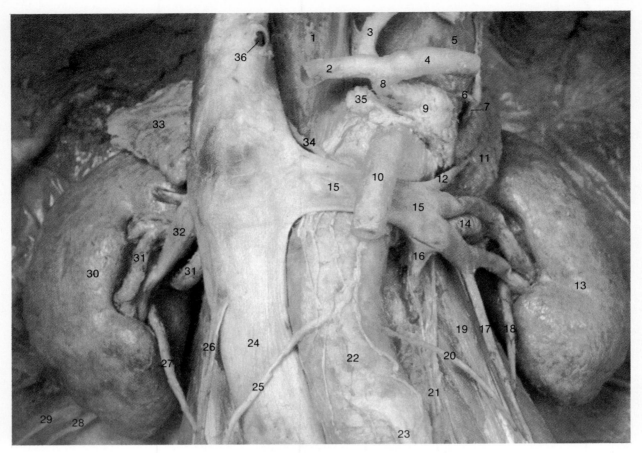

Kidneys and suprarenal glands.

1	Right crus of diaphragm	19	Left psoas major
2	Common hepatic artery	20	Left gonadal artery
3	Left gastric artery	21	Left sympathetic trunk
4	Splenic artery	22	Abdominal aorta and aortic plexus
5	Left crus of diaphragm	23	Inferior mesenteric artery
6	Left inferior phrenic artery	24	Inferior vena cava
7	Left inferior phrenic vein	25	Right gonadal artery
8	Celiac trunk	26	Right gonadal vein
9	Left celiac ganglion	27	Right ureter
10	Superior mesenteric artery	28	Right ilio-inguinal nerve
11	Left suprarenal gland	29	Right iliohypogastric nerve
12	Left suprarenal vein	30	Right kidney
13	Left kidney	31	Right renal artery
14	Left renal artery	32	Right renal vein
15	Left renal vein	33	Right suprarenal gland
16	Lumbar tributary of renal vein	34	Right inferior phrenic artery
17	Left gonadal vein	35	Right celiac ganglion
18	Left ureter	36	A hepatic vein

Kidney.
Internal structure
in longitudinal section.

1	Cortex	6	Minor calyx
2	Medulla	7	Major calyx
3	Renal column	8	Renal pelvis
4	Renal papilla	9	Hilum
5	Medullary pyramid	10	Ureter

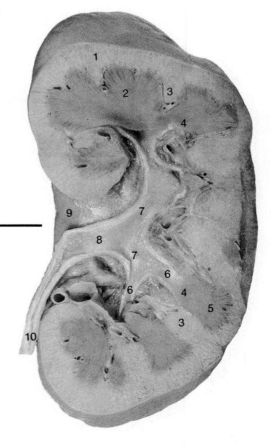

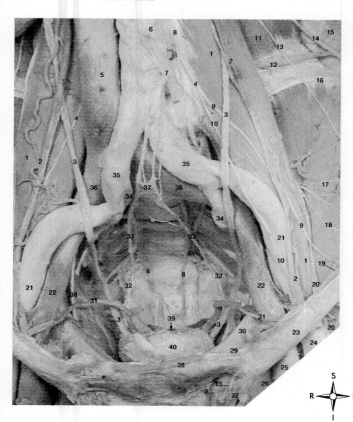

Posterior abdominal and pelvic walls.

1   Psoas major
2   Testicular vessels
3   Ureter
4   Genitofemoral nerve
5   Inferior vena cava
6   Aorta and aortic plexus
7   Inferior mesenteric artery and plexus
8   Sympathetic trunk and ganglia
9   Femoral ⎫ branch of genito-
10 Genital ⎭ femoral nerve
11 Quadratus lumborum
12 Fourth lumbar artery
13 Ilio-inguinal nerve
14 Iliohypogastric nerve
15 Lumbar part of thoracolumbar fascia
16 Iliolumbar ligament
17 Iliacus and branches from femoral nerve and iliolumbar artery
18 Lateral femoral cutaneous nerve arising from femoral nerve

19 Deep circumflex iliac artery
20 Femoral nerve
21 External iliac artery
22 External iliac vein
23 Inguinal ligament
24 Femoral artery
25 Femoral vein
26 Position of femoral canal
27 Spermatic cord
28 Rectus abdominis
29 Lacunar ligament
30 Pectineal ligament
31 Ductus deferens
32 Inferior hypogastric (pelvic) plexus and pelvic splanchnic nerves
33 Hypogastric nerve
34 Internal iliac artery
35 Common iliac artery
36 Common iliac vein
37 Superior hypogastric plexus
38 Obturator nerve and vessels
39 Rectum (cut edge)
40 Bladder

Right half of a midline sagittal section of the male pelvis.

1   Rectus abdominis
2   Extraperitoneal fat
3   Sigmoid colon
4   Promontory of sacrum
5   Rectum
6   Coccyx
7   Anococcygeal body
8   External anal sphincter
9   Anal canal with anal columns of mucous membrane
10 Perineal body
11 Ductus deferens
12 Epididymis
13 Testis
14 Spongy part of urethra and corpus spongiosum
15 Corpus cavernosum
16 Bulbospongiosus
17 Perineal membrane
18 Sphincter urethrae
19 Membranous part of urethra
20 Pubic symphysis
21 Prostate
22 Prostatic part of urethra
23 Seminal colliculus
24 Bristle in ejaculatory duct
25 Internal urethral orifice
26 Bladder
27 Bristle passing up into right ureteral orifice
28 Rectovesical pouch
29 Puborectalis fibers of levator ani

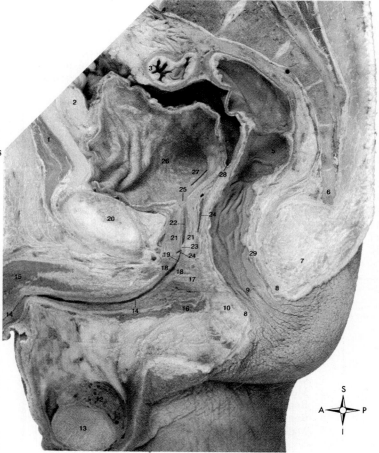

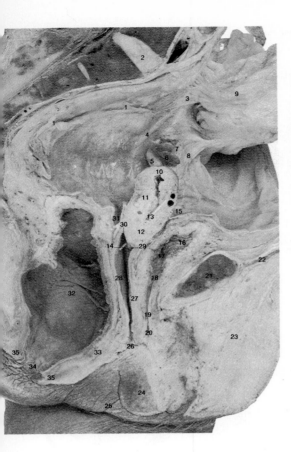

Left half of a midline sagittal section of the female pelvis.

1 Line of attachment of right limb of sigmoid mesocolon
2 Fifth lumbar intervertebral disk
3 Apex of sigmoid mesocolon
4 Ureter underlying peritoneum
5 Ovary
6 Uterine tube
7 Suspensory ligament of ovary containing ovarian vessels
8 Left limb of sigmoid mesocolon overlying external iliac vessels
9 Sigmoid colon (reflected to left and upwards)
10 Fundus ⎫
11 Body ⎬ of uterus
12 Cervix ⎭
13 Marker in internal os
14 Marker in external os
15 Vesico-uterine pouch
16 Bladder

17 Marker in left ureteral orifice
18 Internal urethral orifice
19 Urethra
20 External urethral orifice
21 Pubic symphysis
22 Rectus abdominis (turned forwards)
23 Fat of mons pubis
24 Labium minus
25 Labium majus
26 Vestibule ⎫
27 Anterior wall ⎪
28 Posterior wall ⎬ of vagina
29 Anterior fornix ⎪
30 Posterior fornix ⎭
31 Recto-uterine pouch
32 Rectum
33 Perineal body
34 Anal canal
35 External anal sphincter

Left hip joint, coronal section, from the front.

1 External iliac artery
2 Psoas major
3 Iliacus
4 Iliac crest
5 Gluteus medius
6 Gluteus minimus
7 Greater trochanter
8 Vastus lateralis
9 Shaft of femur
10 Vastus medialis
11 Profunda femoris vessels
12 Adductor longus
13 Pectineus
14 Medial circumflex femoral vessels
15 Capsule of hip joint
16 Neck of femur
17 Zona orbicularis of capsule
18 Head of femur
19 Acetabular labrum
20 Rim of acetabulum
21 Hyaline cartilage of head
22 Hyaline cartilage of acetabulum

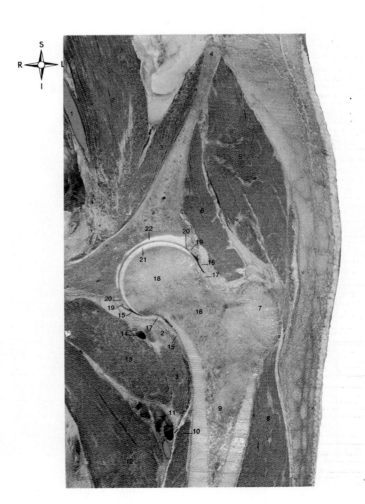

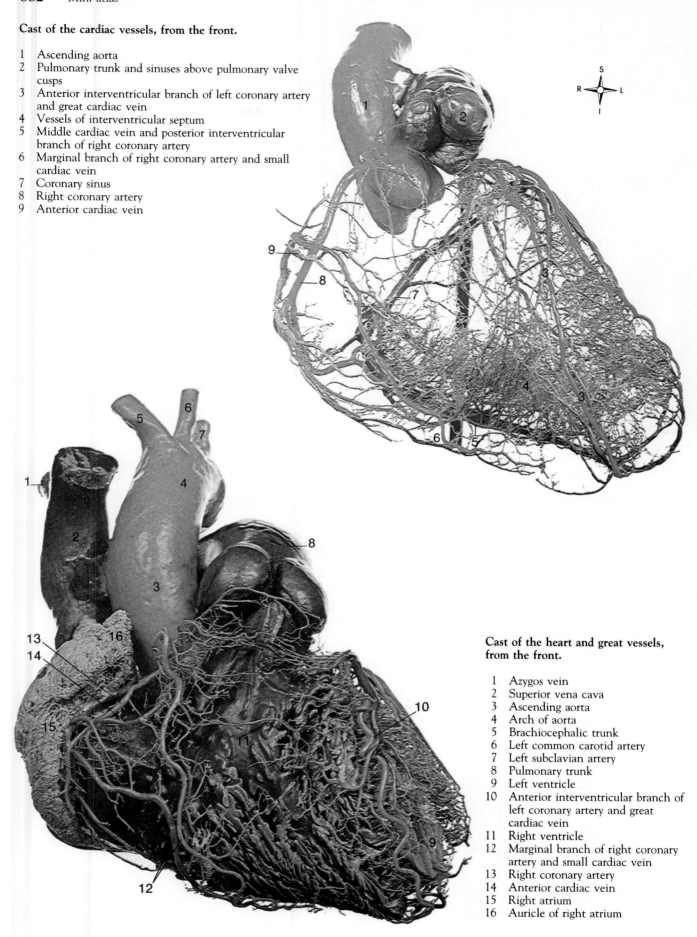

Cast of the cardiac vessels, from the front.

1 Ascending aorta
2 Pulmonary trunk and sinuses above pulmonary valve
   cusps
3 Anterior interventricular branch of left coronary artery
   and great cardiac vein
4 Vessels of interventricular septum
5 Middle cardiac vein and posterior interventricular
   branch of right coronary artery
6 Marginal branch of right coronary artery and small
   cardiac vein
7 Coronary sinus
8 Right coronary artery
9 Anterior cardiac vein

**Cast of the heart and great vessels,
from the front.**

1 Azygos vein
2 Superior vena cava
3 Ascending aorta
4 Arch of aorta
5 Brachiocephalic trunk
6 Left common carotid artery
7 Left subclavian artery
8 Pulmonary trunk
9 Left ventricle
10 Anterior interventricular branch of
   left coronary artery and great
   cardiac vein
11 Right ventricle
12 Marginal branch of right coronary
   artery and small cardiac vein
13 Right coronary artery
14 Anterior cardiac vein
15 Right atrium
16 Auricle of right atrium

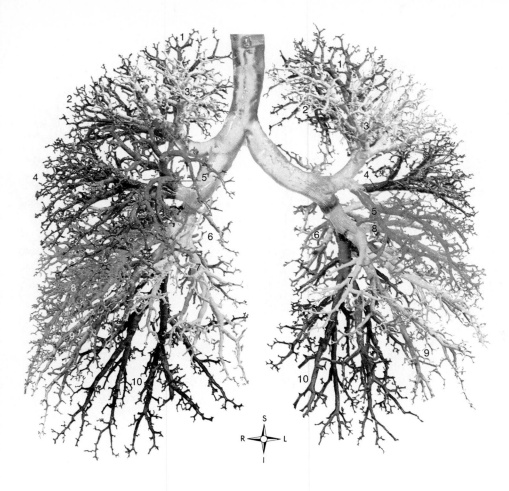

Cast of the bronchial tree.

RIGHT LUNG

Superior lobe
1 Apical
2 Posterior
3 Anterior

Middle lobe
4 Lateral
5 Medial

Inferior lobe
6 Apical (superior)
7 Medial basal
8 Anterior basal
9 Lateral basal
10 Posterior basal

LEFT LUNG

Superior lobe
1 Apical
2 Posterior
3 Anterior
4 Superior lingular
5 Inferior lingular

Inferior lobe
6 Apical (superior)
7 Medial basal (cardiac)
8 Anterior basal
9 Lateral basal
10 Posterior basal

Cast of the duodenum, liver, biliary tract, and associated vessels, from the front.

1 Right branch of portal vein and hepatic artery and right hepatic duct
2 Gallbladder
3 Bile duct
4 Hepatic artery
5 Portal vein
6 Left branch of portal vein and hepatic artery and left hepatic duct
7 Left gastric artery
8 Left gastric vein
9 Splenic artery
10 Splenic vein
11 Short gastric vessels
12 Left gastroepiploic vessels
13 Vessels of left kidney
14 Pancreatic duct
15 Duodenojejunal flexure
16 Superior mesenteric artery
17 Superior mesenteric vein
18 Horizontal (third) part of duodenum
19 Right gastroepiploic vessels
20 Pyloric canal
21 Pylorus
22 Superior (first) part of duodenum
23 Right gastric vessels
24 Branches of superior and inferior pancreaticoduodenal vessels
25 Descending (second) part of duodenum
26 Vessels of right kidney

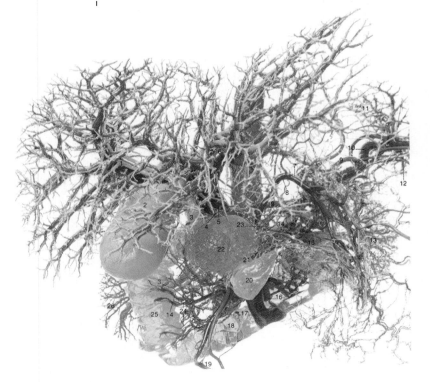

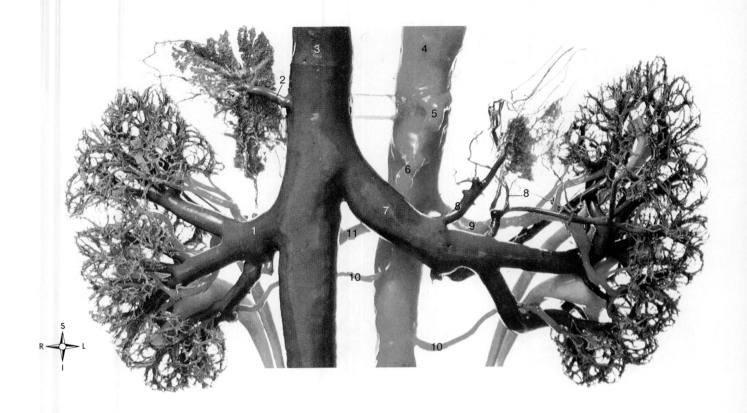

**Cast of the kidneys and great
vessels, from the front.**

1 Right renal vein
2 Right suprarenal vein
3 Inferior vena cava
4 Aorta
5 Celiac trunk
6 Superior mesenteric artery

7 Left renal vein
8 Left suprarenal veins
9 Left renal artery
10 Accessory renal arteries
11 Right renal artery

Appendix A
Metric Measurements and Their Equivalents

Units of measurement

Basic Unit	Metric	English	English/Metric
Time	second	second	same
Length	meter (m)	yard	1.09 yards/1 meter
Volume	liter (l or L)	quart	1.06 quarts/1 liter
Mass	gram (g)	ounce	.035 ounce/1 gram
Temperature	degree Celsius (°C)	degree Fahrenheit (°F)	1.8 °F/1 °C

Prefixes

Less Than One Basic Unit		
nano-	one billionth	.000000001
micro-	one millionth	.000001
milli-	one thousandth	.001
centi-	one hundredth	.01
deci-	one tenth	.1
More Than One Basic Unit		
deka-	ten	10.00
hecto-	one hundred	100.00
kilo-	one thousand	1000.00
mega-	one million	1000000.00

Common conversions

Multiply	By	To Get
Time		
seconds	1000	milliseconds
seconds	.00167	minutes
minutes	60	seconds
milliseconds	.001	seconds
Length		
meters	1.09	yards
meters	3.28	feet
meters	100	centimeters
meters	1000	millimeters
centimeters	.01	meters
centimeters	10	millimeters
centimeters	100000	micrometers
millimeters	.001	meters
millimeters	.1	centimeters

Multiply	By	To Get
Volume		
liters	1.06	quarts
liters	.26	gallons
liters	1000	milliliters
liters	100	centiliters
liters	10	deciliters
centiliters	.01	liters
centiliters	10	milliliters
centiliters	.1	deciliters
deciliters	.1	liters
deciliters	10	centiliters
Mass		
grams	.035	ounces
grams	.001	kilograms
grams	1000	milligrams
milligrams	.001	grams
kilograms	1000	grams
kilograms	2.21	pounds

Temperature conversions

As you can see from the Units of Measurement table, one "degree" in Celsius is a larger unit than one degree Fahrenheit. In fact, a Celsius degree is $\frac{9}{5}$ (1.8) times the size of a Fahrenheit degree. When you convert temperature readings from one form to another, this discrepancy in size must be taken into account. The 32° is added in the conversion to Fahrenheit to account for the fact that, in the Fahrenheit scale, the freezing point of water is 32°, not 0°, as it is in Celsius.

To convert °C to °F

Multiply °C by $\frac{9}{5}$ and add 32

_____°C × $\frac{9}{5}$ + 32 = _____°F

For example, to convert 35°C to °F:
35°C × $\frac{9}{5}$ + 32 = 95°F

To convert °F to °C

Subtract 32 from °F and multiply by $\frac{5}{9}$

(_____°F − 32) × $\frac{5}{9}$ = _____°C

For example, to convert 101°F to °C:
(101°F − 32) × $\frac{5}{9}$ = 38.3°C

Body temperatures in °Celsius and °Fahrenheit

°C	°F	°C	°F
35.0	95.0	37.8	100.0
35.1	95.2	37.9	100.2
35.2	95.4	38.0	100.4
35.3	95.6	38.1	100.6
35.4	95.8	38.2	100.8
35.5	96.0	38.3	101.0
35.7	96.2	38.4	101.2
35.8	96.4	38.6	101.4
35.9	96.6	38.7	101.6
36.0	96.8	38.8	101.8
36.1	97.0	38.9	102.0
36.2	97.2	39.0	102.2
36.3	97.4	39.1	102.4
36.4	97.6	39.2	102.6
36.6	97.8	39.3	102.8
36.7	98.0	39.4	103.0
36.8	98.2	39.6	103.2
36.9	98.4	39.7	103.4
37.0	98.6	39.8	103.6
37.1	98.8	39.9	103.8
37.2	99.0	40.0	104.0
37.3	99.2	40.1	104.2
37.4	99.4	40.2	104.4
37.6	99.6	40.3	104.6
37.7	99.8	40.4	104.8

Appendix B

Determining the Potential Osmotic Pressure of a Solution

Potential osmotic pressure is the maximum osmotic pressure that could develop in a solution if it were separated from distilled water by a selectively permeable membrane. (Actual osmotic pressure is a pressure that already has developed, not just one that could develop.) What determines a solution's potential osmotic pressure? Answer: the number of solute particles in a unit volume of solution directly determines its potential osmotic pressure—the more solute particles per unit volume, the greater the potential osmotic pressure.

If the solute is a nonelectrolyte, the number of solute particles in a liter of solution is determined, as Figure A indicates, solely by the **molar concentration** of the solution. (To calculate molar concentration, divide the number of grams of solute in a liter of solution by the molecular weight of the solute.) If the solute is an electrolyte, the number of solute particles per liter is determined by two factors: the molar concentration of the solution and the number of ions formed from each molecule of the electrolyte (Figure B).

A

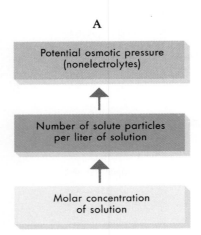

B

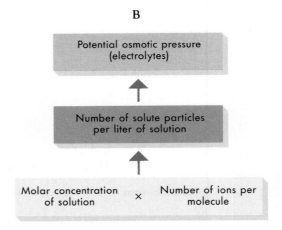

Since it is the number of solute particles per unit volume that directly determines a solution's potential osmotic pressure, one might jump to the conclusion that all solutions having the same percent concentration also have the same potential osmotic pressure. Obviously it is true that all solutions containing the same percent concentration of the same solute do also have the same potential osmotic pressures. All 5% glucose solutions, for example, have a potential osmotic pressure at body temperature of somewhat more than 5,300 mm Hg pressure. But all solutions with the same percent concentrations of different solutes do not have the same molar concentrations. And since it is molar concentration, not percent concentration, that determines potential osmotic pressure, solutions with the same percent concentration of different solutes have different potential osmotic pressures. For example, 5% NaCl at body temperature has a potential osmotic pressure of approximately 33,000 mm Hg—quite different from 5% glucose's potential osmotic pressure of about 5,300 mm Hg. You can calculate these potential osmotic pressures using the formulas given in the table.

Formulas for determining osmotic pressure

$$\text{Potential osmotic pressure of nonelectrolyte (in mm Hg)} = \text{Molar concentration of solution} \times 19{,}300*$$

$$\text{Potential osmotic pressure of electrolyte (in mm Hg)} = \text{Molar concentration of solution} \times \text{Number ions per molecule} \times 19{,}300$$

*Experimentation has shown that a solution with a 1.0 molar concentration of any nonelectrolyte has a potential osmotic pressure of 19,300 mm Hg pressure (at body temperature, 37° C).

$$\text{Molar concentration} = \text{Grams solute in 1 liter solution divided by molecular weight of solute}$$

Example: Two solutions commonly used in hospitals are 0.85% NaCl and 5% glucose. What is the potential osmotic pressure of 0.85% NaCl at body temperature? (0.85 NaCl = 8.5 gm NaCl in 1 liter solution) (Molecular weight of NaCl = 58 [NaCl yields two ions per molecule in solution]).

Using the formula given for computing potential osmotic pressure of an electrolyte:

$$\text{Potential osmotic pressure of 0.85\% NaCl} = \frac{8.5}{58} \times 2 \times 19{,}300 = 5{,}658.6 \; mm \; Hg \; pressure$$

Problem: What is potential osmotic pressure of 5% glucose solution at body temperature? Molecular weight glucose = 180. Glucose does not ionize. It is a nonelectrolyte.†

†5% glucose solution has potential osmotic pressure of 5,359.6 mm Hg pressure.

Appendix C
Clinical and Laboratory Values

TABLE 1   Blood, plasma, and serum values

Test	Normal Values*	Significance of a Change
Acid phosphatase	*Women:* 0.01-0.56 sigma U/ml *Men:* 0.13-0.63 sigma U/ml	↑ in prostate cancer ↑ in kidney disease ↑ after trauma and in fever
Alkaline phosphatase	*Adult:* 13-39 IU/l *Child:* up to 104 IU/l	↑ in bone disorders ↑ in liver disease ↑ during pregnancy ↑ in hypothyroidism
Bicarbonate	22-26 mEq/L	↑ in metabolic alkalosis ↑ in respiratory alkalosis ↑ in metabolic acidosis ↑ in respiratory alkalosis
Blood urea nitrogen (BUN)	5-25 mg/100 ml	↑ with increased protein intake ↓ in kidney failure
Blood volume	*Women:* 65 ml/kg body weight *Men:* 69 ml/kg body weight	↓ during hemorrhage
Calcium	8.4-10.5 mg/100 ml	↑ in hypervitaminosis D ↑ in hyperparathyroidism ↑ in bone cancer and other bone diseases ↓ in severe diarrhea ↓ in hypoparathyroidism ↓ in avitaminosis D (rickets and osteomalacia)
Carbon dioxide content	24-32 mEq/L	↑ in severe vomiting ↑ in respiratory disorders ↑ in obstruction of intestines ↓ in acidosis ↓ in severe diarrhea ↓ in kidney disease
Chloride	96-107 mEq/L	↑ in hyperventilation ↑ in kidney disease ↑ in Cushing's syndrome ↓ in diabetic acidosis ↓ in severe diarrhea ↓ in severe burns ↓ in Addison's disease
Clotting time	5-10 minutes	↓ in hemophilia ↓ (occasionally) in other clotting disorders
Copper	100-200 μg/100 ml	↑ in some liver disorders
Creatine phosphokinase (CPK)	*Women:* 5-35 mU/ml *Men:* 5-55 mU/ml	↑ in Duchenne's muscular dystrophy ↑ during myocardial infarction ↑ in muscle trauma

*Values vary with the analysis method used; 100 ml = 1 dl.

Continued.

TABLE 1   Blood, plasma, and serum values—cont'd

Test	Normal Values*	Significance of a Change
Creatine	0.6-1.5 mg/100 ml	↑ in some kidney disorders
Glucose	70-110 mg/100 ml (fasting)	↑ in diabetes mellitus ↑ in kidney disease ↑ in liver disease ↑ during pregnancy ↑ in hyperthyroidism ↓ in hypothyroidism ↓ in Addison's disease ↓ in hyperinsulinism
Hematocrit (packed cell volume)	*Women:* 38%-47% *Men:* 40%-54%	↑ in polycythemia ↑ in severe dehydration ↓ in anemia ↓ in leukemia ↓ in hyperthyroidism ↓ in cirrhosis of liver
Hemoglobin	*Women:* 12-16g/100ml *Men:* 13-18g/100ml *Newborn:* 14-20g/100ml	↑ in polycythemia ↑ in chronic obstructive pulmonary disease ↑ in congestive heart failure ↓ in anemia ↓ in hyperthyroidism ↓ in cirrhosis of liver
Iron	50-150 μg/100 ml (can be higher in men)	↑ in liver disease ↑ in anemia (some forms) ↓ in iron-deficiency anemia
Lactic dehydrogenase (LDH)	60-120 U/ml	↑ during myocardial infarction ↑ in anemia (several forms) ↑ liver disease ↑ in acute leukemia and other cancers

Lipids—total
  Cholesterol—total
    High-density lipoprotein (HDL)
    Low-density lipoprotein (LDL)
 	
  Triglycerides
  Phospholipids
  Fatty acids
	450-1,000 mg/100 ml
	120-220 mg/100 ml
	>40 mg/100 ml
	<180 mg/100 ml
	
	40-150 mg/100 ml
	145-200 mg/100 ml
	190-420 mg/100 ml
		↑ (total) in diabetes mellitus
		↑ (total) in kidney disease
		↑ (total) in hypothyroidism
		↓ (total) in hyperthyroidism
		↑ in inherited hypercholesterolemia
		↑ (cholesterol) in chronic hepatitis
		↓ (cholesterol) in acute hepatitis
		↑ (HDL) with regular exercise

Mean corpuscular volume	82-98 μl	↑ or ↓ in various forms of anemia
Osmolality	285-295 mOsm/L	↑ or ↓ in fluid and electrolyte imbalances
Pco_2	35-43 mm Hg	↑ in severe vomiting ↑ in respiratory disorders ↑ in obstruction of intestines ↓ in acidosis ↓ in severe diarrhea ↓ in kidney disease
pH	7.35-7.45	↑ during hyperventilation ↑ in Cushing's syndrome ↓ during hypoventilation ↓ in acidosis ↓ in Addison's disease
Phosphorus	2.5-4.5 mg/100 ml	↑ in hypervitaminosis D ↑ in kidney disease ↑ in hypoparathyroidism ↑ in acromegaly ↓ hyperparathyroidism ↓ in hypovitaminosis D (rickets and osteomalacia)

TABLE 1   Blood, plasma, and serum values—cont'd

Test	Normal Values*	Significance of a Change
Plasma volume	*Women:* 40 ml/kg body weight *Men:* 39 ml/kg body weight	↑ or ↓ in fluid and electrolyte imbalances ↓ during hemorrhage
Platelet count	150,000-400,000/mm³	↑ in heart disease ↑ in cancer ↑ in cirrhosis of liver ↑ after trauma ↓ in anemia (some forms) ↓ during chemotherapy ↓ in some allergies
Po_2	75-100 mm Hg (breathing standard air)	↑ in polycythemia ↓ in anemia ↓ in chronic obstructive pulmonary disease
Potassium	3.8 to 5.1 mEq/L	↑ in hypoaldosteronism ↑ in acute kidney failure ↓ in vomiting or diarrhea ↓ in starvation

Protein—total
  Albumin
  Globulin
	6-8.4 g/100 ml
	3.5-5 g/100 ml
	2.3-3.5 g/100 ml
		↑ (total) in severe dehydration
		↓ (total) during hemorrhage
		↓ (total) in starvation

Red blood cell count	*Women:* 4.2-5.4 million/mm³ *Men:* 4.5-6.2 million/mm³	↑ in polycythemia ↑ in dehydration ↓ in anemia (several forms) ↓ in Addison's disease ↓ in systemic lupus erythematosus
Reticulocyte count	25,000-75,000/mm³ (0.5%-1.5% of RBC count)	↑ in hemolytic anemia ↑ in leukemia and metastatic carcinoma ↓ in pernicious anemia ↓ in iron-deficiency anemia ↓ during radiation therapy
Sodium	136-145 mEq/L	↑ in dehydration ↑ in trauma or disease of the central nervous system ↑ or ↓ in kidney disorders ↓ in excessive sweating, vomiting, diarrhea ↓ in burns (sodium shift into cells)
Specific gravity	1.058	↑ or ↓ in fluid imbalances
Transaminase	10-40 U/ml	↑ during myocardial infarction ↑ in liver disease
Uric acid	*Women:* 2.5-7.5 mg/100 ml *Men:* 3-9 mg/100 ml	↑ in gout ↑ in toxemia of pregnancy ↑ during trauma
Viscosity	1.4-1.8 times the viscosity of water	↑ in polycythemia ↑ in dehydration

White blood cell count
  Total
  Neutrophils
  Eosinophils
  Basophils
  Lymphocytes
  Monocytes
	
	4,500-11,000/mm³
	60%-70% of total
	2%-4% of total
	0.5%-1% of total
	20%-25% of total
	3%-8% of total
		↑ (total) in acute infections
		↑ (total) in trauma
		↑ (total) some cancers
		↓ (total) in anemia (some forms)
		↓ (total) during chemotherapy
		↑ (neutrophil) in acute infection
		↑ (eosinophil) in allergies
		↓ (basophil) in severe allergies
		↑ (lymphocyte) during antibody reactions
		↑ (monocyte) in chronic infections

TABLE 2   Urine components

Test	Normal Values*	Significance of a Change
Routine urinalysis		
Acetone and acetoacetate	0	↑ during fasting ↑ in diabetic acidosis
Albumin	0-trace	↑ in hypertension ↑ in kidney disease ↑ after strenuous exercise (temporary)
Ammonia	20-70 mEq/L	↑ in liver disease ↑ in diabetes mellitus
Bile and bilirubin	—	↑ during obstruction of the bile ducts
Calcium	<150 mg/day	↑ in hyperparathyroidism ↓ in hypoparathyroidism
Color	Transparent yellow, straw-colored, or amber	Abnormal color or cloudiness may indicate blood in urine, bile, bacteria, drugs, food pigments, or high solute concentration
Odor	Characteristic slight odor	Acetone odor in diabetes mellitus (diabetic ketosis)
Osmolality	500-800 mOsm/L	↑ in dehydration ↑ in heart failure ↓ in diabetes insipidus ↓ in aldosteronism
pH	4.6-8.0	↑ in alkalosis ↑ during urinary infections ↓ in acidosis ↓ in dehydration ↓ in emphysema
Potassium	25-100 mEq/L	↑ in dehydration ↑ in chronic kidney failure ↓ in diarrhea or vomiting ↓ in adrenal insufficiency
Sodium	75-200 mg/day	↑ in starvation ↑ in dehydration ↓ in acute kidney failure ↓ in Cushing's syndrome
Creatinine clearance	100-140 ml/min	↑ in kidney disease
Creatine	1-2 g/day	↑ in infections ↓ in some kidney diseases ↓ in anemia (some forms)
Glucose	0	↑ in diabetes mellitus ↑ in hyperthyroidism ↑ in hypersecretion of adrenal cortex
Urea clearance	>40 ml blood cleared per min	↑ in some kidney diseases
Urea	25-35 g/day	↑ in some liver diseases ↑ in hemolytic anemia ↓ during obstruction of bile ducts ↓ in severe diarrhea
Uric acid	0.6-1.0 g/day	↑ in gout ↓ in some kidney diseases
Microscopic examination		
Bacteria	<10,000/ml	↑ during urinary infections
Blood cells (RBC)	0-trace	↑ in pyelonephritis ↑ from damage by calculi ↑ in infection ↑ in cancer
Blood cells (WBC)	0-trace	↑ in infections
Blood cell casts (RBC)	0-trace	↑ in pyelonephritis
Blood cell casts (WBC)	0-trace	↑ in infection

*Values vary with the analysis method used.

TABLE 2   Urine components—cont'd

Test	Normal Values*	Significance of a Change
Crystals	0-trace	↑ in urinary retention Very large crystaline masses are calculi
Epithelial casts	0-trace	↑ in some kidney disorders ↑ in heavy metal toxicity
Granular casts	0-trace	↑ in some kidney disorders
Hyaline casts	0-trace	↑ in some kidney disorders ↑ in fever

Appendix D
Conversion Factors to International System of Units (SI Units)

CONVERSION FACTORS (SI UNITS)

Component	Normal Range in Units as Customarily Reported	Conversion Factor	Normal Range in SI Units, Molecular Units, International Units, or Decimal Fractions
Biochemical Components of Blood*			
Acetoacetic acid (S)	0.2-1.0 mg/dL	98	19.6-98.0 μmol/L
Acetone (S)	0.3-2.0 mg/dL	172	51.6-344.0 μmol/L
Albumin (S)	3.2-4.5 g/dL	10	32-45 g/L
Ammonia (P)	20-120 μg/dL	0.588	11.7-70.5 μmol/L
Amylase (S)	60-160 Somogyi units/dL	1.85	111-296 U/L
Base, total (S)	145-160 mEq/L	1	145-160 mmol/L
Bicarbonate (P)	21-28 mEq/L	1	21-28 mmol/L
Bile acids (S)	0.3-3.0 mg/dL	10 2.547	3-30 mg/L 0.8-7.6 μmol/L
Bilirubin, direct (S)	Up to 0.3 mg/dL	17.1	Up to 5.1 μmol/L
Bilirubin, indirect (S)	0.1-1.0 mg/dL	17.1	1.7-17.1 μmol/L
Blood gases (B) P_{CO_2} arterial P_{O_2} arterial	 35-40 mm Hg 95-100 mm Hg	 0.133 0.133	 4.66-5.32 kPa 12.64-13.30 kPa
Calcium (S)	8.5-10.5 mg/dL	0.25	2.1-2.6 mmol/L
Chloride (S)	95-103 mEq/L	1	95-103 mmol/L
Creatine (S)	0.1-0.4 mg/dL	76.3	7.6-30.5 μmol/L
Creatinine (S)	0.6-1.2 mg/dL	88.4	53-106 μmol/L
Creatinine clearance (P)	107-139 mL/min	0.0167	1.78-2.32 mL/s
Fatty acids (total) (S)	8-20 mg/dL	0.01	0.08-2.00 mg/L
Fibrinogen (P)	200-400 mg/dL	0.01	2.00-4.00 g/L
Gamma globulin (S)	0.5-1.6 g/dL	10	5-16 g/L
Globulins (total) (S)	2.3-3.5 g/dL	10	23-35 g/L
Glucose (fasting) (S)	70-110 mg/dL	0.055	3.85-6.05 mmol/L

Continued.

CONVERSION FACTORS (SI UNITS)

Component	Normal Range in Units as Customarily Reported	Conversion Factor	Normal Range in SI Units, Molecular Units, International Units, or Decimal Fractions
Insulin (radioimmunoassay) (P)	4-24 μIU/ml	0.0417	0.17-1.00 μg/L
	0.20-0.84 μg/L	172.2	35-145 pmol/L
Iodine, BEI (S)	3.5-6.5 μg/dL	0.079	0.28-0.51 μmol/L
Iodine, PBI (S)	4.0-8.0 μg/dL	0.079	0.32-0.63 μmol/L
Iron, total (S)	60-150 μg/dL	0.179	11-27 μmol/L
Iron-binding capacity (S)	300-360 μg/dL	0.179	54-64 μmol/L
17-Ketosteroids (P)	25-125 μg/dL	0.01	0.25-1.25 mg/L
Lactic dehydrogenase (S)	80-120 units at 30 °C Lactate → pyruvate	0.48	38-62 U/L at 30 °C
	100-190 U/L at 37 °C	1	100-190 U/L at 37 °C
Lipase (S)	0-1.5 U/ml (Cherry-Crandall)	278	0-417 U/L
Lipids (total) (S)	400-800 mg/dL	0.01	4.00-8.00 g/L
Cholesterol	150-250 mg/dL	0.026	3.9-6.5 mmol/L
Triglycerides	75-165 mg/dL	0.0114	0.85-1.89 mmol/L
Phospholipids	150-380 mg/dL	0.01	1.50-380 g/L
Free fatty acids	9.0-15.0 mM/L	1	9.0-15.0 mmol/L
Nonprotein nitrogen (S)	20-35 mg/dL	0.714	14.3-25.0 mmol/L
Phosphatase (P)			
Acid (units/dL)	Cherry-Crandall	2.77	0-5.5 U/L
	King-Armstrong	1.77	0-5.5 U/L
	Bodansky	5.37	0-5.5 U/L
Alkaline (units/dL)	King-Armstrong	1.77	30-120 U/L
	Bodansky	5.37	30-120 U/L
	Bessey-Lowry-Brock	16.67	30-120 U/L
Phosphorus, inorganic (S)	3.0-4.5 mg/dL	0.323	0.97-1.45 mmol/L
Potassium (P)	3.8-5.0 mEq/L	1	3.8-5.0 mmol/L
Proteins, total (S)	6.0-7.8 g/dL	10	60-78 g/L
Albumin	3.2-4.5 g/dL	10	32-45 g/L
Globulin	2.3-3.5 g/dL	10	23-35 g/L
Sodium (P)	136-142 mEq/L	1	136-142 mmol/L
Testosterone: Male (S)	300-1,200 ng/dL	0.035	10.5-42.0 nmol/L
Female	30-95 ng/dL	0.035	1.0-3.3 nmol/L
Thyroid tests (S)			
Thyroxine (T_4)	4-11 μg/dL	12.87	51-142 nmol/L
T_4 expressed as iodine	3.2-7.2 μg/dL	79.0	253-569 nmol/L
T_3 resin uptake	25%-38% relative uptake	0.01	0.25%-0.38% relative uptake
TSH (S)	10 μU/mL	1	$<10^{-3}$ IU/L
Urea nitrogen (S)	8-23 mg/dL	0.357	2.9-8.2 mmol/L
Uric acid (S)	2-6 mg/dL	59.5	0.120-0.360 mmol/L
Vitamin B_{12} (S)	160-195 pg/mL	0.74	118-703 pmol/L
Hematology Values*			
Red cell volume (male)	25-35 mL/kg body weight	0.001	0.025-0.035 L/kg body weight
Hematocrit	40%-50%	0.01	0.40-0.50
Hemoglobin	13.5-18.0 g/dL	10	135-180 g/L
Hemoglobin	13.5-18.0 g/dL	0.155	2.09-2.79 mmol/L
RBC count	$4.5\text{-}6 \times 10^6/\mu L$	1	$4.6\text{-}6 \times 10^{12}/L$
WBC count	$4.5\text{-}10 \times 10^3/\mu L$	1	$4.5\text{-}10 \times 10^9/L$
Mean corpuscular volume	80-96 μm³	1	80-96 fL

*The International Committee for Standardization in Hematology recommends that the numbers remain the same but that the units change, so that hemoglobin is expressed as grams per deciliter (g/dL) even though other measurements are expressed as units per liter (U/L).

Glossary

A

abducens nerve (ab-DOO-sens nerv) cranial nerve VI; motor nerve; controls movement of the eye and proprioception

abduction (ab-DUK-shun) moving away from the midline of the body, opposite motion of adduction

abruptio placentae (ab-RUP-shee-o plah-SEN-tah) separation of normally positioned placenta from the uterine wall; may result in hemorrhage and death of the fetus and/or mother

abscess (AB-ses) cavity, often pus filled, formed by disintegration of tissues

absolute refractory period (AB-so-loot ree-FRAK-to-ree PE-ree-od) time during which the local area of the membrane has surpassed the threshold potential and will not respond to any stimulus

absorption (ab-SORP-shun) passage of a substance through a membrane such as skin or mucosa; often refers to passage of nutrients into blood

accessory gland (ak-SES-o-ree gland) a gland that assists organs in accomplishing their functions

accessory nerve (ak-SES-o-ree nerve) cranial nerve XI; motor nerve

accessory organ (ak-SES-o-ree OR-gan) an organ that assists other organs in accomplishing their functions

accommodation (ah-kom-o-DAY-shun) mechanism that allows the normal eye to focus on objects closer than 20 feet

acetylcholinesterase (as-e-til-ko-lin-ES-ter-ase) enzyme that rapidly inactivates acetylcholine bound to postsynaptic receptors

acid (AS-id) substance that ionizes in water to release hydrogen ions; substance with a pH of less than 7.0

acidophil (AS-id-o-fil) cell that stains easily with acid dyes

acini (AS-i-nee) pancreatic cells that secrete serous fluid with digestive enzymes

acne (AK-nee) see acne vulgaris

acne vulgaris (AK-nee vul-GAR-is) inflammatory skin condition affecting sebaceous gland ducts; occurs most frequently in adolescent years; see comedo

acoustic nerve (ah-KOOS-tik nerv) cranial nerve VIII; see vestibulocochlear nerve

acquired immunity (ah-KWIRED i-MYOO-ni-tee) immunity that is obtained after birth through exposure to a harmful agent

acromegaly (ak-ro-MEG-ah-lee) condition caused by hypersecretion of growth hormone after puberty, resulting in enlargement of facial features (for example, jaw, nose), fingers, and toes

acrosome (AK-ro-sohm) specialized structure on the sperm head containing enzymes that break down the covering of the ovum to facilitate conception

actin (AK-tin) contractile protein found in the thin myofilaments of skeletal muscle; see sliding filament theory

action potential (AK-shun po-TEN-shal) nerve impulse; membrane potential of an active neuron

active transport (AK-tiv TRANS-port) movement of a substance into or out of a living cell requiring the use of cellular energy

acuity (ah-KYOO-i-tee) sharpness of visual perception

acute (ah-KYOOT) intense; short in duration—as in acute disease

adaptation (ad-ap-TAY-shun) condition of many sensory receptors in which the magnitude of a receptor potential decreases over a period of time in response to a continuous stimulus

adduction (ah-DUK-shun) moving toward the midline of the body, opposite motion of abduction

adenine (AD-e-neen) one of the nitrogenous bases in RNA and DNA

adenocarcinoma (ad-e-no-kar-si-NO-mah) cancer of glandular epithelium

adenofibroma (ad-e-no-fi-BRO-mah) benign neoplasm formed by epithelial and connective tissues

adenohypophysis (ad-e-no-hye-POF-i-sis) anterior pituitary gland, which has the structure of an endocrine gland

adenoid (AD-e-noyd) literally, glandlike; adenoids, or pharyngeal tonsils, are paired lymphoid structures in the nasopharynx

adenoma (ad-e-NO-mah) benign tumor of glandular epithelium

adenosine triphosphate (ATP) (ah-DEN-o-sen try-FOS-fate) chemical compound that provides energy for use by body cells

adenyl cyclase (AD-e-nil SYE-klas) enzyme that promotes ATP change into cyclic AMP

adipose tissue (AD-i-pose TISH-yoo) fat tissue

adolescence (ad-o-LES-ens) period between puberty, the onset of reproductive maturity, and adulthood

adrenal (ah-DREE-nal) near the kidney, as in adrenal gland

adrenal cortex (ah-DREE-nal KOR-teks) outer portion of adrenal gland that secretes hormones called corticoids

adrenal gland (ah-DREE-nal gland) endocrine gland that rests on the top of each kidney; made up of cortex and medulla regions

adrenal medulla (ah-DREE-nal me-DUL-ah) inner portion of adrenal gland that secretes epinephrine and norepinephrine

adrenaline (ah-DREN-ah-len) see epinephrine

adrenergic (adre-NER-jik) describes axons whose terminals release norepinephrine and epinephrine

adrenocorticotropic hormone (ACTH) (ah-dree-no-kor-te-ko-TRO-pic HOR-mone) hormone that stimulates the adrenal cortex to secrete larger amounts of hormones

adulthood (ah-DULT-hood) period after adolescence

adult respiratory distress syndrome (ARDS) (ah-DULT re-SPY-rah-tor-ee di-STRESS SIN-drome) syndrome resulting from impairment or removal of surfactant in the alveoli

aerobic respiration (air-O-bik res-pi-RAY-shun) catabolic process; the stage of cellular respiration requiring oxygen

aerobic training (air-O-bik TRAIN-ing) continuous vigorous exercise requiring the body to increase its consumption of oxygen and develop the muscles' ability to sustain activity over a long period of time; also known as endurance training

afferent impulse (AF-fer-ent IM-pulse) impulse traveling toward the central nervous system

afferent nervous system (AF-fer-ent NER-vus sis-tem) subdivision of PNS; consists of all incoming sensory nerves

afferent neuron (AF-fer-ent NOO-ron) neuron that carries impulses toward the CNS from the periphery; sensory neuron

agglutinogen (ag-loo-TIN-o-jen) substance that stimulates agglu-tination, particularly of red blood cells; antigens present on red blood cell membranes

agranulocyte (ah-GRAN-u-lo-site) white blood cells without cytoplasmic granules

alarm reaction (ah-LARM ree-AK-shun) the initial response to stress

albinism (AL-bi-nizm) recessive, inherited condition characterized by a lack of the dark brown pigment melanin in the skin and eyes, resulting in vision problems and susceptibility to sunburn and skin cancer; ocular albinism is a lack of pigment in the layers of the eyeball

albumin (al-BYOO-min) plasma protein that aids in the regulation of the osmotic concentration of the blood

aldosterone (AL-doe-ste-rone) hormone that stimulates the kidney to retain sodium, ions, and water; only physiologically important mineralocorticoid

aldosteronism (al-do-STER-on-izm) hypersecretion of aldosterone

alimentary canal (al-e-MEN-tar-ee kah-NAL) the digestive tract as a whole

alkaline (AL-kah-lin) substance with a pH of greater than 7.0

all or none response, principle of when stimulated, a muscle fiber will contract fully or not at all; whether a contraction occurs depends on whether the stimulus reaches the required threshold

alpha cell (AL-fah-sell) pancreatic islet cell that secretes glucagon; also called A cell

alpha particle (AL-fah PAR-ti-kul) radioactive particle consisting of two protons plus two neutrons

alpha receptor (AL-fah ree-SEP-tor) adrenergic receptor for norepinephrine

alveolar cell (al-VEE-o-lar sell) milk-producing cell that releases secretions into ducts of the breast

alveolar duct (al-VEE-o-lar dukt) airway that branches from the smallest bronchioles; alveolar sacs arise from alveolar ducts

alveolar ventilation (al-VEE-o-lar ven-ti-LAY-shun) volume of inspired air that actually reaches the alveoli

alveolus (al-VEE-o-lus) literally, a small cavity; alveoli of lungs are microscopic saclike dilations of terminal bronchioles; gas exchange in the lungs occurs

across the membranes of the alveoli

amenorrhea (ay-men-o-REE-ah) absence of normal menstruation

amine (AM-een) organic compound containing nitrogen; neurotransmitter synthesized from amino acid molecules

amino acid (ah-MEE-no AS-id) structural units from which proteins are built

amino acid derivative hormone (ah-MEE-no AS-id de-RIV-ahtiv HOR-mone) category of nonsteroid hormones; each hormone is derived from a single amino acid molecule

amniocentesis (am-nee-o-sen-TEE-sis) procedure in which a sample of amniotic fluid is removed with a syringe for use in genetic testing, often to produce a karyotype of the fetus

amniotic cavity (am-nee-OT-ik KAV-i-tee) cavity within the blastocyst that eventually becomes a fluid-filled sac in which the embryo will float during development

amphiarthrosis (am-fee-ar-THRO-sis) slightly moveable joint such as the joint joining the two pubic bones

ampulla (am-PUL-ah) saclike dilation of a tube or duct; found at end of each semicircular canal, contains crista ampullaris

amygdaloid nucleus (ah-MIG-dah-loid NOO-klee-us) basal ganglia at the tip of the caudate nucleus

amylase (AM-i-lase) enzyme that digests carbohydrates

anabolic hormone (an-ah-BOL-ik HOR-mone) hormone that stimulates anabolism in the target organ

anabolism (ah-NAB-o-lizm) cells making complex molecules (for example, hormones) from simpler compounds (for example, amino acids); opposite of catabolism

anaerobic (an-air-O-bik) metabolic process that does not require the presence of oxygen

anaerobic respiration (an-air-O-bik respi-RAY-shun) catabolic process; stage of cellular respiration not requiring oxygen

anal triangle (AY-nal TRY-angul) area surrounding anus

anaphase (AN-ah-faze) latter stage of mitosis; duplicate chromosomes move to poles of dividing cell

anaplasia (an-ah-PLAY-zee-ah) growth of abnormal (undifferentiated) cells, as in a tumor or neoplasm

anastomosis (ah-nas-toe-MOE-sis) connection between vessels that allows collateral circulation

anatomical dead space (an-ah-TOM-i-kal ded space) air passageways that contain air that does not reach the alveoli

anatomical position (an-ah-TOM-i-kal po-ZISH-un) the standard reference position for the body, which gives meaning to directional terms

anatomy (ah-NAT-o-mee) study of the structure of an organism and the relationships of its parts

anemia (ah-NEE-mee-ah) deficient number of red blood cells, or deficient hemoglobin

anesthetic (an-es-THET-ik) substance that reduces or eliminates the sensation of pain

aneurysm (AN-yoo-rizm) abnormal widening of the arterial wall; aneurysms promote formation of thrombi and also tend to burst

angina pectoris (an-JYE-nah PECK-tor-is) severe chest pain resulting when the myocardium is deprived of sufficient oxygen

angioplasty (AN-jee-o-plas-tee) medical procedure in which vessels occluded by arteriosclerosis are opened (that is, the channel for blood flow is widened)

angiotensin I (an-jee-o-TEN-sin I) formed by conversion of angiotensinogen by renin; causes vasoconstriction and an increase in blood pressure

angiotensin II (an-jee-o-TEN-sin II) formed in the lungs by enzyme conversion of angiotensin I; ultimately stimulates secretion of aldosterone; causes vasoconstriction

angiotensinogen (an-jee-o-TEN-sin-o-jen) normal constituent of blood, it is a precursor to angiotensin

anion (AN-eye-on) negatively charged molecule

annulus fibrosis (AN-yoo-lus fye-BRO-sis) tough outer edge of intervertebral disk; surrounds nucleus pulposus

anorexia (an-o-REK-see-ah) loss of appetite (a symptom, rather than a distinct disorder)

anorexia nervosa (an-o-REK-see-ah ner-VO-sah) behavioral eating disorder characterized by chronic refusal to eat, often related to an abnormal fear of becoming obese

antagonism (an-TAG-o-nizm) action in which one hormone produces the opposite effect of another hormone

antagonist muscle (an-TAG-o-nist MUS-el) muscle that directly opposes prime movers; for example, muscles that flex the upper arm are antagonists to muscles that extend it

anterior (an-TEER-ee-or) front, or ventral; opposite of posterior, or dorsal

anterior cavity (of eye) (an-TEER-ee-or KAV-i-tee) entire space located in front of the lens of the eye; divided into anterior and posterior chambers

antibody (AN-ti-bod-ee) substance produced by the body

that destroys or inactivates a specific substance (antigen) that has entered the body

antibody-mediated immunity (AN-ti-bod-ee MEE-dee-ate-ed i-MYOO-ni-tee) immunity that is produced when antibodies render antigens unable to harm the body

antidiuresis (an-ti-dye-yoo-REE-sis) opposing the production of a large urine volume

antidiuretic hormone (ADH) (an-ti-dye-yoo-RET-ik HOR-mone) hormone produced in the posterior pituitary gland to regulate the balance of water in the body by accelerating reabsorption of water in the kidney tubules

antigen (AN-ti-jen) substance that, when introduced into the body, causes formation of antibodies against it

aortic baroreceptor (ay-OR-tik bar-o-ree-SEP-tor) stretch receptor located in the aorta that is sensitive to changes in blood pressure

aortic body (ay-OR-tik BOD-ee) small cluster of chemosensitive cells that respond to changes in blood levels of carbon dioxide and oxygen

aortic semilunar valve (ay-OR-tic sem-i-LOO-nar valve) valve between the aorta and left ventricle that prevents blood from flowing back into the ventricle

apatite (AP-ah-tite) crystals of calcium and phosphate that contribute to the hardness of bone

aphasia (ah-FAY-zee-ah) language deficit due to damage in speech centers of the brain

apnea (AP-nee-ah) temporary cessation of breathing

apneusis (ap-NYOO-sis) cessation of breathing in the inspiratory position

apneustic center (ap-NYOO-stik SEN-ter) located in the pons; stimulates the inspiratory center to increase the length and depth of inspiration

apocrine glands (AP-o-krin glands) glands that collect their secretions near the apex of the cell and then release them by pinching off the distended end; for example, mammary glands

apocrine sweat glands (AP-o-krin swet glands) sweat glands located in the axilla and genital regions; these glands enlarge and begin to function at puberty

aponeurosis (ap-o-nyoo-RO-sis) broad, flat sheet of connective tissue

appendicitis (ah-pen-di-SYE-tis) inflammation of the vermiform appendix

appendicular skeleton (a-pen-DIK-yoo-lar SKEL-e-ton) bones of the upper and lower extremities of the body

appetite center (AP-e-tite SEN-ter) cluster of neurons in the lateral hypothalmus whose impulses cause an increase in appetite

appositional growth (ap-o-ZISH-un-al growth) process by which a flat bone or cartilage grows in size

aqueous humor (AY-kwee-us HYOO-mor) watery fluid that fills the anterior chamber of the eye, in front of the lens

arachnoid (ah-RAK-noyd) web-like; particularly the middle layer of the meninges

arbor vitae (AR-bor VI-tay) internal white matter of the cerebellum

areolar (ah-REE-o-lar) a type of connective tissue consisting of fibers and a variety of cells embedded in a loose matrix of soft, sticky gel

arrector pili muscle (ah-REK-tor PYE-lie MUS-el) smooth muscles of the skin, attached to hair follicles; when contraction occurs, the hair stands up, resulting in "goose flesh"

arrhythmia (ah-RITH-mee-ah) term referring to any abnormality of cardiac rhythm

arteriole (ar-TEER-ee-ole) small branch of an artery

arteriosclerosis (ar-tee-ree-o-skle-RO-sis) hardening of arteries; materials such as lipids (as in atherosclerosis) accumulate in arterial walls, often becoming hardened via calcification

artery (AR-ter-ee) vessel carrying blood away from the heart

arthritis (ar-THRY-tis) inflammatory joint disease

articular cartilage (ar-TIK-yoo-lar KAR-ti-lij) layer of hyaline cartilage covering the joint surfaces of epiphyses

articulation (ar-tik-yoo-LAY-shun) joint

artificial immunity (ar-ti-FISH-al i-MYOO-ni-tee) immunization; deliberate exposure to potentially harmful antigens; as opposed to natural immunity

artificial pacemaker (ar-ti-FISH-al PAYS-may-ker) an electrical device that is implanted into the heart to treat problems with heart conduction

ascending tract (ah-SEND-ing tract) spinal cord tract that conducts impulses up the cord to the brain

ascites (ah-SYE-tees) effusion in the abdominal cavity; abdominal bloating

association tract (ah-so-see-AY-shun trakt) most common cerebral tract; extends from one convolution to another in the same hemisphere

asthma (AZ-mah) obstructive pulmonary disorder characterized by recurring spasms of muscles in bronchial walls accompanied

by edema and mucus production, making breathing difficult

astigmatism (ah-STIG-mah-tizm) irregular curvature of the cornea or lens that impairs formation of a well-focused image in the eye

astrocyte (AS-tro-site) star shaped neuroglial cell

atherosclerosis (ath-er-o-skle-RO-sis) type of "hardening of the arteries" in which lipids and other substances build up on the inside wall of blood vessels

atom (AT-om) smallest particle of a chemical element that retains the properties of that element; particles that combine to form molecules (chemical building blocks)

atomic number (ah-TOM-ik NUM-ber) number of protons in an atom of an element

atomic weight (ah-TOM-ik wate) number of protons plus the number of neutrons in an atom of an element

atrial fibrillation (AY-tree-al fibri-LAY-shun) frequent premature contractions of the atrium

atrioventricular bundle (ay-tree-o-ven-TRIK-yoo-lar BUN-dul) bundle of specialized cardiac muscle fibers that extend from the AV node to the Purkinje fibers; involved in coordination of heart muscle contraction

atrioventricular node (ay-tree-o-ven-TRIK-yoo-lar node) a small mass of special cardiac muscle tissue; part of the conduction system of heart

atrioventricular (AV) valves (ay-tree-o-ven-TRIK-yoo-lar valves) two valves that separate the atrial chambers from the ventricles

atrium (AY-tree-um) one of the upper chambers of the heart; receives blood from either the systemic or pulmonary circulation

atrophy (AT-ro-fee) wasting away of tissue; decrease in size of a part; sometimes referred to as disuse atrophy

auditory nerve (AW-di-toe-ree nerv) cranial nerve VIII; *see* vestibulocochlear nerve

auditory ossicles (AW-di-toe-ree OS-si-kuls) tiny bones in middle ear: malleus, incus, and stapes; they function to amplify soundwaves passing from the eardrum to the membranes of inner ear

auditory tube (AW-di-toe-ree tube) tube that connects the throat with the middle ear and serves to equalize pressure between the middle ear and the exterior; also known as eustachian tube

auricle (AW-ri-kul) part of the ear attached to the side of the head; earlike appendage of each atrium of the heart

autonomic effector (aw-toe-NOM-ik ef-FEK-tor) tissues to

which autonomic neurons conduct impulses

autonomic nervous system (ANS) (aw-toe-NOM-ik NER-vus SIS-tem) subdivision of the peripheral nervous system, it regulates involuntary actions

autosome (AW-toe-sohm) one of the 44 (22 pairs) chromosomes in the human genome besides the two sex chromosomes; means "same body," referring to the fact that each member of a pair of autosomes matches the other in size and other structural features

avascular (ah-VAS-kyoo-lar) free of blood vessels

avitaminosis (ay-vye-tah-mi-NO-sis) general name for any condition resulting from a vitamin deficiency

axial skeleton (AK-see-al SKEL-e-ton) the bones of the head, neck, and torso

axillary tail of Spence (AK-si-lar-ee tail of Spens) an extension of breast tissue that is in physical contact with several very large lymph nodes

axon (AK-son) in a neuron, the single process that extends from the axon hillock and transmits impulses away from the cell body

axon collateral (AK-son ko-LAT-er-al) one or more side branches from the axon

axon hillock (AK-son HIL-ok) portion of the cell body from which the axon extends

B

bacterium (bak-TEE-ree-um) microbe capable of causing disease; a primitive, single-celled organism without membranous organelles

ball and socket joint (ball and SOK-et joynt) spheroid joint, such as shoulder or hip joint; most moveable type of joint

baroreflex (bar-o-REE-fleks) baroreceptors in blood vessels, operating in feedback loops to maintain homeostasis of blood pressure; also called pressoreflex

basal cell carcinoma (BAY-sal sell kar-si-NO-mah) one of the most common forms of skin cancer; usually occurs on the upper face

basal ganglia (BAY-sal GANG-glee-ah) islands of gray matter located in the cerebral cortex that are responsible for automatic movements and postures

basal lamina (BAY-sal LAM-i-nah) glycoprotein material secreted by epithelial cells

basal metabolic rate (BMR) (BAY-sal met-ah-BOL-ik rate) number of calories of heat that must be produced per hour by catabolism to keep the body alive, awake, and comfortably warm

base (base) substance that ionizes

in water to decrease the number of hydrogen ions; also known as alkaline

basement membrane (BASE-ment MEM-brane) the connective tissue layer of the serous membrane that holds and supports epithelial cells

base pair (base pare) adenine-thymine or cytosine-guanine; occurs when complementary bases from each helical chain of DNA are held together by hydrogen bonds

basilar membrane (BAS-i-lar MEM-brane) floor of the cochlear duct, this structure's vibrating response to sound frequencies leads to the transduction of sound waves into nerve impulses

basophil (BAY-so-fil) white blood cell that stains readily with basic dyes

benign (be-NINE) refers to a tumor or neoplasm that does not metastasize (spread to different tissues)

beta blocker (BAY-ta BLOCK-er) drug that blocks beta receptors and therefore prevents dilation of blood vessels and increased contraction of heart muscle

beta cell (BAY-tah sell) pancreatic islet cell that secretes insulin; also called B cell

beta particle (BAY-tah PAR-ti-kul) electrons formed in a radioactive atom's nucleus by a neutron breaking down into a proton and electron

beta receptor (BAY-tah ree-SEP-tor) adrenergic receptor that, when stimulated, causes vessels to dilate and heart muscle to contract faster and stronger

bicarbonate ion (bye-KAR-bo-nate EYE-on) HCO_3^-; serves an important role in maintaining normal blood pH

bilateral symmetry (bye-LAT-er-al SIM-e-tree) right and left sides of the body are mirror images of each other

biochemistry (bye-o-KEM-is-tree) science of chemistry of living organisms

biofeedback (bye-o-FEED-bak) method of learning to consciously control autonomic effectors

biological clock (bye-o-LOJ-i-kal klok) internal clock that governs hunger, sleeping, reproduction, and behavior

biopsy (BYE-op-see) procedure in which living tissue is removed from a patient for laboratory examination, as in determining the presence of cancer cells

Biot's breathing (BYE-ots BREETH-ing) breathing pattern characterized by repeated sequences of deep gasps and apnea due to increased intracranial pressure

bipolar neuron (bye-PO-lar NOO-ron) neuron with only one dendrite and only one axon

blackhead (BLACK-hed) when sebum accumulates, darkens, and enlarges some of the ducts of the sebaceous glands, as in the case of acne; also known as a comedo

blastocyst (BLAS-toe-sist) stage of developing embryo that implants in uterine wall; consists of hollow ball of cells plus an inner cell mass

blind spot (blind spot) point on retina, where blood vessels and nerves exit the eyeball; "blind" portion of visual field because there are no photoreceptors present in that portion of retina; also known as optic disk

blister (BLIS-ter) fluid-filled skin lesion; *see* vesicle

blood-brain barrier (blud-brane BAR-ee-er) structural and functional barrier formed by astrocytes and blood vessel walls in the brain; it prevents some substances from diffusing from the blood into brain tissue

blood coagulation (blud ko-ag-yoo-LAY-shun) process by which ruptured vessels are plugged up to stop bleeding and prevent loss of precious body fluids

blood serum (blud SE-rum) in a sample of blood, the pale yellowish liquid left after a clot forms

blood type (blud tipe) one of the four different types of blood; identified by certain antigens on red blood cells (A, B, AB, and O)

B lymphocyte (B LIM-fo-site) immune system cell that produces antibodies against specific antigens

body composition (BOD-ee kom-po-ZISH-un) percentages of the body made of lean tissue and fat tissue

Bohr effect (BOR e-FECT) when increased P_{CO_2} decreases the affinity between hemoglobin and oxygen

bolus (BOW-lus) rounded mass of food that is swallowed

bone (bone) highly specialized connective tissue whose matrix is hard and calcified

bone marking (bone MARK-ing) specific feature on an individual bone

bone seeking isotope (bone SEEK-ing EYE-so-tope) radioactive element that will substitute for calcium in apatite crystals of bone; causes damage to red marrow and other tissues by radioactive emissions

Bowman's capsule (BO-mens KAP-sul) in the kidney, the cup-shaped top of a nephron that surrounds the glomerulus

brachial plexus (BRAY-kee-al PLEK-sus) plexus located deep

in the shoulder that innervates the lower part of the shoulder and the entire arm

bradycardia (bray-de-KAR-dee-ah) slow heart rhythm (below 50 beats/min)

brain (brane) part of central nervous system contained within the cranium; consists of medulla oblongata, cerebellum, and cerebrum

brain stem (brane stem) part of brain containing the midbrain, pons, and medulla oblongata

brain waves (brane waves) fluctuating electrical activity occurring in the brain

bronchial tree (BRONG-kee-al tree) the trachea, two primary bronchi, and their many branches

bronchiole (BRONG-kee-ole) small branch of a bronchus

bronchitis (brong-KYE-tis) inflammation of the bronchi of the lungs, characterized by edema and excessive mucus production that causes coughing and difficulty in breathing (especially expiration)

brush border (brush BOR-der) lining of small intestine that resembles a brush; formed by microvilli

buccal cavity (BUK-al KAV-i-tee) oral cavity

buffer (BUF-er) compound that combines with an acid or with a base to form a weaker acid or base, thereby lessening the change in hydrogen-ion concentration that would occur without the buffer; often operates as buffer pairs

buffy coat (BUF-ee kote) in a centrifuged sample of blood, the thin layer of leukocytes and platelets located at the interface between packed red cells and plasma

bursa (BER-sah) small, cushion-like sacs found between moving body parts, making movement easier

bursitis (ber-SYE-tis) inflammation of a bursa

C

cachexia (kah-KEKS-e-ah) syndrome involving loss of weight, loss of appetite, and general weakness; usually associated with cancer

cadaver (kah-DAV-er) corpse preserved for anatomical study

calcitonin (CT) (kal-si-TOE-nin) hormone secreted by the thyroid that decreases calcium levels in the blood

callus (KAL-us) bony tissue that forms a sort of collar around the broken ends of fractured bone during the healing process; in the skin, abnormally thick stratum corneum found at points of friction

calorie (c) (KAL-or-ree) heat unit; the amount of heat needed to raise the temperature of 1 g of water 1° C

Calorie (C) (KAL-or-ree) heat unit; kilocalorie; the amount of heat needed to raise the tempeature of 1 kilogram of water 1° C

calyx (KAY-liks) cup-shaped division of the renal pelvis

canaliculus (kan-ah-LIK-yoo-lus) an extremely narrow tubular passage or channel in compact bone; radiates from lacunae and connects with other lacunae and the haversian canal

canal of Schlemm (kah-NAL of shlem) a ring-shaped venous sinus located deep within the anterior portion of the sclera

cancellous bone (KAN-se-lus bone) bone containing trabeculae; also known as spongy bone

cancer (KAN-cer) malignant cellular neoplasm (tumor) that invades other cells and often metastasizes to many parts of the body

capacitation (kah-pas-i-TAY-shun) complex process needed for a mature sperm to become capable of fertilizing an ovum

capillary (KAP-i-lair-ee) tiny vessels that connect arterioles and venules; gas exchange from blood to tissues occurs in capillaries

carbohydrate (kar-bo-HYE-drate) organic compounds containing carbon, hydrogen, and oxygen in certain specific proportions; for example, sugars, starches, and cellulose

carbonic acid (kar-BON-ik AS-id) product of the reaction between carbon dioxide and water

carbonic anhydrase (kar-BON-ik an-HYE-drays) the enzyme that converts carbon dioxide into carbonic acid

carbuncle (KAR-bung-kul) a mass of connected boils, pus-filled lesions associated with hair follicle infections; *see* furuncle

carcinogen (kar-SIN-o-jen) substance that promotes the development of cancer

carcinoma (kar-si-NO-mah) malignant tumors that arise from epithelial tissue

cardiac cycle (KAR-dee-ak SYE-kul) complete heartbeat consisting of diastole and systole of both atria and both ventricles

cardiac muscle (KAR-dee-ak MUS-el) specialized muscle that makes up the heart

cardiac output (KAR-dee-ak OUT-put) volume of blood pumped by one ventricle per minute

cardiac plexus (KAR-dee-ak PLEK-sus) combination of sympathetic and parasympathetic nerves that are located near the aortic arch

cardiac sphincter (KAR-dee-ak SFINGK-ter) a ring of muscle between the stomach and esophagus that prevents food from reentering the esophagus when the stomach contracts

cardiac tamponade (KAR-dee-ak tam-po-NAHD) compression of the heart caused by fluid buildup in the pericardial space, as in pericarditis or mechanical damage to the pericardium

cardiomyopathy (kar-dee-o-my-OP-ah-thee) general term for disease of the myocardium (heart muscle)

cardiopulmonary resuscitation (CPR) (kar-dee-o-PUL-mo-nair-ree ree-sus-i-TAY-shun) combined external cardiac (heart) massage and artificial respiration

caries (KARE-eez) decay of teeth or of bone

carotene (KAR-o-teen) yellowish pigment; contributes to skin color

carpal tunnel syndrome (KAR-pal TUN-el SIN-drome) muscle weakness, pain, and tingling in the radial side (thumb side) of the wrist, hand, and fingers—perhaps radiating to the forearm and shoulder; caused by compression of the median nerve within the carpal tunnel (a passage along the ventral concavity of the wrist)

carrier (KARE-ee-er) in genetics, a person who possesses the gene for a recessive trait, but who does not actually exhibit the trait

cartilage (KAR-ti-lij) a specialized, fibrous connective tissue that has the consistency of a firm plastic or gristlelike gel

catabolism (kah-TAB-o-lizm) breakdown of food compounds or cytoplasmic constituents into simpler compounds; opposite of anabolism

catalyst (KAT-ah-list) chemical that speeds up reactions without being changed itself

cataract (KAT-ah-rakt) opacity of the lens of the eye

catecholamines (kat-e-kol-AM-eens) norepinephrine, epinephrine, and dopamine

catheterization (kath-e-ter-i-ZAY-shun) passage of a flexible tube (catheter) into the body through an exterior opening (as into the bladder through the urethra for the withdrawal of urine)

cation (KAT-eye-on) positively charged particle

cauda equina (KAW-dah e-KWINE-ah) "horse's tail"; lower end of spinal cord with its attached spinal nerve roots

celiac ganglion (SE-lee-ak GANG-gle-on) solar plexus; collateral ganglion that lies just below the diaphragm

cell (sell) basic biological and structural unit of the body consisting of a nucleus surrounded by cytoplasm and enclosed by a membrane

cellulose (SEL-yoo-lose) dietary fiber; major component of most plant tissue; nondigestable by humans

central nervous system (CNS) (SEN-tral NER-vus SIS-tem) the brain and spinal cord

central sulcus (fissure of Rolando) (SEN-tral SUL-kus (FISH-ur of rol-LAN-do)) groove between frontal and parietal lobes of the cerebrum

centriole (SEN-tree-ol) one of a pair of tiny cylinders in the centrosome of a cell; believed to be involved with the spindle fibers formed during mitosis

centromere (SEN-tro-meer) a beadlike structure that attaches one chromatid to another during the early stages of mitosis

centrosome (SEN-tro-some) area of the cytoplasm near the nucleus that coordinates the building and breaking of microtubules in the cell

cephalic phase (se-FAL-ik faze) stage in regulation of stomach secretion in which mental factors stimulate gastric juice secretion

cerebellum (sair-e-BELL-um) second largest part of the human brain; plays an essential role in the production of normal movements

cerebral cortex (se-REE-bral KOR-teks) thin layer of gray matter made up of neuron dendrites and cell bodies that compose the surface of the cerebrum

cerebral hemisphere (se-REE-bral HEM-i-sfer) either of the two right and left halves of the cerebrum

cerebrospinal fluid (CSF) (se-ree-bro-SPY-nal FLOO-id) plasma-like fluid that fills the subarachnoid space in the brain and spinal cord and in the cerebral ventricles

cerebrovascular accident (CVA) (se-ree-bro-VAS-kyoo-lar AK-si-dent) a hemorrhage or cessation of blood flow through cerebral blood vessels resulting in destruction of neurons; commonly called a stroke

cerebrum (se-REE-brum) largest and uppermost part of the human brain that controls consciousness, memory, sensations, emotions, and voluntary movements

cerumen (se-ROO-men) ear wax

ceruminous gland (se-ROO-mi-nus gland) gland that produces a waxy substance called cerumen (ear wax)

cervical plexus (SER-vi-kal PLEK-sus) plexus located deep within the neck; innervates muscles and skin of the neck, upper shoulder, and part of the head

cervix (SER-viks) neck; particularly the inferior necklike portion of the uterus

cesarean section (se-SAIR-ee-an SEK-shun) surgical removal of a fetus, often through an incision of the skin and uterine wall; also called C-section

chemical bond (KEM-i-kal bond) energy relationship joining two or more atoms; involves sharing or exchange of electrons

chemical digestion (KEM-i-kal di-JEST-chun) changes in chemical composition of food as it passes through the alimentary canal

chemoreceptors (kee-mo-ree-SEP-tors) receptors that respond to chemicals; responsible for taste and smell and monitoring concentration of specific chemicals in the blood

chemotaxis (kee-mo-TAK-sis) process in which a substance attracts cells or organisms into (or away from) its vicinity; for example, when white blood cells move toward the source of inflammation mediators

chemotherapy (kee-mo-THER-ah-pee) technique of using chemicals to treat disease (for example, infections, cancer)

Cheyne-Stokes respiration (CSR) (chain stokes res-pi-RAY-shun) pattern of breathing associated with critical conditions such as brain injury or drug overdose and characterized by cycles of apnea and hyperventilation

childhood (CHILD-hood) period of human development from infancy to puberty

chloride shift (KLOR-ide shift) diffusion of chloride ions into red blood cells as bicarbonate ions diffuse out; maintains electrical neutrality of red blood cells

cholecystectomy (ko-lee-sis-TEK-to-me) surgical removal of the gall bladder

cholecystitis (koh-lee-sis-TYE-tis) inflammation of the gallbladder

cholecystokinin (CCK) (ko-le-sis-toe-KYE-nin) hormone secreted from the intestinal mucosa of the duodenum that stimulates contraction of the gall bladder, resulting in bile flowing into the duodenum

cholesterol (ko-LES-ter-ol) steroid lipid found in many body tissues and in animal fats

cholinergic (ko-lin-ER-jik) having to do with acetylcholine

chondral fracture (KON-dral FRAK-chur) fracture of an articular surface in a synovial joint

chondrocyte (KON-dro-site) cartilage cell

chondroma (kon-DRO-mah) benign tumor of cartilage

chondromalacia patellae (kon-dro-may-LAY-she-ah pa-TEL-ah) degenerative process that results in a softening of the articular surface of the patella

chordae tendineae (KOR-dee ten-DIN-ee) stringlike structures that attach the AV valves to the wall of the heart

chorion (KO-ree-on) outermost fetal membrane; contributes to tissues in the placenta

chorionic villi (ko-re-ON-ik VIL-i) connection between blood vessels of the chorion and those of the placenta

chorionic villus sampling (ko-ree-ON-ik VIL-lus SAM-pling) procedure in which a tube is inserted through the (uterine) cervical opening and a sample of the chorionic tissue surrounding a developing embryo is removed for karyotyping

choroid (KO-royd) middle layer of the eyeball; contains a dark pigment to prevent the scattering of incoming light rays

choroid plexus (KO-royd PLEK-sus) specialized group of capillaries in ventricles of the brain that secrete cerebrospinal fluid

chromatid (KRO-mah-tid) either of the two DNA strands joined by a centromere existing after DNA has replicated (prior to cell division) but before the centromere has divided

chromatin (KRO-mah-tin) threadlike genetic material in the nucleus

chromosome (KRO-mo-sohm) barlike bodies of chromatin (DNA) that have coiled to form a compact mass during mitosis or meiosis; each chromosome is composed of regions called genes, each of which transmits hereditary information

chronic obstructive pulmonary disease (COPD) (KRON-ik ob-STRUK-tiv PUL-mo-nair-ee di-ZEEZ) general term referring to a group of disorders characterized by progressive, irreversible obstruction of air flow in the lungs; see bronchitis, emphysema, asthma

chyle (kile) milky fluid; the fat-containing lymph in the lymphatics of the intestine

chylomicron (ki-lo-MY-kron) small fat droplet

chyme (kime) partially digested food mixture leaving the stomach

cilia (SIL-ee-ah) hairlike projections of cells

ciliary body (SIL-e-air-e BOD-ee) thickening of the choroid that is located between the anterior margin of the retina and the posterior margin of the iris

ciliary muscle (SIL-e-air-e MUS-el) smooth muscle in the ciliary body of the eye that suspends the lens and functions in accommodation

circulatory system (SER-kyoo-lah-tor-ee SIS-tem) system composed of the heart, blood vessels, and lymphatic vessels, which permits transportation of material to and from all the cells of the body

circumcision (ser-kum-SIZH-un) surgical removal of the foreskin, or prepuce, of the penis

circumduction (ser-kum-DUK-shun) moving a part so its distal end moves in a circle

cirrhosis (sir-RO-sis) degeneration of liver tissue characterized by the replacement of damaged liver tissue with fibrous or fatty connective tissue

cisterna chyli (sis-TER-nah KYE-lye) an enlarged pouch on the thoracic duct that serves as a storage area for lymph moving toward its point of entry in to the venous system

cisternae (sis-TER-nay) tiny, membranous sacs that comprise the Golgi apparatus

citric acid cycle (SIT-rik AS-id SYE-kul) second series of chemical reactions in the process of glucose metabolism where carbon dioxide is formed and energy is released; it is an aerobic process; see Krebs cycle

cleavage line (KLEEV-ij line) see Langer's lines

cleft palate (kleft PAL-ate) facial deformity that is an X-linked inherited condition; when the palatine bones fail to unite completely

clitoris (KLIT-o-ris) small, erectile body located within the vestibule of the vagina

clonal selection theory (KLOE-nal se-LEK-shun) process through which a B or T cell, once sensitized through contact with an antigen, divides rapidly to create a colony of clones that destroys the "selecting" antigen

clone (klone) any of a family of many identical cells descended from a single "parent" cell

cochlea (KOKE-lee-ah) snail shell-like structure in the inner ear that houses the organ of Corti responsible for sense of hearing

cochlear duct (KOKE-lee-ar dukt) membranous tube within the bony cochlea; only part of the internal ear concerned with hearing

cochlear nerve (KOKE-kee-ar nerv) part of vestibulocochlear nerve (cranial nerve VIII); sensory nerve responsible for hearing

co-dominance (ko-DOM-i-nance) in genetics, a form of dominance in which two dominant versions of a trait are both expressed in the same individual

codon (KO-don) in RNA, a triplet of three base pairs that codes for a particular amino acid

co-enzyme (co-EN-zime) organic, nonprotein catalyst that acts as molecule carrier

colitis (ko-LYE-tis) any inflammatory condition of the colon and/or rectum

collagen (KOL-ah-jen) principal organic constituent of connective tissue

collateral ganglion (ko-LAT-er-al GANG-glee-on) sympathetic, prevertebral ganglion; named for nearby blood vessels

collecting duct (ko-LEK-ting dukt) in the kidney, straight tubule joined by the distal tubules of several nephrons

colloid (KOL-oyd) dissolved particles that resist separation from the gas, liquid, or solid medium in which they are dissolved

columnar (ko-LUM-nar) cell classification by shape in which cells are higher than they are wide

coma (KO-mah) altered state of consciousness from which an individual cannot be aroused

comedo (KOM-e-doe) inflamed, plugged sebaceous gland duct, common in acne conditions

commisural tract (kom-MIS-yoo-ral trakt) nerve tissue that connects the left and right hemispheres of the brain; see corpus callosum

common bile duct (KOM-on bile dukt) duct from the liver that empties into the duodenum; made up of the merging of the hepatic duct with the cystic duct

compact bone (kom-PAKT bone) dense bone; contains structural units called haversian systems

complement (KOM-ple-ment) any of several proteins normally present in blood serum that, when activated, kill foreign cells by puncturing them

complementary base pairing (kom-ple-MEN-ta-ree base PAIR-ing) bonding purines and pyridines in DNA; adenine always binds with thymine, and cytosine always binds with guanine

complete fracture (kom-PLEET FRAK-chur) fracture that totally divides a bone into separate pieces

compound fracture (kom-pound FRAK-chur) fracture in which the broken ends of the bone protrude through the skin

concentration gradient (kon-sen-TRAY-shun GRAY-dee-ent) measureable difference in concentration from one area to another

conduction (kon-DUK-shun) in human anatomy, the transfer of heat energy to the skin and then the external environment

conductivity (kon-duk-TIV-i-tee) ability of living cells and tissues to selectively transmit a wave of excitation from one point to another within the body

condyloid joint (KON-di-loid) see ellipsoidal joint

cone (cone) receptor cell located in the retina that is stimulated by bright light

congenital disorder (kon-JEN-i-tal dis-OR-der) term that refers to a condition present at birth; congenital conditions may be inherited or may be acquired in the womb or during delivery

congestive heart failure (CHF) (kon-JES-tiv hart FAIL-yoor) left heart failure; inability of the left ventricle to pump effectively, resulting in congestion in the systemic and pulmonary circulations

conjunctiva (kon-junk-TYE-vah) mucous membrane that lines the eyelids and covers the sclera (white portion of the eye)

conjunctivitis (kon-junk-ti-VYE-tis) inflammation of the conjunctiva, usually due to irritation, infection, or allergy

connective tissue (ko-NEK-tiv TISH-yoo) most abundant and widely distributed tissue in the body

consciousness (KON-shus-ness) state of awareness of one's self and environment and other beings

contact dermatitis (KON-takt der-mah-TYE-tis) local skin inflammation lasting a few hours or days after being exposed to an antigen

contact digestion (KON-tact di-JEST-chun) when substrates bind onto enzymes located on the surface of the brush border and complete carbohydrate digestion

contractility (kon-trak-TIL-i-tee) ability, as of muscle cell, to contract or shorten to produce movement

contralateral (kon-trah-LAT-er-al) on the opposite side of the body

contralateral reflex arc (kon-trak-LAT-er-al REE-fleks ark) reflex arc whose receptor and effectors are located on opposite sides of the body

convection (kon-VEK-shun) in human anatomy, transfer of heat energy to air that is flowing away from the skin

convergence (kon-VER-jens) movement of the two eyeballs inward so that their visual axes come together at the object viewed; also, when more than one presynaptic axon synapses with a single postsynaptic neuron

convulsion (con-VUL-shun) abnormal, uncoordinated tetanic contractions of varying groups of muscles

cornea (KOR-nee-ah) transparent, anterior portion of the sclera

coronal plane (ko-RO-nal plane) frontal plane; divides the body into front and back halves

coronary bypass surgery (KOR-o-nair-ee BYE-pass SER-jer-ee) surgery to relieve severely restricted coronary blood flow; veins are taken from other parts of the body and grafted in to bypass the blockage

coronary sinus (KOR-o-nair-ee SYE-nus) area that receives deoxygenated blood from the coronary veins and empties into the right atrium

cor pulmonale (kor pul-mo-NA-lee) failure of the right atrium and ventricle to pump blood effectively, resulting from obstruction of pulmonary blood flow

corpora cavernosa (KOR-por-ah kav-er-NO-sah) two columns of erectile tissue found in the shaft of the penis

corpora quadrigemina (KOR-por-ah kwod-ri-JEM-i-nah) midbrain landmark comprised of inferior and superior colliculi

corpus albicans (KOR-pus AL-bi-kans) white scar on ovary that replaces the degenerated corpus luteum

corpus callosum (KOR-pus kal-LO-sum) nerve tissue connecting the right and left cerebral hemispheres

corpus luteum (KOR-pus LOO-tee-um) a hormone-secreting glandular structure formed after ovulation at the site of the ruptured follicle; it secretes chiefly progesterone with some estrogen

corpus spongiosum (KOR-pus spun-jee-O-sum) a column of erectile tissue surrounding the urethra in the penis

cortex (KOR-teks) outer part of an internal organ; for example, the outer part of the cerebrum or the outer portion of the kidneys

corticosteroid (kor-ti-ko-STE-roid) glucocorticoid secreted by zona fasciculata of the adrenal cortex

cortisol (KOR-ti-sol) glucocorticoid secreted by zona fasciculata of the adrenal cortex; also known as hydrocortisone

cortisone (KOR-ti-sone) glucocorticoid secreted by zona fasciculata of the adrenal cortex

costal cartilage (KOS-tal KAR-ti-lij) cartilage that attaches ribs two through ten to the body of the sternum

coumarin (KOO-mar-in) compound that retards blood coagulation, often given as medication to prevent heart attacks

covalent bond (ko-VAL-ent bond) chemical bond formed by two atoms sharing one or more pairs of electrons

cramps (kramps) painful muscle spasms (involuntary twitches) that result from irritating stimuli, as in mild inflammation, or from ion imbalances

cranial nerve (KRAY-nee-al nerv) any of twelve pairs of nerves that attach to the undersurface of the brain and conduct impulses between the brain and structures in the head, neck, and thorax

craniosacral division (kray-nee-o-SAY-kral di-vi-shun) parasympathetic division of the autonomic nervous system

cranium (KRAY-nee-um) bony vault, made up of eight bones, that encases the brain

cremaster muscle (kre-MAS-ter MUS-el) muscle responsible for elevating the testes

cretinism (KREE-tin-izm) dwarfism caused by hyposecretion of the thyroid gland

cribriform plate (KRIB-ri-form plate) perforated portion of ethmoid bone that separates the nasal and cranial cavities

crista ampullaris (KRIS-tah am-pyoo-LAIR-is) specialized receptor organ located within the ampulla of the semicircular canals; detects head movements

cristae (KRIS-tee) folds of the inner membrane of a mitochondria

crossing-over (KROS-ing O-ver) phenomenon that occurs during meiosis when pairs of homologous chromosomes synapse and exchange genes

cuboidal (KYOO-boyd-al) cell classification by shape in which cells resemble a cube

cupula (KYOO-pyoo-lah) within the crista ampularis of the semicircular canal; the gelatinous cap in which the hair cells are imbedded

Cushing's syndrome (KOOSH-ings SIN-drome) condition caused by hypersecretion of glucocorticoids from the adrenal cortex

cutaneous membrane (ku-TANE-ee-us MEM-brane) primary organ of the integumentary system; the skin

cuticle (KYOO-ti-kul) skin fold covering the root of the nail

cyanosis (sye-ah-NO-sis) condition of blueness, particularly of the skin due to inadequate oxygenation of the blood

cyclic AMP (SIK-lik A M P) one of several second messengers that delivers information inside the cell and thus regulates the cell's activity

cystic duct (SIS-tik dukt) joins with the common hepatic duct to form the common bile duct

cystic fibrosis (SIS-tik fye-BRO-sis) inherited disease involving abnormal chloride ion (Cl^-) transport; causes secretion of abnormally thick mucus and other problems

cytology (sye-TOL-o-jee) study of cells

cytolysis (sye-TOL-i-sis) literally, "cell bursting"; often oc-

curs when ions and water rush into a cell, causing it to burst

cytoplasm (SYE-toe-plazm) gel-like substance of a cell exclusive of the nucleus and other organelles

cytoskeleton (sye-to-SKEL-e-ton) cell's internal supporting framework

cytotoxic T cell (sye-toe-TOK-sik T sell) "cell killing" T cell

D

deamination (dee-am-i-NAY-shun) removal of an amino group from an amino acid to form a molecule of ammonia and one of keto acid; occurs in the liver as first step in protein catabolism

decarboxylase (dee-kar-BOK-si-lase) enzyme that removes carbon dioxide

deciduous teeth (de-SID-yoo-us teeth) commonly referred to as baby teeth; teeth that shed at a certain age prior to development of permanent teeth

decomposition reaction (de-kom-po-ZISH-un ree-AK-shun) chemical reaction that breaks down a substance into two or more simpler substances

decubitus ulcer (de-KYOO-bi-tus UL-ser) area of destroyed tissue resulting from inadequate blood supply that often develops when lying in one position for prolonged periods

deep (deep) farther away from the body's surface, as opposed to superficial

defecation (def-e-KAY-shun) expelling feces from the digestive tract

defibrillation (de-fib-ri-LAY-shun) application of an electric shock to force cardiac muscle fibers to contract in unison

deglutition (deg-loo-TISH-un) swallowing

dehydration (dee-hye-DRAY-shun) clinical term that refers to an abnormal loss of fluid from the body's internal environment

dehydration synthesis (dee-hye-DRAY-shun SIN-the-sis) anabolic process by which molecules are joined to form larger molecules

delta cell (DEL-tah sell) pancreatic islet cell that secretes somatostatin; also called D cell

dendrite (DEN-drite) branching or treelike nerve cell process that receives input from other neurons and transmits impulses toward the cell body

dentate nuclei (DEN-tate NOO-klee-i) paired cerebellar nuclei that are connected by tracts with the thalamus, as well as motor areas of the cerebral cortex

deoxyribose (dee-ok-se-RYE-bose) sugar in DNA whose molecules contain only five carbons

deoxyribonucleic acid (DNA) (dee-ok-see-rye-bo-noo-KLEE-ik AS-id) genetic material of the cell that carries the chemical "blueprint" of the body

depolarization (dee-po-lar-i-ZAY-shun) electrical activity that triggers a contraction of the heart muscle

dermal papillae (DER-mal pah-PIL-ee) upper region of the dermis that forms part of the dermal-epidermal junction and forms the ridges and grooves of fingerprints

dermatitis (der-mah-TIE-tis) general term referring to any inflammation of the skin

dermatology (der-mah-TOL-o-jee) study of the integument and its diseases

dermatome (DER-mah-tohm) skin surface areas supplied by a single spinal nerve

dermatosis (der-mah-TOE-sis) general term meaning "skin condition"

dermis (DER-mis) the deeper of the two major layers of the skin, composed of dense fibrous connective tissue interspersed with glands, nerve endings, and blood vessels

descending tract (de-SEND-ing trakt) bundle of axons in the spinal cord that conduct impulses down the cord from the brain

desmosomes (DES-mo-somes) specialized junctions that hold adjacent cells together; consist of dense plate at point of adhesion plus extracellular cementing material

desquamation (des-kwah-MAY-shun) shedding of epithelial elements from the skin surface

detrusor muscle (dee-TROO-sor MUS-el) smooth muscle tissue making up the wall of the bladder

developmental biology (de-vel-op-MEN-tal bye-OL-o-jee) branch of biology that studies the process of change over the life cycle

deviated septum (DEE-vee-ay-ted SEP-tum) abnormal condition in which the nasal septum is far from its normal position, possibly obstructing normal nasal breathing

diabetes insipidus (dye-ah-BEE-teez in-SIP-i-dus) condition resulting from hyposecretion of ADH in which large volumes of urine are formed and, if left untreated, may cause serious health problems

diabetes mellitus (dye-ah-BEE-teez mell-EYE-tus) condition resulting when the pancreatic islets secrete too little insulin, resulting in increased levels of blood glucose

diad (dye-AD) double structure of T tubules in cardiac muscle fiber

dialysis (dye-AL-i-sis) separation of smaller (diffusible) particles from larger (nondiffusible) particles through a semipermeable membrane

diapedesis (dye-ah-pee-DEE-sis) passage of any formed elements within blood through the vessel wall, as in movement of white cells into the area of injury and infection

diaphragm (DYE-ah-fram) the flat muscular sheet that separates the thorax and abdomen; a major muscle of respiration

diaphysis (dye-AF-i-sis) shaft of a long bone

diarrhea (dye-ah-REE-ah) abnormally frequent defecation of more or less liquid feces

diarthrosis (dye-ar-THRO-sis) freely movable joint

diastasis (dye-AS-tah-sis) reduced ventricular filling of the heart

diastole (dye-AS-toe-lee) relaxation of the heart (especially the ventricles), during which it fills with blood; opposite of systole

diastolic pressure (dye-ah-STOL-ik PRESH-ur) blood pressure in arteries during diastole (relaxation) of the heart; clinically more important than systolic pressure

diencephalon (dye-en-SEF-ah-lon) "between" brain; parts of the brain between the cerebral hemispheres and the mesencephalon, or midbrain

differential count (dif-er-EN-shal kownt) percentage of total white blood cell count of the different types of leukocytes

differentiation (dif-er-EN-she-AY-shun) process of the development of cell specialization

diffusion (di-FYOO-shun) spreading; natural tendency of small particles to spread out evenly within any given space; for example, scattering of dissolved particles

digestion (di-JEST-chun) breakdown of food materials either mechanically (that is, through chewing) or chemically (that is, via digestive enzymes)

diploid number (DIP-loid NUM-ber) normal number of chromosomes per somatic cell (46 in humans)

diplopia (di-PLO-pe-ah) double vision

dissection (di-SEK-shun) cutting technique used to separate body parts for study

dissociate (di-SO-see-ate) when a compound breaks apart in solution forming ions that are surrounded by solvent molecules

distal (DIS-tal) toward the end of a structure; opposite of proximal

distal tubule (DIS-tal TOOB-yool) in the kidney, the part of the tubule distal to the ascending limb of the loop of Henle

that terminates in a collecting tubule; also known as distal convoluted tubule

divergence (dye-VER-jens) when a single presynaptic axon synapses with more than one different posynaptic neuron

diverticulitis (dye-ver-tik-yoo-LIE-tis) inflammation of diverticula (abnormal outpouchings) of the large intestine, possibly causing constipation

dominant (DOM-i-nant) in genetics, term referring to genes that have effects that appear in the offspring (dominant forms of a gene are often represented by upper case letters); *see* recessive

dopamine (DOE-pah-meen) chemical neurotransmitter

dorsal ramus (DOR-sal RAY-mus) branch of spinal nerve that supplies somatic motor and sensory fibers to several smaller nerves

dorsal root ganglion (DOR-sal root GANG-gle-on) small region of gray matter in dorsal nerve root made up of cell bodies of unipolar, sensory neurons

dorsiflexion (dor-si-FLEK-shun) when the top of the foot is elevated (brought toward the front of the lower leg) with toes pointing upward

double bond (DUB-ul bond) covalent chemical bond in which two pairs of electrons are shared

Down's syndrome (downz SIN-drome) group of symptoms usually caused by trisomy of chromosome 21; characterized by mental retardation and multiple structural defects, including facial, skeletal, and cardiovascular abnormalities

ductus arteriosus (DUK-tus ar-teer-ee-O-sus) in the developing fetus, it connects the aorta and the pulmonary artery, allowing most blood to bypass the fetus' developing lungs

ductus deferens (DUK-tus DEF-er-enz) *see* vas deferens

ductus venosus (DUK-tus ve-NO-sus) continuation of the umbilical vein that shunts blood returning from the placenta past the fetus' developing liver directly into the inferior vena cava

duodenum (doo-o-DEE-num) first subdivision of the small intestine; where most chemical digestion occurs

dura mater (DOO-rah MAH-ter) literally "strong or hard mother"; outermost layer of the meninges

dural sinuses (DOO-ral SYE-nus) name for large veins of cranial cavity

dwarfism (DWARF-izm) condition of abnormally small stature, sometimes resulting from hyposecretion of growth hormone

dynamic equilibrium (dye-NAM-ik e-kwi-LIB-ree-um) maintaining balance when the

head or body is rotated or suddenly moved

dyspnea (DISP-nee-ah) difficult or labored breathing

E

eccrine sweat gland (EK-rin swet gland) water-producing exocrine sweat glands widely dispersed throughout the skin

eclampsia (e-KLAMP-see-ah) potentially fatal condition associated with toxemia of pregnancy; characterized by convulsions and coma

ectoderm (EK-toe-derm) outermost of the primary germ layers that develops early in the first trimester of pregnancy; gives rise to the skin and the nervous system

ectomorph (EK-toe-morf) thin, lean body type

ectopic pregnancy (ek-TOP-ik PREG-nan-see) pregnancy in which the fertilized ovum develops in some place other than in the uterus (literally, "out of place")

eczema (EK-ze-mah) inflammatory skin condition associated with various diseases and characterized by erythema, papules, vesicles, and crusts

edema (e-DEE-mah) accumulation of fluid in a tissue, as in inflammation; swelling

effector (ee-FEK-ter) organ, gland, or muscle that responds to a nerve stimulus

efferent ductule (EF-fer-ent DUKT-yools) series of sperm ducts that drain the rete testis and pierce the tunica albuginea

efferent nervous system (EF-fer-ent NER-vus SIS-tem) subdivision of the peripheral nervous system (PNS) that consists of all outgoing motor nerves

efferent neuron (EF-fer-ent NOO-ron) neuron that transmits nerve impulses away from the central nervous system to muscles or glands (effectors)

ejaculation (ee-jak-yoo-LAY-shun) sudden discharging of semen from the body

ejaculatory duct (ee-JAK-yoo-lah-toe-ree dukt) duct formed by the joining of the ductus deferens and the duct from the seminal vesicle that allows sperm to enter the urethra

elastic cartilage (e-LAS-tik KAR-ti-lij) cartilage with elastic, as well as collagenous, fibers; provides elasticity and firmness, as in, for example, the cartilage of the external ear

elastin (e-LAS-tin) protein found in elastic fiber

electrocardiogram (ECG) (e-lek-tro-KAR-dee-o-gram) graphic record of the heart's action potentials

electrocardiograph (e-lek-tro-KAR-dee-o-graf) machine that produces electrocardiograms

electroencephalogram (EEG) (e-lek-tro-en-SEF-lo-gram) graphic representation of voltage changes in brain tissue used to evaluate nerve tissue function

electrolyte (e-LEK-tro-lite) substance that ionizes in solution, rendering the solution capable of conducting an electric current

electromyography (e-lek-tro-my-OG-rah-fee) recording electrical impulses from muscles as they contract

electron (e-LEK-tron) small, negatively charged subatomic particle

electron transport system (e-LEK-tron TRANS-port SIS-tem) carrier molecules imbedded in the inner membrane of the mitochondria that take high energy electrons from the citric acid cycle and form water and energy for oxidative phosphorylation

electrophoresis (e-lek-tro-fo-REE-sis) laboratory procedure in which different types of charged molecules are separated according to molecular weight using a weak electric current

element (EL-e-ment) substance composed of only one type of atom that cannot be broken into simpler constituents by chemical means

elephantiasis (el-e-fan-TIE-ah-sis) extreme lymphedema (swelling due to lymphatic blockage) in the limbs caused by a parasitic worm infestation; so called because the limbs swell to "elephant proportions"

elimination (e-lim-i-NAY-shun) *see* defecation

embryo (EM-bree-o) animal in early stages of intrauterine development; in humans, the first 8 weeks after conception

embryology (em-bree-OL-o-gee) study of the development of an individual from conception to birth

embryonic disk (em-bree-ON-ik disk) cells of the early embryo that differentiate into the three primary germ layers

emesis (EM-e-sis) vomiting

emission (e-MISH-un) reflex movement of spermatozoa and secretions from genital ducts and accessory glands into prostatic urethra; preceeds ejaculation

emmetropic (em-e-TROP-ik) relaxed normal eye

emphysema (em-fi-SEE-mah) abnormal condition characterized by trapping of air in alveoli of the lung that causes them to rupture and fuse to other alveoli

endocardium (en-doe-KAR-dee-um) thin layer of very smooth tissue lining each chamber of the heart

endochondral ossification (en-doe-KON-dral os-i-fi-KAY-shun) process by which bones are formed by replacement of cartilage models

endocrine (EN-doe-krin) secreting into blood or tissue fluid rather than into a duct; opposite of exocrine

endocrine reflex (EN-doe-krin REE-fleks) response that results from feedback loops within the endocrine system

endocrinology (en-doe-krin-OL-o-jee) study of the endocrine glands and their hormones

endocytosis (en-doe-sye-TOE-sis) process that allows extracellular material to enter the cell without actually passing through the plasma membrane

endoderm (EN-doe-derm) innermost layer of the primary germ layers that develops early in the first trimester of pregnancy; gives rise to digestive and urinary structures, as well as many other glands and organ parts

endogenous growth (en-DOJ-e-nus growth) *see* interstitial growth

endolymph (EN-doe-limf) clear potassium-rich fluid that fills the membranous labyrinth of the inner ear

endometrium (en-doe-MEE-tree-um) mucous membrane lining the uterus

endomorph (EN-doe-morf) body type characterized by excessive fat

endomysium (en-doe-ME-see-um) delicate connective tissue membrane covering the highly specialized skeletal muscle fibers

endoneurium (en-doe-NOO-ree-um) thin wrapping of fibrous connective tissue that surrounds each axon in a nerve

endoplasmic reticulum (ER) (endoe-PLAS-mik re-TIK-yoo-lum) network of tubules and vesicles in cytoplasm that contributes to cellular protein manufacture (via attached ribosomes) and distribution

endorphin (en-DOR-fin) chemical in central nervous system that influences pain perception; a natural painkiller

endosteum (en-DOS-tee-um) fibrous membrane that lines the medullary cavity of long bones

endothelium (en-doe-THEE-lee-um) squamous epithelial cells that line the inner surface of the entire circulatory system and the vessels of the lymphatic system

endotracheal intubation (en-doe-TRAY-kee-al in-too-BAY-shun) placing a tube in the trachea to ensure an open airway

endurance training (en-DOOR-ance TRAIN-ing) continuous vigorous exercise requiring the body to increase its consumption of oxygen and developing the muscles' ability to sustain activity over a prolonged period of time

enkephalin (en-KEF-ah-lin) peptide chemical in the central nervous system that acts as a natural painkiller

enterogastric reflex (en-ter-o-GAS-trik REE-fleks) nervous reflex causing inhibition of gastric peristalsis in response to the presence of acid and distention of duodenal mucosa; may also inhibit gastric secretion

enterogastrone (en-ter-o GAS-trone) hormone theorized to be involved with decreasing gastric peristalsis

enzyme (EN-zime) biochemical catalyst allowing chemical reactions to take place; functional proteins that regulate various metabolic pathways of the body

eosinophil (ee-o-SIN-o-fil) white blood cell, readily stained by eosin

ependymal cell (e-PEN-di-mal sell) cells that line the ventricles of the brain and the central canal of the spinal cord

epicardium (ep-i-KAR-dee-um) inner layer of the pericardium that covers the surface of the heart; it is also called the visceral pericardium

epidermis (ep-i-DER-mis) outermost layer of the skin; sometimes called the "false" skin

epidural space (ep-i-DOO-ral space) in the brain, the space above the dura mater

epiglottis (ep-i-GLOT-is) lidlike cartilage overhanging the entrance to the larynx

epimysium (ep-i-MIS-e-um) coarse sheet of connective tissue that covers a muscle as a whole

epinephrine (ep-i-NEF-rin) adrenaline; secretion of the adrenal medulla

epineurium (ep-i-NOO-ree-um) tough fibrous sheath that covers the whole nerve

epiphyseal fracture (ep-i-FEEZ-ee-al FRAK-chur) when the epiphyseal plate is separated from the epiphysis or diaphysis; this type of fracture can disrupt normal growth of the bone

epiphyseal plate (ep-i-FEEZ-ee-al plate) cartilage plate that is between the epiphysis and the diaphysis and allows growth to occur; sometimes referred to as a growth plate

epiphysis (e-PIF-i-sis) end of a long bone; also, the pineal body of the brain

episiotomy (e-piz-ee-OT-o-mee) surgical procedure used during birth to prevent a laceration of the mother's perineum or the vagina

epistaxis (ep-i-STAK-sis) clinical term referring to a bloody nose

epithalamus (ep-i-THAL-a-mus) small nuclei located outside the thalamus and hypothalamus; considered to be one of the structures of the diencephalon

epithelial membrane (ep-i-THEE-lee-al MEM-brane) membrane composed of epithelial tissue with an underlying layer of specialized connective tissue

epithelial tissue (ep-i-THEE-lee-al TISH-yoo) covers the body and its parts; lines various parts of the body; forms continuous sheets that contain no blood vessels; classified according to shape and arrangement

epitope (EP-i-tope) specific portion of an antigen that elicits an immune response

equatorial plate (e-kwah-TO-ree-al plate) plane at the "equator" of a cell during metaphase where the chromosomes align

equilibration (e-kwi-li-BRAY-shun) balance between opposing elements

erection (e-REK-shun) condition of erectile tissue when filled with blood; often refers to the penis' enlargement during sexual arousal

erythema (er-i-THEE-mah) reddening of the skin

erythroblastosis fetalis (e-rith-ro-blas-TOE-sis fe-TAL-sis) condition of a fetus or infant caused by the mother's Rh antibodies reacting with the baby's Rh-positive RBCs, characterized by massive agglutination of the blood and resulting life-threatening circulatory problems for the infant

erythrocytes (e-RITH-ro-sites) red blood cells

erythropoiesis (e-rith-ro-poi-EE-sis) process of red blood cell formation

erythropoietin (e-rith-ro-POI-e-tin) glycoprotein secreted to increase RBC production in response to oxygen deficiency

esophagus (e-SOF-ah-gus) muscular, mucus-lined tube that connects the pharynx with the stomach; also known as the foodpipe

essential fatty acid (ee-SEN-shal FAH-tee AS-id) unsaturated fatty acid that must be provided by the diet; serves as a source within the body for prostaglandin synthesis

essential reproductive organs (ee-SEN-shal ree-pro-DUK-tiv OR-gans) reproductive organs that must be present for reproduction to occur; the gonads

estrogen (ES-tro-jen) sex hormone secreted by the ovary that causes development and maintenance of female secondary sex characteristics and stimulates growth of the epithelial cells lining the uterus

eupnea (YOOP-nee-ah) normal respiration

eustachian tube (yoo-STAY-shun toob) *see* auditory tube

evaporation (ee-vap-o-RAY-shun) in anatomy and physiology, heat lost from the body by vaporization of liquid (sweat) from the skin

exchange reaction (eks-CHANJ ree-AK-shun) chemical reaction that breaks down a compound and then synthesizes a new compound by switching portions of the molecules; for example, AB + CD → AD + BC.

excitability (ek-site-ah-BIL-i-tee) ability of a muscle to be stimulated; also known as irritability

excitation (ek-sye-TAY-shun) occurs when a neuron is stimulated and additional Na$^+$ channels open

excitatory neurotransmitter (ek-SYE-tah-to-ree noo-ro-TRANS-mit-er) neurotransmitter that causes excitation (and thus, depolarization) of the postsynaptic neuron

excitatory postsynaptic potential (EPSP) (ek-SYE-tah-to-ree post-si-NAP-tik po-TEN-shal) temporary depolarization of postsynaptic membrane following stimulation

excretion (eks-KREE-shun) removal of waste products produced during body functions; occurs through defecation, urination, respiration, and through the skin

exocrine (EK-so-krin) secreting into a duct, as in glands that secrete their products via ducts onto a surface or into a cavity; opposite of endocrine

exocytosis (eks-o-sye-TOE-sis) process that allows large molecules to leave the cell without actually passing through the plasma membrane

exogenous growth (eks-OJ-e-nus growth) see appositional growth

expiration (eks-pi-RAY-shun) exhaling

expiratory center (eks-PYE-rah-tor-ee SEN-ter) one of the two most important respiratory control centers, located in the medulla

expiratory muscles (eks-PYE-rah-tor-ee MUS-els) muscles that allow more forceful expiration to increase the rate and depth of ventilation; internal intercostals and abdominal muscles

expiratory reserve volume (ERV) (eks-PYE-rah-tor-ee re-ZERV VOL-yoom) amount of air that can be forcibly exhaled after expiring the tidal volume (TV)

extensibility (eks-ten-si-BIL-i-tee) ability of a muscle to extend or stretch and return to resting length

extension (ek-STEN-shun) increasing the angle between two bones at a joint; as opposed to flexion

external auditory meatus (eks-TER-nal AW-di-tor-ee me-AY-tus) ear canal; a curved tube

(approximately 2.5 cm) extending from the auricle into the temporal bone, ending at the tympanic membrane

external ear (eks-TER-nal ear) outer part ear: auricle and external auditory canal

external genital (eks-TER-nal JEN-i-tal) external reproductive organs; penis and scrotum in males; vagina, vulva, and related structures in females

exteroceptor (eks-ter-o-SEP-tor) somatic sense receptor located on the body surface

extracellular fluid (ECF) (eks-trah-SELL-yoo-lar FLOO-id) water found outside of cells, located in two compartments: between cells (interstitial fluid) and in blood (plasma); lymph, cerebrospinal fluid, and joint fluids are also extracellular fluids

extrapyramidal tract (eks-trah-pi-RAM-i-dal trakt) motor tract from the brain to the spinal cord anterior horn motor neurons except the corticospinal tract

extrinsic eye muscle (eks-TRIN-sik eye MUS-el) voluntary muscle that attaches the eyeball to the socket and produces movement of the eye ball

extrinsic factor (eks-TRIN-sik FAK-tor) substance secreted in the stomach that allows vitamin B$_{12}$ to be absorbed by the body

F

facial nerve (FAY-shal nerv) cranial nerve VII, mixed nerve

facilitated diffusion (fah-SIL-i-tay-ted di-FYOO-shun) special type of diffusion; when movement of a molecule is made more efficient by action of carrier mechanisms in the plasma membrane

fallopian tubes (fal-LO-pee-an toobs) uterine tubes; pair of tubes that conduct the ovum from the ovary to the uterus

falx cerebelli (falks cer-e-BEL-li) small fold in the dura mater in the posterior cranial fossa

falx cerebri (falks CER-e-bri) fold in the dura mater that separates the two cerebral hemispheres

farsightedness (far-SITE-ed-ness) see hyperopia

fascicle (FAS-i-kl) small bundle or cluster, especially of groups of skeletal muscle fibers bound together by perimysium

fast muscle fiber (fast MUS-el FYE-ber) white muscle fiber; primarily relies on anaerobic respiration to produce ATP; responds quickly

fatty acid (FAH-tee AS-id) product of fat digestion; building block of fat molecules

fauces (FAW-seez) opening from the mouth into the oropharynx

febrile (FEB-ril) referring to fever

feces (FEE-seez) waste material discharged from the intestines

feedback control loop (FEED-bak kon-TROL loop) highly complex and integrated communication control network, classified as negative or positive; negative feedback loops are the most important and numerous homeostatic control mechanisms

fenestrations (fen-es-TRAY-shuns) perforations in the glomerular endothelium

fertilization (FER-ti-li-ZAY-shun) union of an ovum and a sperm; conception

fetal alcohol syndrome (FAS) (FEE-tal AL-ko-hol SIN-drome) a condition that may cause congenital abnormalities in a baby; it results from a woman consuming alcohol during pregnancy

fetus (FEE-tus) unborn young, especially in the later stages; in human beings, from the third month of the intrauterine period until birth

fibrillation (fi-bri-LAY-shun) condition in which individual muscle fibers, or small groups of fibers, contract asynchronously (out of time) with other muscle fibers in an organ (especially the heart), producing no effective movement

fibrin (FYE-brin) insoluble protein in clotted blood

fibrinogen (fye-BRIN-o-jen) soluble blood protein that is converted to insoluble fibrin during clotting

fibrinolysis (fye-brin-OL-i-sis) physiological mechanism that dissolves clots

fibroblast (FYE-bro-blast) connective tissue cell that synthesizes interstitial fibers and gels

fibrocartilage (fye-bro-KAR-ti-lij) cartilage with greatest number of collagenous fibers; strongest and most durable type of cartilage

fibrosarcoma (fye-bro-sar-KO-mah) cancer of fibrous connective tissue

fibrous connective tissue (FYE-brus ko-NEK-tiv TISH-yoo) strong, nonstretchable, white collagen fibers that compose tendons

fight-or-flight reaction (fite-or-flite ree-AK-shun) changes produced by increased sympathetic impulses allowing the body to deal with any type of stress

filtration (fil-TRAY-shun) movement of water and solutes through a membrane due to a higher hydrostatic pressure on one side

filum terminale (FYE-lum ter-mi-NAL-ee) slender filament formed by the pia mater that blends with the dura mater and then the periosteum of the coccyx

fimbriae (FIM-bree-ee) fringe

first-degree burn (first-de-GREE

bern) partial-thickness burn, actual tissue destruction is minimal

fissure (FISH-ur) groove

fixator muscle (fik-SAY-tor MUS-el) muscle that functions as a joint stabilizer

fixed-membrane-receptor hypothesis (fixed-MEM-brane-ree-SEP-tor hye-POTH-e-sis) see second-messenger hypothesis

flaccid muscle (FLAK-sid MUS-el) muscle with less tone than normal

flagellum (flah-JEL-um) single projection extending from the cell surface; only example in human is the "tail" of the male sperm

flavine adenine dinucleotide (FAD) (FLAY-vin AD-e-neen dye-NOO-klee-o-tide) molecule that serves as an electron carrier in the electron transport system

flexion (FLEK-shun) act of bending; decreasing the angle between two bones at the joint; opposite of extension

fluid balance (FLOO-id BAL-ans) homeostasis of fluids; the volumes of interstitial fluid, intracellular fluid, and plasma and total volume of water remain relatively constant

fluid compartments (FLOO-id kom-PART-ments) areas in the body where fluid is located; interstitial fluid, plasma, and intracellular fluid

fluid mosaic model (FLOO-id mo-ZAY-ik MO-del) theory of plasma membrane composition in which molecules of the membrane are bound tightly enough to form a continuous layer but loosely enough so molecules can slip past one another

follicle (FOL-li-kul) ovarian structure consisting of oocyte surrounded by numerous supporting cells (follicle cells); also, specialized structures required for hair growth; also, a small hollow sphere with a wall of simple cuboidal glandular epithelium, found in thyroid tissue

follicle-stimulating hormone (FSH) (FOL-li-kul STIM-yoo-lay-ting HOR-mone) hormone present in males and females; in males, FSH stimulates the production of sperm; in females, FSH stimulates the ovarian follicles to mature and follicle cells to secrete estrogen

follicular cell (fo-LIK-yoo-lar sell) cell that produces thyroid colloid

foramen ovale (fo-RAY-men o-VAL-ee) in the developing fetus, opening that shunts blood from the right atrium directly into the left atrium, allowing most blood to bypass the baby's developing lungs

formed elements (formed EL-e-ments) red blood cells, white blood cells, and platelets in blood

fornix (FOR-niks) corner of the vagina where it meets the cervix of the uterus

fovea centralis (FO-vee-ah sen-TRAL-is) small depression in macula lutea where cones are most densely packed; vision is sharpest where light rays focus on the fovea

free fatty acid (FFA) (free FAH-tee AS-id) fatty acid combined with albumin to be transported by the blood to other cells

free nerve endings (free nerv END-ings) specialized receptors in the skin that respond to pain

frontal plane (FRON-tal plane) lengthwise plane running from side to side, dividing the body into anterior and posterior portions

full-thickness burn (full-THIK-nes bern) third-degree burn; skin is severely damaged, nerve endings destroyed

functional residual capacity (FUNK-shun-al re-ZID-yoo-al kah-PAH-si-tee) amount of air left in the lungs at the end of a normal expiration

furuncle (FUR-un-kl) boil; pus-filled cavity formed by some hair follicle infections

G

gametes (GAM-eets) sex cells; spermatozoa and ova

gamma ray (GAM-ah ray) electromagnetic radiations; more penetrating than alpha beta particles

ganglia (GANG-gle-ah) in peripheral nerves; regions of gray matter made up of unmyelinated fibers

gangrene (GANG-green) tissue death (necrosis) that involves decay of tissue

gap junctions (gap JUNK-shun) cell connection formed when membrane channels of adjacent plasma membranes adhere to each other

gastric juice (GAS-trik joos) stomach secretion containing acid and enzymes that aids in the digestion of food

gastrin (GAS-trin) GI hormone that plays an important regulatory role in the digestive process by stimulating gastric secretion

gastroenteritis (gas-tro-en-ter-EYE-tis) inflammation of the stomach and intestines

gastroenterology (gas-tro-en-ter-OL-o-jee) study of the stomach and intestines and their diseases

gastrointestinal tract (GI tract) (gas-tro-in-TES-ti-nal trakt) alimentary canal; tube formed by the major organs of digestion

gastrulation (gas-troo-LAY-shun) process by which blastocyst cells move and then differentiate into the three primary germ layers

gated channel (GATE-ed CHA-nel) channels in the plasma membrane that can be opened and closed to alter membrane permeability

gene (jean) one of many segments of a chromosome (DNA molecule); each gene contains the genetic code for synthesizing a protein molecule such as an enzyme or hormone

gene linkage (jean LIN-kij) when a whole group of genes stay together during the crossing over process

general adaptation syndrome (GAS) (JEN-er-al a-dap-TAY-shun SIN-drome) group of changes that make the presence of stress in the body known

gene therapy (jeen THER-ah-pee) manipulation of genes to cure genetic problems; most forms of gene therapy have not yet proven effective in humans

genetic counseling (je-NET-ik KOWN-sel-ing) professional consultations with families regarding genetic diseases

genetic mutation (je-NET-ik myoo-TAY-shun) change in the genetic material within a genome; may occur spontaneously or due to mutagens

genetic predisposition (je-NET-ik pre-dis-po-ZISH-un) likelihood due to inherited genes of developing a condition even though the condition itself may not be solely caused by genetic mechanisms

genetics (je-NET-iks) scientific study of heredity and the genetic code

genitalia (jen-i-TAIL-yah) reproductive organs

genome (JEE-nome) entire set of chromosomes in a cell; the *human genome* refers to the entire set of human chromosomes

genotype (JEN-o-tipe) alleles present at one or more specific loci on a chromosome of a given individual; *see* phenotype

germinal epithelium (JER-mi-nal ep-i-THEE-lee-um) small epithelial cells that are on the surface of the ovaries

germinal matrix (JER-mi-nal MAY-triks) cap-shaped cluster of cells at the bottom of a hair follicle

gerontology (jair-on-TAHL-o-jee) study of the aging process

gestation period (jes-TAY-shun PEER-i-od) length of pregnancy, approximately 9 months in humans

gigantism (jye-GAN-tizm) condition produced by hypersecretion of growth hormone during the early years of life; results in a child who grows to gigantic size

gingivitis (jin-ji-VYE-tis) inflammation of the gum (gingiva), often as a result of poor oral hygiene

glans penis (glans PEE-nis) slightly bulging structure formed by the distal end of the corpus spongiosum; covered by the foreskin

glaucoma (glaw-KO-mah) disorder characterized by elevated pressure in the eye; can lead to permanent blindness

glomerular-capsular membrane (glu-MER-yoo-lar-KAP-soo-lar MEM-brane) membrane made up of glomerular endothelium, basement membrane, and visceral layer of Bowman's capsule; function is filtration

glomerulonephritis (glo-mer-yoo-lo-ne-FRY-tis) inflammatory disease of the glomerular-capsular membranes of the kidney

glomerulus (glo-MARE-yoo-lus) compact cluster, particularly when referring to the tuft of capillaries forming part of the nephron

glossopharyngeal nerve (glos-o-fah-RIN-jee-al nerv) cranial nerve IX; mixed nerve

glucagon (GLOO-kah-gon) hormone secreted by alpha cells of the pancreatic islets; increases activity of phosphorylase

glucocorticoid (GC) (gloo-ko-KOR-ti-koyds) hormones that influence food metabolism; secreted by the adrenal cortex

gluconeogenesis (gloo-ko-nee-o-JEN-e-sis) formulation of glucose or glycogen from protein or fat compounds

glucose (GLOO-kose) monosaccharide, or simple sugar; principal blood sugar used by cells

glucose phosphorylation (GLOO-kose fos-for-i-LAY-shun) converting glucose to glucose-6-phosphate; prepares glucose for further metabolic reactions

glutamate (GLOO-tah-mate) glutamic acid; amino acid believed to be responsible for up to 75% of the excitatory signals in the brain

glycerol (GLIS-er-ol) product of fat digestion

glycine (GLYE-sine) amino acid; most widely distributed inhibitory neurotransmitter in the spinal cord

glycogen (GLYE-ko-jen) polysaccharide; main carbohydrate stored in animal cells

glycogenesis (glye-ko-JEN-e-sis) anabolic pathway of glycogen formation; formation of glycogen from glucose or from other monosaccharides, fructose, or galactose

glycogenolysis (glye-ko-je-NOL-i-sis) hydrolysis of glycogen to glucose-6-phosphate or to glucose

glycolysis (glye-KOL-i-sis) first series of chemical reactions in carbohydrate catabolism; changes glucose to pyruvic acid in a series of anaerobic reactions

glycosuria (glye-ko-SOO-ree-ah) glucose in urine; a sign of diabetes mellitus

Golgi apparatus (GOL-jee ap-ah-RA-tus) organelle consisting of small sacs stacked on one another near the nucleus that makes carbohydrate compounds, combines them with protein molecules, and packages the product for distribution from cell

Golgi tendon receptors (GOL-jee TEN-don ree-SEP-tors) sensors that are responsible for proprioception; stimulated by excessive muscle contraction

gomphosis (gom-FO-sis) fibrous joint where a process is inserted into a socket; for example, the joint between the tooth and mandible

gonadotropins (go-na-do-TROE-pin) hormones (FSH and LSH) produced by the anterior pituitary that stimulate growth and maintenance of the testes or ovaries

gonads (GO-nads) sex glands in which reproductive cells are formed; ovaries in women, testes in men

gout (gowt) condition characterized by excessive levels of uric acid in the blood that are deposited in the joints

Graafian follicle (GRAF-ee-an FOL-li-kul) a secondary follicle; consists of a mature ovum surrounded by granulosa cells at boundary of fluid-filled antrum

graded potential (GRAY-ded po-TEN-shal) local potentials that vary according to strength of stimulus and distance on membrane from source

graded strength principle (GRAY-ded strength PRIN-si-pul) skeletal muscles contract with varying degrees of strength at different times

gradient (GRAY-dee-ent) measurable difference between two points of a given variable such as molecular concentration, pressure, or electrical charge

granulocyte (GRAN-yoo-lo-site) leukocyte with granules in cytoplasm

Graves' disease (gravz di-ZEEZ) inherited, possibly immune endocrine disorder characterized by hyperthyroidism accompanied by exophthalmos (protruding eyes)

gray fibers (gray FYE-bers) unmyelinated nerve fibers; gray matter

gray ramus (gray RAY-mus) short branch by which some postganglionic axons return to a spinal nerve

greater omentum (GRATE-er o-MEN-tum) pouchlike extension of the visceral peritoneum extending from greater curvature of the stomach to the transverse colon; often called "the lace apron"

greater vestibular glands (GRATE-er ves-TIB-yoo-lar glands) glands on each side of the vaginal orifice that secrete a

lubricating fluid; also called Bartholin's glands

greenstick fracture (green stick FRAK-chur) incomplete fracture most frequently seen in children due to resilience of their bones

gross anatomy (gross ah-NAT-o-mee) study of body parts visible to the naked eye

ground substance (ground SUB-stans) organic matrix of bone and cartilage

growth (growth) normal increase in size or number of cells

growth hormone (GH) (growth HOR-mone) hormone secreted by the anterior pituitary gland that controls the rate of skeletal and visceral growth

gustatory (GUS-tah-tor-ee) refers to taste

gustatory cell (GUS-tah-tor-ee sell) chemoreceptors in tongue that sense taste

gustatory hair (GUS-tah-tor-ee hair) cilia-like structures projecting from gustatory cells and into taste pores

gynecology (gye-ne-KOL-o-jee) study of the female reproductive system

gyrus (JYE-rus) convoluted ridge, usually refers to rounded elevations of the brain surface

H

hair cells (hair sells) specialized mechanoreceptors in the ear that are responsible for balance and hearing

hair follicle (hair FOL-li-kul) small blind-end tube extending from the dermis through the epidermis that contains the hair root and where hair growth occurs; sebaceous and apocrine skin glands have ducts leading into the follicle

hair papilla (hair pah-PIL-ah) small, cap-shaped cluster of cells located at the base of the follicle where hair growth begins

haploid number (HAP-loid NUM-ber) halved number of chromosomes in gametes due to meiosis; in humans, 23 chromosomes per sex cell

haversian canal (ha-VER-shun kah-NAL) canal in the haversian system of bone that runs parallel to the long axis; contains blood vessels and nerves

haversian system (hah-VER-shun SIS-tem) circular arrangements of calcified matrix and cells that give microscopic bone its characteristic appearance; also called an osteon

heart (hart) organ of circulatory system that pumps the blood; composed of cardiac muscle tissue

heart block (hart blok) blockage of impulse conduction from atria to ventricles so that the heart beats at a slower rate than normal

heart failure (hart FAYL-yoor) inability of the heart to pump returned blood sufficiently

heart murmur (hart MUR-mur) abnormal heart sound that may indicate valvular insufficiency or stenosing of the valve

heat exhaustion (heet ek-ZAWS-chun) condition caused by fluid loss resulting from activity of thermoregulatory mechanisms in a warm external environment

heat stroke (heet stroke) life-threatening condition characterized by high body temperature; failure of thermoregulatory mechanisms to maintain homeostasis in a very warm external environment

helper T cells (HEL-per t sells) immune system cells that help B cells differentiate into antibody-secreting plasma cells; also help coordinate cellular immunity through direct contact with other immune cells

hematocrit (he-MAT-o-krit) volume percent of blood cells in whole blood; packed cell volume

hematopoiesis (hem-ah-to-poi-EE-sis) process of blood cell formation

hematopoietic tissue (he-MAT-o-poy-et-ik TISH-yoo) specialized connective tissue that is responsible for formation of blood cells and lymphatic system cells; in red bone marrow, spleen, tonsils, and lymph nodes

hemocytoblast (he-mo-SYE-toe blast) bone marrow stem cell from which all formed elements of blood arise

hemodynamics (he-mo-dye-NAM-iks) mechanisms that influence dynamic circulation of blood

hemoglobin (he-mo-GLO-bin) iron-containing protein in red blood cells responsible for their oxygen-carrying capacity

hemorrhoid (HEM-o-royd) varicose vein in the rectum; also called piles

hemostasis (hee-mo-STAY-sis) stoppage of blood flow

heparin (HEP-ah-rin) substance obtained from the liver; inhibits blood clotting

hepatic duct (he-PAT-ik dukt) one of two ducts that drains bile out of the liver

hepatic lobule (he-PAT-ik LOB-yool) anatomical units of the liver

hepatitis (hep-ah-TIE-tis) inflammation of the liver; may be caused by toxins, viruses (for example, hepatitis A, hepatitis B), bacteria, or parasites

heredity (he-RED-i-tee) transmission of characteristics from parents to offspring

Hering-Breuer reflexes (HER-ing BRYOO-er REE-fleks-es) control respirations—especially rate and rhythmicity

herniated disk (HER-nee-ate-ed disk) condition of the vertebral disk when the annulus fibrosis becomes disrupted, allowing the nucleus pulposis to come out

herpes zoster (HER-peez ZOS-ter) "shingles," viral infection that affects the skin of a single dermatone

heterozygous (het-er-o-ZYE-gus) genotype with two different forms of a trait

hexose (HEK-sose) monosaccharide molecule with six carbons

hiatal hernia (hye-AY-tal HER-nee-ah) condition in which a portion of the stomach is pushed through the hiatus (opening) of the diaphragm, often weakening or expanding the cardiac sphincter at the inferior end of the esophagus

hiccup (HIK-up) involuntary spasmodic contraction of the diaphragm

hilum (HYE-lum) slit on medial surface of each lung where primary bronchi and pulmonary vessels enter; slit on medial surface of each kidney where blood vessels and other structures enter the kidney

histamine (HIS-tah-mine) inflammatory chemical

histogenesis (his-toe-JEN-e-sis) formation of tissues from primary germ layers of embryo

histology (his-TOL-o-jee) branch of microscopic anatomy that studies tissues; biology of tissues

histophysiology (his-toe-fiz-e-OL-o-jee) study of microscopic anatomy of cells and tissues and how that correlates with function

Hodgkin's lymphoma (HOJ-kinz lim-FO-mah) type of lymphoma (malignant lymph tumor) characterized by painless swelling of lymph nodes in the neck, progressing to other regions

holocrine gland (HO-lo-krin gland) gland that collects secretory product inside its cells which then rupture completely to release it; for example, sebaceous glands of the skin

homeostasis (ho-me-o-STAY-sis) relative constancy of the normal body's internal environment

homeostatic control mechanisms (ho-me-o-STA-tik kon-TROL MEK-ah-nism) devices for maintaining or restoring homeostasis

homozygous (ho-mo-ZYE-gus) genotype with two identical forms of a trait

horizontal fissure (hor-i-ZON-tal FISH-ur) separates the superior lobe of the right lung from the middle lobe

horizontal plane (hor-i-ZON-tal plane) transverse plane; divides the body into superior and inferior portions

hormone (HOR-mone) substance secreted by an endocrine gland

into the bloodstream that acts on a specific target tissue to produce a given response

hormone-receptor complex (HOR-mone-ree-SEP-tor KOM-pleks) when a steroid hormone combines to a receptor site in the target cell and then moves into the nucleus of the target cell

human chorionic gonadotropin (HCG) (HYOO-man KO-ri-on-ic go-na-do-TRO-pin) hormone secreted early in pregnancy by the placenta that serves to maintain the uterine lining

humoral immunity (HYOO-mo-ral i-MYOO-ni-tee) antibody-mediated immunity occurring within blood plasma and other body fluids

hyaline cartilage (HYE-ah-lin KAR-ti-lij) most common type of cartilage; appears gelatinous and glossy

hybridoma (hye-brid-O-ma) hybrid cell formed by fusion of cancerous cell with a lymphocyte to mass-produce a specific antibody (monoclonal antibodies)

hydrase (HYE-drase) enzyme that adds water to a molecule without splitting it

hydrocortisone (hye-dro-KOR-ti-zone) hormone secreted by the adrenal cortex; cortisol; compound F

hydrogen bond (HYE-dro-jen bond) weak chemical bond that occurs between the partial positive charge on a hydrogen atom covalently bound to a nitrogen or oxygen and the partial negative charge of another polar molecule

hydrogen ion (HYE-dro-jen EYE-on) a proton or a hydrogen atom without its electron; in water and water solutions; produces an acidic solution; has a positive charge

hydrolase (HYE-dro-lase) hydrolyzing enzyme

hydrolysis (hye-DROL-i-sis) chemical process in which a compound is split by addition of H^+ and OH^- portions of a water molecule

hydrophilic (hye-dro-FIL-ik) "water-loving"

hydrophobic (hye-dro-FOE-bik) "water-fearing"

hydrostatic pressure (hye-dro-STAT-ik PRESH-ur) force of a fluid pushing against some surface

hymen (HYE-men) Greek for "membrane"; mucous membrane that may partially or entirely occlude the vaginal outlet

hypercapnia (hey-per-KAP-nee-ah) excessive carbon dioxide in the blood

hyperextension (hye-per-ex-TEN-shun) stretching an ex-

tended part beyond its anatomical position

hyperglycemia (hye-per-glye-SEE-mee-ah) higher than normal blood glucose concentration

hyperkalemia (hye-per-kah-LEE-me-ah) excessive potassium in the blood

hyperkeratosis (hye-per-ker-ah-TOE-sis) thickening of the horny layer of the skin

hyperopia (hye-per-O-pee-ah) refractive disorder of the eye caused by a shorter than normal eyeball; results in ability to see objects at a distance better than objects nearer; also called farsightedness

hyperplasia (hye-per-PLAY-zee-ah) growth of an abnormally large number of cells at a local site, as in a neoplasm or tumor

hyperpnea (hye-PERP-nee-ah) abnormal increase in respiratory rate and depth

hyperpolarization (hye-per-pol-lar-i-ZAY-shun) increase in electrical charges separated by the cell membrane; causes change away from 0 mV

hypersecretion (hye-per-se-KREE-shun) too much secretion of a substance

hypertension (hye-per-TEN-shun) abnormally high blood pressure

hypertonic (hye-per-TON-ik) solution containing a higher level of salt (NaCl) than is found in a living red blood cell (above 0.9% NaCl); causes cells to shrink

hypertrophy (hye-PER-tro-fee) increased size of an organ or part caused by an increase in the size of its cells

hyperventilation (hye-per-ven-ti-LAY-shun) very rapid, deep respirations

hypervitaminosis (hye-per-vye-tah-mi-NO-sis) general name for any condition resulting from an abnormally high intake of vitamins

hypervolemia (hye-per-vo-LEE-me-ah) abnormally increased blood volume

hypochloremia (hye-po-klo-REE-me-ah) abnormally low levels of chloride in the blood associated with potassium loss

hypodermis (hye-po-DER-mis) loose layer of skin, rich in fat and areolar tissue located beneath the dermis

hypoglossal (hye-po-GLOS-al) under the tongue

hypoglossal nerve (hye-po-GLOS-al nerv) cranial nerve XII; motor nerve; responsible for tongue movement

hypoglycemia (hye-po-glye-SEE-mee-ah) lower than normal blood glucose concentration

hypophyseal portal system (hye-po-FIZ-e-al POR-tal SIS-tem) complex of small blood vessels through which releasing hormones travel from the hypothalamus to the pituitary

hypophysis (hye-POF-i-sis) pituitary gland

hypopotassemia (hye-po-po-tah-SEE-me-ah) potassium deficit

hyposecretion (hye-po-se-KREE-shun) too little secretion of a substance

hypothalamus (hye-po-THAL-ah-mus) important autonomic and neuroendocrine control center located inferior to the thalamus in the brain

hypothermia (hye-po-THER-mee-ah) failure of thermoregulatory mechanisms to maintain homeostasis in a very cold external environment; results in abnormally—sometimes life-threatening—low body temperature

hypotonic (hye-po-TON-ik) solution containing a lower level of salt (NaCl) than is found in a living red blood cell (below 0.9% NaCl); causes cells to swell and perhaps burst

hypoventilation (hye-po-ven-ti-LAY-shun) slow and shallow respirations

hypoxia (hye-POK-see-ah) deficiency of oxygen in the blood

I

ileum (IL-ee-um) distal portion of the small intestine

immune system (i-MYOON SIS-tem) body's defense system against disease

immunization (i-myoo-ni-ZAY-shun) deliberate artificial exposure to disease to produce acquired immunity

immunoglobulin (i-myoo-no-GLOB-yoo-lin) antibodies

immunology (i-myoo-NOL-o-jee) study of immune system functions and mechanisms

immunotherapy (i-myoo-no-THER-ah-pee) therapeutic technique that bolsters a person's immune system in an attempt to control a disease

impetigo (im-pe-TIE-go) highly contagious bacterial skin infection that occurs most often in children

impulse (IM-puls) an electrical signal

inborn immunity (IN-born i-MYOO-ni-tee) inherited immunity to disease

incomplete tetanus (in-kom-PLEET TET-ah-nus) tetanus with very short periods of relaxation occurring between peaks of muscle tension

incontinence (in-KON-ti-nens) involuntary voiding of urine

infancy (IN-fan-see) period of human development from birth to about 18 months of age

infant respiratory distress syndrome (IRDS) (IN-fant re-SPY-rah-toe-ree di-STRESS SIN-drome) leading cause of death in premature babies, due to a lack of surfactant in the alveolar air sacs

inferior (in-FEER-ee-or) lower; opposite of superior

inflammation (in-flah-MAY-shun) group of responses to a tissue irritant marked by signs of redness, heat, swelling, and pain

inflammatory response (in-FLAM-ah-to-ree re-SPONSE) specific process involving tissues and blood vessels in response to injury

infundibulum (in-fun-DIB-yoo-lum) stalk that connects the pituitary gland to the hypothalamus

ingestion (in-JEST-chun) taking in of complex foods, usually by mouth

inherited immunity (in-HAIR-i-ted i-MYOO-ni-tee) inborn immunity; occurs when immune mechanisms are put in place by genetic mechanisms during the early stages of human development

inhibitory (in-HIB-i-tor-ee) something that slows or stops a process

inhibitory neurotransmitter (in-HIB-i-tor-ee noo-ro-TRANS-mit-er) neurotransmitter that causes hyperpolarization of the postsynaptic membrane, thereby decreasing chances of impulse propagation

inhibitory postsynaptic potential (IPSP) (in-HIB-i-tor-ee post-si-NAP-tik po-TEN-shal) temporary hyperpolarization that makes the inside of the membrane even more negative than at the resting potential

inorganic compounds (in-or-GAN-ik KOM-pownds) chemical constituents that do not contain both carbon and hydrogen; for example, water, carbon dioxide, and oxygen

insertion (in-SER-shun) attachment of a muscle to the bone that it moves when contraction occurs (as distinguished from its origin)

inspiratory capacity (IC) (in-SPY-rah-tor-ee kah-PAS-i-tee) maximal amount of air an individual can inspire following a normal expiration

inspiratory center (in-SPY-rah-tor-ee SEN-ter) one of the two most important control centers located in the medulla; the other is the expiratory center

inspiratory reserve volume (IRV) (in-SPY-rah-tor-ee re-SERV VOL-yoom) amount of air that can be forcibly inspired over and above a normal respiration

insulin (IN-suh-lin) hormone secreted by beta cells of the pancreatic islets that increases the uptake of glucose and amino acids by most body cells

integument (in-TEG-yoo-ment) skin; the body's largest organ

integumentary system (in-teg-yoo-MEN-tar-ee SIS-tem) skin and its related structures

intercalated disk (in-TER-kah-lay-ted disk) cross striations and unique dark bands in cardiac muscle fibers

interferon (in-ter-FEER-on) small protein produced by the immune system that inhibits virus multiplication

intermediate fiber (in-ter-MEE-de-it FYE-ber) muscle fiber exhibiting characteristics between fast and slow fibers

interneuron (in-ter-NOO-ron) in a three-neuron reflex arc, nerve cell that conducts impulses from a sensory neuron to a motor neuron

interphase (IN-ter-faze) mitotic phase immediately before visible condensation of the chromosomes during which the DNA of each chromosome replicates itself

interstitial cell (in-ter-STISH-al sell) small, specialized cells in the testes that secrete the male sex hormone, testosterone

interstitial fluid (IF) (in-ter-STISH-al FLOO-id) fluid located in microscopic spaces between cells

intestinal flora (in-TES-ti-nal FLOR-ah) community of various bacterial population that normally inhabit the colon

intestine (in-TES-tin) part of the digestive tract into which food passes after it leaves the stomach; separated into two segments, the small and the large

intracellular fluid (ICF) compartment (in-tra-SELL-yoo-lar FLOO-id kom-PART-ment) fluid located within cells; largest fluid compartment

intramembranous ossification (in-trah-MEM-brah-nus os-i-fi-KAY-shun) process by which most flat bones are formed within connective tissue membranes

intrinsic eye muscles (in-TRIN-sik eye MUS-els) involuntary muscles located within the eye; responsible for size of the iris and shape of the lens

intrinsic rhythm (in-TRIN-sik RITH-m) specialized cells in the sinoatrial (SA) node that produce regular impulses without stimulation from the brain or spinal cord

involuntary muscle (in-VOL-un-tair-ee MUS-el) smooth muscle not under conscious control and found in organs such as the stomach, small intestine, and ureters

ion (EYE-on) electrically charged atom or group of atoms

ionic bond (eye-ON-ik bond) electrocovalent bond; bond formed by transferring of electrons from one atom to another

ipsilateral (ip-si-LAT-er-al) on the same side

ipsilateral reflex arc (ip-si-LAT-er-al REE-fleks ark) reflex arc whose receptors and effectors are located on the same side of the body

iris (EYE-ris) colored portion of the eye

irritability (ir-i-tah-BIL-i-tee) ability of a muscle to be stimulated; see excitability

ischemia (is-KEE-me-ah) reduced flow of blood to tissue resulting in impairment of cell function

islets of Langerhans (EYE-lets of lahn-GER-hans) see pancreatic islets

isometric contraction (eye-so-MET-rik kon-TRAK-shun) type of muscle contraction in which muscle does not alter the distance between two bones; see isotonic contraction

isotonic (eye-so-TON-ik) two fluids that have the same potential osmotic pressure

isotonic contraction (eye-so-TON-ik kon-RAK-shun) type of muscle contraction in which the muscle sustains the same tension or pressure and there is a change in the distance between two bones

isotope (EYE-so-tope) atoms with the same atomic number but different atomic weights

J

jaundice (JAWN-dis) abnormal yellowing of skin, mucous membranes, and white of eyes

jejunum (je-JOO-num) middle third of the small intestine

joint (joynt) junction between two or more bones; articulation

joint capsule (joynt KAP-sool) sleeve-like extension of the periosteum of each of the bones at an articulation

joint cavity (joynt KAV-i-tee) small space between articulating surfaces of the two bones of the joint

juxtaglomerular apparatus (juks-tah-glo-MER-yoo-lar ap-ah-RAH-tus) in the nephron, the complex of cells from the distal tubule and the afferent arteriole, which helps regulate blood pressure by secreting renin in response to blood pressure changes in the kidney; located near the glomerulus

K

Kaposi's sarcoma (KAH-po-seez sar-KO-mah) rare malignant neoplasm of the skin that often spreads to lymph nodes and internal organs; Kaposi's sarcoma is often found in AIDS patients

karyotype (KAIR-ee-o-type) ordered arrangement of photographs of chromosomes from a single cell used in genetic counseling to identify chromosomal disorders such as trisomy or monosomy

keloid (KEE-loid) unusually thick fibrous scar on the skin

keratin (KER-ah-tin) tough, fibrous protein substance in hair, nails, outer skin cells, and horny tissues

keratinization (ker-ah-tin-i-ZAY-shun) process by which cells of the stratum corneum become fitted with keratin and move to the surface

keratinocyte (ke-RAT-i-no-site) epidermal cell responsible for synthesizing keratin

ketogenesis (kee-toe-JEN-e-sis) process that produces ketone bodies

ketone body (KEY-tone BOD-ee) acid produced during fat catabolism

ketosis (key-TOE-sis) large amount of ketone bodies present in the blood of a person with uncontrolled diabetes mellitus

kidney (KID-nee) one of the two organs that cleanses the blood of waste products continually produced by metabolism; the kidneys produce urine

killer T cells (KIL-er T sells) see cytotoxic T cells

kilocalorie (Kcal) (KIL-o-kal-o-ree) 1000 calories; see Calorie

kinase (KYE-nase) substance that converts proenzymes to active enzymes

kinesthesia (kin-es-THEE-zee-ah) "muscle sense"; that is, sense of position and movement of body parts

Klinefelter's syndrome (KLINE-fel-ter SIN-drome) genetic disorder caused by the presence of two or more X chromosomes in a male (typically trisomy XXY); characterized by long legs, enlarged breasts, low intelligence, small testes, sterility, and chronic pulmonary disease

Krause's end bulb (KROWZ end bulb) skin receptor that detects sensations of touch, low frequency vibration, and texture differences

Krebs cycle (krebs SY-kul) see citric acid cycle; also called tricarboxylic acid (TCA) cycle

Kupffer cell (KOOP-fer sell) macrophage found in spaces between liver cells

kyphosis (kye-FO-sis) abnormally exaggerated thoracic curvature of the vertebral column

L

labia majora (LAY-bee-ah ma-JO-rah) "large lips" of the vulva; singular, labium majus

labia minora (LAY-bee-ah mi-NO-rah) "small lips" of the vulva; singular, labium minus

labor (LAY-bor) process of expulsion of the fetus and the placenta; childbirth

labyrinth (LAB-i-rinth) bony cavities and membranes of the inner ear

lacrimal apparatus (LAK-ri-mal app-a-RAT-us) in the eye, the tear (lacrimal) gland plus associated ducts that form tears

lactase (LAK-tase) enzyme needed to digest lactose

lactation (lak-TAY-shun) milk production

lacteal (LAK-tee-al) lymphatic vessel located in each villus of the intestine; serves to absorb fat materials from chyme passing through the small intestine

lactic acid (LAK-tik AS-id) product of anaerobic respiration that accumulates in muscle tissue during exercise and causes a burning sensation

lactose intolerance (LAK-tose in-TOL-er-ans) condition in which one lacks the enzyme lactase, resulting in an inability to digest lactose (a disaccharide in milk and dairy products)

lacuna (lah-KOO-nah) space or cavity; for example, lacunae in bone contain bone cells

lamella (lah-MEL-ah) thin layer, as of bone

Langerhans' cell (lahn-GER-hanz sell) cell that plays a limited role in immunological reactions that affect the skin; serves as defense mechanism for the body

Langer's lines (LANG-erz lines) pattern of dense bundles of white collagenous fibers that characterize the reticular layers of dermis

lanugo (lah-NOO-go) extremely fine and soft hair coat on developing fetus

laryngitis (lar-in-JYE-tis) inflammation of the mucous tissues of the larynx (voice box)

laryngopharynx (lah-ring-go-FAIR-inks) lowest part of the pharynx

larynx (LAIR-inks) voice box located just below the pharynx; the largest piece of cartilage making up the larynx is the thyroid cartilage, commonly known as the Adam's apple

latent period (LAY-tent PEE-re-od) period between time a muscle fiber is stimulated and when it contracts; preceeds the contraction phase

lateral (LAT-er-al) of or toward the side; opposite of medial

lateral fissure (fissure of Sylvius) (LAT-er-al FISH-ur [FISH-ur of SIL-vi-us]) deep groove between the temporal lobe below and the frontal and parietal lobes of the brain

length-tension relationship (lenth-TEN-shun ree-LAY-shun-ship) maximum strength that a muscle can develop bears a direct relationship to the initial length of its fibers

lens (lenz) refracting mechanism of the eye that is located directly behind the pupil

leukocyte (LOO-ko-site) white blood cell

leukocytosis (loo-ko-sye-TOE-sis) abnormally high white blood cell numbers in the blood

leukopenia (loo-ko-PEE-nee-ah) abnormally low white blood cell numbers in the blood

ligament (LIG-ah-ment) band of white fibrous tissue connecting bones to other bones

limbic system (LIM-bik SIS-tem) parts of the brain involved in emotions and sense of smell; plays key role in coupling sensory inputs to short and long term memory; consists of the hippocampus, the hypothalamus, and several other structures

lingual tonsil (LING-gwal TON-sil) tonsil located at the base of the tongue

lipid (LIP-id) class of organic compounds that includes fats, oils, and related substances

lipogenesis (lip-o-JEN-e-sis) formation of body fat from food sources

lipoma (li-PO-mah) benign tumor of adipose (fat) tissue

lipoprotein (lip-o-PRO-teen) substance that is part lipid and part protein; produced mainly in the liver

litmus (LIT-mus) pigment used to test for acidity and alkalinity

lobar (LO-bar) referring to a lobe

lobectomy (lo-BEK-toe-mee) surgical removal of a single lobe of an organ, as in the removal of one lobe of a lung

local potential (LO-kal po-TEN-shal) slight shift from resting membrane potential in a specific region of the plasma membrane

lock-and-key model (lok-and-key MO-del) description of how a specific enzyme will fit into only a specific substrate, like a key fits into a lock

loop of Henle (loop of HEN-lee) extension of the proximal tubule of the nephron

lordosis (lor-DOE-sis) abnormally exaggerated lumbar curvature of the vertebral column

lumbar plexus (LUM-bar PLEK-sus) spinal nerve plexus located in the low back

lumbar puncture (LUM-bar PUNK-chur) clinical procedure in which some cerebrospinal fluid is withdrawn from the subarachnoid space in the lumbar region of the spinal cord for analysis

lunula (LOO-nyoo-lah) crescent-shaped white area under the proximal nail bed

luteinization (loo-te-ni-ZAY-shun) formation of a golden body (corpus luteum) in the ruptured follicle

luteinizing hormone (LH) (LOO-te-nye-zing HOR-mone) in females, acts in conjunction with follicle-stimulating hormone (FSH) to stimulate follicle and ovum maturation, release of

estrogen, and ovulation; known as the ovulating hormone; in males, causes testes to develop and secrete testosterone

lymph (limf) specialized fluid formed in the tissue spaces that returns excess fluid and protein molecules to the blood via the lymphatic vessels

lymph node (limf node) performs biological filtration of lymph on its way to the circulatory system

lymphangitis (lim-fan-JYE-tis) inflammation of lymph vessels, usually due to infection, characterized by fine red streaks extending from the site of infection; may progress to *septicemia* (blood infection)

lymphatic vessels (lim-FAT-ik VES-els) system of blind-ended vessels that collect lymph and deliver it to the circulatory system via the thoracic duct and the right lymphatic duct

lymphedema (lim-fah-DEE-mah) swelling due to lymphatic vessel blockage

lymphocyte (LIM-fo-site) one type of white blood cell; *see* B and T lymphocytes

lymphokine (LIM-fo-kine) chemical compounds released by antigen-bound sensitized T cells

lymphoma (lim-FO-mah) cancer of lymphatic tissue

lymphotoxin (lim-fo-TOK-sin) powerful poison that quickly kills any cells it attacks

lysosome (LYE-so-sohm) membranous organelle containing various enzymes that can dissolve most cellular compounds; hence called digestive bags or suicide bags of cell

M

macromolecule (mak-ro-MOL-e-kyool) large, complex chemical made of combinations of molecules

macrophage (MAK-ro-faje) phagocytic cell in the immune system

macula (MAK-yoo-lah) highly specialized strip of epithelium in the utricle and saccule; provides information related to head position or acceleration

macula lutea (MAK-yoo-lah LOO-tee-ah) yellowish area near center of the retina where cones are densely distributed

magnetic resonance imaging (MRI) (mag-NET-ik REZ-o-nans IM-ah-jing) scanning technique that uses a magnetic field to induce tissues to emit radio waves that can be used by computer to construct a sectional view of a patient's body

malignant (mah-LIG-nant) refers to a tumor or neoplasm that is capable of metastasizing, or spreading, to new tissues; cancer

maltase (MAL-tase) enzyme that catalyzes the final steps of carbohydrate digestion

mammary gland (MAM-er-ee gland) milk-producing glands of the breasts; classified as external accessory sex organs in females

mammography (mam-OG-rah-fee) x-ray photography of the breast

mandibular nerve (man-DIB-yoo-lar nerv) part of the trigeminal nerve

marrow cavity (MAR-o KAV-i-tee) *see* medullary cavity

mast cell (mast sell) immune system cell to which antibodies become attached in early stages of inflammation

mastectomy (mas-TEK-toe-mee) surgical removal of the breast

mastication (mas-ti-KAY-shun) chewing

mastitis (mas-TYE-tis) breast inflammation

mastoiditis (mas-toy-DYE-tis) inflammation of the air cells within the mastoid portion of the temporal bone; usually caused by infection

matrix (MAY-triks) intracellular substance of a tissue; for example, the matrix of bone is calcified, whereas that of blood is liquid

maxillary nerve (MAK-si-lair-ee nerv) branch of the trigeminal nerve

mechanical digestion (me-KAN-i-kal di-JEST-chun) process through which food is broken into smaller portions through chewing and movements of the alimentary canal; enables enzymes to act on a larger surface area to accomplish chemical digestion

mechanoreceptor (mek-an-o-ree-SEP-tor) receptors that respond to physical movement in the environment, such as sound waves; for example, equilibrium and balance sensors in the ears

medial (MEE-dee-al) of or toward the middle; opposite of lateral

medial lemniscal system (MEE-dee-al lem-NIS-kal SIS-tem) posterior white columns of the spinal cord plus the medial lemniscus; functions in touch sensations and conscious proprioception

medulla (me-DUL-ah) Latin for "marrow"; hence the inner portion of an organ in contrast to the outer portion, or cortex

medulla oblongata (me-DUL-ah ob-long-GAH-tah) lowest part of the brain stem; an enlarged extension of the spinal cord; the vital centers are located within this area

medullary cavity (MED-yoo-lair-ee KAV-i-tee) hollow area inside the diaphysis of the bone that contains yellow bone marrow

medullary rhythmicity area (MED-yoo-lair-ee rith-MI-si-tee AIR-ee-ah) area in the brain

stem that generates the basic rhythm of the respiratory cycle of inspiration and expiration

megakaryocytes (meg-ah-KAIR-ee-o-sites) large cells in the spleen that contribute to formation of platelets

meiosis (my-O-sis) nuclear division in which the number of chromosomes is reduced to half their original number through separation of homologous pairs; produces gametes

Meissner's corpuscle (MIZS-ners KOR-pus-ul) sensory receptor located in the skin close to the surface that detects light touch

melanin (MEL-ah-nin) brown pigment primarily in skin and hair

melanocyte (me-LAN-o-site) specialized cell in the stratum basale of the skin that produces melanin

melanocyte-stimulating hormone (MSH) (me-LAN-o-site-STIM-yoo-lay-ting HOR-mone) hormone secreted by the pituitary that increases production of melanin, leading to darker skin color

melanoma (mel-ah-NO-mah) cancer of pigmented epithelial cells

melatonin (mel-ah-TOE-nin) important hormone produced by the pineal gland; it is believed to regulate onset of puberty and the menstrual cycle; also referred to as the third eye because it responds to levels of light and is thought to be involved with the body's internal clock

membrane (MEM-brane) thin, sheetlike structure

membrane channel (MEM-brane CHAN-el) pores within the cell membrane through which specific ions or other small, water-soluble molecules can pass

membrane potential (MEM-brane po-TEN-shal) difference in electrical charge between inside and outside of the plasma membrane

membrane protein (MEM-brane PRO-teen) protein imbedded in the phospholipid bilayer of the membrane; contributes to cell transport, cell-cell recognition, and other cell functions

membranous organelles (MEM-brah-nus or-gah-NELS) organelles made of the same membrane material as the plasma membrane; includes endoplasmic reticulum, Golgi apparatus, lysosomes, peroxisomes, mitochondria, and the nucleus

menarche (me-NAR-kee) first menses occurring at the onset of puberty

meninges (me-NIN-jeez) fluid-containing membranes surrounding the brain and spinal cord

meningitis (men-in-JYE-tis) inflammation of the meninges caused by various factors, in-

cluding bacterial infection, mycosis, viral infection, and tumors

meniscus (me-NIS-kus) articular cartilage disk

menopause (MEN-o-pawz) termination of menstrual cycles

menses (MEN-seez) periodic shedding of endometrial lining in uterus; occurs in cycles of about 28 days

menstrual cycle (MEN-stroo-al sye-kul) series of changes that regularly occurs in sexually mature, nonpregnant women that results in shedding of the uterine lining approximately once a month

menstruation (men-stroo-AY-shun) menses; regular event of the female reproductive cycle that allows the endometrium to renew itself

merocrine gland (MER-o-krin gland) gland that discharges secretions directly through the cell or plasma membrane

mesentary (MEZ-en-tair-ee) large double fold of peritoneal tissue that anchors the loops of the digestive tract to the posterior wall of the abdominal cavity

mesoderm (MEZ-o-derm) middle layer of the primary germ layers; gives rise to such structures as muscle, bones, and blood vessels

mesomorph (MEZ-o-morf) body type characterized by a muscular build

messenger RNA (mRNA) (MES-en-jer RNA) duplicate copy of a gene sequence on the DNA that passes from the nucleus to the cytoplasm; used by ribosomes to create specific proteins

metabolic acidosis (met-ah-BOL-ik ah-si-DOE-sis) disturbances affecting the bicarbonate element of the bicarbonate-carbonic acid buffer pair in the blood; bicarbonate deficit

metabolic alkalosis (met-ah-BOL-ik al-kah-LO-sis) disturbance affecting the bicarbonate element of the bicarbonate-carbonic acid buffer pair in the blood; bicarbonate excess

metabolic rate (met-ah-BOL-ik rate) amount of energy released by catabolism in the body over a given period of time

metabolism (me-TAB-o-lizm) complex process by which food is used by a living organism; *see* catabolism, and anabolism

metaphase (MET-ah-faze) second stage of mitosis, during which the nuclear membrane and nucleolus disappear and the chromosomes align on the equatorial plane

metastasis (me-TAS-tah-sis) process characteristic of cancer by which malignant tumor cells separate from a primary tumor and then migrate to a new tissue to initiate a secondary tumor

micelles (my-SELLS) droplet of lipid surrounded by bile salts, which makes the lipid temporarily water soluble

microcephaly (my-kro-SEF-ah-lee) congenital abnormality in which an infant is born with an abnormally small head

microcirculation (my-kro-sir-kyoo-LAY-shun) flow of blood through the capillary bed

microfilament (my-kro-FIL-ah-ment) smallest cell fibers; "cellular muscles"

microtubule (my-kro-TOOB-yool) thick cell fiber (compared to microfilament); hollow tube responsible for movement of substances within the cell or movement of the cell itself

microvillus (my-kro-VIL-us) brushlike border made up of epithelial cells on each villus in the small intestine; increases the surface area for absorption of nutrients

micturition (mik-too-RISH-un) urination, voiding

midbrain (MID-brain) region of the brain stem between the pons and the diencephalon

middle ear (MID-ul eer) tiny and very thin epithelium-lined cavity in the temporal bone that houses the ossicles; in the middle ear, sound waves are amplified

midsagittal plane (mid-SAJ-i-tal plane) cut, or plane, that divides the body or any of its parts into two equal (mirror image) halves

milliequivalent (mil-i-e-KWIV-ah-lent) number of ionic charges or electrocovalent bonds in a solution; abbreviated mEq

mineralocorticoid (MC) (min-er-al-o-KOR-ti-koyd) hormone that influences mineral salt metabolism; secreted by adrenal cortex; aldosterone is the chief mineralocorticoid

minerals (MIN-er-als) inorganic elements or salts naturally in the earth, many of which are vital to proper functioning of the body and must be obtained in the diet

minute volume (MIN-ut VOL-yoom) volume of blood circulating through the body per minute

miscarriage (mis-KAR-ij) loss of an embryo or fetus before the twentieth week of pregnancy; after 20 weeks, the event is termed a *stillbirth*

mitochondrial DNA (mDNA) (my-toe-KON-dree-al DNA) DNA specifically in the mitochondrion that has the only genetic code for several important enzymes

mitochondrion (my-toe-KON-dree-on) organelle in which ATP generation occurs; often termed *"powerhouse of cell"*

mitosis (my-TOE-sis) complex process in which a cell's DNA is replicated and divided equally between two daughter cells

mitral valve (MY-tral valv) located between the left atrium and ventricle, this valve prevents backflow of blood into the left atrium; named for its resemblance to a bishop's hat (miter); also known as the bicuspid valve

mitral valve prolapse (MVP) (MY-tral valv PRO-laps) condition in which the bicuspid (mitral) valve extends into the left atrium, causing incompetence (leaking) of the valve

mittelschmerz (MIT-el-shmerts) abdominal discomfort experienced by many women at the time of ovulation

mixed nerves (mixt nervs) nerves with axons of both sensory and motor neurons

mobile-receptor hypothesis (MO-bil hye-POTH-e-sis ree-SEP-tor) theory according to which the hormone-receptor complex migrates into the nucleus of the target cell, activating the beginning of mRNA transcription

modiolus (mo-DI-o-lus) cone-shaped core of bone around which the tube of the cochlea is wound

molecule (MOL-e-kyool) formed when two or more atoms join

monoamine oxidase (MAO) (mon-o-ah-MEEN OK-si-dase) enzyme located in synaptic knobs of postganglionic neurons; responsible for breaking down neurotransmitters

monoclonal antibody (mon-o-KLONE-al AN-ti-bod-ee) specific antibody produced from a population of identical cells; *see* hybridoma

monocyte (MON-o-site) large white blood cell; an agranular leukocyte

monosaccharide (mon-o-SAK-ah-ride) simple sugar, such as glucose or fructose; building block of carbohydrates

monosomy (MON-o-so-mee) abnormal genetic condition in which cells have only one chromosome instead of a pair; usually caused by nondisjunction (failure of chromosome pairs to separate) during gamete production

mons pubis (monz PYOO-bis) skin-covered pad of fat over the symphysis pubis in the female

morula (MOR-yoo-lah) solid mass of cells formed by the divisions of a fertilized egg

motility (mo-TIL-i-tee) ability to move spontaneously

motor endplate (MO-tor END-plate) point at which motor neurons connect to the sarcolemma to form the neuromuscular junction

motor unit (MO-tor YOO-nit) functional unit composed of a single motor neuron with the muscle cells it innervates

motorneuron (mo-tor-NOO-ron) transmits nerve impulses from the brain and spinal cord to muscles and glandular epithelial tissues

mucous membrane (MYOO-kus MEM-brane) epithelial membrane that lines body surfaces opening directly to the exterior and secretes mucus

mucus (MYOO-kus) thick, slippery material secreted by mucous membranes that keeps the membrane moist

multiple neurofibromatosis (MUL-ti-pul noo-ro-fye-bro-mah-TOE-sis) disorder characterized by multiple, sometimes disfiguring, benign tumors of the Schwann cells (neuroglia) that surround nerve fibers

multiple sclerosis (MS) (MUL-ti-pul skle-RO-sis) most common primary disease of the central nervous system, MS leads to demyelination of nerves, which frequently causes problems with vision, muscle control, and incontinence

multiple wave summation (MUL-ti-pul wave sum-MAY-shun) condition in which a series of stimuli come in rapid enough succession that the muscle is not able to relax completely, and sustained or more forceful contraction results

multipolar neuron (mul-ti-POL-ar NOO-ron) neuron with only one axon but several dendrites

mumps (mumps) acute viral disease characterized by swelling of the parotid salivary glands

muscarinic receptor (mus-kah-RIN-ik ree-SEP-tor) type of cholinergic receptor responding to muscarine, as well as acetylcholine

muscle fatigue (MUS-el fah-TEEG) state of exhaustion produced by strenous muscular activity

muscle spindle (MUS-el SPIN-dul) stretch receptor in muscle cells involved in maintaining muscle tone

muscle strain (MUS-el strane) muscle injury resulting from overexertion or trauma and involving overstretching or tearing of muscle fibers

muscle tissue (MUS-el TISH-yoo) specialized tissue type that produces movement

muscle tone (MUS-el tone) tonic contraction; characteristic of muscle of a normal individual who is awake

muscularis (mus-kyoo-LAIR-is) two layers of muscle surrounding the digestive tube that produce wavelike, rhythmic contractions, called peristalsis, that move food material

mutagen (MYOO-tah-jen) agent capable of causing mutation (alteration) of DNA

mutation (myoo-TAY-shun) change in genetic material within a cell

myelin (MY-e-lin) lipoprotein substance in the myelin sheath around many nerve fibers that contributes to high-speed conductivity of impulses

myelin sheath (MY-e-lin sheeth) whitish, fatty sheath surrounding nerve fibers; substance produced by Schwann cells

myeloid tissue (MY-e-loyd TISH-yoo) red bone marrow; specialized type of soft, diffuse connective tissue; the site of hematopoeisis

myocardial infarction (MI) (my-o-KAR-dee-al in-FARK-shun) death of cardiac muscle cells resulting from inadequate blood supply, as in coronary thrombosis

myocardium (my-o-KAR-dee-um) muscle of the heart

myofibril (my-o-FYE-bril) very fine longitudinal fibers found in skeletal muscle cells; composed of thick and thin filaments

myofilament (my-o-FIL-ah-ment) ultramicroscopic, threadlike structures found in myofibrils; composed of myosin (thick) and actin (thin)

myoglobin (my-o-GLO-bin) large protein molecule in the sarcoplasm of muscle cells that attracts oxygen and holds it temporarily

myography (my-OG-rah-fee) procedure in which the contraction of an isolated muscle is recorded

myometrium (my-o-MEE-tree-um) middle muscle layer in the uterus

myopia (my-O-pee-ah) refractive disorder of the eye caused by an elongated eyeball; nearsightedness

myosin (MY-o-sin) contractile protein found in the thick filaments of skeletal muscle myofilaments

myxedema (mik-se-DEE-mah) condition caused by deficiency of thyroid hormone in adults

N

nail (nail) multilayered, protective structure composed of epithelial cells containing hard keratin; located at the distal ends of fingers and toes

nail body (nail BOD-ee) visible part of the nail

nail root (nail root) part of the nail hidden by the cuticle

nasolacrimal duct (nay-zo-LAK-ri-mal dukt) tube that extends from lacrimal sac into the inferior meatus of the nose

nasopharynx (nay-zo-FAIR-inks) uppermost portion of the tube just behind the nasal cavities

natural immunity (NAT-yoo-ral i-MYOO-ni-tee) acquired immunity due to exposure to disease-causing agents in the course of daily living

natural killer (NK) cells (NAT-yoo-ral KILL-er sells) group of lymphocytes that kills many types of tumor cells

nausea (NAW-see-ah) unpleasant sensation of the gastrointestinal tract that commonly precedes the urge to vomit; upset stomach

near reflex (near REE-fleks) constriction of the pupil for near vision

necrosis (ne-KRO-sis) death of cells in a tissue, often resulting from ischemia (reduced blood flow)

negative feedback (NEG-ah-tiv FEED-bak) feedback control system in which the level of a variable is changed in the direction opposite to that of the initial stimulus

neonatal period (nee-o-NAY-tal PEE-re-od) period of development immediately after an infant is first born

neonatology (nee-o-nay-TOL-o-jee) diagnosis and treatment of disorders of the newborn infant

nephron (NEF-ron) anatomical and functional unit of the kidney, consisting of the renal corpuscle and the renal tubule

nerve (nerv) bundle nerve fibers, plus surrounding connective tissue, located outside the brain or spinal cord

nerve impulse (nerv IM-puls) self-propagating wave of electrical depolarization carries information along nerves; also called action potential

nervous system (NER-vus SIS-tem) brain, spinal cord, and nerves

nervous tissue (NER-vus TISH-yoo) specialized tissue type consisting of neurons and neuroglia that provides rapid communication and control of body function

neuralgia (noo-RAL-jee-ah) general term referring to nerve pain

neurilemma (noo-ri-LEM-mah) nerve sheath

neuroblastoma (noo-ro-blas-TOE-mah) malignant tumor of sympathetic nervous tissue, found mainly in young children

neuroendocrine system (noo-ro-EN-doe-krin SIS-tem) endocrine and nervous systems working in concert to perform communication, integration, and control within the body

neurofibril (noo-ro-FYE-bril) fine strands extending through the cytoplasm of each neuron; formed by cell's cytoskeleton

neuroglia (noo-ROG-lee-ah) nonexcitable supporting cells of nervous tissue

neurohypophysis (noo-ro-hye-POF-i-sis) posterior pituitary gland

neurology (noo-ROL-o-jee) branch of science that deals with the nervous system and its disorders

neuromodulator (noo-ro-MOD-yoo-lay-tor) "cotransmitter" that regulates the effects of neurotransmitter(s) released along with it

neuromuscular junction (noo-ro-MUS-kyoo-lar JUNK-shun) point of contact between nerve endings and muscle fibers; *see* motor endplate

neuron (NOO-ron) nerve cell, including its processes (axons and dendrites)

neuropeptide (noo-roe-PEP-tide) neurotransmitter with short strands of polypeptides

neurosecretory tissue (non-roe-SEK-ro-tor-ee TISH-yoo) modified neurons that secrete chemical messengers that diffuse into the bloodstream rather than across a synapse; for example, in the hypothalamus

neurotransmitter (noo-ro-TRANS-mit-ter) chemicals by which neurons communicate; the substance is released by a neuron, diffuses across the synapse, and binds to the postsynaptic neuron

neutron (NOO-tron) neutral subatomic particle located in the nucleus of an atom

neutrophil (NOO-tro-fil) white blood cell that stains readily with neutral dyes

nevus (NEE-vus) small, pigmented benign tumor of the skin; a mole

nicotinamide adenine dinucleotide (NAD) (nik-o-TIN-ah-mide AD-e-neen dye-NOO-klee-o-tide) molecule that serves as an electron carrier in the electron transport system

Nissl bodies (NIS-l BOD-ees) in rough endoplasmic reticulum, provide protein molecules needed for transmission of nerve impulses from one neuron to another

nociceptor (no-se-SEP-tor) receptor activated by intense stimuli of any type that results in tissue damage; pain receptor

node of Ranvier (node of rahn-vee-AY) short space in the myelin sheath between adjacent Schwann cells

nondisjunction (non-dis-JUNK-shun) occurs during meiosis when a pair of chromosomes fails to separate

nonelectrolyte (non-e-LEK-tro-lite) compound that does not dissociate in solution; for example, glucose

nonself antigens (NON-self AN-ti-jens) foreign and tumor cell markers that a body's immune cells can recognize as being different from the self

nonspecific immunity (non-spe-SIF-ik i-MYOO-ni-tee) mechanisms that resist various threatening agents or conditions

nonsteroid hormone (non-STE-roid HOR-mone) hormone synthesized primarily from amino acids rather than from cholesterol

norepinephrine (nor-ep-i-NEF-rin) hormone secreted by adrenal medulla that increases cardiac output; neurotransmitter released by sympathetic nervous system cells

nuclear magnetic resonance (NMR) (NOO-klee-ar mag-NET-ik RE-son-ans) *see* magnetic resonance imaging

nuclear membrane (NOO-klee-ar MEM-brane) two-layered membrane that surrounds the cell nucleus

nuclei (NOO-kle-i) distinct regions of gray matter in the central nervous system; clusters of neuron cell bodies

nucleolus (noo-KLEE-o-lus) dense, well-defined but membraneless body within the nucleus; critical to protein formation because it "programs" the formation of ribosomes in the nucleus

nucleus (NOO-klee-us) membranous organelle that contains most of the genetic material of the cell; also, group of neuron cell bodies in the brain or spinal cord

nucleus pulposus (NOO-klee-us pul-PO-sus) gel-like center of the intervertebral disk; surrounded by annulus fibrosis

O

octet rule (ok-TET rool) general principle in chemistry whereby atoms usually form bonds in ways that will provide each atom with an outer shell of eight electrons

oculomotor nerve (ok-yoo-loe-MO-tor nerv) cranial nerve III; motor nerve; controls eye movements

olfactory nerve (ol-FAK-tor-ee nerv) cranial nerve I; sensory nerve; responsible for the sense of smell

oligodendrocyte (ol-i-go-DEN-dro-site) small astrocyte with few cell processes; helps to form myelin sheaths around axons within the central nervous system

oncogene (ON-ko-jeen) gene (DNA segment) thought to be responsible for the development of a cancer

oncologist (ong-KOL-o-jist) cancer specialist

oocyte (O-o-site) developing female sex cell contained within ovarian follicles

oogenesis (o-o-JEN-e-sis) production of female gametes

oogonia (o-o-GO-nee-ah) primitive cell from which oocytes derive meiosis

ophthalmic nerve (op-THAL-mik nerv) part of the trigeminal nerve

opsin (OP-sin) protein produced by the breakdown of rhodopsin in rods and cones of the retina; involved in a chemical reaction that initiates an impulse that results in interpretation of light energy as vision

optic chiasma (OP-tik ki-AS-mah) region where right and left optic nerves enter the brain and cross each other, exchanging fibers

optic disc (OP-tic disk) area in the retina where the optic nerve fibers exit the eye and where therefore there are no rods or cones; also known as a blind spot

optic nerve (OP-tik nerv) cranial nerve II; sensory nerve; nerves that conduct visual information to the brain

optic tract (OP-tik trakt) bundles of fibers formed after the optic nerves pass through the optic chiasma

organ (OR-gan) group of several tissue types that together perform a special function

organ of Corti (OR-gan of KOR-tee) organ of hearing located in the cochlea; consists of membranes and imbedded hair cells capable of converting mechanical motion of soundwaves into neural impulses

organelle (or-gah-NELL) cell organ; for example, the ribosome

organic (or-GAN-ik) referring to chemicals that contain covalently bound carbon and hydrogen atoms and are involved in metabolic reactions

organism (OR-gah-niz-im) any living entity considered as a whole; may be unicellular or composed of many different body systems working together to maintain life

organogenesis (or-gah-no-JEN-e-sis) formation of organs from the primary germ layers of the embryo

origin (OR-i-jin) attachment of a muscle to the bone, which does not move when contraction occurs; *see* insertion

oropharynx (o-ro-FAIR-inks) portion of the pharynx that is located behind the mouth

orthopnea (or-THOP-nee-ah) dyspnea (difficulty in breathing) that is relieved after moving into an upright or sitting position

osmolality (os-mo-LAL-i-tee) osmotic concentration of a solution; the number of moles of a substance per kilogram times the

number of particles into which the solute dissociates

osmoreceptor (os-mo-ree-SEP-tor) special receptors near the supraoptic nucleus that detect decreased osmotic pressure of blood when the body dehydrates

osmosis (os-MO-sis) movement of a fluid (usually water) through a semipermeable membrane from an area of lesser solute concentration to an area of greater concentration

osmotic pressure (os-MOT-ik PRESH-ur) water pressure that develops in a solution across a semipermeable membrane as a result of osmosis

osseous tissue (OS-ee-us TISH-yoo) bone tissue

osteitis fibrosa cystica (os-tee-EYE-tis fye-BRO-sah SIS-ti-kah) bone disease caused by hypercalcemia

osteoarthritis (os-tee-o-ar-THRY-tis) degenerative joint disease; a noninflammatory disorder of a joint characterized by degeneration of articular cartilage

osteoblast (OS-tee-o-blast) bone-forming cell

osteoclast (OS-tee-o-klast) bone-absorbing cell

osteocyte (OS-tee-o-site) bone cell

osteogenesis (os-tee-o-JEN-e-sis) combined action of osteoblasts and osteoclasts to mold bones into adult shape

osteogenesis imperfecta (os-tee-o-JEN-e-sis im-per-FEK-tah) dominant, inherited disorder of connective tissue characterized by imperfect skeletal development, resulting in brittle bones

osteoid (OS-tee-oid) organic matrix of bone

osteoma (os-tee-O-mah) benign bone tumor

osteomyelitis (os-tee-o-my-e-LIE-tis) bacterial (usually staphylococcus) infection of bone tissue

osteon (OS-tee-on) see haversian system

osteoporosis (os-tee-o-po-RO-sis) bone disorder characterized by loss of minerals and collagen from bone matrix, reducing the volume and strength of skeletal bone

osteosarcoma (os-tee-o-sar-KO-mah) bone cancer

otitis media (o-TIE-tis MEE-dee-ah) a middle ear infection

otoliths (O-toe-liths) tiny "ear stones" composed of protein and calcium carbonate in the maculae of the ear, which, by responding to gravity and changes in body position, trigger hair cells that initiate impulses resulting in sense of balance

ova (O-vah) female sex cells (singular ovum)

ovarian follicle (o-VARE-ee-an FOL-i-kul) spherical configuration of cells in the ovary that contains a single oocyte

ovaries (O-var-ees) female gonads that produce ova (sex cells)

oviducts (O-vi-dukts) see uterine or fallopian tubes

ovulation (o-vyoo-LAY-shun) release of an ovum from the ovary at the end of oogenesis

oxidative phosphorylation (ok-si-DAY-tiv fos-for-i-LAY-shun) reaction that joins a phosphate group to ADP to form ATP

oxygen debt (OK-si-jen det) oxygen required for ATP synthesis to remove excess lactic acid following anaerobic exercise

oxyhemoglobin (ok-see-hee-mo-GLO-bin) hemoglobin combined with oxygen

oxytocin (OT) (ok-see-TOE-sin) hormone secreted by the posterior pituitary gland before and after delivering a baby; thought to initiate and maintain labor, as well as cause the release of breast milk into ducts of the mammary glands

P

pacinian corpuscle (pah-SIN-ee-an KOR-pus-ul) receptor deep in the dermis that detects pressure on the skin surface

packed cell volume (PCV) (pakd sell VOL-yoom) volume percent of red blood cells in whole blood; see hematocrit

palatine tonsil (PAL-ah-tine TON-sil) tonsils located behind and below the pillars of the fauces

palpebrae (PAL-pe-bree) eyelids

palpebral fissure (PAL-pe-bral FISH-ur) opening between the two eyelids

pancreas (PAN-kree-as) endocrine gland located in the abdominal cavity; contains pancreatic islets that secrete glucagon and insulin

pancreatic islets of Langerhans (pan-kree-AT-ik eye-LETS) endocrine portion of the pancreas; made up of alpha and beta cells, among others; source of insulin and glucagon

pancreatitis (pan-kre-ah-TYE-tis) inflammation of the pancreas

papilloma (pap-i-LO-mah) benign skin tumor characterized by fingerlike projections (for example, a wart)

paralysis (pah-RAL-i-sis) loss of power of motion, especially voluntary motion

parasympathetic division of (ANS) (par-ah-sim-pah-THE-tic di-VI-shun) part of the autonomic nervous system; ganglia are connected to the brain stem and the sacral segments of the spinal cord; controls many visceral effectors under normal ("rest and repair") conditions

parathyroid gland (pair-ah-THYE-royd gland) endocrine gland located in the neck on the posterior aspect of the thyroid gland; secretes parathyroid hormone (parahormone)

parathyroid hormone (PTH) (pair-ah-THYE-royd HOR-mone) hormone released by the parathyroid gland that increases concentration of calcium in the blood

parenteral therapy (pah-REN-ter-al THER-al-pee) administration of nutrients, special fluids, and/or electrolytes by injection

parietal (pah-RYE-i-tal) refers to the walls of an organ or cavity

parietal membrane (pah-RYE-i-tal MEM-brane) portion of the serous membrane that lines the wall of a body cavity

parietal peritoneum (pah-RYE-i-tal pair-i-toe-NEE-um) serous membrane that covers organs of the abdomen, pelvis, and thorax

Parkinson's disease (PAR-kin-sunz di-ZEEZ) nervous disorder characterized by abnormally low levels of the neurotransmitter dopamine in parts of the brain that control voluntary movement—victims usually exhibit involuntary trembling and muscle rigidity

parotid glands (pah-ROT-id glands) largest of the paired salivary glands

pars anterior (parz an-TEER-ee-or) major part of the anterior pituitary gland

pars intermedia (PARS in-ter-MEE-dee-ah) forms a small part of the anterior pituitary gland

partial pressure (PAR-shal PRES-ur) pressure exerted by any one gas in a mixture of gases or in a liquid

partial-thickness burn (PAR-shal THIK-ness bern) first or second degree burn

parturition (par-too-RISH-un) act of giving birth

pathogen (PATH-o-jen) microorganism that causes disease

pathology (path-o-lo-jee) study of diseased body structures

pedicel (PED-i-cel) branches of the podocytes of the Bowman's capsule, packed tightly together with only filtration slits between them

pedigree (PED-i-gree) chart used in genetic counseling to illustrate genetic relationships over several generations

pentose (PEN-tose) five-carbon sugar such as ribose or deoxyribose

peptide bond (PEP-tide bond) bond that forms between the amino group of one amino acid and the carboxyl group of another

peptide hormone (PEP-tide HOR-mone) major category of nonsteroid hormones; smaller than protein hormones; comprised of a short chain of amino acids; for example, antidiuretic hormone (ADH)

pericardial fluid (pair-i-KAR-de-al FLOO-id) lubricating fluid secreted by the serous membrane; in pericardial space

pericardial space (pair-i-KAR-dee-al space) space between the visceral layer and the parietal layer surrounding the heart, filled with pericardial fluid

pericardium (pair-i-KAR-dee-um) membrane that surrounds the heart

perichondrium (pair-i-KON-dree-um) fibrous covering of cartilage structures

perikaryon (pair-i-KAR-e-on) cell body

perilymph (PAIR-i-limf) watery fluid that fills the bony labyrinth of the ear

perimysium (pair-i-ME-see-um) tough, connective tissue surrounding fascicles

perineum (pair-i-NEE-um) area between anus and genitals

perineurium (pair-i-NOO-ree-um) connective tissue that encircles a bundle of nerve fibers within a nerve

periodontitis (pair-ee-o-don-TIE-tis) inflammation of the periodontal membrane that anchors teeth to jaw bone; common cause of tooth loss among adults

periosteum (pair-i-OS-tee-um) tough, connective tissue covering the bone

peripheral nervous system (PNS) (pe-RIF-er-al NER-vus SIS-tem) nerves connecting brain and spinal cord to other parts of the body

peripheral resistance (pe-RIF-er-al ree-SIS-tance) resistance to blood flow caused by friction of blood passing through blood vessels

peristalsis (pair-i-STAL-sis) wavelike, rhythmic contractions of the stomach and intestines that move food material along the digestive tract

peritoneum (pair-i-toe-NEE-um) large, moist, slippery sheet of serous membrane that lines the abdominopelvic cavity (parietal layer) and its organs (visceral layer)

peritonitis (pair-i-toe-NYE-tis) inflammation of the serous membranes in the abdominopelvic cavity; sometimes a serious complication of an infected appendix

peritubular capillary (pair-e-TOOB-yoo-lar KAP-e-lar-ee) capillaries around the tubules in the kidney

permanent teeth (PER-mah-nent teeth) set of teeth that replace deciduous teeth

peroxisome (pe-ROSK-i-some) organelles that detoxify harmful substances that have entered cells

perspiration (per-spi-RAY-shun) transparent, watery liquid released by glands in the skin that

eliminates ammonia and uric acid and helps maintain body temperature; also known as sweat

pH (pH) units by which acid and base concentrations are measured; scale ranges from O (extremely acidic) to 14 (extremely basic, or alkaline)

phagocytes (FAG-o-sites) white blood cells that engulf microbes and digest them

phagocytosis (fag-o-sye-TOE-sis) ingestion and digestion of particles by a cell

pharyngeal tonsil (fah-RIN-jee-al TON-sil) tonsil located in the nasopharynx on its posterior wall; when enlarged, referred to as adenoids

pharyngitis (fair-in-JYE-tis) sore throat; inflammation or infection of the pharynx

phenotype (FEE-no-tipe) overt, observable expression of a genotype

phenylalanine hydroxylase (fen-il-AL-ah-nin hye-DROK-si-lase) enzyme needed to convert phenylalanine into tyrosine; its absence causes phenylketonuria

phenylketonuria (PKU) (fen-il-kee-toe-NOO-ree-ah) recessive, inherited condition characterized by excess of phenylketone in the urine, caused by accumulation of phenylalanine (an amino acid) in the tissues; may cause brain injury and death if phenylalanine intake is not managed properly

phlebitis (fle-BYE-tis) inflammation of a vein

phosphatase (FOS-fay-tase) enzyme that removes phosphate groups

phospholipid (fos-fo-LIP-id) phosphate-containing fat molecule; an important constituent of cell membranes

phosphorylase (fos-FOR-i-lase) enzyme that adds phosphate groups

photopigment (fo-toe-PIG-ment) chemicals in retinal cells that are sensitive to light

photopupil reflex (fo-toe-PYOO-pul REE-fleks) constriction of the pupil in bright light to protect the retina from too intense or too sudden stimulation; also known as pupillary light reflex

photoreceptor (fo-to-ree-SEP-tor) receptor only in the eye; responds to light stimuli if the intensity is great enough to generate a receptor potential

phrenic nerve (FREN-ik nerv) nerve that stimulates the diaphragm to contract

physiological buffer (fiz-ee-o-LOJ-i-kal BUF-er) secondary defense against harmful shifts in pH of body fluids; comes into play after the chemical buffer system

physiological fatigue (fiz-ee-o-LOJ-i-kal fah-TEEG) fatigue caused by a relative lack of ATP, rendering the myosin cross bridges incapable of producing the force required for further muscle contractions

physiological polycythemia (fiz-ee-o-LOJ-i-kal pol-e-sye-THEE-mee-ah) elevated red blood cell numbers and hematocrit values in healthy individuals who live and work in high altitudes

physiology (fiz-ee-OL-o-jee) study of body function

physique (fi-ZEEK) body build

pia mater (PEE-ah MAH-ter) vascular innermost covering (meninx) of the brain and spinal cord

pimple (PIM-pl) pustule on the skin, usually associated with acne

pineal body (PI-nee-al BOD-ee) endocrine gland located in the diencephalon and thought to be involved with regulating the body's biological clock; produces melatonin; also called pineal gland

pineal gland (PI-nee-al gland) *see* pineal body

pinna (PIN-nah) flap of the external ear

pinocytosis (pin-o-sye-TOE-sis) active transport mechanism used to transfer fluids or dissolved substances into cells

pituitary dwarfism (pi-TOO-i-tair-ee DWARF-izm) condition resulting from hyposecretion of growth hormone during the growth years

pituitary gland (pi-TOO-i-tair-ee gland) neuroendocrine gland located near base of the brain that has numerous and important regulatory functions; also called the hypophysis

placenta (plah-SEN-tah) structure that anchors the developing fetus to the uterus and provides a "bridge" for the exchange of nutrients and waste products between the mother and developing baby

placenta previa (plah-SEN-tah PRE-vee-ah) abnormal condition in which a blastocyst implants in the lower uterus, developing a placenta that approaches or covers the cervical opening; placenta previa involves risk of placental separation and hemorrhage

plantar flexion (PLAN-tar FLEK-shun) the bottom of the foot is directed downward; this motion allows a person to stand on tiptoe

plaque (plak) raised skin lesion greater than 1 cm in diameter

plasma (PLAZ-mah) liquid part of the blood

plasma membrane (PLAZ-mah MEM-brane) membrane that separates the contents of a cell from the tissue fluid, encloses the cytoplasm, and forms the outer boundary of the cell

plasmid (PLAS-mid) small circular ring of bacterial DNA

platelet (PLATE-let) specialized cell fragments in the blood; thrombocyte; important component in the clotting mechanism

pleura (PLOOR-ah) serous membrane in the thoracic cavity

pleurisy (PLOOR-i-see) inflammation of the pleura

plexus (PLEK-sus) complex network formed by converging and diverging nerves, blood vessels, or lymphatics

pneumonectomy (noo-mo-NEK-toe-mee) surgical procedure in which an entire lung is removed

pneumonia (noo-MO-nee-ah) abnormal condition characterized by acute inflammation of the lungs in which alveoli and bronchial passages become plugged with thick fluid (exudate)

pneumotaxic center (noo-mo-TAK-sik SEN-ter) group of cells in the pons of the brain that affects the rate of respiration by inhibiting the inspiration center

pneumothorax (noo-mo-THO-raks) abnormal condition in which air is present in the pleural space surrounding the lung, possibly causing collapse of the lung

podocyte (POD-o-site) special epithelial cells making up the visceral layer of the Bowman's capsule

Poiseuille's law (pwah-ZUH-yez law) volume of blood circulated per minute is directly related to mean arterial pressure minus central venous pressure and inversely related to resistance

polar molecule (PO-lar MOL-e-kyool) molecule in which the electrical charge is not evenly distributed, causing one side of the molecule to be more positive or negative than the other

poliomyelitis (po-lee-o-my-e-LYE-tis) viral disease that damages motor nerves, often progressing to paralysis of skeletal muscles

polymer (POL-i-mer) large molecule made up of many identical smaller molecules joined together in sequence

polypeptide (pol-e-PEP-tide) compound made of many amino acids connected by peptide bonds

pons (ponz) part of the brain stem between the medulla oblongata and the midbrain

portal system (POR-tal SIS-tem) arrangement of blood vessels in which blood exiting one tissue is immediately carried to a second tissue before being returned to the heart and lungs for oxygenation and redistribution

positive feedback mechanism (POZ-i-tiv FEED-bak MEK-ah-nizm) feedback control system that is stimulatory; tends to amplify or reinforce a change in the internal environment

posterior (pos-TEER-ee-or) located behind; opposite of anterior

posterior cavity (of eye) (pos-TEER-ee-or KAV-i-tee) all the space posterior to the lens of the eye; contains vitreous humor

posterior chamber (of eye) (pos-TEER-ee-or CHAM-ber) subdivision of the anterior cavity of the eye; small space behind the iris and anterior to the lens

postganglionic neurons (post-gang-glee-ON-ik NOO-rons) autonomic neurons that conduct nerve impulses from a ganglion to effectors such as cardiac or smooth muscle or glandular epithelial tissue

postnatal period (POST-nay-tal PEER-ee-od) period after birth, ending at death

postsynaptic neuron (post-si-NAP-tik NOO-ron) in neuron-to-neuron communication, the neuron that receives a stimulus via an adjacent neuron's transmission of neurotransmitters across the synapse

postsynaptic potential (post-si-NAP-tik po-TEN-shal) local potential produced by opening of ion channels in the postsynaptic membrane

posture (POS-chur) position of the body; often refers to the erect position of the body maintained unconsciously

potential osmotic pressure (po-TEN-shal os-MOT-ik PRESH-ur) maximum osmotic pressure that could develop in a solution when it is separated from pure water by a selectively permeable membrane

precapillary sphincter (pree-CAP-pi-lair-ee SFINGK-ter) smooth muscle cells that guard the entrance to the capillary

preeclampsia (pree-e-KLAMP-see-ah) syndrome of abnormal conditions in pregnancy of uncertain cause; symptoms include hypertension, proteinuria, and edema; also called toxemia of pregnancy, it may progress to eclampsia—severe toxemia that may cause death

preganglionic neurons (pree-gang-glee-ON-ik NOO-rons) autonomic neurons that conduct nerve impulses between the spinal cord and a ganglion

prenatal period (PREE-nay-tal PEER-i-od) developmental period after conception until birth

prepuce (PREE-pus) foreskin, especially covering fold of skin over the glans penis

presbyopia (pres-bee-O-pee-ah) farsightedness of old age

pressoreflex (pres-o-REE-fleks) *see* baroreflex

presynaptic neuron (pree-si-NAP-tik NOO-ron) in neuron-to-neuron communication, the neuron that transmits a signal to an adjacent neuron via release of neurotransmitters that cross the synapse and bind to the postsynaptic neuron

prevertebral ganglion (pree-ver-TEE-bral GANG-gle-on) collateral ganglion; pairs of sympathetic ganglia located a short distance from the spinal cord

primary bronchi (PRYE-mar-ee BRONG-ki) the two branches of the trachea that enter the lungs

primary germ layers (PRYE-mair-ee jerm LAY-ers) three layers of specialized cells that give rise to definite structures as the embryo develops; see ectoderm, endoderm, and mesoderm

primary ossification center (PRYE-mair-ee os-i-fi-KAY-shun SEN-ter) where a blood vessel enters the cartilage model of bone at the midpoint of the diaphysis to initiate bone formation

prime mover (prime MOO-ver) main muscle responsible for producing a particular movement

product (PROD-ukt) molecules or atoms that result from a chemical reaction

proenzyme (pro-EN-zime) inactive form in which many enzymes are synthesized

progesterone (pro-JES-ter-ohn) steroid hormone produced by the ovaries (particularly the corpus luteum) that helps prepare the uterus for implantation; along with estrogen, helps maintain normal uterine and mammary gland function

prohormone (pro-HOR-mone) hormone precursor

projection tract (pro-JEK-shun trakt) extension of the sensory spinothalamic tracts and descending corticospinal tracts in the brain

prolactin (PRL) (pro-LAK-tin) hormone secreted by the anterior pituitary gland during pregnancy to stimulate the breast development needed for lactation

prophase (PRO-faze) first stage of mitosis during which chromosomes become visible

proprioception (pro-pree-o-SEP-shun) perception of movement and position of the body

proprioceptors (pro-pree-o-SEP-tors) receptors located in the muscles, tendons, and joints; allow the body to recognize its position

prostaglandins (PGs) (pross-tah-GLAN-dins) group of naturally occurring lipid-based substances that act in a hormonelike way to affect many body functions, including vasodilation, uterine smooth muscle contraction, and the inflammatory response

prostate gland (PROSS-tate gland) lies just below the bladder; secretes a fluid that constitutes about 30% of the seminal fluid volume; helps activate sperm and helps them maintain motility

protein (PRO-teen) large molecules formed by linkage of amino acids by peptide bonds; one of the basic building blocks of the body

proton (PRO-ton) positively charged subatomic particle

proximal (PROK-si-mal) next or nearest; located nearest the center of the body or the point of attachment of a structure; opposite of distal

proximal convoluted tubule (PROK-si-mal kon-vo-LOO-ted TOOB-yool) second part of the nephron and the first segment of a renal tubule

psoriasis (so-RYE-ah-sis) chronic, inflammatory skin disorder characterized by cutaneous inflammation and scaly plaques

psychological stressor (sye-KOL-o-jee STRESS-or) anything an individual perceives as a threat

psychophysiology (sye-ko-fiz-ee-OL-o-jee) scientific discipline that studies physiological responses of individuals being subjected to psychological stressors

psychosomatic (sye-ko-so-MAT-ik) mind influencing the body

ptosis (TOE-sis) downward displacement of an organ; for example the lowered kidneys of very thin individuals

puerperal fever (pu-ER-per-al FEE-ver) condition caused by bacterial infection in a woman after delivery of an infant, possibly progressing to septicemia and death; also called childbed fever

pulmonary artery (PUL-mo-nair-ee AR-ter-ee) artery that carries deoxygenated blood from the right ventricle of the heart to the lungs

pulmonary circulation (PUL-mo-nair-ee ser-kyoo-LAY-shun) blood flow from the right ventricle to the lung and returning to the left atrium

pulmonary edema (PUL-mo-nair-ee e-DEE-mah) congestion of blood in the pulmonary circulation

pulmonary embolism (PUL-mo-nair-ee EM-bo-lizm) blockage of the pulmonary circulation by a thrombus or other matter; may lead to death if blockage of pulmonary blood flow is significant

pulmonary semilunar valve (PUL-mo-nair-ee sem-i-LOO-nar valv) valve located at the beginning of the pulmonary artery

pulmonary vein (PUL-mo-nair-ee vane) any vein that carries oxygenated blood from the lungs to the left atrium

pulmonary ventilation (PUL-mo-nair-ee ven-ti-LAY-shun) breathing; process that moves air in and out of the lungs

pulp cavity (pulp KAV-i-tee) cavity in the dentin of a tooth that contains connective tissue, blood and lymphatic vessels, and sensory nerves

pulse point (puls point) where the pulse can be palpated; i.e., wherever an artery lies near the surface and over a bone or other firm background; for example, radial artery

pulse pressure (puls PRESH-ur) difference between systolic and diastolic pressure

punctae (PUNK-tay) opening into the lacrimal canals located at the inner canthus of the eye

Punnett square (PUN-it skwair) grid used in genetic counseling to determine the probability of inheriting genetic traits

pupil (PYOO-pil) opening in the center of the iris that regulates the amount of light entering the eye

pupillary light reflex (PYOO-pi-lair-ee lite REE-fleks) see photopupil reflex

purine bases (PYOO-rine bases) one of two types of nitrogenous bases that are vital components of DNA derived from purine; adenine and guanine

Purkinje fibers (pur-KIN-jee FYE-bers) specialized cells located in the walls of the ventricles; relay nerve impulses from the AV node to the ventricles, causing them to contract

pus (pus) accumulation of white blood cells, dead bacterial cells, and damaged tissue cells at site of an infection

pustule (PUS-tyool) small, raised skin lesion filled with pus

pyloric sphincter (pye-LOR-ik SFINGK-ter) sphincter that prevents food from leaving the stomach and entering the duodenum

pyramidal tract (pi-RAM-i-dal trakt) fibers that come together in the medulla to form the pyramids; also called corticospinal tract

pyramids (PEER-ah-mids) triangular-shaped divisions of the medulla of the kidney

pyrimidine bases (pi-RIM-i-deen bases) one of two types of nitrogenous bases that are vital components of DNA derived from pyrimidine; cytosine and thymine

R

radiation (ray-dee-AY-shun) electromagnetic energy, including light, x-rays, heat; in physiology, often refers to flow of excess heat energy away from the body via the blood

radiation therapy (ray-dee-AY-shun THER-ah-pee) treatment often used with cancer in which high-intensity radiation is used to destroy cancer cells; also called radiotherapy

radioactive isotope (ray-dee-o-ak-tiv EYE-so-tope) unstable isotope that spontaneously emits subatomic particles and electromagnetic radiation

radiography (ray-de-OG-rah-fee) imaging technique using x-rays that pass through certain tissues more easily than others, allowing an image of tissues to form on a photographic plate

radiotherapy (ray-de-o-THER-ah-pee) see radiation therapy

ramus (RAY-mus) large branch of a spinal nerve as it emerges from the spinal cavity

reabsorption (ree-ab-SORP-shun) process occurring in the nephrons during urine formation whereby essential materials that were filtered out of the blood in the glomerulus are reabsorbed into the interstitial space

reactant (ree-AK-tant) participant in a chemical reaction that is changed by the reaction

receptor (ree-SEP-tor) portion of a sensory neuron that responds to an external stimulus; also, any molecule on the surface of a cell that binds specifically to other molecules (such as cell markers, hormones, neurotransmitters)

receptor potential (ree-SEP-tor po-TEN-shal) potential that develops in the receptor's membrane when an adequate stimulus has been received

recessive (ree-SES-iv) in genetics, refers to genes that have effects that do not appear in the offspring when they are masked by a dominant gene (recessive forms of a gene are represented by lower case letters); see dominant

recombinant DNA (ree-KOM-bi-nant DNA) joining together of hereditary material into new, biologically functional combinations

red fiber (red FYE-ber) muscle fibers that contain large amounts of myoglobin and have a deep red appearance

red marrow (red MAR-o) bone marrow found in the ends of long bones and in flat bones; so named due to its function in the production of red blood cells

reduction (ree-DUK-shun) proper alignment of a fractured bone; in chemistry, the gain of one or more electrons by a molecule (as in oxidation-reduction reactions)

reduction division (ree-DUK-shun di-VI-shun) meiosis; when the diploid chromosome number (46) is reduced to the haploid number (23)

referred pain (ree-FERD pane) pain from stimulation of nociceptors in deep structures felt on the surface of the body; for example, experiencing pain in the left arm when there is heart pain

reflex (REE-fleks) automatic involuntary reaction to a stimulus resulting from a nerve impulse passing over a reflex arc

reflex arc (REE-fleks ark) impulse conduction route to and from the central nervous system; smallest portion of nervous system that can receive a stimulus and generate a response

refraction (ree-FRAK-shun) bending of a ray of light as it passes from a medium of one density to one of a different density; occurs as light rays pass through the eye

regeneration (ree-jen-er-AY-shun) process of replacing missing tissue with new tissue by means of cell division

relative refractory period (REL-ah-tiv ree-FRAK-tor-ee PEE-re-od) in muscle cell contraction, the few milliseconds after the absolute refractory period; time during which the membrane is repolarizing and restoring the resting membrane potential

relaxation (ree-lak-SAY-shun) state of lessened tension

releasing hormones (ree-LEE-sing HOR-mones) hormone produced by the hypothalamus that causes the pituitary gland to release its hormones

renal artery (REE-nal AR-ter-ee) large branch of the abdominal aorta that brings blood into each kidney

renal clearance (REE-nal KLER-ans) amount of a substance removed from the blood by the kidneys per minute

renal columns (REE-nal KOL-ums) within the kidneys, the cortical tissue in the medulla between the pyramids

renal corpuscle (REE-nal KOR-pus-ul) within the nephron, the glomerulus plus the Bowman's capsule surrounding it

renal cortex (REE-nal KOR-teks) outer portion of the kidney

renal diabetes (REE-nal dye-ah-BEE-teez) when the maximum cotransport capacity of the kidney is greatly reduced and glucose appears in the urine even though the blood sugar level may be normal; also called renal glycosuria

renal glycosuria (REE-nal glye-ko-soo-ree-ah) *see* renal diabetes

renal medulla (REE-nal me-DUL-ah) inner portion of the kidney

renal pelvis (REE-nal PEL-vis) basinlike upper end of the ureter that is located inside the kidney into which the collecting tubules drain

renal pyramids (REE-nal PEER-ah-mids) distinct triangular wedges that make up most of the medullary tissue in the kidney

renal tubule (REE-nal TOOB-yool) a principal part of the nephron, it consists of the proximal tubule, the loop of Henle, and the distal tubule; receives filtrate from Bowman's capsule and transports it to collecting tubules while filtrate is reabsorbed and additional substances are secreted into it in the process of urine formation

renin (RE-nin) enzyme produced by the juxtaglomerular apparatus of the kidney nephrons that catalyzes the formation of angiotensin, a substance that increases blood pressure

renin-angiotensin mechanism (RE-nun-an-jee-o-TEN-sis MEK-ah-nizm) causes changes in blood plasma volume mainly by controlling aldosterone secretion

repolarization (ree-po-lah-ri-ZAY-shun) phase of the action potential in which the membrane potential changes from its maximum degree of depolarization toward the resting state potential

reproduction (ree-pro-DUK-shun) formation of a new individual; formation of new cells in the body to permit growth

residual volume (RV) (re-ZID-yoo-al VOL-yoom) air that remains in the lungs after the most forceful expiration

resorption (re-SORP-shun) bone loss due to osteoclastic activity

respiration (res-pi-RAY-shun) processes that result in the absorption, transport, and utilization or exchange of respiratory gases between an organism and its environment

respiratory acidosis (re-SPY-rah-tor-ee as-i-DOE-sis) retention of carbon dioxide in the blood

respiratory alkalosis (re-SPY-rah-tor-ee al-kah-LOE-sis) excessive loss of carbonic acid from the blood

respiratory arrest (re-SPY-rah-tor-ee ah-REST) cessation of breathing without resumption

respiratory center (re-SPY-rah-tor-ee SEN-ter) center located in the medulla and pons that stimulates muscles of respiration

respiratory distress syndrome (re-SPY-rah-tor-ee di-STRESS SIN-drome) difficulty in breathing caused by absence or failure of the surfactant in fluid lining the alveoli of the lung; IRDS is infant respiratory distress syndrome; ARDS is adult respiratory distress syndrome

respiratory membrane (re-SPY-rah-tor-ee MEM-brane) single layer of cells that makes up the wall of the alveoli

respiratory mucosa (re-SPY-rah-tor-ee myoo-KO-sah) mucus-

covered membrane that lines tubes of the respiratory tree

respiratory tract (re-SPY-rah-tor-ee trakt) organs of the respiratory system, divided into lower and upper respiratory tracts

responsiveness (ree-SPONS-iv-ness) characteristic of life that permits an organism to sense, monitor, and respond to changes in its external environment

resting membrane potential (RMP) (REST-ing MEM-brane po-TEN-shal) membrane potential maintained by a nonconducting neuron's plasma membrane; approximately −70 mV

retention (ree-TEN-shun) inability to void urine even though the bladder contains an excessive amount of urine

reticular activating system (re-TIK-yoo-lar AK-ti-vay-ting SIS-tem) network of relays in the brain responsible for maintaining consciousness

reticular formation (re-TIK-yoo-lar for-MAY-shun) located in the medulla where bits of gray and white matter mix intricately, this structure is involved in regulating input from sensory neurons, arousal, and motor control

reticulin (re-TIK-yoo-lin) specialized type of collagen found in reticular fibers

reticulocyte (re-TIK-yoo-lo-site) immature red blood cells

reticulocyte count (re-TIK-yoo-la-site cownt) medical procedure to determine the rate of erythropoiesis

reticulum (re-TIK-yoo-lum) intricate network of fibers

retina (RET-i-nah) innermost layer of the eyeball; contains rods and cones and continues posteriorly with the optic nerve

retroperitoneal (re-tro-pair-i-toe-NEE-al) area outside of the peritoneum

reversible reaction (ree-VER-si-bl ree-AK-shun) when the products of a chemical reaction can change back to the original reactants; generally, an equilibrium of products and reactants exists

Rh factor (RH FAK-tor) antigen present on the red blood cells of Rh⁺ individuals; so named because it was first studied in the rhesus monkey

rheumatic fever (roo-MAT-ik FEE-ver) delayed inflammatory response to streptococcal infection that, if improperly treated, may allow the cardiac valves to become inflamed

rheumatic heart disease (roo-MAT-ik hart di-ZEEZ) cardiac damage (especially to the endocardium, including the valves) resulting from a delayed inflammatory response to strep-

tococcal infection; *see* rheumatic fever

rhinitis (rye-NYE-tis) inflammation of the nasal mucosa often caused by nasal infections

rhodopsin (roe-DOP-sin) photopigment in rods; *see* opsin

ribose (RYE-bose) five-carbon sugar; the sugar in RNA

ribosome (RYE-bo-sohm) organelle in the cytoplasm of cells that synthesizes proteins; sometimes called "protein factory"

righting reflexes (RITE-ing REE-fleks) muscular responses to restore the body and its parts to their normal position when they have been displaced

rigor mortis (RIG-or MOR-tis) literally "stiffness of death," the permanent contraction of muscle tissue after death caused by the depletion of ATP during the actin-myosin reaction—preventing myosin from releasing actin to allow relaxation of the muscle

rotation (ro-TAY-shun) joint movement around a longitudinal axis; for example, shaking your head "no"

rotator cuff (RO-tay-tor kuf) musculotendinous cuff resulting from fusion of the tendons of the supraspinatus, infraspinatus, teres minor, and subscapularis; adds to the stability of the glenohumeral (shoulder) joint

rubrospinal tract (roo-bro-SPY-nal trakt) spinal cord nerves that transmit impulses that coordinate body movements and maintain posture

Ruffini's corpuscle (roo-FEE-neez KOR-pus-ul) receptor that senses deep pressure and continuous touch; located in dermis of the skin

"rule of nines" (rule of nines) frequently used method to estimate extent of a burn injury in an adult; the body is divided into areas that are multiples/fractions of 9%

S

saccule (SAK-yool) part of the membranous labyrinth in the inner ear; contains sensory structure called a macula, which functions in equilibrium

sacral plexus (SAY-kral PLEKS-us) plexus formed by fibers from the fourth and fifth lumbar nerves and the first four sacral nerves

sagittal plane (SAJ-i-tal plane) longitudinal plane that divides the body or a part into left and right sides

salpingitis (sal-pin-JYE-tis) inflammation of the uterine tubes

saltatory conduction (SAL-tah-tor-ee kon-DUKT-shun) process in which a nerve impulse travels along a myelinated fiber by jumping from one node of Ranvier to the next

sarcolemma (sar-ko-LEM-ah) plasma membrane of a striated muscle fiber

sarcoma (sar-ko-mah) tumor of muscle tissue

sarcomere (SAR-ko-meer) contractile unit of muscle cells; length of a myofibril between two Z bands

sarcoplasm (SAR-ko-plazm) cytoplasm of muscle fibers

sarcoplasmic reticulum (sar-ko-PLAZ-mik re-TIK-yoo-lum) network of tubules and sacs in muscle cells; similar to endoplasmic reticulum of other cells

satiety center (sah-TYE-e-ti SEN-ter) cells in the hypothalamus that send impulses to decrease appetite so that an individual feels satisfied

saturated fat (SACH-e-ray-ted fat) fats containing triglycerides in which fatty acid chains contain no double bonds (because they are "saturated" with hydrogen atoms)

scala tympani (SKAH-lah TIM-pah-nee) lower portion of the cochlear duct lying below the basilar membrane

scala vestibuli (SKAH-lah ves-TIB-yoo-li) upper portion of the cochlear duct

scar (skahr) thickened mass of tissue, usually fibrous connective tissue, that remains after a damaged tissue has been repaired

Schwann cells (shwon sells) large nucleated cells that form myelin around the axons of neurons

sciatica (sye-AT-i-kal) neuralgia of the sciatic nerve

sclera (SKLE-rah) white outer coat of the eyeball

scleroderma (skle-ro-DER-mah) rare disorder affecting vessels and connective tissue of skin and other tissues, characterized by tissue hardening

scoliosis (sko-lee-O-sis) abnormal lateral (side-to-side) curvature of the vertebral column

scrotum (SKRO-tum) pouchlike sac that contains the testes

seasonal affective disorder (SAD) (SEE-son-al ah-FEK-tiv dis-OR-der) mental disorder in which a patient suffers severe depression only in winter; linked to the pineal gland

sebaceous gland (se-BAY-shus gland) oil-producing glands in the skin

sebum (SEE-bum) secretion of sebaceous glands

second messenger hypothesis (SEK-und MES-en-jer) theory of hormone action in which the hormone binds to receptors of the target cell, which then triggers a second molecule within the cell (such as cyclic AMP) to accomplish its function; also called fixed membrane receptor hypothesis

secondary bronchi (SEK-on-dair-ee BRONG-ki) when the primary bronchi enters the lung on its respective side and immediately splits into smaller bronchioles

secondary ossification center (SEK-on-dair-ee os-i-fi-KAY-shun SEN-ter) growth center located in the epiphyses of long-bones; *see* primary ossification center

secondary sexual characteristics (SEK-on-dair-ee SEK-shoo-al kair-ak-ter-IS-tiks) external physical characteristics of sexual maturity resulting from action of sex hormones; they include male and female patterns of body hair and fat distribution, as well as development of external genitals

second-degree burn (SEK-und-de-GREE bern) burn involving deep epidermal layers of the skin but not causing irreparable damage

secretin (se-KREE-tin) gastrointestinal hormone; first hormone discovered

secretion (se-KREE-shun) process by which a substance is released outside the cell

segmental reflex (seg-MEN-tal REE-fleks) reflex whose mediating reflexes enter and leave the same segment of the spinal cord

segmentation (seg-men-TAY-shun) when digestive reflexes cause a forward-and-backward movement within a single region of the GI tract

seizure (SEE-zhur) *see* convulsion

self antigen (self AN-ti-jen) molecule located on the plasma membrane of all body cells that identifies all normal cells of the body for the immune system

semen (SEE-men) ejaculate from the penis that contains spermatozoa plus fluids from the testes, seminal vesicles, bulbourethral glands, and prostate

semicircular canals (sem-i-SIR-kyoo-lar kah-NALS) three tube-like structures located in the inner ear that function in sense of equilibrium

semilunar valves (sem-i-LOO-nar valvs) valves located between each ventricle and the large artery that carries blood away from it; valves in the veins

seminiferous tubule (se-mi-NIF-er-us TOOB-yool) long, coiled structure that forms the bulk of the testicular mass and in which spermatozoa develop

senescence (se-NES-enz) older adulthood; aging

sensation (sen-SAY-shun) interpretation of sensory nerve impulses by the brain as an awareness of an internal or external event; for example, feeling pain

sensory nerves (SEN-sor-ee nervs) nerves that contain primarily sensory neurons

sensory neurons (SEN-sor-ee NOO-rons) neurons that transmit impulses to the spinal cord and brain from all parts of the body

sensory projection (SEN-sor-ee pro-JEK-shun) brain function that pinpoints the area of the body from which a receptor potential was initiated

sensory receptor (SEN-sor-ee ree-SEP-tor) sense organs in the peripheral nervous system that enable the body to respond to stimuli caused by changes in its internal or external environment

septicemia (sep-ti-SEE-me-ah) blood poisoning

septum (SEP-tum) a wall that divides two areas; e.g., nasal septum

serosa (se-RO-sah) outermost covering of the digestive tract; composed of the parietal pleura in the abdominal cavity

serotonin (sair-o-TOE-nin) neurotransmitter that belongs to a group of compounds called catecholamines

serous membrane (SE-rus MEM-brane) two-layer epithelial membrane that lines body cavities and covers surfaces of organs

serous pericardium (SE-rus per-i-KAR-de-um) part of pericardial coverings of the heart; made up of a parietal layer and a visceral layer

Sertoli cell (ser-TOE-lee sell) testis cell found within walls of the seminiferous tubules; provides mechanical support and protection for the developing sperm

sex chromosomes (seks KRO-mo-soms) pair of chromosomes in the human genome that determine gender; normal males have one X chromosome and one Y chromosome (XY); normal females have two X chromosomes (XX)

sex hormones (seks HOR-mones) hormones that target reproductive tissue; examples include estrogen, progesterone, and testosterone

sex-linked trait (seks-linked trait) nonsexual, inherited trait governed by genes located in a sex chromosome (X or Y); most known sex-linked traits are X-linked

shoulder girdle (SHOL-der GIR-dl) clavicle and scapula, which form the only bony attachment of the upper extremity to the trunk

sickle cell anemia (SIK-ul sell ah-NEE-mee-ah) severe, possibly fatal, hereditary disease in which red blood cells become sickle-shaped due to presence of an abnormal type of hemoglobin

sickle cell trait (SIK-ul sell trate) condition that occurs when only one gene for sickle cell is inherited and only a small amount of abnormal hemoglobin is produced

simple fracture (SIM-pl FRAK-chur) classification of a broken bone when the skin remains intact

sinoatrial (SA) node (sye-no-AY-tree-al node) the heart's pacemaker; where the impulse conduction of the heart normally starts; located in the wall of the right atrium near opening of the superior vena cava

sinus (SYE-nus) space or cavity inside some cranial bones

sinus arrhythmia (SYE-nus ah-RITH-me-ah) variation in rhythm of heart rate during the breathing cycle (inspiration and expiration)

sinusitis (sye-nyoo-SYE-tis) sinus infections

skeletal muscle (SKEL-e-tal MUS-el) also known as voluntary or striated voluntary muscle; muscles under willed or voluntary control

skeletal system (SKEL-e-tal SIS-tem) bones, cartilage, and ligaments that provide the body with a rigid framework for support and protection

sliding filament theory (SLY-ding FIL-ah-ment THEE-o-ree) model of muscle contraction in which sliding of thin filaments toward the center of each sarcomere quickly shortens the muscle and thereby the entire muscle

slow muscle fiber (slow MUS-el FYE-ber) red muscle fiber; also called slow twitch muscle fiber

smooth muscle (smooth MUS-el) muscles that are not under conscious control; also known as involuntary or visceral; forms the walls of blood vessels and hollow organs

sodium cotransport (SO-dee-um ko-TRANS-port) complex transport process in which carriers that bind both sodium ions and glucose molecules passively transport the molecules together out of the GI lumen

sodium-potassium pump (SO-dee-um-po-TAS-ee-um pump) active transport pump that operates in the plasma membrane of all human cells; transports both sodium ions and potassium ions but in opposite directions and in a 3:2 ratio, thereby maintaining a gradient across the plasma membrane

soft palate (soft PAH-late) partition between mouth and nasopharynx

solute (SOL-yoot) dissolved particles in solution

solvent (SOL-vent) liquid portion of a solution in which a solute is dissolved

soma (SO-mah) cell body

somatic (so-MA-tik) refers to the body or specifically, all cells of the body except the gametes

somatic motor neuron (so-MA-tik MO-tor NOO-ron) motor neuron that stimulates a muscle fiber

somatic nervous system (so-MA-tik NER-vus SIS-tem) motor neurons that control voluntary actions of skeletal muscles

somatic reflex (so-MA-tik REE-fleks) reflexive contraction of skeletal muscles

somatic sense (so-MA-tik sens) sense that enables an individual to detect sensations such as pain, temperature, and proprioception

somatopsychic (so-mah-toe-SYE-kik) refers to the body affecting the mind; physical disorder that produces mental symptoms; *see* psychosomatic

somatostatin (so-mah-toe-STAT-in) hormone produced by delta cells of the pancreas that inhibits secretion of glucagon, insulin, and pancreatic polypeptide

somatotropin (STH) (so-mah-toe-TRO-pin) growth hormone

somatotype (so-MAT-o-type) classification of body type determined on the basis of certain physical characteristics; *see* ectomorph, endomorph, and mesomorph

spatial summation (SPAY-shal sum-MAY-shun) ability of the postsynaptic neuron to add together the inhibitory and stimulatory input received from numerous different presynaptic neurons and produce an action potential based on that collation of information

species resistance (SPEE-shez ree-SIS-tens) genetic characteristics common to all organisms of a particular species that provide natural inborn immunity to certain disease

specific heat (spe-SIF-ik heat) the amount of heat required to raise one gram of substance one degree centigrade

specific immunity (spe-SI-fik i-MYOON-i-tee) protective mechanisms by which the immune system is able to recognize, remember, and destroy specific types of bacteria or toxins

spermatic cord (sper-MAT-ik cord) cylindrical casings of white fibrous tissue formed by the ductus deferens and located in the inguinal canal between the scrotum and the abdominal cavity

spermatogenesis (sper-mah-toe-JEN-e-sis) production of sperm cells

spermatogonia (sper-mah-toe-GO-nee-ah) stem cell population that gives rise to sperm cells

sphygmomanometer (sfig-mo-mah-NOM-e-ter) device for measuring blood pressure in the arteries of a limb

spinal cord (SPY-nal cord) portion of central nervous system that provides two-way conduction from brain; major reflex center

spinal ganglion (SPY-nal GANG-gle-on) enlarged portion of the dorsal root of the spinal cord, where afferent nerve fibers from sensory receptors synapse with associated sensory neurons on their way to the brain or lower reflex centers

spinal meningitis (SPY-nal men-in-JYE-tis) inflammation of the spinal meninges

spinal nerve (SPY-nal nerv) nerve that connects the spinal cord to peripheral structures such as the skin and skeletal muscles

spinal reflex (SPY-nal REE-fleks) reflex arc whose center is located in the spinal cord

spinal tracts (spy-nal tracts) white columns of the spinal cord that provide conduction paths to and from the brain; ascending tracts carry information to the brain, whereas descending tracts conduct impulses from the brain

spindle fiber (SPIN-dul FYE-ber) network of tubules formed in the cytoplasm between the centrioles as they are moving away from each other during mitosis

spinothalamic pathway (spy-no-thah-LAM-ik PATH-way) pathway that conducts impulses that produce sensations of crude touch and pressure

spinothalamic tract (spy-no-thah-LAM-ik trakt) ascending tracts of the spinal cord that convey information about pain, temperature, deep pressure, and coarse touch; these tracts cross over in the spinal cord

spirometer (spi-ROM-e-ter) instrument used to measure amount of air exchanged in breathing

spleen (spleen) largest lymphoid organ; filters blood, destroys worn out red blood cells, salvages iron from hemoglobin, and serves as a blood reservoir

splenomegaly (sple-no-MEG-ah-lee) condition of enlargement of the spleen

spongy bone (SPUN-jee bone) porous bone in the end of the long bone, which may be filled with marrow

squamous (SKWAY-mus) scale-like

stage of exhaustion (staj of eks-ZAWS-chun) third stage of the general adaptation syndrome; when the body can no longer cope or adapt to stressors

stage of resistance (staj of ree-SIS-tens) second stage of the general adaptation syndrome

staircase phenomenon (STAIR-case fe-NOM-e-non) *see* treppe

Starling's law of the heart (STAR-lings law of the hart) principle stating that as myocardial fibers are stretched, the force of contraction is increased

static equilibrium (STAT-ik e-kwi-LIB-ree-um) sensing the position of the head relative to gravity; compare to dynamic equilibrium

stem cell (stem sell) cells that have the ability to maintain a constant population of newly differentiating cells

stenosed (cardiac) valves (ste-NOST KAR-dee-ak valvs) valves that are narrower than normal, slowing blood flow from a heart chamber

stereognosis (ste-ree-og-NO-sis) awareness of an object's size, shape, and texture

steroid hormones (STE-royd HOR-mones) lipid-soluble hormones that pass intact through the cell membrane of the target cell and influence cell activity by acting on specific genes

stillbirth (STIL-berth) delivery of a dead fetus after 20th week of gestation; before 20 weeks it is termed *a spontaneous abortion*

stimulus (STIM-yoo-lus) excitant or irritating agent that induces a response

stomach (STUM-ak) an expansion of the digestive tract between the esophagus and small intestine, where some protein digestion begins and where food is churned and mixed with gastric juices before entering the small intestine

strabismus (strah-BIS-mus) abnormal condition in which lack of coordination of, or weakness in, the muscles that control the eye causes improper focusing of images on the retina, thus making depth perception difficult

stratum (STRA-tum) layer; plural, stratum

stratum basale (STRA-tum BAY-sah-le) "base layer"; deepest layer of the epidermis; cells in this layer are able to reproduce themselves

stratum compactum (STRA-tum kom-PAKT-um) surface layer of the endometrium in the uterus

stratum corneum (STRA-tum KOR-nee-um) tough outer layer of the epidermis; cells are filled with keratin

stratum germinativum (STRA-tum JER-mi-nah-tiv-um) term for combined stratum basale and stratum spinosum in the epidermis

stratum granulosum (STRA-tum gran-yoo-LOE-sum) "granular layer," layer in which the process of keratinization begins

stratum lucidum (STRA-tum loo-SEE-dum) "clear" layer of the epidermis, in thick skin between the stratum granulosum and the stratum corneum

stratum spinosum (STRA-tum spi-NO-sum) "spiny layer"; layer of epidermis that is rich in RNA to aid in protein synthesis required for keratin production

stratum spongiosum (STRA-tum spon-gee-O-sum) middle layer of the endometrium (of uterus), composed of loose connective tissue

strength training (strength TRAIN-ing) contracting muscles against resistance to enhance muscle hypertrophy; *see* endurance or aerobic training

stress (stress) any stimulus that directly or indirectly stimulates neurons of the hypothalamus to release corticotropin-releasing hormone

stress response (stress re-SPONS) *see* stress syndrome

stress syndrome (stress SIN-drome) signs and symptoms associated with the body's frequently maladaptive response to stressors; diverse changes initiated by stress

stressor (STRESS-or) any agent or stimulus that produces stress

stretch marks (stretch marks) tiny silver-white scars resulting when elastic fibers in the dermis are stretched too much; also called striae

stretch reflex (stretch REE-fleks) automatic response in which the muscle reacts and tries to maintain a constancy of muscle length when a load is applied

striated involuntary muscle (STRYE-ay-ted in-VOL-un-TAIR-ee MUS-el) cardiac muscle

striated muscle (STRYE-ay-ted MUS-el) *see* skeletal muscle

stroke (stroke) cerebrovascular accident (CVA); results from ischemia of brain tissue caused by an embolism or ruptured aneurysm in brain blood vessel

stroke volume (stroke VOL-yoom) amount of blood that is ejected from the ventricles of the heart with each beat

subarachnoid space (sub-ah-RAK-noid space) within the meninges, space under the arachnoid and outside the pia mater

subareolar plexus (sub-ah-REE-o-lar PLEKS-us) where the cutaneous plexus and large lymphatics that drain the secretory tissue and ducts of the breast come together

subatomic particles (sub-ah-TOM-ik PAR-ti-kuls) particles that make up atoms; for example, neutrons, protons, and electrons

subdural space (sub-DOO-ral space) within the meninges;

the space between the dura mater and the arachnoid membrane

sublingual glands (sub-LING-gwal glands) smallest salivary glands; produce a mucous type of saliva

submandibular glands (sub-man-DIB-yoo-lar glands) salivary glands located just below the mandibular angle; contain enzyme and mucus-producing elements

submucosa (sub-myoo-KO-sah) connective tissue layer containing blood vessels and nerves in the wall of the digestive tract

substantia nigra (sub-STAN-shee-ah NYE-grah) midbrain structure that consists of clusters of cell bodies of neurons involved in muscular control

substrate (SUB-strate) substance on which an enzyme acts

sucrase (soo-krays) enzyme that catalyzes the hydrolysis of sucrose and maltose

sulcus (SUL-kus) furrow, or groove, often associated with the cerebral cortex

superficial (soo-per-FISH-al) near the body surface; opposite of deep

superficial fascia (soo-per-FISH-al FAH-sha) hypodermis; subcutaneous layer beneath the dermis

superficial reflex (soo-per-FISH-al REE-fleks) reflexes elicited by stimulation of receptors located in the skin or mucosa

superior (soo-PEER-ee-or) higher; opposite of inferior

suprarenal (soo-prah-REE-nal) situated above a kidney

surface film (SUR-fis film) mixture of residues and secretions from sweat and sebaceous glands with epithelial cells being shed from the epidermis; works as a protective barrier

surfactant (sur-FAK-tant) substance covering the surface of the respiratory membrane inside the alveolus; it reduces surface tension and prevents the alveoli from collapsing

suspensory ligament (sus-PEN-so-see LIG-ah-ment) fibers attached to the capsule of the lens that help hold it in place

suture (SOO-chur) immovable joint, such as those between the bones of the skull

sweat gland (swet gland) gland in the skin that produces a transparent, watery liquid that eliminates ammonia and uric acid and helps maintain body temperature

sympathetic division (sim-pah-THE-tik di-VI-shun) part of the autonomic nervous system; ganglia are connected to the thoracic and lumbar regions of the spinal cord; functions in "fight-or-flight" response

symphysis (SIM-fi-sis) joint characterized by the presence of a pad or disk of fibrocartilage connecting the two bones

synapse (SIN-aps) membrane to membrane junction between a neuron and another neuron, effector cell, or sensory cell; function to propagate action potential (via neurotransmitters)

synaptic cleft (si-NAP-tik kleft) space between a synaptic knob and the plasma membrane of a postsynaptic neuron

synaptic knob (si-NAP-tik nob) tiny bulge at the end of a terminal branch of a presynaptic neuron's axon that contains vesicles with neurotransmitters

synarthrosis (sin-ar-THRO-sis) joint in which fibrous connective tissue joins bones and holds them together tightly; commonly called sutures

synchondrosis (sin-kon-DROE-sis) joint characterized by the presence of hyaline cartilage between articulating bones

syncytium (sin-SIT-e-um) continuous, electrically coupled mass of cardiac fibers; allows an efficient, coordinated pumping action

syndesmosis (sin-des-MOE-sis) fibrous joint

synergism (SIN-er-jizm) with hormones; a combination of hormones has a greater effect on a target than the sum of effects that each would have if acting alone

synergist (SIN-er-jists) muscle that assists a prime mover

synovial fluid (si-NO-vee-al FLOO-id) thick, colorless lubricating fluid secreted by the synovial membrane in synovial joints

synovial membrane (si-NO-vee-al MEM-brane) connective tissue membrane lining spaces between bones and joints that secretes synovial fluid

synthesis reaction (SIN-the-sis ree-AK-shun) reaction that combines two or more reactants to form a more complex structure

systemic anatomy (sis-TEM-ik ah-NAT-o-mee) study of anatomy that focuses on learning about body systems

systemic circulation (sis-TEM-ik ser-kyoo-LAY-shun) blood flow from the left ventricle to all parts of the body and back to the right atrium

systole (SIS-toe-lee) contraction of heart muscle

systolic blood pressure (sis-TOL-ik blud PRESH-ur) force with which blood pushes against artery walls when ventricles contract

systolic discharge (sis-TOL-ik DIS-charj) volume of blood pumped by one contraction

T

T cell (T sell) *see* T lymphocytes

T lymphocytes (T LIM-fo-sites) cells of the immune system that have undergone maturation in the thymus; produce cell-mediated immunity

T tubules (T TOOB-yools) transverse tubules unique to muscle cells; formed by inward extensions of the sarcolemma that allow electrical impulses to move deeper into the cell

tachycardia (tak-e-KAR-dee-ah) rapid heart rhythm (above 100 beats/min)

target cell (TAR-get sell) cell that, when acted on by a particular hormone, responds because it has receptors to which hormone can bind

tarsal plate (TAR-sal plate) border of thick connective tissue at free edge of each eyelid

taste buds (taste buds) chemical receptors in the tongue that generate nerve impulses, resulting in the sense of taste

Tay-Sachs disease (TAY-saks di-ZEEZ) recessive, inherited condition in which abnormal lipids accumulate in the brain and cause tissue damage that leads to death by age 4

tectorial membrane (tek-TO-ree-al MEM-brane) gelatinous membrane of inner ear that tops hair cells in the organ of Corti

telodendria (tel-o-DEN-dre-ah) distal tips of axons that form branches, each branch terminates in a synaptic knob

telophase (TEL-o-faze) last stage of mitosis

temporal summation (TEM-po-ral sum-MAY-shun) when synaptic knobs stimulate a postsynaptic neuron in rapid succession and the effects add up over time to produce an action potential (*see also* spatial summation)

tendon (TEN-don) bands or cords of fibrous connective tissue that attach a muscle to a bone or other structure

tendon reflex (TEN-don REE-fleks) reflex stimulated by tapping on a tendon

tendon sheath (TEN-don sheeth) tube-shaped structure lined with synovial membrane that encloses certain tendons

teratogen (TER-ah-toe-jen) physical or chemical agent that disrupts normal embryonic development and thus causes congenital defects

terminal hair (TER-mi-nal hair) coarse pubic and axillary hair that develops at puberty

testosterone (tes-TOS-te-rone) male sex hormone produced by interstitial cells in the testes; the "masculinizing hormone"

tetanus (TET-ah-nus) smooth, sustained muscular contraction caused by high-frequency stimulation

thalamus (THAL-ah-mus) mass of gray matter located in diencephalon just above the hypothalamus; helps produce sensations, associates sensations with emotions, and plays a part in the arousal mechanism

thermoreceptor (ther-mo-ree-SEP-tor) receptors activated by heat or cold

third-degree burn (third deGREE bern) full thickness burn; most serious of the three degrees of burns

thoracic duct (thor-AS-ik dukt) largest lymphatic vessel in the body

thoracolumbar division (thor-rah-ko-LUM-bar di-VI-shun) sympathetic division of autonomic nervous system

thorax (THOR-aks) chest

threshold potential (THRESH-hold po-TEN-shal) magnitude of voltage across a membrane at which an action potential, or nerve impulse, is produced

threshold stimulus (THRESH-hold STIM-yoo-lus) minimal level of stimulation required to cause a muscle fiber to contract

thrombocytes (THROM-bo-sites) also called platelets; play a role in blood clotting

thrombophlebitis (throm-bo-fle-BYE-tis) vein inflammation (phlebitis) accompanied by clot formation

thromboxane (throm-BOKS-an) prostaglandin-like substance in platelets that plays a role in hemostasis and blood clotting

thymocytes (THY-mo-sites) cells in the thymus that develop into T lymphocytes

thymosin (THY-mo-sin) hormone produced by the thymus that is vital to the development and functioning of the body's immune system

thymus (THY-mus) endocrine gland located in the mediastinum; vital part of the body's immune system

thyroid cartilage (THY-roid KAR-ti-lij) largest cartilage of the larynx; Adam's apple

thyroid gland (THY-royd gland) endocrine gland located in the neck that stores its hormones until needed; thyroid hormones regulate cellular metabolism

thyroid hormone (THY-royd HOR-mone) hormone that accelerates catabolism of glucose

thyroid-stimulating hormone (TSH) (THY-royd STIM-yoo-lay-ting HOR-mone) a tropic hormone secreted by the anterior pituitary gland that stimulates the thyroid gland to increase its secretion of thyroid hormone

thyroxine (T$_4$) (thy-ROK-sin) thyroid hormone that stimulates cellular metabolism

tic douloureux (tik doo-loo-ROO) *see* trigeminal neuralgia

tidal volume (TV) (TIE-dal VOL-yoom) amount of air breathed in and out with each breath

tight junction (tite JUNK-shun) connection between cells in which they are joined by "collars" of tightly fused membrane

tissue (TISH-yoo) group of similar cells that performs a common function

tonic contraction (TON-ik kon-TRAK-shun) special type of skeletal muscle contraction used to maintain posture

tonicity (toe-NIS-i-tee) potential osmotic pressure

tonsillitis (ton-si-LIE-tis) inflammation of the tonsils, usually due to infection

tonsils (TON-sils) masses of lymphoid tissue; protect against bacteria; three types: palatine tonsils, located on each side of the throat; pharyngeal tonsils (adenoids), near the posterior opening of the nasal cavity; and lingual tonsils, near base of the tongue

total lung capacity (TOE-tal lung kah-PA-si-tee) total volume of air a lung can hold

total metabolic rate (TMR) (TOE-tal met-ah-BOL-ik rate) total amount of energy used by the body per day

trabeculae (trah-BEK-yoo-lee) needlelike threads of spongy bone that surround a network of spaces

trace element (trase EL-e-ment) elements in small amounts in the body

trachea (TRAY-kee-ah) windpipe; the tube extending from the larynx to the bronchi

tracheostomy (tray-kee-OS-to-mee) surgical procedure to cut an opening into the trachea

transcription (trans-KRIP-shun) process in which DNA molecule is used as template to form mRNA

transfer RNA (tRNA) (TRANS-fer RNA) RNA involved with protein synthesis; tRNA molecules carry amino acids to the ribosome for placement in the sequence prescribed by mRNA

translation (trans-LAY-shun) process in which mRNA is used by ribosomes in the synthesis of a protein

transverse plane (TRANS-vers plane) horizontal plane that divides the body or any of its parts into upper and lower parts

treppe (TREP-ee) gradual increase in the extent of muscular contraction following rapidly repeated stimulation; staircase phenomenon

triad (TRY-ad) triplet of tubules; allows an electrical impulse traveling along a T tubule to stimu-

late the membranes of adjacent sacs of the sarcoplasmic reticulum

tricarboxylic acid cycle (TAC) (try-kar-bok-SIL-ik AS-id SYE-kul) aerobic metabolic pathway in which acetyl CoA (from other metabolic pathways) is converted into CO_2 and H_2O with the formation of ATP; also known as citric acid cycle, or Krebs cycle

tricuspid valve (try-KUS-pid valve) heart valve located between right atrium and ventricle

trigeminal nerve (try-JEM-i-nal nerv) cranial nerve V; responsible for chewing movements and sensations of the head and face

trigeminal neuralgia (try-JEM-i-nal noo-RAL-jee-ah) pain in one or more (of three) branches of the fifth cranial nerve (trigeminal nerve) that runs along the face; also called tic douloureux

triglyceride (try-GLI-ser-ide) lipid that is synthesized from fatty acids and glycerol or from excess glucose or amino acids; stored mainly in adipose tissue cells

triiodothyronine (T_3) (try-eye-o-doe-THY-ro-nine) thyroid hormone that stimulates cellular metabolism

trimester (TRY-mes-ter) three month segments of the gestation period

trisomy (TRY-so-me) abnormal genetic condition in which cells have three chromosomes (a triplet) where there should be a pair; usually caused by nondisjunction (failure of chromosome pairs to separate) during gamete production

trochlear nerve (TROK-lee-ar nerv) cranial nerve IV; motor nerve; responsible for eye movements

trophoblast (TRO-fo-blast) outer wall of the blastocyst; contributes in formation of the placenta

tropic hormone (TRO-pik HOR-mone) hormone that stimulates another endocrine gland to secrete its hormones

tropomyosin (tro-po-MY-o-sin) in sliding filament theory of muscle cell contraction, the long protein molecule that covers the active sites of myosin; myofilament

troponin (tro-PO-nin) in sliding filament theory of muscle cell contraction, the molecules spaced at intervals along the thin filament that block troponin when the myofilament is at rest

true vocal cords (tru VO-kul kords) lower pair of vocal cords in the larynx, responsible for vocalization

tubal pregnancy (TOO-bal PREG-nan-see) ectopic preg-

nancy that occurs in a uterine tube

tuberculosis (TB) (too-ber-kyoo-LO-sis) chronic bacterial (bacillus) infection of the lungs or other tissues caused by *Mycobacterium tuberculosis* organisms

tumor (TOO-mer) growth of tissues in which cell proliferation is uncontrolled and progressive

tumor suppressor genes (TOO-mer soo-PRESS-or jeans) genes that work against the development of cancerous cells

tunica adventitia (TOO-ni-kah ad-ven-TISH-ah) outermost layer of arteries and veins; made of strong, flexible fibrous connective tissue

tunica albuginea (TOO-ni-kah al-byoo-JIN-ee-ah) tough, whitish membrane that surrounds each testis and enters the gland to divide it into lobules

tunica interna (TOO-ni-kah in-TER-nah) endothelium that lines blood vessels

tunica media (TOO-ni-kah MEE-dee-ah) muscular middle layer in blood vessels; the tunica media of arteries is more muscular than that of veins

Turner syndrome (TUR-ner SIN-drome) genetic disorder caused by monosomy of the X chromosome (XO) in females; characterized by immaturity of sex organs (causing sterility), webbed neck, cardiovascular defects, and learning disorders

twitch contraction (twich con-TRAK-shun) quick jerk of a muscle produced by a single, brief threshold stimulus

tympanic membrane (tim-PAN-ik MEM-brane) eardrum

Type A (tipe A) blood type that has antigen A on red blood cells

Type AB (tipe AB) blood type that has both antigen A and antigen B on red blood cells

Type B (tipe B) blood type that has antigen B on red blood cells

Type O (tipe O) blood type that has neither antigen A or antigen B on red blood cells

U

ulcer (UL-ser) necrotic open sore or lesion

ultrasonography (ul-tra-son-OG-rah-fee) imaging technique in which high-frequency sound waves are reflected off tissue to form an image

umbilical artery (um-BIL-i-kul AR-ter-ee) two small arteries that carry oxygen-poor blood from the developing fetus to the placenta

umbilical cord (um-BIL-i-kul cord) flexible structure connecting the fetus with the placenta; contains umbilical arteries and vein

umbilical vein (um-BIL-i-kul vane) large vein carrying oxy-

gen-rich blood from the placenta to the developing fetus

umbilicus (um-BIL-i-kus) navel

unmyelinated fiber (un-MY-e-lin-ay-ted FYE-ber) nerve fiber that does not have a myelin sheath; also referred to as gray fiber

unsaturated fat (un-SACH-e-ray-ted fat) fat containing fatty acid chains where there are some double bonds, not all sites for hydrogen are filled; usually liquid at room temperature

upper motor neurons (UP-er MO-ter NOO-rons) neurons whose axons lie in either pyramidal or extrapyramidal tracts

upper respiratory tract (UP-er re-SPY-rah-tor-ee trakt) respiratory organs that are not contained within the thorax; includes nasal cavity, pharynx, and associated structures

urea (yoo-REE-ah) nitrogen-containing waste product

uremia (yoo-REE-mee-ah) condition in which blood urea concentration is abnormally elevated, expressed as a high BUN (blood urea nitrogen) value; uremia is often caused by renal failure; also called uremic syndrome

ureter (YOOR-e-ter) long tube that carries urine from kidney to bladder

urethra (yoo-REE-thrah) passageway from bladder to exterior; functions in elimination of urine; in males, also acts as a genital duct that carries sperm to the exterior

urethral stricture (yoo-REE-thral STRIK-chur) narrowing or blockage of the urethra

urethritis (yoo-re-THRY-tis) inflammation or infection of the urethra

urinary bladder (YOOR-i-nair-ee BLAD-er) collapsible sacklike organ that collects urine from the kidneys and stores it prior to elimination

urinary meatus (YOOR-i-nair-ee mee-AY-tus) external opening of the urethra

urinary system (YOOR-i-nair-ee SIS-tem) system responsible for excreting most liquid wastes from the body

urination (yoor-i-NAY-shun) passage of urine from the body; emptying of the bladder; also called micturition

urine (YOOR-in) fluid waste excreted by kidneys

urochrome (YOOR-o-krome) pigments from the breakdown of old red blood cells in the liver and elsewhere that are found in the urine

urodynamics (yoo-ro-dye-NAM-iks) force of urine flow in the urinary tract

urticaria (er-ti-KAIR-ee-ah) hives; allergic or hypersensitive response characterized by raised red lesions

uterine tube (YOO-ter-in toob) fallopian tube

uterus (YOO-ter-us) hollow, muscular organ that nourishes developing offspring until birth

utricle (YOO-tri-kle) part of membranous labyrinth of inner ear; involved with equilibrium

uvula (YOO-vyoo-lah) cone-shaped process hanging from the soft palate that helps prevent food and liquid from entering the nasal cavities

V

vaccination (vak-si-NAY-shun) method used to achieve active immunity by triggering the body to form antibodies against specific pathogens

vaccine (VAK-seen) application of killed or attenuated (weakened) pathogens (or portions of pathogens) to a patient to stimulate immunity against that pathogen

vagina (vah-JYE-nah) internal tube from uterus to vulva

vaginal orifice (VAH-ji-nal OR-i-fis) opening of the vagina to the outside of the body

vaginitis (vaj-i-NYE-tis) inflammation of the vagina

vagus nerve (VAY-gus nerv) cranial nerve X; mixed nerve; sensations and movements of organs

varicose vein (VAIR-i-kose vane) enlarged vein in which blood pools; also called varix (plural, varices)

varix (VAIR-iks) varicose vein; plural, varices

vas deferens (vas DEF-er-enz) testicle duct that extends from the epididymis to the ejaculatory duct; also called ductus deferens

vasa vasorum (VAS-ah va-SOR-um) tiny blood vessels that supply the smooth muscles that surround the walls of larger blood vessels

vasectomy (va-SEK-to-mee) surgical severing of the vas deferens to render a male sterile

vasoactive intestinal peptide (VIP) (vay-so-AK-tiv in-TES-ti-nal PEP-tide) hormone involved with controlling intestinal secretion

vasodilator (vay-so-DYE-lay-tor) drugs that trigger the smooth muscles of arterial walls to relax, causing the arteries to dilate

vasomotor control mechanism (vay-so-MO-tor kon-trol MEK-ah-nizm) factors that control changes in the diameter of arterioles

vasomotor pressoreflex (vay-so-MO-tor pres-o-REE-fleks) reflex that occurs in response to a change in arterial blood pressure

vector (VEK-tor) arthropod that carries an infectious pathogen from one organism to another

vein (vane) vessel carrying blood from capillaries toward the heart

ventral (VEN-tral) of or near the belly; in humans, front or anterior; opposite of dorsal or posterior

ventral nerve root (VEN-tral nerv root) fibers that carry motor information out of the spinal cord

ventral ramus (VEN-tral RAY-mus) large, complex branch of each spinal nerve

ventral root (VEN-tral root) motor branch of a spinal nerve by which it is attached to the spinal cord

ventricle (VEN-tri-kul) a cavity, such as the large, fluid-filled spaces within the brain or the chambers of the heart

ventricular fibrillation (ven-TRIK-yoo-lar fib-ri-LAY-shun) an immediately life-threatening condition caused by the lack of ventricular pumping suddenly stopping flow of blood to vital organs

venule (VEN-yool) small blood vessels that collect blood from capillaries and join to form veins

vermiform appendix (VERM-i-form ah-PEN-diks) tubular structure attached to the cecum

vertebrae (VER-te-bray) bones that make up the spinal column

vertebral column (ver-TEE-bral KOL-um) the spinal column, made up of a series of separate vertebrae that form a flexible, curved rod; made up of the cervical, thoracic, lumbar, sacral, and coccygeal segments

vertebral foramen (ver-TEE-bral for-AY-men) the central opening in the vertebral column that contains the spinal cord

vertigo (VER-ti-go) abnormal sensation of spinning; dizziness

vesicle (VES-i-kl) tiny membranous bubble; clinical term referring to blisters, fluid-filled skin lesions

vestibular membrane (ves-TIB-yoo-lar MEM-brane) roof of the cochlear duct; also called Reissner's membrane

vestibular nerve (ves-TIB-yoo-lar nerv) division of the vestibulocochlear nerve (eighth cranial nerve)

vestibule (VES-ti-byool) located in the inner ear; portion adjacent to the oval window between the semicircular canals and the cochlea

vestibulocochlear nerve (ves-TIB-yoo-lo-kok-lee-ar nerv) cranial nerve VIII; sensory nerve; responsible for hearing and equilibrium

villi (VIL-eye) fingerlike folds covering the plicae of the small intestines

Vincent's infection (VIN-sents in-FEK-shun) bacterial (spirochete) infection of the gum, producing gingivitis; also called Vincent's angina and trench mouth

virus (VYE-rus) microscopic, intracellular parasitic entity consisting of a nucleic acid bound by a protein coat and sometimes a lipoprotein envelope

viscera (VIS-er-ah) refers to internal organs

visceral membrane (VIS-er-al MEM-brane) serous membrane that covers the surface of the viscera

visceral muscle tissue (VIS-er-al MUS-el TISH-yoo) smooth muscle tissue

visceral reflex (VIS-er-al REE-fleks) autonomic reflex; contractions of smooth or cardiac muscles or secretion by glands

visceral pericardium (VIS-er-al pair-i-KAR-dee-um) the pericardium that covers the heart

visceral portion (VIS-er-al POR-shun) serous membrane that covers the surfaces of organs in the body cavity

viseroceptor (vis-er-o-SEP-tor) somatic sense receptors located in the internal visceral organs

vital capacity (VC) (VYE-tal kah-PAS-i-tee) largest amount of air that can be moved in and out of the lungs in one inspiration and expiration

vitamin (VYE-tah-min) organic molecules needed in small quantities to help enzymes operate effectively

vitiligo (vi-ti-LYE-go) acquired condition that results in loss of pigment in certain areas of the skin

vitreous humor (VIT-ree-us HYOO-mor) jellylike fluid in the eye, posterior to the lens

voiding (VOYD-ing) emptying the bladder

volar (VO-lar) palm or sole

Volkmann's canals (VOLK-manz kah-NALz) communicating canals between haversian canals that contain vessels to carry blood from the exterior surface of the bone to the osteons

voluntary muscle (VOL-un-tair-ee MUS-el) *see* skeletal muscle

vulva (VUL-vah) external genitals of the female

vulvitis (vul-VYE-tis) inflammation of the vulva (external female genitals)

W

white fibers (wite FYE-bers) muscle fibers containing little myoglobin; also called fast fibers; nerve fibers with a thick myelin sheath

white matter (wite MAT-ter) nerves covered with white myelin sheath

white ramus (wite RAY-mus) small branch of myelinated sympathetic preganglionic fibers

windpipe (WIND-pipe) trachea

withdrawal reflex (with-DRAW-al REE-fleks) reflex that moves a body part away from an irritating stimulus

X

X chromosome (x KROH-mo-sohm) sex chromosome, which, in females, is paired with another X chromosome; as opposed to a Y chromosome

X-linked trait (x-linkd trate) genetic trait associated with the female (X) chromosome

Y

Y chromosome (y-KROH-mo-sohm) sex chromosome that contains genes that determine maleness

yeast (yeest) single-celled fungus (compared to mold, which is a multicellular fungus)

yellow marrow (YEL-o MAR-o) connective tissue rich in fat that is found in the medullary cavity of the long bones of an adult; inactive in red cell production

Y-linked trait (y-linkd trate) genetic trait associated with the male (Y) chromosome

yolk sac (yoke sak) in humans, involved with production of blood cells in the developing embryo

Z

zona fasciculata (ZO-nah fas-sic-yoo-LAY-tah) middle zone of the adrenal cortex that secretes glucocorticoids

zona glomerulosa (ZO-nah glo-mare-yoo-LO-sah) outer zone of the adrenal cortex that secretes mineralocorticoids

zona reticularis (ZO-nah re-tik-yoo-LAIR-is) inner zone of the adrenal cortex that secretes small amounts of sex hormones

zone of calcification (zone of kal-si-fi-KAY-shun) deepest layer of the epiphyseal plate; composed of cartilage undergoing rapid calcification

zone of hypertrophy (zone of hye-PER-tro-fee) third layer of the epiphyseal plate; composed of older, enlarged cartilage cells that are undergoing degenerative changes associated with calcium deposition

zone of proliferation (zone of pro-LIF-er-ay-shun) second layer of the epiphyseal plate; composed of cartilage cells undergoing active mitosis

zygomaticus (zye-go-MAT-ik-us) muscle that elevates corners of the mouth and lips; also known as the smiling muscle

zygote (ZYE-gote) a fertilized ovum

Credits

Cousins; 12-6, Joan M. Beck; 12-6 (inset), 12-13 (inset), Rolin Graphics; 12-7, Christine Oleksyk; 12-8, courtesy of B. Vidic and F.R. Suarez (from *Photographic Atlas of the Human Body*, 1984, The C.V. Mosby Co.); 12-10, courtesy of R.M.H. McMinn and R.T. Hutchings (from *Color Atlas of Human Anatomy*, ed. 2, 1988, Year Book Medical Publishers); 12-13, 12-16, Bill Ober; 12-14, Michael P. Schenk; 12-17 (inset), Karen Waldo; Brain Studies box 12-17, (CT scan), Silver/Custom Medical Stock Photo; Brain Studies box (PET scan), Peter Arnold; Brain Studies box (MRI), CNRI/Science Photo Library; 12-23, Graphic Works.

Chapter 13

Chapter opener: courtesy of S. Erlandsen and J. Magney (from *Color Atlas of Histology*, 1992, Mosby-Year Book, Inc.)

13-1, Rolin Graphics; 13-2, 13-3, 13-4, 13-6, David J. Mascaro & Associates; 13-7, George J. Wassilchenko; 13-2B, G. David Brown; 13-7 (inset), courtesy of B. Vidic and F.R. Suarez (from *Photographic Atlas of the Human Body*, 1984, The C.V. Mosby Co.); 13-8A, 13-12, Ernest Beck; 13-2B, G. David Brown; 13-8B, 13-9, 13-10, 13-12, Michael P. Schenk; Herpes Zoster box, courtesy of Thomas P. Habif (from *Clinical Dermatology*, ed. 2, 1990, Mosby-Year Book, Inc.); 13-13, 13-15, Barbara Cousins; 13-14, Joan M. Beck; 13-16, Pagecrafters; 13-17, Graphic Works.

Chapter 14

Chapter opener: courtesy of S. Erlandsen and J. Magney (from *Color Atlas of Histology*, 1992, Mosby-Year Book, Inc.)

14-1, 14-5, Cochlear Implants box, Rolin Graphics; 14-2, Joan M. Beck; 14-3A, 14-4, 14-6, 14-7A, 14-8A, 14-10, 14-11B, 14-17, 14-21, Marsha J. Dohrmann; 14-3BC, 14-7B, 14-8B, Kathy Mitchell Grey; 14-3D, Omikron/Science Source; 14-7CD, 14-8CDE, 14-19, G. David Brown; 14-9, George J. Wassilchenko; 14-11A, Christine Oleksyk; 14-11C, Scott Mittman; 14-12, Trent Stephens; 14-13, Ernest W. Beck; 14-14, John V. Hagen; 14-15, Lisa Chuck/Michael P. Schenk; 14-16, courtesy of Frank W. Newell (from *Ophthalmology: Principles & Concepts*, ed. 7, 1992, Mosby-Year Book, Inc.); 14-18, Graphic Works; 14-19, 14-20, Donna Odle; Color Blindness box, courtesy of Isihara (from *Tests for Colour-Blindness*, Tokyo, Japan, 1973, Kanehara Shuppan Co., Ltd., provided by Washington University Dept. of Ophthalmology).

Chapter 15

Chapter opener: courtesy of S. Erlandsen and J. Magney (from *Color Atlas of Histology*, 1992, Mosby-Year Book, Inc.)

Table 15-1, 15-18, 15-21, 15-22 (drawing), Rolin Graphics; 15-1, 15-20, Joan M. Beck; 15-4, Regulation of Target Cell Insensitivity box, 15-9 (inset), Andrew Grivas; 15-5, 15-13, Nadine Sokol; 15-6, 15-11, Kevin A. Somerville; 15-7, Yvonne Wylie Walston; 15-9, George J. Wassilchenko; 15-10, courtesy of S. Erlandsen and J. Magney (from *Color Atlas of Histology*, 1992, Mosby-Year Book, Inc.); 15-12, 15-14, Barbara Stackhouse; 15-15, 15-19, Ernest W. Beck; Thyroid Hormone Abnormalities box (A), courtesy of Paul W. Ladenson, M.D., Johns Hopkins University and Hospital, Baltimore, Maryland; Thyroid Hormone Abnormalities box (B), L.V. Bergman & Associates, Inc., Cold Spring, New York; 15-23, courtesy of Dr. O. Hegre and Dr. R. Sorenson, Dept. of Cell Biology and Neuroanatomy, University of Minnesota Medical School, Minneapolis, Minnesota.

Chapter 16

Chapter opener: courtesy of G. Bevelander and J.A. Ramaley (from *Essentials of Histology*, ed. 8, 1979, The C.V. Mosby Co.)

16-1, Ernest W. Beck; 16-2, 16-17, 16-19, Rolin Graphics; 16-3, Sickle Cell Anemia box, courtesy of G. Bevelander and J.A. Ramaley (from *Essentials of Histology*, ed. 8, 1979, The C.V. Mosby Co.); 16-4, 16-8, Christine Oleksyk; 16-5, 16-14, courtesy of J.B. Miale (from *Laboratory Medicine: Hematology*, ed. 6, 1982, The C.V. Mosby Co.); 16-6, Kathryn A. Born; 16-7, Yvonne Wylie Walston; 16-9 through 16-13, courtesy of S. Erlandsen and J. Magney (from *Color Atlas of Histology*, 1992, Mosby-Year Book, Inc.); 16-15, 16-16, 16-18, Molly Babich/John Daugherty; 16-20, N.F. Rothman; 16-21, Laurie O'Keefe/John Daugherty.

Chapter 17

Chapter opener: courtesy of S. Erlandsen and J. Magney (from *Color Atlas of Histology*, 1992, Mosby-Year Book, Inc.)

17-1, 17-4, 17-5, 17-7BC, 17-8, 17-15, 17-19, George J. Wassilchenko; 17-2, courtesy of Patricia Kane, Radiology Dept., Indiana University Medical Center; 17-3, 17-11A, 17-16, 17-18, Ernest W. Beck; 17-6, 17-14, Christine Oleksyk; 17-7A, 17-9, Rusty Jones; 17-10, Lisa Shoemaker/Joan M. Beck; 17-12, Bill Ober; 17-12 (inset), 17-13, Joan Beck; 17-17, H. Williams; How to Trace the Flow of Blood box, 17-31, Rolin Graphics; 17-20, 17-21, 17-22, 17-23, 17-24, 17-26CD, Karen Waldo; 17-25, Kevin A. Somerville; Fetal Alcohol Syndrome box, courtesy of Claus Simon and Michael Janner (from *Color Atlas of Pediatric Diseases with Differential Diagnosis*, ed. 2, 1990, B.C. Decker, Inc.); 17-26, 17-27, Molly

Babich/John Daugherty; 17-28, courtesy of K.L. McCance and S. Huether (from *Pathophysiology*, 1990, Mosby-Year Book, Inc.); 17-29, Marcia Williams (from *Mosby's Medical and Nursing Dictionary*, ed. 2, 1986, The C.V. Mosby Co.); 17-30, G. David Brown.

Chapter 18

Chapter opener: Ed Reschke

18-1, Joan M. Beck; Artificial Cardiac Pacemakers box (photo), Science Vu-Cardiac Control Systems Visuals Unlimited; Artificial Cardiac Pacemakers box (drawing), George J. Wassilchenko; 18-3, Lisa Shoemaker/Joan M. Beck; 18-4, 18-7, 18-14, 18-19A, Barbara Stackhouse; 18-5, Yvonne Wylie Walston; Cardiac Arrhythmia box, courtesy of Guzzetta and Dossey, 1984; Doppler Ultrasonography box, courtesy of C.M. Rumack, S.R. Wilson and J.W. Charboneau (from *Diagnostic Ultrasound*, 1991, Mosby-Year Book, Inc.); 18-12, 18-13, Christine Oleksyk; 18-15, Ernest W. Beck; 18-16, Nadine Sokol; 18-17, Graphic Works; 18-18, Joan M. Beck/Donna Odle; 18-20, G. David Brown.

Chapter 19

Chapter opener: courtesy of S. Erlandsen and J. Magney (from *Color Atlas of Histology*, 1992, Mosby-Year Book, Inc.)

19-1, 19-4, G. David Brown; 19-2, Barbara Cousins; 19-5, Ernest W. Beck; 19-7, Joan M. Beck; 19-8, George J. Wassilchenko; 19-9, Rolin Graphics; 19-10A, Nadine Sokol; 19-10B, John D. Cunningham/Visuals Unlimited; 19-11A, David J. Mascaro; 19-11B, Kathy Mitchell Grey.

Chapter 20

Chapter opener: Manfred Kage/Peter Arnold, Inc.

20-1, Immunization box, 20-7, Rolin Graphics; 20-3, Joan M. Beck; 20-4, Christine Oleksyk; 20-6, Bill Ober/Pagecrafters; 20-8, Laurie O'Keefe/John Daugherty; Monoclonal Antibodies box, David Scharf/Peter Arnold, Inc.; 20-11, courtesy of Dr. James T. Barrett; 20-11 (inset), Barbara Cousins; 20-12, Nadine Sokol.

Chapter 21

Chapter opener: Earth Scenes

21-1A, J. Gebhardt/The Image Bank; 21-1B, R. Koch/Contrasto-Picture Group; 21-1C, Mickey Gibson/Tom Stack & Associates; 21-1D, NASA; 21-2, 21-3, 21-4, 21-5, Graphic Works.

Chapter 22

Chapter opener: Ellen Dirkson/Visuals Unlimited

22-1, 22-16, Joan M. Beck; 22-2, 22-7, 22-8, 22-9A, 22-18, Ernest W. Beck; 22-3, 22-6, 22-10, 22-19, George J. Wassilchenko; 22-4, courtesy of G. Bevelander and J.A. Ramaley (from *Essentials of Histology*, ed. 8, 1979, The C.V. Mosby Co.); 22-5, Margaret Gerrity; 22-9B, 22-11 (inset), 22-14B, 22-15, courtesy of S. Erlandsen and J. Magney (from *Color Atlas of Histology*, 1992, Mosby-Year Book, Inc.); 22-11, Lisa Shoemaker/Joan M. Beck; 22-13, courtesy of R.M.H. McMinn and R.T. Hutchings (from *Color Atlas of Human Anatomy*, ed. 2, 1988, Year Book Medical Publishers); 22-14A, Barbara Stackhouse; Keeping the Trachea Open box, courtesy of W.D. Glanze, editor (from *Mosby's Medical and Nursing Dictionary*, ed. 2, 1986, The C.V. Mosby Co.); 22-17, courtesy of B. Vidic and F.R. Suarez (from *Photographic Atlas of the Human Body*, 1984 The C.V. Mosby Co.); Pulmonary Radiology box, Specialized Radiographs box, courtesy of S.F. Wilson and J.M. Thompson (from *Respiratory Disorders*, 1990, Mosby-Year Book, Inc.).

Chapter 23

Chapter opener: courtesy of Brenda Russell, Ph.D., University of Illinois at Chicago

23-1, John Daugherty; 23-2, 23-3, Christy Krames; 23-4A, courtesy of N.B. Slomin and L.H. Hamilton (from *Respiratory Physiology*, ed. 4, 1981, The C.V. Mosby Co.); 23-4B, courtesy of S.F. Wilson and J.M. Thompson (from *Respiratory Disorders*, 1990, Mosby-Year Book, Inc.); 23-6B, Barbara Cousins; Pneumothorax box, courtesy of W.D. Glanze, editor (from *Mosby's Medical and Nursing Dictionary*, ed. 2, 1986, The C.V. Mosby Co.); 23-7, 23-8, 23-11, 23-12, 23-17, Rolin Graphics; 23-15, Bill Ober; Heimlich Maneuver box, Joan M. Beck.

Chapter 24

Chapter opener: CNRI/SPL/Photo Researchers

24-1, Lisa Shoemaker; 24-2, Christy Krames; 24-3, 24-4AB, 24-22AB, 24-25A, Rolin Graphics; 24-4C, 24-6A, 24-7, 24-8, 24-13, 24-16, 24-19, 24-20, 24-24, Ernest W. Beck; 24-6B, Kathy Mitchell Grey; 24-10, G. David Brown; 24-11, Nadine Sokol; 24-12B, 24-15, courtesy of S. Erlandsen and J. Magney (from *Color Atlas of Histology*, 1992, Mosby-Year Book, Inc.); Upper GI Diagnosis box, courtesy

of R.H. Daffner (from *Introduction to Clinical Radiology*, 1978, The C.V. Mosby Co.); 24-14, courtesy of Professor Ivan Stotz, Dept. of Veterinary Science, South Dakota State University, Brookings, South Dakota; Barium Enema box, CNRI Science Photo Library/Photo Researchers; 24-17, George J. Wassilchenko; 24-21, David J. Mascaro; 24-11C, 24-25B, Barbara Cousins.

Chapter 25

Chapter opener: courtesy of S. Erlandsen and J. Magney (from *Color Atlas of Histology*, 1992, Mosby-Year Book, Inc.)

25-1, G. David Brown; 25-2, 25-3, 25-6, 25-8, 25-9, 25-11, 25-12, Rolin Graphics; 25-4, Ron J. Ervin; 25-5, Nadine Sokol; 25-7, 25-13, Graphic Works; Table 25-3, 25-14, Laurie O'Keefe/John Daugherty; 25-10, Barbara Stackhouse; 25-10 (inset), courtesy of Dr. Andrew Evan, Indiana University.

Chapter 26

Chapter opener: Rivera Collection/Superstock, Inc.

26-1, Transferring Chemical Energy box, 26-4, 26-5, 26-6, 26-9, 26-14, 26-19, Rolin Graphics; 26-3, 26-8, Pagecrafters/Nadine Sokol; 26-7, Joan M. Beck; 26-10B, courtesy of Brenda Russell, Ph.D., University of Illinois at Chicago.

Chapter 27

Chapter opener: courtesy of S. Erlandsen and J. Magney (from *Color Atlas of Histology*, 1992, Mosby-Year Book, Inc.)

27-1A, 27-2A, 27-3A, 27-4A, 27-8, Ernest W. Beck; 27-1B, Christine Oleksyk; 27-2B, courtesy of B. Vidic and F.R. Suarez (from *Photographic Atlas of the Human Body*, 1984, The C.V. Mosby Co.); 27-3B, 27-19, Yvonne Wylie Walston; Urodynamic Testing box, courtesy of Patricia Kane, University of Indiana Medical School; 27-4B, Artificial Kidney box (B), Barbara Stackhouse; 27-5A, 27-9, Jody L. Fulks; 27-5B, 27-6, courtesy of R.G. Kessel and R.H. Kardon (from *Tissues and Organs: A Text-Atlas of Scanning Electron Microscopy*, 1979, W.H. Freeman and Co. Publishers); 27-7, Courtesy of Dr. Andrew Evan, Indiana University; 27-8 (inset), courtesy of Cleveland P. Hickman; 27-10, 27-12 through 27-16, 27-20, Rolin Graphics; Artificial Kidney box (A), Joan M. Beck/Donna Odle.

Chapter 28

Chapter opener: Tom Stack & Associates

28-1, 28-3, Joan M. Beck; 28-2, Table 28-1, Rolin Graphics; 28-5, Barbara Stackhouse; 28-10, Andrew Grivas.

Chapter 29

Chapter opener: Herb Charles Ohlmeyer/Fran Heyl Associates

29-7, 29-10, Yvonne Wylie Walston.

Chapter 30

Chapter opener: courtesy of S. Erlandsen and J. Magney (from *Color Atlas of Histology*, 1992, Mosby-Year Book, Inc.)

30-1B, David J. Mascaro & Associates; 30-2A, Ernest W. Beck; 30-2B, Lennart Nilsson; 30-4, 30-7, 30-8, 30-10, courtesy of S. Erlandsen and J. Magney (from *Color Atlas of Histology*, 1992, Mosby-Year Book, Inc.); 30-5, Yvonne Wylie Walston; 30-6A, Carolyn B. Coulam/John A. McIntyre; 30-6B, Bill Ober; 30-9, 30-11A, Kevin A. Somerville; 30-11B, courtesy of B. Vidic and F.R. Suarez (from *Photographic Atlas of the Human Body*, 1984, The C.V. Mosby Co.).

Chapter 31

Chapter opener: courtesy of S. Erlandsen and J. Magney (from *Color Atlas of Histology*, 1992, Mosby-Year Book, Inc.)

31-3, 31-7, Ernest W. Beck; 31-6, 31-12, courtesy of S. Erlandsen and J. Magney (from *Color Atlas of Histology*, 1992, Mosby-Year Book, Inc.); 31-9, David J. Mascaro & Associates; 31-11, 31-13, Graphic Works; 31-14, 31-15, Yvonne Wylie Walston.

Chapter 32

Chapter opener: Lennart Nilsson, The Incredible Machine, National Geographic Society

32-1, Barbara Cousins; 32-2, 32-3 (drawing), 32-12 (bottom), 32-15, Rolin Graphics; 32-2 (photo), 32-11, Lennart Nilsson; 32-4, Scott Bodell; In Vitro Fertilization box, AP/Wide World Photos; 32-5, courtesy of Lucinda L. Veeck, Jones Institute for Reproductive Medicine, Norfolk, Virginia; 32-7, 32-8, 32-9, 32-13, 32-16, Ernest W. Beck; 32-10, 32-14, Kevin A. Somerville; 32-12 (top), Marcia Hartsock; Antenatal Diagnosis and Treatment box (drawing), Rebecca S. Montgomery; Antenatal Diagnosis and Treatment box (photo), Carolyn B. Coulam/John A. McIntyre; 32-17, courtesy of Marjorie M. Pyle, for Lifecircle, Costa Mesa, California; 32-18, Ron Edwards; 32-19, Patrick James Watson; 32-20 (left), The Image Bank; 32-20 (center), Joel Gordon 1981; 32-20 (right), Science Vu Visuals Unlimited; 32-21, Marcia Williams (from *Mosby's Medical and Nursing Dictionary*, ed. 2, 1986, The C.V. Mosby Co.).

Chapter 33

Chapter opener: Laurence-Berkeley Laboratory

33-1 (drawing), 33-2, 33-4, 33-6 through 33-9, Maternal and Paternal Genes box, 33-15, 33-17, Rolin Graphics; 33-1 (photo), CNRI/Science Photo Library; 33-3, Kevin A. Somerville; 33-5, 33-14, Pagecrafters/Nadine Sokol; 33-11A, courtesy of Louis McGavran, Denver Children's Hospital; 33-11B, Richard Hutching/Photo Researchers; 33-16, Rebecca S. Montgomery; DNA Analysis box (A), BioProducts; DNA Analysis box (B), Barbara Cousins.

Mini-Atlas

All photos in the atlas are courtesy of R.M.H. McMinn and R.T. Hutchings (from *Color Atlas of Human Anatomy*, ed. 2, 1988, Year Book Medical Publishers).

Appendix D

From Tilkian, S.M., Conover, M.B., and Tilkian, A.G.: Clinical implications of laboratory tests, ed. 3, St. Louis, 1983, The C. V. Mosby Co.

Index

Word Parts Commonly Used as Prefixes

WORD PART	MEANING
a-	Without, not
af-	Toward
an-	Without, not
ante-	Before
anti-	Against; resisting
auto-	Self
bi-	Two; double
circum-	Around
co-, con-	With; together
contra-	Against
de-	Down from, undoing
dia-	Across, through
dipl-	Twofold, double
dys-	Bad; disordered; difficult
ectop-	Displaced
ef-	Away from
em-, en-	In, into
endo-	Within
epi-	Upon
eu-	Good
ex-, exo-	Out of, out from
extra-	Outside of
hapl-	Single
hem-, hemat-	Blood
hemi-	Half
hom(e)o-	Same; equal
hyper-	Over; above
hypo-	Under; below
infra-	Below, beneath
inter-	Between
intra-	Within
iso-	Same, equal
macro-	Large
mega-	Large; million(th)
mes-	Middle
meta-	Beyond, after
micro-	Small; millionth
milli-	Thousandth
mono-	One (single)
neo-	New
non-	Not
oligo-	Few, scanty
ortho-	Straight; correct, normal
para-	By the side of; near
per-	Through
peri-	Around; surrounding
poly-	Many
post-	After
pre-	Before
pro-	First; promoting
quadr-	Four
re-	Back again
retro-	Behind
semi-	Half
sub-	Under
super-, supra-	Over, above, excessive
trans-	Across; through
tri-	Three; triple

Word Parts Commonly Used as Suffixes

WORD PART	MEANING
-al, -ac	Pertaining to
-algia	Pain
-aps, -apt	Fit; fasten
-arche	Beginning; origin
-ase	Signifies an enzyme
-blast	Sprout; make
-centesis	A piercing
-cide	To kill
-clast	Break; destroy
-crine	Release; secrete
-ectomy	A cutting out
-emesis	Vomiting
-emia	Refers to blood condition
-flux	Flow
-gen	Creates; forms
-genesis	Creation, production
-gram	Something written
-graph(y)	To write, draw
-hydrate	Containing H_2O (water)
-ia, -sia	Condition; process
-iasis	Abnormal condition
-ic, -ac	Pertaining to
-in	Signifies a protein
-ism	Signifies "condition of"
-itis	Signifies "inflammation of"
-lemma	Rind; peel
-lepsy	Seizure
-lith	Stone; rock
-logy	Study of
-lunar	Moon; moonlike
-malacia	Softening
-megaly	Enlargement
-metric, -metry	Measurement, length
-oid	Like; in the shape of
-oma	Tumor
-opia	Vision, vision condition
-oscopy	Viewing
-ose	Signifies a carbohydrate (especially sugar)
-osis	Condition, process
-ostomy	Formation of an opening
-otomy	Cut
-penia	Lack
-philic	Loving
-phobic	Fearing
-phragm	Partition
-plasia	Growth, formation
-plasm	Substance, matter
-plasty	Shape; make
-plegia	Paralysis
-pnea	Breath, breathing
-(r)rhage, -(r)rhagia	Breaking out, discharge
-(r)rhaphy	Sew, suture
-(r)rhea	Flow
-some	Body
-tensin, -tension	Pressure
-tonic	Pressure, tension
-tripsy	Crushing
-ule	Small, little
-uria	Refers to urine condition